Geometric Formulas

Conversion between Radians and Degrees: π radians =

Triangle

$A = \frac{1}{2}bh$
$\quad = \frac{1}{2}ab\sin\theta$

Circle

$A = \pi r^2$
$C = 2\pi r$

$A = \frac{1}{2}r^2\theta \quad (\theta \text{ in radians})$
$s = r\theta \quad (\theta \text{ in radians})$

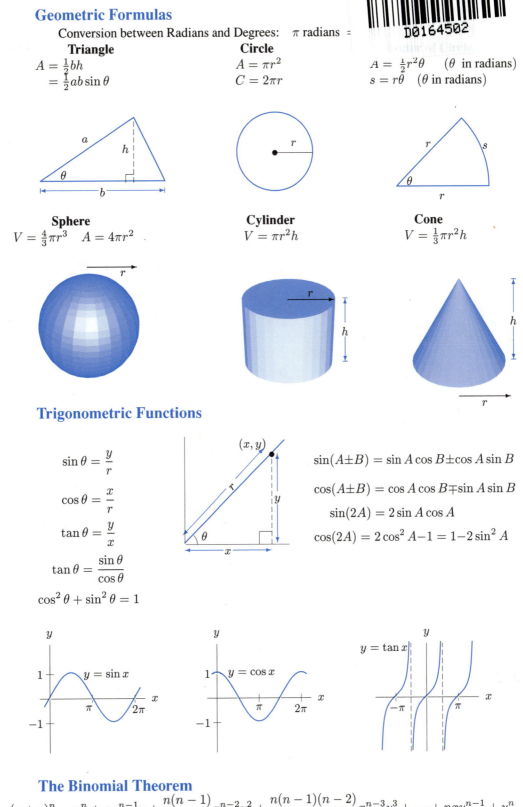

Sphere

$V = \frac{4}{3}\pi r^3 \quad A = 4\pi r^2$

Cylinder

$V = \pi r^2 h$

Cone

$V = \frac{1}{3}\pi r^2 h$

Trigonometric Functions

$\sin\theta = \dfrac{y}{r}$

$\cos\theta = \dfrac{x}{r}$

$\tan\theta = \dfrac{y}{x}$

$\tan\theta = \dfrac{\sin\theta}{\cos\theta}$

$\cos^2\theta + \sin^2\theta = 1$

$\sin(A \pm B) = \sin A\cos B \pm \cos A\sin B$

$\cos(A \pm B) = \cos A\cos B \mp \sin A\sin B$

$\sin(2A) = 2\sin A\cos A$

$\cos(2A) = 2\cos^2 A - 1 = 1 - 2\sin^2 A$

$y = \sin x$

$y = \cos x$

$y = \tan x$

The Binomial Theorem

$$(x + y)^n = x^n + nx^{n-1}y + \frac{n(n-1)}{1\cdot 2}x^{n-2}y^2 + \frac{n(n-1)(n-2)}{1\cdot 2\cdot 3}x^{n-3}y^3 + \cdots + nxy^{n-1} + y^n$$

$$(x - y)^n = x^n - nx^{n-1}y + \frac{n(n-1)}{1\cdot 2}x^{n-2}y^2 - \frac{n(n-1)(n-2)}{1\cdot 2\cdot 3}x^{n-3}y^3 + \cdots \pm nxy^{n-1} \mp y^n$$

CALCULUS

Third Edition

Produced by the Calculus Consortium and initially funded by a National Science Foundation Grant.

Deborah Hughes-Hallett
University of Arizona

William G. McCallum
University of Arizona

Andrew M. Gleason
Harvard University

Brad G. Osgood
Stanford University

Daniel E. Flath
University of South Alabama

Andrew Pasquale
Chelmsford High School

Sheldon P. Gordon
SUNY at Farmingdale

Douglas Quinney
University of Keele

Patti Frazer Lock
St. Lawrence University

Wayne Raskind
University of Southern California

David O. Lomen
University of Arizona

Karen Rhea
University of Michigan

David Lovelock
University of Arizona

Jeff Tecosky-Feldman
Haverford College

David Mumford
Brown University

Joe B. Thrash
University of Southern Mississippi

Thomas W. Tucker
Colgate University

with the assistance of

Otto K. Bretscher
Colby College

Adrian Iovita
University of Washington

Coordinated by

Elliot J. Marks

John Wiley & Sons, Inc.

EXECUTIVE EDITOR	Ruth Baruth
SENIOR DEVELOPMENTAL EDITOR	Nancy Perry
DEVELOPMENTAL EDITOR	Anne Scanlan-Rohrer/Two Ravens Editorial
PROGRAM ASSISTANT	Merillat Saint-Amand
MARKETING MANAGER	Julie Z. Lindstrom
SENIOR PRODUCTION EDITOR	Ken Santor
COVER AND CHAPTER OPENER DESIGNER	Madelyn Lesure
COVER AND CHAPTER OPENING PHOTO	Eddie Hironaka/The Image Bank
OUTSIDE PRODUCTION SERVICES	Publication Services, Inc.

Problems from Calculus: The Analysis of Functions, by Peter D. Taylor (Toronto: Wall & Emerson, Inc., 1992). Reprinted with permission of the publisher.

This book was set in Times Roman by the Consortium using TeX, Mathematica, and the package AsTeX, which was written by Alex Kasman. It was printed and bound by Von Hoffmann Press. The cover was printed by Von Hoffmann Press.

This book is printed on acid-free paper.

The paper in this book was manufactured by a mill whose forest management programs include sustained yield harvesting of its timberlands. Sustained yield harvesting principles ensure that the numbers of trees cut each year does not exceed the amount of new growth.

To order books or for customer service please, call 1(800)-CALL-WILEY (225-5945).

This material is based upon work supported by the National Science Foundation under Grant No. DUE-9352905. Opinions expressed are those of the authors and not necessarily those of the Foundation.

ISBN 0471-40827-1

Printed in the United States of America

10 9 8 7 6 5 4 3 2

PREFACE

Calculus is one of the greatest achievements of the human intellect. Inspired by problems in astronomy, Newton and Leibniz developed the ideas of calculus 300 years ago. Since then, each century has demonstrated the power of calculus to illuminate questions in mathematics, the physical sciences, engineering, and the social and biological sciences.

Calculus has been so successful because of its extraordinary power to reduce complicated problems to simple rules and procedures. Therein lies the danger in teaching calculus: it is possible to teach the subject as nothing but the rules and procedures—thereby losing sight of both the mathematics and of its practical value. This edition of *Calculus* continues our effort to refocus the teaching of calculus on concepts as well as procedures.

A Focused Vision: Conceptual Understanding

Our goal is to provide students with a clear understanding of the ideas of calculus as a solid foundation for subsequent courses in mathematics and other disciplines. When we designed this curriculum we started with a clean slate. We increased the emphasis on some topics, such as differential equations, and omitted some traditional topics whose inclusion we could not justify after discussions with mathematicians, engineers, physicists, chemists, biologists, and economists. We focused on a small number of key concepts, emphasizing depth of understanding rather than breadth of coverage.

The Third Edition: An Evolutionary Curriculum

Since 1992, when we developed the first edition of Calculus, we have helped widen the array of options available to instructors teaching calculus. Our first edition struck a new balance between concepts, modeling, and skills to enable instructors to address all aspects of student learning. Our approach, for which we coined the name "Rule of Four," included developing new problems that probed student understanding in areas often taken for granted. These problems, praised for their creativity and variety, have had an influence far beyond the users of our textbook.

The Third Edition of *Calculus* has the same philosophy as previous editions, but a different balance of topics. Always working in collaboration, consortium members and colleagues in mathematics and client disciplines from all over the world have helped shape this edition. The information we gathered from users and non-users of our text reflects the growing consensus for a curriculum that balances new and traditional topics. Our new Third Edition brings together the best of both.

An example of the balance of traditional and new topics is the chapter on series, which has been extensively rewritten and split into two chapters with more material added. Our experience teaching from both traditional and newer materials put us in a unique position to make this new curriculum meaningful to the students going into mathematics and to those going into science and engineering.

Our book retains the hallmarks of our earlier editions, which have now been adopted as standard features of other calculus textbooks: The Rule of Four, emphasis on modeling, exposition that students can read, and a flexible approach to technology.

As in earlier editions, instructors can add to, subtract from, or rearrange our suggested curriculum. For the Third Edition, we have adopted a more commonly preferred organization in which all chapters on derivatives are together, while still allowing users to follow the Key Concepts approach of the earlier editions. As members of the mathematical community, we have responded to requests for compromise with a book that provides a middle ground. We expect that while some parts may look familiar, some may not. As active classroom teachers ourselves, we enjoy teaching both.

General Changes in the Third Edition

- Certain portions of the Second Edition Focus on Theory sections have been integrated into the body of the text where appropriate.

- All Focus on Theory sections from the Second Edition have been moved to the web site in their entirety for reference.

- Focus on Modeling sections have been integrated into the body of the text.

- All theorems have been numbered for easy reference.

- Exercise sections have been arranged into routine applications and drill problems followed by more challenging/conceptual problems and applications.

- Overall, 15% of the problems are new, including many applications. More easy and mid-level problems have been added where appropriate.

- More problems using computer algebra systems (CAS) have been added.

- Concept tests modeled on those developed by the Harvard physicist Eric Mazur, which test and foster student discussion of basic concepts, are in the Instructor's Manual.

- Check Your Understanding: New true/false questions probe students' understanding.

The Development of Mathematical Thinking

The first stage in the development of mathematical thinking is the acquisition of a clear intuitive picture of the central ideas. In the next stage, the student learns to reason with the intuitive ideas and explain the reasoning clearly in plain English. After this foundation has been laid, there is a choice of direction. Some students, such as mathematics majors, may benefit from a more theoretical approach, while others, such as science and engineering majors, may benefit from further exploration of modeling.

The Development of Mathematical Skills

To use calculus effectively, students need skill in both symbolic manipulation and the use of technology. The exact proportions of each may vary widely, depending on the preparation of the student and the wishes of the instructor. The book is adaptable to many different combinations.

In multivariable calculus, even more than in single variable calculus, computers can help students learn to think mathematically. For example, looking at surface graphs and contour diagrams can promote an understanding of functions of many variables.

The book does not require any specific software or technology. It has been used with graphing calculators, graphing software, and computer algebra systems. Any technology with the ability to graph functions and perform numerical integration will suffice. Students are expected to use their own judgment to determine where technology is useful.

Chapter-by-Chapter Description of Changes from the Second Edition:

- Chapter 1: The chapter has been shortened and reorganized. The Binomial Theorem has been moved to the web site.

- Chapter 2: Sections 2.2, Limits, and 2.7, Continuity and Differentiability, are new sections derived from former Focus on Theory sections.

- Chapter 3: This chapter comes earlier in the third edition. Section 3.8, Parametric Equations, is new and contains material on motion and parametric curves. Section 3.9 includes new material on differentiation and estimating the error in the tangent line approximation. Section 3.10, Using Local Linearity to Find Limits, is a new section, and includes L'Hopital's rule.

- Chapter 4: Section 4.7 is a new section containing the Extreme Value Theorem, Mean Value Theorem, Increasing Function Theorem, Constant Function Theorem, and Racetrack Principle.

- Chapter 5: Section 5.2 includes general Riemann sums. Section 5.3 includes material about integrating rates of change.

- Chapter 6: Section 6.5, The Equations of Motion, is a former Focus on Modeling section.

- Chapter 7: Section 7.4, Algebraic Identities and Trigonometric Substitutions, is a new section including partial fractions and trigonometric substitutions, including completing the square.

- Chapter 8: Section 8.1 includes a more accessible introduction to setting up integrals focusing on basic concepts. The material on center of mass in Section 8.3 has been expanded, and Section 8.4, on applications to physics, has been simplified and made easier to use.

- Chapter 9: In the Third Edition, Chapters 9 and 10 replace Chapter 9 in the Second Edition. The material has been extensively reorganized and rewritten. All sections have new problems. Material is clearly divided between series and convergence (Chapter 9) and approximations of functions (Chapter 10) for users who wish to emphasize one or the other. Section 9.1 had been 9.4 in the second edition. Section 9.2 is a new section derived from a former Focus on Theory Section. Substantial new material on the integral test, ratio test, and alternating series has been added in Sections 9.2 and 9.3. Section 9.4 is material from former 9.2 with substantial new material on intervals and radius of convergence added. Additional theory material is on the web site.

- Chapter 10: Section 10.1 is the first part of former 9.1. Section 10.2 is the second part of former Section 9.1 and Section 9.2; Section 10.3 is former Section 9.3. Section 10.4 is a former Focus on Theory section, substantially rewritten. Section 10.5 is the former 9.5.

- Chapter 11: Section 11.7, Models of Population Growth, which focuses on the logistic equation, has been streamlined. Section 11.8, Systems of Differential Equations, and 11.9, Analyzing the Phase Plane, are former Focus on Modeling sections.

- Chapter 12: Section 12.1 combines former Sections 11.1 and 11.2. Section 12.6 is a new section derived from former Focus on Theory section.

- Chapter 13: This is former Chapter 12.

- Chapter 14: Section 14.7 is from fromer Sections 13.7 and 13.8. Material on partial differential equations is on the web site.

- Chapter 15: This is former Chapter 14.

- Chapter 16: Material on the Monte Carlo Method is on the web site.

- Chapter 17: This chapter is a combination of former Chapters 16 and 17. Sections 17.1 and 17.2 are former Sections 16.1 and 16.2. Sections 17.3 and 17.4 are former Sections 17.1 and 17.2. Section 17.5 is a new sections on parameterized surfaces. There is material on the Implicit Function Theorem and Kepler's Laws on the web site.

- Chapter 18: The proof of Green's Theorem is on the web site.

- Chapter 19: Section 19.3 is a new section on flux through parameterized surfaces.

- Chapter 20: Section 20.5 is a new section derived from a former Focus on Theory section. The proofs of the Divergence and Stokes' Theorems are on the web site.

Content

This content represents our vision of how calculus can be taught. It is flexible enough to accommodate individual course needs and requirements. Topics can easily be added or deleted, or the order changed.

Chapter 1: A Library of Functions

This chapter introduces all the elementary functions to be used in the book. Although the functions are probably familiar, the graphical, numerical, verbal, and modeling approach to them is likely to be new. We introduce exponential functions at the earliest possible stage, since they are fundamental to the understanding of real-world processes.

Chapter 2: Key Concept—The Derivative

The purpose of this chapter is to give the student a practical understanding of the definition of the derivative and its interpretation as an instantaneous rate of change. The power rule is introduced; other rules are introduced in Chapter 3.

Chapter 3: Short-Cuts to Differentiation

The derivatives of all the functions in Chapter 1 are introduced, as well as the rules for differentiating products, quotients, and composite functions.

Chapter 4: Using the Derivative

The aim of this chapter is to enable the student to use the derivative in solving problems, including optimization and graphing. It is not necessary to cover all the sections in this chapter.

Chapter 5: Key Concept—The Definite Integral

The purpose of this chapter is to give the student a practical understanding of the definite integral as a limit of Riemann sums and to bring out the connection between the derivative and the definite integral in the Fundamental Theorem of Calculus.

Chapter 6: Constructing Antiderivatives

This chapter focuses on going backward from a derivative to the original function, first graphically and numerically, then analytically. It introduces the Second Fundamental Theorem of Calculus and the concept of a differential equation.

Chapter 7: Integration

This chapter includes several techniques of integration, including substitution, parts, partial fractions, and trigonometric substitutions; others are included in the table of integrals. There is a discussion of numerical methods and of improper integrals.

Chapter 8: Using the Definite Integral

This chapter emphasizes the idea of subdividing a quantity to produce Riemann sums which, in the limit, yield a definite integral. It shows how the integral is used in geometry, physics, economics, and probability; it is not necessary to cover all the sections in this chapter.

Chapter 9: Series

This chapter focuses on material on series and convergence, including the integral test, ratio test, and alternating series, power series, and intervals of convergence.

Chapter 10: Approximating Functions

This chapter introduces Taylor Series and Fourier Series using the idea of approximating functions by simpler functions.

Chapter 11: Differential Equations

This chapter introduces differential equations. The emphasis is on qualitative solutions, modeling, and interpretation.

Chapter 12: Functions of Several Variables

This chapter introduces functions of many variables from several points of view, using surface graphs, contour diagrams, and tables. We assume throughout that functions of two or more variables are defined on regions with piecewise smooth boundaries. We conclude with a section on continuity.

Chapter 13: A Fundamental Tool: Vectors

This chapter introduces vectors geometrically and algebraically and discusses the dot and cross product.

Chapter 14: Differentiating Functions of Several Variables

Partial derivatives, directional derivatives, gradients, and local linearity are introduced. The chapter also discusses higher order partial derivatives, quadratic Taylor approximations, and differentiability.

Chapter 15: Optimization

The ideas of the previous chapter are applied to optimization problems, both constrained and unconstrained.

Chapter 16: Integrating Functions of Several Variables

This chapter discusses double and triple integrals in Cartesian, polar, cylindrical, and spherical coordinates, as well as the formula for a general change of variable.

Chapter 17: Parameterization and Vector Fields

This chapter discusses parameterized curves and motion, vector fields and flowlines, and parameterized surfaces.

Chapter 18: Line Integrals

This chapter introduces line integrals and shows how to calculate them using parameterizations. Conservative fields, gradient fields, the Fundamental Theorem of Calculus for Line Integrals, and Green's Theorem are discussed.

Chapter 19: Flux Integrals

This chapter introduces flux integrals and shows how to calculate them over surface graphs, portions of cylinders, portions of spheres, and arbitrary parameterized surfaces.

Chapter 20: Calculus of Vector Fields

The purpose of this chapter is to give students a practical understanding of the divergence and curl, as well as of the Divergence and Stokes' Theorems.

Appendices

There are appendices on roots, accuracy, and bounds; polar coordinates; complex numbers; Newton's Method; and determinants.

Supplementary Materials

The following supplementary materials are available for this edition:
- **Instructor's Manual** with Sample Exam Questions containing teaching tips, calculator programs, some overhead transparency masters and test questions arranged according to section.
- **Instructor's Solution Manual** with complete solutions to all problems.
- **Student's Solution Manual** with complete solutions to half the odd-numbered problems.
- **Student's Study Guide** with additional study aids for students that are tied directly to the book.
- **Instructor's Resource CD-ROM** which contains the Instructor's Manual, Instructor's Solutions Manual, additional projects, as well as other valuable resources.

The Faculty Resource Network

The *Faculty Resource Network* is a peer-to-peer network of academic faculty dedicated to the effective use of technology in the classroom. This group can help you apply innovative classroom techniques, implement specific software packages, and tailor the technology experience to the specific needs of each indivicual class. Ask your Wiley representative for more details.

eGrade

eGrade is an on-line assessment system which contains a large bank of skill-building problems, homework problems, and solutions. Instructors can now automate the process of assigning, delivering, grading, and routing all kinds of homework, quizzes, and tests while providing students with immediate scoring and feedback on their work. Wiley *eGrade* "does the math" and much more. For more information, visit www.wiley.com/college/egrade or speak with your Wiley representative.

StudyAgent for Calculus

StudyAgent for Calculus is a web-based, intelligent software package that solves and documents calculus problems in real time. For students, StudyAgent is a step-by-step electronic tutor. As the student works through a particular problem online, StudyAgent provides customized feedback from the text, allowing students to identify and learn from their mistakes. Instructors can use StudyAgent to preview problems and explore functions graphically and analytically. For more information, visit www.wiley.com/college/cch or speak with your Wiley representative.

Supplements for the instructor can be obtained by sending a request on your institutional letterhead to Mathematics Marketing Manager, John Wiley & Sons Inc., 605 Third Avenue, New York, NY 10158-0012, or by contacting your local Wiley representative.

Acknowledgements

First and foremost, we want to express our appreciation to the National Science Foundation for their faith in our ability to produce a revitalized calculus curriculum and, in particular, to our program officers, Louise Raphael, John Kenelly, John Bradley, and James Lightbourne. We also want to thank the members of our Advisory Board, Benita Albert, Lida Barrett, Simon Bernau, Robert Davis, M. Lavinia DeConge-Watson, John Dossey, Ron Douglas, Eli Fromm, William Haver, Seymour Parter and Stephen Rodi for their ongoing guidance and advice.

In addition, a host of other people around the country and abroad deserve our thanks for all that they did to help our project succeed. They include: Huriye Arikan, Ralph Baierlein, Paul Balister, Ruth Baruth, Martin Batey, Frank Beatrous, Melanie Bell, Thomas Bird, Keith Blackwell, Paul Blanchard, Melkana Brakalova, John Bravman, David Bressoud, Dennis W. Brewer, Stephen Boyd, Otto Bretscher, Mark Bridger, Morton Brown, Lucille Buonocore, Marilyn Carlson, Phil Cheifetz, Howie Choset, Jurie Conradie, Sterling G. Crossley, Caspar Curjel, Ehud de Shalit, Ann Davidian, Bob Decker, Dennis De Turck, Tom Dick, Jeff Edmunds, Wade Ellis, Johann Engelbrecht, Hermann Flaschka, G. Martin Forrest, Michael R. Foster, Leonid Fridman, Leonid Friedlander, Amanda Galtman, Avijit Gangopadhgay, Lynn Garner, Tom Gearhart, Howard Georgi, James F. Gibbons, Allan Gleason, Florence Gordon, Danny Goroff, Robin Gottlieb, Biddy Greene, Marty Greenlee, John Hagood, Kenneth Hannsgen, Dudley Herschbach, JoEllen Hillyer, Luke Hunsberger, John Huth, Richard Iltis, Rob Indik, Mary Johenk, Jerry Johnson, Kerry Johnson, Mille Johnson, Matthias Kawski, David Kazhdan, Thomas Kerler, Mike Klucznik, Donna Krawczyk, Terry L. Kohutek, Mark Kunka, Ted Laetsch, Brigitte Lahme, Carl Leinbach, Carl Leinert, Maddy Lesure, David Levermore, Don Lewis, Jim Lewis, Dorian Liepmann, Julie Lindstrom, Howard Littman, Guadalupe Lozano, Bin Lu, Chris Lucas, John Lucas, Tom McMahon, Dan Madden, Lisa A. Mantini, Joe A. Marlin, Barry Mazur, Eric Mazur, Dan McGee, Ansie Meiring, Dave Meredith, Andrew Metrick, Vijay Modi, Lang Moore, Jerry Morris, Don Myers, Bridget Neale-Paris, Alan Newell, Steve Olson, John Orr, Grace Orzech, Arnie Ostebee, Mike Pavloff, Howard Penn, Anthony L. Peressini, Nancy Perry, Tony Phillips, Ago Pisztora, David R. Pitts, John Prados, Ted Pyne, Amy Radunskaya, Wayne Raskind, Gabriella Ratay, Jeffrey Rauch, Janet Ray, Maria Robinson, J. W. Rogers, Jr., George Rublein, Merillat Saint-Amand, John Sanchez, Ken Santor, Anne Scanlan-Rohrer,

Anneliese Schauerte, Wilfried Schmid, Walter Seaman, Pat Shure, Esther Silberstein, Scott Pilzer, Lynne Small, David Smith, Don Snow, Bob Speiser, Howard Stone, Michell Anne Stillman, Steve Strogatz, "Suds" Sudholz, Brian Tankersley, Cliff Taubes, Peter Taylor, Ralph Teixeira, Sally Thomas, Tom Timchek, Alan Tucker, Jerry Uhl, Doug Ulmer, Ignatios Vakalis, Bill Vélez, Charles Walter, Mary Jean Winter, Chris Yetman, Debbie Yoklic, Bruce Yoshiwara, Kathy Yoshiwara, Lee Zia, Paul Zorn.

Reports from the following reviewers were most helpful in shaping the third edition:

Jennifer Aikens, Dean Allison, Cathy Anderson, Richard I. Avery, Janet Beery, Hamid Behmard, Victor Belfi, Tom Bengtson , David Borthwick, Sean Bradley, Carl T. Brezovec, Paul Wayne Britt, Denise Brown, Dave Bush, Don A. Campbell, Jack Carter, Stan Chadick, Ruth Charney, Jon Clauss, Ann Commito, Lawerence J. Cohen, Mary V. Connolly, Randall Crist, Rick Dalrymple, John Davis, Mike Dellens, Matt DeLong, Sylvia Desjardins, Meighan I. Dillon, Dawn Draus, Steven R. Dunbar, Jim Edmondson, Karen Edwards, Murray Eisenberg, Sue Esch, Phil Farmer, Richard E. Faulkenberry,George F. Feissner, Nelson Fong, Paula Castagna, Ben Fusaro, Kris Green, Michael Green, Leah Griffith, Cheryl Grood, David L. Gross, Brenda Guthrie, John Hagood, William Hamill, John Haverhals, Elizabeth Hentges, Curtis Herink, Jim Holmes, John Hoover, Rick Hough, Thomas Hunter, Kathy Ivey, Craig A. Jensen, Sue Jensen, Jerry Johnson, Nancy L. Johnson, Guarav Khanna, Helmut Knaust, Robert Kowalczyk, Anne Landry, James W. Lea, Phyllis Leonard, Tom Linton, Harold Lund, Juan Manfredi, William O. Martin, Alex Martsinkovsky, Anita Maxwell, Barbara McGee, Mary Ann McLoughlin, Douglas B. Meade, Richard Mercer, Raquel Mesa, C.R. Messer, Mark A. Mills, Yair Minsky, Derek Mitchem, Jeff Mock, Doug Mupasiri, Douglas Nelson, Dave Nolan, Sandy Norman, David A. Olson, Amos Ong, Grace Orzech, Wes Ostertag, Jeffrey S. Palmer, Barry Perratt, Peter Philliou, Shirley Pomeranz, G.J. Porter, Rebecca Porter, Vrunda Prabhu, Harald Proppe, Margie Reinhart, Mike Robillard, Donna Fields Rochon, Peter Rojas, Ned I. Rosen, Peter Rosnick, Tom Rousseau, Brenda Rudin, Mark D. Schlatter, Howard Schwesky, Patrick Shanahan, Michael Sheard, Angela B. Shiflet, Mehrdad Simkani, Lynn Small, Katy Snyder, Edward J. Soares, Ernie Solheid, Richard Stout, Robert Styer, James Sullivan, Ted Sundstrom, Stuart Swain, Ray Tennant, Buren Thomas, George Tintera, Dave Trautman, Mark Turner, Marian Van Vleet, Mike Veatch, Rachel Westlake, Bill Waltmann, Michael Ward, Steve Waters, Jack Winn, John Wood, Lisa Woodrum, William Wunderlich, Christos Xenophontos, W. Xie, Deane Yang, Kathie Yoder, Bruce Yoshiwara, Andrzej Zarach, Ed Zeidman, Peter Zekert.

Most of all, to the remarkable team that worked day and night to get the text into the computer (and out again), to get the solutions written and the pictures labeled: we greatly appreciate your ingenuity, energy and dedication. Thanks to: Ebo Bentil, Srdjan Divac, Paul Erickson, Mike Esposito, David Grenda, Alex Kasman, Alex Mallozzi, Ann Ryu, and Xianbao Xu.

In particular, we would like to thank Ray Cannon, C.K. Cheung, Bob Condon, Paul Feehan, and Ernie Solheid for their remarkable contributions.

Deborah Hughes-Hallett

Andrew M. Gleason

William G. McCallum

Daniel E. Flath

Patti Frazer Lock

Sheldon P. Gordon

David O. Lomen

David Lovelock

David Mumford

Brad G. Osgood

Andrew Pasquale

Douglas Quinney

Wayne Raskind

Karen R. Rhea

Jeff Tecosky-Feldman

Joe B. Thrash

Thomas W. Tucker

To Students: How to Learn from this Book

- This book may be different from other math textbooks that you have used, so it may be helpful to know about some of the differences in advance. This book emphasizes at every stage the *meaning* (in practical, graphical or numerical terms) of the symbols you are using. There is much less emphasis on "plug-and-chug" and using formulas, and much more emphasis on the interpretation of these formulas than you may expect. You will often be asked to explain your ideas in words or to explain an answer using graphs.

- The book contains the main ideas of multivariable calculus in plain English. Your success in using this book will depend on your reading, questioning, and thinking hard about the ideas presented. Although you may not have done this with other books, you should plan on reading the text in detail, not just the worked examples.

- There are very few examples in the text that are exactly like the homework problems. This means that you can't just look at a homework problem and search for a similar–looking "worked out" example. Success with the homework will come by grappling with the ideas of calculus.

- Many of the problems that we have included in the book are open-ended. This means that there may be more than one approach and more than one solution, depending on your analysis. Many times, solving a problem relies on common sense ideas that are not stated in the problem but which you will know from everyday life.

- Some problems in this book assume that you have access to a graphing calculator or computer; preferably one that can draw surface graphs, contour diagrams, and vector fields, and can compute multivariable integrals and line integrals numerically. There are many situations where you may not be able to find an exact solution to a problem, but you can use a calculator or computer to get a reasonable approximation.

- This book attempts to give equal weight to three methods for describing functions: graphical (a picture), numerical (a table of values) and algebraic (a formula). Sometimes you may find it easier to translate a problem given in one form into another. For example, if you have to find the maximum of a function, you might use a contour diagram to estimate its approximate position, use its formula to find equations that give the exact position, then use a numerical method to solve the equations. The best idea is to be flexible about your approach: if one way of looking at a problem doesn't work, try another.

- Students using this book have found discussing these problems in small groups very helpful. There are a great many problems which are not cut-and-dried; it can help to attack them with the other perspectives your colleagues can provide. If group work is not feasible, see if your instructor can organize a discussion session in which additional problems can be worked on.

- You are probably wondering what you'll get from the book. The answer is, if you put in a solid effort, you will get a real understanding of one of the most important accomplishments of the millennium – calculus – as well as a real sense of the power of mathematics in the age of technology.

CONTENTS

4 USING THE DERIVATIVE   165

5 KEY CONCEPT: THE DEFINITE INTEGRAL   221

6 CONSTRUCTING ANTIDERIVATIVES   261

7 INTEGRATION   289

8 USING THE DEFINITE INTEGRAL   345

18 LINE INTEGRALS   825

19 FLUX INTEGRALS   863

20 CALCULUS OF VECTOR FIELDS   887

APPENDIX   927

READY REFERENCE   951

ANSWERS TO ODD NUMBERED PROBLEMS   967

INDEX   993

Chapter One

A LIBRARY OF FUNCTIONS

Functions are truly fundamental to mathematics. In everyday language we say, "The fuel needed to launch a rocket is a function of its payload" or "The patient's blood pressure is a function of the drugs prescribed." In each case, the word *function* expresses the idea that knowledge of one fact tells us another. In mathematics, the most important functions are those in which knowledge of one number tells us another number. If we know the length of the side of a square, its area is determined. If the circumference of a circle is known, its radius is determined.

Calculus starts with the study of functions. This chapter lays the foundation for calculus by surveying the behavior of the most common functions, including exponential, logarithmic, and trigonometric functions. We also explore ways of handling the graphs, tables, and formulas that represent these functions.

1.1 FUNCTIONS AND CHANGE

In mathematics, a *function* is used to represent the dependence of one quantity upon another.

Let's look at an example. In December 2000, the temperatures in Chicago were unusually low over winter vacation. The daily high temperatures for December 19–28 are given in Table 1.1.

Table 1.1 *Daily high temperature in Chicago, December 19–28, 2000*

Date (December 2000)	19	20	21	22	23	24	25	26	27	28
High temperature (°F)	20	17	19	7	20	11	17	19	17	20

Although you may not have thought of something so unpredictable as temperature as being a function, the temperature *is* a function of date, because each day gives rise to one and only one high temperature. There is no formula for temperature (otherwise we would not need the weather bureau), but nevertheless the temperature does satisfy the definition of a function: Each date, t, has a unique high temperature, H, associated with it.

We define a function as follows:

> A **function** is a rule that takes certain numbers as inputs and assigns to each a definite output number. The set of all input numbers is called the **domain** of the function and the set of resulting output numbers is called the **range** of the function.

The input is called the *independent variable* and the output is called the *dependent variable*. In the temperature example, the domain is the set of dates $t = \{19, 20, 21, 22, 23, 24, 25, 26, 27, 28\}$ and the range is the set of temperatures $H = \{7, 11, 17, 19, 20\}$. We call the function f and write $H = f(t)$. Notice that a function may have identical outputs for different inputs (December 20, 25, and 27, for example).

Some quantities, such as date, are *discrete*, meaning they take only certain isolated values (dates must be integers). Other quantities, such as time, are *continuous* as they can be any number. For a continuous variable, domains and ranges are often written using interval notation:

The set of numbers t such that $a \leq t \leq b$ is written $[a, b]$.

The set of numbers t such that $a < t < b$ is written (a, b).

Representation of Functions: Tables, Graphs, Formulas, and Words

Functions can be represented by tables, graphs, formulas, and descriptions in words. For example, the function giving the daily high temperatures in Chicago can be represented by the graph in Figure 1.1, as well as by Table 1.1.

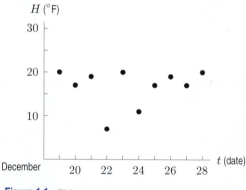

Figure 1.1: Chicago temperatures, December 20000

As another example of a function, consider the snow tree cricket. Surprisingly enough, all such crickets chirp at essentially the same rate if they are at the same temperature. That means that the chirp rate is a function of temperature. In other words, if we know the temperature, we can determine the chirp rate. Even more surprisingly, the chirp rate, C, in chirps per minute, increases steadily with the temperature, T, in degrees Fahrenheit, and can be computed by the formula

$$C = 4T - 160$$

to a fair degree of accuracy. We write $C = f(T)$ to express the fact that we think of C as a function of T and that we have named this function f. The graph of this function is in Figure 1.2.

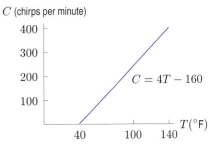

Figure 1.2: Cricket chirp rate versus temperature

Examples of Domain and Range

If the domain of a function is not specified, we usually take it to be the largest possible set of real numbers. For example, we usually think of the domain of the function $f(x) = x^2$ as all real numbers. However, the domain of the function $g(x) = 1/x$ is all real numbers except zero, since we cannot divide by zero.

Sometimes we restrict the domain to be smaller than the largest possible set of real numbers. For example, if the function $f(x) = x^2$ is used to represent the area of a square of side x, we restrict the domain to nonnegative values of x.

Example 1   The function $C = f(T)$ gives chirp rate as a function of temperature. We restrict this function to temperatures for which the predicted chirp rate is positive, and up to the highest temperature ever recorded at a weather station, $136°F$. What is the domain of this function f?

Solution   If we consider the equation

$$C = 4T - 160$$

simply as a mathematical relationship between two variables C and T, any T value is possible. However, if we think of it as a relationship between cricket chirps and temperature, then C cannot be less than 0. Since $C = 0$ leads to $0 = 4T - 160$, and so $T = 40°F$, we see that T cannot be less than $40°F$. (See Figure 1.2.) In addition, we are told that the function is not defined for temperatures above $136°$. Thus, for the function $C = f(T)$ we have

$$\text{Domain} = \text{All } T \text{ values between } 40°F \text{ and } 136°F$$
$$= \text{All } T \text{ values with } 40 \leq T \leq 136$$
$$= [40, 136].$$

Example 2   Find the range of the function f, given the domain from Example 1. In other words, find all possible values of the chirp rate, C, in the equation $C = f(T)$.

Solution   Again, if we consider $C = 4T - 160$ simply as a mathematical relationship, its range is all real C values. However, when thinking of the meaning of $C = f(T)$ for crickets, we see that the function predicts cricket chirps per minute between 0 (at $T = 40°F$) and 384 (at $T = 136°F$). Hence,

$$\text{Range} = \text{All } C \text{ values from 0 to 384}$$
$$= \text{All } C \text{ values with } 0 \leq C \leq 384$$
$$= [0, 384].$$

In using the temperature to predict the chirp rate, we thought of the temperature as the *independent variable* and the chirp rate as the *dependent variable*. However, we could do this backwards, and calculate the temperature from the chirp rate. From this point of view, the temperature is dependent on the chirp rate. Thus, which variable is dependent and which is independent may depend on your viewpoint.

Linear Functions

The chirp-rate function, $C = f(T)$, is an example of a *linear function*. A function is linear if its slope, or rate of change, is the same at every point. The rate of change of a function that is not linear may vary from point to point.

Olympic and World Records

During the early years of the Olympics, the height of the men's winning pole vault increased approximately 8 inches every four years. Table 1.2 shows that the height started at 130 inches in 1900, and increased by the equivalent of 2 inches a year. So the height was a linear function of time from 1900 to 1912. If y is the winning height in inches and t is the number of years since 1900, we can write

$$y = f(t) = 130 + 2t.$$

Since $y = f(t)$ increases with t, we say that f is an *increasing function*. The coefficient 2 tells us the rate, in inches per year, at which the height increases.

Table 1.2   *Men's Olympic pole vault winning height (approximate)*

Year	1900	1904	1908	1912
Height (inches)	130	138	146	154

This rate of increase is the *slope* of the line in Figure 1.3. The slope is given by the ratio

$$\text{Slope} = \frac{\text{Rise}}{\text{Run}} = \frac{146 - 138}{8 - 4} = \frac{8}{4} = 2 \text{ inches/year}.$$

Calculating the slope (rise/run) using any other two points on the line gives the same value.

What about the constant 130? This represents the initial height in 1900, when $t = 0$. Geometrically, 130 is the *intercept* on the vertical axis.

You may wonder whether the linear trend continues beyond 1912. Not surprisingly, it doesn't exactly. The formula $y = 130 + 2t$ predicts that the height in the 2000 Olympics would be 330 inches or 27 feet 6 inches, which is considerably higher than the actual value of 19 feet 4.27 inches. There is clearly a danger in *extrapolating* too far from the given data. You should also observe that the data in Table 1.2 is *discrete*, because it is given only at specific points (every four years). However, we have treated the variable t as though it were *continuous*, because the function $y = 130 + 2t$ makes

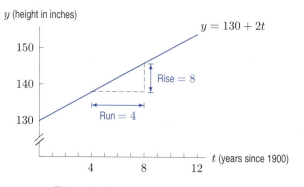

Figure 1.3: Olympic pole vault records

sense for all values of t. The graph in Figure 1.3 is of the continuous function because it is a solid line, rather than four separate points representing the years in which the Olympics were held.

As the pole vault heights have increased over the years, the time to run the mile has decreased. If y is the world record time to run the mile, in seconds, and t is the number of years since 1900, then records show that, approximately,

$$y = g(t) = 260 - 0.4t.$$

The 260 tells us that the world record was 260 seconds in 1900 (at $t = 0$). The slope, -0.4, tells us that the world record decreased by about 0.4 seconds per year. We say that g is a *decreasing function*.

Difference Quotients and Delta Notation

We use the symbol Δ (the Greek letter capital delta) to mean "change in," so Δx means change in x and Δy means change in y.

The slope of a linear function $y = f(x)$ can be calculated from values of the function at two points, given by x_1 and x_2, using the formula

$$m = \frac{\text{Rise}}{\text{Run}} = \frac{\Delta y}{\Delta x} = \frac{f(x_2) - f(x_1)}{x_2 - x_1}.$$

The quantity $(f(x_2) - f(x_1))/(x_2 - x_1)$ is called a *difference quotient* because it is the quotient of two differences. (See Figure 1.4). Since $m = \Delta y/\Delta x$, the units of m are y-units over x-units.

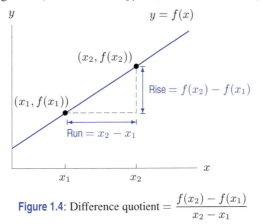

Figure 1.4: Difference quotient $= \dfrac{f(x_2) - f(x_1)}{x_2 - x_1}$

Families of Linear Functions

A **linear function** has the form

$$y = f(x) = b + mx.$$

Its graph is a line such that
- m is the **slope**, or rate of change of y with respect to x.
- b is the **vertical intercept**, or value of y when x is zero.

Notice that if the slope, m, is zero, we have $y = b$, a horizontal line.

To recognize that a table of x and y values comes from a linear function, $y = b + mx$, look for differences in y-values that are constant for equally spaced x-values.

Formulas such as $f(x) = b + mx$, in which the constants m and b can take on various values, give a *family of functions*. All the functions in a family share certain properties—in this case, all the graphs are straight lines. The constants m and b are called *parameters*; their meaning is shown in Figures 1.5 and 1.6. Notice the greater the magnitude of m, the steeper the line.

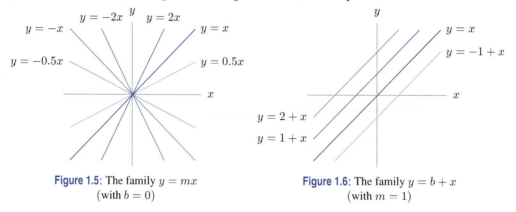

Figure 1.5: The family $y = mx$
(with $b = 0$)

Figure 1.6: The family $y = b + x$
(with $m = 1$)

Increasing versus Decreasing Functions

The terms increasing and decreasing can be applied to other functions, not just linear ones. See Figure 1.7. In general,

> A function f is **increasing** if the values of $f(x)$ increase as x increases.
> A function f is **decreasing** if the values of $f(x)$ decrease as x increases.
>
> The graph of an *increasing* function *climbs* as we move from left to right.
> The graph of a *decreasing* function *falls* as we move from left to right.

Increasing

Decreasing

Figure 1.7: Increasing and decreasing functions

Proportionality

A common functional relationship occurs when one quantity is *proportional* to another. For example, the area, A, of a circle is proportional to the square of the radius, r, because

$$A = f(r) = \pi r^2.$$

> We say y is (directly) **proportional** to x if there is a nonzero constant k such that
>
> $$y = kx.$$
>
> This k is called the constant of proportionality.

We also say that one quantity is *inversely proportional* to another if one is proportional to the reciprocal of the other. For example, the speed, v, at which you make a 50-mile trip is inversely proportional to the time, t, taken, because v is proportional to $1/t$:

$$v = 50 \left(\frac{1}{t} \right) = \frac{50}{t}.$$

Exercises and Problems for Section 1.1

Exercises

1. The population of a city, P, in millions, is a function of t, the number of years since 1950, so $P = f(t)$. Explain the meaning of the statement $f(35) = 12$ in terms of the population of this city.

2. The pollutant PCB (polychlorinated biphenyl) affects the thickness of pelican eggs. Thinking of the thickness, T, of the eggs, in mm, as a function of the concentration, P, of PCBs in ppm (parts per million), we have $T = f(P)$. Explain the meaning of $f(200)$ in terms of thickness of pelican eggs and concentration of PCBs.

3. The value of a car, $V = f(a)$, in thousands of dollars, is a function of the age of the car, a, in years.

   (a) Interpret the statement $f(5) = 6$
   (b) Sketch a possible graph of V against a. Is f an increasing or decreasing function? Explain.
   (c) Explain the significance of the horizontal and vertical intercepts in terms of the value of the car.

4. Match each story to the most appropriate graph in Figure 1.8. Write a story for the remaining graph.

   (a) During my journey, I had a flat tire. Then after fixing the flat tire, I had to speed up to avoid being late.
   (b) My car broke down and I parked it at the road side.
   (c) As soon as I'd dropped off the package, I turned around and drove home.

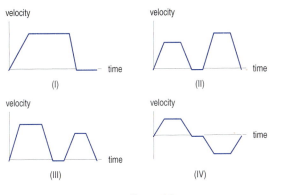

Figure 1.8

For Exercises 5–7, determine the slope and the y-intercept of the line whose equation is given.

5. $7y + 12x - 2 = 0$ 6. $-4y + 2x + 8 = 0$

7. $12x = 6y + 4$

For Exercises 8–10, find the equation of the line that passes through the given points.

8. $(0, 0)$ and $(1, 1)$ 9. $(0, 2)$ and $(2, 3)$

10. $(-2, 1)$ and $(2, 3)$

For Exercises 11–13, use the facts that parallel lines have equal slopes and that the slopes of perpendicular lines are negative reciprocals of one another.

11. Find the equation of the line through the point $(2, 1)$ which is perpendicular to the line $y = 5x - 3$.

12. Find the equations of the lines through the point $(1, 5)$ that are parallel to and perpendicular to the line with equation $y + 4x = 7$.

13. Find the equations of the lines through the point (a, b) that are parallel and perpendicular to the line $y = mx + c$, assuming $m \neq 0$.

14. Match the graphs in Figure 1.9 with the following equations. (Note that the x and y scales may be unequal.)

   (a) $y = x - 5$ (b) $-3x + 4 = y$
   (c) $5 = y$ (d) $y = -4x - 5$
   (e) $y = x + 6$ (f) $y = x/2$

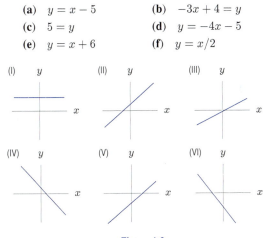

Figure 1.9

15. Match the graphs in Figure 1.10 with the following equations. (Note that the x and y scales may be unequal.)

   (a) $y = -2.72x$ (b) $y = 0.01 + 0.001x$
   (c) $y = 27.9 - 0.1x$ (d) $y = 0.1x - 27.9$
   (e) $y = -5.7 - 200x$ (f) $y = x/3.14$

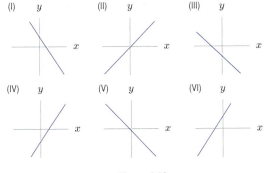

Figure 1.10

For Exercises 16–18, give the approximate domain and range of each function. Assume the entire graph is shown.

16.

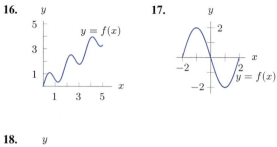

17.

18.

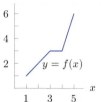

Find domain and range in Exercises 19–20.

19. $y = x^2 + 2$

20. $y = \dfrac{1}{x^2 + 2}$

21. If $f(t) = \sqrt{t^2 - 16}$, find all values of t for which $f(t)$ is a real number. Solve $f(t) = 3$.

22. If $g(x) = (4 - x^2)/(x^2 + x)$, find the domain of $g(x)$. Solve $g(x) = 0$.

In Exercises 23–25, write a formula representing the function.

23. The strength, S, of a beam is proportional to the square of its thickness, h.

24. The energy, E, expended by a swimming dolphin is proportional to the cube of the speed, v, of the dolphin.

25. The number of animal species, N, of a certain body length, l, is inversely proportional to the square of l.

Problems

26. A car starts out slowly and then goes faster and faster until a tire blows out. Sketch a possible graph of the distance the car has traveled as a function of time.

27. An object is put outside on a cold day at time $t = 0$. Its temperature, $H = f(t)$, in °C, is graphed in Figure 1.11.

(a) What does the statement $f(30) = 10$ mean in terms of temperature? Include units for 30 and for 10 in your answer.

(b) Explain what the vertical intercept, a, and the horizontal intercept, b, represent in terms of temperature of the object and time outside.

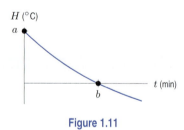

Figure 1.11

28. A flight from Dulles Airport in Washington, DC, to La-Guardia Airport in New York City has to circle La-Guardia several times before being allowed to land. Plot a graph of the distance of the plane from Washington, DC, against time, from the moment of takeoff until landing.

29. You drive at a constant speed from Chicago to Detroit, a distance of 275 miles. About 120 miles from Chicago you pass through Kalamazoo, Michigan. Sketch a graph of your distance from Kalamazoo as a function of time.

30. The monthly charge for a waste collection service is $32 for 100 kg of waste and is $48 for 180 kg of waste.

(a) Find a linear formula for the cost, C, of waste collection as a function of the number of kilograms of waste, w.

(b) What is the slope of the line found in part (a)? Give units with your answer and interpret it in terms of the cost of waste collection.

(c) What is the vertical intercept of the line found in part (a)? Give units with your answer and interpret it in terms of the cost of waste collection.

31. Residents of the town of Maple Grove who are connected to the municipal water supply are billed a fixed amount yearly plus a charge for each cubic foot of water used. A household using 1000 cubic feet was billed $90, while one using 1600 cubic feet was billed $105.

(a) What is the charge per cubic foot?

(b) Write an equation for the total cost of a resident's water as a function of cubic feet of water used.

(c) How many cubic feet of water used would lead to a bill of $130?

32. The graph of Fahrenheit temperature, °F, as a function of Celsius temperature, °C, is a line. You know that 212°F and 100°C both represent the temperature at which water boils. Similarly, 32°F and 0°C both represent water's freezing point.

(a) What is the slope of the graph?

(b) What is the equation of the line?

(c) Use the equation to find what Fahrenheit temperature corresponds to 20°C.

(d) What temperature is the same number of degrees in both Celsius and Fahrenheit?

33. The table gives the average weight, w, in pounds, of American men in their sixties for various heights, h, in inches.[1]

   (a) How do you know that the data in this table could represent a linear function?

   (b) Find weight, w, as a linear function of height, h. What is the slope of the line? What are the units for the slope?

   (c) Find height, h, as a linear function of weight, w. What is the slope of the line? What are the units for the slope?

h (inches)	68	69	70	71	72	73	74	75
w (pounds)	166	171	176	181	186	191	196	201

34. For tax purposes, you may have to report the value of your assets, such as cars or refrigerators. The value you report drops with time. "Straight-line depreciation" assumes that the value is a linear function of time. If a $950 refrigerator depreciates completely in seven years, find a formula for its value as a function of time.

35. The demand function for a certain product, $q = D(p)$, is linear, where p is the price per item in dollars and q is the quantity demanded. If p increases by $5, market research shows that q drops by two items. In addition, 100 items are purchased if the price is $550.

   (a) Find a formula for

      (i) q as a linear function of p

      (ii) p as a linear function of q

   (b) Draw a graph with q on the horizontal axis.

36. For small changes in temperature, the formula for the expansion of a metal rod under a change in temperature is:

$$l - l_0 = a l_0 (t - t_0),$$

where l is the length of the object at temperature t, and l_0 is the length at temperature t_0, and a is a constant which depends on the type of metal.

   (a) Express l as a linear function of t. Find the slope and vertical intercept in terms of l_0, t_0, and a.

   (b) A rod is 100 cm long at $60°$ F and made of a metal with $a = 10^{-5}$. Write an equation giving the length of this rod at temperature t.

   (c) What does the sign of the slope tell you about the expansion of a metal under a change in temperature?

37. When a cold yam is put into a hot oven to bake, the temperature of the yam rises. The rate, R (in degrees per minute), at which the temperature of the yam rises is governed by Newton's Law of Heating, which says that the rate is proportional to the temperature difference between the yam and the oven. If the oven is at $350°$ F and the temperature of the yam is $H°$ F,

   (a) Write a formula giving R as a function of H.

   (b) Sketch the graph of R against H.

38. A body of mass m is falling downward with velocity v. Newton's Second Law of Motion, $F = ma$, says that the net downward force, F, on the body is proportional to its downward acceleration, a. The net force, F, consists of the force due to gravity, F_g, which acts downward, minus the air resistance, F_r, which acts upward. The force due to gravity is mg, where g is a constant. Assume the air resistance is proportional to the velocity of the body.

   (a) Write an expression for the net force, F, as a function of the velocity, v.

   (b) Write a formula giving a as a function of v.

   (c) Sketch a against v.

39. When Galileo was formulating the laws of motion, he considered the motion of a body starting from rest and falling under gravity. He originally thought that the velocity of such a falling body was proportional to the distance it had fallen. What do the experimental data in Table 1.3 tell you about Galileo's hypothesis? What alternative hypothesis is suggested by the two sets of data in Table 1.3 and Table 1.4?

Table 1.3

Distance (ft)	0	1	2	3	4
Velocity (ft/sec)	0	8	11.3	13.9	16

Table 1.4

Time (sec)	0	1	2	3	4
Velocity (ft/sec)	0	32	64	96	128

1.2 EXPONENTIAL FUNCTIONS

Population Growth

The population of Mexico in the early 1980s is given in Table 1.5. To see how the population is growing, we look at the increase in population in the third column. If the population had been growing linearly, all the numbers in the third column would be the same.

---

[1] Adapted from "Average Weight of Americans by Height and Age", *The World Almanac* (New Jersey: Funk and Wagnalls, 1992), p. 956.

Table 1.5 *Population of Mexico (estimated), 1980–1986*

Year	Population (millions)	Change in population (millions)
1980	67.38	
		1.75
1981	69.13	
		1.80
1982	70.93	
		1.84
1983	72.77	
		1.89
1984	74.66	
		1.94
1985	76.60	
		1.99
1986	78.59	

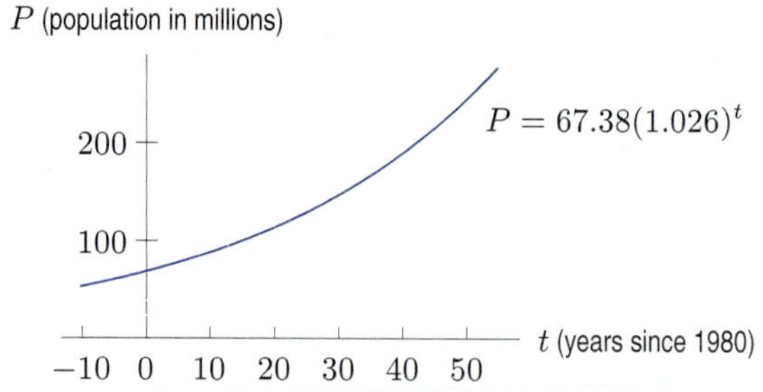

P (population in millions)

$$P = 67.38(1.026)^t$$

Figure 1.12: Population of Mexico (estimated): Exponential growth

Suppose we divide each year's population by the previous year's population. For example,

$$\frac{\text{Population in 1981}}{\text{Population in 1980}} = \frac{69.13 \text{ million}}{67.38 \text{ million}} = 1.026$$

$$\frac{\text{Population in 1982}}{\text{Population in 1981}} = \frac{70.93 \text{ million}}{69.13 \text{ million}} = 1.026.$$

The fact that both calculations give 1.026 shows the population grew by about 2.6% between 1980 and 1981 *and* between 1981 and 1982. Similar calculations for other years show that the population grew by a factor of about 1.026, or 2.6%, every year. Whenever we have a constant growth factor (here 1.026), we have exponential growth. The population t years after 1980 is given by the *exponential* function

$$P = 67.38(1.026)^t.$$

If we assume that the formula holds for 50 years, the population graph has the shape shown in Figure 1.12. Since the population is growing faster and faster as time goes on, the graph is bending upward; we say it is *concave up*. Even exponential functions which climb slowly at first, such as this one, eventually climb extremely quickly.

> To recognize that a table of t and P values comes from an exponential function $P = P_0 a^t$, look for ratios of P values that are constant for equally spaced t values.

Concavity

We have used the term concave up[2] to describe the graph in Figure 1.12. In words:

> The graph of a function is **concave up** if it bends upward as we move left to right; it is **concave down** if it bends downward. (See Figure 1.13.) A line is neither concave up nor concave down.

Concave up

Concave down

Figure 1.13: Concavity of a graph

Elimination of a Drug from the Body

Now we look at a quantity which is decreasing exponentially instead of increasing. When a patient is

---

[2]In Chapter 2 we consider concavity in more depth.

given medication, the drug enters the bloodstream. As the drug passes through the liver and kidneys, it is metabolized and eliminated at a rate that depends on the particular drug. For the antibiotic ampicillin, approximately 40% of the drug is eliminated every hour. A typical dose of ampicillin is 250 mg. Suppose $Q = f(t)$, where Q is the quantity of ampicillin, in mg, in the bloodstream at time t hours since the drug was given. At $t = 0$, we have $Q = 250$. Since every hour the amount remaining is 60% of the previous amount, we have

$$f(0) = 250$$
$$f(1) = 250(0.6)$$
$$f(2) = (250(0.6))(0.6) = 250(0.6)^2,$$

and after t hours,

$$Q = f(t) = 250(0.6)^t.$$

This is an *exponential decay function*. Some values of the function are in Table 1.6; its graph is in Figure 1.14.

Notice the way the function in Figure 1.14 is decreasing. Each hour a smaller quantity of the drug is removed than in the previous hour. This is because as time passes, there is less of the drug in the body to be removed. Compare this to the exponential growth in Figure 1.12, where each step upward is larger than the previous one. Notice, however, that both graphs are concave up.

Table 1.6

t (hours)	Q (mg)
0	250
1	150
2	90
3	54
4	32.4
5	19.4

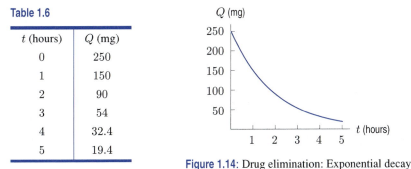

Figure 1.14: Drug elimination: Exponential decay

The General Exponential Function

> We say P is an **exponential function** of t with base a if
>
> $$P = P_0 a^t,$$
>
> where P_0 is the initial quantity (when $t = 0$) and a is the factor by which P changes when t increases by 1.
> If $a > 1$, we have exponential growth; if $0 < a < 1$, we have exponential decay.

Provided $a > 0$, the largest possible domain for the exponential function is all real numbers. The reason we do not want $a \leq 0$ is that, for example, we cannot define $a^{1/2}$ if $a < 0$. Also, we do not usually have $a = 1$, since $P = P_0 1^t = P_0$ is then a constant function.

The value of a is closely related to the percent growth (or decay) rate. For example, if $a = 1.03$, then P is growing at 3%; if $a = 0.94$, then P is decaying at 6%.

Example 1   Suppose that $Q = f(t)$ is an exponential function of t. If $f(20) = 88.2$ and $f(23) = 91.4$:

(a)   Find the base.   (b)   Find the growth rate.   (c)   Evaluate $f(25)$.

Solution   (a) Let

$$Q = Q_0 a^t.$$

Substituting $t = 20, Q = 88.2$ and $t = 23, Q = 91.4$ gives two equations for Q_0 and a:

$$88.2 = Q_0 a^{20} \quad \text{and} \quad 91.4 = Q_0 a^{23}.$$

Dividing the two equations enables us to eliminate Q_0:

$$\frac{91.4}{88.2} = \frac{Q_0 a^{23}}{Q_0 a^{20}} = a^3.$$

Solving for the base, a, gives

$$a = \left(\frac{91.4}{88.2}\right)^{1/3} = 1.012.$$

(b) Since $a = 1.012$, the growth rate is $0.012 = 1.2\%$.

(c) We want to evaluate $f(25) = Q_0 a^{25} = Q_0(1.012)^{25}$. First we find Q_0 from the equation

$$88.2 = Q_0(1.012)^{20}.$$

Solving gives $Q_0 = 69.5$. Thus,

$$f(25) = 69.5(1.012)^{25} = 93.6.$$

Half-Life and Doubling Time

Radioactive substances, such as uranium, decay exponentially. A certain percentage of the mass disintegrates in a given unit of time; the time it takes for half the mass to decay is called the *half-life* of the substance.

A well-known radioactive substance is carbon-14, which is used to date organic objects. When a piece of wood or bone was part of a living organism, it accumulated small amounts of radioactive carbon-14. Once the organism dies, it no longer picks up carbon-14. Using the half-life of carbon-14 (about 5730 years), we can estimate the age of the object. We use the following definitions:

> The **half-life** of an exponentially decaying quantity is the time required for the quantity to be reduced by a factor of one half.
>
> The **doubling time** of an exponentially increasing quantity is the time required for the quantity to double.

The Family of Exponential Functions

The formula $P = P_0 a^t$ gives a family of exponential functions with positive parameters P_0 (the initial quantity) and a (the base, or growth/decay factor). The base tells us whether the function is increasing ($a > 1$) or decreasing ($0 < a < 1$). Since a is the factor by which P changes when t is increased by 1, large values of a mean fast growth; values of a near 0 mean fast decay. (See Figures 1.15 and 1.16.) All members of the family $P = P_0 a^t$ are concave up.

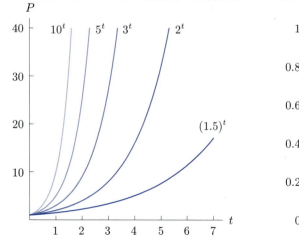

Figure 1.15: Exponential growth: $P = a^t$, for $a > 1$

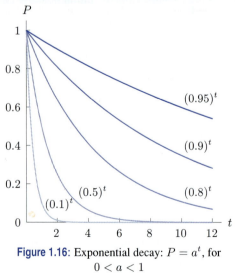

Figure 1.16: Exponential decay: $P = a^t$, for $0 < a < 1$

Example 2   Figure 1.17 is the graph of three exponential functions. What can you say about the values of the six constants, a, b, c, d, p, q?

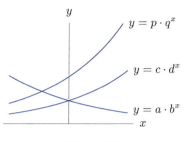

Figure 1.17

Solution   All the constants are positive. Since a, c, p represent y-intercepts, we see that $a = c$ because these graphs intersect on the y-axis. In addition, $a = c < p$, since $y = p \cdot q^x$ crosses the y-axis above the other two.

Since $y = a \cdot b^x$ is decreasing, we have $0 < b < 1$. The other functions are increasing, so $1 < d$ and $1 < q$.

Exponential Functions with Base e

The most frequently used base for an exponential function is the famous number $e = 2.71828\ldots$. This base is used so often that you will find an e^x button on most scientific calculators. At first glance, this is all somewhat mysterious. Why is it convenient to use the base $2.71828\ldots$? The full answer to that question must wait until Chapter 3, where we show that many calculus formulas come out neatly when e is used as the base. We often use the following result:

Any **exponential growth** function can be written, for some $a > 1$ and $k > 0$, in the form

$$P = P_0 a^t \quad \text{or} \quad P = P_0 e^{kt}$$

and any **exponential decay** function can be written, for some $0 < a < 1$ and $k > 0$, as

$$Q = Q_0 a^t \quad \text{or} \quad Q = Q_0 e^{-kt},$$

where P_0 and Q_0 are the initial quantities.

We say that P and Q are growing or decaying at a *continuous*[3] rate of k. (For example, $k = 0.02$ corresponds to a continuous rate of 2%.)

Example 3   Convert the functions $P = e^{0.5t}$ and $Q = 5e^{-0.2t}$ into the form $y = y_0 a^t$. Use the results to explain the shape of the graphs in Figures 1.18 and 1.19.

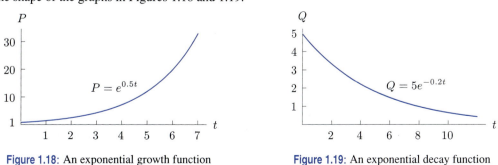

Figure 1.18: An exponential growth function   **Figure 1.19:** An exponential decay function

[3]The reason that k is called the continuous rate is explored in detail in Chapter 11.

Solution   We have
$$P = e^{0.5t} = (e^{0.5})^t = (1.65)^t.$$
Thus, P is an exponential growth function with $P_0 = 1$ and $a = 1.65$. The function is increasing and itas graph is concave up, similar to those in Figure 1.15. Also,
$$Q = 5e^{-0.2t} = 5(e^{-0.2})^t = 5(0.819)^t,$$
so Q is an exponential decay function with $Q_0 = 5$ and $a = 0.819$. The function is decreasing and its graph is concave up, similar to those in Figure 1.16.

Example 4   The quantity, Q, of a drug in a patient's body at time t is represented for positive constants S and k by the function $Q = S(1 - e^{-kt})$. For $t \geq 0$, describe how Q changes with time. What does S represent? How does the graph of Q relate to the exponential decay graph in Figure 1.19?

Solution   The graph of Q is shown in Figure 1.20. Initially none of the drug is present, but the quantity increases with time. Since the graph is concave down, the quantity increases at a decreasing rate. This is realistic because as the quantity of the drug in the body increases, so does the rate at which the body excretes the drug. Thus, we expect the quantity to level off. Figure 1.20 shows that S is the saturation level.

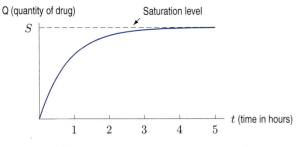

Figure 1.20: Buildup of the quantity of a drug in body

The graph in Figure 1.20 looks like an exponential decay function, but upside down. Since $Q = S - Se^{-kt}$, we have
$$S - Q = Se^{-kt}.$$
Thus, the difference between the saturation level, S, and the quantity in the blood, Q, is decaying exponentially. The graph in Figure 1.20 can be obtained from the graph in Figure 1.19 by reflection about the t-axis and moving up by a distance S.

Exercises and Problems for Section 1.2

Exercises

In Exercises 1–4, decide whether the graph is concave up, concave down, or neither.

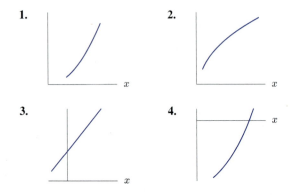

The functions in Exercises 5–8 represent exponential growth or decay. What is the initial quantity? What is the growth rate? (State if the growth rate is continuous.)

5. $P = 5(1.07)^t$

6. $P = 7.7(0.92)^t$

7. $P = 3.2e^{0.03t}$

8. $P = 15e^{-0.06t}$

9. A town has a population of 1000 people at time $t = 0$. In each of the following cases, write a formula for the population, P, of the town as a function of year t.

(a) The population increases by 50 people a year.
(b) The population increases by 5% a year.

10. An air-freshener starts with 30 grams and evaporates. In each of the following cases, write a formula for the quantity, Q grams, of air-freshener remaining t days after the start and sketch a graph of the function. The decrease is:

(a) 2 grams a day   (b) 12% a day

11. Identify the x-intervals on which the function graphed in Figure 1.21 is:

(a) Increasing and concave up
(b) Increasing and concave down
(c) Decreasing and concave up
(d) Decreasing and concave down

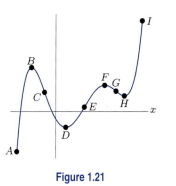

Figure 1.21

Problems

12. When a new product is advertised, more and more people try it. However, the rate at which new people try it slows as time goes on.

(a) Sketch a graph of the total number of people who have tried such a product against time.
(b) What do you know about the concavity of the graph?

13. Sketch reasonable graphs for the following. Pay particular attention to the concavity of the graphs.

(a) The total revenue generated by a car rental business, plotted against the amount spent on advertising.
(b) The temperature of a cup of hot coffee standing in a room, plotted as a function of time.

14. (a) Which (if any) of the functions in the following table could be linear? Find formulas for those functions.
(b) Which (if any) of these functions could be exponential? Find formulas for those functions.

x	$f(x)$	$g(x)$	$h(x)$
-2	12	16	37
-1	17	24	34
0	20	36	31
1	21	54	28
2	18	81	25

15. In 1999, the world's population reached 6 billion and was increasing at a rate of 1.3% per year. Assume that this growth rate remains constant. (In fact, the growth rate has decreased since 1987.)

(a) Write a formula for the world population (in billions) as a function of the number of years since 1999.
(b) Use your formula to estimate the population of the world in the year 2020.
(c) Sketch a graph of world population as a function of years since 1999. Use the graph to estimate the doubling time of the population of the world.

16. A photocopy machine can reduce copies to 80% of their original size. By copying an already reduced copy, further reductions can be made.

(a) If a page is reduced to 80%, what percent enlargement is needed to return it to its original size?
(b) Estimate the number of times in succession that a page must be copied to make the final copy less than 15% of the size of the original.

17. (a) Niki invested $10,000 in the stock market. The investment was a loser, declining in value 10% per year each year for 10 years. How much was the investment worth after 10 years?
(b) After 10 years, the stock began to gain value at 10% per year. After how long will the investment regain its initial value ($10,000)?

Give a possible formula for the functions in Problems 18–21.

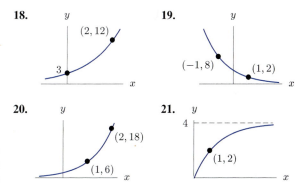

18.

19.

20.

21.

In Problems 22–23, let $f(t) = Q_0 a^t = Q_0(1 + r)^t$.
(a) Find the base, a.
(b) Find the percentage growth rate, r.

22. $f(5) = 75.94$ and $f(7) = 170.86$

23. $f(0.02) = 25.02$ and $f(0.05) = 25.06$

Write the functions in Problems 24–27 in the form $P = P_0 a^t$. Which represent exponential growth and which represent exponential decay?

24. $P = 15e^{0.25t}$

25. $P = 2e^{-0.5t}$

26. $P = P_0 e^{0.2t}$

27. $P = 7e^{-\pi t}$

28. (a) A population, P, grows at a continuous rate of 2% a year and starts at 1 million. Write P in the form $P = P_0 e^{kt}$, with P_0, k constants.
  (b) Plot the population in part (a) against time.

29. Aircrafts require longer takeoff distances, called takeoff rolls, at high altitude airports because of diminished air density. The table shows how the takeoff roll for a certain light airplane depends on the airport elevation. (Takeoff rolls are also strongly influenced by air temperature; the data shown assume a temperature of $0°$ C.) Determine a formula for this particular aircraft that gives the takeoff roll as an exponential function of airport elevation.

Elevation (ft)	Sea level	1000	2000	3000	4000
Takeoff roll (ft)	670	734	805	882	967

30. Table 1.7 shows some values of a linear function f and an exponential function g. Find exact values (not decimal approximations) for each of the missing entries.

Table 1.7

x	0	1	2	3	4
$f(x)$	10	?	20	?	?
$g(x)$	10	?	20	?	?

31. When the Olympic Games were held outside Mexico City in 1968, there was much discussion about the effect the high altitude (7340 feet) would have on the athletes. Assuming air pressure decays exponentially by 0.4% every 100 feet, by what percentage is air pressure reduced by moving from sea level to Mexico City?

32. During 1988, Nicaragua's inflation rate averaged 1.3% a day. This means that, on average, prices went up by 1.3% from one day to the next.

  (a) By what percentage did Nicaraguan prices increase in June of 1988?
  (b) What was Nicaragua's annual inflation rate during 1988?

33. Each of the functions g, h, k in Table 1.8 is increasing, but each increases in a different way. Which of the graphs in Figure 1.22 best fits each function?

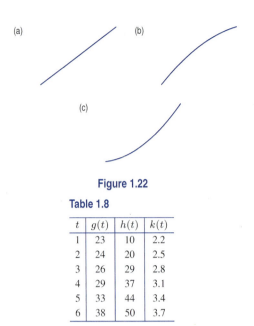

Figure 1.22

Table 1.8

t	$g(t)$	$h(t)$	$k(t)$
1	23	10	2.2
2	24	20	2.5
3	26	29	2.8
4	29	37	3.1
5	33	44	3.4
6	38	50	3.7

34. A certain region has a population of 10,000,000 and an annual growth rate of 2%. Estimate the doubling time by guessing and checking.

35. (a) The half-life of radium-226 is 1620 years. Write a formula for the quantity, Q, of radium left after t years, if the initial quantity is Q_0.
  (b) What percentage of the original amount of radium is left after 500 years?

36. In the early 1960s, radioactive strontium-90 was released during atmospheric testing of nuclear weapons and got into the bones of people alive at the time. If the half-life of strontium-90 is 29 years, what fraction of the strontium-90 absorbed in 1960 remained in people's bones in 1990?

37. Estimate graphically the doubling time of the exponentially growing population shown in Figure 1.23. Check that the doubling time is independent of where you start on the graph. Show algebraically that if $P = P_0 a^t$ doubles between time t and time $t + d$, then d is the same number for any t.

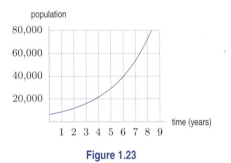

Figure 1.23

38. Match the functions $h(s)$, $f(s)$, and $g(s)$, whose values are in Table 1.9, with the formulas

$$y = a(1.1)^s, \quad y = b(1.05)^s, \quad y = c(1.03)^s,$$

assuming a, b, and c are constants. Note that the function values have been rounded to two decimal places.

Table 1.9

s	$h(s)$	s	$f(s)$	s	$g(s)$
2	1.06	1	2.20	3	3.47
3	1.09	2	2.42	4	3.65
4	1.13	3	2.66	5	3.83
5	1.16	4	2.93	6	4.02
6	1.19	5	3.22	7	4.22

39. The median price, P, of a home rose from \$50,000 in 1970 to \$100,000 in 1990. Let t be the number of years since 1970.

(a) Assume the increase in housing prices has been linear. Give an equation for the line representing price, P, in terms of t. Use this equation to complete column (a) of Table 1.10. Use units of \$1000.

(b) If instead the housing prices have been rising exponentially, determine an equation of the form $P = P_0 a^t$ which would represent the change in housing prices from 1970–1990, and complete column (b) of Table 1.10.

(c) On the same set of axes, sketch the functions represented in column (a) and column (b) of Table 1.10.

(d) Which model for the price growth do you think is more realistic?

Table 1.10

t	(a) Linear growth price in \$1000 units	(b) Exponential growth price in \$1000 units
0	50	50
10		
20	100	100
30		
40		

1.3 NEW FUNCTIONS FROM OLD

Shifts and Stretches

The graph of a constant multiple of a given function is easy to visualize: each y-value is stretched or shrunk by that multiple. For example, consider the function $f(x)$ and its multiples $y = 3f(x)$ and $y = -2f(x)$. Their graphs are shown in Figure 1.24. The factor 3 in the function $y = 3f(x)$ stretches each $f(x)$ value by multiplying it by 3; the factor -2 in the function $y = -2f(x)$ stretches $f(x)$ by multiplying by 2 and reflects it about the x-axis. You can think of the multiples of a given function as a family of functions.

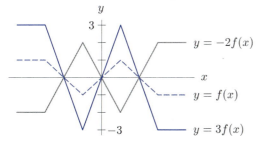

Figure 1.24: Multiples of the function $f(x)$

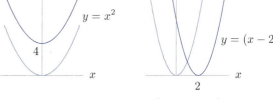

Figure 1.25: Graphs of $y = x^2$ with $y = x^2 + 4$ and $y = (x-2)^2$

It is also easy to create families of functions by shifting graphs. For example, $y - 4 = x^2$ is the same as $y = x^2 + 4$, which is the graph of $y = x^2$ shifted up by 4. Similarly, $y = (x-2)^2$ is the graph of $y = x^2$ shifted right by 2. (See Figure 1.25.)

- Multiplying a function by a constant, c, stretches the graph vertically (if $c > 1$) or shrinks the graph vertically (if $0 < c < 1$). A negative sign (if $c < 0$) reflects the graph about the x-axis, in addition to shrinking or stretching.
- Replacing y by $(y - k)$ moves a graph up by k (down if k is negative).
- Replacing x by $(x - h)$ moves a graph to the right by h (to the left if h is negative).

Composite Functions

If oil is spilled from a tanker, the area of the oil slick grows with time. Suppose that the oil slick is always a perfect circle. Then the area, A, of the oil slick is a function of its radius, r:

$$A = f(r) = \pi r^2.$$

The radius is also a function of time, because the radius increases as more oil spills. Thus, the area, being a function of the radius, is also a function of time. If, for example, the radius is given by

$$r = g(t) = 1 + t,$$

then the area is given as a function of time by substitution:

$$A = \pi r^2 = \pi(1 + t)^2.$$

We are thinking of A as a *composite function* or a "function of a function," which is written

$$A = \underbrace{f(g(t))} = \pi(g(t))^2 = \pi(1 + t)^2.$$

Composite function;
f is outside function,
g is inside function

To calculate A using the formula $\pi(1 + t)^2$, the first step is to find $1 + t$, and the second step is to square and multiply by π. The first step corresponds to the inside function $g(t) = 1 + t$, and the second step corresponds to the outside function $f(r) = \pi r^2$.

Example 1   If $f(x) = x^2$ and $g(x) = x + 1$, find each of the following:
(a) $f(g(2))$ (b) $g(f(2))$ (c) $f(g(x))$ (d) $g(f(x))$

Solution   (a) Since $g(2) = 3$, we have $f(g(2)) = f(3) = 9$.
(b) Since $f(2) = 4$, we have $g(f(2)) = g(4) = 5$. Notice that $f(g(2)) \neq g(f(2))$.
(c) $f(g(x)) = f(x + 1) = (x + 1)^2$.
(d) $g(f(x)) = g(x^2) = x^2 + 1$. Again, notice that $f(g(x)) \neq g(f(x))$.

Example 2   Express each of the following functions as a composition:

(a) $h(t) = (1 + t^3)^{27}$ (b) $k(y) = e^{-y^2}$ (c) $l(y) = -(e^y)^2$

Solution   In each case think about how you would calculate a value of the function. The first stage of the calculation gives you the inside function, and the second stage gives you the outside function.
(a) For $(1 + t^3)^{27}$, the first stage is cubing and adding 1, so an inside function is $g(t) = 1 + t^3$. The second stage is taking the 27th power, so an outside function is $f(y) = y^{27}$. Then

$$f(g(t)) = f(1 + t^3) = (1 + t^3)^{27}.$$

In fact, there are lots of different answers: $g(t) = t^3$ and $f(y) = (1 + y)^{27}$ is another possibility.
(b) To calculate e^{-y^2} we square y, take its negative, and then take e to that power. So if $g(y) = -y^2$ and $f(z) = e^z$, then we have

$$f(g(y)) = e^{-y^2}.$$

(c) To calculate $-(e^y)^2$, we find e^y, square it, and take the negative. Using the same definitions of f and g as in part (b), the composition is

$$g(f(y)) = -(e^y)^2.$$

Since parts (b) and (c) give different answers, we see the order in which functions are composed is important.

Odd and Even Functions: Symmetry

There is a certain symmetry apparent in the graphs of $f(x) = x^2$ and $g(x) = x^3$ in Figure 1.26. For each point (x, x^2) on the graph of f, the point $(-x, x^2)$ is also on the graph; for each point (x, x^3) on the graph of g, the point $(-x, -x^3)$ is also on the graph. The graph of $f(x) = x^2$ is symmetric about the y-axis, whereas the graph of $g(x) = x^3$ is symmetric about the origin. The graph of any polynomial involving only even powers of x has symmetry about the y-axis, while polynomials with only odd powers of x are symmetric about the origin. Consequently, any functions with these symmetry properties are called *even* and *odd*, respectively.

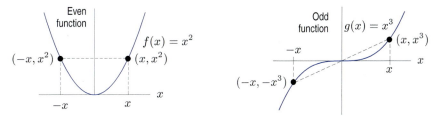

Figure 1.26: Symmetry of even and odd functions

> For any function f,
> f is an **even** function if $f(-x) = f(x)$ for all x.
> f is an **odd** function if $f(-x) = -f(x)$ for all x.

For example, $g(x) = e^{x^2}$ is even and $h(x) = x^{1/3}$ is odd. However, many functions do not have any symmetry and are neither even nor odd.

Inverse Functions

On August 18, 1989, the runner Arturo Barrios of Mexico set a world record for the 10,000-meter race. His times, in seconds, at 2000-meter intervals are recorded in Table 1.11, where $t = f(d)$ is the number of seconds Barrios took to complete the first d meters of the race. For example, Barrios ran the first 4000 meters in 650.10 seconds, so $f(4000) = 650.10$. The function f was useful to athletes planning to compete with Barrios.

Let us now change our point of view and ask for distances rather than times. If we ask how far Barrios ran during the first 650.10 seconds of his race, the answer is clearly 4000 meters. Going backwards in this way from numbers of seconds to numbers of meters gives f^{-1}, the *inverse function*[4] of f. We write $f^{-1}(650.10) = 4000$. Thus, $f^{-1}(t)$ is the number of meters that Barrios ran during the first t seconds of his race. See Table 1.12 which contains values of f^{-1}.

The independent variable for f is the dependent variable for f^{-1}, and vice versa. The domains and ranges of f and f^{-1} are also interchanged. The domain of f is all distances d such that $0 \leq d \leq 10000$, which is the range of f^{-1}. The range of f is all times t, such that $0 \leq t \leq 1628.23$, which is the domain of f^{-1}.

Table 1.11   *Barrios's running time*

d (meters)	$t = f(d)$ (seconds)
0	0.00
2000	325.90
4000	650.10
6000	975.50
8000	1307.00
10000	1628.23

Table 1.12   *Distance run by Barrios*

t (seconds)	$d = f^{-1}(t)$ (meters)
0.00	0
325.90	2000
650.10	4000
975.50	6000
1307.00	8000
1628.23	10000

[4]The notation f^{-1} represents the inverse function, which is not the same as the reciprocal, $1/f$.

Which Functions Have Inverses?

If a function has an inverse, we say it is *invertible*. Let's look at a function which is not invertible. Consider the flight of the Mercury spacecraft *Freedom 7*, which carried Alan Shepard, Jr. into space in May 1961. Shepard was the first American to journey into space. After launch, his spacecraft rose to an altitude of 116 miles, and then came down into the sea. The function $f(t)$ giving the altitude in miles t minutes after lift-off does not have an inverse. To see why not, try to decide on a value for $f^{-1}(100)$, which should be the time when the altitude of the spacecraft was 100 miles. However, there are two such times, one when the spacecraft was ascending and one when it was descending. (See Figure 1.27.)

The reason the altitude function does not have an inverse is that the altitude has the same value for two different times. The reason the Barrios time function did have an inverse is that each running time, t, corresponds to a unique distance, d.

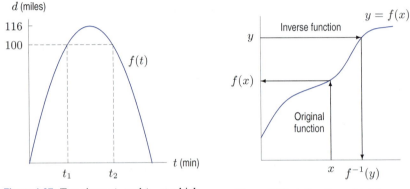

Figure 1.27: Two times, t_1 and t_2, at which altitude of spacecraft is 100 miles

Figure 1.28: A function which has an inverse

Figure 1.28 suggests when an inverse exists. The original function, f, takes us from an x-value to a y-value, as shown in Figure 1.28. Since having an inverse means there is a function going from a y-value to an x-value, the crucial question is whether we can get back. In other words, does each y-value correspond to a unique x-value? If so, there's an inverse; if not, there isn't. This principle may be stated geometrically, as follows:

> A function has an inverse if (and only if) its graph intersects any horizontal line at most once.

For example, the function $f(x) = x^2$ does not have an inverse because many horizontal lines intersect the parabola twice.

Definition of an Inverse Function

If the function f is invertible, its inverse is defined as follows:

$$f^{-1}(y) = x \quad \text{means} \quad y = f(x).$$

Formulas for Inverse Functions

If a function is defined by a formula, it is sometimes possible to find a formula for the inverse function. In Section 1.1, we looked at the snow tree cricket, whose chirp rate, C, in chirps per minute, is approximated at the temperature, T, in degrees Fahrenheit, by the formula

$$C = f(T) = 4T - 160.$$

So far we have used this formula to predict the chirp rate from the temperature. But it is also possible to use this formula backwards to calculate the temperature from the chirp rate.

Example 3 Find the formula for the function giving temperature in terms of the number of cricket chirps per minute; that is, find the inverse function f^{-1} such that

$$T = f^{-1}(C).$$

Solution Since C is an increasing function, f is invertible. We know $C = 4T - 160$. We solve for T, giving

$$T = \frac{C}{4} + 40,$$

so

$$f^{-1}(C) = \frac{C}{4} + 40.$$

Graphs of Inverse Functions

The function $f(x) = x^3$ is increasing everywhere and so has an inverse. To find the inverse, we solve

$$y = x^3$$

for x, giving

$$x = y^{1/3}.$$

The inverse function is

$$f^{-1}(y) = y^{1/3}$$

or, if we want to call the independent variable x,

$$f^{-1}(x) = x^{1/3}.$$

The graphs of $y = x^3$ and $y = x^{1/3}$ are shown in Figure 1.29. Notice that these graphs are the reflections of one another about the line $y = x$. For example, $(8, 2)$ is on the graph of $y = x^{1/3}$ because $2 = 8^{1/3}$, and $(2, 8)$ is on the graph of $y = x^3$ because $8 = 2^3$. The points $(8, 2)$ and $(2, 8)$ are reflections of one another about the line $y = x$. In general, if the x- and y-axes have the same scales:

> The graph of f^{-1} is the reflection of the graph of f about the line $y = x$.

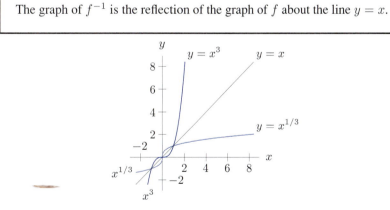

Figure 1.29: Graphs of inverse functions, $y = x^3$ and $y = x^{1/3}$, are reflections about the line $y = x$

Exercises and Problems for Section 1.3

Exercises

1. For $g(x) = x^2 + 2x + 3$, find and simplify:

    (a) $g(2 + h)$ **(b)** $g(2)$

    (c) $g(2 + h) - g(2)$

2. If $f(x) = x^2 + 1$, find and simplify:

    (a) $f(t + 1)$ **(b)** $f(t^2 + 1)$

    (c) $f(2)$ **(d)** $2f(t)$

    (e) $[f(t)]^2 + 1$

Values of $\log x$ are in Table 1.13. Because no power of 10 gives 0, $\log 0$ is undefined. The graph of $y = \log x$ is shown in Figure 1.33. The domain of $y = \log x$ is positive real numbers; the range is all real numbers. In contrast, the inverse function $y = 10^x$ has domain all real numbers and range all positive real numbers. The graph of $y = \log x$ has a vertical asymptote at $x = 0$, whereas $y = 10^x$ has a horizontal asymptote at $y = 0$.

One big difference between $y = 10^x$ and $y = \log x$ is that the exponential function grows extremely quickly whereas the log function grows extremely slowly. However, $\log x$ does go to infinity, albeit slowly, as x increases. Since $y = \log x$ and $y = 10^x$ are inverse functions, the graphs of the two functions are reflections of one another about the line $y = x$, provided the scales along the x- and y-axes are equal.

Table 1.13 *Values for $\log x$ and 10^x*

x	$\log x$	x	10^x
0	undefined	0	1
1	0	1	10
2	0.3	2	100
3	0.5	3	10^3
4	0.6	4	10^4
$\vdots$	$\vdots$	$\vdots$	$\vdots$
10	1	10	10^{10}

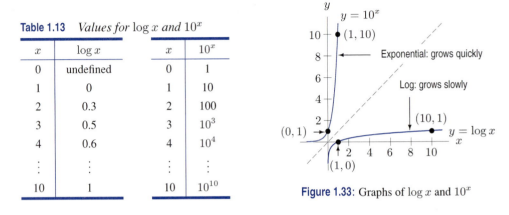

Figure 1.33: Graphs of $\log x$ and 10^x

The graph of $y = \ln x$ in Figure 1.34 has roughly the same shape as the graph of $y = \log x$. The x-intercept is $x = 1$, since $\ln 1 = 0$. The graph of $y = \ln x$ also climbs very slowly as x increases.

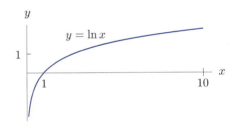

Figure 1.34: Graph of the natural logarithm

The following properties of logarithms may be deduced from the properties of exponents:

Properties of Logarithms

Note that $\log x$ and $\ln x$ are not defined when x is negative or 0.

1. $\log(AB) = \log A + \log B$
2. $\log\left(\dfrac{A}{B}\right) = \log A - \log B$
3. $\log(A^p) = p \log A$
4. $\log(10^x) = x$
5. $10^{\log x} = x$

1. $\ln(AB) = \ln A + \ln B$
2. $\ln\left(\dfrac{A}{B}\right) = \ln A - \ln B$
3. $\ln(A^p) = p \ln A$
4. $\ln e^x = x$
5. $e^{\ln x} = x$

In addition, $\log 1 = 0$ because $10^0 = 1$, and $\ln 1 = 0$ because $e^0 = 1$.

Solving Equations Using Logarithms

Logs are frequently useful when we have to solve for unknown exponents, as in the next examples.

Example 1   Find t such that $2^t = 7$.

Solution   First, notice that we expect t to be between 2 and 3 (because $2^2 = 4$ and $2^3 = 8$). To calculate t, we take logs to base 10. (Natural logs could also be used.)

$$\log(2^t) = \log 7.$$

Then use the third property of logs, which says $\log(2^t) = t \log 2$, and get:

$$t \log 2 = \log 7.$$

Using a calculator to find the logs gives

$$t = \frac{\log 7}{\log 2} \approx 2.81.$$

Example 2   Find when the population of Mexico reaches 200 million by solving $200 = 67.38(1.026)^t$.

Solution   Dividing both sides of the equation by 67.38, we get

$$\frac{200}{67.38} = (1.026)^t.$$

Now take logs of both sides:

$$\log\left(\frac{200}{67.38}\right) = \log(1.026^t).$$

Using the fact that $\log(A^t) = t \log A$, we get

$$\log\left(\frac{200}{67.38}\right) = t \log(1.026).$$

Solving this equation using a calculator to find the logs, we get

$$t = \frac{\log(200/67.38)}{\log(1.026)} \approx 42.4 \text{ years}$$

which is between $t = 42$ and $t = 43$. This value of t corresponds to the year 2022.

Example 3   The release of chlorofluorocarbons used in air conditioners and in household sprays (hair spray, shaving cream, etc.) destroys the ozone in the upper atmosphere. Currently, the amount of ozone, Q, is decaying exponentially at a continuous rate of 0.25% per year. What is the half-life of ozone?

Solution   We want to find how long it takes for half the ozone to disappear. If Q_0 is the initial quantity of ozone, then

$$Q = Q_0 e^{-0.0025t}.$$

We want to find T, the value of t making $Q = Q_0/2$, that is,

$$Q_0 e^{-0.0025T} = \frac{Q_0}{2}.$$

Dividing by Q_0 and then taking natural logs yields

$$\ln\left(e^{-0.0025T}\right) = -0.0025T = \ln\left(\frac{1}{2}\right) \approx -0.6931,$$

so

$$T \approx 277 \text{ years.}$$

The half-life of ozone is about 277 years.

Let's consider $y = 2^x$ and $y = x^3$. The close-up view in Figure 1.56(a) shows that between $x = 2$ and $x = 4$, the graph of $y = 2^x$ lies below the graph of $y = x^3$. The far-away view in Figure 1.56(b) shows that the exponential function $y = 2^x$ eventually overtakes $y = x^3$. Figure 1.56(c), which gives a very far–away view, shows that, for large x, the value of x^3 is insignificant compared to 2^x. Indeed, 2^x is growing so much faster than x^3 that the graph of 2^x appears almost vertical in comparison to the more leisurely climb of x^3.

We say that Figure 1.56(a) gives a *local* view of the functions' behavior, whereas Figure 1.56(c) gives a *global* view.

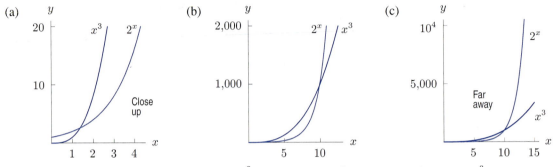

Figure 1.56: Comparison of $y = 2^x$ and $y = x^3$: Notice that $y = 2^x$ eventually dominates $y = x^3$

In fact, *every* exponential growth function eventually dominates *every* power function. Although an exponential function may be below a power function for some values of x, if we look at large enough x-values, a^x (with $a > 1$) will eventually dominate x^n, no matter what n is.

Polynomials

Polynomials are the sums of power functions with nonnegative integer exponents:

$$y = p(x) = a_n x^n + a_{n-1} x^{n-1} + \cdots + a_1 x + a_0.$$

Here n is a nonnegative integer called the *degree* of the polynomial, and $a_n, a_{n-1}, \ldots, a_1, a_0$ are constants, with leading coefficient $a_n \neq 0$. An example of a polynomial of degree $n = 3$ is

$$y = p(x) = 2x^3 - x^2 - 5x - 7.$$

In this case $a_3 = 2, a_2 = -1, a_1 = -5$, and $a_0 = -7$. The shape of the graph of a polynomial depends on its degree; typical graphs are shown in Figure 1.57. These graphs correspond to a positive coefficient for x^n; a negative leading coefficient turns the graph upside down. Notice that the quadratic "turns around" once, the cubic "turns around" twice, and the quartic (fourth degree) "turns around" three times. An nth degree polynomial "turns around" at most $n - 1$ times (where n is a positive integer), but there may be fewer turns.

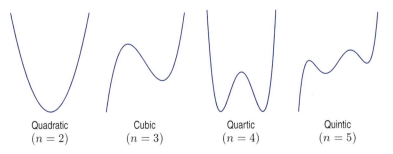

Quadratic $(n = 2)$ Cubic $(n = 3)$ Quartic $(n = 4)$ Quintic $(n = 5)$

Figure 1.57: Graphs of typical polynomials of degree n

Example 1 Find possible formulas for the polynomials whose graphs are in Figure 1.58.

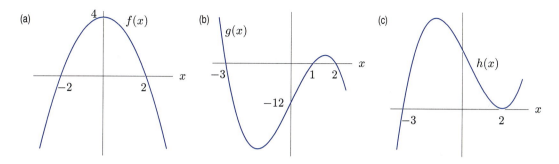

Figure 1.58: Graphs of polynomials

Solution (a) This graph appears to be a parabola, turned upside down, and moved up by 4, so

$$f(x) = -x^2 + 4.$$

The negative sign turns the parabola upside down and the $+4$ moves it up by 4. Notice that this formula does give the correct x-intercepts since $0 = -x^2 + 4$ has solutions $x = \pm 2$. These values of x are called *zeros* of f.

We can also solve this problem by looking at the x-intercepts first, which tell us that $f(x)$ has factors of $(x + 2)$ and $(x - 2)$. So

$$f(x) = k(x + 2)(x - 2).$$

To find k, use the fact that the graph has a y-intercept of 4, so $f(0) = 4$, giving

$$4 = k(0 + 2)(0 - 2),$$

or $k = -1$. Therefore, $f(x) = -(x + 2)(x - 2)$, which multiplies out to $-x^2 + 4$.

Note that $f(x) = 4 - x^4/4$ also has the same basic shape, but is flatter near $x = 0$. There are many possible answers to these questions.

(b) This looks like a cubic with factors $(x + 3)$, $(x - 1)$, and $(x - 2)$, one for each intercept:

$$g(x) = k(x + 3)(x - 1)(x - 2).$$

Since the y-intercept is -12, we have

$$-12 = k(0 + 3)(0 - 1)(0 - 2).$$

So $k = -2$, and we get the cubic polynomial

$$g(x) = -2(x + 3)(x - 1)(x - 2).$$

(c) This also looks like a cubic with zeros at $x = 2$ and $x = -3$. Notice that at $x = 2$ the graph of $h(x)$ touches the x-axis but doesn't cross it, whereas at $x = -3$ the graph crosses the x-axis. We say that $x = 2$ is a *double zero*, but that $x = -3$ is a single zero.

To find a formula for $h(x)$, imagine the graph of $h(x)$ to be slightly lower down, so that the graph has one x-intercept near $x = -3$ and two near $x = 2$, say at $x = 1.9$ and $x = 2.1$. Then a formula would be

$$h(x) \approx k(x + 3)(x - 1.9)(x - 2.1).$$

Now move the graph back to its original position. The zeros at $x = 1.9$ and $x = 2.1$ move toward $x = 2$, giving

$$h(x) = k(x + 3)(x - 2)(x - 2) = k(x + 3)(x - 2)^2.$$

The double zero leads to a repeated factor, $(x - 2)^2$. Notice that when $x > 2$, the factor $(x - 2)^2$ is positive, and when $x < 2$, the factor $(x - 2)^2$ is still positive. This reflects the fact that $h(x)$ doesn't change sign near $x = 2$. Compare this with the behavior near the single zero at $x = -3$, where h does change sign.

We cannot find k, as no coordinates are given for points off of the x-axis. Any positive value of k stretches the graph vertically but does not change the zeros, so any positive k works.

Example 2   Using a calculator or computer, sketch graphs of $y = x^4$ and $y = x^4 - 15x^2 - 15x$ for $-4 \leq x \leq 4$ and for $-20 \leq x \leq 20$. Set the y range to $-100 \leq y \leq 100$ for the first domain, and to $-100 \leq y \leq 200{,}000$ for the second. What do you observe?

Solution   From the graphs in Figure 1.59 we see that close up ($-4 \leq x \leq 4$) the graphs look different; from far away, however, they are almost indistinguishable. The reason is that the leading terms (those with the highest power of x) are the same, namely x^4, and for large values of x, the leading term dominates the other terms.

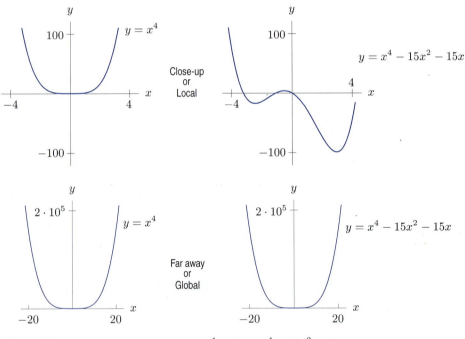

Figure 1.59: Local and global views of $y = x^4$ and $y = x^4 - 15x^2 - 15x$

When $x = \pm 20$, the differences in the values of the two functions, although large, are tiny compared with the vertical scale (-100 to $200{,}000$). See Table 1.15. So these differences can't be seen on the graph.

Table 1.15   *Numerical values of $y = x^4$ and $y = x^4 - 15x^2 - 15x$*

x	$y = x^4$	$y = x^4 - 15x^2 - 15x$	Difference
-20	160,000	154,300	5700
-15	50,625	47,475	3150
15	50,625	47,025	3600
20	160,000	153,700	6300

Rational Functions

Rational functions are ratios of polynomials, p and q:

$$f(x) = \frac{p(x)}{q(x)}.$$

Example 3   Look at a graph and explain the behavior of $y = \dfrac{1}{x^2 + 4}$.

Solution   The function is even, so the graph is symmetric about the y-axis. As x gets larger, the denominator gets larger, making the value of the function closer to 0. Thus the graph gets arbitrarily close to the x-axis as x increases without bound. See Figure 1.60.

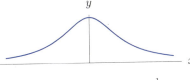

Figure 1.60: Graph of $y = \frac{1}{x^2+4}$

In the previous example, we say that $y = 0$ (i.e. the x-axis) is a *horizontal asymptote*. Writing "$\to$" to mean "tends to," we have $y \to 0$ as $x \to \infty$ and $y \to 0$ as $x \to -\infty$.

If the graph of $y = f(x)$ approaches a horizontal line $y = L$ as $x \to \infty$ or $x \to -\infty$, then the line $y = L$ is called a **horizontal asymptote**.[6] This occurs when

$$f(x) \to L \quad \text{as} \quad x \to \infty \qquad \text{or} \qquad f(x) \to L \quad \text{as} \quad x \to -\infty.$$

If the graph of $y = f(x)$ approaches the vertical line $x = K$ as $x \to K$ from one side or the other, that is, if

$$y \to \infty \quad \text{or} \quad y \to -\infty \quad \text{when} \quad x \to K,$$

then the line $x = K$ is called a **vertical asymptote**.

The graphs of rational functions may have vertical asymptotes where the denominator is zero. For example, the function in Example 3 has no vertical asymptotes as the denominator is never zero. The function in Example 4 has two vertical asymptotes corresponding to the two zeros in the denominator.

Rational functions have horizontal asymptotes if $f(x)$ approaches a finite number as $x \to \infty$ or $x \to -\infty$. We call the behavior of a function as $x \to \pm\infty$ its *end behavior*.

Example 4   Look at a graph and explain the behavior of $y = \dfrac{3x^2 - 12}{x^2 - 1}$, including end behavior.

Solution   Factoring gives

$$y = \frac{3x^2 - 12}{x^2 - 1} = \frac{3(x + 2)(x - 2)}{(x + 1)(x - 1)}$$

so $x = \pm 1$ are vertical asymptotes. If $y = 0$, then $3(x + 2)(x - 2) = 0$ or $x = \pm 2$; these are the x-intercepts. Note that zeros of the denominator give rise to the vertical asymptotes, whereas zeros of the numerator give rise to x-intercepts. Substituting $x = 0$ gives $y = 12$; this is the y-intercept. The function is even, so the graph is symmetric about the y-axis.

[6]We are assuming that $f(x)$ gets arbitrarily close to L as $x \to \infty$.

CAS Challenge Problems

43. (a) Factor $f(x) = x^4 + bx^3 - cx^3 - a^2x^2 - bcx^2 - a^2bx + a^2cx + a^2bc$ using a computer algebra system.

(b) Assuming a, b, c are constants with $0 < a < b < c$, use your answer to part (a) to make a hand sketch of the graph of f. Explain how you know its shape.

44. (a) Using a computer algebra system, factor $f(x) = -x^5 + 11x^4 - 46x^3 + 90x^2 - 81x + 27$.

(b) Use your answer to part (a) to make a hand sketch of the graph of f. Explain how you know its shape.

45. Let $f(x) = e^{6x} + e^{5x} - 2e^{4x} - 10e^{3x} - 8e^{2x} + 16e^x + 16$.

(a) What happens to the value of $f(x)$ as $x \to \infty$? As $x \to -\infty$? Explain your answer.

(b) Using a computer algebra system, factor $f(x)$ and predict the number of zeros of the function $f(x)$.

(c) What are the exact values of the zeros? What is the relationship between successive zeros?

46. Let $f(x) = x^2 - x$.

(a) Find the polynomials $f(f(x))$ and $f(f(f(x)))$ in expanded form.

(b) What do you expect to be the degree of the polynomial $f(f(f(f(f(f(x))))))$? Explain.

47. (a) Use a computer algebra system to rewrite the rational function

$$f(x) = \frac{x^3 - 30}{x - 3}$$

in the form

$$f(x) = p(x) + \frac{r(x)}{q(x)},$$

where $p(x), q(x), r(x)$ are polynomials and the degree of $r(x)$ is less than the degree of $q(x)$.

(b) What is the vertical asymptote of f? Use your answer to part (a) to write the formula for a function whose graph looks like the graph of f for x near the vertical asymptote.

(c) Use your answer to part (a) to write the formula for a function whose graph looks like the graph of f for $x \to \infty$ and $x \to -\infty$.

(d) Using graphs, confirm the asymptote you found in part (b) and the formula you found in part (c).

We say that a function can be written as a polynomial in $\sin x$ (or $\cos x$) if it is of the form $p(\sin x)$ (or $p(\cos x)$) for some polynomial $p(x)$. For example, $\cos 2x$ can be written as a polynomial in $\sin x$ because $\cos(2x) = 1 - 2\sin^2 x = p(\sin x)$, where $p(x) = 1 - 2x^2$.

48. Use the trigonometric capabilities of your computer algebra system to express $\sin(5x)$ as a polynomial in $\sin x$.

49. Use the trigonometric capabilities of your computer algebra system to express $\cos(4x)$ as a polynomial in

(a) $\sin x$

(b) $\cos x$.

CHECK YOUR UNDERSTANDING

Are the statements in Problems 1–25 true or false? Give an explanation for your answer.

1. For any two points in the plane, there is a linear function whose graph passes through them.

2. The graph of $f(x) = 100(10^x)$ is a horizontal shift of the graph of $g(x) = 10^x$.

3. The graph of $f(x) = \ln x$ is concave down.

4. The graph of $g(x) = \log(x - 1)$ crosses the x-axis at $x = 1$.

5. Every polynomial of odd degree has at least one zero.

6. The function $y = 2 + 3e^{-t}$ has a y-intercept of $y = 3$.

7. The function $y = 5 - 3e^{-4t}$ has a horizontal asymptote of $y = 5$.

8. If $y = f(x)$ is a linear function, then increasing x by 1 unit changes the corresponding y by m units, where m is the slope.

9. If $y = f(x)$ is an exponential function and if increasing x by 1 increases y by a factor of 5, then increasing x by 2 increases y by a factor of 10.

10. If $y = Ab^x$ and increasing x by 1 increases y by a factor of 3, then increasing x by 2 increases y by a factor of 9.

11. The function $f(\theta) = \cos\theta - \sin\theta$ is increasing on $0 \le \theta \le \pi/2$.

12. The function $f(t) = \sin(0.05\pi t)$ has period 0.05.

13. If t is in seconds, $g(t) = \cos(200\pi t)$ executes 100 cycles in one second.

14. The function $f(\theta) = \tan(\theta - \pi/2)$ is not defined at $\theta = \pi/2, 3\pi/2, 5\pi/2 \ldots$.

15. The function $f(x) = \sin(x^2)$ is periodic, with period 2π.

16. The function $g(\theta) = e^{\sin\theta}$ is periodic.

17. The function $f(x) = e^{-x^2}$ is decreasing for all x.

18. The inverse function of $y = \log x$ is $y = 1/\log x$.

19. If f is an increasing function, then f^{-1} is an increasing function.

20. The function $f(x) = |\sin x|$ is even.

21. If a and b are positive constants, $b \neq 1$, then $y = a + ab^x$ has a horizontal asymptote.

22. If a and b are positive constants, then $y = \ln(ax + b)$ has no vertical asymptote.

23. The function $y = 20/(1 + 2e^{-kt})$ with $k > 0$, has a horizontal asymptote at $y = 20$.

24. If $g(x)$ is an even function then $f(g(x))$ is even for every function $f(x)$.

25. If $f(x)$ is an even function then $f(g(x))$ is even for every function $g(x)$.

In Problems 26–31, give an example of a function with the specified properties. Express your answer using formulas.

26. Continuous on $[0, 1]$ but not continuous on $[1, 3]$.

27. Increasing but not continuous on $[0, 10]$.

28. Has a vertical asymptote at $x = -7\pi$.

29. Has exactly 17 vertical asymptotes.

30. Has a vertical asymptote which is crossed by a horizontal asymptote.

31. Two functions $f(x)$ and $g(x)$ such that moving the graph of f to the left 2 units gives the graph of g and moving the graph of f up 3 also gives the graph of g.

Suppose f is an increasing function and g is a decreasing function. In Problems 32–35, give an example for f and g for which the statement is true, or say why such an example is impossible.

32. $f(x) + g(x)$ is decreasing for all x.

33. $f(x) - g(x)$ is decreasing for all x.

34. $f(x)g(x)$ is decreasing for all x.

35. $f(g(x))$ is increasing for all x.

Are the statements in Problems 36–42 true or false? If a statement is true, explain how you know. If a statement is false, give a counterexample.

36. Every rational function that is not a polynomial has a vertical asymptote.

37. If a function is increasing on an interval, then it is concave up on that interval.

38. If y is a linear function of x, then the ratio y/x is constant for all points on the graph at which $x \neq 0$.

39. An exponential function can be decreasing.

40. If $y = f(x)$ is a linear function, then increasing x by 2 units adds $m + 2$ units to the corresponding y, where m is the slope.

41. If a function is not continuous at a point, then it is not defined at that point.

42. There is a function which is both even and odd.

43. Which of the following statements is a direct consequence of the statement: "If f and g are continuous at $x = a$ and $g(a) \neq 0$ then f/g is continuous at $x = a$"?

   (a) If f and g are continuous at $x = a$ and $f(a) \neq 0$ then g/f is continuous at $x = a$.

   (b) If f and g are continuous at $x = a$ and $g(a) = 0$, then f/g is not continuous at $x = a$.

   (c) If f, g, are continuous at $x = a$, but f/g is not continuous at $x = a$, then $g(a) = 0$.

   (d) If f and f/g are continuous at $x = a$ and $g(a) \neq 0$, then g is continuous at $x = a$.

PROJECTS FOR CHAPTER ONE

1. Matching Functions to Data

From the data in Table 1.20, determine a possible formula for each function.[10] Write an explanation of your reasoning.

Table 1.20

x	$f(x)$	$g(x)$	$h(x)$	$F(x)$	$G(x)$	$H(x)$
-5	-10	20	25	0.958924	0.544021	2.958924
-4.5	-9	19	20.25	0.97753	-0.412118	2.97753
-4	-8	18	16	0.756802	-0.989358	2.756802
-3.5	-7	17	12.25	0.350783	-0.656987	2.350783
-3	-6	16	9	-0.14112	0.279415	1.85888
-2.5	-5	15	6.25	-0.598472	0.958924	1.401528
-2	-4	14	4	-0.909297	0.756802	1.090703
-1.5	-3	13	2.25	-0.997495	-0.14112	1.002505
-1	-2	12	1	-0.841471	-0.909297	1.158529
-0.5	-1	11	0.25	-0.479426	-0.841471	1.520574
0	0	10	0	0	0	2
0.5	1	9	0.25	0.479426	0.841471	2.479426
1	2	8	1	0.841471	0.909297	2.841471
1.5	3	7	2.25	0.997495	0.14112	2.997495
2	4	6	4	0.909297	-0.756802	2.909297
2.5	5	5	6.25	0.598472	-0.958924	2.598472
3	6	4	9	0.14112	-0.279415	2.14112
3.5	7	3	12.25	-0.350783	0.656987	1.649217
4	8	2	16	-0.756802	0.989358	1.243198
4.5	9	1	20.25	-0.97753	0.412118	1.02247
5	10	0	25	-0.958924	-0.544021	1.041076

2. Compound Interest

The newspaper article below is from *The New York Times*, May 27, 1990. Fill in the three blanks. (For the first blank, assume that daily compounding is essentially the same as continuous compounding. For the last blank, assume the interest has been compounded yearly, and give your answer in dollars. Ignore the occurrence of leap years.)

213 Years After Loan, Uncle Sam Is Dunned

By LISA BELKIN

Special to The New York Times

SAN ANTONIO, May 26 — More than 200 years ago, a wealthy Pennsylvania merchant named Jacob DeHaven lent $450,000 to the Continental Congress to rescue the troops at Valley Forge. That loan was apparently never repaid.

So Mr. DeHaven's descendants are taking the United States Government to court to collect what they believe they are owed.

The total: ____ in today's dollars if the interest is compounded daily at 6 percent, the going rate at the time. If compounded yearly, the bill is only ____.

Family Is Flexible

The descendants say that they are willing to be flexible about the amount of a settlement and that they might even accept a heartfelt thank you or perhaps a DeHaven statue. But they also note that interest is accumulating at ____ a second.

---

[10]Based on a problem by Lee Zia

Chapter Two

KEY CONCEPT: THE DERIVATIVE

We begin this chapter by investigating the problem of speed: How can we measure the speed of a moving object at a given instant in time? Or, more fundamentally, what do we mean by the term *speed*? We'll come up with a definition of speed that has wide-ranging implications — not just for the speed problem, but for measuring the rate of change of any quantity. Our journey will lead us to the key concept of *derivative*, which forms the basis for our study of calculus.

The derivative can be interpreted geometrically as the slope of a curve, and physically as a rate of change. Because derivatives can be used to represent everything from fluctuations in interest rates to the rates at which fish populations vary and gas molecules move, they have applications throughout the sciences.

2.1   HOW DO WE MEASURE SPEED?

The speed of an object at an instant in time is surprisingly difficult to define precisely. Consider the statement "At the instant it crossed the finish line, the horse was traveling at 42 mph." How can such a claim be substantiated? A photograph taken at that instant will show the horse motionless—it is no help at all. There is some paradox in trying to study the horse's motion at a particular instant in time, since by focusing on a single instant we stop the motion!

Problems of motion were of central concern to Zeno and other philosophers as early as the fifth century B.C. The modern approach, made famous by Newton's calculus, is to stop looking for a simple notion of speed at an instant, and instead to look at speed over small time intervals containing the instant. This method sidesteps the philosophical problems mentioned earlier but introduces new ones of its own.

We illustrate the ideas discussed above by an idealized example, called a thought experiment. It is idealized in the sense that we assume that we can make measurements of distance and time as accurately as we wish.

A Thought Experiment: Average and Instantaneous Velocity

We look at the speed of a small object (say, a grapefruit) that is thrown straight upward into the air at $t = 0$ seconds. The grapefruit leaves the thrower's hand at high speed, slows down until it reaches its maximum height, and then speeds up in the downward direction and finally, "Splat!" (See Figure 2.1.)

Suppose that we want to be more precise in our determination of the speed, say, at $t = 1$ second. We think of the height, y, of the grapefruit above the ground as a function of time. (See Table 2.1.) "Splat!" comes sometime between 6 and 7 seconds. The numbers in the table show that during the first second the grapefruit travels $90 - 6 = 84$ feet, and during the second second it travels only $142 - 90 = 52$ feet. Hence the grapefruit traveled faster over the first interval, $0 \leq t \leq 1$, than the second interval, $1 \leq t \leq 2$.

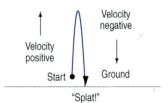

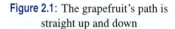

Figure 2.1: The grapefruit's path is straight up and down

Table 2.1   *Height of the grapefruit above the ground*

t (sec)	0	1	2	3	4	5	6
y (feet)	6	90	142	162	150	106	30

Velocity versus Speed

From now on, we will make a distinction between velocity and speed. Suppose an object moves along a line. We pick one direction to be positive and say that the *velocity* is positive if it is in the same direction, and negative if it is in the opposite direction. For the grapefruit, upward is positive and downward is negative. (See Figure 2.1.) *Speed* is the magnitude of the velocity and so is always positive or zero.

If $s(t)$ is the position of an object at time t, then the **average velocity** of the object over the interval $a \leq t \leq b$ is

$$\text{Average velocity} = \frac{\text{Change in position}}{\text{Change in time}} = \frac{s(b) - s(a)}{b - a}.$$

In words, the **average velocity** of an object over an interval is the net change in position during the interval divided by the change in time.

Example 1 Compute the average velocity of the grapefruit over the interval $4 \leq t \leq 5$. What is the significance of the sign of your answer?

Solution During this interval, the grapefruit moves $(106 - 150) = -44$ feet. Therefore the average velocity is -44 ft/sec. The negative sign means the height is decreasing and the grapefruit is moving downward.

Example 2 Compute the average velocity of the grapefruit over the interval $1 \leq t \leq 3$.

Solution Average velocity $= (162 - 90)/(3 - 1) = 72/2 = 36$ ft/sec.

The average velocity is a useful concept since it gives a rough idea of the behavior of the grapefruit: if two grapefruits are hurled into the air, and one has an average velocity of 10 ft/sec over the interval $0 \leq t \leq 1$ while the second has an average velocity of 100 ft/sec over the same interval, the second one is moving faster.

But average velocity over an interval doesn't solve the problem of measuring the velocity of the grapefruit at *exactly* $t = 1$ second. To get closer to an answer to that question, we have to look at what happens near $t = 1$ in more detail. The data[1] in Figure 2.2 shows the average velocity over small intervals on either side of $t = 1$.

Notice that the average velocity before $t = 1$ is slightly more than the average velocity after $t = 1$. We expect to define the velocity *at* $t = 1$ to be between these two average velocities. As the size of the interval shrinks, the values of the velocity before $t = 1$ and the velocity after $t = 1$ get closer together. In the smallest interval in Figure 2.2, both velocities are 68.0 ft/sec (to one decimal place), so we define the velocity at $t = 1$ to be 68.0 ft/sec (to one decimal place).

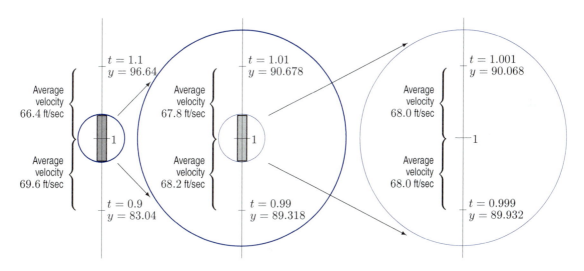

Figure 2.2: Average velocities over intervals on either side of $t = 1$: Showing successively smaller intervals

Of course, if we calculate to more decimal places, the average velocities before and after $t = 1$ would no longer agree. To calculate the velocity at $t = 1$ to more decimal places of accuracy, we take smaller and smaller intervals on either side of $t = 1$ until the average velocities agree to the number of decimal places we want. In this way, we can estimate the velocity at $t = 1$ to any accuracy.

[1]The data is in fact calculated from the formula $y = 6 + 100t - 16t^2$.

Defining Instantaneous Velocity Using Limit Notation

When we take smaller and smaller intervals, it turns out that the average velocities are always just above or just below 68 ft/sec. It seems natural, then, to define velocity at the instant $t = 1$ to be 68 ft/sec. This is called the *instantaneous velocity* at this point. Its definition depends on our being convinced that smaller and smaller intervals will provide average speeds that come arbitrarily close to 68. This process is called *taking the limit*.

Notice how we have replaced the original difficulty of computing velocity at a point by a search for an argument to convince ourselves that the average velocities do approach a number as the time intervals shrink in size. In a sense, we have traded one hard question for another, since we don't yet have any idea how to be certain what number the average velocities are approaching. In the thought experiment, the number seems to be exactly 68, but what if it were 68.000001? How can we be sure that we have taken small enough intervals? Showing that the limit is exactly 68 requires more precise knowledge of the limiting process.

We now define instantaneous velocity at an arbitrary point $t = a$. We use the same method as for $t = 1$: we look at small intervals of size h around $t = a$. Then, over the interval $a \leq t \leq a + h$,

$$\text{Average velocity} = \frac{s(a + h) - s(a)}{h}.$$

The same formula holds when $h < 0$. The instantaneous velocity is the number that the average velocities approach as the intervals decrease in size, that is, as h becomes smaller. So we define

$$\text{Instantaneous velocity} = \text{Limit, as } h \text{ approaches 0, of } \frac{s(a + h) - s(a)}{h}.$$

This is written more compactly using limit notation, as follows:

Let $s(t)$ give the position at time t. Then the **instantaneous velocity** at $t = a$ is defined as

$$\begin{matrix}\text{Instantaneous velocity} \\ \text{at } t = a\end{matrix} = \lim_{h \to 0} \frac{s(a + h) - s(a)}{h}.$$

In words, the **instantaneous velocity** of an object at time $t = a$ is given by the limit of the average velocity over an interval, as the interval shrinks around a.

This expression forms the foundation of the rest of calculus. This limit refers to the number that the average velocities approach as the intervals shrink. To estimate the limit, we look at intervals of smaller and smaller, but never zero, length.

Visualizing Velocity: Slope of Curve

Now visualize velocity using a graph of height. Let's go back to the grapefruit. Figure 2.3 shows the height of the grapefruit plotted against time. (Note that this is not a picture of the grapefruit's path, which is straight up and down.)

How can we visualize the average velocity on this graph? Suppose $y = s(t)$. We consider the interval $1 \leq t \leq 2$ and the expression

$$\text{Average velocity} = \frac{\text{Change in position}}{\text{Change in time}} = \frac{s(2) - s(1)}{2 - 1} = \frac{142 - 90}{1} = 52 \text{ ft/sec.}$$

Now $s(2) - s(1)$ is the change in position over the interval, and it is marked vertically in Figure 2.3. The 1 in the denominator is the time elapsed and is marked horizontally in Figure 2.3. Therefore,

$$\text{Average velocity} = \frac{\text{Change in position}}{\text{Change in time}} = \text{Slope of line joining } B \text{ and } C.$$

(See Figure 2.3.) A similar argument shows the following:

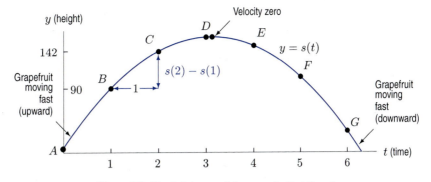

Figure 2.3: The height, y, of the grapefruit at time t

The **average velocity** over any time interval $a \leq t \leq b$ is the slope of the line joining the points on the graph of $s(t)$ corresponding to $t = a$ and $t = b$.

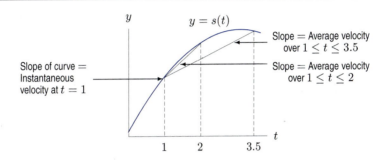

Figure 2.4: Average velocities over small intervals

To visualize the instantaneous velocity, we think about how we calculated it. We took average velocities over small intervals starting at $t = 1$. Two such velocities are represented by the slopes of the lines in Figure 2.4. As the length of the interval shrinks, the slope of the line gets closer to the slope of the curve at $t = 1$.

The cornerstone of the idea is the fact that, on a very small scale, most functions look almost like straight lines. Imagine taking the graph of a function near a point and "zooming in" to get a close-up view. (See Figure 2.5.) The more we zoom in, the more the curve appears to be a straight line. We call the slope of this line the *slope of the curve* at the point. Therefore, the slope of the magnified line is the instantaneous velocity. Thus, we can say:

The **instantaneous velocity** is the slope of the curve at a point.

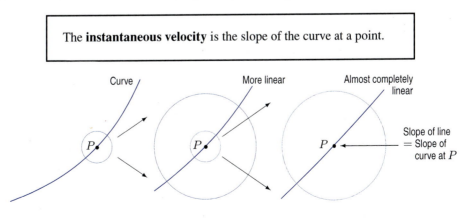

Figure 2.5: Estimating the slope of the curve at the point by "zooming in"

Look back at the graph of the grapefruit's height as a function of time in Figure 2.3 to see how the grapefruit's velocity varies during its journey. At points A and B the curve has a large positive slope, indicating that the grapefruit is traveling up rapidly. Point D is almost at the top: the grapefruit is slowing down. At the peak, the slope of the curve is zero: the fruit has slowed to zero velocity for an instant in preparation for its return to earth. At point E the curve has a small negative slope, indicating a slow velocity of descent. Finally, the slope of the curve at point G is large and negative, indicating a large downward velocity that is responsible for the "Splat."

The Idea of a Limit

In the process of defining the instantaneous velocity, we observed average velocities as time intervals shrank around a point and introduced the notation for a limit. Now we look a bit more at the idea of a limit, which is considered in more detail in Section 2.2. We talk about the *limit* of the function at the point c.

> We write $\lim_{x \to c} f(x)$ to represent the number L approached by $f(x)$ as x approaches c.

Example 3   Investigate $\lim_{x \to 2} x^2$.

Solution   Notice that we can make x^2 as close to 4 as we like by taking x sufficiently close to 2. (Look at the values of 1.9^2, 1.99^2, 1.999^2, and 2.1^2, 2.01^2, 2.001^2; they seem to be approaching 4 in Table 2.2.) We write

$$\lim_{x \to 2} x^2 = 4,$$

which is read "the limit, as x approaches 2, of x^2 is 4." Notice that the limit doesn't ask what happens *at* $x = 2$, so it is not sufficient to substitute 2 to find the answer. The limit describes behavior of a function *near* a point, not *at* the point.

Table 2.2   *Values of x^2 near $x = 2$*

x	1.9	1.99	1.999	2.001	2.01	2.1
x^2	3.61	3.96	3.996	4.004	4.04	4.41

Example 4   Use a graph to estimate $\lim_{\theta \to 0} \dfrac{\sin \theta}{\theta}$. (Use radians.)

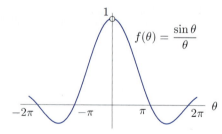

Figure 2.6: Find the limit as $\theta \to 0$

Solution   Figure 2.6 shows that as θ approaches 0 from either side, the value of $\dfrac{\sin \theta}{\theta}$ appears to approach 1, suggesting that $\lim_{\theta \to 0} \dfrac{\sin \theta}{\theta} = 1$. Zooming in on the graph near $\theta = 0$ provides further support for this conclusion. Notice that $\dfrac{\sin \theta}{\theta}$ is undefined at $\theta = 0$.

Example 5   Estimate $\lim_{h \to 0} \dfrac{(3 + h)^2 - 9}{h}$ numerically.

Solution   The limit is the value approached by this expression as h approaches 0. The values in Table 2.3 seem to be converging to 6 as $h \to 0$. So it is a reasonable guess that

$$\lim_{h \to 0} \frac{(3+h)^2 - 9}{h} = 6.$$

However, we cannot be sure that the limit is *exactly* 6 by looking at the table. To calculate the limit exactly requires algebra.

Table 2.3   *Values of $\left((3+h)^2 - 9\right)/h$ near $h = 0$*

h	-0.1	-0.01	-0.001	0.001	0.01	0.1
$\left((3+h)^2 - 9\right)/h$	5.9	5.99	5.999	6.001	6.01	6.1

Example 6   Use algebra to find $\displaystyle\lim_{h \to 0} \frac{(3+h)^2 - 9}{h}$.

Solution   Expanding the numerator gives

$$\frac{(3+h)^2 - 9}{h} = \frac{9 + 6h + h^2 - 9}{h} = \frac{6h + h^2}{h}.$$

Since taking the limit as $h \to 0$ means looking at values of h near, but not equal, to 0, we can cancel h, giving

$$\lim_{h \to 0} \frac{(3+h)^2 - 9}{h} = \lim_{h \to 0}(6 + h).$$

As h approaches 0, the values of $(6 + h)$ approach 6, so

$$\lim_{h \to 0} \frac{(3+h)^2 - 9}{h} = \lim_{h \to 0}(6 + h) = 6.$$

Exercises and Problems for Section 2.1

Exercises

1. The distance, s, a car has traveled on a trip is shown in the table as a function of the time, t, since the trip started. Find the average velocity between $t = 2$ and $t = 5$.

t (hours)	0	1	2	3	4	5
s (km)	0	45	135	220	300	400

Slope	Point
-3	
-1	
0	
$1/2$	
1	
2	

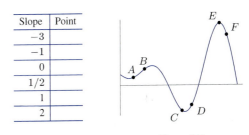

Figure 2.7

2. In a time of t seconds, a particle moves a distance of s meters from its starting point, where $s = 3t^2$.

   (a) Find the average velocity between $t = 1$ and $t = 1 + h$ if:

      (i) $h = 0.1$,   (ii) $h = 0.01$, (iii) $h = 0.001$.

   (b) Use your answers to part (a) to estimate the instantaneous velocity of the particle at time $t = 1$.

3. Match the points labeled on the curve in Figure 2.7 with the given slopes.

4. For the function shown in Figure 2.8, at what labeled points is the slope of the graph positive? Negative? At which labeled point does the graph have the greatest (i.e., most positive) slope? The least slope (i.e., negative and with the largest magnitude)?

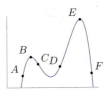

Figure 2.8

Estimate the limits in Exercises 5–8 by substituting smaller and smaller values of h. For trigonometric functions, use radians. Give answers to one decimal place.

5. $\lim\limits_{h\to 0} \dfrac{(3+h)^3 - 27}{h}$

6. $\lim\limits_{h\to 0} \dfrac{\cos h - 1}{h}$

7. $\lim\limits_{h\to 0} \dfrac{7^h - 1}{h}$

8. $\lim\limits_{h\to 0} \dfrac{e^{1+h} - e}{h}$

Use a graph to estimate each of the limits in Exercises 9–12. Use radians unless degrees are indicated by $\theta°$.

9. $\lim\limits_{\theta\to 0} \dfrac{\sin(2\theta)}{\theta}$

10. $\lim\limits_{\theta\to 0} \dfrac{\cos\theta - 1}{\theta}$

11. $\lim\limits_{\theta\to 0} \dfrac{\sin\theta°}{\theta°}$

12. $\lim\limits_{\theta\to 0} \dfrac{\theta}{\tan(3\theta)}$

Problems

13. A car is driven at a constant speed. Sketch a graph of the distance the car has traveled as a function of time.

14. A car is driven at an increasing speed. Sketch a graph of the distance the car has traveled as a function of time.

15. A car starts at a high speed, and its speed then decreases slowly. Sketch a graph of the distance the car has traveled as a function of time.

16. For the graph $y = f(x)$ in Figure 2.9, arrange the following numbers from smallest to largest:

- The slope of the graph at A.
- The slope of the graph at B.
- The slope of the graph at C.
- The slope of the line AB.
- The number 0.
- The number 1.

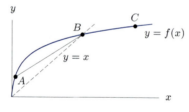

Figure 2.9

17. The population of the world reached 1 billion in 1804, 2 billion in 1927, 3 billion in 1960, 4 billion in 1974, 5 billion in 1987 and 6 billion in 1999. Find the average rate of change of the population of the world, in people per minute, during each of these intervals. [That is, from 1804 to 1927, 1927 to 1960, etc.]

18. The graph of $f(t)$ in Figure 2.10 gives the position of a particle at time t. List the following quantities in order, smallest to largest.

- A, average velocity between $t = 1$ and $t = 3$,
- B, average velocity between $t = 5$ and $t = 6$,
- C, instantaneous velocity at $t = 1$,
- D, instantaneous velocity at $t = 3$,
- E, instantaneous velocity at $t = 5$,
- F, instantaneous velocity at $t = 6$.

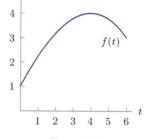

Figure 2.10

19. A particle moves at varying velocity along a line and $s = f(t)$ represents the particle's distance from a point as a function of time, t. Sketch a possible graph for f if the average velocity of the particle between $t = 2$ and $t = 6$ is the same as the instantaneous velocity at $t = 5$.

20. The notation $\lim\limits_{x\to 0^+}$ means that we only consider values of x greater than 0. Estimate the limit

$$\lim\limits_{x\to 0^+} x^x,$$

either by evaluating x^x for smaller and smaller positive values of x (say $x = 0.1, 0.01, 0.001, \ldots$) or by zooming in on the graph of $y = x^x$ near $x = 0$.

2.2   LIMITS

In this section we develop a formal definition of limit from the intuitive idea introduced in the previous section.

Definition of Limit

By the beginning of the 19th century, calculus had proved its worth, and there was no doubt about the correctness of its answers. However, it was not until the work of the French mathematician Augustin Cauchy (1789–1857) that a formal definition of the limit was given, similar to the following:

Suppose a function f is defined on an interval around c, except perhaps not at the point $x = c$. We define the **limit** of the function $f(x)$ as x approaches c, written $\lim_{x \to c} f(x)$, to be a number L (if one exists) such that $f(x)$ is as close to L as we please whenever x is sufficiently close to c (but $x \neq c$). If L exists, we write

$$\lim_{x \to c} f(x) = L.$$

Shortly, we will see how "as close as we please" and "sufficiently close" can be given a precise meaning using inequalities. First, we look at $\lim_{\theta \to 0}(\sin \theta / \theta)$ more closely (see Example 4 on page 60).

Example 1   By graphing $y = (\sin \theta)/\theta$ in an appropriate window, find how close θ should be to 0 in order to make $(\sin \theta)/\theta$ within 0.01 of 1.

Solution   Since we want $(\sin \theta)/\theta$ to be within 0.01 of 1, we set the y-range on the graphing window to go from 0.99 to 1.01. Our first attempt with $-0.5 \leq \theta \leq 0.5$ yields the graph in Figure 2.11. Since we want the y-values to stay within the range $0.99 < y < 1.01$, we do not want the graph to leave the window through the top or bottom. By trial and error, we find that changing the θ-range to $-0.2 \leq \theta \leq 0.2$ gives the graph in Figure 2.12. Thus, the graph suggests that $(\sin \theta)/\theta$ will be within 0.01 of 1 whenever θ is within 0.2 of 0. Proving this requires an analytical argument, not just graphs from a calculator.

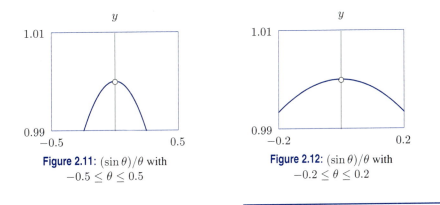

Figure 2.11: $(\sin \theta)/\theta$ with $-0.5 \leq \theta \leq 0.5$

Figure 2.12: $(\sin \theta)/\theta$ with $-0.2 \leq \theta \leq 0.2$

When we say "$f(x)$ is as close to L as we please," we mean that we can specify a maximum distance between $f(x)$ and L. We express the distance using absolute values:

$$|f(x) - L| = \text{Distance between } f(x) \text{ and } L.$$

Using ϵ (the Greek letter epsilon) to stand for the distance we have specified, we write

$$|f(x) - L| < \epsilon$$

to indicate that the distance between $f(x)$ and L is less than ϵ. In Example 1 we used $\epsilon = 0.01$. In a similar manner we interpret "x is sufficiently close to c" as specifying a maximum distance between x and c:

$$|x - c| < \delta,$$

where δ (the Greek letter delta) tells us how close x should be to c. In Example 1 we found $\delta = 0.2$.

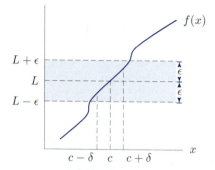

Figure 2.13: What the definition of the limit means practically

If $\lim\limits_{x \to c} f(x) = L$, we know that no matter how narrow the band determined by ϵ in Figure 2.13, there's always a δ which makes the graph stay within the band for $c - \delta < x < c + \delta$.

Thus we restate the definition of a limit, using symbols:

Definition of Limit

We define $\lim\limits_{x \to c} f(x)$ to be the number L (if one exists) such that for any $\epsilon > 0$ (as small as we want), there is a $\delta > 0$ (sufficiently small) such that if $|x - c| < \delta$ and $x \neq c$, then $|f(x) - L| < \epsilon$.

Realize that the point of this definition is that for any ϵ we are given, we need to show that a corresponding δ exists. One way to do this is to give an explicit expression for δ in terms of ϵ.

Example 2 Use algebra to find a maximum distance between x and 2 which ensures that x^2 is within 0.1 of 4. Use a similar argument to show that $\lim\limits_{x \to 2} x^2 = 4$.

Solution We write $x = 2 + h$. We want to find the values of h making x^2 within 0.1 of 4. We know

$$x^2 = (2 + h)^2 = 4 + 4h + h^2,$$

so x^2 differs from 4 by $4h + h^2$. Since we want x^2 to be within 0.1 of 4, we need

$$\left| x^2 - 4 \right| = \left| 4h + h^2 \right| = |h| \cdot |4 + h| < 0.1.$$

Assuming $0 < |h| < 1$, we know $|4 + h| < 5$, so that

$$\left| x^2 - 4 \right| < 5 |h|.$$

Hence we want

$$5 |h| < 0.1.$$

Thus, if we choose h such that $0 < |h| < 0.1/5 = 0.02$, then x^2 is less than 0.1 from 4.

An analogous argument using any small ϵ instead of 0.1 shows that if we take $\delta = \epsilon/5$, then

$$\left| x^2 - 4 \right| < \epsilon \quad \text{for all} \quad |x - 2| < \epsilon/5.$$

Thus, we have used the definition to show that

$$\lim_{x \to 2} x^2 = 4.$$

It is important to understand that the ϵ, δ definition by itself does not make it easier to calculate limits. The advantage of the ϵ, δ definition is that it makes it possible to put calculus on a rigorous foundation. From this foundation, we can prove the following properties. See Problems 46–48.

Theorem 2.1: Properties of Limits

Assuming all the limits on the right hand side exist:

1. If b is a constant, then $\displaystyle\lim_{x \to c} (bf(x)) = b \left(\lim_{x \to c} f(x) \right)$.

2. $\displaystyle\lim_{x \to c} (f(x) + g(x)) = \lim_{x \to c} f(x) + \lim_{x \to c} g(x)$.

3. $\displaystyle\lim_{x \to c} (f(x)g(x)) = \left(\lim_{x \to c} f(x) \right) \left(\lim_{x \to c} g(x) \right)$.

4. $\displaystyle\lim_{x \to c} \frac{f(x)}{g(x)} = \frac{\lim_{x \to c} f(x)}{\lim_{x \to c} g(x)}$, provided $\displaystyle\lim_{x \to c} g(x) \neq 0$.

5. For any constant k, $\displaystyle\lim_{x \to c} k = k$.

6. $\displaystyle\lim_{x \to c} x = c$.

These properties underlie many limit calculations, though we seldom acknowledge them explicitly.

Example 3   Explain how the limit properties are used in the following calculation:

$$\lim_{x \to 3} \frac{x^2 + 5x}{x + 9} = \frac{3^2 + (5)(3)}{3 + 9} = 2.$$

Solution   We calculate this limit in stages, using the limit properties to justify each step:

$$\lim_{x \to 3} \frac{x^2 + 5x}{x + 9} = \frac{\displaystyle\lim_{x \to 3} (x^2 + 5x)}{\displaystyle\lim_{x \to 3} (x + 9)} \qquad \text{(Property 4, since } \lim_{x \to 3}(x + 9) \neq 0)$$

$$= \frac{\displaystyle\lim_{x \to 3} (x^2) + \lim_{x \to 3} (5x)}{\displaystyle\lim_{x \to 3} x + \lim_{x \to 3} 9} \qquad \text{(Property 2)}$$

$$= \frac{\left(\displaystyle\lim_{x \to 3} x\right)^2 + 5 \left(\displaystyle\lim_{x \to 3} x\right)}{\displaystyle\lim_{x \to 3} x + \lim_{x \to 3} 9} \qquad \text{(Properties 1 and 3)}$$

$$= \frac{3^2 + (5)(3)}{3 + 9} = 2. \qquad \text{(Properties 5 and 6)}$$

One- and Two-Sided Limits

When we write

$$\lim_{x \to 2} f(x),$$

we mean the number that $f(x)$ approaches as x approaches 2 *from both sides*. We examine values of $f(x)$ as x approaches 2 through values greater than 2 (such as 2.1, 2.01, 2.003) and values less than 2 (such as 1.9, 1.99, 1.994). If we want x to approach 2 only through values greater than 2, we write

$$\lim_{x \to 2^+} f(x)$$

for the number that $f(x)$ approaches (assuming such a number exists). Similarly,

$$\lim_{x \to 2^-} f(x)$$

denotes the number (if it exists) obtained by letting x approach 2 through values less than 2. We call $\lim\limits_{x \to 2^+} f(x)$ a *right-hand limit* and $\lim\limits_{x \to 2^-} f(x)$ a *left-hand limit*. Problems 25 and 26 ask for formal definitions of left and right-hand limits.

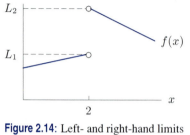

Figure 2.14: Left- and right-hand limits at $x = 2$

For the function graphed in Figure 2.14, we have

$$\lim_{x \to 2^-} f(x) = L_1 \qquad \lim_{x \to 2^+} f(x) = L_2.$$

If the left- and right-hand limits were equal, that is, if $L_1 = L_2$, then it could easily be proved that $\lim\limits_{x \to 2} f(x)$ exists and $\lim\limits_{x \to 2} f(x) = L_1 = L_2$. Since $L_1 \neq L_2$ in Figure 2.14, we see that $\lim\limits_{x \to 2} f(x)$ does not exist in this case.

When Limits Do Not Exist

Whenever there is no number L such that $\lim\limits_{x \to c} f(x) = L$, we say $\lim\limits_{x \to c} f(x)$ does not exist. Here are three examples in which limits fail to exist.

Example 4   Explain why $\lim\limits_{x \to 2} \dfrac{|x - 2|}{x - 2}$ doesn't exist.

Solution   Figure 2.15 shows the problem: The right-hand limit and the left-hand limit are different. For $x > 2$, we have $|x - 2| = x - 2$, so as x approaches 2 from the right,

$$\lim_{x \to 2^+} \frac{|x - 2|}{x - 2} = \lim_{x \to 2^+} \frac{x - 2}{x - 2} = \lim_{x \to 2^+} 1 = 1.$$

Similarly, if $x < 2$, then $|x - 2| = 2 - x$ so

$$\lim_{x \to 2^-} \frac{|x - 2|}{x - 2} = \lim_{x \to 2^-} \frac{2 - x}{x - 2} = \lim_{x \to 2^-} (-1) = -1.$$

So if $\lim\limits_{x \to 2} \dfrac{|x - 2|}{x - 2} = L$ then L would have to be both 1 and -1. Since L cannot have two different values, the limit does not exist.

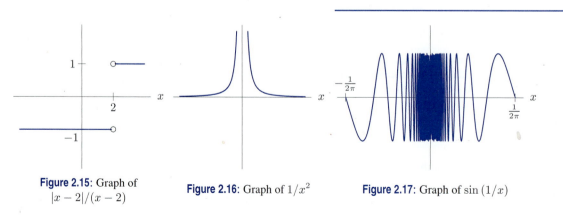

Figure 2.15: Graph of $|x - 2|/(x - 2)$

Figure 2.16: Graph of $1/x^2$

Figure 2.17: Graph of $\sin(1/x)$

Example 5   Explain why $\lim\limits_{x \to 0} \dfrac{1}{x^2}$ doesn't exist.

Solution   As x approaches zero, $1/x^2$ becomes arbitrarily large, so it can't stay close to any finite number L. See Figure 2.16. Therefore we say $1/x^2$ has no limit as $x \to 0$.

Example 6   Explain why $\lim\limits_{x \to 0} \sin\left(\dfrac{1}{x}\right)$ doesn't exist.

Solution   The sine function has values between -1 and 1. The graph in Figure 2.17 oscillates more and more rapidly as $x \to 0$. There are x-values as close to 0 as we like where $\sin(1/x) = 0$. There are also x-values as close to 0 as we like where $\sin(1/x) = 1$. So if the limit existed, it would have to be both 0 and 1. Thus, the limit does not exist.

If $\lim_{x \to a} f(x)$ doesn't exist because $f(x)$ gets arbitrarily large on both sides of a, we also say $\lim_{x \to a} f(x) = \infty$. So in Example 5 we could say $\lim_{x \to 0} 1/x^2 = \infty$.

Limits at Infinity

Sometimes we want to know what happens to $f(x)$ as x gets large, that is, the end behavior of f.

> If $f(x)$ gets as close to a number L as we please when x gets sufficiently large, then we write
> $$\lim_{x \to \infty} f(x) = L.$$
> Similarly, if $f(x)$ approaches L when x is negative and has a sufficiently large absolute value, then we write
> $$\lim_{x \to -\infty} f(x) = L.$$

The symbol ∞ does not represent a number. Writing $x \to \infty$ means that we consider arbitrarily large values of x. If the limit of $f(x)$ as $x \to \infty$ or $x \to -\infty$ is L, we say that the graph of f has $y = L$ as a *horizontal asymptote*. Problem 27 asks for a formal definition of $\lim_{x \to \infty} f(x)$.

Example 7   Investigate $\lim\limits_{x \to \infty} \dfrac{1}{x}$ and $\lim\limits_{x \to -\infty} \dfrac{1}{x}$.

Solution   A graph of $f(x) = 1/x$ in a large window shows $1/x$ approaching zero as x increases in either the positive or the negative direction (See Figure 2.18). This is as we would expect, since dividing 1 by larger and larger numbers yields answers which are smaller and smaller. This suggests that
$$\lim_{x \to \infty} \frac{1}{x} = \lim_{x \to -\infty} \frac{1}{x} = 0,$$
and that $f(x) = 1/x$ has $y = 0$ as a horizontal asymptote as $x \to \pm\infty$.

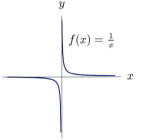

Figure 2.18: The end behavior of $f(x) = 1/x$

Exercises and Problems for Section 2.2

Exercises

1. Use Figure 2.19 to give approximate values for the following limits (if they exist).

   (a) $\lim_{x \to -2} f(x)$ (b) $\lim_{x \to 0} f(x)$

   (c) $\lim_{x \to 2} f(x)$ (d) $\lim_{x \to 4} f(x)$

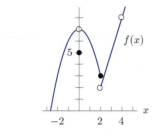

Figure 2.19

2. Using Figures 2.20 and 2.21, estimate

   (a) $\lim_{x \to 1^-} (f(x) + g(x))$ (b) $\lim_{x \to 1^+} (f(x) + 2g(x))$

   (c) $\lim_{x \to 1^-} f(x)g(x)$ (d) $\lim_{x \to 1^+} \dfrac{f(x)}{g(x)}$

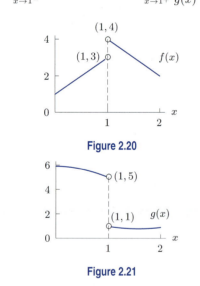

Figure 2.20

Figure 2.21

For the functions in Exercises 3–10, do the following:

(a) Make a table of values of $f(x)$ for $x = 0.1, 0.01, 0.001, 0.0001, -0.1, -0.01, -0.001,$ and -0.0001.

(b) Make a conjecture about the value of $\lim_{x \to 0} f(x)$.

(c) Graph the function to see if it is consistent with your answers to parts (a) and (b).

(d) Find an interval for x near 0 such that the difference between your conjectured limit and the value of the function is less than 0.01. (In other words, find a window of height 0.02 such that the graph exits the sides of the window and not the top or bottom of the window.)

3. $f(x) = 3x + 1$ 4. $f(x) = x^2 - 1$

5. $f(x) = \sin 2x$ 6. $f(x) = \sin 3x$

7. $f(x) = \dfrac{\sin 2x}{x}$ 8. $f(x) = \dfrac{\sin 3x}{x}$

9. $f(x) = \dfrac{e^x - 1}{x}$ 10. $f(x) = \dfrac{e^{2x} - 1}{x}$

For the functions in Exercises 11–14, do the following:

(a) Make a table of values of $f(x)$ for $x = a+0.1, a+0.01, a + 0.001, a + 0.0001, a - 0.1, a - 0.01, a - 0.001,$ and $a - 0.0001$.

(b) Make a conjecture about the value of $\lim_{x \to a} f(x)$.

(c) Graph the function to see if it is consistent with your answers to parts (a) and (b).

(d) Find an interval for x containing a such that the difference between your conjectured limit and the value of the function is less than 0.01 on that interval. (In other words, find a window of height 0.02 such that the graph exits the sides of the window and not the top or bottom of the window.)

11. $f(x) = \dfrac{x^2 - 4}{x - 2}, \quad a = 2$

12. $f(x) = \dfrac{x^2 - 9}{x - 3}, \quad a = 3$

13. $f(x) = \dfrac{\sin x - 1}{x - \pi/2}, \quad a = \dfrac{\pi}{2}$

14. $f(x) = \dfrac{e^{2x-2} - 1}{x - 1}, \quad a = 1$

Use algebra to evaluate the limits in Exercises 15–19.

15. $\lim_{h \to 0} \dfrac{(2 + h)^2 - 4}{h}$ 16. $\lim_{h \to 0} \dfrac{1/(1 + h) - 1}{h}$

17. $\lim_{h \to 0} \dfrac{1/(1 + h)^2 - 1}{h}$

18. $\lim_{h \to 0} \dfrac{\sqrt{4 + h} - 2}{h}$ [Hint: Multiply by $\sqrt{4 + h} + 2$ in numerator and denominator.]

19. $\lim_{h \to 0} \dfrac{1/\sqrt{4 + h} - 1/2}{h}$

For the functions in Exercises 20–22, use algebra to evaluate the limits $\lim_{x \to a+} f(x)$, $\lim_{x \to a-} f(x)$, and $\lim_{x \to a} f(x)$ if they exist. Sketch a graph to confirm your answers.

20. $a = 4, \quad f(x) = \dfrac{|x - 4|}{x - 4}$

21. $a = 2, \quad f(x) = \dfrac{|x - 2|}{x}$

22. $a = 3$, $f(x) = \begin{cases} x^2 - 2, & 0 < x < 3 \\ 2, & x = 3 \\ 2x + 1, & 3 < x \end{cases}$

23. Consider the function $(\sin \theta)/\theta$. Estimate how close θ should be to 0 to make $(\sin \theta)/\theta$ stay within 0.001 of 1.

Problems

In Problems 25–27, modify the definition of limit on page 64 to give a definition of each of the following.

25. A right-hand limit

26. A left-hand limit

27. $\lim\limits_{x \to \infty} f(x) = L$

28. If $p(x)$ is the function on page 45 giving the price of mailing a first-class letter, explain why $\lim\limits_{x \to 1} p(x)$ does not exist.

29. Investigate $\lim\limits_{h \to 0} (1 + h)^{1/h}$ numerically.

Assuming that limits as $x \to \infty$ have the properties listed for limits as $x \to c$ on page 65, use algebraic manipulations to evaluate $\lim\limits_{x \to \infty}$ for the functions in Problems 30–36.

30. $f(x) = \dfrac{x + 3}{2 - x}$

31. $f(x) = \dfrac{x^2 + 2x - 1}{3 + 3x^2}$

32. $f(x) = \dfrac{x^2 + 4}{x + 3}$

33. $f(x) = \dfrac{2x^3 - 16x^2}{4x^2 + 3x^3}$

34. $f(x) = \dfrac{x^4 + 3x}{x^4 + 2x^5}$

35. $f(x) = \dfrac{3e^x + 2}{2e^x + 3}$

36. $f(x) = \dfrac{2e^{-x} + 3}{3e^{-x} + 2}$

In Problems 37–41, find a value of the constant k such that the limit exists.

37. $\lim\limits_{x \to 4} \dfrac{x^2 - k^2}{x - 4}$

38. $\lim\limits_{x \to 1} \dfrac{x^2 - kx + 4}{x - 1}$

39. $\lim\limits_{x \to -2} \dfrac{x^2 + 4x + k}{x + 2}$

40. $\lim\limits_{x \to \infty} \dfrac{x^2 + 3x + 5}{4x + 1 + x^k}$

41. $\lim\limits_{x \to -\infty} \dfrac{e^{2x} - 5}{e^{kx} + 3}$

For each value of ϵ in Problems 42–43, find a positive value of δ such that the graph of the function leaves the window $a - \delta < x < a + \delta$, $b - \epsilon < y < b + \epsilon$ by the sides and not through the top or bottom.

42. $f(x) = -2x + 3$; $a = 0$; $b = 3$; $\epsilon = 0.2, 0.1, 0.02,$ $0.01, 0.002, 0.001$.

43. $g(x) = -x^3 + 2$; $a = 0$; $b = 2$; $\epsilon = 0.1, 0.01, 0.001$.

44. Show that $\lim\limits_{x \to 0} (-2x + 3) = 3$. [Hint: Use Problem 42.]

24. Write the definition of the following statement both in words and in symbols:
$$\lim_{h \to a} g(h) = K.$$

45. Consider the function $f(x) = \sin(1/x)$.

(a) Find a sequence of x-values that approach 0 such that $\sin(1/x) = 0$.
[Hint: Use the fact that $\sin(\pi) = \sin(2\pi) = \sin(3\pi) = \ldots = \sin(n\pi) = 0$.]

(b) Find a sequence of x-values that approach 0 such that $\sin(1/x) = 1$.
[Hint: Use the fact that $\sin(n\pi/2) = 1$ if $n = 1, 5, 9, \ldots$.]

(c) Find a sequence of x-values that approach 0 such that $\sin(1/x) = -1$.

(d) Explain why your answers to any two of parts (a)–(c) show that $\lim\limits_{x \to 0} \sin(1/x)$ does not exist.

46. This problem suggests a proof of the first property of limits on page 65: $\lim\limits_{x \to c} bf(x) = b \lim\limits_{x \to c} f(x)$.

(a) First, prove the property in the case $b = 0$.

(b) Now suppose that $b \neq 0$. Let $\epsilon > 0$. Show that if $|f(x) - L| < \epsilon/|b|$, then $|bf(x) - bL| < \epsilon$.

(c) Finally, prove that if $\lim\limits_{x \to c} f(x) = L$ then $\lim\limits_{x \to c} bf(x) = bL$. [Hint: Choose δ so that if $|x - c| < \delta$, then $|f(x) - L| < \epsilon/|b|$.]

47. Prove the second property of limits: $\lim\limits_{x \to c} (f(x) + g(x)) = \lim\limits_{x \to c} f(x) + \lim\limits_{x \to c} g(x)$. Assume that the limits on the right exist.

48. This problem suggests a proof of the third property of limits (assuming the limits on the right exist):
$$\lim_{x \to c} (f(x)g(x)) = \left(\lim_{x \to c} f(x) \right) \left(\lim_{x \to c} g(x) \right)$$
Let $L_1 = \lim_{x \to c} f(x)$ and $L_2 = \lim_{x \to c} g(x)$.

(a) First, show that if $\lim\limits_{x \to c} f(x) = \lim\limits_{x \to c} g(x) = 0$, then $\lim\limits_{x \to c} (f(x)g(x)) = 0$.

(b) Show algebraically that $f(x)g(x) = (f(x) - L_1)(g(x) - L_2) + L_1 g(x) + L_2 f(x) - L_1 L_2$.

(c) Use the second limit property (see Problem 47) to explain why
$$\lim_{x \to c} (f(x) - L_1) = \lim_{x \to c} (g(x) - L_2) = 0.$$

(d) Use parts (a) and (c) to explain why $\lim\limits_{x \to c} (f(x) - L_1)(g(x) - L_2) = 0$.

(e) Finally, use parts (b) and (d) and the first and second limit properties to show that
$$\lim_{x \to c} (f(x)g(x)) = \left(\lim_{x \to c} f(x) \right) \left(\lim_{x \to c} g(x) \right).$$

2.3   THE DERIVATIVE AT A POINT

Average Rate of Change

In Section 2.1, we looked at the change in height divided by the change in time, which tells us

$$\text{Average rate of change of height with respect to time} = \frac{s(a + h) - s(a)}{h}.$$

This ratio is called the *difference quotient*. Now we apply the same analysis to any function f, not necessarily a function of time. We say:

$$\text{Average rate of change of } f \text{ over the interval from } a \text{ to } a + h = \frac{f(a + h) - f(a)}{h}.$$

The numerator, $f(a + h) - f(a)$, measures the change in the value of f over the interval from a to $a + h$. So, the difference quotient is the change in f divided by the change in x. See Figure 2.22.

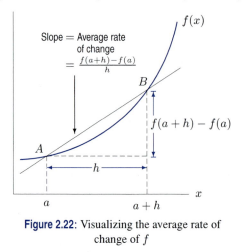

Figure 2.22: Visualizing the average rate of change of f

Although the interval is no longer necessarily a time interval, we still talk about the *average rate of change* of f over the interval. If we want to emphasize the independent variable, we talk about the average rate of change of f *with respect to x*.

Average Rate of Change versus Absolute Change

The average rate of change of a function over an interval is not the same as the absolute change. Absolute change is just the difference in the values of f at the ends of the interval:

$$f(a + h) - f(a).$$

The average rate of change is the absolute change divided by the size of the interval:

$$\frac{f(a + h) - f(a)}{h}.$$

The average rate of change tells how quickly (or slowly) the function changes from one end of the interval to the other, relative to the size of the interval. It is often more useful to know the rate of change than the absolute change. For example, if someone offers you a $100 salary, you will want to know how long to work to make that money. Just knowing the absolute change in your money, $100, is not enough, but knowing the rate of change (i.e., $100 divided by the time it takes to make it) helps you decide whether or not to accept the salary.

Blowing Up a Balloon

Consider the function which gives the radius of a sphere in terms of its volume. For example, think of blowing air into a balloon. You've probably noticed that a balloon seems to blow up faster at the start and then slows down as you blow more air into it. What you're seeing is variation in the rate of change of the radius with respect to volume.

Example 1   The volume, V, of a sphere of radius r is given by $V = 4\pi r^3/3$. Solving for r in terms of V gives

$$r = f(V) = \left(\frac{3V}{4\pi}\right)^{1/3}.$$

Calculate the average rate of change of r with respect to V over the intervals $0.5 \leq V \leq 1$ and $1 \leq V \leq 1.5$.

Solution   Using the formula for the average rate of change gives

$$\begin{array}{l} \text{Average rate of change} \\ \text{of radius for } 0.5 \leq V \leq 1 \end{array} = \frac{f(1) - f(0.5)}{0.5} = 2\left(\left(\frac{3}{4\pi}\right)^{1/3} - \left(\frac{1.5}{4\pi}\right)^{1/3}\right) \approx 0.26$$

$$\begin{array}{l} \text{Average rate of change} \\ \text{of radius for } 1 \leq V \leq 1.5 \end{array} = \frac{f(1.5) - f(1)}{0.5} = 2\left(\left(\frac{4.5}{4\pi}\right)^{1/3} - \left(\frac{3}{4\pi}\right)^{1/3}\right) \approx 0.18.$$

So we see that the rate decreases as the volume increases.

Instantaneous Rate of Change: The Derivative

We define the *instantaneous rate of change* of a function at a point in the same way that we defined instantaneous velocity: we look at the average rate of change over smaller and smaller intervals. This instantaneous rate of change is called the *derivative of f at a*, denoted by $f'(a)$.

> The **derivative of f at a**, written $f'(a)$, is defined as
>
> $$\begin{array}{l} \text{Rate of change} \\ \text{of } f \text{ at } a \end{array} = f'(a) = \lim_{h \to 0} \frac{f(a+h) - f(a)}{h}.$$
>
> If the limit exists, then f is said to be **differentiable at a**.

To emphasize that $f'(a)$ is the rate of change of $f(x)$ as the variable x changes, we call $f'(a)$ the derivative of f *with respect to x* at $x = a$. When the function $y = s(t)$ represents the position of an object, the derivative $s'(t)$ is the velocity.

Example 2   By choosing small values for h, estimate the instantaneous rate of change of the radius, r, of a sphere with respect to change in volume at $V = 1$.

Solution   The formula for $r = f(V)$ is given in Example 1. With $h = 0.01$ and $h = -0.01$, we have the difference quotients

$$\frac{f(1.01) - f(1)}{0.01} \approx 0.2061 \quad \text{and} \quad \frac{f(0.99) - f(1)}{-0.01} \approx 0.2075.$$

With $h = 0.001$ and $h = -0.001$,

$$\frac{f(1.001) - f(1)}{0.001} \approx 0.2067 \quad \text{and} \quad \frac{f(0.999) - f(1)}{-0.001} \approx 0.2069.$$

The values of these difference quotients suggest that the limit is between 0.2067 and 0.2069. We conclude that the value is about 0.207; taking smaller h values confirms this. So we say

$$f'(1) = \begin{array}{c} \text{Instantaneous rate of change of radius} \\ \text{with respect to volume at } V = 1 \end{array} \approx 0.207.$$

In this example we found an approximation to the instantaneous rate of change, or derivative, by substituting in smaller and smaller values of h. Now we see how to visualize the derivative.

Visualizing the Derivative: Slope of Curve and Slope of Tangent

As with velocity, we can visualize the derivative $f'(a)$ as the slope of the graph of f at $x = a$. In addition, there is another way to think of $f'(a)$. Consider the difference quotient $(f(a+h) - f(a))/h$. The numerator, $f(a+h) - f(a)$, is the vertical distance marked in Figure 2.23 and h is the horizontal distance, so

$$\text{Average rate of change of } f = \frac{f(a+h) - f(a)}{h} = \text{Slope of line } AB.$$

As h becomes smaller, the line AB approaches the tangent line to the curve at A. (See Figure 2.24.) We say

$$\begin{array}{c} \text{Instantaneous rate of change} \\ \text{of } f \text{ at } a \end{array} = \lim_{h \to 0} \frac{f(a+h) - f(a)}{h} = \text{Slope of tangent at } A.$$

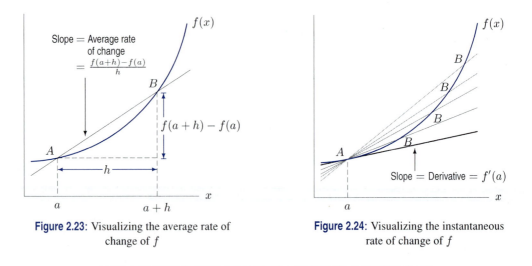

Figure 2.23: Visualizing the average rate of change of f

Figure 2.24: Visualizing the instantaneous rate of change of f

> The derivative at point A can be interpreted as:
> - The slope of the curve at A.
> - The slope of the tangent line to the curve at A.

The slope interpretation is often useful in gaining rough information about the derivative, as the following examples show.

Example 3   Is the derivative of $\sin x$ at $x = \pi$ positive or negative?

Solution   Looking at a graph of $\sin x$ in Figure 2.25 (remember, x is in radians), we see that a tangent line drawn at $x = \pi$ has negative slope. So the derivative at this point is negative.

Figure 2.25: Tangent line to $\sin x$ at $x = \pi$

Recall that if we zoom in on the graph of a function $y = f(x)$ at the point $x = a$, we usually find that the graph looks like a straight line with slope $f'(a)$.

Example 4   By zooming in on the point $(0,0)$ on the graph of the sine function, estimate the value of the derivative of $\sin x$ at $x = 0$, with x in radians.

Solution   Figure 2.26 shows graphs of $\sin x$ with smaller and smaller scales. On the interval $-0.1 \leq x \leq 0.1$, the graph looks like a straight line of slope 1. Thus, the derivative of $\sin x$ at $x = 0$ is about 1.

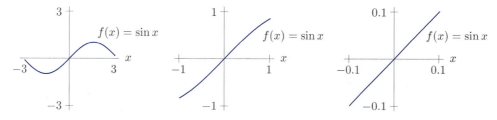

Figure 2.26: Zooming in on the graph of $\sin x$ near $x = 0$ shows the derivative is about 1 at $x = 0$

Later we will show that the derivative of $\sin x$ at $x = 0$ is exactly 1. (See page 131 in Section 3.5.) From now on we will assume that this is so.

Example 5   Use the tangent line at $x = 0$ to estimate values of $\sin x$ near $x = 0$.

Solution   In the previous example we see that near $x = 0$, the graph of $y = \sin x$ looks like the straight line $y = x$; we can use this line to estimate values of $\sin x$ when x is close to 0. For example, the point on the straight line $y = x$ with x coordinate 0.32 is $(0.32, 0.32)$. Since the line is close to the graph of $y = \sin x$, we estimate that $\sin 0.32 \approx 0.32$. (See Figure 2.27.) Checking on a calculator, we find that $\sin 0.32 \approx 0.3146$, so our estimate is quite close. Notice that the graph suggests that the real value of $\sin 0.32$ is slightly less than 0.32.

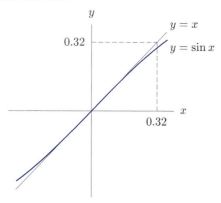

Figure 2.27: Approximating $y = \sin x$ by the line $y = x$

Why Do We Use Radians and Not Degrees?

After Example 4 we stated that the derivative of $\sin x$ at $x = 0$ is 1, when x is in radians. This is the reason we choose to use radians. If we had done Example 4 in degrees, the derivative of $\sin x$ would have turned out to be a much messier number. (See Problem 29, page 77.)

Estimating the Derivative of an Exponential Function

Example 6   Estimate the value of the derivative of $f(x) = 2^x$ at $x = 0$ graphically and numerically.

Solution   Graphically: Figure 2.28 indicates that the graph is concave up. Assuming this, the slope at A is between the slope of BA and the slope of AC. Since

$$\text{Slope of line } BA = \frac{(2^0 - 2^{-1})}{(0 - (-1))} = \frac{1}{2} \quad \text{and} \quad \text{Slope of line } AC = \frac{(2^1 - 2^0)}{(1 - 0)} = 1,$$

we know that at $x = 0$ the derivative of 2^x is between $1/2$ and 1.

Numerically: To estimate the derivative at $x = 0$, we look at values of the difference quotient

$$\frac{f(0 + h) - f(0)}{h} = \frac{2^h - 2^0}{h} = \frac{2^h - 1}{h}$$

for small h. Table 2.4 shows some values of 2^h together with values of the difference quotients. (See Problem 35 on page 78 for what happens for very small values of h.)

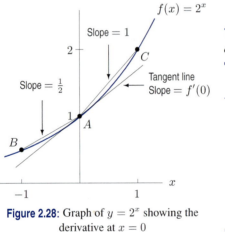

Figure 2.28: Graph of $y = 2^x$ showing the derivative at $x = 0$

Table 2.4   *Numerical values for difference quotient of 2^x at $x = 0$*

h	2^h	Difference quotient: $\frac{2^h - 1}{h}$
-0.0003	0.999792078	0.693075
-0.0002	0.999861380	0.693099
-0.0001	0.999930688	0.693123
0	1	
0.0001	1.00006932	0.693171
0.0002	1.00013864	0.693195
0.0003	1.00020797	0.693219

The concavity of the curve tells us that difference quotients calculated with negative h's are smaller than the derivative, and those calculated with positive h's are larger. From Table 2.4 we see that the derivative is between 0.693123 and 0.693171. To three decimal places, $f'(0) = 0.693$.

Example 7   Find an approximate equation for the tangent line to $f(x) = 2^x$ at $x = 0$.

Solution   From the previous example, we know the slope of the tangent line is about 0.693. Since the tangent line has y-intercept 1, its equation is

$$y = 0.693x + 1.$$

Computing the Derivative of x^2

In previous examples we got an approximation to the derivative by using smaller and smaller values of h. We now see how to find the derivative of x^2 exactly.

Example 8   Find the derivative of the function $f(x) = x^2$ at the point $x = 1$.

Solution   The derivative is the limit of the difference quotient, so we look at

$$f'(1) = \lim_{h \to 0} \frac{f(1+h) - f(1)}{h}.$$

Using the formula for f, we have

$$f'(1) = \lim_{h \to 0} \frac{(1+h)^2 - 1^2}{h} = \lim_{h \to 0} \frac{(1 + 2h + h^2) - 1}{h} = \lim_{h \to 0} \frac{2h + h^2}{h}.$$

Since the limit only examines values of h close to, but not equal to zero, we can cancel h in the expression $(2h + h^2)/h$. We get

$$f'(1) = \lim_{h \to 0} \frac{h(2+h)}{h} = \lim_{h \to 0} (2 + h).$$

This limit is 2, so $f'(1) = 2$. At $x = 1$ the rate of change of x^2 is 2.

Since the derivative is the rate of change, $f'(1) = 2$ means that for small changes in x near $x = 1$, the change in $f(x) = x^2$ is about twice as big as the change in x. As an example, if x changes from 1 to 1.1, a net change of 0.1, then $f(x)$ changes by about 0.2. Figure 2.29 shows this geometrically. Near $x = 1$ the function is approximately linear with slope of 2.

Table 2.5 shows the derivative of $f(x) = x^2$ numerically. Notice that near $x = 1$, every time the value of x increases by 0.001, the value of x^2 increases by approximately 0.002.

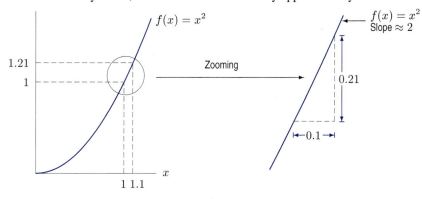

Figure 2.29: Graph of $f(x) = x^2$ near $x = 1$ has slope ≈ 2

Table 2.5   *Values of $f(x) = x^2$ near $x = 1$*

x	x^2	Difference in successive x^2 values
0.998	0.996004	
		0.001997
0.999	0.998001	
		0.001999
1.000	1.000000	
		0.002001
1.001	1.002001	
		0.002003
1.002	1.004004	
↑		↑
x increments of 0.001		All approximately 0.002

Exercises and Problems for Section 2.3

Exercises

1. The table shows values of $f(x) = x^3$ near $x = 2$ (to three decimal places). Use it to estimate $f'(2)$.

x	1.998	1.999	2.000	2.001	2.002
x^3	7.976	7.988	8.000	8.012	8.024

2. Sketch a rough graph of $f(x) = \sin x$, and use the graph to decide whether the derivative of $f(x)$ at $x = 3\pi$ is positive or negative. Give reasons for your decision.

3. (a) Make a table of values rounded to two decimal places for the function $f(x) = e^x$ for $x = 1, 1.5, 2, 2.5,$ and 3. Then use the table to answer parts (b) and (c).
(b) Find the average rate of change of $f(x)$ between $x = 1$ and $x = 3$.
(c) Use average rates of change to approximate the instantaneous rate of change of $f(x)$ at $x = 2$.

4. (a) Make a table of values, rounded to two decimal places, for $f(x) = \log x$ (that is, log base 10) with $x = 1, 1.5, 2, 2.5, 3$. Then use this table to answer parts (b) and (c).
(b) Find the average rate of change of $f(x)$ between $x = 1$ and $x = 3$.
(c) Use average rates of change to approximate the instantaneous rate of change of $f(x)$ at $x = 2$.

5. For the function $f(x) = \log x$, estimate $f'(1)$. From the graph of $f(x)$, would you expect your estimate to be greater than or less than $f'(1)$?

6. Estimate $f'(2)$ for $f(x) = 3^x$. Explain your reasoning.

7. Figure 2.30 shows the graph of f. Match the derivatives in the table with the points a, b, c, d, e.

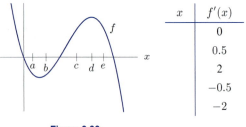

x	$f'(x)$
	0
	0.5
	2
	−0.5
	−2

Figure 2.30

8. Label points $A, B, C, D, E,$ and F on the graph of $y = f(x)$ in Figure 2.31.

(a) Point A is a point on the curve where the derivative is negative.
(b) Point B is a point on the curve where the value of the function is negative.
(c) Point C is a point on the curve where the derivative is largest.

(d) Point D is a point on the curve where the derivative is zero.
(e) Points E and F are different points on the curve where the derivative is about the same.

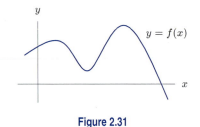

Figure 2.31

9. The graph of $y = f(x)$ is shown in Figure 2.32. Which is larger in each of the following pairs?

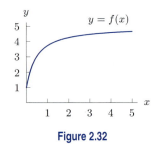

Figure 2.32

(a) Average rate of change: Between $x = 1$ and $x = 3$? Or between $x = 3$ and $x = 5$?
(b) $f(2)$ or $f(5)$?
(c) $f'(1)$ or $f'(4)$?

Find the derivatives in Exercises 10–15 algebraically.

10. $f(x) = 5x^2$ at $x = 10$ **11.** $f(x) = x^3$ at $x = -2$

12. $g(t) = 3t^2 + 5t$ at $t = -1$ **13.** $f(x) = x^3 + 5$ at $x = 1$

14. $g(x) = 1/x$ at $x = 2$ **15.** $g(z) = z^{-2}$, find $g'(2)$

For Exercises 16–19, find the equation of the line tangent to the function at the given point.

16. $f(x) = 5x^2$ at $x = 10$ **17.** $f(x) = x^3$ at $x = -2$

18. $f(x) = x$ at $x = 20$ **19.** $f(x) = 1/x^2$ at $(1, 1)$

Problems

20. Suppose that $f(x)$ is a function with $f(100) = 35$ and $f'(100) = 3$. Estimate $f(102)$.

21. The function graphed in Figure 2.33 has $f(4) = 25$ and $f'(4) = 1.5$. Find the coordinates of the points A, B, C.

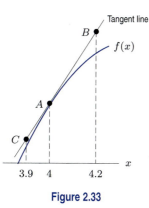

Figure 2.33

22. Use Figure 2.34 to fill in the blanks in the following statements about the function g at point B.

(a) $g(\underline{}) = \underline{}$ (b) $g'(\underline{}) = \underline{}$

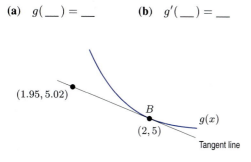

Figure 2.34

23. Show how you can represent the following on a copy of Figure 2.35.

(a) $f(4)$ (b) $f(4) - f(2)$

(c) $\dfrac{f(5) - f(2)}{5 - 2}$ (d) $f'(3)$

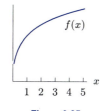

Figure 2.35

24. Consider the function $y = f(x)$ shown in Figure 2.35. For each of the following pairs of numbers, decide which is larger. Explain your answer.

(a) $f(3)$ or $f(4)$?

(b) $f(3) - f(2)$ or $f(2) - f(1)$?

(c) $\dfrac{f(2) - f(1)}{2 - 1}$ or $\dfrac{f(3) - f(1)}{3 - 1}$?

(d) $f'(1)$ or $f'(4)$?

25. With the function f given by Figure 2.35, arrange the following quantities in ascending order:

$$0, \quad f'(2), \quad f'(3), \quad f(3) - f(2)$$

26. Suppose $y = f(x)$ graphed in Figure 2.35 represents the cost of manufacturing x kilograms of a chemical. Then $f(x)/x$ represents the average cost of producing 1 kilogram when x kilograms are made. This problem asks you to visualize these averages graphically.

(a) Show how to represent $f(4)/4$ as the slope of a line.

(b) Which is larger, $f(3)/3$ or $f(4)/4$?

27. Consider the function shown in Figure 2.36.

(a) Write an expression involving f for the slope of the line joining A and B.

(b) Draw the tangent line at C. Compare its slope to the slope of the line in part (a).

(c) Are there any other points on the curve at which the slope of the tangent line is the same as the slope of the tangent line at C? If so, mark them on the graph. If not, why not?

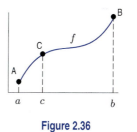

Figure 2.36

28. Estimate the instantaneous rate of change of the function $f(x) = x \ln x$ at $x = 1$ and at $x = 2$. What do these values suggest about the concavity of the function between 1 and 2?

29. (a) Estimate $f'(0)$ if $f(x) = \sin x$, with x in degrees.

(b) In Example 4 on page 73, we found that the derivative of $\sin x$ at $x = 0$ was 1. Why do we get a different result here? (This problem shows why radians are almost always used in calculus.)

30. Estimate the derivative of $f(x) = x^x$ at $x = 2$.

31. For $y = f(x) = 3x^{3/2} - x$, use your calculator to construct a graph of $y = f(x)$, for $0 \leq x \leq 2$. From your graph, estimate $f'(0)$ and $f'(1)$.

32. Let $f(x) = \ln(\cos x)$. Use your calculator to approximate the instantaneous rate of change of f at the point $x = 1$. Do the same thing for $x = \pi/4$. (Note: Be sure that your calculator is set in radians.)

33. The population, P, of China, in billions, can be approximated by the function

$$P = 1.15(1.014)^t,$$

where t is the number of years since the start of 1993. According to this model, how fast is the population growing at the start of 1993 and at the start of 1995? Give your answers in millions of people per year.

34. (a) Sketch graphs of the functions $f(x) = \frac{1}{2}x^2$ and $g(x) = f(x) + 3$ on the same set of axes. What can you say about the slopes of the tangent lines to the two graphs at the point $x = 0$? $x = 2$? Any point $x = x_0$?

(b) Explain why adding a constant value, C, to any function does not change the value of the slope of its graph at any point. [Hint: Let $g(x) = f(x) + C$, and calculate the difference quotients for f and g.]

35. Suppose Table 2.4 on page 74 is continued with smaller values of h. A particular calculator gives the results in Table 2.6. (Your calculator may give slightly different results.) Comment on the values of the difference quotient in Table 2.6. In particular, why is the last value of $(2^h - 1)/h$ zero? What do you expect the calculated value of $(2^h - 1)/h$ to be when $h = 10^{-20}$?

Table 2.6   *Questionable values of difference quotients of 2^x near $x = 0$*

h	Difference quotient: $(2^h - 1)/h$
10^{-4}	0.6931712
10^{-6}	0.693147
10^{-8}	0.6931
10^{-10}	0.69
10^{-12}	0

2.4   THE DERIVATIVE FUNCTION

In the previous section we looked at the derivative of a function at a fixed point. Now we consider what happens at a variety of points. The derivative generally takes on different values at different points and is itself a function.

First, remember that the derivative of a function at a point tells us the rate at which the value of the function is changing at that point. Geometrically, we can think of the derivative as the slope of the curve or of the tangent line at the point.

Example 1   Estimate the derivative of the function $f(x)$ graphed in Figure 2.37 at $x = -2, -1, 0, 1, 2, 3, 4, 5$.

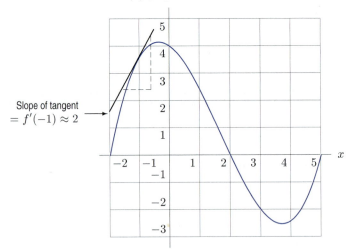

Figure 2.37: Estimating the derivative graphically as the slope of the tangent line

Solution   From the graph we estimate the derivative at any point by placing a straightedge so that it forms the tangent line at that point, and then using the grid squares to estimate the slope of the straightedge. For example, the tangent at $x = -1$ is drawn in Figure 2.37, and has a slope of about 2, so $f'(-1) \approx 2$. Notice that the slope at $x = -2$ is positive and fairly large; the slope at $x = -1$ is positive but smaller. At $x = 0$, the slope is negative, by $x = 1$ it has become more negative, and so on.

Table 2.7   *Estimated values of derivative of function in Figure 2.37*

x	-2	-1	0	1	2	3	4	5
$f'(x)$	6	2	-1	-2	-2	-1	1	4

Some estimates of the derivative are listed in Table 2.7. You should check these values. Are they reasonable? Is the derivative positive where you expect? Negative?

Notice that for every x-value, there's a corresponding value of the derivative. Therefore, the derivative is itself a function of x.

For any function f, we define the **derivative function**, f', by

$$f'(x) = \text{Rate of change of } f \text{ at } x = \lim_{h \to 0} \frac{f(x+h) - f(x)}{h}.$$

For every x-value for which this limit exists, we say f is *differentiable at* that x-value. If the limit exists for all x in the domain of f, we say f is *differentiable everywhere*. Most functions we meet are differentiable at every point in their domain, except perhaps for a few isolated points.

The Derivative Function: Graphically

Example 2   Sketch the graph of the derivative of the function shown in Figure 2.37.

Solution   We plot the values of this derivative given in Table 2.7. We obtain Figure 2.38, which shows a graph of the derivative (the black curve), along with the original function (color).

You should check that this graph of f' makes sense. The values of f' are positive where f is increasing ($x < -0.3$ or $x > 3.8$) and negative where f is decreasing. Notice that at the points where f has large positive slope, such as $x = -2$, the graph of the derivative is far above the x-axis, as it should be, since the value of the derivative is large there. At points where the slope is gentler, such as $x = -1$, the graph of f' is closer to the x-axis, since the derivative is smaller.

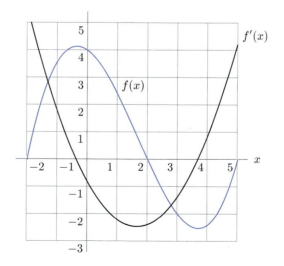

Figure 2.38: Function (colored) and derivative (black) from Example 1

What Does the Derivative Tell Us Graphically?

Where f' is positive, the tangent line to f is sloping up; where f' is negative, the tangent line to f is sloping down. If $f' = 0$ everywhere, then the tangent line to f is horizontal everywhere, and f is constant. We see that the sign of f' tells us whether f is increasing or decreasing.

> If $f' > 0$ on an interval, then f is *increasing* over that interval.
> If $f' < 0$ on an interval, then f is *decreasing* over that interval.

Moreover, the magnitude of the derivative gives us the magnitude of the rate of change; so if f' is large (positive or negative), then the graph of f is steep (up or down), whereas if f' is small the graph of f slopes gently. With this in mind, we can learn about the behavior of a function from the behavior of its derivative.

The Derivative Function: Numerically

If we are given values of a function instead of its graph, we can estimate values of the derivative.

Example 3   Table 2.8 gives values of $c(t)$, the concentration (μg/cm^3) of a drug in the bloodstream at time t (min). Construct a table of estimated values for $c'(t)$, the rate of change of $c(t)$ with respect to time.

Table 2.8   *Concentration as a function of time*

t (min)	0	0.1	0.2	0.3	0.4	0.5	0.6	0.7	0.8	0.9	1.0
$c(t)$ (μg/cm^3)	0.84	0.89	0.94	0.98	1.00	1.00	0.97	0.90	0.79	0.63	0.41

Solution   We estimate values of c' using the values in the table. To do this, we have to assume that the data points are close enough together that the concentration doesn't change wildly between them. From the table, we see that the concentration is increasing between $t = 0$ and $t = 0.4$, so we expect a positive derivative there. However, the increase is quite slow, so we expect the derivative to be small. The concentration doesn't change between 0.4 and 0.5, so we expect the derivative to be roughly 0 there. From $t = 0.5$ to $t = 1.0$, the concentration starts to decrease, and the rate of decrease gets larger and larger, so we expect the derivative to be negative and of greater and greater magnitude.

Using the data in the table, we estimate the derivative using the difference quotient:

$$c'(t) \approx \frac{c(t + h) - c(t)}{h}.$$

Since the data points are 0.1 apart, we use $h = 0.1$, giving

$$c'(0) \approx \frac{c(0.1) - c(0)}{0.1} = \frac{0.89 - 0.84}{0.1} = 0.5 \ \mu\text{g/cm}^3/\text{min}$$

$$c'(0.1) \approx \frac{c(0.2) - c(0.1)}{0.1} = \frac{0.94 - 0.89}{0.1} = 0.5 \ \mu\text{g/cm}^3/\text{min}$$

$$c'(0.2) \approx \frac{c(0.3) - c(0.2)}{0.1} = \frac{0.98 - 0.94}{0.1} = 0.4 \ \mu\text{g/cm}^3/\text{min}$$

$$c'(0.3) \approx \frac{c(0.4) - c(0.3)}{0.1} = \frac{1.00 - 0.98}{0.1} = 0.2 \ \mu\text{g/cm}^3/\text{min}$$

$$c'(0.4) \approx \frac{c(0.5) - c(0.4)}{0.1} = \frac{1.00 - 1.00}{0.1} = 0.0 \ \mu\text{g/cm}^3/\text{min}$$

and so on. These values are tabulated in Table 2.9. Notice that the derivative has small positive values up until $t = 0.4$, where it is roughly 0, and then it gets more and more negative, as we expected. The slopes are shown on the graph of $c(t)$ in Figure 2.39.

Table 2.9 *Estimated derivative of concentration*

t	$c'(t)$
0	0.5
0.1	0.5
0.2	0.4
0.3	0.2
0.4	0.0
0.5	−0.3
0.6	−0.7
0.7	−1.1
0.8	−1.6
0.9	−2.2

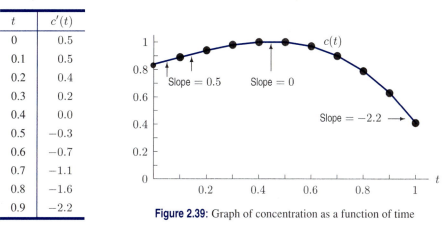

Figure 2.39: Graph of concentration as a function of time

Improving Numerical Estimates for the Derivative

In the previous example, the estimate for the derivative at 0.2 used the interval to the right; we found the average rate of change between $t = 0.2$ and $t = 0.3$. However, we could equally well have gone to the left and used the rate of change between $t = 0.1$ and $t = 0.2$ to approximate the derivative at 0.2. For a more accurate result, we could average these slopes and say

$$c'(0.2) \approx \frac{1}{2}\left(\begin{matrix}\text{Slope to left} \\ \text{of } 0.2\end{matrix} + \begin{matrix}\text{Slope to right} \\ \text{of } 0.2\end{matrix}\right) = \frac{0.5 + 0.4}{2} = 0.45.$$

In general, averaging the slopes leads to a more accurate answer.

Derivative Function: From a Formula

If we are given a formula for f, can we come up with a formula for f'? We often can, as shown in the next example. Indeed, much of the power of calculus depends on our ability to find formulas for the derivatives of all the functions we described earlier. This is done systematically in Chapter 3.

Derivative of a Constant Function

The graph of a constant function $f(x) = k$ is a horizontal line, with a slope of 0 everywhere. Therefore, its derivative is 0 everywhere. (See Figure 2.40.)

> If $f(x) = k$, then $f'(x) = 0$.

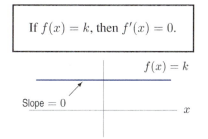

Figure 2.40: A constant function

Derivative of a Linear Function

We already know that the slope of a straight line is constant. This tells us that the derivative of a linear function is constant.

> If $f(x) = b + mx$, then $f'(x) = \text{Slope} = m$.

Derivative of a Power Function

Example 4   Find a formula for the derivative of $f(x) = x^2$.

Solution   Before computing the formula for $f'(x)$ algebraically, let's try to guess the formula by looking for a pattern in the values of $f'(x)$. Table 2.10 contains values of $f(x) = x^2$ (rounded to three decimals), which we can use to estimate the values of $f'(1)$, $f'(2)$, and $f'(3)$.

Table 2.10   *Values of $f(x) = x^2$ near $x = 1$, $x = 2$, $x = 3$ (rounded to three decimals)*

x	x^2	x	x^2	x	x^2
0.999	0.998	1.999	3.996	2.999	8.994
1.000	1.000	2.000	4.000	3.000	9.000
1.001	1.002	2.001	4.004	3.001	9.006
1.002	1.004	2.002	4.008	3.002	9.012

Near $x = 1$, the value of x^2 increases by about 0.002 each time x increases by 0.001, so

$$f'(1) \approx \frac{0.002}{0.001} = 2.$$

Similarly, near $x = 2$ and $x = 3$, the value of x^2 increases by about 0.004 and 0.006, respectively, when x increases by 0.001. So

$$f'(2) \approx \frac{0.004}{0.001} = 4 \quad \text{and} \quad f'(3) \approx \frac{0.006}{0.001} = 6.$$

Knowing the value of f' at specific points can never tell us the formula for f', but it certainly can be suggestive: Knowing $f'(1) \approx 2$, $f'(2) \approx 4$, $f'(3) \approx 6$ suggests that $f'(x) = 2x$.

The derivative is calculated by forming the difference quotient and taking the limit as h goes to zero. The difference quotient is

$$\frac{f(x+h) - f(x)}{h} = \frac{(x+h)^2 - x^2}{h} = \frac{x^2 + 2xh + h^2 - x^2}{h} = \frac{2xh + h^2}{h}.$$

Since h never actually reaches zero, we can cancel it in the last expression to get $2x + h$. The limit of this as h goes to zero is $2x$, so

$$f'(x) = \lim_{h \to 0} (2x + h) = 2x.$$

Example 5   Calculate $f'(x)$ if $f(x) = x^3$.

Solution   We look at the difference quotient

$$\frac{f(x+h) - f(x)}{h} = \frac{(x+h)^3 - x^3}{h}.$$

Multiplying out gives $(x+h)^3 = x^3 + 3x^2 h + 3xh^2 + h^3$, so

$$f'(x) = \lim_{h \to 0} \frac{x^3 + 3x^2 h + 3xh^2 + h^3 - x^3}{h} = \lim_{h \to 0} \frac{3x^2 h + 3xh^2 + h^3}{h}.$$

Since in taking the limit as $h \to 0$, we consider values of h near, but not equal to, zero, we can cancel h giving

$$f'(x) = \lim_{h \to 0} \frac{3x^2 h + 3xh^2 + h^3}{h} = \lim_{h \to 0} (3x^2 + 3xh + h^2).$$

As $h \to 0$, the value of $(3xh + h^2) \to 0$ so

$$f'(x) = \lim_{h \to 0} (3x^2 + 3xh + h^2) = 3x^2.$$

The previous two examples show how to compute the derivatives of power functions of the form $f(x) = x^n$, when n is 2 or 3. We can use the Binomial Theorem to show the *power rule* for a positive integer n:

$$\text{If } f(x) = x^n \text{ then } f'(x) = nx^{n-1}.$$

This result is in fact valid for any real value of n.

Exercises and Problems for Section 2.4

Exercises

For Exercises 1–9, sketch a graph of the derivative function of each of the given functions.

1.
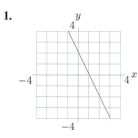

2.
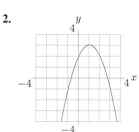

9.
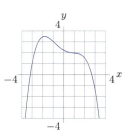

10. For $f(x) = \ln x$, construct tables, rounded to four decimals, near $x = 1$, $x = 2$, $x = 5$, and $x = 10$. Use the tables to estimate $f'(1)$, $f'(2)$, $f'(5)$, and $f'(10)$. Then guess a general formula for $f'(x)$.

3.
4.
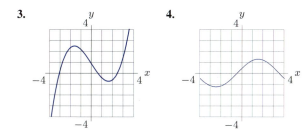

11. (a) Estimate $f'(2)$ using the values of f in the table.
(b) For what values of x does $f'(x)$ appear to be positive? Negative?

x	0	2	4	6	8	10	12
$f(x)$	10	18	24	21	20	18	15

12. Find approximate values for $f'(x)$ at each of the x-values given in the following table.

x	0	5	10	15	20
$f(x)$	100	70	55	46	40

5.
6.
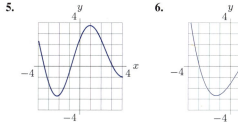

In Exercises 13–14, find a formula for the derivative using the power rule. Confirm it using difference quotients.

13. $k(x) = 1/x$ **14.** $l(x) = 1/x^2$

Find a formula for the derivatives of the functions in Exercises 15–16 using difference quotients.

15. $g(x) = 2x^2 - 3$ **16.** $m(x) = 1/(x+1)$

7.
8.
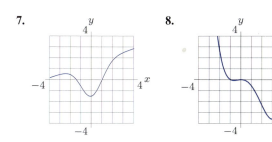

For Exercises 17–20, sketch the graph of $f(x)$, and use this graph to sketch the graph of $f'(x)$.

17. $f(x) = x^2$ **18.** $f(x) = x(x-1)$

19. $f(x) = \cos x$ **20.** $f(x) = \log x$

Problems

21. Given the numerical values shown, find approximate values for the derivative of $f(x)$ at each of the x-values given. Where is the rate of change of $f(x)$ positive? Where is it negative? Where does the rate of change of $f(x)$ seem to be greatest?

x	0	1	2	3	4	5	6	7	8
$f(x)$	18	13	10	9	9	11	15	21	30

22. Values of x and $g(x)$ are given in the table. For what value of x is $g'(x)$ closest to 3?

x	2.7	3.2	3.7	4.2	4.7	5.2	5.7	6.2
$g(x)$	3.4	4.4	5.0	5.4	6.0	7.4	9.0	11.0

For Problems 23–32, sketch the graph of $f'(x)$.

23.

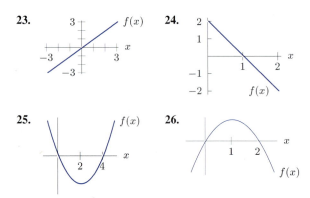

24.

25.

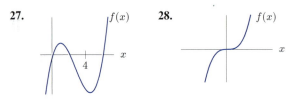

26.

27.

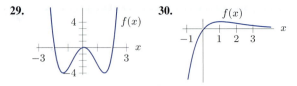

28.

29.

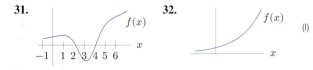

30.

31.

32.

33. Draw the graph of a continuous function $y = f(x)$ that satisfies the following three conditions.

- $f'(x) > 0$ for $x < -2$,
- $f'(x) < 0$ for $-2 < x < 2$,
- $f'(x) = 0$ for $x > 2$.

34. Draw the graph of a continuous function $y = f(x)$ that satisfies the following three conditions.

- $f'(x) > 0$ for $-\frac{\pi}{2} < x < \frac{\pi}{2}$,
- $f'(x) < 0$ for $-\pi < x < -\frac{\pi}{2}$ and $\frac{\pi}{2} < x < \pi$,
- $f'(x) = 0$ at $x = -\frac{\pi}{2}$ and $x = \frac{\pi}{2}$.

35. In the graph of f in Figure 2.41, at which of the labeled x-values is

(a) $f(x)$ greatest?   (b) $f(x)$ least?

(c) $f'(x)$ greatest?   (d) $f'(x)$ least?

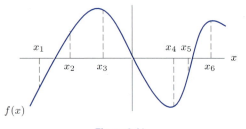

Figure 2.41

36. Figure 2.42 shows a graph of voltage across an electrical capacitor as a function of time. The current is proportional to the derivative of the voltage; the constant of proportionality is positive. Sketch a graph of the current as a function of time.

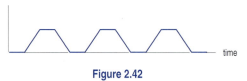

Figure 2.42

37. A vehicle moving along a straight road has distance $f(t)$ from its starting point at time t. Which of the graphs in Figure 2.43 could be $f'(t)$ for the following scenarios? (Assume the scales on the vertical axes are all the same.)

(a) A bus on a popular route, with no traffic

(b) A car with no traffic and all green lights

(c) A car in heavy traffic conditions

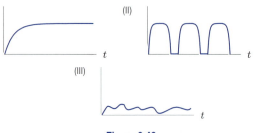

Figure 2.43

38. A child inflates a balloon, admires it for a while and then lets the air out at a constant rate. If $V(t)$ gives the volume of the balloon at time t, then Figure 2.44 shows $V'(t)$ as a function of t. At what time does the child:

(a) Begin to inflate the balloon?
(b) Finish inflating the balloon?
(c) Begin to let the air out?
(d) What would the graph of $V'(t)$ look like if the child had alternated between pinching and releasing the open end of the balloon, instead of letting the air out at a constant rate?

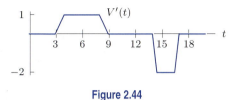

Figure 2.44

39. The population of a herd of deer is modeled by

$$P(t) = 4000 + 500 \sin\left(2\pi t - \frac{\pi}{2}\right)$$

where t is measured in years from January 1.

(a) How does this population vary with time? Sketch a graph of $P(t)$ for one year.

(b) Use the graph to decide when in the year the population is a maximum. What is that maximum? Is there a minimum? If so, when?
(c) Use the graph to decide when the population is growing fastest. When is it decreasing fastest?
(d) Estimate roughly how fast the population is changing on the first of July.

40. Figure 2.45 is the graph of f', the derivative of a function f. On what interval(s) is the function f

(a)   Increasing?   **(b)**   Decreasing?

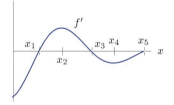

Figure 2.45: Graph of f', not f

41. Looking at the graph, explain why if $f(x)$ is an even function, then $f'(x)$ is odd.

42. Looking at the graph, explain why if $g(x)$ is an odd function, then $g'(x)$ is even.

2.5   INTERPRETATIONS OF THE DERIVATIVE

We have seen the derivative interpreted as a slope and as a rate of change. In this section, we see other interpretations. The purpose of these examples is not to make a catalog of interpretations but to illustrate the process of obtaining them.

An Alternative Notation for the Derivative

So far we have used the notation f' to stand for the derivative of the function f. An alternative notation for derivatives was introduced by the German mathematician Wilhelm Gottfried Leibniz (1646–1716). If the variable y depends on the variable x, that is, if

$$y = f(x),$$

then he wrote dy/dx for the derivative, so

$$\frac{dy}{dx} = f'(x).$$

Leibniz's notation is quite suggestive if we think of the letter d in dy/dx as standing for "small difference in . . . ." The notation dy/dx reminds us that the derivative is a limit of ratios of the form

$$\frac{\text{Difference in } y\text{-values}}{\text{Difference in } x\text{-values}}.$$

The notation dy/dx suggests the units for the derivative: the units for y divided by the units for x. The separate entities dy and dx officially have no independent meaning: they are all part of one notation. In fact, a good formal way to view the notation dy/dx is to think of d/dx as a single symbol meaning "the derivative with respect to x of ...". So dy/dx can be viewed as

$$\frac{d}{dx}(y), \quad \text{meaning "the derivative with respect to } x \text{ of } y\text{."}$$

On the other hand, many scientists and mathematicians think of dy and dx as separate entities representing "infinitesimally" small differences in y and x, even though it is difficult to say exactly how small "infinitesimal" is. Although not formally correct, it can be helpful to think of dy/dx as a small change in y divided by a small change in x.

For example, recall that if $s = f(t)$ is the position of a moving object at time t, then $v = f'(t)$ is the velocity of the object at time t. Writing

$$v = \frac{ds}{dt}$$

reminds us that v is a velocity, since the notation suggests a distance, ds, over a time, dt, and we know that distance over time is velocity. Similarly, we recognize

$$\frac{dy}{dx} = f'(x)$$

as the slope of the graph of $y = f(x)$ since slope is vertical rise, dy, over horizontal run, dx.

The disadvantage of Leibniz's notation is that it is awkward to specify the x-value at which we are evaluating the derivative. To specify $f'(2)$, for example, we have to write

$$\left.\frac{dy}{dx}\right|_{x=2}.$$

Using Units to Interpret the Derivative

The following examples illustrate how useful units can be in suggesting interpretations of the derivative. We use the fact that the units of the instantaneous and the average rate of change are the same.

For example, suppose $s = f(t)$ gives the distance, in meters, of a body from a fixed point as a function of time, t, in seconds. Then knowing that

$$\left.\frac{ds}{dt}\right|_{t=2} = f'(2) = 10 \text{ meters/sec}$$

tells us that when $t = 2$ seconds, the body is moving at an instantaneous velocity of 10 meters/sec. This means that if the body continued to move at this speed for a whole second, it would move 10 meters. In practice, however, the velocity of the body may not remain 10 meters/sec for long. Notice that the units of instantaneous velocity and of average velocity are the same.

Example 1   The cost C (in dollars) of building a house A square feet in area is given by the function $C = f(A)$. What is the practical interpretation of the function $f'(A)$?

Solution   In the alternative notation,

$$f'(A) = \frac{dC}{dA}.$$

This is a cost divided by an area, so it is measured in dollars per square foot. You can think of dC as the extra cost of building an extra dA square feet of house. Then you can think of dC/dA as the additional cost per square foot. So if you are planning to build a house roughly A square feet in area, $f'(A)$ is the cost per square foot of the *extra* area involved in building a slightly larger house, and is called the *marginal cost*. The marginal cost is probably smaller than the average cost per square foot for the entire house, since once you are already set up to build a large house, the cost of adding a few square feet is likely to be small.

Example 2   The cost of extracting T tons of ore from a copper mine is $C = f(T)$ dollars. What does it mean to say that $f'(2000) = 100$?

Solution   In the alternative notation,

$$f'(2000) = \left.\frac{dC}{dT}\right|_{T=2000} = 100.$$

Since C is measured in dollars and T is measured in tons, dC/dT must be measured in dollars per ton. So the statement $f'(2000) = 100$ says that when 2000 tons of ore have already been extracted from the mine, the cost of extracting the next ton is approximately $100.

Example 3   If $q = f(p)$ gives the number of pounds of sugar produced when the price per pound is p dollars, then what are the units and the meaning of the statement $f'(3) = 50$?

Solution   Since $f'(3)$ is the limit as $h \to 0$ of the difference quotient

$$\frac{f(3+h) - f(3)}{h},$$

the units of $f'(3)$ and the difference quotient are the same. Since $f(3+h) - f(3)$ is in pounds and h is in dollars, the units of the difference quotient and $f'(3)$ are pounds/dollar. The statement

$$f'(3) = 50 \text{ pounds/dollar}$$

tells us that the instantaneous rate of change of q with respect to p is 50 when $p = 3$. In other words, when the price is $3, the quantity produced is increasing at 50 pounds/dollar. Thus, if the price increased by a dollar, the quantity produced would increase by approximately 50 pounds.

Example 4   You are told that water is flowing through a pipe at a constant rate of 10 cubic feet per second. Interpret this rate as the derivative of some function.

Solution   You might think at first that the statement has something to do with the velocity of the water, but in fact a flow rate of 10 cubic feet per second could be achieved either with very slowly moving water through a large pipe, or with very rapidly moving water through a narrow pipe. If we look at the units—cubic feet per second—we realize that we are being given the rate of change of a quantity measured in cubic feet. But a cubic foot is a measure of volume, so we are being told the rate of change of a volume. One way to visualize this is to imagine all the water that is flowing through the pipe ending up in a tank somewhere. Let $V(t)$ be the volume of water in the tank at time t. Then we are being told that the rate of change of $V(t)$ is 10, or

$$V'(t) = \frac{dV}{dt} = 10.$$

Example 5   Suppose $P = f(t)$ is the population of Mexico in millions, where t is the number of years since 1980. Explain the meaning of the statements:

(a) $f'(6) = 2$ (b) $f^{-1}(95.5) = 16$ (c) $(f^{-1})'(95.5) = 0.46$

Solution   (a) The units of P are millions of people, the units of t are years, so the units of $f'(t)$ are millions of people per year. Therefore the statement $f'(6) = 2$ tells us that at $t = 6$ (that is, in 1986), the population of Mexico was increasing at 2 million people per year.

(b) The statement $f^{-1}(95.5) = 16$ tells us that the year when the population was 95.5 million was $t = 16$ (that is, in 1996).

(c) The units of $(f^{-1})'(P)$ are years per million of population. The statement $(f^{-1})'(95.5) = 0.46$ tells us that when the population was 95.5 million, it took about 0.46 years for the population to increase by 1 million.

Exercises and Problems for Section 2.5

Exercises

1. The temperature, H, in degrees Celsius, of a cup of coffee placed on the kitchen counter is given by $H = f(t)$, where t is in minutes since the coffee was put on the counter.

   (a) Is $f'(t)$ positive or negative? Give a reason for your answer.

   (b) What are the units of $f'(20)$? What is its practical meaning in terms of the temperature of the coffee?

2. The temperature, T, in degrees Fahrenheit, of a cold yam placed in a hot oven is given by $T = f(t)$, where t is the time in minutes since the yam was put in the oven.

   (a) What is the sign of $f'(t)$? Why?

   (b) What are the units of $f'(20)$? What is the practical meaning of the statement $f'(20) = 2$?

3. The cost, C (in dollars) to produce g gallons of ice cream can be expressed as $C = f(g)$. Using units, explain the meaning of the following statements in terms of ice cream.

   (a) $f(200) = 350$ **(b)** $f'(200) = 1.4$

4. The time for a chemical reaction, T (in minutes), is a function of the amount of catalyst present, a (in milliliters), so $T = f(a)$.

   (a) If $f(5) = 18$, what are the units of 5? What are the units of 18? What does this statement tell us about the reaction?

   (b) If $f'(5) = -3$, what are the units of 5? What are the units of -3? What does this statement tell us?

5. After investing $1000 at an annual interest rate of 7% compounded continuously for t years, your balance is $\$B$, where $B = f(t)$. What are the units of dB/dt? What is the financial interpretation of dB/dt?

6. Investing $1000 at an annual interest rate of $r\%$, compounded continuously, for 10 years gives you a balance of $\$B$, where $B = g(r)$. Give a financial interpretation of the statements:

   (a) $g(5) \approx 1649$.

   (b) $g'(5) \approx 165$. What are the units of $g'(5)$?

7. Suppose $C(r)$ is the total cost of paying off a car loan borrowed at an annual interest rate of $r\%$. What are the units of $C'(r)$? What is the practical meaning of $C'(r)$? What is its sign?

8. Suppose $P(t)$ is the monthly payment, in dollars, on a mortgage which will take t years to pay off. What are the units of $P'(t)$? What is the practical meaning of $P'(t)$? What is its sign?

9. Let $f(x)$ be the elevation in feet of the Mississippi river x miles from its source. What are the units of $f'(x)$? What can you say about the sign of $f'(x)$?

10. An economist is interested in how the price of a certain item affects its sales. At a price of $\$p$, a quantity, q, of the item is sold. If $q = f(p)$, explain the meaning of each of the following statements:

   (a) $f(150) = 2000$ **(b)** $f'(150) = -25$

Problems

11. A laboratory study investigating the relationship between diet and weight in adult humans found that the weight of a subject, W, in pounds, was a function, $W = f(c)$, of the average number of Calories per day, c, consumed by the subject.

   (a) Interpret the statements $f(1800) = 155$, $f'(2000) = 0$, and $f^{-1}(162) = 2200$ in terms of diet and weight.

   (b) What are the units of $f'(c) = dW/dc$?

12. A city grew in population throughout the 1980s. The population was at its largest in 1990, and then shrank throughout the 1990s. Let $P = f(t)$ represent the population of the city t years since 1980. Sketch graphs of $f(t)$ and $f'(t)$, labeling the units on the axes.

13. If t is the number of years since 1993, the population, P, of China, in billions, can be approximated by the function

$$P = f(t) = 1.15(1.014)^t,$$

Estimate $f(6)$ and $f'(6)$, giving units. What do these two numbers tell you about the population of China?

14. For some painkillers, the size of the dose, D, given depends on the weight of the patient, W. Thus, $D = f(W)$, where D is in milligrams and W is in pounds.

   (a) Interpret the statements $f(140) = 120$ and $f'(140) = 3$ in terms of this painkiller.

   (b) Use the information in the statements in part (a) to estimate $f(145)$.

15. Let $f(t)$ be the number of centimeters of rainfall that has fallen since midnight, where t is the time in hours. Interpret the following in practical terms, giving units.

   (a) $f(10) = 3.1$ **(b)** $f^{-1}(10) = 16$

   (c) $f'(8) = 0.4$ **(d)** $(f^{-1})'(5) = 2$

16. Let $p(h)$ be the pressure in dynes per cm^2 on a diver at a depth of h meters below the surface of the ocean. What do each of the following quantities mean to the diver? Give units for the quantities.

   (a) $p(100)$ **(b)** h such that $p(h) = 1.2 \cdot 10^6$

   (c) $p(h) + 20$ **(d)** $p(h + 20)$

   (e) $p'(100)$ **(f)** h such that $p'(h) = 20$

17. If $g(v)$ is the fuel efficiency, in miles per gallon, of a car going at v miles per hour, what are the units of $g'(90)$? What is the practical meaning of the statement $g'(55) = -0.54$?

18. (a) If you jump out of an airplane without a parachute, you fall faster and faster until wind resistance causes you to approach a steady velocity, called the *terminal* velocity. Sketch a graph of your velocity against time.
   (b) Explain the concavity of your graph.
   (c) Assuming wind resistance to be negligible at $t = 0$, what natural phenomenon is represented by the slope of the graph at $t = 0$?

19. A company's revenue from car sales, C (in thousands of dollars), is a function of advertising expenditure, a, in thousands of dollars, so $C = f(a)$.

   (a) What does the company hope is true about the sign of f'?
   (b) What does the statement $f'(100) = 2$ mean in practical terms? How about $f'(100) = 0.5$?
   (c) Suppose the company plans to spend about $100,000 on advertising. If $f'(100) = 2$, should the company spend more or less than $100,000 on advertising? What if $f'(100) = 0.5$?

20. Let $P(x)$ be the number of people of height $\leq x$ inches in the US. What is the meaning of $P'(66)$? What are its units? Estimate $P'(66)$ (using common sense). Is $P'(x)$ ever negative? [Hint: You may want to approximate $P'(66)$ by a difference quotient, using $h = 1$. Also,

you may assume the US population is about 250 million, and note that 66 inches = 5 feet 6 inches.]

21. When you breathe, a muscle (called the diaphragm) reduces the pressure around your lungs and they expand to fill with air. The table shows the volume of a lung as a function of the reduction in pressure from the diaphragm. Pulmonologists (lung doctors) define the *compliance* of the lung as the derivative of this function.[2]

   (a) What are the units of compliance?
   (b) Estimate the maximum compliance of the lung.
   (c) Explain why the compliance gets small when the lung is nearly full (around 1 liter).

Pressure reduction (cm of water)	Volume (liters)
0	0.20
5	0.29
10	0.49
15	0.70
20	0.86
25	0.95
30	1.00

2.6 THE SECOND DERIVATIVE

Since the derivative is itself a function, we can consider its derivative. For a function f, the derivative of its derivative is called the *second derivative*, and written f'' (read "f double-prime"). If $y = f(x)$, the second derivative can also be written as $\dfrac{d^2y}{dx^2}$, which means $\dfrac{d}{dx}\left(\dfrac{dy}{dx}\right)$, the derivative of $\dfrac{dy}{dx}$.

What Do Derivatives Tell Us?

Recall that the derivative of a function tells you whether a function is increasing or decreasing:
- If $f' > 0$ on an interval, then f is *increasing* over that interval.
- If $f' < 0$ on an interval, then f is *decreasing* over that interval.

Since f'' is the derivative of f',
- If $f'' > 0$ on an interval, then f' is *increasing* over that interval.
- If $f'' < 0$ on an interval, then f' is *decreasing* over that interval.

What does it mean for f' to be increasing or decreasing? An example in which f' is increasing is shown in Figure 2.46, where the curve is bending upward, or is *concave up*. In the example shown in Figure 2.47, in which f' is decreasing, the graph is bending downward, or is *concave down*. These figures suggest

---

[2]Adapted from John B. West, *Respiratory Physiology* 4th Ed. (New York: Williams and Wilkins, 1990).

If $f'' > 0$ on an interval, then f' is increasing, so the graph of f is concave up there.
If $f'' < 0$ on an interval, then f' is decreasing, so the graph of f is concave down there.

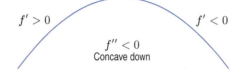

Figure 2.46: Meaning of f'': The slope increases from left to right, so f'' is positive and f is concave up

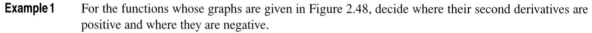

Figure 2.47: Meaning of f'': The slope decreases from left to right, so f'' is negative and f is concave down

Example 1   For the functions whose graphs are given in Figure 2.48, decide where their second derivatives are positive and where they are negative.

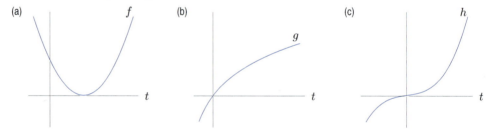

Figure 2.48: What signs do the second derivatives have?

Solution   From the graphs it appears that:

(a) $f'' > 0$ everywhere, because the graph of f is concave up everywhere.

(b) $g'' < 0$ everywhere, because the graph is concave down everywhere.

(c) $h'' < 0$ for $t < 0$, because the graph of h is concave down there; $h'' > 0$ for $t > 0$, because the graph of h is concave up there.

Interpretation of the Second Derivative as a Rate of Change

If we think of the derivative as a rate of change, then the second derivative is a rate of change of a rate of change. If the second derivative is positive, the rate of change of f is increasing; if the second derivative is negative, the rate of change of f is decreasing.

The second derivative can be a matter of practical concern. In 1985 a newspaper headline reported the Secretary of Defense as saying that Congress had cut the defense budget. As his opponents pointed out, however, Congress had merely cut the rate at which the defense budget was increasing.[3] In other words, the derivative of the defense budget was still positive (the budget was increasing), but the second derivative was negative (the budget's rate of increase had slowed).

---

[3] In the *Boston Globe*, March 13, 1985, Representative William Gray (D–Pa.) was reported as saying: "It's confusing to the American people to imply that Congress threatens national security with reductions when you're really talking about a reduction in the increase."

Example 2   A population, P, growing in a confined environment often follows a *logistic* growth curve, like that shown in Figure 2.49. Use d^2P/dt^2 to describe how the rate at which the population is increasing changes over time. What are practical interpretations of t_0 and L?

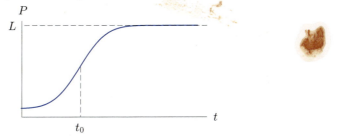

Figure 2.49: Logistic growth curve

Solution   Initially, the population is increasing, and at an increasing rate. So, initially dP/dt is increasing and so $d^2P/dt^2 > 0$. At t_0, the rate at which the population is increasing is a maximum. In other words, at time t_0 the population is growing fastest. Beyond t_0, the rate at which the population is growing is decreasing, and so $d^2P/dt^2 < 0$. At t_0, the curve changes from concave up to concave down, and $d^2P/dt^2 = 0$ there.

 The quantity L represents the limiting value of the population as $t \to \infty$. Biologists call L the *carrying capacity* of the environment.

Example 3   Tests on the C5 Chevy Corvette sports car gave the results[4] in Table 2.11.

(a) Estimate dv/dt for the time intervals shown.

(b) What can you say about the sign of d^2v/dt^2 over the period shown?

Table 2.11   *Velocity of C5 Chevy Corvette*

Time, t (sec)	0	3	6	9	12
Velocity, v (meters/sec)	0	20	33	43	51

Solution   (a) For each time interval we can calculate the average rate of change of velocity. For example, from $t = 0$ to $t = 3$ we have

$$\frac{dv}{dt} \approx \text{Average rate of change of velocity} = \frac{20 - 0}{3 - 0} = 6.67 \; \frac{\text{m/sec}}{\text{sec}}.$$

Estimated values of dv/dt are in Table 2.12.

(b) Since the values of dv/dt are decreasing, $d^2v/dt^2 < 0$. The graph of v against t in Figure 2.50 confirms this; it is concave down. The fact that $dv/dt > 0$ tells us that the car is speeding up; the fact that $d^2v/dt^2 < 0$ tells us that the rate of increase decreased over this time period.

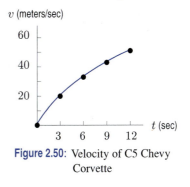

Figure 2.50: Velocity of C5 Chevy Corvette

Table 2.12   *Estimates for dv/dt (meters/sec/sec)*

Time interval (sec)	$0 - 3$	$3 - 6$	$6 - 9$	$9 - 12$
Average rate of change (dv/dt)	6.67	4.33	3.33	2.67

[4]Adapted from the report in *Car and Driver*, February 1997.

Velocity and Acceleration

The car in Example 3 is speeding up; we say that it is accelerating. We define *acceleration* as the rate of change of velocity with respect to time. If $v(t)$ is the velocity of an object at time t, we have

$$\text{Average acceleration from } t \text{ to } t+h = \frac{v(t+h) - v(t)}{h},$$

$$\text{Instantaneous acceleration} = v'(t) = \lim_{h \to 0} \frac{v(t+h) - v(t)}{h}.$$

If the term velocity or acceleration is used alone, it is assumed to be instantaneous. Since velocity is the derivative of position, acceleration is the second derivative of position. Summarizing:

If $y = s(t)$ is the position of an object at time t, then

- Velocity: $v(t) = \dfrac{dy}{dt} = s'(t).$

- Acceleration: $a(t) = \dfrac{d^2y}{dt^2} = s''(t) = v'(t).$

Example 4   A particle is moving along a straight line. If its distance, s, to the right of a fixed point is given by Figure 2.51, estimate:

(a) When the particle is moving to the right and when it is moving to the left.
(b) When the particle has positive acceleration and when it has negative acceleration.

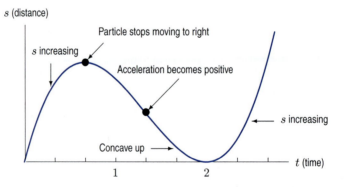

Figure 2.51: Distance of particle to right of a fixed point

Solution   (a) The particle is moving to the right whenever s is increasing. From the graph, this appears to be for $0 < t < \frac{2}{3}$ and for $t > 2$. For $\frac{2}{3} < t < 2$, the value of s is decreasing, so the particle is moving to the left.

(b) The particle has positive acceleration whenever the curve is concave up, which appears to be for $t > \frac{4}{3}$. The particle has negative acceleration when the curve is concave down, for $t < \frac{4}{3}$.

Exercises and Problems for Section 2.6

Exercises

1. For the function graphed in Figure 2.52, are the following quantities positive or negative?

   (a) $f(2)$ **(b)** $f'(2)$ **(c)** $f''(2)$

   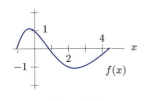

   Figure 2.52

2. The graph of a function $f(x)$ is shown in Figure 2.53. On a copy of the table indicate whether f, f', f'' at each marked point is positive, negative, or zero.

   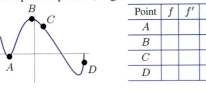

Point	f	f'	f''
A			
B			
C			
D			

   Figure 2.53

3. At which of the labeled points on the graph in Figure 2.54 are both dy/dx and d^2y/dx^2 positive?

   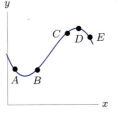

   Figure 2.54

4. The distance of a car from its initial position t minutes after setting out is given by $s(t) = 5t^2 + 3$ kilometers. What are the car's velocity and acceleration at time t? Give units.

5. Sketch the graph of a function whose first and second derivatives are everywhere positive.

6. Sketch the graph of a function whose first derivative is everywhere negative and whose second derivative is positive for some x-values and negative for other x-values.

7. Sketch the graph of the height of a particle against time if velocity is positive and acceleration is negative.

For Exercises 8–13, give the signs of the first and second derivatives for each of the following functions.

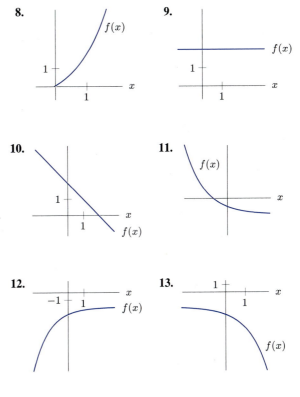

8.

9.

10.

11.

12.

13.

Problems

14. The table gives the number of passenger cars, $C = f(t)$, in millions, in the US in the year t.

    (a) Do $f'(t)$ and $f''(t)$ appear to be positive or negative during the period 1940–1980?
    (b) Estimate $f'(1975)$. Using units, interpret your answer in terms of passenger cars.

t (year)	1940	1950	1960	1970	1980
C (cars, in millions)	27.5	40.3	61.7	89.3	121.6

15. An accelerating sports car goes from 0 mph to 60 mph in five seconds. Its velocity is given in the following table, converted from miles per hour to feet per second, so that all time measurements are in seconds. (Note: 1 mph is 22/15 ft/sec.) Find the average acceleration of the car over each of the first two seconds.

Time, t (sec)	0	1	2	3	4	5
Velocity, $v(t)$ (ft/sec)	0	30	52	68	80	88

16. A function f satisfies the following conditions: $f(5) = 20$, $f'(5) = 2$, and $f''(x) < 0$, for $x \geq 5$. Which of the following are possible values for $f(7)$ and which are impossible?

   (a)   26   **(b)**   24   **(c)**   22

17. Let $P(t)$ represent the price of a share of stock of a corporation at time t. What does each of the following statements tell us about the signs of the first and second derivatives of $P(t)$?

   (a) "The price of the stock is rising faster and faster."
   (b) "The price of the stock is close to bottoming out."

18. In economics, *total utility* refers to the total satisfaction from consuming some commodity. According to the economist Samuelson:[5]

> As you consume more of the same good, the total (psychological) utility increases. However, ...with successive new units of the good, your total utility will grow at a slower and slower rate because of a fundamental tendency for your psychological ability to appreciate more of the good to become less keen.

   (a) Sketch the total utility as a function of the number of units consumed.
   (b) In terms of derivatives, what is Samuelson saying?

19. IBM-Peru uses second derivatives to assess the relative success of various advertising campaigns. They assume that all campaigns produce some increase in sales. If a graph of sales against time shows a positive second derivative during a new advertising campaign, what does this suggest to IBM management? Why? What does a negative second derivative suggest?

20. A continuous function defined for all x has the following properties:
   • f is increasing   • f is concave down
   • $f(5) = 2$ • $f'(5) = \frac{1}{2}$

   (a) Sketch a possible graph for f.   **(b)** How many zeros does f have?
   (c) What can you say about the location of the zeros?
   (d) What is $\lim\limits_{x \to -\infty} f(x)$?
   (e) Is it possible that $f'(1) = 1$?   **(f)** Is it possible that $f'(1) = \frac{1}{4}$?

21. An industry is being charged by the Environmental Protection Agency (EPA) with dumping unacceptable levels of toxic pollutants in a lake. Over a period of several months, an engineering firm makes daily measurements of the rate at which pollutants are being discharged into the lake. The engineers produce a graph similar to either Figure 2.55(a) or Figure 2.55(b). For each case, give an idea of what argument the EPA might make in court against the industry and of the industry's defense.

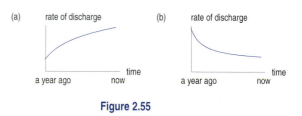

Figure 2.55

22. The graph of f' (not f) is given in Figure 2.56. At which of the marked values of x is

   (a) $f(x)$ greatest?   **(b)** $f(x)$ least?
   (c) $f'(x)$ greatest?   **(d)** $f'(x)$ least?
   (e) $f''(x)$ greatest?   **(f)** $f''(x)$ least?

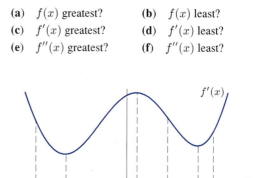

Figure 2.56: Graph of f', not f

23. Which of the points labeled by letters in the graph of f in Figure 2.57 have

   (a) f' and f'' nonzero and of the same sign?
   (b)   At least two of f, f', and f'' equal to zero?

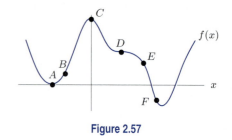

Figure 2.57

---

[5]From Paul A. Samuelson, *Economics*, 11th edition (New York: McGraw-Hill, 1981).

2.7   CONTINUITY AND DIFFERENTIABILITY

Definition of Continuity

Recall that the idea of continuity rules out breaks, jumps, or holes by demanding that the behavior of a function *near* a point be consistent with its behavior *at* the point:

> The function f is **continuous** at $x = c$ if f is defined at $x = c$ and
>
> $$\lim_{x \to c} f(x) = f(c).$$
>
> In other words, $f(x)$ is as close as we please to $f(c)$ provided x is close enough to c. The function is **continuous on an interval** $[a, b]$ if it is continuous at every point in the interval.[6]

Constant functions and $f(x) = x$ are continuous. Using the continuity of sums and products, we can show that any polynomial is continuous. Proving that $\sin x$, $\cos x$, and e^x are continuous is more difficult. The following theorem, based on the properties of limits on page 65, makes it easier to decide whether a given function is continuous.

> ## Theorem 2.2: Continuity of Sums, Products, and Quotients of Functions
>
> Suppose that f and g are continuous on an interval and that b is a constant. Then, on that same interval,
>
> 1. $bf(x)$ is continuous
> 2. $f(x) + g(x)$ is continuous
> 3. $f(x)g(x)$ is continuous
> 4. $f(x)/g(x)$ is continuous, provided $g(x) \neq 0$ on the interval.

We prove the first of these properties.

Proof   To prove that $bf(x)$ is continuous, pick any point c in the interval. We must show that $\lim_{x \to c} bf(x) = bf(c)$. Since $f(x)$ is continuous, we already know that $\lim_{x \to c} f(x) = f(c)$. So, by the first property of limits in Theorem 2.1,

$$\lim_{x \to c} (bf(x)) = b \left(\lim_{x \to c} f(x) \right) = bf(c).$$

Since c was chosen arbitrarily, we have shown that $bf(x)$ is continuous at every point in the interval.

> ## Theorem 2.3: Continuity of Composite Functions
>
> If f and g are continuous, and if the composite function $f(g(x))$ is defined on an interval, then $f(g(x))$ is continuous on that interval.

Assuming the continuity of $\sin x$ and e^x, Theorem 2.3 shows us, for example, that $\sin(e^x)$ and $e^{\sin x}$ are both continuous.

---

[6]If c is an endpoint of the interval, we define continuity at $x = c$ using one-sided limits at c.

How Can We Recognize Whether a Function Is Differentiable?

A function is differentiable at a point if it has a derivative there. Its graph must have a nonvertical tangent line at a point at which it is differentiable; the slope of the tangent line is the derivative.

Occasionally we meet a function which fails to have a derivative at a few points. A function fails to be differentiable at a point if:

- The function is not continuous at the point.
- The graph has a sharp corner at that point.
- The graph has a vertical tangent line.

Figure 2.58 shows a function which appears to be differentiable at all points except $x = a$ and $x = b$. There is no tangent at A because the graph has a corner there. As x approaches a from the left, the slope of the line joining P to A converges to some positive number. As x approaches a from the right, the slope of the line joining P to A converges to some negative number. Thus the slopes approach different numbers as we approach $x = a$ from different sides. Therefore the function is not differentiable at $x = a$. At B, the graph has a vertical tangent. As x approaches b, the slope of the line joining B to Q does not approach a limit; it just keeps growing larger and larger. Again, the limit defining the derivative does not exist and the function is not differentiable at $x = b$.

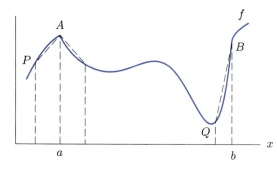

Figure 2.58: A function which is not differentiable at A or B

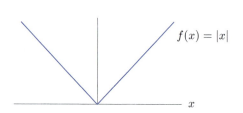

Figure 2.59: Graph of absolute value function, showing point of non-differentiability at $x = 0$

Examples of Nondifferentiable Functions

An example of a function whose graph has a corner is the *absolute value* function defined as follows:

$$f(x) = |x| = \begin{cases} x & \text{if } x \geq 0, \\ -x & \text{if } x < 0. \end{cases}$$

This function is called *piecewise linear* because each part of it is linear. Its graph is in Figure 2.59. Near $x = 0$, even close-up views of the graph of $f(x)$ look the same, so this is a corner which can't be straightened out by zooming in.

Example 1   Try to compute the derivative of the function $f(x) = |x|$ at $x = 0$. Is f differentiable there?

Solution   To find the derivative at $x = 0$, we want to look at

$$\lim_{h \to 0} \frac{f(h) - f(0)}{h} = \lim_{h \to 0} \frac{|h| - 0}{h} = \lim_{h \to 0} \frac{|h|}{h}.$$

As h approaches 0 from the right, h is positive, so $|h| = h$, and the ratio is always 1. As h approaches 0 from the left, h is negative, so $|h| = -h$, and the ratio is -1. Since the limits are different from each side, the limit of the difference quotient does not exist. Thus, the absolute value function is not differentiable at $x = 0$. The limits of 1 and -1 correspond to the fact that the slope of the right-hand part of the graph is 1, and the slope of the left-hand part is -1.

Example 2   Investigate the differentiability of $f(x) = x^{1/3}$ at $x = 0$.

Solution   This function is smooth at $x = 0$ (no sharp corners) but appears to have a vertical tangent there. (See Figure 2.60.) Looking at the difference quotient at $x = 0$, we see

$$\lim_{h \to 0} \frac{(0 + h)^{1/3} - 0^{1/3}}{h} = \lim_{h \to 0} \frac{h^{1/3}}{h} = \lim_{h \to 0} \frac{1}{h^{2/3}}.$$

As $h \to 0$ the denominator becomes small, so the fraction grows without bound. Hence, the function fails to have a derivative at $x = 0$.

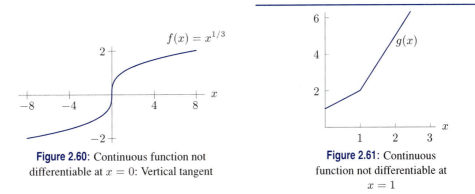

Figure 2.60: Continuous function not differentiable at $x = 0$: Vertical tangent

Figure 2.61: Continuous function not differentiable at $x = 1$

Example 3   Consider the function given by the formulas

$$g(x) = \begin{cases} x + 1 & \text{if} \quad x \le 1 \\ 3x - 1 & \text{if} \quad x > 1. \end{cases}$$

Draw the graph of g. Is g continuous? Is g differentiable at $x = 1$?

Solution   The graph in Figure 2.61 has no breaks in it, so the function is continuous. However, the graph has a corner at $x = 1$ which no amount of magnification will remove. To the left of $x = 1$, the slope is 1; to the right of $x = 1$, the slope is 3. Thus, the difference quotient at $x = 1$ has no limit, so the function g is not differentiable at $x = 1$.

Differentiability and Continuity

The fact that a function which is differentiable at a point has a tangent line suggests that the function is continuous there, as the next theorem shows.

Theorem 2.4: A Differentiable Function Is Continuous

If $f(x)$ is differentiable at a point $x = a$, then $f(x)$ is continuous at $x = a$.

Proof   We assume f is differentiable at $x = a$. Then we know that $f'(a)$ exists where

$$f'(a) = \lim_{x \to a} \frac{f(x) - f(a)}{x - a}.$$

To show that f is continuous at $x = a$, we want to show that $\lim_{x \to a} f(x) = f(a)$. We calculate $\lim_{x \to a}(f(x) - f(a))$, hoping to get 0. By algebra, we know that for $x \ne a$,

$$f(x) - f(a) = (x - a) \cdot \frac{f(x) - f(a)}{x - a}.$$

37. The population of a herd of deer is modeled by

$$P(t) = 4000 + 400 \sin\left(\frac{\pi}{6}t\right) + 180 \sin\left(\frac{\pi}{3}t\right)$$

where t is measured in months from the first of April.

(a) Use a calculator or computer to sketch a graph showing how this population varies with time.

Use the graph to answer the following questions.

(b) When is the herd largest? How many deer are in it at that time?

(c) When is the herd smallest? How many deer are in it then?

(d) When is the herd growing the fastest? When is it shrinking the fastest?

(e) How fast is the herd growing on April 1?

38. The number of hours, H, of daylight in Madrid is a function of t, the number of days since the start of the year. Figure 2.66 shows a one-month portion of the graph of H.

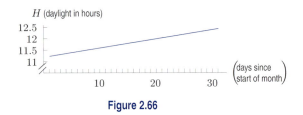

Figure 2.66

(a) Comment on the shape of the graph. Why does it look like a straight line?

(b) What month does this graph show? How do you know?

(c) What is the approximate slope of this line? What does the slope represent in practical terms?

CAS Challenge Problems

42. Use a computer algebra system to find the derivative of $f(x) = \sin^2 x + \cos^2 x$ and simplify your answer. Explain your result.

43. (a) Use a computer algebra system to find the derivative of $f(x) = 2 \sin x \cos x$.

(b) Simplify $f(x)$ and $f'(x)$ using double angle formulas. Write down the derivative formula that you get after doing this simplification.

44. (a) Use a computer algebra system to find the second derivative of $g(x) = e^{-ax^2}$ with respect to x.

(b) Graph $g(x)$ and $g''(x)$ on the same axes for $a = 1, 2, 3$ and describe the relation between the two graphs.

(c) Explain your answer to part (b) in terms of concavity.

39. Suppose you put a yam in a hot oven, maintained at a constant temperature of $200°$C. As the yam picks up heat from the oven, its temperature rises.[7]

(a) Draw a possible graph of the temperature T of the yam against time t (minutes) since it is put into the oven. Explain any interesting features of the graph, and in particular explain its concavity.

(b) Suppose that, at $t = 30$, the temperature T of the yam is $120°$ and increasing at the (instantaneous) rate of $2°$/min. Using this information, plus what you know about the shape of the T graph, estimate the temperature at time $t = 40$.

(c) Suppose in addition you are told that at $t = 60$, the temperature of the yam is $165°$. Can you improve your estimate of the temperature at $t = 40$?

(d) Assuming all the data given so far, estimate the time at which the temperature of the yam is $150°$.

40. You are given the following values for the error function, $\text{erf}(x)$.

$$\text{erf}(0) = 0 \quad \text{erf}(1) = 0.84270079$$
$$\text{erf}(0.1) = 0.11246292 \quad \text{erf}(0.01) = 0.01128342.$$

(a) Use all this information to determine your best estimate for $\text{erf}'(0)$. (Give only those digits of which you feel reasonably certain.)

(b) Suppose you find that $\text{erf}(0.001) = 0.00112838$. How does this extra information change your answer to part (a)?

41. (a) Use your calculator to approximate the derivative of the hyperbolic sine function (written $\sinh x$) at the points $0, 0.3, 0.7$, and 1.

(b) Can you find a relation between the values of this derivative and the values of the hyperbolic cosine (written $\cosh x$)?

45. (a) Use a computer algebra system to find the derivative of $f(x) = \ln(x)$, $g(x) = \ln(2x)$, and $h(x) = \ln(3x)$. What is the relationship between the answers?

(b) Use the properties of logarithms to explain what you see in part (a).

46. (a) Use a computer algebra system to find the derivative of $(x^2 + 1)^2$, $(x^2 + 1)^3$, and $(x^2 + 1)^4$.

(b) Conjecture a formula for the derivative of $(x^2 + 1)^n$ that works for any integer n. Check your formula using the computer algebra system.

47. (a) Use a computer algebra system to find the derivatives of $\sin x$, $\cos x$ and $\sin x \cos x$.

(b) Is the derivative of a product of two functions always equal to the product of their derivatives?

[7] From Peter D. Taylor, *Calculus: The Analysis of Functions* (Toronto: Wall & Emerson, Inc., 1992).

CHECK YOUR UNDERSTANDING

Are the statements in Problems 1–20 true or false? Give an explanation for your answer.

1. If a car is going 50 miles per hour at 2 pm and 60 miles per hour at 3 pm then it travels between 50 and 60 miles during the hour between 2 pm and 3 pm.

2. If a car travels 80 miles between 2 a0nd 4 pm, then its velocity is close to 40 mph at 2 pm.

3. If the time interval is short enough, then the average velocity of a car over the time interval and the instantaneous velocity at a time in the interval can be expected to be close.

4. If an object moves with the same average velocity over every time interval, then its average velocity equals its instantaneous velocity at any time.

5. The formula Distance traveled $=$ Average velocity $\times$ Time is valid for every moving object for every time interval.

6. By definition, the instantaneous velocity of an object equals a difference quotient.

7. If $\lim_{h \to 0} f(h) = L$, then $f(0.0001)$ is closer to L than is $f(0.01)$.

8. If $f(x)$ is concave up, then
$f'(a) < (f(b) - f(a))/(b - a)$ for $a < b$.

9. You cannot be sure of the exact value of a derivative of a function at a point using only the information in a table of values of the function. The best you can do is find an approximation.

10. If $f'(x)$ is increasing, then $f(x)$ is also increasing.

11. If $f(a) \neq g(a)$, then $f'(a) \neq g'(a)$.

12. The derivative of a linear function is constant.

13. If $g(x)$ is a vertical shift of $f(x)$, then $f'(x) = g'(x)$.

14. If $f(x)$ is defined for all x but $f'(0)$ is not defined, then the graph of $f(x)$ must have a corner at the point where $x = 0$.

15. If $y = f(x)$, then $\left. \dfrac{dy}{dx} \right|_{x=a} = f'(a)$.

16. If you zoom in (with your calculator) on the graph of $y = f(x)$ in a small interval around $x = 10$ and see a straight line, then the slope of that line equals the derivative $f'(10)$.

17. If $f''(x) > 0$ then $f'(x)$ is increasing.

18. The instantaneous acceleration of a moving particle at time t is the limit of difference quotients.

19. If $f(t)$ is the quantity in grams of a chemical produced after t minutes and $g(t)$ is the same quantity in kilograms, then $f'(t) = 1000g'(t)$.

20. If $f(t)$ is the quantity in kilograms of a chemical produced after t minutes and $g(t)$ is the quantity in kilograms produced after t seconds, then $f'(t) = 60g'(t)$.

Are the statements in Problems 21–26 true or false? If a statement is true, give an example illustrating it. If a statement is false, give a counterexample.

21. There is a function which is continuous on $[1, 5]$ but not differentiable at $x = 3$.

22. If a function is differentiable, then it is continuous.

23. If a function is continuous, then it is differentiable.

24. If a function is not continuous, then it is not differentiable.

25. If a function is not differentiable, then it is not continuous.

26. If $\lim_{x \to c^+} (f(x) + g(x))$ exists, then $\lim_{x \to c^+} f(x)$ and $\lim_{x \to c^+} g(x)$ must exist.

Suppose that $\lim_{x \to 3} f(x) = 7$. Are the statements in Problems 27–33 true or false? If a statement is true, explain how you know. If a statement is false, give a counterexample.

27. $\lim_{x \to 3} (x f(x)) = 21$.

28. If $g(3) = 4$, then $\lim_{x \to 3} (f(x)g(x)) = 28$.

29. If $\lim_{x \to 3} g(x) = 5$, then $\lim_{x \to 3} (f(x) + g(x)) = 12$.

30. If $\lim_{x \to 3} (f(x) + g(x)) = 12$, then $\lim_{x \to 3} g(x) = 5$.

31. $f(2.99)$ is closer to 7 than $f(2.9)$ is.

32. If $f(3.1) > 0$, then $f(3.01) > 0$.

33. If $\lim_{x \to 3} g(x)$ does not exist, then $\lim_{x \to 3} (f(x)g(x))$ does not exist.

Which of the statements in Problems 34–38 are true about every function $f(x)$ such that $\lim_{x \to c} f(x) = L$? Give a reason for your answer.

34. If $f(x)$ is within 10^{-3} of L, then x is within 10^{-3} of c.

35. There is a positive ϵ such that, provided x is within 10^{-3} of c, and $x \neq c$, we can be sure $f(x)$ is within ϵ of L.

36. For any positive ϵ, we can find a positive δ such that, provided x is within δ of c, and $x \neq c$, we can be sure that $f(x)$ is within ϵ of L.

37. For each $\epsilon > 0$, there is a $\delta > 0$ such that if x is not within δ of c, then $f(x)$ is not within ϵ of L.

38. For each $\epsilon > 0$, there is some $\delta > 0$ such that if $f(x)$ is within ϵ of L, then we can be sure that x is within δ of c.

39. Which of the following would be a counterexample to the statement: "If f is differentiable at $x = a$ then f is continuous at $x = a$"?

   (a) A function which is not differentiable at $x = a$ but is continuous at $x = a$.
   (b) A function which is not continuous at $x = a$ but is differentiable at $x = a$.
   (c) A function which is both continuous and differentiable at $x = a$.
   (d) A function which is neither continuous nor differentiable at $x = a$.

PROJECTS FOR CHAPTER TWO

1. **Hours of Daylight as a Function of Latitude**

   Let $S(x)$ be the number of sunlight hours on a cloudless June 21, as a function of latitude, x, measured in degrees.

   (a) What is $S(0)$?

   (b) Let x_0 be the latitude of the Arctic Circle ($x_0 \approx 66°30'$). In the northern hemisphere, $S(x)$ is given, for some constants a and b, by the formula:

   $$S(x) = \begin{cases} a + b \arcsin\left(\dfrac{\tan x}{\tan x_0}\right) & \text{for } 0 \le x < x_0 \\ 24 & \text{for } x_0 \le x \le 90. \end{cases}$$

   Find a and b so that $S(x)$ is continuous.

   (c) Calculate $S(x)$ for Tucson, Arizona ($x = 32°13'$) and Walla Walla, Washington ($46°4'$).

   (d) Graph $S(x)$, for $0 \le x \le 90$.

   (e) Does $S(x)$ appear to be differentiable?

2. **US Population**

   Census figures for the US population (in millions) are listed in Table 2.13. Let f be the function such that $P = f(t)$ is the population (in millions) in year t.

   Table 2.13 *US population (in millions), 1790–1990*

Year	Population	Year	Population	Year	Population	Year	Population
1790	3.9	1850	23.1	1910	92.0	1970	205.0
1800	5.3	1860	31.4	1920	105.7	1980	226.5
1810	7.2	1870	38.6	1930	122.8	1990	248.7
1820	9.6	1880	50.2	1940	131.7		
1830	12.9	1890	62.9	1950	150.7		
1840	17.1	1900	76.0	1960	179.0		

   (a) **(i)** Estimate the rate of change of the population for the years 1900, 1945, and 1990.

   (ii) When, approximately, was the rate of change of the population greatest?

   (iii) Estimate the US population in 1956.

   (iv) Based on the data from the table, what would you predict for the census in the year 2000?

   (b) Assume that f is increasing (as the values in the table suggest). Then f is invertible.

   (i) What is the meaning of $f^{-1}(100)$?

   (ii) What does the derivative of $f^{-1}(P)$ at $P = 100$ represent? What are its units?

   (iii) Estimate $f^{-1}(100)$.

   (iv) Estimate the derivative of $f^{-1}(P)$ at $P = 100$.

   (c) **(i)** Usually we think the US population $P = f(t)$ as a smooth function of time. To what extent is this justified? What happens if we zoom in at a point of the graph? What about events such as the Louisiana Purchase? Or the moment of your birth?

   (ii) What do we in fact mean by the rate of change of the population at a particular time t?

   (iii) Give another example of a real-world function which is not smooth but is usually treated as such.

Chapter Three

SHORT-CUTS TO DIFFERENTIATION

In Chapter 2, we defined the derivative function

$$f'(x) = \lim_{h \to 0} \frac{f(x+h) - f(x)}{h}$$

and saw how the derivative represents a slope and a rate of change. We learned how to approximate the derivative of a function given graphically (by estimating the slope of the tangent at each point) and numerically (by finding the average rate of change of the function between data values). We calculated the derivatives of x^2 and x^3 exactly using the definition.

In this chapter we make a systematic study of the derivatives of functions given by formulas. These functions include powers, polynomials, exponential, logarithmic, and trigonometric functions. The chapter also contains general rules, such as the product, quotient, and chain rules, which allow us to differentiate combinations of functions.

Useful Notation: We write $\frac{d}{dx}(x^3)$, for example, to mean the derivative of x^3 with respect to x. Similarly, $\frac{d}{d\theta}\left(\sin(\theta^2)\right)$ denotes the derivative of $\sin(\theta^2)$ with θ as the variable.

3.1 POWERS AND POLYNOMIALS

Derivative of a Constant Times a Function

Figure 3.1 shows the graph of $y = f(x)$ and of three multiples: $y = 3f(x)$, $y = \frac{1}{2}f(x)$, and $y = -2f(x)$. What is the relationship between the derivatives of these functions? In other words, for a particular x-value, how are the slopes of these graphs related?

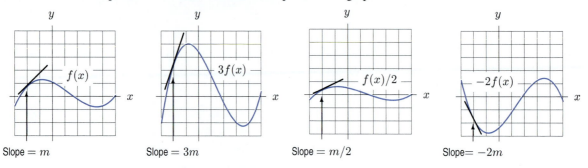

Figure 3.1: A function and its multiples: Derivative of multiple is multiple of derivative

Multiplying by a constant stretches or shrinks the graph (and reflects it in the x-axis if the constant is negative). This changes the slope of the curve at each point. If the graph has been stretched, the "rises" have all been increased by the same factor, whereas the "runs" remain the same. Thus, the slopes are all steeper by the same factor. If the graph has been shrunk, the slopes are all smaller by the same factor. If the graph has been reflected in the x-axis, the slopes will all have their signs reversed. In other words, if a function is multiplied by a constant, c, so is its derivative:

Theorem 3.1: Derivative of a Constant Multiple

If f is differentiable and c is a constant, then

$$\frac{d}{dx}[cf(x)] = cf'(x).$$

Proof Although the graphical argument makes the theorem plausible, to prove it we must use the definition of the derivative:

$$\frac{d}{dx}[cf(x)] = \lim_{h \to 0} \frac{cf(x+h) - cf(x)}{h} = \lim_{h \to 0} c\,\frac{f(x+h) - f(x)}{h}$$

$$= c \lim_{h \to 0} \frac{f(x+h) - f(x)}{h} = cf'(x).$$

We can take c across the limit sign by the properties of limits (part 1 of Theorem 2.1 on page 65).

Derivatives of Sums and Differences

Suppose we have two functions, $f(x)$ and $g(x)$, with the values listed in Table 3.1. Values of the sum $f(x) + g(x)$ are given in the same table.

Table 3.1   *Sum of Functions*

x	$f(x)$	$g(x)$	$f(x) + g(x)$
0	100	0	100
1	110	0.2	110.2
2	130	0.4	130.4
3	160	0.6	160.6
4	200	0.8	200.8

We see that adding the increments of $f(x)$ and the increments of $g(x)$ gives the increments of $f(x) + g(x)$. For example, as x increases from 0 to 1, $f(x)$ increases by 10 and $g(x)$ increases by 0.2, while $f(x) + g(x)$ increases by $110.2 - 100 = 10.2$. Similarly, as x increases from 3 to 4, $f(x)$ increases by 40 and $g(x)$ by 0.2, while $f(x) + g(x)$ increases by $200.8 - 160.6 = 40.2$.

From this example, we see that the rate at which $f(x) + g(x)$ is increasing is the sum of the rates at which $f(x)$ and $g(x)$ are increasing. Similar reasoning applies to the difference, $f(x) - g(x)$. In terms of derivatives:

Theorem 3.2: Derivative of Sum and Difference

If f and g are differentiable, then

$$\frac{d}{dx}[f(x) + g(x)] = f'(x) + g'(x) \quad \text{and} \quad \frac{d}{dx}[f(x) - g(x)] = f'(x) - g'(x).$$

Proof Using the definition of the derivative:

$$\frac{d}{dx}[f(x) + g(x)] = \lim_{h \to 0} \frac{[f(x + h) + g(x + h)] - [f(x) + g(x)]}{h}$$

$$= \lim_{h \to 0} \left[\underbrace{\frac{f(x + h) - f(x)}{h}}_{\text{Limit of this is } f'(x)} + \underbrace{\frac{g(x + h) - g(x)}{h}}_{\text{Limit of this is } g'(x)} \right]$$

$$= f'(x) + g'(x).$$

We have used the fact that the limit of a sum is the sum of the limits, part 2 of Theorem 2.1 on page 65. The proof for $f(x) - g(x)$ is similar.

Powers of x

In Chapter 2 we showed that

$$f'(x) = \frac{d}{dx}(x^2) = 2x \quad \text{and} \quad g'(x) = \frac{d}{dx}(x^3) = 3x^2.$$

The graphs of $f(x) = x^2$ and $g(x) = x^3$ and their derivatives are shown in Figures 3.2 and 3.3. Notice $f'(x) = 2x$ has the behavior we expect. It is negative for $x < 0$ (when f is decreasing), zero for $x = 0$, and positive for $x > 0$ (when f is increasing). Similarly, $g'(x) = 3x^2$ is zero when $x = 0$, but positive everywhere else, as g is increasing everywhere else.

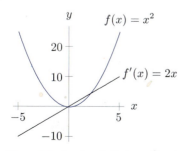

Figure 3.2: Graphs of $f(x) = x^2$ and its derivative $f'(x) = 2x$

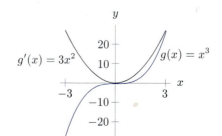

Figure 3.3: Graphs of $g(x) = x^3$ and its derivative $g'(x) = 3x^2$

These examples are special cases of the power rule which we justify for any positive integer n on page 109:

The Power Rule

For any constant real number n,

$$\frac{d}{dx}(x^n) = nx^{n-1}$$

Problem 62 asks you to show that this rule holds for negative integral powers. In Section 3.6 we indicate how to justify it for powers of the form $1/n$.

Example 1   Use the power rule to differentiate (a) $\dfrac{1}{x^3}$,   (b) $x^{1/2}$,   (c) $\dfrac{1}{\sqrt[3]{x}}$.

Solution

(a) For $n = -3$: $\dfrac{d}{dx}\left(\dfrac{1}{x^3}\right) = \dfrac{d}{dx}(x^{-3}) = -3x^{-3-1} = -3x^{-4} = -\dfrac{3}{x^4}.$

(b) For $n = 1/2$: $\dfrac{d}{dx}\left(x^{1/2}\right) = \dfrac{1}{2}x^{(1/2)-1} = \dfrac{1}{2}x^{-1/2} = \dfrac{1}{2\sqrt{x}}.$

(c) For $n = -1/3$: $\dfrac{d}{dx}\left(\dfrac{1}{\sqrt[3]{x}}\right) = \dfrac{d}{dx}\left(x^{-1/3}\right) = -\dfrac{1}{3}x^{(-1/3)-1} = -\dfrac{1}{3}x^{-4/3} = -\dfrac{1}{3x^{4/3}}.$

Example 2   Use the definition of the derivative to justify the power rule for $n = -2$: Show $\dfrac{d}{dx}(x^{-2}) = -2x^{-3}$.

Solution   Provided $x \neq 0$, we have

$$\frac{d}{dx}\left(x^{-2}\right) = \frac{d}{dx}\left(\frac{1}{x^2}\right) = \lim_{h \to 0}\left(\frac{\frac{1}{(x+h)^2} - \frac{1}{x^2}}{h}\right) = \lim_{h \to 0}\frac{1}{h}\left[\frac{x^2 - (x+h)^2}{(x+h)^2 x^2}\right] \quad \text{(Combining fractions over a common denominator)}$$

$$= \lim_{h \to 0}\frac{1}{h}\left[\frac{x^2 - (x^2 + 2xh + h^2)}{(x+h)^2 x^2}\right] \quad \text{(Multiplying out)}$$

$$= \lim_{h \to 0}\frac{-2xh - h^2}{h(x+h)^2 x^2} \quad \text{(Simplifying numerator)}$$

$$= \lim_{h \to 0}\frac{-2x - h}{(x+h)^2 x^2} \quad \text{(Dividing numerator and denominator by } h\text{)}$$

$$= \frac{-2x}{x^4} \quad \text{(Letting } h \to 0\text{)}$$

$$= -2x^{-3}.$$

The graphs of x^{-2} and its derivative, $-2x^{-3}$, are shown in Figure 3.4. Does the graph of the derivative have the features you expect?

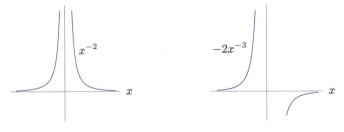

Figure 3.4: Graphs of x^{-2} and its derivative, $-2x^{-3}$

Justification of $\frac{d}{dx}(x^n) = nx^{n-1}$, for n a Positive Integer

To calculate the derivatives of x^2 and x^3, we had to expand $(x+h)^2$ and $(x+h)^3$. To calculate the derivative of x^n, we must expand $(x+h)^n$. Let's look back at the previous expansions:

$$(x+h)^2 = x^2 + 2xh + h^2, \qquad (x+h)^3 = x^3 + 3x^2h + 3xh^2 + h^3,$$

and multiply out a few more examples:

$$(x+h)^4 = x^4 + 4x^3h + 6x^2h^2 + 4xh^3 + h^4,$$
$$(x+h)^5 = x^5 + 5x^4h + \underbrace{10x^3h^2 + 10x^2h^3 + 5xh^4 + h^5}.$$
$$\text{Terms involving } h^2 \text{ and higher powers of } h$$

In general, we can say

$$(x+h)^n = x^n + nx^{n-1}h + \underbrace{\cdots\cdots + h^n}.$$
$$\text{Terms involving } h^2 \text{ and higher powers of } h$$

We have just seen this is true for $n = 2, 3, 4, 5$. It can be proved in general using the Binomial Theorem (see the online theory supplement). Now to find the derivative,

$$\frac{d}{dx}(x^n) = \lim_{h \to 0} \frac{(x+h)^n - x^n}{h}$$

$$= \lim_{h \to 0} \frac{(x^n + nx^{n-1}h + \cdots + h^n) - x^n}{h}$$
$$\text{Terms involving } h^2 \text{ and higher powers of } h$$

$$= \lim_{h \to 0} \frac{nx^{n-1}h + \overbrace{\cdots + h^n}}{h}.$$

When we factor out h from terms involving h^2 and higher powers of h, each term will still have an h in it. Factoring and dividing, we get:

$$\text{Terms involving } h \text{ and higher powers of } h$$
$$\frac{d}{dx}(x^n) = \lim_{h \to 0} \frac{h(nx^{n-1} + \cdots + h^{n-1})}{h} = \lim_{h \to 0}(nx^{n-1} + \overbrace{\cdots + h^{n-1}}).$$

But as $h \to 0$, all terms involving an h will go to 0, so

$$\frac{d}{dx}(x^n) = \lim_{h \to 0}(nx^{n-1} + \underbrace{\cdots + h^{n-1}}) = nx^{n-1}.$$
$$\text{These terms go to 0}$$

Derivatives of Polynomials

Now that we know how to differentiate powers, constant multiples, and sums, we can differentiate any polynomial.

Example 3 Find the derivatives of (a) $5x^2 + 3x + 2$, (b) $\sqrt{3}x^7 - \dfrac{x^5}{5} + \pi$.

Solution (a)

$$\frac{d}{dx}(5x^2 + 3x + 2) = 5\frac{d}{dx}(x^2) + 3\frac{d}{dx}(x) + \frac{d}{dx}(2)$$
$$= 5 \cdot 2x + 3 \cdot 1 + 0 \qquad \text{(Since the derivative of a constant, } \frac{d}{dx}(2)\text{, is zero.)}$$
$$= 10x + 3.$$

(b)

$$\frac{d}{dx}\left(\sqrt{3}x^7 - \frac{x^5}{5} + \pi\right) = \sqrt{3}\frac{d}{dx}(x^7) - \frac{1}{5}\frac{d}{dx}(x^5) + \frac{d}{dx}(\pi)$$

$$= \sqrt{3} \cdot 7x^6 - \frac{1}{5} \cdot 5x^4 + 0 \qquad \text{(Since } \pi \text{ is a constant, } d\pi/dx = 0.\text{)}$$

$$= 7\sqrt{3}x^6 - x^4.$$

We can also use the rules we have seen so far to differentiate expressions which are not polynomials.

Example 4   Differentiate (a) $5\sqrt{x} - \dfrac{10}{x^2} + \dfrac{1}{2\sqrt{x}}$ (b) $0.1x^3 + 2x^{\sqrt{2}}$

Solution   (a) $\dfrac{d}{dx}\left(5\sqrt{x} - \dfrac{10}{x^2} + \dfrac{1}{2\sqrt{x}}\right) = \dfrac{d}{dx}\left(5x^{1/2} - 10x^{-2} + \dfrac{1}{2}x^{-1/2}\right)$

$$= 5 \cdot \frac{1}{2}x^{-1/2} - 10(-2)x^{-3} + \frac{1}{2}\left(-\frac{1}{2}\right)x^{-3/2}$$

$$= \frac{5}{2\sqrt{x}} + \frac{20}{x^3} - \frac{1}{4x^{3/2}}.$$

(b) $\dfrac{d}{dx}(0.1x^3 + 2x^{\sqrt{2}}) = 0.1\dfrac{d}{dx}(x^3) + 2\dfrac{d}{dx}(x^{\sqrt{2}}) = 0.3x^2 + 2\sqrt{2}x^{\sqrt{2}-1}.$

Example 5   Find the second derivative and interpret its sign for
(a) $f(x) = x^2$,   (b) $g(x) = x^3$,   (c) $k(x) = x^{1/2}$.

Solution   (a) If $f(x) = x^2$, then $f'(x) = 2x$, so $f''(x) = \dfrac{d}{dx}(2x) = 2$. Since f'' is always positive, f is concave up, as expected for a parabola opening upwards. (See Figure 3.5.)

(b) If $g(x) = x^3$, then $g'(x) = 3x^2$, so $g''(x) = \dfrac{d}{dx}(3x^2) = 3\dfrac{d}{dx}(x^2) = 3 \cdot 2x = 6x$. This is positive for $x > 0$ and negative for $x < 0$, which means x^3 is concave up for $x > 0$ and concave down for $x < 0$. (See Figure 3.6.)

(c) If $k(x) = x^{1/2}$, then $k'(x) = \frac{1}{2}x^{(1/2)-1} = \frac{1}{2}x^{-1/2}$, so

$$k''(x) = \frac{d}{dx}\left(\frac{1}{2}x^{-1/2}\right) = \frac{1}{2} \cdot (-\frac{1}{2})x^{-(1/2)-1} = -\frac{1}{4}x^{-3/2}.$$

Now k' and k'' are only defined on the domain of k, that is, $x \geq 0$. When $x > 0$, we see that $k''(x)$ is negative, so k is concave down. (See Figure 3.7.)

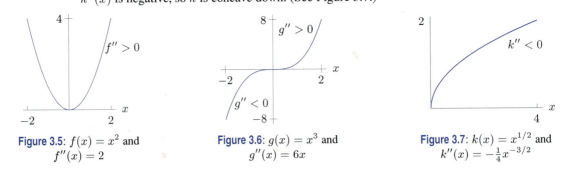

Figure 3.5: $f(x) = x^2$ and $f''(x) = 2$

Figure 3.6: $g(x) = x^3$ and $g''(x) = 6x$

Figure 3.7: $k(x) = x^{1/2}$ and $k''(x) = -\frac{1}{4}x^{-3/2}$

Example 6   If the position of a body, in meters, is given as a function of time t, in seconds, by

$$s = -4.9t^2 + 5t + 6,$$

find the velocity and acceleration of the body at time t.

Solution

The velocity, v, is the derivative of the position:

$$v = \frac{ds}{dt} = \frac{d}{dt}(-4.9t^2 + 5t + 6) = -9.8t + 5,$$

and the acceleration, a, is the derivative of the velocity:

$$a = \frac{dv}{dt} = \frac{d}{dt}(-9.8t + 5) = -9.8.$$

Notice that v is in meters/second and a is in meters/second^2.

Example 7   Figure 3.8 shows the graph of a cubic polynomial. Both graphically and algebraically, describe the behavior of the derivative of this cubic.

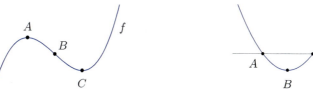
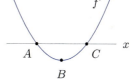

Figure 3.8: The cubic of Example 7   **Figure 3.9:** Derivative of the cubic of Example 7

Solution

Graphical approach: Suppose we move along the curve from left to right. To the left of A, the slope is positive; it starts very positive and decreases until the curve reaches A, where the slope is 0. Between A and C the slope is negative. Between A and B the slope is decreasing (getting more negative); it is most negative at B. Between B and C the slope is negative but increasing; at C the slope is zero. From C to the right, the slope is positive and increasing. The graph of the derivative function is shown in Figure 3.9.

Algebraic approach: f is a cubic that goes to $+\infty$ as $x \to +\infty$, so

$$f(x) = ax^3 + bx^2 + cx + d$$

with $a > 0$. Hence,

$$f'(x) = 3ax^2 + 2bx + c,$$

whose graph is a parabola opening upward, as in Figure 3.9.

Exercises and Problems for Section 3.1

Exercises

1. Let $f(x) = 7$. Using the definition of the derivative, show that $f'(x) = 0$ for all values of x.

2. Let $f(x) = 17x + 11$. Use the definition of the derivative to calculate $f'(x)$.

For Exercises 3–35, find the derivatives of the given functions. Assume that a, b, c, and k are constants.

3. $y = x^{11}$

4. $y = x^{12}$

5. $y = -x^{-11}$

6. $y = x^{3.2}$

7. $y = x^{-12}$

8. $y = x^{4/3}$

9. $y = x^{3/4}$

10. $y = x^{-3/4}$

11. $f(x) = \dfrac{1}{x^4}$

12. $f(x) = \sqrt[4]{x}$

13. $f(x) = x^e$

14. $y = 4x^{3/2} - 5x^{1/2}$

15. $f(t) = 3t^2 - 4t + 1$

16. $y = 17x + 24x^{1/2}$

17. $g(t) = \dfrac{t^3 + k}{t}$

18. $f(x) = 5x^4 + \dfrac{1}{x^2}$

19. $h(w) = -2w^{-3} + 3\sqrt{w}$

20. $y = 6x^3 + 4x^2 - 2x$

21. $y = 3t^5 - 5\sqrt{t} + \dfrac{7}{t}$

22. $y = 3t^2 + \dfrac{12}{\sqrt{t}} - \dfrac{1}{t^2}$

23. $y = z^2 + \dfrac{1}{2z}$

24. $y = \dfrac{x^2 + 1}{x}$

25. $f(z) = \dfrac{z^2 + 1}{3z}$

26. $f(t) = \dfrac{t^2 + t^3 - 1}{t^4}$

27. $y = \dfrac{\theta - 1}{\sqrt{\theta}}$

28. $j(x) = \dfrac{x^3}{a} + \dfrac{a}{b}x^2 - cx$

33. $g(x) = -\frac{1}{2}(x^5 + 2x - 9)$

34. $y = -3x^4 - 4x^3 - 6x + 2$

29. $\dfrac{dV}{dr}$ if $V = \frac{4}{3}\pi r^2 b$

30. $\dfrac{dw}{dq}$ if $w = 3ab^2 q$

35. $g(z) = \dfrac{z^7 + 5z^6 - z^3}{z^2}$

31. $\dfrac{dy}{dx}$ if $y = ax^2 + bx + c$

32. $\dfrac{dP}{dt}$ if $P = a + b\sqrt{t}$

Problems

For Problems 36–41, determine if the derivative rules from this section apply. If they do, find the derivative. If they don't apply, indicate why.

36. $y = (x + 3)^{1/2}$

37. $y = 3^x$

38. $g(x) = x^\pi - x^{-\pi}$

39. $y = 3x^2 + 4$

40. $y = \dfrac{1}{3x^2 + 4}$

41. $y = \dfrac{1}{3z^2} + \dfrac{1}{4}$

42. If $f(t) = 2t^3 - 4t^2 + 3t - 1$, find $f'(t)$ and $f''(t)$.

43. If $f(x) = 4x^3 + 6x^2 - 23x + 7$, find the intervals on which $f'(x) \geq 1$.

44. On what intervals is the function $f(x) = x^4 - 4x^3$ both decreasing and concave up?

45. For what values of x is the graph of $y = x^5 - 5x$ both increasing and concave up?

46. If $f(x) = x^3 - 6x^2 - 15x + 20$, find analytically all values of x for which $f'(x) = 0$. Show your answers on a graph of f.

47. If $f(x) = 13 - 8x + \sqrt{2}x^2$ and $f'(r) = 4$, find r.

48. **(a)** Find the *eighth* derivative of $f(x) = x^7 + 5x^5 - 4x^3 + 6x - 7$. Think ahead!
(The nth derivative, $f^{(n)}(x)$, is the result of differentiating $f(x)$ n times.)
(b) Find the seventh derivative of $f(x)$.

49. Find the equation of the line tangent to the graph of f at $(1, 1)$, where f is given by $f(x) = 2x^3 - 2x^2 + 1$.

50. **(a)** Find the equation of the tangent line to $f(x) = x^3$ at the point where $x = 2$.
(b) Graph the tangent line and the function on the same axes. If the tangent line is used to estimate values of the function, will the estimates be overestimates or underestimates?

51. Using a graph to help you, find the equations of all lines through the origin tangent to the parabola
$$y = x^2 - 2x + 4.$$
Sketch the lines on the graph.

52. Show that for any power function $f(x) = x^n$, we have $f'(1) = n$.

53. Given a power function of the form $f(x) = ax^n$, with $f'(2) = 3$ and $f'(4) = 24$, find n and a.

54. Is there a value of n which makes $y = x^n$ a solution to the equation $13x(dy/dx) = y$? If so, what value?

55. The height of a sand dune (in centimeters) is represented by $f(t) = 700 - 3t^2$, where t is measured in years since 1995. Find $f(5)$ and $f'(5)$. Using units, explain what each means in terms of the sand dune.

56. A ball is dropped from the top of the Empire State building to the ground below. The height, y, of the ball above the ground (in feet) is given as a function of time, t, (in seconds) by
$$y = 1250 - 16t^2.$$
(a) Find the velocity of the ball at time t. What is the sign of the velocity? Why is this to be expected?
(b) Show that the acceleration of the ball is a constant. What are the value and sign of this constant?
(c) When does the ball hit the ground, and how fast is it going at that time? Give your answer in feet per second and in miles per hour (1 ft/sec = 15/22 mph).

57. At a time t seconds after it is thrown up in the air, a tomato is at a height of $f(t) = -4.9t^2 + 25t + 3$ meters.
(a) What is the average velocity of the tomato during the first 2 seconds? Give units.
(b) Find (exactly) the instantaneous velocity of the tomato at $t = 2$. Give units.
(c) What is the acceleration at $t = 2$?
(d) How high does the tomato go?
(e) How long is the tomato in the air?

58. The gravitational attraction, F, between the earth and a satellite of mass m at a distance r from the center of the earth is given by
$$F = \frac{GMm}{r^2},$$
where M is the mass of the earth, and G is a constant. Find the rate of change of force with respect to distance.

59. The period, T, of a pendulum is given in terms of its length, l, by
$$T = 2\pi\sqrt{\frac{l}{g}},$$
where g is the acceleration due to gravity (a constant).
(a) Find dT/dl.
(b) What is the sign of dT/dl? What does this tell you about the period of pendulums?

60. (a) Use the formula for the area of a circle of radius r, $A = \pi r^2$, to find dA/dr.

(b) The result from part (a) should look familiar. What does dA/dr represent geometrically?

(c) Use the difference quotient to explain the observation you made in part (b).

61. What is the formula for V, the volume of a sphere of ra-

dius r? Find dV/dr. What is the geometrical meaning of dV/dr?

62. Using the definition of derivative, justify the formula $d(x^n)/dx = nx^{n-1}$.

(a) For $n = -1$; for $n = -3$.

(b) For any negative integer n.

3.2 THE EXPONENTIAL FUNCTION

What do we expect the graph of the derivative of the exponential function $f(x) = a^x$ to look like? The graph of the exponential function is shown in Figure 3.10. The function increases slowly for $x < 0$ and more rapidly for $x > 0$, so the values of f' are small for $x < 0$ and larger for $x > 0$. Since the function is increasing for all values of x, the graph of the derivative must lie above the x-axis. In fact, it appears that the graph of f' must resemble the graph of f itself. We confirm this observation for $f(x) = 2^x$ and $g(x) = 3^x$.

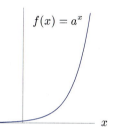

$f(x) = a^x$

x

Figure 3.10: $f(x) = a^x$, with $a > 1$

The Derivatives of 2^x and 3^x

In Chapter 2, we saw that the derivative of $f(x) = 2^x$ at $x = 0$ is given by

$$f'(0) = \lim_{h \to 0} \frac{2^h - 2^0}{h} = \lim_{h \to 0} \frac{2^h - 1}{h} \approx 0.6931.$$

The estimate for this limit is obtained by evaluating $(2^h - 1)/h$ for small values of h. Similarly, we can estimate the derivative at $x = 1$ and $x = 2$:

$$f'(1) = \lim_{h \to 0} \frac{2^{1+h} - 2^1}{h} \approx 1.3863,$$

$$f'(2) = \lim_{h \to 0} \frac{2^{2+h} - 2^2}{h} \approx 2.7726.$$

Can you see the relationship between these values of the derivative? If we notice that $1.3863 \approx 2(0.6931)$ and $2.7726 \approx 4(0.6931)$, we see

$$f'(0) \approx 0.6931 = 0.6931 \cdot 2^0,$$
$$f'(1) \approx 1.3863 \approx 0.6931 \cdot 2^1,$$
$$f'(2) \approx 2.7726 \approx 0.6931 \cdot 2^2.$$

It looks as though we have $f'(x) \approx 0.6931 \cdot 2^x$, which is in fact true. Let's look at the derivative function to see why this happens:

$$f'(x) = \lim_{h \to 0} \left(\frac{2^{x+h} - 2^x}{h} \right) = \lim_{h \to 0} \left(\frac{2^x 2^h - 2^x}{h} \right) = \lim_{h \to 0} 2^x \left(\frac{2^h - 1}{h} \right)$$

$$= 2^x \lim_{h \to 0} \left(\frac{2^h - 1}{h} \right) \qquad \text{(Since x and 2^x are fixed during this calculation)}$$

$$= f'(0) 2^x.$$

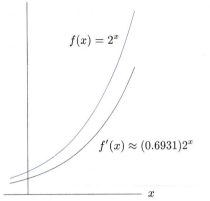

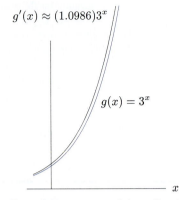
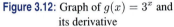

Figure 3.11: Graph of $f(x) = 2^x$ and its derivative

Figure 3.12: Graph of $g(x) = 3^x$ and its derivative

We have already estimated $f'(0) \approx 0.6931$, so we get

$$\frac{d}{dx}(2^x) = f'(x) \approx (0.6931)2^x.$$

Figure 3.11 shows that the graphs of $f(x) = 2^x$ and $f'(x) \approx (0.6931)2^x$ do indeed resemble one another.

Example 1   Find the derivative of $g(x) = 3^x$ and plot g and g' on the same axes.

Solution   As before

$$g'(x) = \lim_{h \to 0} \frac{3^{x+h} - 3^x}{h} = \lim_{h \to 0} \frac{3^x 3^h - 3^x}{h} = 3^x \lim_{h \to 0} \left(\frac{3^h - 1}{h} \right).$$

Using a calculator gives

$$\lim_{h \to 0} \left(\frac{3^h - 1}{h} \right) \approx 1.0986,$$

so

$$g'(x) \approx (1.0986)3^x.$$

The graphs of g and g' are shown in Figure 3.12; they resemble one another closely.

Notice that for both $f(x) = 2^x$ and $g(x) = 3^x$, *the derivative is proportional to the original function.* For $f(x) = 2^x$, we have $f'(x) \approx (0.6931)2^x$, so the constant of proportionality is less than 1, and the graph of the derivative is below the graph of the original function. For $g(x) = 3^x$, we have $g'(x) \approx (1.0986)3^x$, so the constant is greater than 1 and the graph of the derivative is above that of the original function.

Note on Round-Off Error and Limits

If we try to evaluate $(2^h - 1)/h$ on a calculator by taking smaller and smaller values of h, the values of $(2^h - 1)/h$ will at first get closer to 0.6931. However, they will eventually move away from the correct value 0.6931... because of the *round-off error* (i.e., errors introduced by the fact that the calculator can only hold a certain number of digits).

As we try smaller and smaller values of h, how do we know when to stop? Unfortunately, there is no fixed rule. A calculator can only suggest the value of a limit, but can never confirm that this value is correct. In this case, it looks like the limit is about 0.6931 because the values of $(2^h - 1)/h$ hover around 0.6931 for a while. To be sure this is correct, we would have to find the limit by theoretical means.

The Derivative of a^x and the Definition of e

The calculation of the derivative of $f(x) = a^x$, for $a > 0$, is similar to that of 2^x and 3^x:

$$f'(x) = \lim_{h \to 0} \frac{a^{x+h} - a^x}{h} = a^x \lim_{h \to 0} \frac{a^h - 1}{h}.$$

The quantity $\lim_{h \to 0} (a^h - 1)/h$ doesn't depend on x, and so is a constant for any particular a. Therefore the derivative is again proportional to the original function, with constant of proportionality

$$\lim_{h \to 0} \frac{a^h - 1}{h}.$$

We can't use a calculator to evaluate this limit without knowing the value of a. However, when $a = 2$, we know that the limit (0.6931) is less than 1, and the derivative is smaller than the original function. When $a = 3$, the limit (1.0986) is more than 1, and the derivative is greater than the original function. Is there an in-between case, when derivative and function are exactly equal? In other words:

$$\text{Is there a value of } a \text{ that makes } \frac{d}{dx}(a^x) = a^x?$$

If so, we have found a function with the remarkable property that it is equal to its own derivative.
So let's look for such an a. This means we want to find a such that

$$\lim_{h \to 0} \frac{a^h - 1}{h} = 1, \qquad \text{or, for small } h, \qquad \frac{a^h - 1}{h} \approx 1.$$

Solving for a suggests that we can calculate a as follows:

$$a^h - 1 \approx h, \qquad \text{or} \qquad a^h \approx 1 + h, \qquad \text{so} \qquad a \approx (1+h)^{1/h}.$$

Taking small values of h, as in Table 3.2, we can see $a \approx 2.718\ldots$, which looks like the number e introduced in Chapter 1. In fact, it can be shown that

$$e = \lim_{h \to 0} (1+h)^{1/h} = 2.718\ldots \qquad \text{and} \qquad \lim_{h \to 0} \frac{e^h - 1}{h} = 1.$$

This means that e^x is its own derivative:

$$\boxed{\frac{d}{dx}(e^x) = e^x.}$$

It turns out that the constants involved in the derivatives of 2^x and 3^x are natural logarithms. In fact, since $0.6931 \approx \ln 2$ and $1.0986 \approx \ln 3$, we (correctly) guess that

Table 3.2

h	$(1+h)^{1/h}$
0.001	2.7169239
0.0001	2.7181459
0.00001	2.7182682

$$\frac{d}{dx}(2^x) = (\ln 2)2^x \qquad \text{and} \qquad \frac{d}{dx}(3^x) = (\ln 3)3^x.$$

In Section 3.6 we show that, in general,

$$\boxed{\frac{d}{dx}(a^x) = (\ln a)a^x.}$$

Figure 3.13 shows the graph of the derivative of 2^x below the graph of the function, and the graph of the derivative of 3^x above the graph of the function. With $e \approx 2.718$, the function e^x and its derivative are identical.

Since $\ln a$ is a constant, the derivative of a^x is proportional to a^x. Many quantities have rates of change which are proportional to themselves; for example, the simplest model of population growth has this property. The fact that the constant of proportionality is 1 when $a = e$ makes e a particularly useful base for exponential functions.

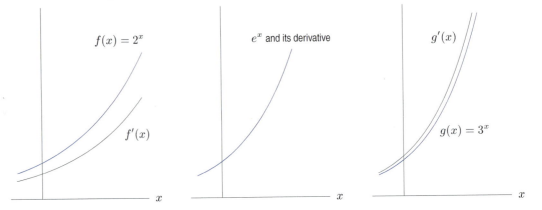

Figure 3.13: Graphs of the functions 2^x, e^x, and 3^x and their derivatives

Example 2   Differentiate $2 \cdot 3^x + 5e^x$.

Solution

$$\frac{d}{dx}(2 \cdot 3^x + 5e^x) = 2\frac{d}{dx}(3^x) + 5\frac{d}{dx}(e^x) = 2\ln 3 \cdot 3^x + 5e^x \approx (2.1972)3^x + 5e^x.$$

Exercises and Problems for Section 3.2

Exercises

Find the derivatives of the functions in Exercises 1–26. Assume that a, b, c, and k are constants.

1. $f(x) = 2e^x + x^2$

2. $y = 5t^2 + 4e^t$

3. $y = 5^x + 2$

4. $f(x) = 2^x + 2 \cdot 3^x$

5. $y = 5x^2 + 2^x + 3$

6. $f(x) = 12e^x + 11^x$

7. $y = 4 \cdot 10^x - x^3$

8. $y = 3x - 2 \cdot 4^x$

9. $y = \dfrac{3^x}{3} + \dfrac{33}{\sqrt{x}}$

10. $f(x) = e^2 + x^e$

11. $f(x) = e^{1+x}$

12. $f(t) = e^{t+2}$

13. $y = e^{\theta - 1}$

14. $z = (\ln 4)e^x$

15. $z = (\ln 4)4^x$

16. $f(t) = (\ln 3)^t$

17. $f(x) = x^3 + 3^x$

18. $y = 5 \cdot 5^t + 6 \cdot 6^t$

19. $y = \pi^2 + \pi^x$

20. $h(z) = (\ln 2)^z$

21. $f(x) = e^\pi + \pi^x$

22. $f(x) = \pi^x + x^\pi$

23. $y(x) = a^x + x^a$.

24. $f(x) = x^{\pi^2} + (\pi^2)^x$

25. $f(z) = (\ln 3)z^2 + (\ln 4)e^z$

26. $g(x) = 2x - \dfrac{1}{\sqrt[3]{x}} + 3^x - e$

Which of the functions in Exercises 27–35 can be differentiated using the rules we have developed so far? Differentiate if you can; otherwise, indicate why the rules discussed so far do not apply.

27. $y = x^2 + 2^x$

28. $y = \sqrt{x} - (\frac{1}{2})^x$

29. $y = x^2 \cdot 2^x$

30. $y = \dfrac{2^x}{x}$

31. $y = e^{x+5}$

32. $y = e^{5x}$

33. $y = 4^{(x^2)}$

34. $f(z) = (\sqrt{4})^z$

35. $f(\theta) = 4^{\sqrt{\theta}}$

Problems

36. Since January 1, 1960, the population of Slim Chance has been described by the formula

$$P = 35{,}000(0.98)^t,$$

where P is the population of the city t years after the start of 1960. At what rate was the population changing on January 1, 1983?

37. With a yearly inflation rate of 5%, prices are described by

$$P = P_0(1.05)^t,$$

where P_0 is the price in dollars when $t = 0$ and t is time in years. Suppose $P_0 = 1$. How fast (in cents/year) are prices rising when $t = 10$?

38. The population of the world in billions of people can be modeled by the function $f(t) = 5.3(1.018)^t$, where t is years since 1990. Find $f(0)$ and $f'(0)$. Find $f(30)$ and $f'(30)$. Using units, explain what each answer tells you about the population of the world.

39. Certain pieces of antique furniture increased very rapidly in price in the 1970s and 1980s. For example, the value of a particular rocking chair is well approximated by

$$V = 75(1.35)^t,$$

where V is in dollars and t is the number of years since 1975. Find the rate, in dollars per year, at which the price is increasing.

40. The value of a certain automobile purchased in 1997 can be approximated by the function $V(t) = 25(0.85)^t$, where t is the time, in years, from the date of purchase, and V is the value, in thousands of dollars.

(a) Evaluate and interpret $V(4)$.
(b) Find an expression for $V'(t)$, including units.
(c) Evaluate and interpret $V'(4)$.
(d) Use $V(t)$, $V'(t)$, and any other considerations you think are relevant to write a paragraph in support of or in opposition to the following statement: "From a monetary point of view, it is best to keep this vehicle as long as possible."

41. At any time, t, a population, $P(t)$, is growing at a rate proportional to the population at that moment.

(a) Using derivatives, write an equation representing the growth of the population. Let k be the constant of proportionality.
(b) Show that the function $P(t) = Ae^{kt}$ satisfies the equation in part (a) for any constant A.

42. (a) Find the slope of the graph of $f(x) = 1 - e^x$ at the point where it crosses the x-axis.
(b) Find the equation of the tangent line to the curve at this point.
(c) Find the equation of the line which is perpendicular to the tangent line at this point. (This line is known as the *normal* line.)

43. Find the value of c in Figure 3.14, where the line l tangent to the graph of $y = 2^x$ at $(0, 1)$ intersects the x-axis.

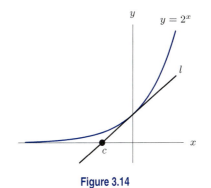

Figure 3.14

44. Find the quadratic polynomial $g(x) = ax^2 + bx + c$ which best fits the function $f(x) = e^x$ at $x = 0$, in the sense that

$$g(0) = f(0), \quad \text{and} \quad g'(0) = f'(0), \quad \text{and} \quad g''(0) = f''(0).$$

Using a computer or calculator, sketch graphs of f and g on the same axes. What do you notice?

45. Using the equation of the tangent line to the graph of e^x at $x = 0$, show that

$$e^x \geq 1 + x$$

for all values of x. A sketch may be helpful.

46. Find all solutions of the equation

$$2^x = 2x.$$

How do you know that you found all solutions?

47. For what value(s) of a are $y = a^x$ and $y = 1 + x$ tangent at $x = 0$? Explain.

3.3   THE PRODUCT AND QUOTIENT RULES

We now know how to find derivatives of powers and exponentials, and of sums and constant multiples of functions. This section shows how to find the derivatives of products and quotients.

Using Δ Notation

To express the difference quotients of general functions, some additional notation is helpful. We write Δf, read "delta f," for a small change in the value of f,

$$\Delta f = f(x+h) - f(x).$$

In this notation, the derivative is the limit of the ratio $\Delta f/h$:

$$f'(x) = \lim_{h \to 0} \frac{\Delta f}{h}.$$

The Product Rule

Suppose we know the derivatives of $f(x)$ and $g(x)$ and want to calculate the derivative of the product, $f(x)g(x)$. The derivative of the product is calculated by taking the limit, namely,

$$\frac{d[f(x)g(x)]}{dx} = \lim_{h \to 0} \frac{f(x+h)g(x+h) - f(x)g(x)}{h}.$$

To picture the quantity $f(x+h)g(x+h) - f(x)g(x)$, imagine the rectangle with sides $f(x+h)$ and $g(x+h)$ in Figure 3.15, where $\Delta f = f(x+h) - f(x)$ and $\Delta g = g(x+h) - g(x)$.

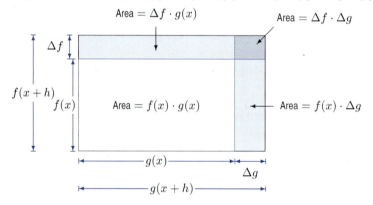

Figure 3.15: Illustration for the product rule (with Δf, Δg positive)

Then

$$f(x+h)g(x+h) - f(x)g(x) = (\text{Area of whole rectangle}) - (\text{Unshaded area})$$
$$= \text{Area of the three shaded rectangles}$$
$$= \Delta f \cdot g(x) + f(x) \cdot \Delta g + \Delta f \cdot \Delta g.$$

Now divide by h:

$$\frac{f(x+h)g(x+h) - f(x)g(x)}{h} = \frac{\Delta f}{h} \cdot g(x) + f(x) \cdot \frac{\Delta g}{h} + \frac{\Delta f \cdot \Delta g}{h}.$$

To evaluate the limit as $h \to 0$, we examine the three terms on the right separately. Notice that

$$\lim_{h \to 0} \frac{\Delta f}{h} \cdot g(x) = f'(x)g(x) \quad \text{and} \quad \lim_{h \to 0} f(x) \cdot \frac{\Delta g}{h} = f(x)g'(x).$$

In the third term we multiply the top and bottom by h to get $\dfrac{\Delta f}{h} \cdot \dfrac{\Delta g}{h} \cdot h$. Then,

$$\lim_{h \to 0} \frac{\Delta f \cdot \Delta g}{h} = \lim_{h \to 0} \frac{\Delta f}{h} \cdot \frac{\Delta g}{h} \cdot h = \lim_{h \to 0} \frac{\Delta f}{h} \cdot \lim_{h \to 0} \frac{\Delta g}{h} \cdot \lim_{h \to 0} h = f'(x) \cdot g'(x) \cdot 0 = 0.$$

Therefore, we conclude that

$$\lim_{h \to 0} \frac{f(x+h)g(x+h) - f(x)g(x)}{h} = \lim_{h \to 0} \left(\frac{\Delta f}{h} \cdot g(x) + f(x) \cdot \frac{\Delta g}{h} + \frac{\Delta f \cdot \Delta g}{h} \right)$$

$$= \lim_{h \to 0} \frac{\Delta f}{h} \cdot g(x) + \lim_{h \to 0} f(x) \cdot \frac{\Delta g}{h} + \lim_{h \to 0} \frac{\Delta f \cdot \Delta g}{h}$$

$$= f'(x)g(x) + f(x)g'(x).$$

Thus we have proved the following rule:

Theorem 3.3: The Product Rule

If $u = f(x)$ and $v = g(x)$ are differentiable, then

$$(fg)' = f'g + fg'.$$

The product rule can also be written

$$\frac{d(uv)}{dx} = \frac{du}{dx} \cdot v + u \cdot \frac{dv}{dx}.$$

In words:

  The derivative of a product is the derivative of the first times the second plus the first times the derivative of the second.

Another justification of the product rule is given in Problem 16 on page 154.

Example 1 Differentiate (a) $x^2 e^x$,   (b) $(3x^2 + 5x)e^x$,   (c) $\dfrac{e^x}{x^2}$.

Solution (a)
$$\frac{d(x^2 e^x)}{dx} = \left(\frac{d(x^2)}{dx} \right) e^x + x^2 \frac{d(e^x)}{dx} = 2xe^x + x^2 e^x = (2x + x^2)e^x.$$

(b)
$$\frac{d((3x^2 + 5x)e^x)}{dx} = \left(\frac{d(3x^2 + 5x)}{dx} \right) e^x + (3x^2 + 5x)\frac{d(e^x)}{dx}$$

$$= (6x + 5)e^x + (3x^2 + 5x)e^x = (3x^2 + 11x + 5)e^x.$$

(c) First we must write $\dfrac{e^x}{x^2}$ as the product $x^{-2}e^x$:

$$\frac{d}{dx}\left(\frac{e^x}{x^2} \right) = \frac{d(x^{-2}e^x)}{dx} = \left(\frac{d(x^{-2})}{dx} \right) e^x + x^{-2}\frac{d(e^x)}{dx}$$

$$= -2x^{-3}e^x + x^{-2}e^x = (-2x^{-3} + x^{-2})e^x.$$

The Quotient Rule

Suppose we want to differentiate a function of the form $Q(x) = f(x)/g(x)$. (Of course, we have to avoid points where $g(x) = 0$.) We want a formula for Q' in terms of f' and g'.

Assuming that $Q(x)$ is differentiable,[1] we can use the product rule on $f(x) = Q(x)g(x)$:

$$f'(x) = Q'(x)g(x) + Q(x)g'(x)$$

$$= Q'(x)g(x) + \frac{f(x)}{g(x)}g'(x).$$

Solving for $Q'(x)$ gives

$$Q'(x) = \frac{f'(x) - \dfrac{f(x)}{g(x)}g'(x)}{g(x)}.$$

---

[1]The method in Example 6 on page 125 can be used to explain why $Q(x)$ must be differentiable.

Multiplying the top and bottom by $g(x)$ to simplify gives

$$\left(\frac{f(x)}{g(x)}\right)' = \frac{f'(x)g(x) - f(x)g'(x)}{(g(x))^2}.$$

So we have the following rule:

Theorem 3.4: The Quotient Rule

If $u = f(x)$ and $v = g(x)$ are differentiable, then

$$\left(\frac{f}{g}\right)' = \frac{f'g - fg'}{g^2},$$

or equivalently,

$$\frac{d}{dx}\left(\frac{u}{v}\right) = \frac{\dfrac{du}{dx} \cdot v - u \cdot \dfrac{dv}{dx}}{v^2}.$$

In words:

> The derivative of a quotient is the derivative of the numerator times the denominator minus the numerator times the derivative of the denominator, all over the denominator squared.

Example 2   Differentiate (a) $\dfrac{5x^2}{x^3 + 1}$,   (b) $\dfrac{1}{1 + e^x}$,   (c) $\dfrac{e^x}{x^2}$.

Solution   (a)

$$\frac{d}{dx}\left(\frac{5x^2}{x^3 + 1}\right) = \frac{\left(\dfrac{d}{dx}(5x^2)\right)(x^3 + 1) - 5x^2 \dfrac{d}{dx}(x^3 + 1)}{(x^3 + 1)^2} = \frac{10x(x^3 + 1) - 5x^2(3x^2)}{(x^3 + 1)^2}$$

$$= \frac{-5x^4 + 10x}{(x^3 + 1)^2}.$$

(b)

$$\frac{d}{dx}\left(\frac{1}{1 + e^x}\right) = \frac{\left(\dfrac{d}{dx}(1)\right)(1 + e^x) - 1\dfrac{d}{dx}(1 + e^x)}{(1 + e^x)^2} = \frac{0(1 + e^x) - 1(0 + e^x)}{(1 + e^x)^2}$$

$$= \frac{-e^x}{(1 + e^x)^2}.$$

(c) This is the same as part (c) of Example 1, but this time we do it by the quotient rule.

$$\frac{d}{dx}\left(\frac{e^x}{x^2}\right) = \frac{\left(\dfrac{d(e^x)}{dx}\right)x^2 - e^x\left(\dfrac{d(x^2)}{dx}\right)}{(x^2)^2} = \frac{e^x x^2 - e^x(2x)}{x^4}$$

$$= e^x\left(\frac{x^2 - 2x}{x^4}\right) = e^x\left(\frac{x - 2}{x^3}\right).$$

This is, in fact, the same answer as before, although it looks different. Can you show that it is the same?

Exercises and Problems for Section 3.3

Exercises

1. If $f(x) = x^2(x^3 + 5)$, find $f'(x)$ two ways: by using the product rule and by multiplying out before taking the derivative. Do you get the same result? Should you?

2. If $f(x) = 2^x \cdot 3^x$, find $f'(x)$ two ways: by using the product rule and by using the fact that $2^x \cdot 3^x = 6^x$. Do you get the same result?

For Exercises 3–29, find the derivative. It may be to your advantage to simplify first. Assume that a, b, c, and k are constants.

3. $f(x) = xe^x$

4. $y = x \cdot 2^x$

5. $y = \sqrt{x} \cdot 2^x$

6. $f(x) = (x^2 - \sqrt{x})3^x$

7. $z = (s^2 - \sqrt{s})(s^2 + \sqrt{s})$

8. $y = (t^2 + 3)e^t$

9. $y = (t^3 - 7t^2 + 1)e^t$

10. $f(x) = \dfrac{x}{e^x}$

11. $g(x) = \dfrac{25x^2}{e^x}$

12. $g(w) = \dfrac{w^{3.2}}{5^w}$

13. $q(r) = \dfrac{3r}{5r + 2}$

14. $g(t) = \dfrac{t - 4}{t + 4}$

15. $z = \dfrac{3t + 1}{5t + 2}$

16. $z = \dfrac{t^2 + 5t + 2}{t + 3}$

17. $z = \dfrac{t^2 + 3t + 1}{t + 1}$

18. $f(x) = \dfrac{x^2 + 3}{x}$

19. $w = \dfrac{y^3 - 6y^2 + 7y}{y}$

20. $y = \dfrac{\sqrt{t}}{t^2 + 1}$

21. $f(z) = \dfrac{z^2 + 1}{\sqrt{z}}$

22. $g(t) = \dfrac{4}{3 + \sqrt{t}}$

23. $h(r) = \dfrac{r^2}{2r + 1}$

24. $f(z) = \dfrac{3z^2}{5z^2 + 7z}$

25. $w(x) = \dfrac{17e^x}{2^x}$

26. $h(p) = \dfrac{1 + p^2}{3 + 2p^2}$

27. $f(x) = \dfrac{1 + x}{2 + 3x + 4x^2}$

28. $f(x) = \dfrac{ax + b}{cx + k}$

29. $w = (t^3 + 5t)(t^2 - 7t + 2)$

Problems

30. If $f(x) = (3x + 8)(2x - 5)$, find $f'(x)$ and $f''(x)$.

31. For what intervals is $f(x) = xe^{-x}$ concave down?

32. For what intervals is $g(x) = \dfrac{1}{x^2 + 1}$ concave down?

33. Find the equation of the tangent line to the graph of $f(x) = \dfrac{2x - 5}{x + 1}$ at the point at which $x = 0$.

34. Differentiate $f(t) = e^{-t}$ by writing it as $f(t) = \dfrac{1}{e^t}$.

35. Differentiate $f(x) = e^{2x}$ by writing it as $f(x) = e^x \cdot e^x$.

36. Differentiate $f(x) = e^{3x}$ by writing it as $f(x) = e^x \cdot e^{2x}$ and using the result of Problem 35.

37. Based on your answers to Problems 35 and 36, guess the derivative of e^{4x}.

38. (a) Differentiate $y = \dfrac{e^x}{x}$, $y = \dfrac{e^x}{x^2}$, and $y = \dfrac{e^x}{x^3}$.

   (b) What do you anticipate the derivative of $y = \dfrac{e^x}{x^n}$ will be? Confirm your guess.

39. Using the product rule and the fact that $\dfrac{d(x)}{dx} = 1$, show that $\dfrac{d(x^2)}{dx} = 2x$ and $\dfrac{d(x^3)}{dx} = 3x^2$.

40. Use the product rule to show that $\dfrac{d}{dx}(x^{1/2}) = \dfrac{1}{2x^{1/2}}$.
   [Hint: Write $x = x^{1/2}x^{1/2}$.]

41. Suppose f and g are differentiable functions with the values shown in the following table. For each of the following functions h, find $h'(2)$.

   (a) $h(x) = f(x) + g(x)$ (b) $h(x) = f(x)g(x)$

   (c) $h(x) = \dfrac{f(x)}{g(x)}$

x	$f(x)$	$g(x)$	$f'(x)$	$g'(x)$
2	3	4	5	-2

42. If $H(3) = 1$, $H'(3) = 3$, $F(3) = 5$, $F'(3) = 4$, find:

   (a) $G'(3)$ if $G(z) = F(z) \cdot H(z)$
   (b) $G'(3)$ if $G(w) = F(w)/H(w)$

43. Find a possible formula for a function $y = f(x)$ such that $f'(x) = 10x^9 e^x + x^{10} e^x$.

44. The quantity, q, of a certain skateboard sold depends on the selling price, p, in dollars, so we write $q = f(p)$. You are given that $f(140) = 15{,}000$ and $f'(140) = -100$.

   (a) What do $f(140) = 15{,}000$ and $f'(140) = -100$ tell you about the sales of skateboards?
   (b) The total revenue, R, earned by the sale of skateboards is given by $R = pq$. Find $\left.\dfrac{dR}{dp}\right|_{p=140}$.

(c) What is the sign of $\left.\dfrac{dR}{dp}\right|_{p=140}$? If the skateboards are currently selling for $140, what happens to revenue if the price is increased to $141?

45. When an electric current passes through two resistors with resistance r_1 and r_2, connected in parallel, the combined resistance, R, can be calculated from the equation

$$\frac{1}{R} = \frac{1}{r_1} + \frac{1}{r_2}.$$

Find the rate at which the combined resistance changes with respect to changes in r_1. Assume that r_2 is constant.

46. A museum has decided to sell one of its paintings and to invest the proceeds. If the picture is sold between the years 2000 and 2020 and the money from the sale is invested in a bank account earning 5% annual interest compounded once a year, then $B(t)$, the balance in the year 2020, depends on the year, t, in which the painting is sold and the sale price $P(t)$. If t is measured from the year 2000 so that $0 < t < 20$ then

$$B(t) = P(t)(1.05)^{20-t}.$$

(a) Explain why $B(t)$ is given by this formula.
(b) Show that the formula for $B(t)$ is equivalent to

$$B(t) = (1.05)^{20}\frac{P(t)}{(1.05)^t}.$$

(c) Find $B'(10)$, given that $P(10) = 150,000$ and $P'(10) = 5000$.

47. Let $f(v)$ be the gas consumption (in liters/km) of a car going at velocity v (in km/hr). In other words, $f(v)$ tells you how many liters of gas the car uses to go one kilometer, if it is going at velocity v. You are told that

$$f(80) = 0.05 \text{ and } f'(80) = 0.0005.$$

(a) Let $g(v)$ be the distance the same car goes on one liter of gas at velocity v. What is the relationship between $f(v)$ and $g(v)$? Find $g(80)$ and $g'(80)$.
(b) Let $h(v)$ be the gas consumption in liters per hour. In other words, $h(v)$ tells you how many liters of gas the car uses in one hour if it is going at velocity v. What is the relationship between $h(v)$ and $f(v)$? Find $h(80)$ and $h'(80)$.
(c) How would you explain the practical meaning of the values of these functions and their derivatives to a driver who knows no calculus?

48. The function $f(x) = e^x$ has the properties

$$f'(x) = f(x) \text{ and } f(0) = 1.$$

Explain why $f(x)$ is the only function with both these properties. [Hint: Assume $g'(x) = g(x)$, and $g(0) = 1$,

for some function $g(x)$. Define $h(x) = g(x)/e^x$, and compute $h'(x)$. Then use the fact that a function with a derivative of 0 must be a constant function.]

49. Find $f'(x)$ for the following functions with the product rule, rather than by multiplying out.

(a) $f(x) = (x-1)(x-2)$.
(b) $f(x) = (x-1)(x-2)(x-3)$.
(c) $f(x) = (x-1)(x-2)(x-3)(x-4)$.

50. Use the answer from Problem 49 to guess $f'(x)$ for the following function:

$$f(x) = (x - r_1)(x - r_2)(x - r_3) \cdots (x - r_n)$$

where $r_1, r_2, \ldots, r_n$ are any real numbers.

51. (a) Provide a three dimensional analogue for the geometrical demonstration of the formula for the derivative of a product, given in Figure 3.15 on page 118. In other words, find a formula for the derivative of $F(x) \cdot G(x) \cdot H(x)$ using Figure 3.16.
(b) Confirm your results by writing $F(x) \cdot G(x) \cdot H(x)$ as $[F(x) \cdot G(x)] \cdot H(x)$ and using the product rule twice.
(c) Generalize your result to n functions: what is the derivative of

$$f_1(x) \cdot f_2(x) \cdot f_3(x) \cdots f_n(x)?$$

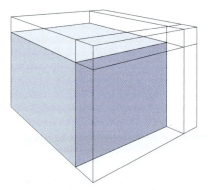

Figure 3.16: A graphical representation of the 3-dimensional product rule

52. We call $x = a$ a double zero of the polynomial $P(x)$ if it appears twice as a factor, that is, if $P(x) = (x - a)^2 Q(x)$, where $Q(x)$ is a polynomial.

(a) If $x = a$ is a double zero of a polynomial $P(x)$, show that $P(a) = P'(a) = 0$.
(b) If $P(x)$ is a polynomial and $P(a) = P'(a) = 0$, show that $x = a$ is a double zero of $P(x)$.

3.4   THE CHAIN RULE

Composite functions such as $\sin(3t)$ or e^{-x^2} occur frequently in practice. In this section we see how to differentiate such functions.

The Derivative of a Composition of Functions

Suppose $f(g(x))$ is a composite function, with f being the outside function and g being the inside. Let us write

$$z = g(x) \quad \text{and} \quad y = f(z), \quad \text{so} \quad y = f(g(x)).$$

Then a small change in x, called Δx, generates a small change in z, called Δz. In turn, Δz generates a small change in y called Δy. Provided Δx and Δz are not zero, we can say:

$$\frac{\Delta y}{\Delta x} = \frac{\Delta y}{\Delta z} \cdot \frac{\Delta z}{\Delta x}.$$

Since $\dfrac{dy}{dx} = \lim\limits_{\Delta x \to 0} \dfrac{\Delta y}{\Delta x}$, this suggests that in the limit as Δx, Δy, and Δz get smaller and smaller, we have:

The Chain Rule

$$\frac{dy}{dx} = \frac{dy}{dz} \cdot \frac{dz}{dx}.$$

Since $\dfrac{dy}{dz} = f'(z)$ and $\dfrac{dz}{dx} = g'(x)$, we can also write

$$\frac{d}{dx} f(g(x)) = f'(z) \cdot g'(x).$$

Substituting $z = g(x)$, we can rewrite this as follows:

Theorem 3.5: The Chain Rule

If f and g are differentiable functions, then

$$\frac{d}{dx} f(g(x)) = f'(g(x)) \cdot g'(x).$$

In words:
   The derivative of a composite function is the product of the derivatives of the outside and inside functions. The derivative of the outside function must be evaluated at the inside function.

A justification of the chain rule is given in Problem 17 on page 154.

Example 1   The length, L, in cm, of a steel bar depends on the air temperature, $H°C$, and the temperature H depends on time, t, measured in hours. If the length increases by 2 cm for every degree increase in temperature, and the temperature is increasing at $3°C$ per hour, how fast is the length of the bar increasing? What are the units for your answer?

Solution   We expect the rate at which the length is increasing to be in cm/hr. We are told that

$$\text{Rate length increasing with respect to temperature} = \frac{dL}{dH} = 2 \text{ cm/}°\text{C}$$

$$\text{Rate temperature increasing with respect to time} = \frac{dH}{dt} = 3°\text{C/hr.}$$

We want to calculate the rate at which the length is increasing with respect to time, or dL/dt. We think of L as a function of H, and H as a function of t. By the chain rule we know that

$$\frac{dL}{dt} = \frac{dL}{dH} \cdot \frac{dH}{dt} = \left(2\frac{\text{cm}}{\text{°C}}\right) \cdot \left(3\frac{\text{°C}}{\text{hr}}\right) = 6 \text{ cm/hr.}$$

Thus, the length is increasing at 6 cm/hr.

Example 1 above shows us how to interpret the chain rule in practical terms. The next examples show how it is used to compute derivatives of functions given by formulas.

Example 2   Find the derivatives of the following functions: (a) $(4x^2 + 1)^7$ (b) e^{3x}

Solution   (a) Here $z = g(x) = 4x^2 + 1$ is the inside function; $f(z) = z^7$ is the outside function. Since $g'(x) = 8x$ and $f'(z) = 7z^6$, we have

$$\frac{d}{dx}\left[(4x^2 + 1)^7\right] = 7z^6 \cdot 8x = 7(4x^2 + 1)^6 \cdot 8x = 56x(4x^2 + 1)^6.$$

(b) Let $z = g(x) = 3x$ and $f(z) = e^z$. Then $g'(x) = 3$ and $f'(z) = e^z$, so

$$\frac{d}{dx}\left(e^{3x}\right) = e^z \cdot 3 = 3e^{3x}.$$

Example 3   Differentiate

(a) $(x^2+1)^{100}$ (b) $\sqrt{3x^2 + 5x - 2}$ (c) $\dfrac{1}{x^2 + x^4}$ (d) $\sqrt{e^x + 1}$ (e) e^{x^2}

Solution   (a) Here $z = g(x) = x^2 + 1$ is the inside function; $f(z) = z^{100}$ is the outside function. Now $g'(x) = 2x$ and $f'(z) = 100z^{99}$, so

$$\frac{d[(x^2 + 1)^{100}]}{dx} = 100z^{99} \cdot 2x = 100(x^2 + 1)^{99} \cdot 2x = 200x(x^2 + 1)^{99}.$$

(b) Here $z = g(x) = 3x^2 + 5x - 2$ and $f(z) = \sqrt{z}$, so $g'(x) = 6x + 5$ and $f'(z) = \dfrac{1}{2\sqrt{z}}$. Hence

$$\frac{d(\sqrt{3x^2 + 5x - 2})}{dx} = \frac{1}{2\sqrt{z}} \cdot (6x + 5) = \frac{1}{2\sqrt{3x^2 + 5x - 2}} \cdot (6x + 5).$$

(c) Let $z = g(x) = x^2 + x^4$ and $f(z) = 1/z$, so $g'(x) = 2x + 4x^3$ and $f'(z) = -z^{-2} = -\dfrac{1}{z^2}$. Then

$$\frac{d}{dx}\left(\frac{1}{x^2 + x^4}\right) = -\frac{1}{z^2}(2x + 4x^3) = -\frac{2x + 4x^3}{(x^2 + x^4)^2}.$$

We could have done this problem using the quotient rule. Try it and see that you get the same answer!

(d) Let $z = g(x) = e^x + 1$ and $f(z) = \sqrt{z}$. Therefore $g'(x) = e^x$ and $f'(z) = \dfrac{1}{2\sqrt{z}}$. We get

$$\frac{d(\sqrt{e^x + 1})}{dx} = \frac{1}{2\sqrt{z}}e^x = \frac{e^x}{2\sqrt{e^x + 1}}.$$

(e) To figure out which is the inside function and which is the outside, notice that to evaluate e^{x^2} we first evaluate x^2 and then take e to that power. This tells us that the inside function is $z = g(x) = x^2$ and the outside function is $f(z) = e^z$. Therefore, $g'(x) = 2x$, and $f'(z) = e^z$, giving

$$\frac{d(e^{x^2})}{dx} = e^z \cdot 2x = e^{x^2} \cdot 2x = 2xe^{x^2}.$$

Example 4   Find the derivative of e^{2x} by the chain rule and by the product rule.

Solution   Chain rule: Let the inside function be $z = g(x) = 2x$ and the outside function be $f(z) = e^z$. Then

$$\frac{d(e^{2x})}{dx} = f'(g(x)) \cdot g'(x) = e^{2x} \cdot 2 = 2e^{2x}.$$

Product rule: Write $e^{2x} = e^x \cdot e^x$. Then

$$\frac{d(e^{2x})}{dx} = \frac{d(e^x e^x)}{dx} = \left(\frac{d(e^x)}{dx}\right)e^x + e^x\left(\frac{d(e^x)}{dx}\right) = e^x \cdot e^x + e^x \cdot e^x = 2e^{2x}.$$

The chain rule can often be used to find rates, as in the following example.

Example 5   A circular oil spill is spreading. If the radius is increasing at 0.2 km/hr when the radius is 3 km, find the rate at which the area is increasing at that time. Give units in your answer.

Solution   If A is the area of the spill in km^2 and r is the radius in km, then

$$A = \pi r^2.$$

Differentiating using the chain rule gives

$$\frac{dA}{dt} = \frac{dA}{dr} \cdot \frac{dr}{dt} = 2\pi r \cdot \frac{dr}{dt}.$$

We know that $dr/dt = 0.2$ km/hr when $r = 3$, so

$$\frac{dA}{dt} = (2\pi \cdot 3 \text{ km})(0.2 \text{ km/hr}) = 1.2\pi \approx 3.77 \text{ km}^2/\text{hr}.$$

Notice that the units of dA/dt are (km)(km/hr)=km^2/hr, which is area per unit time, as expected.

Using the Product and Chain Rules to Differentiate a Quotient

If you prefer, you can differentiate a quotient by the product and chain rules, instead of by the quotient rule. The resulting formulas may look different, but they will be equivalent.

Example 6   Find $k'(x)$ if $k(x) = \dfrac{x}{x^2 + 1}$.

Solution   One way is to use the quotient rule:

$$k'(x) = \frac{1 \cdot (x^2 + 1) - x \cdot (2x)}{(x^2 + 1)^2}$$

$$= \frac{1 - x^2}{(x^2 + 1)^2}.$$

Alternatively, we can write the original function as a product,

$$k(x) = x\frac{1}{x^2 + 1} = x \cdot (x^2 + 1)^{-1},$$

and use the product rule:

$$k'(x) = 1 \cdot (x^2 + 1)^{-1} + x \cdot \frac{d}{dx}\left[(x^2 + 1)^{-1}\right].$$

Now use the chain rule to differentiate $(x^2 + 1)^{-1}$. Let $z = x^2 + 1$, and $f(z) = z^{-1}$, so

$$\frac{d}{dx}\left[(x^2 + 1)^{-1}\right] = -z^{-2} \cdot 2x = -(x^2 + 1)^{-2} \cdot 2x = \frac{-2x}{(x^2 + 1)^2}.$$

Therefore,

$$k'(x) = \frac{1}{x^2 + 1} + x \cdot \frac{-2x}{(x^2 + 1)^2} = \frac{1}{x^2 + 1} - \frac{2x^2}{(x^2 + 1)^2}.$$

Putting these two fractions over a common denominator gives the same answer as the quotient rule.

Exercises and Problems for Section 3.4

Exercises

Find the derivatives of the functions in Exercises 1–48. Assume that a, b, c, and k are constants.

1. $f(x) = (x+1)^{99}$

2. $f(x) = \sqrt{1-x^2}$

3. $w = (t^2+1)^{100}$

4. $w = (t^3+1)^{100}$

5. $w = (\sqrt{t}+1)^{100}$

6. $f(t) = e^{3t}$

7. $h(w) = (w^4 - 2w)^5$

8. $w(r) = \sqrt{r^4+1}$

9. $g(x) = e^{\pi x}$

10. $f(\theta) = 2^{-\theta}$

11. $y = \pi^{(x+2)}$

12. $g(x) = 3^{(2x+7)}$

13. $k(x) = (x^3 + e^x)^4$

14. $f(x) = e^{2x}\left(x^2 + 5^x\right)$

15. $v(t) = t^2 e^{-ct}$

16. $p(t) = e^{4t+2}$

17. $g(t) = e^{(1+3t)^2}$

18. $z(x) = \sqrt[3]{2x+5}$

19. $z = 2^{5t-3}$

20. $w = \sqrt{(x^2 \cdot 5^x)^3}$

21. $y = e^{3w/2}$

22. $y = e^{-4t}$

23. $y = \sqrt{s^3+1}$

24. $w = e^{\sqrt{s}}$

25. $y = te^{-t^2}$

26. $f(z) = \sqrt{z}e^{-z}$

27. $f(z) = \dfrac{\sqrt{z}}{e^z}$

28. $y = \dfrac{\sqrt{z}}{2^z}$

29. $f(t) = te^{5-2t}$

30. $y = \left(\dfrac{x^2+2}{3}\right)^2$

31. $h(x) = \sqrt{\dfrac{x^2+9}{x+3}}$

32. $y = \dfrac{e^{2x}}{x^2+1}$

33. $y = \dfrac{1}{e^{3x}+x^2}$

34. $h(z) = \left(\dfrac{b}{a+z^2}\right)^4$

35. $h(x) = 2e^{3x}$

36. $f(z) = \dfrac{1}{(e^z+1)^2}$

37. $f(\theta) = \dfrac{1}{1+e^{-\theta}}$

38. $f(x) = 6e^{5x} + e^{-x^2}$

39. $f(w) = (5w^2+3)e^{w^2}$

40. $w = (t^2+3t)(1-e^{-2t})$

41. $f(y) = \sqrt{10^{(5-y)}}$

42. $f(x) = e^{-(x-1)^2}$

43. $f(y) = e^{e^{(y^2)}}$

44. $f(t) = 2 \cdot e^{-2e^{2t}}$

45. $f(t) = ae^{bt}$

46. $f(x) = (ax^2+b)^3$

47. $f(x) = axe^{-bx}$

48. $f(x) = (3x^2+\pi)(e^x - 4)$

Problems

49. Find the equation of the tangent line to $f(x) = (x-1)^3$ at the point where $x = 2$.

50. Find the equation of the line tangent to $y = f(x)$ at $x = 1$, where $f(x)$ is the function in Problem 38.

51. For what values of x is the graph of $y = e^{-x^2}$ concave down?

52. Suppose $f(x) = (2x+1)^{10}(3x-1)^7$. Find a formula for $f'(x)$. Decide on a reasonable way to simplify your result, and find a formula for $f''(x)$.

53. Given $F(2) = 1$, $F'(2) = 5$, $F(4) = 3$, $F'(4) = 7$ and $G(4) = 2$, $G'(4) = 6$, $G(3) = 4$, $G'(3) = 8$, find:

(a) $H(4)$ if $H(x) = F(G(x))$
(b) $H'(4)$ if $H(x) = F(G(x))$
(c) $H(4)$ if $H(x) = G(F(x))$
(d) $H'(4)$ if $H(x) = G(F(x))$
(e) $H'(4)$ if $H(x) = F(x)/G(x)$

54. Given $y = f(x)$ with $f(1) = 4$ and $f'(1) = 3$, find

(a) $g'(1)$ if $g(x) = \sqrt{f(x)}$.
(b) $h'(1)$ if $h(x) = f(\sqrt{x})$.

55. Suppose f and g are differentiable functions with the values given in the following table. For each of the following functions h, find $h'(2)$.

(a) $h(x) = f(g(x))$ (b) $h(x) = g(f(x))$
(c) $h(x) = f(f(x))$

x	$f(x)$	$g(x)$	$f'(x)$	$g'(x)$
2	5	5	e	$\sqrt{2}$
5	2	8	π	7

56. If the derivative of $y = k(x)$ equals 2 when $x = 1$, what is the derivative of

(a) $k(2x)$ when $x = \dfrac{1}{2}$?
(b) $k(x+1)$ when $x = 0$?
(c) $k\left(\dfrac{1}{4}x\right)$ when $x = 4$?

57. Is $x = \sqrt[3]{2t+5}$ a solution to the equation $3x^2\dfrac{dx}{dt} = 2$? Why or why not?

58. Find a possible formula for a function $m(x)$ such that $m'(x) = x^5 \cdot e^{(x^6)}$.

59. The balance, B, in a bank account t years after a deposit of $5000 is given by $B = 5000e^{0.08t}$. At what rate is the balance in the account changing at $t = 5$ years? Use units to interpret your answer in financial terms.

60. At a time t hours after it was administered, the concentration of a drug in the body is $f(t) = 27e^{-0.14t}$ ng/ml. What is the concentration 4 hours after it was administered? At what rate is the concentration changing at that time?

61. The world's population is about $f(t) = 6e^{0.013t}$ billion, where t is time in years since 1999. Find $f(0)$, $f'(0)$, $f(10)$, and $f'(10)$. Using units, interpret your answers in terms of population.

62. One gram of radioactive carbon-14 decays according to the formula

$$Q = e^{-0.000121t}$$

where Q is the number of grams of carbon-14 remaining after t years.

(a) Find the rate at which carbon-14 is decaying (in grams/year).

(b) Sketch the rate you found in part (a) against time.

63. The temperature, H, in degrees Fahrenheit ($°F$), of a can of soda that is put into a refrigerator to cool is given as a function of time, t, in hours, by

$$H = 40 + 30e^{-2t}.$$

(a) Find the rate at which the temperature of the soda is changing (in $°F$/hour).

(b) What is the sign of dH/dt? Explain.

(c) When, for $t \geq 0$, is the magnitude of dH/dt largest? In terms of the can of soda, why is this?

64. If you invest P dollars in a bank account at an annual interest rate of $r\%$, then after t years you will have B dollars, where

$$B = P\left(1 + \frac{r}{100}\right)^t.$$

(a) Find dB/dt, assuming P and r are constant. In terms of money, what does dB/dt represent?

(b) Find dB/dr, assuming P and t are constant. In terms of money, what does dB/dr represent?

65. A pebble is dropped in still water, forming a circular ripple whose radius is expanding at a constant rate of 10 cm/sec. Find a formula giving the area enclosed by the ripple as a function of time. When the radius is 20 cm, how fast is the area enclosed by the ripple increasing?

66. The theory of relativity predicts that an object whose mass is m_0 when it is at rest will appear heavier when moving at speeds near the speed of light. When the object is moving at speed v, its mass m is given by

$$m = \frac{m_0}{\sqrt{1 - (v^2/c^2)}}, \qquad \text{where } c \text{ is the speed of light.}$$

(a) Find dm/dv.

(b) In terms of physics, what does dm/dv tell you?

67. The charge, Q, on a capacitor which starts discharging at time $t = 0$ is given by

$$Q = \begin{cases} Q_0 & \text{for } t \leq 0 \\ Q_0 e^{-t/RC} & \text{for } t > 0, \end{cases}$$

where R and C are positive constants depending on the circuit and Q_0 is the charge at $t = 0$, where $Q_0 \neq 0$. The current, I, flowing in the circuit is given by $I = dQ/dt$.

(a) Find the current I for $t < 0$ and for $t > 0$.

(b) Is it possible to define I at $t = 0$?

(c) Is the function Q differentiable at $t = 0$?

68. An electric charge is decaying exponentially according to the formula

$$Q = Q_0 e^{-t/RC}.$$

Show that the charge, Q, and the electric current, $I = dQ/dt$, have the same time constant. (The time constant is the time to decay to $1/e$ times the original value.)

69. A particle is moving on the x-axis, where x is in centimeters. Its velocity, v, in cm/sec, when it is at the point with coordinate x is given by

$$v = x^2 + 3x - 2.$$

Find the acceleration of the particle when it is at the point $x = 2$. Give units in your answer.

70. The rate of change of a population depends on the current population, P, and is given by

$$\frac{dP}{dt} = kP(L - P) \qquad \text{for positive constants } k, L.$$

(a) For what nonnegative values of P is the population increasing? Decreasing? For what values of P does the population remain constant?

(b) Find d^2P/dt^2 as a function of P.

71. A function f is said to have a *zero of multiplicity m* at $x = a$ if

$$f(x) = (x - a)^m h(x), \qquad \text{with } h(a) \neq 0.$$

Explain why a function having a zero of multiplicity m at $x = a$ satisfies $f^{(p)}(a) = 0$, for $p = 1, 2, \ldots m - 1$. [Note: $f^{(p)}$ is the pth derivative.]

3.5   THE TRIGONOMETRIC FUNCTIONS

Derivatives of the Sine and Cosine

Since the sine and cosine functions are periodic, their derivatives must be periodic also. (Why?) Let's look at the graph of $f(x) = \sin x$ in Figure 3.17 and estimate the derivative function graphically.

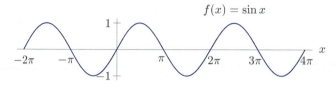

$$f(x) = \sin x$$

Figure 3.17: The sine function

First we might ask where the derivative is zero. (At $x = \pm\pi/2, \pm 3\pi/2, \pm 5\pi/2$, etc.) Then ask where the derivative is positive and where it is negative. (Positive for $-\pi/2 < x < \pi/2$; negative for $\pi/2 < x < 3\pi/2$, etc.) Since the largest positive slopes are at $x = 0, 2\pi$, and so on, and the largest negative slopes are at $x = \pi, 3\pi$, and so on, we get something like the graph in Figure 3.18.

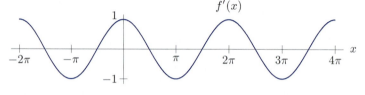

$$f'(x)$$

Figure 3.18: Derivative of $f(x) = \sin x$

The graph of the derivative in Figure 3.18 looks suspiciously like the graph of the cosine function. This might lead us to conjecture, quite correctly, that the derivative of the sine is the cosine.

Of course, we cannot be sure, just from the graphs, that the derivative of the sine really is the cosine. However, for now we'll assume that the derivative of the sine *is* the cosine and confirm the result at the end of the section.

One thing we can do now is to check that the derivative function in Figure 3.18 has amplitude 1 (as it ought to if it is the cosine). That means we have to convince ourselves that the derivative of $f(x) = \sin x$ is 1 when $x = 0$. The next example suggests that this is true when x is in radians.

Example 1   Using a calculator set in radians, estimate the derivative of $f(x) = \sin x$ at $x = 0$.

Solution   Since $f(x) = \sin x$,

$$f'(0) = \lim_{h \to 0} \frac{\sin(0+h) - \sin 0}{h} = \lim_{h \to 0} \frac{\sin h}{h}.$$

Table 3.3 contains values of $(\sin h)/h$ which suggest that this limit is 1, so we estimate

$$f'(0) = \lim_{h \to 0} \frac{\sin h}{h} = 1.$$

Table 3.3

h (radians)	−0.1	−0.01	−0.001	−0.0001	0.0001	0.001	0.01	0.1
$(\sin h)/h$	0.99833	0.99998	1.0000	1.0000	1.0000	1.0000	0.99998	0.99833

Warning: It is important to notice that in the previous example h was in *radians*; any conclusions we have drawn about the derivative of $\sin x$ are valid *only* when x is in radians.

Example 2   Starting with the graph of the cosine function, sketch a graph of its derivative.

Solution
The graph of $g(x) = \cos x$ is in Figure 3.19(a). Its derivative is 0 at $x = 0, \pm\pi, \pm 2\pi$, and so on; it is positive for $-\pi < x < 0$, $\pi < x < 2\pi$, and so on; and it is negative for $0 < x < \pi$, $2\pi < x < 3\pi$, and so on. The derivative is in Figure 3.19(b).

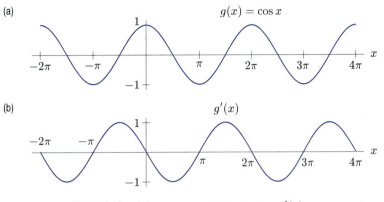

Figure 3.19: $g(x) = \cos x$ and its derivative, $g'(x)$

As we did with the sine, we use the graphs to make a conjecture. The derivative of the cosine in Figure 3.19(b) looks exactly like the graph of sine, except reflected about the x-axis. But how can we be sure that the derivative is $-\sin x$?

Example 3   Use the relation $\dfrac{d}{dx}(\sin x) = \cos x$ to show that $\dfrac{d}{dx}(\cos x) = -\sin x$.

Solution
Since the cosine function is the sine function shifted to the left by $\pi/2$ (that is, $\cos x = \sin(x + \pi/2)$), we expect the derivative of the cosine to be the derivative of the sine, shifted to the left by $\pi/2$. Differentiating using the chain rule:

$$\frac{d}{dx}(\cos x) = \frac{d}{dx}\left(\sin\left(x + \frac{\pi}{2}\right)\right) = \cos\left(x + \frac{\pi}{2}\right).$$

But $\cos(x + \pi/2)$ is the cosine shifted to the left by $\pi/2$, which gives a sine curve reflected about the x-axis. So we have

$$\frac{d}{dx}(\cos x) = \cos\left(x + \frac{\pi}{2}\right) = -\sin x.$$

For x in radians, $\dfrac{d}{dx}(\sin x) = \cos x$ and $\dfrac{d}{dx}(\cos x) = -\sin x.$

Example 4   Differentiate (a) $2\sin(3\theta)$,   (b) $\cos^2 x$,   (c) $\cos(x^2)$,   (d) $e^{-\sin t}$.

Solution
Use the chain rule:

(a) $\dfrac{d}{d\theta}(2\sin(3\theta)) = 2\dfrac{d}{d\theta}(\sin(3\theta)) = 2(\cos(3\theta))\dfrac{d}{d\theta}(3\theta) = 2(\cos(3\theta))3 = 6\cos(3\theta).$

(b) $\dfrac{d}{dx}(\cos^2 x) = \dfrac{d}{dx}\left((\cos x)^2\right) = 2(\cos x) \cdot \dfrac{d}{dx}(\cos x) = 2(\cos x)(-\sin x) = -2\cos x \sin x.$

(c) $\dfrac{d}{dx}\left(\cos(x^2)\right) = -\sin(x^2) \cdot \dfrac{d}{dx}(x^2) = -2x\sin(x^2).$

(d) $\dfrac{d}{dt}(e^{-\sin t}) = e^{-\sin t}\dfrac{d}{dt}(-\sin t) = -(\cos t)e^{-\sin t}.$

Derivative of the Tangent Function

Since $\tan x = \sin x / \cos x$, we differentiate $\tan x$ using the quotient rule. Writing $(\sin x)'$ for $d(\sin x)/dx$, we have:

$$\frac{d}{dx}(\tan x) = \frac{d}{dx}\left(\frac{\sin x}{\cos x}\right) = \frac{(\sin x)'(\cos x) - (\sin x)(\cos x)'}{\cos^2 x} = \frac{\cos^2 x + \sin^2 x}{\cos^2 x} = \frac{1}{\cos^2 x}.$$

> For x in radians, $\qquad \dfrac{d}{dx}(\tan x) = \dfrac{1}{\cos^2 x}.$

The graphs of $f(x) = \tan x$ and $f'(x) = 1/\cos^2 x$ are in Figure 3.20. Is it reasonable that f' is always positive? Are the asymptotes of f' where we expect?

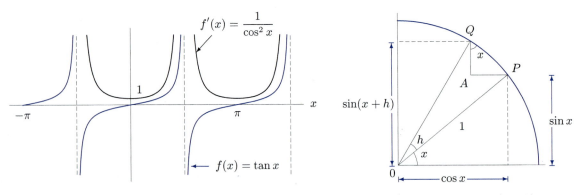

Figure 3.20: The function $\tan x$ and its derivative

Figure 3.21: Unit circle showing $\sin(x + h)$ and $\sin x$

Example 5   Differentiate (a) $2\tan(3t)$,   (b) $\tan(1 - \theta)$,   (c) $\dfrac{1 + \tan t}{1 - \tan t}$.

Solution   (a) Use the chain rule:

$$\frac{d}{dt}(2\tan(3t)) = 2\frac{1}{\cos^2(3t)}\frac{d}{dt}(3t) = \frac{6}{\cos^2(3t)}.$$

(b) Use the chain rule:

$$\frac{d}{d\theta}(\tan(1 - \theta)) = \frac{1}{\cos^2(1 - \theta)} \cdot \frac{d}{d\theta}(1 - \theta) = \frac{-1}{\cos^2(1 - \theta)}.$$

(c) Use the quotient rule:

$$\frac{d}{dt}\left(\frac{1 + \tan t}{1 - \tan t}\right) = \frac{\left(\dfrac{d(1 + \tan t)}{dt}\right)(1 - \tan t) - (1 + \tan t)\dfrac{d(1 - \tan t)}{dt}}{(1 - \tan t)^2}$$

$$= \frac{\dfrac{1}{\cos^2 t}(1 - \tan t) - (1 + \tan t)\left(-\dfrac{1}{\cos^2 t}\right)}{(1 - \tan t)^2}$$

$$= \frac{2}{\cos^2 t \cdot (1 - \tan t)^2}.$$

Informal Justification of $\frac{d}{dx}(\sin x) = \cos x$

Consider the unit circle in Figure 3.21. To find the derivative of $\sin x$, we need to estimate

$$\frac{\sin(x+h) - \sin x}{h}.$$

In Figure 3.21, the quantity $\sin(x+h) - \sin x$ is represented by the length QA. The arc QP is of length h, so

$$\frac{\sin(x+h) - \sin x}{h} = \frac{QA}{\text{Arc } QP}.$$

Now, if h is small, QAP is approximately a right triangle because the arc QP is almost a straight line. Furthermore, using geometry, we can show that angle $AQP \approx x$. For small h, we have

$$\frac{\sin(x+h) - \sin x}{h} = \frac{QA}{\text{Arc } QP} \approx \cos x.$$

As $h \to 0$, the approximation gets better, so

$$\frac{d}{dx}(\sin x) = \lim_{h \to 0} \frac{\sin(x+h) - \sin x}{h} = \cos x.$$

Other derivations of this result are given in Problems 53 and 54 on page 132.

Exercises and Problems for Section 3.5

Exercises

1. Construct a table of values for $\cos x$, $x = 0, 0.1, 0.2, \ldots, 0.6$. Using the difference quotient, estimate the derivative at these points (use $h = 0.001$), and compare it with $(-\sin x)$.

Find the derivatives of the functions in Exercises 2–41. Assume a is a constant.

2. $r(\theta) = \sin \theta + \cos \theta$
3. $s(\theta) = \cos \theta \sin \theta$
4. $z = \cos(4\theta)$
5. $f(x) = \sin(3x)$
6. $g(x) = \sin(2 - 3x)$
7. $R(x) = 10 - 3\cos(\pi x)$
8. $g(\theta) = \sin^2(2\theta) - \pi\theta$
9. $f(x) = x^2 \cos x$
10. $w = \sin(e^t)$
11. $f(x) = e^{\cos x}$
12. $f(y) = e^{\sin y}$
13. $z = \theta e^{\cos \theta}$
14. $R(\theta) = e^{\sin(3\theta)}$
15. $g(\theta) = \sin(\tan \theta)$
16. $w(x) = \tan(x^2)$
17. $f(x) = \sqrt{1 - \cos x}$
18. $f(x) = \cos(\sin x)$
19. $f(x) = \tan(\sin x)$
20. $k(x) = \sqrt{(\sin(2x))^3}$
21. $f(x) = 2x\sin(3x)$
22. $y = e^\theta \sin(2\theta)$
23. $f(x) = e^{-2x} \cdot \sin x$
24. $z = \sqrt{\sin t}$
25. $y = \sin^5 \theta$
26. $g(z) = \tan(e^z)$
27. $z = \tan(e^{-3\theta})$
28. $w = e^{-\sin \theta}$
29. $h(t) = t\cos t + \tan t$
30. $f(\alpha) = \cos \alpha + 3\sin \alpha$
31. $k(\alpha) = \sin^5 \alpha \cos^3 \alpha$
32. $t(\theta) = \frac{\cos \theta}{\sin \theta}$
33. $f(x) = \sqrt{\frac{1 - \sin x}{1 - \cos x}}$
34. $G(x) = \frac{\sin^2 x + 1}{\cos^2 x + 1}$
35. $r(y) = \frac{y}{\cos y + a}$
36. $h(x) = 2^{\sin x}$
37. $w = 2^{2\sin x + e^x}$
38. $f(x) = \sin(2x) \cdot \sin(3x)$
39. $f(\theta) = \theta^2 \sin \theta + 2\theta \cos \theta - 2\sin \theta$
40. $f(x) = \sin(\sin x + \cos x)$
41. $f(w) = \cos^2 w + \cos(w^2)$

Notice that although the equation $y^3 - xy = -6$ leads to a curve which is difficult to deal with algebraically, it still looks like a straight line locally.

Example 2   Find all points where the tangent line to $y^3 - xy = -6$ is either horizontal or vertical.

Solution   From the previous example, $\dfrac{dy}{dx} = \dfrac{y}{3y^2 - x}$. The tangent is horizontal when the numerator of dy/dx equals 0, so $y = 0$. Since we also must satisfy $y^3 - xy = -6$, we get $0^3 - x \cdot 0 = -6$, which is impossible. We conclude that there are no points on the curve where the tangent line is horizontal.

The tangent is vertical when the denominator of dy/dx is 0, giving $3y^2 - x = 0$. Thus, $x = 3y^2$ at any point with a vertical tangent line. Again, we must also satisfy $y^3 - xy = -6$, so

$$y^3 - (3y^2)y = -6,$$
$$-2y^3 = -6,$$
$$y = \sqrt[3]{3} \approx 1.442.$$

We can then find x by substituting $y = \sqrt[3]{3}$ in $y^3 - xy = -6$. We get $3 - x(\sqrt[3]{3}) = -6$, so $x = 9/(\sqrt[3]{3}) \approx 6.240$. So the tangent line is vertical at $(6.240, 1.442)$.

Using implicit differentiation and the expression for dy/dx to locate the points where the tangent is vertical or horizontal, as in the previous example, is a first step in obtaining an overall picture of the curve $y^3 - xy = -6$. However, filling in the rest of the graph, even roughly, by using the sign of dy/dx to tell us where the curve is increasing or decreasing can be difficult.

Exercises and Problems for Section 3.7

Exercises

For Exercises 1–15, find dy/dx. Assume a, b, c are constants.

1. $x^2 + y^2 = \sqrt{7}$

2. $x^2 + xy - y^3 = xy^2$

3. $\sqrt{x} = 5\sqrt{y}$

4. $\sqrt{x} + \sqrt{y} = 25$

5. $xy - x - 3y - 4 = 0$

6. $6x^2 + 4y^2 = 36$

7. $ax^2 - by^2 = c^2$

8. $\ln x + \ln(y^2) = 3$

9. $e^{x^2} + \ln y = 0$

10. $\arctan(x^2 y) = xy^2$

11. $x \ln y + y^3 = \ln x$

12. $\sin(xy) = 2x + 5$

13. $x^{2/3} + y^{2/3} = a^{2/3}$

14. $e^{\cos y} = x^3 \arctan y$

15. $\cos^2 y + \sin^2 y = y + 2$

In Exercises 16–19, find the slope of the tangent to the curve at the point specified.

16. $x^2 + y^2 = 1$ at $(0, 1)$

17. $\sin(xy) = x$ at $(1, \pi/2)$

18. $x^3 + 5x^2 y + 2y^2 = 4y + 11$ at $(1, 2)$

19. $x^3 + 2xy + y^2 = 4$ at $(1, 1)$

For Exercises 20–23, find the equations of the tangent lines to the following curves at the indicated points.

20. $xy^2 = 1$ at $(1, -1)$

21. $\ln(xy) = 2x$ at $(1, e^2)$

22. $y^2 = \dfrac{x^2}{xy - 4}$ at $(4, 2)$

23. $x^{2/3} + y^{2/3} = a^{2/3}$ at $(a, 0)$

Problems

24. (a) Find dy/dx given the equation $x^2 + y^2 - 4x + 7y = 15$.

  (b) Under what conditions on x and/or y is the tangent line to this curve horizontal? Vertical?

25. (a) Find the slope of the tangent line to the ellipse $\dfrac{x^2}{25} + \dfrac{y^2}{9} = 1$ at the point (x, y).

  (b) Are there any points where the slope is not defined?

26. (a) Find the equations of the tangent lines to the circle $x^2 + y^2 = 25$ at the points where $x = 4$.

  (b) Find the equations of the normal lines to this circle at the same points. (The normal line to a curve at a point is perpendicular to the tangent line at that point.)

  (c) At what point do the two normal lines intersect?

27. Consider the equation $x^3 + y^3 - xy^2 = 5$.

 (a) Find dy/dx by implicit differentiation.
 (b) Give a table of approximate y-values near $x = 1, y = 2$ for $x = 0.96, 0.98, 1, 1.02, 1.04$.
 (c) Find the y-value for $x = 0.96$ by substituting $x = 0.96$ in the equation and solving for y using a computer or calculator. Compare with your answer in part (b).
 (d) Find all points where the tangent line is horizontal or vertical.

28. Find the equation of the tangent line to the curve $y = x^2$ at $x = 1$. Show that this line is also a tangent to a circle centered at $(8, 0)$ and find the equation of this circle.

29. Sketch the circles $y^2 + x^2 = 1$ and $y^2 + (x - 3)^2 = 4$. There is a line with positive slope that is tangent to both circles. Determine the points at which this tangent line touches each circle.

30. Show that the power rule for derivatives applies to rational powers of the form $y = x^{m/n}$ by raising both sides to the nth power and using implicit differentiation.

3.8 PARAMETRIC EQUATIONS

Representing Motion in the Plane

To represent the motion of a particle in the xy-plane we use two equations, one for the x-coordinate of the particle, $x = f(t)$, and another for the y-coordinate, $y = g(t)$. Thus at time t the particle is at the point $(f(t), g(t))$. The equation for x describes the right-left motion; the equation for y describes the up-down motion. The two equations for x and y are called *parametric equations* with *parameter t*.

Example 1   Describe the motion of the particle whose coordinates at time t are $x = \cos t$ and $y = \sin t$.

Solution   Since $(\cos t)^2 + (\sin t)^2 = 1$, we have $x^2 + y^2 = 1$. That is, at any time t the particle is at a point (x, y) on the unit circle $x^2 + y^2 = 1$. We plot points at different times to see how the particle moves on the circle. (See Figure 3.31 and Table 3.4.) The particle moves at a uniform speed, completing one full trip counterclockwise around the circle every 2π units of time. Notice how the x-coordinate goes repeatedly back and forth from -1 to 1 while the y-coordinate goes repeatedly up and down from -1 to 1. The two motions combine to trace out a circle.

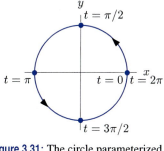

Figure 3.31: The circle parameterized by $x = \cos t, y = \sin t$

Table 3.4   *Points on the circle with $x = \cos t$, $y = \sin t$*

t	x	y
0	1	0
$\pi/2$	0	1
π	-1	0
$3\pi/2$	0	-1
2π	1	0

Example 2   Figure 3.32 shows the graphs of two functions, $f(t)$ and $g(t)$. Describe the motion of the particle whose coordinates at time t are $x = f(t)$ and $y = g(t)$.

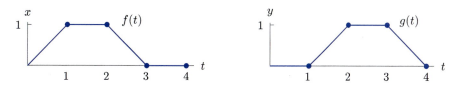

Figure 3.32: Graphs of $x = f(t)$ and $y = g(t)$ used to trace out the path $(f(t), g(t))$ in Figure 3.33

Solution   Between times $t = 0$ and $t = 1$, the x-coordinate goes from 0 to 1, while the y-coordinate stays fixed at 0. So the particle moves along the x-axis from $(0,0)$ to $(1,0)$. Then, between times $t = 1$ and $t = 2$, the x-coordinate stays fixed at $x = 1$, while the y-coordinate goes from 0 to 1. Thus, the particle moves along the vertical line from $(1,0)$ to $(1,1)$. Similarly, between times $t = 2$ and $t = 3$, it moves horizontally back to $(0,1)$, and between times $t = 3$ and $t = 4$ it moves down the y-axis to $(0,0)$. Thus, it traces out the square in Figure 3.33.

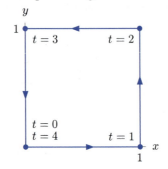

Figure 3.33: The square parameterized by $(f(t), g(t))$

Different Motions Along the Same Path

Example 3   Describe the motion of the particle whose x and y coordinates at time t are given by the equations
$$x = \cos(3t), \quad y = \sin(3t).$$

Solution   Since $(\cos(3t))^2 + (\sin(3t))^2 = 1$, we have $x^2 + y^2 = 1$, giving motion around the unit circle. But from Table 3.5, we see that the particle in this example is moving three times as fast as the particle in Example 1. (See Figure 3.34.)

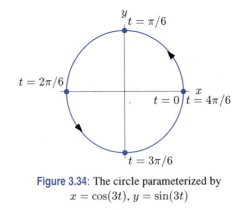

Figure 3.34: The circle parameterized by
$x = \cos(3t), y = \sin(3t)$

Table 3.5   *Points on circle with*
$x = \cos(3t), y = \sin(3t)$

t	x	y
0	1	0
$\pi/6$	0	1
$2\pi/6$	-1	0
$3\pi/6$	0	-1
$4\pi/6$	1	0

Example 3 is obtained from Example 1 by replacing t by $3t$; this is called a *change in parameter*. If we make a change in parameter, the particle traces out the same curve (or a part of it) but at a different speed or in a different direction.

Example 4   Describe the motion of the particle whose x and y coordinates at time t are given by
$$x = \cos(e^{-t^2}), \quad y = \sin(e^{-t^2}).$$

Solution   As in Examples 1 and 3, we have $x^2 + y^2 = 1$ so the motion lies on the unit circle. As time t goes from $-\infty$ (way back in the past) to 0 (the present) to ∞ (way off in the future), e^{-t^2} goes from near 0 to 1 back to near 0. So $(x, y) = (\cos(e^{-t^2}), \sin(e^{-t^2}))$ goes from near $(1, 0)$ to $(\cos 1, \sin 1)$ and back to near $(1, 0)$. The particle does not actually reach the point $(1, 0)$. (See Figure 3.35 and Table 3.6.)

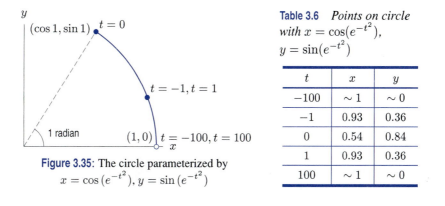

Figure 3.35: The circle parameterized by $x = \cos(e^{-t^2})$, $y = \sin(e^{-t^2})$

Table 3.6   *Points on circle with* $x = \cos(e^{-t^2})$, $y = \sin(e^{-t^2})$

t	x	y
-100	~ 1	~ 0
-1	0.93	0.36
0	0.54	0.84
1	0.93	0.36
100	~ 1	~ 0

Motion in a Straight Line

An object moves with constant speed along a straight line through the point (x_0, y_0). Both the x- and y-coordinates have a constant rate of change. Let $a = dx/dt$ and $b = dy/dt$. Then at time t the object has coordinates $x = x_0 + at$, $y = y_0 + bt$. (See Figure 3.36.) Notice that a represents the horizontal run in one unit of time, and b represents the vertical rise. Thus the line has slope $m = b/a$.

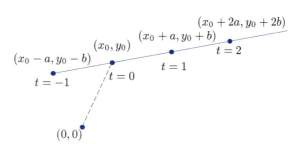

Figure 3.36: The line $x = x_0 + at$, $y = y_0 + bt$

This yields the following:

Parametric Equations for a Straight Line

An object moving along a line through the point (x_0, y_0), with $dx/dt = a$ and $dy/dt = b$, has parametric equations

$$x = x_0 + at, \quad y = y_0 + bt.$$

The slope of the line is $m = b/a$.

Example 5   Find parametric equations for:

(a) The line passing through the points $(2, -1)$ and $(-1, 5)$.
(b) The line segment from $(2, -1)$ to $(-1, 5)$.

Solution   (a) Imagine an object moving with constant speed along a straight line from $(2, -1)$ to $(-1, 5)$, making the journey from the first point to the second in one unit of time. Then $dx/dt = ((-1) - 2)/1 = -3$ and $dy/dt = (5 - (-1))/1 = 6$. Thus the parametric equation are

$$x = 2 - 3t, \quad y = -1 + 6t.$$

(b) In the parameterization in part (a), $t = 0$ corresponds to the point $(2, -1)$ and $t = 1$ corresponds to the point $(-1, 5)$. So the parameterization of the segment is

$$x = 2 - 3t, \quad y = -1 + 6t, \quad 0 \le t \le 1.$$

There are many other possible parametric equations for this line.

Speed and Velocity

An object moves along a straight line at a constant speed, with $dx/dt = a$ and $dy/dt = b$. In one unit of time, the object moves a units horizontally and b units vertically. Thus, by the Pythagorean Theorem, it travels a distance $\sqrt{a^2 + b^2}$. So its speed is

$$\text{Speed} = \frac{\text{Distance traveled}}{\text{Time taken}} = \frac{\sqrt{a^2 + b^2}}{1} = \sqrt{a^2 + b^2}.$$

For general motion along a curve with varying speed, we make the following definition:

The **instantaneous speed** of a moving object is defined to be

$$v = \sqrt{\left(\frac{dx}{dt}\right)^2 + \left(\frac{dy}{dt}\right)^2}.$$

The quantity $v_x = dx/dt$ is the **instantaneous velocity** in the x-direction; $v_y = dy/dt$ is the **instantaneous velocity** in the y-direction.

The quantities v_x and v_y are called the *components* of the velocity in the x- and y-directions.

Example 6   A particle moves in the xy-plane with $x = 2t^3 - 9t^2 + 12t$ and $y = 3t^4 - 16t^3 + 18t^2$, where t is time.

(a) At what times is the particle
  (i)   Stopped   (ii)   Moving parallel to the x- or y- axis?
(b) Find the speed of the particle at time t. Is the speed ever 0?

Solution   (a) Differentiating gives

$$\frac{dx}{dt} = 6t^2 - 18t + 12 \quad \frac{dy}{dt} = 12t^3 - 48t^2 + 36t.$$

We are interested in the points at which $dx/dt = 0$ or $dy/dt = 0$. Solving gives

$$\frac{dx}{dt} = 6(t^2 - 3t + 2) = 6(t - 1)(t - 2) \quad \text{so} \quad \frac{dx}{dt} = 0 \quad \text{if } t = 1 \text{ or } t = 2.$$

$$\frac{dy}{dt} = 12t(t^2 - 4t + 3) = 12t(t - 1)(t - 3) \quad \text{so} \quad \frac{dy}{dt} = 0 \quad \text{if } t = 0, t = 1, \text{ or } t = 3.$$

(i) The particle is stopped if both dx/dt and dy/dt are 0, which occurs at $t = 1$.

(ii) The particle is moving parallel to the x-axis if $dy/dt = 0$ but $dx/dt \neq 0$. This occurs at $t = 0$ and $t = 3$. The particle is moving parallel to the y-axis if $dx/dt = 0$ but $dy/dt \neq 0$. This occurs at $t = 2$.

(b) We have

$$\text{Speed} = \sqrt{\left(\frac{dx}{dt}\right)^2 + \left(\frac{dy}{dt}\right)^2} = \sqrt{(6t^2 - 18t + 12)^2 + (12t^3 - 48t^2 + 36t)^2}$$

$$= 6\sqrt{4t^6 - 32t^5 + 89t^4 - 102t^3 + 49t^2 - 12t + 4}.$$

The speed is zero when $dx/dt = dy/dt = 0$, so $t = 1$.

Example 7   A child is sitting on a ferris wheel of diameter 10 meters, making one revolution every 2 minutes. Find the speed of the child

(a) Using geometry.   (b) Using a parameterization of the motion.

Solution   (a) The child moves at a constant speed around a circle of radius 5 meters, completing one revolution every 2 minutes. One revolution around a circle of radius 5 is a distance of 10π, so the child's speed is $10\pi/2 = 5\pi \approx 15.7$ m/min. See Figure 3.37.

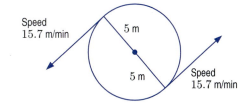

Figure 3.37: Motion of a child on a ferris wheel at two different times is represented by the arrows. The direction of each arrow is the direction of motion at that time.

(b) The ferris wheel has radius 5 meters and completes 1 revolution counterclockwise every 2 minutes. If the origin is at the center of the circle and we measure x and y in meters, the motion is parameterized by equations of the form

$$x = 5\cos(\omega t), \quad y = 5\sin(\omega t),$$

where ω is chosen to make the period 2 minutes. Since the period of $\cos(\omega t)$ and $\sin(\omega t)$ is $2\pi/\omega$, we must have

$$\frac{2\pi}{\omega} = 2, \quad \text{so} \quad \omega = \pi.$$

Thus, for t in minutes, the motion is described by the equations

$$x = 5\cos(\pi t), \quad y = 5\sin(\pi t).$$

So the speed is given by

$$v = \sqrt{\left(\frac{dx}{dt}\right)^2 + \left(\frac{dy}{dt}\right)^2}$$

$$= \sqrt{(-5\pi)^2 \sin^2(\pi t) + (5\pi)^2 \cos^2(\pi t)} = 5\pi\sqrt{\sin^2(\pi t) + \cos^2(\pi t)} = 5\pi \approx 15.7 \text{ m/min},$$

which agrees with the speed we calculated in part (a).

Tangent Lines

To find the tangent line at a point (x_0, y_0) to a curve given parametrically, we find the straight line motion through (x_0, y_0) with the same velocity in the x and y directions as the curve.

Example 8   Find the tangent line at the point $(1, 2)$ to the curve defined by the parametric equation

$$x = t^3, \quad y = 2t.$$

Solution   At time $t = 1$ the particle is at the point $(1, 2)$. The velocity in the x-direction at time t is $v_x = dx/dt = 3t^2$, and the velocity in the y-direction is $v_y = dy/dt = 2$. So at $t = 1$ the velocity in the x-direction is 3 and the velocity in the y-direction is 2. Thus the tangent line has parametric equations

$$x = 1 + 3t, \quad y = 2 + 2t.$$

Parametric Representations of Curves in the Plane

Sometimes we are more interested in the curve traced out by the particle than we are in the motion itself. In that case we will call the parametric equations a *parameterization* of the curve. As we can see by comparing Examples 1 and 3, two different parameterizations can describe the same curve in the xy-plane. Though the parameter, which we usually denote by t, may not have physical meaning, it is often helpful to think of it as time.

Example 9   Give a parameterization of the semicircle of radius 1 shown in Figure 3.38.

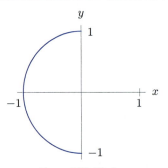

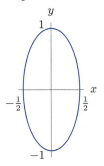

Figure 3.38: Find a parameterization of this semicircle

Figure 3.39: Find a parameterization of the ellipse $4x^2 + y^2 = 1$

Solution   We can use the equations $x = \cos t$ and $y = \sin t$ for counterclockwise motion in a circle, from Example 1 on page 141. The particle passes $(0, 1)$ at $t = \pi/2$, moves counterclockwise around the circle, and reaches $(0, -1)$ at $t = 3\pi/2$. So a parameterization is

$$x = \cos t, \ y = \sin t, \quad \frac{\pi}{2} \le t \le \frac{3\pi}{2}.$$

To find the xy-equation of a curve given parametrically, we eliminate the parameter t in the parametric equations. In the previous example, we use the Pythagorean identity, so

$$\cos^2 t + \sin^2 t = 1 \quad \text{gives} \quad x^2 + y^2 = 1.$$

Example 10   Give a parameterization of the ellipse $4x^2 + y^2 = 1$ shown in Figure 3.39.

Solution   Since $(2x)^2 + y^2 = 1$, we adapt the parameterization of the circle in Example 1. Replacing x by $2x$ gives the equations $2x = \cos t, y = \sin t$. A parameterization of the ellipse is thus

$$x = \tfrac{1}{2}\cos t, \quad y = \sin t, \quad 0 \le t \le 2\pi.$$

We usually require that the parameterization of a curve go from one end of the curve to the other without retracing any portion of the curve. This is different from parameterizing the motion of a particle, where, for example, a particle may move around the same circle many times.

Parameterizing the Graph of a Function

The graph of any function $y = f(x)$ can be parameterized by letting the parameter t be x:

$$x = t, \quad y = f(t).$$

Example 11   Give parametric equations for the curve $y = x^3 - x$. In which direction does this parameterization trace out the curve?

Solution   Let $x = t$, $y = t^3 - t$. Thus, $y = t^3 - t = x^3 - x$. Since $x = t$, as time increases the x-coordinate moves from left to right, so the particle traces out the curve $y = x^3 - x$ from left to right.

Curves Given Parametrically

Some complicated curves can be graphed more easily using parametric equations; the next example shows such a curve.

Example 12   Assume t is time in seconds. Sketch the curve traced out by the particle whose motion is given by

$$x = \cos(3t), \quad y = \sin(5t).$$

Solution   The x-coordinate oscillates back and forth between 1 and -1, completing 3 oscillations every 2π seconds. The y-coordinate oscillates up and down between 1 and -1, completing 5 oscillations every 2π seconds. Since both the x- and y-coordinates return to their original values every 2π seconds, the curve is retraced every 2π seconds. The result is a pattern called a Lissajous figure. (See Figure 3.40.) Problems 36–39 concern Lissajous figures $x = \cos(at)$, $y = \sin(bt)$ for other values of a and b.

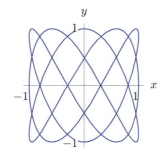

Figure 3.40: A Lissajous figure: $x = \cos(3t), y = \sin(5t)$

Slopes of Parametric Curves

Suppose we have a curve traced out by the parametric equations $x = f(t)$, $y = g(t)$. To find the slope at a point on the curve, we could, in theory, eliminate the parameter t and then differentiate the function we obtain. However, the chain rule gives us an easier way.

Suppose the curve traced out by the parametric equations is represented by $y = h(x)$. (It may be represented by an implicit function.) Thinking of x and y as functions of t, the chain rule gives

$$\frac{dy}{dt} = \frac{dy}{dx} \cdot \frac{dx}{dt},$$

so we obtain the slope of the curve as a function of t:

$$\boxed{\text{Slope of curve,} \quad \frac{dy}{dx} = \frac{dy/dt}{dx/dt}.}$$

Example 13   If $x = \cos t$, $y = \sin t$, find the point corresponding to $t = \pi/4$ and the slope of the curve through the point.

Solution   The point corresponding to $t = \pi/4$ is $(\cos(\pi/4), \sin(\pi/4)) = (1/\sqrt{2}, 1/\sqrt{2})$. We have

$$\frac{dy}{dx} = \frac{dy/dt}{dx/dt} = \frac{\cos t}{-\sin t},$$

so when $t = \pi/4$,

$$\text{Slope} = \frac{\cos(\pi/4)}{-\sin(\pi/4)} = -1.$$

Thus, the curve has slope -1 at the point $(1/\sqrt{2}, 1/\sqrt{2})$. This is as we would expect, since the curve traced out is the circle of Example 8.

Exercises and Problems for Section 3.8

Exercises

For Exercises 1–4, use the graphs of f and g to describe the motion of a particle whose position at time t is given by $x = f(t)$, $y = g(t)$.

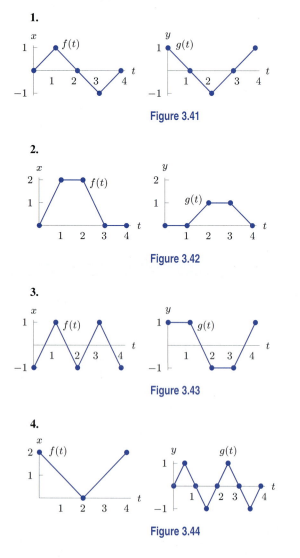

1.

Figure 3.41

2.

Figure 3.42

3.

Figure 3.43

4.

Figure 3.44

Exercises 5–10 give parameterizations of the unit circle or a part of it. Describe in words how the circle is traced out, including when and where the particle is moving clockwise and when and where the particle is moving counterclockwise.

5. $x = \cos t$, $y = -\sin t$

6. $x = \sin t$, $y = \cos t$

7. $x = \cos(t^2)$, $y = \sin(t^2)$

8. $x = \cos(t^3 - t)$, $y = \sin(t^3 - t)$

9. $x = \cos(\ln t)$, $y = \sin(\ln t)$

10. $x = \cos(\cos t)$, $y = \sin(\cos t)$

In Exercises 11–17, write a parameterization for the curves in the xy-plane.

11. A circle of radius 3 centered at the origin and traced out clockwise.

12. A vertical line through the point $(-2, -3)$.

13. A circle of radius 5 centered at the point $(2, 1)$ and traced out counterclockwise.

14. A circle of radius 2 centered at the origin traced clockwise starting from $(-2, 0)$ when $t = 0$.

15. The line through the points $(2, -1)$ and $(1, 3)$.

16. An ellipse centered at the origin and crossing the x-axis at ± 5 and the y-axis at ± 7.

17. An ellipse centered at the origin, crossing the x-axis at ± 3 and the y-axis at ± 7. Start at the point $(-3, 0)$ and trace out the ellipse counterclockwise.

In Exercises 18–20, find an equation of the tangent line to the curve for the given value of t.

18. $x = t^3 - t$, $y = t^2$ when $t = 2$

19. $x = \sin(3t)$, $y = \sin(4t)$ when $t = \pi$

20. $x = t^2 - 2t$, $y = t^2 + 2t$ when $t = 1$

For Exercises 21–24, find the speed for the given motion of a particle. Find any times when the particle comes to a stop.

21. $x = t^2, \quad y = t^3$

22. $x = \cos(t^2), \quad y = \sin(t^2)$

23. $x = \cos 2t, \quad y = \sin t$

24. $x = t^2 - 2t, y = t^3 - 3t$

25. Find parametric equations for the tangent line at $t = 2$ for Problem 21.

Problems

26. (a) Explain how you know that the following two pairs of equations parameterize the same line:

$$x = 2 + t, y = 4 + 3t \text{ and } x = 1 - 2t, y = 1 - 6t.$$

(b) What are the slope and y intercept of this line?

27. A line is parameterized by $x = 10 + t$ and $y = 2t$.

(a) What part of the line do we get by restricting t to $t < 0$?

(b) What part of the line do we get by restricting t to $0 \leq t \leq 1$?

28. A line is parameterized by $x = 2 + 3t$ and $y = 4 + 7t$.

(a) What part of the line is obtained by restricting t to nonnegative numbers?

(b) What part of the line is obtained if t is restricted to $-1 \leq t \leq 0$?

(c) How should t be restricted to give the part of the line to the left of the y-axis?

29. (a) Sketch the parameterized curve $x = t \cos t, \ y = t \sin t$ for $0 \leq t \leq 4\pi$.

(b) By calculating the position at $t = 2$ and $t = 2.01$, estimate the speed at $t = 2$.

(c) Use derivatives to calculate the speed at $t = 2$ and compare your answer to part (b).

30. For x and y in meters, the motion of the particle given by the parametric equations

$$x = t^3 - 3t, \quad y = t^2 - 2t,$$

where the y-axis is vertical and the x-axis is horizontal.

(a) Does the particle ever come to a stop? If so, when and where?

(b) Is the particle ever moving straight up or down? If so, when and where?

(c) Is the particle ever moving straight horizontally right or left? If so, when and where?

31. Describe the similarities and differences among the motions in the plane given by the following three pairs of parametric equations:

(a) $x = t, \quad y = t^2$ **(b)** $x = t^2, \quad y = t^4$
(c) $x = t^3, \quad y = t^6$.

32. Suppose $a, b, c, d, m, n, p, q > 0$. Match each pair of parametric equations with one of the lines l_1, l_2, l_3, l_4 in Figure 3.45.

I. $\begin{cases} x = a + ct, \\ y = -b + dt. \end{cases}$ II. $\begin{cases} x = m + pt, \\ y = n - qt. \end{cases}$

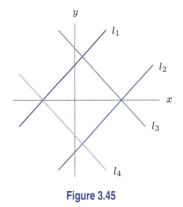

Figure 3.45

33. What can you say about the values of a, b and k if the equations

$$x = a + k \cos t, \quad y = b + k \sin t, \quad 0 \leq t \leq 2\pi,$$

trace out the following circles in Figure 3.46?
(a) C_1 **(b)** C_2 **(c)** C_3

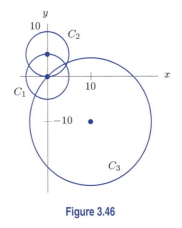

Figure 3.46

34. Describe in words the curve represented by the parametric equations

$$x = 3 + t^3, \quad y = 5 - t^3.$$

35. A hypothetical moon orbits a planet which in turn orbits a star. Suppose that the orbits are circular and that the moon orbits the planet 12 times in the time it takes for the planet to orbit the star once. In this problem we will investigate whether the moon could come to a stop at some instant. (See Figure 3.47.)

(a) Suppose the radius of the moon's orbit around the planet is 1 unit and the radius of the planet's orbit around the star is R units. Explain why the motion of the moon relative to the star can be described by the parametric equations

$$x = R\cos t + \cos(12t), \quad y = R\sin t + \sin(12t).$$

(b) Find values for R and t such that the moon stops relative to the star at time t.

(c) On a graphing calculator, plot the path of the moon for the value of R you obtained in part (b). Experiment with other values for R.

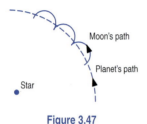

Moon's path

Planet's path

Star

Figure 3.47

Graph the Lissajous figures in Problems 36–39 using a calculator or computer.

36. $x = \cos 2t, \quad y = \sin 5t$

37. $x = \cos 3t, \quad y = \sin 7t$

38. $x = \cos 2t, \quad y = \sin 4t$

39. $x = \cos 2t, \quad y = \sin \sqrt{3}t$

40. To find the second derivative d^2y/dx^2 for a parametrically defined curve, we use

$$\frac{d^2y}{dx^2} = \frac{dw}{dx} = \frac{dw/dt}{dx/dt} \quad \text{where} \quad w = \frac{dy}{dx}.$$

(a) Find d^2y/dx^2 for $x = t^3 + t, y = t^2$. Is the curve concave up or down at $t = 1$?

(b) Derive the general formula:

$$\frac{d^2y}{dx^2} = \frac{(dx/dt)(d^2y/dt^2) - (dy/dt)(d^2x/dt^2)}{(dx/dt)^3}.$$

3.9 LINEAR APPROXIMATION AND THE DERIVATIVE

The Tangent Line Approximation

When we zoom in on the graph of a differentiable function, it looks like a straight line. In fact, the graph is not exactly a straight line when we zoom in; however, its deviation from straightness is so small that it can't be detected by the naked eye. Let's examine what this means. The straight line that we think we see when we zoom in on the graph of $f(x)$ at $x = a$ has slope equal to the derivative, $f'(a)$, so the equation is

$$y = f(a) + f'(a)(x - a).$$

The fact that the graph looks like a line means that y is a good approximation to $f(x)$. (See Figure 3.48.) This suggests the following definition:

> ### The Tangent Line Approximation
>
> Suppose f is differentiable at a. Then, for values of x near a, the tangent line approximation to $f(x)$ is
>
> $$f(x) \approx f(a) + f'(a)(x - a).$$
>
> The expression $f(a) + f'(a)(x - a)$ is called the *local linearization* of f near $x = a$. We are thinking of a as fixed, so that $f(a)$ and $f'(a)$ are constant.
> The **error**, $E(x)$, in the approximation is defined by
>
> $$E(x) = f(x) - f(a) - f'(a)(x - a).$$

It can be shown that the tangent line approximation is the best linear approximation to f near a. See Problem 18.

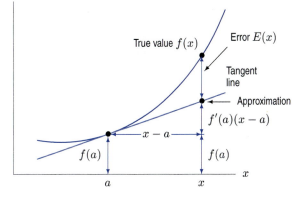

Figure 3.48: The tangent line approximation and its error

Example 1   What is the tangent line approximation for $f(x) = \sin x$ near $x = 0$?

Solution   The tangent line approximation of f near $x = 0$ is

$$f(x) \approx f(0) + f'(0)(x - 0).$$

If $f(x) = \sin x$, then $f'(x) = \cos x$, so $f(0) = \sin 0 = 0$ and $f'(0) = \cos 0 = 1$, and the approximation is

$$\sin x \approx x.$$

This means that, near $x = 0$, the function $f(x) = \sin x$ is well approximated by the function $y = x$. If we zoom in on the graphs of the functions $\sin x$ and x near the origin, we won't be able to tell them apart. (See Figure 3.49.)

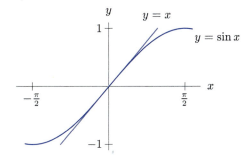

Figure 3.49: Tangent line approximation to $y = \sin x$

Example 2   What is the local linearization of e^{kx} near $x = 0$?

Solution   If $f(x) = e^{kx}$, then $f(0) = 1$ and, by the chain rule, $f'(x) = ke^{kx}$, so $f'(0) = ke^{k \cdot 0} = k$. Thus

$$f(x) \approx f(0) + f'(0)(x - 0)$$

becomes

$$e^{kx} \approx 1 + kx.$$

This is the tangent line approximation to e^{kx} near $x = 0$. In other words, if we zoom in on the functions $f(x) = e^{kx}$ and $y = 1 + kx$ near the origin, we won't be able to tell them apart.

Estimating the Error in the Approximation

Let us look at the error, $E(x)$, which is the difference between $f(x)$ and the local linearization. (Look back at Figure 3.48.) The fact that the graph of f looks like a line as we zoom in means that not only is $E(x)$ small for x near a, but also that $E(x)$ is small relative to $(x - a)$. To demonstrate this, we prove the following theorem about the ratio $E(x)/(x - a)$.

Theorem 3.6: Differentiability and Local Linearity

Suppose f is differentiable at $x = a$ and $E(x)$ is the error in the tangent line approximation, that is:

$$E(x) = f(x) - f(a) - f'(a)(x - a).$$

Then

$$\lim_{x \to a} \frac{E(x)}{x - a} = 0.$$

Proof Using the definition of $E(x)$, we have

$$\frac{E(x)}{x - a} = \frac{f(x) - f(a) - f'(a)(x - a)}{x - a} = \frac{f(x) - f(a)}{x - a} - f'(a).$$

Taking the limit as $x \to a$ and using the definition of the derivative, we see that

$$\lim_{x \to a} \frac{E(x)}{x - a} = \lim_{x \to a} \left(\frac{f(x) - f(a)}{x - a} - f'(a) \right) = f'(a) - f'(a) = 0.$$

Why Differentiability Makes A Graph Look Straight

We can use the error $E(x)$ to understand why differentiability makes a graph look straight when we zoom in.

Example 3 Consider the graph of $f(x) = \sin x$ near $x = 0$, and its linear approximation computed in Example 1. Show that there is an interval around 0 with the property that the distance from $f(x) = \sin x$ to the linear approximation is less than $0.1|x|$ for all x in the interval.

Solution The linear approximation of $f(x) = \sin x$ near 0 is $y = x$, so we write

$$\sin x = x + E(x).$$

Since $\sin x$ is differentiable at $x = 0$, Theorem 3.6 tells us that

$$\lim_{x \to 0} \frac{E(x)}{x} = 0.$$

If we take $\epsilon = 1/10$, then the definition of limit guarantees that there is a $\delta > 0$ such that

$$\left| \frac{E(x)}{x} \right| < 0.1 \quad \text{for all} \quad |x| < \delta.$$

In other words, for x in the interval $(-\delta, \delta)$, we have $|x| < \delta$, so

$$|E(x)| < 0.1|x|.$$

(See Figure 3.50.)

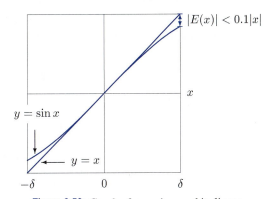

Figure 3.50: Graph of $y = \sin x$ and its linear approximation $y = x$, showing a window in which the magnitude of the error, $|E(x)|$, is less than $0.1|x|$ for all x in the window

We can generalize from this example to explain why differentiability makes the graph of f look straight when viewed over a small graphing window. Suppose f is differentiable at $x = a$. Then we know $\lim\limits_{x \to a} \left| \dfrac{E(x)}{x - a} \right| = 0$. So, for any $\epsilon > 0$, we can find a δ small enough so that

$$\left| \frac{E(x)}{x - a} \right| < \epsilon, \quad \text{for} \quad a - \delta < x < a + \delta.$$

So, for any x in the interval $(a - \delta, a + \delta)$, we have

$$|E(x)| < \epsilon |x - a|.$$

Thus, the error, $E(x)$, is less than ϵ times $|x - a|$, the distance between x and a. So, as we zoom in on the graph by choosing smaller ϵ, the deviation, $|E(x)|$, of f from its tangent line shrinks, even relative to the scale on the x-axis. So, zooming makes a differentiable function look straight.

Exercises and Problems for Section 3.9

Exercises

1. Find the tangent line approximation for $\sqrt{1 + x}$ near $x = 0$.

2. What is the tangent line approximation to e^x near $x = 0$?

3. Find tangent line approximation to $1/x$ near $x = 1$.

4. Show that $1 - x/2$ is the tangent line approximation to $1/\sqrt{1 + x}$ near $x = 0$.

5. Show that $e^{-x} \approx 1 - x$ near $x = 0$.

6. What is the local linearization of e^{x^2} near $x = 1$?

Problems

7. (a) Show that $1 + kx$ is the local linearization of $(1 + x)^k$ near $x = 0$.
   (b) Someone claims that the square root of 1.1 is about 1.05. Without using a calculator, do you think that this estimate is about right?
   (c) Is the actual number above or below 1.05?

8. Multiply the local linearization of e^x near $x = 0$ by itself to obtain an approximation for e^{2x}. Compare this with the actual local linearization of e^{2x}. Explain why these two approximations are consistent, and discuss which one is more accurate.

9. (a) Show that $1 - x$ is the local linearization of $\dfrac{1}{1 + x}$ near $x = 0$.
   (b) From your answer to part (a), show that near $x = 0$,

$$\frac{1}{1 + x^2} \approx 1 - x^2.$$

   (c) Without differentiating, what do you think the derivative of $\dfrac{1}{1 + x^2}$ is at $x = 0$?

10. From the local linearizations of e^x and $\sin x$ near $x = 0$, write down the local linearization of the function $e^x \sin x$. From this result, write down the derivative of $e^x \sin x$ at $x = 0$. Using this technique, write down the derivative of $e^x \sin x/(1 + x)$ at $x = 0$.

11. The acceleration due to gravity, g, is given by

$$g = \frac{GM}{r^2},$$

where M is the mass of the earth, r is the distance from the center of the earth, and G is the universal gravitational constant.

(a) Show that when r changes by Δr, the change in the acceleration due to gravity, Δg, is given by

$$\Delta g \approx -2g\frac{\Delta r}{r}.$$

(b) What is the significance of the negative sign?
(c) What is the percent change in g when moving from sea level to the top of Pike's Peak (4.315 km)? Assume the radius of the earth is 6400 km.

12. Writing g for the acceleration due to gravity, the period, T, of a pendulum of length l is given by

$$T = 2\pi\sqrt{\frac{l}{g}}.$$

(a) Show that if the length of the pendulum changes by Δl, the change in the period, ΔT, is given by

$$\Delta T \approx \frac{T}{2l}\Delta l.$$

(b) If the length of the pendulum increases by 2%, by what percent does the period change?

13. Suppose now the length of the pendulum in Problem 12 remains constant, but that the acceleration due to gravity changes.

(a) Use the method of the preceding problem to relate ΔT approximately to Δg, the change in g.
(b) If g increases by 1%, find the percent change in T.

14. Suppose f has a continuous positive second derivative for all x. Which is larger, $f(1+\Delta x)$ or $f(1)+f'(1)\Delta x$? Explain.

15. Suppose $f'(x)$ is a differentiable decreasing function for all x. In each of the following pairs, which number is the larger? Give a reason for your answer.

(a) $f'(5)$ and $f'(6)$
(b) $f''(5)$ and 0
(c) $f(5 + \Delta x)$ and $f(5) + f'(5)\Delta x$

16. Use local linearization to derive the product rule,

$$[f(x)g(x)]' = f'(x)g(x) + f(x)g'(x).$$

[Hint: Use the definition of the derivative and the local linearizations $f(x+h) \approx f(x)+f'(x)h$ and $g(x+h) \approx g(x) + g'(x)h$.]

17. Derive the chain rule using local linearization. [Hint: In other words, differentiate $f(g(x))$, using $g(x + h) \approx g(x) + g'(x)h$ and $f(z + k) \approx f(z) + f'(z)k$.]

18. Consider a function f and a point a. Suppose there is a number L such that the linear function g

$$g(x) = f(a) + L(x - a)$$

is a good approximation to f. By good approximation, we mean that

$$\lim_{x \to a} \frac{E_L(x)}{x - a} = 0,$$

where $E_L(x)$ is the approximation error defined by

$$f(x) = g(x) + E_L(x) = f(a) + L(x - a) + E_L(x).$$

Show that f is differentiable at $x = a$ and that $f'(a) = L$. Thus the tangent line approximation is the only good linear approximation.

3.10   USING LOCAL LINEARITY TO FIND LIMITS

Suppose we want to calculate the exact value of the limit

$$\lim_{x \to 0} \frac{e^{2x} - 1}{x}.$$

Substituting $x = 0$ gives us $0/0$, which is undefined:

$$\frac{e^{2(0)} - 1}{0} = \frac{1 - 1}{0} = \frac{0}{0}.$$

Substituting values of x near 0 gives us an approximate value for the limit.

However, the limit can be calculated exactly using local linearity. Suppose we let $f(x)$ be the numerator, so $f(x) = e^{2x} - 1$, and $g(x)$ be the denominator, so $g(x) = x$. Then $f(0) = 0$ and $f'(x) = 2e^{2x}$, so $f'(0) = 2$. When we zoom in on the graph of $f(x) = e^{2x} - 1$ near the origin, we see its tangent line $y = 2x$ shown in Figure 3.51. We are interested in the ratio $f(x)/g(x)$, which is approximately the ratio of the y-values in Figure 3.51. This ratio of y-values is just the ratio of the slopes of the lines. So

$$\frac{f(x)}{g(x)} = \frac{e^{2x}-1}{x} \approx \frac{2}{1} = \frac{f'(0)}{g'(0)}.$$

As $x \to 0$, this approximation gets better, and we have

$$\lim_{x\to 0}\frac{e^{2x}-1}{x} = 2.$$

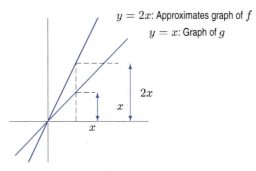

Figure 3.51: Ratio $(e^{2x}-1)/x$ is approximated by ratio of slopes as we zoom in near the origin

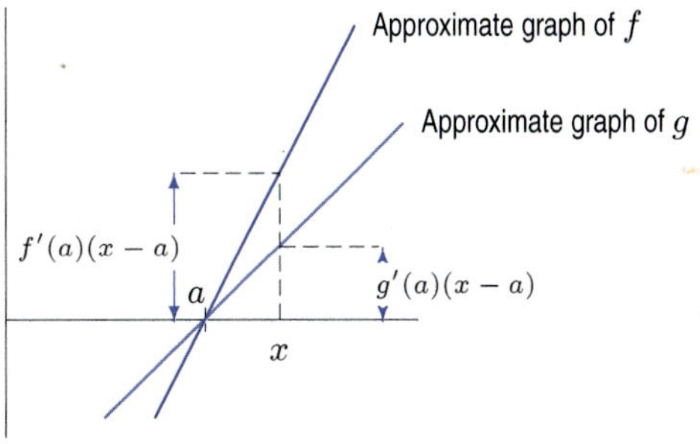

Figure 3.52: Ratio $f(x)/g(x)$ is approximated by ratio of slopes, $f'(a)/g'(a)$, as we zoom in at a

L'Hopital's Rule

If $f(a) = g(a) = 0$, we can use the same method to investigate limits of the form

$$\lim_{x\to a}\frac{f(x)}{g(x)}.$$

As in the previous case, we zoom in on the graphs of $f(x)$ and $g(x)$. Figure 3.52 shows that both graphs cross the x-axis at $x = a$. This suggests that the limit of $f(x)/g(x)$ as $x \to a$ is the ratio of slopes, giving the following result:

L'Hopital's rule: If f and g are differentiable, $f(a) = g(a) = 0$, and $g'(a) \neq 0$, then

$$\lim_{x\to a}\frac{f(x)}{g(x)} = \frac{f'(a)}{g'(a)}.$$

To justify this result, let us assume $g'(a) \neq 0$ and consider the quantity $f'(a)/g'(a)$. Using the definition of the derivative and the fact that $f(a) = g(a) = 0$, we have

$$\frac{f'(a)}{g'(a)} = \frac{\displaystyle\lim_{h\to 0}\frac{f(a+h)-f(a)}{h}}{\displaystyle\lim_{h\to 0}\frac{g(a+h)-g(a)}{h}} = \frac{\displaystyle\lim_{h\to 0}\frac{f(a+h)}{h}}{\displaystyle\lim_{h\to 0}\frac{g(a+h)}{h}}$$

$$= \lim_{h\to 0}\frac{f(a+h)}{g(a+h)} = \lim_{x\to a}\frac{f(x)}{g(x)}.$$

Note that if $f'(a) \neq 0$ and $g'(a) = 0$, the limit of $f(x)/g(x)$ does not exist.

Example 1 Use l'Hopital's rule to confirm that $\lim\limits_{x \to 0} \dfrac{\sin x}{x} = 1$.

Solution Let $f(x) = \sin x$ and $g(x) = x$. Then $f(0) = g(0) = 0$ and $f'(x) = \cos x$ and $g'(x) = 1$. Thus,

$$\lim_{x \to 0} \frac{\sin x}{x} = \frac{\cos 0}{1} = 1.$$

If we also have $f'(a) = g'(a) = 0$, then we can use the following result:

More general form of **l'Hopital's rule:** If f and g are differentiable and $f(a) = g(a) = 0$, then

$$\lim_{x \to a} \frac{f(x)}{g(x)} = \lim_{x \to a} \frac{f'(x)}{g'(x)},$$

provided the limit on the right exists.

Example 2 Calculate $\lim\limits_{t \to 0} \dfrac{e^t - 1 - t}{t^2}$.

Solution Let $f(t) = e^t - 1 - t$ and $g(t) = t^2$. Then $f(0) = e^0 - 1 - 0 = 0$ and $g(0) = 0$, and $f'(t) = e^t - 1$ and $g'(t) = 2t$. So

$$\lim_{t \to 0} \frac{e^t - 1 - t}{t^2} = \lim_{t \to 0} \frac{e^t - 1}{2t}.$$

Since $f'(0) = g'(0) = 0$, the ratio $f'(0)/g'(0)$ is not defined. So we use l'Hopital's rule again:

$$\lim_{t \to 0} \frac{e^t - 1 - t}{t^2} = \lim_{t \to 0} \frac{e^t - 1}{2t} = \lim_{t \to 0} \frac{e^t}{2} = \frac{1}{2}.$$

The following forms of l'Hopital's rule apply to limits involving infinity.

L'Hopital's rule can also be used in the following cases, provided f and g are differentiable:
- When $\lim_{x \to a} f(x) = \pm\infty$ and $\lim_{x \to a} g(x) = \pm\infty$,

or

- When $a = \infty$ (or $a = -\infty$) and $\lim\limits_{x \to \infty} f(x) = \lim\limits_{x \to \infty} g(x) = 0$ or $\lim\limits_{x \to \infty} f(x) = \pm\infty$ and $\lim\limits_{x \to \infty} g(x) = \pm\infty$.

It can be shown that under these circumstances:

$$\lim_{x \to a} \frac{f(x)}{g(x)} = \lim_{x \to a} \frac{f'(x)}{g'(x)}$$

(where a may be $\pm\infty$), provided the limit on the right-hand side exists.

Notice that we cannot evaluate $f'(x)/g'(x)$ directly when $a = \pm\infty$. The next example shows how this version of l'Hopital's rule is used.

Example 3   Calculate $\lim\limits_{x \to \infty} \dfrac{5x + e^{-x}}{7x}$.

Solution   Let $f(x) = 5x + e^{-x}$ and $g(x) = 7x$. Then $\lim\limits_{x \to \infty} f(x) = \lim\limits_{x \to \infty} g(x) = \infty$, and $f'(x) = 5 - e^{-x}$ and $g'(x) = 7$, so

$$\lim_{x \to \infty} \frac{5x + e^{-x}}{7x} = \lim_{x \to \infty} \frac{(5 - e^{-x})}{7} = \frac{5}{7}.$$

We can also use l'Hopital's rule to calculate some limits of the form $\lim\limits_{x \to \infty} f(x)g(x)$, providing we rewrite them appropriately.

Example 4   Calculate $\lim\limits_{x \to \infty} xe^{-x}$.

Solution   Since $\lim\limits_{x \to \infty} x = \infty$ and $\lim\limits_{x \to \infty} e^{-x} = 0$, we see that

$$xe^{-x} \to 0 \cdot \infty \qquad \text{as } x \to \infty.$$

Since $0 \cdot \infty$ is undefined, we rewrite the function xe^{-x} as

$$xe^{-x} = \frac{x}{e^x}.$$

Now we use l'Hopital's rule, since

$$xe^{-x} = \frac{x}{e^x} \to \frac{\infty}{\infty} \qquad \text{as } x \to \infty.$$

Taking $f(x) = x$ and $g(x) = e^x$ gives $f'(x) = 1$ and $g'(x) = e^x$, so

$$\lim_{x \to \infty} xe^{-x} = \lim_{x \to \infty} \frac{x}{e^x} = \lim_{x \to \infty} \frac{1}{e^x} = 0.$$

Dominance: Powers, Polynomials, Exponentials, and Logarithms

In Chapter 1, we saw that some functions were much larger than others as $x \to \infty$. We say that g *dominates* f as $x \to \infty$ if $\lim\limits_{x \to \infty} \dfrac{f(x)}{g(x)} = 0$. L'Hopital's rule gives us an easy way of checking this.

Example 5   Check that $x^{1/2}$ dominates $\ln x$ as $x \to \infty$.

Solution   We apply l'Hopital's rule to $(\ln x)/x^{1/2}$:

$$\lim_{x \to \infty} \frac{\ln x}{x^{1/2}} = \lim_{x \to \infty} \frac{1/x}{\frac{1}{2}x^{-1/2}}.$$

To evaluate this limit, we simplify and get

$$\lim_{x \to \infty} \frac{1/x}{\frac{1}{2}x^{-1/2}} = \lim_{x \to \infty} \frac{2x^{1/2}}{x} = \lim_{x \to \infty} \frac{2}{x^{1/2}} = 0.$$

Therefore we have

$$\lim_{x \to \infty} \frac{\ln x}{x^{1/2}} = 0,$$

which tells us that $x^{1/2}$ dominates $\ln x$ as $x \to \infty$.

Example 6   Check any exponential function of the form e^{kx} (with $k > 0$) dominates any power function of the form Ax^p (with A and p positive) as $x \to \infty$.

Solution   We apply l'Hopital's rule repeatedly to Ax^p/e^{kx}:

$$\lim_{x\to\infty} \frac{Ax^p}{e^{kx}} = \lim_{x\to\infty} \frac{Apx^{p-1}}{ke^{kx}} = \lim_{x\to\infty} \frac{Ap(p-1)x^{p-2}}{k^2 e^{kx}} = \cdots$$

Keep applying l'Hopital's rule until the power of x is no longer positive. Then the limit of the numerator must be a finite number, while the limit of the denominator must be ∞. Therefore we have

$$\lim_{x\to\infty} \frac{Ax^p}{e^{kx}} = 0,$$

so e^{kx} dominates Ax^p.

Exercises and Problems for Section 3.10

Exercises

For Exercises 1–4, find the sign of $\lim\limits_{x\to a} \dfrac{f(x)}{g(x)}$ from the figure.

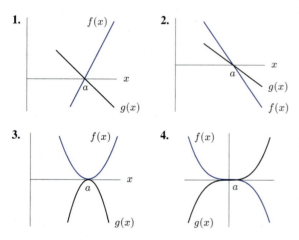

Based on your knowledge of the behavior of the numerator and denominator, predict the value of the limits in Exercises 5–8. Then find each limit using l'Hopital's rule.

5. $\lim\limits_{x\to 0} \dfrac{x^2}{\sin x}$

6. $\lim\limits_{x\to 0} \dfrac{\sin^2 x}{x}$

7. $\lim\limits_{x\to 0} \dfrac{\sin x}{x^{1/3}}$

8. $\lim\limits_{x\to 0} \dfrac{x}{(\sin x)^{1/3}}$

In Exercises 9–12, which function dominates as $x \to \infty$?

9. x^5 and $0.1x^7$

10. $0.01x^3$ and $50x^2$

11. $\ln(x+3)$ and $x^{0.2}$

12. x^{10} and $e^{0.1x}$

Problems

13. Evaluate $\lim\limits_{x\to 0^+} x \ln x$. [Hint: Write $x \ln x = \dfrac{\ln x}{1/x}$.]

14. (a) What is the slope of $f(x) = \sin(3x)$ at $x = 0$?
   (b) What is the slope of $g(x) = 5x$ at $x = 0$?
   (c) Use the results of parts (a) and (b) to calculate
   $$\lim_{x\to 0} \frac{\sin(3x)}{5x}.$$

Explain why l'Hopital's rule cannot be used to calculate the limits in Problems 15–17. Then evaluate the limit if it exists.

15. $\lim\limits_{x\to 1} \dfrac{\sin(2x)}{x}$

16. $\lim\limits_{x\to 0} \dfrac{\cos x}{x}$

17. $\lim\limits_{x\to\infty} \dfrac{e^{-x}}{\sin x}$

18. Find the horizontal asymptote of $f(x) = \dfrac{2x^3 + 5x^2}{3x^3 - 1}$.

19. The functions f and g and their tangent lines at $(4, 0)$ are shown in Figure 3.53. Find $\lim\limits_{x\to 4} \dfrac{f(x)}{g(x)}$.

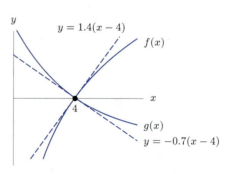

Figure 3.53

CHAPTER SUMMARY

- **Derivatives of elementary functions**
  Powers, polynomials, rational functions, exponential functions, logarithms, trigonometric functions, inverse trigonometric functions.

- **Derivatives of sums, differences, and constant multiples**

- **Product and quotient rules**

- **Chain rule**

Differentiation of implicitly defined functions, inverse functions.

- **Working with the derivative**
  Tangent line approximation, local linearity, l'Hopital's rule.

- **Parametric equations**
  Motion of a particle in the plane, parametric equations for lines and other curves, velocity, tangent lines.

REVIEW EXERCISES AND PROBLEMS FOR CHAPTER THREE

Exercises

Find derivatives for the functions in Exercises 1–52. Assume $a, b, c,$ and k are constants.

1. $f(t) = 2te^t - \dfrac{1}{\sqrt{t}}$

2. $w = \dfrac{5 - 3z}{5 + 3z}$

3. $f(y) = \ln\left(\ln(2y^3)\right)$

4. $f(x) = \dfrac{x^3}{9}(3\ln x - 1)$

5. $g(x) = x^k + k^x$

6. $z = \sin^3\theta$

7. $f(t) = \cos^2(3t + 5)$

8. $M(\alpha) = \tan^2(2 + 3\alpha)$

9. $s(\theta) = \sin^2(3\theta - \pi)$

10. $h(t) = \ln\left(e^{-t} - t\right)$

11. $p(\theta) = \dfrac{\sin(5 - \theta)}{\theta^2}$

12. $w(\theta) = \dfrac{\theta}{\sin^2\theta}$

13. $g(x) = \dfrac{x^2 + \sqrt{x} + 1}{x^{3/2}}$

14. $s(x) = \arctan(2 - x)$

15. $r(\theta) = e^{(e^\theta + e^{-\theta})}$

16. $m(n) = \sin(e^n)$

17. $k(\alpha) = e^{\tan(\sin\alpha)}$

18. $g(t) = t\cos(\sqrt{t}e^t)$

19. $f(r) = (\tan 2 + \tan r)^e$

20. $y = e^{-\pi} + \pi^{-e}$

21. $h(x) = xe^{\tan x}$

22. $y = e^{2x}\sin^2(3x)$

23. $g(x) = \tan^{-1}(3x^2 + 1)$

24. $g(w) = \dfrac{1}{2^w + e^w}$

25. $y = 2^{\sin x}\cos x$

26. $h(x) = \ln e^{ax}$

27. $k(x) = \ln e^{ax} + \ln b$

28. $f(\theta) = e^{k\theta} - 1$

29. $f(t) = e^{-4kt}\sin t$

30. $f(x) = a^{5x}$

31. $f(x) = \dfrac{a^2 - x^2}{a^2 + x^2}$

32. $w(r) = \dfrac{ar^2}{b + r^3}$

33. $f(s) = \dfrac{a^2 - s^2}{\sqrt{a^2 + s^2}}$

34. $H(t) = (at^2 + b)e^{-ct}$

35. $g(\theta) = \sqrt{a^2 - \sin^2\theta}$

36. $y = \arctan\left(\dfrac{2}{x}\right)$

37. $r(t) = \ln\left(\sin\left(\dfrac{t}{k}\right)\right)$

38. $g(u) = \dfrac{e^{au}}{a^2 + b^2}$

39. $g(w) = \dfrac{5}{(a^2 - w^2)^2}$

40. $y = \dfrac{e^x - e^{-x}}{e^x + e^{-x}}$

41. $y = \dfrac{e^{ax} - e^{-ax}}{e^{ax} + e^{-ax}}$

42. $y = \sqrt{\cos(5\theta) + \sin^2(6\theta)}$

43. $r(\theta) = \sin\left((3\theta - \pi)^2\right)$

44. $y = (x^2 + 5)^3(3x^3 - 2)^2$

45. $N(\theta) = \tan(\arctan(k\theta))$

46. $h(t) = e^{kt}(\sin at + \cos bt)$

47. $f(x) = (2 - 4x - 3x^2)(6x^e - 3\pi)$

48. $f(t) = (\sin(2t) - \cos(3t))^4$

49. $s(y) = \sqrt[3]{(\cos^2 y + 3 + \sin^2 y)}$

50. $f(x) = (4 - x^2 + 2x^3)(6 - 4x + x^7)$

51. $h(x) = \left(\dfrac{1}{x} - \dfrac{1}{x^2}\right)(2x^3 + 4)$

52. $f(z) = \sqrt{5z} + 5\sqrt{z} + \dfrac{5}{\sqrt{z}} - \sqrt{\dfrac{5}{z}} + \sqrt{5}$

53. Find the slope of the curve $x^2 + 3y^2 = 7$ at the point $(2, -1)$.

54. Assume y is a differentiable function of x and that $y + \sin y + x^2 = 9$. Find dy/dx at the point $x = 3, y = 0$.

For Exercises 55–57, assume that y is a differentiable function of x and find dy/dx.

55. $x^2y - 2y + 5 = 0$

56. $x^3 + y^3 - 4x^2y = 0$

57. $\sin(ay) + \cos(bx) = xy$

58. Find the equations for the lines tangent to the graph of $xy + y^2 = 4$ where $x = 3$.

Problems

59. Suppose W is proportional to r^3. The derivative dW/dr is proportional to what power of r?

60. Using the information in the table about f and g, find:

  (a) $h(4)$ if $h(x) = f(g(x))$
  (b) $h'(4)$ if $h(x) = f(g(x))$
  (c) $h(4)$ if $h(x) = g(f(x))$
  (d) $h'(4)$ if $h(x) = g(f(x))$
  (e) $h'(4)$ if $h(x) = g(x)/f(x)$
  (f) $h'(4)$ if $h(x) = f(x)g(x)$

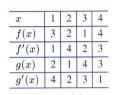

x	1	2	3	4
$f(x)$	3	2	1	4
$f'(x)$	1	4	2	3
$g(x)$	2	1	4	3
$g'(x)$	4	2	3	1

61. Given: $r(2) = 4$, $s(2) = 1$, $s(4) = 2$, $r'(2) = -1$, $s'(2) = 3$, $s'(4) = 3$. Compute the following derivatives, or state what additional information you would need to be able to do so.

  (a) $H'(2)$ if $H(x) = r(x) \cdot s(x)$
  (b) $H'(2)$ if $H(x) = \sqrt{r(x)}$
  (c) $H'(2)$ if $H(x) = r(s(x))$
  (d) $H'(2)$ if $H(x) = s(r(x))$

62. If $g(2) = 3$ and $g'(2) = -4$, find $f'(2)$ for the following:

  (a) $f(x) = x^2 - 4g(x)$ **(b)** $f(x) = \dfrac{x}{g(x)}$
  (c) $f(x) = x^2 g(x)$ **(d)** $f(x) = (g(x))^2$
  (e) $f(x) = x \sin(g(x))$ **(f)** $f(x) = x^2 \ln(g(x))$

63. For parts (a)–(f) of Problem 62, determine the equation of the line tangent to f at $x = 2$.

64. Imagine you are zooming in on the graphs of the following functions near the origin:

$$y = \arcsin x \qquad y = \sin x - \tan x \qquad y = x - \sin x$$

$$y = \arctan x \qquad y = \frac{\sin x}{1 + \sin x} \qquad y = \frac{x^2}{x^2+1}$$

$$y = \frac{1 - \cos x}{\cos x} \qquad y = \frac{x}{x^2+1} \qquad y = \frac{\sin x}{x} - 1$$

$$y = -x \ln x \qquad y = e^x - 1 \qquad y = x^{10} + \sqrt[10]{x}$$

$$y = \frac{x}{x+1}$$

Which of them look the same? Group together those functions which become indistinguishable, and give the equation of the line they look like. [Note: $(\sin x)/x - 1$ and $-x \ln x$ never quite make it to the origin.]

65. The graphs of $\sin x$ and $\cos x$ intersect once between 0 and $\pi/2$. What is the angle between the two curves at the point where they intersect? (You need to think about how the angle between two curves should be defined.)

In Problems 66–67, show that the curves meet at least once and determine whether the curves are perpendicular at the point of intersection.

66. $y = 1 + x - x^2$ and $y = 1 - x + x^2$

67. $y = 1 - x^3/3$ and $y = x - 1$

68. For some constant b and $x > 0$, let $y = x \ln x - bx$. Find the equation of the tangent line to this graph at the point at which the graph crosses the x-axis.

69. The acceleration due to gravity, g, at a distance r from the center of the earth is given by

$$g = \frac{GM}{r^2},$$

where M is the mass of the earth and G is a constant.

  (a) Find dg/dr.
  (b) What is the practical interpretation (in terms of acceleration) of dg/dr? Why would you expect it to be negative?
  (c) You are told that $M = 6 \cdot 10^{24}$ and $G = 6.67 \cdot 10^{-20}$ where M is in kilograms and r in kilometers. What is the value of dg/dr at the surface of the earth ($r = 6400$ km) ?
  (d) What does this tell you about whether or not it is reasonable to assume g is constant near the surface of the earth?

70. In 1990, the population of Mexico was about 84 million and growing at 2.6% annually, while the population of the US was about 250 million and growing at 0.7% annually. Which population was growing faster, if we measure growth rates in people/year? Explain your answer.

71. The distance, s, of a moving body from a fixed point is given as a function of time, t, by $s = 20e^{t/2}$.

  (a) Find the velocity, v, of the body as a function of t.
  (b) Find a relationship between v and s, then show that s satisfies the differential equation $s' = \frac{1}{2}s$.

72. Air pressure at sea level is 30 inches of mercury. At an altitude of h feet above sea level, the air pressure, P, in inches of mercury, is given by

$$P = 30e^{-3.23 \times 10^{-5} h}$$

  (a) Sketch a graph of P against h.
  (b) Find the equation of the tangent line at $h = 0$.
  (c) A rule of thumb used by travelers is that air pressure drops about 1 inch for every 1000-foot increase in height above sea level. Write a formula for the air pressure given by this rule of thumb.

(d) What is the relation between your answers to parts (b) and (c)? Explain why the rule of thumb works.

(e) Are the predictions made by the rule of thumb too large or too small? Why?

73. The depth of the water, y, in meters, in the Bay of Fundy, Canada, is given as a function of time, t, in hours after midnight, by the function

$$y = 10 + 7.5 \cos(0.507t).$$

How quickly is the tide rising or falling (in meters/hour) at each of the following times?

(a) 6:00 am   **(b)** 9:00 am

(c) Noon   **(d)** 6:00 pm

74. A yam is put in a hot oven, maintained at a constant temperature $200°$C. Suppose that at time $t = 30$ minutes, the temperature T of the yam is $120°$ and is increasing at an (instantaneous) rate of $2°$/min. Newton's law of cooling (or, in our case, warming) implies that the temperature at time t will be given by a formula of the form

$$T(t) = 200 - ae^{-bt}.$$

Find a and b.

75. An object is oscillating at the end of a spring. Its position, in centimeters, relative to a fixed point, is given as a function of time, t, in seconds, by

$$y = y_0 \cos(2\pi\omega t), \quad \text{with } \omega \text{ a constant.}$$

(a) Find an expression for the velocity and acceleration of the object.

(b) How do the amplitudes of the position, velocity, and acceleration functions compare? How do the periods of these functions compare?

(c) Show that the function y satisfies the differential equation

$$\frac{d^2y}{dt^2} + 4\pi^2\omega^2 y = 0.$$

76. The total number of people, N, who have contracted a disease by a time t days after its outbreak is given by

$$N = \frac{1{,}000{,}000}{1 + 5{,}000e^{-0.1t}}.$$

(a) In the long run, how many people get the disease?

(b) Is there any day on which more than a million people fall sick? Half a million? Quarter of a million? (Note: You do not have to try to find out on what days these things happen.)

77. A spherical balloon is inflated so that its radius is increasing at a constant rate of 1 cm per second. At what rate is air being blown into the balloon when its radius is 5 cm?

78. When the growth of a spherical cell depends on the flow of nutrients through the surface, it is reasonable to assume that the growth rate, dV/dt, is proportional to the surface area, S. Assume that for a particular cell $dV/dt = \frac{1}{3} \cdot S$. At what rate is its radius r increasing?

79. A horizontal disk of radius a centered at the origin in the xy-plane is rotating about a vertical axis through the center. The angle between the positive x-axis and a radial line painted on the disk is θ radians.

(a) What does $d\theta/dt$ represent?

(b) What is the relationship between $d\theta/dt$ and the speed v of a point on the rim?

80. A radio navigation system used by aircraft gives a cockpit readout of the distance, s, in miles, between a fixed ground station and the aircraft. The system also gives a readout of the instantaneous rate of change, ds/dt, of this distance in miles/hour. An aircraft on a straight flight path at a constant altitude of 10,560 feet (2 miles) has passed directly over the ground station and is now flying away from it. What is the speed of this aircraft along its constant altitude flight path when the cockpit readouts are $s = 4.6$ miles and $ds/dt = 210$ miles/hour?

Problems 81–82 involve Boyle's Law which states that for a fixed quantity of gas at a constant temperature, the pressure, P, and the volume, V, are inversely related. Thus, for some constant k

$$PV = k.$$

81. A fixed quantity of gas is allowed to expand at constant temperature. Find the rate of change of pressure with respect to volume.

82. A certain quantity of gas occupies 10 cm³ at a pressure of 2 atmospheres. The pressure is increased, while keeping the temperature constant.

(a) Does the volume increase or decrease?

(b) If the pressure is increasing at a rate of 0.05 atmospheres/minute when the pressure is 2 atmospheres, find the rate at which the volume is changing at that moment. What are the units of your answer?

83. Find the nth derivative of the following functions:

(a) $\ln x$ **(b)** xe^x **(c)** $e^x \cos x$

84. The derivative f' gives the (absolute) rate of change of a quantity f, and f'/f gives the relative rate of change of the quantity. In this problem, we show that the product rule is equivalent to an additive rule for relative rates of change. Assume $h = f \cdot g$ with $f \neq 0$ and $g \neq 0$.

(a) Show that the additive rule

$$\frac{f'}{f} + \frac{g'}{g} = \frac{h'}{h}$$

implies the product rule, by multiplying through by h and using the fact that $h = f \cdot g$.

(b) Show that the product rule implies the additive rule in part (a), by starting with the product rule and dividing through by $h = f \cdot g$.

85. The relative rate of change of a function f is defined to be f'/f. Find an expression for the relative rate of change of a quotient f/g in terms of the relative rates of change of the functions f and g.

CAS Challenge Problems

86. (a) Use a computer algebra system to differentiate $(x+1)^x$ and $(\sin x)^x$.
  (b) Conjecture a rule for differentiating $(f(x))^x$, where f is any differentiable function.
  (c) Apply your rule to $g(x) = (\ln x)^x$. Does your answer agree with the answer given by the computer algebra system?
  (d) Prove your conjecture by rewriting $(f(x))^x$ in the form $e^{h(x)}$.

For Problems 87–89,
  (a) Use a computer algebra system to find and simplify the derivative of the given function.
  (b) Without a computer algebra system, use differentiation rules to calculate the derivative. Make sure that the answer simplifies to the same answer as in part (a).
  (c) Explain how you could have predicted the derivative by using algebra before taking the derivative.

87. $f(x) = \sin(\arcsin x)$

88. $g(r) = 2^{-2r}4^r$

89. $h(t) = \ln(1 - 1/t) + \ln(t/(t-1))$

CHECK YOUR UNDERSTANDING

Are the statements in Problems 1–10 true or false? Give an explanation for your answer.

1. The derivative of a polynomial is always a polynomial.

2. The derivative of π/x^2 is $-\pi/x$.

3. The derivative of $\tan \theta$ is periodic.

4. The graph of $\ln(x^2)$ is concave up for $x > 0$

5. If $f'(x)$ is defined for all x, then $f(x)$ is defined for all x.

6. If $f'(2) = 3.1$ and $g'(2) = 7.3$, then the graph of $f(x) + g(x)$ has slope 10.4 at $x = 2$.

7. Let f and g be two functions whose second derivatives are defined. Then

$$(fg)'' = fg'' + f''g.$$

8. If a function is periodic, with period c, then so is its derivative.

9. If y satisfies the equation $y^2 + xy - 1 = 0$, then dy/dx exists everywhere.

10. If a curve is given parametrically by $x = \cos(t^2)$, $y = \sin(t^2)$, then its slope is $\tan(t^2)$.

Are the statements in Problems 11–17 true or false? If a statement is true, explain how you know. If a statement is false, give a counterexample.

11. If $f(x)$ is defined for all x, then $f'(x)$ is defined for all x.

12. If $f(x)$ is increasing, then $f'(x)$ is increasing.

13. The only functions whose fourth derivatives are equal to $\cos t$ are of the form $\cos t + C$, where C is any constant.

14. If $f(x)$ has an inverse function, $g(x)$, then the derivative of $g(x)$ is $1/f'(x)$.

15. $(fg)'(x)$ is never equal to $f'(x)g'(x)$.

16. If the function $f(x)/g(x)$ is defined but not differentiable at $x = 1$, then either $f(x)$ or $g(x)$ is not differentiable at $x = 1$.

17. If the derivative of $f(g(x))$ is equal to the derivative of $f(x)$ for all x, then $g(x) = x$ for all x.

Suppose that f'' and g'' exist and that f and g are concave up for all x. Are the statements in Problems 18–21 true or false for all such f and g? If a statement is true, explain how you know. If a statement is false, give a counterexample.

18. $f(x) + g(x)$ is concave up for all x.

19. $f(x)g(x)$ is concave up for all x.

20. $f(x) - g(x)$ cannot be concave up for all x.

21. $f(g(x))$ is concave up for all x.

22. Let f be a differentiable function and let L be the linear function $L(x) = f(a) + k(x - a)$ for some constant a. Decide whether the following statements are true or false for all constants k. Explain your answer.

  (a) L is the local linearization for f near $x = a$,
  (b) If $\lim_{x \to a} (f(x) - L(x)) = 0$, then L is the local linearization for f near $x = a$.

23. Which of the following would be a counterexample to the product rule?

  (a) Two differentiable functions f and g satisfying $(fg)' = f'g'$.
  (b) A differentiable function f such that $(xf(x))' = xf'(x) + f(x)$.
  (c) A differentiable function f such that $(f(x)^2)' = 2f(x)$.
  (d) Two differentiable functions f and g such that $f'(a) = 0$ and $g'(a) = 0$ and fg has positive slope at $x = a$.

PROJECTS FOR CHAPTER THREE

1. **Rule of 70**

   The "Rule of 70" is a rule of thumb to estimate how long it takes money in a bank to double. Suppose the money is in an account earning $i\%$ annual interest, compounded yearly. The Rule of 70 says that the time it takes the amount of money to double is approximately $70/i$ years, assuming i is small. Find the local linearization of $\ln(1 + x)$, and use it to explain why this rule works.

2. **Newton's Method**

   Read about how to find roots using bisection and Newton's method in Appendices A and D.

   (a) What is the smallest positive zero of the function $f(x) = \sin x$? Apply Newton's method, with initial guess $x_0 = 3$, to see how fast it converges to $\pi = 3.1415926536\ldots$.

      (i) Compute the first two approximations, x_1 and x_2; compare x_2 with π.

      (ii) Newton's method works very well here. Explain why. To do this, you will have to outline the basic idea behind Newton's method.

      (iii) Estimate the location of the zero using bisection, starting with the interval $[3, 4]$. How does bisection compare to Newton's method in terms of accuracy?

   (b) Newton's method can be very sensitive to your initial estimate, x_0. For example, consider finding a zero of $f(x) = \sin x - \frac{2}{3}x$.

      (i) Use Newton's method with the following initial estimates to find a zero:

   $$x_0 = 0.904, \quad x_0 = 0.905, \quad x_0 = 0.906.$$

      (ii) What happens?

Chapter Four

USING THE DERIVATIVE

In Chapter 2, we introduced the derivative and some of its interpretations. In Chapter 3, we saw how to differentiate all of the standard functions, including powers, exponentials, logarithms, and trigonometric functions. Now we use first and second derivatives to analyze the behavior of families of functions and to solve optimization problems.

4.1 USING FIRST AND SECOND DERIVATIVES

What Derivatives Tell Us About a Function and its Graph

As we saw in Chapter 2, the connection between derivatives of a function and the function itself is given by the following:

- If $f' > 0$ on an interval, then f is increasing on that interval.
- If $f' < 0$ on an interval, then f is decreasing on that interval.
- If $f'' > 0$ on an interval, then the graph of f is concave up on that interval.
- If $f'' < 0$ on an interval, then the graph of f is concave down on that interval.

We can do more with these principles now than we could in Chapter 2 because we now have formulas for the derivatives of the elementary functions.

When we graph a function on a computer or calculator, we often see only part of the picture. Information given by the first and second derivatives can help identify regions with interesting behavior.

Example 1   Use a computer or calculator to sketch a useful graph of the function $f(x) = x^3 - 9x^2 - 48x + 52$.

Solution   Since f is a cubic polynomial, we expect a graph that is roughly S-shaped. Graphing this function with $-10 \le x \le 10$, $-10 \le y \le 10$, gives the two nearly vertical lines in Figure 4.1. We know that there is more going on than this, but how do we know where to look?

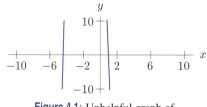

Figure 4.1: Unhelpful graph of
$f(x) = x^3 - 9x^2 - 48x + 52$

We use the derivative to determine where the function is increasing and where it is decreasing. The derivative of f is

$$f'(x) = 3x^2 - 18x - 48.$$

To find where $f' > 0$ or $f' < 0$, we first find where $f' = 0$, that is, where $3x^2 - 18x - 48 = 0$. Factoring, we get $3(x - 8)(x + 2) = 0$, so $x = -2$ or $x = 8$. Since $f' = 0$ *only* at $x = -2$ and $x = 8$, and since f' is continuous, f' cannot change sign on any of the three intervals $x < -2$, or $-2 < x < 8$, or $8 < x$. How can we tell the sign of f' on each of these intervals? The easiest way is to pick a point and substitute into f'. For example, since $f'(-3) = 33 > 0$, we know f' is positive for $x < -2$, so f is increasing for $x < -2$. Similarly, since $f'(0) = -48$ and $f'(10) = 72$, we know that f decreases between $x = -2$ and $x = 8$ and increases for $x > 8$. Summarizing:

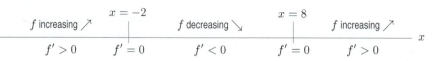

We find that $f(-2) = 104$ and $f(8) = -396$. Hence on the interval $-2 < x < 8$ the function decreases from a high of 104 to a low of -396. (Now we see why not much showed up in our first calculator graph.) One more point on the graph is easy to get: the y intercept, $f(0) = 52$. With just these three points we can get a much more helpful graph. By setting the plotting window to

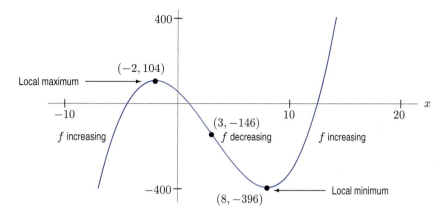

Figure 4.2: Useful graph of $f(x) = x^3 - 9x^2 - 48x + 52$

$-10 \leq x \leq 20$ and $-400 \leq y \leq 400$, we get Figure 4.2, which gives much more insight into the behavior of $f(x)$ than the graph in Figure 4.1.

In Figure 4.2, we see that part of the graph is concave up and part is concave down. We can use the second derivative to analyze concavity. We have

$$f''(x) = 6x - 18.$$

So, $f''(x) < 0$ when $x < 3$ and $f''(x) > 0$ when $x > 3$, and so the graph of f is concave down for $x < 3$ and concave up for $x > 3$. At $x = 3$, we have $f''(x) = 0$. Summarizing:

$$x = 3$$

f concave down $\cap$	f concave up $\cup$
$f'' < 0$	$f'' = 0$ $\quad$ $f'' > 0$

Local Maxima and Minima

We are often interested in points such as those marked local maximum and local minimum in Figure 4.2. We have the following definition:

> Suppose p is a point in the domain of f:
> - f has a **local minimum** at p if $f(p)$ is less than or equal to the values of f for points near p.
> - f has a **local maximum** at p if $f(p)$ is greater than or equal to the values of f for points near p.

We use the adjective "local" because we are describing only what happens near p.

How Do We Detect a Local Maximum or Minimum?

In the preceding example, the points $x = -2$ and $x = 8$, where $f'(x) = 0$, played a key role in leading us to local maxima and minima. We give a name to such points:

> For any function f, a point p in the domain of f where $f'(p) = 0$ or $f'(p)$ is undefined is called a **critical point** of the function. In addition, the point $(p, f(p))$ on the graph of f is also called a critical point. A **critical value** of f is the value, $f(p)$, at a critical point, p.

Notice that "critical point of f" can refer either to points in the domain of f or to points on the graph of f. You will know which meaning is intended from the context.

Geometrically, at a critical point where $f'(p) = 0$, the line tangent to the graph of f at p is horizontal. At a critical point where $f'(p)$ is undefined, there is no horizontal tangent to the graph—there's either a vertical tangent or no tangent at all. (For example, $x = 0$ is a critical point for the absolute value function $f(x) = |x|$.) However, most of the functions we work with are differentiable everywhere, and therefore most of our critical points are of the $f'(p) = 0$ variety.

The critical points divide the domain of f into intervals on which the sign of the derivative remains the same, either positive or negative. Therefore, if f is defined on the interval between two successive critical points, its graph cannot change direction on that interval; it is either increasing or decreasing. We have the following result, which is proved on page 207, after Theorem 4.2.

Theorem 4.1: Local Extrema and Critical Points

Suppose f is defined on an interval and has a local maximum or minimum at the point $x = a$, which is not an endpoint of the interval. If f is differentiable at $x = a$, then $f'(a) = 0$.

Warning! The sign of f' doesn't *have* to change at a critical point. Consider $f(x) = x^3$, which has a critical point at $x = 0$. (See Figure 4.3.) The derivative , $f'(x) = 3x^2$, is positive on both sides of $x = 0$, so f increases on both sides of $x = 0$, and there is neither a local maximum nor a local minimum at $x = 0$. In other words, not every critical point is a local maximum or local minimum.

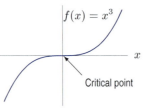

Figure 4.3: Critical point which is not a local maximum or minimum

Testing For Local Maxima and Minima

If f' has different signs on either side of a critical point p, with $f'(p) = 0$, then the graph changes direction at p and looks like one of those in Figure 4.4. So we have the following criterion:

The First-Derivative Test for Local Maxima and Minima

Suppose p is a critical point of a continuous function f.
- If f' changes from negative to positive at p, then f has a local minimum at p.
- If f' changes from positive to negative at p, then f has a local maximum at p.

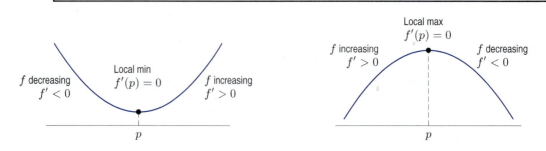

Figure 4.4: Changes in direction at a critical point, p: Local maxima and minima

---

Example 2   Use a graph of the function $f(x) = \dfrac{1}{x(x-1)}$ to observe its local maxima and minima. Confirm your observation analytically.

Solution   The graph in Figure 4.5 suggests that this function has no local minima but that there is a local maximum at about $x = \frac{1}{2}$. Confirming this analytically means using the formula for the derivative to show that what we expect is true. Since $f(x) = (x^2 - x)^{-1}$, we have

$$f'(x) = -1(x^2 - x)^{-2}(2x - 1) = -\frac{2x - 1}{(x^2 - x)^2}.$$

So $f'(x) = 0$ where $2x - 1 = 0$. Thus, the only critical point in the domain of f is $x = \frac{1}{2}$.

Furthermore, $f'(x) > 0$ where $0 < x < 1/2$, and $f'(x) < 0$ where $1/2 < x < 1$. Thus, f increases for $0 < x < 1/2$ and decreases for $1/2 < x < 1$. According to the first derivative test, the critical point $x = 1/2$ is a local maximum.

For $-\infty < x < 0$ or $1 < x < \infty$, there are no critical points and no local maxima or minima. Although $1/(x(x-1)) \to 0$ both as $x \to \infty$ and as $x \to -\infty$, we don't say 0 is a local minimum because $1/(x(x-1))$ never actually *equals* 0.

Notice that although $f' > 0$ everywhere that it is defined for $x < \frac{1}{2}$, the function f is not increasing throughout this interval. The problem is that f and f' are not defined at $x = 0$, so we cannot conclude that f is increasing when $x < 1/2$.

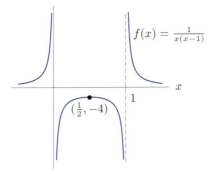

$f(x) = \frac{1}{x(x-1)}$

$\left(\frac{1}{2}, -4\right)$

Figure 4.5: Find local maxima and minima

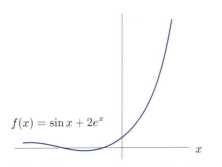

$f(x) = \sin x + 2e^x$

Figure 4.6: Explain the absence of local maxima and minima for $x \geq 0$

Example 3   The graph of $f(x) = \sin x + 2e^x$ is in Figure 4.6. Using the derivative, explain why there are no local maxima or minima for $x \geq 0$.

Solution   Local maxima and minima can occur only at critical points. Now $f'(x) = \cos x + 2e^x$, which is defined for all x. We know $\cos x$ is always between -1 and 1, and $2e^x \geq 2$ for $x \geq 0$, so $f'(x)$ cannot be 0 for any $x \geq 0$. Therefore there are no local maxima or minima there.

The Second-Derivative Test for Local Maxima and Minima

Knowing the concavity of a function can also be useful in testing if a critical point is a local maximum or a local minimum. Suppose p is a critical point of f, with $f'(p) = 0$, so that the graph of f has a horizontal tangent line at p. If the graph is concave up at p, then f has a local minimum at p. Likewise, if the graph is concave down, f has a local maximum. (See Figure 4.7.) This suggests:

Concave up   Local max

Local min   Concave down

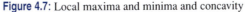

Figure 4.7: Local maxima and minima and concavity

> ## The Second-Derivative Test for Local Maxima and Minima
> - If $f'(p) = 0$ and $f''(p) > 0$ then f has a local minimum at p.
> - If $f'(p) = 0$ and $f''(p) < 0$ then f has a local maximum at p.
> - If $f'(p) = 0$ and $f''(p) = 0$ then the test tells us nothing.

Example 4   Classify as local maxima or local minima the critical points of $f(x) = x^3 - 9x^2 - 48x + 52$.

Solution   As we saw in Example 1 on page 166,

$$f'(x) = 3x^2 - 18x - 48$$

and the critical points of f are $x = -2$ and $x = 8$. We have

$$f''(x) = 6x - 18.$$

Thus $f''(8) = 30 > 0$, so f has a local minimum at $x = 8$. Since $f''(-2) = -30 < 0$, f has a local maximum at $x = -2$.

Warning! The second-derivative test does not tell us anything if both $f'(p) = 0$ and $f''(p) = 0$. For example, if $f(x) = x^3$ and $g(x) = x^4$, both $f'(0) = g'(0) = 0$ and $f''(0) = g''(0) = 0$. The point $x = 0$ is a minimum for g but is neither a maximum nor a minimum for f. However, the first-derivative test is still useful. For example, g' changes sign from negative to positive at $x = 0$, so we know g has a local minimum there.

Concavity and Inflection Points

We have studied points where the slope changes sign, which led us to critical points. Now we look at points where the concavity changes.

> A point at which the graph of a function changes concavity is called an **inflection point** of f.

The words "inflection point of f" can refer either to a point in the domain of f or to a point on the graph of f. The context of the problem will tell you which is meant.

How Do We Locate an Inflection Point?

Since the concavity changes at an inflection point, the sign of f'' changes there. It is positive on one side of the inflection point, and negative on the other; so at the inflection point, f'' is zero or undefined.

Example 5   Find the critical and inflection points for $g(x) = xe^{-x}$ and sketch the graph of g for $x \geq 0$.

Solution   Taking derivatives and simplifying, we have

$$g'(x) = (1 - x)e^{-x} \quad \text{and} \quad g''(x) = (x - 2)e^{-x}.$$

So $x = 1$ is a critical point, and $g' > 0$ for $x < 1$ and $g' < 0$ for $x > 1$. Hence g increases to a local maximum at $x = 1$ and then decreases. Also, $g(x) \to 0$ as $x \to \infty$. There is an inflection point at $x = 2$ since $g'' < 0$ for $x < 2$ and $g'' > 0$ for $x > 2$. The graph is sketched in Figure 4.8.

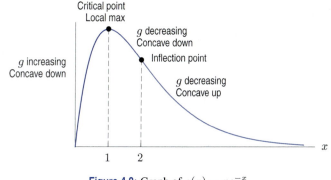

Figure 4.8: Graph of $g(x) = xe^{-x}$

Warning! Not every point x where $f''(x) = 0$ (or f'' is undefined) is an inflection point (just as not every point where $f' = 0$ is a local maximum or minimum). For instance $f(x) = x^4$ has $f''(x) = 12x^2$ so $f''(0) = 0$, but $f'' > 0$ when $x > 0$ and when $x < 0$, so there is *no* change in concavity at $x = 0$. See Figure 4.9.

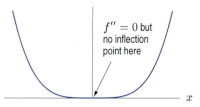

Figure 4.9: Graph of $f(x) = x^4$

Inflection Points and Local Maxima and Minima of the Derivative

Inflection points can also be interpreted in terms of first derivatives. If p is an inflection point, then $f''(p) = 0$ (or $f''(p)$ is undefined) and hence p is a critical point of the derivative function f'. If f' is continuous, this critical point is a local maximum or minimum of f', since f'' changes sign at p.

> A function f with a continuous derivative has an inflection point at p if either of the following conditions hold:
> - f' has a local minimum or a local maximum at p.
> - f'' changes sign at p.

Figure 4.10 shows two inflection points. Notice that the curve crosses the tangent line at these points and that the slope of the curve is a maximum or a minimum there.

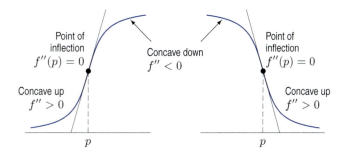

Figure 4.10: Change in concavity at p: Points of inflection

Example 6   Graph $f(x) = x + \sin x$, and determine where f is increasing most rapidly, and least rapidly.

Solution   The graph of $f(x) = x + \sin x$ is in Figure 4.11 and the graph of $f'(x) = 1 + \cos x$ is in Figure 4.12.

Where is f increasing most rapidly? At the points $x = \ldots, -2\pi, 0, 2\pi, 4\pi, 6\pi, \ldots$, because these points are local maxima for f', and f' has the same value at each of them. Likewise f is increasing least rapidly at the points $x = \ldots, -3\pi, -\pi, \pi, 3\pi, 5\pi, \ldots$, since these points are local minima for f'. Notice that the points where f is increasing most rapidly and the points where it is increasing least rapidly are inflection points of f.

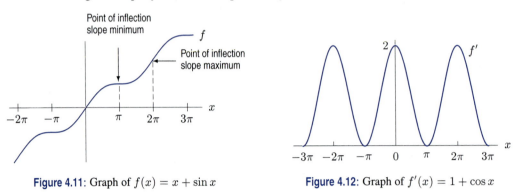

Figure 4.11: Graph of $f(x) = x + \sin x$ Figure 4.12: Graph of $f'(x) = 1 + \cos x$

Example 7   Water is being poured into the vase in Figure 4.13 at a constant rate, measured in volume per unit time. Graph $y = f(t)$, the depth of the water against time, t. Explain the concavity and indicate the inflection points.

Solution   At first the water level, y, rises slowly because the base of the vase is wide, and it takes a lot of water to make the depth increase. However, as the vase narrows, the rate at which the water is rising increases. Thus, y is increasing at an increasing rate and the graph is concave up. The rate of increase in the water level is at a maximum when the water reaches the middle of the vase, where the diameter is smallest; this is an inflection point. After that, the rate at which y increases decreases again, so the graph is concave down. (See Figure 4.14.)

Figure 4.13: A vase

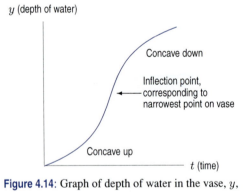

Figure 4.14: Graph of depth of water in the vase, y, against time, t

Exercises and Problems for Section 4.1

Exercises

1. Sketch the graph of a function with exactly two critical points, one of which is a local minimum and the other is neither a local maximum nor a local minimum.

2. Sketch the graph of a function which has exactly one critical point, at $x = 2$, and exactly one inflection point, at $x = 4$.

3. Indicate all critical points on the graph of f in Figure 4.15 and determine which correspond to local maxima of f, which to local minima, and which to neither.

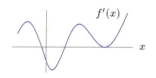

Figure 4.15

4. Indicate on the graph of the derivative function f' in Figure 4.16 the x-values that are critical points of the function f itself. At which critical points does f have local maxima, local minima, or neither?

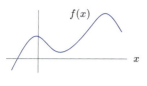

Figure 4.16

5. (a) Use a graph of $f(x) = e^{-x^2}$ to estimate roughly the x-values of any critical points and inflection points of f.
   (b) Use derivatives to find the x-values of any critical points and inflection points exactly.

6. Find all critical points of $f(x) = 10.2x^2 e^{-0.4x}$.

Classify the critical points of the functions in Problems 7–8 as local maxima or local minima.

7. $g(x) = xe^{-x}$ 8. $h(x) = x + 1/x$

9. Suppose f has a continuous derivative whose values are given in the following table.

   (a) Estimate the x-coordinates of critical points of f for $0 \le x \le 10$.
   (b) For each critical point, indicate if it is a local maximum of f, local minimum, or neither.

x	0	1	2	3	4	5	6	7	8	9	10
$f'(x)$	5	2	1	-2	-5	-3	-1	2	3	1	-1

10. Indicate on the graph of the derivative f' in Figure 4.17 the x-values that are inflection points of the function f.

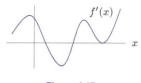

Figure 4.17

11. Indicate on the graph of the second derivative f'' in Figure 4.18 the x-values that are inflection points of the function f.

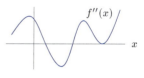

Figure 4.18

12. Find the inflection points of $f(x) = x^4 + x^3 - 3x^2 + 2$.

Using a calculator or computer, graph the functions in Exercises 13–20. Describe briefly in words the interesting features of the graph including the location of the critical points and where the function is increasing/decreasing. Then use the derivative and algebra to explain the shape of the graph.

13. $f(x) = x^3 - 6x + 1$ 14. $f(x) = x^3 + 6x + 1$

15. $f(x) = 3x^5 - 5x^3$ 16. $f(x) = x + 2\sin x$

17. $f(x) = e^x - 10x$ 18. $f(x) = e^x + \sin x$

19. $f(x) = xe^{-x^2}$ 20. $f(x) = x \ln x, \quad x > 0$

21. Use the graphs you drew in Problems 13–20 to describe in words the concavity of each graph, including approximate x-coordinates for all points of inflection. Then use f'' and algebra to explain what you see.

Problems

22. Give examples of graphs of functions with no, one, and infinitely many critical points.

23. (a) The following table gives values of the differentiable function $y = f(x)$. Estimate the x-values of critical points of $f(x)$ on the interval $0 < x < 10$. Classify each critical point as a local maximum, local minimum, or neither.

   (b) Now assume that the table gives values of the continuous function $y = f'(x)$ (instead of $f(x)$). Estimate and classify critical points of the function $f(x)$.

x	0	1	2	3	4	5	6	7	8	9	10
y	1	2	1	-2	-5	-3	-1	2	3	1	-1

24. Find values of a and b so that the function $f(x) = x^2 + ax + b$ has a local minimum at the point $(6, -5)$.

25. Find the value of a so that the function $f(x) = xe^{ax}$ has a critical point at $x = 3$.

26. Find constants a and b in the function $f(x) = axe^{bx}$ such that $f(\frac{1}{3}) = 1$ and the function has a local maximum at $x = \frac{1}{3}$.

27. You might think the graph of $f(x) = x^2 + \cos x$ should look like a parabola with some waves on it. Sketch the actual graph of $f(x)$ using a calculator or computer. Explain what you see using $f''(x)$.

28. The rabbit population on a small Pacific island is approximated by
$$P(t) = \frac{2000}{1 + e^{(5.3 - 0.4t)}}$$
with t measured in years since 1774, when Captain James Cook left 10 rabbits on the island.

(a) Using a calculator, graph P. Does the population level off?

(b) Estimate when the rabbit population grew most rapidly. How large was the population at that time?

(c) What natural causes could lead to the shape of the graph of P?

29. (a) Water is flowing at a constant rate into a cylindrical container standing vertically. Sketch a graph showing the depth of water against time.

(b) Water is flowing at a constant rate into a cone-shaped container standing on its point. Sketch a graph showing the depth of the water against time.

30. If water is flowing at a constant rate (i.e., constant volume per unit time) into the Grecian urn in Figure 4.19, sketch a graph of the depth of the water against time. Mark on the graph the time at which the water reaches the widest point of the urn.

Figure 4.19

31. If water is flowing at a constant rate (i.e., constant volume per unit time) into the vase in Figure 4.20, sketch a graph of the depth of the water against time. Mark on the graph the time at which the water reaches the corner of the vase.

Figure 4.20

32. Sketch the graph of f given that:

- $f'(x) = 0$ at $x = 2$, $f'(x) < 0$ for $x < 2$, $f'(x) > 0$ for $x > 2$,
- $f''(x) = 0$ at $x = 4$, $f''(x) > 0$ for $x < 4$, $f''(x) < 0$ for $x > 4$.

33. Sketch graphs of two continuous functions f and g, each of which has exactly five critical points, the points A–E in Figure 4.21, and which satisfy the following conditions:

(a) $\lim\limits_{x \to -\infty} f(x) = \infty$ and $\lim\limits_{x \to \infty} f(x) = \infty$

(b) $\lim\limits_{x \to -\infty} g(x) = -\infty$ and $\lim\limits_{x \to \infty} g(x) = 0$

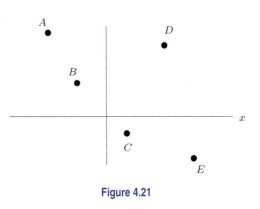

Figure 4.21

34. Assume the function f is differentiable everywhere and has just one critical point, at $x = 3$. In parts (a) through (d), you are given additional conditions. In each case decide whether $x = 3$ is a local maximum, a local minimum, or neither. Explain your reasoning. Also, sketch possible graphs for all four cases.

(a) $f'(1) = 3$ and $f'(5) = -1$

(b) $\lim\limits_{x \to \infty} f(x) = \infty$ and $\lim\limits_{x \to -\infty} f(x) = \infty$

(c) $f(1) = 1$, $f(2) = 2$, $f(4) = 4$, $f(5) = 5$

(d) $f'(2) = -1$, $f(3) = 1$, $\lim\limits_{x \to \infty} f(x) = 3$

35. Assume that the polynomial f has exactly two local maxima and one local minimum, and that these are the only critical points of f.

(a) Sketch a possible graph of f.

(b) What is the largest number of zeros f could have?

(c) What is the least number of zeros f could have?

(d) What is the least number of inflection points f could have?

(e) What is the smallest degree f could have?

(f) Find a possible formula for $f(x)$.

Problems 36–37 show graphs of the three functions f, f', f''. Identify which is which.

36.

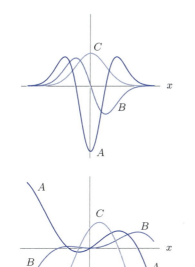

37.

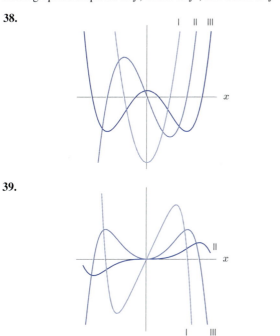

Problems 38–39 show graphs of f, f', f''. Each of these three functions is either odd or even. Decide which functions are odd and which are even. Use this information to identify which graph corresponds to f, which to f', and which to f''.

38.

39.

For Problems 40–43, sketch a possible graph of $y = f(x)$, using the given information about the derivatives $y' = f'(x)$ and $y'' = f''(x)$. Assume that the function is defined and continuous for all real x.

40.

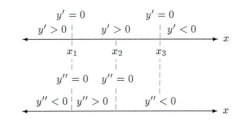

41.

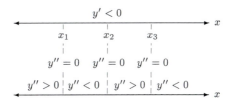

42.

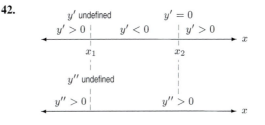

43.

44. Let f be a function with $f(x) > 0$ for all x. Set $g = 1/f$.

(a) If f is increasing in an interval around x_0, what about g?

(b) If f has a local maximum at x_1, what about g?

(c) If f is concave down at x_2, what about g?

45. What happens to concavity when functions are added?

(a) If $f(x)$ and $g(x)$ are concave up for all x, is $f(x) + g(x)$ concave up for all x?

(b) If $f(x)$ is concave up for all x and $g(x)$ is concave down for all x, what can you say about the concavity of $f(x) + g(x)$? For example, what happens if $f(x)$ and $g(x)$ are both polynomials of degree 2?

(c) If $f(x)$ is concave up for all x and $g(x)$ is concave down for all x, is it possible for $f(x) + g(x)$ to change concavity infinitely often?

4.2 FAMILIES OF CURVES

We saw in Chapter 1 that knowledge of one function can provide knowledge of the graphs of many others. The shape of the graph of $y = x^2$ also tells us, indirectly, about the graphs of $y = x^2 + 2$, $y = (x + 2)^2$, $y = 2x^2$, and countless other functions. We say that all functions of the form $y = a(x + b)^2 + c$ form a *family of functions*; their graphs are like that of $y = x^2$, except for shifts and stretches determined by the values of $a, b,$ and c. The constants a, b, c are called *parameters*. Different values of the parameters give different members of the family.

One reason for studying families of functions is their use in mathematical modeling. Confronted with the problem of modeling some phenomenon, a crucial first step involves recognizing families of functions which might fit the available data.

Motion Under Gravity: $y = -4.9t^2 + v_0t + y_0$

The position of an object moving vertically under the influence of gravity can be described by a function in the two-parameter family

$$y = -4.9t^2 + v_0t + y_0$$

where t is time in seconds and y is the distance in meters above the ground. Why do we need the parameters v_0 and y_0 to describe all such motions? Notice that at time $t = 0$ we have $y = y_0$. Thus the parameter y_0 gives the height above ground of the object at time $t = 0$. Since $dy/dt = -9.8t + v_0$, the parameter v_0 gives the velocity of the object at time $t = 0$.

Curves of the Form $y = A\sin(Bx)$

This family is used to model a wave. We saw in Section 1.5 that $|A|$ is the amplitude of the wave and that $2\pi/|B|$ is its period. Figures 4.22 and 4.23 illustrate these facts.

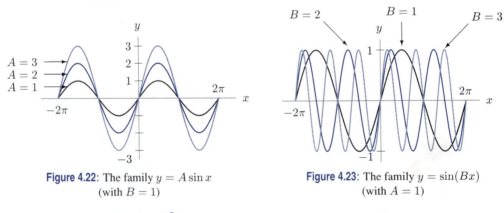

Figure 4.22: The family $y = A\sin x$
(with $B = 1$)

Figure 4.23: The family $y = \sin(Bx)$
(with $A = 1$)

Curves of the Form $y = e^{-(x-a)^2}$

The role of the parameter a in this family is to shift the graph of $y = e^{-x^2}$ to the right or left. Notice that the value of y is always positive. Since $y \to 0$ as $x \to \pm\infty$, the x-axis is a horizontal asymptote. Thus $y = e^{-(x-a)^2}$ is the family of horizontal shifts of the bell-shaped curve shown in Figure 4.24.

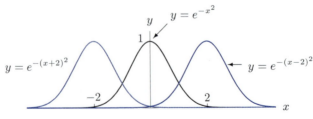

Figure 4.24: The family $y = e^{-(x-a)^2}$

Curves of the Form $y = e^{-(x-a)^2/b}$

This family is related to the *normal density* function, used in probability and statistics.[1] We assume that $b > 0$. The graph of a typical member of the family is a bell-shaped curve as in Figure 4.24. We know that the parameter a shifts the graph horizontally. We now consider the role of the parameter b by studying the family with $a = 0$:

$$y = e^{-x^2/b}.$$

To investigate the critical points and points of inflection, we calculate

$$\frac{dy}{dx} = -\frac{2x}{b}e^{-x^2/b}$$

and, using the product rule, we get

$$\frac{d^2y}{dx^2} = -\frac{2}{b}e^{-x^2/b} - \frac{2x}{b}\left(-\frac{2x}{b}e^{-x^2/b}\right) = \frac{2}{b}\left(\frac{2x^2}{b} - 1\right)e^{-x^2/b}.$$

Critical points occur where $dy/dx = 0$, that is, where

$$\frac{dy}{dx} = -\frac{2x}{b}\,e^{-x^2/b} = 0.$$

Since $e^{-x^2/b}$ is never zero, the only critical point is $x = 0$. At that point, $y = 1$ and $d^2y/dx^2 < 0$. Hence, by the second derivative test, there is a local maximum at $x = 0$.

Inflection points occur where the second derivative changes sign; thus, we start by finding values of x for which $d^2y/dx^2 = 0$. Since $e^{-x^2/b}$ is never zero, $d^2y/dx^2 = 0$ when

$$\frac{2x^2}{b} - 1 = 0.$$

Solving for x gives

$$x = \pm\sqrt{\frac{b}{2}}.$$

Looking at the expression for d^2y/dx^2, we see that d^2y/dx^2 is negative for $x = 0$, and positive as $x \to \pm\infty$. Therefore the concavity changes at $x = -\sqrt{b/2}$ and at $x = \sqrt{b/2}$, so we have inflection points there. Returning to the two-parameter family $y = e^{-(x-a)^2/b}$, we conclude that there is a maximum at $x = a$, obtained by horizontally shifting the maximum at $x = 0$ of $y = e^{-x^2/b}$ by a units. There are inflection points at $x = a \pm \sqrt{b/2}$ obtained by shifting the inflection points $x = \pm\sqrt{b/2}$ of $y = e^{-x^2/b}$ by a units. (See Figure 4.25.) At these points $y = e^{-1/2} \approx 0.6$.

With this information we can see the effect of the parameters. The parameter a determines the location of the center of the bell and the parameter b determines how narrow or wide the bell is. (See Figure 4.26.) If b is small, then the inflection points are close to a and the bell is sharply peaked near a; if b is large, the inflection points are farther away from a and the bell is spread out.

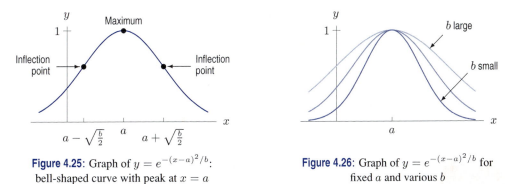

Figure 4.25: Graph of $y = e^{-(x-a)^2/b}$: bell-shaped curve with peak at $x = a$

Figure 4.26: Graph of $y = e^{-(x-a)^2/b}$ for fixed a and various b

[1] Probabilists divide our function by a constant, $\sqrt{\pi b}$, to get the normal density.

Curves of the Form $y = a(1 - e^{-bx})$

This is a two-parameter family. We consider $a, b > 0$. The graph of one member, with $a = 2$ and $b = 1$, is in Figure 4.27. Such a graph represents a quantity which is increasing but leveling off. For example, a body dropped in a thick fluid speeds up initially, but its velocity levels off as it approaches a terminal velocity. Similarly, if a pollutant pouring into a lake builds up toward a saturation level, its concentration may be described in this way. The graph might also represent the temperature of an object in an oven.

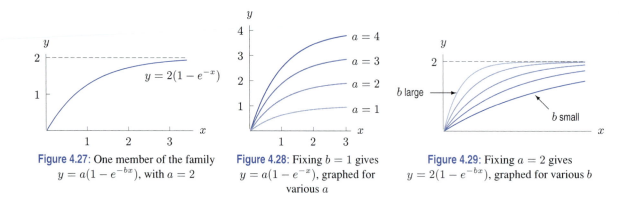

Figure 4.27: One member of the family $y = a(1 - e^{-bx})$, with $a = 2$

Figure 4.28: Fixing $b = 1$ gives $y = a(1 - e^{-x})$, graphed for various a

Figure 4.29: Fixing $a = 2$ gives $y = 2(1 - e^{-bx})$, graphed for various b

We examine the effect on the graph of varying a. Fix b at some positive number, say $b = 1$. Substitute different values for a and look at the graphs in Figure 4.28. We see that as x gets larger, y approaches a from below. The reason is $e^{-bx} \to 0$ as $x \to \infty$. Physically, the value of a represents the terminal velocity of a falling body or the saturation level of the pollutant in the lake.

Now examine the effect of varying b on the graph. Fix a at some positive number, say $a = 2$. Substitute different values for b and look at the graphs in Figure 4.29. The parameter b determines how sharply the curve rises and how quickly it gets close to the line $y = a$.

Let's confirm this last observation analytically. For $y = a(1 - e^{-bx})$, we have $dy/dx = abe^{-bx}$, so the slope of the tangent to the curve at $x = 0$ is ab. For larger b, the curve rises more rapidly at $x = 0$. How long does it take the curve to climb halfway up from $y = 0$ to $y = a$? When $y = a/2$, we have

$$a(1 - e^{-bx}) = \frac{a}{2}, \qquad \text{which leads to} \qquad x = \frac{\ln 2}{b}.$$

If b is large then $(\ln 2)/b$ is small, so in a short distance the curve is already half way up to a. If b is small, then $(\ln 2)/b$ is large and we have to go a long way out to get up to $a/2$. See Figure 4.30.

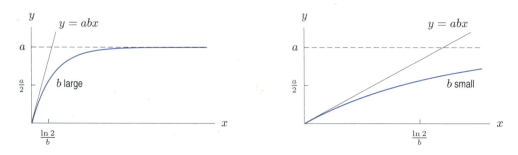

Figure 4.30: Tangent at $x = 0$ to $y = a(1 - e^{-bx})$, with fixed a, and large and small b

Exercises and Problems for Section 4.2

Exercises

Find formulas for the functions described in Exercises 1–11.

1. A parabola opening downward with its vertex at $(2, 5)$.

2. The top half of a circle with center $(-1, 2)$ and radius 3.

3. A function of the form $y = a(1 - e^{-bx})$ with $a, b > 0$ and a horizontal asymptote of $y = 5$.

4. A function of the form $y = be^{-(x-a)^2/2}$ with its maximum at the point $(0, 3)$.

5. A rational function of the form $y = ax/(x + b)$ with a vertical asymptote at $x = 2$ and a horizontal asymptote of $y = -5$.

6. A cubic polynomial having x-intercepts at 1, 5, 7.

7. A function of the form $y = A\sin(Bx) + C$ with a maximum at $(5, 2)$, a minimum at $(15, 1.5)$, and no critical points between these two points.

8. A cubic polynomial with a local maximum at $x = 1$, a local minimum at $x = 3$, a y-intercept of 5, and an x^3 term whose coefficient is 1.

9. A quartic polynomial whose graph is symmetric about the y-axis and has local maxima at $(-1, 4)$ and $(1, 4)$ and a y-intercept of 3.

10. A curve of the form $y = e^{-(x-a)^2/b}$ for $b > 0$ with a local maximum at $x = 2$ and points of inflection at $x = 1$ and $x = 3$.

11. A function of the form $y = ax^b \ln x$, where a and b are nonzero constants, which has a local maximum at the point $(e^2, 6e^{-1})$.

Problems

12. Let $p(x) = x^3 - ax$, where a is constant.

(a) If $a < 0$, show that $p(x)$ is always increasing.
(b) If $a > 0$, show that $p(x)$ has a local maximum and a local minimum.
(c) Sketch and label typical graphs for the cases when $a < 0$ and when $a > 0$.

13. Let $p(x) = x^3 - ax$, where a is constant and $a > 0$.

(a) Find the local maxima and minima of p.
(b) What effect does increasing the value of a have on the positions of the maxima and minima?
(c) On the same axes, sketch and label the graphs of p for three positive values of a.

14. What effect does increasing the value of a have on the graph of $f(x) = x^2 + 2ax$? Consider roots, maxima and minima, and both positive and negative values of a.

15. The number, N, of people who have heard a rumor spread by mass media at time, t, is given by

$$N(t) = a(1 - e^{-kt}).$$

There are 200,000 people in the population who hear the rumor eventually. If 10% of them heard it the first day, find a and k, assuming t is measured in days.

16. The temperature, T, in $°$ C, of a yam put into a $200°$C oven is given as a function of time, t, in minutes, by

$$T = a(1 - e^{-kt}) + b.$$

(a) If the yam starts at $20°$C, find a and b.
(b) If the temperature of the yam is initially increasing at $2°$C per minute, find k.

17. Consider the family of functions $y = f(x) = x - k\sqrt{x}$, with k a positive constant and $x \geq 0$. Show that the graph of $f(x)$ has a local minimum at a point whose x-coordinate is $1/4$ of the way between its x-intercepts.

18. Sketch graphs of $y = xe^{-bx}$ for $b = 1, 2, 3, 4$. Describe the graphical significance of b.

19. Find the coordinates of the critical point of $y = xe^{-bx}$ and use it to confirm your answer to Problem 18.

20. (a) Find all critical points of $f(x) = x^4 + ax^2 + b$.
(b) Under what conditions on a and b does this function have exactly one critical point? What is the one critical point, and is it a local maximum, a local minimum, or neither?
(c) Under what conditions on a and b does this function have exactly three critical points? What are they? Which are local maxima and which are local minima?
(d) Is it ever possible for this function to have two critical points? No critical points? More than three critical points? Give an explanation in each case.

21. If $a > 0$, $b > 0$, confirm analytically that $f(x) = a(1 - e^{-bx})$ is everywhere increasing and everywhere concave down.

22. A family of functions is given by

$$r(x) = \frac{1}{a + (x - b)^2}.$$

(a) For what values of a and b does the graph of r have a vertical asymptote? Where are the vertical asymptotes in this case?
(b) Find values of a and b so that the function r has a local maximum at the point $(3, 5)$.

23. For any constant a, let $f(x) = ax - x \ln x$ for $x > 0$.

    (a) What is the x-intercept of the graph of $f(x)$?

    (b) Sketch the graph of $f(x)$ for $a = -1$ and $a = 1$.

    (c) For what values of a does $f(x)$ have a critical point for $x > 0$? Find the coordinates of the critical point and decide if it is a local maximum, a local minimum, or neither.

24. Let $f(x) = x^2 + \cos(kx)$, for $k > 0$.

    (a) Using a calculator or computer, sketch the graph of f for $k = 0.5, 1, 3, 5$. Find the smallest number k at which you see points of inflection in the graph of f.

    (b) Explain why the graph of f has no points of inflection if $k \leq \sqrt{2}$, and infinitely many points of inflection if $k > \sqrt{2}$.

    (c) Explain why f has only a finite number of critical points, no matter what the value of k.

25. Let $f(x) = e^x - kx$, for $k > 0$.

    (a) Using a calculator or computer, sketch the graph of f for $k = 1/4, 1/2, 1, 2, 4$. Describe what happens as k changes.

    (b) Show that f has a local minimum at $x = \ln k$.

    (c) Find the value of k for which the local minimum is the largest.

26. Let $y = Ae^{-Bx^2}$ for positive A, B. Analyze the effect of varying A and B on the shape of the curve. Illustrate your answer with sketches.

27. Consider the surge function $y = axe^{-bx}$ for $a, b > 0$.

    (a) Find the local maxima, local minima, and points of inflection.

    (b) How does varying a and b affect the shape of the graph?

    (c) On one set of axes, graph this function for several values of a and b.

28. Sketch several members of the family $y = e^{-ax} \sin bx$ for $b = 1$, and describe the graphical significance of the parameter a.

29. Sketch several members of the family $e^{-ax} \sin bx$ for $a = 1$, and describe the graphical significance of the parameter b.

30. Consider the family

$$y = \frac{A}{x + B}.$$

    (a) If $B = 0$, what is the effect of varying A on the graph?

    (b) If $A = 1$, what is the effect of varying B?

    (c) On one set of axes, graph the function for several values of A and B.

31. For positive a, b, the potential energy, U, of a particle is given by

$$U = b\left(\frac{a^2}{x^2} - \frac{a}{x}\right) \quad \text{for } x > 0.$$

    (a) Find the intercepts and asymptotes.

    (b) Compute the local maxima and minima.

    (c) Sketch the graph.

32. The force, F, on a particle with potential energy U is given by

$$F = -\frac{dU}{dx}.$$

Using the expression for U in Problem 31, graph F and U on the same axes, labeling intercepts and local maxima and minima.

33. For positive A, B, the force between two atoms is a function of the distance, r, between them:

$$f(r) = -\frac{A}{r^2} + \frac{B}{r^3} \quad r > 0.$$

    (a) What are the zeros and asymptotes of f?

    (b) Find the coordinates of the critical points and inflection points of f.

    (c) Sketch a graph of f.

    (d) Illustrating your answers with a sketch, describe the effect on the graph of f of:

        (i) Increasing B, holding A fixed

        (ii) Increasing A, holding B fixed

34. The x-axis runs along the water surface in a straight, horizontal canal. At time t, the displacement, y, of the surface of the water at position x is given by

$$y = \sin(x - t)$$

where x and y are in meters and t is in seconds.

    (a) Graph and label y as a function of x for $t = 0, 0.5, 1, 1.5, 2$.

    (b) What does the function $y = \sin(x - t)$ represent

        (i) For fixed t?   **(ii)** For fixed x?

    (c) What does the derivative dy/dx represent for fixed t?

    (d) What does the derivative dy/dt represent for fixed x?

4.3 OPTIMIZATION

It is often important to find the largest or smallest value of some quantity. For example, automobile engineers want to construct a car that uses the least amount of fuel, scientists want to calculate which wavelength carries the maximum radiation at a given temperature, and urban planners want to design traffic patterns to minimize delays. Such problems belong to the field of mathematics called *optimization*. The next three sections show how the derivative provides an efficient way of solving many optimization problems.

Global Maxima and Minima

The single greatest (or least) value of a function f over a specified domain is called the *global maximum* (or *minimum*) of f. Recall that the local maxima and minima tell us where a function is locally largest or smallest. Now we are interested in where the function is absolutely largest or smallest in a given domain. We define

- f has a **global minimum** at p if $f(p)$ is less than or equal to all values of f.
- f has a **global maximum** at p if $f(p)$ is greater than or equal to all values of f.

Global maxima and minima are sometimes called *extrema* or *optimal values*.

How Do We Find Global Maxima and Minima?

If f is a continuous function defined on a closed interval $a \le x \le b$ (that is, an interval containing its endpoints), Figure 4.31 illustrates that the global maximum or minimum of f occurs either at a local maximum or a local minimum, or at one of the endpoints of the interval, $x = a$ or $x = b$.

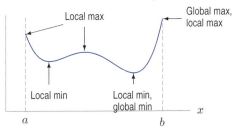

Figure 4.31: Global maximum and minimum on a closed interval $a \le x \le b$

To find the global maximum and minimum of a continuous function on a closed interval: Compare values of the function at all the critical points in the interval and at the endpoints.

If the function is defined on an open interval $a < x < b$ (that is, an interval not including its endpoints) or on all real numbers, there may or may not be a global maximum or minimum. For example, there is no global maximum in Figure 4.32 because the function has no actual largest value. The global minimum in Figure 4.32 coincides with the local minimum. There is a global minimum but no global maximum in Figure 4.33.

To find the global maximum and minimum of a continuous function on an open interval or on all real numbers: Find the value of the function at all the critical points and sketch a graph. Look at the function values when x approaches the endpoints of the interval, or approaches $\pm\infty$, as appropriate.

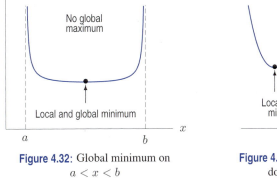

Figure 4.32: Global minimum on $a < x < b$

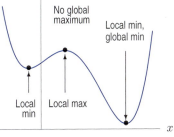

Figure 4.33: Global minimum when the domain is all real numbers

Example 1   Find the global maxima and minima of $f(x) = x^3 - 9x^2 - 48x + 52$ on the following intervals:

(a) $-5 \leq x \leq 12$ (b) $-5 \leq x \leq 14$ (c) $-5 \leq x < \infty$.

Solution   (a) We have previously obtained the critical points $x = -2$ and $x = 8$ using

$$f'(x) = 3x^2 - 18x - 48 = 3(x + 2)(x - 8).$$

We evaluate f at the critical points and the endpoints of the interval:

$$f(-5) = (-5)^3 - 9(-5)^2 - 48(-5) + 52 = -58$$
$$f(-2) = 104$$
$$f(8) = -396$$
$$f(12) = -92.$$

Comparing these function values, we see that the global maximum on $[-5, 12]$ is 104 and occurs at $x = -2$, and the global minimum on $[-5, 12]$ is -396 and occurs at $x = 8$.

(b) For the interval $[-5, 14]$, we compare

$$f(-5) = -58, \quad f(-2) = 104, \quad f(8) = -396, \quad f(14) = 360.$$

The global maximum is now 360 and occurs at $x = 14$, and the global minimum is still -396 and occurs at $x = 8$. Since the function is increasing for $x > 8$, changing the right-hand end of the interval from $x = 12$ to $x = 14$ alters the global maximum but not the global minimum. See Figure 4.34.

(c) Figure 4.34 shows that for $-5 \leq x < \infty$ there is no global maximum, because we can make $f(x)$ as large as we please by choosing x large enough. The global minimum remains -396 at $x = 8$.

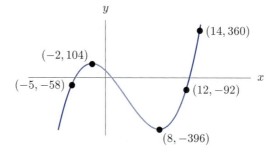

Figure 4.34: Graph of $f(x) = x^3 - 9x^2 - 48x + 52$

Example 2   When an arrow is shot into the air, its range, R, is defined as the horizontal distance from the archer to the point where the arrow hits the ground. If the ground is horizontal and we neglect air resistance, it can be shown that

$$R = \frac{v_0{}^2 \sin(2\theta)}{g},$$

where v_0 is the initial velocity of the arrow, g is the (constant) acceleration due to gravity, and θ is the angle above horizontal, so $0 \leq \theta \leq \pi/2$. (See Figure 4.35.) What initial angle maximizes R?

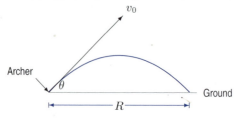

Figure 4.35: Arrow's path

Solution   We can find the maximum of this function without using calculus. The maximum value of R occurs when $\sin(2\theta) = 1$, so $\theta = \arcsin(1)/2 = \pi/4$, giving $R = v_0^2/g$.

Let's see how we can do the same problem with calculus. We want to find the global maximum of R for $0 \leq \theta \leq \pi/2$. First we look for critical points:

$$\frac{dR}{d\theta} = 2\frac{v_0^2\cos(2\theta)}{g}.$$

Setting $dR/d\theta$ equal to 0, we get

$$0 = \cos(2\theta), \quad \text{or} \quad 2\theta = \pm\frac{\pi}{2}, \pm\frac{3\pi}{2}, \pm\frac{5\pi}{2}, \ldots$$

so $\pi/4$ is the only critical point in the interval $0 \leq \theta \leq \pi/2$. The range at $\theta = \pi/4$ is $R = v_0^2/g$.

Now we must check the value of R at the endpoints $\theta = 0$ and $\theta = \pi/2$. Since $R = 0$ at each endpoint, the critical point $\theta = \pi/4$ gives both a local and a global maximum on $0 \leq \theta \leq \pi/2$. Therefore, the arrow goes farthest if shot at an angle of $\pi/4$, or $45°$.

A Graphical Example: Minimizing Gas Consumption

Next we look at an example in which a function is given graphically and the optimum values are read from a graph. You already know how to estimate the optimum values of $f(x)$ from a graph of $f(x)$—read off the highest and lowest values. In this example, we see how to estimate the optimum value of the quantity $f(x)/x$ from a graph of $f(x)$ against x.

The question we investigate is how to set driving speeds to maximize fuel efficiency.[2] We assume that gas consumption, g (in gallons/hour), as a function of velocity, v (in mph) is as shown in Figure 4.36. We want to minimize the gas consumption per *mile*, not the gas consumption per hour. Let $G = g/v$ represent the average gas consumption per mile. (The units of G are gallons/mile.)

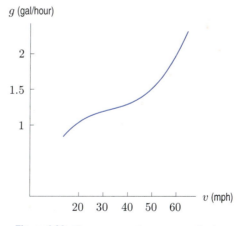

Figure 4.36: Gas consumption versus velocity

Example 3   Using Figure 4.36, estimate the velocity which minimizes G.

Solution   We want to find the minimum value of $G = g/v$ when g and v are related by the graph in Figure 4.36. We could use Figure 4.36 to sketch a graph of G against v and estimate a critical point. But there is an easier way.

[2]Adapted from Peter D. Taylor, *Calculus: The Analysis of Functions* (Toronto: Wall & Emerson, 1992).

Figure 4.37 shows that g/v is the slope of the line from the origin to the point P. Where on the curve should P be to make the slope a minimum? From the possible positions of the line shown in Figure 4.37, we see that the slope of the line is both a local and global minimum when the line is tangent to the curve. From Figure 4.38, we can see that the velocity at this point is about 50 mph. Thus to minimize gas consumption per mile, we should drive about 50 mph.

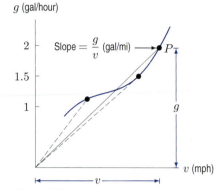

Figure 4.37: Graphical representation of gas consumption per mile, $G = g/v$

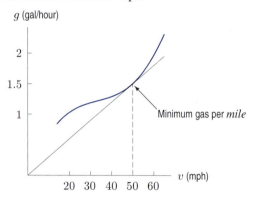

Figure 4.38: Velocity for maximum fuel efficiency

Finding Upper and Lower Bounds

A problem which is closely related to finding maxima and minima is finding the *bounds* of a function. In Example 1, the value of $f(x)$ on the interval $[-5, 12]$ ranges from -396 to 104. Thus

$$-396 \leq f(x) \leq 104,$$

and we say that -396 is a *lower bound* for f and 104 is an *upper bound* for f on $[-5, 12]$. (See Appendix A for more on bounds.) Of course, we could also say that

$$-400 \leq f(x) \leq 150,$$

so that f is also bounded below by -400 and above by 150 on $[-5, 12]$. However, we consider the -396 and 104 to be the *best possible bounds* because they describe more accurately how the function $f(x)$ behaves on $[-5, 12]$.

Example 4   An object on a spring oscillates about its equilibrium position at $y = 0$. Its distance from equilibrium is given as a function of time, t, by

$$y = e^{-t} \cos t.$$

Find the greatest distance the object goes above and below the equilibrium for $t \geq 0$.

Solution   We are looking for the bounds of the function. What does the graph of the function look like? We can think of it as a cosine curve with a decreasing amplitude of e^{-t}; in other words, it is a cosine curve squashed between the graphs of $y = e^{-t}$ and $y = -e^{-t}$, forming a wave with lower and lower crests and shallower and shallower troughs. (See Figure 4.39.)

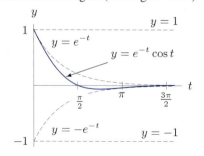

Figure 4.39: $f(t) = e^{-t} \cos t$ for $t \geq 0$

From the graph we can see that for $t \geq 0$, the graph lies between the horizontal lines $y = -1$ and $y = 1$. This means that -1 and 1 are bounds:

$$-1 \leq e^{-t} \cos t \leq 1.$$

The line $y = 1$ is the best possible upper bound because the graph does come up that high (at $t = 0$). However, we can find a better lower bound if we find the global minimum value of f for $t \geq 0$; this minimum occurs in the first trough between $t = \pi/2$ and $t = 3\pi/2$ because later troughs are squashed closer to the t-axis. At the minimum, $dy/dt = 0$. The product rule gives

$$\frac{dy}{dt} = (-e^{-t}) \cos t + e^{-t}(-\sin t) = -e^{-t}(\cos t + \sin t) = 0.$$

Since e^{-t} is never 0, we must have

$$\cos t + \sin t = 0, \quad \text{so} \quad \frac{\sin t}{\cos t} = -1.$$

Hence

$$\tan t = -1, \quad \text{giving} \quad t = \frac{3\pi}{4}.$$

Thus, the global minimum we see on the graph occurs at $t = 3\pi/4$. The value of y at that minimum is

$$y = e^{-3\pi/4} \cos\left(\frac{3\pi}{4}\right) \approx -0.067.$$

Rounding down so that the inequalities still hold for all $t \geq 0$ gives

$$-0.07 < e^{-t} \cos t \leq 1.$$

Notice how much smaller in magnitude the lower bound is than the upper. This is a reflection of how quickly the factor e^{-t} causes the oscillation to die out.

Exercises and Problems for Section 4.3

Exercises

For Exercises 1–2, indicate all critical points on the given graphs. Determine which correspond to local minima, local maxima, global minima, global maxima, or none of these. (Note that the graphs are on closed intervals.)

1.

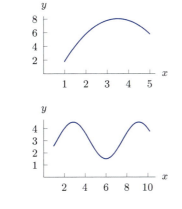

2.

3. For $y = f(x) = x^{10} - 10x$, and $0 \leq x \leq 2$, find the value(s) of x for which:

    (a) $f(x)$ has a local maximum or local minimum. Indicate which ones are maxima and which are minima.

    (b) $f(x)$ has a global maximum or global minimum.

4. For $f(x) = x - \ln x$, and $0.1 \leq x \leq 2$, find the value(s) of x for which:

    (a) $f(x)$ has a local maximum or local minimum. Indicate which ones are maxima and which are minima.

    (b) $f(x)$ has a global maximum or global minimum.

5. For $f(x) = \sin^2 x - \cos x$, and $0 \leq x \leq \pi$, find, to two decimal places, the value(s) of x for which:

    (a) $f(x)$ has a local maximum or local minimum. Indicate which ones are maxima and which are minima.

    (b) $f(x)$ has a global maximum or global minimum.

Problems

6. A function $y = h(x)$ has a second derivative shown in Figure 4.40 for $-2 \leq x \leq 1$. If $h'(-1) = 0$ and $h(-1) = 2$,

   (a) Explain why $h'(x)$ is never negative on this interval.
   (b) Explain why $h(x)$ must have a global maximum at $x = 1$.
   (c) Sketch a possible graph of $h(x)$ for $-2 \leq x \leq 1$.

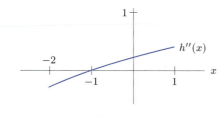

Figure 4.40

7. A grapefruit is tossed straight up with an initial velocity of 50 ft/sec. The grapefruit is 5 feet above the ground when it is released. Its height at time t is given by

$$y = -16t^2 + 50t + 5.$$

How high does it go before returning to the ground?

8. For some positive constant C, a patient's temperature change, T, due to a dose, D, of a drug is given by

$$T = \left(\frac{C}{2} - \frac{D}{3}\right)D^2.$$

   (a) What dosage maximizes the temperature change?
   (b) The sensitivity of the body to the drug is defined as dT/dD. What dosage maximizes sensitivity?

9. When you cough, your windpipe contracts. The speed, v, with which air comes out depends on the radius, r, of your windpipe. If R is the normal (rest) radius of your windpipe, then for $r \leq R$, the speed is given by:

$$v = a(R - r)r^2 \quad \text{where } a \text{ is a positive constant.}$$

What value of r maximizes the speed?

10. A chemical reaction converts a substance A to a substance Y; the presence of Y catalyzes the reaction. At the start of the reaction, the quantity of A present is a grams. At time t seconds later, the quantity of Y present is y grams. The rate of the reaction, in grams/sec, is given by

$$\text{Rate} = ky(a - y), \quad k \text{ is a positive constant.}$$

   (a) For what values of y is the rate nonnegative? Sketch a graph of the rate against y.
   (b) For what values of y is the rate a maximum?

11. In a chemical reaction, substance A combines with substance B to form substance Y. At the start of the reaction, the quantity of A present is a grams, and the quantity of B present is b grams. Assume $a < b$. At time t seconds after the start of the reaction, the quantity of Y present is y grams. For certain types of reactions, the rate of the reaction, in grams/sec, is given by

$$\text{Rate} = k(a - y)(b - y), \quad k \text{ is a positive constant.}$$

   (a) For what values of y is the rate nonnegative? Sketch a graph of the rate against y.
   (b) Use your graph to find the value of y at which the rate of the reaction is fastest.

12. A warehouse selling cement has to decide how often and in what quantities to reorder. It is cheaper, on average, to place large orders, because this reduces the ordering cost per unit. On the other hand, larger orders mean higher storage costs. The warehouse always reorders cement in the same quantity, q. The total weekly cost, C, of ordering and storage is given by

$$C = \frac{a}{q} + bq, \quad \text{where } a, b \text{ are positive constants.}$$

   (a) Which of the terms, a/q and bq, represents the ordering cost and which represents the storage cost?
   (b) What value of q gives the minimum total cost?

13. For positive constants A and B, the force between two atoms in a molecule is given by

$$f(r) = -\frac{A}{r^2} + \frac{B}{r^3},$$

where $r > 0$ is the distance between the molecules. What value of r minimizes the force between the atoms?

14. The potential energy, U, of a particle moving along the x-axis is given by

$$U = b\left(\frac{a^2}{x^2} - \frac{a}{x}\right),$$

where a and b are positive constants and $x > 0$. What value of x minimizes the potential energy?

15. The bending moment M of a beam, supported at one end, at a distance x from the support is given by

$$M = \tfrac{1}{2}wLx - \tfrac{1}{2}wx^2,$$

where L is the length of the beam, and w is the uniform load per unit length. Find the point on the beam where the moment is greatest.

16. The efficiency of a screw, E, is given by

$$E = \frac{(\theta - \mu\theta^2)}{\mu + \theta}, \quad \theta > 0,$$

where θ is the angle of pitch of the thread and μ is the coefficient of friction of the material, a (positive) constant. What value of θ maximizes E?

17. A woman pulls a sled which, together with its load, has a mass of m kg. If her arm makes an angle of θ with her body (assumed vertical) and the coefficient of friction (a positive constant) is μ, the least force, F, she must exert to move the sled is given by

$$F = \frac{mg\mu}{\sin\theta + \mu\cos\theta}.$$

If $\mu = 0.15$, find the maximum and minimum values of F for $0 \leq \theta \leq \pi/2$. Give answers as multiples of mg.

18. A circular ring of wire of radius r_0 lies in a plane perpendicular to the x-axis and is centered at the origin. The ring has a positive electric charge spread uniformly over it. The electric field in the x-direction, E, at the point x on the axis is given by

$$E = \frac{kx}{(x^2 + r_0^2)^{3/2}} \quad \text{for} \quad k > 0.$$

At what point on the x-axis is the field greatest? Least?

19. An electric current, I, in amps, is given by

$$I = \cos(wt) + \sqrt{3}\sin(wt),$$

where $w \neq 0$ is a constant. What are the maximum and minimum values of I?

20. When an electric current passes through two resistors with resistance r_1 and r_2, connected in parallel, the combined resistance, R, can be calculated from the equation

$$\frac{1}{R} = \frac{1}{r_1} + \frac{1}{r_2},$$

where R, r_1, and r_2 are positive. Assume that r_2 is constant.

(a) Show that R is an increasing function of r_1.
(b) Where on the interval $a \leq r_1 \leq b$ does R take its maximum value?

Find the best possible bounds for each of the functions in Problems 21–26.

21. e^{-x^2}, for $|x| \leq 0.3$

22. $\ln(1 + x)$, for $x \geq 0$

23. $\ln(1 + x^2)$, for $-1 \leq x \leq 2$

24. $x^3 - 4x^2 + 4x$, for $0 \leq x \leq 4$

25. $x + \sin x$, for $0 \leq x \leq 2\pi$

26. $x^3 e^{-x}$ for $x \geq 0$

27. The distance, s, traveled by a cyclist, who starts at 1 pm, is given in Figure 4.41. Time, t, is in hours since noon.

(a) Explain why the quantity, s/t, is represented by the slope of a line from the origin to the point (t, s) on the graph.

(b) Estimate the time at which the quantity s/t is a maximum.

(c) What is the relationship between the quantity s/t and the instantaneous speed of the cyclist at the time you found in part (b)?

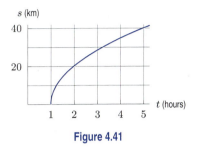

Figure 4.41

28. When birds lay eggs, they do so in clutches of several at a time. When the eggs hatch, each clutch gives rise to a brood of baby birds. We want to determine the clutch size which maximizes the number of birds surviving to adulthood per brood. If the clutch is small, there are few baby birds in the brood; if the clutch is large, there are so many baby birds to feed that most die of starvation. The number of surviving birds per brood as a function of clutch size is shown by the benefit curve in Figure 4.42.[3]

(a) Estimate the clutch size which maximizes the number of survivors per brood.

(b) Suppose also that there is a biological cost to having a larger clutch: the female survival rate is reduced by large clutches. This cost is represented by the dotted line in Figure 4.42. If we take cost into account by assuming that the optimal clutch size in fact maximizes the vertical distance between the curves, what is the new optimal clutch size?

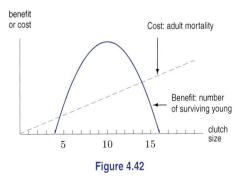

Figure 4.42

[3]Data from C. M. Perrins and D. Lack reported by J. R. Krebs and N. B. Davies in *An Introduction to Behavioural Ecology* (Oxford: Blackwell, 1987).

29. Let $f(v)$ be the amount of energy consumed by a flying bird, measured in joules per second (a joule is a unit of energy), as a function of its speed v (in meters/sec). Let $a(v)$ be the amount of energy consumed by the same bird, measured in joules *per meter*.

(a) Suggest a reason (in terms of the way birds fly) for the shape of the graph of $f(v)$ in Figure 4.43.
(b) What is the relationship between $f(v)$ and $a(v)$?
(c) Where is $a(v)$ a minimum?
(d) Should the bird try to minimize $f(v)$ or $a(v)$ when it is flying? Why?

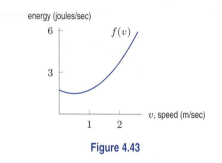

Figure 4.43

30. The forward motion of an aircraft in level flight is reduced by two kinds of forces, known as *induced drag* and *parasite drag*. Induced drag is a consequence of the downward deflection of air as the wings produce lift. Parasite drag results from friction between the air and the entire surface of the aircraft. Induced drag is inversely proportional to the square of speed and parasite drag is directly proportional to the square of speed. The sum of induced drag and parasite drag is called total drag. The graph in Figure 4.44 shows a certain aircraft's induced drag and parasite drag functions.

(a) Sketch of the total drag as a function of airspeed.
(b) Estimate two different airspeeds which each result in a total drag of 1000 pounds. Does the total drag function have an inverse? What about the induced and parasite drag functions?
(c) Fuel consumption (in gallons per hour) is roughly proportional to total drag. Suppose you are low on fuel and the control tower has instructed you to enter a circular holding pattern of indefinite duration to await the passage of a storm at your landing field. At what airspeed should you fly the holding pattern? Why?

drag (thousands of lbs)

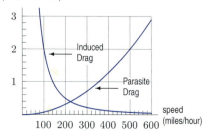

Figure 4.44

31. Let $f(v)$ be the fuel consumption, in gallons per hour, of a certain aircraft as a function of its airspeed, v, in miles per hour. A graph of $f(v)$ is given in Figure 4.45.

(a) Let $g(v)$ be the fuel consumption of the same aircraft, but measured in gallons per mile instead of gallons per hour. What is the relationship between $f(v)$ and $g(v)$?
(b) For what value of v is $f(v)$ minimized?
(c) For what value of v is $g(v)$ minimized?
(d) Should a pilot try to minimize $f(v)$ or $g(v)$?

fuel consumption
(gallons/hour)

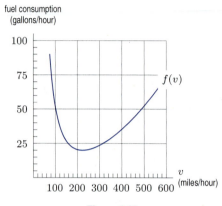

Figure 4.45

32. The function $y = t(x)$ is positive and continuous with a global maximum at the point $(3, 3)$. Sketch a possible graph of $t(x)$ if $t'(x)$ and $t''(x)$ have the same sign for $x < 3$, but opposite signs for $x > 3$.

33. A function $y = g(x)$ has a derivative that is given in Figure 4.46 for $-2 \le x \le 2$.

(a) Write a few sentences describing the behavior of $g(x)$ on this interval.
(b) Does the graph of $g(x)$ have any inflection points? If so, give the approximate x-coordinates of their locations. Explain your reasoning.
(c) What are the global maxima and minima of g on $[-2, 2]$?
(d) If $g(-2) = 5$, what do you know about $g(0)$ and $g(2)$? Explain.

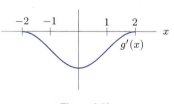

Figure 4.46

34. (a) Show that $x > 2 \ln x$ for all $x > 0$. [Hint: Find the minimum of $f(x) = x - 2 \ln x$.]
(b) Use the above result to show that $e^x > x^2$ for all positive x.
(c) Is $x > 3 \ln x$ for all positive x?

4.4 APPLICATIONS TO MARGINALITY

Management decisions within a particular business usually aim at maximizing profit for the company. In this section we will see how the derivative can be used to maximize profit. Profit depends on both production cost and revenue (or income) from sales. We begin by looking at the cost and revenue functions.

> The **cost function**, $C(q)$, gives the total cost of producing a quantity q of some good.

What sort of function do we expect C to be? The more goods that are made, the higher the total cost, so C is an increasing function. In fact, cost functions usually have the general shape shown in Figure 4.47. The intercept on the C-axis represents the *fixed costs*, which are incurred even if nothing is produced. (This includes, for instance, the machinery needed to begin production.) The cost function increases quickly at first and then more slowly because producing larger quantities of a good is usually more efficient than producing smaller quantities—this is called *economy of scale*. At still higher production levels, the cost function starts to increase faster again as resources become scarce, and sharp increases may occur when new factories have to be built. Thus, the graph of $C(q)$ may start out concave down and become concave up later on.

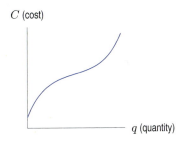

Figure 4.47: Cost as a function of quantity

> The **revenue function**, $R(q)$, gives the total revenue received by a firm from selling a quantity q of some good.

Revenue is income obtained from sales. If the price per item is p, and the quantity sold is q, then

$$\text{Revenue} = \text{Price} \times \text{Quantity}, \quad \text{so} \quad R = pq.$$

If the price per item does not depend on the quantity sold, then the graph of $R(q)$ is a straight line through the origin with slope equal to the price p. See Figure 4.48. In practice, for large values of q, the market may become glutted, causing the price to drop, giving $R(q)$ the shape in Figure 4.49.

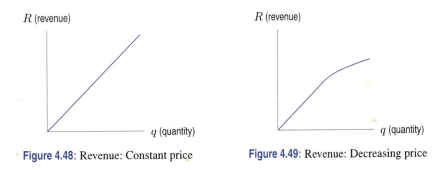

Figure 4.48: Revenue: Constant price   **Figure 4.49**: Revenue: Decreasing price

The profit is usually written as π. (Economists use π to distinguish it from the price, p; this π has nothing to do with the area of a circle, and merely stands for the Greek equivalent of the letter "p.") The profit resulting from producing and selling q items is defined by

$$\text{Profit} = \text{Revenue} - \text{Cost}, \quad \text{so} \quad \pi(q) = R(q) - C(q).$$

Example 1 If cost, C, and revenue, R, are given by the graph in Figure 4.50, for what production quantities, q, does the firm make a profit? Approximately what production level maximizes profit?

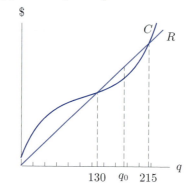

Figure 4.50: Costs and revenues for Example 1

Solution The firm makes a profit whenever revenues are greater than costs, that is, when $R > C$. The graph of R is above the graph of C approximately when $130 < q < 215$. Production between $q = 130$ units and $q = 215$ units generates a profit. The vertical distance between the cost and revenue curves is largest at q_0, so q_0 units gives maximum profit.

Marginal Analysis

Many economic decisions are based on an analysis of the costs and revenues "at the margin." Let's look at this idea through an example.

Suppose we are running an airline and we are trying to decide whether to offer an additional flight. How should we decide? We'll assume that the decision is to be made purely on financial grounds: if the flight will make money for the company, it should be added. Obviously we need to consider the costs and revenues involved. Since the choice is between adding this flight and leaving things the way they are, the crucial question is whether the *additional costs* incurred are greater or smaller than the *additional revenues* generated by the flight. These additional costs and revenues are called the *marginal costs* and *marginal revenues*.

Suppose $C(q)$ is the function giving the total cost of running q flights. If the airline had originally planned to run 100 flights, its costs would be $C(100)$. With the additional flight, its costs would be $C(101)$. Therefore,

$$\text{Marginal cost} = C(101) - C(100).$$

Now

$$C(101) - C(100) = \frac{C(101) - C(100)}{101 - 100},$$

and this quantity is the average rate of change of cost between 100 and 101 flights. In Figure 4.51 the average rate of change is the slope of the line joining the $C(100)$ and $C(101)$ points on the graph. If the graph of the cost function is not curving fast near the point, the slope of this line is close to the slope of the tangent line there. Therefore, the average rate of change is close to the instantaneous rate of change. Since these rates of change are not very different, many economists choose to define marginal cost, MC, as the instantaneous rate of change of cost with respect to quantity:

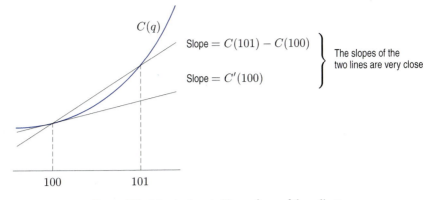

Figure 4.51: Marginal cost: Slope of one of these lines

$$\text{Marginal cost} = MC = C'(q).$$

Similarly if the revenue generated by q flights is $R(q)$, the additional revenue generated by increasing the number of flights from 100 to 101 is

$$\text{Marginal revenue} = R(101) - R(100).$$

Now $R(101) - R(100)$ is the average rate of change of revenue between 100 and 101 flights. As before, the average rate of change is usually almost equal to the instantaneous rate of change, so economists often define:

$$\text{Marginal revenue} = MR = R'(q).$$

We often refer to total cost and total revenue to distinguish them from marginal cost and marginal revenue. If the words cost and revenue are used alone, they are understood to mean total cost and total revenue.

Example 2   If $C(q)$ and $R(q)$ for the airline are given in Figure 4.52, should the company add the 101st flight?

Solution   The marginal revenue is the slope of the revenue curve, and the marginal cost is the slope of the cost curve at the point 100. From Figure 4.52, you can see that the slope at the point A is smaller than the slope at B, so $MC < MR$. This means that the airline will make more in extra revenue than it will spend in extra costs if it runs another flight, so it should go ahead and run the 101st flight.

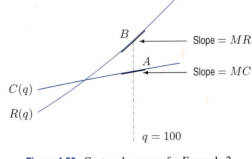

Figure 4.52: Cost and revenue for Example 2

Since MC and MR are derivative functions, they can be estimated from the graphs of total cost and total revenue.

Example 3   If R and C are given by the graphs in Figure 4.53, sketch graphs of $MR = R'(q)$ and $MC = C'(q)$.

Figure 4.53: Total revenue and total cost for Example 3

Solution   The revenue graph is a line through the origin, with equation

$$R = pq$$

where p is the price, which is a constant. The slope is p and

$$MR = R'(q) = p.$$

The total cost is increasing, so the marginal cost is always positive (above the q-axis). For small q values, the total cost curve is concave down, so the marginal cost is decreasing. For larger q, say $q > 100$, the total cost curve is concave up and the marginal cost is increasing. Thus the marginal cost has a minimum at about $q = 100$. (See Figure 4.54.)

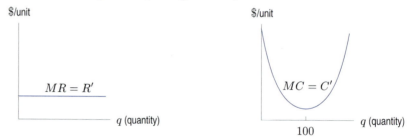

Figure 4.54: Marginal revenue and costs for Example 3

Maximizing Profit

Now let's look at how to maximize total profit, given functions for total revenue and total cost.

Example 4   Find the maximum profit if the total revenue and total cost are given, for $0 \leq q \leq 200$, by the curves R and C in Figure 4.55.

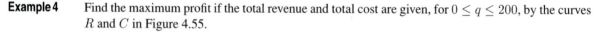

Figure 4.55: Maximum profit at $q = 140$

Solution   The profit is represented by the vertical difference between the curves and is marked by the vertical arrows on the graph. When revenue is below cost, the company is taking a loss; when revenue is above cost, the company is making a profit. We can see that the profit is largest at about $q = 140$, so this is the production level we're looking for. To be sure that the local maximum is a global maximum, we need to check the endpoints. At $q = 0$ and $q = 200$, the profit is negative, so the global maximum is indeed at $q = 140$.

To find the actual maximum profit, we estimate the vertical distance between the curves at $q = 140$. This gives a maximum profit of $\$80{,}000 - \$60{,}000 = \$20{,}000$.

Suppose we wanted to find the minimum profit. In this example, we must look at the endpoints, when $q = 0$ or $q = 200$. We see the minimum profit is negative (a loss), and it occurs at $q = 0$.

Maximum Profit Occurs Where $MR = MC$

In Example 4, observe that at $q = 140$ the slopes of the two curves in Figure 4.55 are equal. To the left of $q = 140$, the revenue curve has a larger slope than the cost curve, and the profit increases as q increases. The company will make more money by producing more units, so production should increase toward $q = 140$. To the right of $q = 140$, the slope of the revenue curve is less than the slope of the cost curve, and the profit is decreasing. The company will make more money by producing fewer units so production should decrease toward $q = 140$. At the point where the slopes are equal, the profit has a local maximum; otherwise the profit could be increased by moving toward that point. Since the slopes are equal at $q = 140$, we have $MR = MC$ there.

Now let's look at the general situation. To maximize or minimize profit over an interval, we optimize the profit, π, where

$$\pi(q) = R(q) - C(q).$$

We know that global maxima and minima can only occur at critical points or at endpoints of an interval. To find critical points of π, look for zeros of the derivative:

$$\pi'(q) = R'(q) - C'(q) = 0.$$

So

$$R'(q) = C'(q),$$

that is, the slopes of the revenue and cost curves are equal. This is the same observation that we made in the previous example. In economic language,

> The maximum (or minimum) profit can occur where
>
> Marginal cost = Marginal revenue.

Of course, maximal or minimal profit does not *have* to occur where $MR = MC$; there are also the endpoints to consider.

Example 5   Find the quantity q which maximizes profit if the total revenue, $R(q)$, and total cost, $C(q)$, are given in dollars by

$$R(q) = 5q - 0.003q^2$$
$$C(q) = 300 + 1.1q,$$

where $0 \leq q \leq 800$ units. What production level gives the minimum profit?

Solution

We look for production levels that give marginal revenue = marginal cost:

$$MR = R'(q) = 5 - 0.006q$$
$$MC = C'(q) = 1.1.$$

So $5 - 0.006q = 1.1$, giving

$$q = 3.9/0.006 = 650 \text{ units}.$$

Does this value of q represent a local maximum or minimum of π? We can tell by looking at production levels of 649 units and 651 units. When $q = 649$ we have $MR = \$1.106$, which is greater than the (constant) marginal cost of $\$1.1$. This means that producing one more unit will bring in more revenue than its cost, so profit will increase. When $q = 651$, $MR = \$1.094$, which is *less* than MC, so it is not profitable to produce the 651st unit. We conclude that $q = 650$ is a local maximum for the profit function π. The profit earned by producing and selling this quantity is $\pi(650) = R(650) - C(650) = \$967.50$.

To check for global maxima we need to look at the endpoints. If $q = 0$, the only cost is $\$300$ (the fixed costs) and there is no revenue, so $\pi(0) = -\$300$. At the upper limit of $q = 800$, we have $\pi(800) = \$900$. Therefore, the maximum profit is at the production level of 650 units, where $MR = MC$. The minimum profit (a loss) occurs when $q = 0$ and there is no production at all.

Exercises and Problems for Section 4.4

Exercises

1. Total cost and revenue are approximated by the functions $C = 5000 + 2.4q$ and $R = 4q$, both in dollars. Identify the fixed cost, marginal cost per item, and the price at which this commodity is sold.

2. (a) Fixed costs are $\$3$ million; variable costs are $\$0.4$ million per item. Write a formula for total cost as a function of quantity, q.
   (b) The item in part (a) is sold for $\$0.5$ million each. Write a formula for revenue as a function of q.
   (c) Write a formula for the profit function for this item.

3. The revenue from selling q items is $R(q) = 500q - q^2$, and the total cost is $C(q) = 150 + 10q$. Write a function that gives the total profit earned, and find the quantity which maximizes the profit.

4. If $C'(500) = 75$ and $R'(500) = 100$, should the quantity produced be increased or decreased from $q = 500$ in order to increase profits?

5. Figure 4.56 gives cost and revenue. What are fixed costs? What quantity maximizes profit, and what is the maximum profit earned?

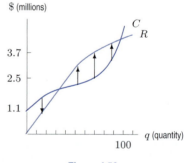

Figure 4.56

Problems

6. Using the cost and revenue graphs in Figure 4.57, sketch the following functions. Label the points q_1 and q_2.

   (a) Total profit   (b) Marginal cost
   (c) Marginal revenue

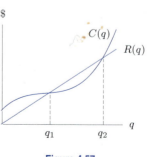

Figure 4.57

7. A manufacturer's cost of producing a product is given in Figure 4.57. The manufacturer can sell the product for a price p each (regardless of the quantity sold), so that the total revenue from selling a quantity q is $R(q) = pq$.

   (a) The difference $\pi(q) = R(q) - C(q)$ is the total profit. For which quantity q_0 is the profit a maximum? Mark your answer on a sketch of the graph.
   (b) What is the relationship between p and $C'(q_0)$? Explain your result both graphically and analytically. What does this mean in terms of economics? (Note that p is the slope of the line $R(q) = pq$. Note also that $\pi(q)$ has a maximum at $q = q_0$, so $\pi'(q_0) = 0$.)
   (c) Graph $C'(q)$ and p (as a horizontal line) on the same axes. Mark q_0 on the q-axis.

8. Let $C(q)$ be the total cost of producing a quantity q of a certain product. See Figure 4.58.

   (a) What is the meaning of $C(0)$?
   (b) Describe in words how the marginal cost changes as the quantity produced increases.
   (c) Explain the concavity of the graph (in terms of economics).
   (d) Explain the economic significance (in terms of marginal cost) of the point at which the concavity changes.
   (e) Do you expect the graph of $C(q)$ to look like this for all types of products?

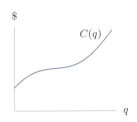

Figure 4.58

9. The marginal revenue and marginal cost for a certain item are graphed in Figure 4.59. Do the following quantities maximize profit for the company? Explain your answer.

   (a) $q = a$ (b) $q = b$

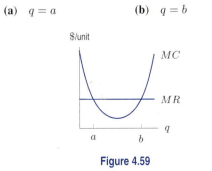

Figure 4.59

10. The total cost $C(q)$ of producing q goods is given by:

$$C(q) = 0.01q^3 - 0.6q^2 + 13q.$$

   (a) What is the fixed cost?
   (b) What is the maximum profit if each item is sold for $7? (Assume you sell everything you produce.)
   (c) Suppose exactly 34 goods are produced. They all sell when the price is $7 each, but for each $1 increase in price, 2 fewer goods are sold. Should the price be raised, and if so by how much?

11. (a) A cruise line offers a trip for $1000 per passenger. If at least 100 passengers sign up, the price is reduced for *all* the passengers by $5 for every additional passenger (beyond 100) who goes on the trip. The boat can accommodate up to 250 passengers. What number of passengers maximizes the cruise line's total revenue? What price does each passenger pay then?
    (b) The cost to the cruise line for q passengers is $40{,}000 + 200q$. What is the maximum profit that the cruise line can make on one trip? How many passengers must sign up for the maximum to be reached and what price will each pay?

12. A company manufactures only one product. The quantity, q, of this product produced per month depends on the amount of capital, K, invested (i.e., the number of machines the company owns, the size of its building, and so on) and the amount of labor, L, available each month. We assume that q can be expressed as a *Cobb-Douglas production function*:

$$q = cK^\alpha L^\beta$$

where c, α, β are positive constants, with $0 < \alpha < 1$ and $0 < \beta < 1$. In this problem we will see how the Russian government could use a Cobb-Douglas function to estimate how many people a newly privatized industry might employ. A company in such an industry has only a small amount of capital available to it and needs to use all of it, so K is fixed. Suppose L is measured in man-hours per month, and that each man-hour costs the company w rubles (a ruble is the unit of Russian currency). Suppose the company has no other costs besides labor, and that each unit of the good can be sold for a fixed price of p rubles. How many man-hours of labor per month should the company use in order to maximize its profit?

13. An agricultural worker in Uganda is planting clover to increase the number of bees making their home in the region. There are 100 bees in the region naturally, and for every acre put under clover, 20 more bees are found in the region.

    (a) Draw a graph of the total number, $N(x)$, of bees as a function of x, the number of acres devoted to clover.
    (b) Explain, both geometrically and algebraically, the shape of the graph of:
       (i) The marginal rate of increase of the number of bees with acres of clover, $N'(x)$.
       (ii) The average number of bees per acre of clover, $N(x)/x$.

14. If you invest x dollars in a certain project, your return is $R(x)$. You want to choose x to maximize your return per dollar invested,[4] which is

$$r(x) = \frac{R(x)}{x}.$$

(a) The graph of $R(x)$ is in Figure 4.60, with $R(0) = 0$. Illustrate on the graph that the maximum value of $r(x)$ is reached at a point at which the line from the origin to the point is tangent to the graph of $R(x)$.

(b) Also, the maximum of $r(x)$ occurs at a point at which the slope of the graph of $r(x)$ is zero. On the same axes as part (a), sketch $r(x)$. Illustrate that the maximum of $r(x)$ occurs where its slope is 0.

(c) Show, by taking the derivative of the formula for $r(x)$, that the conditions in part (a) and (b) are equivalent: the x-value at which the line from the origin is tangent to the graph of R is the same as the x-value at which the graph of r has zero slope.

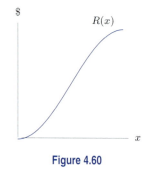

Figure 4.60

Problems 15–17 involve the *average cost* of manufacturing a quantity q of a good, which is defined to be

$$a(q) = \frac{C(q)}{q}.$$

15. Figure 4.61 shows the cost of production, $C(q)$, as a function of quantity produced, q.

(a) For some q_0, sketch a line whose slope is the marginal cost, MC, at that point.

(b) For the same q_0, explain why the average cost $a(q_0)$ can be represented by the slope of the line from that point on the curve to the origin.

(c) Use the method of Example 3 on page 183 to explain why at the value of q which minimizes $a(q)$, the average and marginal costs are equal.

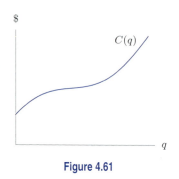

Figure 4.61

16. The average cost per item to produce q items is given by

$$a(q) = 0.01q^2 - 0.6q + 13, \quad \text{for} \quad q > 0.$$

(a) What is the total cost, $C(q)$, of producing q goods?

(b) What is the minimum marginal cost? What is the practical interpretation of this result?

(c) At what production level is the average cost a minimum? What is the lowest average cost?

(d) Compute the marginal cost at $q = 30$. How does this relate to your answer to part (c)? Explain this relationship both analytically and in words.

17. A reasonably realistic model of a firm's costs is given by the *short-run Cobb-Douglas cost curve*

$$C(q) = Kq^{1/a} + F,$$

where a is a positive constant, F is the fixed cost, and K measures the technology available to the firm.

(a) Show that C is concave down if $a > 1$.

(b) Assuming that $a < 1$ and that average cost is minimized when average cost equals marginal cost, find what value of q minimizes the average cost.

4.5   OPTIMIZATION AND MODELING

Finding global maxima and minima is made much easier by having a formula for the function to be maximized or minimized. The process of translating a problem into a function whose formula we know is called *mathematical modeling*. By working through the examples that follow, you will get the flavor of some kinds of modeling.

Example 1   What are the dimensions of an aluminum can that holds 40 in³ of juice and that uses the least material (i.e., aluminum)? Assume that the can is cylindrical, and is capped on both ends.

---
[4]From Peter D. Taylor, *Calculus: The Analysis of Functions* (Toronto: Wall & Emerson, 1992).

Solution   It is often a good idea to think about a problem in general terms before trying to solve it. Since we're trying to use as little material as possible, why not make the can very small, say, the size of a peanut? We can't, since the can must hold 40 in^3. If we make the can short, to try to use less material in the sides, we'll have to make it fat as well, so that it can hold 40 in^3. Saving aluminum in the sides might actually use more aluminum in the top and bottom in a short, fat can. See Figure 4.62(a).

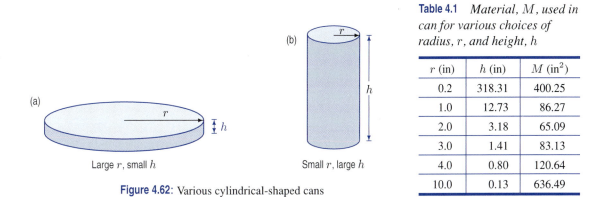

Figure 4.62: Various cylindrical-shaped cans

(a) Large r, small h

(b) Small r, large h

Table 4.1 *Material, M, used in can for various choices of radius, r, and height, h*

r (in)	h (in)	M (in^2)
0.2	318.31	400.25
1.0	12.73	86.27
2.0	3.18	65.09
3.0	1.41	83.13
4.0	0.80	120.64
10.0	0.13	636.49

If we try to save material by making the top and bottom small, the can has to be tall to accommodate the 40 in^3 of juice. So any savings we get by using a small top and bottom might be outweighed by the height of the sides. See Figure 4.62(b). This is, in fact, true, as Table 4.1 shows.

The table gives the amount of material used in the can for some choices of the radius, r, and the height, h. You can see that r and h change in opposite directions, and that more material is used at the extremes (very large or very small r and h) than in the middle. From the table it appears that the optimal radius for the can lies somewhere in $1.0 \leq r \leq 3.0$. If we consider the material used, M, as a function of the radius, r, a graph of this function looks like Figure 4.63. The graph shows that the global minimum we want is at a critical point.

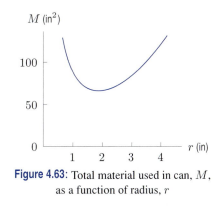

Figure 4.63: Total material used in can, M, as a function of radius, r

Both the table and the graph were obtained from a mathematical model, which in this case is a formula for the material used in making the can. Finding such a formula depends on knowing the geometry of a cylinder, in particular its area and volume. We have

$$M = \text{Material used in the can} = \text{Material in ends} + \text{Material in the side}$$

where

$$\text{Material in ends} = 2 \cdot \text{Area of a circle with radius } r = 2 \cdot \pi r^2,$$
$$\text{Material in the side} = \text{Surface area of cylinder with height } h \text{ and radius } r = 2\pi rh.$$

However, h is not independent of r: if r grows, h shrinks, and vice-versa. To find the relationship, we use the fact that the volume of the cylinder, $\pi r^2 h$, is equal to the constant 40 in^3:

$$\text{Volume of can } = \pi r^2 h = 40, \quad \text{giving} \quad h = \frac{40}{\pi r^2}.$$

This means

$$\text{Material in the side} = 2\pi r h = 2\pi r \frac{40}{\pi r^2} = \frac{80}{r}.$$

Thus we obtain the formula for the total material, M, used in a can of radius r:

$$M = 2\pi r^2 + \frac{80}{r}.$$

The domain of this function is all $r > 0$.

Now we use calculus to find the minimum of M. We look for critical points:

$$\frac{dM}{dr} = 4\pi r - \frac{80}{r^2} = 0 \quad \text{at a critical point,} \quad \text{so} \quad 4\pi r = \frac{80}{r^2}.$$

Therefore,

$$\pi r^3 = 20, \quad \text{so} \quad r = \left(\frac{20}{\pi}\right)^{1/3} \approx 1.85 \text{ inches,}$$

which agrees with our graph. We also have

$$h = \frac{40}{\pi r^2} \approx \frac{40}{\pi (1.85)^2} \approx 3.7 \text{ inches.}$$

Thus, the material used, $M = 2\pi (1.85)^2 + 80/1.85 = 64.7 \text{ in}^2$.

Practical Tips for Modeling Optimization Problems

1. Make sure that you know what quantity or function is to be optimized.

2. If possible, make several sketches showing how the elements that vary are related. Label your sketches clearly by assigning variables to quantities which change.

3. Try to obtain a formula for the function to be optimized in terms of the variables that you identified in the previous step. If necessary, eliminate from this formula all but one variable. Identify the domain over which this variable varies.

4. Find the critical points and evaluate the function at these points and the endpoints to find the global maxima and minima.

The next example, another problem in geometry, illustrates this approach.

Example 2   Alaina wants to get to the bus stop as quickly as possible. The bus stop is across a grassy park, 2000 feet west and 600 feet north of her starting position. Alaina can walk west along the edge of the park on the sidewalk at a speed of 6 ft/sec. She can also travel through the grass in the park, but only at a rate of 4 ft/sec. What path will get her to the bus stop the fastest?

Solution

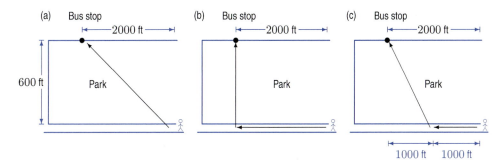

Figure 4.64: Three possible paths to the bus stop

We might first think that she should take a path that is the shortest distance. Unfortunately, the path that follows the shortest distance to the bus stop is entirely in the park, where her speed is slow. (See Figure 4.64(a).) That distance is $\sqrt{2000^2 + 600^2} \approx 2100$ feet, which takes her about 525 seconds to traverse. She could instead walk quickly the entire 2000 feet along the sidewalk, which leaves her just the 600-foot northward journey through the park. (See Figure 4.64(b).) This method would take $2000/6 + 600/4 \approx 483$ seconds total walking time.

But can she do even better? Perhaps another combination of sidewalk and park gives a shorter travel time. For example, what is the travel time if she walks 1000 feet west along the sidewalk and the rest of the way through the park? (See Figure 4.64(c).) The answer is about 458 seconds.

We make a model for this problem. We label the distance that Alaina walks west along the sidewalk x and the distance she walks through the park y, as in Figure 4.65. Then the total time, t, is

$$t = t_{\text{sidewalk}} + t_{\text{park}}.$$

Since

$$\text{Time} = \text{Distance}/\text{Speed},$$

and she can walk 6 ft/sec on the sidewalk and 4 ft/sec in the park, we have that

$$t = \frac{x}{6} + \frac{y}{4}.$$

Now, by the Pythagorean Theorem, $y = \sqrt{(2000 - x)^2 + 600^2}$. Therefore

$$t = \frac{x}{6} + \frac{\sqrt{(2000 - x)^2 + 600^2}}{4}.$$

We can find the critical points of this function analytically. (See Exercise 1 on page 201.) Alternatively, we can graph the function on a calculator and estimate the critical point, which is $x \approx 1463$ feet. This gives a minimum total time of about 445 seconds.

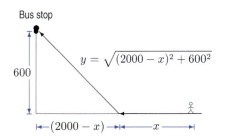

Figure 4.65: Modeling time to bus stop

Example 3   One hallway which is 4 feet wide meets another hallway which is 8 feet wide in a right angle. (See Figure 4.66.) What is the length of the longest ladder which can be carried horizontally around the corner?

Solution   We imagine the ladder being carried on its side and ignore its width. To allow the longest ladder possible we carry the ladder around the corner so that it just touches both walls (at A and C) and just touches the corner at B. Let's draw some lines that do this. (See Figure 4.66.) The length of the line $\overline{ABC}$ decreases as the corner is turned, then increases again. The minimum such length would be the length of the longest ladder that could make it around the corner. A smaller ladder would still work (it wouldn't touch A, B, and C simultaneously), but a larger one would not fit.

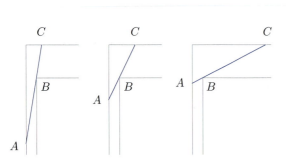

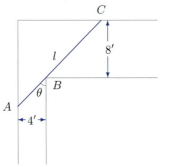

Figure 4.66: Various ladders that touch both walls and corner   **Figure 4.67**: Ladder and hallway

We want the *smallest* length of the line $\overline{ABC}$. We express the length, l, as a function of θ, the angle between the line and the wall of the narrow hall. (See Figure 4.67.) We have

$$l = \overline{AB} + \overline{BC}.$$

Now $\overline{AB} = 4/\sin\theta$ and $\overline{BC} = 8/\cos\theta$, so

$$l = \frac{4}{\sin\theta} + \frac{8}{\cos\theta}.$$

Here the domain of θ is $0 < \theta < \pi/2$.

Next we differentiate:

$$\frac{dl}{d\theta} = -\frac{4}{(\sin\theta)^2}(\cos\theta) - \frac{8}{(\cos\theta)^2}(-\sin\theta).$$

To minimize l, we solve $dl/d\theta = 0$:

$$-4\frac{\cos\theta}{(\sin\theta)^2} + 8\frac{\sin\theta}{(\cos\theta)^2} = 0,$$

so

$$2(\sin\theta)^3 = (\cos\theta)^3, \quad \text{or} \quad \frac{(\sin\theta)^3}{(\cos\theta)^3} = \frac{1}{2}.$$

This means

$$\tan\theta = \sqrt[3]{0.5} \approx 0.79, \quad \text{so} \quad \theta \approx 0.67 \text{ radians.}$$

Thus $\theta \approx 0.67$ is a critical point, and an investigation of what happens for θ near 0 and θ near $\pi/2$ shows that l has a global minimum at $\theta = 0.67$. The minimum value of l is therefore

$$l = \frac{4}{\sin(0.67)} + \frac{8}{\cos(0.67)} \approx 16.65 \text{ feet,}$$

so the ladder can be at most 16.65 feet long and still make it around the corner.

Exercises and Problems for Section 4.5

Exercises

1. Find analytically the exact critical point of the function which represents the time, t, to walk to the bus stop in Example 2. Recall that t is given by

$$t = \frac{x}{6} + \frac{\sqrt{(2000 - x)^2 + 600^2}}{4}.$$

2. A smokestack deposits soot on the ground with a concentration inversely proportional to the square of the distance from the stack. With two smokestacks 20 miles apart, the concentration of the combined deposits on the line joining them, at a distance x from one stack, is given by

$$S = \frac{k_1}{x^2} + \frac{k_2}{(20 - x)^2}$$

where k_1 and k_2 are positive constants which depend on the quantity of smoke each stack is emitting. If $k_1 = 7k_2$, find the point on the line joining the stacks where the concentration of the deposit is a minimum.

3. A wave of wavelength λ traveling in deep water has speed, v, given by

$$v = k\sqrt{\frac{\lambda}{c} + \frac{c}{\lambda}},$$

where c and k are positive constants. As λ varies, does such a wave have a maximum or minimum velocity? If so, what is it? Explain.

Problems

4. Figure 4.68 shows the curves $y = \sqrt{x}$, $x = 9$, $y = 0$, and a rectangle with its sides parallel to the axes and its left end at $x = a$. Find the dimensions of the rectangle having the maximum possible area.

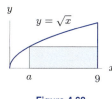

Figure 4.68

5. A closed box has a fixed surface area A and a square base with side x.

   (a) Find a formula for its volume, V, as a function of x.
   (b) Sketch a graph of V against x.
   (c) Find the maximum value of V.

6. If you have 100 feet of fencing and want to enclose a rectangular area up against a long, straight wall, what is the largest area you can enclose?

7. A rectangular beam is cut from a cylindrical log of radius 30 cm. The strength of a beam of width w and height h is proportional to wh^2. (See Figure 4.69.) Find the width and height of the beam of maximum strength.

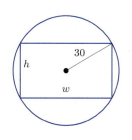

Figure 4.69

8. A landscape architect plans to enclose a 3000 square foot rectangular region in a botanical garden. She will use shrubs costing $25 per foot along three sides and fencing costing $10 per foot along the fourth side. Find the minimum total cost.

9. A rectangular swimming pool is to be built with an area of 1800 square feet. The owner wants 5-foot wide decks along either side and 10-foot wide decks at the two ends. Find the dimensions of the smallest piece of property on which the pool can be built satisfying these conditions.

10. A square-bottomed box with no top has a fixed volume, V. What dimensions minimize the surface area?

11. A light is suspended at a height h above the floor. (See Figure 4.70.) The illumination at the point P is inversely proportional to the square of the distance from the point P to the light and directly proportional to the cosine of the angle θ. How far from the floor should the light be to maximize the illumination at the point P?

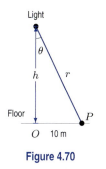

Figure 4.70

12. Which point on the parabola $y = x^2$ is nearest to $(1, 0)$? Find the coordinates to two decimals. [Hint: Minimize the square of the distance—this avoids square roots.]

13. Find the coordinates of the point on the parabola $y = x^2$ which is closest to the point (3,0).

14. The cross-section of a tunnel is a rectangle of height h surmounted by a semicircular roof section of radius r (See Figure 4.71 on page 202). If the cross-sectional area is A, determine the dimensions of the cross section which minimize the perimeter.

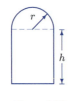

Figure 4.71

15. Of all rectangles with given area, A, which has the shortest diagonals?

16. You run a small furniture business. You sign a deal with a customer to deliver up to 400 chairs, the exact number to be determined by the customer later. The price will be $90 per chair up to 300 chairs, and above 300, the price will be reduced by $0.25 per chair (on the whole order) for every additional chair over 300 ordered. What are the largest and smallest revenues your company can make under this deal?

17. The cost of fuel to propel a boat through the water (in dollars per hour) is proportional to the cube of the speed. A certain ferry boat uses $100 worth of fuel per hour when cruising at 10 miles per hour. Apart from fuel, the cost of running this ferry (labor, maintenance, and so on) is $675 per hour. At what speed should it travel so as to minimize the cost *per mile* traveled?

18. **(a)** For which positive number x is $x^{1/x}$ largest? Justify your answer.
[Hint: You may want to write $x^{1/x} = e^{\ln(x^{1/x})}$.]
(b) For which positive integer n is $n^{1/n}$ largest? Justify your answer.
(c) Use your answer to parts (a) and (b) to decide which is larger: $3^{1/3}$ or $\pi^{1/\pi}$.

19. The *arithmetic mean* of two numbers a and b is defined as $(a+b)/2$; the *geometric mean* of two positive numbers a and b is defined as $\sqrt{ab}$.

(a) For two positive numbers, which of the two means is larger? Justify your answer.
[Hint: Define $f(x) = (a+x)/2 - \sqrt{ax}$ for fixed a.]
(b) For three positive numbers a, b, c, the arithmetic and geometric mean are $(a+b+c)/3$ and $\sqrt[3]{abc}$, respectively. Which of the two means of three numbers is larger?
[Hint: Redefine $f(x)$ for fixed a *and* b.]

20. A bird such as a starling feeds worms to its young. To collect worms, the bird flies to a site where worms are to be found, picks up several in its beak, and flies back to its nest. The *loading curve* in Figure 4.72 shows how the number of worms (the load) a starling collects depends on the time it has been searching for them.[5] The curve is concave down because the bird can pick up worms more efficiently when its beak is empty; when its beak is partly full, the bird becomes much less efficient. The traveling time (from nest to site and back) is represented by the distance PO in Figure 4.72. The bird wants to maximize the rate at which it brings worms to the nest, where

$$\text{Rate worms arrive} = \frac{\text{Load}}{\text{Traveling time} + \text{Searching time}}$$

(a) Draw a line in Figure 4.72 whose slope is this rate.
(b) Using the graph, estimate the load which maximizes this rate.
(c) If the traveling time is increased, does the optimal load increase or decrease? Why?

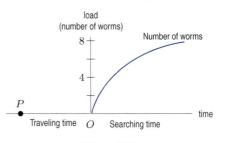

Figure 4.72

21. On the same side of a straight river are two towns, and the townspeople want to build a pumping station, S. See Figure 4.73. The pumping station is to be at the river's edge with pipes extending straight to the two towns. Where should the pumping station be located to minimize the total length of pipe?

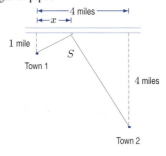

Figure 4.73

22. A pigeon is released from a boat (point B in Figure 4.74) floating on a lake. Because of falling air over the cool water, the energy required to fly one meter over the lake is twice the corresponding energy e required for flying over the bank ($e = 3$ joule/meter). To minimize the energy required to fly from B to the loft, L, the pigeon heads to a point P on the bank and then flies along the bank to L. The distance $\overline{AL}$ is 2000 m, and $\overline{AB}$ is 500 m.

(a) Express the energy required to fly from B to L via P as a function of the angle θ (the angle BPA).

[5]Alex Kacelnick (1984). Reported by J. R. Krebs and N. B. Davis, *An Introduction to Behavioural Ecology* (Oxford: Blackwell, 1987).

(b) What is the optimal angle θ?

(c) Does your answer change if $\overline{AL}$, $\overline{AB}$, and e have different numerical values?

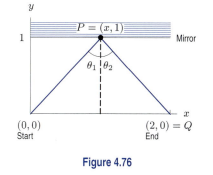

Figure 4.76

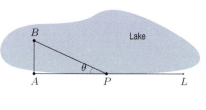

Figure 4.74

23. To get the best view of the Statue of Liberty in Figure 4.75, you should be at the position where θ is a maximum. If the statue stands 92 meters high, including the pedestal, which is 46 meters high, how far from the base should you be? [Hint: Find a formula for θ in terms of your distance from the base. Use this function to maximize θ, noting that $0 \leq \theta \leq \pi/2$.]

Figure 4.75

24. A light ray starts at the origin and is reflected off a mirror along the line $y = 1$ to the point $(2, 0)$. See Figure 4.76. Fermat's Principle[6] says that light's path minimizes the time of travel. The speed of light is a constant.

(a) Using Fermat's principle, find the optimal position of P.

(b) Using your answer to part (a), derive the Law of Reflection, that $\theta_1 = \theta_2$.

25. When a ray of light travels from one medium to another (for example, from air to water), it changes direction. This phenomenon is known as refraction. In Figure 4.77, light is traveling from A to B. The amount of refraction depends on the velocities, v_1 and v_2, of light in the two media and on Fermat's Principle which states that the light's time of travel, T, from A to B, is a minimum.

(a) Find an expression for T in terms of x and the constants a, b, v_1, v_2, and c.

(b) Show that if R is chosen so that the time of travel is minimized, then

$$\frac{\sin \theta_1}{\sin \theta_2} = \frac{v_1}{v_2}.$$

This result is known as Snell's Law and the ratio v_1/v_2 is called the index of refraction of the second medium with respect to the first medium.

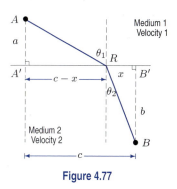

Figure 4.77

26. Show that when the value of x in Figure 4.77 is chosen according to Snell's Law, the time taken by the light ray is a minimum.

4.6 HYPERBOLIC FUNCTIONS

There are two combinations of e^x and e^{-x} which are used so often in engineering that they are given their own name. They are the *hyperbolic sine*, abbreviated sinh, and the *hyperbolic cosine*, abbreviated cosh. They are defined as follows:

---

[6]See, for example, D. Halliday, R. Resnik, K. Kane, *Physics* Vol 2, 4th edn, p. 909 (New York: Wiley, 1992).

Hyperbolic Functions

$$\cosh x = \frac{e^x + e^{-x}}{2} \qquad \sinh x = \frac{e^x - e^{-x}}{2}$$

Properties of Hyperbolic Functions

The graphs of $\cosh x$ and $\sinh x$ are given in Figures 4.78 and 4.79 together with the graphs of multiples of e^x and e^{-x}. The graph of $\cosh x$ is called a *catenary*; it is the shape of a hanging cable.

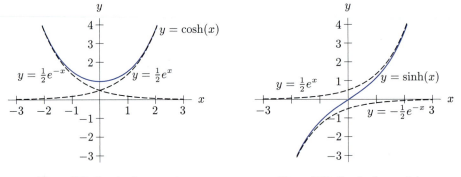

Figure 4.78: Graph of $y = \cosh x$ Figure 4.79: Graph of $y = \sinh x$

The graphs suggest that the following results hold:

$$\cosh 0 = 1 \qquad \sinh 0 = 0$$

$$\cosh(-x) = \cosh x \qquad \sinh(-x) = -\sinh x$$

To show that the hyperbolic functions really do have these properties, we use their formulas.

Example 1   Show that (a) $\cosh(0) = 1$ (b) $\cosh(-x) = \cosh x$

Solution   (a) Substituting $x = 0$ into the formula for $\cosh x$ gives

$$\cosh 0 = \frac{e^0 + e^{-0}}{2} = \frac{1+1}{2} = 1.$$

(b) Substituting $-x$ for x gives

$$\cosh(-x) = \frac{e^{-x} + e^{-(-x)}}{2} = \frac{e^{-x} + e^x}{2} = \cosh x.$$

Thus, we know that $\cosh x$ is an even function.

Example 2   Describe and explain the behavior of $\cosh x$ as $x \to \infty$ and $x \to -\infty$.

Solution   From Figure 4.78, it appears that as $x \to \infty$, the graph of $\cosh x$ resembles the graph of $\frac{1}{2}e^x$. Similarly, as $x \to -\infty$, the graph of $\cosh x$ resembles the graph of $\frac{1}{2}e^{-x}$. This behavior is explained by using the formula for $\cosh x$ and the facts that $e^{-x} \to 0$ as $x \to \infty$ and $e^x \to 0$ as $x \to -\infty$:

$$\text{As } x \to \infty, \qquad \cosh x = \frac{e^x + e^{-x}}{2} \to \frac{1}{2}e^x.$$

$$\text{As } x \to -\infty, \qquad \cosh x = \frac{e^x + e^{-x}}{2} \to \frac{1}{2}e^{-x}.$$

Identities Involving cosh x and sinh x

The reason the hyperbolic functions have names that remind us of the trigonometric functions is that they share similar properties. A familiar identity for trigonometric functions is

$$(\cos x)^2 + (\sin x)^2 = 1.$$

To discover an analogous identity relating $(\cosh x)^2$ and $(\sinh x)^2$, we first calculate

$$(\cosh x)^2 = \left(\frac{e^x + e^{-x}}{2}\right)^2 = \frac{e^{2x} + 2e^x e^{-x} + e^{-2x}}{4} = \frac{e^{2x} + 2 + e^{-2x}}{4}$$

$$(\sinh x)^2 = \left(\frac{e^x - e^{-x}}{2}\right)^2 = \frac{e^{2x} - 2e^x e^{-x} + e^{-2x}}{4} = \frac{e^{2x} - 2 + e^{-2x}}{4}.$$

If we add these expressions, the resulting right-hand side contains terms involving both e^{2x} and e^{-2x}. If, however, we subtract the expressions for $(\cosh x)^2$ and $(\sinh x)^2$, we obtain a simple result:

$$(\cosh x)^2 - (\sinh x)^2 = \frac{e^{2x} + 2 + e^{-2x}}{4} - \frac{e^{2x} - 2 + e^{-2x}}{4} = \frac{4}{4} = 1.$$

Thus, writing $\cosh^2 x$ for $(\cosh x)^2$ and $\sinh^2 x$ for $(\sinh x)^2$, we have the identity

$$\boxed{\cosh^2 x - \sinh^2 x = 1}$$

The Hyperbolic Tangent

Extending the analogy to the trigonometric functions, we define

$$\tanh x = \frac{\sinh x}{\cosh x} = \frac{e^x - e^{-x}}{e^x + e^{-x}}$$

Derivatives of Hyperbolic Functions

We calculate the derivatives using the fact that $\frac{d}{dx}(e^x) = e^x$. The results are again reminiscent of the trigonometric functions. For example,

$$\frac{d}{dx}(\cosh x) = \frac{d}{dx}\left(\frac{e^x + e^{-x}}{2}\right) = \frac{e^x - e^{-x}}{2} = \sinh x.$$

We find $\frac{d}{dx}(\sinh x)$ similarly, giving the following results:

$$\boxed{\frac{d}{dx}(\cosh x) = \sinh x \qquad \frac{d}{dx}(\sinh x) = \cosh x}$$

Example 3 Compute the derivative of $\tanh x$.

Solution Using the quotient rule gives

$$\frac{d}{dx}(\tanh x) = \frac{d}{dx}\left(\frac{\sinh x}{\cosh x}\right) = \frac{(\cosh x)^2 - (\sinh x)^2}{(\cosh x)^2} = \frac{1}{\cosh^2 x}.$$

Exercises and Problems for Section 4.6

Exercises

Find the derivatives of the functions in Exercises 1–8.

1. $y = \cosh(2x)$

2. $y = \sinh(3z + 5)$

3. $f(t) = \cosh(\sinh t)$

4. $f(t) = t^3 \sinh t$

5. $g(t) = \cosh^2 t$

6. $f(t) = \cosh(e^{t^2})$

7. $f(y) = \sinh(\sinh(3y))$

8. $g(\theta) = \ln(\cosh(1 + \theta))$

9. Show that $\sinh 0 = 0$.

10. Show that $\sinh(-x) = -\sinh(x)$.

11. Show that $d(\sinh x)/dx = \cosh x$.

Problems

12. Describe and explain the behavior of $\sinh x$ as $x \to \infty$ and as $x \to -\infty$.

13. Is there an identity analogous to $\sin(2x) = 2 \sin x \cos x$ for the hyperbolic functions? Explain.

14. Is there an identity analogous to $\cos(2x) = \cos^2 x - \sin^2 x$ for the hyperbolic functions? Explain.

15. **(a)** Find $\tanh 0$.
   (b) For what values of x is $\tanh x$ positive? Negative? Explain your answer algebraically.
   (c) On what intervals is $\tanh x$ increasing? Decreasing? Use derivatives to explain your answer.
   (d) Find $\lim_{x \to \infty} \tanh x$ and $\lim_{x \to -\infty} \tanh x$. Show this information on a graph.
   (e) Does $\tanh x$ have an inverse? Justify your answer using derivatives.

16. Consider the family of functions $y = a \cosh(x/a)$ for $a > 0$. Sketch graphs for $a = 1, 2, 3$. Describe in words the effect of increasing a.

17. **(a)** Using a calculator or computer, sketch the graph of $y = 2e^x + 5e^{-x}$ for $-3 \le x \le 3, 0 \le y \le 20$. Observe that it looks like the graph of $y = \cosh x$. Approximately where is its minimum?
   (b) Show algebraically that $y = 2e^x + 5e^{-x}$ can be written in the form $y = A \cosh(x - c)$. Calculate the values of A and c. Explain what this tells you about the graph in part (a).

18. The following problem is a generalization of Problem 17. Show that any function of the form

$$y = Ae^x + Be^{-x}, \quad A > 0, \ B > 0,$$

can be written, for some K and c, in the form

$$y = K \cosh(x - c).$$

What does this tell you about the graph of $y = Ae^x + Be^{-x}$?

19. Let $y = Ae^x + Be^{-x}$ for any constants A, B.

   (a) Sketch the function for
   (i) $A = 1, \ B = 1$
   (ii) $A = 1, \ B = -1$
   (iii) $A = 2, \ B = 1$
   (iv) $A = 2, \ B = -1$
   (v) $A = -2, \ B = -1$
   (vi) $A = -2, \ B = 1$

   (b) Describe in words the general shape of the graph if A and B have the same sign. What effect does the sign of A have on the graph?
   (c) Describe in words the general shape of the graph if A and B have different signs. What effect does the sign of A have on the graph?
   (d) For what values of A and B does the function have a local maximum? A local minimum? Justify your answer using derivatives.

20. The cable between the two towers of a suspension bridge hangs in the shape of the curve

$$y = \frac{T}{w} \cosh\left(\frac{wx}{T}\right),$$

where T is the tension in the cable at its lowest point and w is the weight of the cable per unit length. This curve is called a *catenary*.

   (a) Suppose the cable stretches between the points $x = -T/w$ and $x = T/w$. Find an expression for the "sag" in the cable. (That is, find the difference between the height of the cable at the highest and lowest points.)
   (b) Show that the shape of the cable satisfies the equation

$$\frac{d^2 y}{dx^2} = \frac{w}{T}\sqrt{1 + \left(\frac{dy}{dx}\right)^2}.$$

21. The Saint Louis arch can be approximated by using a function of the form $y = b - a \cosh(x/a)$. Putting the origin on the ground in the center of the arch and the y-axis upward, find an approximate equation for the arch given the dimensions shown in Figure 4.80. (In other words, find a and b.)

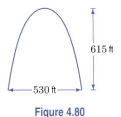

615 ft

530 ft

Figure 4.80

4.7 THEOREMS ABOUT CONTINUOUS AND DIFFERENTIABLE FUNCTIONS

In this chapter we have used some basic facts without proof: for example, that a continuous function has a maximum on a bounded, closed interval, or that a function whose derivative is positive on an interval is increasing on that interval.

From a geometric point of view, these facts seem obvious. If we draw the graph of a continuous function, starting at one end of a bounded, closed interval and going to the other, it seems obvious that we must pass a highest point on the way. If the derivative of a function is positive, then its graph must be sloping up, so the function has to be increasing.

However, this sort of graphical reasoning is not a rigorous proof, for two reasons. First, no matter how many pictures we imagine, we can't be sure we have covered all possibilities. Second, our pictures often depend on the theorems we are trying to prove.

A Continuous Function on a Closed Interval Has a Maximum

On page 181 we gave a method for finding global maxima and minima of a continuous function on a closed interval: Compare values of the function at all the critical points in the interval and at the endpoints. This method depends on knowing that global maxima and minima exist, and on knowing that they occur at critical points or endpoints. The first of these is guaranteed by the following theorem.

Theorem 4.2: The Extreme Value Theorem

If f is continuous on the interval $[a, b]$, then f has a global maximum and a global minimum on that interval.

For a proof of Theorem 4.2, see the online theory supplement for this book. To prove the second fact, we notice that any global maximum (or minimum) is a local maximum (or minimum). Thus, the second fact is guaranteed by Theorem 4.1 on page 168, which tells us that inside an interval, local maxima (and minima) only occur at critical points, where the derivative is either zero or undefined.

We now prove Theorem 4.1. Suppose that f has a local maximum at $x = a$. Assuming that $f'(a)$ is defined, the definition of the derivative gives

$$f'(a) = \lim_{h \to 0} \frac{f(a+h) - f(a)}{h}.$$

Since this is a two-sided limit, we have

$$f'(a) = \lim_{h \to 0^-} \frac{f(a+h) - f(a)}{h} = \lim_{h \to 0^+} \frac{f(a+h) - f(a)}{h}.$$

By the definition of local maximum, $f(a+h) \leq f(a)$ for all sufficiently small h. Thus $f(a+h) - f(a) \leq 0$ for sufficiently small h. The denominator, h, is positive when we take the limit from the right and negative when we take the limit from the left. Thus

$$\lim_{h \to 0^-} \frac{f(a+h) - f(a)}{h} \geq 0 \quad \text{and} \quad \lim_{h \to 0^+} \frac{f(a+h) - f(a)}{h} \leq 0.$$

Since both these limits are equal to $f'(a)$, we have $f'(a) \geq 0$ and $f'(a) \leq 0$, so we must have $f'(a) = 0$. The proof for a local minimum at $x = a$ is similar.

A Relationship Between Local and Global: The Mean Value Theorem

We often want to infer a global conclusion (for example, f is increasing on an interval) from local information (f' is positive.) For example, the fact that f is increasing when f' is positive and decreasing when f' is negative lies behind the First Derivative Test for local maxima and minima on page 168. The following theorem relates the average rate of change of a function on an interval (global information) to the instantaneous rate of change at a point in the interval (local information).

Theorem 4.3: The Mean Value Theorem

If f is continuous on $[a, b]$ and differentiable on (a, b), then there exists a number c, with $a < c < b$, such that

$$f'(c) = \frac{f(b) - f(a)}{b - a}.$$

In other words, $f(b) - f(a) = f'(c)(b - a)$.

To understand what this theorem is saying geometrically, consider the graph in Figure 4.81. Join the points on the curve where $x = a$ and $x = b$ with a line and observe that the slope of this secant line AB is given by

$$m = \frac{f(b) - f(a)}{b - a}.$$

Now consider the tangent line drawn to the curve at each point between $x = a$ and $x = b$. In general, these lines have different slopes. For the curve shown in Figure 4.81, the tangent line at $x = a$ is flatter than the secant line from A to B. Similarly, the tangent line at $x = b$ is steeper than the secant line. However, there is at least one point between a and b where the slope of the tangent line to the curve is precisely the same as the slope of the secant line. Suppose this occurs at $x = c$. Then

$$f'(c) = m = \frac{f(b) - f(a)}{b - a}.$$

The Mean Value Theorem tells us that the point $x = c$ exists, but it does not tell us how to find c.

Problems 30 and 31 show how the Mean Value Theorem can be deduced from the Extreme Value Theorem.

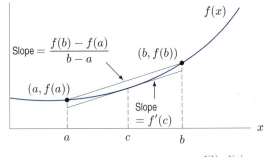

Figure 4.81: The point c with $f'(c) = \frac{f(b)-f(a)}{b-a}$

The Increasing Function Theorem

We say that a function f is *increasing* on an interval if, for any two numbers x_1 and x_2 in the interval such that $x_1 < x_2$, we have $f(x_1) < f(x_2)$. If instead we have $f(x_1) \leq f(x_2)$, we say f is *nondecreasing*.

Theorem 4.4: The Increasing Function Theorem

Suppose that f is continuous on $[a, b]$ and differentiable on (a, b).
- If $f'(x) > 0$ on (a, b), then f is increasing on $[a, b]$.
- If $f'(x) \geq 0$ on (a, b), then f is nondecreasing on $[a, b]$.

Proof Suppose $a \le x_1 < x_2 \le b$. By the Mean Value Theorem, there is a number c, with $x_1 < c < x_2$, such that

$$f(x_2) - f(x_1) = f'(c)(x_2 - x_1).$$

If $f'(c) > 0$, this says $f(x_2) - f(x_1) > 0$, which means f is increasing. If $f'(c) \ge 0$, this says $f(x_2) - f(x_1) \ge 0$, which means f is nondecreasing.

It may seem that something as simple as the Increasing Function Theorem should follow immediately from the definition of the derivative, and that the use of the Mean Value Theorem (which in turn depends on the Extreme Value Theorem) is surprising.

The Constant Function Theorem

If f is constant on an interval, then we know that $f'(x) = 0$ on the interval. The following theorem is the converse.

Theorem 4.5: The Constant Function Theorem

Suppose that f is continuous on $[a, b]$ and differentiable on (a, b). If $f'(x) = 0$ on (a, b), then f is constant on $[a, b]$.

Proof The proof is the same as for the Increasing Function Theorem, only in this case $f'(c) = 0$ so $f(x_2) - f(x_1) = 0$. Thus $f(x_2) = f(x_1)$ for $a \le x_1 < x_2 \le b$, so f is constant.

A proof of the Constant Function Theorem using the Increasing Function Theorem is given in Problems 17 and 25.

The Racetrack Principle

Theorem 4.6: The Racetrack Principle[7]

Suppose that g and h are continuous on $[a, b]$ and differentiable on (a, b), and that $g'(x) \le h'(x)$ for $a < x < b$.
- If $g(a) = h(a)$, then $g(x) \le h(x)$ for $a \le x \le b$.
- If $g(b) = h(b)$, then $g(x) \ge h(x)$ for $a \le x \le b$.

The Racetrack Principle has the following interpretation. We can think of $g(x)$ and $h(x)$ as the positions of two racehorses at time x, with horse h always moving faster than horse g. If they start together, horse h is ahead during the whole race. If they finish together, horse g was ahead during the whole race.

Proof Consider the function $f(x) = h(x) - g(x)$. Since $f'(x) = h'(x) - g'(x) \ge 0$, we know that f is nondecreasing by the Increasing Function Theorem. So $f(x) \ge f(a) = h(a) - g(a) = 0$. Thus $g(x) \le h(x)$ for $a \le x \le b$. This proves the first part of the Racetrack Principle. Problem 24 asks for a proof of the second part.

Example 1 Explain graphically why $e^x \ge 1 + x$ for all values of x. Then use the Racetrack Principle to prove the inequality.

---

[7]Based on the Racetrack Principle in *Calculus&Mathematica*, by William Davis, Horacio Porta, Jerry Uhl (Reading: Addison Wesley, 1994).

Solution   The graph of the function $y = e^x$ is concave up everywhere and the equation of its tangent line at the point $(0, 1)$ is $y = x + 1$. (See Figure 4.82.) Since the graph always lies above its tangent, we have the inequality

$$e^x \geq 1 + x.$$

Now we prove the inequality using the Racetrack Principle. Let $g(x) = 1 + x$ and $h(x) = e^x$. Then $g(0) = h(0) = 1$. Furthermore, $g'(x) = 1$ and $h'(x) = e^x$. Hence $g'(x) \leq h'(x)$ for $x \geq 0$. So by the Racetrack Principle, with $a = 0$, we have $g(x) \leq h(x)$, that is, $1 + x \leq e^x$.

For $x \leq 0$ we have $h'(x) \leq g'(x)$. So by the Racetrack Principle, with $b = 0$, we have $g(x) \leq h(x)$, that is, $1 + x \leq e^x$.

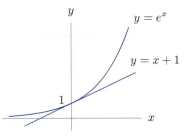

Figure 4.82: Graph showing that $e^x \geq 1 + x$

Exercises and Problems for Section 4.7

Exercises

Decide if the statements in Exercises 1–8 are true or false. Give an explanation for your answer.

1. A critical point of f must be a local maximum or minimum of f.

2. Let $f(x) = [x]$, the greatest integer less than or equal to x. Then $f'(x) = 0$, so $f(x)$ is constant by the Constant Function Theorem.

3. Since the function $f(x) = 1/x$ is continuous for $x > 0$ and the interval $(0, 1)$ is bounded, f has a maximum on the interval $(0, 1)$.

4. The Extreme Value Theorem says that only continuous functions have global maxima and minima on every closed, bounded interval.

5. The Racetrack Principle can be used to justify the statement that if two horses start a race at the same time, the horse that wins must have been moving faster than the other throughout the race.

6. Two horses start a race at the same time and one runs slower than the other throughout the race. The Racetrack Principle can be used to justify the fact that the slower horse loses the race.

7. If $a < b$ and $f'(x)$ is positive on $[a, b]$ then $f(a) < f(b)$.

8. If $f(x)$ is increasing and differentiable on the interval $[a, b]$, then $f'(x) > 0$ on $[a, b]$.

Do the functions graphed in Exercises 9–12 appear to satisfy the hypotheses of the Mean Value Theorem on the interval $[a, b]$? Do they satisfy the conclusion?

9.

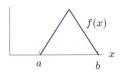

10.

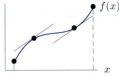

11.

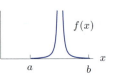

12.

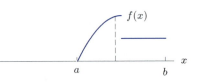

Problems

13. Use the Racetrack Principle and the fact that $\sin 0 = 0$ to show that $\sin x \leq x$ for all $x \geq 0$.

14. Use the Racetrack Principle to show that $\ln x \leq x - 1$.

15. Use the fact that $\ln x$ and e^x are inverse functions to show that the inequalities $e^x \geq 1 + x$ and $\ln x \leq x - 1$ are equivalent for $x > 0$.

16. Show that if every continuous function on an interval $[a, b]$ has a global maximum, then every continuous function has a global minimum as well. [Hint: Consider $-f$.]

17. State a Decreasing Function Theorem, analogous to the Increasing Function Theorem. Deduce your theorem from the Increasing Function Theorem. [Hint: Apply the Increasing Function Theorem to $-f$.]

Use one of the theorems in this section to prove the statements in Problems 18–22.

18. If $f'(x) \leq 1$ for all x and $f(0) = 0$, then $f(x) \leq x$ for all $x \geq 0$.

19. If $f''(t) \leq 3$ for all t and $f(0) = f'(0) = 0$, then $f(t) \leq \frac{3}{2}t^2$ for all $t \geq 0$.

20. If $f'(x) = g'(x)$ for all x and $f(5) = g(5)$, then $f(x) = g(x)$ for all x.

21. If f is differentiable and $f(0) < f(1)$, then there is a number c, with $0 < c < 1$, such that $f'(c) > 0$.

22. If $f''(x)$ is continuous and $f(x)$ has exactly two critical points, then $f(x)$ must have an inflection point.

23. The position of a particle on the x-axis is given by $s = f(t)$; its initial position and velocity are $f(0) = 3$ and $f'(0) = 4$. The acceleration is bounded by $5 \leq f''(t) \leq 7$ for $0 \leq t \leq 2$. What can we say about the position $f(2)$ of the particle at $t = 2$?

24. Suppose that g and h are continuous on $[a, b]$ and differentiable on (a, b). Prove that if $g'(x) \leq h'(x)$ for $a < x < b$ and $g(b) = h(b)$, then $h(x) \leq g(x)$ for $a \leq x \leq b$.

25. Deduce the Constant Function Theorem from the Increasing Function Theorem and the Decreasing Function Theorem (see problem 17).

26. Prove that if $f'(x) = g'(x)$ for all x in (a, b), then there is a constant C such that $f(x) = g(x) + C$ on (a, b). [Hint: Apply the Constant Function Theorem to $h(x) = f(x) - g(x)$.]

27. Suppose that $f'(x) = f(x)$ for all x. Prove that $f(x) = Ce^x$ for some constant C. [Hint: Consider $f(x)/e^x$.]

28. Suppose that f is continuous on $[a, b]$ and differentiable on (a, b) and that $m \leq f'(x) \leq M$ on (a, b). Use the Racetrack Principle to prove that $f(x) - f(a) \leq M(x - a)$ for all x in $[a, b]$, and that $m(x - a) \leq f(x) - f(a)$ for all x in $[a, b]$. Conclude that $m \leq (f(b) - f(a))/(b - a) \leq M$. This is called the Mean Value Inequality. In words: If the instantaneous rate of change of f is between m and M on an interval, so is the average rate of change of f over the interval.

29. Suppose that $f''(x) \geq 0$ for all x in (a, b). We will show the graph of f lies above the tangent line at $(c, f(c))$ for any c with $a < c < b$.

(a) Use the Increasing Function Theorem to prove that $f'(c) \leq f'(x)$ for $c \leq x < b$ and that $f'(x) \leq f'(c)$ for $a < x \leq c$.

(b) Use (a) and the Racetrack Principle to conclude that $f(c) + f'(c)(x - c) \leq f(x)$, for $a < x < b$.

30. In this problem we prove a special case of the Mean Value Theorem where $f(a) = f(b) = 0$. This special case is called Rolle's Theorem: If f is continuous on $[a, b]$ and differentiable on (a, b), and if $f(a) = f(b) = 0$, then there is a number c, with $a < c < b$, such that

$$f'(c) = 0.$$

By the Extreme Value Theorem, f has a global maximum and a global minimum on $[a, b]$.

(a) Prove Rolle's theorem in the case that both the global maximum and the global minimum are at endpoints of $[a, b]$. [Hint: $f(x)$ must be a very simple function in this case.]

(b) Prove Rolle's theorem in the case that either the global maximum or the global minimum is not at an endpoint. [Hint: Think about local maxima and minima.]

31. Use Rolle's Theorem to prove the Mean Value Theorem. Suppose that $f(x)$ is continuous on $[a, b]$ and differentiable on (a, b). Let $g(x)$ be the difference between $f(x)$ and the y-value on the secant line joining $(a, f(a))$ to $(b, f(b))$, so

$$g(x) = f(x) - f(a) - \frac{f(b) - f(a)}{b - a}(x - a).$$

(a) Show $g(x)$ on a sketch of $f(x)$.

(b) Use Rolle's Theorem (Problem 30) to show that there must be a point c in (a, b) such that $g'(c) = 0$.

(c) Show that if c is the point in part (b), then

$$f'(c) = \frac{f(b) - f(a)}{b - a}.$$

CHAPTER SUMMARY

- **Local extrema**
  Maximum, minimum, critical point, tests for local maxima/minima.

- **Using the second derivative**
  Concavity, inflection point.

- **Families of curves**

- **Optimization**
  Global extremum, modeling problems, graphical optimization, upper and lower bounds.

- **Marginality**
  Cost/revenue functions, marginal cost/marginal revenue functions.

- **Hyperbolic functions**

- **Theorems about continuous and differentiable functions**
  Extreme value theorem, mean value theorem, increasing function theorem, racetrack principle.

REVIEW EXERCISES AND PROBLEMS FOR CHAPTER FOUR

Exercises

For Exercises 1–2, indicate all critical points on the given graphs. Which correspond to local minima, local maxima, global maxima, global minima, or none of these? (Note that the graphs are on closed intervals.)

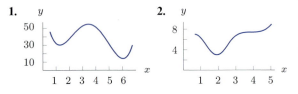

1.   **2.**

For the graphs of f' in Exercises 3–6 decide:
  (a) Over what intervals is f increasing? Decreasing?
  (b) Does f have maxima or minima? If so, which, and where?

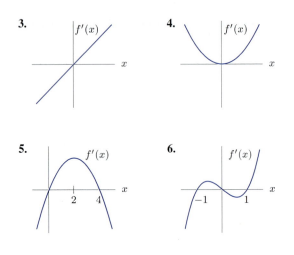

3.   **4.**

5.   **6.**

For each of the functions in Exercises 7–10, do the following:

(a) Find f' and f''.
(b) Find the critical points of f.
(c) Find any inflection points.
(d) Evaluate f at the critical points and the endpoints. Identify the global maxima and minima of f.
(e) Sketch f. Indicate clearly where f is increasing or decreasing, and its concavity.

7. $f(x) = x^3 - 3x^2 \quad (-1 \le x \le 3)$

8. $f(x) = x + \sin x \quad (0 \le x \le 2\pi)$

9. $f(x) = e^{-x} \sin x \quad (0 \le x \le 2\pi)$

10. $f(x) = x^{-2/3} + x^{1/3} \quad (1.2 \le x \le 3.5)$

For each of the functions in Exercises 11–13, find the limits as x tends to $+\infty$ and $-\infty$, and then proceed as in Exercises 7–10 (That is, find f', etc.).

11. $f(x) = 2x^3 - 9x^2 + 12x + 1$

12. $f(x) = \dfrac{4x^2}{x^2 + 1}$

13. $f(x) = xe^{-x}$

For each of the functions in Exercises 14–19, use derivatives to identify local maxima and minima and points of inflection. Sketch a graph of the function. Confirm your answers using a calculator or computer.

14. $f(x) = x^3 + 3x^2 - 9x - 15$

15. $f(x) = x^5 - 15x^3 + 10$

16. $f(x) = x - 2\ln x \quad$ for $x > 0$

17. $f(x) = x^2 e^{5x}$

18. $f(x) = e^{-x^2}$

19. $f(x) = \dfrac{x^2}{x^2 + 1}$

Problems

20. Find values of a and b so that the function $y = axe^{-bx}$ has a local maximum at the point $(2, 10)$.

21. A drug is injected into a patient at a rate given by $r(t) = ate^{-bt}$ ml/sec, where t is in seconds since the injection started, $0 \le t \le 5$, and a and b are constants. The maximum rate of 0.3 ml/sec occurs half a second after the injection starts. Find a formula for a and b.

22. On the graph of the derivative function f' in Figure 4.83, indicate the x-values that are critical points of the function f itself. Are they local maxima, local minima, or neither?

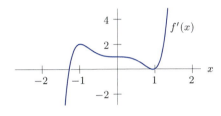

Figure 4.83: Graph of f', not f

23. Figure 4.84 is the graph of f', the derivative of a function f. At which of the points $0, x_1, x_2, x_3, x_4, x_5$, is the function f:

(a) At a local maximum value?
(b) At a local minimum value?
(c) Climbing fastest?
(d) Falling most steeply?

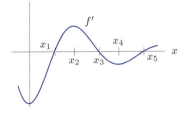

Figure 4.84: Graph of f' not f

24. For the function, f, given in the graph in Figure 4.85:

(a) Sketch $f'(x)$.
(b) Where does $f'(x)$ change its sign?
(c) Where does $f'(x)$ have local maxima or minima?

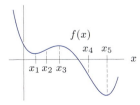

Figure 4.85

25. Using your answer to Problem 24 as a guide, write a short paragraph (using complete sentences) which describes the relationships between the following features of a function f:

(a) The local maxima and minima of f.
(b) The points at which the graph of f changes concavity.
(c) The sign changes of f'.
(d) The local maxima and minima of f'.

26. You are given the n numbers $a_1, a_2, a_3, \cdots, a_n$. For a variable x, consider the expression

$$D = (x - a_1)^2 + (x - a_2)^2 + (x - a_3)^2 + \cdots + (x - a_n)^2.$$

Show that D is a minimum when x is the average of $a_1, a_2, a_3, \cdots, a_n$.

27. A square-bottomed box with a top has a fixed volume, V. What dimensions minimize the surface area?

28. A ship is steaming due north at 12 knots (1 knot = 1.85 kilometers/hour) and sights a large tanker 3 kilometers away northwest steaming at 15 knots due east. For reasons of safety, the ships want to maintain a distance of at least 100 meters between them. Use a calculator or computer to determine the shortest distance between them if they remain on their current headings, and hence decide if they need to change course.

29. A polystyrene cup is in the shape of a frustum (the part of a cone between two parallel planes cutting the cone), has top radius $2r$, base radius r and height h. The surface area S of such a cup is given by $S = 3\pi r \sqrt{r^2 + h^2}$ and its volume V by $V = 7\pi r^2 h / 3$. If the cup is to hold 200 ml, use a calculator or a computer to estimate the value of r that minimizes its surface area.

30. A business sells an item at a constant rate of r units per month. It reorders in batches of q units, at a cost of $a + bq$ dollars per order. Storage costs are k dollars per item per month, and, on average, $q/2$ items are in storage, waiting to be sold. [Assume r, a, b, k are positive constants.]

(a) How often does the business reorder?
(b) What is the average monthly cost of reordering?
(c) What is the total monthly cost, C of ordering and storage?
(d) Obtain Wilson's lot size formula, the optimal batch size which minimizes cost.

31. Boise, Idaho, is about 300 miles inland from the nearest point on the Pacific coast; San Diego is about 1000 miles south of that point down the coast. (See Figure 4.86.) Assuming the coast is a straight line going north-south, C is the point along the coast directly west of Boise. It costs 2 cents per mile to transport a ton of potatoes by truck and 1 cent per mile to transport them by ship. The Idaho Potato Company wants to find the point, P, on the Pacific coast to which it should truck its potatoes before loading them aboard a cargo ship in order to minimize the total cost of transporting them from Boise to San Diego.

(a) Set up the function which The Idaho Potato Company must minimize.
(b) Find the position of P which minimizes cost.

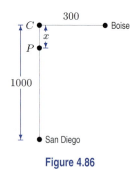

Figure 4.86

Find the best possible bounds for the functions in Problems 32–34.

32. $e^{-x} \sin x$, for $x \geq 0$

33. $x \sin x$, for $0 \leq x \leq 2\pi$

34. $x^3 - 6x^2 + 9x + 5$ for $0 \leq x \leq 5$

35. Sketch several members of the family $y = x^3 - ax^2$ on the same axes. Show that the critical points lie on the curve $y = -\frac{1}{2}x^3$.

36. For $a > 0$, the line

$$a(a^2 + 1)y = a - x$$

forms a triangle in the first quadrant with the x- and y-axes.

(a) Find, in terms of a, the x- and y-intercepts of the line.
(b) Find the area of the triangle, as a function of a.
(c) Find the value of a making the area a maximum.
(d) What is this greatest area?
(e) If you want the triangle to have area $1/5$, what choices do you have for a?

37. Suppose $g(t) = (\ln t)/t$ for $t > 0$.

(a) Does g have either a global maximum or a global minimum on $0 < t < \infty$? If so, where, and what are their values?

(b) What does your answer to part (a) tell you about the number of solutions to the equation

$$\frac{\ln x}{x} = \frac{\ln 5}{5}?$$

(Note: There are many ways to investigate the number of solutions to this equation. We are asking you to draw a conclusion from your answer to part (a).)

(c) Estimate the solution(s).

38. The vase in Figure 4.87 is filled with water at a constant rate (i.e., constant volume per unit time).

(a) Graph $y = f(t)$, the depth of the water, against time, t. Show on your graph the points at which the concavity changes.
(b) At what depth is $y = f(t)$ growing most quickly? Most slowly? Estimate the ratio between the growth rates at these two depths.

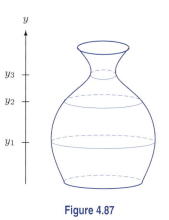

Figure 4.87

39. Any body radiates energy at various wavelengths. The power of the radiation (per meter^2 of surface) and the distribution of the radiation among the wavelengths varies with temperature.

The function graphed in Figure 4.88 represents the intensity of the radiation of a black body at a temperature $T = 3000°$ Kelvin, as a function of the wavelength. The intensity of the radiation is highest in the infrared range, that is, at wavelengths longer than that of visible light (0.4–0.7μm). Max Planck's radiation law, announced to the Berlin Physical Society on October 19, 1900, states that

$$r(\lambda) = \frac{a}{\lambda^5 (e^{b/\lambda} - 1)},$$

where a and b are empirical constants chosen to best fit the experimental data. Find a and b so that the formula fits the graph.

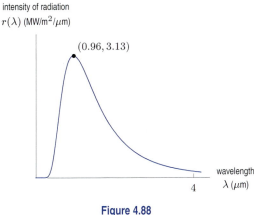

intensity of radiation
$r(\lambda)$ (MW/m^2/μm)

$(0.96, 3.13)$

wavelength
λ (μm)

4

Figure 4.88

(Later in 1900 Planck was able to derive his radia-

tion law entirely from theory. He found that $a = 2\pi c^2 h$ and $b = \frac{hc}{Tk}$ where c = speed of light, h = Planck's constant, and k = Boltzmann's constant.)

40. A piece of wire of length L cm is cut into two pieces. One piece, of length x cm, is made into a circle; the rest is made into a square.

(a) Find the value of x that makes the sum of the areas of the circle and square a minimum. Find the value of x giving a maximum.

(b) For the values of x found in part (a), show that the ratio of the length of wire in the square to the length of wire in the circle is equal to the ratio of the area of the square to the area of the circle.[8]

(c) Are the values of x found in part (a) the only values of x for which the ratios in part (b) are equal?

CAS Challenge Problems

41. A population, P, in a restricted environment may grow with time, t, according to the *logistic function*

$$P = \frac{L}{1 + Ce^{-kt}}$$

where L is called the carrying capacity and L, C and k are positive constants.

(a) Find $\lim_{t \to \infty} P$. Explain why L is called the carrying capacity.

(b) Using a computer algebra system, show that the graph of P has an inflection point at $P = L/2$.

42. For positive a, consider the family of functions

$$y = \arctan\left(\frac{\sqrt{x} + \sqrt{a}}{1 - \sqrt{ax}}\right), \qquad x > 0.$$

(a) Graph several curves in the family and describe how the graph changes as a varies.

(b) Use a computer algebra system to find dy/dx, and graph the derivative for several values of a. What do you notice?

(c) Do your observations in part (b) agree with the answer to part (a)? Explain. [Hint: Use the fact that $\sqrt{ax} = \sqrt{a}\sqrt{x}$ for $a > 0$, $x > 0$.]

43. The function arcsinh x is the inverse function of sinh x.

(a) Use a computer algebra system to find a formula for the derivative of arcsinh x.

(b) Derive the formula by hand by differentiating both sides of the equation

$$\sinh(\text{arcsinh } x) = x.$$

[Hint: Use the identity $\cosh^2 x - \sinh^2 x = 1$.]

44. The function arccosh x, for $x \geq 0$, is the inverse function of cosh x, for $x \geq 0$.

(a) Use a computer algebra system to find the derivative of arccosh x.

(b) Derive the formula by hand by differentiating both sides of the equation

$$\cosh(\text{arccosh } x) = x, \quad x \geq 1.$$

[Hint: Use the identity $\cosh^2 x - \sinh^2 x = 1$.]

45. Consider the family of functions

$$f(x) = \frac{\sqrt{a + x}}{\sqrt{a} + \sqrt{x}}, \qquad x \geq 0, \quad \text{for positive } a.$$

(a) Using a computer algebra system, find the local maxima and minima of f.

(b) On one set of axes, graph this function for several values of a. How does varying a affect the shape of the graph? Explain your answer in terms of the answer to part (a).

(c) Use your computer algebra system to find the inflection points of f when $a = 2$.

46. (a) Use a computer algebra system to find the derivative of

$$y = \arctan\left(\sqrt{\frac{1 - \cos x}{1 + \cos x}}\right).$$

(b) Graph the derivative. Does the graph agree with the answer you got in part (a)? Explain using the identity $\cos(x) = \cos^2(x/2) - \sin^2(x/2)$.

---

[8] From Sally Thomas.

47. In 1696, the first calculus textbook was published by the Marquis de l'Hopital. The following problem is a simplified version of a problem from this text.

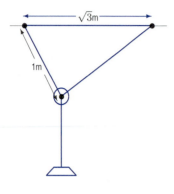

Figure 4.89

In Figure 4.89, two ropes are attached to the ceiling at points $\sqrt{3}$ meters apart. The rope on the left is 1 meter long and has a pulley attached at its end. The rope on the right is 3 meters long; it passes through the pulley and has a weight tied to its end. When the weight is released, the ropes and pulley arrange themselves so that the distance from the weight to the ceiling is maximized.

(a) Show that the maximum distance occurs when the weight is exactly halfway between the the points where the ropes are attached to the ceiling. [Hint: Write the vertical distance from the weight to the ceiling in terms of its horizontal distance to the point at which the left rope is tied to the ceiling. A computer algebra system will be useful.]

(b) Does the weight always end up halfway between the ceiling anchor points no matter how long the left-hand rope is? Explain.

CHECK YOUR UNDERSTANDING

Are the statements in Problems 1–12 true or false for a function f whose domain is all real numbers? If a statement is true, explain how you know. If a statement is false, give a counterexample.

1. A local minimum of f occurs at a critical point of f.

2. If $x = p$ is not a critical point of f, then $x = p$ is not a local maximum of f.

3. A local maximum of f occurs at a point where $f'(x) = 0$.

4. If $x = p$ is not a local maximum of f, then $x = p$ is not a critical point of f.

5. If $f'(p) = 0$, then $f(x)$ has a local minimum or local maximum at $x = p$.

6. If $f'(x)$ is continuous and $f(x)$ has no critical points, then f is everywhere increasing or everywhere decreasing.

7. If $f''(p) = 0$, then the graph of f has an inflection point at $x = p$.

8. If $f''(x)$ is continuous and the graph of f has an inflection point at $x = p$, then $f''(p) = 0$.

9. If $f'(x) \geq 0$ for all x, then $f(a) \leq f(b)$ whenever $a \leq b$.

10. If $f'(x) \leq g'(x)$ for all x, then $f(x) \leq g(x)$ for all x.

11. If $f'(x) = g'(x)$ for all x, then $f(x) = g(x)$ for all x.

12. If $f'(x) \leq 1$ for all x and $f(0) = 0$, then $f(x) \leq x$ for all x.

In Problems 13–17, give an example of function(s) with the given properties.

13. A family of functions, $f(x)$, depending on a parameter a, such that each member of the family has exactly one critical point.

14. A family of functions, $g(x)$, depending on two parameters, a and b, such that each member of the family has exactly two critical points and one inflection point. You may want to restrict a and b.

15. A continuous function f on the interval $[-1, 1]$ that does not satisfy the conclusion of the Mean Value Theorem.

16. A function f that is differentiable on the interval $(0, 2)$, but does not satisfy the conclusion of the Mean Value Theorem on the interval $[0, 2]$.

17. A function that is differentiable on $(0, 1)$ and not continuous on $[0, 1]$, but which satisfies the conclusion of the Mean Value Theorem.

18. Let $f(x) = x^2$. Decide if the following statements are true or false. Explain your answer.

(a) f has an upper bound on the interval $(0, 2)$.
(b) f has a global maximum on the interval $(0, 2)$.
(c) f does not have a global minimum on the interval $(0, 2)$.
(d) f does not have a global minimum on any interval (a, b).
(e) f has a global minimum on any interval $[a, b]$.

19. Which of the following statements is implied by the statement "If f is continuous on $[a, b]$ then f has a global maximum on $[a, b]$"?

(a) If f has a global maximum on $[a, b]$ then f must be continuous on $[a, b]$.
(b) If f is not continuous on $[a, b]$ then f does not have a global maximum on $[a, b]$.
(c) If f does not have a global maximum on $[a, b]$ then f is not continuous on $[a, b]$.

In Problems 20–25, give an example of a function f that makes the statement true, or say why such an example is impossible. Assume that f'' exists everywhere.

20. f is concave up and $f(x)$ is positive for all x

21. f is concave down and $f(x)$ is positive for all x

22. f is concave down and $f(x)$ is negative for all x

23. f is concave up and $f(x)$ is negative for all x

24. $f(x)f''(x) < 0$ for all x

25. $f(x)f'(x)f''(x)f'''(x) < 0$ for all x.[9]

PROJECTS FOR CHAPTER FOUR

1. **Building a Greenhouse**

   Your parents are going to knock out the bottom of the entire length of the south wall of their house and turn it into a greenhouse by replacing the bottom portion of the wall with a huge sloped piece of glass (which is expensive). They have already decided they are going to spend a certain fixed amount. The triangular ends of the greenhouse will be made of various materials they already have lying around.[10]

   The floor space in the greenhouse is only considered usable if they can both stand up in it, so part of it will be unusable. They want to choose the dimensions of the greenhouse to get the most usable floor space. What should the dimensions of the greenhouse be and how much usable space will your parents get?

2. **Fitting a Line to Data**

   (a) The line which best fits the data points $(x_1, y_1), (x_{,2}, y_2) \ldots (x_n, y_n)$ is the one which minimizes the sum of the squares of the vertical distances from the points to the line. These are the distances marked in Figure 4.90. Find the best fitting line of the form $y = mx$ for the points $(2, 3.5), (3, 6.8), (5, 9.1)$.

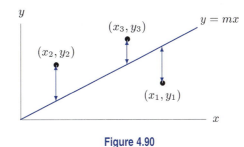

Figure 4.90

   (b) A cone with height and radius both equal to r has volume, V, proportional to r^3; that is, $V = kr^3$ for some constant k. A lab experiment is done to measure the volume of several cones; the results are in the following table. Using the method of part (a), determine the best value of k. [Note: Since the volumes were determined experimentally, the values may not be accurate. Assume that the radii were measured accurately.]

Radius (cm)	2	5	7	8
Volume (cm^3)	8.7	140.3	355.8	539.2

   (c) Using the method of part (a), show that the best fitting line of the form $y = mx$ for the points $(x_1, y_1), (x_2, y_2) \ldots (x_n, y_n)$ has

$$m = \frac{x_1 y_1 + x_2 y_2 + \cdots + x_n y_n}{x_1^2 + x_2^2 + \cdots + x_n^2}.$$

[9]From the 1998 William Lowell Putnam Mathematical Competition, by permission of the Mathematical Association of America.

[10]Adapted from M. Cohen, E. Gaughan, A. Knoebel, D. Kurtz, D. Pengelley, *Student Research Projects in Calculus* (Washington DC: Mathematical Association of America, 1992).

3. Interstate Trucking

In this project we investigate the effect of the price of diesel fuel and drivers' wages on a trucker's optimal driving speed. We analyze the relationship between fuel cost and wages and the 55 mph speed limit. [11]

After an extensive study, the Interstate Commerce Commission has concluded that the primary variable expenses for any over-the-road freight hauler are the wages for the driver and the cost of fuel. The study concluded that maintenance and replacement costs for vehicles, although considerable, did not vary significantly from carrier to carrier. On the other hand, the fuel costs and wages did vary. Given this result, your project is to determine how these two variables affect the cost of transporting freight. In particular, you are to determine if there is an optimal wage to fuel cost relationship for which most haulers would encourage their drivers to abide by a 55 mile per hour speed limit.

Reading the study tells you that under ideal conditions, an interstate freight hauling vehicle (18 wheeler) gets 6 miles per gallon of fuel. This mileage is affected by the speed of the vehicle and its weight. In general, the miles per gallon of fuel decreases by 0.2 for each increase of 10,000 pounds in the weight of the truck and freight over 25,000 pounds. Also, the miles per gallon of fuel decreases by 0.1 mile per gallon for each mile per hour the truck averages above 45 miles per hour.

(a) Using this information, create an expression for the cost per mile of driving, taking into account only the drivers' wages and fuel cost.

(b) Suppose that the national average for diesel fuel is $1.25 per gallon, that drivers on the average earn $15.00 per hour, and that the average weight of a loaded truck is 75,000 pounds. What is the optimal average speed under these conditions?

(c) Compute the cost per mile for 55 and 60 miles per hour. Suppose you hauled produce from California to Maine. What average speed would you choose?

(d) In solving parts (a)-(c), you found an equation relating cost, wages, fuel cost, weight of the truck, and the average speed of the truck. This equation is a mathematical model of interstate freight hauling costs. Using this model, determine the fuel cost that makes 55 miles per hour the optimum average speed for trucking companies.

(e) Each year drivers' wages change as a result of contract renewals and inflation. The cost of fuel also fluctuates, but can be adjusted by surcharges and fuel taxes. Thus, the relationship between wages and the cost of fuel can be altered by various government agencies. Assuming that 75,000 pounds remains the national average for the weight of a truck involved in interstate freight hauling, what should be the relationship between fuel cost and wages to maintain 55 miles per hour as the optimal speed?

(f) There have been some suggestions of changing the road use tax to lower the average weight of over-the-road freight haulers. Assuming the relationship of wages to fuel cost found in part (e) is maintained, how would lowering the average weight affect the optimal average speed?

4. Firebreaks

The summer of 2000 was devastating for forests in the western US: over 3.5 million acres of trees were lost to fires, making this the worst fire season in 30 years. This project studies a fire management technique called *firebreaks*, which reduce the damage done by forest fires. A firebreak is a strip where trees have been removed in a forest so that a fire started on one side of the strip will not spread to the other side. Having many firebreaks helps confine a fire to a small area. On the other hand, having too many firebreaks involves removing large swaths of trees. [12]

(a) A forest in the shape of a 50 km by 50 km square has firebreaks in rectangular strips 50 km by 0.01 km. The trees between two firebreaks are called a stand of trees. All firebreaks in this forest are parallel to each other and to one edge of the forest, with the first firebreak at

---

[11] Adapted from L. Carl Leinbach, *Calculus Laboratories Using Derive*, (Belmont, CA: Wadsworth, 1991). Reprinted by permission of Brooks/Cole Publishing Company, a division of International Thompson Publishing, Inc.

[12] Adapted from D. Quinney and R. Harding, *Calculus Connections* (New York: John Wiley & Sons, 1996).

the edge of the forest. The firebreaks are evenly spaced throughout the forest. (For example, Figure 4.91 shows four firebreaks.) The total area lost in the case of a fire is the area of the stand of trees in which the fire started plus the area of all the firebreaks.

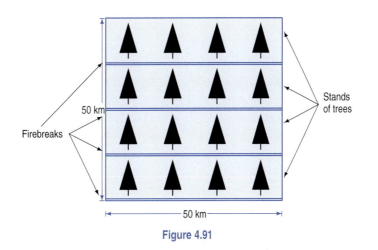

Figure 4.91

(i) Find the number of firebreaks that minimizes the total area lost to the forest in the case of a fire.

(ii) If a firebreak is 50 km by b km, find the optimal number of firebreaks as a function of b. If the width, b, of a firebreak is quadrupled, how does the optimal number of firebreaks change?

(b) Now suppose firebreaks are arranged in two equally spaced sets of parallel lines, as shown in Figure 4.92 The forest is a 50 km by 50 km square, and each firebreak is a rectangular strip 50 km by 0.01 km. Find the number of firebreaks in each direction that minimizes the total area lost to the forest in the case of a fire.

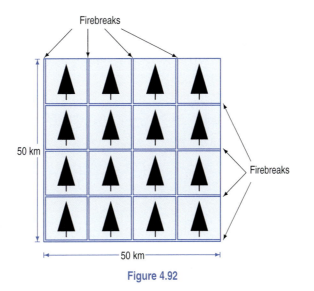

Figure 4.92

Chapter Five

KEY CONCEPT: THE DEFINITE INTEGRAL

We started Chapter 2 by calculating the velocity from the distance traveled. This led us to the notion of the derivative, or rate of change, of a function. Now we consider the reverse problem: given the velocity, how can we calculate the distance traveled? This leads to our second key concept, the *definite integral*, which computes the total change in a function from its rate of change. We then see that the definite integral can be used to calculate not only distance but also other quantities, such as the area under a curve and the average value of a function.

This chapter concludes with the Fundamental Theorem of Calculus, which tells us that we can use a definite integral to get information about a function from its derivative. Calculating derivatives and calculating definite integrals are, in a sense, reverse processes.

5.1   HOW DO WE MEASURE DISTANCE TRAVELED?

If the velocity of a moving object is a constant, we can find the distance it travels using the formula

$$\text{Distance} = \text{Velocity} \times \text{Time}.$$

In this section we see how to estimate the distance when the velocity is not a constant.

A Thought Experiment: How Far Did the Car Go?

Velocity Data Every Two Seconds

A car is moving with increasing velocity. We measure the car's velocity every two seconds and obtain the data in Table 5.1:

Table 5.1   *Velocity of car every two seconds*

Time (sec)	0	2	4	6	8	10
Velocity (ft/sec)	20	30	38	44	48	50

How far has the car traveled? Since we don't know how fast the car is moving at every moment, we can't calculate the distance exactly, but we can make an estimate. The velocity is increasing, so the car is going at least 20 ft/sec for the first two seconds. Since Distance = Velocity × Time, the car goes at least $(20)(2) = 40$ feet during the first two seconds. Likewise, it goes at least $(30)(2) = 60$ feet during the next two seconds, and so on. During the ten-second period it goes at least

$$(20)(2) + (30)(2) + (38)(2) + (44)(2) + (48)(2) = 360 \text{ feet}.$$

Thus, 360 feet is an underestimate of the total distance traveled during the ten seconds.

To get an overestimate, we can reason this way: During the first two seconds, the car's velocity is at most 30 ft/sec, so it moved at most $(30)(2) = 60$ feet. In the next two seconds it moved at most $(38)(2) = 76$ feet, and so on. Therefore, over the ten-second period it moved at most

$$(30)(2) + (38)(2) + (44)(2) + (48)(2) + (50)(2) = 420 \text{ feet}.$$

Therefore,

$$360 \text{ feet} \leq \text{Total distance traveled} \leq 420 \text{ feet}.$$

There is a difference of 60 feet between the upper and lower estimates.

Velocity Data Every One Second

What if we want a more accurate estimate? Then we make more frequent velocity measurements, say every second. The data is in Table 5.2.

As before, we get a lower estimate for each second by using the velocity at the beginning of that second. During the first second the velocity is at least 20 ft/sec, so the car travels at least $(20)(1) = 20$ feet. During the next second the car moves at least 26 feet, and so on. We have

$$\text{New lower estimate} = (20)(1) + (26)(1) + (30)(1) + (35)(1) + (38)(1)$$
$$+ (42)(1) + (44)(1) + (46)(1) + (48)(1) + (49)(1)$$
$$= 378 \text{ feet}.$$

Table 5.2   *Velocity of car every second*

Time (sec)	0	1	2	3	4	5	6	7	8	9	10
Velocity (ft/sec)	20	26	30	35	38	42	44	46	48	49	50

Notice that this lower estimate is greater than the old lower estimate of 360 feet.

We get a new upper estimate by considering the velocity at the end of each second. During the first second the velocity is at most 26 ft/sec, so the car moves at most $(26)(1) = 26$ feet; in the next second it moves at most 30 feet, and so on.

$$\begin{aligned}
\text{New upper estimate} = {} & (26)(1) + (30)(1) + (35)(1) + (38)(1) + (42)(1) \\
& + (44)(1) + (46)(1) + (48)(1) + (49)(1) + (50)(1) \\
= {} & 408 \text{ feet.}
\end{aligned}$$

This is less than the old upper estimate of 420 feet. Now we know that

$$378 \text{ feet} \le \text{Total distance traveled} \le 408 \text{ feet.}$$

The difference between upper and lower estimates is now 30 feet, half of what it was before. By halving the interval of measurement, we have halved the difference between the upper and lower estimates.

Visualizing Distance on the Velocity Graph: Two Second Data

We can represent both upper and lower estimates on a graph of the velocity. On such a graph we can also see how changing the time interval between velocity measurements changes the accuracy of our estimates.

The velocity can be graphed by plotting the two-second data in Table 5.1 and drawing a curve through the data points. (See Figure 5.1.) The area of the first dark rectangle is $(20)(2) = 40$, the lower estimate of the distance moved during the first two seconds. The area of the second dark rectangle is $(30)(2) = 60$, the lower estimate for the distance moved in the next two seconds. The total area of the dark rectangles represents the lower estimate for the total distance moved during the ten seconds.

If the dark and light rectangles are considered together, the first area is $(30)(2) = 60$, the upper estimate for the distance moved in the first two seconds. The second area is $(38)(2) = 76$, the upper estimate for the next two seconds. The upper estimate for the total distance is represented by the sum of the areas of the dark and light rectangles. Therefore, the area of the light rectangles alone represents the difference between the two estimates.

To visualize the difference between the two estimates, look at Figure 5.1 and imagine the light rectangles all pushed to the right and stacked on top of each other. This gives a rectangle of width 2 and height 30. The height, 30, is the difference between the initial and final values of the velocity: $30 = 50 - 20$. The width, 2, is the time interval between velocity measurements.

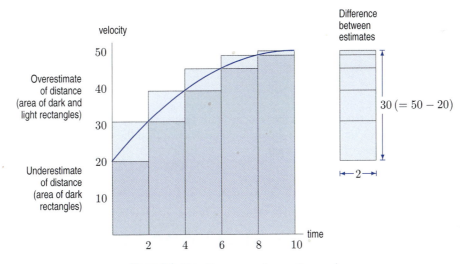

Figure 5.1: Velocity measured every 2 seconds

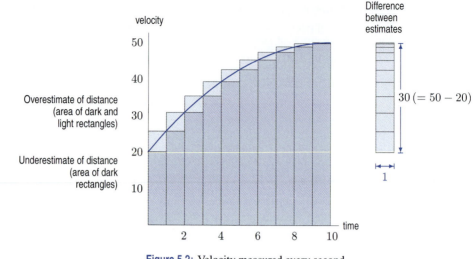

Figure 5.2: Velocity measured every second

Visualizing Distance on the Velocity Graph: One Second Data

The data for the velocities measured every second is graphed in Figure 5.2. The area of the dark rectangles again represents the lower estimate, and the dark and light rectangles together represent the upper estimate. As before, the difference between the two estimates is represented by the area of the light rectangles. This difference can be calculated by stacking the light rectangles vertically, giving a rectangle of the same height as before but of half the width. Its area is therefore half what it was before. Again, the height of this stack is $50 - 20 = 30$, but its width is the time interval between measurements, which is 1.

Example 1   What would be the difference between the upper and lower estimates if the velocity were given every tenth of a second? Every hundredth of a second? Every thousandth of a second?

Solution   Every tenth of a second: Difference between estimates $= (50 - 20)(1/10) = 3$ feet.
Every hundredth of a second: Difference between estimates $= (50 - 20)(1/100) = 0.3$ feet.
Every thousandth of a second: Difference between estimates $= (50 - 20)(1/1000) = 0.03$ feet.

Example 2   How frequently must the velocity be recorded in order to estimate the total distance traveled to within 0.1 feet?

Solution   The difference between the velocity at the beginning and end of the observation period is $50 - 20 = 30$. If the time between the measurements is h, then the difference between the upper and lower estimates is $(30)h$. We want

$$(30)h < 0.1,$$

or

$$h < \frac{0.1}{30} \approx 0.0033.$$

So if the measurements are made less than 0.0033 seconds apart, the distance estimate is accurate to within 0.1 feet.

Making the Estimates for Distance Precise

We now obtain an exact expression for the total distance traveled. We express the exact total distance traveled as a limit of upper or lower estimates.

We want to know the distance traveled by a moving object over the time interval $a \leq t \leq b$. Let the velocity at time t be given by the function $v = f(t)$. We take measurements of $f(t)$ at equally spaced times $t_0, t_1, t_2, \ldots, t_n$, with time $t_0 = a$ and time $t_n = b$. The time interval between any two consecutive measurements is

$$\Delta t = \frac{b-a}{n},$$

where Δt means the change, or increment, in t.

During the first time interval, the velocity can be approximated by $f(t_0)$, so the distance traveled is approximately

$$f(t_0)\Delta t.$$

During the second time interval, the velocity is about $f(t_1)$, so the distance traveled is about

$$f(t_1)\Delta t.$$

Continuing in this way and adding up all the estimates, we get an estimate for the total distance. In the last interval, the velocity is approximately $f(t_{n-1})$, so the last term is $f(t_{n-1})\Delta t$:

$$\text{Total distance traveled} \atop \text{between } t = a \text{ and } t = b \quad \approx f(t_0)\Delta t + f(t_1)\Delta t + f(t_2)\Delta t + \cdots + f(t_{n-1})\Delta t.$$

This is called a *left-hand sum* because we used the value of velocity from the left end of each time interval. It is represented by the sum of the areas of the rectangles in Figure 5.3.

We can also calculate a *right-hand sum* by using the value of the velocity at the right end of each time interval. In that case the estimate for the first interval is $f(t_1)\Delta t$, for the second interval it is $f(t_2)\Delta t$, and so on. The estimate for the last interval is now $f(t_n)\Delta t$, so

$$\text{Total distance traveled} \atop \text{between } t = a \text{ and } t = b \quad \approx f(t_1)\Delta t + f(t_2)\Delta t + f(t_3)\Delta t + \cdots + f(t_n)\Delta t.$$

The right-hand sum is represented by the area of the rectangles in Figure 5.4.

If f is an increasing function, as in Figures 5.3 and 5.4, the left-hand sum is an underestimate of the total distance traveled. Similarly, if f is increasing, the right-hand sum is an overestimate. If f is decreasing, as in Figure 5.5, then the roles of the two sums are reversed.

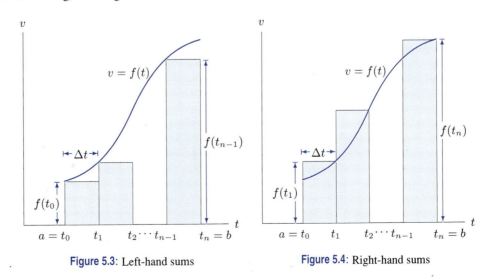

Figure 5.3: Left-hand sums   Figure 5.4: Right-hand sums

For either increasing or decreasing functions, the exact value of the distance traveled lies somewhere between the two estimates. Thus the accuracy of our estimate depends on how close these two sums are. For a function which is increasing throughout or decreasing throughout the interval $[a, b]$:

$$\left| \begin{array}{c} \text{Difference between} \\ \text{upper and lower estimates} \end{array} \right| = \left| \begin{array}{c} \text{Difference between} \\ f(a) \text{ and } f(b) \end{array} \right| \cdot \Delta t = |f(b) - f(a)| \cdot \Delta t.$$

(Absolute values are used to make the difference nonnegative.) In Figure 5.5, the area of the light rectangles is the difference between estimates. By making the time interval, Δt, between measurements small enough, we can make this difference between lower and upper estimates as small as we like.

In the car example, as n increased, the overestimates of the distance traveled decreased and the underestimates increased, trapping the exact distance between them. To find exactly the total distance traveled between $t = a$ and $t = b$, we take the limit of the sums, as n goes to infinity.

$$\begin{array}{rl} \text{Total distance traveled} \\ \text{between } t = a \text{ and } t = b = & \lim_{n \to \infty} \text{ (Left-hand sum)} \\ = & \lim_{n \to \infty} [f(t_0)\Delta t + f(t_1)\Delta t + \cdots + f(t_{n-1})\Delta t] \\ = & \text{Area under curve } f(t) \text{ from } t = a \text{ to } t = b \end{array}$$

and

$$\begin{array}{rl} \text{Total distance traveled} \\ \text{between } t = a \text{ and } t = b = & \lim_{n \to \infty} \text{ (Right-hand sum)} \\ = & \lim_{n \to \infty} [f(t_1)\Delta t + f(t_2)\Delta t + \cdots + f(t_n)\Delta t] \\ = & \text{Area under curve } f(t) \text{ from } t = a \text{ to } t = b. \end{array}$$

Provided f is continuous, the limits of the left and the right sums are both equal to the total distance traveled. This method of calculating the distance by taking the limit of a sum works even if the velocity is not increasing throughout, or decreasing throughout, the interval.

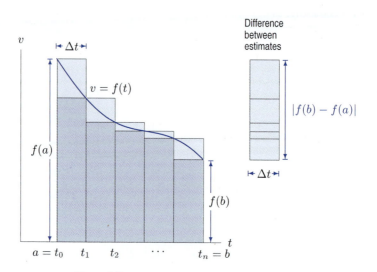

Figure 5.5: Left and right sums if f is decreasing

Exercises and Problems for Section 5.1

Exercises

1. A village wishes to measure the quantity of water that is piped to a factory during a typical morning. A gauge on the water line gives the flowrate (in cubic meters per hour) at any instant. The flowrate is about 100 m^3/hr at 6 am and increases steadily to about 280 m^3/hr at 9 am.

   (a) Using only this information, give your best estimate of the total volume of water used by the factory between 6 am and 9 am.

   (b) How often should the flowrate gauge be read to obtain an estimate of this volume to within 6 m^3?

2. A car comes to a stop six seconds after the driver applies the brakes. While the brakes are on, the following velocities are recorded:

Time since brakes applied (sec)	0	2	4	6
Velocity (ft/sec)	88	45	16	0

   (a) Give lower and upper estimates for the distance the car traveled after the brakes were applied.

   (b) On a sketch of velocity against time, show the lower and upper estimates of part (a).

3. Roger runs a marathon. His friend Jeff rides behind him on a bicycle and clocks his speed every 15 minutes. Roger starts out strong, but after an hour and a half he is so exhausted that he has to stop. Jeff's data follow:

Time since start (min)	0	15	30	45	60	75	90
Speed (mph)	12	11	10	10	8	7	0

   (a) Assuming that Roger's speed is never increasing, give upper and lower estimates for the distance Roger ran during the first half hour.

   (b) Give upper and lower estimates for the distance Roger ran in total during the entire hour and a half.

   (c) How often would Jeff have needed to measure Roger's speed in order to find lower and upper estimates within 0.1 mile of the actual distance he ran?

4. Coal gas is produced at a gasworks. Pollutants in the gas are removed by scrubbers, which become less and less efficient as time goes on. The following measurements, made at the start of each month, show the rate at which pollutants are escaping (in tons/month) in the gas:

Time (months)	0	1	2	3	4	5	6
Rate pollutants escape	5	7	8	10	13	16	20

   (a) Make an overestimate and an underestimate of the total quantity of pollutants that escaped during the first month.

   (b) Make an overestimate and an underestimate of the total quantity of pollutants that escaped during the six months.

   (c) How often would measurements have to be made to find overestimates and underestimates which differ by less than 1 ton from the exact quantity of pollutants that escaped during the first six months?

Problems

5. Two cars start at the same time and travel in the same direction along a straight road. Figure 5.6 gives the velocity, v, of each car as a function of time, t. Which car:

   (a) Attains the larger maximum velocity?
   (b) Stops first?
   (c) Travels farther?

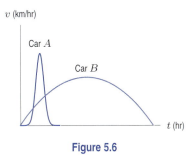

Figure 5.6

6. Two cars travel in the same direction along a straight road. Figure 5.7 shows the velocity, v, of each car at time t. Car B starts 2 hours after car A and car B reaches a maximum velocity of 50 km/hr.

   (a) For approximately how long does each car travel?
   (b) Estimate car A's maximum velocity.
   (c) Approximately how far does each car travel?

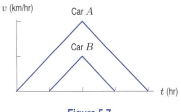

Figure 5.7

7. A student is speeding down Route 11 in his fancy red Porsche when his radar system warns him of an obstacle 400 feet ahead. He immediately applies the brakes, starts to slow down, and spots a skunk in the road directly ahead of him. The "black box" in the Porsche records the car's speed every two seconds, producing the following table. The speed decreases throughout the 10 seconds it takes to stop, although not necessarily at a uniform rate.

Time since brakes applied (sec)	0	2	4	6	8	10
Speed (ft/sec)	100	80	50	25	10	0

(a) What is your best estimate of the total distance that the student's car traveled before coming to rest?

(b) Which one of the following statements can you justify from the information given?

    (i) The car stopped before getting to the skunk.

    (ii) The "black box" data is inconclusive. The skunk may or may not have been hit.

    (iii) The skunk was hit by the car.

8. At time, t, in seconds, your velocity, v, in meters/second, is given by

$$v(t) = 1 + t^2 \quad \text{for} \quad 0 \le t \le 6.$$

Use $\Delta t = 2$ to estimate the distance traveled during this time. Find the left- and right-hand sums, and then average the two.

9. For time, t, in hours, $0 \le t \le 1$, a bug is crawling at a velocity, v, in meters/hour given by

$$v = \frac{1}{1+t}.$$

Use $\Delta t = 0.2$ to estimate the distance that the bug crawls during this hour. Find an overestimate and an underestimate. Then average the two to get a new estimate.

10. In Figure 5.8, use the grid to get upper and a lower estimates of the area of the region bounded by the curve, the horizontal axis and the vertical lines $x = 3$ and $x = -3$.

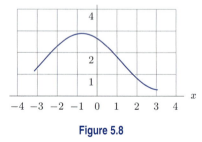

Figure 5.8

11. Figure 5.9 shows the graph of the velocity, v, of an object (in meters/sec). Estimate the total distance the object traveled between $t = 0$ and $t = 6$.

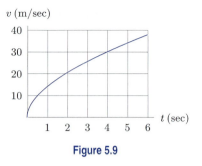

Figure 5.9

12. In Figure 5.10 estimate the shaded area with an error of at most 0.1.

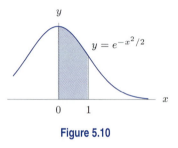

Figure 5.10

13. You jump out of an airplane. Before your parachute opens you fall faster and faster, but your acceleration decreases as you fall because of air resistance. The table gives your acceleration, a (in m/sec^2), after t seconds.

t	0	1	2	3	4	5
a	9.81	8.03	6.53	5.38	4.41	3.61

(a) Give upper and lower estimates of your speed at $t = 5$.

(b) Get a new estimate by taking the average of your upper and lower estimates. What does the concavity of the graph of acceleration tell you about your new estimate?

14. A car speeds up at a constant rate from 10 to 70 mph over a period of half an hour. Its fuel efficiency (in miles per gallon) at various speeds is shown in the table. Make lower and upper estimates of the quantity of fuel used during the half hour.

Speed (mph)	10	20	30	40	50	60	70
Fuel efficiency (mpg)	15	18	21	23	24	25	26

5.2 THE DEFINITE INTEGRAL

In Section 5.1 we saw how distance traveled can be approximated by sums and expressed exactly as the limit of a sum. In this section we construct these sums for any function f, whether or not it represents a velocity. Suppose $f(t)$ is a continuous function for $a \leq t \leq b$. We divide the interval from a to b into n equal subdivisions, and we call the width of an individual subdivision Δt, so

$$\Delta t = \frac{b - a}{n}.$$

Let $t_0, t_1, t_2, \ldots, t_n$ be endpoints of the subdivisions, as in Figures 5.11 and 5.12. As before, we construct two sums:

$$\text{Left-hand sum} = f(t_0)\Delta t + f(t_1)\Delta t + \cdots + f(t_{n-1})\Delta t$$

and

$$\text{Right-hand sum} = f(t_1)\Delta t + f(t_2)\Delta t + \cdots + f(t_n)\Delta t.$$

These sums represent the areas of the rectangles in Figures 5.11 and 5.12, provided $f(t) \geq 0$. In Figure 5.11, the first rectangle has width Δt and height $f(t_0)$, since the top of its left edge just touches the curve, and hence it has area $f(t_0)\Delta t$. The second rectangle has width Δt and height $f(t_1)$, and hence has area $f(t_1)\Delta t$, and so on. The sum of all these areas is the left-hand sum. The right-hand sum, shown in Figure 5.12, is constructed in the same way, except that each rectangle touches the curve on its right edge instead of its left.

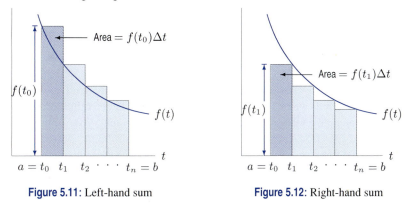

Figure 5.11: Left-hand sum   Figure 5.12: Right-hand sum

Writing Left and Right Sums Using Sigma Notation

Both the left-hand and right-hand sums can be written more compactly using *sigma*, or summation, notation. The symbol $\sum$ is a capital sigma, or Greek letter "S." We write

$$\text{Right-hand sum} = \sum_{i=1}^{n} f(t_i)\Delta t = f(t_1)\Delta t + f(t_2)\Delta t + \cdots + f(t_n)\Delta t.$$

The $\sum$ tells us to add terms of the form $f(t_i)\Delta t$. The "$i = 1$" at the base of the sigma sign tells us to start at $i = 1$, and the "n" at the top tells us to stop at $i = n$.

In the left-hand sum we start at $i = 0$ and stop at $i = n - 1$, so we write

$$\text{Left-hand sum} = \sum_{i=0}^{n-1} f(t_i)\Delta t = f(t_0)\Delta t + f(t_1)\Delta t + \cdots + f(t_{n-1})\Delta t.$$

Taking the Limit to Obtain the Definite Integral

Now we take the limit of these sums as n goes to infinity, as we did in the previous section. If f is continuous for $a \leq t \leq b$, the limits of the left- and right-hand sums exist and are equal. The *definite integral* is the limit of these sums. A formal definition of the definite integral is given in the online supplement to the text.

Suppose f is continuous for $a \le t \le b$. The **definite integral** of f from a to b, written

$$\int_a^b f(t) \, dt,$$

is the limit of the left-hand or right-hand sums with n subdivisions of $[a, b]$ as n gets arbitrarily large. In other words,

$$\int_a^b f(t) \, dt = \lim_{n \to \infty} (\text{Left-hand sum}) = \lim_{n \to \infty} \left(\sum_{i=0}^{n-1} f(t_i) \Delta t \right)$$

and

$$\int_a^b f(t) \, dt = \lim_{n \to \infty} (\text{Right-hand sum}) = \lim_{n \to \infty} \left(\sum_{i=1}^{n} f(t_i) \Delta t \right).$$

Each of these sums is called a *Riemann sum*, f is called the *integrand*, and a and b are called the *limits of integration*.

The "$\int$" notation comes from an old-fashioned "S," which stands for "sum" in the same way that $\sum$ does. The "dt" in the integral comes from the factor Δt. Notice that the limits on the $\sum$ symbol are 0 and $n-1$ for the left-hand sum, and 1 and n for the right-hand sum, whereas the limits on the $\int$ sign are a and b.

Computing a Definite Integral

In practice, we often approximate definite integrals numerically using a calculator or computer. They use programs which compute sums for larger and larger values of n, and eventually give a value for the integral. Different calculators and computers may give slightly different estimates, owing to round-off error and the fact that they may use different approximation methods. Some (but not all) definite integrals can be computed exactly. However, any definite integral can be approximated numerically.

In the next example, we see how numerical approximation works. For each value of n, we calculate an over- and an under-estimate for the integral. As we increase the value of n the over- and under-estimates get closer together, trapping the value of the integral between them. By increasing the value of n sufficiently, we can calculate the integral to any desired accuracy.

Example 1   Calculate the left-hand and right-hand sums with $n = 2$ and $n = 10$ for $\int_1^2 \dfrac{1}{t} \, dt$. How do the values of these sums compare with the exact value of the integral?

Solution   Here $a = 1$ and $b = 2$, so for $n = 2$, $\Delta t = (2 - 1)/2 = 0.5$. Therefore, $t_0 = 1$, $t_1 = 1.5$ and $t_2 = 2$. (See Figure 5.13.) We have

$$\text{Left-hand sum} = f(1)\Delta t + f(1.5)\Delta t = 1(0.5) + \frac{1}{1.5}(0.5) \approx 0.8333,$$

$$\text{Right-hand sum} = f(1.5)\Delta t + f(2)\Delta t = \frac{1}{1.5}(0.5) + \frac{1}{2}(0.5) \approx 0.5833.$$

From Figure 5.13 we see that the left-hand sum is bigger than the area under the curve and the right-hand sum is smaller. So the area under the curve $f(t) = 1/t$ from $t = 1$ to $t = 2$ is between 0.5833 and 0.8333 :

$$0.5833 < \int_1^2 \frac{1}{t} \, dt < 0.8333.$$

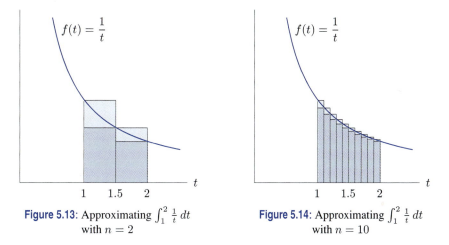

Figure 5.13: Approximating $\int_1^2 \frac{1}{t}\, dt$
with $n = 2$

Figure 5.14: Approximating $\int_1^2 \frac{1}{t}\, dt$
with $n = 10$

When $n = 10$, we have $\Delta t = (2 - 1)/10 = 0.1$ (see Figure 5.14), so

$$\text{Left sum} = f(1)\Delta t + f(1.1)\Delta t + \cdots + f(1.9)\Delta t = \left(1 + \frac{1}{1.1} + \cdots + \frac{1}{1.9}\right) 0.1 \approx 0.7188,$$

$$\text{Right sum} = f(1.1)\Delta t + f(1.2)\Delta t + \cdots + f(2)\Delta t = \left(\frac{1}{1.1} + \frac{1}{1.2} + \cdots + \frac{1}{2}\right) 0.1 \approx 0.6688.$$

From Figure 5.14 we see that the left-hand sum is larger than the area under the curve, and the right-hand sum smaller, so

$$0.6688 < \int_1^2 \frac{1}{t}\, dt < 0.7188.$$

The left- and right-hand sums trap the exact value of the integral between them. As the subdivisions become finer, the left- and right-hand sums get closer together.

Example 2   Use left and right sums with $n = 250$ for $\int_1^2 \frac{1}{t}\, dt$ to estimate the value of the integral.

Solution   Using a calculator or computer, we see that

$$0.6921 < \int_1^2 \frac{1}{t}\, dt < 0.6941.$$

So, to two decimal places, we can say that

$$\int_1^2 \frac{1}{t}\, dt \approx 0.69.$$

The exact value is known to be $\int_1^2 \frac{1}{t}\, dt = \ln 2 = 0.693147\ldots$.

The Definite Integral as an Area

If $f(x)$ is positive we can interpret each term $f(x_0)\Delta x, f(x_1)\Delta x, \ldots$ in a left- or right-hand Riemann sum as the area of a rectangle. See Figure 5.15. As the width Δx of the rectangles approaches zero, the rectangles fit the curve of the graph more exactly, and the sum of their areas gets closer and closer to the area under the curve shaded in Figure 5.16. This suggests that:

When $f(x)$ is positive and $a < b$:

$$\text{Area under graph of } f \atop \text{between } a \text{ and } b = \int_a^b f(x)\,dx.$$

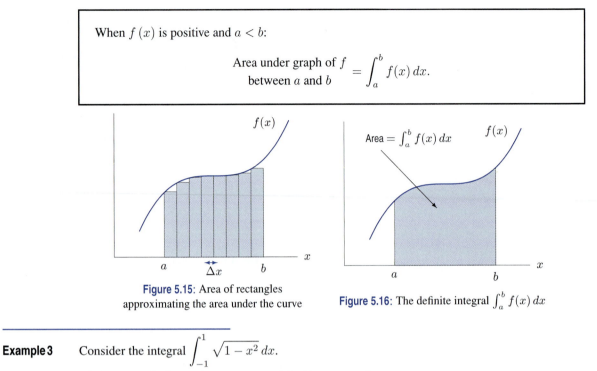

Figure 5.15: Area of rectangles
approximating the area under the curve

Figure 5.16: The definite integral $\int_a^b f(x)\,dx$

Example 3   Consider the integral $\int_{-1}^1 \sqrt{1-x^2}\,dx$.

(a) Interpret the integral as an area, and find its exact value.
(b) Estimate the integral using a calculator or computer. Compare your answer to the exact value.

Solution   (a) The integral is the area under the graph of $y = \sqrt{1-x^2}$ between -1 and 1. See Figure 5.17. Rewriting this equation as $x^2 + y^2 = 1$, we see that the graph is a semicircle of radius 1 and area $\pi/2$.
(b) A calculator gives the value of the integral as $1.5707963\ldots$.

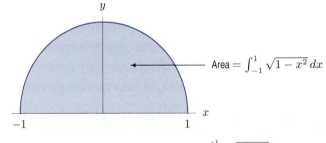

Figure 5.17: Area interpretation of $\int_{-1}^1 \sqrt{1-x^2}\,dx$

When $f(x)$ is Not Positive

We have assumed in drawing Figure 5.16 that the graph of $f(x)$ lies above the x-axis. If the graph lies below the x-axis, then each value of $f(x)$ is negative, so each $f(x)\Delta x$ is negative, and the area gets counted negatively. In that case, the definite integral is the negative of the area.

When $f(x)$ is positive for some x values and negative for others, and $a < b$:

$\int_a^b f(x)\,dx$ is the sum of the areas above the x-axis, counted positively, and the areas below the x-axis, counted negatively.

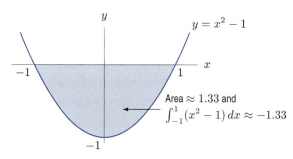

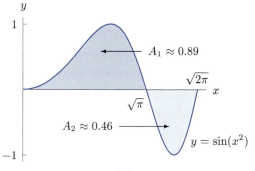
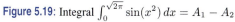

Figure 5.18: Integral $\int_{-1}^{1}(x^2 - 1)\, dx$ is negative of shaded area

Figure 5.19: Integral $\int_{0}^{\sqrt{2\pi}} \sin(x^2)\, dx = A_1 - A_2$

Example 4   How does the definite integral $\displaystyle\int_{-1}^{1}(x^2 - 1)\, dx$ relate to the area between the parabola $y = x^2 - 1$ and the x-axis?

Solution   A calculator gives $\int_{-1}^{1}(x^2 - 1)\, dx \approx -1.33$. The parabola lies below the axis between $x = -1$ and $x = 1$. (See Figure 5.18.) So the area between the parabola and the x-axis is approximately 1.33.

Example 5   Interpret the definite integral $\displaystyle\int_{0}^{\sqrt{2\pi}} \sin(x^2)\, dx$ in terms of areas.

Solution   The integral is the area above the x-axis, A_1, minus the area below the x-axis, A_2. See Figure 5.19. Approximating the integral with a calculator gives

$$\int_{0}^{\sqrt{2\pi}} \sin(x^2)\, dx \approx 0.43.$$

The graph of $y = \sin(x^2)$ crosses the x-axis where $x^2 = \pi$, that is, at $x = \sqrt{\pi}$. The next crossing is at $x = \sqrt{2\pi}$. Breaking the integral into two parts and calculating each one separately gives

$$\int_{0}^{\sqrt{\pi}} \sin(x^2)\, dx \approx 0.89 \quad\text{and}\quad \int_{\sqrt{\pi}}^{\sqrt{2\pi}} \sin(x^2)\, dx \approx -0.46.$$

So $A_1 \approx 0.89$ and $A_2 \approx 0.46$. Then, as we would expect,

$$\int_{0}^{\sqrt{2\pi}} \sin(x^2)\, dx = A_1 - A_2 \approx 0.89 - 0.46 = 0.43.$$

More General Riemann Sums

Left- and right-hand sums are special cases of Riemann sums. For a general Riemann sum we allow subdivisions to have different lengths. Also, instead of evaluating f only at the left or right endpoint of each subdivision, we allow it to be evaluated anywhere in the subdivision. Thus, a general Riemann sum has the form

$$\sum_{i=1}^{n} \text{Value of } f \text{ at some point in } i\text{-th subdivision} \cdot \text{Length of } i\text{-th subdivision}.$$

(See Figure 5.20.) As before, we let $x_0, x_1, \ldots, x_n$ be the endpoints of the subdivisions, so the length of the i-th subdivision is $\Delta x_i = x_i - x_{i-1}$. For each i we choose a point c_i in the i-th subinterval at which to evaluate f, leading to the following definition:

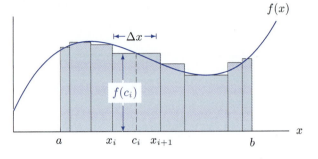

Figure 5.20: A general Riemann sum approximating $\int_a^b f(x)\,dx$

A general Riemann sum for f on the interval $[a, b]$ is a sum of the form

$$\sum_{i=1}^{n} f(c_i)\Delta x_i,$$

where $a = x_0 < x_1 < \cdots < x_n = b$, and, for $i = 1, \ldots, n$, $\Delta x_i = x_i - x_{i-1}$, and $x_{i-1} \le c_i \le x_i$.

If f is continuous, we can make a general Riemann sum as close as we like to the value of the definite integral by making the interval lengths small enough. Thus, in approximating definite integrals or in proving theorems about them, we can use general Riemann sums rather than left- or right-hand sums. The online theory supplement shows how general Riemann sums are used in proving theorems.

Exercises and Problems for Section 5.2

Exercises

1. Using Figure 5.21, draw rectangles representing each of the following Riemann sums for the function f on the interval $0 \le t \le 8$. Calculate the value of each sum.

   (a) Left-hand sum with $\Delta t = 4$
   (b) Right-hand sum with $\Delta t = 4$
   (c) Left-hand sum with $\Delta t = 2$
   (d) Right-hand sum with $\Delta t = 2$

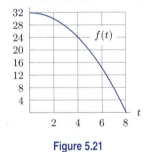

Figure 5.21

2. The graph of a function $f(t)$ is given in Figure 5.22. Which of the following four numbers could be an estimate of $\int_0^1 f(t)\,dt$ accurate to two decimal places? Explain how you chose your answer.

   (a) -98.35 **(b)** 71.84
   (c) 100.12 **(d)** 93.47

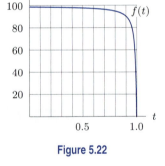

Figure 5.22

3. Use Figure 5.23 to estimate $\int_0^3 f(x)\,dx$.

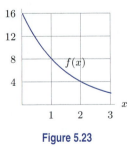

Figure 5.23

4. Use Figure 5.24 to estimate $\int_{-10}^{15} f(x)dx$.

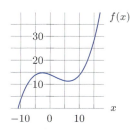

Figure 5.24

5. Use Figure 5.25 to estimate $\int_{-15}^{20} f(x)\,dx$.

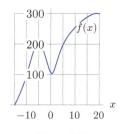

Figure 5.25

6. Use the table to estimate $\int_0^{40} f(x)dx$. What value of n and what Δx did you use?

x	0	10	20	30	40
$f(x)$	350	410	435	450	460

7. Use the table to estimate $\int_0^{12} f(x)\,dx$.

x	0	3	6	9	12
$f(x)$	32	22	15	11	9

8. Use the table to estimate $\int_0^{15} f(x)\,dx$.

x	0	3	6	9	12	15
$f(x)$	50	48	44	36	24	8

In Exercises 9–11, use a calculator or a computer to find the value of the definite integral to four decimal places.

9. $\int_0^3 2^x\,dx$ **10.** $\int_0^1 \sin(t^2)dt$ **11.** $\int_{-1}^1 e^{-x^2}\,dx$

For Exercises 12–14, construct left- and right-hand sums with 2, 10, 50, and 250 subdivisions. Observe the limit to which your sums are tending as the number of subdivisions gets larger, and hence estimate the value of the definite integral.

12. $\int_0^1 x^3\,dx$ **13.** $\int_0^{\pi/2} \cos x\,dx$

14. $\int_2^3 \sin(t^2)\,dt$

Estimate the area of the regions in Exercises 15–21.

15. Under $y = \ln x$ for $1 \le x \le 4$.

16. Under the curve $y = \cos t$ for $0 \le t \le \pi/2$.

17. Under the curve $y = 7 - x^2$ and above the x-axis.

18. Under the curve $y = \cos\sqrt{x}$ for $0 \le x \le 2$.

19. Under the curve $y = e^x$ and above the line $y = 1$ for $0 \le x \le 2$.

20. Between $y = x^2$ and $y = x^3$ for $0 \le x \le 1$.

21. Between $y = x^{1/2}$ and $y = x^{1/3}$ for $0 \le x \le 1$.

Problems

22. (a) On a sketch of $y = \ln x$, represent the left Riemann sum with $n = 2$ approximating $\int_1^2 \ln x\,dx$. Write out the terms in the sum, but do not evaluate it.
   (b) On another sketch, represent the right Riemann sum with $n = 2$ approximating $\int_1^2 \ln x\,dx$. Write out the terms in the sum, but do not evaluate it.
   (c) Which sum is an overestimate? Which sum is an underestimate?

23. (a) Use a calculator or computer to find $\int_0^6 (x^2 + 1)\,dx$. Represent this value as the area under a curve.
   (b) Estimate $\int_0^6 (x^2 + 1)\,dx$ using a left-hand sum with $n = 3$. Represent this sum graphically on a sketch of $f(x) = x^2 + 1$. Is this sum an overestimate or underestimate of the true value found in part (a)?
   (c) Estimate $\int_0^6 (x^2 + 1)\,dx$ using a right-hand sum with $n = 3$. Represent this sum on your sketch. Is this sum an overestimate or underestimate?

24. Estimate $\int_1^2 x^2\,dx$ using left- and right-hand sums with four subdivisions. How far from the true value of the integral could your estimate be?

25. In Exercises 20 and 21 we calculated the areas between $y = x^2$ and $y = x^3$ and between $y = x^{1/2}$ and $y = x^{1/3}$ on $0 \le x \le 1$. Explain why you would expect these two areas to be equal.

26. (a) Sketch a graph of $f(x) = x(x + 2)(x - 1)$.
   (b) Find the total area between the graph and the x-axis between $x = -2$ and $x = 1$.
   (c) Find $\int_{-2}^1 f(x)\,dx$ and interpret it in terms of areas.

27. Compute the definite integral $\int_0^4 \cos\sqrt{x}\,dx$ and interpret the result in terms of areas.

28. Without computing the integral, decide if $\int_0^{2\pi} e^{-x} \sin x\,dx$ is positive or negative, and explain your decision. [Hint: Sketch $e^{-x} \sin x$.]

29. Suppose that we use $n = 500$ subintervals to approximate $\int_{-1}^{1}(2x^3 + 4)\,dx$. Without computing the Riemann sums, find the difference between the right- and left-hand Riemann sums.

30. Use Figure 5.26 to find the values of

(a) $\displaystyle\int_{a}^{b} f(x)\,dx$ (b) $\displaystyle\int_{b}^{c} f(x)\,dx$

(c) $\displaystyle\int_{a}^{c} f(x)\,dx$ (d) $\displaystyle\int_{a}^{c} |f(x)|\,dx$

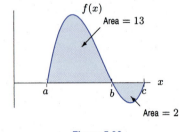

Figure 5.26

31. Given $\int_{-2}^{0} f(x)\,dx = 4$ and Figure 5.27, estimate:

(a) $\int_{0}^{2} f(x)\,dx$ (b) $\int_{-2}^{2} f(x)\,dx$

(c) The total shaded area

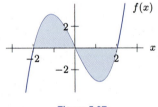

Figure 5.27

32. (a) Using Figure 5.28, find $\int_{-3}^{0} f(x)\,dx$.

(b) If the area of the shaded region is A, estimate $\int_{-3}^{4} f(x)\,dx$.

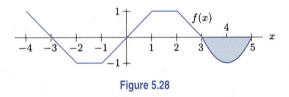

Figure 5.28

33. Three terms of a left-hand sum used to approximate a definite integral $\int_{a}^{b} f(x)\,dx$ are as follows.

$$\left(2 + 0 \cdot \frac{4}{3}\right)^2 \cdot \frac{4}{3} + \left(2 + 1 \cdot \frac{4}{3}\right)^2 \cdot \frac{4}{3} + \left(2 + 2 \cdot \frac{4}{3}\right)^2 \cdot \frac{4}{3}.$$

Find possible values for a and b and a possible formula for $f(x)$.

34. Consider the integral $\int_{1}^{2}(1/t)\,dt$. In Example 1, by dividing the interval $1 \le t \le 2$ into 10 equal parts, we showed that

$$0.1\left[\frac{1}{1.1} + \frac{1}{1.2} + \ldots + \frac{1}{2}\right] \le \int_{1}^{2}\frac{1}{t}\,dt$$

and

$$\int_{1}^{2}\frac{1}{t}\,dt \le 0.1\left[\frac{1}{1} + \frac{1}{1.1} + \ldots + \frac{1}{1.9}\right].$$

(a) Now divide the interval $1 \le t \le 2$ into n equal parts to show that

$$\sum_{r=1}^{n}\frac{1}{n+r} \quad < \quad \int_{1}^{2}\frac{1}{t}\,dt \quad < \quad \sum_{r=0}^{n-1}\frac{1}{n+r}.$$

(b) Show that the difference between the upper and lower sums in part (a) is $1/(2n)$.

(c) The exact value of $\int_{1}^{2}(1/t)\,dt$ is $\ln 2$. How large should n be to approximate $\ln 2$ with an error of at most $5 \cdot 10^{-6}$, using one of the sums in part (a)?

35. Write a few sentences in support of or in opposition to the following statement:

"If a left-hand sum underestimates a definite integral by a certain amount, then the corresponding right-hand sum will overestimate the integral by the same amount."

36. Using the graph of $2 + \cos x$, for $0 \le x \le 4\pi$, list the following quantities in increasing order: the value of the integral $\int_{0}^{4\pi}(2 + \cos x)\,dx$, the left sum with $n = 2$ subdivisions, and the right sum with $n = 2$ subdivisions.

37. Sketch the graph of a function f (you do not need to give a formula for f) on an interval $[a, b]$ with the property that with $n = 2$ subdivisions,

$$\int_{a}^{b} f(x)\,dx < \text{Left-hand sum} < \text{Right-hand sum.}$$

5.3 INTERPRETATIONS OF THE DEFINITE INTEGRAL

The Definite Integral of a Rate Gives Total Change

We have seen how the definite integral of a velocity function can be interpreted as total distance traveled. If $v(t)$ is velocity and $s(t)$ is position, then $v(t) = s'(t)$ and we know that

$$\text{Total change in position} = s(b) - s(a) = \int_a^b s'(t)\, dt.$$

Generalizing, in this section we see that the integral of the rate of change of any quantity gives the total change in that quantity.

Suppose $F'(t)$ is the rate of change of some quantity $F(t)$ with respect to time, and that we are interested in the total change in $F(t)$ between $t = a$ and $t = b$. We divide the interval $a \leq t \leq b$ into n subintervals, each of length Δt. For each small interval, we estimate the change in $F(t)$, written ΔF, and then add all these up. In each subinterval we assume the rate of change of $F(t)$ is approximately constant, so that we can say

$$\Delta F \approx \text{Rate of change of } F \times \text{Time elapsed}.$$

For the first subinterval, from t_0 to t_1, the rate of change of $F(t)$ is approximately $F'(t_0)$, so

$$\Delta F \approx F'(t_0)\, \Delta t.$$

Similarly, for the second interval

$$\Delta F \approx F'(t_1)\, \Delta t.$$

Summing over all the subintervals, we get

$$\begin{array}{c} \text{Total change in } F(t) \\ \text{between } t = a \text{ and } t = b \end{array} = \sum_{i=0}^{n-1} \Delta F \approx \sum_{i=0}^{n-1} F'(t_i)\, \Delta t.$$

We have approximated the change in $F(t)$ as a left-hand sum.

However, the total change in $F(t)$ between the times $t = a$ and $t = b$ is simply $F(b) - F(a)$. Taking the limit as n goes to infinity converts the Riemann sum to a definite integral and suggests the following result:[1]

$$F(b) - F(a) = \begin{array}{c} \text{Total change in } F(t) \\ \text{between } t = a \text{ and } t = b \end{array} = \int_a^b F'(t)\, dt.$$

This result is an interpretation of the Fundamental Theorem of Calculus, which we discuss in Section 5.4. It is used to find the total change, $F(b) - F(a)$, from the rate, $F'(t)$. If we also know $F(a)$, the result enables us to reconstruct the function F from its derivative F'.

Example 1   Let $F(t)$ represent a bacteria population which is 5 million at time $t = 0$. After t hours, the population is growing at an instantaneous rate of 2^t million bacteria per hour. Estimate the total increase in the bacteria population during the first hour, and the population at $t = 1$.

Solution   Since the rate at which the population is growing is $F'(t) = 2^t$, we have

$$\text{Change in population} = F(1) - F(0) = \int_0^1 2^t\, dt.$$

Using a calculator to evaluate the integral,

$$\text{Change in population} = \int_0^1 2^t\, dt \approx 1.44 \text{ million bacteria}.$$

Since $F(0) = 5$, the population at $t = 1$ is given by

$$\text{Population} = F(1) = F(0) + \int_0^1 2^t\, dt \approx 5 + 1.44 = 6.44 \text{ million}.$$

---

[1]We could equally well have used a right-hand sum, since the definite integral is their common limit.

The Notation and Units for the Definite Integral

Just as the Leibniz notation dy/dx for the derivative reminds us that the derivative is the limit of a ratio of differences, the notation for the definite integral helps us recall the meaning of the integral. The symbol

$$\int_a^b f(x)\, dx$$

reminds us that an integral is a limit of sums (the integral sign is an old-fashioned S) of terms of the form "$f(x)$ times a small difference of x." Officially, dx is not a separate entity, but a part of the whole integral symbol. Just as one thinks of d/dx as a single symbol meaning "the derivative with respect to x of...," one can think of $\int_a^b \ldots dx$ as a single symbol meaning "the integral of ... with respect to x."

However, many scientists and mathematicians informally think of dx as an "infinitesimally" small bit of x multiplied by $f(x)$. This viewpoint is often the key to interpreting the meaning of a definite integral. For example, if $f(t)$ is the velocity of a moving particle at time t, then $f(t)\, dt$ may by thought of informally as velocity × time, giving the distance traveled by the particle during a small bit of time dt. The integral $\int_a^b f(t)\, dt$ may then be thought of as the sum of all these small distances, giving us the net change in position of the particle between $t = a$ and $t = b$. The notation for the integral suggests units for the value of the integral. Since the terms being added up are products of the form "$f(x)$ times a difference in x," the unit of measurement for $\int_a^b f(x)\, dx$ is the product of the units for $f(x)$ and the units for x. For example, if $f(t)$ is velocity measured in meters/second and t is time measured in seconds, then

$$\int_a^b f(t)\, dt$$

has units of (meters/sec)×(sec) = meters. This is what we expect, since the value of this integral represents change in position.

As another example, graph $y = f(x)$ with the same units of measurement of length along the x- and y-axes, say cm. Then $f(x)$ and x are measured in the same units, so

$$\int_a^b f(x)\, dx$$

is measured in square units of cm × cm = cm^2. Again, this is what we would expect since in this context the integral represents an area.

The Definite Integral as an Average

We know how to find the average of n numbers: Add them and divide by n. But how do we find the average value of a continuously varying function? Let us consider an example. Suppose $f(t)$ is the temperature at time t, measured in hours since midnight, and that we want to calculate the average temperature over a 24-hour period. One way to start is to average the temperatures at n equally spaced times, $t_1, t_2, \ldots, t_n$, during the day.

$$\text{Average temperature} \approx \frac{f(t_1) + f(t_2) + \cdots + f(t_n)}{n}.$$

The larger we make n, the better the approximation. We can rewrite this expression as a Riemann sum over the interval $0 \leq t \leq 24$ if we use the fact that $\Delta t = 24/n$, so $n = 24/\Delta t$:

$$\begin{aligned}
\text{Average temperature} &\approx \frac{f(t_1) + f(t_2) + \cdots + f(t_n)}{24/\Delta t} \\
&= \frac{f(t_1)\Delta t + f(t_2)\Delta t + \cdots + f(t_n)\Delta t}{24} \\
&= \frac{1}{24} \sum_{i=1}^{n} f(t_i)\Delta t.
\end{aligned}$$

As $n \to \infty$, the Riemann sum tends towards an integral, and $1/24$ of the sum also approximates the average temperature better. It makes sense, then, to write

$$\text{Average temperature} = \lim_{n \to \infty} \frac{1}{24} \sum_{i=1}^{n} f(t_i) \Delta t = \frac{1}{24} \int_0^{24} f(t) \, dt.$$

We have found a way of expressing the average temperature over an interval in terms of an integral. Generalizing for any function f, if $a < b$, we define

$$\text{Average value of } f \text{ from } a \text{ to } b = \frac{1}{b-a} \int_a^b f(x) \, dx.$$

How to Visualize the Average on a Graph

The definition of average value tells us that

$$(\text{Average value of } f) \cdot (b - a) = \int_a^b f(x) \, dx.$$

Let's interpret the integral as the area under the graph of f. Then the average value of f is the height of a rectangle whose base is $(b - a)$ and whose area is the same as the area under the graph of f. (See Figure 5.29.)

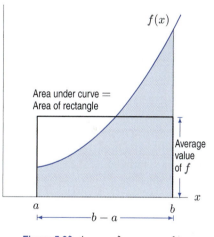

Figure 5.29: Area and average value

Example 2   Suppose that $C(t)$ represents the daily cost of heating your house, measured in dollars per day, where t is time measured in days and $t = 0$ corresponds to January 1, 2001. Interpret $\int_0^{90} C(t) \, dt$ and $\frac{1}{90 - 0} \int_0^{90} C(t) \, dt$.

Solution   The units for the integral $\int_0^{90} C(t) \, dt$ are (dollars/day) × (days) = dollars. The integral represents the total cost in dollars to heat your house for the first 90 days of 2001, namely the months of January, February, and March. The second expression is measured in (1/days)(dollars) or dollars per day, the same units as $C(t)$. It represents the average cost per day to heat your house during the first 90 days of 2001.

Example 3   On page 10, we saw that the population of Mexico could be modeled by the function

$$P = f(t) = 67.38(1.026)^t,$$

where P is in millions of people and t is in years since 1980. Use this function to predict the average population of Mexico between the years 2000 and 2020.

Solution   We want the average value of $f(t)$ between $t = 20$ and $t = 40$. This is given by

$$\text{Average population} = \frac{1}{40 - 20} \int_{20}^{40} f(t)\, dt \approx \frac{1}{20}(2942.66) = 147.1.$$

We used a calculator to evaluate the integral. We see that the average population of Mexico between 2000 and 2020 is predicted to be about 147 million people.

Applications of the Definite Integral

The total distance traveled by a moving object in a given time interval may be represented by a definite integral of the velocity. The following examples show how representing a quantity as a definite integral, and thereby as an area, can be helpful even if we don't evaluate the integral.

Example 4   Two cars start from rest at a traffic light and accelerate for several minutes. Figure 5.30 shows their velocities as a function of time.   (a) Which car is ahead after one minute?   (b) Which car is ahead after two minutes?

Solution   (a) For the first minute car 1 goes faster than car 2, and therefore car 1 must be ahead at the end of one minute.

(b) At the end of two minutes the situation is less clear, since car 1 was going faster for the first minute and car 2 for the second. However, if $v = f(t)$ is the velocity of a car after t minutes, then we know that

$$\text{Distance traveled in two minutes} = \int_0^2 f(t)\, dt,$$

since the integral of velocity is distance traveled. This definite integral may also be interpreted as the area under the graph of f between 0 and 2. Since the area representing the distance traveled by car 2 is clearly larger than the area for car 1 (see Figure 5.30), we know that car 2 has traveled farther than car 1.

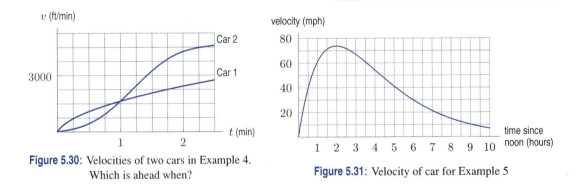

Figure 5.30: Velocities of two cars in Example 4. Which is ahead when?

Figure 5.31: Velocity of car for Example 5

Example 5   A car starts at noon and travels with the velocity shown in Figure 5.31. A truck starts at 1 pm from the same place and travels at a constant velocity of 50 mph.

(a) How far away is the car when the truck starts?

(b) During the period when the car is ahead of the truck, when is the distance between them greatest, and what is that greatest distance?

(c) When does the truck overtake the car, and how far have both traveled then?

Solution   To find distances from the velocity graph, we use the fact that if t is the time measured from noon, and v is the velocity, then

$$\text{Distance traveled by car up to time } T = \int_0^T v\, dt = \text{Area under velocity graph between 0 and } T.$$

The truck's motion can be represented on the same graph by the horizontal line $v = 50$, starting at $t = 1$. The distance traveled by the truck is then the rectangular area under this line, and the distance between the two vehicles is the difference between these areas. Note each small rectangle on the graph corresponds to moving at 10 mph for a half hour (i.e., to a distance of 5 miles).

(a) The distance traveled by the car when the truck starts is represented by the shaded area in Figure 5.32, which totals about seven rectangles or about 35 miles.

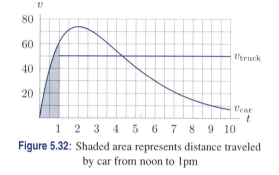

Figure 5.32: Shaded area represents distance traveled by car from noon to 1pm

(b) The car starts ahead of the truck, and the distance between them increases as long as the velocity of the car is greater than the velocity of the truck. Later, when the truck's velocity exceeds the car's, the truck starts to gain on the car. In other words, the distance between the car and the truck will increase as long as $v_{\text{car}} > v_{\text{truck}}$, and it will decrease when $v_{\text{car}} < v_{\text{truck}}$. Therefore, the maximum distance occurs when $v_{\text{car}} = v_{\text{truck}}$, that is, when $t \approx 4.3$ hours (at about 4:20 pm). (See Figure 5.33.) The distance traveled by the car is the area under the v_{car} graph between $t = 0$ and $t = 4.3$; the distance traveled by the truck is the area under the v_{truck} line between $t = 1$ (when it started) and $t = 4.3$. So the distance between the car and truck is represented by the shaded area in Figure 5.33, which is approximately

$$35 \text{ miles} + 50 \text{ miles} = 85 \text{ miles}.$$

(c) The truck overtakes the car when both have traveled the same distance. This occurs when the area under the curve (v_{car}) up to that time equals the area under the line (v_{truck}) up to that time. Since the areas under v_{car} and v_{truck} overlap (see Figure 5.34), they are equal when the lightly shaded area equals the heavily shaded area (which we know is about 85 miles). This happens when $t \approx 8.3$ hours, or about 8:20 pm. At this time, each has traveled about 365 miles.

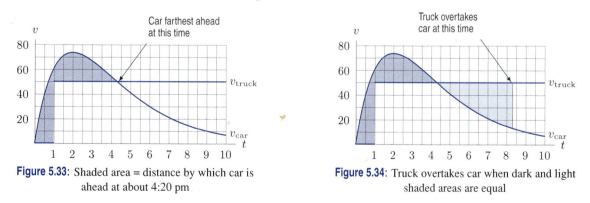

Figure 5.33: Shaded area = distance by which car is ahead at about 4:20 pm

Figure 5.34: Truck overtakes car when dark and light shaded areas are equal

Exercises and Problems for Section 5.3

Exercises

In Exercises 1–3, find the average value of the function over the given interval.

1. $g(t) = 1 + t$ over $[0, 2]$ **2.** $g(t) = e^t$ over $[0, 10]$

3. $f(x) = 4x + 7$ over $[1, 3]$

4. If $f(t)$ is measured in meters/second^2 and t is measured in seconds, what are the units of $\int_a^b f(t)\,dt$?

5. If $f(t)$ is measured in dollars per year and t is measured in years, what are the units of $\int_a^b f(t)\,dt$?

6. If $f(x)$ is measured in pounds and x is measured in feet, what are the units of $\int_a^b f(x)\,dx$?

7. For the two cars in Example 4, page 240, estimate:

(a) The distances moved by car 1 and car 2 during the first minute.
(b) The time at which the two cars have gone the same distance.

8. Consider the car and the truck in Example 5, page 240.

(a) How fast is the distance between the car and the truck increasing or decreasing at 3 pm?
(b) What is the practical significance (in terms of the distance between the car and the truck) of the fact that the car's velocity is maximized at about 2 pm?

9. Consider the car and the truck in Example 5, page 240, but suppose the truck starts at noon. (Everything else remains the same.)

(a) Sketch a new graph showing the velocities of both car and truck against time.
(b) How many times do the two graphs intersect? What does each intersection mean in terms of the distance between the two?

10. Oil leaks out of a tanker at a rate of $r = f(t)$ gallons per minute, where t is in minutes. Write a definite integral expressing the total quantity of oil which leaks out of the tanker in the first hour.

Problems

11. Water is leaking out of a tank at a rate of $R(t)$ gallons/hour, where t is measured in hours.

(a) Write a definite integral that expresses the total amount of water that leaks out in the first two hours.
(b) In Figure 5.35, shade the region whose area represents the total amount of water that leaks out in the first two hours.
(c) Give an upper and lower estimate of the total amount of water that leaks out in the first two hours.

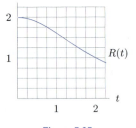

Figure 5.35

12. (a) Using Figure 5.36, find $\int_1^6 f(x)\,dx$.
(b) What is the average value of f on $[1, 6]$?

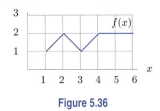

Figure 5.36

^{2}*Statistical Abstract of the US*, 1992.

13. (a) Using Figure 5.37, estimate $\int_{-3}^3 f(x)\,dx$.
(b) Which of the following average values of $f(x)$ is larger?
  (i) Between $x = -3$ and $x = 3$
  (ii) Between $x = 0$ and $x = 3$

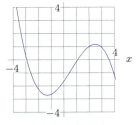

Figure 5.37

14. The following table gives the emissions, E, of nitrogen oxides in millions of metric tons per year in the US.^{2} Let t be the number of years since 1940 and $E = f(t)$.

(a) What are the units and meaning of $\int_0^{50} f(t)\,dt$?
(b) Estimate $\int_0^{50} f(t)\,dt$.

Year	1940	1950	1960	1970	1980	1990
E	6.9	9.4	13.0	18.5	20.9	19.6

15. The amount of waste a company produces, W, in metric tons per week, is approximated by $W = 3.75e^{-0.008t}$, where t is in weeks since January 1, 2000. Waste removal for the company costs $15/ton. How much does the company pay for waste removal during the year 2000?

16. A two-day environmental clean up started at 9 am on the first day. The number of workers fluctuated as shown in Figure 5.38. If the workers were paid $10 per hour, how much was the total personnel cost of the clean up?

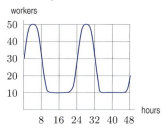

Figure 5.38

17. Suppose in Problem 16 that the workers were paid $10 per hour for work during the time period 9 am to 5 pm and were paid $15 per hour for work during the rest of the day. What would the total personnel costs of the clean up have been under these conditions?

18. A warehouse charges its customers $5 per day for every 10 cubic feet of space used for storage. Figure 5.39 records the storage used by one company over a month. How much will the company have to pay?

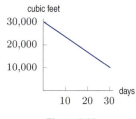

Figure 5.39

19. (a) Without computing any integrals, explain why the average value of $f(x) = \sin x$ on $[0, \pi]$ must be between 0.5 and 1.
   (b) Compute this average.

20. (a) What is the average value of $f(x) = \sqrt{1 - x^2}$ over the interval $0 \leq x \leq 1$?
   (b) How can you tell whether this average value is more or less than 0.5 without doing any calculations?

21. How do the units for the average value relate to the units for $f(x)$ and the units for x?

22. A bar of metal is cooling from $1000°$C to room temperature, $20°$C. The temperature, H, of the bar t minutes after it starts cooling is given, in $°$C, by

$$H = 20 + 980e^{-0.1t}.$$

   (a) Find the temperature of the bar at the end of one hour.
   (b) Find the average value of the temperature over the first hour.
   (c) Is your answer to part (b) greater or smaller than the average of the temperatures at the beginning and the end of the hour? Explain this in terms of the concavity of the graph of H.

23. The value, V, of a Tiffany lamp, worth $225 in 1965, increases at 15% per year. Its value in dollars t years after 1965 is given by

$$V = 225(1.15)^t.$$

Find the average value of the lamp over the period 1965–2000.

24. The number of hours, H, of daylight in Madrid as a function of date is approximated by the formula

$$H = 12 + 2.4 \sin[0.0172(t - 80)],$$

where t is the number of days since the start of the year. Find the average number of hours of daylight in Madrid:

   (a) in January   (b) in June   (c) over a year

   (d) Explain why the relative magnitudes of your answers to parts (a), (b), and (c) are reasonable.

25. A news broadcast in early 1993 said the average American's annual income is changing at a rate of $r(t) = 40(1.002)^t$ dollars per month, where t is in months from January 1, 1993. How much did the average American's income change during 1993?

26. A bicyclist pedals along a straight road with velocity, v, given in Figure 5.40. She starts 5 miles from a lake; positive velocities take her away from the lake and negative velocities take her towards the lake. When is the cyclist farthest from the lake, and how far away is she then?

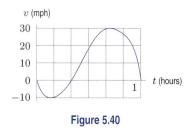

Figure 5.40

27. Figure 5.41 shows the rate, $f(x)$, in thousands of algae per hour, at which a population of algae is growing, where x is in hours.

   (a) Estimate the average value of the rate of growth of this population over the interval $x = -1$ to $x = 3$.
   (b) Estimate the total change in the algae population over the interval $x = -3$ to $x = 3$.

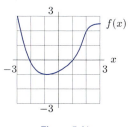

Figure 5.41

28. Height velocity graphs are used by endocrinologists to follow the progress of children with growth deficiencies. Figure 5.42 shows the height velocity curves of an average boy and an average girl between age 3 and 18.

   (a) Which curve is for girls and which is for boys? Explain how you can tell.

   (b) About how much does the average boy grow between ages 3 and 10?

   (c) The growth spurt associated with adolescence and the onset of puberty occurs between ages 12 and 15 for the average boy and between ages 10 and 12.5 for the average girl. Estimate the height gained by each average child during this growth spurt.

   (d) When fully grown, about how much taller is the average man than the average woman? (The average boy and girl are about the same height at age 3.)

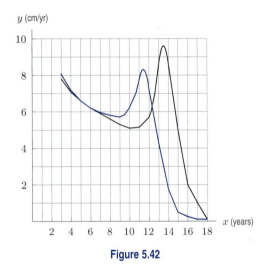

Figure 5.42

29. A force F parallel to the x-axis is given by the graph in Figure 5.43. Estimate the work, W, done by the force, where $W = \int_0^{16} F(x)\, dx$.

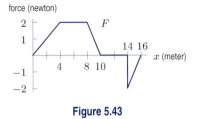

Figure 5.43

30. For the even function f in Figure 5.44, write an expression involving one or more definite integrals that denotes:

   (a) The average value of f for $0 \le x \le 5$.

   (b) The average value of $|f|$ for $0 \le x \le 5$.

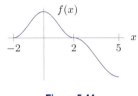

Figure 5.44

31. For the even function f in Figure 5.44, consider the average value of f over the following intervals:

   I. $0 \le x \le 1$ II. $0 \le x \le 2$

   III. $0 \le x \le 5$ IV. $-2 \le x \le 2$

   (a) For which interval is the average value of f least?

   (b) For which interval is the average value of f greatest?

   (c) For which pair of intervals are the average values equal?

32. A woman drives 10 miles, accelerating uniformly from rest to 60 mph. Sketch a graph of her velocity versus time. How long does it take for her to reach 60 mph?

5.4 THEOREMS ABOUT DEFINITE INTEGRALS

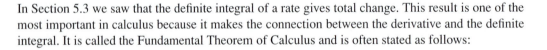

In Section 5.3 we saw that the definite integral of a rate gives total change. This result is one of the most important in calculus because it makes the connection between the derivative and the definite integral. It is called the Fundamental Theorem of Calculus and is often stated as follows:

Theorem 5.1: The Fundamental Theorem of Calculus

If f is continuous on the interval $[a, b]$ and $f(t) = F'(t)$, then

$$\int_a^b f(t)\, dt = F(b) - F(a).$$

In words:
   The definite integral of a rate of change gives total change.

What Is Involved in Proving the Fundamental Theorem?

The argument we gave in Section 5.3 makes the Fundamental Theorem plausible. It expresses the total change as a sum of small changes:

$$\text{Total change in } F = \sum_{i=0}^{n-1} \Delta F \approx \sum_{i=0}^{n-1} F'(t_i)\, \Delta t.$$

However, it is not a mathematical proof. Each of the approximations

$$\Delta F \approx F'(t_i)\, \Delta t, \quad i = 1, \ldots, n,$$

involves a small error. To prove that the total change in F is well-approximated by the Riemann sum $\sum_{i=0}^{n-1} F'(t_i)\, \Delta t$, we must show that the sum of all the errors is small (in fact, that it can be made as small as we like by choosing n large enough). We will not give the details of this argument. However, the key idea brings out an important fact about local linearization: the error in the local linearization $\Delta F \approx F'(t_i)\, \Delta t$ is not merely small, it is small relative to the size of Δt. This follows from the definition of the derivative: if we choose Δt small enough, we can ensure that the error in the approximation

$$F'(t_i) \approx \frac{\Delta F}{\Delta t}$$

is as small as we like. In other words,

$$F'(t_i) = \frac{\Delta F}{\Delta t} + \epsilon,$$

where ϵ can be made as small as we like. Multiplying through by Δt, we get an approximation

$$F'(t_i)\Delta t = \Delta F + \epsilon \Delta t.$$

Thus, in the approximation $F'(t_i)\Delta t \approx \Delta F$, the error is $\epsilon \Delta t$, which is small even relative to Δt. When we add these small errors, we still get a small total error. For a proof of the Fundamental Theorem from another point of view, see the online theory supplement.

Using the Fundamental Theorem to Compute Integrals

The Fundamental Theorem provides a precise way of computing certain definite integrals.

Example 1   Compute $\displaystyle\int_1^3 2x\, dx$ by two different methods.

Solution   Using left- and right-hand sums, we can approximate this integral as accurately as we want. With $n = 100$, for example, the left-sum is 7.96 and the right sum is 8.04. Using $n = 500$ we learn

$$7.992 < \int_1^3 2x\, dx < 8.008.$$

The Fundamental Theorem, on the other hand, allows us to compute the integral exactly. We take $f(x) = 2x$. By Example 4, on page 82, we know that if $F(x) = x^2$, then $F'(x) = 2x$. So we use $f(x) = 2x$ and $F(x) = x^2$ and obtain

$$\int_1^3 2x\, dx = F(3) - F(1) = 3^2 - 1^2 = 8.$$

Using Symmetry to Evaluate Integrals

Symmetry can be useful in evaluating definite integrals. An even function is symmetric about the y-axis. An odd function is symmetric about the origin. Figures 5.45 and 5.46 suggest the following results:

$$\text{If } f \text{ is even, then} \quad \int_{-a}^{a} f(x)\,dx = 2\int_{0}^{a} f(x)\,dx. \quad \text{If } g \text{ is odd, then} \quad \int_{-a}^{a} g(x)\,dx = 0.$$

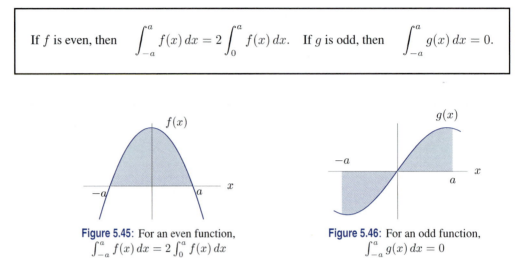

Figure 5.45: For an even function,
$\int_{-a}^{a} f(x)\,dx = 2\int_{0}^{a} f(x)\,dx$

Figure 5.46: For an odd function,
$\int_{-a}^{a} g(x)\,dx = 0$

Properties of the Definite Integral

For the definite integral $\int_{a}^{b} f(x)\,dx$, we have so far only considered the case $a < b$. We now allow $a \geq b$. We still set $x_0 = a$, $x_n = b$, and $\Delta x = (b-a)/n$. As before, we have $\int_{a}^{b} f(x)\,dx = \lim_{n\to\infty} \sum_{i=1}^{n} f(x_i)\Delta x$.

Theorem 5.2: Properties of Limits of Integration

If a, b, and c are any numbers and f is a continuous function, then

1. $\displaystyle\int_{b}^{a} f(x)\,dx = -\int_{a}^{b} f(x)\,dx.$

2. $\displaystyle\int_{a}^{c} f(x)\,dx + \int_{c}^{b} f(x)\,dx = \int_{a}^{b} f(x)\,dx.$

In words:

1. The integral from b to a is the negative of the integral from a to b.

2. The integral from a to c plus the integral from c to b is the integral from a to b.

By interpreting the integrals as areas, we can justify these results for $f \geq 0$. In fact, they are true for all functions for which the integrals make sense.

Why is $\int_{b}^{a} f(x)\,dx = -\int_{a}^{b} f(x)\,dx$?

By definition, both integrals are approximated by sums of the form $\sum f(x_i)\Delta x$. The only difference in the sums for $\int_{b}^{a} f(x)\,dx$ and $\int_{a}^{b} f(x)\,dx$ is that in the first $\Delta x = (a-b)/n = -(b-a)/n$ and in the second $\Delta x = (b-a)/n$. Since everything else about the sums is the same, we must have $\int_{b}^{a} f(x)\,dx = -\int_{a}^{b} f(x)\,dx$.

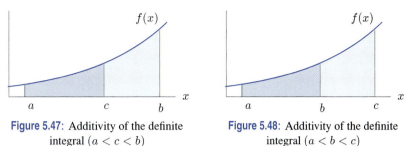

Figure 5.47: Additivity of the definite integral $(a < c < b)$

Figure 5.48: Additivity of the definite integral $(a < b < c)$

Why is $\int_a^c f(x)\,dx + \int_c^b f(x)\,dx = \int_a^b f(x)\,dx$?

Suppose $a < c < b$. Figure 5.47 suggests that $\int_a^c f(x)\,dx + \int_c^b f(x)\,dx = \int_a^b f(x)\,dx$ since the area under f from a to c plus the area under f from c to b together make up the whole area under f from a to b.

This property holds for all numbers a, b, and c, not just those satisfying $a < c < b$. (See Figure 5.48.) For example, the area under f from 3 to 6 is equal to the area from 3 to 8 *minus* the area from 6 to 8, so

$$\int_3^6 f(x)\,dx = \int_3^8 f(x)\,dx - \int_6^8 f(x)\,dx = \int_3^8 f(x)\,dx + \int_8^6 f(x)\,dx.$$

Example 2   Suppose $\int_0^{1.25} \cos(x^2)\,dx = 0.98$ and $\int_0^1 \cos(x^2)\,dx = 0.90$. (See Figure 5.49.) What are the values of the following integrals?

(a) $\displaystyle\int_1^{1.25} \cos(x^2)\,dx$ (b) $\displaystyle\int_{-1}^1 \cos(x^2)\,dx$ (c) $\displaystyle\int_{1.25}^{-1} \cos(x^2)\,dx$

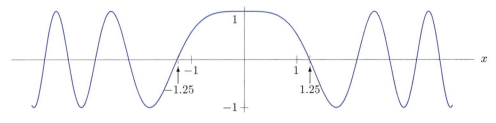

Figure 5.49: Graph of $f(x) = \cos(x^2)$

Solution

(a) Since $\int_0^{1.25} \cos(x^2)\,dx = \int_0^1 \cos(x^2)\,dx + \int_1^{1.25} \cos(x^2)\,dx$ by the additivity property, we get $0.98 = 0.90 + \int_1^{1.25} \cos(x^2)\,dx$, so $\int_1^{1.25} \cos(x^2)\,dx = 0.08$.

(b) $\int_{-1}^1 \cos(x^2)\,dx = \int_{-1}^0 \cos(x^2)\,dx + \int_0^1 \cos(x^2)\,dx$.

By the symmetry of $\cos(x^2)$ about the y-axis, $\int_{-1}^0 \cos(x^2)\,dx = \int_0^1 \cos(x^2)\,dx = 0.90$.

So $\int_{-1}^1 \cos(x^2)\,dx = 0.90 + 0.90 = 1.80$.

(c) $\int_{1.25}^{-1} \cos(x^2)\,dx = -\int_{-1}^{1.25} \cos(x^2)\,dx = -(\int_{-1}^0 \cos(x^2)\,dx + \int_0^{1.25} \cos(x^2)\,dx) = -(0.90 + 0.98) = -1.88$.

Theorem 5.3: Properties of Sums and Constant Multiples of the Integrand

Let f and g be continuous functions and let c be a constant.

1. $\displaystyle \int_a^b (f(x) \pm g(x))\, dx = \int_a^b f(x)\, dx \pm \int_a^b g(x)\, dx.$

2. $\displaystyle \int_a^b cf(x)\, dx = c \int_a^b f(x)\, dx.$

In words:

1. The integral of the sum (or difference) of two functions is the sum (or difference) of their integrals.

2. The integral of a constant times a function is that constant times the integral of the function.

Why Do these Properties Hold?

Both can be visualized by thinking of the definite integral as the limit of a sum of areas of rectangles.

For property 1, suppose that f and g are positive on the interval $[a, b]$ so that the area under $f(x)+g(x)$ is approximated by the sum of the areas of rectangles like the one shaded in Figure 5.50. The area of this rectangle is

$$[f(x_i) + g(x_i)]\Delta x = f(x_i)\Delta x + g(x_i)\Delta x.$$

Since $f(x_i)\Delta x$ is the area of a rectangle under the graph of f, and $g(x_i)\Delta x$ is the area of a rectangle under the graph of g, the area under $f(x) + g(x)$ is the sum of the areas under $f(x)$ and $g(x)$.

For property 2, notice that multiplying a function by c stretches or flattens the graph in the vertical direction by a factor of c. Thus, it stretches or flattens the height of each approximating rectangle by c, and hence multiplies the area by c.

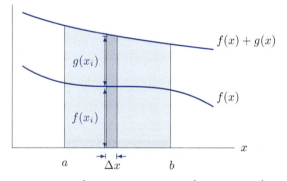

Figure 5.50: Area $= \int_a^b [f(x) + g(x)]\, dx = \int_a^b f(x)\, dx + \int_a^b g(x)\, dx$

Example 3   Evaluate the definite integral $\displaystyle \int_0^2 (1 + 3x)\, dx$ exactly.

Solution   We can break this integral up as follows:

$$\int_0^2 (1 + 3x)\, dx = \int_0^2 1\, dx \ + \ \int_0^2 3x\, dx = \int_0^2 1\, dx \ + \ 3\int_0^2 x\, dx.$$

This expresses our original integral in terms of two simpler integrals. From the area interpretation of the integral, we see that

$$\int_0^2 1\, dx = 2,$$

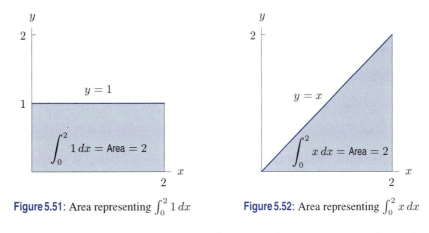

Figure 5.51: Area representing $\int_0^2 1\,dx$ Figure 5.52: Area representing $\int_0^2 x\,dx$

since it represents the area under the horizontal line $y = 1$ between $x = 0$ and $x = 2$ (see Figure 5.51). Similarly,

$$\int_0^2 x\,dx = \frac{1}{2} \cdot 2 \cdot 2 = 2$$

because it is the area of the triangle in Figure 5.52. Therefore,

$$\int_0^2 (1 + 3x)\,dx = \int_0^2 1\,dx + 3\int_0^2 x\,dx = 2 + 3(2) = 8.$$

Comparing Integrals

Suppose we have constants m and M such that $m \leq f(x) \leq M$ for $a \leq x \leq b$. We say f is *bounded above* by M and *bounded below* by m. Then the graph of f lies between the horizontal lines $y = m$ and $y = M$. So the definite integral lies between $m(b - a)$ and $M(b - a)$. See Figure 5.53.

Suppose $f(x) \leq g(x)$ for $a \leq x \leq b$, as in Figure 5.54. By a similar argument, the definite integral of f is less than or equal to the definite integral of g. This leads us to the following results:

Theorem 5.4: Comparison of Definite Integrals

Let f and g be continuous functions.

1. If $m \leq f(x) \leq M$ for $a \leq x \leq b$, then $m(b - a) \leq \int_a^b f(x)\,dx \leq M(b - a)$.

2. If $f(x) \leq g(x)$ for $a \leq x \leq b$, then $\int_a^b f(x)\,dx \leq \int_a^b g(x)\,dx$.

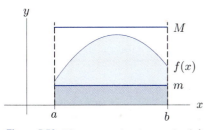

Figure 5.53: The area under the graph of f lies between the areas of the rectangles

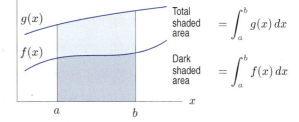
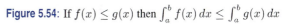

Figure 5.54: If $f(x) \leq g(x)$ then $\int_a^b f(x)\,dx \leq \int_a^b g(x)\,dx$

Example 4   Explain why $\int_0^{\sqrt{\pi}} \sin(x^2)\,dx \leq \sqrt{\pi}$.

Solution   Since $\sin(x^2) \leq 1$ for all x (see Figure 5.55), part 2 of Theorem 5.4 gives

$$\int_0^{\sqrt{\pi}} \sin(x^2)\,dx \leq \int_0^{\sqrt{\pi}} 1\,dx = \sqrt{\pi}.$$

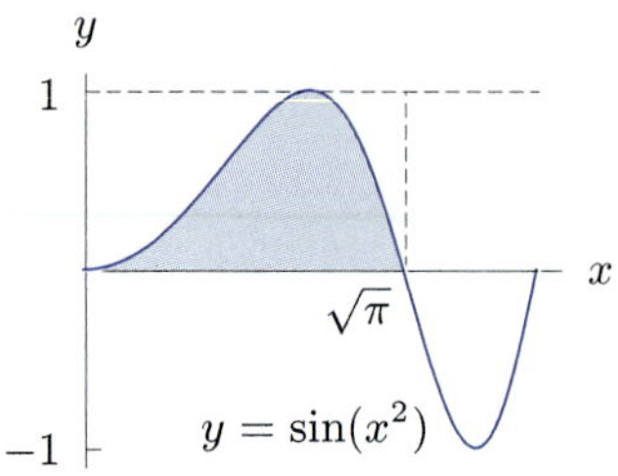

Figure 5.55: Graph showing that $\int_0^{\sqrt{\pi}} \sin(x^2)\,dx < \sqrt{\pi}$

Example 5   Show that $2 \leq \int_0^2 \sqrt{1+x^3}\,dx \leq 6$.

Solution   Notice that $f(x) = \sqrt{1+x^3}$ is increasing for $0 \leq x \leq 2$, since x^3 gets bigger as x increases. This means that $f(0) \leq f(x) \leq f(2)$. For this function, $f(0) = 1$ and $f(2) = 3$. Thus we can imagine the area under $f(x)$ as lying between the area under the line $y = 1$ and the area under the line $y = 3$ on the interval $0 \leq x \leq 2$. That is,

$$1(2-0) \leq \int_0^2 \sqrt{1+x^3}\,dx \leq 3(2-0).$$

Exercises and Problems for Section 5.4

Exercises

1. The graph of a derivative $f'(x)$ is shown in Figure 5.56. Fill in the table of values for $f(x)$ given that $f(0) = 2$.

x	0	1	2	3	4	5	6
$f(x)$	2						

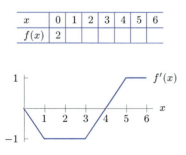

Figure 5.56: Graph of f', not f

2. Figure 5.57 shows the graph of f. If $F' = f$ and $F(0) = 0$, find $F(b)$ for $b = 1, 2, 3, 4, 5, 6$.

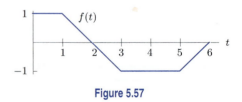

Figure 5.57

3. (a) Suppose $f'(x) = \sin(x^2)$ and $f(0) = 2$. Use a graph of $f'(x)$ to decide which is larger:

  (i) $f(0)$ or $f(1)$ (ii) $f(2)$ or $f(2.5)$

  (b) Estimate $f(b)$ for $b = 0, 1, 2, 3$.

Problems

Suppose $\int_a^b f(x)\,dx = 8$, $\int_a^b (f(x))^2\,dx = 12$, $\int_a^b g(t)\,dt = 2$, and $\int_a^b (g(t))^2\,dt = 3$. Find the integrals in Problems 4–9.

4. $\int_a^b (f(x) + g(x))\,dx$ **5.** $\int_a^b cf(z)\,dz$

6. $\int_a^b \left((f(x))^2 - (g(x))^2\right)\,dx$

7. $\int_a^b (f(x))^2\,dx - (\int_a^b f(x)\,dx)^2$

8. $\int_a^b \left(c_1 g(x) + (c_2 f(x))^2\right)\,dx$

9. $\int_{a+5}^{b+5} f(x-5)\,dx$

10. Using Figure 5.58, write $\int_0^3 f(x)\,dx$ in terms of $\int_{-1}^1 f(x)\,dx$ and $\int_1^3 f(x)\,dx$.

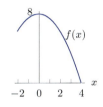

Figure 5.58

11. (a) Let $\int_0^3 f(x)dx = 6$. What is the average value of $f(x)$ on the interval $x = 0$ to $x = 3$?
   (b) If $f(x)$ is even, what is $\int_{-3}^3 f(x)dx$? What is the average value of $f(x)$ on the interval $x = -3$ to $x = 3$?
   (c) If $f(x)$ is odd, what is $\int_{-3}^3 f(x)dx$? What is the average value of $f(x)$ on the interval $x = -3$ to $x = 3$?

12. (a) Is $\int_{-1}^1 e^{x^2}\,dx$ positive, negative, or zero? Explain.
   (b) Explain why $0 < \int_0^1 e^{x^2}\,dx < 3$.

13. Without calculating the integral, explain why the following statements are false.

   (a) $\int_{-2}^{-1} e^{x^2}\,dx = -3$

   (b) $\int_{-1}^1 \left|\dfrac{\cos(x+2)}{1 + \tan^2 x}\right|\,dx = 0$

Problems 14–15 concern the graph of f' in Figure 5.59.

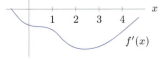

Figure 5.59: Graph of f', not f

14. Which is greater, $f(0)$ or $f(1)$?

15. List the following in increasing order:
$$\dfrac{f(4) - f(2)}{2}, \quad f(3) - f(2), \quad f(4) - f(3).$$

16. Without doing any computations, find the values of

   (a) $\displaystyle\int_{-2}^2 \sin x\,dx$,   **(b)** $\displaystyle\int_{-\pi}^\pi x^{113}\,dx$.

For Problems 17–20, mark the following quantities on a copy of the graph of f in Figure 5.60.

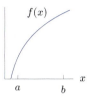

Figure 5.60

17. A length representing $f(b) - f(a)$.

18. A slope representing $\dfrac{f(b) - f(a)}{b - a}$.

19. An area representing $F(b) - F(a)$, where $F' = f$.

20. A length roughly approximating
$$\dfrac{F(b) - F(a)}{b - a}, \text{ where } F' = f.$$

21. The function for the *standard normal distribution*, which is often used in statistics, has the formula
$$\dfrac{1}{\sqrt{2\pi}} e^{-x^2/2}$$

and the graph in Figure 5.61. Statistics books usually contain tables such as the one below, showing only the area under the curve from 0 to b, for different values of b.

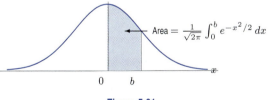

Figure 5.61

b	$\frac{1}{\sqrt{2\pi}} \int_0^b e^{-x^2/2}\,dx$
1	0.3413
2	0.4772
3	0.4987
4	0.5000

Use the information given in the table and the symmetry of the standard normal curve about the y-axis to find:

   (a) $\dfrac{1}{\sqrt{2\pi}} \displaystyle\int_1^3 e^{-x^2/2}\,dx$.   **(b)** $\dfrac{1}{\sqrt{2\pi}} \displaystyle\int_{-2}^3 e^{-x^2/2}\,dx$.

22. Use the property $\int_b^a f(x)\,dx = -\int_a^b f(x)\,dx$ to show that $\int_a^a f(x)\,dx = 0$.

23. The average value of $y = v(x)$ equals 4 for $1 \leq x \leq 6$, and equals 5 for $6 \leq x \leq 8$. What is the average value of $v(x)$ for $1 \leq x \leq 8$?

CHAPTER SUMMARY

- **Definite integral as limit of right-hand or left-hand sums**
- **Interpretations of the definite integral**
  Total change from rate of change, change in position given velocity, area, $(b - a)\cdot$ average value.
- **Properties of the definite integral**

- Properties involving integrand, properties involving limits, comparison between integrals.
- **Fundamental Theorem of Calculus**
- **Working with the definite integral**
  Estimate definite integral from graph, table of values, or formula.

REVIEW EXERCISES AND PROBLEMS FOR CHAPTER FIVE

Exercises

1. A car going 80 ft/sec (about 55 mph) brakes to a stop in 8 seconds. Its velocity is recorded every 2 seconds and is given in the following table.

(a) Give your best estimate of the distance traveled by the car during the 8 seconds.
(b) To estimate the distance traveled accurate to within 20 feet, how often should you record the velocity?

t (seconds)	0	2	4	6	8
$v(t)$ (ft/sec)	80	52	28	10	0

2. Use Figure 5.62 to estimate $\int_0^{20} f(x)\,dx$.

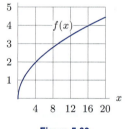

Figure 5.62

3. A car accelerates smoothly from 0 to 60 mph in 10 seconds with the velocity given in Figure 5.63. Estimate how far the car travels during the 10-second period.

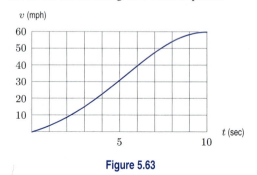

Figure 5.63

4. Using Figure 5.64, estimate $\int_{-3}^{5} f(x)\,dx$.

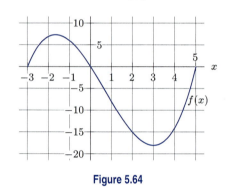

Figure 5.64

5. Using the table, estimate $\int_0^{100} f(t)\,dt$.

t	0	20	40	60	80	100
$f(t)$	1.2	2.8	4.0	4.7	5.1	5.2

6. (a) What is the area between the graph of $f(x)$ in Figure 5.65 and the x-axis, between $x = 0$ and $x = 5$?
(b) What is $\int_0^5 f(x)\,dx$?

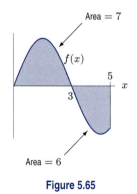

Area = 7

Area = 6

Figure 5.65

7. Your velocity is $v(t) = \ln(t^2 + 1)$ for $0 \le t \le 3$. Estimate the distance traveled during this time.

8. Your velocity is $v(t) = \sin(t^2)$ for $0 \le t \le 1.1$. Estimate the distance traveled during this time.

Find the area of the regions in Exercises 9–13.

9. Between $y = x^2 - 9$ and the x-axis.

10. Under one arch of the curve $y = \sin x$.

11. Between the parabola $y = 4 - x^2$ and the x-axis.

12. Between the line $y = 1$ and one arch of the curve $y = \sin \theta$.

13. Between $y = -x^2 + 5x - 4$ and the x-axis, $0 \le x \le 3$.

Problems

14. If $\int_2^5 (2f(x) + 3)\,dx = 17$, find $\int_2^5 f(x)\,dx$.

15. Annual coal production in the US (in quadrillion BTU per year) is given in the table.[3] Estimate the total amount of coal produced in the US between 1960 and 1990. If $r = f(t)$ is the rate of coal production t years since 1960, write an integral to represent the 1960–1990 coal production.

Year	1960	1965	1970	1975	1980	1985	1990
Rate	10.82	13.06	14.61	14.99	18.60	19.33	22.46

16. An old rowboat has sprung a leak. Water is flowing into the boat at a rate given in the following table.

t minutes	0	5	10	15
$r(t)$ liters/min	12	20	24	16

   (a) Compute upper and lower estimates for the volume of water that has flowed into the boat during the 15 minutes.
   (b) Draw a graph to illustrate the lower estimate.

17. Figure 5.66 gives your velocity during a trip starting from home. Positive velocities take you away from home and negative velocities take you toward home. Where are you at the end of the 5 hours? When are you farthest from home? How far away are you at that time?

18. A bicyclist is pedaling along a straight road for one hour with a velocity v shown in Figure 5.67. She starts out five kilometers from the lake and positive velocities take her towards the lake. [Note: The vertical lines on the graph are at 10 minute (1/6 hour) intervals.]

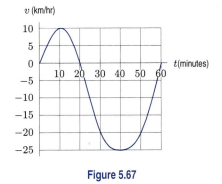

Figure 5.67

   (a) Does the cyclist ever turn around? If so, at what time(s)?
   (b) When is she going the fastest? How fast is she going then? Toward the lake or away?
   (c) When is she closest to the lake? Approximately how close to the lake does she get?
   (d) When is she farthest from the lake? Approximately how far from the lake is she then?

19. The graph in Figure 5.68 shows the rate of change of the quantity of water in a water tower, in liters per day, during the month of April. If the tower had 12,000 liters of water in it on April 1, estimate the quantity of water in the tower on April 30.

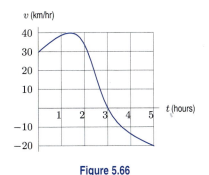

Figure 5.66

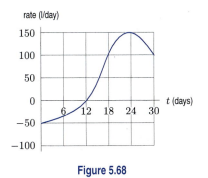

Figure 5.68

---

[3]*World Almanac, 1995.*

20. The graph of a continuous function f is given in Figure 5.69. Rank the following integrals in ascending numerical order. Explain your reasons.

(i) $\int_0^2 f(x)\,dx$ (ii) $\int_0^1 f(x)\,dx$

(iii) $\int_0^2 (f(x))^{1/2}\,dx$ (iv) $\int_0^2 (f(x))^2\,dx$.

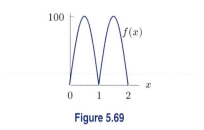

Figure 5.69

21. **(a)** Sketch a graph of $f(x) = \sin(x^2)$ and mark on it the points $x = \sqrt{\pi}, \sqrt{2\pi}, \sqrt{3\pi}, \sqrt{4\pi}$.
    (b) Use your graph to decide which of the four numbers

$$\int_0^{\sqrt{n\pi}} \sin(x^2)\,dx \quad n = 1, 2, 3, 4$$

is largest. Which is smallest? How many of the numbers are positive?

22. Find $\int_{-1}^1 |x|\,dx$ geometrically.

23. Using Figure 5.70, list the following integrals in increasing order (from smallest to largest). Which integrals are negative, which are positive? Give reasons.

I. $\int_a^b f(x)\,dx$ II. $\int_a^c f(x)\,dx$ III. $\int_a^e f(x)\,dx$
IV. $\int_b^e f(x)\,dx$ V. $\int_b^c f(x)\,dx$

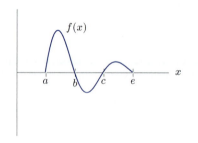

Figure 5.70

24. For the even function f graphed in Figure 5.71:

(a) Suppose you know $\int_0^2 f(x)\,dx$. What is $\int_{-2}^2 f(x)\,dx$?

(b) Suppose you know $\int_0^5 f(x)\,dx$ and $\int_2^5 f(x)\,dx$. What is $\int_0^2 f(x)\,dx$?

(c) Suppose you know $\int_{-2}^5 f(x)\,dx$ and $\int_{-2}^2 f(x)\,dx$. What is $\int_0^5 f(x)\,dx$?

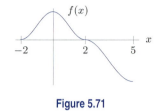

Figure 5.71

25. For the even function f graphed in Figure 5.71:

(a) Suppose you know $\int_{-2}^2 f(x)\,dx$ and $\int_0^5 f(x)\,dx$. What is $\int_2^5 f(x)\,dx$?

(b) Suppose you know $\int_{-2}^5 f(x)\,dx$ and $\int_{-2}^0 f(x)\,dx$. What is $\int_2^5 f(x)\,dx$?

(c) Suppose you know $\int_2^5 f(x)\,dx$ and $\int_{-2}^5 f(x)\,dx$. What is $\int_0^2 f(x)\,dx$?

26. Assume w, b_1, and b_2 are positive constants, with $b_2 > b_1$. Compute the following integral using its geometric interpretation.

$$\int_0^w \left(b_1 + \frac{b_2 - b_1}{w}x\right) dx$$

27. A cup of coffee at $90°C$ is put into a $20°C$ room when $t = 0$. The coffee's temperature is changing at a rate of $r(t) = -7e^{-0.1t}$ °C per minute, with t in minutes. Estimate the coffee's temperature when $t = 10$.

28. The rate at which the world's oil is being consumed is continuously increasing. Suppose the rate of oil consumption (in billions of barrels per year) is given by the function $r = f(t)$, where t is measured in years and $t = 0$ is the start of 1990.

(a) Write a definite integral which represents the total quantity of oil used between the start of 1990 and the start of 1995.
(b) Suppose $r = 32e^{0.05t}$. Using a left-hand sum with five subdivisions, find an approximate value for the total quantity of oil used between the start of 1990 and the start of 1995.
(c) Interpret each of the five terms in the sum from part (b) in terms of oil consumption.

29. Water is pumped out of a holding tank at a rate of $5 - 5e^{-0.12t}$ liters/minute, where t is in minutes since the pump is started. If the holding tank contains 1000 liters of water when the pump is started, how much water does it hold one hour later?

30. Two trains travel along parallel tracks. The velocity, v, of the trains as functions of time t are shown in Figure 5.72.

  (a) Describe in words the trips taken by each train.
  (b) Estimate the ratio of the following quantities for Train A to Train B:

  (i) Maximum velocity   (ii)   Time traveled
  (iii) Distance traveled

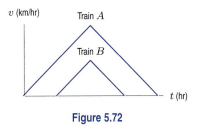

Figure 5.72

31. The graphs in Figure 5.73 represent the velocity, v, of a particle moving along the x-axis for time $0 \le t \le 5$. The vertical scales of all graphs are the same. Identify the graph showing which particle:

  (a) Has a constant acceleration.
  (b) Ends up farthest to the left of where it started.
  (c) Ends up the farthest from its starting point.
  (d) Experiences the greatest initial acceleration.
  (e) Has the greatest average velocity.
  (f) Has the greatest average acceleration.

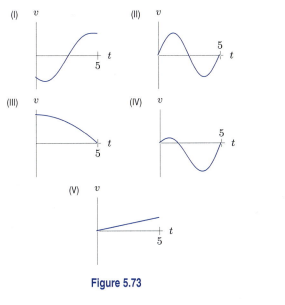

Figure 5.73

32. Water is run into a large tank through a hose at a constant rate. After 5 minutes a hole is opened in the bottom of the tank, and water starts to flow out. Initially the flow rate through the hole is twice as great as the rate through the hose, but as the water level in the tank goes down, the flow rate through the hole decreases; after another 10 minutes the water level in the tank appears to be constant. Plot graphs of the flow rates through the hose and through the hole against time on the same pair of axes. Show how the volume of water in the tank at any time can be interpreted as an area (or the difference between two areas) on the graph. In particular, interpret the steady-state volume of water in the tank.[4]

33. The Glen Canyon Dam at the top of the Grand Canyon prevents natural flooding. In 1996, scientists decided an artificial flood was necessary to restore the environmental balance. Water was released through the dam at a controlled rate[5] shown in Figure 5.74. The figure also shows the rate of flow of the last natural flood in 1957.

  (a) At what rate was water passing through the dam in 1996 before the artificial flood?
  (b) At what rate was water passing down the river in the pre-flood season in 1957?
  (c) Estimate the maximum rates of discharge for the 1996 and 1957 floods.
  (d) Approximately how long did the 1996 flood last? How long did the 1957 flood last?
  (e) Estimate how much additional water passed down the river in 1996 as a result of the artificial flood.
  (f) Estimate how much additional water passed down the river in 1957 as a result of the flood.

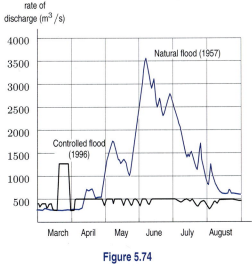

Figure 5.74

---

[4]From *Calculus: The Analysis of Functions*, by Peter D. Taylor (Toronto: Wall & Emerson, Inc., 1992)
[5]Adapted from M. Collier, R. Webb, E. Andrews "Experimental Flooding in Grand Canyon" in *Scientific American* (January 1997).

34. The Montgolfier brothers (Joseph and Etienne) were eighteenth-century pioneers in the field of hot-air ballooning. Had they had the appropriate instruments, they might have left us a record, like that shown in Figure 5.75, of one of their early experiments. The graph shows their vertical velocity, v, with upward as positive.

    (a) Over what intervals was the acceleration positive? Negative?
    (b) What was the greatest altitude achieved, and at what time?
    (c) At what time was the upward acceleration greatest?
    (d) At what time was the deceleration greatest?
    (e) What might have happened during this flight to explain the answer to part (d)?
    (f) This particular flight ended on top of a hill. How do you know that it did, and what was the height of the hill above the starting point?

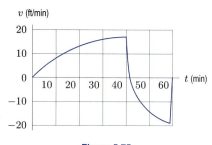

Figure 5.75

35. A mouse moves back and forth in a straight tunnel, attracted to bits of cheddar cheese alternately introduced to and removed from the ends (right and left) of the tunnel. The graph of the mouse's velocity, v, is given in Figure 5.76, with positive velocity corresponding to motion toward the right end. Assuming that the mouse starts $(t = 0)$ at the center of the tunnel, use the graph to estimate the time(s) at which:

    (a) The mouse changes direction.
    (b) The mouse is moving most rapidly to the right; to the left.
    (c) The mouse is farthest to the right of center; farthest to the left.
    (d) The mouse's speed (i.e., the magnitude of its velocity) is decreasing.
    (e) The mouse is at the center of the tunnel.

CAS Challenge Problems

38. Consider the definite integral $\int_0^1 x^4 dx$.

    (a) Write an expression for a right-hand Riemann sum approximation for this integral using n subdivisions. Express each x_i, $i = 1,2,\dots,n$, in terms of i.
    (b) Use a computer algebra system to obtain a formula for the sum you wrote in part (a) in terms of n.

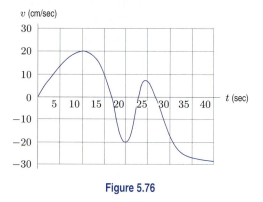

Figure 5.76

36. When an aircraft attempts to climb as rapidly as possible, its climb rate decreases with altitude. (This occurs because the air is less dense at higher altitudes.) The table shows performance data for a single-engine aircraft.

Altitude (1000 ft)	0	1	2	3	4	5
Climb rate (ft/min)	925	875	830	780	730	685
Altitude (1000 ft)	6	7	8	9	10	
Climb rate (ft/min)	635	585	535	490	440	

    (a) Calculate upper and lower estimates for the time required for this aircraft to climb from sea level to 10,000 ft.
    (b) If climb rate data were available in increments of 500 ft, what would be the difference between a lower and upper estimate of climb time based on 20 subdivisions?

37. Pollution is being dumped into a lake at a rate which is increasing at a constant rate from 10 kg/year to 50 kg/year until a total of 270 kg has been dumped. Sketch a graph of the rate at which pollution is being dumped in the lake against time. How long does it take until 270 kg has been dumped?

    (c) Take the limit of this expression for the sum as $n \to \infty$, thereby finding the exact value of this integral.

39. Repeat Problem 38, using the definite integral $\int_0^1 x^5 dx$.

For Problems 40–42, you will write a Riemann sum approximating a definite integral and use a computer algebra system to find a formula for the Riemann sum. By evaluating the limit of this sum as the number of subdivisions approaches infinity, you will obtain the definite integral.

40. **(a)** Using summation notation, write the left-hand Riemann sum with n subdivisions for $\int_1^2 t\, dt$.
   (b) Use a computer algebra system to find a formula for the Riemann sum.
   (c) Evaluate the limit of the sum as n approaches infinity.
   (d) Calculate directly the area under the graph of $y = t$ between $t = 1$ and $t = 2$, and compare it with your answer to part (c).

41. **(a)** Using summation notation, write the left-hand Riemann sum with n subdivisions for $\int_1^2 t^2\, dt$.
   (b) Use a computer algebra system to find a formula for the Riemann sum.
   (c) Evaluate the limit of the sum as n approaches infinity.
   (d) What is the area under the graph of $y = t^2$ between $t = 1$ and $t = 2$?

42. **(a)** Using summation notation, write the right-hand Riemann sum with n subdivisions for $\int_0^\pi \sin x\, dx$.
   (b) Use a computer algebra system to find a formula for the Riemann sum. [Note: Not all computer algebra systems can evaluate this sum.]
   (c) Use a computer algebra system to evaluate the limit of the sum as n approaches infinity.
   (d) Confirm your answer to part (c) by calculating the definite integral with the computer algebra system.

In Problems 43–44:

   (a) Use a computer algebra system to compute the given definite integral.

   (b) From your answer to part (a) and the Fundamental Theorem of Calculus, guess a function whose derivative is the integrand. Check your guess using the computer algebra system.

[Hint: Make sure that the constants a, b, and c do not have previously assigned values in your computer algebra system.]

43. $\displaystyle \int_a^b \sin(cx)\, dx$

44. $\displaystyle \int_a^c \frac{x}{1 + bx^2}\, dx, \quad b > 0$

CHECK YOUR UNDERSTANDING

In Problems 1–17, are the statements true for all continuous functions $f(x)$ and $g(x)$? Give an explanation for your answer.

1. If $\int_0^2 (f(x) + g(x))dx = 10$ and $\int_0^2 f(x)dx = 3$, then $\int_0^2 g(x)dx = 7$.

2. If $\int_0^2 (f(x) + g(x))dx = 10$, then $\int_0^2 f(x)dx = 3$ and $\int_0^2 g(x)dx = 7$.

3. If $\int_0^2 f(x)dx = 6$, then $\int_0^4 f(x)dx = 12$.

4. If $\int_0^2 f(x)dx = 6$ and $g(x) = 2f(x)$, then $\int_0^2 g(x)dx = 12$.

5. If $\int_0^2 f(x)dx = 6$ and $h(x) = f(5x)$, then $\int_0^2 h(x)dx = 30$.

6. If $a = b$, then $\int_a^b f(x)\, dx = 0$.

7. If $a \neq b$, then $\int_a^b f(x)\, dx \neq 0$.

8. $\int_1^2 f(x)\, dx + \int_2^3 g(x)\, dx = \int_1^3 (f(x) + g(x))\, dx$.

9. $\int_{-1}^1 f(x)\, dx = 2\int_0^1 f(x)\, dx$.

10. $\int_0^2 f(x)dx \leq \int_0^3 f(x)dx$.

11. $\int_0^2 f(x)dx = \int_0^2 f(t)dt$.

12. If $\int_2^6 f(x)\, dx \leq \int_2^6 g(x)\, dx$, then $f(x) \leq g(x)$ for $2 \leq x \leq 6$.

13. If $f(x) \leq g(x)$ on the interval $[a, b]$, then the average value of f is less than or equal to the average value of g on the interval $[a, b]$.

14. The average value of f on the interval $[0, 10]$ is the average of the average value of f on $[0, 5]$ and the average value of f on $[5, 10]$.

15. If $a < c < d < b$, then the average value of f on the interval $[c, d]$ is less than the average value of f on the interval $[a, b]$.

16. Suppose that A is the average value of f on the interval $[1, 4]$ and B is the average value of f on the interval $[4, 9]$. Then the average value of f on $[1, 9]$ is the weighted average $(3/8)A + (5/8)B$.

17. On the interval $[a, b]$, the average value of $f(x) + g(x)$ is the average value of $f(x)$ plus the average value of $g(x)$.

18. Which of the following statements follow directly from the rule

$$\int_a^b (f(x) + g(x))\, dx = \int_a^b f(x)\, dx + \int_a^b g(x)\, dx?$$

   (a) If $\int_a^b (f(x) + g(x))\, dx = 5 + 7$, then $\int_a^b f(x)\, dx = 5$ and $\int_a^b g(x)\, dx = 7$.
   (b) If $\int_a^b f(x)\, dx = \int_a^b g(x)\, dx = 7$, then $\int_a^b (f(x) + g(x))\, dx = 14$.
   (c) If $h(x) = f(x) + g(x)$, then $\int_a^b (h(x) - g(x))\, dx = \int_a^b h(x)\, dx - \int_a^b g(x)\, dx$.

PROJECTS FOR CHAPTER FIVE

1. An Orbiting Satellite

A NASA satellite orbits the earth every 90 minutes. During an orbit, the satellite's electric power comes either from solar array wings, when these are illuminated by the sun, or from batteries. The batteries discharge whenever the satellite uses more electricity than the solar array can provide or whenever the satellite is in the shadow of the earth (where the solar array cannot be used). If the batteries are overused, however, they can be damaged. [6]

You are to determine whether the batteries could be damaged in either of the following operations. You are told that the battery capacity is 50 ampere-hours. If the total battery discharge does not exceed 40% of battery capacity, the batteries will not be damaged.

(a) Operation 1 is performed by the satellite while orbiting the earth. At the beginning of a given 90-minute orbit, the satellite performs a 15-minute maneuver which requires more current than the solar array can deliver, causing the batteries to discharge. The maneuver causes a sinusoidally varying battery discharge of period 30 minutes with a maximum discharge of ten amperes at 7.5 minutes. For the next 45 minutes the solar array meets the total satellite current demand, and the batteries do not discharge. During the last 30 minutes, the satellite is in the shadow of the earth and the batteries supply the total current demand of 30 amperes.

  (i) The battery current in amperes is a function of time. Plot the function, showing the current in amperes as a function of time for the 90-minute orbit. Write a formula (or formulas) for the battery current function.

  (ii) Calculate the total battery discharge (in units of ampere-hours) for the 90-minute orbit for Operation 1.

  (iii) What is your recommendation regarding the advisability of Operation 1?

(b) Operation 2 is simulated at NASA's laboratory in Houston. The following graph was produced by the laboratory simulation of the current demands on the battery during the 90-minute orbit required for Operation 2.

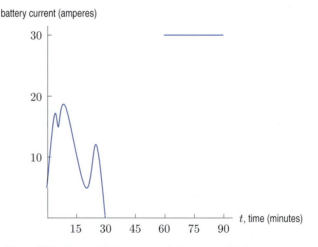

Figure 5.77: Battery discharge simulation graph for Operation 2

  (i) Calculate the total battery discharge (in units of ampere-hours) for the 90-minute orbit for Operation 2.

  (ii) What is your recommendation regarding the advisability of Operation 2?

---

[6]Adapted from Amy C. R. Gerson, "Electrical Engineering: Space Systems," in *She Does Math! Real Life Problems from Women on the Job*, ed. Marla Parker, p. 61 (Washington, DC: Mathematical Association of America, 1995).

2. Gas Tanks

Gasoline is stored at a gas station in an underground tank which is a circular cylinder lying on its side. You may have seen a gas station attendant dropping a long measuring stick into the tank to measure the amount of gasoline remaining. Suppose the remaining fuel comes to a height of h inches and the diameter of the tank is d inches. It might be tempting to guess that the tank is h/d full. The purpose of this project is to see how far off such a guess is.

Assume a cylindrical gas tank has diameter d and length l, as in Figure 5.78.

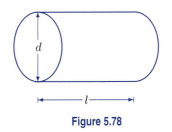

Figure 5.78

(a) Suppose another gas tank has the shape of a w by d by l rectangular box. What is w if this tank holds the same amount as the cylindrical tank? Does h/d give an exact measure of how full the rectangular tank is?

(b) When computing the accuracy of the stick method for determining what fraction of the tank is full, why can we ignore the length of the tank?

(c) Draw cross-sections of both tanks on one set of axes, with the x-axis measuring height and with both tanks centered at the origin.

(d) Let $E(h)$ be the error, as a function of h, when using h/d to measure how full the cylindrical tank is. Use the picture you drew in part (c) to write $E(h)$ as an integral.

(e) For what value of h is the error maximal?

Chapter Six

CONSTRUCTING ANTIDERIVATIVES

In Chapter 2, we saw how to calculate velocity given position, and in Chapter 5, we saw how to reconstruct distance from velocity.

In this chapter we look in more detail at how to reconstruct a function from its derivative using the Fundamental Theorem of Calculus. We introduce antiderivatives and show how to construct them numerically, graphically, and analytically. We introduce differential equations and the Second Fundamental Theorem of Calculus.

6.1 ANTIDERIVATIVES GRAPHICALLY AND NUMERICALLY

The Family of Antiderivatives

If the derivative of F is f, we call F an *antiderivative* of f. For example, since the derivative of x^2 is $2x$, we say that

$$x^2 \text{ is an antiderivative of } 2x.$$

Notice that $2x$ has many antiderivatives, since $x^2 + 1$, $x^2 + 2$, and $x^2 + 3$, all have derivative $2x$. In fact, if C is any constant, we have

$$\frac{d}{dx}(x^2 + C) = 2x + 0 = 2x,$$

so any function of the form $x^2 + C$ is an antiderivative of $2x$. The function $f(x) = 2x$ has a *family of antiderivatives*.

Let us look at another example. If v is the velocity of a car and s is its position, then $v = ds/dt$ and s is an antiderivative of v. As before, $s + C$ is an antiderivative of v for any constant C. In terms of the car, adding C to s is equivalent to adding C to the odometer reading. Adding a constant to the odometer reading simply means measuring distance from a different point, which doesn't alter the car's velocity.

Visualizing Antiderivatives Using Slopes

Suppose we have the graph of f', and we want to sketch an approximate graph of f. We are looking for the graph of f whose slope at any point is equal to the value of f' there. Where f' is above the x-axis, f is increasing; where f' is below the x-axis, f is decreasing. If f' is increasing, f is concave up; if f' is decreasing, f is concave down.

Example 1   The graph of f' is given in Figure 6.1. Sketch a graph of f in the cases when $f(0) = 0$ and $f(0) = 1$.

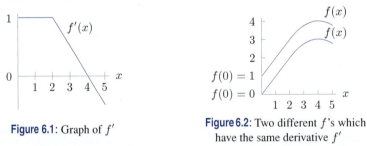

Figure 6.1: Graph of f'

Figure 6.2: Two different f's which have the same derivative f'

Solution   For $0 \leq x \leq 2$, the function f has a constant slope of 1, so the graph of f is a straight line. For $2 \leq x \leq 4$, the function f is increasing but more and more slowly; it has a maximum at $x = 4$ and decreases thereafter. (See Figure 6.2.) The solutions with $f(0) = 0$ and $f(0) = 1$ start at different points on the vertical axis but have the same shape.

Example 2   Sketch a graph of the antiderivative F of $f(x) = e^{-x^2}$ satisfying $F(0) = 0$.

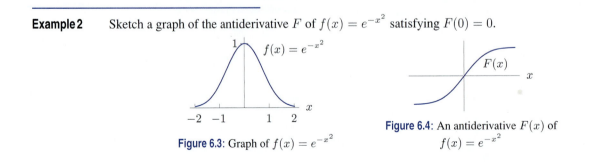

Figure 6.3: Graph of $f(x) = e^{-x^2}$

Figure 6.4: An antiderivative $F(x)$ of $f(x) = e^{-x^2}$

Solution   The graph of $f(x) = e^{-x^2}$ is shown in Figure 6.3. The slope of the antiderivative $F(x)$ is given by $f(x)$. Since $f(x)$ is always positive, the antiderivative $F(x)$ is always increasing. Since $f(x)$ is increasing for negative x, we know that $F(x)$ is concave up for negative x. Since $f(x)$ is decreasing for positive x, we know that $F(x)$ is concave down for positive x. Since $f(x) \to 0$ as $x \to \pm\infty$, the graph of $F(x)$ levels off at both ends. See Figure 6.4.

Example 3   For the function f' given in Figure 6.5, sketch a graph of three antiderivative functions f, one with $f(0) = 0$, one with $f(0) = 1$, and one with $f(0) = 2$.

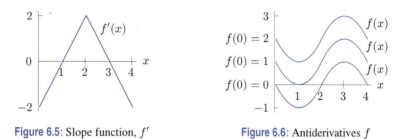

Figure 6.5: Slope function, f' **Figure 6.6:** Antiderivatives f

Solution   To graph f, start at the point on the vertical axis specified by the initial condition and move with slope given by the value of f' in Figure 6.5. Different initial conditions lead to different graphs for f, but for a given x-value they all have the same slope (because the value of f' is the same for each). Thus, the different f curves are obtained from one another by a vertical shift. See Figure 6.6.

- Where f' is positive $(1 < x < 3)$, we see f is increasing; where f' is negative $(0 < x < 1$ or $3 < x < 4)$, we see f is decreasing.
- Where f' is increasing $(0 < x < 2)$, we see f is concave up; where f' is decreasing $(2 < x < 4)$, we see f is concave down.
- Where $f' = 0$, we see f has a local maximum at $x = 3$ and a local minimum at $x = 1$.
- Where f' has a maximum $(x = 2)$, we see f has a point of inflection.

Computing Values of an Antiderivative Using the Fundamental Theorem

A graph of f' shows where f is increasing and where f is decreasing. We can calculate the actual value of the function f by using the Fundamental Theorem of Calculus (Theorem 5.1 on page 244): If f' is continuous, then

$$\int_a^b f'(x)\,dx = f(b) - f(a).$$

Example 4   Figure 6.7 is the graph of the derivative $f'(x)$ of a function $f(x)$. It is given that $f(0) = 100$. Sketch the graph of $f(x)$, showing all critical points and inflection points of f and giving their coordinates.

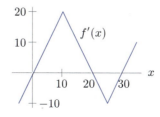

Figure 6.7: Graph of derivative

Solution   The critical points of f occur at $x = 0$, $x = 20$, and $x = 30$, where $f'(x) = 0$. The inflection points of f occur at $x = 10$ and $x = 25$, where $f'(x)$ has a maximum or minimum. To find the coordinates of the critical points and inflection points of f, we evaluate $f(x)$ for $x = 0, 10, 20, 25, 30$. Using the Fundamental Theorem, we can express the values of $f(x)$ in terms of definite integrals. We evaluate the definite integrals using the areas of triangular regions under the graph of $f'(x)$, remembering that areas below the x-axis are subtracted. (See Figure 6.8.)

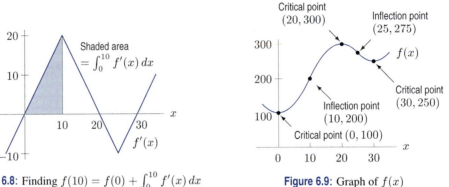

Figure 6.8: Finding $f(10) = f(0) + \int_0^{10} f'(x)\,dx$ **Figure 6.9**: Graph of $f(x)$

Since $f(0) = 100$, the Fundamental Theorem gives us the following values of f, which are marked in Figure 6.9.

$$f(10) = f(0) + \int_0^{10} f'(x)\,dx = 100 + \text{(Shaded area in Figure 6.8)} = 100 + \frac{1}{2}(10)(20) = 200,$$

$$f(20) = f(10) + \int_{10}^{20} f'(x)\,dx = 200 + \frac{1}{2}(10)(20) = 300,$$

$$f(25) = f(20) + \int_{20}^{25} f'(x)\,dx = 300 - \frac{1}{2}(5)(10) = 275,$$

$$f(30) = f(25) + \int_{25}^{30} f'(x)\,dx = 275 - \frac{1}{2}(5)(10) = 250.$$

Example 5   Suppose $F'(t) = t \cos t$ and $F(0) = 2$. Find $F(b)$ at the points $b = 0, 0.1, 0.2, \ldots, 1.0$.

Solution   We apply the Fundamental Theorem with $f(t) = t \cos t$ and $a = 0$ to get values for $F(b)$:

$$F(b) - F(0) = \int_0^b F'(t)\,dt = \int_0^b t \cos t\,dt.$$

Since $F(0) = 2$, we have

$$F(b) = 2 + \int_0^b t \cos t\,dt.$$

Calculating the definite integral $\int_0^b t \cos t\,dt$ numerically for $b = 0, 0.1, 0.2, \ldots, 1.0$ gives the values for F in Table 6.1:

Table 6.1   *Approximate values for F*

b	0	0.1	0.2	0.3	0.4	0.5	0.6	0.7	0.8	0.9	1.0
$F(b)$	2.000	2.005	2.020	2.044	2.077	2.117	2.164	2.216	2.271	2.327	2.382

Notice that $F(b)$ appears to be increasing between $b = 0$ and $b = 1$. This could have been predicted from the fact that $t \cos t$, the derivative of $F(t)$, is positive for t between 0 and 1.

Exercises and Problems for Section 6.1

Exercises

In Exercises 1–4, sketch two functions F such that $F' = f$. In one case let $F(0) = 0$ and in the other, let $F(0) = 1$.

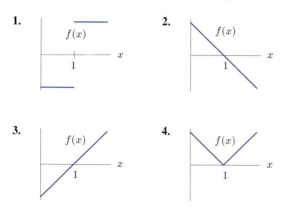

1.

2.

3.

4.

5. Given the values of the derivative $f'(x)$ in the table and that $f(0) = 100$, estimate $f(x)$ for $x = 2, 4, 6$.

x	0	2	4	6
$f'(x)$	10	18	23	25

6. Estimate $f(x)$ for $x = 2,\ 4,\ 6$, using the given values of $f'(x)$ and the fact that $f(0) = 50$.

x	0	2	4	6
$f'(x)$	17	15	10	2

7. (a) Using Figure 6.10, estimate $\int_0^7 f(x)dx$.
   (b) If F is an antiderivative of the same function f and $F(0) = 25$, estimate $F(7)$.

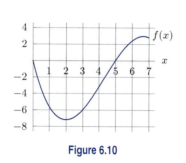

Figure 6.10

8. Use Figure 6.11 and the fact that $P = 2$ when $t = 0$ to find values of P when $t = 1, 2, 3, 4$ and 5.

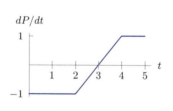

Figure 6.11

Problems

In Problems 9–10, a graph of f is given. Let $F'(x) = f(x)$.
(a) What are the critical points of $F(x)$?
(b) Which critical points are local maxima, which are local minima, and which are neither?
(c) Sketch a possible graph of $F(x)$.

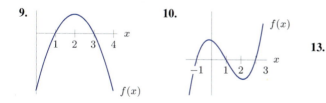

9.

10.

11.

12.

13.

In Problems 11–13, sketch two functions F where $F'(x) = f(x)$. In one case, let $F(0) = 0$, and in the other, let $F(0) = 1$. Mark x_1, x_2, and x_3 on the x-axis of your graph. Identify local maxima, minima, and inflection points of $F(x)$.

14. A particle moves back and forth along the x-axis. Figure 6.12 approximates the velocity of the particle as a function of time. Positive velocities represent movement to the right and negative velocities represent movement to the left. The particle starts at the point $(5, 0)$. Graph the distance of the particle from the origin, with distance measured in kilometers and time in hours.

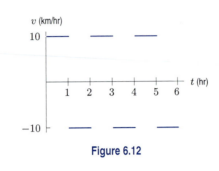

Figure 6.12

15. Assume f' is given by the graph in Figure 6.13. Suppose f is continuous and that $f(0) = 0$.

(a) Find $f(3)$ and $f(7)$.
(b) Find all x with $f(x) = 0$.
(c) Sketch a graph of f over the interval $0 \le x \le 7$.

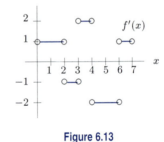

Figure 6.13

16. Use Figure 6.14 and the fact that $F(2) = 3$ to sketch the graph of $F(x)$. Label the values of at least four points.

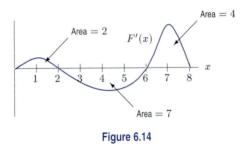

Figure 6.14

17. Using Figure 6.15, sketch a graph of an antiderivative $G(t)$ of $g(t)$ satisfying $G(0) = 5$. Label each critical point of $G(t)$ with its coordinates.

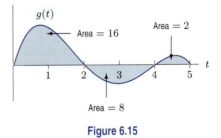

Figure 6.15

18. Using the graph of g' in Figure 6.16 and the fact that $g(0) = 50$, sketch the graph of $g(x)$. Give the coordinates of all critical points and inflection points of g.

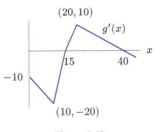

Figure 6.16

19. The vertical velocity of a cork bobbing up and down on the waves in the sea is given by Figure 6.17. Upward is considered positive. Describe the motion of the cork at each of the labeled points. At which point(s), if any, is the acceleration zero? Sketch a graph of the height of the cork above the sea floor as a function of time.

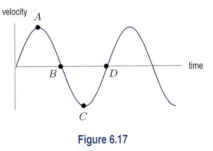

Figure 6.17

20. Figure 6.18 shows the rate of change of the concentration of adrenaline, in micrograms per milliliter per minute, in a person's body. Sketch a graph of the concentration of adrenaline, in micrograms per milliliter, in the body as a function of time, in minutes.

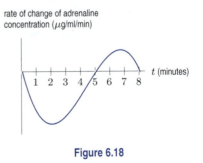

Figure 6.18

21. Urologists are physicians who specialize in the health of the bladder. In a common diagnostic test, urologists monitor the emptying of the bladder using a device that produces two graphs. In one of the graphs the flow rate (in milliliters per second) is measured as a function of time (in seconds). In the other graph, the volume emptied from the bladder is measured (in milliliters) as a function of time (in seconds). See Figure 6.19.

(a) Which graph is the flow rate and which is the volume?

(b) Which one of these graphs is an antiderivative of the other?

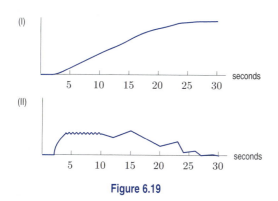

Figure 6.19

22. Use a graph of $f(x) = 2\sin(x^2)$ to determine where an antiderivative, F, of this function reaches its maximum on $0 \leq x \leq 3$. If $F(1) = 5$, find the maximum value attained by F.

23. The graph of f'' is given in Figure 6.20. Draw graphs of f and f', assuming both go through the origin, and use them to decide at which of the labeled x-values:

(a) $f(x)$ is greatest.
(b) $f(x)$ is least.
(c) $f'(x)$ is greatest.
(d) $f'(x)$ is least.
(e) $f''(x)$ is greatest.
(f) $f''(x)$ is least.

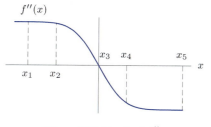

Figure 6.20: Graph of f''

24. Two functions, $f(x)$ and $g(x)$, are shown in Figure 6.21. Let F and G be antiderivatives of f and g, respectively. On the same axes, sketch graphs of the antiderivatives $F(x)$ and $G(x)$ satisfying $F(0) = 0$ and $G(0) = 0$. Compare F and G, including a discussion of zeros and x- and y-coordinates of critical points.

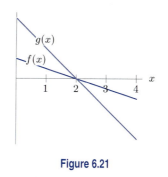

Figure 6.21

25. The Quabbin Reservoir in the western part of Massachusetts provides most of Boston's water. The graph in Figure 6.22 represents the flow of water in and out of the Quabbin Reservoir throughout 1993.

(a) Sketch a possible graph for the quantity of water in the reservoir, as a function of time.

(b) When, in the course of 1993, was the quantity of water in the reservoir largest? Smallest? Mark and label these points on the graph you drew in part (a).

(c) When was the quantity of water increasing most rapidly? Decreasing most rapidly? Mark and label these times on both graphs.

(d) By July 1994 the quantity of water in the reservoir was about the same as in January 1993. Draw plausible graphs for the flow into and the flow out of the reservoir for the first half of 1994. Explain your graphs.

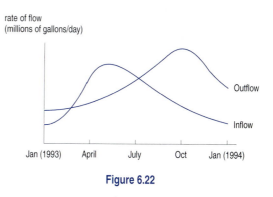

Figure 6.22

6.2 CONSTRUCTING ANTIDERIVATIVES ANALYTICALLY

What Is an Antiderivative of $f(x) = 0$?

A function whose derivative is zero everywhere on an interval must have a horizontal tangent line at every point of its graph, and the only way this can happen is if the function is constant. Alternatively, if we think of the derivative as a velocity, and if the velocity is always zero, then the object is standing still; the position function is constant. A rigorous proof of this result using the definition of the derivative is surprisingly subtle. (See the Constant Function Theorem on page 209.)

If $F'(x) = 0$ on an interval, then $F(x) = C$ on this interval.

What Is the Most General Antiderivative of f ?

We know that if a function f has an antiderivative F, then it has a family of antiderivatives of the form $F(x) + C$, where C is any constant. You might wonder if there are any others. To decide, suppose that we have two functions F and G with $F' = f$ and $G' = f$: that is, F and G are both antiderivatives of the same function f. Since $F' = G'$ we have $(F-G)' = 0$. But this means that we must have $F - G = C$, so $F(x) = G(x) + C$, where C is a constant. Thus, any two antiderivatives of the same function differ only by a constant.

If F and G are both antiderivatives of f, then $F(x) = G(x) + C$ for some constant C.

The Indefinite Integral

All antiderivatives of $f(x)$ are of the form $F(x) + C$. We introduce a notation for the general antiderivative that looks like the definite integral without the limits and is called the *indefinite integral*:

$$\int f(x)\,dx = F(x) + C.$$

It is important to understand the difference between

$$\int_a^b f(x)\,dx \quad \text{and} \quad \int f(x)\,dx.$$

The first is a number and the second is a family of *functions*. The word "integration" is frequently used for the process of finding the antiderivative as well as of finding the definite integral. The context usually makes clear which is intended.

What is an Antiderivative of $f(x) = k$?

If k is a constant, the derivative of kx is k, so we have

$$\text{An antiderivative of } k \text{ is } kx.$$

Using the indefinite integral notation, we have

If k is constant,

$$\int k\,dx = kx + C.$$

Finding Antiderivatives

Finding antiderivatives of functions is like taking square roots of numbers: if we pick a number at random, such as 7 or 493, we may have trouble finding its square root without a calculator. But if we happen to pick a number such as 25 or 64, which we know is a perfect square, then we can find its square root exactly. Similarly, if we pick a function which we recognize as a derivative, then we can find its antiderivative easily.

For example, to find an antiderivative of $f(x) = x$, notice that $2x$ is the derivative of x^2; this tells us that x^2 is an antiderivative of $2x$. If we divide by 2, then we guess that

$$\text{An antiderivative of } x \text{ is } \frac{x^2}{2}.$$

To check this statement, take the derivative of $x^2/2$:

$$\frac{d}{dx}\left(\frac{x^2}{2}\right) = \frac{1}{2} \cdot \frac{d}{dx}x^2 = \frac{1}{2} \cdot 2x = x.$$

What about an antiderivative of x^2? The derivative of x^3 is $3x^2$, so the derivative of $x^3/3$ is $3x^2/3 = x^2$. Thus,

$$\text{An antiderivative of } x^2 \text{ is } \frac{x^3}{3}.$$

The pattern looks like

$$\text{An antiderivative of } x^n \text{ is } \frac{x^{n+1}}{n+1}.$$

(We assume $n \neq -1$, or we would have $x^0/0$, which doesn't make sense.) It is easy to check this formula by differentiation:

$$\frac{d}{dx}\left(\frac{x^{n+1}}{n+1}\right) = \frac{(n+1)x^n}{n+1} = x^n.$$

In indefinite integral notation, we have shown that

$$\int x^n \, dx = \frac{x^{n+1}}{n+1} + C, \quad n \neq -1.$$

What about when $n = -1$? In other words, what is an antiderivative of $1/x$? Fortunately, we know a function whose derivative is $1/x$, namely, the natural logarithm. Thus, since

$$\frac{d}{dx}(\ln x) = \frac{1}{x},$$

we know that

$$\int \frac{1}{x} \, dx = \ln x + C, \quad \text{for } x > 0.$$

If $x < 0$, then $\ln x$ is not defined, so it can't be an antiderivative of $1/x$. In this case, we can try $\ln(-x)$:

$$\frac{d}{dx}\ln(-x) = (-1)\frac{1}{-x} = \frac{1}{x}$$

so

$$\int \frac{1}{x} \, dx = \ln(-x) + C, \quad \text{for } x < 0.$$

This means $\ln x$ is an antiderivative of $1/x$ if $x > 0$, and $\ln(-x)$ is an antiderivative of $1/x$ if $x < 0$. Since $|x| = x$ when $x > 0$ and $|x| = -x$ when $x < 0$ we can collapse these two formulas into:

An antiderivative of $\dfrac{1}{x}$ is $\ln|x|$.

Therefore

$$\int \frac{1}{x}\,dx = \ln|x| + C.$$

Since the exponential function is its own derivative, it is also its own antiderivative; thus

$$\int e^x\,dx = e^x + C.$$

Also, antiderivatives of the sine and cosine are easy to guess. Since

$$\frac{d}{dx}\sin x = \cos x \qquad \text{and} \qquad \frac{d}{dx}\cos x = -\sin x,$$

we get

$$\int \cos x\,dx = \sin x + C \qquad \text{and} \qquad \int \sin x\,dx = -\cos x + C.$$

Example 1 Find $\displaystyle\int (3x + x^2)\,dx$.

Solution We know that $x^2/2$ is an antiderivative of x and that $x^3/3$ is an antiderivative of x^2, so we expect

$$\int (3x + x^2)\,dx = 3\left(\frac{x^2}{2}\right) + \frac{x^3}{3} + C.$$

You should always check your antiderivatives by differentiation—it's easy to do. Here

$$\frac{d}{dx}\left(\frac{3}{2}x^2 + \frac{x^3}{3} + C\right) = \frac{3}{2}\cdot 2x + \frac{3x^2}{3} = 3x + x^2.$$

The preceding example illustrates that the sum and constant multiplication rules of differentiation work in reverse:

Theorem 6.1: Properties of Antiderivatives: Sums and Constant Multiples

In indefinite integral notation,

1. $\displaystyle\int [f(x) \pm g(x)]\,dx = \int f(x)\,dx \pm \int g(x)\,dx$

2. $\displaystyle\int cf(x)\,dx = c\int f(x)\,dx.$

In words,

1. An antiderivative of the sum (or difference) of two functions is the sum (or difference) of their antiderivatives.

2. An antiderivative of a constant times a function is the constant times an antiderivative of the function.

Example 2 Find $\displaystyle\int (\sin x + 3\cos x)\,dx$.

Solution   We break the antiderivative into two terms:

$$\int (\sin x + 3 \cos x)\, dx = \int \sin x\, dx + 3 \int \cos x\, dx = -\cos x + 3 \sin x + C.$$

Check by differentiating:

$$\frac{d}{dx}(-\cos x + 3 \sin x + C) = \sin x + 3 \cos x.$$

Using Antiderivatives to Compute Definite Integrals

The Fundamental Theorem of Calculus gives us a new way of calculating definite integrals. The theorem says that if $F' = f$, then

$$\int_a^b f(x)\, dx = F(b) - F(a),$$

To find $\int_a^b f(x)\, dx$, we first try to find F, and then calculate $F(b) - F(a)$. This method of computing definite integrals gives an exact answer. However, the method only works in situations where we can find the antiderivative $F(x)$. This is not always easy; for example, none of the functions we have encountered so far is an antiderivative of $\sin(x^2)$.

Example 3   Compute $\int_1^2 3x^2\, dx$ using the Fundamental Theorem.

Solution   Since $F(x) = x^3$ is an antiderivative of $f(x) = 3x^2$, we have

$$\int_1^2 3x^2\, dx = F(2) - F(1) = 2^3 - 1^3 = 7.$$

Notice in this example we used the antiderivative x^3, but $x^3 + C$ works just as well because the constant cancels out:

$$\int_1^2 3x^2\, dx = F(2) - F(1) = (2^3 + C) - (1^3 + C) = 7.$$

It is helpful to introduce a shorthand notation for $F(b) - F(a)$: we write it as $F(x)\Big|_a^b$. For example:

$$\int_1^2 3x^2\, dx = x^3 \Big|_1^2 = 2^3 - 1^3 = 7.$$

Exercises and Problems for Section 6.2

Exercises

In Exercises 1–21, find an antiderivative.

1. $f(x) = 5$

2. $f(x) = 5x$

3. $f(x) = x^2$

4. $g(t) = t^2 + t$

5. $h(t) = \cos t$

6. $g(z) = \sqrt{z}$

7. $h(z) = \dfrac{1}{z}$

8. $r(t) = \dfrac{1}{t^2}$

9. $g(z) = \dfrac{1}{z^3}$

10. $f(z) = e^z$

11. $g(t) = \sin t$

12. $f(t) = 2t^2 + 3t^3 + 4t^4$

13. $p(t) = t^3 - \dfrac{t^2}{2} - t$

14. $q(y) = y^4 + \dfrac{1}{y}$

15. $p(t) = \cos t + \dfrac{1}{\cos^2 t}$

16. $f(t) = \dfrac{t^2 + 1}{t}$

17. $p(\theta) = 2 \sin(2\theta)$

18. $r(t) = e^t + 5e^{5t}$

19. $q(t) = (t + 1)^2$

20. $f(x) = 5^x$.

21. $f(x) = 5x - \sqrt{x}$

In Exercises 22–33, find the general antiderivative.

22. $f(t) = 6t$

23. $h(x) = x^3 - x$

24. $f(x) = x^2 - 4x + 7$

25. $r(t) = t^3 + 5t - 1$

26. $f(z) = z + e^z$

27. $g(t) = \sqrt{t}$

28. $g(x) = \sin x + \cos x$

29. $h(x) = 4x^3 - 7$

30. $p(t) = 2 + \sin t$

31. $p(t) = \dfrac{1}{\sqrt{t}}$

32. $g(x) = \dfrac{5}{x^3}$

33. $f(x) = \dfrac{1}{x^2}$

For Exercises 34–41, find an antiderivative $F(x)$ with $F'(x) = f(x)$ and $F(0) = 0$. Is there only one possible solution in each case?

34. $f(x) = 3$

35. $f(x) = 2x$

36. $f(x) = -7x$

37. $f(x) = \dfrac{1}{4}x$

38. $f(x) = x^2$

39. $f(x) = \sqrt{x}$

40. $f(x) = 2 + 4x + 5x^2$

41. $f(x) = \sin x$

Find the indefinite integrals in Exercises 42–60.

42. $\displaystyle\int 5x\, dx$

43. $\displaystyle\int x^3\, dx$

44. $\displaystyle\int \sin\theta\, d\theta$

45. $\displaystyle\int (x^3 - 2)\, dx$

46. $\displaystyle\int \left(t^2 + \dfrac{1}{t^2}\right) dt$

47. $\displaystyle\int 4\sqrt{w}\, dw$

48. $\displaystyle\int (x^2 + 5x + 8)\, dx$

49. $\displaystyle\int \dfrac{4}{t^2}\, dt$

50. $\displaystyle\int (4t + 7)\, dt$

51. $\displaystyle\int \cos\theta\, d\theta$

52. $\displaystyle\int 5e^z\, dz$

53. $\displaystyle\int \left(x + \dfrac{1}{\sqrt{x}}\right) dx$

54. $\displaystyle\int \sin t\, dt$

55. $\displaystyle\int (\pi + x^{11})\, dx$

56. $\displaystyle\int \left(t\sqrt{t} + \dfrac{1}{t\sqrt{t}}\right) dt$

57. $\displaystyle\int \cos(x + 1)\, dx$

58. $\displaystyle\int e^{2r}\, dr$

59. $\displaystyle\int \dfrac{1}{e^z}\, dz$

60. $\displaystyle\int \left(\dfrac{y^2 - 1}{y}\right)^2 dy$

In Exercises 61–73, evaluate the definite integrals exactly [as in $\ln(3\pi)$], using the Fundamental Theorem, and numerically $[\ln(3\pi) \approx 2.243]$:

61. $\displaystyle\int_0^3 (x^2 + 4x + 3)\, dx$

62. $\displaystyle\int_1^3 \dfrac{1}{t}\, dt$

63. $\displaystyle\int_0^{\pi/4} \sin x\, dx$

64. $\displaystyle\int_0^2 3e^x\, dx$

65. $\displaystyle\int_2^5 (x^3 - \pi x^2)\, dx$

66. $\displaystyle\int_0^1 \sin\theta\, d\theta$

67. $\displaystyle\int_1^2 \dfrac{1 + y^2}{y}\, dy$

68. $\displaystyle\int_0^2 \left(\dfrac{x^3}{3} + 2x\right) dx$

69. $\displaystyle\int_0^{\pi/4} (\sin t + \cos t)\, dt$

70. $\displaystyle\int_{-3}^{-1} \dfrac{2}{r^3}\, dr$

71. $\displaystyle\int_0^1 2e^x\, dx$

72. $\displaystyle\int_0^{\pi/4} \dfrac{1}{\cos^2 x}\, dx$

73. $\displaystyle\int_{-1}^1 2^x\, dx$

Problems

74. Use the Fundamental Theorem to find the area under $f(x) = x^2$ between $x = 1$ and $x = 4$.

75. Calculate the exact area between the x-axis and the graph of $y = 7 - 8x + x^2$.

76. Calculate the exact area above the graph of $y = \sin\theta$ and below the graph of $y = \cos\theta$ for $0 \le \theta \le \pi/4$.

77. Find the exact value of the area between the graphs of $y = \cos x$ and $y = e^x$ for $0 \le x \le 1$.

78. Use the Fundamental Theorem to determine the value of b if the area under the graph of $f(x) = 8x$ between $x = 1$ and $x = b$ is equal to 192. Assume $b > 1$.

79. Find the exact positive value of c if the area between the graph of $y = x^2 - c^2$ and the x-axis is 36.

80. Use the Fundamental Theorem to find the average value of $f(x) = x^2 + 1$ on the interval $x = 0$ to $x = 10$. Illustrate your answer on a graph of $f(x)$.

81. The average value of the function $v(x) = 6/x^2$ on the interval $[1, c]$ is equal to 1. Find the value of c.

82. (a) What is the average value of $f(t) = \sin t$ over $0 \le t \le 2\pi$? Why is this a reasonable answer?
(b) Find the average of $f(t) = \sin t$ over $0 \le t \le \pi$.

83. In drilling an oil well, the total cost, C, consists of fixed costs (independent of the depth of the well) and marginal costs, which depend on depth; drilling becomes more expensive, per meter, deeper into the earth. Suppose the fixed costs are 1,000,000 riyals (the riyal is the unit of currency of Saudi Arabia), and the marginal costs are

$$C'(x) = 4000 + 10x \text{ riyals/meter},$$

where x is the depth in meters. Find the total cost of drilling a well x meters deep.

84. One of the earliest pollution problems brought to the attention of the Environmental Protection Agency (EPA) was the case of the Sioux Lake in eastern South Dakota. For years a small paper plant located nearby had been discharging waste containing carbon tetrachloride (CCl_4) into the waters of the lake. At the time the EPA learned of the situation, the chemical was entering at a rate of 16 cubic yards/year.

The agency ordered the installation of filters designed to slow (and eventually stop) the flow of CCl_4 from the mill. Implementation of this program took exactly three years, during which the flow of pollutant was steady at 16 cubic yards/year. Once the filters were installed, the flow declined. If t is time measured in years since the EPA learned of the situation, the rate of flow was well approximated by

$$\text{Rate (in cubic yards/year)} = t^2 - 14t + 49,$$

between the time the filters were installed and the time the flow stopped.

(a) Draw a graph showing the rate of CCl_4 flow into the lake as a function of time, beginning at the time the EPA first learned of the situation.

(b) How many years elapsed between the time the EPA learned of the situation and the time the pollution flow stopped entirely?

(c) How much CCl_4 entered the waters during the time shown in the graph in part (a)?

6.3 DIFFERENTIAL EQUATIONS

In Chapter 2 we saw that velocity is the derivative of distance and that acceleration is the derivative of velocity. In this section we analyze the motion of an object falling freely under the influence of gravity. This involves going "backward" from acceleration to velocity to position.

Motion With Constant Velocity

Let's briefly consider a familiar problem: An object moving in a straight line with constant velocity. Imagine a car moving at 50 mph. How far does it go in a given time? The answer is given by

$$\text{Distance} = \text{Rate} \times \text{Time}$$

or

$$s = 50t$$

where s is the distance of the car (in miles) from a fixed reference point and t is the time in hours. Alternatively, we can describe the motion by writing the equation

$$\frac{ds}{dt} = 50.$$

This is called a *differential equation* for the *function s*. The solution to this equation is the antiderivative

$$s = 50t + C.$$

The equation $s = 50t + C$ tells us that $s = C$ when $t = 0$. Thus, the constant C represents the initial distance, s_0, of the car from the reference point.

Uniformly Accelerated Motion

Now we consider an object moving with *constant acceleration* along a straight line, or *uniformly accelerated motion*. It has been known since Galileo's time that an object moving under the influence of gravity (ignoring air resistance) has constant acceleration, g. In the most frequently used units, its value is approximately

$$g = 9.8 \text{ m/sec}^2, \quad \text{or} \quad g = 32 \text{ ft/sec}^2.$$

20. An object is shot vertically upward from the ground with a velocity of 160 ft/sec.

   (a) Sketch a graph of the velocity of the object (with upward as positive) against time.

   (b) Mark on the graph the points when the object reaches its highest point and when it lands.

   (c) Find the maximum height reached by the object by considering an area on the graph.

   (d) Now express velocity as a function of time, and find the greatest height by antidifferentiation.

21. A stone thrown upward from the top of a 320-foot cliff at 128 ft/sec eventually falls to the beach below.

   (a) How long does the stone take to reach its highest point?

   (b) What is its maximum height?

   (c) How long before the stone hits the beach?

   (d) What is the velocity of the stone on impact?

22. On the moon, the acceleration due to gravity is about 1.6 m/sec^2 (compared to $g \approx 9.8$ m/sec^2 on earth). If you drop a rock on the moon (with initial velocity 0), find formulas for:

   (a) Its velocity, $v(t)$, at time t.

   (b) The distance, $s(t)$, it falls in time t.

23. **(a)** Imagine throwing a rock straight up in the air. What is its initial velocity if the rock reaches a maximum height of 100 feet above its starting point?

   (b) Now imagine being transplanted to the moon and throwing a moon rock vertically upward with the same velocity as in part (a). How high will it go? (On the moon, $g = 5$ ft/sec^2.)

24. A cat, walking along the window ledge of a New York apartment, knocks off a flower pot, which falls to the street 200 feet below. How fast is the flower pot traveling when it hits the street? (Give your answer in ft/sec and in mph, given that 1 ft/sec $= 15/22$ mph.)

25. An Acura NSX going at 70 mph stops in 157 feet. Find the acceleration, assuming it is constant.

6.4 SECOND FUNDAMENTAL THEOREM OF CALCULUS

Suppose f is an elementary function, that is, a combination of constants, powers of x, $\sin x$, $\cos x$, e^x, and $\ln x$. Then we have to be lucky to find an antiderivative F which is also an elementary function. But if we can't find F as an elementary function, how can we be sure that F exists at all? In this section we learn to use the definite integral to construct antiderivatives.

Construction of Antiderivatives Using the Definite Integral

Consider the function $f(x) = e^{-x^2}$. We would like to find a way of calculating values of its antiderivative, F, which is not an elementary function. However, we know from the Fundamental Theorem of Calculus that

$$F(b) - F(a) = \int_a^b e^{-t^2}\, dt.$$

Setting $a = 0$ and replacing b by x, we have

$$F(x) - F(0) = \int_0^x e^{-t^2}\, dt.$$

Suppose we want the antiderivative that satisfies $F(0) = 0$. Then we get

$$F(x) = \int_0^x e^{-t^2}\, dt.$$

This is a formula for F. For any value of x, there is a unique value for $F(x)$, so F is a function. For any fixed x, we can calculate $F(x)$ numerically. For example,

$$F(2) = \int_0^2 e^{-t^2}\, dt = 0.88208\ldots.$$

Notice that our expression for F is not an elementary function; we have *created* a new function using the definite integral. The next theorem says that this method of constructing antiderivatives works in general. This means that if we define F by

$$F(x) = \int_a^x f(t)\, dt$$

then F must be an antiderivative of f.

Theorem 6.2: Construction Theorem for Antiderivatives

(Second Fundamental Theorem of Calculus) If f is a continuous function on an interval, and if a is any number in that interval, then the function F defined as follows is an antiderivative of f:

$$F(x) = \int_a^x f(t)\, dt.$$

Proof   Our task is to show that F, defined by this integral, is an antiderivative of f. We want to show that $F'(x) = f(x)$. By the definition of the derivative,

$$F'(x) = \lim_{h \to 0} \frac{F(x+h) - F(x)}{h}.$$

To gain some geometric insight, let's suppose f is positive and h is positive. Then we can visualize

$$F(x) = \int_a^x f(t)\, dt \quad \text{and} \quad F(x+h) = \int_a^{x+h} f(t)\, dt$$

as areas, which leads to representing

$$F(x+h) - F(x) = \int_x^{x+h} f(t)\, dt$$

as a difference of two areas. From Figure 6.24, we see that $F(x+h) - F(x)$ is roughly the area of a rectangle of height $f(x)$ and width h (shaded darker in Figure 6.24), so we have

$$F(x+h) - F(x) \approx f(x)h,$$

hence

$$\frac{F(x+h) - F(x)}{h} \approx f(x).$$

More precisely, we can use Theorem 5.4 on comparing integrals on page 249 to conclude that

$$mh \le \int_x^{x+h} f(t)\, dt \le Mh,$$

where m is the greatest lower bound for f on the interval from x to $x+h$ and M is the least upper bound on that interval. (See Figure 6.25.) Hence

$$mh \le F(x+h) - F(x) \le Mh,$$

so

$$m \le \frac{F(x+h) - F(x)}{h} \le M.$$

Since f is continuous, both m and M approach $f(x)$ as h approaches zero. Thus

$$f(x) \le \lim_{h \to 0} \frac{F(x+h) - F(x)}{h} \le f(x).$$

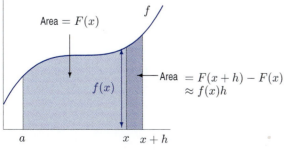

Figure 6.24: $F(x+h) - F(x)$ is area of roughly rectangular region

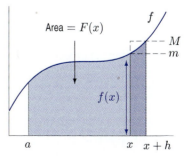

Figure 6.25: Upper and lower bounds for $F(x+h) - F(x)$

Thus both inequalities must actually be equalities, so we have the result we want:

$$f(x) = \lim_{h \to 0} \frac{F(x+h) - F(x)}{h} = F'(x).$$

Notice that if we already know an antiderivative of f, which we may call G, then the Fundamental Theorem of Calculus tells us that

$$G(x) - G(a) = \int_a^x f(t)\, dt.$$

By the definition of F, we have

$$F(x) = \int_a^x f(t)\, dt.$$

Therefore,

$$G(x) - G(a) = F(x),$$

so F and G differ by a constant—namely $G(a)$—so F and G are *both* antiderivatives of f.

Using the Construction Theorem for Antiderivatives

The construction theorem enables us to write down antiderivatives of functions that do not have elementary antiderivatives. For example, an antiderivative of $(\sin x)/x$ is

$$F(x) = \int_0^x \frac{\sin t}{t}\, dt.$$

Notice that F is a function; we can calculate its values to any degree of accuracy. This function already has a name: it is called the *sine-integral*, and it is denoted $\mathrm{Si}(x)$.

Example 1   Construct a table of values of $\mathrm{Si}(x)$ for $x = 0, 1, 2, 3$.

Solution   Using numerical methods, we calculate the values of $\mathrm{Si}(x) = \int_0^x \sin t/t\, dt$ given in Table 6.2. Since the integrand is undefined at $t = 0$, we took the lower limit as 0.00001 instead of 0.

Table 6.2   *A table of values of* $\mathrm{Si}(x)$

x	0	1	2	3
$\mathrm{Si}(x)$	0	0.95	1.61	1.85

The reason the sine-integral has a name is that some scientists and engineers use it all the time (for example, in optics). For them, it is just another common function like sine or cosine. Its derivative is given by

$$\frac{d}{dx}\, \mathrm{Si}(x) = \frac{\sin x}{x}.$$

Example 2   Find the derivative of $x\,\mathrm{Si}(x)$.

Solution   Using the product rule,

$$\frac{d}{dx}\left(x\,\mathrm{Si}(x)\right) = \left(\frac{d}{dx}\, x\right)\mathrm{Si}(x) + x\left(\frac{d}{dx}\, \mathrm{Si}(x)\right)$$

$$= 1 \cdot \mathrm{Si}(x) + x\frac{\sin x}{x}$$

$$= \mathrm{Si}(x) + \sin x.$$

Exercises and Problems for Section 6.4

Exercises

In Exercises 1–3, let $F(x) = \int_0^x f(t)\,dt$. Graph $F(x)$ as a function of x.

1.
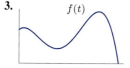

2.
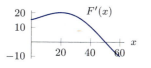

3.
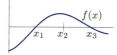

4. For $x = 0, 0.5, 1.0, 1.5,$ and 2.0, make a table of values for $I(x) = \int_0^x \sqrt{t^4 + 1}\,dt$.

5. Assume that $F'(t) = \sin t \cos t$ and $F(0) = 1$. Find $F(b)$ for $b = 0, 0.5, 1, 1.5, 2, 2.5,$ and 3.

6. (a) Continue the table of values for $\mathrm{Si}(x) = \int_0^x (\sin t/t)\,dt$ on page 280 for $x = 4$ and $x = 5$.
   (b) Why is $\mathrm{Si}(x)$ decreasing between $x = 4$ and $x = 5$?

In Exercises 7–9, write an expression for the function, $f(x)$, with the given properties.

7. $f'(x) = \sin(x^2)$ and $f(0) = 7$

8. $f'(x) = (\sin x)/x$ and $f(1) = 5$

9. $f'(x) = \mathrm{Si}(x)$ and $f(0) = 2$

Problems

10. Use Figure 6.26 to sketch a graph of $F(x) = \int_0^x f(t)\,dt$. Label the points x_1, x_2, x_3.

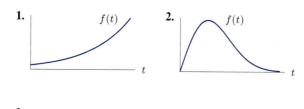

Figure 6.26

11. The graph of the derivative F' of some function F is given in Figure 6.27. If $F(20) = 150$, estimate the maximum value attained by F.

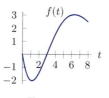

Figure 6.27

In Problems 12–13, find the value of the function with the given properties.

12. $F(1)$, where $F'(x) = e^{-x^2}$ and $F(0) = 2$

13. $G(-1)$, where $G'(x) = \cos(x^2)$ and $G(0) = -3$

Find the derivatives in Problems 14–19.

14. $\dfrac{d}{dx}\displaystyle\int_0^x \cos(t^2)\,dt$

15. $\dfrac{d}{dx}\displaystyle\int_1^x (1+t)^{200}\,dt$

16. $\dfrac{d}{dx}\displaystyle\int_{0.5}^x \arctan(t^2)\,dt$

17. $\dfrac{d}{dt}\displaystyle\int_t^\pi \cos(z^3)\,dz$

18. $\dfrac{d}{dx}\displaystyle\int_x^1 \ln t\,dt$

19. $\dfrac{d}{dx}\left[\mathrm{Si}(x^2)\right]$

20. Let $g(x) = \int_0^x f(t)\,dt$. Using Figure 6.28, find
   (a) $g(0)$ **(b)** $g'(1)$
   (c) The interval where g is concave up.
   (d) The value of x where g takes its maximum on the interval $0 \le x \le 8$.

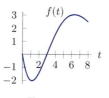

Figure 6.28

21. Let $F(x) = \int_0^x \sin(2t)\,dt$.
   (a) Evaluate $F(\pi)$.
   (b) Draw a sketch to explain geometrically why the answer to part (a) is correct.
   (c) For what values of x is $F(x)$ positive? negative?

22. Let $F(x) = \int_2^x (1/\ln t)\,dt$ for $x \ge 2$.
   (a) Find $F'(x)$.
   (b) Is F increasing or decreasing? What can you say about the concavity of its graph?
   (c) Sketch a graph of $F(x)$.

In Problems 23–26, find the given quantities. The *error function*, $\mathrm{erf}(x)$, is defined by

$$\mathrm{erf}(x) = \frac{2}{\sqrt{\pi}}\int_0^x e^{-t^2}\,dt.$$

23. $\dfrac{d}{dx}\left(x\,\mathrm{erf}(x)\right)$

24. $\dfrac{d}{dx}\left(\mathrm{erf}(\sqrt{x})\right)$

25. $\dfrac{d}{dx}\left(\displaystyle\int_0^{x^3} e^{-t^2}\,dt\right)$

26. $\dfrac{d}{dx}\left(\displaystyle\int_x^{x^3} e^{-t^2}\,dt\right)$

6.5 THE EQUATIONS OF MOTION

The problem of a body moving freely under the influence of gravity near the surface of the earth intrigued mathematicians and philosophers from Greek times onward and was finally solved by Galileo and Newton. The question to be answered was: How do the velocity and the position of the body vary with time? We define s to be the position, or height, of the body above a fixed point (often ground level); v then is the velocity of the body measured upward. We assume that the acceleration of the body is a constant, $-g$ (the negative sign means that the acceleration is downward), so

$$\text{Acceleration } = \frac{dv}{dt} = -g.$$

Thus, velocity is the antiderivative of $-g$:

$$v = -gt + C.$$

If the initial velocity is v_0, then $C = v_0$, so

$$v = -gt + v_0.$$

How about the position? We know that

$$\frac{ds}{dt} = v = -gt + v_0.$$

Therefore, we can find s by antidifferentiating again, giving:

$$s = -\frac{gt^2}{2} + v_0 t + C.$$

If the initial position is s_0, then we must have

$$s = -\frac{gt^2}{2} + v_0 t + s_0.$$

Our derivation of the formulas for the velocity and the position of the body took little effort. It hides an almost 2000-year struggle to understand the mechanics of falling bodies, from Aristotle's *Physics* to Galileo's *Dialogues Concerning Two New Sciences*.

Though it is an oversimplification of his ideas, we can say that Aristotle's conception of motion was primarily in terms of *change of position*. This seems entirely reasonable; it is what we commonly observe, and this view dominated discussions of motion for centuries. But it misses a subtlety that was brought to light by Descartes, Galileo, and, with a different emphasis, by Newton. That subtlety is now usually referred to as the *principle of inertia*.

This principle holds that a body traveling undisturbed at constant velocity in a straight line will continue in this motion indefinitely. Stated another way, it says that one cannot distinguish in any absolute sense (that is, by performing an experiment), between being at rest and moving with constant velocity in a straight line. If you are reading this book in a closed room and have no external reference points, there is no experiment that will tell you, one way or the other, whether you are at rest or whether you, the room, and everything in it are moving with constant velocity in a straight line. Therefore, as Newton saw, an understanding of motion should be based on *change of velocity* rather than change of position. Since acceleration is the rate of change of velocity, it is acceleration that must play a central role in the description of motion.

How does acceleration come about? How does the velocity change? Through the action of *forces*. Newton placed a new emphasis on the importance of forces. Newton's laws of motion do not say what a force *is*, they say how it *acts*. His first law is the principle of inertia, which says what happens in the *absence* of a force—there is no change in velocity. His second law says that a force acts to produce a change in velocity, that is, an acceleration. It states that $F = ma$, where m is the mass of the object, F is the net force, and a is the acceleration produced by this force.

Let's return to Galileo. He demonstrated that a body falling under the influence of gravity does so with constant acceleration. Furthermore, assuming we can neglect air resistance, this constant acceleration is independent of the mass of the body. This last fact was the outcome of Galileo's famous observation that a heavy ball and a light ball dropped off the Leaning Tower of Pisa hit the ground at the same time. Whether or not he actually performed this experiment, Galileo presented a very clear thought experiment in the *Dialogues* to prove the same point. (This point was counter to Aristotle's more common sense notion that the heavier ball would reach the ground first.) Galileo showed that the mass of the object did not appear as a variable in the equation of motion. Thus, the same constant acceleration equation applies to all bodies falling under the influence of gravity.

Nearly a hundred years after Galileo's experiment, Newton formulated his laws of motion and gravity, which led to a differential equation describing the motion of a falling body. According to Newton, acceleration is caused by force, and in the case of falling bodies, that force is the force of gravity. Newton's law of gravity says that the gravitational force between two bodies is attractive and given by

$$F = \frac{GMm}{r^2},$$

where G is the gravitational constant, m and M are the masses of the two bodies, and r is the distance between them. This is the famous *inverse square law*. For a falling body, we take M to be the mass of the earth and r to be the distance from the body to the center of the earth. So, actually, r changes as the body falls, but for anything we can easily observe (say, a ball dropped from the Tower of Pisa), it won't change significantly over the course of the motion. Hence, as an approximation, it is reasonable to assume that the force is constant. According to Newton's second law,

$$\text{Force} = \text{Mass} \times \text{Acceleration}.$$

Since the gravitational force is acting downwards

$$-\frac{GMm}{r^2} = m\frac{d^2 s}{dt^2}.$$

Hence,

$$\frac{d^2 s}{dt^2} = -\frac{GM}{r^2} = \text{Constant}.$$

If we define $g = GM/r^2$, then

$$\frac{d^2 s}{dt^2} = -g.$$

The fact that the mass cancels out of Newton's equations of motion reflects Galileo's experimental observation that the acceleration due to gravity is independent of the mass of the body.

Exercises and Problems for Section 6.5

Exercises

1. An object is thrown upward at time $t = 0$. After t seconds, its height is $y = -4.9t^2 + 7t + 1.5$ meters above the ground.

   (a) From what height was the object thrown?
   (b) What is the initial velocity of the object?
   (c) What is the acceleration due to gravity?

2. At time $t = 0$, a stone is thrown off a 250-meter cliff with velocity 20 meters/sec downward. Express its height, $h(t)$, in meters above the ground as a function of time, t, in seconds.

Problems

3. An object is dropped from a 400-foot tower. When does it hit the ground and how fast is it going at the time of the impact?

4. The object in Problem 3 falls off the same 400-foot tower. What would the acceleration due to gravity have to be to make it reach the ground in half the time?

5. A ball that is dropped from a window hits the ground in five seconds. How high is the window? (Give your answer in feet.)

6. On the moon the acceleration due to gravity is 5 ft/sec². An astronaut jumps into the air with an initial upward velocity of 10 ft/sec. How high does he go? How long is the astronaut off the ground?

7. Galileo was the first person to show that the distance traveled by a body falling from rest is proportional to the square of the time it has traveled, and independent of the mass of the body. Derive this result from the fact that the acceleration due to gravity is a constant.

8. While attempting to understand the motion of bodies under gravity, Galileo stated that:

> The time in which any space is traversed by a body starting at rest and uniformly accelerated is equal to the time in which that same space would be traversed by the same body moving at a uniform speed whose value is the mean of the highest velocity and the velocity just before acceleration began.

(a) Write Galileo's statement in symbols, defining all the symbols you use.

(b) Check Galileo's statement for a body dropped off a 100-foot building accelerating from rest under gravity until it hits the ground.

(c) Show why Galileo's statement is true in general.

9. In his *Dialogues Concerning Two New Sciences*, Galileo wrote:

> The distances traversed during equal intervals of time by a body falling from rest stand to one another in the same ratio as the odd numbers beginning with unity.

Assume, as is now believed, that $s = -(gt^2)/2$, where s is the total distance traveled in time t, and g is the acceleration due to gravity.

(a) How far does a falling body travel in the first second (between $t = 0$ and $t = 1$)? During the second second (between $t = 1$ and $t = 2$)? The third second? The fourth second?

(b) What do your answers tell you about the truth of Galileo's statement?

10. The acceleration due to gravity 2 meters from the ground is 9.8 m/sec². What is the acceleration due to gravity 100 meters from the ground? At 100,000 meters? (The radius of the earth is $6.4 \cdot 10^6$ meters.)

CHAPTER SUMMARY

- **Constructing antiderivatives**
  Graphically, numerically, analytically.
- **The family of antiderivatives**
  The indefinite integral.
- **Differential equations**

Initial value problems, uniform motion.

- **Construction theorem (Second Fundamental Theorem of Calculus)**
  Constructing antiderivatives using definite integrals.

REVIEW EXERCISES AND PROBLEMS FOR CHAPTER SIX

Exercises

Find the indefinite integrals in Exercises 1–21.

1. $\int (5x + 7)\, dx$

2. $\int \left(4t + \dfrac{1}{t}\right) dt$

3. $\int (2 + \cos t)\, dt$

4. $\int 7e^x\, dx$

5. $\int (3e^x + 2\sin x)\, dx$

6. $\int (x + 3)^2\, dx$

7. $\int \dfrac{8}{\sqrt{x}}\, dx$

8. $\int \left(\dfrac{3}{t} - \dfrac{2}{t^2}\right) dt$

9. $\int (e^x + 5)\, dx$

10. $\int \left(\sqrt{x^3} - \dfrac{2}{x}\right) dx$

11. $\int \dfrac{1}{\cos^2 x}\, dx$

12. $\int 2^x\, dx$

13. $\int (x + 1)^2\, dx$

14. $\int (x + 1)^3\, dx$

15. $\int (x + 1)^9\, dx$

16. $\int \left(\dfrac{x + 1}{x}\right) dx$

17. $\int \left(\dfrac{x^2 + x + 1}{x}\right) dx$

18. $\int \left(3\cos t + 3\sqrt{t}\right) dt$

19. $\int (3\cos x - 7\sin x)\, dx$

20. $\int \left(\dfrac{2}{x} + \pi \sin x\right) dx$

21. $\int (2e^x - 8\cos x)\, dx$

Find antiderivatives for the functions in Exercises 22–29. Check by differentiation.

22. $p(t) = \dfrac{1}{t}$

23. $f(x) = \cos x$

24. $f(x) = \dfrac{1}{x^2}$

25. $g(x) = \sin x$

26. $f(x) = 5e^x$

27. $h(t) = \dfrac{5}{t}$

28. $f(t) = t + \dfrac{1}{t}$

29. $f(x) = e^x - 1$

Problems

37. Use the Fundamental Theorem to find the area under $f(x) = x^2$ between $x = 0$ and $x = 3$.

38. Find the exact area of the region bounded by the x-axis and the graph of $y = x^3 - x$.

39. Calculate the exact area above the graph of $y = \frac{1}{2}\left(\frac{3}{\pi}x\right)^2$ and below the graph of $y = \cos x$. The curves intersect at $x = \pm\pi/3$.

40. Find the exact area of the shaded region in Figure 6.29 between $y = 3x^2 - 3$ and the x-axis.

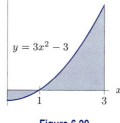

$y = 3x^2 - 3$

Figure 6.29

41. (a) Find the exact area between $f(x) = x^3 - 7x^2 + 10x$, the x-axis, $x = 0$, and $x = 5$.
   (b) Find $\int_0^5 (x^3 - 7x^2 + 10x)\, dx$ exactly and interpret this integral in terms of areas.

42. The area under $1/\sqrt{x}$ on the interval $1 \le x \le b$ is equal to 6. Find the value of b using the Fundamental Theorem.

43. Find the exact positive value of c which makes the area under the graph of $y = c(1 - x^2)$ and above the x-axis equal to 1.

44. Sketch the parabola $y = x(x - \pi)$ and the curve $y = \sin x$, showing their points of intersection. Find the exact area between the two graphs.

45. Find the exact average value of $f(x) = \sqrt{x}$ on the interval $0 \le x \le 9$. Illustrate your answer on a graph of $f(x) = \sqrt{x}$.

46. For t in years, $0 \le t \le 3$, the rate of discharge of a pollutant is estimated to be $t^2 - 14t + 49$ cubic meters per year. What is the total amount discharged during the three years?

For Exercises 30–35, find an antiderivatve $F(x)$ with $F'(x) = f(x)$ and $F(0) = 4$.

30. $f(x) = x^2$

31. $f(x) = x^3 + 6x^2 - 4$

32. $f(x) = \sqrt{x}$

33. $f(x) = e^x$

34. $f(x) = \sin x$

35. $f(x) = \cos x$

36. Use the Fundamental Theorem of Calculus to evaluate $\int_1^3 (6x^2 + 8x + 5)\,dx$.

47. For a function f, you are given the graph of the derivative f' in Figure 6.30 and that $f(0) = 50$.
   (a) On the interval $0 \le t \le 5$, at what value of t does f appear to reach its maximum value? Its minimum value?
   (b) Estimate these maximum and minimum values.
   (c) Estimate $f(5) - f(0)$.

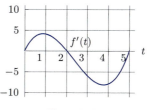

Figure 6.30

48. Assume f' is given by the graph in Figure 6.31. Suppose f is continuous and that $f(3) = 0$.
   (a) Sketch a graph of f.
   (b) Find $f(0)$ and $f(7)$.
   (c) Find $\int_0^7 f'(x)\, dx$ in two different ways.

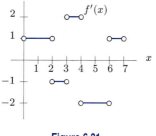

Figure 6.31

For Problems 49–50 the graph of $f'(x)$ is given. Sketch a possible graph for $f(x)$. Mark the points $x_1 \ldots x_4$ on your graph and label local maxima, local minima and points of inflection.

49.

50.

51. Let $F(x) = \int_{\pi/2}^{x} (\sin t / t)\, dt$. Find the value(s) of x between $\pi/2$ and $3\pi/2$ for which $F(x)$ has a global maximum and global minimum.

52. The graphs of three functions are given in Figure 6.32. Determine which is f, which is f', and which is $\int_0^x f(t)\, dt$. Explain your answer.

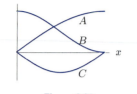

Figure 6.32

53. Write an expression for the function which is an antiderivative of e^{x^2} and which passes through the point
(a) $(0, 3)$ (b) $(-1, 5)$

54. An object thrown in the air on a planet in a distant galaxy is at height $s = -25t^2 + 72t + 40$ feet at time t seconds after it is thrown. What is the acceleration due to gravity on this planet? With what velocity was the object thrown? From what height?

55. An object is attached to a coiled spring which is suspended from the ceiling of a room. The function $h(t)$ gives the height of the object above the floor of the room at time, t. The graph of $h'(t)$ is given in Figure 6.33. Sketch a possible graph of $h(t)$.

Figure 6.33

56. The acceleration, a, of a particle as a function of time is shown in Figure 6.34. Sketch graphs of velocity and position against time. The particle starts at rest at the origin.

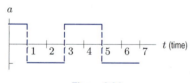

Figure 6.34

57. The angular speed of a car engine increases from 1100 revs/min to 2500 revs/min in 6 sec.

(a) Assuming that it is constant, find the angular acceleration in revs/min^2.
(b) How many revolutions does the engine make in this time?

58. A helicopter rotor slows down at a constant rate from 350 revs/min to 260 revs/min in 1.5 minutes.

(a) Find the angular acceleration during this time interval. What are the units of this acceleration?
(b) Assuming the angular acceleration remains constant, how long does it take for the rotor to stop? (Measure time from the moment when speed was 350 revs/min.)
(c) How many revolutions does the rotor make between the time the angular speed was 350 revs/min and stopping?

59. An object is thrown vertically upward with a velocity of 80 ft/sec.

(a) Make a table showing its velocity every second.
(b) When does it reach its highest point? When does it hit the ground?
(c) Using your table, write left and right sums which under- and overestimate the height the object attains.
(d) Use antidifferentiation to find the greatest height it reaches.

60. A car, initially moving at 60 mph, has a constant deceleration and stops in a distance of 200 feet. What is its deceleration? (Give your answer in ft/sec^2. Note that 1 mph $= 22/15$ ft/sec.)

61. The birth rate, B, in births per hour, of a bacteria population is given in Figure 6.35. The curve marked D gives the death rate, in deaths per hour, of the same population.

(a) Explain what the shape of each of these graphs tells you about the population.
(b) Use the graphs to find the time at which the net rate of increase of the population is at a maximum.
(c) At time $t = 0$ the population has size N. Sketch the graph of the total number born by time t. Also sketch the graph of the number alive at time t. Estimate the time at which the population is a maximum.

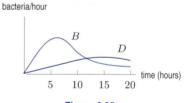

Figure 6.35

62. Water flows at a constant rate into the left side of the W-shaped container in Figure 6.36. Sketch a graph of the height, H, of the water in the left side of the container as a function of time, t. The container starts empty.

Figure 6.36

63. A particle of mass, m, acted on by a force, F, moves in a straight line. Its acceleration, a, is given by Newton's Law:

$$F = ma.$$

The work, W, done by a constant force when the particle moves through a displacement, d, is

$$W = Fd.$$

The velocity, v, of the particle as a function of time, t, is given in Figure 6.37. What is the sign of the work done during each of the time intervals: $[0, t_1], [t_1, t_2], [t_2, t_3], [t_3, t_4], [t_2, t_4]$?

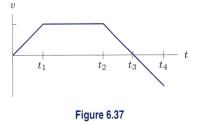

Figure 6.37

CAS Challenge Problems

64. (a) Set up a right-hand Riemann sum for $\int_a^b x^3 \, dx$ using n subdivisions. What is Δx? Express each x_i, for $i = 1, 2, \ldots, n$, in terms of i.

(b) Use a computer algebra system to find an expression for the Riemann sum in part (a); then find the limit of this expression as $n \to \infty$.

(c) Simplify the final expression and compare the result to that obtained using the Fundamental Theorem of Calculus.

65. (a) Use a computer algebra system to find $\int e^{2x} \, dx$, $\int e^{3x} \, dx$, and $\int e^{3x+5} \, dx$.

(b) Using your answers to part (a), conjecture a formula for $\int e^{ax+b} \, dx$, where a and b are constants.

(c) Check your formula by differentiation. Explain which differentiation rules you are using.

66. (a) Use a computer algebra system to find $\int \sin(3x) \, dx$, $\int \sin(4x) \, dx$, and $\int \sin(3x - 2) \, dx$.

(b) Using your answers to part (a), conjecture a formula for $\int \sin(ax + b) \, dx$, where a and b are constants.

(c) Check your formula by differentiation. Explain which differentiation rules you are using.

67. (a) Use a computer algebra system to find

$$\int \frac{x-2}{x-1} \, dx, \quad \int \frac{x-3}{x-1} \, dx, \quad \text{and} \quad \int \frac{x-1}{x-2} \, dx.$$

(b) If a and b are constants, use your answers to part (a) to conjecture a formula for

$$\int \frac{x-a}{x-b} \, dx.$$

(c) Check your formula by differentiation. Explain which rules of differentiation you are using.

68. (a) Use a computer algebra system to find

$$\int \frac{1}{(x-1)(x-3)} \, dx, \quad \int \frac{1}{(x-1)(x-4)} \, dx$$

and

$$\int \frac{1}{(x-1)(x+3)} \, dx.$$

(b) If a and b are constants, use your answers to part (a) to conjecture a formula for

$$\int \frac{1}{(x-a)(x-b)} \, dx.$$

(c) Check your formula by differentiation. Explain which rules of differentiation you are using.

CHECK YOUR UNDERSTANDING

Are the statements in Problems 1–16 true or false? Give an explanation for your answer.

1. A function $f(x)$ has at most one derivative.

2. An antiderivative of $3\sqrt{x+1}$ is $2(x+1)^{3/2}$.

3. An antiderivative of $3x^2$ is $x^3 + \pi$.

4. An antiderivative of $1/x$ is $\ln|x| + \ln 2$.

5. An antiderivative of e^{-x^2} is $-e^{-x^2}/2x$.

6. $\int f(x) \, dx = (1/x) \int x f(x) \, dx$.

7. If $F(x)$ is an antiderivative of $f(x)$ and $G(x) = F(x) + 2$, then $G(x)$ is an antiderivative of $f(x)$.

8. If $F(x)$ is an antiderivative of $f(x)$, then $y = F(x)$ is a solution to the differential equation $dy/dx = f(x)$.

9. If $y = F(x)$ is a solution to the differential equation $dy/dx = f(x)$, then $F(x)$ is an antiderivative of $f(x)$.

10. If an object has constant nonzero acceleration, then the position of the object as a function of time is a quadratic polynomial.

11. $\int_0^x \sin(t^2) \, dt$ is an antiderivative of $\sin(x^2)$.

12. If $F(x) = \int_0^x f(t) \, dt$, then $F(5) - F(3) = \int_3^5 f(t) \, dt$.

13. If $F(x) = \int_0^x f(t)dt$, then $F(x)$ must be increasing.

14. If $F(x) = \int_0^x f(t)dt$ and $G(x) = \int_2^x f(t)dt$, then $F(x) = G(x) + C$.

15. If $F(x) = \int_0^x f(t)dt$ and $G(x) = \int_2^x f(t)dt$, then $F(x) = G(x)$.

16. If $F(x) = \int_0^x f(t)dt$ and $G(x) = \int_0^x g(t)dt$, then $F(x) + G(x) = \int_0^x (f(t) + g(t))dt$.

PROJECTS FOR CHAPTER SIX

1. Distribution of Resources

Whether a resource is distributed evenly among members of a population is often an important political or economic question. How can we measure this? How can we decide if the distribution of wealth in this country is becoming more or less equitable over time? How can we measure which country has the most equitable income distribution? This problem describes a way of making such measurements. Suppose the resource is distributed evenly. Then any 20% of the population will have 20% of the resource. Similarly, any 30% will have 30% of the resource and so on. If, however, the resource is not distributed evenly, the poorest $p\%$ of the population (in terms of this resource) will not have $p\%$ of the goods. Suppose $F(x)$ represents the fraction of the resources owned by the poorest fraction x of the population. Thus $F(0.4) = 0.1$ means that the poorest 40% of the population owns 10% of the resource.

(a) What would F be if the resource were distributed evenly?

(b) What must be true of any such F? What must $F(0)$ and $F(1)$ equal? Is F increasing or decreasing? Is the graph of F concave up or concave down?

(c) Gini's index of inequality, G, is one way to measure how evenly the resource is distributed. It is defined by

$$G = 2 \int_0^1 [x - F(x)]\, dx.$$

Show graphically what G represents.

2. Yield from an Apple Orchard

Figure 6.38 is a graph of the annual yield, $y(t)$, in bushels per year, from an orchard t years after planting. The trees take about 10 years to get established, but for the next 20 years they give a substantial yield. After about 30 years, however, age and disease start to take their toll, and the annual yield falls off.[1]

(a) Represent on a sketch of Figure 6.38 the total yield, $F(M)$, up to M years, with $0 \leq M \leq 60$. Write an expression for $F(M)$ in terms of $y(t)$.

(b) Sketch a graph of $F(M)$ against M for $0 \leq M \leq 60$.

(c) Write an expression for the average annual yield, $a(M)$, up to M years.

(d) When should the orchard be cut down and replanted? Assume that we want to maximize average revenue per year, and that fruit prices remain constant, so that this is achieved by maximizing average annual yield. Use the graph of $y(t)$ to estimate the time at which the average annual yield is a maximum. Explain your answer geometrically and symbolically.

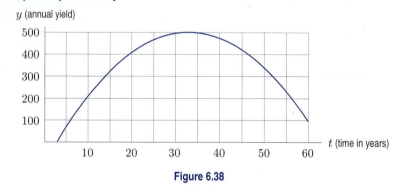

Figure 6.38

[1] From Peter D. Taylor, *Calculus: The Analysis of Functions*, (Toronto: Wall & Emerson, Inc., 1992).

Chapter Seven

INTEGRATION

In Chapter 5, we introduced the definite integral, and in Chapter 6, we introduced antiderivatives. In this chapter, we focus on methods of finding antiderivatives and definite integrals. We first see additional ways to evaluate antiderivatives and definite integrals exactly, and we then develop more efficient numerical methods for approximating definite integrals. Finally, we introduce improper integrals and discuss methods of evaluating them.

7.1 INTEGRATION BY SUBSTITUTION

In Chapter 3, we learned rules to differentiate any function obtained by combining constants, powers of x, $\sin x$, $\cos x$, e^x, and $\ln x$, using addition, multiplication, division, or composition of functions. Such functions are called *elementary*.

In the next few sections, we introduce two methods of antidifferentiation: substitution and integration by parts, which reverse the chain and product rules, respectively. However, there is a great difference between looking for derivatives and looking for antiderivatives. Every elementary function has elementary derivatives, but many elementary functions do not have elementary antiderivatives. Some examples are $\sqrt{x^3 + 1}$, $(\sin x)/x$, and e^{-x^2}. These are not exotic functions, but ordinary functions that arise naturally.

The Guess-and-Check Method

A good strategy for finding simple antiderivatives is to *guess* an answer (using knowledge of differentiation rules) and then *check* the answer by differentiating it. If we get the expected result, then we're done; otherwise, we revise the guess and check again.

The method of guess-and-check is useful in reversing the chain rule. According to the chain rule,

$$\frac{d}{dx}(f(g(x))) = \underbrace{f'}_{\text{Derivative of outside}} \overbrace{(g(x))}^{\text{Inside}} \cdot \underbrace{g'(x)}_{\text{Derivative of inside}}.$$

Thus, any function which is the result of applying the chain rule is the product of two factors: the "derivative of the outside" and the "derivative of the inside." If a function has this form, its antiderivative is $f(g(x))$.

Example 1   Find $\displaystyle\int 3x^2 \cos(x^3)\,dx$.

Solution   The function $3x^2 \cos(x^3)$ looks like the result of applying the chain rule: there is an "inside" function x^3 and its derivative $3x^2$ appears as a factor. Since the outside function is a cosine which has a sine as an antiderivative, we guess $\sin(x^3)$ for the antiderivative. Differentiating to check gives

$$\frac{d}{dx}(\sin(x^3)) = \cos(x^3) \cdot (3x^2).$$

Since this is what we began with, we know that

$$\int 3x^2 \cos(x^3)\,dx = \sin(x^3) + C.$$

The basic idea of this method is to try to find an inside function whose derivative appears as a factor. This works even when the derivative is missing a constant factor, as in the next example.

Example 2   Find $\displaystyle\int t e^{(t^2+1)}\,dt$.

Solution   It looks like $t^2 + 1$ is an inside function. So we guess $e^{(t^2+1)}$ for the antiderivative, since taking the derivative of an exponential results in the reappearance of the exponential together with other terms from the chain rule. Now we check:

$$\frac{d}{dt}\left(e^{(t^2+1)}\right) = \left(e^{(t^2+1)}\right) \cdot 2t.$$

The original guess was too large by a factor of 2. We change the guess to $\frac{1}{2}e^{(t^2+1)}$ and check again:

$$\frac{d}{dt}\left(\frac{1}{2}e^{(t^2+1)}\right) = \frac{1}{2}e^{(t^2+1)}\cdot 2t = e^{(t^2+1)}\cdot t.$$

Thus, we know that

$$\int te^{(t^2+1)}\,dt = \frac{1}{2}e^{(t^2+1)} + C.$$

Example 3 Find $\int x^3\sqrt{x^4+5}\,dx.$

Solution Here the inside function is x^4+5, and its derivative appears as a factor, with the exception of a missing 4. Thus, the integrand we have is more or less of the form

$$g'(x)\sqrt{g(x)},$$

with $g(x) = x^4+5$. Since $x^{3/2}/(3/2)$ is an antiderivative of the outside function $\sqrt{x}$, we might guess that an antiderivative is

$$\frac{(g(x))^{3/2}}{3/2} = \frac{(x^4+5)^{3/2}}{3/2}.$$

Let's check and see:

$$\frac{d}{dx}\left(\frac{(x^4+5)^{3/2}}{3/2}\right) = \frac{3}{2}\frac{(x^4+5)^{1/2}}{3/2}\cdot 4x^3 = 4x^3(x^4+5)^{1/2},$$

so $\dfrac{(x^4+5)^{3/2}}{3/2}$ is too big by a factor of 4. The correct antiderivative is

$$\frac{1}{4}\frac{(x^4+5)^{3/2}}{3/2} = \frac{1}{6}(x^4+5)^{3/2}.$$

Thus

$$\int x^3\sqrt{x^4+5}\,dx = \frac{1}{6}(x^4+5)^{3/2} + C.$$

As a final check:

$$\frac{d}{dx}\left(\frac{1}{6}(x^4+5)^{3/2}\right) = \frac{1}{6}\cdot\frac{3}{2}(x^4+5)^{1/2}\cdot 4x^3 = x^3(x^4+5)^{1/2}.$$

As we have seen in the preceding examples, antidifferentiating a function often involves "correcting for" constant factors: if differentiation produces an extra factor of 4, antidifferentiation will require a factor of $\frac{1}{4}$.

The Method of Substitution

When the integrand is complicated, it helps to formalize this guess-and-check method as follows:

To Make a Substitution

Let w be the "inside function" and $dw = w'(x)\,dx = \dfrac{dw}{dx}dx.$

Let's redo the first example using a substitution.

Example 4 Find $\int 3x^2\cos(x^3)\,dx.$

Solution   As before, we look for an inside function whose derivative appears—in this case x^3. We let $w = x^3$. Then $dw = w'(x)\, dx = 3x^2\, dx$. The original integrand can now be completely rewritten in terms of the new variable w:

$$\int 3x^2 \cos(x^3)\, dx = \int \cos \underbrace{(x^3)}_{w} \cdot \underbrace{3x^2\, dx}_{dw} = \int \cos w\, dw = \sin w + C = \sin(x^3) + C.$$

By changing the variable to w, we can simplify the integrand. We now have $\cos w$, which can be antidifferentiated more easily. The final step, after antidifferentiating, is to convert back to the original variable, x.

Why Does Substitution Work?

The substitution method makes it look as if we can treat dw and dx as separate entities, even canceling them in the equation $dw = (dw/dx)dx$. Let's see why this works. Suppose we have an integral of the form $\int f(g(x))g'(x)\, dx$, where $g(x)$ is the inside function and $f(x)$ is the outside function. If F is an antiderivative of f, then $F' = f$, and by the chain rule $\frac{d}{dx}(F(g(x))) = f(g(x))g'(x)$. Therefore,

$$\int f(g(x))g'(x)\, dx = F(g(x)) + C.$$

Now write $w = g(x)$ and $dw/dx = g'(x)$ on both sides of this equation:

$$\int f(w)\frac{dw}{dx}\, dx = F(w) + C.$$

On the other hand, knowing that $F' = f$ tells us that

$$\int f(w)\, dw = F(w) + C.$$

Thus, the following two integrals are equal:

$$\int f(w)\frac{dw}{dx}\, dx = \int f(w)\, dw.$$

Substituting w for the inside function and writing $dw = w'(x)dx$ leaves the indefinite integral unchanged.

    Let's revisit the second example that we did by guess-and-check.

Example 5   Find $\displaystyle\int te^{(t^2+1)}\, dt$.

Solution   Here the inside function is $t^2 + 1$, with derivative $2t$. Since there is a factor of t in the integrand, we try

$$w = t^2 + 1.$$

Then

$$dw = w'(t)\, dt = 2t\, dt.$$

Notice, however, that the original integrand has only $t\, dt$, not $2t\, dt$. We therefore write

$$\frac{1}{2}\, dw = t\, dt$$

and then substitute:

$$\int te^{(t^2+1)}\, dt = \int e^{\overbrace{(t^2 + 1)}^{w}} \cdot \underbrace{t\, dt}_{\frac{1}{2}dw} = \int e^w \frac{1}{2}\, dw = \frac{1}{2}\int e^w\, dw = \frac{1}{2}e^w + C = \frac{1}{2}e^{(t^2+1)} + C.$$

This gives the same answer as we found using guess-and-check.

Why didn't we put $\frac{1}{2} \int e^w \, dw = \frac{1}{2} e^w + \frac{1}{2} C$ in the preceding example? Since the constant C is arbitrary, it doesn't really matter whether we add C or $\frac{1}{2}C$. The convention is always to add C to whatever antiderivative we have calculated.

Now let's redo the third example that we solved previously by guess-and-check.

Example 6   Find $\int x^3 \sqrt{x^4 + 5} \, dx$.

Solution   The inside function is $x^4 + 5$, with derivative $4x^3$. The integrand has a factor of x^3, and since the only thing missing is a constant factor, we try

$$w = x^4 + 5.$$

Then

$$dw = w'(x) \, dx = 4x^3 \, dx,$$

giving

$$\frac{1}{4} dw = x^3 \, dx.$$

Thus,

$$\int x^3 \sqrt{x^4 + 5} \, dx = \int \sqrt{w} \, \frac{1}{4} dw = \frac{1}{4} \int w^{1/2} \, dw = \frac{1}{4} \cdot \frac{w^{3/2}}{3/2} + C = \frac{1}{6}(x^4 + 5)^{3/2} + C.$$

Once again, we get the same result as with guess-and-check.

Warning

We saw in the preceding examples that we can apply the substitution method when a *constant* factor is missing from the derivative of the inside function. However, we may not be able to use substitution if anything other than a constant factor is missing. For example, setting $w = x^4 + 5$ to find

$$\int x^2 \sqrt{x^4 + 5} \, dx$$

does us no good because $x^2 \, dx$ is not a constant multiple of $dw = 4x^3 \, dx$. In order to use substitution, the integrand must contain the derivative of the inside function, *to within a constant factor*.

Some people prefer the substitution method over guess-and-check since it is more systematic, but both methods achieve the same result. For simple problems, guess-and-check can be faster.

Example 7   Find $\int e^{\cos \theta} \sin \theta \, d\theta$.

Solution   We let $w = \cos \theta$ since its derivative is $-\sin \theta$ and there is a factor of $\sin \theta$ in the integrand. This gives

$$dw = w'(\theta) \, d\theta = -\sin \theta \, d\theta,$$

so

$$-dw = \sin \theta \, d\theta.$$

Thus

$$\int e^{\cos \theta} \sin \theta \, d\theta = \int e^w \, (-dw) = (-1) \int e^w \, dw = -e^w + C = -e^{\cos \theta} + C.$$

Example 8   Find $\displaystyle\int \frac{e^t}{1+e^t}\, dt$.

Solution   Observing that the derivative of $1+e^t$ is e^t, we see $w=1+e^t$ is a good choice. Then $dw=e^t\, dt$, so that

$$\int \frac{e^t}{1+e^t}\, dt = \int \frac{1}{1+e^t}\, e^t\, dt = \int \frac{1}{w}\, dw = \ln|w| + C$$
$$= \ln|1+e^t| + C$$
$$= \ln(1+e^t) + C \qquad \text{(Since } (1+e^t) \text{ is always positive.)}$$

Since the numerator is $e^t\, dt$, we might also have tried $w=e^t$. This substitution leads to the integral $\int (1/(1+w))\, dw$, which is better than the original integral but requires another substitution, $u = 1+w$, to finish. There are often several different ways of doing an integral by substitution.

Notice the pattern in the previous example: having a function in the denominator and its derivative in the numerator leads to a natural logarithm. The next example follows the same pattern.

Example 9   Find $\displaystyle\int \tan\theta\, d\theta$.

Solution   Recall that $\tan\theta = (\sin\theta)/(\cos\theta)$. If $w=\cos\theta$, then $dw = -\sin\theta\, d\theta$, so

$$\int \tan\theta\, d\theta = \int \frac{\sin\theta}{\cos\theta}\, d\theta = \int \frac{-dw}{w} = -\ln|w| + C = -\ln|\cos\theta| + C.$$

Definite Integrals by Substitution

Example 10   Compute $\displaystyle\int_0^2 x e^{x^2}\, dx$.

Solution   To evaluate this definite integral using the Fundamental Theorem of Calculus, we first need to find an antiderivative of $f(x)=xe^{x^2}$. The inside function is x^2, so we let $w=x^2$. Then $dw=2x\, dx$, so $\frac{1}{2}\, dw = x\, dx$. Thus,

$$\int xe^{x^2}\, dx = \int e^w \frac{1}{2}\, dw = \frac{1}{2}e^w + C = \frac{1}{2}e^{x^2} + C.$$

Now we find the definite integral

$$\int_0^2 xe^{x^2}\, dx = \frac{1}{2}e^{x^2}\Big|_0^2 = \frac{1}{2}(e^4 - e^0) = \frac{1}{2}(e^4 - 1).$$

There is another way to look at the same problem. After we established that

$$\int xe^{x^2}\, dx = \frac{1}{2}e^w + C,$$

our next two steps were to replace w by x^2, and then x by 2 and 0. We could have directly replaced the original limits of integration, $x=0$ and $x=2$, by the corresponding w limits. Since $w=x^2$, the w limits are $w=0^2=0$ (when $x=0$) and $w=2^2=4$ (when $x=2$), so we get

$$\int_{x=0}^{x=2} xe^{x^2}\, dx = \frac{1}{2}\int_{w=0}^{w=4} e^w\, dw = \frac{1}{2}e^w\Big|_0^4 = \frac{1}{2}\left(e^4 - e^0\right) = \frac{1}{2}(e^4 - 1).$$

As we would expect, both methods give the same answer.

> ## To Use Substitution to Find Definite Integrals
>
> Either
> - Compute the indefinite integral, expressing an antiderivative in terms of the original variable, and then evaluate the result at the original limits,
>
> or
> - Convert the original limits to new limits in terms of the new variable and do not convert the antiderivative back to the original variable.

Example 11   Evaluate $\displaystyle\int_0^{\pi/4} \frac{\tan^3\theta}{\cos^2\theta}\, d\theta$.

Solution   To use substitution, we must decide what w should be. There are two possible inside functions, $\tan\theta$ and $\cos\theta$. Now

$$\frac{d}{d\theta}(\tan\theta) = \frac{1}{\cos^2\theta} \quad \text{and} \quad \frac{d}{d\theta}(\cos\theta) = -\sin\theta,$$

and since the integral contains a factor of $1/\cos^2\theta$ but not of $\sin\theta$, we try $w = \tan\theta$. Then $dw = (1/\cos^2\theta)d\theta$. When $\theta = 0$, $w = \tan 0 = 0$, and when $\theta = \pi/4$, $w = \tan(\pi/4) = 1$, so

$$\int_0^{\pi/4} \frac{\tan^3\theta}{\cos^2\theta}\, d\theta = \int_0^{\pi/4} (\tan\theta)^3 \cdot \frac{1}{\cos^2\theta}\, d\theta = \int_0^1 w^3\, dw = \frac{1}{4}w^4\Big|_0^1 = \frac{1}{4}.$$

Example 12   Evaluate $\displaystyle\int_1^3 \frac{dx}{5-x}$.

Solution   Let $w = 5 - x$, so $dw = -dx$. When $x = 1$, $w = 4$, and when $x = 3$, $w = 2$, so

$$\int_1^3 \frac{dx}{5-x} = \int_4^2 \frac{-dw}{w} = -\ln|w|\Big|_4^2 = -(\ln 2 - \ln 4) = \ln\left(\frac{4}{2}\right) = \ln 2 \approx 0.69.$$

Notice that we write the limit $w = 4$ at the bottom, even though it is larger than $w = 2$, because $w = 4$ corresponds to the lower limit $x = 1$.

More Complex Substitutions

In the examples of substitution presented so far, we guessed an expression for w and hoped to find dw (or some constant multiple of it) in the integrand. What if we are not so lucky? It turns out that it often works to let w be some messy expression contained inside, say, a cosine or under a root, even if we cannot see immediately how such a substitution helps.

Example 13   Find $\displaystyle\int \sqrt{1 + \sqrt{x}}\, dx$.

Solution   This time, the derivative of the inside function is nowhere to be seen. Nevertheless, we try $w = 1 + \sqrt{x}$. Then $w - 1 = \sqrt{x}$, so $(w-1)^2 = x$. Therefore $2(w-1)\, dw = dx$. We have

$$\int \sqrt{1 + \sqrt{x}}\, dx = \int \sqrt{w}\, 2(w-1)\, dw = 2\int w^{1/2}(w-1)\, dw$$

$$= 2\int (w^{3/2} - w^{1/2})\, dw = 2\left(\frac{2}{5}w^{5/2} - \frac{2}{3}w^{3/2}\right) + C$$

$$= 2\left(\frac{2}{5}(1+\sqrt{x})^{5/2} - \frac{2}{3}(1+\sqrt{x})^{3/2}\right) + C.$$

Notice that the substitution in the preceding example again converts the inside of the messiest function into something simple. In addition, since the derivative of the inside function is not waiting for us, we have to solve for x so that we can get dx entirely in terms of w and dw.

Example 14   Find $\int (x+7)\sqrt[3]{3-2x}\,dx$.

Solution   Here, instead of the derivative of the inside function (which is -2), we have the factor $(x+7)$. However, substituting $w = 3-2x$ turns out to help anyway. Then $dw = -2\,dx$, so $(-1/2)\,dw = dx$. Now we must convert everything to w, including $x + 7$. If $w = 3 - 2x$, then $2x = 3 - w$, so $x = 3/2 - w/2$, and therefore we can write $x + 7$ in terms of w. Thus

$$\int (x+7)\sqrt[3]{3-2x}\,dx = \int \left(\frac{3}{2} - \frac{w}{2} + 7\right)\sqrt[3]{w}\left(-\frac{1}{2}\right)dw$$

$$= -\frac{1}{2}\int \left(\frac{17}{2} - \frac{w}{2}\right)w^{1/3}\,dw$$

$$= -\frac{1}{4}\int (17 - w)w^{1/3}\,dw$$

$$= -\frac{1}{4}\int (17w^{1/3} - w^{4/3})\,dw$$

$$= -\frac{1}{4}\left(17\frac{w^{4/3}}{4/3} - \frac{w^{7/3}}{7/3}\right) + C$$

$$= -\frac{1}{4}\left(\frac{51}{4}(3-2x)^{4/3} - \frac{3}{7}(3-2x)^{7/3}\right) + C.$$

Looking back over the solution, the reason this substitution works is that it converts $\sqrt[3]{3-2x}$, the messiest part of the integrand, to $\sqrt[3]{w}$, which can be combined with the other term and then integrated.

Exercises and Problems for Section 7.1

Exercises

1. (a) Find the derivatives of $\sin(x^2 + 1)$ and $\sin(x^3 + 1)$.
   (b) Use your answer to part (a) to find antiderivatives of:

   (i) $x\cos(x^2 + 1)$ (ii) $x^2\cos(x^3 + 1)$

   (c) Find the general antiderivatives of:

   (i) $x\sin(x^2 + 1)$ (ii) $x^2\sin(x^3 + 1)$

2. Use substitution to express each of the following integrals as a multiple of $\int_a^b (1/w)\,dw$ for some a and b. Then evaluate the integrals.

   (a) $\displaystyle\int_0^1 \frac{x}{1+x^2}\,dx$ (b) $\displaystyle\int_0^{\pi/4} \frac{\sin x}{\cos x}\,dx$

Find the integrals in Exercises 3–40. Check your answers by differentiation.

3. $\displaystyle\int e^{3x}\,dx$

4. $\displaystyle\int 25e^{-0.2t}\,dt$

5. $\displaystyle\int \sin(2x)\,dx$

6. $\displaystyle\int t\cos(t^2)\,dt$

7. $\displaystyle\int y(y^2 + 5)^8\,dy$

8. $\displaystyle\int t^2(t^3 - 3)^{10}\,dt$

9. $\displaystyle\int y^2(1 + y)^2\,dy$

10. $\displaystyle\int x^2(1 + 2x^3)^2\,dx$

11. $\displaystyle\int x(x^2 - 4)^{7/2}\,dx$

12. $\displaystyle\int x(x^2 + 3)^2\,dx$

13. $\displaystyle\int \frac{1}{\sqrt{4-x}}\,dx$

14. $\displaystyle\int \frac{dy}{y + 5}$

15. $\displaystyle\int (2t - 7)^{73}\,dt$

16. $\displaystyle\int (x^2 + 3)^2\,dx$

17. $\displaystyle\int \sin\theta(\cos\theta + 5)^7\,d\theta$

18. $\displaystyle\int \sqrt{\cos 3t}\sin 3t\,dt$

19. $\displaystyle\int xe^{-x^2}\,dx$

20. $\displaystyle\int \sin^6\theta \cos\theta\,d\theta$

21. $\displaystyle\int x^2 e^{x^3+1}\,dx$

22. $\displaystyle\int \sin^6(5\theta)\cos(5\theta)\,d\theta$

23. $\displaystyle\int \sin^3\alpha\cos\alpha\,d\alpha$

24. $\displaystyle\int \tan(2x)\,dx$

25. $\displaystyle\int \frac{(\ln z)^2}{z}\,dz$

26. $\displaystyle\int \frac{e^t+1}{e^t+t}\,dt$

27. $\displaystyle\int \frac{y}{y^2+4}\,dy$

28. $\displaystyle\int \frac{\cos\sqrt{x}}{\sqrt{x}}\,dx$

29. $\displaystyle\int \frac{e^{\sqrt{y}}}{\sqrt{y}}\,dy$

30. $\displaystyle\int \frac{1+e^x}{\sqrt{x+e^x}}\,dx$

31. $\displaystyle\int \frac{e^x}{2+e^x}\,dx$

32. $\displaystyle\int \frac{x+1}{x^2+2x+19}\,dx$

33. $\displaystyle\int \frac{t}{1+3t^2}\,dt$

34. $\displaystyle\int \frac{e^x-e^{-x}}{e^x+e^{-x}}\,dx$

35. $\displaystyle\int \frac{(t+1)^2}{t^2}\,dt$

36. $\displaystyle\int \frac{x\cos(x^2)}{\sqrt{\sin(x^2)}}\,dx$

37. $\displaystyle\int \cosh x\,dx$

38. $\displaystyle\int \sinh 3t\,dt$

39. $\displaystyle\int (\sinh z)e^{\cosh z}\,dz$

40. $\displaystyle\int \cosh(2w+1)\,dw$

For the functions in Exercises 41–48, find the general antiderivative. Check your answers by differentiation.

41. $p(t)=\pi t^3+4t$

42. $f(x)=\sin 3x$

43. $f(x)=2x\cos(x^2)$

44. $r(t)=12t^2\cos(t^3)$

45. $f(x)=\sin(2-5x)$

46. $f(x)=e^{\sin x}\cos x$

47. $f(x)=\dfrac{x}{x^2+1}$

48. $f(x)=\dfrac{1}{3\cos^2(2x)}$

For Exercises 49–56, use the Fundamental Theorem to calculate the definite integrals.

49. $\displaystyle\int_0^\pi \cos(x+\pi)\,dx$

50. $\displaystyle\int_0^{1/2} \cos(\pi x)\,dx$

51. $\displaystyle\int_0^{\pi/2} e^{-\cos\theta}\sin\theta\,d\theta$

52. $\displaystyle\int_1^2 2xe^{x^2}\,dx$

53. $\displaystyle\int_1^8 \frac{e^{\sqrt[3]{x}}}{\sqrt[3]{x^2}}\,dx$

54. $\displaystyle\int_{-1}^{e-2} \frac{1}{t+2}\,dt$

55. $\displaystyle\int_1^4 \frac{\cos\sqrt{x}}{\sqrt{x}}\,dx$

56. $\displaystyle\int_0^2 \frac{x}{(1+x^2)^2}\,dx$

For Exercises 57–62, evaluate the definite integrals. Whenever possible, use the Fundamental Theorem of Calculus, perhaps after a substitution. Otherwise, use numerical methods.

57. $\displaystyle\int_{-1}^3 (x^3+5x)\,dx$

58. $\displaystyle\int_{-1}^1 \frac{1}{1+y^2}\,dy$

59. $\displaystyle\int_1^3 \frac{1}{x}\,dx$

60. $\displaystyle\int_1^3 \frac{dt}{(t+7)^2}$

61. $\displaystyle\int_{-1}^2 \sqrt{x+2}\,dx$

62. $\displaystyle\int_1^2 \frac{\sin t}{t}\,dt$

Problems

63. If appropriate, evaluate the following integrals by substitution. If substitution is not appropriate, say so, and do not evaluate.

(a) $\displaystyle\int x\sin(x^2)\,dx$

(b) $\displaystyle\int x^2\sin x\,dx$

(c) $\displaystyle\int \frac{x^2}{1+x^2}\,dx$

(d) $\displaystyle\int \frac{x}{(1+x^2)^2}\,dx$

(e) $\displaystyle\int x^3 e^{x^2}\,dx$

(f) $\displaystyle\int \frac{\sin x}{2+\cos x}\,dx$

64. (a) Calculate exactly: $\displaystyle\int_{-\pi}^\pi \cos^2\theta\sin\theta\,d\theta$.
   (b) Calculate the exact area under the curve
   $y=\cos^2\theta\sin\theta$ between $\theta=0$ and $\theta=\pi$.

65. Use the Fundamental Theorem to find the area under the graph of $f(x)=1/(x+1)$ between $x=0$ and $x=2$.

66. Find the exact area under the graph of $f(x)=xe^{x^2}$ between $x=0$ and $x=2$.

67. Find the exact area enclosed by the hyperbola $xy=4$, the x-axis and the lines $x=2$ and $x=4$.

68. Find the exact value of the area under one arch of the curve $V(t)=V_0\sin(\omega t)$, where $V_0>0$ and $\omega>0$.

69. Find the average value of $f(x)=1/(x+1)$ on the interval $x=0$ to $x=2$. Sketch a graph showing the function and the average value.

70. Find $\int 4x(x^2+1)\,dx$ using two methods:

(a) Do the multiplication first, and then antidifferentiate.
(b) Use the substitution $w=x^2+1$.

(c) Explain how the expressions from parts (a) and (b) are different. Are they both correct?

71. (a) Find $\int \sin \theta \cos \theta \, d\theta$.

(b) You probably solved part (a) by making the substitution $w = \sin \theta$ or $w = \cos \theta$. (If not, go back and do it that way.) Now find $\int \sin \theta \cos \theta \, d\theta$ by making the *other* substitution.

(c) There is yet another way of finding this integral which involves the trigonometric identities

$$\sin(2\theta) = 2 \sin \theta \cos \theta$$
$$\cos(2\theta) = \cos^2 \theta - \sin^2 \theta.$$

Find $\int \sin \theta \cos \theta \, d\theta$ using one of these identities and then the substitution $w = 2\theta$.

(d) You should now have three different expressions for the indefinite integral $\int \sin \theta \cos \theta \, d\theta$. Are they really different? Are they all correct? Explain.

72. Suppose $\int_0^1 f(t) \, dt = 3$. Calculate the following:

(a) $\int_0^{0.5} f(2t) \, dt$ **(b)** $\int_0^1 f(1 - t) \, dt$

(c) $\int_1^{1.5} f(3 - 2t) \, dt$

73. If t is in years since 1990, the population, P, of the world in billions can be modeled by $P = 5.3e^{0.014t}$.

(a) What does this model give for the world population in 1990? In 2000?

(b) Use the Fundamental Theorem to find the average population of the world during the 1990s.

74. Oil is leaking out of a ruptured tanker at the rate of $r(t) = 50e^{-0.02t}$ thousand liters per minute.

(a) At what rate, in liters per minute, is oil leaking out at $t = 0$? At $t = 60$?

(b) How many liters leak out during the first hour?

75. Throughout much of the 20th century, the yearly consumption of electricity in the US increased exponentially at a continuous rate of 7% per year. Assume this trend continues and that the electrical energy consumed in 1900 was 1.4 million megawatt-hours.

(a) Write an expression for yearly electricity consumption as a function of time, t, in years since 1900.

(b) Find the average yearly electrical consumption throughout the 20th century.

(c) During what year was electrical consumption closest to the average for the century?

(d) Without doing the calculation for part (c), how could you have predicted which half of the century the answer would be in?

76. If we assume that wind resistance is proportional to velocity, then the downward velocity, v, of a body of mass m falling vertically is given by

$$v = \frac{mg}{k} \left(1 - e^{-kt/m} \right),$$

where g is the acceleration due to gravity and k is a constant. Find the height, h, above the surface of the earth as a function of time. Assume the body starts at height h_0.

77. If we assume that wind resistance is proportional to the square of velocity, then the downward velocity, v, of a falling body is given by

$$v = \sqrt{\frac{g}{k}} \left(\frac{e^{t\sqrt{gk}} - e^{-t\sqrt{gk}}}{e^{t\sqrt{gk}} + e^{-t\sqrt{gk}}} \right).$$

Use the substitution $w = e^{t\sqrt{gk}} + e^{-t\sqrt{gk}}$ to find the height, h, of the body above the surface of the earth as a function of time. Assume the body starts at a height h_0.

78. (a) During the 1970s, ACME Widgets sold at a continuous rate of $R = R_0 e^{0.15t}$ widgets per year, where t is time in years since January 1, 1970. Suppose they were selling widgets at a rate of 1000 per year on the first day of the decade. How many widgets did they sell during the decade? How many did they sell if the rate on January 1, 1970 was 150,000,000 widgets per year?

(b) In the first case above (1000 widgets per year on January 1, 1970), how long did it take for half the widgets in the 1970s to be sold? In the second case (150,000,000 widgets per year on January 1, 1970), when had half the widgets in the 1970s been sold?

(c) In 1980 ACME began an advertising campaign claiming that half the widgets it had sold in the previous ten years were still in use. Based on your answer to part (b), about how long must a widget last in order to justify this claim?

7.2 INTEGRATION BY PARTS

The method of substitution reverses the chain rule. Now we introduce *integration by parts*, which is based on the product rule.

Example 1   Find $\int xe^x \, dx$.

Solution   We are looking for a function whose derivative is xe^x. The product rule might lead us to guess xe^x, because we know that the derivative has two terms, one of which is xe^x:

$$\frac{d}{dx}(xe^x) = \frac{d}{dx}(x)e^x + x\frac{d}{dx}(e^x) = e^x + xe^x.$$

Of course, our guess is wrong because of the extra e^x. But we can adjust our guess by subtracting e^x; this leads us to try $xe^x - e^x$. Let's check it:

$$\frac{d}{dx}(xe^x - e^x) = \frac{d}{dx}(xe^x) - \frac{d}{dx}(e^x) = e^x + xe^x - e^x = xe^x.$$

It works, so $\int xe^x \, dx = xe^x - e^x + C$.

Example 2   Find $\int \theta \cos \theta \, d\theta$.

Solution   We guess the antiderivative is $\theta \sin \theta$ and use the product rule to check:

$$\frac{d}{d\theta}(\theta \sin \theta) = \frac{d(\theta)}{d\theta} \sin \theta + \theta \frac{d}{d\theta}(\sin \theta) = \sin \theta + \theta \cos \theta.$$

To correct for the extra $\sin \theta$ term, we must subtract from our original guess something whose derivative is $\sin \theta$. Since $\frac{d}{d\theta}(\cos \theta) = -\sin \theta$, we try:

$$\frac{d}{d\theta}(\theta \sin \theta + \cos \theta) = \frac{d}{d\theta}(\theta \sin \theta) + \frac{d}{d\theta}(\cos \theta) = \sin \theta + \theta \cos \theta - \sin \theta = \theta \cos \theta.$$

Thus, $\int \theta \cos \theta \, d\theta = \theta \sin \theta + \cos \theta + C$.

The General Formula for Integration by Parts

We can formalize the process illustrated in the last two examples in the following way. We begin with the product rule:

$$\frac{d}{dx}(uv) = u'v + uv'$$

where u and v are functions of x with derivatives u' and v', respectively. We rewrite this as:

$$uv' = \frac{d}{dx}(uv) - u'v$$

and then integrate both sides:

$$\int uv' \, dx = \int \frac{d}{dx}(uv) \, dx - \int u'v \, dx.$$

Since an antiderivative of $\frac{d}{dx}(uv)$ is just uv, we get the following formula:

Integration by Parts

$$\int uv' \, dx = uv - \int u'v \, dx.$$

This formula is useful when the integrand can be viewed as a product and when the integral on the right-hand side is simpler than that on the left. In effect, we were using integration by parts in the previous two examples. In Example 1, we let $xe^x = (x) \cdot (e^x) = uv'$, and choose $u = x$ and $v' = e^x$. Thus, $u' = 1$ and $v = e^x$, so

$$\int \underbrace{(x)}_{u}\underbrace{(e^x)}_{v'} \, dx = \underbrace{(x)}_{u}\underbrace{(e^x)}_{v} - \int \underbrace{(1)}_{u'}\underbrace{(e^x)}_{v} \, dx = xe^x - e^x + C.$$

So uv represents our first guess, and $\int u'v \, dx$ represents the correction to our guess.

Notice what would have happened if we took $v = e^x + C_1$. Then

$$\int xe^x \, dx = x(e^x + C_1) - \int (e^x + C_1) \, dx$$
$$= xe^x + C_1 x - e^x - C_1 x + C$$
$$= xe^x - e^x + C,$$

as before. Thus, it is not necessary to include an arbitrary constant in the antiderivative for v; any antiderivative will do.

What would have happened if we had picked u and v' the other way around? If $u = e^x$ and $v' = x$, then $u' = e^x$ and $v = x^2/2$. The formula for integration by parts then gives

$$\int xe^x \, dx = \frac{x^2}{2}e^x - \int \frac{x^2}{2} \cdot e^x \, dx,$$

which is true but not helpful since the integral on the right is worse than the one on the left. To use this method, we must choose u and v' to make the integral on the right easier to find than the integral on the left.

How to Choose u and v'

- Whatever you let v' be, you need to be able to find v.
- It helps if u' is simpler than u (or at least no more complicated than u).
- It helps if v is simpler than v' (or at least no more complicated than v').

If we pick $v' = x$ in Example 1, then $v = x^2/2$, which is certainly "worse" than v'.

There are some examples which don't look like good candidates for integration by parts because they don't appear to involve products, but for which the method works well. Such examples often involve $\ln x$ or the inverse trigonometric functions. Here is one:

Example 3   Find $\displaystyle\int_2^3 \ln x \, dx$.

Solution   This doesn't look like a product unless we write $\ln x = (1)(\ln x)$. Then we might say $u = 1$ so $u' = 0$, which certainly makes things simpler. But if $v' = \ln x$, what is v? If we knew, we wouldn't need integration by parts. Let's try the other way: if $u = \ln x$, $u' = 1/x$ and if $v' = 1$, $v = x$, so

$$\int_2^3 \underbrace{(\ln x)}_{u}\, \underbrace{(1)}_{v'} \; dx = \underbrace{(\ln x)}_{u} \underbrace{(x)}_{v} \Big|_2^3 - \int_2^3 \underbrace{\left(\frac{1}{x}\right)}_{u'} \cdot \underbrace{(x)}_{v} \, dx$$

$$= x \ln x \Big|_2^3 - \int_2^3 1 \, dx = (x \ln x - x)\Big|_2^3$$

$$= 3\ln 3 - 3 - 2\ln 2 + 2 = 3\ln 3 - 2\ln 2 - 1.$$

Notice that when doing a definite integral by parts, we must remember to put the limits of integration (here 2 and 3) on the uv term (in this case $x \ln x$) as well as on the integral $\int u'v \, dx$.

Example 4   Find $\displaystyle\int x^6 \ln x \, dx$.

Solution   View $x^6 \ln x$ as uv' where $u = \ln x$ and $v' = x^6$. Then $v = \frac{1}{7}x^7$ and $u' = 1/x$, so integration by parts gives us:

$$\int x^6 \ln x \, dx = \int (\ln x) x^6 \, dx = (\ln x)\left(\frac{1}{7}x^7\right) - \int \frac{1}{7}x^7 \cdot \frac{1}{x} \, dx$$

$$= \frac{1}{7}x^7 \ln x - \frac{1}{7}\int x^6 \, dx$$

$$= \frac{1}{7}x^7 \ln x - \frac{1}{49}x^7 + C.$$

In Example 4 we did not choose $v' = \ln x$, because it is not immediately clear what v would be. In fact, we used integration by parts in Example 3 to find out what the antiderivative of $\ln x$ was. Also, using $u = \ln x$, as we have done, gives $u' = 1/x$, which most people would consider simpler than $u = \ln x$. This shows that u does not have to be the first factor in the integrand (here x^6).

Example 5   Find $\int x^2 \sin 4x \, dx$.

Solution   If we let $v' = \sin 4x$, then $v = -\frac{1}{4}\cos 4x$, which is no worse than v'. Also letting $u = x^2$, we get $u' = 2x$, which is simpler than $u = x^2$. Using integration by parts:

$$\int x^2 \sin 4x \, dx = x^2\left(-\frac{1}{4}\cos 4x\right) - \int 2x\left(-\frac{1}{4}\cos 4x\right) \, dx$$

$$= -\frac{1}{4}x^2 \cos 4x + \frac{1}{2}\int x \cos 4x \, dx.$$

The trouble is we still have to grapple with $\int x \cos 4x \, dx$. This can be done by using integration by parts again with a new u and v, namely $u = x$ and $v' = \cos 4x$:

$$\int x \cos 4x \, dx = x\left(\frac{1}{4}\sin 4x\right) - \int 1 \cdot \frac{1}{4}\sin 4x \, dx$$

$$= \frac{1}{4}x \sin 4x - \frac{1}{4} \cdot \left(-\frac{1}{4}\cos 4x\right) + C$$

$$= \frac{1}{4}x \sin 4x + \frac{1}{16}\cos 4x + C.$$

Thus,

$$\int x^2 \sin 4x \, dx = -\frac{1}{4}x^2 \cos 4x + \frac{1}{2}\int x \cos 4x \, dx$$

$$= -\frac{1}{4}x^2 \cos 4x + \frac{1}{2}\left(\frac{1}{4}x \sin 4x + \frac{1}{16}\cos 4x + C\right)$$

$$= -\frac{1}{4}x^2 \cos 4x + \frac{1}{8}x \sin 4x + \frac{1}{32}\cos 4x + C.$$

Notice that, in this example, each time we used integration by parts, the exponent of x went down by 1. In addition, when the arbitrary constant C is multiplied by $\frac{1}{2}$, it is still represented by C.

Example 6   Find $\int \cos^2 \theta \, d\theta$.

Solution   Using integration by parts with $u = \cos \theta$, $v' = \cos \theta$ gives $u' = -\sin \theta$, $v = \sin \theta$, so we get

$$\int \cos^2 \theta \, d\theta = \cos \theta \sin \theta + \int \sin^2 \theta \, d\theta$$

Substituting $\sin^2 \theta = 1 - \cos^2 \theta$ leads to

$$\int \cos^2 \theta \, d\theta = \cos \theta \sin \theta + \int (1 - \cos^2 \theta) \, d\theta$$

$$= \cos \theta \sin \theta + \int 1 \, d\theta - \int \cos^2 \theta \, d\theta.$$

Looking at the right side, we see that the original integral has reappeared. If we move it to the left, we get

$$2 \int \cos^2 \theta \, d\theta = \cos \theta \sin \theta + \int 1 \, d\theta = \cos \theta \sin \theta + \theta + C.$$

Dividing by 2 gives

$$\int \cos^2 \theta \, d\theta = \frac{1}{2} \cos \theta \sin \theta + \frac{1}{2}\theta + C.$$

Problem 39 asks you to do this integral by another method.

The previous example illustrates a useful technique : Use integration by parts to transform the integral into an expression containing another copy of the same integral, possibly multiplied by a coefficient, then solve for the original integral. It may be necessary to apply integration by parts twice, as the next example shows.

Example 7   Use integration by parts twice to find $\int e^{2x} \sin(3x) \, dx$.

Solution   Using integration by parts with $u = e^{2x}$ and $v' = \sin(3x)$ gives $u' = 2e^{2x}$, $v = -\frac{1}{3}\cos(3x)$, so we get

$$\int e^{2x} \sin(3x) \, dx = -\frac{1}{3}e^{2x} \cos(3x) + \frac{2}{3} \int e^{2x} \cos(3x) \, dx.$$

On the right side we have an integral similar to the original one, with the sine replaced by a cosine. Using integration by parts on that integral in the same way gives

$$\int e^{2x} \cos(3x) \, dx = \frac{1}{3}e^{2x} \sin(3x) - \frac{2}{3} \int e^{2x} \sin(3x) \, dx.$$

Substituting this into the expression we obtained for the original integral gives

$$\int e^{2x} \sin(3x) \, dx = -\frac{1}{3}e^{2x} \cos(3x) + \frac{2}{3} \left(\frac{1}{3}e^{2x} \sin(3x) - \frac{2}{3} \int e^{2x} \sin(3x) \, dx \right)$$

$$= -\frac{1}{3}e^{2x} \cos(3x) + \frac{2}{9}e^{2x} \sin(3x) - \frac{4}{9} \int e^{2x} \sin(3x) \, dx$$

The right side now has a copy of the original integral, multiplied by $-4/9$. Moving it to the left, we get

$$\left(1 + \frac{4}{9}\right) \int e^{2x} \sin(3x) \, dx = -\frac{1}{3}e^{2x} \cos(3x) + \frac{2}{9}e^{2x} \sin(3x).$$

Dividing through by the coefficient on the left, $(1 + 4/9) = 13/9$, we get

$$\int e^{2x} \sin(3x) \, dx = \frac{9}{13} \left(-\frac{1}{3}e^{2x} \cos(3x) + \frac{2}{9}e^{2x} \sin(3x) \right)$$

$$= \frac{1}{13}e^{2x} \left(2 \sin(3x) - 3 \cos(3x) \right) + C.$$

Exercises and Problems for Section 7.2

Exercises

1. Write $\arctan x = 1 \cdot \arctan x$ to find $\int \arctan x \, dx$.

Find the integrals in Exercises 2–27.

2. $\displaystyle \int te^{5t} \, dt$

3. $\displaystyle \int t^2 e^{5t} \, dt$

4. $\displaystyle \int pe^{-0.1p} \, dp$

5. $\displaystyle \int t \sin t \, dt$

6. $\displaystyle \int y \ln y \, dy$

7. $\displaystyle \int x^3 \ln x \, dx$

8. $\displaystyle \int (z+1)e^{2z} \, dz$

9. $\displaystyle \int t^2 \sin t \, dt$

10. $\displaystyle \int \theta^2 \cos 3\theta \, d\theta$

11. $\displaystyle \int \sin^2 \theta \, d\theta$

12. $\displaystyle \int \cos^2(3\alpha + 1) \, d\alpha$

13. $\displaystyle \int q^5 \ln 5q \, dq$

14. $\displaystyle \int y\sqrt{y + 3} \, dy$

15. $\displaystyle \int (\ln t)^2 \, dt$

16. $\displaystyle \int (t+2)\sqrt{2 + 3t} \, dt$

17. $\displaystyle \int (\theta + 1) \sin(\theta + 1) \, d\theta$

18. $\displaystyle \int \frac{z}{e^z} \, dz$

19. $\displaystyle \int \frac{\ln x}{x^2} \, dx$

20. $\displaystyle \int \frac{y}{\sqrt{5 - y}} \, dy$

21. $\displaystyle \int \frac{t + 7}{\sqrt{5 - t}} \, dt$

22. $\displaystyle \int x(\ln x)^4 \, dx$

23. $\displaystyle \int \arcsin w \, dw$

24. $\displaystyle \int \arctan 7z \, dz$

25. $\displaystyle \int x \arctan x^2 \, dx$

26. $\displaystyle \int x^3 e^{x^2} \, dx$

27. $\displaystyle \int x^5 \cos x^3 \, dx$

Evaluate the integrals in Exercises 28–35 both exactly [e.g. $\ln(3\pi)$] and numerically [e.g. $\ln(3\pi) \approx 2.243$]:

28. $\displaystyle \int_1^5 \ln t \, dt$

29. $\displaystyle \int_3^5 x \cos x \, dx$

30. $\displaystyle \int_0^{10} ze^{-z} \, dz$

31. $\displaystyle \int_1^3 t \ln t \, dt$

32. $\displaystyle \int_0^1 \arctan y \, dy$

33. $\displaystyle \int_0^5 \ln(1 + t) \, dt$

34. $\displaystyle \int_0^1 \arcsin z \, dz$

35. $\displaystyle \int_0^1 u \arcsin u^2 \, du$

Problems

36. For each of the following integrals, indicate whether integration by substitution or integration by parts is more appropriate. Do not evaluate the integrals.

(a) $\displaystyle \int x \sin x \, dx$

(b) $\displaystyle \int \frac{x^2}{1 + x^3} \, dx$

(c) $\displaystyle \int xe^{x^2} \, dx$

(d) $\displaystyle \int x^2 \cos(x^3) \, dx$

(e) $\displaystyle \int \frac{1}{\sqrt{3x + 1}} \, dx$

(f) $\displaystyle \int x^2 \sin x \, dx$

(g) $\displaystyle \int \ln x \, dx$

37. Find the exact value of the area under the first arch of $f(x) = x \sin x$.

38. In Problem 11, you evaluated $\int \sin^2 \theta \, d\theta$ using integration by parts. (If you didn't do it by parts, do so now!) Redo this integral using the identity $\sin^2 \theta = (1 - \cos 2\theta)/2$. Explain any differences in the form of the answer obtained by the two methods.

39. Compute $\int \cos^2 \theta \, d\theta$ in two different ways and explain any differences in the form of your answers. (The identity $\cos^2 \theta = (1 + \cos 2\theta)/2$ may be useful.)

40. Use integration by parts twice to find $\int e^x \sin x \, dx$.

41. Use integration by parts twice to find $\int e^\theta \cos \theta \, d\theta$.

42. Use the results from Problems 40 and 41 and integration by parts to find $\int xe^x \sin x \, dx$.

43. Use the results from Problems 40 and 41 and integration by parts to find $\int \theta e^\theta \cos \theta \, d\theta$.

In Problems 44–47, derive the given formulas.

44. $\displaystyle \int x^n e^x \, dx = x^n e^x - n \int x^{n-1} e^x \, dx$

45. $\displaystyle \int x^n \cos ax \, dx = \frac{1}{a} x^n \sin ax - \frac{n}{a} \int x^{n-1} \sin ax \, dx$

46. $\displaystyle \int x^n \sin ax \, dx = -\frac{1}{a} x^n \cos ax + \frac{n}{a} \int x^{n-1} \cos ax \, dx$

47. $\displaystyle \int \cos^n x \, dx = \frac{1}{n} \cos^{n-1} x \sin x + \frac{n-1}{n} \int \cos^{n-2} x \, dx$

48. Integrating $e^{ax} \sin bx$ by parts twice yields a result of the form

$$\int e^{ax} \sin bx \, dx = e^{ax}(A \sin bx + B \cos bx) + C.$$

(a) Find the constants A and B in terms of a and b. [Hint: Don't actually perform the integration by parts.]

(b) Evaluate $\int e^{ax} \cos bx \, dx$ by modifying the result in part (a). [Again, it is not necessary to perform integration by parts, as the result is of the same form as that in part (a).]

49. Estimate $\int_0^{10} f(x)g'(x) \, dx$ if $f(x) = x^2$ and g has the values in the following table.

x	0	2	4	6	8	10
$g(x)$	2.3	3.1	4.1	5.5	5.9	6.1

50. Let f be twice differentiable with $f(0) = 6$, $f(1) = 5$, and $f'(1) = 2$. Evaluate the integral $\int_0^1 xf''(x)dx$.

51. Let $F(a)$ be the area under the graph of $y = x^2 e^{-x}$ between $x = 0$ and $x = a$, for $a > 0$.

(a) Find a formula for $F(a)$.
(b) Is F an increasing or decreasing function?
(c) Is F concave up or concave down for $0 < a < 2$?

52. The concentration, C, in ng/ml, of a drug in the blood as a function of the time, t, in hours since the drug was administered is given by $C = 15te^{-0.2t}$. The area under the concentration curve is a measure of the overall effect of the drug on the body, called the bioavailability. Find the bioavailability of the drug between $t = 0$ and $t = 3$.

53. The voltage, V, in an electric circuit is given as a function of time, t, by

$$V = V_0 \cos(\omega t + \phi).$$

Suppose each of the positive constants, V_0, ω, ϕ is increased (while the other two are held constant). What is the effect of each increase on the following quantities:

(a) The maximum value of V?
(b) The maximum value of dV/dt?
(c) The average value of V^2 over one period?

54. During a surge in the demand for electricity, the rate, r, at which energy is used can be approximated by

$$r = te^{-at},$$

where t is the time in hours and a is a positive constant.

(a) Find the total energy, E, used in the first T hours. Give your answer as a function of a.
(b) What happens to E as $T \to \infty$?

55. In describing the behavior of an electron, we use wave functions $\Psi_1, \Psi_2, \Psi_3, \ldots$ of the form

$$\Psi_n(x) = C_n \sin(n\pi x) \qquad \text{for } n = 1, 2, 3, \ldots$$

where x is the distance from a fixed point and C_n is a positive constant.

(a) Find C_1 so that Ψ_1 satisfies

$$\int_0^1 (\Psi_1(x))^2 \, dx = 1.$$

This is called normalizing the wave function Ψ_1.
(b) For any integer n, find C_n so that Ψ_n is normalized.

7.3 TABLES OF INTEGRALS

Since so few functions have elementary antiderivatives, they have been compiled in a list called a table of integrals.[1] A short table of indefinite integrals is given inside the back cover of this book. The key to using these tables is being able to recognize the general class of function that you are trying to integrate, so you can know in what section of the table to look.

Warning: This section involves long division of polynomials and completing the square. You may want to review these topics!

Using the Table of Integrals

Part I of the table inside the back cover gives the antiderivatives of the basic functions x^n, a^x, $\ln x$, $\sin x$, $\cos x$, and $\tan x$. (The antiderivative for $\ln x$ is found using integration by parts and is a special case of the more general formula III-13.) Most of these are already familiar.

Part II of the table contains antiderivatives of functions involving products of e^x, $\sin x$, and $\cos x$. All of these antiderivatives were obtained using integration by parts.

---

[1] See, for example, *CRC Standard Mathematical Tables* (Boca Raton, Fl: CRC Press). Many computer programs and calculators can compute antiderivatives as well.

Example 1   Find $\int \sin 7z \sin 3z \, dz$.

Solution   Since the integrand is the product of two sines, we should use II-10 in the table,

$$\int \sin 7z \sin 3z \, dz = -\frac{1}{40}(7 \cos 7z \sin 3z - 3 \cos 3z \sin 7z) + C.$$

Part III of the table contains antiderivatives for products of a polynomial and e^x, $\sin x$, or $\cos x$. It also has an antiderivative for $x^n \ln x$, which can easily be used to find the antiderivatives of the product of a general polynomial and $\ln x$. Each *reduction formula* is used repeatedly to reduce the degree of the polynomial until a zero degree polynomial is obtained.

Example 2   Find $\int (x^5 + 2x^3 - 8)e^{3x} \, dx$.

Solution   Since $p(x) = x^5 + 2x^3 - 8$ is a polynomial multiplied by e^{3x}, this is of the form in III-14. Now $p'(x) = 5x^4 + 6x^2$ and $p''(x) = 20x^3 + 12x$, and so on, giving

$$\int (x^5 + 2x^3 - 8)e^{3x} \, dx = e^{3x} \left[\frac{1}{3}(x^5 + 2x^3 - 8) - \frac{1}{9}(5x^4 + 6x^2) + \frac{1}{27}(20x^3 + 12x) \right.$$
$$\left. - \frac{1}{81}(60x^2 + 12) + \frac{1}{243}(120x) - \frac{1}{729} \cdot 120 \right] + C.$$

Here we have the successive derivatives of the original polynomial $x^5 + 2x^3 - 8$, occurring with alternating signs and multiplied by successive powers of 1/3.

Part IV of the table contains reduction formulas for the antiderivatives of $\cos^n x$ and $\sin^n x$, which can be obtained by integration by parts. When n is a positive integer, formulas IV-17 and IV-18 can be used repeatedly to reduce the power n until it is 0 or 1.

Example 3   Find $\int \sin^6 \theta \, d\theta$.

Solution   Use IV-17 repeatedly:

$$\int \sin^6 \theta \, d\theta = -\frac{1}{6} \sin^5 \theta \cos \theta + \frac{5}{6} \int \sin^4 \theta \, d\theta$$
$$\int \sin^4 \theta \, d\theta = -\frac{1}{4} \sin^3 \theta \cos \theta + \frac{3}{4} \int \sin^2 \theta \, d\theta$$
$$\int \sin^2 \theta \, d\theta = -\frac{1}{2} \sin \theta \cos \theta + \frac{1}{2} \int 1 \, d\theta.$$

Calculate $\int \sin^2 \theta \, d\theta$ first, and use this to find $\int \sin^4 \theta \, d\theta$; then calculate $\int \sin^6 \theta \, d\theta$. Putting this all together, we get

$$\int \sin^6 \theta \, d\theta = -\frac{1}{6} \sin^5 \theta \cos \theta - \frac{5}{24} \sin^3 \theta \cos \theta - \frac{15}{48} \sin \theta \cos \theta + \frac{15}{48}\theta + C.$$

The last item in **Part IV** of the table is not a formula: it is advice on how to antidifferentiate products of integer powers of $\sin x$ and $\cos x$. There are various techniques to choose from, depending on the nature (odd or even, positive or negative) of the exponents.

Example 4   Find $\int \cos^3 t \sin^4 t \, dt$.

Solution   Here the exponent of $\cos t$ is odd, so IV-23 recommends making the substitution $w = \sin t$. Then $dw = \cos t \, dt$. To make this work, we'll have to separate off one of the cosines to be part of dw. Also, the remaining even power of $\cos t$ can be rewritten in terms of $\sin t$ by using $\cos^2 t = 1 - \sin^2 t = 1 - w^2$, so that

$$\int \cos^3 t \sin^4 t \, dt = \int \cos^2 t \sin^4 t \cos t \, dt$$

$$= \int (1 - w^2) w^4 \, dw = \int (w^4 - w^6) \, dw$$

$$= \frac{1}{5} w^5 - \frac{1}{7} w^7 + C = \frac{1}{5} \sin^5 t - \frac{1}{7} \sin^7 t + C.$$

Example 5   Find $\int \cos^2 x \sin^4 x \, dx$.

Solution   In this example, both exponents are even. The advice given in IV-23 is to convert to all sines or all cosines. We'll convert to all sines by substituting $\cos^2 x = 1 - \sin^2 x$, and then we'll multiply out the integrand:

$$\int \cos^2 x \sin^4 x \, dx = \int (1 - \sin^2 x) \sin^4 x \, dx = \int \sin^4 x \, dx - \int \sin^6 x \, dx.$$

In Example 3 we found $\int \sin^4 x \, dx$ and $\int \sin^6 x \, dx$. Put them together to get

$$\int \cos^2 x \sin^4 x \, dx = -\frac{1}{4} \sin^3 x \cos x - \frac{3}{8} \sin x \cos x + \frac{3}{8} x$$

$$- \left(-\frac{1}{6} \sin^5 x \cos x - \frac{5}{24} \sin^3 x \cos x - \frac{15}{48} \sin x \cos x + \frac{15}{48} x \right) + C$$

$$= \frac{1}{6} \sin^5 x \cos x - \frac{1}{24} \sin^3 x \cos x - \frac{3}{48} \sin x \cos x + \frac{3}{48} x + C.$$

The last two parts of the table are concerned with quadratic functions: **Part V** has expressions with quadratic denominators; **Part VI** contains square roots of quadratics. The quadratics that appear in these formulas are of the form $x^2 \pm a^2$ or $a^2 - x^2$, or in factored form $(x - a)(x - b)$, where a and b are different constants. Quadratics can be converted to these forms by factoring or completing the square.

Preparing to Use the Table: Transforming the Integrand

To use the integral table, we often need to manipulate or reshape integrands to fit entries in the table. The manipulations that tend to be useful are factoring, long division, completing the square, and substitution.

Using Factoring

Example 6   Find $\int \dfrac{3x + 7}{x^2 + 6x + 8} \, dx$.

Solution   In this case we factor the denominator to get it into a form in the table:

$$x^2 + 6x + 8 = (x + 2)(x + 4).$$

Now in V-27 we let $a = -2$, $b = -4$, $c = 3$, and $d = 7$, to obtain

$$\int \frac{3x + 7}{x^2 + 6x + 8} \, dx = \frac{1}{2} (\ln |x + 2| - (-5) \ln |x + 4|) + C.$$

Long Division

Example 7   Find $\displaystyle\int \frac{x^2}{x^2+4}\,dx$.

Solution   A good rule of thumb when integrating a rational function whose numerator has a degree greater than or equal to that of the denominator is to start by doing *long division*. This results in a polynomial plus a simpler rational function as a remainder. Performing long division here, we obtain:

$$\frac{x^2}{x^2+4} = 1 - \frac{4}{x^2+4}.$$

Then, by V-24 with $a = 2$, we obtain:

$$\int \frac{x^2}{x^2+4}\,dx = \int 1\,dx - 4\int \frac{1}{x^2+4}\,dx = x - 4\cdot\frac{1}{2}\arctan\frac{x}{2} + C = x - 2\arctan\frac{x}{2} + C.$$

Completing the Square to Rewrite the Quadratic in the Form $w^2 + a^2$

Example 8   Find $\displaystyle\int \frac{1}{x^2+6x+14}\,dx$.

Solution   By completing the square, we can get this integrand into a form in the table:

$$x^2 + 6x + 14 = (x^2 + 6x + 9) - 9 + 14$$
$$= (x+3)^2 + 5.$$

Let $w = x + 3$. Then $dw = dx$ and so the substitution gives

$$\int \frac{1}{x^2+6x+14}\,dx = \int \frac{1}{w^2+5}\,dw = \frac{1}{\sqrt{5}}\arctan\frac{w}{\sqrt{5}} + C = \frac{1}{\sqrt{5}}\arctan\frac{x+3}{\sqrt{5}} + C,$$

where the antidifferentiation uses V-24 with $a^2 = 5$.

Substitution

Example 9   Find $\displaystyle\int e^t \sin(5t+7)\,dt$.

Solution   This looks similar to II-8. To make the correspondence more complete, let's try the substitution $w = 5t + 7$. Then $dw = 5\,dt$, so $dt = \frac{1}{5}\,dw$. Also, $t = (w-7)/5$. Then the integral becomes

$$\int e^t \sin(5t+7)\,dt = \int e^{(w-7)/5}\sin w\,\frac{dw}{5}$$

$$= \frac{e^{-7/5}}{5}\int e^{w/5}\sin w\,dw \qquad \text{(Since } e^{(w-7)/5} = e^{w/5}e^{-7/5} \text{ and } e^{-7/5} \text{ is a constant)}$$

Now we can use II-8 with $a = \frac{1}{5}$ and $b = 1$ to write

$$\int e^{w/5}\sin w\,dw = \frac{1}{(\frac{1}{5})^2 + 1^2}e^{w/5}\left(\frac{\sin w}{5} - \cos w\right) + C,$$

so

$$\int e^t \sin(5t+7)\,dt = \frac{e^{-7/5}}{5}\left(\frac{25}{26}e^{(5t+7)/5}\left(\frac{\sin(5t+7)}{5} - \cos(5t+7)\right)\right) + C$$

$$= \frac{5e^t}{26}\left(\frac{\sin(5t+7)}{5} - \cos(5t+7)\right) + C.$$

Exercises and Problems for Section 7.3

Exercises

For Exercises 1–35, antidifferentiate using the table of integrals. You may need to transform the integrand first.

1. $\int e^{-3\theta} \cos\theta \, d\theta$

2. $\int x^5 \ln x \, dx$

3. $\int x^3 \sin 5x \, dx.$

4. $\int (x^2 + 3) \ln x \, dx.$

5. $\int (x^3 + 5)^2 \, dx.$

6. $\int \sin w \cos^4 w \, dw$

7. $\int \sin^4 x \, dx$

8. $\int \frac{1}{3 + y^2} \, dy$

9. $\int \frac{1}{\cos^3 x} \, dx$

10. $\int x^3 e^{2x} \, dx$

11. $\int \sin 3\theta \cos 5\theta \, d\theta$

12. $\int \sin 3\theta \sin 5\theta \, d\theta$

13. $\int x^2 e^{3x} \, dx$

14. $\int x^2 e^{x^3} \, dx$

15. $\int x^4 e^{3x} \, dx$

16. $\int u^5 \ln(5u) \, du$

17. $\int \frac{t^2 + 1}{t^2 - 1} \, dt$

18. $\int x^3 \sin x^2 \, dx$

19. $\int \cos 2y \cos 7y \, dy$

20. $\int y^2 \sin 2y \, dy$

21. $\int e^{5x} \sin 3x \, dx$

22. $\int \frac{1}{\cos^5 x} \, dx.$

23. $\int \frac{1}{\sin^2 2\theta} \, d\theta$

24. $\int \frac{1}{\sin^3 3\theta} \, d\theta$

25. $\int \frac{1}{\cos^4 7x} \, dx$

26. $\int \frac{1}{x^2 + 4x + 3} \, dx$

27. $\int \tan^4 x \, dx$

28. $\int \frac{dz}{z(z-3)}$

29. $\int \frac{dy}{4 - y^2}$

30. $\int \frac{1}{1 + (z+2)^2} \, dz$

31. $\int \frac{1}{y^2 + 4y + 5} \, dy$

32. $\int \frac{1}{x^2 + 4x + 4} \, dx$

33. $\int \sin^3 x \, dx$

34. $\int \sin^3 3\theta \cos^2 3\theta \, d\theta$

35. $\int z e^{2z^2} \cos(2z^2) \, dz$

Problems

36. Show that for all integers m and n, with $m \neq \pm n$, $\int_{-\pi}^{\pi} \sin m\theta \sin n\theta \, d\theta = 0$.

37. Show that for all integers m and n, with $m \neq \pm n$, $\int_{-\pi}^{\pi} \cos m\theta \cos n\theta \, d\theta = 0$.

38. The voltage, V, in an electrical outlet is given as a function of time, t, by the function $V = V_0 \cos(120\pi t)$, where V is in volts and t is in seconds, and V_0 is a positive constant representing the maximum voltage.

  (a) What is the average value of the voltage over 1 second?
  (b) Engineers do not use the average voltage. They use the root mean square voltage defined by $\overline{V} = \sqrt{\text{average of } (V^2)}$. Find $\overline{V}$ in terms of V_0. (Take the average over 1 second.)
  (c) The standard voltage in an American house is 110 volts, meaning that $\overline{V} = 110$. What is V_0?

39. For some constants A and B, the rate of production, $R(t)$, of oil in a new oil well is modeled by:

$$R(t) = A + Be^{-t} \sin(2\pi t)$$

where t is the time in years, A is the equilibrium rate, and B is the "variability" coefficient.

  (a) Find the total amount of oil produced in the first N years of operation. (Take N to be an integer.)
  (b) Find the average amount of oil produced per year over the first N years (where N is an integer).
  (c) From your answer to part (b), find the average amount of oil produced per year as $N \to \infty$.
  (d) Looking at the function $R(t)$, explain how you might have predicted your answer to part (c) without doing any calculations.
  (e) Do you think it is reasonable to expect this model to hold over a very long period? Why or why not?

40. Suppose n is a positive integer and $\Psi_n = C_n \sin(n\pi x)$ is the wave function used in describing the behavior of an electron. If n and m are different integers, find

$$\int_0^1 \Psi_n(x) \cdot \Psi_m(x) \, dx.$$

7.4 ALGEBRAIC IDENTITIES AND TRIGONOMETRIC SUBSTITUTIONS

Although not all functions have elementary antiderivatives, many do. In this section we introduce two powerful methods of integration which show that large classes of functions have elementary antiderivatives. The first is the method of partial fractions, which depends on an algebraic identity, and allows us to integrate rational functions. The second is the method of trigonometric substitutions, which allows us to handle expressions involving the square root of a quadratic polynomial. Rather than giving a complete description of these methods, we illustrate their power in a variety of examples. Some of the formulas in the table of integrals can be derived using the techniques of this section.

Method of Partial Fractions

The integral of some rational functions can be obtained by splitting the integrand into *partial fractions*. For example, to find

$$\int \frac{1}{(x-2)(x-5)}\, dx,$$

the integrand is split into partial fractions with denominators $(x-2)$ and $(x-5)$. We write

$$\frac{1}{(x-2)(x-5)} = \frac{A}{x-2} + \frac{B}{x-5},$$

where A and B are constants that need to be found. Multiplying by $(x-2)(x-5)$ gives the identity

$$1 = A(x-5) + B(x-2)$$

so

$$1 = (A+B)x - 5A - 2B.$$

Since this equation holds for all x, the constant terms on both sides must be equal. Similarly, the coefficients of x on both sides must be equal. So

$$-5A - 2B = 1$$
$$A + B = 0.$$

Solving these equations gives $A = -1/3$, $B = 1/3$. Thus,

$$\frac{1}{(x-2)(x-5)} = \frac{-1/3}{x-2} + \frac{1/3}{x-5}.$$

Example 1   Use partial fractions to integrate $\displaystyle\int \frac{1}{(x-2)(x-5)}\, dx$.

Solution   We split the integrand into partial fractions, each of which can be integrated:

$$\int \frac{1}{(x-2)(x-5)}\, dx = \int \left(\frac{-1/3}{x-2} + \frac{1/3}{x-5} \right) dx = -\frac{1}{3}\ln|x-2| + \frac{1}{3}\ln|x-5| + C.$$

You can check that using formula V-26 in the integral table gives the same result.

This method can be used to derive formulas V-26 and V-27 in the integral table. A similar method works whenever the denominator of the integrand factors into distinct linear factors and the numerator has degree less than the denominator.

Example 2   Find $\int \dfrac{x+2}{x^2+x}\, dx$.

Solution   We factor the denominator and split the integrand into partial fractions:

$$\frac{x+2}{x^2+x} = \frac{x+2}{x(x+1)} = \frac{A}{x} + \frac{B}{x+1}.$$

Multiplying by $x(x+1)$ gives the identity

$$\begin{aligned} x+2 &= A(x+1) + Bx \\ &= (A+B)x + A. \end{aligned}$$

Equating constant terms and coefficients of x gives $A = 2$ and $A + B = 1$, so $B = -1$. Then we split the integrand into two parts and integrate:

$$\int \frac{x+2}{x^2+x}\, dx = \int \left(\frac{2}{x} - \frac{1}{x+1} \right) dx = 2 \ln |x| - \ln |x+1| + C.$$

The next example illustrates what to do if there is a repeated factor in the denominator.

Example 3   Calculate $\int \dfrac{10x - 2x^2}{(x-1)^2(x+3)}\, dx$ using partial fractions of the form $\dfrac{A}{x-1}, \dfrac{B}{(x-1)^2}, \dfrac{C}{x+3}$.

Solution   We are given that the squared factor, $(x-1)^2$, leads to partial fractions of the form:

$$\frac{10x - 2x^2}{(x-1)^2(x+3)} = \frac{A}{x-1} + \frac{B}{(x-1)^2} + \frac{C}{x+3}$$

Multiplying through by $(x-1)^2(x+3)$ gives

$$\begin{aligned} 10x - 2x^2 &= A(x-1)(x+3) + B(x+3) + C(x-1)^2 \\ &= (A+C)x^2 + (2A + B - 2C)x - 3A + 3B + C. \end{aligned}$$

Equating the coefficients of x^2 and x and the constant terms, we get the simultaneous equations:

$$\begin{aligned} A + C &= -2 \\ 2A + B - 2C &= 10 \\ -3A + 3B + C &= 0 \end{aligned}$$

Solving gives $A = 1, B = 2, C = -3$. Thus, we obtain three integrals which can be evaluated:

$$\begin{aligned} \int \frac{10x - 2x^2}{(x-1)^2(x+3)}\, dx &= \int \left(\frac{1}{x-1} + \frac{2}{(x-1)^2} - \frac{3}{x+3} \right) dx \\ &= \ln |x-1| - \frac{2}{(x-1)} - 3 \ln |x+3| + K. \end{aligned}$$

For the second integral, we use the fact that $\int 2/(x-1)^2\, dx = 2 \int (x-1)^{-2}\, dx = -2(x-1)^{-1} + K$.

If there is a quadratic in the denominator which cannot be factored, we must allow a numerator of the form $Ax + B$ in the numerator, as the next example shows.

Example 4   Find $\displaystyle\int \frac{2x^2 - x - 1}{(x^2 + 1)(x - 2)}\,dx$ using partial fractions of the form $\dfrac{Ax + B}{x^2 + 1}$ and $\dfrac{C}{x - 2}$.

Solution   We are given that the quadratic denominator, $(x^2 + 1)$, which cannot be factored further, has a numerator of the form $Ax + B$, so we have

$$\frac{2x^2 - x - 1}{(x^2 + 1)(x - 2)} = \frac{Ax + B}{x^2 + 1} + \frac{C}{x - 2}.$$

Multiplying by $(x^2 + 1)(x - 2)$ gives

$$
\begin{aligned}
2x^2 - x - 1 &= (Ax + B)(x - 2) + C(x^2 + 1) \\
&= (A + C)x^2 + (B - 2A)x + C - 2B
\end{aligned}
$$

Equating the coefficients of x^2 and x and the constant terms gives the simultaneous equations

$$
\begin{aligned}
A + C &= 2 \\
B - 2A &= -1 \\
C - 2B &= -1.
\end{aligned}
$$

Solving gives $A = B = C = 1$, so we rewrite the integral as follows:

$$\int \frac{2x^2 - x - 1}{(x^2 + 1)(x - 2)}\,dx = \int \left(\frac{x + 1}{x^2 + 1} + \frac{1}{x - 2}\right)\,dx.$$

This identity is useful provided we can perform the integration on the right-hand side. The first integral can be done if it is split into two; the second integral is similar to those in the previous examples. We have

$$\int \frac{2x^2 - x - 1}{(x^2 + 1)(x - 2)}\,dx = \int \frac{x}{x^2 + 1}\,dx + \int \frac{1}{x^2 + 1}\,dx + \int \frac{1}{x - 2}\,dx.$$

To calculate $\int (x/(x^2 + 1))\,dx$, substitute $w = x^2 + 1$, or guess and check. The final result is

$$\int \frac{2x^2 - x - 1}{(x^2 + 1)(x - 2)}\,dx = \frac{1}{2}\ln|x^2 + 1| + \arctan x + \ln|x - 2| + K.$$

Finally, the next example shows what to do if the numerator has degree larger than the denominator.

Example 5   Calculate $\displaystyle\int \frac{x^3 - 7x^2 + 10x + 1}{x^2 + 7x + 10}\,dx$ using long division.

Solution   The degree of the numerator is greater than the degree of the denominator, so we divide first:

$$\frac{x^3 - 7x^2 + 10x + 1}{x^2 + 7x + 10} = \frac{x(x^2 - 7x + 10) + 1}{x^2 - 7x + 10} = x + \frac{1}{x^2 - 7x + 10}.$$

The remainder, in this case $1/(x^2 - 7x + 10)$, is a rational function on which we try to use partial fractions. We have

$$\frac{1}{x^2 - 7x + 10} = \frac{1}{(x - 2)(x - 5)}$$

so in this case we use the result of Example 1 to obtain

$$\int \frac{x^3 - 7x^2 + 10x + 1}{x^2 - 7x + 10}\,dx = \int \left(x + \frac{1}{(x - 2)(x - 5)}\right)\,dx = \frac{x^2}{2} - \frac{1}{3}\ln|x - 2| + \frac{1}{3}\ln|x - 5| + C.$$

Many, though not all, rational functions can be integrated by the strategy suggested by the previous examples.

Strategy for Integrating a Rational Function, $\frac{P(x)}{Q(x)}$

- If degree of $P(x) \geq$ degree of $Q(x)$, try long division and the method of partial fractions on the remainder.
- If $Q(x)$ is the product of distinct linear factors, use partial fractions of the form

$$\frac{A}{(x - c)}.$$

- If $Q(x)$ contains a repeated linear factor, $(x - c)^n$, use partial fractions of the form

$$\frac{A_1}{(x - c)} + \frac{A_2}{(x - c)^2} + \cdots + \frac{A_n}{(x - c)^n}.$$

- If $Q(x)$ contains an unfactorable quadratic $q(x)$, try a partial fraction of the form

$$\frac{Ax + B}{q(x)}.$$

To use this method, we need to be able to integrate each partial fraction. We know how to integrate terms of the form $A/(x - c)^n$ using the power rule (if $n > 1$) and logarithms (if $n = 1$). Next we see how to integrate terms of the form $(Ax + B)/q(x)$, where $q(x)$ is an unfactorable quadratic.

Trigonometric Substitutions

Section 7.1 showed how substitutions could be used to transform complex integrands. Now we see how substitution of $\sin \theta$ or $\tan \theta$ can be used for integrands involving square roots of quadratics or unfactorable quadratics.

Sine Substitutions

Substitutions involving $\sin \theta$ make use of the Pythagorean identity, $\cos^2 \theta + \sin^2 \theta = 1$, to simplify an integrand involving $\sqrt{a^2 - x^2}$.

Example 6   Find $\displaystyle\int \frac{1}{\sqrt{1 - x^2}}\, dx$ using the substitution $x = \sin \theta$.

Solution   If $x = \sin \theta$, then $dx = \cos \theta\, d\theta$, and substitution gives

$$\int \frac{1}{\sqrt{1 - x^2}}\, dx = \int \frac{1}{\sqrt{1 - \sin^2 \theta}} \cos \theta\, d\theta = \int \frac{\cos \theta}{\sqrt{\cos^2 \theta}}\, d\theta.$$

Now either $\sqrt{\cos^2 \theta} = \cos \theta$ or $\sqrt{\cos^2 \theta} = -\cos \theta$ depending on the values taken by θ. If we choose $-\pi/2 \leq \theta \leq \pi/2$, then $\cos \theta \geq 0$, so $\sqrt{\cos^2 \theta} = \cos \theta$. Then

$$\int \frac{\cos \theta}{\sqrt{\cos^2 \theta}}\, d\theta = \int \frac{\cos \theta}{\cos \theta}\, d\theta = \int 1\, d\theta = \theta + C = \arcsin x + C.$$

The last step uses the fact that $\theta = \arcsin x$ if $x = \sin \theta$ and $-\pi/2 \leq \theta \leq \pi/2$.

From now on, when we substitute $\sin \theta$, we assume that the interval $-\pi/2 \leq \theta \leq \pi/2$ has been chosen. Notice that the previous example is the case $a = 1$ of VI-28 in the table of integrals. The next example illustrates how to choose the substitution when $a \neq 1$.

Example 7   Use a trigonometric substitution to find $\displaystyle\int \frac{1}{\sqrt{4 - x^2}}\, dx$.

Solution   This time we choose $x = 2 \sin \theta$, with $-\pi/2 \leq \theta \leq \pi/2$, so that $4 - x^2$ becomes a perfect square:

$$\sqrt{4 - x^2} = \sqrt{4 - 4 \sin^2 \theta} = 2\sqrt{1 - \sin^2 \theta} = 2\sqrt{\cos^2 \theta} = 2 \cos \theta.$$

Then $dx = 2 \cos \theta \, d\theta$, so substitution gives

$$\int \frac{1}{\sqrt{4 - x^2}} \, dx = \int \frac{1}{2 \cos \theta} 2 \cos \theta \, d\theta = \int 1 \, d\theta = \theta + C = \arcsin\left(\frac{x}{2}\right) + C.$$

The general rule for choosing a sine substitution is:

> To simplify $\sqrt{a^2 - x^2}$, for constant a, try $x = a \sin \theta$, with $-\pi/2 \leq \theta \leq \pi/2$.

Notice $\sqrt{a^2 - x^2}$ is only defined on the interval $[-a, a]$. Assuming that the domain of the integrand is $[-a, a]$, the substitution $x = a \sin \theta$, with $-\pi/2 \leq \theta \leq \pi/2$, is valid for all x in the domain, because its range is $[-a, a]$ and it has an inverse $\theta = \arcsin(x/a)$ on $[-a, a]$.

Example 8   Find the area of the ellipse $4x^2 + y^2 = 9$.

Solution   Solving for y shows that $y = \sqrt{9 - 4x^2}$ gives the upper half of the ellipse. From Figure 7.1, we see that

$$\text{Area} = 4 \int_0^{3/2} \sqrt{9 - 4x^2} \, dx.$$

To decide which trigonometric substitution to use, we write the integrand as

$$\sqrt{9 - 4x^2} = 2\sqrt{\frac{9}{4} - x^2} = 2\sqrt{\left(\frac{3}{2}\right)^2 - x^2}.$$

This suggests that we should choose $x = (3/2) \sin \theta$, so that $dx = (3/2) \cos \theta \, d\theta$ and

$$\sqrt{9 - 4x^2} = 2\sqrt{\left(\frac{3}{2}\right)^2 - \left(\frac{3}{2}\right)^2 \sin^2 \theta} = 2\left(\frac{3}{2}\right)\sqrt{1 - \sin^2 \theta} = 3 \cos \theta.$$

When $x = 0$, $\theta = 0$, and when $x = 3/2$, $\theta = \pi/2$, so

$$4 \int_0^{3/2} \sqrt{9 - 4x^2} \, dx = 4 \int_0^{\pi/2} 3 \cos \theta \left(\frac{3}{2}\right) \cos \theta \, d\theta = 18 \int_0^{\pi/2} \cos^2 \theta \, d\theta.$$

Using Example 6 on page 301 or table of integrals IV-18, we find

$$\int \cos^2 \theta \, d\theta = \frac{1}{2} \cos \theta \sin \theta + \frac{1}{2} \theta + C.$$

So we have

$$\text{Area} = 4 \int_0^{3/2} \sqrt{9 - 4x^2} \, dx = \frac{18}{2} (\cos \theta \sin \theta + \theta)\Big|_0^{\pi/2} = 9\left(0 + \frac{\pi}{2}\right) = \frac{9\pi}{2}.$$

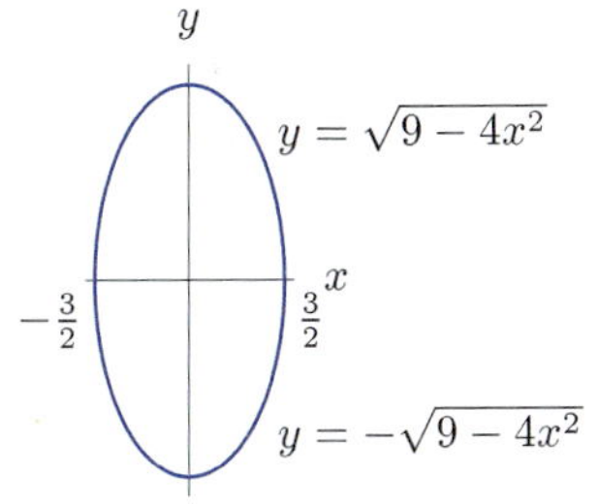

Figure 7.1: The ellipse $4x^2 + y^2 = 9$

Tangent Substitutions

Integrals involving $a^2 + x^2$ may be simplified by a substitution involving $\tan \theta$ and the trigonometric identities $\tan \theta = \sin \theta / \cos \theta$ and $\cos^2 \theta + \sin^2 \theta = 1$.

Example 9   Find $\displaystyle \int \frac{1}{x^2 + 9} \, dx$ using the substitution $x = 3 \tan \theta$.

Solution   If $x = 3 \tan \theta$, then $dx = (3 / \cos^2 \theta) \, d\theta$, so

$$\int \frac{1}{x^2 + 9} \, dx = \int \left(\frac{1}{9 \tan^2 \theta + 9} \right) \left(\frac{3}{\cos^2 \theta} \right) d\theta = \frac{1}{3} \int \frac{1}{\left(\frac{\sin^2 \theta}{\cos^2 \theta} + 1 \right) \cos^2 \theta} \, d\theta$$

$$= \frac{1}{3} \int \frac{1}{\sin^2 \theta + \cos^2 \theta} \, d\theta = \frac{1}{3} \int 1 \, d\theta = \frac{1}{3} \theta + C = \frac{1}{3} \arctan \left(\frac{x}{3} \right) + C.$$

To simplify $a^2 + x^2$ or $\sqrt{a^2 + x^2}$, for constant a, try $x = a \tan \theta$, with $-\pi/2 < \theta < \pi/2$.

Note that $a^2 + x^2$ and $\sqrt{a^2 + x^2}$ are defined on $(-\infty, \infty)$. Assuming that the domain of the integrand is $(-\infty, \infty)$, the substitution $x = a \tan \theta$, with $-\pi/2 < \theta < \pi/2$, is valid for all x in the domain, because its range is $(-\infty, \infty)$ and it has an inverse $\theta = \arctan(x/a)$ on $(-\infty, \infty)$.

Example 10   Use a tangent substitution to show that the following two integrals are equal:

$$\int_0^1 \sqrt{1 + x^2} \, dx = \int_0^{\pi/4} \frac{1}{\cos^3 \theta} \, d\theta.$$

What area do these integrals represent?

Solution   We put $x = \tan \theta$, with $-\pi/2 < \theta < \pi/2$, so that $dx = (1 / \cos^2 \theta) \, d\theta$, and

$$\sqrt{1 + x^2} = \sqrt{1 + \frac{\sin^2 \theta}{\cos^2 \theta}} = \sqrt{\frac{\cos^2 \theta + \sin^2 \theta}{\cos^2 \theta}} = \frac{1}{\cos \theta}.$$

When $x = 0$, $\theta = 0$, and when $x = 1$, $\theta = \pi/4$, so

$$\int_0^1 \sqrt{1 + x^2} \, dx = \int_0^{\pi/4} \left(\frac{1}{\cos \theta} \right) \left(\frac{1}{\cos^2 \theta} \right) d\theta = \int_0^{\pi/4} \frac{1}{\cos^3 \theta} \, d\theta.$$

The left-hand integral represents the area under the hyperbola $y^2 - x^2 = 1$ in Figure 7.2.

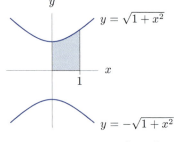

Figure 7.2: The hyperbola $y^2 - x^2 = 1$

Completing the Square to Use a Trigonometric Substitution

To make a trigonometric substitution, we may first need to complete the square.

Example 11   Find $\displaystyle\int \frac{3}{\sqrt{2x-x^2}}\, dx$.

Solution   To use a sine or tangent substitution, the expression under the square root sign should be in the form $a^2 + x^2$ or $a^2 - x^2$. Completing the square, we get

$$2x - x^2 = 1 - (x-1)^2.$$

This suggests we substitute $x - 1 = \sin\theta$, or $x = \sin\theta + 1$. Then $dx = \cos\theta\, d\theta$, and

$$\int \frac{3}{\sqrt{2x-x^2}}\, dx = \int \frac{3}{\sqrt{1-(x-1)^2}}\, dx = \int \frac{3}{\sqrt{1-\sin^2\theta}}\cos\theta\, d\theta$$

$$= \int \frac{3}{\cos\theta}\cos\theta\, d\theta = \int 3\, d\theta = 3\theta + C.$$

Since $x - 1 = \sin\theta$, we have $\theta = \arcsin(x-1)$, so

$$\int \frac{3}{\sqrt{2x-x^2}}\, dx = 3\arcsin(x-1) + C.$$

Example 12   Find $\displaystyle\int \frac{1}{x^2+x+1}\, dx$.

Solution   Completing the square, we get

$$x^2 + x + 1 = \left(x + \frac{1}{2}\right)^2 + \frac{3}{4} = \left(x + \frac{1}{2}\right)^2 + \left(\frac{\sqrt{3}}{2}\right)^2.$$

This suggests we substitute $x + 1/2 = (\sqrt{3}/2)\tan\theta$, or $x = -1/2 + (\sqrt{3}/2)\tan\theta$. Then $dx = (\sqrt{3}/2)(1/\cos^2\theta)\, d\theta$, so

$$\int \frac{1}{x^2+x+1}\, dx = \int \left(\frac{1}{(x+\frac{1}{2})^2 + \frac{3}{4}}\right)\left(\frac{\sqrt{3}}{2}\frac{1}{\cos^2\theta}\right) d\theta$$

$$= \frac{\sqrt{3}}{2}\int \left(\frac{1}{\frac{3}{4}\tan^2\theta + \frac{3}{4}}\right)\left(\frac{1}{\cos^2\theta}\right) d\theta = \frac{2}{\sqrt{3}}\int \frac{1}{(\tan^2\theta + 1)\cos^2\theta}\, d\theta$$

$$= \frac{2}{\sqrt{3}}\int \frac{1}{\sin^2\theta + \cos^2\theta}\, d\theta = \frac{2}{\sqrt{3}}\int 1\, d\theta = \frac{2}{\sqrt{3}}\theta + C.$$

Since $x + 1/2 = (\sqrt{3}/2)\tan\theta$, we have $\theta = \arctan((2/\sqrt{3})x + 1/\sqrt{3})$, so

$$\int \frac{1}{x^2+x+1}\, dx = \frac{2}{\sqrt{3}}\arctan\left(\frac{2}{\sqrt{3}}x + \frac{1}{\sqrt{3}}\right) + C.$$

Exercises and Problems for Section 7.4

Exercises

Split the functions in Exercises 1–4 into partial fractions.

1. $\dfrac{20}{25-x^2}$

2. $\dfrac{x+1}{6x+x^2}$

3. $\dfrac{8}{y^3-4y}$

4. $\dfrac{2y}{y^3-y^2+y-1}$

5. Integrate the function in Exercise 1 with respect to x.

6. Integrate the function in Exercise 2 with respect to x.

7. Integrate the function in Exercise 3 with respect to y.

8. Integrate the function in Exercise 4 with respect to y.

9. Which of the following integrals are best done by a trigonometric substitution, and what substitution?

   (a) $\displaystyle\int \sqrt{9-x^2}\, dx$ **(b)** $\displaystyle\int x\sqrt{9-x^2}\, dx$

In Exercises 10–14, evaluate the integral.

10. $\int \dfrac{3x^2 - 8x + 1}{x^3 - 4x^2 + x + 6} \, dx$; use $\dfrac{A}{x-2} + \dfrac{B}{x+1} + \dfrac{C}{x-3}$.

11. $\int \dfrac{dx}{x^3 - x^2}$; use $\dfrac{A}{x} + \dfrac{B}{x^2} + \dfrac{C}{x-1}$.

12. $\int \dfrac{10x + 2}{x^3 - 5x^2 + x - 5} \, dx$; use $\dfrac{A}{x-5} + \dfrac{Bx + C}{x^2 + 1}$.

13. $\int \dfrac{x^4 + 12x^3 + 15x^2 + 25x + 11}{x^3 + 12x^2 + 11x} \, dx$;

use division and $\dfrac{A}{x} + \dfrac{B}{x+1} + \dfrac{C}{x+11}$.

14. $\int \dfrac{x^4 + 3x^3 + 2x^2 + 1}{x^2 + 3x + 2} \, dx$; use division.

Use the given substitution in Exercises 15–17.

15. $\int \dfrac{1}{\sqrt{9 - 4x^2}} \, dx$, $\quad x = \dfrac{3}{2} \sin t$

16. $\int \dfrac{1}{x^2 + 4x + 5} \, dx$, $\quad x = \tan t - 2$

17. $\int \dfrac{1}{\sqrt{4x - 3 - x^2}} \, dx$, $\quad x = \sin t + 2$

18. Give a substitution (not necessarily trigonometric) which could be used to compute the following integrals:

(a) $\int \dfrac{x}{\sqrt{x^2 + 10}} \, dx$ (b) $\int \dfrac{1}{\sqrt{x^2 + 10}} \, dx$

Problems

Complete the square and give a substitution (not necessarily trigonometric) which could be used to compute the integrals in Problems 19–26.

19. $\int \dfrac{1}{x^2 + 6x + 25} \, dx$

20. $\int \dfrac{dy}{y^2 + 3y + 3}$

21. $\int \dfrac{1}{x^2 + 2x + 2} \, dx$

22. $\int \dfrac{x + 1}{x^2 + 2x + 2} \, dx$

23. $\int \dfrac{4}{\sqrt{2z - z^2}} \, dz$

24. $\int \dfrac{z - 1}{\sqrt{2z - z^2}} \, dz$

25. $\int (t + 2) \sin(t^2 + 4t + 7) \, dt$

26. $\int (2 - \theta) \cos(\theta^2 - 4\theta) \, d\theta$

Calculate the integrals in Problems 27–36.

27. $\int \dfrac{1}{(x - 5)(x - 3)} \, dx$

28. $\int \dfrac{y + 2}{2y^2 + 3y + 1} \, dy$

29. $\int \dfrac{x + 1}{x^3 + x} \, dx$

30. $\int \dfrac{x - 2}{x^2 + x^4} \, dx$

31. $\int \dfrac{x^2}{\sqrt{9 - x^2}} \, dx$

32. $\int \dfrac{y^2}{25 + y^2} \, dy$

33. $\int \dfrac{dt}{t^2 \sqrt{1 + t^2}}$

34. $\int \dfrac{dz}{(4 - z^2)^{3/2}}$

35. $\int \dfrac{x}{x^2 - 3x + 2} \, dx$

36. $\int \dfrac{1}{x^2 + 4x + 13} \, dx$

Find the exact area of the regions in Problems 37–42.

37. Bounded by $3x/((x - 1)(x - 4))$, $y = 0$, $x = 2$, $x = 3$.

38. Bounded by $y = (3x^2 + x)/((x^2 + 1)(x + 1))$, $y = 0$, $x = 0$, $x = 1$.

39. Bounded by $y = x^2/\sqrt{1 - x^2}$, $y = 0$, $x = 0$, $x = 1/2$.

40. Bounded by $y = x^3/\sqrt{4 - x^2}$, $y = 0$, $x = 0$, $x = \sqrt{2}$.

41. Bounded by $y = 1/\sqrt{x^2 + 9}$, $y = 0$, $x = 0$, $x = 3$.

42. Bounded by $y = 1/(x\sqrt{x^2 + 9})$, $y = 0$, $x = \sqrt{3}$, $x = 3$.

43. (a) Show $\int \dfrac{1}{\sin^2 \theta} \, d\theta = -\dfrac{1}{\tan \theta} + C$.

(b) Calculate $\int \dfrac{dy}{y^2 \sqrt{5 - y^2}}$.

44. Calculate the integral $\int \dfrac{dx}{1 - x^2}$ both by partial fractions and by using a trigonometric substitution. Show that the results you get are the same.

45. The Law of Mass Action tells us that the time, T, taken by a chemical to create a quantity x_0 of the product (in molecules) is given by

$$T = \int_0^{x_0} \dfrac{k \, dx}{(a - x)(b - x)}$$

where a and b are initial quantities of the two ingredients used to make the product, and k is a positive constant. Suppose $0 < a < b$.

(a) Find the time taken to make a quantity $x_0 = a/2$ of the product.

(b) What happens to T as $x_0 \to a$?

46. A population, P, is said to be growing logistically if the time, T, taken for it to increase from P_1 to P_2 is given by

$$T = \int_{P_1}^{P_2} \dfrac{k \, dP}{P(L - P)},$$

where k and L are positive constants and $P_1 < P_2 < L$.

(a) Calculate the time taken for the population to grow from $P_1 = L/4$ to $P_2 = L/2$.

(b) What happens to T as $P_2 \to L$?

7.5 APPROXIMATING DEFINITE INTEGRALS

The methods of the last few sections allow us to get exact answers for definite integrals in a variety of cases. However, many functions do not have elementary antiderivatives. To evaluate the definite integrals of such functions, we cannot use the Fundamental Theorem; we must use numerical methods.

We already know how to approximate a definite integral numerically using left- and right-hand Riemann sums. In the next two sections we introduce better methods for approximating definite integrals—better in the sense that they give more accurate results with less work than that required to find the left- and right-hand sums.

The Midpoint Rule

In the left- and right-hand Riemann sums, the heights of the rectangles are found using the left-hand or right-hand endpoints, respectively, of the subintervals. For the *midpoint rule*, we use the midpoint of each of the subintervals. For example, in approximating $\int_1^2 f(x)\,dx$ by a Riemann sum with two subdivisions, we first divide the interval $1 \leq x \leq 2$ into two pieces. The midpoint of the first subinterval is 1.25 and the midpoint of the second is 1.75. The heights of the two rectangles are $f(1.25)$ and $f(1.75)$, respectively. (See Figure 7.3.) The Riemann sum is

$$f(1.25)0.5 + f(1.75)0.5.$$

Figure 7.3 shows that evaluating f at the midpoint of each subdivision usually gives a better approximation to the area under the curve than evaluating f at either end. For this particular f, it appears that each rectangle is partly above and partly below the graph on each subinterval. Furthermore, the area under the curve which is not under the rectangle appears to be nearly equal to the area under the rectangle which is above the curve. In fact, this new midpoint Riemann sum is generally a better approximation to the definite integral than the left- or right-hand sum with the same number of subdivisions, n.

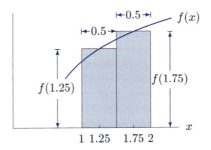

Figure 7.3: Midpoint rule with two subdivisions

So far, we have three ways of estimating an integral using a Riemann sum:

1. The **left rule** uses the left endpoint of each subinterval.
2. The **right rule** uses the right endpoint of each subinterval.
3. The **midpoint rule** uses the midpoint of each subinterval.

We write $\text{LEFT}(n)$, $\text{RIGHT}(n)$, and $\text{MID}(n)$ to denote the results obtained by using these rules with n subdivisions.

Example 1 For $\int_1^2 \dfrac{1}{x}\,dx$, compute $\text{LEFT}(2)$, $\text{RIGHT}(2)$ and $\text{MID}(2)$, and compare your answers with the exact value of the integral.

Solution For $n = 2$ subdivisions of the interval $[1, 2]$, we use $\Delta x = 0.5$. Then

$$\text{LEFT}(2) = f(1)(0.5) + f(1.5)(0.5) = \frac{1}{1}(0.5) + \frac{1}{1.5}(0.5) = 0.8333\ldots$$

$$\text{RIGHT}(2) = f(1.5)(0.5) + f(2)(0.5) = \frac{1}{1.5}(0.5) + \frac{1}{2}(0.5) = 0.5833\ldots$$

$$\text{MID}(2) = f(1.25)(0.5) + f(1.75)(0.5) = \frac{1}{1.25}(0.5) + \frac{1}{1.75}(0.5) = 0.6857\ldots$$

All three Riemann sums in this example are approximating

$$\int_1^2 \frac{1}{x}\, dx = \ln x \Big|_1^2 = \ln 2 - \ln 1 = \ln 2 = 0.6931\ldots.$$

With only two subdivisions, the left and right rules give quite poor approximations but the midpoint rule is already fairly close to the exact answer. Figure 7.4(a) illustrates why the left and right rules are so inaccurate. Since $f(x) = 1/x$ is decreasing from 1 to 2, the left rule overestimates on each subdivision while the right rule underestimates. However, the midpoint rule approximates with rectangles on each subdivision that are each partly above and partly below the graph, so the errors tend to balance out. (See Figure 7.4(b).)

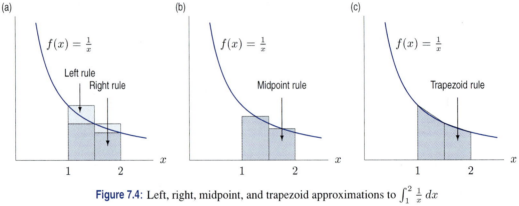

Figure 7.4: Left, right, midpoint, and trapezoid approximations to $\int_1^2 \frac{1}{x}\, dx$

The Trapezoid Rule

We have just seen how the midpoint rule can have the effect of balancing out the errors of the left and right rules. There is another way of balancing these errors: we average the results from the left and right rules. This approximation is called the *trapezoid rule*:

$$\text{TRAP}(n) = \frac{\text{LEFT}(n) + \text{RIGHT}(n)}{2}.$$

The trapezoid rule averages the values of f at the left and right endpoints of each subinterval and multiplies by Δx. This is the same as approximating the area under the graph of f in each subinterval by a trapezoid (see Figure 7.5).

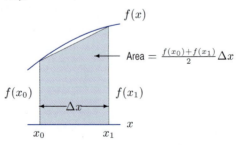

Figure 7.5: Area used in the trapezoid rule

Example 2   For $\int_1^2 \frac{1}{x}\,dx$, compare the trapezoid rule with two subdivisions with the left, right, and midpoint rules.

Solution   In the previous example we got $\text{LEFT}(2) = 0.8333\ldots$ and $\text{RIGHT}(2) = 0.5833\ldots$. The trapezoid rule is the average of these, so $\text{TRAP}(2) = 0.7083\ldots$. (See Figure 7.4(c).) The exact value of the integral is $0.6931\ldots$, so the trapezoid rule is better than the left or right rules. The midpoint rule is still the best, however, since $\text{MID}(2) = 0.6857\ldots$.

Is the Approximation an Over- or Underestimate?

It is useful to know when a rule is producing an overestimate and when it is producing an underestimate. In Chapter 5 we saw that the following relationship holds.

If f is increasing on $[a, b]$, then

$$\text{LEFT}(n) \leq \int_a^b f(x)\,dx \leq \text{RIGHT}(n).$$

If f is decreasing on $[a, b]$, then

$$\text{RIGHT}(n) \leq \int_a^b f(x)\,dx \leq \text{LEFT}(n).$$

The Trapezoid Rule

If the graph of the function is concave down on $[a, b]$, then each trapezoid lies below the graph and the trapezoid rule underestimates. If the graph is concave up on $[a, b]$, the trapezoid rule overestimates. (See Figure 7.6.)

f concave down:
Trapezoid underestimates

f concave up:
Trapezoid overestimates

Figure 7.6: Error in the trapezoid rule

The Midpoint Rule

To understand the relationship between the midpoint rule and concavity, take a rectangle whose top intersects the curve at the midpoint of a subinterval. Draw a tangent to the curve at the midpoint; this gives a trapezoid. See Figure 7.7. (This is *not* the same trapezoid as in the trapezoid rule.) The midpoint rectangle and the new trapezoid have the same area, because the shaded triangles in Figure 7.7 are congruent. Hence, if the graph of the function is concave down, the midpoint rule overestimates; if the graph is concave up, the midpoint rule underestimates. (See Figure 7.8.)

If the graph of f is concave down on $[a, b]$, then

$$\text{TRAP}(n) \leq \int_a^b f(x)\,dx \leq \text{MID}(n).$$

If the graph of f is concave up on $[a, b]$, then

$$\text{MID}(n) \leq \int_a^b f(x)\,dx \leq \text{TRAP}(n).$$

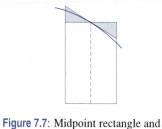

Figure 7.7: Midpoint rectangle and trapezoid with same area

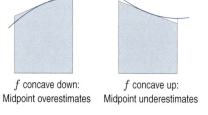

f concave down: f concave up:
Midpoint overestimates   Midpoint underestimates

Figure 7.8: Error in the midpoint rule

Exercises and Problems for Section 7.5

Exercises

In Exercises 1–4, sketch the area given by the following approximations to $\int_a^b f(x)\,dx$. Identify each approximation as an overestimate or an underestimate.

(a) LEFT(2)　　　　　**(b)** RIGHT(2)
(c) TRAP(2)　　　　　**(d)** MID(2)

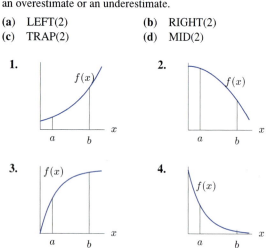

1.

$f(x)$

a　b　x

2.

$f(x)$

a　b　x

3.

$f(x)$

a　b　x

4.

$f(x)$

a　b　x

5. Calculate the following approximations to the integral $\int_0^6 x^2\,dx$.

(a) LEFT(2)
(b) RIGHT(2)
(c) TRAP(2)
(d) MID(2)

6. (a) Find LEFT(2) and RIGHT(2) for $\int_0^4 (x^2 + 1)\,dx$.
(b) Illustrate your answers to part (a) graphically. Is each approximation an underestimate or overestimate?

7. (a) Find MID(2) and TRAP(2) for $\int_0^4 (x^2 + 1)\,dx$.
(b) Illustrate your answers to part (a) graphically. Is each approximation an underestimate or overestimate?

Problems

8. (a) Estimate $\int_0^1 1/(1 + x^2)\,dx$ by subdividing the interval into eight parts using:

(i) The left Riemann sum.

(ii) The right Riemann sum.

(iii) The trapezoidal rule.

(b) Since the exact value of the integral is $\pi/4$, you can estimate the value of π using part(a). Explain why your first estimate is too large and your second estimate too small.

9. Using the table, estimate the total distance traveled from time $t = 0$ to time $t = 6$ using LEFT, RIGHT, and TRAP.

Time, t	0	1	2	3	4	5	6
Velocity, v	3	4	5	4	7	8	11

10. Using Figure 7.9, order the following approximations to the integral $\int_0^3 f(x)dx$ and its exact value from smallest to largest:
LEFT(n), RIGHT(n), MID(n), TRAP(n), Exact value.

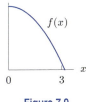

Figure 7.9

11. The results from the left, right, trapezoid, and midpoint rules used to approximate $\int_0^1 g(t)\,dt$, with the same number of subdivisions for each rule, are as follows:
0.601, 0.632, 0.633, 0.664.

(a) Using Figure 7.10, match each rule with its approximation.
(b) Between which two consecutive approximations does the true value of the integral lie?

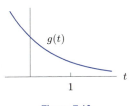

Figure 7.10

For the functions in Problems 12–15, pick which approximation — left, right, trapezoid, or midpoint — is guaranteed to give an overestimate for $\int_0^5 f(x)\,dx$, and which is guaranteed to give an underestimate. (There may be more than one.)

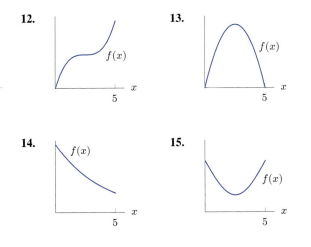

16. Using a fixed number of subdivisions, we approximate the integrals of f and g on the interval in Figure 7.11.

(a) For which function, f or g, is LEFT more accurate? RIGHT? Explain.
(b) For which function, f or g, is TRAP more accurate? MID? Explain.

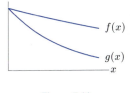

Figure 7.11

17. (a) Values for $f(x)$ are in the table. Which of the four approximation methods in this section is most likely to give the best estimate of $\int_0^{12} f(x)\,dx$? Estimate the integral using this method.
(b) Assume $f(x)$ is continuous with no critical points or points of inflection on the interval $0 \le x \le 12$. Is the estimate found in part (a) an over- or underestimate? Explain.

x	0	3	6	9	12
$f(x)$	100	97	90	78	55

18. (a) Find the exact value of $\int_0^{2\pi} \sin\theta\,d\theta$.
(b) Explain, using pictures, why the MID(1) and MID(2) approximations to this integral give the exact value.
(c) Does MID(3) give the exact value of this integral? How about MID(n)? Explain.

19. (a) Show geometrically why $\int_0^1 \sqrt{2-x^2}\,dx = \frac{\pi}{4} + \frac{1}{2}$.
[Hint: Break up the area under $y = \sqrt{2-x^2}$ from $x = 0$ to $x = 1$ into two pieces: a sector of a circle and a right triangle.]
(b) Approximate $\int_0^1 \sqrt{2-x^2}\,dx$ for $n = 5$ using the left, right, trapezoid, and midpoint rules. Compute the error in each case using the answer to part (a), and compare the errors.

20. The width, in feet, at various points along the fairway of a hole on a golf course is given in Figure 7.12. If one pound of fertilizer covers 200 square feet, estimate the amount of fertilizer needed to fertilize the fairway.

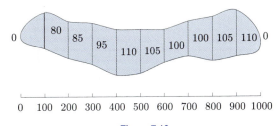

Figure 7.12

Problems 21–25 involve approximating $\int_a^b f(x)\,dx$.

21. Show $\text{RIGHT}(n) = \text{LEFT}(n) + f(b)\Delta x - f(a)\Delta x$.

22. Show $\text{TRAP}(n) = \text{LEFT}(n) + \frac{1}{2}\left(f(b) - f(a)\right)\Delta x$.

23. Show $\text{LEFT}(2n) = \frac{1}{2}\left(\text{LEFT}(n) + \text{MID}(n)\right)$.

24. Check that the equations in Problems 21 and 22 hold for $\int_1^2 (1/x)\,dx$ when $n = 10$.

25. Suppose that $a = 2$, $b = 5$, $f(2) = 13$, $f(5) = 21$ and that $\text{LEFT}(10) = 3.156$ and $\text{MID}(10) = 3.242$. Use Problems 21–23 to compute $\text{RIGHT}(10)$, $\text{TRAP}(10)$, $\text{LEFT}(20)$, $\text{RIGHT}(20)$, and $\text{TRAP}(20)$.

7.6   APPROXIMATION ERRORS AND SIMPSON'S RULE

When we compute an approximation, we are always concerned about the error, namely the difference between the exact answer and the approximation. We usually do not know the exact error; if we did, we would also know the exact answer. Often the best we can get is an upper bound on the error and some idea of how much work is involved in making the error smaller. The study of numerical approximations is really the study of errors. The errors for some methods are much smaller than those for others. The errors for the midpoint and trapezoid rules are related to each other in a way that suggests an even better method, called Simpson's rule. We work with the example $\int_1^2 (1/x)\,dx$ because we know the exact value of this integral ($\ln 2$) and we can investigate the behavior of the errors.

Error in Left and Right Rules

For any approximation, we take

$$\text{Error} = \text{Actual value} - \text{Approximate value}.$$

Let us see what happens to the error in the left and right rules as we increase n. We increase n each time by a factor of 5 starting at $n = 2$. The results are in Table 7.1. A positive error indicates that the Riemann sum is less than the exact value, $\ln 2$. Notice that the errors for the left and right rules have opposite signs but are approximately equal in magnitude. (See Figure 7.13.) The best way to try to get the errors to cancel is to average the left and right rules; this average is the trapezoid rule. If we had not already thought of the trapezoid rule, we might have been led to invent it by this observation.

There is another pattern to the errors in Table 7.1. If we compute the *ratio* of the errors in Table 7.2, we see that the error[2] in both the left and right rules decreases by a factor of about 5 as n increases by a factor of 5.

There is nothing special about the number 5; the same holds for any factor. To get one extra digit of accuracy in any calculation, we must make the error $1/10$ as big, so we must increase n by a factor of 10. In fact, *for the left or right rules, each extra digit of accuracy requires about 10 times the work.* The calculator used to produce these tables took about half a second to compute the left rule approximation for $n = 50$, and this yields $\ln 2$ to two digits. To get three correct digits, n would need to be around 500 and the time would be about 5 seconds. Four digits requires $n = 5000$ and 50 seconds. Ten digits requires $n = 5 \cdot 10^9$ and $5 \cdot 10^7$ seconds, which is more than a year! Clearly, the errors for the left and right rules do not decrease fast enough as n increases for practical use.

Table 7.1   *Errors for the left and right rule approximation to* $\int_1^2 \frac{1}{x}\,dx = \ln 2 \approx 0.6931471806$

n	Error in left rule	Error in right rule
2	−0.1402	0.1098
10	−0.0256	0.0244
50	−0.0050	0.0050
250	−0.0010	0.0010

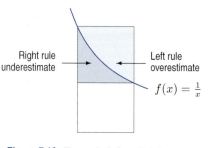

Right rule underestimate

Left rule overestimate

$f(x) = \frac{1}{x}$

Figure 7.13: Errors in left and right sums

---

[2]The values in Table 7.1 are rounded to 4 decimal places; those in Table 7.2 were computed using more decimal places and then rounded.

Table 7.2 *Ratio of the errors as n increases for $\int_1^2 \frac{1}{x}\,dx$*

	Ratio of errors in left rule	Ratio of errors in right rule
Error(2)/Error(10)	5.47	4.51
Error(10)/Error(50)	5.10	4.90
Error(50)/Error(250)	5.02	4.98

Error in Trapezoid and Midpoint Rules

Table 7.3 shows that the trapezoid and midpoint rules produce much better approximations to $\int_1^2 (1/x)\,dx$ than the left and right rules.

Again there is a pattern to the errors. For each n, the midpoint rule is noticeably better than the trapezoid rule; the error for the midpoint rule, in absolute value, seems to be about half the error of the trapezoid rule. To see why, compare the shaded areas in Figure 7.14. Also, notice in Table 7.3 that the errors for the two rules have opposite signs; this is due to concavity.

We are interested in how the errors behave as n increases. Table 7.4 gives the ratios of the errors for each rule. For each rule, we see that as n increases by a factor of 5, the error decreases by a factor of about $25 = 5^2$. In fact, it can be shown that this squaring relationship holds for any factor, so increasing n by a factor of 10 will decrease the error by a factor of about $100 = 10^2$. Reducing the error by a factor of 100 is equivalent to adding two more decimal places of accuracy to the result.

Table 7.3 *The errors for the trapezoid and midpoint rules for $\int_1^2 \frac{1}{x}\,dx$*

n	Error in trapezoid rule	Error in midpoint rule
2	−0.0152	0.0074
10	−0.00062	0.00031
50	−0.0000250	0.0000125
250	−0.0000010	0.0000005

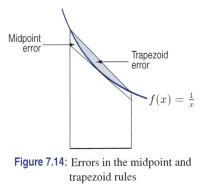

Figure 7.14: Errors in the midpoint and trapezoid rules

In other words: *In the trapezoid or midpoint rules, each extra 2 digits of accuracy requires about 10 times the work.*

This result shows the advantage of the midpoint and trapezoid rules over the left and right rules: less additional work needs to be done to get another decimal place of accuracy. The calculator used to produce these tables again took about half a second to compute the midpoint rule for $\int_1^2 \frac{1}{x}\,dx$ with $n = 50$, and this gets 4 digits correct. Thus to get 6 digits would take $n = 500$ and 5 seconds, to get 8 digits would take 50 seconds, and to get 10 digits would take 500 seconds, or about 10 minutes. That is still not great, but it is certainly better than the 1 year required by the left or right rule.

Table 7.4 *Ratios of the errors as n increases for $\int_1^2 \frac{1}{x}\,dx$*

	Ratio of errors in trapezoid rule	Ratio of errors in midpoint rule
Error(2)/Error(10)	24.33	23.84
Error(10)/Error(50)	24.97	24.95
Error(50)/Error(250)	25.00	25.00

Simpson's Rule

Still more improvement is possible. Observing that the trapezoid error has the opposite sign and about twice the magnitude of the midpoint error, we may guess that a weighted average of the two rules, with the midpoint rule weighted twice the trapezoid rule, will have a much smaller error. This approximation is called *Simpson's rule*[3]:

$$\text{SIMP}(n) = \frac{2 \cdot \text{MID}(n) + \text{TRAP}(n)}{3}.$$

Table 7.5 gives the errors for Simpson's rule. Notice how much smaller the errors are than the previous errors. Of course, it is a little unfair to compare Simpson's rule at $n = 50$, say, with the previous rules, because Simpson's rule must compute the value of f at both the midpoint and the endpoints of each subinterval and hence involves evaluating the function at twice as many points. We know by our previous analysis, however, that even if we did compute the other rules at $n = 100$ to compare with Simpson's rule at $n = 50$, the other errors would only decrease by a factor of 2 for the left and right rules and by a factor of 4 for the trapezoid and midpoint rules.

We see in Table 7.5 that as n increases by a factor of 5, the errors decrease by a factor of about 600, or about 5^4. Again this behavior holds for any factor, so increasing n by a factor of 10 decreases the error by a factor of about 10^4. In other words: *In Simpson's rule, each extra 4 digits of accuracy requires about 10 times the work.*

Table 7.5   *The errors for Simpson's rule and the ratios of the errors*

n	Error	Ratio
2	−0.0001067877	
		550.15
10	−0.0000001940	
		632.27
50	−0.0000000003	

This is a great improvement over either the midpoint or trapezoid rules, which only give two extra digits of accuracy when n is increased by a factor of 10. Simpson's rule is so efficient that we get 9 digits correct with $n = 50$ in about 1 second on our calculator. Doubling n will decrease the error by a factor of about $2^4 = 16$ and hence will give the tenth digit. The total time is 2 seconds, which is pretty good.

In general, Simpson's rule achieves a reasonable degree of accuracy when using relatively small values of n, and is a good choice for an all-purpose method for estimating definite integrals.

Analytical View of the Trapezoid and Simpson's Rules

Our approach to approximating $\int_a^b f(x)\, dx$ numerically has been empirical: try a method, see how the error behaves, and then try to improve it. We can also develop the various rules for numerical integration by making better and better approximations to the integrand, f. The left, right, and midpoint rules are all examples of approximating f by a constant (flat) function on each subinterval. The trapezoid rule is obtained by approximating f by a linear function on each subinterval. Simpson's rule can, in the same spirit, be obtained by approximating f by quadratic functions. The details are given in Problems 11 and 12 on page 326.

How the Error Depends on the Integrand

Other factors besides the size of n affect the size of the error in each of the rules. Instead of looking at how the error behaves as we increase n, let's leave n fixed and imagine trying our approximation methods on different functions. We observe that the error in the left or right rule depends on how steeply the graph of f rises or falls. A steep curve makes the triangular regions missed by the left or right rectangles tall and hence large in area. This observation suggests that the error in the left or right rules depends on the size of the derivative of f (see Figure 7.15).

---

[3]Some books and computer programs use slightly different terminology for Simpson's rule; what we call $n = 50$, they call $n = 100$.

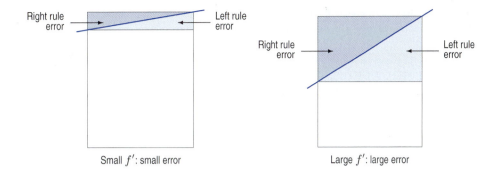

Figure 7.15: The error in the left and right rules depends on the steepness of the curve

From Figure 7.16 it appears that the errors in the trapezoid and midpoint rules depend on how much the curve is bent up or down. In other words, the concavity, and hence the size of the second derivative of f, has an effect on the errors of these two rules. Finally, it can be shown[4] that the error in Simpson's rule depends on the size of the *fourth* derivative of f, written $f^{(4)}$.

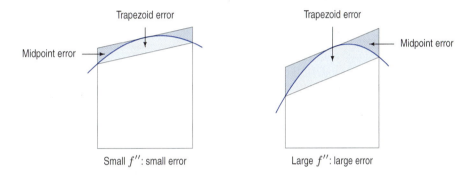

Figure 7.16: The error in the trapezoid and midpoint rules depends on how bent the curve is

Exercises and Problems for Section 7.6

Exercises

1. Estimate $\int_0^6 x^2\,dx$ using SIMP(2).

2. (a) Using the result of Problem 7 on page 320, compute SIMP(2) for $\int_0^4 (x^2 + 1)\,dx$.

   (b) Use the Fundamental Theorem of Calculus to find $\int_0^4 (x^2 + 1)\,dx$ exactly.

   (c) What is the error in SIMP(2) for this integral?

Problems

3. In this problem you will investigate the behavior of the errors in the approximation of the integral

$$\int_1^2 \frac{1}{x}\,dx \approx 0.6931471806\ldots.$$

   (a) For $n = 2, 4, 8, 16, 32, 64, 128$ subdivisions, find the left and right approximations and the errors in each.

   (b) What are the signs of the errors in the left and right approximations? How do the errors change if n is

   doubled?

   (c) For the values of n in part (a), compute the midpoint and trapezoid approximations and the errors in each.

   (d) What are the signs of the errors in the midpoint and trapezoid approximations? How do the errors change if n is doubled?

   (e) For $n = 2, 4, 8, 16, 32$, compute Simpson's rule approximation and the error in each. How do these errors change as n doubles?

[4]See Kendall E. Atkinson, *An Introduction to Numerical Analysis* (New York: John Wiley and Sons, 1978).

4. (a) What is the exact value of $\int_0^2 (x^3 + 3x^2)\, dx$?
   (b) Find $\text{SIMP}(n)$ for $n = 2, 4, 100$. What do you notice?

5. (a) What is the exact value of $\int_0^4 e^x\, dx$?
   (b) Find $\text{LEFT}(2)$, $\text{RIGHT}(2)$, $\text{TRAP}(2)$, $\text{MID}(2)$, and $\text{SIMP}(2)$. Compute the error for each.
   (c) Repeat part (b) with $n = 4$ (instead of $n = 2$).
   (d) For each rule in part (b), as n goes from $n = 2$ to $n = 4$, does the error go down approximately as you would expect? Explain.

6. The approximation to a definite integral using $n = 10$ is 2.346; the exact value is 4.0. If the approximation was found using each of the following rules, use the same rule to estimate the integral with $n = 30$.

   (a) LEFT   **(b)** TRAP   **(c)** SIMP

7. (a) A certain computer takes two seconds to compute a certain definite integral accurate to 4 digits to the right of the decimal point, using the left rectangle rule. How long (in years) will it take to get 8 digits correct using the left rectangle rule? How about 12 digits? 20 digits?
   (b) Repeat part(a) but this time assume that the trapezoidal rule is being used throughout.

8. For a certain definite integral, $\text{LEFT}(10) = 0.38745$ and $\text{LEFT}(20) = 0.36517$. Estimate the actual error for $\text{LEFT}(10)$ (and thus the actual value of the integral) by assuming that the error is reduced by a factor of 2 in going from $\text{LEFT}(10)$ to $\text{LEFT}(20)$.

9. For a certain definite integral, $\text{MID}(10) = 35.619$ and $\text{MID}(20) = 35.415$. Estimate the actual error for $\text{MID}(10)$ assuming that the error for $\text{MID}(10)$ is reduced by a factor of 4 in going to $\text{MID}(20)$.

10. For a certain definite integral, $\text{TRAP}(10) = 12.676$ and $\text{TRAP}(30) = 10.420$. Estimate the actual value of the integral.

Problems 11–12 show how Simpson's rule can be obtained by approximating the integrand, f, by quadratic functions.

11. Suppose that $a < b$ and that m is the midpoint $m = (a + b)/2$. Let $h = b - a$. The purpose of this problem is to show that if f is a quadratic function, then

$$\int_a^b f(x)\, dx = \frac{h}{3}\left(\frac{f(a)}{2} + 2f(m) + \frac{f(b)}{2}\right).$$

   (a) Show that this equation holds for the functions $f(x) = 1$, $f(x) = x$, and $f(x) = x^2$.
   (b) Use part (a) and the properties of the integral on page 248 to show that the equation holds for any quadratic function, $f(x) = Ax^2 + Bx + C$.

12. Consider the following method for approximating $\int_a^b f(x)\, dx$. Divide the interval $[a, b]$ into n equal subintervals. On each subinterval approximate f by a quadratic function that agrees with f at both endpoints and at the midpoint of the subinterval.

   (a) Explain why the integral of f on the subinterval $[x_i, x_{i+1}]$ is approximately equal to the expression

$$\frac{h}{3}\left(\frac{f(x_i)}{2} + 2f(m_i) + \frac{f(x_{i+1})}{2}\right),$$

   where m_i is the midpoint of the subinterval, $m_i = (x_i + x_{i+1})/2$. (See Problem 11.)
   (b) Show that if we add up these approximations for each subinterval, we get Simpson's rule:

$$\int_a^b f(x)\,dx \approx \frac{2 \cdot \text{MID}(n) + \text{TRAP}(n)}{3}.$$

7.7 IMPROPER INTEGRALS

Our original discussion of the definite integral $\int_a^b f(x)\, dx$ assumed that the interval $a \le x \le b$ was of finite length and that f was continuous. Integrals that arise in applications don't necessarily have these nice properties. In this section we investigate a class of integrals, called *improper* integrals, in which one limit of integration is infinite or the integrand is unbounded. As an example, to estimate the mass of the earth's atmosphere, we might calculate an integral which sums the mass of the air up to different heights. In order to represent the fact that the atmosphere doesn't end at a specific height, we let the upper limit of integration get larger and larger, or tend to infinity.

We will usually consider only improper integrals with positive integrands since they are the most common.

One Type of Improper Integral: When the Limit of Integration Is Infinite

Here is an example of an improper integral:

$$\int_1^\infty \frac{1}{x^2}\, dx.$$

To evaluate this integral, we first compute the definite integral $\int_1^b (1/x^2)\, dx$:

$$\int_1^b \frac{1}{x^2}\, dx = -x^{-1}\Big|_1^b = -\frac{1}{b} + \frac{1}{1}.$$

Now take the limit as $b \to \infty$. Since

$$\lim_{b \to \infty} \int_1^b \frac{1}{x^2}\, dx = \lim_{b \to \infty}\left(-\frac{1}{b} + 1\right) = 1,$$

we say that the improper integral $\int_1^\infty (1/x^2)\, dx$ *converges* to 1.

If we think in terms of areas, the integral $\int_1^\infty (1/x^2)\, dx$ represents the area under $f(x) = 1/x^2$ from $x = 1$ extending infinitely far to the right. (See Figure 7.17(a).) It may seem strange that this region has finite area. What our limit computations are saying is that

$$\text{When } b = 10: \quad \int_1^{10} \frac{1}{x^2}\, dx = -\frac{1}{x}\Big|_1^{10} = -\frac{1}{10} + 1 = 0.9$$

$$\text{When } b = 100: \quad \int_1^{100} \frac{1}{x^2}\, dx = -\frac{1}{100} + 1 = 0.99$$

$$\text{When } b = 1000: \quad \int_1^{1000} \frac{1}{x^2}\, dx = -\frac{1}{1000} + 1 = 0.999$$

and so on. In other words, as b gets larger and larger, the area between $x = 1$ and $x = b$ tends to 1. See Figure 7.17(b). Thus, it does make sense to declare that $\int_1^\infty (1/x^2)\, dx = 1$.

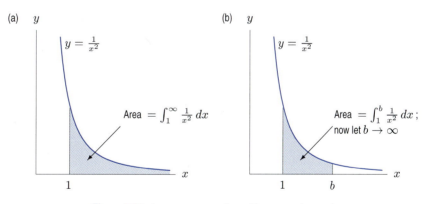

Figure 7.17: Area representation of improper integral

Of course, in another example, we might not get a finite limit as b gets larger and larger. In that case we say the improper integral *diverges*.

Suppose $f(x)$ is positive for $x \geq a$.

If $\displaystyle\lim_{b \to \infty} \int_a^b f(x)\, dx$ is a finite number, we say that $\displaystyle\int_a^\infty f(x)\, dx$ **converges** and define

$$\int_a^\infty f(x)\, dx = \lim_{b \to \infty} \int_a^b f(x)\, dx.$$

Otherwise, we say that $\displaystyle\int_a^\infty f(x)\, dx$ **diverges**. We define $\displaystyle\int_{-\infty}^b f(x)\, dx$ similarly.

Example 1   Does the improper integral $\int_1^\infty \frac{1}{\sqrt{x}}\,dx$ converge or diverge?

Solution   We consider

$$\int_1^b \frac{1}{\sqrt{x}}\,dx = \int_1^b x^{-1/2}\,dx = 2x^{1/2}\Big|_1^b = 2b^{1/2} - 2.$$

We see that $\int_1^b (1/\sqrt{x})\,dx$ grows without bound as $b \to \infty$. Thus we say the integral $\int_1^\infty (1/\sqrt{x})\,dx$ *diverges*. We have shown that the area under the curve in Figure 7.18 is not finite.

Notice that $f(x) \to 0$ as $x \to \infty$ does not guarantee convergence of $\int_a^\infty f(x)\,dx$.

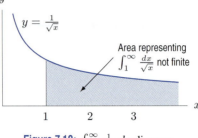

Figure 7.18: $\int_1^\infty \frac{1}{\sqrt{x}}\,dx$ diverges

What is the difference between the functions $1/x^2$ and $1/\sqrt{x}$ that makes the area under the graph of $1/x^2$ approach 1 as $x \to \infty$, whereas the area under $1/\sqrt{x}$ grows very large? Both functions approach 0 as x grows, so as b grows larger, smaller bits of area are being added to the definite integral. The difference between the functions is subtle: the values of the function $1/\sqrt{x}$ don't *shrink fast enough* for the integral to have a finite value. Of the two functions, $1/x^2$ drops to 0 much faster than $1/\sqrt{x}$, and this feature keeps the area under $1/x^2$ from growing beyond 1.

Example 2   Find $\int_0^\infty e^{-5x}\,dx$.

Solution   First we consider $\int_0^b e^{-5x}\,dx$:

$$\int_0^b e^{-5x}\,dx = -\frac{1}{5}e^{-5x}\Big|_0^b = -\frac{1}{5}e^{-5b} + \frac{1}{5}.$$

Since $e^{-5b} = \frac{1}{e^{5b}}$, this term tends to 0 as b approaches infinity, so $\int_0^\infty e^{-5x}\,dx$ converges. Its value is

$$\int_0^\infty e^{-5x}\,dx = \lim_{b\to\infty} \int_0^b e^{-5x}\,dx = \lim_{b\to\infty}\left(-\frac{1}{5}e^{-5b} + \frac{1}{5}\right) = 0 + \frac{1}{5} = \frac{1}{5}.$$

Since e^{5x} grows very rapidly, we expect that e^{-5x} will approach 0 rapidly. The fact that the area approaches $1/5$ instead of growing without bound is a consequence of the speed with which the integrand e^{-5x} approaches 0.

Example 3   Determine for which values of the exponent, p, the improper integral $\int_1^\infty \frac{1}{x^p}\,dx$ diverges.

Solution   For $p \neq 1$,

$$\int_1^b x^{-p}\,dx = \frac{1}{-p+1}x^{-p+1}\Big|_1^b = \left(\frac{1}{-p+1}b^{-p+1} - \frac{1}{-p+1}\right).$$

The important question is whether the exponent of b is positive or negative. If it is negative, then as b approaches infinity, b^{-p+1} approaches 0. If the exponent is positive, then b^{-p+1} grows without bound as b approaches infinity.

What happens if $p = 1$? In this case we get

$$\int_1^\infty \frac{1}{x}\, dx = \lim_{b\to\infty} \ln x \Big|_1^b = \lim_{b\to\infty} \ln b - \ln 1.$$

Since $\ln b$ becomes arbitrarily large as b approaches infinity, the integral grows without bound. We conclude that $\int_1^\infty (1/x^p)\, dx$ diverges precisely when $p \leq 1$. For $p > 1$ the integral has the value

$$\int_1^\infty \frac{1}{x^p}\, dx = \lim_{b\to\infty} \int_1^b \frac{1}{x^p}\, dx = \lim_{b\to\infty} \left(\frac{1}{-p+1} b^{-p+1} - \frac{1}{-p+1} \right) = -\left(\frac{1}{-p+1} \right) = \frac{1}{p-1}.$$

Application of Improper Integrals to Energy

The energy, E, required to separate two charged particles, originally a distance a apart, to a distance b, is given by the integral

$$E = \int_a^b \frac{kq_1 q_2}{r^2}\, dr$$

where q_1 and q_2 are the magnitudes of the charges and k is a constant. If q_1 and q_2 are in coulombs, a and b are in meters, and E is in joules, the value of the constant k is $9 \cdot 10^9$.

Example 4

A hydrogen atom consists of a proton and an electron, with opposite charges of magnitude $1.6 \cdot 10^{-19}$ coulombs. Find the energy required to take a hydrogen atom apart (that is, to move the electron from its orbit to an infinite distance from the proton). Assume that the initial distance between the electron and the proton is the Bohr radius, $R_B = 5.3 \cdot 10^{-11}$ meter.

Solution

Since we are moving from an initial distance of R_B to a final distance of ∞, the energy is represented by the improper integral

$$E = \int_{R_B}^\infty k\frac{q_1 q_2}{r^2}\, dr = kq_1 q_2 \lim_{b\to\infty} \int_{R_B}^b \frac{1}{r^2}\, dr$$

$$= kq_1 q_2 \lim_{b\to\infty} -\frac{1}{r} \Big|_{R_B}^b = kq_1 q_2 \lim_{b\to\infty} \left(-\frac{1}{b} + \frac{1}{R_B} \right) = \frac{kq_1 q_2}{R_B}.$$

Substituting numerical values, we get

$$E = \frac{(9 \cdot 10^9)(1.6 \cdot 10^{-19})^2}{5.3 \cdot 10^{-11}} \approx 4.35 \cdot 10^{-18} \text{ joules.}$$

This is about the amount of energy needed to lift a speck of dust 0.000000025 inch off the ground. (In other words, not much!)

What happens if the limits of integration are $-\infty$ and ∞? In this case, we break the integral at any point and write the original integral as a sum of two new improper integrals.

We can use any (finite) number c to define

$$\int_{-\infty}^\infty f(x)\, dx = \int_{-\infty}^c f(x)\, dx + \int_c^\infty f(x)\, dx.$$

If *either* of the two new improper integrals diverges, we say the original integral diverges. Only if both of the new integrals have a finite value do we add the values to get a finite value for the original integral.

It is not hard to show that the preceding definition does not depend on the choice for c.

Another Type of Improper Integral: When the Integrand Becomes Infinite

There is another way for an integral to be improper. The interval may be finite but the function may be unbounded near some points in the interval. For example, consider $\int_0^1 (1/\sqrt{x})\, dx$. Since the graph of $y = 1/\sqrt{x}$ has a vertical asymptote at $x = 0$, the region between the graph, the x-axis, and the lines $x = 0$ and $x = 1$ is unbounded. Instead of extending to infinity in the horizontal direction as in the previous improper integrals, this region extends to infinity in the vertical direction. See Figure 7.19(a). We handle this improper integral in a similar way as before: we compute $\int_a^1 (1/\sqrt{x})\, dx$ for values of a slightly larger than 0 and look at what happens as a approaches 0 from the positive side. (This is written as $a \to 0^+$.)

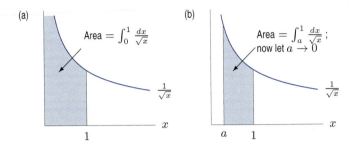

Figure 7.19: Area representation of improper integral

First we compute the integral:

$$\int_a^1 \frac{1}{\sqrt{x}}\, dx = 2x^{1/2} \Big|_a^1 = 2 - 2a^{1/2}.$$

Now we take the limit:

$$\lim_{a \to 0^+} \int_a^1 \frac{1}{\sqrt{x}}\, dx = \lim_{a \to 0^+} (2 - 2a^{1/2}) = 2.$$

Since the limit is finite, we say the improper integral converges, and that

$$\int_0^1 \frac{1}{\sqrt{x}}\, dx = 2.$$

Geometrically, what we have done is to calculate the finite area between $x = a$ and $x = 1$ and take the limit as a tends to 0 from the right. See Figure 7.19(b). Since the limit exists, the integral converges to 2. If the limit did not exist, we would say the improper integral diverges.

Example 5   Investigate the convergence of $\displaystyle\int_0^2 \frac{1}{(x-2)^2}\, dx$.

Solution   This is an improper integral since the integrand tends to infinity as x approaches 2, and is undefined at $x = 2$. Since the trouble is at the right endpoint, we replace the upper limit by b, and let b tend to 2 from the left. This is written $b \to 2^-$, with the "$-$" signifying that 2 is approached from below. See Figure 7.20.

$$\int_0^2 \frac{1}{(x-2)^2}\, dx = \lim_{b \to 2^-} \int_0^b \frac{1}{(x-2)^2}\, dx = \lim_{b \to 2^-} (-1)(x-2)^{-1} \Big|_0^b = \lim_{b \to 2^-} \left(-\frac{1}{(b-2)} - \frac{1}{2} \right).$$

Therefore, since $\displaystyle\lim_{b \to 2^-} \left(-\frac{1}{b-2} \right)$ does not exist, the integral diverges.

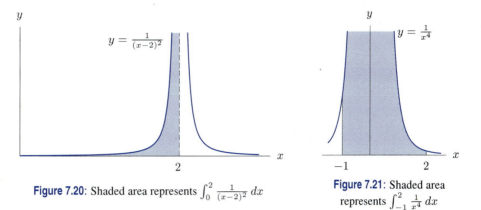

Figure 7.20: Shaded area represents $\int_0^2 \frac{1}{(x-2)^2}\, dx$

Figure 7.21: Shaded area represents $\int_{-1}^2 \frac{1}{x^4}\, dx$

Suppose $f(x)$ is positive and continuous on $a \le x < b$ and tends to infinity as $x \to b$. If $\displaystyle\lim_{c \to b^-} \int_a^c f(x)\, dx$ is a finite number, we say that $\displaystyle\int_a^b f(x)\, dx$ **converges** and define

$$\int_a^b f(x)\, dx = \lim_{c \to b^-} \int_a^c f(x)\, dx.$$

Otherwise, we say that $\displaystyle\int_a^b f(x)\, dx$ **diverges**.

When $f(x)$ tends to infinity as x approaches a, we define convergence in a similar way. In addition, an integral can be improper because the integrand tends to infinity *inside* the interval of integration rather than at an endpoint. In this case, we break the given integral into two (or more) improper integrals so that the integrand tends to infinity only at endpoints.

Suppose that $f(x)$ is positive and continuous on $[a, b]$ except at the point c. If $f(x)$ tends to infinity as $x \to c$, then we define

$$\int_a^b f(x)\, dx = \int_a^c f(x)\, dx + \int_c^b f(x)\, dx.$$

If *either* of the two new improper integrals diverges, we say the original integral diverges. Only if *both* of the new integrals have a finite value do we add the values to get a finite value for the original integral.

Example 6   Investigate the convergence of $\displaystyle\int_{-1}^2 \frac{1}{x^4}\, dx$.

Solution   See the graph in Figure 7.21. The trouble spot is $x = 0$, rather than $x = -1$ or $x = 2$. To handle this situation, we break the given improper integral into two other improper integrals each of which have $x = 0$ as one of the endpoints:

$$\int_{-1}^2 \frac{1}{x^4}\, dx = \int_{-1}^0 \frac{1}{x^4}\, dx + \int_0^2 \frac{1}{x^4}\, dx.$$

We can now use the previous technique to evaluate the new integrals, if they converge. Since

$$\int_0^2 \frac{1}{x^4}\, dx = \lim_{a \to 0^+} -\frac{1}{3} x^{-3} \Big|_a^2 = \lim_{a \to 0^+} \left(-\frac{1}{3}\right)\left(\frac{1}{8} - \frac{1}{a^3}\right)$$

the integral $\int_0^2 (1/x^4)\,dx$ diverges. Thus, the original integral diverges. A similar computation shows that $\int_{-1}^0 (1/x^4)\,dx$ also diverges.

It is easy to miss an improper integral when the integrand tends to infinity inside the interval. For example, it is fundamentally incorrect to say that $\int_{-1}^2 (1/x^4)\,dx = -\frac{1}{3}x^{-3}\Big|_{-1}^2 = -\frac{1}{24} - \frac{1}{3} = -\frac{3}{8}$.

Example 7   Find $\displaystyle\int_0^6 \frac{1}{(x-4)^{2/3}}\,dx$.

Solution   Figure 7.22 shows that the trouble spot is at $x = 4$, so we break the integral at $x = 4$ and consider the separate parts.

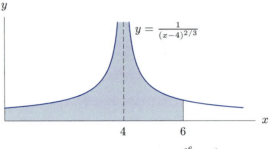

Figure 7.22: Shaded area represents $\int_0^6 \frac{1}{(x-4)^{2/3}}\,dx$

We have

$$\int_0^4 \frac{1}{(x-4)^{2/3}}\,dx = \lim_{b \to 4^-} 3(x-4)^{1/3}\Big|_0^b = \lim_{b \to 4^-} \left(3(b-4)^{1/3} - 3(-4)^{1/3}\right) = 3(4)^{1/3}.$$

Similarly,

$$\int_4^6 \frac{1}{(x-4)^{2/3}}\,dx = \lim_{a \to 4^+} 3(x-4)^{1/3}\Big|_a^6 = \lim_{a \to 4^+} \left(3 \cdot 2^{1/3} - 3(a-4)^{1/3}\right) = 3(2)^{1/3}.$$

Since both of these integrals converge, the original integral converges:

$$\int_0^6 \frac{1}{(x-4)^{2/3}}\,dx = 3(4)^{1/3} + 3(2)^{1/3} \approx 8.54.$$

Finally, there is a question of what to do when an integral is improper at both endpoints. In this case, we just break the integral at any interior point of the interval. The original integral diverges if either or both of the new integrals diverge.

Example 8   Investigate the convergence of $\displaystyle\int_0^\infty \frac{1}{x^2}\,dx$.

Solution   This integral is improper both because the upper limit is ∞ and because the function is undefined at $x = 0$. We break the integral into two parts at, say, $x = 1$. We know by Example 3 that $\int_1^\infty (1/x^2)\,dx$ has a finite value. However, the other part, $\int_0^1 (1/x^2)\,dx$, diverges since:

$$\int_0^1 \frac{1}{x^2}\,dx = \lim_{a \to 0^+} -x^{-1}\Big|_a^1 = \lim_{a \to 0^+} \left(\frac{1}{a} - 1\right).$$

Therefore $\displaystyle\int_0^\infty \frac{1}{x^2}\,dx$ diverges as well.

Exercises and Problems for Section 7.7

Exercises

1. Shade the area represented by:

(a) $\int_1^\infty (1/x^2)\, dx$ (b) $\int_0^1 (1/\sqrt{x})\, dx$

2. Evaluate the improper integral $\int_0^\infty e^{-0.4x} dx$ and sketch the area it represents.

3. (a) Use a calculator or computer to estimate $\int_0^b xe^{-x} dx$ for $b = 5, 10, 20$.

(b) Use your answers to part (a) to estimate the value of $\int_0^\infty xe^{-x} dx$, assuming it is finite.

4. (a) Sketch the the area represented by the improper integral $\int_{-\infty}^\infty e^{-x^2} dx$.

(b) Use a calculator or computer to estimate $\int_{-a}^a e^{-x^2} dx$ for $a = 1,2,3,4,5$.

(c) Use the answers to part (b) to estimate the value of $\int_{-\infty}^\infty e^{-x^2} dx$, assuming it is finite.

Calculate the integrals in Exercises 5–32, if they converge.

5. $\int_1^\infty \dfrac{1}{5x+2}\, dx$

6. $\int_1^\infty \dfrac{1}{(x+2)^2}\, dx$

7. $\int_0^\infty xe^{-x^2}\, dx$

8. $\int_1^\infty e^{-2x}\, dx$

9. $\int_0^\infty \dfrac{x}{e^x}\, dx$

10. $\int_1^\infty \dfrac{x}{4+x^2}\, dx$

11. $\int_{-\infty}^0 \dfrac{e^x}{1+e^x}\, dx$

12. $\int_{-\infty}^\infty \dfrac{dz}{z^2+25}$

13. $\int_0^4 \dfrac{dx}{\sqrt{16-x^2}}$

14. $\int_{\pi/4}^{\pi/2} \dfrac{\sin x}{\sqrt{\cos x}}\, dx$

15. $\int_{-1}^1 \dfrac{1}{v}\, dv$

16. $\int_0^1 \dfrac{x^4+1}{x}\, dx$

17. $\int_1^\infty \dfrac{1}{x^2+1}\, dx$

18. $\int_1^\infty \dfrac{1}{\sqrt{x^2+1}}\, dx$

19. $\int_0^4 \dfrac{1}{u^2-16}\, du$

20. $\int_1^\infty \dfrac{y}{y^4+1}\, dy$

21. $\int_2^\infty \dfrac{dx}{x\ln x}$

22. $\int_0^1 \dfrac{\ln x}{x}\, dx$

23. $\int_{16}^{20} \dfrac{1}{y^2-16}\, dy$

24. $\int_1^2 \dfrac{dx}{x\ln x}$

25. $\int_0^\pi \dfrac{1}{\sqrt{x}} e^{-\sqrt{x}}\, dx$

26. $\int_3^\infty \dfrac{dx}{x(\ln x)^2}$

27. $\int_0^2 \dfrac{1}{\sqrt{4-x^2}}\, dx$

28. $\int_4^\infty \dfrac{dx}{(x-1)^2}$

29. $\int_4^\infty \dfrac{dx}{x^2-1}$

30. $\int_7^\infty \dfrac{dy}{\sqrt{y-5}}$

31. $\int_{-3}^3 \dfrac{y\, dy}{\sqrt{9-y^2}}$

32. $\int_3^6 \dfrac{d\theta}{(4-\theta)^2}$

Problems

33. Find the area under the curve $y = xe^{-x}$ for $x \geq 0$.

34. Find the area under the curve $y = 1/\cos^2 t$ between $t = 0$ and $t = \pi/2$.

35. Given that $\int_{-\infty}^\infty e^{-x^2}\, dx = \sqrt{\pi}$, calculate the exact value of

$$\int_{-\infty}^\infty e^{-(x-a)^2/b}\, dx.$$

For what values of p do the integrals in Problems 36–37 converge or diverge? What is the value of the integral when it converges?

36. $\int_e^\infty x^p \ln x\, dx$

37. $\int_0^e x^p \ln x\, dx$

38. The gamma function is defined for all $x > 0$ by the rule

$$\Gamma(x) = \int_0^\infty t^{x-1} e^{-t}\, dt.$$

(a) Find $\Gamma(1)$ and $\Gamma(2)$.

(b) Integrate by parts with respect to t to show that, for positive n,

$$\Gamma(n+1) = n\Gamma(n).$$

(c) Find a simple expression for $\Gamma(n)$ for positive integers n.

39. The rate, r, at which people get sick during an epidemic of the flu can be approximated by $r = 1000te^{-0.5t}$, where r is measured in people/day and t is measured in days since the start of the epidemic.

(a) Sketch a graph of r as a function of t.

(b) When are people getting sick fastest?

(c) How many people get sick altogether?

40. Find the energy required to separate opposite electric charges of magnitude 1 coulomb. The charges are initially 1 meter apart and one is moved infinitely far from the other. (The definition of energy is on page 329.)

7.8 COMPARISON OF IMPROPER INTEGRALS

Making Comparisons

Sometimes it is difficult to find the exact value of an improper integral by antidifferentiation, but it may be possible to determine whether an integral converges or diverges. The key is to *compare* the given integral to one whose behavior we already know. Let's look at an example.

Example 1 Determine whether $\int_1^\infty \dfrac{1}{\sqrt{x^3 + 5}}\, dx$ converges.

Solution First, let's see what this integrand does as $x \to \infty$. For large x, the 5 becomes insignificant compared with the x^3, so
$$\frac{1}{\sqrt{x^3 + 5}} \approx \frac{1}{\sqrt{x^3}} = \frac{1}{x^{3/2}}.$$

Since
$$\int_1^\infty \frac{1}{\sqrt{x^3}}\, dx = \int_1^\infty \frac{1}{x^{3/2}}\, dx = \lim_{b \to \infty} \int_1^b \frac{1}{x^{3/2}}\, dx = \lim_{b \to \infty} -2x^{-1/2}\Big|_1^b = \lim_{b \to \infty} \left(2 - 2b^{-1/2}\right) = 2,$$

the integral $\int_1^\infty (1/x^{3/2})\, dx$ converges. So we expect our integral to converge as well. In order to confirm this, we observe that for $0 \le x^3 \le x^3 + 5$, we have
$$\frac{1}{\sqrt{x^3 + 5}} \le \frac{1}{\sqrt{x^3}}.$$

and so for $b \ge 1$,
$$\int_1^b \frac{1}{\sqrt{x^3 + 5}}\, dx \le \int_1^b \frac{1}{\sqrt{x^3}}\, dx.$$

(See Figure 7.23.) Since $\int_1^b (1/\sqrt{x^3 + 5})\, dx$ increases as b approaches infinity but is always smaller than $\int_1^b (1/x^{3/2})\, dx < \int_1^\infty (1/x^{3/2})\, dx = 2$, we know $\int_1^\infty (1/\sqrt{x^3 + 5})\, dx$ must have a finite value less than 2. Thus,
$$\int_1^\infty \frac{dx}{\sqrt{x^3 + 5}} \quad \text{converges to a value less than 2.}$$

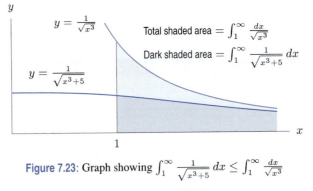

Figure 7.23: Graph showing $\int_1^\infty \dfrac{1}{\sqrt{x^3+5}}\, dx \le \int_1^\infty \dfrac{dx}{\sqrt{x^3}}$

A little more work is required to estimate the value of a convergent improper integral.

Example 2 Estimate the value of $\int_1^\infty \dfrac{1}{\sqrt{x^3 + 5}}\, dx$ with an error of less than 0.01 using the approximation
$$\int_1^\infty \frac{1}{\sqrt{x^3 + 5}}\, dx \approx \int_1^b \frac{1}{\sqrt{x^3 + 5}}\, dx.$$

Solution   We must figure out how large a value of b to take. Since

$$\int_1^\infty \frac{1}{\sqrt{x^3+5}}\,dx = \int_1^b \frac{1}{\sqrt{x^3+5}}\,dx + \int_b^\infty \frac{1}{\sqrt{x^3+5}}\,dx,$$

we find b such that the tail of the integral satisfies the inequality

$$\left| \int_b^\infty \frac{1}{\sqrt{x^3+5}}\,dx \right| < 0.01.$$

From the solution of Example 1, we have

$$0 < \int_b^\infty \frac{1}{\sqrt{x^3+5}}\,dx < \int_b^\infty \frac{1}{\sqrt{x^3}}\,dx = \frac{2}{\sqrt{b}}.$$

We choose b such that $2/\sqrt{b} < 0.01$, which means that $b > 40,000$. With an error of less than 0.01, we have

$$\int_1^\infty \frac{1}{\sqrt{x^3+5}}\,dx \approx \int_1^{50,000} \frac{1}{\sqrt{x^3+5}}\,dx = 1.699.$$

Notice that we first looked at the behavior of the integrand as $x \to \infty$. This is useful because the convergence or divergence of the integral is determined by what happens as $x \to \infty$.

The Comparison Test for $\displaystyle\int_a^\infty f(x)\,dx$

Assume $f(x)$ is positive. Making a comparison involves two stages:
1. Guess, by looking at the behavior of the integrand for large x, whether the integral converges or not. (This is the "behaves like" principle.)
2. Confirm the guess by comparison:
   - If $0 \le f(x) \le g(x)$ and $\int_a^\infty g(x)\,dx$ converges, then $\int_a^\infty f(x)\,dx$ converges.
   - If $0 \le g(x) \le f(x)$ and $\int_a^\infty g(x)\,dx$ diverges, then $\int_a^\infty f(x)\,dx$ diverges.

Example 3   Decide whether $\displaystyle\int_4^\infty \frac{dt}{(\ln t)-1}$ converges or diverges.

Solution   Since $\ln t$ grows without bound as $t \to \infty$, the -1 is eventually going to be insignificant in comparison to $\ln t$. Thus, as far as convergence is concerned,

$$\int_4^\infty \frac{1}{(\ln t)-1}\,dt \quad \text{behaves like} \quad \int_4^\infty \frac{1}{\ln t}\,dt.$$

Does $\int_4^\infty (1/\ln t)\,dt$ converge or diverge? Since $\ln t$ grows very slowly, $1/\ln t$ goes to zero very slowly, and so the integral probably doesn't converge. We know that $(\ln t) - 1 < \ln t < t$ for all positive t. So, provided $t > e$, we take reciprocals:

$$\frac{1}{(\ln t)-1} > \frac{1}{\ln t} > \frac{1}{t}.$$

Since $\int_4^\infty (1/t)\,dt$ diverges, we conclude that

$$\int_4^\infty \frac{1}{(\ln t)-1}\,dt \quad \text{diverges.}$$

How Do We Know What To Compare With?

In Examples 1 and 3, we investigated the convergence of an integral by comparing it with an easier integral. How did we pick the easier integral? This is a matter of trial and error, guided by any information we get by looking at the original integrand as $x \to \infty$. We want the comparison integrand to be easy and, in particular, to have a simple antiderivative.

Useful Integrals for Comparison

- $\displaystyle \int_1^\infty \frac{1}{x^p}\,dx$ converges for $p > 1$ and diverges for $p \leq 1$.

- $\displaystyle \int_0^1 \frac{1}{x^p}\,dx$ converges for $p < 1$ and diverges for $p \geq 1$.

- $\displaystyle \int_0^\infty e^{-ax}\,dx$ converges for $a > 0$.

Of course, we can use any function for comparison, provided we can determine its behavior.

Example 4   Investigate the convergence of $\displaystyle \int_1^\infty \frac{(\sin x) + 3}{\sqrt{x}}\,dx$.

Solution   Since it looks difficult to find an antiderivative of this function, we try comparison. What happens to this integrand as $x \to \infty$? Since $\sin x$ oscillates between -1 and 1,

$$\frac{2}{\sqrt{x}} = \frac{-1+3}{\sqrt{x}} \leq \frac{(\sin x) + 3}{\sqrt{x}} \leq \frac{1+3}{\sqrt{x}} = \frac{4}{\sqrt{x}},$$

the integrand oscillates between $2/\sqrt{x}$ and $4/\sqrt{x}$. (See Figure 7.24.)

What do $\int_1^\infty (2/\sqrt{x})\,dx$ and $\int_1^\infty (4/\sqrt{x})\,dx$ do? As far as convergence is concerned, they certainly do the same thing, and whatever that is, the original integral does it too. It is important to notice that $\sqrt{x}$ grows very slowly. This means that $1/\sqrt{x}$ gets small slowly, which means that convergence is unlikely. Since $\sqrt{x} = x^{1/2}$, the result in the preceding box (with $p = \frac{1}{2}$) tells us that $\int_1^\infty (1/\sqrt{x})\,dx$ diverges. So the comparison test tells us that the original integral diverges.

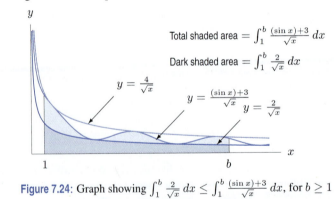

Figure 7.24: Graph showing $\int_1^b \frac{2}{\sqrt{x}}\,dx \leq \int_1^b \frac{(\sin x)+3}{\sqrt{x}}\,dx$, for $b \geq 1$

Notice that there are two possible comparisons we could have made in Example 4:

$$\frac{2}{\sqrt{x}} \leq \frac{(\sin x) + 3}{\sqrt{x}} \qquad \text{or} \qquad \frac{(\sin x) + 3}{\sqrt{x}} \leq \frac{4}{\sqrt{x}}.$$

Since both $\int_1^\infty (2/\sqrt{x})\,dx$ and $\int_1^\infty (4/\sqrt{x})\,dx$ diverge, only the first comparison is useful. Knowing that an integral is *smaller* than a divergent integral is of no help whatsoever!

The next example shows what to do if the comparison does not hold throughout the interval of integration.

Example 5   Show $\int_1^\infty e^{-x^2/2}\, dx$ converges to a finite value.

Solution   We know that $e^{-x^2/2}$ goes very rapidly to zero as $x \to \infty$, so we expect this integral to converge. Hence we look for some larger integrand which has a convergent integral. One possibility is $\int_1^\infty e^{-x}\, dx$, because e^{-x} has an elementary antiderivative and $\int_1^\infty e^{-x}\, dx$ converges. What is the relationship between $e^{-x^2/2}$ and e^{-x}? We know that for $x \geq 2$,

$$x \leq \frac{x^2}{2} \quad \text{so} \quad -\frac{x^2}{2} \leq -x,$$

and so, for $x \geq 2$

$$e^{-x^2/2} \leq e^{-x}.$$

Since this inequality holds only for $x \geq 2$, we split the interval of integration into two pieces:

$$\int_1^\infty e^{-x^2/2}\, dx = \int_1^2 e^{-x^2/2}\, dx + \int_2^\infty e^{-x^2/2}\, dx.$$

Now $\int_1^2 e^{-x^2/2}\, dx$ is finite (it is not improper) and $\int_2^\infty e^{-x^2/2}\, dx$ is finite by comparison with $\int_2^\infty e^{-x}\, dx$. Therefore, $\int_1^\infty e^{-x^2/2}\, dx$ is the sum of two finite pieces and therefore must be finite.

Exercises and Problems for Section 7.8

Exercises

In Exercises 1–9, use the box on page 336 and the behavior of rational and exponential functions as $x \to \infty$ to predict whether the integrals converge or diverge.

1. $\int_1^\infty \frac{x^2}{x^4 + 1}\, dx$

2. $\int_2^\infty \frac{x^3}{x^4 - 1}\, dx$

3. $\int_1^\infty \frac{x^2 + 1}{x^3 + 3x + 2}\, dx$

4. $\int_1^\infty \frac{1}{x^2 + 5x + 1}\, dx$

5. $\int_1^\infty \frac{x}{x^2 + 2x + 4}\, dx$

6. $\int_1^\infty \frac{x^2 - 6x + 1}{x^2 + 4}\, dx$

7. $\int_1^\infty \frac{5x + 2}{x^4 + 8x^2 + 4}\, dx$

8. $\int_1^\infty \frac{1}{e^{5t} + 2}\, dt$

9. $\int_1^\infty \frac{x^2 + 4}{x^4 + 3x^2 + 11}\, dx$

In Exercises 10–25 decide if the improper integral converges or diverges. Explain your reasoning.

10. $\int_{50}^\infty \frac{dz}{z^3}$

11. $\int_1^\infty \frac{dx}{1 + x}$

12. $\int_1^\infty \frac{dx}{x^3 + 1}$

13. $\int_5^8 \frac{6}{\sqrt{t - 5}}\, dt$

14. $\int_0^1 \frac{1}{x^{19/20}}\, dx$

15. $\int_{-1}^5 \frac{dt}{(t + 1)^2}$

16. $\int_{-\infty}^\infty \frac{du}{1 + u^2}$

17. $\int_1^\infty \frac{du}{u + u^2}$

18. $\int_1^\infty \frac{d\theta}{\sqrt{\theta^2 + 1}}$

19. $\int_2^\infty \frac{d\theta}{\sqrt{\theta^3 + 1}}$

20. $\int_0^1 \frac{d\theta}{\sqrt{\theta^3 + \theta}}$

21. $\int_0^\infty \frac{dy}{1 + e^y}$

22. $\int_1^\infty \frac{2 + \cos\phi}{\phi^2}\, d\phi$

23. $\int_0^\infty \frac{dz}{e^z + 2^z}$

24. $\int_0^\pi \frac{2 - \sin\phi}{\phi^2}\, d\phi$

25. $\int_4^\infty \frac{3 + \sin\alpha}{\alpha}\, d\alpha$

Estimate the values of the integrals in Exercises 26–27 to two decimal places by integrating the functions on your calculator or computer for large values of the upper limit of integration.

26. $\int_1^\infty e^{-x^2}\, dx$

27. $\int_0^\infty e^{-x^2} \cos^2 x\, dx$

Problems

28. The graphs of $y = 1/x, y = 1/x^2$ and the functions $f(x), g(x), h(x)$, and $k(x)$ are shown in Figure 7.25.

    (a) Is the area between $y = 1/x$ and $y = 1/x^2$ on the interval from $x = 1$ to ∞ finite or infinite? Explain.

    (b) Using the graph, decide whether the integral of each of the functions $f(x), g(x), h(x)$ and $k(x)$ on the interval from $x = 1$ to ∞ converges, diverges, or whether it is impossible to tell.

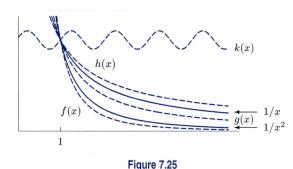

Figure 7.25

For what values of p do the integrals in Problems 29–30 converge or diverge?

29. $\displaystyle\int_2^\infty \frac{dx}{x(\ln x)^p}$
    30. $\displaystyle\int_1^2 \frac{dx}{x(\ln x)^p}$

31. Find the value of a (to three decimal places) that makes

$$\int_{-\infty}^\infty ae^{-x^2/2}\, dx = 1.$$

32. (a) The function $g(x) = ae^{-(x-k)^2/2}$ is used by statisticians. To three decimal places, what value of a should be chosen to ensure that

$$\int_{-\infty}^\infty g(x)\, dx = 1?$$

    (b) Is your answer the same as or different from your answer to Problem 31? Why?

33. (a) Find an upper bound for

$$\int_3^\infty e^{-x^2}\, dx.$$

    [Hint: $e^{-x^2} \le e^{-3x}$ for $x \ge 3$.]

    (b) For any positive n, generalize the result of part (a) to find an upper bound for

$$\int_n^\infty e^{-x^2}\, dx$$

    by noting that $nx \le x^2$ for $x \ge n$.

34. In Planck's Radiation Law, we encounter the integral

$$\int_1^\infty \frac{dx}{x^5(e^{1/x} - 1)}.$$

    (a) Explain why a graph of the tangent line to e^t at $t = 0$ tells us that for all t

$$1 + t \le e^t.$$

    (b) Substituting $t = 1/x$, show that for all x

$$e^{1/x} - 1 > \frac{1}{x}.$$

    (c) Use the comparison test to show that the original integral converges.

CHAPTER SUMMARY

- **Integration techniques**
  Substitution, parts, partial fractions, trigonometric substitution, using tables.
- **Numerical approximations**

  Riemann sums (left, right, midpoint), trapezoid rule, Simpson's rule, approximation errors.

- **Improper integrals**
  Convergence/divergence, comparison test for integrals.

REVIEW EXERCISES AND PROBLEMS FOR CHAPTER SEVEN

Exercises

For Exercises 1–114, evaluate the following integrals. Assume a, b, c, and k are constants. Evaluate the definite integrals using the Fundamental Theorem of Calculus, if possible, and check your answers numerically. Exercises 1–74 can be done without an integral table, as can some of the later problems.

1. $\displaystyle\int \sin t\, dt$
    2. $\displaystyle\int \cos 2t\, dt$

3. $\displaystyle\int e^{5z}\,dz$

4. $\displaystyle\int (3w+7)\,dw$

43. $\displaystyle\int \tan\theta\,d\theta$

44. $\displaystyle\int \sin(5\theta)\cos(5\theta)\,d\theta$

5. $\displaystyle\int \sin 2\theta\,d\theta$

6. $\displaystyle\int (x^3-1)^4 x^2\,dx$

45. $\displaystyle\int \frac{x}{x^2+1}\,dx$

46. $\displaystyle\int \frac{dz}{1+z^2}$

7. $\displaystyle\int \left(x^{3/2}+x^{2/3}\right)dx$

8. $\displaystyle\int (e^x+3^x)\,dx$

47. $\displaystyle\int \frac{dz}{1+4z^2}$

48. $\displaystyle\int \cos^3 2\theta \sin 2\theta\,d\theta$

9. $\displaystyle\int (r+1)^3\,dr$

10. $\displaystyle\int \left(\frac{4}{x^2}-\frac{3}{x^3}\right)dx$

49. $\displaystyle\int \sin 5\theta \cos^3 5\theta\,d\theta$

50. $\displaystyle\int \sin^3 z\cos^3 z\,dz$

11. $\displaystyle\int \frac{x^3+x+1}{x^2}dx$

12. $\displaystyle\int \frac{(1+\ln x)^2}{x}\,dx$

51. $\displaystyle\int t(t-10)^{10}\,dt$

52. $\displaystyle\int \cos\theta\sqrt{1+\sin\theta}\,d\theta$

13. $\displaystyle\int te^{t^2}\,dt$

14. $\displaystyle\int x\cos x\,dx$

53. $\displaystyle\int xe^x\,dx$

54. $\displaystyle\int t^3 e^t\,dt$

15. $\displaystyle\int x^2 e^{2x}\,dx$

16. $\displaystyle\int x\sqrt{1-x}\,dx$

55. $\displaystyle\int_1^3 x(x^2+1)^{70}\,dx$

56. $\displaystyle\int (3z+5)^3\,dz$

17. $\displaystyle\int x\ln x\,dx$

18. $\displaystyle\int y\sin y\,dy$

57. $\displaystyle\int \frac{du}{9+u^2}$

58. $\displaystyle\int \frac{\cos w}{1+\sin^2 w}\,dw$

19. $\displaystyle\int (\ln x)^2\,dx$

20. $\displaystyle\int \ln(x^2)\,dx$

59. $\displaystyle\int \frac{1}{x}\tan(\ln x)\,dx$

60. $\displaystyle\int \frac{1}{x}\sin(\ln x)\,dx$

21. $\displaystyle\int e^{0.5-0.3t}\,dt$

22. $\displaystyle\int \sin^2\theta\cos\theta\,d\theta$

61. $\displaystyle\int \frac{dx}{\sqrt{1-4x^2}}$

62. $\displaystyle\int \frac{w\,dw}{\sqrt{16-w^2}}$

23. $\displaystyle\int x\sqrt{4-x^2}\,dx$

24. $\displaystyle\int \frac{(u+1)^3}{u^2}\,du$

63. $\displaystyle\int \frac{e^{2y}+1}{e^{2y}}\,dy$

64. $\displaystyle\int \frac{\sin w\,dw}{\sqrt{1-\cos w}}$

25. $\displaystyle\int \frac{\cos\sqrt{y}}{\sqrt{y}}\,dy$

26. $\displaystyle\int \frac{1}{\cos^2 z}\,dz$

65. $\displaystyle\int \frac{dx}{x\ln x}$

66. $\displaystyle\int \frac{du}{3u+8}$

27. $\displaystyle\int \cos^2\theta\,d\theta$

28. $\displaystyle\int t^{10}(t-10)\,dt$

67. $\displaystyle\int \frac{x\cos\sqrt{x^2+1}}{\sqrt{x^2+1}}\,dx$

68. $\displaystyle\int \frac{t^3}{\sqrt{1+t^2}}\,dt$

29. $\displaystyle\int \tan(2x-6)\,dx$

30. $\displaystyle\int_1^3 \ln(x^3)\,dx$

69. $\displaystyle\int ue^{ku}\,du$

70. $\displaystyle\int (w+5)^4 w\,dw$

31. $\displaystyle\int_1^e (\ln x)^2\,dx$

32. $\displaystyle\int_{-\pi}^{\pi} e^{2x}\sin 2x\,dx$

71. $\displaystyle\int e^{\sqrt{2x+3}}\,dx$

72. $\displaystyle\int r(\ln r)^2\,dr$

33. $\displaystyle\int_0^{10} ze^{-z}\,dz$

34. $\displaystyle\int_{-\pi/3}^{\pi/4} \sin^3\theta\cos\theta\,d\theta$

73. $\displaystyle\int (e^x+x)^2\,dx$

74. $\displaystyle\int u^2\ln u\,du$

35. $\displaystyle\int_{-\pi/4}^{\pi/4} x^3\cos x^2\,dx$

36. $\displaystyle\int_1^4 \frac{e^{\sqrt{x}}}{\sqrt{x}}\,dx$

75. $\displaystyle\int \frac{5x+6}{x^2+4}\,dx$

76. $\displaystyle\int \frac{1}{\sin^3(2x)}\,dx$

37. $\displaystyle\int_0^1 \frac{dx}{x^2+1}$

38. $\displaystyle\int \frac{(\ln x)^2}{x}\,dx$

77. $\displaystyle\int \frac{dr}{r^2-100}$

78. $\displaystyle\int y^2\sin(cy)\,dy$

39. $\displaystyle\int \frac{(t+2)^2}{t^3}\,dt$

40. $\displaystyle\int \left(x^2+2x+\frac{1}{x}\right)dx$

79. $\displaystyle\int e^{-ct}\sin kt\,dt$

80. $\displaystyle\int e^{5x}\cos(3x)\,dx$

41. $\displaystyle\int \frac{t+1}{t^2}\,dt$

42. $\displaystyle\int te^{t^2+1}\,dt$

81. $\displaystyle\int \left(x^{\sqrt{k}}+\sqrt{k}^{\,x}\right)dx$

82. $\displaystyle\int \sqrt{3+12x^2}\,dx.$

83. $\displaystyle\int \frac{dx}{x^2 + 5x + 4}$

84. $\displaystyle\int \frac{1}{\sqrt{x^2 - 3x + 2}}\, dx$

85. $\displaystyle\int \frac{x^3}{x^2 + 3x + 2}\, dx$

86. $\displaystyle\int \frac{x^2 + 1}{x^2 - 3x + 2}\, dx$

87. $\displaystyle\int \frac{dx}{ax^2 + bx}$

88. $\displaystyle\int \frac{ax + b}{ax^2 + 2bx + c}\, dx$

89. $\displaystyle\int \frac{dz}{z^2 + z}$

90. $\displaystyle\int \left(\frac{x}{3} + \frac{3}{x}\right)^2 dx$

91. $\displaystyle\int \frac{2^t}{2^t + 1}\, dt$

92. $\displaystyle\int 10^{1-x}\, dx$

93. $\displaystyle\int (x^2 + 5)^3\, dx$

94. $\displaystyle\int v \arcsin v\, dv$

95. $\displaystyle\int \sqrt{4 - x^2}\, dx$

96. $\displaystyle\int \frac{z^3}{z - 5}\, dz$

97. $\displaystyle\int \frac{\sin w \cos w}{1 + \cos^2 w}\, dw$

98. $\displaystyle\int \frac{1}{\tan(3\theta)}\, d\theta$

99. $\displaystyle\int \frac{x}{\cos^2 x}\, dx$

100. $\displaystyle\int \frac{x + 1}{\sqrt{x}}\, dx$

101. $\displaystyle\int \frac{x}{\sqrt{x + 1}}\, dx$

102. $\displaystyle\int \frac{\sqrt{\sqrt{x} + 1}}{\sqrt{x}}\, dx$

103. $\displaystyle\int \frac{e^{2y}}{e^{2y} + 1}\, dy$

104. $\displaystyle\int \frac{z}{(z^2 - 5)^3}\, dz$

105. $\displaystyle\int \frac{z}{(z - 5)^3}\, dz$

106. $\displaystyle\int \frac{(1 + \tan x)^3}{\cos^2 x}\, dx$

107. $\displaystyle\int \frac{(2x - 1)e^{x^2}}{e^x}\, dx$

108. $\displaystyle\int (2x + 1)e^{x^2} e^x\, dx$

109. $\displaystyle\int \sin x (\sqrt{2 + 3\cos x})\, dx$

110. $\displaystyle\int (x^2 - 3x + 2)e^{-4x}\, dx$

111. $\displaystyle\int \sin^2(2\theta)\cos^3(2\theta)\, d\theta$

112. $\displaystyle\int \cos(2\sin x)\cos x\, dx$

113. $\displaystyle\int (x + \sin x)^3 (1 + \cos x)\, dx$

114. $\displaystyle\int (2x^3 + 3x + 4)\cos(2x)\, dx$

For a, b positive constants, evaluate the integrals in Exercises 115–126 using partial fractions or a trigonometric substitution.

115. $\displaystyle\int \frac{1}{(x - 2)(x + 2)}\, dx$

116. $\displaystyle\int \frac{1}{\sqrt{25 - x^2}}\, dx,$

117. $\displaystyle\int \frac{1}{x(x + 5)}\, dx$

118. $\displaystyle\int \frac{dP}{3P - 3P^2}$

119. $\displaystyle\int \frac{1}{\sqrt{1 - 9x^2}}\, dx$

120. $\displaystyle\int \frac{2x + 3}{x(x + 2)(x - 1)}\, dx$

121. $\displaystyle\int \frac{3x + 1}{x(x^2 - 1)}\, dx$

122. $\displaystyle\int \frac{1 + x^2}{x(1 + x)^2}\, dx$

123. $\displaystyle\int \frac{1}{x^2 + 2x + 2}\, dx$

124. $\displaystyle\int \frac{1}{x^2 + 4x + 5}\, dx$

125. $\displaystyle\int \frac{3x + 1}{x^2 - 3x + 2}\, dx$

126. $\displaystyle\int \frac{1}{\sqrt{a^2 - (bx)^2}}\, dx$

For Exercises 127–140 decide if the integral converges or diverges. If the integral converges, find its value or give a bound on its value.

127. $\displaystyle\int_4^\infty \frac{dt}{t^{3/2}}$

128. $\displaystyle\int_{10}^\infty \frac{dx}{x \ln x}$

129. $\displaystyle\int_0^\infty we^{-w}\, dw$

130. $\displaystyle\int_{-1}^1 \frac{1}{x^4}\, dx$

131. $\displaystyle\int_{-\pi/4}^{\pi/4} \tan\theta\, d\theta$

132. $\displaystyle\int_2^\infty \frac{1}{4 + z^2}\, dz$

133. $\displaystyle\int_{10}^\infty \frac{1}{z^2 - 4}\, dz$

134. $\displaystyle\int_{-5}^{10} \frac{dt}{\sqrt{t + 5}}$

135. $\displaystyle\int_0^{\pi/2} \frac{1}{\sin\phi}\, d\phi$

136. $\displaystyle\int_0^{\pi/4} \tan 2\theta\, d\theta$

137. $\displaystyle\int_1^\infty \frac{x}{x + 1}\, dx$

138. $\displaystyle\int_0^\infty \frac{\sin^2\theta}{\theta^2 + 1}\, d\theta$

139. $\displaystyle\int_0^\pi \tan^2\theta\, d\theta$

140. $\displaystyle\int_0^1 (\sin x)^{-3/2}\, dx$

Problems

141. A function is defined by $f(t) = t^2$ for $0 \le t \le 1$ and $f(t) = 2 - t$ for $1 < t \le 2$. Compute $\int_0^2 f(t)\, dt$.

142. The curves $y = \sin x$ and $y = \cos x$ cross each other infinitely often. What is the area of the region bounded by these two curves between two consecutive crossings?

143. Find the average (vertical) height of the shaded area in Figure 7.26.

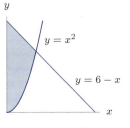

Figure 7.26

144. Find the average (horizontal) width of the shaded area in Figure 7.26.

145. (a) Find the average value of the following functions over one cycle:

(i) $f(t) = \cos t$

(ii) $g(t) = |\cos t|$

(iii) $k(t) = (\cos t)^2$

(b) Write the averages you have just found in ascending order. Using words and graphs, explain why the averages come out in the order they do.

146. Using Figure 7.27, put the following approximations to the integral $\int_a^b f(x)\,dx$ and its exact value in order from smallest to largest: LEFT(5), LEFT(10), RIGHT(5), RIGHT(10), MID(10), TRAP(10), Exact value

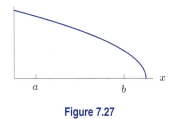

Figure 7.27

147. For a certain definite integral, TRAP(10) = 4.6891 and TRAP(50) = 4.6966. Estimate the actual error for TRAP(10) (and thus the actual value of the integral) by assuming that the error is reduced by a factor of roughly 25 in going from TRAP(10) to TRAP(50).

148. In 1987 the average per capita income in the US was $26,000 and increasing continuously at a rate of $r(t) = 480(1.024)^t$ dollars per year, where t is the number of years since 1987.

(a) Estimate the average per capita income in 1995.

(b) Find a formula for the average per capita income as a function of time after 1987.

149. A patient is given an injection of imitrex, a migraine medicine, at a rate of $r(t) = 2te^{-2t}$ ml/sec, where t is the number of seconds since the injection started.

(a) By letting $t \to \infty$, estimate the total quantity of imitrex injected.

(b) What fraction of this dose has the patient received at the end of 5 seconds?

150. In 1990 humans generated $1.4 \cdot 10^{20}$ joules of energy from petroleum. At the time, it was estimated that all of the earth's petroleum would generate approximately 10^{22} joules. Assuming the use of energy generated by petroleum increases by 2% each year, how long will it be before all of our petroleum resources are used up?

CAS Challenge Problems

151. (a) Use a computer algebra system to find $\int \dfrac{(\ln x)}{x}\,dx$, $\int \dfrac{(\ln x)^2}{x}\,dx$, and $\int \dfrac{(\ln x)^3}{x}\,dx$.

(b) Guess a formula for $\int \dfrac{\ln(x)^n}{x}\,dx$ that works for any positive integer n.

(c) Use a substitution to check your formula.

152. (a) Using a computer algebra system, find $\int (\ln x)^n\,dx$ for $n = 1, 2, 3, 4$.

(b) There is a formula relating $\int (\ln x)^n\,dx$ to $\int (\ln x)^{n-1}\,dx$ for any positive integer n. Guess this formula using your answer to part (a). Check your guess using integration by parts.

In Problems 153–155:

(a) Use a computer algebra system to find the indefinite integral of the given function.

(b) Use the computer algebra system again to differentiate the result of part (a). Do not simplify.

(c) Use algebra to show that the result of part (b) is the same as the original function. Show all the steps in your calculation.

153. $\sin^3 x$

154. $\sin x \cos x \cos(2x)$

155. $\dfrac{x^4}{(1+x^2)^2}$

CHECK YOUR UNDERSTANDING

The left and right Riemann sums of a function f on the interval $[2, 6]$ are denoted by LEFT(n) and RIGHT(n), respectively, when the interval is divided into n equal parts. In Problems 1–9, decide whether the statements are true for all continuous functions, f. Give an explanation for your answer.

1. If $n = 10$, then the subdivision size is $\Delta x = 1/10$.

2. If we double the value of n, we make Δx half as large.

3. LEFT$(10) \leq$ RIGHT(10)

4. As n approaches infinity, LEFT(n) approaches 0.

5. LEFT$(n) -$ RIGHT$(n) = (f(2) - f(6))\Delta x$.

6. Doubling n decreases the difference LEFT$(n) -$ RIGHT(n) by exactly the factor $1/2$.

7. If LEFT$(n) =$ RIGHT(n) for all n, then f is a constant function.

8. The trapezoid estimate TRAP$(n) = ($LEFT$(n) +$ RIGHT$(n))/2$ is always closer to $\int_2^6 f(x)dx$ than LEFT(n) or RIGHT(n).

9. $\int_2^6 f(x)dx$ lies between LEFT(n) and RIGHT(n).

In Problems 10–23, decide whether the statements are true or false. Give an explanation for your answer.

10. $\int f'(x)\cos(f(x))\,dx = \sin(f(x)) + C$

11. $\int (1/f(x))\,dx = \ln|f(x)| + C$

12. $\int t\sin(5 - t^2)\,dt$ can be evaluated using substitution.

13. $\int \sin^7\theta\cos^6\theta\,d\theta$ can be written as a polynomial with $\cos\theta$ as the variable.

14. $\int 1/(x^2 + 4x + 5)\,dx$ involves a natural logarithm.

15. $\int 1/(x^2 + 4x - 5)\,dx$ involves an arctangent.

16. $\int x^{-1}((\ln x)^2 + (\ln x)^3)\,dx$ is a polynomial with $\ln x$ as the variable.

17. $\int t\sin(5 - t)\,dt$ can be evaluated by parts.

18. If f is continuous for all x and $\int_0^\infty f(x)dx$ converges, then so does $\int_a^\infty f(x)dx$ for all positive a.

19. If $f(x)$ is a positive periodic function, then $\int_0^\infty f(x)dx$ diverges.

20. If $f(x)$ is continuous and positive for $x > 0$ and if $\lim_{x\to\infty} f(x) = 0$, then $\int_0^\infty f(x)dx$ converges.

21. If $f(x)$ is continuous and positive for $x > 0$ and if $\lim_{x\to\infty} f(x) = \infty$, then $\int_0^\infty (1/f(x))dx$ converges.

22. If $\int_0^\infty f(x)dx$ and $\int_0^\infty g(x)dx$ both converge, then $\int_0^\infty (f(x) + g(x))dx$ converges.

23. If $\int_0^\infty f(x)dx$ and $\int_0^\infty g(x)dx$ both diverge, then $\int_0^\infty (f(x) + g(x))dx$ diverges.

Suppose that f is continuous for all real numbers and that $\int_0^\infty f(x)\,dx$ converges. Let a be any positive number. Decide which of the following statements are true and which are false. Give an explanation for your answer.

24. $\int_0^\infty af(x)\,dx$ converges.

25. $\int_0^\infty f(ax)\,dx$ converges.

26. $\int_0^\infty f(a + x)\,dx$ converges.

27. $\int_0^\infty (a + f(x))\,dx$ converges.

PROJECTS FOR CHAPTER SEVEN

1. **Taylor Polynomial Inequalities**

   (a) Use the fact that $e^x \geq 1 + x$ for all values of x and the formula

   $$e^x = 1 + \int_0^x e^t \, dt$$

   to show that

   $$e^x \geq 1 + x + \frac{x^2}{2}$$

   for all positive values of x. Generalize this idea to get inequalities involving higher-degree polynomials.

   (b) Use the fact that $\cos x \leq 1$ for all x and repeated integration to show that

   $$\cos x \leq 1 - \frac{x^2}{2!} + \frac{x^4}{4!}.$$

2. **Slope Fields**

   Suppose we want to sketch the antiderivative, F, of the function f. To get an accurate graph of F, we must be careful about making F have the right slope at every point. The slope of F at any point (x, y) on its graph should be $f(x)$, since $F'(x) = f(x)$. We arrange this as follows: at the point (x, y) in the plane, draw a small line segment with slope $f(x)$. Do this at many points. We call such a diagram a *slope field*. If $f(x) = x$, we get the slope field in Figure 7.28.

   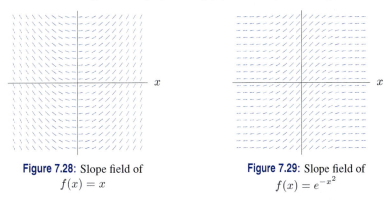

   Figure 7.28: Slope field of $f(x) = x$ **Figure 7.29:** Slope field of $f(x) = e^{-x^2}$

   Notice how the lines in Figure 7.28 seem to be arranged in a parabolic pattern. This is because the general antiderivative of x is $x^2/2 + C$, so the lines are all the tangent lines to the family of parabolas $y = x^2/2 + C$. This suggests a way of finding antiderivatives graphically even if we can't write down a formula for them: plot the slopes, and see if they suggest the graph of an antiderivative. For example, if you do this with $f(x) = e^{-x^2}$, which is one of the functions that does not have an elementary antiderivative, you get Figure 7.29.

   You can see the ghost of the graph of a function lurking behind the slopes in Figure 7.29; in fact there is a whole stack of them. If you move across the plane in the direction suggested by the slope field at every point, you will trace out a curve. The slope field is tangent to the curve everywhere, so this is the graph of an antiderivative of e^{-x^2}.

   (a) (i) Sketch a graph of $f(t) = \dfrac{\sin t}{t}$.

   (ii) What does your graph tell you about the behavior of

   $$\text{Si}(x) = \int_0^x \frac{\sin(t)}{t} \, dt$$

   for $x > 0$? Is $\text{Si}(x)$ always increasing or always decreasing? Does $\text{Si}(x)$ cross the x-axis for $x > 0$?

(iii) By drawing the slope field for $f(t) = \dfrac{\sin t}{t}$, decide whether $\lim\limits_{x \to \infty} \mathrm{Si}(x)$ exists.

(b) (i) Use your calculator or computer to sketch a graph of $y = x^{\sin x}$ for $0 < x \leq 20$.

(ii) Using your answer to part (a), sketch by hand a graph of the function F, where

$$F(x) = \int_0^x t^{\sin t}\, dt.$$

(iii) Use a slope field program to check your answer to part (b).

(c) Let $F(x)$ be the antiderivative of $\sin(x^2)$ satisfying $F(0) = 0$.

(i) Describe any general features of the graph of F that you can deduce by looking at the graph of $\sin(x^2)$ in Figure 7.30.

(ii) By drawing a slope field (using a calculator or computer), sketch a graph of F. Does F ever cross the x-axis in the region $x > 0$? Does $\lim\limits_{x \to \infty} F(x)$ exist?

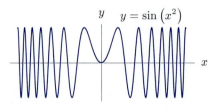

Figure 7.30

Chapter Eight

USING THE DEFINITE INTEGRAL

In Chapter 5, we saw how a definite integral can be used to compute an area, an average value, or a total change.

There are other quantities which can be expressed as definite integrals. To do this, we chop the quantity into small pieces and approximate it by a Riemann sum. In this chapter we use definite integrals to solve problems in geometry, physics, economics, and probability.

8.1 AREAS AND VOLUMES

In Chapter 5, we calculated areas using definite integrals. We obtained the integral by slicing up the region, calculating a Riemann sum, and then taking a limit.

In this section, we calculate volumes using definite integrals. To obtain the integral, we again slice up the region and construct a Riemann sum. The first two examples show how slicing can be used to find both areas and volumes.

Example 1 Use horizontal slices to set up a definite integral to calculate the area of the isosceles triangle in Figure 8.1.

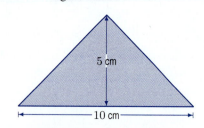

Figure 8.1: Isosceles triangle

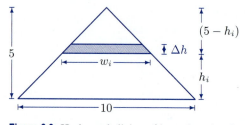

Figure 8.2: Horizontal slicing of isosceles triangle

Solution Notice that we can find the area of a triangle without using an integral; we will use this to check the result from integration:

$$\text{Area} = \frac{1}{2} \text{ Base } \cdot \text{ Height } = 25 \text{ cm}^2.$$

To calculate the area using horizontal slices, see Figure 8.2. A typical strip is approximately a rectangle of length w_i and width Δh, so

$$\text{Area of strip } \approx w_i \Delta h \text{ cm}^2.$$

To get w_i in terms of h_i, the height above the base, use the similar triangles in Figure 8.2:

$$\frac{w_i}{10} = \frac{5 - h_i}{5}$$
$$w_i = 2(5 - h_i) = 10 - 2h_i.$$

Summing the areas of the strips gives the Riemann sum approximation

$$\text{Area of triangle } \approx \sum_{i=1}^{n} w_i \Delta h = \sum_{i=1}^{n} (10 - 2h_i) \Delta h \text{ cm}^2.$$

Taking the limit as $n \to \infty$ and $\Delta h \to 0$ gives the integral:

$$\text{Area of triangle } = \lim_{n \to \infty} \sum_{i=1}^{n} (10 - 2h_i) \Delta h = \int_0^5 (10 - 2h) \, dh \text{ cm}^2.$$

Evaluating the integral gives

$$\text{Area of triangle } = \int_0^5 (10 - 2h) \, dh = (10h - h^2) \Big|_0^5 = 25 \text{ cm}^2.$$

Notice that the limits in the definite integral are the limits for the variable h. Once we decided to slice the triangle horizontally, we knew that a typical slice has thickness Δh, so h is the variable in our definite integral, and the limits must be values of h.

Finding Volumes by Slicing

To calculate the volume of a solid using Riemann sums, we chop the solid into slices whose volumes we can estimate.

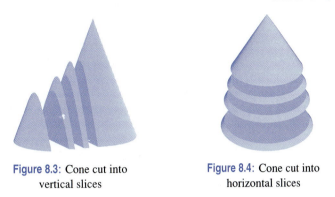

Figure 8.3: Cone cut into vertical slices

Figure 8.4: Cone cut into horizontal slices

Let's see how we might slice a cone standing with the vertex uppermost. We could divide the cone vertically into arch-shaped slices; see Figure 8.3. We could also divide the cone horizontally, giving coin-shaped slices; see Figure 8.4.

To calculate the volume of the cone, we choose the circular slices because it is easier to estimate the volumes of the coin-shaped slices.

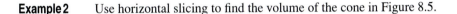

Example 2   Use horizontal slicing to find the volume of the cone in Figure 8.5.

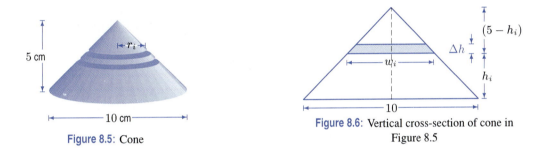

Figure 8.5: Cone

Figure 8.6: Vertical cross-section of cone in Figure 8.5

Solution   Each slice is a circular disk of thickness Δh. See Figure 8.5. The disk at height h_i above the base has radius $r_i = \frac{1}{2} w_i$. From Figure 8.6 and the previous example, we have

$$w_i = 10 - 2h_i \quad \text{so} \quad r_i = 5 - h_i.$$

Each slice is approximately a cylinder of radius r_i and thickness Δh, so

$$\text{Volume of slice} \approx \pi r_i^2 \Delta h = \pi (5 - h_i)^2 \Delta h \text{ cm}^3.$$

Summing over all slices, we have

$$\text{Volume of cone} \approx \sum_{i=1}^{n} \pi (5 - h_i)^2 \Delta h \text{ cm}^3.$$

Taking the limit as $n \to \infty$, so $\Delta h \to 0$, gives

$$\text{Volume of cone} = \lim_{n \to \infty} \sum_{i=1}^{n} \pi (5 - h_i)^2 \Delta h = \int_0^5 \pi (5 - h)^2 \, dh \text{ cm}^3.$$

The integral can be evaluated using the substitution $u = 5 - h$ or by multiplying out $(5 - h)^2$. Using the substitution, we have

$$\text{Volume of cone} = \int_0^5 \pi (5 - h)^2 dh = -\frac{\pi}{3}(5 - h)^3 \bigg|_0^5 = \frac{125}{3} \pi \text{ cm}^3.$$

Note that the sum represented by the $\sum$ sign is over all the strips. In the future, we will not write down the points, $h_0, h_1, \ldots h_n$, that are needed to express the Riemann sum precisely. This is unnecessary if all we want is the final expression for the definite integral. We now calculate the area of a semicircle and the volume of a hemisphere by slicing.

Example 3   Use horizontal slices to set up a definite integral representing the area of the semicircle of radius 7 cm in Figure 8.7.

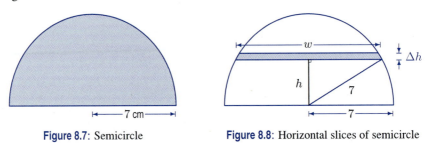

Figure 8.7: Semicircle   **Figure 8.8**: Horizontal slices of semicircle

Solution   As in Example 1, to calculate the area using horizontal slices, we divide the region into strips; see Figure 8.8. A typical strip at height h above the base has width w and thickness Δh, so

$$\text{Area of strip} \approx w\Delta h \text{ cm}^2.$$

To get w in terms of h, we use the Pythagorean Theorem in Figure 8.8, giving

$$h^2 + \left(\frac{w}{2}\right)^2 = 7^2,$$

so

$$w = \sqrt{4(7^2 - h^2)} = 2\sqrt{49 - h^2}.$$

Summing the areas of the strips gives the Riemann sum approximation

$$\text{Area of semicircle} \approx \sum w\Delta h = \sum 2\sqrt{49 - h^2}\Delta h \text{ cm}^2.$$

Taking the limit as $\Delta h \to 0$ gives the integral:

$$\text{Area of semicircle} = \lim_{\Delta h \to 0} \sum 2\sqrt{49 - h^2}\Delta h = 2\int_0^7 \sqrt{49 - h^2} \, dh \text{ cm}^2.$$

Using the table of integrals VI-28 and VI-30, or a calculator or computer, gives

$$\text{Area of semicircle} = 2\cdot\frac{1}{2}\left(h\sqrt{49 - h^2} + 49\arcsin\left(\frac{h}{7}\right)\right)\Bigg|_0^7 = 49\arcsin 1 = \frac{49}{2}\pi = 76.97 \text{ cm}^2.$$

As a check, notice that the area of the whole circle of radius 7 is $\pi \cdot 7^2 = 49\pi \text{ cm}^2$.

Example 4   Set up and evaluate an integral giving the volume of the hemisphere of radius 7 cm in Figure 8.9.

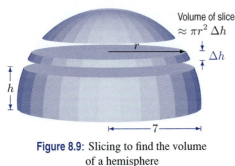

Volume of slice
$\approx \pi r^2 \, \Delta h$

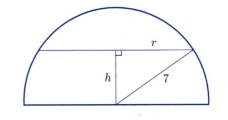

Figure 8.9: Slicing to find the volume   **Figure 8.10**: Vertical cut through center of hemisphere
of a hemisphere   showing relation between r_i and h_i

Solution   We will not use the formula $\frac{4}{3}\pi r^3$ for the volume of a sphere. However our approach can be used to derive that formula.

Divide the hemisphere into horizontal slices of thickness Δh cm. (See Figure 8.9.) Each slice is a circular slab. Let r be the radius of the slice at height h, so

$$\text{Volume of slice} \approx \pi r^2 \Delta h \text{ cm}^3.$$

We want to express r in terms of h. From Figure 8.10, we have $r^2 = 7^2 - h^2$, so

$$\text{Volume of slice} \approx \pi r^2 \Delta h = \pi(7^2 - h^2)\Delta h \text{ cm}^3.$$

Summing the volumes of all slices gives:

$$\text{Volume} \approx \sum \pi r^2 \Delta h = \sum \pi(7^2 - h^2)\Delta h \text{ cm}^3.$$

As the thickness of each slab tends to zero, the sum becomes a definite integral. Since the radius of the hemisphere is 7, we know that h varies from 0 to 7, so these are the limits of integration:

$$\text{Volume} = \int_0^7 \pi(7^2 - h^2)\,dh = \pi\left(7^2 h - \frac{1}{3}h^3\right)\Big|_0^7 = \frac{2}{3}\pi 7^3 = 718.4 \text{ cm}^3.$$

Notice that the volume of the hemisphere is half of $\frac{4}{3}\pi 7^3$ cm^3, as we expected.

We now use slicing to find the volume of a pyramid. You may already know the formula $V = \frac{1}{3}b^2 \cdot h$ for the volume V of a pyramid with height h and a square base of length b. We will not use the formula, but our approach can be used to derive it.

Example 5   Compute the volume, in cubic feet, of the Great Pyramid of Egypt, whose base is a square 755 feet by 755 feet and whose height is 410 feet.

Solution   We slice the pyramid horizontally. Each slice is a square slab with thickness Δh. The bottom layer is a square slab 755 feet by 755 feet and volume about $(755)^2\Delta h$ ft^3. As we move up the pyramid, the layers have shorter side lengths. We divide the height into n subintervals of length Δh. See Figure 8.11. Let s be the side length of the slice at height h, then

$$\text{Volume of slice} \approx s^2\,\Delta h \text{ ft}^3.$$

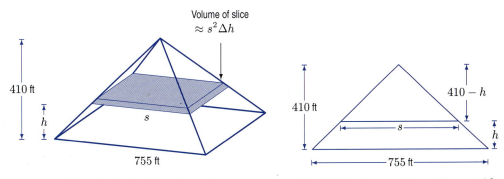

Figure 8.11: The Great Pyramid   **Figure 8.12:** Cross-section relating s and h

We express s as a function of h using the vertical cross-section in Figure 8.12. By similar triangles, we get

$$\frac{s}{755} = \frac{(410 - h)}{410}.$$

Thus,

$$s = \left(\frac{755}{410}\right)(410 - h),$$

and the total volume, V, is approximated by adding the volumes of the n layers:

$$V \approx \sum s^2 \, \Delta h = \sum \left[\left(\frac{755}{410} \right) (410 - h) \right]^2 \Delta h \text{ ft}^3.$$

As the thickness of each slice tends to zero, the sum becomes a definite integral. Finally, since h varies from 0 to 410, the height of the pyramid, we have

$$V = \int_0^{410} \left[\left(\frac{755}{410} \right) (410 - h) \right]^2 dh = \left(\frac{755}{410} \right)^2 \int_0^{410} (410 - h)^2 \, dh$$

$$= \left(\frac{755}{410} \right)^2 \left[-\frac{(410 - h)^3}{3} \right] \Bigg|_0^{410} = \left(\frac{755}{410} \right)^2 \frac{(410)^3}{3} = \frac{1}{3} (755)^2 (410) \approx 78 \text{ million ft}^3.$$

Note that $V = \frac{1}{3} (755)^2 (410) = \frac{1}{3} b^2 \cdot h$, as expected.

Exercises and Problems for Section 8.1

Exercises

In Exercises 1–8, write a Riemann sum and then a definite integral representing the area of the region, using the strip shown. Evaluate the integral exactly.

1.

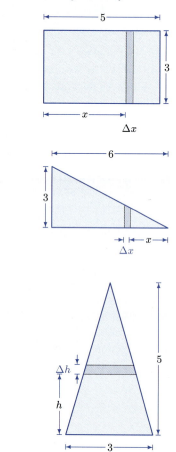

2.

3.

4.

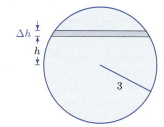

5.

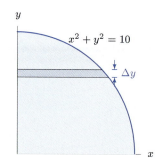

6.

7.

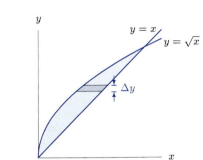

8.

11.

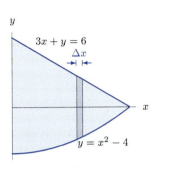

12.

13.

In Exercises 9–14, write a Riemann sum and then a definite integral representing the volume of the region, using the slice shown. Evaluate the integral exactly. (Regions are parts of cones, cylinders, spheres, and pyramids.)

9.

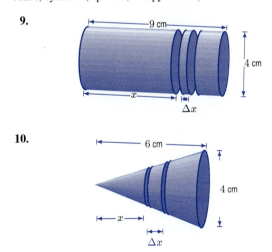

10.

14.

Problems

15. The origin and the point (a, a) are at opposite corners of a square. Calculate the ratio of the areas of the two parts into which the curve $\sqrt{x} + \sqrt{y} = \sqrt{a}$ divides the square.

16. If A_n is the area between the curves $y = x$ and $y = x^n$, show that $A_n \to \frac{1}{2}$ as $n \to \infty$ and explain this result graphically.

The integrals in Problems 17–20 represent the area of either a triangle or part of a circle, and the variable of integration mea-

sures a distance. In each case, say which shape is represented, and give the radius of the circle or the base and height of the triangle. Make a sketch to support your answer showing the variable and all other relevant quantities.

17. $\int_0^1 3x \, dx$

18. $\int_{-9}^9 \sqrt{81 - x^2} \, dx$

19. $\int_0^{\sqrt{15}} \sqrt{15 - h^2} \, dh$

20. $\int_0^7 5\left(1 - \frac{h}{7}\right) dh$

The integrals in Problems 21–24 represent the volume of either a hemisphere or a cone, and the variable of integration measures a length. In each case, say which shape is represented, and give the radius of the hemisphere or the radius and height of the cone. Make a sketch to support your answer showing the variable and all other relevant quantities.

21. $\int_0^{12} \pi(144 - h^2)\, dh$ **22.** $\int_0^{12} \pi(x/3)^2 dx$

23. $\int_0^6 \pi(3 - y/2)^2 dy$ **24.** $\int_0^2 \pi(2^2 - (2-y)^2)\, dy$

25. Find the volume of a sphere of radius r by slicing.

26. Find, by slicing, a formula for the volume of a cone of height h and base radius r.

27. Figure 8.13 shows a solid with both rectangular and triangular cross sections.

   (a) Slice the solid parallel to the triangular faces. Sketch one slice and calculate its volume in terms of x, the distance of the slice from one end. Then write and evaluate an integral giving the volume of the solid.

   (b) Repeat part (a) for horizontal slices. Instead of x, use h, the distance of a slice from the top.

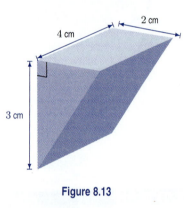

Figure 8.13

28. A rectangular lake is 150 km long and 3 km wide. The vertical cross-section through the lake in Figure 8.14 shows that the lake is 0.2 km deep at the center. (These are the approximate dimensions of Lake Mead, the largest reservoir in the US, which provides water to California, Nevada, and Arizona.) Set up and evaluate a definite integral giving the total volume of the lake.

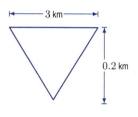

Figure 8.14: Not to scale

29. A dam has a rectangular base 1400 meters long and 160 meters wide. Its cross-section is shown in Figure 8.15. (The Grand Coulee Dam in Washington state is roughly this size.) By slicing horizontally, set up and evaluate a definite integral giving the volume of material used to build this dam.

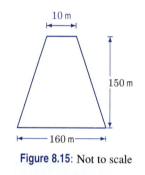

Figure 8.15: Not to scale

8.2   APPLICATIONS TO GEOMETRY

In Section 8.1, we calculated volumes using slicing and definite integrals. In this section, we use the same method to calculate the volumes of more complicated regions as well as the length of a curve. The method is summarized in the following steps:

To Compute a Volume or Length Using an Integral

- Divide the solid (or curve) into small pieces whose volume (or length) we can easily approximate;

- Add the contributions of all the pieces, obtaining a Riemann sum that approximates the total volume (or length);

- Take the limit as the number of terms in the sum tends to infinity, giving a definite integral for the total volume (or total length).

In the previous section, all the slices we created were disks or squares. We now look at different ways of generating volumes whose cross-sections include circles, rings, and squares.

Volumes of Revolution

One way to create a solid having circular cross-sections is to revolve a region in the plane around a line, giving a *solid of revolution,* as in the following examples.

Example 1 The region bounded by the curve $y = e^{-x}$ and the x-axis between $x = 0$ and $x = 1$ is revolved around the x-axis. Find the volume of this solid of revolution.

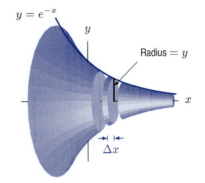

Figure 8.16: A thin strip rotated around the x-axis to form a circular slice

Solution We slice the region perpendicular to the x-axis, giving circular disks of thickness Δx. See Figure 8.16. The radius of the disk is $y = e^{-x}$, so

$$\text{Volume of the slice} \approx \pi y^2 \,\Delta x = \pi (e^{-x})^2 \,\Delta x,$$

$$\text{Total volume} \approx \sum \pi y^2 \,\Delta x = \sum \pi \left(e^{-x}\right)^2 \,\Delta x.$$

As the thickness of each slice tends to zero, we get

$$\text{Total volume} = \int_0^1 \pi (e^{-x})^2 \, dx = \pi \int_0^1 e^{-2x} \, dx = \pi \left(-\frac{1}{2}\right) e^{-2x} \Big|_0^1$$

$$= \pi \left(-\frac{1}{2}\right) (e^{-2} - e^0) = \frac{\pi}{2}(1 - e^{-2}) \approx 1.36.$$

Example 2 A table leg in Figure 8.17 has a circular cross section with radius r cm at a height of y cm above the ground given by $r = 3 + \cos(\pi y/25)$. Find the volume of the table leg.

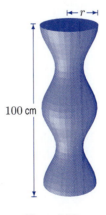

100 cm

Figure 8.17

Solution   The table leg is formed by rotating the curve $r = 3 + \cos(\pi y/25)$ around the y-axis. Slicing the table leg horizontally gives circular disks of thickness Δy and radius $r = 3 + \cos(\pi y/25)$.

To set up a definite integral for the volume, we find the volume of a typical slice:

$$\text{Volume of slice} \approx \pi r^2 \Delta y = \pi \left(3 + \cos\left(\frac{\pi}{25}y\right)\right)^2 \Delta y.$$

Summing over all slices gives the Riemann sum approximation:

$$\text{Total volume} = \sum \pi \left(3 + \cos\left(\frac{\pi}{25}y\right)\right)^2 \Delta y.$$

Taking the limit as $\Delta y \to 0$ gives the definite integral

$$\text{Total volume} = \lim_{\Delta y \to 0} \sum \pi \left(3 + \cos\left(\frac{\pi}{25}y\right)\right)^2 \Delta y = \int_0^{100} \pi \left(3 + \cos\left(\frac{\pi}{25}y\right)\right)^2 dy.$$

Evaluating the integral numerically gives

$$\text{Total volume} = \int_0^{100} \pi \left(3 + \cos\left(\frac{\pi}{25}y\right)\right)^2 dy = 2984.5 \text{ cm}^3.$$

Example 3   The region bounded by the curves $y = x$ and $y = x^2$ is rotated about the line $y = 3$. Compute the volume of the resulting solid.

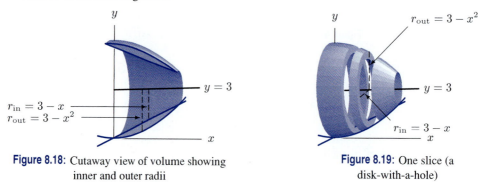

Figure 8.18: Cutaway view of volume showing inner and outer radii

Figure 8.19: One slice (a disk-with-a-hole)

Solution   The solid is shaped like a bowl with the base removed. See Figure 8.18. To compute the volume, we divide the area in the xy-plane into thin vertical strips of thickness Δx, as in Figure 8.20.

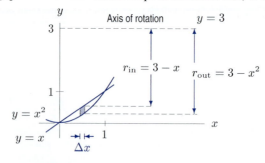

Figure 8.20: The region for Example 3

As each strip is rotated around the line $y = 3$, it sweeps out a slice shaped like a circular disk with a hole in it. See Figure 8.19. This disk-with-a-hole has an inner radius of $r_{\text{in}} = 3 - x$ and an outer radius of $r_{\text{out}} = 3 - x^2$. Think of the slice as a circular disk of radius r_{out} from which has been removed a smaller disk of radius r_{in}. Then,

$$\text{Volume of slice} \approx \pi r_{\text{out}}^2 \, \Delta x - \pi r_{\text{in}}^2 \, \Delta x = \pi (3 - x^2)^2 \, \Delta x - \pi (3 - x)^2 \, \Delta x.$$

Adding the volumes of all the slices, we have

$$\text{Total volume} = V \approx \sum \left(\pi r_{\text{out}}^2 - \pi r_{\text{in}}^2 \right) \Delta x = \sum \left(\pi(3 - x^2)^2 - \pi(3 - x)^2 \right) \Delta x.$$

We let Δx, the thickness of each slice, tend to zero to obtain a definite integral. Since the curves $y = x$ and $y = x^2$ intersect at $x = 0$ and $x = 1$, these are the limits of integration:

$$V = \int_0^1 \left(\pi(3 - x^2)^2 - \pi(3 - x)^2 \right) dx = \pi \int_0^1 \left((9 - 6x^2 + x^4) - (9 - 6x + x^2) \right) dx$$

$$= \pi \int_0^1 (6x - 7x^2 + x^4) \, dx = \pi \left(3x^2 - \frac{7x^3}{3} + \frac{x^5}{5} \right) \Big|_0^1 \approx 2.72.$$

Volumes of Regions of Known Cross-Section

We now calculate the volume of a solid constructed by a different method. Starting with a region in the xy-plane as a base, the solid is built by standing squares, semicircles, or triangles vertically on edge in this region.

Example 4   Find the volume of the solid whose base is the region in the xy-plane bounded by the curves $y = x^2$ and $y = 8 - x^2$ and whose cross-sections perpendicular to the x-axis are squares with one side in the xy-plane. (See Figure 8.21.)

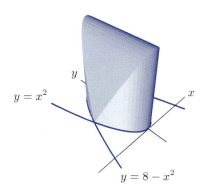

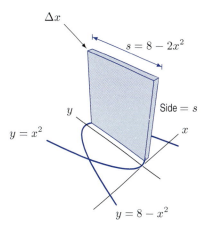

Figure 8.21: The solid for Example 4   **Figure 8.22**: A slice of the solid for Example 4

Solution   We view the solid as a loaf of bread sitting on the xy-plane and made up of square slices. A typical slice of thickness Δx is shown in Figure 8.22. The side length, s, of the square is the distance (in the y direction) between the two curves, so $s = (8 - x^2) - x^2 = 8 - 2x^2$ giving

$$\text{Volume of slice} \approx s^2 \, \Delta x = (8 - 2x^2)^2 \, \Delta x.$$

Thus

$$\text{Total volume} = V \approx \sum s^2 \, \Delta x = \sum (8 - 2x^2)^2 \, \Delta x.$$

As the thickness Δx of each slice tends to zero, the sum becomes a definite integral. Since the curves $y = x^2$ and $y = 8 - x^2$ intersect at $x = -2$ and $x = 2$, these are the limits of integration. We have

$$V = \int_{-2}^2 (8 - 2x^2)^2 \, dx = \int_{-2}^2 (64 - 32x^2 + 4x^4) \, dx$$

$$= 64x - \frac{32}{3}x^3 + \frac{4}{5}x^5 \Big|_{-2}^2 = \frac{2048}{15} \approx 136.5.$$

Arc Length

A definite integral can be used to compute the *arc length*, or length, of a curve. To compute the length of the curve $y = f(x)$ from $x = a$ to $x = b$, where $a < b$, we divide the curve into small pieces, each one approximately straight.

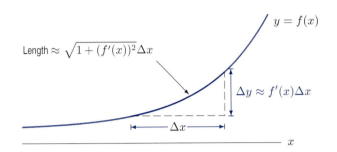

Length $\approx \sqrt{1 + (f'(x))^2} \, \Delta x$

$y = f(x)$

$\Delta y \approx f'(x) \Delta x$

Δx

Figure 8.23: Length of a small piece of curve approximated using Pythagoras' theorem

Figure 8.23 shows that a small change Δx corresponds to a small change $\Delta y \approx f'(x) \, \Delta x$. The length of the piece of the curve is approximated by

$$\text{Length} \approx \sqrt{(\Delta x)^2 + (\Delta y)^2} \approx \sqrt{(\Delta x)^2 + (f'(x) \, \Delta x)^2} = \sqrt{1 + (f'(x))^2} \, \Delta x.$$

Thus, the arc length of the entire curve is approximated by a Riemann sum:

$$\text{Arc length} \approx \sum \sqrt{1 + (f'(x))^2} \, \Delta x.$$

Since x varies between a and b, as we let Δx tend to zero, the sum becomes the definite integral:

For $a < b$, the arc length of the curve $y = f(x)$ from $x = a$ to $x = b$ is given by

$$\text{Arc length} = \int_a^b \sqrt{1 + (f'(x))^2} \, dx.$$

Example 5 Set up and evaluate an integral to compute the length of the curve $y = x^3$ from $x = 0$ to $x = 5$.

Solution If $f(x) = x^3$, then $f'(x) = 3x^2$, so

$$\text{Arc length} = \int_0^5 \sqrt{1 + (3x^2)^2} \, dx.$$

Although the formula for the arc length of a curve is easy to apply, the integrands it generates often do not have elementary antiderivatives. Evaluating the integral numerically, we find the arc length to be 125.68. To check that the answer is reasonable, notice that the curve starts at $(0, 0)$ and goes to $(5, 125)$, so its length must be at least the length of a straight line between these points, or $\sqrt{5^2 + 125^2} = 125.10$. (See Figure 8.24.)

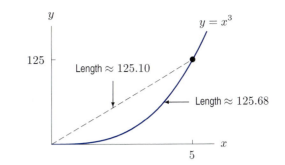

Figure 8.24: Arc length of $y = x^3$ (Note: The picture is distorted because the scales on the two axes are quite different)

Arc Length of a Parametric Curve

A particle moving along a curve in the plane given by the parametric equations $x = f(t)$, $y = g(t)$, where t is time, has speed given by:

$$v(t) = \sqrt{\left(\frac{dx}{dt}\right)^2 + \left(\frac{dy}{dt}\right)^2}.$$

We can find the distance traveled by a particle along a curve between $t = a$ and $t = b$ by integrating its speed. Thus,

$$\text{Distance traveled} = \int_a^b v(t)\, dt.$$

If the particle never stops or reverses its direction as it moves along the curve, the distance it travels is the same as the length of the curve. This suggests the following formula:

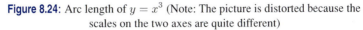

> If a curve is given parametrically for $a \leq t \leq b$ by differentiable functions and if the velocity $v(t)$ is not 0 for $a < t < b$, then
>
> $$\text{Arc length of curve} = \int_a^b v(t)\, dt = \int_a^b \sqrt{\left(\frac{dx}{dt}\right)^2 + \left(\frac{dy}{dt}\right)^2}\, dt.$$

Example 6   Find the circumference of the ellipse given by the parametric equations

$$x = 2\cos t, \quad y = \sin t, \quad 0 \leq t \leq 2\pi.$$

Solution   The circumference of this curve is given by an integral which must be calculated numerically:

$$\text{Circumference} = \int_0^{2\pi} \sqrt{\left(\frac{dx}{dt}\right)^2 + \left(\frac{dy}{dt}\right)^2}\, dt = \int_0^{2\pi} \sqrt{(-2\sin t)^2 + (\cos t)^2}\, dt$$

$$= \int_0^{2\pi} \sqrt{4\sin^2 t + \cos^2 t}\, dt = 9.69.$$

Since the ellipse is inscribed in a circle of radius 2 and circumscribes a circle of radius 1, we expect the length of the ellipse to be between $2\pi(2) \approx 12.57$ and $2\pi(1) \approx 6.28$, so the value of 9.69 is reasonable.

Exercises and Problems for Section 8.2

Exercises

In Exercises 1–9, the region is rotated around the x-axis. Find the volume.

1. Bounded by $y = x^2, y = 0, x = 0, x = 1$.

2. Bounded by $y = (x + 1)^2, y = 0, x = 1, x = 2$.

3. Bounded by $y = e^x, y = 0, x = -1, x = 1$.

4. Bounded by $y = \sqrt{x + 1}, y = 0, x = -1, x = 1$.

5. Bounded by $y = 4 - x^2, y = 0, x = -2, x = 0$.

6. Bounded by $y = 1/(x + 1), y = 0, x = 0, x = 1$.

7. Bounded by $y = \cos x, y = 0, x = 0, x = \pi/2$.

8. Bounded by $y = x^2, y = x, x = 0, x = 1$.

9. Bounded by $y = e^{3x}, y = e^x, x = 0, x = 1$.

For Exercises 10–11, find the arc length of the given function from $x = 0$ to $x = 2$.

10. $f(x) = \sqrt{4 - x^2}$

11. $f(x) = \sqrt{x^3}$

Find the length of the parametric curves in Exercises 12–14.

12. $x = 3 + 5t, y = 1 + 4t$ for $1 \le t \le 2$. Explain why your answer is reasonable.

13. $x = \cos(e^t), y = \sin(e^t)$ for $0 \le t \le 1$. Explain why your answer is reasonable.

14. $x = \cos(3t), y = \sin(5t)$ for $0 \le t \le 2\pi$.

Problems

15. Consider the hyperbola $x^2 - y^2 = 1$ in Figure 8.25.

    (a) The shaded region $2 \le x \le 3$ is rotated around the x-axis. What is the volume generated?

    (b) What is the arc length with $y \ge 0$ from $x = 2$ to $x = 3$?

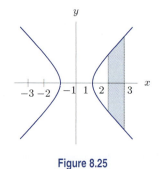

Figure 8.25

16. Rotating the ellipse $x^2/a^2 + y^2/b^2 = 1$ about the x-axis generates an ellipsoid. Compute its volume.

For Problems 17–19, sketch the solid obtained by rotating each region around the indicated axis. Using the sketch, show how to approximate the volume of the solid by a Riemann sum, and hence find the volume.

17. Bounded by $y = x^3, x = 1, y = -1$. Axis: $y = -1$.

18. Bounded by $y = \sqrt{x}, x = 1, y = 0$. Axis: $x = 1$.

19. Bounded by the first arch of $y = \sin x, y = 0$. Axis: x axis.

For Problems 20–26 consider the region bounded by $y = e^x$, the x-axis, and the lines $x = 0$ and $x = 1$. Find the volume of the following solids.

20. The solid obtained by rotating the region about the x-axis.

21. The solid obtained by rotating the region about the horizontal line $y = -3$.

22. The solid obtained by rotating the region about the horizontal line $y = 7$.

23. The solid whose base is the given region and whose cross-sections perpendicular to the x-axis are squares.

24. The solid whose base is the given region and whose cross-sections perpendicular to the x-axis are semicircles.

25. **(a)** A pie dish is 9 inches across the top, 7 inches across the bottom, and 3 inches deep. See Figure 8.26. Compute the volume of this dish.

    (b) Make a rough estimate of the volume in cubic inches of a single cut-up apple, and estimate the number of apples that is needed to make an apple pie that fills this dish.

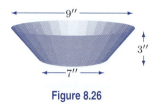

Figure 8.26

26. A 100 cm long gutter is made of three strips of metal, each 5 cm wide; Figure 8.27 shows a cross-section.

    (a) Find the volume of water in the gutter when the depth is h cm.

    (b) What is the maximum value of h?

    (c) What is the maximum volume of water that the gutter can hold?

(d) If the gutter is filled with half the maximum volume of water, is the depth larger or smaller than half of the answer to part (b)? Explain how you can answer without any calculation.

(e) Find the depth of the water when the gutter contains half the maximum possible volume.

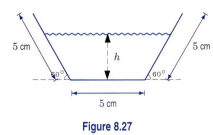

Figure 8.27

27. The design of boats is based on Archimedes' Principle, which states that the buoyant force on an object in water is equal to the weight of the water displaced. Suppose you want to build a sailboat whose hull is parabolic with cross section $y = ax^2$, where a is a constant. Your boat will have length L and its maximum draft (the maximum vertical depth of any point of the boat beneath the water line) will be H. See Figure 8.28. Every cubic meter of water weighs 10,000 newtons. What is the maximum possible weight for your boat and cargo?

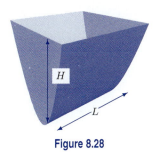

Figure 8.28

28. The circumference of a tree at different heights above the ground is given in the table below. Assume that all horizontal cross-sections of the tree are circles. Estimate the volume of the tree.

Height (inches)	0	20	40	60	80	100	120
Circumference (inches)	31	28	21	17	12	8	2

29. A bowl has the shape of the graph of $y = x^4$ between the points $(1, 1)$ and $(-1, 1)$ rotated about the y-axis. When the bowl contains water to a depth of h units, it flows out through a hole in the bottom at a rate (volume/time) proportional to $\sqrt{h}$, with constant of proportionality 6.

(a) Show that the water level falls at a constant rate.

(b) Find how long it takes to empty the bowl if it is originally full to the brim.

30. The hull of a boat has widths given by the following table. Reading across a row of the table gives widths at points 0, 10, ..., 60 feet from the front to the back at a certain level below waterline. Reading down a column of the table gives widths at levels 0, 2, 4, 6, 8 feet below waterline at a certain distance from the front. Use the trapezoidal rule to estimate the volume of the hull below waterline.

		Front of boat $\longrightarrow$ Back of boat						
		0	10	20	30	40	50	60
	0	2	8	13	16	17	16	10
Depth	2	1	4	8	10	11	10	8
below	4	0	3	4	6	7	6	4
waterline	6	0	1	2	3	4	3	2
(in feet)	8	0	0	1	1	1	1	1

31. (a) Write an integral which represents the circumference of a circle of radius r.

(b) Evaluate the integral, and show that you get the answer you expect.

32. Compute the perimeter of the region used for the base of the solids in Problems 20–24.

33. The catenary $\cosh x = \frac{1}{2}(e^x + e^{-x})$ represents the shape of a hanging cable. Find the exact length of this catenary between $x = -1$ and $x = 1$.

34. Set up an integral for the circumference of an ellipse with semi-major axis $a = 2$ and semi-minor axis $b = 1$. Evaluating this integral numerically (i.e., using a calculator or computer) poses a problem. Describe this problem. Formulate a way to solve it.

35. There are very few elementary functions $y = f(x)$ for which arc length can be computed in elementary terms using the formula

$$\int_a^b \sqrt{1 + \left(\frac{dy}{dx}\right)^2}\, dx.$$

You have seen some such functions f in Problems 10, 11, and 33, namely, $f(x) = \sqrt{4 - x^2}$, $f(x) = \sqrt{x^3}$, and $f(x) = \frac{1}{2}(e^x + e^{-x})$. Try to find some other function that "works," that is, a function whose arc length you can find using this formula and antidifferentiation.

36. After doing Problem 35, you may wonder what sort of functions can represent arc length. If $g(0) = 0$ and g is differentiable and increasing, then can $g(x)$, $x \geq 0$, represent arc length? That is, can we find a function $f(t)$ such that

$$\int_0^x \sqrt{1 + (f'(t))^2}\, dt = g(x)?$$

(a) Show that $f(x) = \int_0^x \sqrt{(g'(t))^2 - 1}\, dt$ works as long as $g'(x) \geq 1$. In other words, show that the arc length of the graph of f from 0 to x is $g(x)$.

(b) Show that if $g'(x) < 1$ for some x, then $g(x)$ cannot represent the arc length of the graph of any function.

(c) Find a function f whose arc length from 0 to x is $2x$.

8.3   DENSITY AND CENTER OF MASS

Density and How to Slice a Region

The examples in this section involve the idea of *density*. For example,
- A population density is measured in, say, people per mile (along the edge of a road), or people per unit area (in a city), or bacteria per cubic centimeter (in a test tube).

- The density of a substance (e.g. air, wood, or metal), is the mass of a unit volume of the substance and is measured in, say, grams per cubic centimeter.

Suppose we want to calculate the total mass or total population, but the density is not constant over a region.

> **To find total quantity from density:** Divide the region into small pieces in such a way that the density is approximately constant on each piece, and add the contributions of the pieces.

Example 1   The Massachusetts Turnpike ("the Pike") starts in the middle of Boston and heads west. The number of people living next to it varies as it gets farther from the city. Suppose that, x miles out of town, the population density adjacent to the Pike is $P = f(x)$ people/mile. Express the total population living next to the Pike within 5 miles of Boston as a definite integral.

Solution   Divide the Pike up into segments of length Δx. The population density at the center of Boston is $f(0)$; let's use that density for the first segment. This gives an estimate of

$$\text{People living in first segment} \approx f(0) \text{ people/ mile} \cdot \Delta x \text{ mile} = f(0)\Delta x \text{ people.}$$

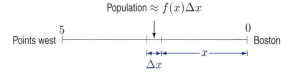

Figure 8.29: Population along the Massachusetts Turnpike

Similarly, the population in a typical segment x miles from the center of Boston is the population density times the length of the interval, or roughly $f(x)\,\Delta x$. (See Figure 8.29.) The sum of all these estimates gives the estimate

$$\text{Total population} \approx \sum f(x)\,\Delta x.$$

Letting $\Delta x \to 0$ gives

$$\text{Total population} = \lim_{\Delta x \to 0} \sum f(x)\,\Delta x = \int_0^5 f(x)\,dx.$$

The 5 and 0 in the limits of the integral are the upper and lower limits of the interval over which we are integrating.

Example 2   The air density h meters above the earth's surface is $f(h)$ kg/m^3. Find the mass of a cylindrical column of air 2 meters in diameter and 25 kilometers high, with base on the surface of the earth.

Solution   The column of air is a circular cylinder 2 meters in diameter and 25 kilometers, or 25,000 meters, high. First we must decide how we are going to slice this column. Since the air density varies with altitude but remains constant horizontally, we take horizontal slices of air. That way, the density will be more or less constant over the whole slice, being close to its value at the bottom of the slice. (See Figure 8.30.)

Figure 8.30: Slicing a column of air horizontally

A slice is a cylinder of height Δh and diameter 2 m, so its radius is 1 m. We find the approximate mass of the slice by multiplying its volume by its density. If the thickness of the slice is Δh, then its volume is $\pi r^2 \cdot \Delta h = \pi 1^2 \cdot \Delta h = \pi \Delta h$ m^3. The density of the slice is given by $f(h)$. Thus,

$$\text{Mass of slice} \approx \text{Volume} \cdot \text{Density} = (\pi \Delta h \text{ m}^3)(f(h) \text{ kg/m}^3) = \pi \Delta h \cdot f(h) \text{ kg.}$$

Adding these slices up yields a Riemann sum:

$$\text{Total mass} \approx \sum \pi f(h) \, \Delta h \text{ kg.}$$

As $\Delta h \to 0$, this sum approximates the definite integral:

$$\text{Total mass} = \int_0^{25,000} \pi f(h) \, dh \text{ kg.}$$

In order to get a numerical value for the mass of air, we need an explicit formula for the density as a function of height, as in the next example.

Example 3   Find the mass of the column of air in Example 2 if the density of air at height h is given by

$$P = f(h) = 1.28 e^{-0.000124h} \text{ kg/m}^3.$$

Solution   Using the result of the previous example, we have

$$\text{Mass} = \int_0^{25,000} \pi 1.28 e^{-0.000124h} \, dh = \frac{-1.28\pi}{0.000124} \left(e^{-0.000124h} \Big|_0^{25,000} \right)$$

$$= \frac{1.28\pi}{0.000124} \left(e^0 - e^{-0.000124(25,000)} \right) \approx 31,000 \text{ kg.}$$

It requires some thought to figure out how to slice a region. The key point is that you want the density to be nearly constant within each piece.

Example 4   The population density in Ringsburg is a function of the distance from the city center. At r miles from the center, the density is $P = f(r)$ people per square mile. Ringsburg is circular with radius 5 miles. Write a definite integral that expresses the total population of Ringsburg.

Solution   We want to slice Ringsburg up and estimate the population of each slice. If we were to take straight-line slices, the population density would vary on each slice, since it depends on the distance from the city center. We want the population density to be pretty close to constant on each slice. We therefore take slices that are thin rings around the center. (See Figure 8.31.) Since the ring is very thin, we can approximate its area by straightening it into a thin rectangle. (See Figure 8.32.) The width of the rectangle is Δr miles, and its length is approximately equal to the circumference of the ring, $2\pi r$ miles, so its area is about $2\pi r\,\Delta r$ mi^2. Since

$$\text{Population on ring} \approx \text{Density} \cdot \text{Area},$$

we get

$$\text{Population on ring} \approx (f(r)\,\text{people/mi}^2)(2\pi r\Delta r\,\text{mi}^2) = f(r) \cdot 2\pi r\,\Delta r\,\text{people}.$$

Adding the contributions from each ring, we get

$$\text{Total population} \approx \sum 2\pi r f(r)\,\Delta r\,\text{people}.$$

So

$$\text{Total population} = \int_0^5 2\pi r f(r)\,dr\,\text{people}.$$

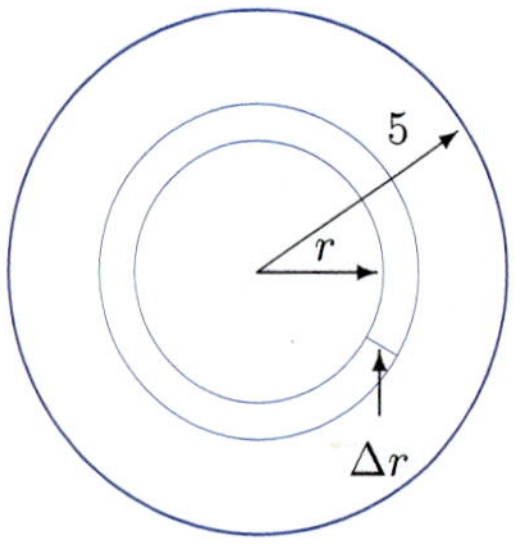

Width $= \Delta r$

$2\pi r$

Figure 8.31: Ringsburg 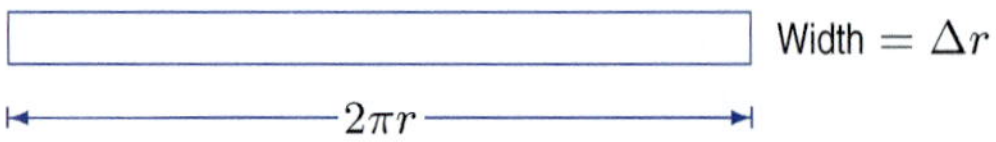   **Figure 8.32:** Ring from Ringsburg (straightened out)

Note: You may wonder what happens if we calculate the area of the ring by subtracting the area of the inner circle (πr^2) from the area of the outer circle $[\pi(r + \Delta r)^2]$, giving

$$\text{Area} = \pi(r + \Delta r)^2 - \pi r^2.$$

Multiplying out and subtracting, we get

$$\begin{aligned}\text{Area} &= \pi[r^2 + 2r\,\Delta r + (\Delta r)^2] - \pi r^2 \\ &= 2\pi r\,\Delta r + \pi(\Delta r)^2.\end{aligned}$$

This expression differs from the one we used before by the $\pi(\Delta r)^2$ term. However, as Δr becomes very small, $\pi(\Delta r)^2$ becomes much, much smaller. We say its smallness is of *second order*, since the power of the small factor, Δr, is 2. In the limit as $\Delta r \to 0$, we can ignore $\pi(\Delta r)^2$.

Center of Mass

The center of mass of a mechanical system is important for studying its behavior when in motion. For example, some sport utility vehicles and light trucks tend to tip over in accidents, because of their high centers of mass.

In this section, we first define the center of mass for a system of point masses on a line. Then we use the definite integral to extend this definition.

Point Masses

Two children on a seesaw, one twice the weight of the other, will balance if the lighter child is twice as far from the pivot as the heavier child. Thus, the balance point is 2/3 of the way from the lighter child and 1/3 of the way from the heavier child. This balance point is the *center of mass* of the mechanical system consisting of the masses of the two children (we ignore the mass of the seesaw itself). See Figure 8.33.

To find the balance point, we use the *displacement* (signed distance) of each child from the pivot to calculate the *moment*, where

$$\text{Moment of mass about pivot} = \text{Mass} \times \text{Displacement from pivot}.$$

A moment represents the tendency of a child to turn the system about the pivot point; the seesaw balances if the total moment is zero. Thus, the center of mass is the point about which the total moment is zero.

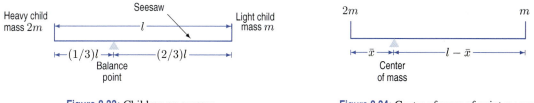

Figure 8.33: Children on seesaw   **Figure 8.34**: Center of mass of point masses

Example 5   Calculate the position of the center of mass of the children in Figure 8.33 using moments.

Solution   Suppose the center of mass in Figure 8.34 is at a distance of $\bar{x}$ from the left end. The moment of the left mass about the center of mass is $-2m\bar{x}$ (it is negative since it is to the left of the center of mass); the moment of the right mass about the center of mass is $m(l - \bar{x})$. The system balances if

$$-2m\bar{x} + m(l - \bar{x}) = 0 \quad \text{or} \quad ml - 3m\bar{x} = 0 \quad \text{so} \quad \bar{x} = \frac{1}{3}l.$$

Thus, the center of mass is $l/3$ from the left end.

We use the same method to calculate the center of mass, $\bar{x}$, of the system in Figure 8.35. The sum of the moments of the three masses about $\bar{x}$ is 0, so

$$m_1(x_1 - \bar{x}) + m_2(x_2 - \bar{x}) + m_3(x_3 - \bar{x}) = 0.$$

Solving for $\bar{x}$, we get

$$m_1\bar{x} + m_2\bar{x} + m_3\bar{x} = m_1x_1 + m_2x_2 + m_3x_3$$

$$\bar{x} = \frac{m_1x_1 + m_2x_2 + m_3x_3}{m_1 + m_2 + m_3} = \frac{\sum_{i=1}^{3} m_i x_i}{\sum_{i=1}^{3} m_i}$$

Generalizing leads to the following formula:

The **center of mass** of a system of n point masses $m_1, m_2, \ldots, m_n$ located at positions $x_1, x_2, \ldots, x_n$ along the x-axis is given by

$$\bar{x} = \frac{\sum x_i m_i}{\sum m_i}.$$

The numerator is the sum of the moments of the masses about the origin; the denominator is the total mass of the system.

Figure 8.35: Center of mass of discrete masses, m_1, m_2, m_3

Example 6   Show that the definition of $\bar{x}$ gives the same answer as we found in Example 5.

Solution   Suppose the origin is at the left end of the seesaw in Figure 8.33. The total mass of the system is $2m + m = 3m$. We compute

$$\bar{x} = \frac{\sum x_i m_i}{\sum m_i} = \frac{1}{3m}(2m \cdot 0 + m \cdot l) = \frac{ml}{3m} = \frac{l}{3}.$$

Continuous Mass Density

Instead of discrete masses arranged along the x-axis, suppose we have an object lying on the x-axis between $x = a$ and $x = b$. At point x, suppose the object has mass density (mass per unit length) of $\delta(x)$. To calculate the center of mass of such an object, divide it into n pieces, each of length Δx. On each piece, the density is nearly constant, so the mass of the piece is given by density times length. See Figure 8.36. Thus, if x_i is a point in the ith piece,

$$\text{Mass of the } i^{\text{th}} \text{ piece, } m_i \approx \delta(x_i)\Delta x.$$

Then the formula for the center of mass, $\bar{x} = \sum x_i m_i / \sum m_i$, applied to the n pieces of the object gives

$$\bar{x} = \frac{\sum x_i \delta(x_i)\Delta x}{\sum \delta(x_i)\Delta x}.$$

In the limit as $n \to \infty$ we have the following formula:

> The **center of mass** $\bar{x}$ of an object lying along the x-axis between $x = a$ and $x = b$ is
>
> $$\bar{x} = \frac{\int_a^b x\delta(x)\,dx}{\int_a^b \delta(x)\,dx},$$
>
> where $\delta(x)$ is the density (mass per unit length) of the object.

As in the discrete case, the denominator is the total mass of the object.

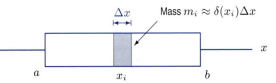

Figure 8.36: Calculating the center of mass of an object of variable density, $\delta(x)$

Example 7   Find the center of mass of a 2-meter rod lying on the x-axis with its left end at the origin if:
(a)   The density is constant and the total mass is 5 kg.   (b)   The density is $\delta(x) = 15x^2$ kg/m.

Solution   (a)   Since the density is constant along the rod, we expect the balance point to be in the middle, that is, $\bar{x} = 1$. To check this, we compute $\bar{x}$. The density is the total mass divided by the length, so $\delta(x) = 5/2$ kg/m. Then

$$\bar{x} = \frac{\text{Moment}}{\text{Mass}} = \frac{\int_0^2 x \cdot \frac{5}{2}\,dx}{5} = \frac{1}{5} \cdot \frac{5}{2} \cdot \frac{x^2}{2}\bigg|_0^2 = 1 \text{ meter.}$$

(b)   Since more of the mass of the rod is closer to its right end (the density is greatest there), we expect the center of mass to be in the right half of the rod, that is, between $x = 1$ and $x = 2$.

We have

$$\text{Total mass} = \int_0^2 15x^2\, dx = 5x^3 \Big|_0^2 = 40 \text{ kg.}$$

Thus,

$$\bar{x} = \frac{\text{Moment}}{\text{Mass}} = \frac{\int_0^2 x \cdot 15x^2 dx}{40} = \frac{15}{40} \cdot \frac{x^4}{4}\Big|_0^2 = 1.5 \text{ meter.}$$

Two- and Three-Dimensional Regions

For a system of masses that lies in the plane, the center of mass is a point with coordinates $(\bar{x}, \bar{y})$. In three dimensions, the center of mass is a point with coordinates $(\bar{x}, \bar{y}, \bar{z})$. To compute the center of mass in three dimensions, we use the following formulas in which $A_x(x)$ is the area of a slice perpendicular to the x-axis at x, and $A_y(y)$ and $A_z(z)$ are defined similarly. In two dimensions, we use the same formulas for $\bar{x}$ and $\bar{y}$, but we interpret $A_x(x)$ and $A_y(y)$ as the lengths of strips perpendicular to the x- and y-axes, respectively.

For a region of constant density δ, the center of mass is given by

$$\bar{x} = \frac{\int x\delta A_x(x)\, dx}{\text{Mass}} \qquad \bar{y} = \frac{\int y\delta A_y(y)\, dy}{\text{Mass}} \qquad \bar{z} = \frac{\int z\delta A_z(z)\, dz}{\text{Mass}}$$

The expression $\delta A_x(x)\Delta x$ is the moment of a slice perpendicular to the x-axis. Thus, these formulas are extensions of that on page 364. In the two and three dimensional case, we are assuming that the density δ is constant. If the density is not constant, finding the center of mass may require a double or triple integral from multivariable calculus.

Example 8   Find the coordinates of the center of mass of the isosceles triangle in Figure 8.37. The triangle has constant density and mass m.

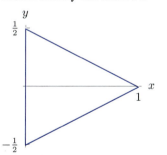

Figure 8.37: Find center of mass of this triangle

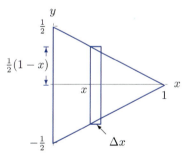

Figure 8.38: Sliced triangle

Solution   Because the mass of the triangle is symmetrically distributed with respect to the x-axis, $\bar{y} = 0$. We expect $\bar{x}$ to be closer to $x = 0$ than to $x = 1$, since the triangle is wider near the origin.

The area of the triangle is $\frac{1}{2} \cdot 1 \cdot 1 = \frac{1}{2}$. Thus Density = Mass/Area = $2m$. If we slice the triangle into strips of width Δx, then the strip at position x has length $A_x(x) = 2 \cdot \frac{1}{2}(1-x) = (1-x)$. (See Figure 8.38.) So

$$\text{Area of strip} = A_x(x)\Delta x \approx (1-x)\Delta x.$$

Since the density is $2m$, the center of mass is given by

$$\bar{x} = \frac{\int x\delta A_x(x)\, dx}{\text{Mass}} = \frac{\int_0^1 2mx(1-x)\, dx}{m} = 2\left(\frac{x^2}{2} - \frac{x^3}{3}\right)\Big|_0^1 = \frac{1}{3}.$$

So the center of mass of this triangle is at the point $(\bar{x}, \bar{y}) = (1/3, 0)$.

Example 9   Find the center of mass of a hemisphere of radius 7 cm and constant density δ.

Solution   Stand the hemisphere with its base horizontal in the xy-plane, with the center at the origin. Symmetry tells us that its center of mass lies directly above the center of the base, so $\bar{x} = \bar{y} = 0$. Since the hemisphere is wider near its base, we expect the center of mass to be nearer to the base than the top.

To calculate the center of mass, slice the hemisphere into horizontal disks as in Figure 8.9 on page 353. A disk of thickness Δz at height z above the base has

$$\text{Volume of disk } = A_z(z)\Delta z \approx \pi(7^2 - z^2)\Delta z \text{ cm}^3.$$

So, since the density is δ,

$$\bar{z} = \frac{\int z\delta A_z(z)\,dz}{\text{Mass}} = \frac{\int_0^7 z\delta\pi(7^2 - z^2)\,dz}{\text{Mass}}.$$

Since the total mass of the hemisphere is $\left(\frac{2}{3}\pi 7^3\right)\delta$, we get

$$\bar{z} = \frac{\delta\pi\int_0^7(7^2z - z^3)\,dz}{\text{Mass}} = \frac{\delta\pi\left(7^2z^2/2 - z^4/4\right)\big|_0^7}{\text{Mass}} = \frac{\frac{7^4}{4}\delta\pi}{\frac{2}{3}\pi 7^3\delta} = \frac{21}{8} = 2.625 \text{ cm}.$$

The center of mass of the hemisphere is 2.625 cm above the center of its base. As expected, it is closer to the base of the hemisphere than its top.

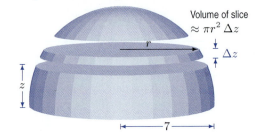

Figure 8.39: Slicing to find the center of mass of a hemisphere

Exercises and Problems for Section 8.3

Exercises

1. Find the mass of a rod of length 10 cm with density $\delta(x) = e^{-x}$ gm/cm at a distance of x cm from the left end.

2. A plate occupying the region $0 \le x \le 2, 0 \le y \le 3$ has density $\delta = 5$ gm/cm^2. Set up two integrals giving the mass of the plate, one corresponding to strips in the x-direction and one corresponding to strips in the y-direction.

3. A rod has length 2 meters. At a distance x meters from its left end, the density of the rod is given by

$$\delta(x) = 2 + 6x \text{ gm/m}.$$

(a) Write a Riemann sum approximating the total mass of the rod.

(b) Find the exact mass by converting the sum into an integral.

4. If a rod lies along the x-axis between a and b, the moment of the rod is $\int_a^b x\delta(x)\,dx$, where $\delta(x)$ is its density in grams/meter at a position x meters. Find the moment and center of mass of the rod in Problem 3.

5. The density of cars (in cars per mile) down a 20-mile stretch of the Pennsylvania Turnpike is approximated by

$$\delta(x) = 300\left(2 + \sin\left(4\sqrt{x + 0.15}\right)\right),$$

at a distance x miles from the Breezewood toll plaza.

(a) Sketch a graph of this function for $0 \le x \le 20$.

(b) Write a Riemann sum that approximates the total number of cars on this 20-mile stretch.

(c) Find the total number of cars on the 20-mile stretch.

6. (a) Find a Riemann sum which approximates the total mass of a 3×5 rectangular sheet, whose density per unit area at a distance x from one of the sides of length 5 is $1/(1 + x^4)$.

(b) Calculate the mass.

7. A point mass of 2 grams located 3 centimeters to the left of the origin and a point mass of 5 grams located 4 centimeters to the right of the origin are connected by a thin, light rod. Find the center of mass of the system.

8. Find the center of mass of a system containing three point masses of 5 gm, 3 gm, and 1 gm located respectively at $x = -10, x = 1$, and $x = 2$.

Problems

9. Find the total mass of the triangular region in Figure 8.40, which has density $\delta(x) = 1 + x$ grams/cm^2.

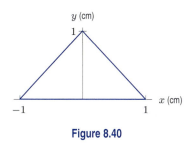

Figure 8.40

10. The density of oil in a circular oil slick on the surface of the ocean at a distance r meters from the center of the slick is given by $\delta(r) = 50/(1 + r)$ kg/m^2.

   (a) If the slick extends from $r = 0$ to $r = 10,000$ m, find a Riemann sum approximating the total mass of oil in the slick.
   (b) Find the exact value of the mass of oil in the slick by turning your sum into an integral and evaluating it.
   (c) Within what distance r is half the oil of the slick contained?

11. The soot produced by a garbage incinerator spreads out in a circular pattern. The depth, $H(r)$, in millimeters, of the soot deposited each month at a distance r kilometers from the incinerator is given by $H(r) = 0.115e^{-2r}$.

   (a) Write a definite integral giving the total volume of soot deposited within 5 kilometers of the incinerator each month.
   (b) Evaluate the integral you found in part (a), giving your answer in cubic meters.

12. A cardboard figure has the shape shown in Figure 8.41. The region is bounded on the left by the line $x = a$, on the right by the line $x = b$, above by $f(x)$, and below by $g(x)$. If the density $\delta(x)$ gm/cm^2 varies only with x, find an expression for the total mass of the figure, in terms of $f(x)$, $g(x)$, and $\delta(x)$.

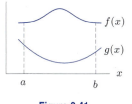

Figure 8.41

13. The storage shed in Figure 8.42 is the shape of a half-cylinder of radius r and length l.

   (a) What is the volume of the shed?
   (b) Suppose the shed is filled with sawdust whose density (mass/unit volume) at any point is proportional to the distance of that point from the floor. The constant of proportionality is k. Calculate the total mass of sawdust in the shed.

Figure 8.42

14. The following table gives the density D (in gm/cm^3) of the earth at a depth x km below the earth's surface. The radius of the earth is about 6370 km. Find an upper and a lower bound for the earth's mass such that the upper bound is less than twice the lower bound. Explain your reasoning; in particular, what assumptions have you made about the density?

x	0	1000	2000	2900	3000	4000	5000	6000	6370
D	3.3	4.5	5.1	5.6	10.1	11.4	12.6	13.0	13.0

15. Water leaks out of a tank through a square hole with 1-inch sides. At time t (in seconds) the velocity of water flowing through the hole is $v = g(t)$ ft/sec. Write a definite integral that represents the total amount of water lost in the first minute.

16. An exponential model for the density of the earth's atmosphere says that if the temperature of the atmosphere were constant, then the density of the atmosphere as a function of height, h (in meters), above the surface of the earth would be given by

$$\delta(h) = 1.28e^{-0.000124h} \text{ kg/m}^3.$$

   (a) Write (but do not evaluate) a sum that approximates the mass of the portion of the atmosphere from $h = 0$ to $h = 100$ m (i.e., the first 100 meters above sea level). Assume the radius of the earth is 6400 km.
   (b) Find the exact answer by turning your sum in part (a) into an integral. Evaluate the integral.

17. Three point masses of 4 gm each are placed at $x = -6, 1$ and 3. Where should a point fourth mass of 4 gm be placed to make the center of mass at the origin?

18. A rod of length 3 meters with density $\delta(x) = 1 + x^2$ grams/meter is positioned along the positive x-axis, with its left end at the origin. Find the total mass and the center of mass of the rod.

19. A rod with density $\delta(x) = 2 + \sin x$ lies on the x-axis between $x = 0$ and $x = \pi$. Find the center of mass of the rod.

20. A rod of length 1 meter has density $\delta(x) = 1 + kx^2$ grams/meter, where k is a positive constant. The rod is lying on the positive x-axis with one end at the origin.

   (a) Find the center of mass as a function of k.
   (b) Show that the center of mass of the rod satisfies $0.5 < \bar{x} < 0.75$.

21. A rod of length 2 meters and density $\delta(x) = 3 - e^{-x}$ kilograms per meter is placed on the x-axis with its ends at $x = \pm 1$.

   (a) Will the center of mass of the rod be on the left or right of the origin? Explain.
   (b) Find the coordinate of the center of mass.

22. A metal plate, with constant density 2 gm/cm^2, has a shape bounded by the curve $y = x^2$ and the x-axis, with $0 \le x \le 1$ and x, y in cm.

   (a) Find the total mass of the plate.
   (b) Sketch the plate, and decide, on the basis of the shape, whether $\bar{x}$ is less than or greater than $1/2$.
   (c) Find $\bar{x}$.

23. A metal plate, with constant density 5 gm/cm^2, has a shape bounded by the curve $y = \sqrt{x}$ and the x-axis, with $0 \le x \le 1$ and x, y in cm.

   (a) Find the total mass of the plate.
   (b) Find $\bar{x}$ and $\bar{y}$.

24. An isosceles triangle with uniform density, altitude a, and base b is placed in the xy-plane as in Figure 8.43. Show that the center of mass is at $\bar{x} = a/3$, $\bar{y} = 0$. Hence show that the center of mass is independent of the triangle's base.

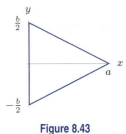

Figure 8.43

25. Find the center of mass of a cone of height 5 cm and base diameter 10 cm with constant density δ gm/cm^3.

26. A solid is formed by rotating the region bounded by the curve $y = e^{-x}$ and the x-axis between $x = 0$ and $x = 1$, around the x-axis. It was shown in Example 1 on page 353 that the volume of this solid is $\pi(1 - e^{-2})/2$. Assuming the solid has constant density δ, find $\bar{x}$ and $\bar{y}$.

27. (a) Find the mass of a pyramid of constant density δ gm/cm^3 with a square base of side 40 cm and height 10 cm. [That is, the vertex is 10 cm above the center of the base.]
   (b) Find the center of mass of the pyramid.

8.4 APPLICATIONS TO PHYSICS

Although geometric problems were a driving force for the development of the calculus in the seventeenth century, it was Newton's spectacularly successful applications of the calculus to physics that most clearly demonstrated the power of this new mathematics.

Work

In physics the word "work" has a technical meaning which is different from its everyday meaning. Physicists say that if a constant force, F, is applied to some object to move it a distance, d, then the force has done work on the object. The force must be parallel to the motion (in the same or the opposite direction). We make the following definition:

$$\text{Work done} = \text{Force} \cdot \text{Distance} \qquad \text{or} \qquad W = F \cdot d$$

Notice that if we walk across a room holding a book, we do no work on the book, since the force we exert on the book is vertical, but the motion of the book is horizontal. On the other hand, if we lift the book from the floor to a table, we accomplish work.

There are several sets of units in common use. To measure work, we will generally use the two sets of units, International (SI) and British, in the following table.

	Force	Distance	Work	Conversions
International (SI) units	newton (nt)	meter (m)	joule (j)	1 lb = 4.45 nt
British units	pound (lb)	foot (ft)	foot-pound (ft-lb)	1 ft = 0.305 m
				1 ft-lb = 1.36 joules

One joule of work is done when a force of 1 newton moves an object through 1 meter, so 1 joule = 1 newton-meter.

Example 1   Calculate the work done on an object when

(a)   A force of 2 newtons moves it 12 meters.   (b)   A 3-lb force moves it 4 feet.

Solution   (a)   Work done = $2 \text{ nt} \cdot 12 \text{ m} = 24$ joules.   (b)   Work done = $3 \text{ lb} \cdot 4 \text{ ft} = 12$ ft-lb.

In the previous example, the force was constant and we calculated the work by multiplication. In the next example, the force varies, so we need an integral. We divide up the distance moved and sum to get a definite integral representing the work.

Example 2   Hooke's Law says that the force, F, required to compress the spring in Figure 8.44 by a distance x, in meters, is given by $F = kx$, for some constant k. Find the work done in compressing the spring by 0.1 m if $k = 8$ nt/m.

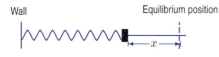

Figure 8.44: Compression of spring: Force is kx

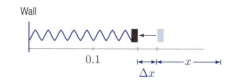

Figure 8.45: Work done in compressing spring a small distance Δx is $kx\Delta x$

Solution   Since k is in newtons/meter and x is in meters, we have $F = 8x$ newtons. Since the force varies with x, we divide the distance moved into small increments, Δx, as in Figure 8.45. Then

$$\text{Work done in moving through an increment} \approx F\Delta x = 8x\Delta x \text{ joules.}$$

So, summing over all increments gives the Riemann sum approximation

$$\text{Total work done} \approx \sum 8x\Delta x.$$

Taking the limit as $\Delta x \to \infty$ gives

$$\text{Total work done} = \int_0^{0.1} 8x \, dx = 4x^2 \Big|_0^{0.1} = 0.04 \text{ joules.}$$

In general, if force is a function $F(x)$ of position x, then in moving from $x = a$ to $x = b$,

$$\boxed{\text{Work done} = \int_a^b F(x) \, dx}$$

The Force Due to Gravity: Mass versus Weight

When an object is lifted, work is done against the force exerted by gravity on the object. By Newton's Second Law, the downward gravitational force acting on a mass m is mg, where g is the acceleration due to gravity. To lift the object, we need to exert a force equal to the gravitational force but in the opposite direction.

In International units, $g \approx 9.8$ m/sec^2, and we will usually measure mass, m, in *kilograms*. In British units, mass is seldom used. Instead, we usually talk about the *weight* of an object, which is the force exerted by gravity on the object. Roughly speaking, the mass represents the quantity of matter in an object, whereas the weight represents the force of gravity on it. The mass of an object is the same everywhere, whereas the weight can vary if, for example, the object is moved to outer space where gravitational forces are smaller.

When we are given the weight of an object, we do not multiply by g to find the gravitational force as it has already been done. In British units, a *pound* is a unit of weight. In International units, a kilogram is a unit of mass, and the unit of weight is a newton, where 1 newton $= 1$ kg $\cdot$ m/sec^2.

Example 3   How much work is done in lifting

(a)   A 5-pound book 3 feet off the floor?   (b)   A 1.5-kilogram book 2 meters off the floor?

Solution   (a) The force due to gravity is 5 lb, so $W = F \cdot d = (5 \text{ lb})(3 \text{ ft}) = 15$ foot-pounds.
(b) The force due to gravity is $mg = (1.5 \text{ kg})(g \text{ m/sec}^2)$, so

$$W = F \cdot d = [(1.5 \text{ kg})(9.8 \text{ m/sec}^2)] \cdot (2 \text{ m}) = 29.4 \text{ joules.}$$

In the previous example, work is found by multiplication. In the next example, different parts of the object move different distances, so an integral is needed.

Example 4   A 28-meter uniform chain with a mass 2 kilograms per meter is dangling from the roof of a building. How much work is needed to pull the chain up to the top of the building?

Solution   Since 1 meter of the chain has mass density 2 kg, the gravitational force per meter of chain is $(2 \text{ kg})(9.8 \text{ m/sec}^2) = 19.6$ newtons. Let's divide the chain into small sections of length Δy, each requiring a force of $19.6 \, \Delta y$ newtons to move it against gravity. See Figure 8.46. If Δy is small, all of this piece is hauled up approximately the same distance, namely y, so

$$\text{Work done on the small piece } \approx (19.6 \, \Delta y \text{ newtons})(y \text{ meters}) = 19.6y \, \Delta y \text{ joules.}$$

The work done on the entire chain is given by the total of the work done on each piece:

$$\text{Work done } \approx \sum 19.6y \, \Delta y \text{ joules.}$$

As $\Delta y \to 0$, we obtain a definite integral. Since y varies from 0 to 28 meters, the total work is

$$\text{Work done } = \int_0^{28} (19.6y) \, dy = 9.8y^2 \Big|_0^{28} = 7683.2 \text{ joules.}$$

Top of building

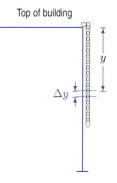

y

Δy

Figure 8.46: Chain for Example 4

Example 5   Calculate the work done in pumping oil from the cone-shaped tank in Figure 8.47 to the rim. The oil has density 800 kg/m^3 and its vertical depth is 10 m.

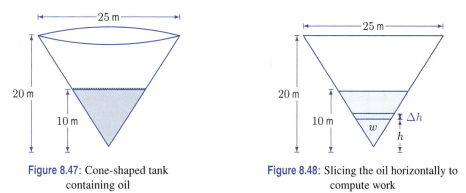

Figure 8.47: Cone-shaped tank
containing oil

Figure 8.48: Slicing the oil horizontally to
compute work

Solution   We slice the oil horizontally because each part of such a slice moves the same vertical distance. Each slice is approximately a circular disk with radius $w/2$ m, so

$$\text{Volume of slice} \approx \pi \left(\frac{w}{2}\right)^2 \Delta h = \frac{\pi}{4} w^2 \Delta h \text{ m}^3.$$

$$\text{Force of gravity on slice} = \text{Density} \cdot g \cdot \text{Volume} = 800g\frac{\pi}{4}w^2\Delta h = 200\pi gw^2\Delta h \text{ nt.}$$

Since each part of the slice has to move a vertical distance of $(20 - h)$ m, we have

$$\text{Work done on slice} \approx \text{Force} \cdot \text{Distance} = 200\pi gw^2\Delta h \text{ nt} \cdot (20 - h) \text{ m}$$
$$= 200\pi gw^2(20 - h)\Delta h \text{ joules.}$$

To find w in terms of h, we use the similar triangles in Figure 8.48:

$$\frac{w}{h} = \frac{25}{20} \quad \text{so} \quad w = \frac{5}{4}h = 1.25h.$$

Thus,

$$\text{Work done on strip} \approx 200\pi g(1.25h)^2(20 - h)\Delta h = 312.5\pi gh^2(20 - h)\Delta h \text{ joules.}$$

Summing and taking the limit as $\Delta h \to 0$ gives an integral with upper limit $h = 10$, the depth of the oil.

$$\text{Total work} = \lim_{\Delta h \to 0} \sum 312.5\pi gh^2(20 - h)\Delta h = \int_0^{10} 312.5\pi gh^2(20 - h)\, dh \text{ joules.}$$

Evaluating the integral using $g = 9.8$ m/sec^2 gives

$$\text{Total work} = 312.5\pi g \left(20\frac{h^3}{3} - \frac{h^4}{4}\right)\Bigg|_0^{10} = 1,302,083\pi g \approx 4.0 \cdot 10^7 \text{ joules.}$$

In the following example, information is given in British units about the weight of the pyramid, so we do not need to multiply by g to find the gravitational force.

Example 6   It is reported that the Great Pyramid of Egypt was built in 20 years. If the stone making up the pyramid has density 200 pounds per cubic foot, find the total amount of work done in building the pyramid. The pyramid is 410 feet high and has a square base 755 feet by 755 feet. Estimate how many workers were needed to build the pyramid.

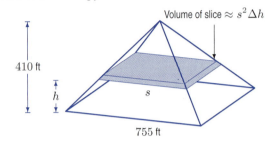

Volume of slice $\approx s^2 \Delta h$

410 ft

h

s

755 ft

Figure 8.49: Pyramid for Example 6

Solution   We assume that the stones were originally located at the approximate height of the construction site. Imagine the pyramid built in layers as we did in Example 5 on page 349.

By similar triangles, the layer at height h has a side length $s = \frac{755}{410}(410 - h)$ ft. (See Figure 8.49.) The layer at height h has a volume of $s^2 \Delta h$ ft^3, so its weight is $200 s^2 \Delta h$ lb. This layer is lifted through a height of h, so

$$\text{Work to lift layer} = (200 s^2 \Delta h \text{ lb})(h \text{ ft}) = 200 s^2 h \Delta h \, ft - lb.$$

Summing over all layers gives

$$\text{Total work} \approx \sum 200 s^2 h \Delta h = \sum 200 \left(\frac{755}{410}\right)^2 (410 - h)^2 h \Delta h \text{ ft-lb.}$$

Since h varies from 0 to 410, as $\Delta h \to 0$, we obtain

$$\text{Total work} = \int_0^{410} 200 \left(\frac{755}{410}\right)^2 (410 - h)^2 h \, dh \approx 1.6 \cdot 10^{12} \text{ foot-pounds.}$$

We have calculated the total work done in building the pyramid; now we want to estimate the total number of workers needed. Let's assume every laborer worked 10 hours a day, 300 days a year, for 20 years. Assume that a typical worker lifted ten 50 pound blocks a distance of 4 feet every hour, thus performing 2000 foot-pounds of work per hour (this is a very rough estimate). Then each laborer performed $(10)(300)(20)(2000) = 1.2 \cdot 10^8$ foot-pounds of work over a twenty year period. Thus, the number of workers needed was about $(1.6 \cdot 10^{12})/(1.2 \cdot 10^8)$, or about 13,000.

Force and Pressure

We can use the definite integral to compute the force exerted by a liquid on a surface, for example, the force of water on a dam. The idea is to get the force from the *pressure*. The pressure in a liquid is the force per unit area exerted by the liquid. Two things you need to know about pressure are:

- At any point, pressure is exerted equally in all directions—up, down, sideways.

- Pressure increases with depth. (That is one of the reasons why deep sea divers have to take much greater precautions than scuba divers.)

At a depth of h meters, the pressure, p, exerted by the liquid, measured in newtons per square meter, is given by computing the total weight of a column of liquid h meters high with a base of 1 square meter. The volume of such a column of liquid is just h cubic meters. If the liquid has density δ (mass per unit volume), then its weight per unit volume is δg, where g is the acceleration due to gravity. The weight of the column of liquid is δgh, so

$$\text{Pressure} = \text{Mass density} \cdot g \cdot \text{Depth} \quad \text{or} \quad p = \delta gh$$

Provided the pressure is constant over a given area, we also have the following relation:

$$\text{Force} = \text{Pressure} \cdot \text{Area}$$

The units and data we will generally use are given in the following table:

	Force	Area	Pressure	Density of water
International (SI) units	newton	meter^2	nt/m^2, called pascal	1000 kg/m^3 (mass)
British units	pound	foot^2	lb/ft^2	62.4 lb/ft^3 (weight)

In International units, the mass density of water is 1000 kg/m^3, so the pressure at a depth of h meters is $\delta gh = 1000 \cdot 9.8h = 9800h$ nt/m^2. See Figure 8.50.

In British units, the density of the liquid is usually given as a weight per unit volume, rather than a mass per unit volume. In that case, we do not need to multiply by g because it has already been done. For example, water weighs 62.4 lb/ft^3, so the pressure at depth h feet is $62.4h$ lb/ft^2. See Figure 8.51.

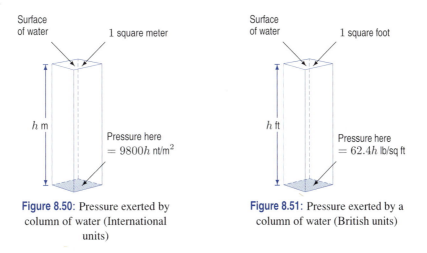

Figure 8.50: Pressure exerted by column of water (International units)

Figure 8.51: Pressure exerted by a column of water (British units)

If the pressure is constant over a surface, we calculate the force on the surface by multiplying the pressure by the area of the surface. If the pressure is not constant, we divide the surface into small pieces *in such a way that the pressure is nearly constant on each one* to obtain a definite integral for the force on the surface. Since the pressure varies with depth, we divide the surface into horizontal strips, each of which is at an approximately constant depth.

Example 7 In 1912, the ocean liner Titanic sank to the bottom of the Atlantic, 12,500 feet (nearly 2.5 miles) below the surface. Find the force on one side of a 100-foot square plate at the depth of the Titanic if the plate is: (a) Lying horizontally   (b) Standing vertically.

Solution   (a) When the plate is horizontal, the pressure is the same at every point on the plate, so

$$\text{Pressure} = 62.4 \text{ lb/ft}^3 \cdot 12{,}500 \text{ ft} = 780{,}000 \text{ lb/ft}^2.$$

To imagine this pressure, convert to pounds per square inch, giving $780{,}000/144 \approx 5400 \text{ lb/in}^2$. For the horizontal plate

$$\text{Force} = 780{,}000 \text{ lb/ft}^2 \cdot 100^2 \text{ ft}^2 = 7.8 \cdot 10^9 \text{ pounds.}$$

(b) When the plate is vertical, only the bottom is at 12,500 feet; the top is at 12,400 feet. Dividing into horizontal strips of width Δh, as in Figure 8.52, we have

$$\text{Area of strip} = 100\Delta h \text{ ft}^2.$$

Since the pressure on a strip at a depth of h feet is $62.4h$ lb/ft^2,

$$\text{Force on strip} \approx 62.4h \cdot 100\Delta h = 6240h\Delta h \text{ pounds.}$$

Summing over all strips and taking the limit as $\Delta h \to 0$ gives a definite integral. The strips vary in depth from 12,400 to 12,500 feet, so

$$\text{Total force} = \lim_{\Delta h \to 0} \sum 6240h\Delta h = \int_{12{,}400}^{12{,}500} 6240h \, dh \text{ pounds.}$$

Evaluating the integral gives

$$\text{Total force} = 6240\frac{h^2}{2}\Big|_{12{,}400}^{12{,}500} = 3120(12{,}500^2 - 12{,}400^2) = 7.77 \cdot 10^9 \text{ pounds.}$$

Notice that the answer to part (b) is smaller than the answer to part (a). This is because part of the plate is at a smaller depth in part (b) than in part (a).

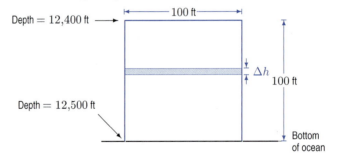

Figure 8.52: Square plate at bottom of ocean; h measured from the surface of water

Example 8   Figure 8.53 shows a dam approximately the size of Hoover Dam, which stores water for California, Nevada, and Arizona. Calculate: (a)   The water pressure at the base of the dam.   (b)   The total force of the water on the dam.

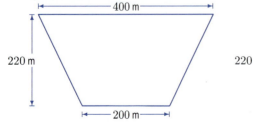

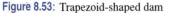

Figure 8.53: Trapezoid-shaped dam

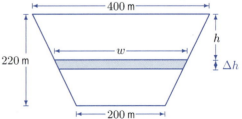

Figure 8.54: Dividing dam into horizontal strips

Solution   (a) Since the density of water is $\delta = 1000$ kg/m^3, at the base of the dam,

$$\text{Water pressure } = \delta g h = 1000 \cdot 9.8 \cdot 220 = 2.156 \cdot 10^6 \text{ nt/m}^2.$$

(b) To calculate the force on the dam, we divide the dam into horizontal strips because the pressure along each strip is approximately constant. See Figure 8.54. Since each strip is approximately rectangular,

$$\text{Area of strip } \approx w \Delta h \text{ m}^2.$$

The pressure at a depth of h meters is $\delta g h = 9800h$ nt/m^2. Thus,

$$\text{Force on strip } \approx \text{ Pressure } \cdot \text{ Area } = 9800 h w \Delta h \text{ nt.}$$

To find w in terms of h, we use the fact that w decreases linearly from $w = 400$ when $h = 0$ to $w = 200$ when $h = 220$. Thus w is a linear function of h, with slope $(200 - 400)/220 = -10/11$, so

$$w = 400 - \frac{10}{11}h.$$

Thus

$$\text{Force on strip } \approx 9800 h \left(400 - \frac{10}{11}h \right) \Delta h \text{ nt.}$$

Summing over all strips and taking the limit as $\Delta h \to 0$ gives

$$\text{Total force on dam } = \lim_{\Delta h \to 0} \sum 9800 h \left(400 - \frac{10}{11}h \right) \Delta h$$

$$= \int_0^{220} 9800 h \left(400 - \frac{10}{11}h \right) dh \text{ newtons.}$$

Evaluating the integral gives

$$\text{Total force } = 9800 \left(200 h^2 - \frac{10}{33} h^3 \right) \Big|_0^{220} = 6.32 \cdot 10^{10} \text{ newtons.}$$

In fact, Hoover Dam is not flat, as the problem assumed, but arched, to better withstand the pressure.

Exercises and Problems for Section 8.4

Exercises

In Exercises 1–3, the force, F, required to compress a spring by a distance x meters is given by $F = 3x$ newtons.

1. Find the work done in compressing the spring from $x = 1$ to $x = 2$.

2. Find the work done to compress the spring to $x = 3$, starting at the equilibrium position, $x = 0$.

3. (a) Find the work done in compressing the spring from $x = 0$ to $x = 1$ and in compressing the spring from $x = 4$ to $x = 5$.
(b) Which of the two answers is larger? Why?

4. The gravitational force on a 1 kg object at a distance r meters from the center of the earth is $F = 4 \cdot 10^{14}/r^2$ newtons. Find the work done in moving the object from the surface of the earth to a height of 10^6 meters above the surface. The radius of the earth is $6.4 \cdot 10^6$ meters.

5. How much work is required to lift a 1000-kg satellite to an altitude of $2 \cdot 10^6$ m above the surface of the earth? The gravitational force is $F = GMm/r^2$, where M is the mass of the earth, m is the mass of the satellite, and r is the distance between them. The radius of the earth is $6.4 \cdot 10^6$ m, its mass is $6 \cdot 10^{24}$ kg, and in these units the gravitational constant, G, is $6.67 \cdot 10^{-11}$.

Problems

6. A worker on a scaffolding 75 ft above the ground needs to lift a 500 lb bucket of cement from the ground to a point 30 ft above the ground by pulling on a rope weighing 5 lb/ft. How much work is required?

7. A 1000-lb weight is being lifted to a height 10 feet off the ground. It is lifted using a rope which weighs 4 lb per foot and which is being pulled up by construction workers standing on a roof 30 feet off the ground. Find the work done to lift the weight.

8. A bucket of water of mass 20 kg is pulled at constant velocity up to a platform 40 meters above the ground. This takes 10 minutes, during which time 5 kg of water drips out at a steady rate through a hole in the bottom. Find the work needed to raise the bucket to the platform.

9. A rectangular water tank has length 20 ft, width 10 ft, and depth 15 ft. If the tank is full, how much work does it take to pump all the water out?

10. A water tank is in the form of a right circular cylinder with height 20 ft and radius 6 ft. If the tank is half full of water, find the work required to pump all of it over the top rim.

11. Suppose the tank in Problem 10 is full of water. Find the work required to pump all of it to a point 10 ft above the top of the tank.

12. A fuel oil tank is an upright cylinder, buried so that its circular top is 10 feet beneath ground level. The tank has a radius of 5 feet and is 15 feet high, although the current oil level is only 6 feet deep. Calculate the work required to pump all of the oil to the surface. Oil weighs 50 lb/ft^3.

13. A gas station stores its gasoline in a tank under the ground. The tank is a cylinder lying horizontally on its side. (In other words, the tank is not standing vertically on one of its flat ends.) If the radius of the cylinder is 4 feet, its length is 12 feet, and its top is 10 feet under the ground, find the total amount of work needed to pump the gasoline out of the tank. (Gasoline weighs 42 lb/ft^3.)

14. A cylindrical barrel, standing upright on its circular end, contains muddy water. The top of the barrel, which has diameter 1 meter, is open. The height of the barrel is 1.8 meter and it is filled to a depth of 1.5 meter. The density of the water at a depth of h meters below the surface is given by $\delta(h) = 1 + kh$ kg/m^3, where k is a positive constant. Find the total work done to pump the muddy water to the top rim of the barrel. (You can leave π, k, and g in your answer.)

15. (a) A reservoir has a dam at one end. The dam is a rectangular wall, 1000 feet long and 50 feet high. Approximate the total force of the water on the dam by a Riemann sum.
    (b) Write an integral which represents the force, and evaluate it.

16. Before any water was pumped out, what was the total force on the bottom and each side of the tank in Problem 9?

17. A lobster tank in a restaurant is 4 ft long by 3 ft wide by 2 ft deep. Find the water force on the bottom and on each of the four sides.

18. The Three Gorges Dam is currently being built in China. When it is finished in 2009, it will be the largest dam in the world: about 2000 m long and 180 m high, creating a lake the length of Lake Superior. Assume the dam is rectangular in shape.
    (a) Estimate the water pressure at the base of the dam.
    (b) Set up and evaluate a definite integral giving the total force of the water on the dam.

19. On August 12, 2000, the Russian submarine Kursk sank to the bottom of the sea, 350 feet below the surface. Find the following at the depth of the Kursk.
    (a) The water pressure in pounds per square foot and pounds per square inch.
    (b) The force on a 5-foot square metal sheet held
        (i) Horizontally.
        (ii) Vertically.

20. The ocean liner Titanic lies under 12,500 feet of water at the bottom of the Atlantic Ocean.
    (a) What is the water pressure at the Titanic? Give your answer in pounds per square foot and pounds per square inch.
    (b) Set up and calculate an integral giving the total force on a circular porthole (window) of diameter 6 feet standing vertically with its center at the depth of the Titanic.

21. Set up and calculate a definite integral giving the total pressure on the dam shown in Figure 8.55, which is about the size of the Aswan Dam in Egypt.

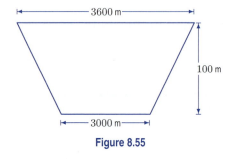

Figure 8.55

22. We define the electric potential at a distance r from an electric charge q by q/r. The electric potential of a charge distribution is obtained by adding up the potential from each point. Electric charge is sprayed (with constant density σ in units of charge/unit area) on to a circular

disk of radius a. Consider the axis perpendicular to the disk and through its center. Find the electric potential at the point P on this axis at a distance R from the center. (See Figure 8.56.)

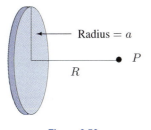

Figure 8.56

For Problems 23–24, find the kinetic energy of the rotating body. Use the fact that the kinetic energy of a particle of mass m moving at a speed v is $\frac{1}{2}mv^2$. Slice the object into pieces in such a way that the velocity is approximately constant on each piece.

23. Find the kinetic energy of a rod of mass 10 kg and length 6 m rotating about an axis perpendicular to the rod at its midpoint, with an angular velocity of 2 radians per second. (Imagine a helicopter blade of uniform thickness.)

24. Find the kinetic energy of a phonograph record of uniform density, mass 50 gm and radius 10 cm rotating at $33\frac{1}{3}$ revolutions per minute.

For Problems 25–27, find the gravitational force between two objects. Use the fact that the gravitational attraction between particles of mass m_1 and m_2 at a distance r apart is Gm_1m_2/r^2. Slice the objects into pieces, use this formula for the pieces, and sum using a definite integral.

25. What is the force of gravitational attraction between a thin uniform rod of mass M and length l and a particle of mass m lying in the same line as the rod at a distance a from one end?

26. Two long, thin, uniform rods of lengths l_1 and l_2 lie on a straight line with a gap between them of length a. Suppose their masses are M_1 and M_2, respectively, and the constant of the gravitation is G. What is the force of attraction between the rods? (Use the result of Problem 25.)

27. Find the gravitational force exerted by a thin uniform ring of mass M and radius a on a particle of mass m lying on a line perpendicular to the ring through its center. Assume m is at a distance y from the center of the ring.

28. A uniform, thin, circular disk of radius a and mass M lies on a horizontal plane. The point P lies a distance y directly above O, the center of the disk. Calculate the gravitational force on a mass m at the point P. (See Figure 8.57.) Use the fact that the gravitational force exerted on the mass m by a thin horizontal ring of radius r, mass μ, and center O is toward O and given by

$$F = \frac{G\mu my}{(r^2 + y^2)^{3/2}}, \quad \text{where } G \text{ is constant.}$$

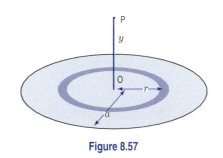

Figure 8.57

8.5   APPLICATIONS TO ECONOMICS

Present and Future Value

Many business deals involve payments in the future. If you buy a car or furniture, for example, you may buy it on credit and pay over a period of time. If you are going to accept payment in the future under such a deal, you obviously need to know how much you should be paid. Being paid $100 in the future is clearly worse than being paid $100 today for many reasons. If you are given the money today, you can do something else with it—for example, put it in the bank, invest it somewhere, or spend it. Thus, even without considering inflation, if you are to accept payment in the future, you would expect to be paid more to compensate for this loss of potential earnings. The question we will consider now is, how much more?

To simplify matters, we consider only what we would lose by not earning interest; we will not consider the effect of inflation. Let's look at some specific numbers. Suppose you deposit $100 in an account which earns 7% interest compounded annually, so that in a year's time you will have $107. Thus, $100 today will be worth $107 a year from now. We say that the $107 is the *future value* of the $100, and that the $100 is the *present value* of the $107. Observe that the present value is smaller than the future value. In general, we say the following:

- The **future value**, B, of a payment, P, is the amount to which the P would have grown if deposited in an interest bearing bank account.
- The **present value**, P, of a future payment, B, is the amount which would have to be deposited in a bank account today to produce exactly B in the account at the relevant time in the future.

With an interest rate of r, compounded annually, and a time period of t years, a deposit of P grows to a future balance of B, where

$$B = P(1+r)^t, \quad \text{or equivalently,} \quad P = \frac{B}{(1+r)^t}.$$

Note that for a 7% interest rate, $r = 0.07$. If instead of annual compounding, we have continuous compounding, we get the following result:

$$B = Pe^{rt}, \quad \text{or equivalently,} \quad P = \frac{B}{e^{rt}} = Be^{-rt}.$$

Example 1   You win the lottery and are offered the choice between $1 million in four yearly installments of $250,000 each, starting now, and a lump-sum payment of $920,000 now. Assuming a 6% interest rate, compounded continuously, and ignoring taxes, which should you choose?

Solution   We will do the problem in two ways. First, we assume that you pick the option with the largest present value. The first of the four $250,000 payments is made now, so

$$\text{Present value of first payment} = \$250{,}000.$$

The second payment is made one year from now, so

$$\text{Present value of second payment} = \$250{,}000e^{-0.06(1)}.$$

Calculating the present value of the third and fourth payments similarly, we find:

$$\text{Total present value} = \$250{,}000 + \$250{,}000e^{-0.06(1)} + \$250{,}000e^{-0.06(2)} + \$250{,}000e^{-0.06(3)}$$
$$\approx \$250{,}000 + \$235{,}441 + \$221{,}730 + \$208{,}818$$
$$= \$915{,}989.$$

Since the present value of the four payments is less than $920,000, you are better off taking the $920,000 right now.

Alternatively, we can compare the future values of the two pay schemes. The scheme with the highest future value is the best from a purely financial point of view. We calculate the future value of both schemes three years from now, on the date of the last $250,000 payment. At that time,

$$\text{Future value of the lump sum payment} = \$920{,}000e^{0.06(3)} \approx \$1{,}101{,}440.$$

Now we calculate the future value of the first $250,000 payment:

$$\text{Future value of the first payment} = \$250{,}000e^{0.06(3)}.$$

Calculating the future value of the other payments similarly, we find:

$$\text{Total future value} = \$250{,}000e^{0.06(3)} + \$250{,}000e^{0.06(2)} + \$250{,}000e^{0.06(1)} + \$250{,}000$$
$$\approx \$299{,}304 + \$281{,}874 + \$265{,}459 + \$250{,}000$$
$$= \$1{,}096{,}637.$$

The future value of the $920,000 payment is greater, so you are better off taking the $920,000 right now. Of course, since the present value of the $920,000 payment is greater than the present value of the four separate payments, you would expect the future value of the $920,000 payment to be greater than the future value of the four separate payments.

(Note: If you read the fine print, you will find that many lotteries do not make their payments right away, but often spread them out, sometimes far into the future. This is to reduce the present value of the payments made, so that the value of the prizes is much less than it might first appear!)

Income Stream

When we consider payments made to or by an individual, we usually think of *discrete* payments, that is, payments made at specific moments in time. However, we may think of payments made by a company as being *continuous*. The revenues earned by a huge corporation, for example, come in essentially all the time and can be represented by a continuous *income stream*, written

$$P(t) \text{ dollars/year.}$$

Notice that $P(t)$ is the *rate* at which deposits are made (its units are dollars per year, for example) and that this rate may vary with time, t.

Present and Future Values of an Income Stream

Just as we can find the present and future values of a single payment, so we can find the present and future values of a stream of payments. We will assume that interest is compounded continuously.

Suppose that we want to calculate the present value of the income stream described by a rate of $P(t)$ dollars per year, and that we are interested in the period from now until M years in the future. We divide the stream into many small deposits, each of which is made at approximately one instant. We divide the interval $0 \leq t \leq M$ into subintervals, each of length Δt:

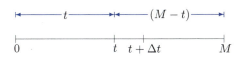

Assuming Δt is small, the rate, $P(t)$, at which deposits are being made will not vary much within one subinterval. Thus, between t and $t + \Delta t$:

$$\text{Amount deposited} \approx \text{Rate of deposits} \times \text{Time}$$
$$\approx (P(t) \text{ dollars/year})(\Delta t \text{ years})$$
$$= P(t)\Delta t \text{ dollars.}$$

Measured from the present, the deposit of $P(t)\Delta t$ is made t years in the future. Thus,

$$\begin{matrix} \text{Present value of money deposited} \\ \text{in interval } t \text{ to } t + \Delta t \end{matrix} \approx P(t)\Delta t e^{-rt}.$$

Summing over all subintervals gives

$$\text{Total present value} \approx \sum P(t)e^{-rt}\Delta t \text{ dollars.}$$

In the limit as $\Delta t \to 0$, we get the following integral:

$$\boxed{\text{Present value} = \int_0^M P(t)e^{-rt}dt \text{ dollars.}}$$

In computing future value, the deposit of $P(t)\Delta t$ has a period of $(M - t)$ years to earn interest, and therefore

$$\begin{matrix} \text{Future value of money deposited} \\ \text{in interval } t \text{ to } t + \Delta t \end{matrix} \approx [P(t)\Delta t]\, e^{r(M-t)}.$$

Summing over all subintervals, we get:

$$\text{Total future value} \approx \sum P(t)\Delta t e^{r(M-t)} \text{ dollars.}$$

As the length of the subdivisions tends toward zero, the sum becomes an integral:

$$\text{Future value} = \int_0^M P(t)e^{r(M-t)}dt \text{ dollars.}$$

In addition, by writing $e^{r(M-t)} = e^{rM} \cdot e^{-rt}$ and factoring out e^{rM}, we see that

$$\text{Future value} = e^{rM} \cdot \text{Present value.}$$

Example 2   Find the present and future values of a constant income stream of \$1000 per year over a period of 20 years, assuming an interest rate of 10% compounded continuously.

Solution   Using $P(t) = 1000$ and $r = 0.1$, we have

$$\text{Present value} = \int_0^{20} 1000e^{-0.1t}dt = 1000\left(-\frac{e^{-0.1t}}{0.1}\right)\Big|_0^{20} = 10{,}000(1-e^{-2}) \approx 8646.65 \text{ dollars.}$$

There are two ways to compute the future value. Using the present value of \$8646.65, we have

$$\text{Future value} = 8646.65e^{0.1(20)} = 63{,}890.58 \text{ dollars.}$$

Alternatively, we can use the integral formula:

$$\text{Future value} = \int_0^{20} 1000e^{0.1(20-t)}dt = \int_0^{20} 1000e^2 e^{-0.1t}dt$$

$$= 1000e^2\left(-\frac{e^{-0.1t}}{0.1}\right)\Big|_0^{20} = 10{,}000e^2(1 - e^{-2}) \approx 63890.58 \text{ dollars.}$$

Notice that the total amount deposited is \$1000 per year for 20 years, or \$20,000. The additional \$43,895.58 of the future value comes from interest earned.

Supply and Demand Curves

In a free market, the quantity of a certain item produced and sold can be described by the supply and demand curves of the item. The *supply curve* shows the quantity of the item the producers will supply at different price levels. It is usually assumed that as the price increases, the quantity supplied will increase. The consumers' behavior is reflected in the *demand curve*, which shows what quantity of goods are bought at various prices. An increase in price is usually assumed to cause a decrease in the quantity purchased. See Figure 8.58.

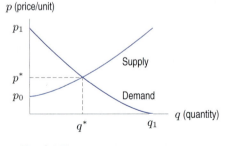

Figure 8.58: Supply and demand curves

It is assumed that the market settles to the *equilibrium price* and *quantity*, p^* and q^*, where the graphs cross. At equilibrium, a quantity q^* of an item is produced and sold for a price of p^* each.

Consumer and Producer Surplus

Notice that at equilibrium, a number of consumers have bought the item at a lower price than they would have been willing to pay. (For example, there are some consumers who would have been willing to pay prices up to p_1.) Similarly, there are some suppliers who would have been willing to produce the item at a lower price (down to p_0, in fact). We define the following terms:

> - The **consumer surplus** measures the consumers' gain from trade. It is the total amount gained by consumers by buying the item at the current price rather than at the price they would have been willing to pay.
> - The **producer surplus** measures the suppliers' gain from trade. It is the total amount gained by producers by selling at the current price, rather than at the price they would have been willing to accept.
>
> In the absence of price controls, the current price is assumed to be the equilibrium price.

Both consumers and producers are richer for having traded. The consumer and producer surplus measure how much richer they are.

Suppose that all consumers buy the good at the maximum price they are willing to pay. Divide the interval from 0 to q^* into subintervals of length Δq. Figure 8.59 shows that a quantity Δq of items are sold at a price of about p_1, another Δq are sold for a slightly lower price of about p_2, the next Δq for a price of about p_3, and so on. Thus,

$$\text{Consumers' total expenditure} \approx p_1 \Delta q + p_2 \Delta q + p_3 \Delta q + \cdots = \sum p_i \Delta q.$$

If D is the demand function given by $p = D(q)$, and if all consumers who were willing to pay more than p^* paid as much as they were willing, then as $\Delta q \to 0$, we would have

$$\text{Consumer expenditure} = \int_0^{q^*} D(q)dq = \begin{array}{l}\text{Area under demand}\\\text{curve from 0 to } q^*.\end{array}$$

Now if all goods are sold at the equilibrium price, the consumers' actual expenditure is $p^* q^*$, the area of the rectangle between the axes and the lines $q = q^*$ and $p = p^*$. Thus, if p^* and q^* are equilibrium price and quantity, the consumer surplus is calculated as follows:

$$\text{Consumer surplus} = \left(\int_0^{q^*} D(q)dq \right) - p^* q^* = \begin{array}{l}\text{Area under demand}\\\text{curve above } p = p^*.\end{array}$$

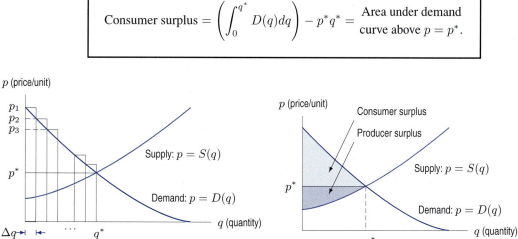

Figure 8.59: Calculation of consumer surplus

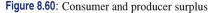
Figure 8.60: Consumer and producer surplus

See Figure 8.60. Similarly, if the supply curve is given by the function $p = S(q)$ and p^* and q^* are equilibrium price and quantity, the producer surplus is calculated as follows:

$$\text{Producer surplus} = p^*q^* - \left(\int_0^{q^*} S(q)dq \right) = \frac{\text{Area between supply}}{\text{curve and line } p = p^*.}$$

Exercises and Problems for Section 8.5

Exercises

1. Find the future value of an income stream of $1000 per year, deposited into an account paying 8% interest, compounded continuously, over a 10-year period.

2. Find the present and future values of an income stream of $3000 per year over a 15-year period, assuming a 6% annual interest rate compounded continuously.

3. Find the present and future values of an income stream of $2000 a year, for a period of 5 years, if the continuous interest rate is 8%.

4. A person deposits money into a retirement account, which pays 7% interest compounded continuously, at a rate of $1000 per year for 20 years. Calculate:

   (a) The balance in the account at the end of the 20 years.
   (b) The amount of money actually deposited into the account.
   (c) The interest earned during the 20 years.

Problems

5. Draw a graph, with time in years on the horizontal axis, of what an income stream might look like for a company that sells sunscreen in the northeast United States.

6. On April 15, 1999, Maria Grasso won the largest lottery amount ever awarded up to that date. She was given her choice between $197 million, paid out continuously over 26 years, or a lump sum of $104 million, paid immediately.

   (a) Which option is better if the interest rate is 6%, compounded continuously ? An interest rate of 5%?
   (b) The winner chose the lump sum option. What assumption was she making about interest rates?

7. (a) A bank account earns 10% interest compounded continuously. At what (constant, continuous) rate must a parent deposit money into such an account in order to save $100,000 in 10 years for a child's college expenses?
   (b) If the parent decides instead to deposit a lump sum now in order to attain the goal of $100,000 in 10 years, how much must be deposited now?

8. (a) If you deposit money continuously at a constant rate of $1000 per year into a bank account that earns 5% interest, how many years will it take for the balance to reach $10,000?
   (b) How many years would it take if the account had $2000 in it initially?

9. A business associate who owes you $3000 offers to pay you $2800 now, or else pay you three yearly installments of $1000 each, with the first installment paid now. If you

use only financial reasons to make your decision, which option should you choose? Justify your answer, assuming a 6% market interest rate, compounded continuously.

10. Big Tree McGee is negotiating his rookie contract with a professional basketball team. They have agreed to a three-year deal which will pay Big Tree a fixed amount at the end of each of the three years, plus a signing bonus at the beginning of his first year. They are still haggling about the amounts and Big Tree must decide between a big signing bonus and fixed payments per year, or a smaller bonus with payments increasing each year. The two options are summarized in the table. All values are payments in millions of dollars.

	Signing bonus	Year 1	Year 2	Year 3
Option #1	6.0	2.0	2.0	2.0
Option #2	1.0	2.0	4.0	6.0

   (a) Big Tree decides to invest all income in stock funds which he expects to grow at a rate of 10% per year, compounded continuously. He would like to choose the contract option which gives him the greater future value at the end of the three years when the last payment is made. Which option should he choose?
   (b) Calculate the present value of each contract offer.

11. Sales of Version 6.0 of a computer software package start out high and decrease exponentially. At time t, in years, the sales are $s(t) = 50e^{-t}$ thousands of dollars per

year. After two years, Version 7.0 of the software is released and replaces Version 6.0. Assume that all income from software sales is immediately invested in government bonds which pay interest at a 6% rate compounded continuously, calculate the total value of sales of Version 6.0 over the two year period.

12. The value of good wine increases with age. Thus, if you are a wine dealer, you have the problem of deciding whether to sell your wine now, at a price of $\$P$ a bottle, or to sell it later at a higher price. Suppose you know that the amount a wine-drinker is willing to pay for a bottle of this wine t years from now is $\$P(1 + 20\sqrt{t})$. Assuming continuous compounding and a prevailing interest rate of 5% per year, when is the best time to sell your wine?

13. An oil company discovered an oil reserve of 100 million barrels. For time $t > 0$, in years, the company's extraction plan is a linear declining function of time as follows:

$$q(t) = a - bt,$$

where $q(t)$ is the rate of extraction of oil in millions of barrels per year at time t and $b = 0.1$ and $a = 10$.

(a) How long does it take to exhaust the entire reserve?
(b) The oil price is a constant $\$20$ per barrel, the extraction cost per barrel is a constant $\$10$, and the market interest rate is 10% per year, compounded continuously. What is the present value of the company's profit?

14. In 1980 West Germany made a loan of 20 billion Deutsche Marks to the Soviet Union, to be used for the construction of a natural gas pipeline connecting Siberia to Western Russia, and continuing to West Germany (Urengoi–Uschgorod–Berlin). Assume that the deal was as follows: In 1985, upon completion of the pipeline, the Soviet Union would deliver natural gas to West Germany, at a constant rate, for all future times. Assuming a constant price of natural gas of 0.10 Deutsche Mark per cubic meter, and assuming West Germany expects 10% annual interest on its investment (compounded continuously), at what rate does the Soviet Union have to deliver the gas, in billions of cubic meters per year? Keep in mind that delivery of gas could not begin until the pipeline was completed. Thus, West Germany received no return on its investment until after five years had passed. (Note: A more complex deal of this type was actually made between the two countries.)

15. In May 1991 *Car and Driver* described a Jaguar that sold for $\$980,000$. At that price only 50 have been sold. It is estimated that 350 could have been sold if the price had been $\$560,000$. Assuming that the demand curve is a straight line, and that $\$560,000$ and 350 are the equilibrium price and quantity, find the consumer surplus at the equilibrium price.

16. Using Riemann sums, explain the economic significance of $\int_0^{q^*} S(q)\,dq$ to the producers.

17. Using Riemann sums, give an interpretation of producer surplus, $\int_0^{q^*} (p^* - S(q))\,dq$ analogous to the interpretation of consumer surplus.

18. In Figure 8.60, page 381, mark the regions representing the following quantities and explain their economic meaning:

(a) $p^* q^*$
(b) $\int_0^{q^*} D(q)\,dq$
(c) $\int_0^{q^*} S(q)\,dq$
(d) $\int_0^{q^*} D(q)\,dq - p^* q^*$
(e) $p^* q^* - \int_0^{q^*} S(q)\,dq$
(f) $\int_0^{q^*} (D(q) - S(q))\,dq$

19. The dairy industry is an example of cartel pricing: the government has set milk prices artificially high. On a supply and demand graph, label p^+, a price above the equilibrium price. Using the graph, describe the effect of forcing the price up to p^+ on:

(a) The consumer surplus.
(b) The producer surplus.
(c) The total gains from trade (Consumer surplus + Producer surplus).

20. Rent controls on apartments are an example of price controls on a commodity. They keep the price of the good artificially low (below the equilibrium price). Sketch a graph of supply and demand curves, and label on it a price p^- below the equilibrium price. What effect does forcing the price down to p^- have on:

(a) The producer surplus?
(b) The consumer surplus?
(c) The total gains from trade (Consumer surplus + Producer surplus)?

8.6   DISTRIBUTION FUNCTIONS

Understanding the distribution of various quantities through the population is important to decision makers. For example, the income distribution gives useful information about the economic structure of a society. In this section we will look at the distribution of ages in the US. To allocate funding for education, health care, and social security, the government needs to know how many people are in each age group. We will see how to represent such information by a density function.

US Age Distribution

Table 8.1   *Distribution of ages in the US in 1995*

Age group	Fraction of total population
0–20	$29\% = 0.29$
20–40	$31\% = 0.31$
40–60	$24\% = 0.24$
60–80	$13\% = 0.13$
80–100	$3\% = 0.03$

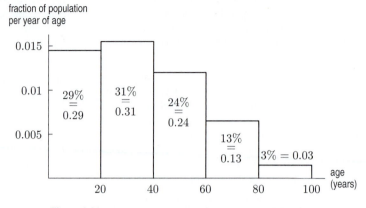

Figure 8.61: How ages were distributed in the US in 1995

The data in Table 8.1 shows how the ages of the US population were distributed in 1995. To represent this information graphically we use a type[1] of *histogram*, putting a vertical bar above each age group in such a way that the *area* of each bar represents the fraction of the population in that age group. The total area of all the rectangles is $100\% = 1$. We only consider people who are less than 100 years old.[2] For the 0–20 age group, the base of the rectangle is 20, and we want the area to be 0.29, so the height must be $0.29/20 = 0.0145$. We treat ages as though they were continuously distributed. The category 0–20, for example, contains people who are just one day short of their twentieth birthday. (See Figure 8.61.)

Example 1   In 1995, estimate what fraction of the US population was:

(a) Between 20 and 60 years old.   (b) Less than 10 years old.
(c) Between 75 and 80 years old.   (d) Between 80 and 85 years old.

Solution

(a) We add the fractions, so $0.31 + 0.24 = 0.55$; that is, 55% of the US population was in this age group.

(b) To find the fraction less than 10 years old, we could assume, for example, that the population was distributed evenly over the 0–20 group. (This means we are assuming that babies were born at a fairly constant rate over the last 20 years, which is probably reasonable.) If we make this assumption, then we can say that the population less than 10 years old was about half that in the 0–20 group, that is, 0.145 of the total population. Notice that we get the same result by computing the area of the rectangle from 0 to 10. (See Figure 8.62.)

(c) To find the population between 75 and 80 years old, since 0.13 of Americans in 1990 were in the 60-80 group, we might apply the same reasoning and say that $\frac{1}{4}(0.13) = 0.0325$ of the population was in this age group. This result is represented as an area in Figure 8.62. The assumption that the population was evenly distributed is not a good one here; certainly there were more people between the ages of 60 and 65 than between 75 and 80. Thus, the estimate of 0.0325 is certainly too high.

(d) Again using the (faulty) assumption that ages in each group were distributed uniformly, we would find that the fraction between 80 and 85 was $\frac{1}{4}(0.03) = 0.0075$. (See Figure 8.62.) This estimate is also poor—there were certainly more people in the 80–85 group than, say, the 95–100 group, and so the 0.0075 estimate is too low.

---

[1]There are other types of histogram which have frequency on the vertical axis.
[2]In fact, 0.02% of the population is over 100, but this is too small to be visible on the histogram.

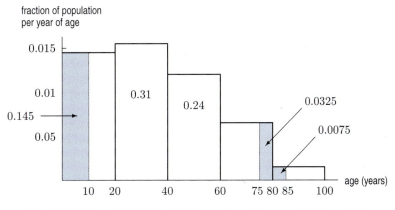

Figure 8.62: Ages in the US in 1995 — various subgroups (for Example 1)

Smoothing Out the Histogram

We could get better estimates if we had smaller age groups (each age group in Figure 8.61 is 20 years, which is quite large). The more detailed data in Table 8.2 leads to the new histogram in Figure 8.63. As we get more detailed information, the upper silhouette of the histogram becomes smoother, but the area of any of the bars still represents the percentage of the population in that age group. Imagine, in the limit, replacing the upper silhouette of the histogram by a smooth curve in such a way that area under the curve above one age group is the same as the area in the corresponding rectangle. The total area under the whole curve is again 100% = 1. (See Figure 8.63.)

The Age Density Function

If t is age in years, we define $p(t)$, the age *density function*, to be a function which "smooths out" the age histogram. This function has the property that

$$\text{Fraction of population between ages } a \text{ and } b = \text{Area under graph of } p \text{ between } a \text{ and } b = \int_a^b p(t)\,dt.$$

If a and b are the smallest and largest possible ages (say, $a = 0$ and $b = 100$), so that the ages of all of the population are between a and b, then

$$\int_a^b p(t)\,dt = \int_0^{100} p(t)\,dt = 1.$$

Table 8.2 *Ages in the US in 1995 (more detailed)*

Age group	Fraction of total population
0–10	15% = 0.15
10–20	14% = 0.14
20–30	14% = 0.14
30–40	17% = 0.17
40–50	14% = 0.14
50–60	10% = 0.10
60–70	8% = 0.08
70–80	5% = 0.05
80–90	2% = 0.02
90–100	1% = 0.01

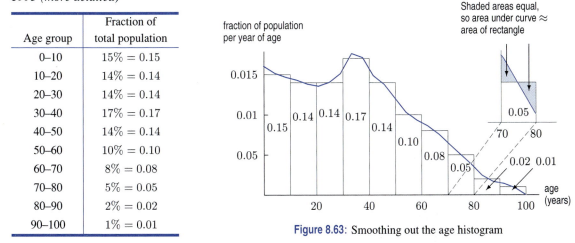

Figure 8.63: Smoothing out the age histogram

What does the age density function p tell us? Notice that we have not talked about the meaning of $p(t)$ itself, but *only* of the integral $\int_a^b p(t)\, dt$. Let's look at this in a bit more detail. Suppose, for example, that $p(10) = 0.015$ per year. This is *not* telling us that 0.015 of the population is precisely 10 years old (where 10 years old means exactly 10, not $10\frac{1}{2}$, not $10\frac{1}{4}$, not 10.1). However, $p(10) = 0.015$ does tell us that for some small interval Δt around 10, the fraction of the population with ages in this interval is approximately $p(10)\,\Delta t = 0.015\,\Delta t$.

The Density Function

Suppose we are interested in how a certain characteristic, x, is distributed through a population. For example, x might be height or age if the population is people, or might be wattage for a population of light bulbs. Then we define a general density function with the following properties:

> The function, $p(x)$, is a **density function** if
>
> $$\begin{array}{c}\text{Fraction of population for which} \\ x \text{ is between } a \text{ and } b\end{array} = \begin{array}{c}\text{Area under graph of } p \\ \text{between } a \text{ and } b\end{array} = \int_a^b p(x)\,dx.$$
>
> $$\int_{-\infty}^{\infty} p(x)\,dx = 1 \quad \text{and} \quad p(x) \geq 0 \quad \text{for all } x.$$

The density function must be nonnegative because its integral always gives a fraction of the population. Also, the fraction of the population with x between $-\infty$ and ∞ is 1 because the entire population has the characteristic x between $-\infty$ and ∞. The function p that was used to smooth out the age histogram satisfies this definition of a density function. We do not assign a meaning to the value $p(x)$ directly, but rather interpret $p(x)\,\Delta x$ as the fraction of the population with the characteristic in a short interval of length Δx around x.

The density function is often approximated by formulas, as in the next example.

Example 2   Find formulas to approximate the density function, p, for the US age distribution. To reflect Figure 8.63, use a continuous function, constant at 0.015 up to age 40 and then dropping linearly.

Solution   We have $p(t) = 0.015$ for $0 \leq t < 40$. For $t \geq 40$, we need a linear function sloping downward. Because p is continuous, we have $p(40) = 0.015$. Because p is a density function we have $\int_0^{100} p(t)dt = 1$. Suppose b is as in Figure 8.64 then

$$\int_0^{100} p(t)dt = \int_0^{40} p(t)dt + \int_{40}^{100} p(t)dt = 40(0.015) + \frac{1}{2}(0.015)b = 1,$$

where $\int_{40}^{100} p(t)dt$ is given by the area of the triangle. This gives

$$\frac{0.015}{2}b = 0.4, \quad \text{and so} \quad b \approx 53.3.$$

Thus the slope of the line is $-0.015/53.3 \approx -0.00028$, so for $40 \leq t \leq 40 + 53.3 = 93.3$, we have

$$p(t) - 0.015 = -0.00028(t - 40),$$
$$p(t) = 0.0262 - 0.00028t.$$

According to this way of smoothing the data, there is no one over 93.3 years old, so $p(t) = 0$ for $t > 93.3$.

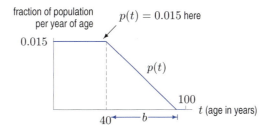

Figure 8.64: Age density function

Cumulative Distribution Function for Ages

Another way of showing how ages are distributed in the US is by using the *cumulative distribution function* $P(t)$, defined by

$$P(t) = \frac{\text{Fraction of population}}{\text{of age less than } t} = \int_0^t p(x)dx.$$

Thus, P is the antiderivative of p with $P(0) = 0$, and $P(t)$ gives the area under the density curve between 0 and t.

Notice that the cumulative distribution function is nonnegative and increasing (or at least nondecreasing), since the number of people younger than age t increases as t increases. Another way of seeing this is to notice that $P' = p$, and p is positive (or nonnegative). Thus the cumulative age distribution is a function which starts with $P(0) = 0$ and increases as t increases. $P(t) = 0$ for $t < 0$ because, when $t < 0$, there is no one whose age is less than t. The limiting value of P, as $t \to \infty$, is 1 since as t becomes very large (100 say), everyone is younger than age t, so the fraction of people with age less than t tends toward 1. (See Figure 8.65.) For t less than 40, the graph of P is a straight line, because p is constant there. For $t > 40$, the graph of P levels off as p tends to 0.

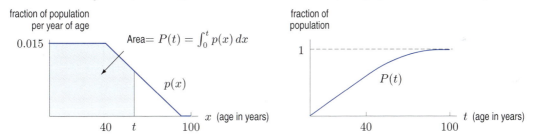

Figure 8.65: $P(t)$, the cumulative age distribution function, and its relation to $p(x)$, the age density function

Cumulative Distribution Function

A **cumulative distribution function**, $P(t)$, of a density function p, is defined by

$$P(t) = \int_{-\infty}^t p(x)\, dx = \frac{\text{Fraction of population having}}{\text{values of } x \text{ below } t.}$$

Thus, P is an antiderivative of p, that is, $P' = p$.
Any cumulative distribution has the following properties:
- P is increasing (or nondecreasing).
- $\lim_{t \to \infty} P(t) = 1$ and $\lim_{t \to -\infty} P(t) = 0$.
- $\frac{\text{Fraction of population having}}{\text{values of } x \text{ between } a \text{ and } b} = \int_a^b p(x)\, dx = P(b) - P(a).$

Exercises and Problems for Section 8.6

Exercises

In Exercises 1–3, sketch graphs of a density function and a cumulative distribution function which could represent the distribution of income through a population with the given characteristics.

1. A large middle class.

2. Small middle and upper classes and many poor people.

3. Small middle class, many poor and many rich people.

Decide if the function graphed in Exercises 4–9 is a probability density function (pdf) or a cumulative distribution function (cdf). Give reasons. Find the value of c. Sketch and label the other function. (That is, sketch and label the cdf if the problem shows a pdf, and the pdf if the problem shows a cdf.)

4.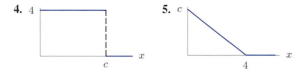

6.

7.

8.

9.

Problems

10. A large number of people take a standardized test, receiving scores described by the density function p graphed in Figure 8.66. Does the density function imply that most people receive a score near 50? Explain why or why not.

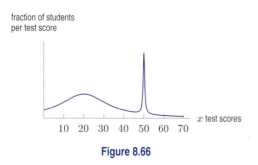

Figure 8.66

11. An experiment is done to determine the effect of two new fertilizers A and B on the growth of a species of peas. The cumulative distribution functions of the heights of the mature peas without treatment and treated with each of A and B are graphed in Figure 8.67.

  (a) About what height are most of the unfertilized plants?

  (b) Explain in words the effect of the fertilizers A and B on the mature height of the plants.

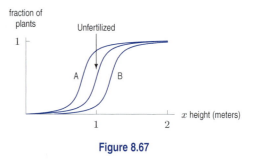

Figure 8.67

12. Suppose $F(x)$ is the cumulative distribution function for heights (in meters) of trees in a forest.

  (a) Explain in terms of trees the meaning of the statement $F(7) = 0.6$.

  (b) Which is greater, $F(6)$ or $F(7)$? Justify your answer in terms of trees.

13. Suppose that $p(x)$ is the density function for heights of American men, in inches. What is the meaning of the statement $p(68) = 0.2$?

14. Suppose $P(t)$ is the fraction of the US population of age less than t. Using Table 8.2 on page 385, make a table of values for $P(t)$.

15. Figure 8.68 shows a density function and the corresponding cumulative distribution function.[3]

[3]Adapted from *Calculus*, by David A. Smith and Lawerence C. Moore (Lexington: D.C. Heath, 1994).

(a) Which curve represents the density function and which represents the cumulative distribution function? Give a reason for your choice.

(b) Put reasonable values on the tick marks on each of the axes.

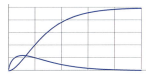

Figure 8.68

16. The density function and cumulative distribution function of heights of grass plants in a meadow are in Figures 8.69 and 8.70, respectively.

(a) There are two species of grass in the meadow, a short grass and a tall grass. Explain how the graph of the density function reflects this fact.

(b) Explain how the graph of the cumulative distribution functions reflects the fact that there are two species of grass in the meadow.

(c) About what percentage of the grasses in the meadow belong to the short grass species?

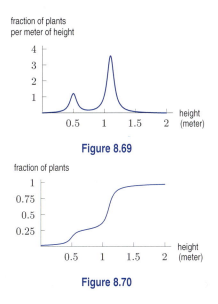

Figure 8.69

Figure 8.70

17. After measuring the duration of many telephone calls, the telephone company found their data was well-approximated by the density function $p(x) = 0.4e^{-0.4x}$, where x is the duration of a call, in minutes.

(a) What percentage of calls last between 1 and 2 minutes?

(b) What percentage of calls last 1 minute or less?

(c) What percentage of calls last 3 minutes or more?

(d) Find the cumulative distribution function.

18. Students at the University of California were surveyed and asked their grade point average . (The GPA ranges from 0 to 4, where 2 is just passing.) The distribution of GPAs is shown in Figure 8.71.[4]

(a) Roughly what fraction of students are passing?

(b) Roughly what fraction of the students have honor grades (GPAs above 3)?

(c) Why do you think there is a peak around 2?

(d) Sketch the cumulative distribution function.

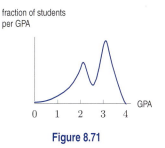

Figure 8.71

19. Figure 8.72[4] shows the distribution of elevation, in miles, across the earth's surface. Positive elevation denotes land above sea level; negative elevation shows land below sea level (i.e., the ocean floor).

(a) Describe in words the elevation of most of the earth's surface.

(b) Approximately what fraction of the earth's surface is below sea level?

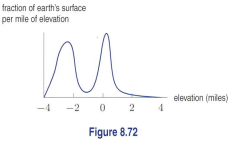

Figure 8.72

20. Consider a population of individuals with a disease. Suppose that t is the number of years since the onset of the disease. The death density function, $f(t) = cte^{-kt}$, approximates the fraction of the sick individuals who die in the time interval $[t, t + \Delta t]$ as follows:

$$\text{Fraction who die} \approx f(t)\Delta t = cte^{-kt}\Delta t$$

where c and k are positive constants whose values depend on the particular disease.

(a) Find the value of c in terms of k.

(b) If 40% of the population dies within 5 years, find c and k.

(c) Find the cumulative death distribution function, $C(t)$. Give your answer in terms of k.

[4]Adapted from *Statistics*, by Freedman, Pisani, Purves, and Adikhari (New York: Norton, 1991).

8.7   PROBABILITY, MEAN, AND MEDIAN

Probability

Suppose we pick a member of the US population at random and ask what is the probability that the person is between, say, the ages of 70 and 80. We saw in Table 8.2 on page 385 that $5\% = 0.05$ of the population is in this age group. We say that the probability, or chance, that the person is between 70 and 80 is 0.05. Using any age density function $p(t)$, we can define probabilities as follows:

$$
\begin{array}{ccc}
\text{Probability that a person is} & = & \text{Fraction of population} \\
\text{between ages } a \text{ and } b & & \text{between ages } a \text{ and } b
\end{array}
= \int_a^b p(t)\, dt.
$$

Since the cumulative distribution function gives the fraction of the population younger than age t, the cumulative distribution can also be used to calculate the probability that a randomly selected person is in a given age group.

$$
\begin{array}{ccc}
\text{Probability that a person is} & = & \text{Fraction of population} \\
\text{younger than age } t & & \text{younger than age } t
\end{array}
= P(t) = \int_0^t p(x)\, dx.
$$

In the next example, both a density function and a cumulative distribution function are used to describe the same situation.

Example 1   Suppose you want to analyze the fishing industry in a small town. Each day, the boats bring back at least 2 tons of fish, and never more than 8 tons.

(a) Using the density function describing the daily catch in Figure 8.73, find and graph the corresponding cumulative distribution function and explain its meaning.

(b) What is the probability that the catch will be between 5 and 7 tons?

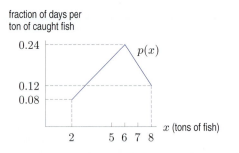

Figure 8.73: Density function of daily catch

Solution   (a) The cumulative distribution function $P(t)$ is equal to the fraction of days on which the catch is less than t tons of fish. Since the catch is never less than 2 tons, we have $P(t) = 0$ for $t \le 2$. Since the catch is always less than 8 tons, we have $P(t) = 1$ for $t \ge 8$. For t in the range $2 < t < 8$, we must evaluate the integral

$$
P(t) = \int_{-\infty}^t p(x)\,dx = \int_2^t p(x)\,dx.
$$

This integral equals the area under the graph of $p(x)$ between $x = 2$ and $x = t$. It can be calculated by noting that $p(x)$ is given by the formula

$$p(x) = \begin{cases} 0.04x & \text{for } 2 \le x \le 6 \\ -0.06x + 0.6 & \text{for } 6 < x \le 8 \end{cases}$$

and $p(x) = 0$ for $x < 2$ or $x > 8$. Thus, for $2 \le t \le 6$,

$$P(t) = \int_2^t 0.04x \, dx = 0.04 \frac{x^2}{2} \Big|_2^t = 0.02t^2 - 0.08.$$

And for $6 \le t \le 8$,

$$P(t) = \int_2^t p(x) \, dx = \int_2^6 p(x) \, dx + \int_6^t p(x) \, dx$$

$$= 0.64 + \int_6^t (-0.06x + 0.6) \, dx = 0.64 + \left(-0.06 \frac{x^2}{2} + 0.6x \right) \Big|_6^t$$

$$= -0.03t^2 + 0.6t - 1.88.$$

Thus

$$P(t) = \begin{cases} 0.02t^2 - 0.08 & \text{for } 2 \le t \le 6 \\ -0.03t^2 + 0.6t - 1.88 & \text{for } 6 < t \le 8. \end{cases}$$

In addition $P(t) = 0$ for $t < 2$ and $P(t) = 1$ for $8 < t$. (See Figure 8.74.)

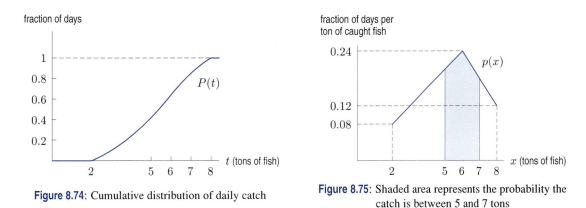

Figure 8.74: Cumulative distribution of daily catch

Figure 8.75: Shaded area represents the probability the catch is between 5 and 7 tons

(b) The probability that the catch is between 5 and 7 tons can be found using either the density function, p, or the cumulative distribution function, P. If we use the density function, this probability can be represented by the shaded area in Figure 8.75, which is about 0.43.

$$\text{Probability catch is between 5 and 7 tons} = \int_5^7 p(x) \, dx = 0.43.$$

The probability can be found from the cumulative distribution as follows:

$$\text{Probability catch is between 5 and 7 tons} = P(7) - P(5) = 0.85 - 0.42 = 0.43.$$

The Median and Mean

It is often useful to be able to give an "average" value for a distribution. Two measures that are in common use are the *median* and the *mean*.

The Median

> A **median** of a quantity x distributed through a population is a value T such that half the population has values of x less than (or equal to) T, and half the population has values of x greater than (or equal to) T. Thus, a median T satisfies
>
> $$\int_{-\infty}^{T} p(x)\, dx = 0.5,$$
>
> where p is the density function. In other words, half the area under the graph of p lies to the left of T.

Example 2   Find the median age in the US in 1995, using the age density function given by

$$p(t) = \begin{cases} 0.015 & \text{for } 0 \le t \le 40 \\ 0.0262 - 0.00028t & \text{for } 40 < t \le 93.3. \end{cases}$$

Solution   We want to find the value of T such that

$$\int_{-\infty}^{T} p(t)\, dt = \int_{0}^{T} p(t)\, dt = 0.5.$$

Since $p(t) = 0.015$ up to age 40, we have

$$\text{Median} = T = \frac{0.5}{0.015} \approx 33 \text{ years.}$$

(See Figure 8.76.)

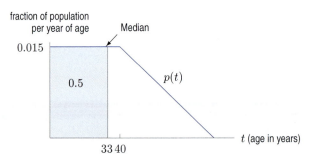

Figure 8.76: Median of age distribution

The Mean

Another commonly used average value is the *mean*. To find the mean of N numbers, you add the numbers and divide the sum by N. For example, the mean of the numbers 1, 2, 7, and 10 is $(1 + 2 + 7 + 10)/4 = 5$. The mean age of the entire US population is therefore defined as

$$\text{Mean age} = \frac{\sum \text{Ages of all people in the US}}{\text{Total number of people in the US}}.$$

Calculating the sum of all the ages directly would be an enormous task; we will approximate the sum by an integral. The idea is to "slice up" the age axis and consider the people whose age is between t and $t + \Delta t$. How many are there?

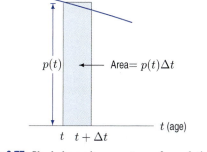

Figure 8.77: Shaded area is percentage of population with
age between t and $t + \Delta t$

The fraction of the population between t and $t + \Delta t$ is the area under the graph of p between these points, which is well approximated by the area of the rectangle, $p(t)\Delta t$. (See Figure 8.77.) If the total number of people in the population is N, then

$$\text{Number of people with age between } t \text{ and } t + \Delta t \approx p(t)\Delta t N.$$

The age of all of these people is approximately t:

$$\text{Sum of ages of people between age } t \text{ and } t + \Delta t \approx tp(t)\Delta t N.$$

Therefore, adding and factoring out an N gives us

$$\text{Sum of ages of all people} \approx \left(\sum tp(t)\Delta t\right) N.$$

In the limit, as we allow Δt to shrink to 0, the sum becomes an integral, so

$$\text{Sum of ages of all people} = \left(\int_0^{100} tp(t)dt\right) N.$$

Therefore, with N equal to the total number of people in the US, and assuming no person is over 100 years old,

$$\text{Mean age} = \frac{\text{Sum of ages of all people in US}}{N} = \int_0^{100} tp(t)dt.$$

We can give the same argument for any[5] density function $p(x)$.

If a quantity has density function $p(x)$,

$$\textbf{Mean value} \text{ of the quantity} = \int_{-\infty}^{\infty} xp(x)\, dx.$$

It can be shown that the mean is the point on the horizontal axis where the region under the graph of the density function, if it were made out of cardboard, would balance.

Example 3   Find the mean age of the US population, using the density function of Example 2.

Solution   The formula for p is

$$p(t) = \begin{cases} 0 & \text{for } t < 0 \\ 0.015 & \text{for } 0 \leq t \leq 40 \\ 0.0262 - 0.00028t & \text{for } 40 < t \leq 93.3 \\ 0 & \text{for } t > 93.3. \end{cases}$$

---

[5]Provided all the relevant improper integrals converge.

Using these formulas, we compute

$$\text{Mean age} = \int_0^{100} tp(t)dt = \int_0^{40} t(0.015)dt + \int_{40}^{93.3} t(0.0262 - 0.00028t)dt$$

$$= 0.015\frac{t^2}{2}\Big|_0^{40} + 0.0262\frac{t^2}{2}\Big|_{40}^{93.3} - 0.00028\frac{t^3}{3}\Big|_{40}^{93.3} \approx 35 \text{ years.}$$

The mean is shown is Figure 8.78.

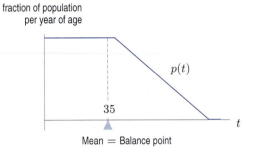

fraction of population
per year of age

Mean = Balance point

Figure 8.78: Mean of age distribution

Normal Distributions

How much rain do you expect to fall in your home town this year? If you live in Anchorage, Alaska, the answer is something close to 15 inches (including the snow). Of course, you don't expect exactly 15 inches. Some years there are more than 15 inches, and some years there is less. Most years, however, the amount of rainfall is close to 15 inches; only rarely is it well above or well below 15 inches. What does the density function for the rainfall look like? To answer this question, we look at rainfall data over many years. Records show that the distribution of rainfall is well-approximated by a *normal distribution*. Its graph is a bell-shaped curve which peaks at 15 inches and slopes downward approximately symmetrically on either side.

Normal distributions are frequently used to model real phenomena, from grades on an exam to the number of airline passengers on a particular flight. A normal distribution is characterized by its *mean*, μ, and its *standard deviation*, σ. The mean tells us the location of the central peak. The standard deviation tells us how closely the data is clustered around the mean. A small value of σ tells us that the data is close to the mean; a large σ tells us the data is spread out. In the following formula for a normal distribution, the factor of $1/(\sigma\sqrt{2\pi})$ makes the area under the graph equal to 1.

A **normal distribution** has a density function of the form

$$p(x) = \frac{1}{\sigma\sqrt{2\pi}}e^{-(x-\mu)^2/(2\sigma^2)},$$

where μ is the mean of the distribution and σ is the standard deviation, with $\sigma > 0$.

To model the rainfall in Anchorage, we use a normal distribution with $\mu = 15$ and $\sigma = 1$. (See Figure 8.79.)

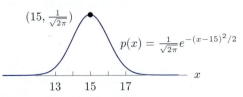

$\left(15, \frac{1}{\sqrt{2\pi}}\right)$

$p(x) = \frac{1}{\sqrt{2\pi}}e^{-(x-15)^2/2}$

Figure 8.79: Normal distribution with $\mu = 15$ and $\sigma = 1$

Example 4   For Anchorage's rainfall, use the normal distribution with the density function with $\mu = 15$ and $\sigma = 1$ to compute the fraction of the years with rainfall between
(a)   14 and 16 inches,   (b)   13 and 17 inches,   (c)   12 and 18 inches.

Solution   (a) The fraction of the years with annual rainfall between 14 and 16 inches is $\int_{14}^{16} \frac{1}{\sqrt{2\pi}} e^{-(x-15)^2/2} dx$.
Since there is no elementary antiderivative for $e^{-(x-15)^2/2}$, we find the integral numerically. Its value is about 0.68.

$$\begin{array}{l} \text{Fraction of years with rainfall} \\ \text{between 14 and 16 inches} \end{array} = \int_{14}^{16} \frac{1}{\sqrt{2\pi}} e^{-(x-15)^2/2} \, dx \approx 0.68.$$

(b) Finding the integral numerically again:

$$\begin{array}{l} \text{Fraction of years with rainfall} \\ \text{between 13 and 17 inches} \end{array} = \int_{13}^{17} \frac{1}{\sqrt{2\pi}} e^{-(x-15)^2/2} \, dx \approx 0.95.$$

(c)

$$\begin{array}{l} \text{Fraction of years with rainfall} \\ \text{between 12 and 18 inches} \end{array} = \int_{12}^{18} \frac{1}{\sqrt{2\pi}} e^{-(x-15)^2/2} \, dx \approx 0.997.$$

Since 0.95 is so close to 1, we expect that most of the time the rainfall will be between 13 and 17 inches a year.

Among the normal distributions, the one having $\mu = 0$, $\sigma = 1$ is called the *standard normal distribution*. Values of the corresponding cumulative distribution function are published in tables.

Exercises and Problems for Section 8.7

Exercises

1. Consider the fishing data given in Example 1 on page 390. Show that the area under the density function in Figure 8.73 is 1. Why is this to be expected?

2. Find the mean daily catch for the fishing data in Figure 8.73, page 390.

3. (a) Using a calculator or computer, sketch graphs of the density function of the normal distribution
$$p(x) = \frac{1}{\sigma\sqrt{2\pi}} e^{-(x-\mu)^2/(2\sigma^2)}.$$

(i) For fixed μ (say, $\mu = 5$) and varying σ (say, $\sigma = 1, 2, 3$).

(ii) For varying μ (say, $\mu = 4, 5, 6$) and fixed σ (say, $\sigma = 1$).

(b) Explain how the graphs confirm that μ is the mean of the distribution and that σ is a measure of how closely the data is clustered around the mean.

Problems

4. The probability of a transistor failing between $t = a$ months and $t = b$ months is given by $c \int_a^b e^{-ct} dt$, for some constant c.

(a) If the probability of failure within the first six months is 10%, what is c?

(b) Given the value of c in part (a), what is the probability the transistor fails within the second six months?

5. Suppose that x measures the time (in hours) it takes for a student to complete an exam. Assume that all students

are done within two hours and the density function for x is given by
$$p(x) = \begin{cases} x^3/4 & \text{if } 0 < x < 2 \\ 0 & \text{otherwise.} \end{cases}$$

(a) What proportion of students take between 1.5 and 2.0 hours to finish the exam?

(b) What is the mean time for students to complete the exam?

(c) Compute the median of this distribution.

6. In 1950 an experiment was done observing the time gaps between successive cars on the Arroyo Seco Freeway.[6] The data show that the density function of these time gaps was given approximately by

$$p(x) = ae^{-0.122x}$$

where x is the time in seconds and a is a constant.

(a) Find a.
(b) Find P, the cumulative distribution function.
(c) Find the median and mean time gap.
(d) Sketch rough graphs of p and P.

7. Consider a group of people who have received treatment for a disease such as cancer. Let t be the *survival time*, the number of years a person lives after receiving treatment. The density function giving the distribution of t is $p(t) = Ce^{-Ct}$ for some positive constant C.

(a) What is the practical meaning for the cumulative distribution function $P(t) = \int_0^t p(x)\,dx$?
(b) The survival function, $S(t)$, is the probability that a randomly selected person survives for at least t years. Find $S(t)$.
(c) Suppose a patient has a 70% probability of surviving at least two years. Find C.

8. While taking a walk along the road where you live, you accidentally drop your glove. You don't know where you dropped it. Suppose the probability density $p(x)$ for having dropped the glove x kilometers from home (along the road) is

$$p(x) = 2e^{-2x} \quad \text{for } x \geq 0.$$

(a) What is the probability that you dropped it within 1 kilometer of home?
(b) At what distance y from home is the probability that you dropped it within y km of home equal to 0.95?

9. The distribution of IQ scores is often modeled by the normal distribution with mean 100 and standard deviation 15.

(a) Write a formula for the density distribution of IQ scores.
(b) Estimate the fraction of the population with IQ between 115 and 120.

10. The speeds of cars on a road are approximately normally distributed with a mean $\mu = 58$ km/hr and standard deviation $\sigma = 4$ km/hr.

(a) What is the probability that a randomly selected car is going between 60 and 65 km/hr?
(b) What fraction of all cars are going slower than 52 km/hr?

11. Consider the normal distribution, $p(x)$.

(a) Show that $p(x)$ is a maximum when $x = \mu$. What is that maximum value?

(b) Show that $p(x)$ has points of inflection where $x = \mu + \sigma$ and $x = \mu - \sigma$.
(c) Describe in your own words what μ and σ tell you about the distribution.

12. For a normal population of mean 0, show that the fraction of the population within one standard deviation of the mean does not depend on the standard deviation.
[Hint: Use the substitution $w = x/\sigma$.]

13. Which of the following functions makes the most sense as a model for the probability density representing the time (in minutes, starting from $t = 0$) that the next customer walks into a store?

(a) $p(t) = \begin{cases} \cos t & 0 \leq t \leq 2\pi \\ e^{t-2\pi} & t \geq 2\pi \end{cases}$
(b) $p(t) = 3e^{-3t}$ for $t \geq 0$
(c) $p(t) = e^{-3t}$ for $t \geq 0$
(d) $p(t) = 1/4$ for $0 \leq t \leq 4$

14. Let $P(x)$ be the cumulative distribution function for the income distribution in the US in 1973 (income is measured in thousands of dollars). Some values of $P(x)$ are in the following table:

Income x (thousands)	1	4.4	7.8	12.6	20	50
$P(x)$ (%)	1	10	25	50	75	99

(a) What fraction of the population made between $20,000 and $50,000?
(b) What was the median income?
(c) Sketch a density function for this distribution. Where, approximately, does your density function have a maximum? What is the significance of this point, in terms of income distribution? How can you recognize this point on the graph of the density function and on the graph of the cumulative distribution?

15. If we think of an electron as a particle, the function

$$P(r) = 1 - (2r^2 + 2r + 1)e^{-2r}$$

is the cumulative distribution function of the distance, r, of the electron in a hydrogen atom from the center of the atom. The distance is measured in Bohr radii. (1 Bohr radius = 5.29×10^{-11} m. Niels Bohr (1885–1962) was a Danish physicist.)

For example, $P(1) = 1 - 5e^{-2} \approx 0.32$ means that the electron is within 1 Bohr radius from the center of the atom 32% of the time.

(a) Find a formula for the density function of this distribution. Sketch the density function and the cumulative distribution function.
(b) Find the median distance and the mean distance. Near what value of r is an electron most likely to be found?
(c) The Bohr radius is sometimes called the "radius of the hydrogen atom." Why?

---

[6]Reported by Daniel Furlough and Frank Barnes.

CHAPTER SUMMARY

- **Geometry**
  Area, volume, arc length.
- **Density**
  Finding total quantity from density, center of mass.
- **Physics**
  Work, force and pressure.

- **Economics**
  Present and future value of income stream, consumer and producer surplus.
- **Probability**
  Density function, cumulative distribution function, mean, median, normal distribution.

REVIEW EXERCISES AND PROBLEMS FOR CHAPTER EIGHT

Exercises

For each region in Exercises 1–3, write a definite integral which represents its area. Evaluate the integral to derive a formula for the area.

1. A rectangle with base b and height h:

2. A circle of radius r:

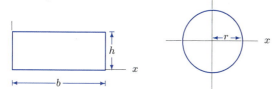

3. A right triangle of base b and height h:

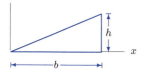

4. Imagine a hard-boiled egg lying on its side cut into thin slices. First think about vertical slices and then horizontal ones. What would these slices look like? Sketch them.

5. Find the volume of the region in Figure 8.80, given that the radius, r of the circular slice at h is $r = \sqrt{h}$.

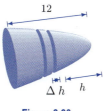

Figure 8.80

6. Figure 8.81 shows thrust-time curves for two model rockets. The thrust or force, F, of the engine (in newtons) is plotted against time, t, (in seconds). The *total impulse* of the rocket's engine is defined as the definite integral of F with respect to t. The total impulse is a measure of the strength of the engine.

(a) For approximately how many seconds is the thrust of rocket B greater than 10 newtons?
(b) Estimate the total impulse for model rocket A.
(c) What are the units for the total impulse calculated in part (b)?
(d) Which rocket has the largest total impulse?
(e) Which rocket has the largest maximum thrust?

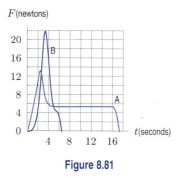

Figure 8.81

For the curves described in Exercises 7–8, write the integral that gives the exact length of the curve; do not evaluate it.

7. One arch of the sine curve, from $x = 0$ to $x = \pi$.

8. The ellipse with equation $(x^2/a^2) + (y^2/b^2) = 1$.

Problems

9. (a) Sketch the solid obtained by rotating the region bounded by $y = \sqrt{x}$, $x = 1$, and $y = 0$ around the line $y = 0$.
(b) Approximate its volume by Riemann sums, showing the volume represented by each term in your sum on the sketch.
(c) Now find the volume of this solid using an integral.

10. Using the region of Problem 9, find the volume when it is rotated around
(a) The line $y = 1$.   (b) The y-axis.

11. Find, by slicing, the volume of a cone whose height is 3 cm and whose base radius is 1 cm. Slice the cone as shown in Figure 8.4 on page 347.

12. The reflector behind a car headlight is made in the shape of the parabola, $x = \frac{4}{9}y^2$, with a circular cross-section, as shown in Figure 8.82.

   (a) Find a Riemann sum approximating the volume contained by this headlight.

   (b) Find the volume exactly.

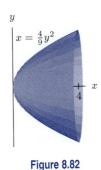

Figure 8.82

13. In this problem, you will derive the formula for the volume of a right circular cone with height l and base radius b by rotating the line $y = ax$ from $x = 0$ to $x = l$ around the x-axis. See Figure 8.83.

   (a) What value should you choose for a such that the cone will have height l and base radius b?

   (b) Given this value of a, find the volume of the cone.

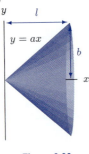

Figure 8.83

14. Figure 8.84 shows a cross section through an apple. (Scale: One division = 1/2 inch.)

   (a) Give a rough estimate for the volume of this apple (in cubic inches).

   (b) The density of these apples is about 0.03 lb/in^3 (a little less than the density of water—as you might expect, since apples float). Estimate how much this apple would cost. (They go for 80 cents a pound.)

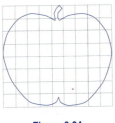

Figure 8.84

15. The circle $x^2 + y^2 = 1$ is rotated about the line $y = 3$ forming a torus (a doughnut-shaped figure). Find the volume of this torus.

16. Find the center of mass of a system containing four identical point masses of 3 gm, located at $x = -5, -3, 2, 7$.

17. A metal plate, with constant density 2 gm/cm^2, has a shape bounded by the two curves $y = x^2$ and $y = \sqrt{x}$, with $0 \le x \le 1$, and x, y in cm.

   (a) Find the total mass of the plate.

   (b) Because of the symmetry of the plate about the line $y = x$, we have $\bar{x} = \bar{y}$. Sketch the plate, and decide, on the basis of the shape, whether $\bar{x}$ is less than or greater than $1/2$.

   (c) Find $\bar{x}$ and $\bar{y}$.

18. A 200-lb weight is hoisted by a rope from the ground to a point 20 ft above the ground. The rope weighs 2 lb/ft. Find the work done in lifting the weight.

19. Water is raised from a well 40 ft deep by a bucket attached to a rope. When the bucket is full, it weighs 30 lb. However, a leak in the bucket causes it to lose water at a rate of 1/4 lb for each foot that the bucket is raised. Neglecting the weight of the rope, find the work done in raising the bucket to the top.

20. A water tank is in the shape of a right circular cone with height 18 ft and radius 12 ft at the top. If it is filled with water to a depth of 15 ft, find the work done in pumping all of the water over the top of the tank. (The density of water is $\delta = 62.4$ lb/ft^3.)

21. The dam in Hannawa Falls, NY, on the Raquette River is approximately 60 feet across and 25 feet high. Find the water force on the dam.

22. (a) Find the present and future values of a constant income stream of $100 per year over a period of 20 years, assuming a 10% annual interest rate compounded continuously.

   (b) How many years will it take for the balance to reach $5000?

23. You are manufacturing a particular item. After t years, the rate at which you earn a profit on the item is $(2-0.1t)$ thousand dollars per year. (A negative profit represents a loss.) Interest is 10%, compounded continuously,

   (a) Write a Riemann sum approximating the present value of the total profit earned up to a time M years in the future.

   (b) Write an integral representing the present value in part (a). (You need not evaluate this integral.)

   (c) For what M is the present value of the stream of profits on this item maximized? What is the present value of the total profit earned up to that time?

24. A nuclear power plant produces strontium-90 at a rate of 3 kg/yr. How much of the strontium produced since 1971 (when the plant opened) was still around in 1992? (The half-life of strontium-90 is 28 years.)

25. Mt. Shasta is a cone-like volcano whose radius at an elevation of h feet above sea level is approximately $(3.5 \cdot 10^5)/\sqrt{h + 600}$ feet. Its bottom is 400 feet above sea level, and its top is 14,400 feet above sea level. See Figure 8.85. (Note: Mt. Shasta is in northern California, and for some time was thought to be the highest point in the US outside Alaska.)

(a) Give a Riemann sum approximating the volume of Mt. Shasta.

(b) Find the volume in cubic feet.

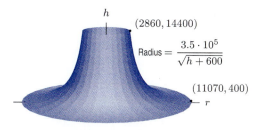

Figure 8.85: Mt. Shasta

26. Figure 8.86 shows an ancient Greek water clock called a clepsydra, which is designed so that the depth of the water decreases at a constant rate as the water runs out a hole in the bottom. This design allows the hours to be marked by a uniform scale. The tank of the clepsydra is a volume of revolution about a vertical axis. According to Torricelli's law, the exit speed of the water flowing through the hole is proportional to the square root of the depth of the water. Use this to find the formula $y = f(x)$ for this profile, assuming that $f(1) = 1$.

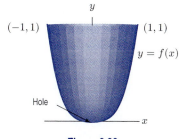

Figure 8.86

27. A radiation detector is a circular disk which registers photons which hit it. The probability that a photon hitting the disk at a distance r from the center is actually detected is given by $S(r)$. A radiation detector of radius R is bombarded by constant radiation of N photons per second per unit area. Write an integral representing the number of photons per second registered by the detector.

28. Find the volume of the snowman in Figure 8.87. If x and y are in meters, and the origin is on the ground, the x-axis is horizontal and the y-axis is vertical, then the body is approximated by rotating the curve $y = 1 - 4x^2$ about

the y-axis. The neck is a cylinder of radius 0.1 meter and length 0.15 meter; the head is spherical with radius 0.2 meter.

Figure 8.87

29. A blood vessel is cylindrical with radius R and length l. The blood near the boundary moves slowly; blood at the center moves the fastest. The velocity, v, of the blood at a distance r from the center of the artery is given by

$$v = \frac{P}{4\eta l}(R^2 - r^2)$$

where P is the pressure difference between the ends of the blood vessel and η is the viscosity of blood.

(a) Find the rate at which the blood is flowing down the blood vessel. (Give your answer as a volume per unit time.)

(b) Show that your result agrees with Poiseuille's Law which says that the rate at which blood is flowing down the blood vessel is proportional to the radius of the blood vessel to the fourth power.

30. A car moving at a speed of v mph achieves $25 + 0.1v$ mpg (miles per gallon) for v between 20 and 60 mph. Your speed as a function of time, t, in hours, is given by

$$v = 50\frac{t}{t + 1} \text{ mph.}$$

How many gallons of gas do you consume between $t = 2$ and $t = 3$?

31. A bowl is made by rotating the curve $y = ax^2$ around the y-axis (a is a constant).

(a) The bowl is filled with water to depth h. What is the volume of water in the bowl? (Your answer will contain a and h.)

(b) What is the area of the surface of the water if the bowl is filled to depth h? (Your answer will contain a and h.)

(c) Water is evaporating from the surface of the bowl at a rate proportional to the surface area, with proportionality constant k. Find a differential equation satisfied by h as a function of time, t. (That is, find an equation for dh/dt.)

(d) If the water starts at depth h_0, find the time taken for all the water to evaporate.

32. A cylindrical centrifuge of radius 1 m and height 2 m is filled with water to a depth of 1 meter (see Figure 8.88(I)). As the centrifuge accelerates, the water level rises along the wall and drops in the center; the cross-section will be a parabola. (See Figure 8.88(II).)

(a) Find the equation of the parabola in Figure 8.88(II) in terms of h, the depth of the water at its lowest point.

(b) As the centrifuge rotates faster and faster, either water will be spilled out the top, as in Figure 8.88(III), or the bottom of the centrifuge will be exposed, as in Figure 8.88(IV). Which happens first?

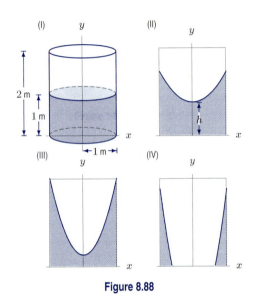

Figure 8.88

In Problems 33–34, you are given two objects which have the same mass M, the same radius R, and the same angular velocity about the indicated axes (say, one revolution per minute). For each problem, determine which of the two objects has the greater kinetic energy. (The kinetic energy of a particle of mass m with speed v is $\frac{1}{2}mv^2$.) Don't compute the kinetic energy of the objects to do this; just use reasoning.

33.

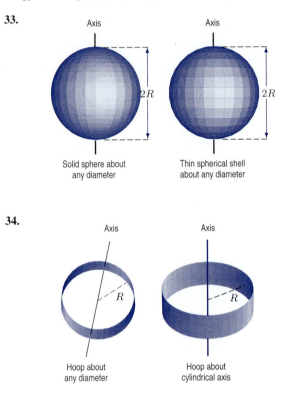

Solid sphere about any diameter   Thin spherical shell about any diameter

34.

Axis   Axis

Hoop about any diameter   Hoop about cylindrical axis

CAS Challenge Problems

35. For a positive constant a, consider the curve

$$y = \sqrt{\frac{x^3}{a-x}}, \qquad 0 \le x < a.$$

(a) Using a computer algebra system, show that for $0 \le t < \pi/2$, the point with coordinates (x, y) lies on the curve if:

$$x = a\sin^2 t, \quad y = \frac{a\sin^3 t}{\cos t}.$$

(b) A solid is obtained by rotating the curve about its asymptote at $x = a$. Use horizontal slicing to write an integral in terms of x and y that represents the volume of this solid.

(c) Use part (a) to substitute in the integral for both x and y in terms of t. Use a computer algebra system or trigonometric identities to calculate the volume of the solid.

For Problems 36–37, define $A(t)$ to be the arc length of the graph of $y = f(x)$ from $x = 0$ to $x = t$, for $t \ge 0$.

(a) Use the integral expression for arc length and a computer algebra system to obtain a formula for $A(t)$.

(b) Graph $A(t)$ for $0 \le t \le 10$. What simple function does $A(t)$ look like? What does this tell you about the approximate value of $A(t)$ for large t?

(c) In order to estimate arc length visually, you need the same scales on both axes, so that the lengths aren't distorted in one direction. Draw a graph of $f(x)$ with viewing window $0 \le x \le 100$, $0 \le y \le 100$. Explain what you noticed in part (b) in terms of this graph.

36. $f(x) = x^2$

37. $f(x) = \sqrt{x}$

38. A bead is formed by drilling a cylindrical hole of circular cross section and radius a through a sphere of radius

$r > a$, the axis of the hole passing through the center of the sphere.

(a) Write a definite integral expressing the volume of

the bead.

(b) Find a formula for the bead by evaluating the definite integral in part (a).

CHECK YOUR UNDERSTANDING

In Problems 1–21, are the statements true or false? Give an explanation for your answer.

1. The integral $\int_{-3}^{3} \pi(9 - x^2)\, dx$ represents the volume of a sphere of radius 3.

2. The integral $\int_0^h \pi(r - y)\, dy$ gives the volume of a cone of radius r and height h.

3. The integral $\int_0^r \pi\sqrt{r^2 - y^2}\, dy$ gives the volume of a hemisphere of radius r.

4. A cylinder of radius r and length l is lying on its side. Horizontal slicing tells us that the volume is given by $\int_{-r}^{r} 2l\sqrt{r^2 - y^2}\, dy$.

5. If a region in the xy-plane lies below the x-axis, then revolving that region around the x-axis gives a solid of negative volume.

6. To find the total population in a circular city, we always slice it into concentric rings, no matter what the population density function.

7. A city occupies a region in the xy-plane, with population density $\delta(y) = 1 + y$. To set up an integral representing the total population in the city, we should slice the region parallel to the y-axis.

8. The population density in a circular city of radius 2 depends on the distance r from the center by $f(r) = 10 - 3r$, so that the density is greatest at the center. Then the population of the inner city, $0 \le r \le 1$, is greater than the population of the suburbs, $1 \le r \le 2$.

9. The location of the center of mass of a system of three masses on the x-axis does not change if all the three masses are doubled.

10. The center of mass of a region in the plane cannot be outside the region.

11. Particles are shot at a circular target. The density of particles hitting the target decreases with the distance from the center. To set up a definite integral to calculate the total number of particles hitting the target, we should slice the region into concentric rings.

12. A metal rod of density $f(x)$ lying along the x-axis from $x = 0$ to $x = 4$ has its center of mass at $x = 2$. Then the two halves of the rod on either side of $x = 2$ have equal mass.

13. It takes more work to lift a 20 lb weight 10 ft slowly than to lift it the same distance quickly.

14. Work can be negative or positive.

15. The force on a rectangular dam is doubled if its length stays the same and its depth is doubled.

16. To find the force of water on a vertical wall, we always slice the wall horizontally, no matter what the shape of the wall.

17. The force of a liquid on a wall can be negative or positive.

18. If the average value of the force $F(x)$ is 7 on the interval $1 \le x \le 4$, then the work done by the force in moving from $x = 1$ to $x = 4$ is 21.

19. The present value of an income stream is always less than its future value.

20. If $p(x) = xe^{-x^2}$ for all x, then $p(x)$ is a probability density function.

21. If $p(x) = xe^{-x^2}$ for all $x > 0$ and $p(x) = 0$ for $x \le 0$, then $p(x)$ is a probability density function.

Are the statements in Problems 22–25 true or false? If a statement is true, explain how you know. If a statement is false, give a counterexample

22. Of two solids of revolution, the one with the greater volume is obtained by revolving the region in the plane with the greater area.

23. If f is differentiable on the interval $[0, 10]$, then the arc length of the graph of f on the interval $[0, 1]$ is less than the arc length of the graph of f on the interval $[1, 10]$.

24. If f is concave up for all x and $f'(0) = 3/4$, then the arc length of the graph of f on the interval $[0, 4]$ is at least 5.

25. If f is concave down for all x and $f'(0) = 3/4$, then the arc length of the graph of f on the interval $[0, 4]$ is at most 5.

In Problems 26–30, a quantity x is distributed through a population with probability density function $p(x)$ and cumulative distribution function $P(x)$. Decide if the statements in Problems 26–30 are true or false. Give an explanation for your answer.

26. If $p(10) = 1/2$, then half the population has $x < 10$.

27. If $P(10) = 1/2$, then half the population has $x < 10$.

28. If $p(10) = 1/2$, then the fraction of the population lying between $x = 9.98$ and $x = 10.04$ is about 0.03.

29. If $p(10) = p(20)$, then none of the population has x values lying between 10 and 20.

30. If $P(10) = P(20)$, then none of the population has x values lying between 10 and 20.

PROJECTS FOR CHAPTER EIGHT

1. Volume Enclosed by Two Cylinders

Two cylinders are inscribed in a cube of side length 2, as shown in Figure 8.89. What is the volume of the solid that the two cylinders enclose? [Hint: Use horizontal slices.] Note: The solution was known to Archimedes. The Chinese mathematician Liu Hui (third century A.D.) tried to find this volume, but he failed; he wrote a poem about his efforts calling the enclosed volume a "box-lid":

Look inside the cube
And outside the box-lid;
Though the dimension increases,
It doesn't quite fit.
The marriage preparations are complete;
But square and circle wrangle,
Thick and thin are treacherous plots,
They are incompatible.
I wish to give my humble reflections,
But fear that I will miss the correct principle;
I dare to let the doubtful points stand,
Waiting
For one who can expound them.

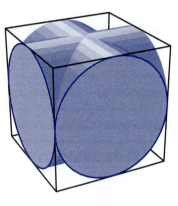

Figure 8.89

2. Length of a Hanging Cable

The distance between the towers of the main span of the Golden Gate Bridge is about 1280 m; the sag of the cable halfway between the towers on a cold winter day is about 143 m. See Figure 8.90.

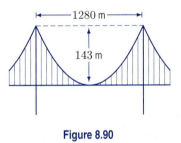

Figure 8.90

(a) How long is the cable, assuming it has an approximately parabolic shape? (Represent the cable as a parabola of the form $y = kx^2$ and determine k to at least 1 decimal place.)

(b) On a hot summer day the cable is about 0.05% longer, due to thermal expansion. By how much does the sag increase? Assume no movement of the towers.

3. Surface Area of an Unpaintable Can of Paint

This project introduces the formula for the surface area of a volume of revolution and shows that it is possible to have a solid with finite volume but infinite surface area.

We know that the arclength of the curve $y = f(x)$ from a to b can be found using the integral

$$\text{Arc length} = \int_a^b \sqrt{1 + (f'(x))^2}\, dx.$$

Figure 8.91 shows the corresponding arc length of a small piece of the curve. Similarly, it can be shown that the surface area from a to b of the solid obtained by revolving $y = f(x)$ around the x-axis is given by

$$\text{Surface area } = 2\pi \int_a^b f(x)\sqrt{1+(f'(x))^2}\,dx.$$

We can see this why this might be true by looking at Figure 8.92. We approximate the small piece of the surface by a slanted cylinder of radius y and "height" equal to the arclength of the curve, so that

$$\text{Surface area of edge of slice } \approx 2\pi y\sqrt{1+(f'(x))^2}\,\Delta x.$$

Integrating this expression from $x = a$ to $x = b$ gives the surface area of the solid.

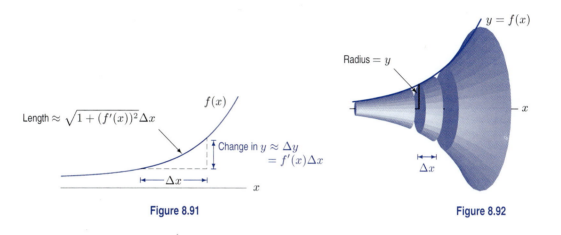

Figure 8.91   **Figure 8.92**

(a) Calculate the surface area of a sphere of radius r.
(b) Calculate the surface area of a cone of radius r and height h.
(c) Rotate the curve $y = 1/x$ for $x \geq 1$ around the x-axis. Find the volume of this solid.
(d) Show that the surface area of the solid in part (c) is infinite. [Hint: You might not be able to find an antiderivative of the integrand in the surface area formula; can you get a lower bound on the integral?]
(e) (Optional. Requires Chapter 11.) Find a curve such that when the portion of the curve from $x = a$ to $x = b$ is rotated around the x-axis (for any a and b), the volume of the solid of revolution is equal to its surface area. You may assume $dy/dx \geq 0$.

4. Maxwell's Distribution of Molecular Velocities

Let v be the speed, in meters/second, of an oxygen molecule, and let $p(v)$ be the density function of the speed distribution of oxygen molecules at room temperature. Maxwell showed that

$$p(v) = av^2 e^{-mv^2/(2kT)},$$

where $k = 1.4 \times 10^{-23}$ is the Boltzmann constant, T is the temperature in Kelvin (at room temperature, $T = 293$), and $m = 5 \times 10^{-26}$ is the mass of the oxygen molecule in kilograms.

(a) Find the value of a.
(b) Estimate the median and the mean speed. Find the maximum of $p(v)$.
(c) How do your answers in part (b) for the mean and the maximum of $p(v)$ change as T changes?

Chapter Nine

SERIES

In this chapter we look at infinite sums, called *infinite series*. We start in Section 9.1 with a particular type of series, called a geometric series. Section 9.2 considers general sequences and series of constants and what it means for such series to converge. The tests that allow us to determine convergence are in Sections 9.2 and 9.3.

Section 9.4 introduces power series, in which the terms are constants multiplied by powers of x. These series converge for some x-values and not for others; the radius of convergence is introduced to identify the interval on which the series converges.

For proofs of these properties, see Problems 26–29. As for improper integrals, the convergence of a series is determined by its behavior for large n. (See the "behaves like" principle on page 335.) From Property 2 we see that, if N is a positive integer, then $\sum_{n=1}^{\infty} a_n$ and $\sum_{n=N}^{\infty} a_n$ either both converge or both diverge. Thus, if all we care about is the convergence of a series, we can omit the limits and write $\sum a_n$.

Property 3 often is used to tell us that a series does *not* converge. If the terms do not go to 0, the series has no chance of converging.

Warning: Knowing that $\lim_{n \to \infty} a_n = 0$ is *not* enough to ensure convergence. (See Example 3.)

Example 2   Does the series $\sum (1 - e^{-n})$ converge?

Solution   Since the terms in the series, $a_n = 1 - e^{-n}$ tend to 1, not 0, as $n \to \infty$, the series diverges by Property 3 of Theorem 9.2.

Comparison of Series and Integrals

We investigate the convergence of some series by comparison with an improper integral. The *harmonic series* is the infinite series

$$1 + \frac{1}{2} + \frac{1}{3} + \frac{1}{4} + \cdots + \frac{1}{n} + \cdots .$$

Convergence of this sum would mean that the sequence of partial sums

$$S_1 = 1, \quad S_2 = 1 + \frac{1}{2}, \quad S_3 = 1 + \frac{1}{2} + \frac{1}{3}, \quad \cdots, \quad S_n = 1 + \frac{1}{2} + \frac{1}{3} + \cdots + \frac{1}{n}, \quad \cdots$$

tends to a limit as $n \to \infty$. Let's look at some values:

$$S_1 = 1, \quad S_{10} \approx 2.93, \quad S_{100} \approx 5.19, \quad S_{1000} \approx 7.49, \quad S_{10000} \approx 9.79.$$

The growth of these partial sums is slow, but they do in fact grow without bound, so the harmonic series diverges. This is justified in the following example and in Problem 30.

Example 3   Show that the harmonic series $1 + 1/2 + 1/3 + 1/4 + \ldots$ diverges.

Solution   The idea is to approximate $\int_1^{\infty} (1/x)\, dx$ by a left-hand sum, where the terms $1, 1/2, 1/3, \ldots$ are heights of rectangles of base 1. In Figure 9.4, the sum of the areas of the 3 rectangles is larger than the area under the curve between $x = 1$ and $x = 4$, and the same kind of relationship holds for the first n rectangles. Thus, we have

$$S_n = 1 + \frac{1}{2} + \frac{1}{3} + \ldots + \frac{1}{n} > \int_1^{n+1} \frac{1}{x}\, dx = \ln(n+1).$$

Since $\ln(n+1)$ gets arbitrarily large as $n \to \infty$, so do the partial sums, S_n. Thus, the partial sums have no limit, so the series diverges.

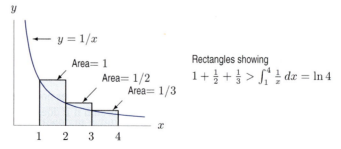

Figure 9.4: Comparing the harmonic series to $\int_1^{\infty} (1/x)\, dx$

Notice that the harmonic series diverges, even though $\lim_{n \to \infty} a_n = \lim_{n \to \infty} (1/n) = 0$.

Example 4   By comparison with the improper integral $\int_1^\infty (1/x^2)\,dx$, show that the following series converges:

$$\sum_{n=1}^\infty \frac{1}{n^2} = 1 + \frac{1}{4} + \frac{1}{9} + \cdots.$$

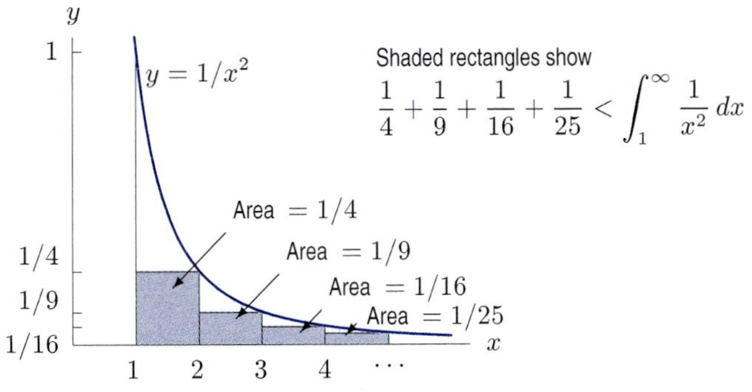

Figure 9.5: Comparing $\sum_{n=1}^\infty 1/n^2$ to $\int_1^\infty (1/x^2)\,dx$

Solution   Since we want to show that $\displaystyle\sum_{n=1}^\infty 1/n^2$ converges, we want to show that the partial sums of this series tend to a limit. We do this by showing that the sequence of partial sums increases and is bounded above, so Theorem 9.1 applies.

Each successive partial sum is obtained from the previous one by adding one more term in the series. Since all the terms are positive, the sequence of partial sums is increasing.

To show that the partial sums of $\displaystyle\sum_{n=1}^\infty 1/n^2$ are bounded, we consider the right-hand sum represented by the area of the rectangles in Figure 9.5. We start at $x = 1$, since the area under the curve is infinite for $0 \leq x \leq 1$. The shaded rectangles in Figure 9.5 suggest that:

$$\frac{1}{4} + \frac{1}{9} + \frac{1}{16} + \cdots + \frac{1}{n^2} \leq \int_1^\infty \frac{1}{x^2}\,dx.$$

The area under the graph is finite, since

$$\int_1^\infty \frac{1}{x^2}\,dx = \lim_{b\to\infty} \int_1^b \frac{1}{x^2}\,dx = \lim_{b\to\infty}\left(\frac{-1}{b} + 1\right) = 1.$$

To get S_n, we add 1 to both sides, giving

$$S_n = 1 + \frac{1}{4} + \frac{1}{9} + \frac{1}{16} + \cdots + \frac{1}{n^2} \leq 1 + \int_1^\infty \frac{1}{x^2}\,dx = 2.$$

Thus, the sequence of partial sums is bounded above by 2. Hence, by Theorem 9.1 the sequence of partial sums converges, so the series converges.

Notice that we have shown that the series in the previous example converges, but we have not found its sum. The integral gives us a bound on the partial sums, but it does not give us the limit of the partial sums. Euler proved the remarkable fact that the sum is $\pi^2/6$.

The method of Examples 3 and 4 can be used to prove the following theorem. See Problem 33.

Theorem 9.3: The Integral Test

Suppose $c \geq 0$ and $f(x)$ is a decreasing positive function, defined for all $x \geq c$, with $a_n = f(n)$ for all n.

- If $\int_c^\infty f(x)\,dx$ converges, then $\sum a_n$ converges.

- If $\int_c^\infty f(x)\,dx$ diverges, then $\sum a_n$ diverges.

The integral test can be used to show that $\sum_{n=1}^\infty 1/n^p$ converges for $p > 1$. See Problem 23.

Exercises and Problems for Section 9.2

Exercises

Exercises 1–8 give expressions for S_n, the general term of a sequence. Find the limit of each sequence, if it exists.

1. $(0.2)^n$

2. 2^n

3. $(-0.3)^n$

4. $3 + e^{-2n}$

5. $\cos(\pi n)$

6. $\dfrac{2^n}{3^n}$

7. $\dfrac{3 + 4n}{5 + 7n}$

8. $\dfrac{2n + (-1)^n 5}{4n - (-1)^n 3}$

Use the integral test to decide whether the series in Exercises 9–12 converge or diverge.

9. $\displaystyle\sum_{n=1}^\infty \frac{1}{n^3}$

10. $\displaystyle\sum_{n=1}^\infty \frac{n}{n^2 + 1}$

11. $\displaystyle\sum_{n=0}^\infty n e^{-n^2}$

12. $\displaystyle\sum_{n=2}^\infty \frac{1}{n(\ln n)^2}$

Problems

Do the series in Problems 13–20 converge or diverge?

13. $\displaystyle\sum_{n=1}^\infty \left(\left(\frac{3}{4}\right)^n + \frac{1}{n} \right)$

14. $\displaystyle\sum_{n=1}^\infty \frac{n}{n + 1}$

15. $\displaystyle\sum_{n=0}^\infty \frac{3}{n + 2}$

16. $\displaystyle\sum_{n=1}^\infty \frac{n + 2^n}{n 2^n}$

17. $\displaystyle\sum_{n=1}^\infty \frac{3}{(2n - 1)^2}$

18. $\displaystyle\sum_{n=0}^\infty \frac{2}{\sqrt{2 + n}}$

19. $\displaystyle\sum_{n=1}^\infty \frac{\ln n}{n}$

20. $\displaystyle\sum_{n=3}^\infty \frac{n + 1}{n^2 + 2n + 2}$

By identifying each one as a left- or right-hand approximation of an improper integral, decide whether the series in Problems 21–22 converge.

21. $1 + \dfrac{1}{5} + \dfrac{1}{9} + \dfrac{1}{13} + \dfrac{1}{17} + \cdots + \dfrac{1}{4n - 3} + \cdots$

22. $1 + \dfrac{1}{2^{3/2}} + \dfrac{1}{3^{3/2}} + \dfrac{1}{4^{3/2}} + \dfrac{1}{5^{3/2}} + \cdots + \dfrac{1}{n^{3/2}} + \cdots$

23. Use the integral test to show that $\displaystyle\sum_{n=1}^\infty 1/n^p$

    (a) Converges if $p > 1$.
    (b) Diverges if $p \leq 1$.

24. Find the sum: $\displaystyle\sum_{n=0}^\infty \frac{3^n + 5}{4^n}$

25. Write a definition for $\lim\limits_{n \to \infty} S_n = L$ similar to the ϵ, δ definition for $\lim\limits_{x \to a} f(x) = L$ in Section 2.2. Instead of δ, you will need N, a value of n.

26. Show that if $\sum a_n$ and $\sum b_n$ converge and if k is a constant, then $\sum (a_n + b_n)$, $\sum (a_n - b_n)$, and $\sum k a_n$ converge.

27. Let N be a positive integer. Show that if $a_n = b_n$ for $n \geq N$, then $\sum a_n$ and $\sum b_n$ either both converge, or both diverge.

28. Show that if $\sum a_n$ converges, then $\lim\limits_{n \to \infty} a_n = 0$. [Hint: Consider $\lim_{n \to \infty}(S_n - S_{n-1})$, where S_n is the nth partial sum.]

29. Show that if $\displaystyle\sum_{n=1}^\infty a_n$ diverges and $k \neq 0$, then $\displaystyle\sum_{n=1}^\infty k a_n$ diverges.

30. Consider the following grouping of terms in the harmonic series:

$$1 + \left(\frac{1}{2}\right) + \left(\frac{1}{3} + \frac{1}{4}\right) + \left(\frac{1}{5} + \frac{1}{6} + \frac{1}{7} + \frac{1}{8}\right) +$$
$$\left(\frac{1}{9} + \frac{1}{10} + \cdots + \frac{1}{16}\right) + \cdots$$

    (a) Show that the sum of each group of fractions is more than $1/2$.
    (b) Explain why this shows that the harmonic series does not converge.

31. Estimate the sum of the first 100,000 terms of the harmonic series,

$$\sum_{k=1}^{100,000} \frac{1}{k},$$

to the closest integer. [Hint: Use left- and right-hand sums of the function $f(x) = 1/x$ on the interval from 1 to 100,000 with $\Delta x = 1$.]

32. Although the harmonic series does not converge, the partial sums grow very, very slowly. Take a right-hand sum approximating the integral of $f(x) = 1/x$ on the interval $[1, n]$, with $\Delta x = 1$, to show that

$$\frac{1}{2} + \frac{1}{3} + \frac{1}{4} + \cdots + \frac{1}{n} < \ln n.$$

If a computer could add a million terms of the harmonic series each second, estimate the sum after one year.

33. In this problem, you will justify the integral test. Suppose $c \geq 0$ and $f(x)$ is a decreasing positive function, defined for all $x \geq c$, with $f(n) = a_n$ for all n.

(a) Suppose $\int_c^\infty f(x)\, dx$ converges. By considering rectangles under the graph of f, show that $\sum a_n$ converges. [Hint: See Example 4 on page 415.]

(b) Suppose that $\int_c^\infty f(x)\, dx$ diverges. By considering rectangles above the graph of f, show that $\sum a_n$ diverges. [Hint: See Example 3 on page 414.]

9.3 TESTS FOR CONVERGENCE

Comparison of Series

In Section 7.8, we compared two integrals to decide whether an improper integral converged. In Theorem 9.3 we compared an integral and a series. Now we compare two series.

Theorem 9.4: Comparison Test

Suppose $0 \leq a_n \leq b_n$ for all n.
- If $\sum b_n$ converges, then $\sum a_n$ converges.
- If $\sum a_n$ diverges, then $\sum b_n$ diverges.

Since $a_n \leq b_n$, the plot of the a_n lies under the plot of the b_n. (See Figure 9.6.) The comparison test says that if the total area for $\sum b_n$ is finite, then the total area for $\sum a_n$ is finite also. If the total area for $\sum a_n$ is not finite, then neither is the total area for $\sum b_n$.

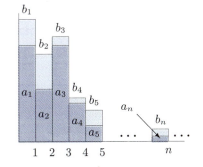

Figure 9.6: Each a_n is represented by the area of a dark rectangle, and each b_n by a dark plus a light rectangle

---

Example 1   Use the comparison test to determine whether the series $\displaystyle\sum_{n=1}^{\infty} 1/n^3$ converges.

Solution   For $n \geq 1$, we know that $n^2 \leq n^3$, so

$$0 \leq \frac{1}{n^3} \leq \frac{1}{n^2}.$$

Thus, every term in the series $\sum_{n=1}^{\infty} 1/n^3$ is less than or equal to the corresponding term in $\sum_{n=1}^{\infty} 1/n^2$. Since we saw that $\sum_{n=1}^{\infty} 1/n^2$ converges from page 415, we know that $\sum_{n=1}^{\infty} 1/n^3$ converges.

Example 2   Use the comparison test to show that if $p < 1$, then $\sum_{n=1}^{\infty} 1/n^p$ diverges.

Solution   Since $p < 1$ and $n \geq 1$, we have $n^p \leq n$, so $1/n^p \geq 1/n$. The harmonic series $\sum 1/n$ diverges, so $\sum 1/n^p$ diverges also.

Example 3   Decide whether the following series converge: (a) $\displaystyle\sum_{n=1}^{\infty} \frac{n-1}{n^3+3}$ (b) $\displaystyle\sum_{n=1}^{\infty} \frac{6n^2+1}{2n^3-1}$.

Solution   (a) Since the convergence is determined by the behavior of the terms for large n, we observe that

$$\frac{n-1}{n^3+3} \quad \text{behaves like} \quad \frac{n}{n^3} = \frac{1}{n^2} \quad \text{as} \quad n \to \infty.$$

Since $\sum 1/n^2$ converges, we guess that $\sum (n-1)/(n^3+3)$ converges. To confirm this, we use the comparison test. Since a fraction increases if its numerator is made larger or its denominator is made smaller, we have

$$0 \leq \frac{n-1}{n^3+3} \leq \frac{n}{n^3} = \frac{1}{n^2} \quad \text{for all} \quad n \geq 1.$$

Thus, the series $\sum (n-1)/(n^3+3)$ converges by comparison with $\sum 1/n^2$.

(b) First, we observe that

$$\frac{6n^2+1}{2n^3-1} \quad \text{behaves like} \quad \frac{6n^2}{2n^3} = \frac{3}{n} \quad \text{as} \quad n \to \infty.$$

Since $\sum 1/n$ diverges, so does $\sum 3/n$, and we guess that $\sum (6n^2+1)/(2n^3-1)$ diverges. To confirm this, we use the comparison test. Since a fraction decreases if its numerator is made smaller or its denominator is made larger, we have

$$0 \leq \frac{6n^2}{2n^3} \leq \frac{6n^2+1}{2n^3-1},$$

so

$$0 \leq \frac{3}{n} \leq \frac{6n^2+1}{2n^3-1}.$$

Thus, the series $\sum (6n^2+1)/(2n^3-1)$ diverges by comparison with $\sum 3/n$.

Series of Both Positive and Negative Terms

If $\sum a_n$ has both positive and negative terms, then its plot has rectangles lying both above and below the horizontal axis. See Figure 9.7. The total area of the rectangles is no longer equal to $\sum a_n$. However, it is still true that if the total area of the rectangles above and below the axis is finite, then the series converges. The area of the nth rectangle is $|a_n|$, so we have:

Theorem 9.5: Convergence of Absolute Values Implies Convergence

If $\sum |a_n|$ converges, then so does $\sum a_n$.

Problem 32 shows how to prove this result.

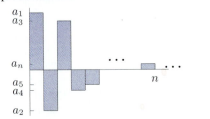

Figure 9.7: Representing a series with positive and negative terms

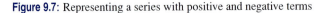

Example 4   Explain how we know that the following series converges

$$\sum_{n=1}^{\infty} \frac{(-1)^{n-1}}{n^2} = 1 - \frac{1}{4} + \frac{1}{9} - \cdots.$$

Solution   Writing $a_n = (-1)^{n-1}/n^2$, we have

$$|a_n| = \left| \frac{(-1)^{n-1}}{n^2} \right| = \frac{1}{n^2}.$$

From Example 4 on page 415, we know that $\sum 1/n^2$ converges, so $\sum (-1)^{n-1}/n^2$ converges.

Comparison with a Geometric Series: The Ratio Test

A geometric series $\sum a_n$ has the property that the ratio a_{n+1}/a_n is constant for all n. For many other series, this ratio, although not constant, tends to a constant as n increases. In some ways, such series behave like geometric series. In particular, a geometric series converges if the ratio $|a_{n+1}/a_n| < 1$. A non-geometric series converges if the ratio $|a_{n+1}/a_n|$ tends to a limit which is less than 1. This idea leads to the following test.

Theorem 9.6: The Ratio Test

For a series $\sum a_n$, suppose the sequence of ratios $|a_{n+1}|/|a_n|$ has a limit:

$$\lim_{n \to \infty} \frac{|a_{n+1}|}{|a_n|} = L.$$

- If $L < 1$, then $\sum a_n$ converges.
- If $L > 1$, or if L is infinite,[3] then $\sum a_n$ diverges.
- If $L = 1$, the test does not tell us anything about the convergence of $\sum a_n$.

Proof   Here are the main steps in the proof. Suppose $\lim_{n \to \infty} \dfrac{|a_{n+1}|}{|a_n|} = L < 1$. Let x be a number between L and 1. Then for all sufficiently large n, say for all $n \geq k$, we have

$$\frac{|a_{n+1}|}{|a_n|} < x.$$

Then,

$$|a_{k+1}| < |a_k|x,$$
$$|a_{k+2}| < |a_{k+1}|x < |a_k|x^2,$$
$$|a_{k+3}| < |a_{k+2}|x < |a_k|x^3,$$

and so on. Thus, writing $a = |a_k|$, we have for $i = 1, 2, 3, \ldots$,

$$|a_{k+i}| < ax^i.$$

Now we can use the comparison test: $\sum |a_{k+i}|$ converges by comparison with the geometric series $\sum ax^i$. Since $\sum |a_{k+i}|$ converges, Theorem 9.5 tells us that $\sum a_{k+i}$ converges. So, by property 2 of Theorem 9.2, we see that $\sum a_n$ converges too.

If $L > 1$, then for sufficiently large n, say $n \geq m$,

$$|a_{n+1}| > |a_n|,$$

---

[3]That is, the sequence $|a_{n+1}|/|a_n|$ grows without bound.

so the sequence $|a_m|, |a_{m+1}|, |a_{m+2}|, \cdots$, is increasing. Thus, $\lim\limits_{n\to\infty} a_n \neq 0$, so $\sum a_n$ diverges (by Theorem 9.2, property 3). The argument in the case that $|a_{n+1}|/|a_n|$ is unbounded is similar.

Example 5   Show that the following series converges:[4]

$$\sum_{n=1}^{\infty} \frac{1}{n!} = 1 + \frac{1}{2!} + \frac{1}{3!} + \cdots.$$

Solution   Since $a_n = 1/n!$ and $a_{n+1} = 1/(n+1)!$, we have

$$\frac{|a_{n+1}|}{|a_n|} = \frac{1/(n+1)!}{1/n!} = \frac{n!}{(n+1)!} = \frac{n(n-1)(n-2)\cdots\cdots 2\cdot 1}{(n+1)n(n-1)\cdots\cdots 2\cdot 1}.$$

We cancel $n(n-1)(n-2)\cdots\cdots 2\cdot 1$, giving

$$\lim_{n\to\infty} \frac{|a_{n+1}|}{|a_n|} = \lim_{n\to\infty} \frac{n!}{(n+1)!} = \lim_{n\to\infty} \frac{1}{n+1} = 0.$$

Because the limit is 0, which is less than 1, the ratio test tells us that $\sum\limits_{n=1}^{\infty} 1/n!$ converges.

Example 6   What does the ratio test tell us about the convergence of the following two series?

$$\sum_{n=1}^{\infty} \frac{1}{n} \quad \text{and} \quad \sum_{n=1}^{\infty} \frac{(-1)^{n-1}}{n}.$$

Solution   Because $|(-1)^n| = 1$, in both cases we have $\lim_{n\to\infty} |a_{n+1}/a_n| = \lim_{n\to\infty} n/(n+1) = 1$. Thus, the ratio test does not tell us anything about the convergence of either series. In fact, the first series is the harmonic series, which diverges. Example 7 will show that the second series converges.

Alternating Series

A series is called an *alternating series* if the terms alternate in sign. For example,

$$\sum_{n=1}^{\infty} \frac{(-1)^{n-1}}{n} = 1 - \frac{1}{2} + \frac{1}{3} - \frac{1}{4} + \cdots + \frac{(-1)^{n-1}}{n} + \cdots.$$

The convergence of an alternating series can often by determined using the following test:

Theorem 9.7: Alternating Series Test

A series of the form

$$\sum_{n=1}^{\infty} (-1)^{n-1} a_n = a_1 - a_2 + a_3 - a_4 + \cdots + (-1)^{n-1} a_n + \cdots$$

converges if

$$0 < a_{n+1} < a_n \quad \text{for all } n \quad \text{and} \quad \lim_{n\to\infty} a_n = 0.$$

Although we do not prove this result, we can see why it is reasonable. The first partial sum, $S_1 = a_1$, is positive. The second, $S_2 = a_1 - a_2$, is still positive, since $a_2 < a_1$, but S_2 is smaller than S_1. (See Figure 9.8.) The next sum, $S_3 = a_1 - a_2 + a_3$, is greater than S_2 but smaller than S_1. The partial sums oscillate back and forth, and since the distance between them tends to 0, they eventually converge.

---

[4]We define 2! to be $2 \cdot 1 = 2$. Similarly, $3! = 3 \cdot 2 \cdot 1 = 6$ and $n! = n \cdot (n-1) \cdots 2 \cdot 1$. We also define 0! = 1.

9.4 POWER SERIES

In Section 9.1 we
otherwise. This s
Chapter 10 shows
and $\ln x$.

A **power serie**

$$C_0 + C$$

We think of
numbers like thos
consider the partia
$a)^2 + \cdots + C_n(x$

For a fixed val
$$\lim_{n \to \infty} S_n(x) =$$

A power serie

Numerical and Graphical V

As an example, co

$$(x - 1) -$$

To investigate the c
ure 9.9. The graph
and 4, we show th
of this series.

At $x = 1.4$, wh

Table 9.1 show
which is inside the
rather slowly. For
larger the value of
fifth term is about 2
shows the interval c

[5]Here we have chose

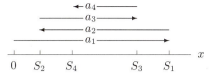

Figure 9.8: Partial sums, S_1, S_2, S_3, S_4 of an alternating series

Example 7 Show that the alternating harmonic series converges

$$\sum_{n=1}^{\infty} \frac{(-1)^{n-1}}{n}.$$

Solution We have $a_n = 1/n$ and $a_{n+1} = 1/(n+1)$. Thus,

$$a_{n+1} = \frac{1}{n+1} < \frac{1}{n} = a_n \quad \text{for all } n, \quad \text{and} \quad \lim_{n \to \infty} 1/n = 0.$$

Thus, the hypothesis of Theorem 9.7 is satisfied, so the alternating harmonic series converges.

Suppose S is the sum of the alternating series, so $S = \lim_{n \to \infty} S_n$. Then S is trapped between any two consecutive partial sums, say S_3 and S_4 or S_4 and S_5 so

$$S_2 < S_4 < \cdots < S < \cdots < S_3 < S_1.$$

Thus, the error in using S_n to approximate the true sum S is less than the distance from S_n to S_{n+1}, which is a_{n+1}. Stated symbolically, we have the following result:

Theorem 9.8: Error Bounds for Alternating Series

Let $S_n = \sum_{i=1}^{n} (-1)^{i-1} a_i$ be the nth partial sum of an alternating series and let $S = \lim_{n \to \infty} S_n$.
Suppose that $0 < a_{n+1} < a_n$ for all n and $\lim_{n \to \infty} a_n = 0$. Then

$$|S - S_n| < a_{n+1}.$$

Thus, the error in using S_n to approximate S is less than the magnitude of the first term of the series which is omitted in the approximation.

Example 8 Estimate the error in approximating the sum of the alternating harmonic series $\sum_{n=1}^{\infty} (-1)^{n-1}/n$ by the sum of the first nine terms.

Solution The ninth partial sum is given by

$$S_9 = 1 - \frac{1}{2} + \frac{1}{3} - \cdots + \frac{1}{9} = 0.7456\ldots.$$

The first term omitted is $-1/10$, with magnitude 0.1. By Theorem 9.8, we know that the true value of the sum differs from $0.7456\ldots$ by less than 0.1.

Exercises and Problems for Section 9.3

Exercises

Use the comparison test to confirm the statements in Exercises 1–3.

1. $\sum_{n=1}^{\infty} \frac{1}{n^2}$ converges, so $\sum_{n=1}^{\infty} \frac{1}{n^2 + 2}$ converges.

2. $\sum_{n=4}^{\infty} \frac{1}{n}$ diverges, so $\sum_{n=4}^{\infty} \frac{1}{n - 3}$ diverges.

3. $\sum_{n=1}^{\infty} \frac{1}{n^2}$ converges, so $\sum_{n=1}^{\infty} \frac{e^{-n}}{n^2}$ converges.

Use the comparison test to d[...]
Exercises 4–11 converge.

4. $\displaystyle\sum_{n=1}^{\infty} \frac{1}{n^3}$

6. $\displaystyle\sum_{n=1}^{\infty} \frac{1}{3^n + 1}$

8. $\displaystyle\sum_{n=1}^{\infty} 2^{-n}\frac{(n+1)}{(n+2)}$

10. $\displaystyle\sum_{n=1}^{\infty} \frac{2^n + 1}{n2^n - 1}$

Problems

Use a computer or calculator [...]
the partial sums of the alternat[...]
Which ones converge? Confir[...]
nating series test. If a series co[...]

18. $1 - 0.1 + 0.01 - 0.001 + $

19. $1 - 2 + 3 - 4 + 5 + \cdots$

20. $1 - \dfrac{1}{1!} + \dfrac{1}{2!} - \dfrac{1}{3!} + \cdots$

Determine which of the series i[...]

21. $\displaystyle\sum_{n=0}^{\infty} e^{-n}$

23. $\displaystyle\sum_{n=1}^{\infty} \frac{(n-1)!}{n^2}$

25. $\displaystyle\sum_{n=1}^{\infty} \frac{(-1)^{n-1}}{\sqrt{3n-1}}$

27. $\displaystyle\sum_{n=1}^{\infty} \frac{n(n+1)}{\sqrt{n^3 + 2n^2}}$

29. (a) Show that if n is even

$$1 - \frac{1}{2} + \frac{1}{3} - \frac{1}{4} + \cdots - \frac{1}{n}$$

(b) Show that $\dfrac{1}{1\cdot 2} + \dfrac{1}{3\cdot}$

(c) Use the results of pa[...]

$$1 - \frac{1}{2} + \frac{1}{3} - \frac{1}{4} + \cdots$$

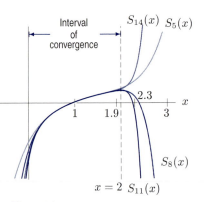

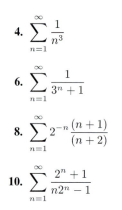

Figure 9.9: Partial sums for series in Example 2 converge for $0 < x < 2$

Table 9.1 *Partial sums for series in Example 2 with $x = 1.9$ inside interval of convergence and $x = 2.3$ outside*

n	$S_n(1.9)$	n	$S_n(2.3)$
2	0.495	2	0.455
5	0.69207	5	1.21589
8	0.61802	8	0.28817
11	0.65473	11	1.71710
14	0.63440	14	-0.70701

Notice that the interval of convergence, $0 < x \le 2$, is centered on $x = 1$. Since the interval extends one unit on either side, we say the *radius of convergence* of this series is 1.

Intervals of Convergence

Each power series falls into one of the three following cases, characterized by its *radius of convergence, R.*

- The series converges only for $x = a$; the **radius of convergence** is defined to be $R = 0$.
- The series converges for all values of x; the **radius of convergence** is defined to be $R = \infty$.
- There is a positive number R, called the **radius of convergence**, such that the series converges for $|x - a| < R$ and diverges for $|x - a| > R$. See Figure 9.10.
- The **interval of convergence** is the interval between $a - R$ and $a + R$, including any endpoint where the series converges.

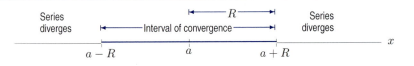

Figure 9.10: Radius of convergence, R, determines an interval, centered at $x = a$, in which the series converges

There are some series whose radius of convergence we already know. For example, the geometric series

$$1 + x + x^2 + \cdots + x^n + \cdots$$

converges for $|x| < 1$ and diverges for $|x| \ge 1$, so its radius of convergence is 1. Similarly, the series

$$1 + \frac{x}{3} + \left(\frac{x}{3}\right)^2 + \cdots + \left(\frac{x}{3}\right)^n + \cdots$$

converges for $|x/3| < 1$ and diverges for $|x/3| \ge 1$, so its radius of convergence is 3.

The next theorem gives a method of computing the radius of convergence for many series. To find the values of x for which the power series $\displaystyle\sum_{n=0}^{\infty} C_n(x - a)^n$ converges, we use the ratio test. Writing $a_n = C_n(x - a)^n$ and assuming $C_n \ne 0$ and $x \ne a$, we have

$$\lim_{n\to\infty} \frac{|a_{n+1}|}{|a_n|} = \lim_{n\to\infty} \frac{|C_{n+1}(x-a)^{n+1}|}{|C_n(x-a)^n|} = \lim_{n\to\infty} \frac{|C_{n+1}||x-a|}{|C_n|} = |x - a| \lim_{n\to\infty} \frac{|C_{n+1}|}{|C_n|}.$$

Case 1. Suppose $\lim\limits_{n\to\infty} |a_{n+1}|/|a_n|$ is infinite. Then the ratio test shows that the power series converges only for $x = a$. The radius of convergence is $R = 0$.

Case 2. Suppose $\lim\limits_{n\to\infty} |a_{n+1}|/|a_n| = 0$. Then the ratio test shows that the power series converges for all x. The radius of convergence is $R = \infty$.

Case 3. Suppose $\lim\limits_{n\to\infty} |a_{n+1}|/|a_n| = K|x - a|$, where $\lim\limits_{n\to\infty} |C_{n+1}|/|C_n| = K$. In Case 1, K does not exist; in Case 2, $K = 0$. Thus, we can assume K exists and $K \neq 0$, and we can define $R = 1/K$. Then we have

$$\lim_{n\to\infty} \frac{|a_{n+1}|}{|a_n|} = K|x - a| = \frac{|x - a|}{R},$$

so the ratio test tells us that the power series:

- Converges for $\dfrac{|x - a|}{R} < 1$; that is, for $|x - a| < R$

- Diverges for $\dfrac{|x - a|}{R} > 1$; that is, for $|x - a| > R$.

The results are summarized in the following theorem.

Theorem 9.9: Method for Computing Radius of Convergence

To calculate the radius of convergence, R, for the power series $\sum\limits_{n=0}^{\infty} C_n(x - a)^n$, use the ratio test with $a_n = C_n(x - a)^n$.

- If $\lim\limits_{n\to\infty} |a_{n+1}|/|a_n|$ is infinite, then $R = 0$.

- If $\lim\limits_{n\to\infty} |a_{n+1}|/|a_n| = 0$, then $R = \infty$.

- If $\lim\limits_{n\to\infty} |a_{n+1}|/|a_n| = K|x - a|$, where K is finite and nonzero, then $R = 1/K$.

Note that the ratio test does not tell us anything if $\lim_{n\to\infty} |a_{n+1}|/|a_n|$ fails to exist, which can occur, for example, if some of the C_ns are zero.

A proof that a power series has a radius of convergence and of Theorem 9.9 is given in the online theory supplement. To understand these facts informally, we can think of a power series as being like a geometric series whose coefficients vary from term to term. The radius of convergence depends on the behavior of the coefficients: if there are constants C and K such that for larger and larger n,

$$|C_n| \approx CK^n,$$

then it is plausible that $\sum C_n x^n$ and $\sum CK^n x^n = \sum C(Kx)^n$ converge or diverge together. The geometric series $\sum C(Kx)^n$ converges for $|Kx| < 1$, that is, for $|x| < 1/K$. We can find K using the ratio test, because

$$\frac{|a_{n+1}|}{|a_n|} = \frac{|C_{n+1}||(x - a)^{n+1}|}{|C_n||(x - a)^n|} \approx \frac{CK^{n+1}|(x - a)^{n+1}|}{CK^n|(x - a)^n|} = K|x - a|.$$

Example 1 Show that the following power series converges for all x:

$$1 + x + \frac{x^2}{2!} + \frac{x^3}{3!} + \cdots + \frac{x^n}{n!} + \cdots.$$

Solution Because $C_n = 1/n!$, none of the C_ns are zero and we can use the ratio test:

$$\lim_{n\to\infty} \frac{|a_{n+1}|}{|a_n|} = |x| \lim_{n\to\infty} \frac{|C_{n+1}|}{|C_n|} = |x| \lim_{n\to\infty} \frac{1/(n+1)!}{1/n!} = |x| \lim_{n\to\infty} \frac{n!}{(n+1)!} = |x| \lim_{n\to\infty} \frac{1}{n+1} = 0.$$

This gives $R = \infty$, so the series converges for all x. We see in Chapter 10 that it converges to e^x.

Example 2 Determine the radius of convergence of the series

$$(x-1) - \frac{(x-1)^2}{2} + \frac{(x-1)^3}{3} - \frac{(x-1)^4}{4} + \cdots + (-1)^{n-1}\frac{(x-1)^n}{n} + \cdots.$$

What does this tell us about the interval of convergence of this series?

Solution The general term of the series is $(x-1)^n/n$ if n is odd and $-(x-1)^n/n$ if n is even, so $C_n = (-1)^{n-1}/n$, and we can use the ratio test. We have

$$\lim_{n\to\infty} \frac{|a_{n+1}|}{|a_n|} = |x-1| \lim_{n\to\infty} \frac{|C_{n+1}|}{|C_n|} = |x-1| \lim_{n\to\infty} \frac{|(-1)^n/(n+1)|}{|(-1)^{n-1}/n|} = |x-1| \lim_{n\to\infty} \frac{n}{n+1} = |x-1|.$$

The radius of convergence is $R = 1$. The power series converges for $|x-1| < 1$ and diverges for $|x-1| > 1$, so the series converges for $0 < x < 2$. Notice that the radius of convergence does not tell us what happens at the endpoints, $x = 0$ and $x = 2$. We see in Chapter 10 that the series converges to $\ln x$ for $0 < x \leq 2$.

The ratio test requires $\lim_{n\to\infty} |a_{n+1}|/|a_n|$ to exist for $a_n = C_n(x-a)^n$. What happens if some of the coefficients C_n are zero? Then we use the fact that an infinite series can be written in several ways and pick one in which the terms are nonzero. For example, we think of the power series

$$x - \frac{x^3}{3!} + \frac{x^5}{5!} - \frac{x^7}{7!} + \cdots + (-1)^{n-1}\frac{x^{2n-1}}{(2n-1)!} + \cdots$$

as the series with $a_1 = x$ and $a_2 = -x^3/3!, \ldots$, so $a_n = (-1)^{n-1}x^{2n-1}/(2n-1)!$. With this choice of a_n, all $a_n \neq 0$, so we can use the ratio test.[6]

Example 3 Find the radius and interval of convergence of the series

$$x - \frac{x^3}{3!} + \frac{x^5}{5!} - \frac{x^7}{7!} + \cdots + (-1)^{n-1}\frac{x^{2n-1}}{(2n-1)!} + \cdots$$

Solution We take

$$a_n = (-1)^{n-1} \frac{x^{2n-1}}{(2n-1)!},$$

so that, replacing n by $n+1$, we have

$$a_{n+1} = (-1)^{(n+1)-1} \frac{x^{2(n+1)-1}}{(2(n+1)-1)!} = (-1)^n \frac{x^{2n+1}}{(2n+1)!}.$$

Thus,

$$\frac{|a_{n+1}|}{|a_n|} = \left| \frac{(-1)^n \frac{x^{2n+1}}{(2n+1)!}}{(-1)^{n-1} \frac{x^{2n-1}}{(2n-1)!}} \right| = \left| \frac{(-1)^n x^{2n+1}(2n-1)!}{(-1)^{n-1}x^{2n-1}(2n+1)!} \right| = \left| \frac{(-1)x^2}{(2n+1)2n} \right| = \frac{x^2}{(2n+1)2n}.$$

Because

$$\lim_{n\to\infty} \frac{|a_{n+1}|}{|a_n|} = \lim_{n\to\infty} \frac{x^2}{(2n+1)2n} = 0,$$

we have $\lim_{n\to\infty} |a_{n+1}|/|a_n| = 0 < 1$ for all x. Thus, the ratio test guarantees that the power series converges for every x. The radius of convergence is infinite and the interval of convergence is all x. We see in Chapter 10 that the series converges to $\sin x$.

---

[6]We do not take $a_1 = x, a_2 = 0, a_3 = -x^3/3!, a_4 = 0, \ldots$ because then $\lim_{n\to\infty} |a_{n+1}|/|a_n|$ does not exist.

What Happens at the Endpoints of the Interval of Convergence?

The ratio test does not tell us whether a power series converges at the endpoints of its interval of convergence, $x = a \pm R$. There is no simple theorem that answers this question. Since substituting $x = a \pm R$ converts the power series to a series of numbers, the tests in Sections 9.2 and 9.3 are often useful. See Examples 4 and 5.

Example 4 Determine the interval of convergence of the series

$$(x - 1) - \frac{(x-1)^2}{2} + \frac{(x-1)^3}{3} - \frac{(x-1)^4}{4} + \cdots + (-1)^{n-1}\frac{(x-1)^n}{n} + \cdots.$$

Solution In Example 2 on page 426, we showed that this series has $R = 1$; it converges for $0 < x < 2$ and diverges for $x < 0$ or $x > 2$. We need to determine whether it converges at the endpoints of the interval of convergence, $x = 0$ and $x = 2$. At $x = 2$, we have the series

$$1 - \frac{1}{2} + \frac{1}{3} - \frac{1}{4} + \cdots + \frac{(-1)^{n-1}}{n} + \cdots.$$

This is an alternating series with $a_n = 1/(n+1)$, so by the alternating series test (Theorem 9.7), it converges. At $x = 0$, we have the series

$$-1 - \frac{1}{2} - \frac{1}{3} - \frac{1}{4} - \cdots - \frac{1}{n} - \cdots.$$

This is the negative of the harmonic series, so it diverges. Therefore, the interval of convergence is $0 < x \le 2$. The right endpoint is included and the left endpoint is not.

Example 5 Find the radius and interval of convergence of the series

$$1 + 2^2 x^2 + 2^4 x^4 + 2^6 x^6 + \cdots + 2^{2n} x^{2n} + \cdots.$$

Solution If we take $a_n = 2^n x^n$ for n even and $a_n = 0$ for n odd, $\lim_{n \to \infty} |a_{n+1}|/|a_n|$ does not exist. Therefore, for this series we take

$$a_n = 2^{2n} x^{2n},$$

so that, replacing n by $n + 1$, we have

$$a_{n+1} = 2^{2(n+1)} x^{2(n+1)} = 2^{2n+2} x^{2n+2}.$$

Thus,

$$\frac{|a_{n+1}|}{|a_n|} = \frac{|2^{2n+2} x^{2n+2}|}{|2^{2n} x^{2n}|} = |2^2 x^2| = 4x^2.$$

Because

$$\lim_{n \to \infty} \frac{|a_{n+1}|}{|a_n|} = 4x^2,$$

we have $\lim |a_{n+1}|/|a_n| = 4x^2$. Thus, the ratio test guarantees that the power series converges if $4x^2 < 1$, that is, if $|x| < \frac{1}{2}$. The radius of convergence is $\frac{1}{2}$. The series converges for $-\frac{1}{2} < x < \frac{1}{2}$ and diverges for $x < \frac{1}{2}$ or $x > -\frac{1}{2}$. At $x = \pm\frac{1}{2}$, all the terms in the series are 1, so the series diverges (by Theorem 9.2 Property 3). Thus, the interval of convergence is $-\frac{1}{2} < x < \frac{1}{2}$.

Exercises and Problems for Section 9.4

Exercises

Which of the series in Exercises 1–4 are power series?

1. $x - x^3 + x^6 - x^{10} + x^{15} - \cdots$

2. $\dfrac{1}{x} + \dfrac{1}{x^2} + \dfrac{1}{x^3} + \dfrac{1}{x^4} + \cdots$

3. $1 + x + (x-1)^2 + (x-2)^3 + (x-3)^4 + \cdots$

4. $x^7 + x + 2$

Find an expression for the general term of the series in Exercises 5–10. Give the starting value of the index (n or k for example).

5. $\dfrac{1}{2}x + \dfrac{1 \cdot 3}{2^2 \cdot 2!}x^2 + \dfrac{1 \cdot 3 \cdot 5}{2^3 \cdot 3!}x^3 + \cdots$

6. $px + \dfrac{p(p-1)}{2!}x^2 + \dfrac{p(p-1)(p-2)}{3!}x^3 + \cdots$

7. $1 - \dfrac{(x-1)^2}{2!} + \dfrac{(x-1)^4}{4!} - \dfrac{(x-1)^6}{6!} + \cdots$

8. $(x-1)^3 - \dfrac{(x-1)^5}{2!} + \dfrac{(x-1)^7}{4!} - \dfrac{(x-1)^9}{6!} + \cdots$

9. $\dfrac{x-a}{1} + \dfrac{(x-a)^2}{2 \cdot 2!} + \dfrac{(x-a)^3}{4 \cdot 3!} + \dfrac{(x-a)^4}{8 \cdot 4!} + \cdots$

10. $2(x+5)^3 + 3(x+5)^5 + \dfrac{4(x+5)^7}{2!} + \dfrac{5(x+5)^9}{3!} + \cdots$

Use the ratio test to find the radius of convergence of the power series in Exercises 11–21.

11. $\displaystyle\sum_{n=0}^{\infty} (5x)^n$

12. $\displaystyle\sum_{n=0}^{\infty} n^3 x^n$

13. $\displaystyle\sum_{n=0}^{\infty} \dfrac{(n+1)x^n}{2^n + n}$

14. $\displaystyle\sum_{n=1}^{\infty} \dfrac{2^n (x-1)^n}{n}$

15. $x + 4x^2 + 9x^3 + 16x^4 + 25x^5 + \cdots$

16. $x - \dfrac{x^2}{4} + \dfrac{x^3}{9} - \dfrac{x^4}{16} + \dfrac{x^5}{25} - \cdots$

17. $1 + 2x + \dfrac{4x^2}{2!} + \dfrac{8x^3}{3!} + \dfrac{16x^4}{4!} + \dfrac{32x^5}{5!} + \cdots$

18. $\dfrac{x}{3} + \dfrac{2x^2}{5} + \dfrac{3x^3}{7} + \dfrac{4x^4}{9} + \dfrac{5x^5}{11} + \cdots$

19. $1 + 2x + \dfrac{4!x^2}{(2!)^2} + \dfrac{6!x^3}{(3!)^2} + \dfrac{8!x^4}{(4!)^2} + \dfrac{10!x^5}{(5!)^2} + \cdots$

20. $3x + \dfrac{5}{2}x^2 + \dfrac{7}{3}x^3 + \dfrac{9}{4}x^4 + \dfrac{11}{5}x^5 + \cdots$

21. $x - \dfrac{x^3}{3} + \dfrac{x^5}{5} - \dfrac{x^7}{7} + \cdots$

Problems

22. (a) Determine the radius of convergence of the series

$$x - \dfrac{x^2}{2} + \dfrac{x^3}{3} - \dfrac{x^4}{4} + \cdots + (-1)^{n-1}\dfrac{x^n}{n} + \cdots.$$

What does this tell us about the interval of convergence of this series?

(b) Investigate convergence at the end points of the interval of convergence of this series.

23. Show that the series $\displaystyle\sum_{n=1}^{\infty} \dfrac{(2x)^n}{n}$ converges for $|x| < 1/2$.

Investigate whether the series converges for $x = 1/2$ and $x = -1/2$.

24. For constant p, find the radius of convergence of the binomial power series:[7]

$$1 + px + \dfrac{p(p-1)x^2}{2!} + \dfrac{p(p-1)(p-2)x^3}{3!} + \cdots.$$

25. Show that if $C_0 + C_1x + C_2x^2 + C_3x^3 + \cdots$ converges for $|x| < R$ with R given by the ratio test, then so does $C_1 + 2C_2x + 3C_3x^2 + \cdots$.

26. Suppose that the power series $\displaystyle\sum_{n=0}^{\infty} C_n x^n$ converges when $x = -4$ and diverges when $x = 7$. Which of the following are true, false or not possible to determine? Give reasons for your answers.

(a) The power series converges when $x = 10$.
(b) The power series converges when $x = 3$.
(c) The power series diverges when $x = 1$.
(d) The power series diverges when $x = 6$.

---

[7]For an explanation of the name, see Section 11.2.

CHAPTER SUMMARY

- **Geometric series**
  Finite sum, infinite sum
- **Convergence of sequences**
- **Harmonic series**
- **Alternating series**

- **Tests for convergence of series**
  Integral test, comparison test, ratio test, alternating series test
- **Power series**
  Interval test for radius of convergence

REVIEW EXERCISES AND PROBLEMS FOR CHAPTER NINE

Exercises

Use the comparison test to confirm the statements in Exercises 1–2.

1. $\displaystyle\sum_{n=1}^{\infty}\left(\frac{1}{3}\right)^n$ converges, so $\displaystyle\sum_{n=1}^{\infty}\left(\frac{n^2}{3n^2+4}\right)^n$ converges.

2. $\displaystyle\sum_{n=1}^{\infty}\frac{1}{n}$ diverges, so $\displaystyle\sum_{n=1}^{\infty}\frac{1}{n\sin^2 n}$ diverges.

Use the comparison test to decide which of the series in Exercises 3–4 converge and which diverge.

3. $\displaystyle\sum_{n=1}^{\infty}\frac{4-n}{n^3+1}$

4. $\displaystyle\sum_{n=1}^{\infty}\frac{3n^2+n+1}{n^5+1}$

Determine which of the series in Exercises 5–11 converge.

5. $\displaystyle\sum_{n=1}^{\infty}\left(\left(\frac{3}{4}\right)^n+\frac{1}{n^2}\right)$

6. $\displaystyle\sum_{n=0}^{\infty}\frac{2+3^n}{5^n}$

7. $\displaystyle\sum_{n=1}^{\infty}\frac{1}{2+\sin n}$

8. $\displaystyle\sum_{n=1}^{\infty}\frac{3^n}{(2n)!}$

9. $\displaystyle\sum_{n=1}^{\infty}\frac{(2n)!}{(n!)^2}$

10. $\displaystyle\sum_{n=1}^{\infty}\frac{(-1)^{n-1}}{\sqrt{n}+1}$

11. $\displaystyle\sum_{k=1}^{\infty}\ln\left(1+\frac{1}{k}\right)$

Find the radius of convergence in Exercises 12–15.

12. $\displaystyle\sum_{n=0}^{\infty}nx^n$

13. $\displaystyle\sum_{n=0}^{\infty}\frac{(2n)!x^n}{(n!)^2}$

14. $\displaystyle\sum_{n=0}^{\infty}(2^n+n^2)x^n$

15. $\displaystyle\sum_{n=0}^{\infty}\frac{x^n}{n!+1}$

Problems

16. A repeating decimal can always be expressed as a fraction. This problem shows how writing a repeating decimal as a geometric series enables you to find the fraction. Consider the decimal $0.232323\ldots$.

(a) Use the fact that $0.232323\ldots = 0.23 + 0.0023 + 0.000023 + \cdots$ to write $0.232323\ldots$ as a geometric series.

(b) Use the formula for the sum of a geometric series to show that $0.232323\ldots = 23/99$.

17. Cephalexin is an antibiotic with a half-life in the body of 0.9 hours, taken in tablets of 250 mg every six hours.

(a) What percentage of the cephalexin in the body at the start of a six-hour period is still there at the end (assuming no tablets are taken during that time)?

(b) Write an expression for Q_1, Q_2, Q_3, Q_4, where Q_n mg, is the amount of cephalexin in the body right after the nth tablet is taken.

(c) Express Q_3, Q_4 in closed-form and evaluate them.

(d) Write an expression for Q_n and put it in closed-form.

(e) If the patient keeps taking the tablets, use your answer to part (d) to find the quantity of cephalexin in the body in the long run, right after taking a pill.

18. Around January 1, 1993, Barbra Streisand signed a contract with Sony Corporation for $2 million a year for 10 years. Suppose the first payment was made on the day of signing and that all other payments are made on the first day of the year. Suppose also that all payments are made into a bank account earning 4% a year, compounded annually.

(a) How much money was in the account

(i) On the night of December 31, 1999?

(ii) On the day the last payment is made?

(b) What was the present value of the contract on the day it was signed?

19. One way of valuing a company is to calculate the present value of all its future earnings. Suppose a farm expects to sell $1000 worth of Christmas trees once a year forever, with the first sale in the immediate future. What is the present value of this Christmas tree business? Assume that the interest rate is 4% per year, compounded continuously.

20. Before World War I, the British government issued what are called *consols*, which pay the owner or his heirs a fixed amount of money every year forever. (Cartoonists of the time described aristocrats living off such payments as "pickled in consols.") What should a person expect to pay for consols which pay £10 a year forever? Assume the first payment is one year from the date of purchase and that interest remains 4% per year, compounded annually. (£ denotes pounds, the British unit of currency.)

Problems 21–23 are about *bonds*, which are issued by a government to raise money. An individual who buys a $1000 bond gives the government $1000 and in return receives a fixed sum of money, called the *coupon*, every six months or every year for the life of the bond. At the time of the last coupon, the individual also gets the $1000, or *principal* back.

21. What is the present value of a $1000 bond which pays $50 a year for 10 years, starting one year from now? Assume interest rate is 6% per year, compounded annually.

22. What is the present value of a $1000 bond which pays $50 a year for 10 years, starting one year from now? Assume the interest rate is 4% per year, compounded annually.

CAS Challenge Problems

In Problems 26–28 we use a computer algebra system to investigate multiplication and composition of power series. Just as a number is represented in a calculator only up to a certain number of decimal places, so a power series is represented in a computer only up to a certain number of terms. Look in the manual for your computer algebra system to find the command for entering a power series up to a certain number of terms. Consider the power series

$$p = \sum_{n=1}^{\infty} n x^n \quad \text{and} \quad q = \sum_{n=0}^{\infty} \frac{x^n}{n!}.$$

Let $S_n(x)$ be the nth partial sum for p and $T_n(x)$ the nth partial sum for q.

26. (a) Calculate and expand the product $S_n(x)T_n(x)$ for $n = 1, 2, 3, 4$.
    (b) What happens to the coefficients of x, x^2 and x^3 as n increases?
    (c) Make a prediction about the coefficient of x^4 in the expansion of $S_5(x)T_5(x)$ and test your prediction.
    (d) Make a general statement about what happens to the

23. (a) What is the present value of a $1000 bond which pays $50 a year for 10 years, starting one year from now? Assume the interest rate is 5% per year, compounded annually.
    (b) Since $50 is 5% of $1000, this bond is often called a 5% bond. What does your answer to part (a) tell you about the relationship between the principal and the present value of this bond when the interest rate is 5%?
    (c) If the interest rate is more than 5% per year, compounded annually, which is larger: the principal or the value of the bond? Why do you think the bond is then described as *trading at discount*?
    (d) If the interest rate is less than 5% per year, compounded annually, why is the bond described as *trading at a premium*?

24. Suppose that the power $\sum_{n=0}^{\infty} C_n(x-2)^n$ converges when $x = 4$ and diverges when $x = 6$. Which of the following are true, false or not possible to determine? Give reasons for your answers.

    (a) The power series converges when $x = 7$.
    (b) The power series diverges when $x = 1$.
    (c) The power series converges when $x = 0.5$.
    (d) The power series diverges when $x = 5$.
    (e) The power series converges when $x = -3$.

25. (a) For a series $\sum a_n$, show that $0 \leq a_n + |a_n| \leq 2|a_n|$.
    (b) Use part (a) to show that if $\sum |a_n|$ converges, then $\sum a_n$ converges.

coefficient of any power of x in $S_n(x)T_n(x)$ as n increases, and explain why your statement is true.

27. (a) Calculate and expand $T_n(S_n(x))$ for $n = 1, 2, 3$.
    (b) What happens to the coefficient of x as n increases. What about the coefficient of x^2?
    (c) Make a prediction about the coefficient of x^3 in the expansion of $T_4(S_4(x))$ and test your prediction.
    (d) State what happens to the coefficient of any power of x in $T_n(S_n(x))$ as n increases, and explain why your statement is true.

28. Consider the geometric series $p = \sum_{n=0}^{\infty}(-1)^n x^n$ and $q = \sum_{n=0}^{\infty} 2^n x^n$.
    (a) What are the radii of convergence of p and q?
    (b) Use a computer algebra system to calculate the first 11 terms of the product pq.
    (c) Use the ratio test to estimate the radius of convergence of the product in part (b).
    (d) Make a conjecture about the relationship between the radii of convergence of two power series and the radius of convergence of their product.

CHECK YOUR UNDERSTANDING

Decide if the statements in Problems 1–25 are true or false. Give an explanation for your answer.

1. A geometric series in x is a power series.

2. $\sum_{n=1}^{\infty} (x-n)^n$ is a power series.

3. If the power series $\sum C_n x^n$ converges for $x = 2$, then it converges for $x = 1$.

4. If the power series $\sum C_n x^n$ converges for $x = 1$, then the power series converges for $x = 2$.

5. If the power series $\sum C_n x^n$ does not converge for $x = 1$, then the power series does not converge for $x = 2$.

6. If $0 \le a_n \le b_n$ and $\sum a_n$ converges, then $\sum b_n$ converges.

7. If $0 \le a_n \le b_n$ and $\sum a_n$ diverges, then $\sum b_n$ diverges.

8. If $b_n \le a_n \le 0$ and $\sum b_n$ converges, then $\sum a_n$ converges.

9. If $\sum a_n$ converges, then $\sum |a_n|$ converges.

10. If $\sum |a_n + b_n|$ converges, then $\sum |a_n|$ and $\sum |b_n|$ converge.

11. If $\sum a_n$ converges, then $\lim_{n \to \infty} |a_{n+1}|/|a_n| \neq 1$.

12. $\sum_{n=0}^{\infty} (-1)^n \cos(\pi n)$ is an alternating series.

13. $\sum_{n=1}^{\infty} (1 + (-1)^n)$ is a series of nonnegative terms.

14. The series $\sum_{n=0}^{\infty} (-1)^n 2^n$ converges.

15. The series $\sum_{n=1}^{\infty} 2^{(-1)^n}$ converges.

16. If $\sum a_n$ converges, then $\sum (-1)^n a_n$ converges.

17. If $\sum |a_n|$ converges, then $\sum (-1)^n |a_n|$ converges.

18. To find the sum of the alternating harmonic series to within 0.01 of the true value, we can sum the first 100 terms.

19. $\sum C_n (x-1)^n$ and $\sum C_n x^n$ have the same radius of convergence.

20. If $\sum C_n x^n$ and $\sum B_n x^n$ have the same radius of convergence, then the coefficients, C_n and B_n, must be equal.

21. If a power series with $R \neq 0$, $R \neq \infty$ converges at one endpoint of its interval of convergence, then it converges at the other endpoint also.

22. If a series $\sum a_n$ converges, then the terms, a_n, tend to zero as n increases.

23. If the terms, a_n, of a series tend to zero as n increases, then the series $\sum a_n$ converges.

24. If $\sum a_n$ does not converge and $\sum b_n$ does not converge, then $\sum a_n b_n$ does not converge.

25. If $\sum a_n b_n$ converges, then $\sum a_n$ and $\sum b_n$ converge.

PROJECTS FOR CHAPTER NINE

1. **Probability of Winning in Sports**

    In certain sports, winning a game requires a lead of two points. That is, if the score is tied you have to score two points in a row to win.

    (a) For some sports (e.g. tennis), a point is scored every play. Suppose your probability of scoring the next point is always p. Then, your opponent's probability of scoring the next point is always $1 - p$.

    (i) What is the probability that you win the next two points?

    (ii) What is the probability that you and your opponent split the next two points, that is, that neither of you wins both points?

    (iii) What is the probability that you split the next two points but you win the two after that?

    (iv) What is the probability that you either win the next two points or split the next two and then win the next two after that?

    (v) Give a formula for your probability w of winning a tied game.

    (vi) Compute your probability of winning a tied game when $p = 0.5$; when $p = 0.6$; when $p = 0.7$; when $p = 0.4$. Comment on your answers.

(b) In other sports (e.g. volleyball), you can score a point only if it is your turn, with turns alternating until a point is scored. Suppose your probability of scoring a point when it is your turn is p, and your opponent's probability of scoring a point when it is her turn is q.

   (i) Find a formula for the probability S that you are the first to score the next point, assuming it is currently your turn.

   (ii) Suppose that if you score a point, the next turn is yours. Using your answers to part (a) and your formula for S, compute the probability of winning a tied game (if you need two points in a row to win).

      • Assume $p = 0.5$ and $q = 0.5$ and it is your turn.

      • Assume $p = 0.6$ and $q = 0.5$ and it is your turn.

2. Quinine

Malaria is a parasitic infection transmitted by mosquito bites, mainly in tropical areas of the world. The disease has existed since ancient times, and currently there are hundreds of millions of cases each year, with millions of deaths. Around 1630, Jesuits in Peru introduced the bark of the cinchona tree to the West as the first treatment for malaria. The drug quinine is the active ingredient in the bark, and it is still used today. However, the concentration in the body of quinine, and of most drugs, must be kept within certain parameters. If the concentration is too low, the drug is ineffective. If the concentration is too high, toxic side effects can result.

Suppose you are a doctor who prescribes quinine for a 70 kg malaria patient. At 8 am each day, she is to receive 50 mg of the drug.[8] To be effective, the average concentration in the body must be at least 0.4 mg/kg. However, concentrations above 3.0 mg/kg can be fatal. The half-life of quinine in the body is 11.5 hours.

(a) What is the continuous rate of decay of quinine (in units of %/min)?

(b) How much quinine is in the patient's system just before and just after the first day's dose? After the second day's dose?

(c) How much is in her system after a few more doses? After the nth dose? Explain what happens in the long run.

(d) Graph the concentration of the drug versus time showing what the steady-state looks like. Find a formula for the repeating function.

(e) What is the average (over time) of the concentration of quinine in the patient's body?

(f) Is this treatment both effective and safe?

(g) Suppose instead that the patient were to receive two doses of 25 mg each day, one at 8 am and one at 8 pm. Is this a safe and effective treatment?

(h) Determine the average value of an exponentially decaying function between two points (x_0, y_0) and (x_1, y_1).

(i) Notice that the average concentration of quinine is the same for the 50 mg once per day and 25 mg twice per day treatments. Use part (h) to explain why. Would the average concentration be the same for a 100 mg dose once every two days?

(j) Suppose the original quinine regimen is stopped. How long after the last dose will the amount of quinine be less than 10^{-10} times the patient's body mass?

---

[8]This is a simplified model; actual treatments involve different drugs and more complicated dosage regimens. See, for example, *The Pharmacological Basis of Therapeutics*, 9th Ed., ed. Joel G. Hardman, Alfred Goodman Gilman, and Lee E. Limbird, (New York: McGraw Hill, 1996).

Chapter Ten

APPROXIMATING FUNCTIONS

In Section 9.1, for $|x| < 1$, we found the sum of a geometric series as a function of the common ratio x:

$$1 + x + x^2 + \cdots + x^n + \cdots = \frac{1}{1-x}.$$

Viewed from another perspective, this formula gives us a series whose sum is the function $f(x) = 1/(1-x)$. We now work in the opposite direction, starting with a function and looking for a series of simpler functions whose sum is that function. We first use polynomials, which lead us to Taylor series, and then trigonometric functions, which lead us to Fourier series.

Taylor approximations use polynomials, which may be considered the simplest functions. They are easy to use because they can be evaluated by simple arithmetic, unlike transcendental functions, such as e^x and $\ln x$.

Fourier approximations use sines and cosines, the simplest periodic functions, instead of polynomials. Taylor approximations are generally good approximations to the function locally (that is, near a specific point), whereas Fourier approximations are generally good approximations over an interval.

10.1 TAYLOR POLYNOMIALS

In this section, we see how to approximate a function by polynomials.

Linear Approximations

We already know how to approximate a function using a degree 1 polynomial, namely the tangent line approximation given in Section 3.9 :

$$f(x) \approx f(a) + f'(a)(x - a).$$

The tangent line and the curve have the same slope at $x = a$. As Figure 10.1 suggests, the tangent line approximation to the function is generally more accurate for values of x close to a.

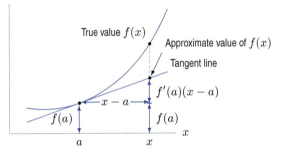

Figure 10.1: Tangent line approximation of $f(x)$ for x near a

We first focus on $a = 0$. The tangent line approximation at $x = 0$ is referred to as the *first Taylor approximation* at $x = 0$, or as follows:

Taylor Polynomial of Degree 1 Approximating $f(x)$ for x near 0

$$f(x) \approx P_1(x) = f(0) + f'(0)x$$

Example 1   Find the Taylor Polynomial of degree 1 for $g(x) = \cos x$, with x in radians, for x near 0.

Solution   The tangent line at $x = 0$ is just the horizontal line $y = 1$, as shown in Figure 10.2, so

$$g(x) = \cos x \approx 1, \quad \text{for } x \text{ near 0.}$$

If we take $x = 0.05$, then

$$g(0.05) = \cos(0.05) = 0.998\ldots,$$

which is quite close to the approximation $\cos x \approx 1$. Similarly, if $x = -0.1$, then

$$g(-0.1) = \cos(-0.1) = 0.995\ldots$$

is close to the approximation $\cos x \approx 1$. However, if $x = 0.4$, then

$$g(0.4) = \cos(0.4) = 0.921\ldots,$$

so the approximation $\cos x \approx 1$ is less accurate. The graph suggests that the farther a point x is away from 0, the worse the approximation is likely to be.

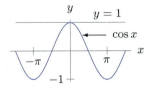

Figure 10.2: Graph of $\cos x$ and its tangent line at $x = 0$

The previous example shows that the Taylor polynomial of degree 1 might actually have degree less than 1.

Quadratic Approximations

To get a more accurate approximation, we use a quadratic function instead of a linear function.

Example 2   Find the quadratic approximation to $g(x) = \cos x$ for x near 0.

Solution   To ensure that the quadratic, $P_2(x)$, is a good approximation to $g(x) = \cos x$ at $x = 0$, we require that $\cos x$ and the quadratic have the same value, the same slope, and the same second derivative at $x = 0$. That is, we require $P_2(0) = g(0)$, $P_2'(0) = g'(0)$, and $P_2''(0) = g''(0)$. We take the quadratic polynomial

$$P_2(x) = C_0 + C_1 x + C_2 x^2,$$

and determine C_0, C_1, and C_2. Since

$$\begin{aligned} P_2(x) &= C_0 + C_1 x + C_2 x^2 && \text{and} & g(x) &= \cos x \\ P_2'(x) &= C_1 + 2C_2 x && & g'(x) &= -\sin x \\ P_2''(x) &= 2C_2 && & g''(x) &= -\cos x, \end{aligned}$$

we have

$$\begin{aligned} C_0 &= P_2(0) = g(0) = \cos 0 = 1 && \text{so} & C_0 &= 1 \\ C_1 &= P_2'(0) = g'(0) = -\sin 0 = 0 && & C_1 &= 0 \\ 2C_2 &= P_2''(0) = g''(0) = -\cos 0 = -1, && & C_2 &= -\tfrac{1}{2}. \end{aligned}$$

Consequently, the quadratic approximation is

$$\cos x \approx P_2(x) = 1 + 0 \cdot x - \frac{1}{2}x^2 = 1 - \frac{x^2}{2}, \quad \text{for } x \text{ near } 0.$$

Figure 10.3 suggests that the quadratic approximation $\cos x \approx P_2(x)$ is better than the linear approximation $\cos x \approx P_1(x)$ for x near 0. Let's compare the accuracy of the two approximations. Recall that $P_1(x) = 1$ for all x. At $x = 0.4$, we have $\cos(0.4) = 0.921\ldots$ and $P_2(0.4) = 0.920$, so the quadratic approximation is a significant improvement over the linear approximation. The magnitude of the error is about 0.001 instead of 0.08.

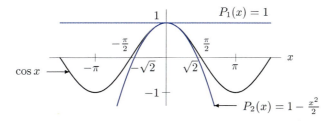

Figure 10.3: Graph of $\cos x$ and its linear, $P_1(x)$, and quadratic, $P_2(x)$, approximations for x near 0

Generalizing the computations in Example 2, we define the *second Taylor approximation* at $x = 0$.

> ## Taylor Polynomial of Degree 2 Approximating $f(x)$ for x near 0
>
> $$f(x) \approx P_2(x) = f(0) + f'(0)x + \frac{f''(0)}{2}x^2$$

Higher-Degree Polynomials

In a small interval around $x = 0$, the quadratic approximation to a function is usually a better approximation than the linear (tangent line) approximation. However, Figure 10.3 shows that the quadratic can still bend away from the original function for large x. We can attempt to fix this by using an approximating polynomial of higher degree. Suppose that we approximate a function $f(x)$ for x near 0 by a polynomial of degree n:

$$f(x) \approx P_n(x) = C_0 + C_1 x + C_2 x^2 + \cdots + C_{n-1} x^{n-1} + C_n x^n.$$

We need to find the values of the constants: $C_0, C_1, C_2, \ldots, C_n$. To do this, we require that the function $f(x)$ and each of its first n derivatives agree with those of the polynomial $P_n(x)$ at the point $x = 0$. In general, the more derivatives there are that agree at $x = 0$, the larger the interval on which the function and the polynomial remain close to each other.

To see how to find the constants, let's take $n = 3$ as an example:

$$f(x) \approx P_3(x) = C_0 + C_1 x + C_2 x^2 + C_3 x^3.$$

Substituting $x = 0$ gives

$$f(0) = P_3(0) = C_0.$$

Differentiating $P_3(x)$ yields

$$P_3'(x) = C_1 + 2C_2 x + 3C_3 x^2,$$

so substituting $x = 0$ shows that

$$f'(0) = P_3'(0) = C_1.$$

Differentiating and substituting again, we get

$$P_3''(x) = 2 \cdot 1 C_2 + 3 \cdot 2 \cdot 1 C_3 x,$$

which gives

$$f''(0) = P_3''(0) = 2 \cdot 1 C_2,$$

so that

$$C_2 = \frac{f''(0)}{2 \cdot 1}.$$

The third derivative, denoted by P_3''', is

$$P_3'''(x) = 3 \cdot 2 \cdot 1 C_3,$$

so

$$f'''(0) = P_3'''(0) = 3 \cdot 2 \cdot 1 C_3,$$

and then

$$C_3 = \frac{f'''(0)}{3 \cdot 2 \cdot 1}.$$

You can imagine a similar calculation starting with $P_4(x)$, using the fourth derivative $f^{(4)}$, which would give

$$C_4 = \frac{f^{(4)}(0)}{4 \cdot 3 \cdot 2 \cdot 1},$$

and so on. Using factorial notation,[1] we write these expressions as

$$C_3 = \frac{f'''(0)}{3!}, \quad C_4 = \frac{f^{(4)}(0)}{4!}.$$

Writing $f^{(n)}$ for the nth derivative of f, we have, for any positive integer n

$$C_n = \frac{f^{(n)}(0)}{n!}.$$

So we define the nth *Taylor approximation* at $x = 0$:

---

[1] Recall that $k! = k(k-1) \cdots 2 \cdot 1$. In addition, $1! = 1$, and $0! = 1$.

> ### Taylor Polynomial of Degree n Approximating $f(x)$ for x near 0
>
> $$f(x) \approx P_n(x)$$
> $$= f(0) + f'(0)x + \frac{f''(0)}{2!}x^2 + \frac{f'''(0)}{3!}x^3 + \frac{f^{(4)}(0)}{4!}x^4 + \cdots + \frac{f^{(n)}(0)}{n!}x^n$$
>
> We call $P_n(x)$ the Taylor polynomial of degree n centered at $x = 0$, or the Taylor polynomial about (or around) $x = 0$.

Example 3

Construct the Taylor polynomial of degree 7 approximating the function $f(x) = \sin x$ for x near 0. Compare the value of the Taylor approximation with the true value of f at $x = \pi/3$.

Solution

We have

$$
\begin{aligned}
f(x) &= \sin x & \text{giving} \quad f(0) &= 0 \\
f'(x) &= \cos x & f'(0) &= 1 \\
f''(x) &= -\sin x & f''(0) &= 0 \\
f'''(x) &= -\cos x & f'''(0) &= -1 \\
f^{(4)}(x) &= \sin x & f^{(4)}(0) &= 0 \\
f^{(5)}(x) &= \cos x & f^{(5)}(0) &= 1 \\
f^{(6)}(x) &= -\sin x & f^{(6)}(0) &= 0 \\
f^{(7)}(x) &= -\cos x, & f^{(7)}(0) &= -1.
\end{aligned}
$$

Using these values, we see that the Taylor polynomial approximation of degree 7 is

$$\sin x \approx P_7(x) = 0 + 1 \cdot x + 0 \cdot \frac{x^2}{2!} - 1 \cdot \frac{x^3}{3!} + 0 \cdot \frac{x^4}{4!} + 1 \cdot \frac{x^5}{5!} + 0 \cdot \frac{x^6}{6!} - 1 \cdot \frac{x^7}{7!}$$
$$= x - \frac{x^3}{3!} + \frac{x^5}{5!} - \frac{x^7}{7!}, \quad \text{for } x \text{ near 0.}$$

Notice that since $f^{(8)}(0) = 0$, the seventh and eighth Taylor approximations to $\sin x$ are the same.

In Figure 10.4 we show the graphs of the sine function and the approximating polynomial of degree 7 for x near 0. They are indistinguishable where x is close to 0. However, as we look at values of x farther away from 0 in either direction, the two graphs move apart. To check the accuracy of this approximation numerically, we see how well it approximates $\sin(\pi/3) = \sqrt{3}/2 = 0.8660254\ldots$.

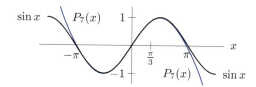

Figure 10.4: Graph of $\sin x$ and its seventh degree Taylor polynomial, $P_7(x)$, for x near 0

When we substitute $\pi/3 = 1.0471976\ldots$ into the polynomial approximation, we obtain $P_7(\pi/3) = 0.8660213\ldots$, which is extremely accurate—to about four parts in a million.

Example 4

Graph the Taylor polynomial of degree 8 approximating $g(x) = \cos x$ for x near 0.

Solution   We find the coefficients of the Taylor polynomial by the method of the preceding example, giving

$$\cos x \approx P_8(x) = 1 - \frac{x^2}{2!} + \frac{x^4}{4!} - \frac{x^6}{6!} + \frac{x^8}{8!}.$$

Figure 10.5 shows that $P_8(x)$ is close to the cosine function for a larger interval of x-values than the quadratic approximation $P_2(x) = 1 - x^2/2$ in Example 2 on page 435.

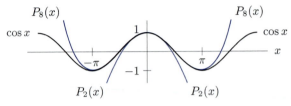

Figure 10.5: $P_8(x)$ approximates $\cos x$ better than $P_2(x)$ for x near 0

Example 5   Construct the Taylor polynomial of degree 10 about $x = 0$ for the function $f(x) = e^x$.

Solution   We have $f(0) = 1$. Since the derivative of e^x is equal to e^x, all the higher-order derivatives are equal to e^x. Consequently, for any $k = 1, 2, \ldots, 10$, $f^{(k)}(x) = e^x$ and $f^{(k)}(0) = e^0 = 1$. Therefore, the Taylor polynomial approximation of degree 10 is given by

$$e^x \approx P_{10}(x) = 1 + x + \frac{x^2}{2!} + \frac{x^3}{3!} + \frac{x^4}{4!} + \cdots + \frac{x^{10}}{10!}, \quad \text{for } x \text{ near } 0.$$

To check the accuracy of this approximation, we use it to approximate $e = e^1 = 2.718281828\ldots$. Substituting $x = 1$ gives $P_{10}(1) = 2.718281801$. Thus, P_{10} yields the first seven decimal places for e. For large values of x, however, the accuracy diminishes because e^x grows faster than any polynomial as $x \to \infty$. Figure 10.6 shows graphs of $f(x) = e^x$ and the Taylor polynomials of degree $n = 0, 1, 2, 3, 4$. Notice that each successive approximation remains close to the exponential curve for a larger interval of x-values.

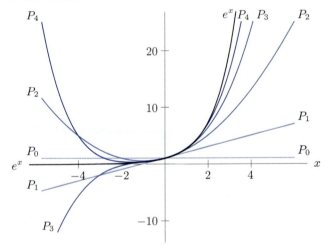

Figure 10.6: For x near 0, the value of e^x is more closely approximated by higher-degree Taylor polynomials

Example 6   Construct the Taylor polynomial of degree n approximating $f(x) = \dfrac{1}{1-x}$ for x near 0.

Solution   Differentiating gives $f(0) = 1$, $f'(0) = 1$, $f''(0) = 2$, $f'''(0) = 3!$, $f^{(4)}(0) = 4!$, and so on. This means

$$\frac{1}{1-x} \approx P_n(x) = 1 + x + x^2 + x^3 + x^4 + \cdots + x^n, \quad \text{for } x \text{ near } 0,$$

Let us compare the Taylor polynomial with the formula obtained from the sum of a finite geometric series on page 408:

$$\frac{1 - x^{n+1}}{1 - x} = 1 + x + x^2 + x^3 + x^4 + \cdots + x^n.$$

If x is close to 0 and x^{n+1} is small enough to neglect, the formula for the sum of a finite geometric series gives us the Taylor approximation of degree n:

$$\frac{1}{1 - x} \approx 1 + x + x^2 + x^3 + x^4 + \cdots + x^n.$$

Taylor Polynomials Around $x = a$

Suppose we want to approximate $f(x) = \ln x$ by a Taylor polynomial. This function has no Taylor polynomial about $x = 0$ because the function is not defined for $x \leq 0$. However, it turns out that we can construct a polynomial centered about some other point, $x = a$.

First, let's look at the equation of the tangent line at $x = a$:

$$y = f(a) + f'(a)(x - a).$$

This gives the first Taylor approximation

$$f(x) \approx f(a) + f'(a)(x - a) \quad \text{for } x \text{ near } a.$$

The $f'(a)(x - a)$ term is a correction term which approximates the change in f as x moves away from a.

Similarly, the Taylor polynomial $P_n(x)$ centered at $x = a$ is set up as $f(a)$ plus correction terms which are zero for $x = a$. This is achieved by writing the polynomial in powers of $(x - a)$ instead of powers of x:

$$f(x) \approx P_n(x) = C_0 + C_1(x - a) + C_2(x - a)^2 + \cdots + C_n(x - a)^n.$$

If we require n derivatives of the approximating polynomial $P_n(x)$ and the original function $f(x)$ to agree at $x = a$, we get the following result for the nth Taylor approximations at $x = a$:

Taylor Polynomial of Degree n Approximating $f(x)$ for x near a

$$f(x) \approx P_n(x)$$
$$= f(a) + f'(a)(x - a) + \frac{f''(a)}{2!}(x - a)^2 + \cdots + \frac{f^{(n)}(a)}{n!}(x - a)^n$$

We call $P_n(x)$ the Taylor polynomial of degree n centered at $x = a$, or the Taylor polynomial about $x = a$.

You can derive the formula for these coefficients in the same way that we did for $a = 0$. (See Problem 28, page 441.)

Example 7   Construct the Taylor polynomial of degree 4 approximating the function $f(x) = \ln x$ for x near 1.

Solution   We have

$$
\begin{array}{lll}
f(x) = & \ln x & \quad \text{so} \quad f(1) = \ln(1) = 0 \\
f'(x) = & 1/x & \quad f'(1) = 1 \\
f''(x) = & -1/x^2 & \quad f''(1) = -1 \\
f'''(x) = & 2/x^3 & \quad f'''(1) = 2 \\
f^{(4)}(x) = & -6/x^4, & \quad f^{(4)}(1) = -6.
\end{array}
$$

The Taylor polynomial is therefore

$$\ln x \approx P_4(x) = 0 + (x - 1) - \frac{(x-1)^2}{2!} + 2\frac{(x-1)^3}{3!} - 6\frac{(x-1)^4}{4!}$$
$$= (x - 1) - \frac{(x-1)^2}{2} + \frac{(x-1)^3}{3} - \frac{(x-1)^4}{4}, \quad \text{for } x \text{ near } 1.$$

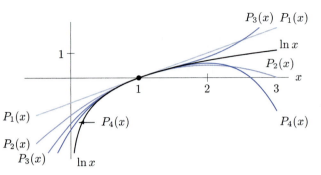

Figure 10.7: Taylor polynomials approximate $\ln x$ closely for x near 1, but not necessarily farther away

Graphs of $\ln x$ and several of its Taylor polynomials are shown in Figure 10.7. Notice that $P_4(x)$ stays reasonably close to $\ln x$ for x near 1, but bends away as x gets farther from 1. Also, note that the Taylor polynomials are defined for $x \leq 0$, but $\ln x$ is not.

The examples in this section suggest that the following results are true for common functions:

- Taylor polynomials centered at $x = a$ give good approximations to $f(x)$ for x near a. Farther away, they may or may not be good.

- The higher the degree of the Taylor polynomial, the larger the interval over which it fits the function closely.

Exercises and Problems for Section 10.1

Exercises

For Exercises 1–10, find the Taylor polynomials of degree n approximating the given functions for x near 0. (Assume p is a constant.)

1. $\dfrac{1}{1+x}$, $n = 4, 6, 8$

2. $\dfrac{1}{1-x}$, $n = 3, 5, 7$

3. $\sqrt{1+x}$, $n = 2, 3, 4$

4. $\cos x$, $n = 2, 4, 6$

5. $\arctan x$, $n = 3, 4$

6. $\tan x$, $n = 3, 4$

7. $\sqrt[3]{1-x}$, $n = 2, 3, 4$

8. $\ln(1+x)$, $n = 5, 7, 9$

9. $\dfrac{1}{\sqrt{1+x}}$, $n = 2, 3, 4$ **10.** $(1+x)^p$, $n = 2, 3, 4$

For Exercises 11–14, find the Taylor polynomial of degree n for x near the given point a.

11. $\sin x$, $a = \pi/2$, $n = 4$

12. $\cos x$, $a = \pi/4$, $n = 3$

13. e^x, $a = 1$, $n = 4$

14. $\sqrt{1+x}$, $a = 1$, $n = 3$

Problems

For Problems 15–18, suppose $P_2(x) = a + bx + cx^2$ is the second degree Taylor polynomial for the function f about $x = 0$. What can you say about the signs of a, b, c if f has the graph given below?

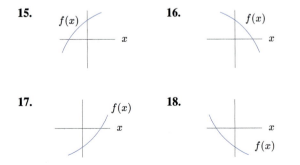

19. Suppose the function $f(x)$ is approximated near $x = 0$ by a sixth degree Taylor polynomial

$$P_6(x) = 3x - 4x^3 + 5x^6.$$

Give the value of

(a) $f(0)$ (b) $f'(0)$ (c) $f'''(0)$

(d) $f^{(5)}(0)$ (e) $f^{(6)}(0)$

20. Suppose g is a function which has continuous derivatives, and that $g(5) = 3, g'(5) = -2, g''(5) = 1, g'''(5) = -3$.

(a) What is the Taylor polynomial of degree 2 for g near 5? What is the Taylor polynomial of degree 3 for g near 5?

(b) Use the two polynomials that you found in part (a) to approximate $g(4.9)$.

21. Find the second-degree Taylor polynomial for $f(x) = 4x^2 - 7x + 2$ about $x = 0$. What do you notice?

22. Find the third-degree Taylor polynomial for $f(x) = x^3 + 7x^2 - 5x + 1$ about $x = 0$. What do you notice?

23. **(a)** Based on your observations in Problems 21–22, make a conjecture about Taylor approximations in the case when f is itself a polynomial.

(b) Show that your conjecture is true.

24. Show how you can use the Taylor approximation $\sin x \approx x - \dfrac{x^3}{3!}$, for x near 0, to explain why $\lim\limits_{x \to 0} \dfrac{\sin x}{x} = 1$.

25. Use the fourth-degree Taylor approximation $\cos x \approx 1 - \dfrac{x^2}{2!} + \dfrac{x^4}{4!}$ for x near 0 to explain why $\lim\limits_{x \to 0} \dfrac{1 - \cos x}{x^2} = \dfrac{1}{2}$.

26. Use a fourth degree Taylor approximation for e^h, for h near 0, to evaluate the following limits. Would your answer be different if you used a Taylor polynomial of higher degree?

(a) $\lim\limits_{h \to 0} \dfrac{e^h - 1 - h}{h^2}$

(b) $\lim\limits_{h \to 0} \dfrac{e^h - 1 - h - \frac{h^2}{2}}{h^3}$

27. If $f(2) = g(2) = h(2) = 0$, and $f'(2) = h'(2) = 0$, $g'(2) = 22$, and $f''(2) = 3$, $g''(2) = 5$, $h''(2) = 7$, calculate the following limits. Explain your reasoning.

(a) $\lim\limits_{x \to 2} \dfrac{f(x)}{h(x)}$ **(b)** $\lim\limits_{x \to 2} \dfrac{f(x)}{g(x)}$

28. Derive the formulas given in the box on page 439 for the coefficients of the Taylor polynomial approximating a function f for x near a.

29. **(a)** Find the Taylor polynomial approximation of degree 4 about $x = 0$ for the function $f(x) = e^{x^2}$.

(b) Compare this result to the Taylor polynomial approximation of degree 2 for the function $f(x) = e^x$ about $x = 0$. What do you notice?

(c) Use your observation in part (b) to write out the Taylor polynomial approximation of degree 20 for the function in part (a).

(d) What is the Taylor polynomial approximation of degree 5 for the function $f(x) = e^{-2x}$?

30. Consider the equations $\sin x = 0.2$ and $x - \dfrac{x^3}{3!} = 0.2$.

(a) How many solutions does each equation have?

(b) Which of the solutions of the two equations are approximately equal? Explain.

31. The integral $\int_0^1 (\sin t / t)\, dt$ is difficult to approximate using, for example, left Riemann sums or the trapezoid rule because the integrand $(\sin t)/t$ is not defined at $t = 0$. However, this integral converges; its value is $0.94608\ldots$. Estimate the integral using Taylor polynomials for $\sin t$ about $t = 0$ of

(a) Degree 3   **(b)** Degree 5

32. One of the two sets of functions, f_1, f_2, f_3, or g_1, g_2, g_3, is graphed in Figure 10.8; the other set is graphed in Figure 10.9. Points A and B each have $x = 0$. Taylor polynomials of degree 2 approximating these functions near $x = 0$ are as follows:

$$f_1(x) \approx 2 + x + 2x^2 \qquad g_1(x) \approx 1 + x + 2x^2$$
$$f_2(x) \approx 2 + x - x^2 \qquad g_2(x) \approx 1 + x + x^2$$
$$f_3(x) \approx 2 + x + x^2 \qquad g_3(x) \approx 1 - x + x^2.$$

(a) Which group of functions, the fs or the gs, is represented by each figure?

(b) What are the coordinates of the points A and B?

(c) Match each function with the graphs (I)–(III) in the appropriate figure.

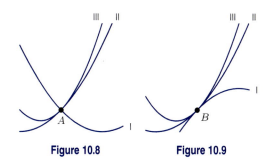

Figure 10.8   **Figure 10.9**

10.2 TAYLOR SERIES

In the previous section we saw how to approximate a function near a point by Taylor polynomials. Now we define a Taylor series, which is a power series that can be thought of as a Taylor polynomial that goes on forever.

Taylor Series for $\cos x$, $\sin x$, e^x

We have the following Taylor polynomials centered at $x = 0$ for $\cos x$:

$$\cos x \approx P_0(x) = 1$$

$$\cos x \approx P_2(x) = 1 - \frac{x^2}{2!}$$

$$\cos x \approx P_4(x) = 1 - \frac{x^2}{2!} + \frac{x^4}{4!}$$

$$\cos x \approx P_6(x) = 1 - \frac{x^2}{2!} + \frac{x^4}{4!} - \frac{x^6}{6!}$$

$$\cos x \approx P_8(x) = 1 - \frac{x^2}{2!} + \frac{x^4}{4!} - \frac{x^6}{6!} + \frac{x^8}{8!}.$$

Here we have a sequence of polynomials, $P_0(x)$, $P_2(x)$, $P_4(x)$, $P_6(x)$, $P_8(x)$, ..., each of which is a better approximation to $\cos x$ than the last, for x near 0. When we go to a higher-degree polynomial (say from P_6 to P_8), we add more terms ($x^8/8!$, for example), but the terms of lower degree don't change. Thus, each polynomial includes the information from all the previous ones. We represent the whole sequence of Taylor polynomials by writing the *Taylor series* for $\cos x$:

$$1 - \frac{x^2}{2!} + \frac{x^4}{4!} - \frac{x^6}{6!} + \frac{x^8}{8!} - \cdots.$$

Notice that the partial sums of this series are the Taylor polynomials, $P_n(x)$.

We define the Taylor series for $\sin x$ and e^x similarly. It turns out that, for these functions, the Taylor series converges to the function for all x, so we can write the following:

$$\sin x = x - \frac{x^3}{3!} + \frac{x^5}{5!} - \frac{x^7}{7!} + \frac{x^9}{9!} - \cdots$$

$$\cos x = 1 - \frac{x^2}{2!} + \frac{x^4}{4!} - \frac{x^6}{6!} + \frac{x^8}{8!} - \cdots$$

$$e^x = 1 + x + \frac{x^2}{2!} + \frac{x^3}{3!} + \frac{x^4}{4!} + \cdots$$

These series are also called *Taylor expansions* of the functions $\sin x$, $\cos x$, and e^x about $x = 0$. The *general term* of a Taylor series is a formula which gives any term in the series. For example, $x^n/n!$ is the general term in the Taylor expansion for e^x, and $(-1)^k x^{2k}/(2k)!$ is the general term in the expansion for $\cos x$. We call n or k the *index*.

Taylor Series in General

Any function f, all of whose derivatives exist at 0, has a Taylor series. However, the Taylor series for f does not necessarily converge to $f(x)$ for all values of x. For the values of x for which the series does converge to $f(x)$, we have the following formula:

Taylor Series for $f(x)$ about $x = 0$

$$f(x) = f(0) + f'(0)x + \frac{f''(0)}{2!}x^2 + \frac{f'''(0)}{3!}x^3 + \cdots + \frac{f^{(n)}(0)}{n!}x^n + \cdots$$

In addition, just as we have Taylor polynomials centered at points other than 0, we can also have a Taylor series centered at $x = a$ (provided all the derivatives of f exist at $x = a$). For the values of x for which the series converges to $f(x)$, we have the following formula:

Taylor Series for $f(x)$ about $x = a$

$$f(x) = f(a) + f'(a)(x - a) + \frac{f''(a)}{2!}(x - a)^2 + \frac{f'''(a)}{3!}(x - a)^3 + \cdots + \frac{f^{(n)}(a)}{n!}(x - a)^n + \cdots$$

The Taylor series is a power series whose partial sums are the Taylor polynomials. As we saw in Section 9.4, power series generally converge on an interval centered at $x = a$. The Taylor series for such a function can be interpreted when x is replaced by a complex number. This extends the domain of the function. See Problem 41.

For a given function f and a given x, even if the Taylor series converges, it might not converge to $f(x)$. However, the Taylor series for most commonly encountered functions, including e^x, $\cos x$, and $\sin x$, do converge to the original function for all x. See Section 10.4.

Intervals of Convergence of Taylor Series

Let us look again at the Taylor polynomial for $\ln x$ about $x = 1$ that we derived in Example 7 on page 439. A similar calculation gives the Taylor Series

$$\ln x = (x-1) - \frac{(x-1)^2}{2} + \frac{(x-1)^3}{3} - \frac{(x-1)^4}{4} + \cdots + (-1)^{n-1}\frac{(x-1)^n}{n} + \cdots.$$

Example 2 on page 426 and Example 4 on page 427 show that this power series has interval of convergence $0 < x \leq 2$. However, although we know that the series converges in this interval, we do not yet know that its sum is $\ln x$. The fact that in Figure 10.10 the polynomials fit the curve well for $0 < x < 2$ suggests that the Taylor series does converge to $\ln x$ for $0 < x \leq 2$. For such x-values, a higher-degree polynomial gives, in general, a better approximation.

However, when $x > 2$, the polynomials move away from the curve and the approximations get worse as the degree of the polynomial increases. Thus, the Taylor polynomials are effective only as approximations to $\ln x$ for values of x between 0 and 2; outside that interval, they should not be used. Inside the interval, but near the ends, 0 or 2, the polynomials converge very slowly. This means we might have to take a polynomial of very high degree to get an accurate value for $\ln x$.

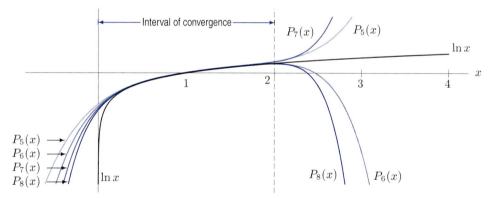

Figure 10.10: Taylor polynomials $P_5(x), P_6(x), P_7(x), P_8(x), \ldots$ converge to $\ln x$ for $0 < x \leq 2$ and diverge outside that interval

To compute the interval of convergence exactly, we first compute the radius of convergence using the method on page 425. Convergence at the endpoints, $x = 0$ and $x = 2$, has to be determined separately. However, proving that the series converges to $\ln x$ on its interval of convergence, as Figure 10.10 suggests, requires the error term introduced in Section 10.4.

Example 1   Find the Taylor series for $\ln(1 + x)$ about $x = 0$, and calculate its interval of convergence.

Solution   Taking derivatives of $\ln(1 + x)$ and substituting $x = 0$ leads to the Taylor series

$$\ln(1 + x) = x - \frac{x^2}{2} + \frac{x^3}{3} - \frac{x^4}{4} + \cdots.$$

Let's substitute the series for $\sin \theta$ for y:

$$e^{\sin \theta} = 1 + \left(\theta - \frac{\theta^3}{3!} + \frac{\theta^5}{5!} - \cdots \right) + \frac{1}{2!} \left(\theta - \frac{\theta^3}{3!} + \frac{\theta^5}{5!} - \cdots \right)^2 + \frac{1}{3!} \left(\theta - \frac{\theta^3}{3!} + \frac{\theta^5}{5!} - \cdots \right)^3 + \cdots .$$

To simplify, we multiply out and collect terms. The only constant term is the 1, and there's only one θ term. The only θ^2 term is the first term we get by multiplying out the square, namely $\theta^2/2!$. There are two contributors to the θ^3 term: the $-\theta^3/3!$ from within the first parentheses, and the first term we get from multiplying out the cube, which is $\theta^3/3!$. Thus the series starts

$$e^{\sin \theta} = 1 + \theta + \frac{\theta^2}{2!} + \left(-\frac{\theta^3}{3!} + \frac{\theta^3}{3!} \right) + \cdots$$

$$= 1 + \theta + \frac{\theta^2}{2!} + 0 \cdot \theta^3 + \cdots \quad \text{for all } \theta.$$

New Series by Differentiation and Integration

Just as we can get new series by substitution, we can also get new series by differentiation and integration. Here again, proofs that this method gives the correct series and that the new series has the same radius of convergence as the original series, can be found in more advanced texts.

Example 3   Find the Taylor Series about $x = 0$ for $\dfrac{1}{(1-x)^2}$ from the series for $\dfrac{1}{1-x}$.

Solution   We know that $\dfrac{d}{dx}\left(\dfrac{1}{1-x}\right) = \dfrac{1}{(1-x)^2}$, so we start with the geometric series

$$\frac{1}{1-x} = 1 + x + x^2 + x^3 + x^4 + \cdots \quad \text{for } -1 < x < 1.$$

Differentiation term by term gives the binomial series

$$\frac{1}{(1-x)^2} = \frac{d}{dx}\left(\frac{1}{1-x}\right) = 1 + 2x + 3x^2 + 4x^3 + \cdots \quad \text{for } -1 < x < 1.$$

Example 4   Find the Taylor series about $x = 0$ for $\arctan x$ from the series for $\dfrac{1}{1+x^2}$.

Solution   We know that $\dfrac{d(\arctan x)}{dx} = \dfrac{1}{1+x^2}$, so we use the series from Example 1 on page 447:

$$\frac{d(\arctan x)}{dx} = \frac{1}{1+x^2} = 1 - x^2 + x^4 - x^6 + x^8 - \cdots \quad \text{for } -1 < x < 1.$$

Antidifferentiating term by term gives

$$\arctan x = \int \frac{1}{1+x^2}\, dx = C + x - \frac{x^3}{3} + \frac{x^5}{5} - \frac{x^7}{7} + \frac{x^9}{9} - \cdots \quad \text{for } -1 < x < 1,$$

where C is the constant of integration. Since $\arctan 0 = 0$, we have $C = 0$, so

$$\arctan x = x - \frac{x^3}{3} + \frac{x^5}{5} - \frac{x^7}{7} + \frac{x^9}{9} - \cdots \quad \text{for } -1 < x < 1.$$

The series for $\arctan x$ was discovered by James Gregory (1638–1675).

Applications of Taylor Series

Example 5   Use the series for $\arctan x$ to estimate the numerical value of π.

Solution   Since $\arctan 1 = \pi/4$, we use the series for $\arctan x$ from Example 4. We assume—as is the case—that the series does converge to $\pi/4$ at $x = 1$, the endpoint of its interval of convergence.

Substituting $x = 1$ into the series for $\arctan x$ gives

$$\pi = 4 \arctan 1 = 4\left(1 - \frac{1}{3} + \frac{1}{5} - \frac{1}{7} + \frac{1}{9} - \cdots\right).$$

Table 10.1 *Approximating π using the series for* $\arctan x$

n	4	5	25	100	500	1000	10,000
S_n	2.895	3.340	3.182	3.132	3.140	3.141	3.141

Table 10.1 shows the value of the nth partial sum, S_n, obtained by summing the nonzero terms from 1 through n. The values of S_n do seem to converge to $\pi = 3.141\ldots$. However, this series converges very slowly, meaning that we have to take a large number of terms to get an accurate estimate for π. So this way of calculating π is not particularly practical. (A better one is given in Problem 1, page 474.) However, the expression for π given by this series is surprising and elegant.

A basic question we can ask about two functions is which one gives larger values. Taylor series can often be used to answer this question over a small interval. If the constant terms of the series for two functions are the same, compare the linear terms; if the linear terms are the same, compare the quadratic terms, and so on.

Example 6

By looking at their Taylor series, decide which of the following functions is largest, and which is smallest, for θ near 0.   (a) $1 + \sin\theta$ (b) e^θ (c) $\dfrac{1}{\sqrt{1 - 2\theta}}$

Solution

The Taylor expansion about $\theta = 0$ for $\sin\theta$ is

$$\sin\theta = \theta - \frac{\theta^3}{3!} + \frac{\theta^5}{5!} - \frac{\theta^7}{7!} + \cdots.$$

So

$$1 + \sin\theta = 1 + \theta - \frac{\theta^3}{3!} + \frac{\theta^5}{5!} - \frac{\theta^7}{7!} + \cdots.$$

The Taylor expansion about $\theta = 0$ for e^θ is

$$e^\theta = 1 + \theta + \frac{\theta^2}{2!} + \frac{\theta^3}{3!} + \frac{\theta^4}{4!} + \cdots.$$

The Taylor expansion about $\theta = 0$ for $1/\sqrt{1 + \theta}$ is

$$\frac{1}{\sqrt{1 + \theta}} = (1 + \theta)^{-1/2} = 1 - \frac{1}{2}\theta + \frac{(-\frac{1}{2})(-\frac{3}{2})}{2!}\theta^2 + \frac{(-\frac{1}{2})(-\frac{3}{2})(-\frac{5}{2})}{3!}\theta^3 + \cdots$$

$$= 1 - \frac{1}{2}\theta + \frac{3}{8}\theta^2 - \frac{5}{16}\theta^3 + \cdots.$$

So, substituting -2θ for θ:

$$\frac{1}{\sqrt{1 - 2\theta}} = 1 - \frac{1}{2}(-2\theta) + \frac{3}{8}(-2\theta)^2 - \frac{5}{16}(-2\theta)^3 + \cdots$$

$$= 1 + \theta + \frac{3}{2}\theta^2 + \frac{5}{2}\theta^3 + \cdots.$$

For θ near 0, we can neglect terms beyond the second degree. We are left with the approximations:

$$1 + \sin\theta \approx 1 + \theta$$

$$e^\theta \approx 1 + \theta + \frac{\theta^2}{2}$$

$$\frac{1}{\sqrt{1 - 2\theta}} \approx 1 + \theta + \frac{3}{2}\theta^2.$$

Since

$$1 + \theta < 1 + \theta + \frac{1}{2}\theta^2 < 1 + \theta + \frac{3}{2}\theta^2,$$

and since the approximations are valid for θ near 0, we conclude that, for θ near 0,

$$1 + \sin\theta < e^\theta < \frac{1}{\sqrt{1 - 2\theta}}.$$

Example 7   Two electrical charges of equal magnitude and opposite signs located near one another are called an electrical dipole. The charges Q and $-Q$ are a distance r apart. (See Figure 10.14.) The electric field, E, at the point P is given by

$$E = \frac{Q}{R^2} - \frac{Q}{(R+r)^2}.$$

Use series to investigate the behavior of the electric field far away from the dipole. Show that when R is large in comparison to r, the electric field is approximately proportional to $1/R^3$.

Figure 10.14: Approximating the electric field at P due to a
dipole consisting of charges Q and $-Q$ a distance r apart

Solution   In order to use a series approximation, we need a variable whose value is small. Although we know that r is much smaller than R, we do not know that r itself is small. The quantity r/R is, however, very small. Hence we expand $1/(R+r)^2$ in powers of r/R so that we can safely use only the first few terms of the Taylor series. First we rewrite using algebra:

$$\frac{1}{(R+r)^2} = \frac{1}{R^2(1 + r/R)^2} = \frac{1}{R^2}\left(1 + \frac{r}{R}\right)^{-2}.$$

Now we use the binomial expansion for $(1 + x)^p$ with $x = r/R$ and $p = -2$:

$$\frac{1}{R^2}\left(1 + \frac{r}{R}\right)^{-2} = \frac{1}{R^2}\left(1 + (-2)\left(\frac{r}{R}\right) + \frac{(-2)(-3)}{2!}\left(\frac{r}{R}\right)^2 + \frac{(-2)(-3)(-4)}{3!}\left(\frac{r}{R}\right)^3 + \cdots\right)$$

$$= \frac{1}{R^2}\left(1 - 2\frac{r}{R} + 3\frac{r^2}{R^2} - 4\frac{r^3}{R^3} + \cdots\right).$$

So, substituting the series into the expression for E, we have

$$E = \frac{Q}{R^2} - \frac{Q}{(R+r)^2} = Q\left[\frac{1}{R^2} - \frac{1}{R^2}\left(1 - 2\frac{r}{R} + 3\frac{r^2}{R^2} - 4\frac{r^3}{R^3} + \cdots\right)\right]$$

$$= \frac{Q}{R^2}\left(2\frac{r}{R} - 3\frac{r^2}{R^2} + 4\frac{r^3}{R^3} - \cdots\right).$$

Since r/R is smaller than 1, the binomial expansion for $(1 + r/R)^{-2}$ converges. We are interested in the electric field far away from the dipole. The quantity r/R is small there, and $(r/R)^2$ and higher powers are smaller still. Thus, we approximate by disregarding all terms except the first, giving

$$E \approx \frac{Q}{R^2}\left(\frac{2r}{R}\right), \quad \text{so} \quad E \approx \frac{2Qr}{R^3}.$$

Since Q and r are constants, this means that E is approximately proportional to $1/R^3$.

In the previous example, we say that E is *expanded in terms of* r/R, meaning that the variable in the expansion is r/R.

Exercises and Problems for Section 10.3

Exercises

Find the first four nonzero terms of the Taylor series about 0 for the functions in Exercises 1–12.

1. $\sqrt{1-2x}$

2. $\cos(\theta^2)$

3. e^{-x}

4. $\dfrac{t}{1+t}$

5. $\ln(1-2y)$

6. $\arcsin x$

7. $\dfrac{1}{\sqrt{1-z^2}}$

8. $\phi^3 \cos(\phi^2)$

9. $\dfrac{z}{e^{z^2}}$

10. $\sqrt{(1+t)}\sin t$

11. $e^t \cos t$

12. $\sqrt{1+\sin\theta}$

Find the Taylor series about 0 for the functions in Exercises 13–15, including the general term.

13. $(1+x)^3$

14. $t\sin(t^2) - t^3$

15. $\dfrac{1}{\sqrt{1-y^2}}$

For Exercises 16–17, expand the quantity about 0 in terms of the variable given. Give four nonzero terms.

16. $\dfrac{1}{2+x}$ in terms of $\dfrac{x}{2}$

17. $\dfrac{a}{\sqrt{a^2+x^2}}$ in terms of $\dfrac{x}{a}$, where $a > 0$

Problems

18. For a, b positive constants, the upper half of an ellipse has equation

$$y = f(x) = b\sqrt{1 - \frac{x^2}{a^2}},$$

(a) Find the second degree Taylor polynomial of $f(x)$ about 0.

(b) Explain how you could have predicted the coefficient of x in the Taylor polynomial from a graph of f.

(c) For x near 0, the ellipse is approximated by a parabola. What is the equation of the parabola? What are its x-intercepts?

(d) Taking $a = 3$ and $b = 2$, estimate the maximum difference between $f(x)$ and the second-degree Taylor polynomial for $-0.1 \le x \le 0.1$.

19. By looking at the Taylor series, decide which of the following functions is largest, and which is smallest, for small positive θ.

(a) $1 + \sin\theta$ (b) $\cos\theta$ (c) $\dfrac{1}{1-\theta^2}$

20. For values of y near 0, put the following functions in increasing order, using their Taylor expansions.

(a) $\ln(1+y^2)$ (b) $\sin(y^2)$ (c) $1 - \cos y$

21. Figure 10.15 shows the graphs of the four functions below for values of x near 0. Use Taylor series to match graphs and formulas.

(a) $\dfrac{1}{1-x^2}$

(b) $(1+x)^{1/4}$

(c) $\sqrt{1+\dfrac{x}{2}}$

(d) $\dfrac{1}{\sqrt{1-x}}$

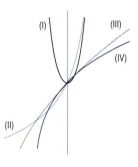

Figure 10.15

22. Consider the functions $y = e^{-x^2}$ and $y = 1/(1+x^2)$.

(a) Write the Taylor expansions for the two functions about $x = 0$. What is similar about the two series? What is different?

(b) Looking at the series, which function do you predict will be greater over the interval $(-1, 1)$? Graph both and see.

(c) Are these functions even or odd? How might you see this by looking at the series expansions?

(d) By looking at the coefficients, explain why it is reasonable that the series for $y = e^{-x^2}$ converges for all values of x, but the series for $y = 1/(1+x^2)$ converges only on $(-1, 1)$.

23. The hyperbolic sine and cosine are differentiable and satisfy the conditions $\cosh 0 = 1$ and $\sinh 0 = 0$, and

$$\frac{d}{dx}(\cosh x) = \sinh x \qquad \frac{d}{dx}(\sinh x) = \cosh x.$$

(a) Using only this information, find the Taylor approximation of degree $n = 8$ about $x = 0$ for $f(x) = \cosh x$.

(b) Estimate the value of $\cosh 1$.

(c) Use the result from part (a) to find a Taylor polynomial approximation of degree $n = 7$ about $x = 0$ for $g(x) = \sinh x$.

24. Padé approximants are rational functions used to approximate more complicated functions. In this problem, you will derive the Padé approximant to the exponential function.

(a) Let $f(x) = (1 + ax)/(1 + bx)$, where a and b are constants. Write down the first three terms of the Taylor series for $f(x)$ about $x = 0$.

(b) By equating the first three terms of the Taylor series about $x = 0$ for $f(x)$ and for e^x, find a and b so that $f(x)$ approximates e^x as closely as possible near $x = 0$.

25. An electric dipole on the x-axis consists of a charge Q at $x = 1$ and a charge $-Q$ at $x = -1$. The electric field, E, at the point $x = R$ on the x-axis is given (for $R > 1$) by

$$E = \frac{kQ}{(R-1)^2} - \frac{kQ}{(R+1)^2}$$

where k is a positive constant whose value depends on the units. Expand E as a series in $1/R$, giving the first two nonzero terms.

26. Assume a is a positive constant. Suppose z is given by the expression

$$z = \sqrt{a^2 + x^2} - \sqrt{a^2 - x^2}.$$

Expand z as a series in x as far as the second nonzero term.

27. The electric potential, V, at a distance R along the axis perpendicular to the center of a charged disc with radius a and constant charge density σ, is given by

$$V = 2\pi\sigma(\sqrt{R^2 + a^2} - R).$$

Show that, for large R,

$$V \approx \frac{\pi a^2 \sigma}{R}.$$

28. One of Einstein's most amazing predictions was that light traveling from distant stars would bend around the sun on the way to earth. His calculations involved solving for ϕ in the equation

$$\sin\phi + b(1 + \cos^2\phi + \cos\phi) = 0$$

where b is a very small positive constant.

(a) Explain why the equation could have a solution for ϕ which is near 0.

(b) Expand the left-hand side of the equation in Taylor series about $\phi = 0$, disregarding terms of order ϕ^2 and higher. Solve for ϕ. (Your answer will involve b.)

29. The Michelson-Morley experiment, which contributed to the formulation of the Theory of Relativity, involved the difference between the two times t_1 and t_2 that light took to travel between two points. If v is the velocity of light; l_1, l_2, and c are constants; and $v < c$, then the times t_1 and t_2 are given by

$$t_1 = \frac{2l_2}{c(1 - v^2/c^2)} - \frac{2l_1}{c\sqrt{1 - v^2/c^2}}$$

$$t_2 = \frac{2l_2}{c\sqrt{1 - v^2/c^2}} - \frac{2l_1}{c(1 - v^2/c^2)}.$$

(a) Find an expression for $\Delta t = t_1 - t_2$, and give its Taylor expansion in terms of v^2/c^2 up to the second nonzero term.

(b) For small v, to what power of v is Δt proportional? What is the constant of proportionality?

30. A hydrogen atom consists of an electron, of mass m, orbiting a proton, of mass M, where m is much smaller than M. The *reduced mass*, μ, of the hydrogen atom is defined by

$$\mu = \frac{mM}{m + M}.$$

(a) Show that $\mu \approx m$.

(b) To get a more accurate approximation for μ, express μ as m times a series in m/M.

(c) The approximation $\mu \approx m$ is obtained by disregarding all but the constant term in the series. The first-order correction is obtained by including the linear term but no higher terms. If $m \approx M/1836$, by what percentage does including the first-order correction change the estimate $\mu \approx m$?

31. A thin disk of radius a and mass M lies horizontally; a particle of mass m is at a height h directly above the center of the disk. The gravitational force, F, exerted by the disk on the mass m is given by

$$F = \frac{2GMmh}{a^2}\left(\frac{1}{h} - \frac{1}{(a^2 + h^2)^{1/2}}\right).$$

Assume $a < h$ and think of F as a function of a, with the other quantities constant.

(a) Expand F as a series in a/h. Give the first two nonzero terms.

(b) Show that the approximation for F obtained by using only the first nonzero term in the series is independent of the radius, a.

(c) If $a = 0.02h$, by what percentage does the approximation in part (a) differ from the approximation in part (b)?

32. When a body is near the surface of the earth, we usually assume that the force due to gravity on it is a constant mg, where m is the mass of the body and g is the acceleration due to gravity at sea level. For a body at a distance h above the surface of the earth, a more accurate expression for the force F is

$$F = \frac{mgR^2}{(R + h)^2}$$

where R is the radius of the earth. We will consider the situation in which the body is close to the surface of the earth so that h is much smaller than R.

(a) Show that $F \approx mg$.

(b) Express F as mg multiplied by a series in h/R.

(c) The first-order correction to the approximation $F \approx mg$ is obtained by taking the linear term in the series but no higher terms. How far above the surface of the earth can you go before the first-order correction changes the estimate $F \approx mg$ by more than 10%? (Assume $R = 6400$ km.)

33. (a) Estimate the value of $\int_0^1 e^{-x^2} dx$ using Riemann sums for both left-hand and right-hand sums with $n = 5$ subdivisions.

(b) Approximate the function $f(x) = e^{-x^2}$ with a Taylor polynomial of degree 6.

(c) Estimate the integral in part (a) by integrating the Taylor polynomial from part (b).

(d) Indicate briefly how you could improve the results in each case.

34. Use Taylor series to explain the patterns in the digits in the following expansions:

(a) $\dfrac{1}{0.98} = 1.020408163264\ldots$

(b) $\left(\dfrac{1}{0.99}\right)^2 = 1.020304050607\ldots$

10.4 THE ERROR IN TAYLOR POLYNOMIAL APPROXIMATIONS

In order to use an approximation intelligently, we need to be able to estimate the size of the error, which is the difference between the exact answer (which we do not know) and the approximate value.

When we use $P_n(x)$, the nth degree Taylor polynomial, to approximate $f(x)$, the error is the difference

$$E_n(x) = f(x) - P_n(x).$$

If E_n is positive, the approximation is smaller than the true value. If E_n is negative, the approximation is too large. Often we are only interested in the magnitude of the error, $|E_n|$.

Recall that we constructed $P_n(x)$ so that its first n derivatives equal the corresponding derivatives of $f(x)$. Therefore, $E_n(0) = 0$, $E_n'(0) = 0$, $E_n''(0) = 0$, $\cdots$, $E_n^{(n)}(0) = 0$. Since $P_n(x)$ is an nth degree polynomial, its $(n+1)^{\text{st}}$ derivative is 0, so $E_n^{(n+1)}(x) = f^{(n+1)}(x)$. In addition, suppose that $\left| f^{(n+1)}(x) \right|$ is bounded by a positive constant M, for all positive values of x near 0, say for $0 \le x \le d$, so that

$$-M \le f^{(n+1)}(x) \le M \quad \text{for } 0 \le x \le d.$$

This means that

$$-M \le E_n^{(n+1)}(x) \le M \quad \text{for } 0 \le x \le d.$$

Writing t for the variable, we integrate this inequality from 0 to x, giving

$$-\int_0^x M \, dt \le \int_0^x E_n^{(n+1)}(t) \, dt \le \int_0^x M \, dt \quad \text{for } 0 \le x \le d,$$

so

$$-Mx \le E_n^{(n)}(x) \le Mx \quad \text{for } 0 \le x \le d.$$

We integrate this inequality again from 0 to x, giving

$$-\int_0^x Mt \, dt \le \int_0^x E_n^{(n)}(t) \, dt \le \int_0^x Mt \, dt \quad \text{for } 0 \le x \le d,$$

so

$$-\frac{1}{2}Mx^2 \le E_n^{(n-1)}(x) \le \frac{1}{2}Mx^2 \quad \text{for } 0 \le x \le d.$$

By repeated integration, we obtain the following estimate:

$$-\frac{1}{(n+1)!}Mx^{n+1} \le E_n(x) \le \frac{1}{(n+1)!}Mx^{n+1} \quad \text{for } 0 \le x \le d,$$

which means that

$$|E_n(x)| = |f(x) - P_n(x)| \le \frac{1}{(n+1)!}Mx^{n+1} \quad \text{for } 0 \le x \le d.$$

When x is to the left of 0, so $-d \leq x \leq 0$, and when the Taylor series is centered at $a \neq 0$, similar calculations lead to the following result:

Theorem 10.1: Bounding the Error in $P_n(x)$

Suppose f and all its derivatives are continuous. If $P_n(x)$ is the nth Taylor approximation to $f(x)$ about a, then

$$|E_n(x)| = |f(x) - P_n(x)| \leq \frac{M}{(n+1)!}|x - a|^{n+1},$$

where $M = \max f^{(n+1)}$ on the interval between a and x.

Using the Error Bound for Taylor Polynomials

Example 1   Give a bound on the error, E_4, when e^x is approximated by its fourth-degree Taylor polynomial for $-0.5 \leq x \leq 0.5$.

Solution   Let $f(x) = e^x$. Then the fifth derivative is $f^{(5)}(x) = e^x$. Since e^x is increasing,

$$|f^{(5)}(x)| \leq e^{0.5} = \sqrt{e} < 2 \quad \text{for} \quad -0.5 \leq x \leq 0.5.$$

Thus

$$|E_4| = |f(x) - P_4(x)| \leq \frac{2}{5!}|x|^5.$$

This means, for example, that on $-0.5 \leq x \leq 0.5$, the approximation

$$e^x \approx 1 + x + \frac{x^2}{2!} + \frac{x^3}{3!} + \frac{x^4}{4!}$$

has an error of at most $\frac{2}{120}(0.5)^5 < 0.0006$.

The error formula for Taylor polynomials can be used to bound the error in a particular numerical approximation, or to see how the accuracy of the approximation depends on the value of x or the value of n. Observe that the error for a Taylor polynomial of degree n depends on the $(n+1)^{\text{st}}$ power of x. That means, for example, with a Taylor polynomial of degree n centered at 0, if we decrease x by a factor of 2, the error bound decreases by a factor of 2^{n+1}.

Example 2   Compare the errors in the approximations

$$e^{0.1} \approx 1 + 0.1 + \frac{1}{2!}(0.1)^2 \quad \text{and} \quad e^{0.05} \approx 1 + (0.05) + \frac{1}{2!}(0.05)^2.$$

Solution   We are approximating e^x by its second-degree Taylor polynomial, first at $x = 0.1$, and then at $x = 0.05$. Since we have decreased x by a factor of 2, the error bound decreases by a factor of about $2^3 = 8$. To see what actually happens to the errors, we compute them:

$$e^{0.1} - \left(1 + 0.1 + \frac{1}{2!}(0.1)^2\right) = 1.105171 - 1.105000 = 0.000171$$

$$e^{0.05} - \left(1 + 0.05 + \frac{1}{2!}(0.05)^2\right) = 1.051271 - 1.051250 = 0.000021$$

Since $(0.000171)/(0.000021) = 8.1$, the error has also decreased by a factor of about 8.

Convergence of Taylor Series for $\cos x$

We have already seen that the Taylor polynomials centered at $x = 0$ for $\cos x$ are good approximations for x near 0. (See Figure 10.16.) In fact, for any value of x, if we take a Taylor polynomial centered at $x = 0$ of high enough degree, its graph is nearly indistinguishable from the graph of the cosine near that point.

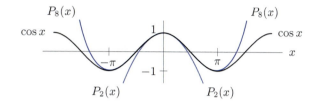

Figure 10.16: Graph of $\cos x$ and two Taylor polynomials for x near 0

Let's see what happens numerically. Let $x = \pi/2$. The successive Taylor polynomial approximations to $\cos(\pi/2) = 0$ about $x = 0$ are

$$
\begin{aligned}
P_2(\pi/2) &= \quad\quad 1 - (\pi/2)^2/2! \quad\quad\quad = -0.23370\ldots \\
P_4(\pi/2) &= 1 - (\pi/2)^2/2! + (\pi/2)^4/4! = \quad 0.01997\ldots \\
P_6(\pi/2) &= \quad\quad\quad \cdots \quad\quad\quad\quad = -0.00089\ldots \\
P_8(\pi/2) &= \quad\quad\quad \cdots \quad\quad\quad\quad = \quad 0.00002\ldots.
\end{aligned}
$$

It appears that the approximations converge to the true value, $\cos(\pi/2) = 0$, very rapidly. Now take a value of x somewhat farther away from 0, say $x = \pi$, then $\cos \pi = -1$ and

$$
\begin{aligned}
P_2(\pi) &= 1 - (\pi)^2/2! = \quad -3.93480\ldots \\
P_4(\pi) &= \quad \cdots \quad\quad = \quad\quad 0.12391\ldots \\
P_6(\pi) &= \quad \cdots \quad\quad = \quad -1.21135\ldots \\
P_8(\pi) &= \quad \cdots \quad\quad = \quad -0.97602\ldots \\
P_{10}(\pi) &= \quad \cdots \quad\quad = \quad -1.00183\ldots \\
P_{12}(\pi) &= \quad \cdots \quad\quad = \quad -0.99990\ldots \\
P_{14}(\pi) &= \quad \cdots \quad\quad = \quad -1.000004\ldots
\end{aligned}
$$

We see that the rate of convergence is somewhat slower; it takes a 14th degree polynomial to approximate $\cos \pi$ as accurately as an 8th degree polynomial approximates $\cos(\pi/2)$. If x were taken still farther away from 0, then we would need still more terms to obtain as accurate an approximation of $\cos x$.

Using the ratio test, we can show the Taylor series for $\cos x$ converges for all values of x. In addition, we will prove that it converges to $\cos x$ using Theorem 10.1. Thus, we are justified in writing the equality:

$$
\cos x = 1 - \frac{x^2}{2!} + \frac{x^4}{4!} - \frac{x^6}{6!} + \frac{x^8}{8!} - \cdots \quad \text{for all } x.
$$

Showing the Taylor Series for $\cos x$ Converges to $\cos x$

The error bound in Theorem 10.1 allows us to see if the Taylor series for a function converges to that function. In the series for $\cos x$, the odd powers are missing, so we assume n is even and write

$$
E_n(x) = \cos x - P_n(x) = \cos x - \left(1 - \frac{x^2}{2!} + \cdots + (-1)^{n/2}\frac{x^n}{n!}\right),
$$

giving

$$\cos x = 1 - \frac{x^2}{2!} + \cdots + (-1)^{n/2}\frac{x^n}{n!} + E_n(x).$$

Thus, for the Taylor series to converge to $\cos x$, we must have $E_n(x) \to 0$ as $n \to \infty$.

Showing $E_n(x) \to 0$ as $n \to \infty$

Proof Since $f(x) = \cos x$, the $(n+1)^{\text{st}}$ derivative $f^{(n+1)}(x)$ is $\pm\cos x$ or $\pm\sin x$, no matter what n is. So for all n, we have $|f^{(n+1)}(x)| \leq 1$ on the interval between 0 and x.

By the error bound formula in Theorem 10.1, we have

$$|E_n(x)| = |\cos x - P_n(x)| \leq \frac{|x|^{n+1}}{(n+1)!} \quad \text{for every } n.$$

To show that the errors go to zero, we must show that for a fixed x,

$$\frac{|x|^{n+1}}{(n+1)!} \to 0 \quad \text{as} \quad n \to \infty.$$

To see why this is true, think about what happens when n is much larger than x. Suppose, for example, that $x = 17.3$. Let's look at the value of the sequence for n more than twice as big as 17.3, say $n = 36$, or $n = 37$, or $n = 38$:

$$\text{For } n = 36: \quad \frac{1}{37!}(17.3)^{37}$$

$$\text{For } n = 37: \quad \frac{1}{38!}(17.3)^{38} = \frac{17.3}{38}\cdot\frac{1}{37!}(17.3)^{37},$$

$$\text{For } n = 38: \quad \frac{1}{39!}(17.3)^{39} = \frac{17.3}{39}\cdot\frac{17.3}{38}\cdot\frac{1}{37!}(17.3)^{37}, \quad \cdots$$

Since $17.3/36$ is less than $\frac{1}{2}$, each time we increase n by 1, the term is multiplied by a number less than $\frac{1}{2}$. No matter what the value of $\frac{1}{37!}(17.3)^{37}$ is, if we keep on dividing it by two, the result gets closer to zero. Thus $\frac{1}{(n+1)!}(17.3)^{n+1}$ goes to 0 as n goes to infinity.

We can generalize this by replacing 17.3 by an arbitrary $|x|$. For $n > 2|x|$, the following sequence converges to 0 because each term is obtained from its predecessor by multiplying by a number less than $\frac{1}{2}$:

$$\frac{x^{n+1}}{(n+1)!}, \quad \frac{x^{n+2}}{(n+2)!}, \quad \frac{x^{n+3}}{(n+3)!}, \ldots$$

Therefore, the Taylor series $1 - x^2/2! + x^4/4! - \cdots$ does converge to $\cos x$.

Problems 16 and 15 ask you to show that the Taylor series for $\sin x$ and e^x converge to the original function for all x. In each case, you again need the following limit:

$$\lim_{n\to\infty}\frac{x^n}{n!} = 0.$$

Exercises and Problems for Section 10.4

Exercises

Use the methods of this section to show how you can estimate the magnitude of the error in approximating the quantities in Exercises 1–4 using a third-degree Taylor polynomial about $x = 0$.

1. $0.5^{1/3}$ **2.** $\ln(1.5)$ **3.** $1/\sqrt{3}$ **4.** $\tan 1$

Problems

5. Suppose you approximate $f(t) = e^t$ by a Taylor polynomial of degree 0 about $t = 0$ on the interval $[0, 0.5]$.

  (a) Reasoning informally, say whether the approximation is an overestimate or an underestimate.

  (b) Using Theorem 10.1, estimate the magnitude of the largest possible error. Check your answer graphically on a computer or calculator.

6. Repeat Problem 5 using the second-degree Taylor approximation, $P_2(t)$, to e^t.

7. Consider the error in using the approximation $\sin\theta \approx \theta$ on the interval $[-1, 1]$.

  (a) Reasoning informally, say where the approximation is an overestimate, and where is it an underestimate.

  (b) Using Theorem 10.1, estimate the magnitude of the largest possible error. Check your answer graphically on a computer or calculator.

8. Repeat Problem 7 for the approximation $\sin\theta \approx \theta - \theta^3/3!$.

9. Use the graphs of $y = \cos x$ and its Taylor polynomials, $P_{10}(x)$ and $P_{20}(x)$, in Figure 10.17 to estimate the following quantities.

  (a) The error in approximating $\cos 6$ by $P_{10}(6)$ and by $P_{20}(6)$.

  (b) The maximum error in approximating $\cos x$ by $P_{20}(x)$ for $|x| \leq 9$.

  (c) If we want to approximate $\cos x$ by $P_{10}(x)$ to an accuracy of within 0.1, what is the largest interval of x-values on which we can work? Give your answer to the nearest integer.

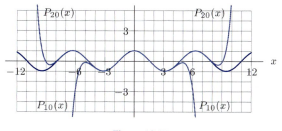

Figure 10.17

10. Give a bound for the maximum possible error for the nth degree Taylor polynomial about $x = 0$ approximating $\cos x$ on the interval $[0, 1]$. What is the bound for $\sin x$?

11. What degree Taylor polynomial about $x = 0$ do you need to calculate $\cos 1$ to four decimal places? To six decimal places? Justify your answer using the results of Problem 10.

12. **(a)** Using a calculator, make a table of the values to four decimal places of $\sin x$ for

$$x = -0.5, -0.4, \ldots, -0.1, 0, 0.1, \ldots, 0.4, 0.5.$$

  (b) Add to your table the values of the error $E_1 = \sin x - x$ for these x-values.

  (c) Using a calculator or computer, draw a graph of the quantity $E_1 = \sin x - x$ showing that

$$|E_1| < 0.03 \quad \text{for} \quad -0.5 \leq x \leq 0.5.$$

13. In this problem, you will investigate the error in the nth degree Taylor approximation to e^x for various values of n.

  (a) Let $E_1 = e^x - P_1(x) = e^x - (1+x)$. Using a calculator or computer, graph E_1 for $-0.1 \leq x \leq 0.1$. What shape is the graph of E_1? Use the graph to confirm that

$$|E_1| \leq x^2 \quad \text{for} \quad -0.1 \leq x \leq 0.1.$$

  (b) Let $E_2 = e^x - P_2(x) = e^x - (1+x+x^2/2)$. Choose a suitable range and graph E_2 for $-0.1 \leq x \leq 0.1$. What shape is the graph of E_2? Use the graph to confirm that

$$|E_2| \leq x^3 \quad \text{for} \quad -0.1 \leq x \leq 0.1.$$

  (c) Explain why the graphs of E_1 and E_2 have the shapes they do.

14. For $|x| \leq 0.1$, graph the error

$$E_0 = \cos x - P_0(x) = \cos x - 1.$$

Explain the shape of the graph, using the Taylor expansion of $\cos x$. Find a bound for $|E_0|$ for $|x| \leq 0.1$.

15. Show that the Taylor series about 0 for e^x converges to e^x for every x. Do this by showing that the error $E_n(x) \to 0$ as $n \to \infty$.

16. Show that the Taylor series about 0 for $\sin x$ converges to $\sin x$ for every x.

10.5 FOURIER SERIES

We have seen how to approximate a function by a Taylor polynomial of fixed degree. Such a polynomial is usually very close to the true value of the function near one point (the point at which the Taylor polynomial is centered), but not necessarily at all close anywhere else. In other words, Taylor polynomials are good approximations of a function *locally*, but not necessarily *globally*. In this section, we take another approach: we approximate the function by trigonometric functions, called *Fourier approximations*. The resulting approximation may not be as close to the original function at some points as the Taylor polynomial. However, the Fourier approximation is, in general, close over a larger interval. In other words, a Fourier approximation can be a better approximation globally. In addition, Fourier approximations are useful even for functions that are not continuous. Unlike Taylor approximations, Fourier approximations are periodic, so they are useful for approximating periodic functions.

Many processes in nature are periodic or repeating, so it makes sense to approximate them by periodic functions. For example, sound waves are made up of periodic oscillations of air molecules. Heartbeats, the movement of the lungs, and the electrical current that powers our homes are all periodic phenomena. Two of the simplest periodic functions are the square wave in Figure 10.18 and the triangular wave in Figure 10.19. Electrical engineers use the square wave as the model for the flow of electricity when a switch is repeatedly flicked on and off.

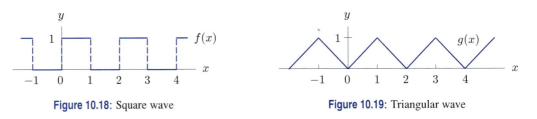

Figure 10.18: Square wave   Figure 10.19: Triangular wave

Fourier Polynomials

We can express the square wave and the triangular wave by the formulas

$$
f(x) = \begin{cases}
\vdots & \vdots \\
0 & -1 \le x < 0 \\
1 & 0 \le x < 1 \\
0 & 1 \le x < 2 \\
1 & 2 \le x < 3 \\
0 & 3 \le x < 4 \\
\vdots & \vdots
\end{cases}
\qquad
g(x) = \begin{cases}
\vdots & \vdots \\
-x & -1 \le x < 0 \\
x & 0 \le x < 1 \\
2 - x & 1 \le x < 2 \\
x - 2 & 2 \le x < 3 \\
4 - x & 3 \le x < 4 \\
\vdots & \vdots
\end{cases}
$$

However, these formulas are not particularly easy to work with. Worse, the functions are not differentiable at various points. Here we show how to approximate such functions by differentiable, periodic functions.

Since the sine and cosine are the simplest periodic functions, they are the building blocks we use. Because they repeat every 2π, we assume that the function f we want to approximate repeats every 2π. (Later, we deal with the case where f has some other period.) We start by considering the square wave in Figure 10.20. Because of the periodicity of all the functions concerned, we only have to consider what happens in the course of a single period; the same behavior repeats in any other period.

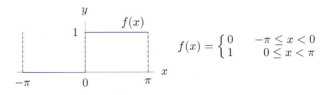

$$
f(x) = \begin{cases}
0 & -\pi \le x < 0 \\
1 & 0 \le x < \pi
\end{cases}
$$

Figure 10.20: Square wave on $[-\pi, \pi]$

We will attempt to approximate f with a sum of trigonometric functions of the form

$$f(x) \approx F_n(x)$$
$$= a_0 + a_1 \cos x + a_2 \cos(2x) + a_3 \cos(3x) + \cdots + a_n \cos(nx)$$
$$+ b_1 \sin x + b_2 \sin(2x) + b_3 \sin(3x) + \cdots + b_n \sin(nx)$$
$$= a_0 + \sum_{k=1}^{n} a_k \cos(kx) + \sum_{k=1}^{n} b_k \sin(kx).$$

$F_n(x)$ is known as a *Fourier polynomial of degree* n, named after the French mathematician Joseph Fourier (1768-1830), who was one of the first to investigate it.[3] The coefficients a_k and b_k are called *Fourier coefficients*. Since each of the component functions $\cos(kx)$ and $\sin(kx)$, $k = 1, 2, \ldots, n$, repeats every 2π, $F_n(x)$ must repeat every 2π and so is a potentially good match for $f(x)$, which also repeats every 2π. The problem is to determine values for the Fourier coefficients that achieve a close match between $f(x)$ and $F_n(x)$. We choose the following values:

The Fourier Coefficients for a Periodic Function f of Period 2π

$$a_0 = \frac{1}{2\pi} \int_{-\pi}^{\pi} f(x)\, dx,$$

$$a_k = \frac{1}{\pi} \int_{-\pi}^{\pi} f(x) \cos(kx)\, dx \quad \text{for } k > 0,$$

$$b_k = \frac{1}{\pi} \int_{-\pi}^{\pi} f(x) \sin(kx)\, dx \quad \text{for } k > 0.$$

Notice that a_0 is just the average value of f over the interval $[-\pi, \pi]$.

For an informal justification for the use of these values, see page 466. In addition, the integrals over $[-\pi, \pi]$ for a_k and b_k can be replaced by integrals over any interval of length 2π.

Example 1   Construct successive Fourier polynomials for the square wave function f, with period 2π, given by

$$f(x) = \begin{cases} 0 & -\pi \leq x < 0 \\ 1 & 0 \leq x < \pi. \end{cases}$$

Solution   Since a_0 is the average value of f on $[-\pi, \pi]$, we suspect from the graph of f that $a_0 = \frac{1}{2}$. We can verify this analytically:

$$a_0 = \frac{1}{2\pi} \int_{-\pi}^{\pi} f(x)\, dx = \frac{1}{2\pi} \int_{-\pi}^{0} 0\, dx + \frac{1}{2\pi} \int_{0}^{\pi} 1\, dx = 0 + \frac{1}{2\pi}(\pi) = \frac{1}{2}.$$

Furthermore,

$$a_1 = \frac{1}{\pi} \int_{-\pi}^{\pi} f(x) \cos x\, dx = \frac{1}{\pi} \int_{0}^{\pi} 1 \cos x\, dx = 0$$

and

$$b_1 = \frac{1}{\pi} \int_{-\pi}^{\pi} f(x) \sin x\, dx = \frac{1}{\pi} \int_{0}^{\pi} 1 \sin x\, dx = \frac{2}{\pi}.$$

Therefore, the Fourier polynomial of degree 1 is given by

$$f(x) \approx F_1(x) = \frac{1}{2} + \frac{2}{\pi} \sin x$$

---

[3]The Fourier polynomials are not polynomials in the usual sense of the word.

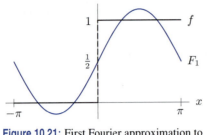

Figure 10.21: First Fourier approximation to the square wave

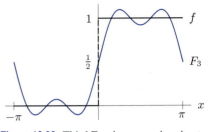

Figure 10.22: Third Fourier approximation to the square wave

and the graphs of the function and the first Fourier approximation are shown in Figure 10.21.

We next construct the Fourier polynomial of degree 2. The coefficients a_0, a_1, b_1 are the same as before. In addition,

$$a_2 = \frac{1}{\pi} \int_{-\pi}^{\pi} f(x) \cos(2x)\,dx = \frac{1}{\pi} \int_0^{\pi} 1 \cos(2x)\,dx = 0$$

and

$$b_2 = \frac{1}{\pi} \int_{-\pi}^{\pi} f(x) \sin(2x)\,dx = \frac{1}{\pi} \int_0^{\pi} 1 \sin(2x)\,dx = 0.$$

Since $a_2 = b_2 = 0$, the Fourier polynomial of degree 2 is identical to the Fourier polynomial of degree 1. Let's look at the Fourier polynomial of degree 3:

$$a_3 = \frac{1}{\pi} \int_{-\pi}^{\pi} f(x) \cos(3x)\,dx = \frac{1}{\pi} \int_0^{\pi} 1 \cos(3x)\,dx = 0$$

and

$$b_3 = \frac{1}{\pi} \int_{-\pi}^{\pi} f(x) \sin(3x)\,dx = \frac{1}{\pi} \int_0^{\pi} 1 \sin(3x)\,dx = \frac{2}{3\pi}.$$

So the approximation is given by

$$f(x) \approx F_3(x) = \frac{1}{2} + \frac{2}{\pi} \sin x + \frac{2}{3\pi} \sin(3x).$$

The graph of F_3 is shown in Figure 10.22. This approximation is better than $F_1(x) = \frac{1}{2} + \frac{2}{\pi} \sin x$, as comparing Figure 10.22 to Figure 10.21 shows.

Without going through the details, we calculate the coefficients for higher-degree Fourier approximations:

$$F_5(x) = \frac{1}{2} + \frac{2}{\pi} \sin x + \frac{2}{3\pi} \sin(3x) + \frac{2}{5\pi} \sin(5x)$$

$$F_7(x) = \frac{1}{2} + \frac{2}{\pi} \sin x + \frac{2}{3\pi} \sin(3x) + \frac{2}{5\pi} \sin(5x) + \frac{2}{7\pi} \sin(7x).$$

Figure 10.23 shows that higher-degree approximations match the step-like nature of the square wave function more and more closely.

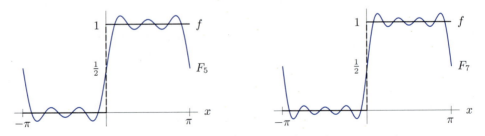

Figure 10.23: Fifth and seventh Fourier approximations to the square wave

We could have used a Taylor series to approximate the square wave, provided we did not center the series at a point of discontinuity. Since the square wave is a constant function on each interval, all its derivatives are zero, and so its Taylor series approximations are the constant functions: 0 or 1, depending on where the Taylor series is centered. They approximate the square wave perfectly on each piece, but they do not do a good job over the whole interval of length 2π. That is what Fourier polynomials succeed in doing: they approximate a curve fairly well everywhere, rather than just near a particular point. The Fourier approximations above look a lot like square waves, so they approximate well *globally*. However, they may not give good values near points of discontinuity. (For example, near $x = 0$, they all give values near $1/2$, which are incorrect.) Thus Fourier polynomials may not be good *local* approximations.

> Taylor polynomials give good *local* approximations to a function;
> Fourier polynomials give good *global* approximations to a function.

Fourier Series

As with Taylor polynomials, the higher the degree of the Fourier approximation, the more accurate it is. Therefore, we carry this procedure on indefinitely by letting $n \to \infty$, and we call the resulting infinite series a *Fourier series*.

> **The Fourier Series for f on $[-\pi, \pi]$**
>
> $$f(x) = a_0 + a_1 \cos x + a_2 \cos 2x + a_3 \cos 3x + \cdots$$
> $$+ b_1 \sin x + b_2 \sin 2x + b_3 \sin 3x + \cdots$$
>
> where a_k and b_k are the Fourier coefficients.

Thus, the Fourier series for the square wave is

$$f(x) = \frac{1}{2} + \frac{2}{\pi} \sin x + \frac{2}{3\pi} \sin 3x + \frac{2}{5\pi} \sin 5x + \frac{2}{7\pi} \sin 7x + \cdots.$$

Harmonics

Let us start with a function $f(x)$ that is periodic with period 2π, expanded in a Fourier series:

$$f(x) = a_0 + a_1 \cos x + a_2 \cos 2x + a_3 \cos 3x + \cdots$$
$$+ b_1 \sin x + b_2 \sin 2x + b_3 \sin 3x + \cdots$$

The function

$$a_k \cos kx + b_k \sin kx$$

is referred to as the kth *harmonic* of f, and it is customary to say that the Fourier series expresses f in terms of its harmonics. The first harmonic, $a_1 \cos x + b_1 \sin x$, is sometimes called the *fundamental harmonic* of f.

Example 2 Find a_0 and the first four harmonics of a *pulse train* function f of period 2π shown in Figure 10.24:

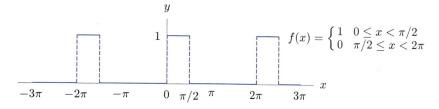

$$f(x) = \begin{cases} 1 & 0 \le x < \pi/2 \\ 0 & \pi/2 \le x < 2\pi \end{cases}$$

Figure 10.24: A train of pulses with period 2π

First- and Second-Order Differential Equations

First, some more definitions. We often write y' to represent the derivative of y. The differential equation

$$y' = 100 - y$$

is called *first-order* because it involves the first derivative, but no higher derivatives. By contrast, if s is the height (in meters) of a body moving under the force of gravity and t is time (in seconds), then

$$\frac{d^2 s}{dt^2} = -9.8.$$

This is a *second-order* differential equation because it involves the second derivative of the unknown function, $s = f(t)$, but no higher derivatives.

Example 1   Show that $y = e^{2t}$ is not a solution to the second-order differential equation

$$\frac{d^2 y}{dt^2} + 4y = 0.$$

Solution   To decide whether the function $y = e^{2t}$ is a solution, substitute it into the differential equation:

$$\frac{d^2 y}{dt^2} + 4y = 2(2e^{2t}) + 4e^{2t} = 8e^{2t}.$$

Since $8e^{2t}$ is not identically zero, $y = e^{2t}$ is not a solution.

How Many Arbitrary Constants Should We Expect in the Family of Solutions?

Since a differential equation involves the derivative of an unknown function, solving it usually involves antidifferentiation, which introduces arbitrary constants. The solution to a first-order differential equation usually involves one antidifferentiation and one arbitrary constant (for example, the C in $y = 100 + Ce^{-t}$). Picking out one particular solution involves knowing one additional piece of information, such as an initial condition. Solving a second-order differential equation generally involves two antidifferentiations and so two arbitrary constants. Consequently, finding a particular solution usually involves two initial conditions.

For example, if s is the height (in meters) of a body above the surface of the earth at time t (in seconds), then

$$\frac{d^2 s}{dt^2} = -9.8.$$

Integrating gives

$$\frac{ds}{dt} = -9.8t + C_1,$$

and integrating again gives

$$s = -4.9t^2 + C_1 t + C_2.$$

Thus the general solution for s involves the two arbitrary constants C_1 and C_2. We can find C_1 and C_2 if we are told, for example, that the initial velocity is 30 meters per second upward and that the initial position is 5 meters above the ground. In this case, $C_1 = 30$ and $C_2 = 5$.

Exercises and Problems for Section 11.1

Exercises

1. Match the graphs in Figure 11.3 with the following descriptions.

(a) The population of a new species introduced onto a tropical island

(b) The temperature of a metal ingot placed in a furnace and then removed

(c) The speed of a car traveling at uniform speed and then braking uniformly

(d) The mass of carbon-14 in a historical specimen

(e) The concentration of tree pollen in the air over the course of a year.

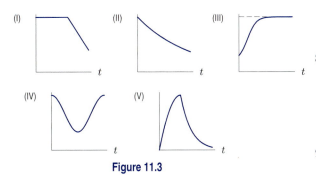

Figure 11.3

2. Fill in the missing values in the table given if you know that $dy/dt = 0.5y$. Assume the rate of growth given by dy/dt is approximately constant over each unit time interval and that the initial value of y is 8.

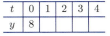

t	0	1	2	3	4
y	8				

3. Is $y = x^3$ a solution to the differential equation

$$xy' - 3y = 0?$$

4. Find the values of k for which $y = x^2 + k$ is a solution to the differential equation

$$2y - xy' = 10.$$

5. For what values of k (if any) does $y = 5 + 3e^{kx}$ satisfy the differential equation:

$$\frac{dy}{dx} = 10 - 2y ?$$

6. Show that, for any constant P_0, the function $P = P_0 e^t$ satisfies the differential equation

$$\frac{dP}{dt} = P.$$

7. Show that $y = A + Ce^{kt}$ is a solution to the differential equation

$$\frac{dy}{dt} = k(y - A).$$

8. Suppose $Q = Ce^{kt}$ satisfies the differential equation

$$\frac{dQ}{dt} = -0.03Q.$$

What (if anything) does this tell you about the values of C and k?

9. Show that $y = \sin 2t$ satisfies

$$\frac{d^2y}{dt^2} + 4y = 0.$$

10. Find the value(s) of ω for which $y = \cos \omega t$ satisfies

$$\frac{d^2y}{dt^2} + 9y = 0.$$

11. Show that any function of the form

$$x = C_1 \cosh \omega t + C_2 \sinh \omega t$$

satisfies the differential equation

$$x'' - \omega^2 x = 0.$$

Problems

12. (a) For what values of C and n (if any) is $y = Cx^n$ a solution to the differential equation:

$$x\frac{dy}{dx} - 3y = 0 ?$$

(b) If the solution satisfies $y = 40$ when $x = 2$, what more (if anything) can you say about C and n?

13. (a) Show that $P = 1/(1 + e^{-t})$ satisfies the logistic equation

$$\frac{dP}{dt} = P(1 - P).$$

(b) What is the limiting value of P as $t \to \infty$?

14. Pick out which functions are solutions to which differential equations. (Note: Functions may be solutions to more than one equation or to none; an equation may have more than one solution.)

(a) $\dfrac{dy}{dx} = -2y$ (I) $y = 2\sin x$

(b) $\dfrac{dy}{dx} = 2y$ (II) $y = \sin 2x$

(c) $\dfrac{d^2y}{dx^2} = 4y$ (III) $y = e^{2x}$

(d) $\dfrac{d^2y}{dx^2} = -4y$ (IV) $y = e^{-2x}$

15. Match the following differential equations and possible solutions. (Note the given functions may satisfy more than one equation or none, and some equations may have more than one solution.)

(a) $y'' = y$
(b) $y' = -y$
(c) $y' = 1/y$
(d) $y'' = -y$
(e) $x^2 y'' - 2y = 0$

(I) $y = \cos x$
(II) $y = \cos(-x)$
(III) $y = x^2$
(IV) $y = e^x + e^{-x}$
(V) $y = \sqrt{2x}$

11.2 SLOPE FIELDS

In this section, we see how to visualize a first-order differential equation. We start with the equation

$$\frac{dy}{dx} = y.$$

Any solution to this differential equation has the property that the slope at any point is equal to the y coordinate at that point. (That's what the equation $dy/dx = y$ is telling us!) If the solution goes through the point $(0, 0.5)$, its slope there is 0.5; if it goes through a point with $y = 1.5$, its slope there is 1.5. See Figure 11.4.

In Figure 11.4 a small line segment is drawn at each of the marked points showing the slope of the curve there. Imagine drawing many of these line segments, but leaving out the curves; this gives the *slope field* for the equation $dy/dx = y$ in Figure 11.5. From this picture, we can see that above the x-axis, the slopes are all positive (because y is positive there), and they increase as we move upward (as y increases). Below the x-axis, the slopes are all negative, and get more so as we move downward. On any horizontal line (where y is constant) the slopes are constant.

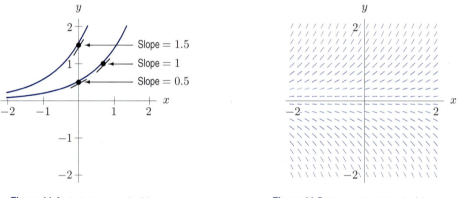

Figure 11.4: Solutions to $dy/dx = y$ **Figure 11.5:** Slope field for $dy/dx = y$

In the slope field we can see the ghost of the solution curve lurking. Start anywhere on the plane and move so that the slope lines are tangent to our path; we trace out one of the solution curves. We think of the slope field as a set of signposts pointing in the direction we should go at each point. In this case, the slope field should trace out exponential curves of the form $y = Ce^x$, the solutions to the differential equation $dy/dx = y$.

Example 1   Figure 11.6 shows the slope field of the differential equation $dy/dx = 2x$.
(a) How does the slope field vary as we move around the xy-plane?
(b) Compare the solution curves sketched on the slope field with the formula for the solutions.

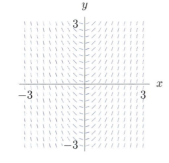

Figure 11.6: Slope field for $dy/dx = 2x$

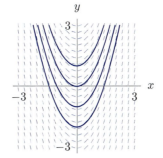

Figure 11.7: Some solutions to $dy/dx = 2x$

Solution   (a) In Figure 11.6 we notice that on a vertical line (where x is constant) the slopes are constant. This is because in this differential equation dy/dx depends on x only. (In the previous example, $dy/dx = y$, the slopes depended on y only.)

(b) The solution curves in Figure 11.7 look like parabolas. By antidifferentiation, we see that the solution to the differential equation $dy/dx = 2x$ is

$$y = \int 2x \, dx = x^2 + C,$$

so the solution curves really are parabolas.

Example 2   Using the slope field, guess the form of the solution curves of the differential equation

$$\frac{dy}{dx} = -\frac{x}{y}.$$

Solution   The slope field is shown in Figure 11.8. On the y-axis, where x is 0, the slope is 0. On the x-axis, where y is 0, the line segments are vertical and the slope is infinite. At the origin the slope is undefined, and there is no line segment.

The slope field suggests that the solution curves are circles centered at the origin. Later we see how to obtain the solution analytically, but even without this, we can check that the circle is a solution. We take the circle of radius r,

$$x^2 + y^2 = r^2,$$

and differentiate implicitly, thinking of y as a function of x. Using the chain rule, we get

$$2x + 2y \cdot \frac{dy}{dx} = 0.$$

Solving for dy/dx gives our differential equation,

$$\frac{dy}{dx} = -\frac{x}{y}.$$

This tells us that $x^2 + y^2 = r^2$ is a solution to the differential equation.

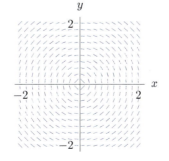

Figure 11.8: Slope field for $dy/dx = -x/y$

The previous example shows that the solution to a differential equation may be an implicit function.

Example 3   The slope fields of $dy/dt = 2 - y$ and $dy/dt = t/y$ are in Figures 11.9 and 11.10.
(a) On each slope field, sketch solution curves with initial conditions
        (i) $y = 1$ when $t = 0$ (ii) $y = 0$ when $t = 1$ (iii) $y = 3$ when $t = 0$
(b) For each solution curve, what can you say about the long-run behavior of y? For example, does $\lim_{t \to \infty} y$ exist? If so, what is its value?

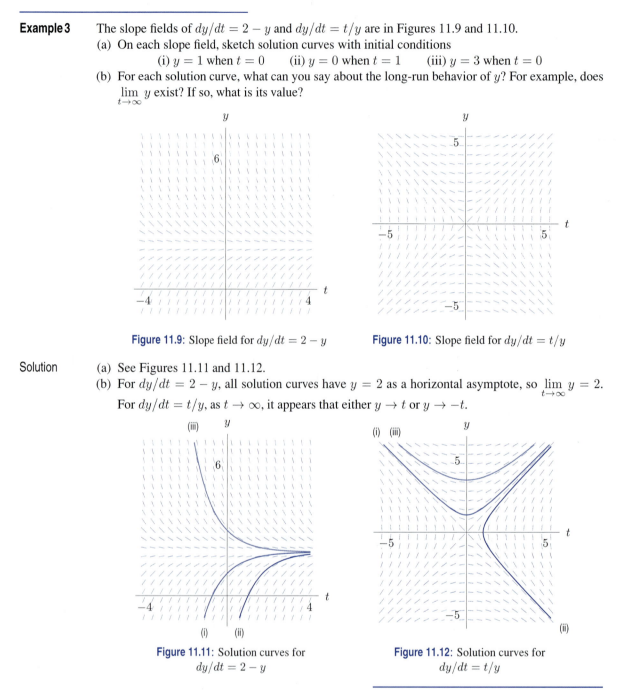

Figure 11.9: Slope field for $dy/dt = 2 - y$ **Figure 11.10:** Slope field for $dy/dt = t/y$

Solution   (a) See Figures 11.11 and 11.12.
(b) For $dy/dt = 2 - y$, all solution curves have $y = 2$ as a horizontal asymptote, so $\lim_{t \to \infty} y = 2$.
For $dy/dt = t/y$, as $t \to \infty$, it appears that either $y \to t$ or $y \to -t$.

Figure 11.11: Solution curves for $dy/dt = 2 - y$ **Figure 11.12:** Solution curves for $dy/dt = t/y$

Existence and Uniqueness of Solutions

Since differential equations are used to model many real situations, the question of whether a solution is unique can have great practical importance. If we know how the velocity of a satellite is changing, can we know its velocity at any future time? If we know the initial population of a city, and we know how the population is changing, can we predict the population in the future? Common sense says yes: if we know the initial value of some quantity and we know exactly how it is changing, we should be able to figure out its future value.

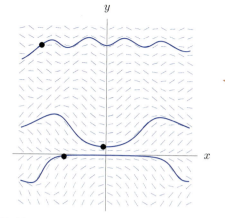

Figure 11.13: There's one and only one solution curve through each
point in the plane for this slope field (Dots represent initial conditions)

In the language of differential equations, an initial-value problem (that is, a differential equation and an initial condition) almost always has a unique solution. One way to see this is by looking at the slope field. Imagine starting at the point representing the initial condition. Through that point there is usually a line segment pointing in the direction of the solution curve. By following these line segments, we trace out the solution curve. See Figure 11.13. In general, at each point there is one line segment and therefore only one direction for the solution curve to go. The solution curve *exists* and is *unique* provided we are given an initial point. Notice that even though we can draw the solution curves, we may have no simple formula for them.

It can be shown that if the slope field is continuous as we move from point to point in the plane, we can be sure that a solution curve exists everywhere. Ensuring that each point has only one solution curve through it requires a slightly stronger condition.

Exercises and Problems for Section 11.2

Exercises

1. Sketch three solution curves for each of the slope fields in Figures 11.14 and 11.15.

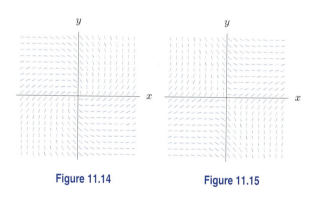

Figure 11.14   **Figure 11.15**

2. The slope field for the equation $y' = x(y-1)$ is shown in Figure 11.16.

(a) Sketch the solutions passing through the points
  (i) $(0, 1)$ (ii) $(0, -1)$ (iii) $(0, 0)$
(b) From your sketch, write down the equation of the solution with $y(0) = 1$.
(c) Check your solution to part (b) by substituting it into the differential equation.

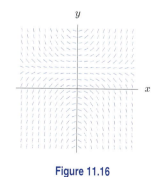

Figure 11.16

3. The slope field for the equation $y' = x + y$ is shown in Figure 11.17.

   (a) Carefully sketch the solutions that pass through the points
      (i) $(0, 0)$ (ii) $(-3, 1)$ (iii) $(-1, 0)$
   (b) From your sketch, write the equation of the solution passing through $(-1, 0)$.
   (c) Check your solution to part (b) by substituting it into the differential equation.

Problems

4. The slope field for the equation $dP/dt = 0.1P(10 - P)$, for $P \geq 0$, is in Figure 11.18.

   (a) Plot the solutions through the following points:
      (i) $(0, 0)$ (ii) $(1, 4)$ (iii) $(4, 1)$
      (iv) $(-5, 1)$ (v) $(-2, 12)$ (vi) $(-2, 10)$
   (b) For which positive values of P are the solutions increasing? Decreasing? What is the limiting value of P as t gets large?

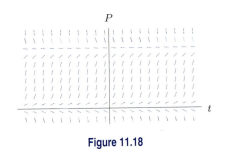

Figure 11.18

5. Figure 11.19 shows the slope field for the equation $y' = (\sin x)(\sin y)$.

   (a) Sketch the solutions that pass through the points:
      (i) $(0, -2)$ (ii) $(0, \pi)$.
   (b) What is the equation of the solution that passes through $(0, n\pi)$, where n is any integer?

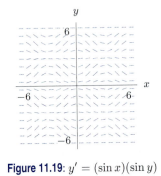

Figure 11.19: $y' = (\sin x)(\sin y)$

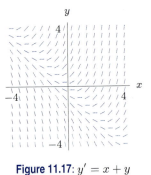

Figure 11.17: $y' = x + y$

6. One of the slope fields in Figure 11.20 has the equation $y' = (x + y)/(x - y)$. Which one?

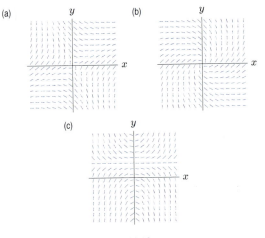

Figure 11.20

7. The slope field for $y' = 0.5(1 + y)(2 - y)$ is shown in Figure 11.21.

   (a) Plot the following points on the slope field:
      (i) the origin   (ii) $(0, 1)$ (iii) $(1, 0)$
      (iv) $(0, -1)$ (v) $(0, -5/2)$ (vi) $(0, 5/2)$
   (b) Plot solution curves through the points in part (a).
   (c) For which regions are all solution curves increasing? For which regions are all solution curves decreasing? When can the solution curves have horizontal tangents? Explain why, using both the slope field and the differential equation.

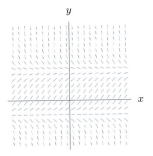

Figure 11.21: Note: x and y scales are equal

8. Match the slope fields in Figure 11.22 with their differential equations:

(a) $y' = -y$ **(b)** $y' = y$ **(c)** $y' = x$
(d) $y' = 1/y$ **(e)** $y' = y^2$

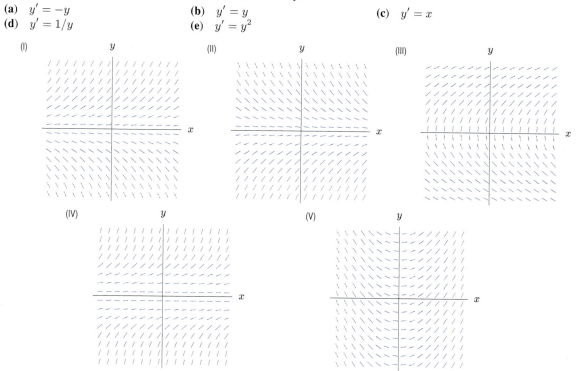

Figure 11.22

9. Match the slope fields in Figure 11.23 with their differential equations:

(a) $y' = 1 + y^2$ **(b)** $y' = x$ **(c)** $y' = \sin x$
(d) $y' = y$ **(e)** $y' = x - y$ **(f)** $y' = 4 - y$

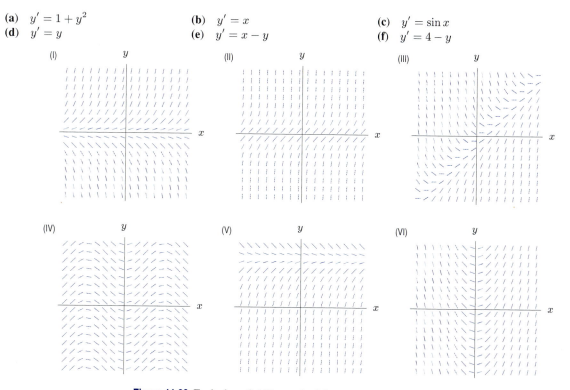

Figure 11.23: Each slope field is graphed for $-5 \le x \le 5$, $-5 \le y \le 5$

10. Match the slope fields in Figure 11.24 to the corresponding differential equations:

(a) $y' = xe^{-x}$

(b) $y' = \sin x$

(c) $y' = \cos x$

(d) $y' = x^2 e^{-x}$

(e) $y' = e^{-x^2}$

(f) $y' = e^{-x}$

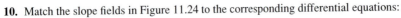

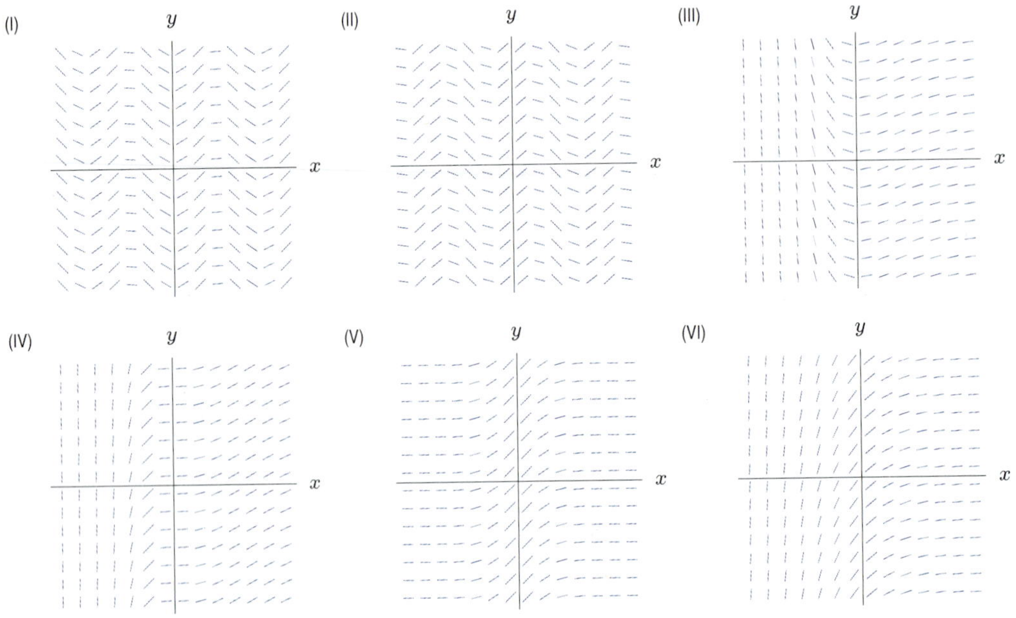

Figure 11.24

11.3   EULER'S METHOD

In the preceding section we saw how to sketch a solution curve to a differential equation using its slope field. In this section we compute points on a solution curve numerically using *Euler's method*. (Leonhard Euler was an eighteenth-century Swiss mathematician.) In Section 11.4 we find formulas for some solution curves.

Here's the concept behind Euler's method. Think of the slope field as a set of signposts directing you across the plane. Pick a starting point (corresponding to the initial value), and calculate the slope at that point using the differential equation. This slope is a signpost telling you the direction to take. Head off a small distance in that direction. Stop and look at the new signpost. Recalculate the slope from the differential equation, using the coordinates of the new point. Change direction to correspond to the new slope, and move another small distance, and so on.

Example 1   Use Euler's method for $dy/dx = y$. Start at the point $P_0 = (0, 1)$ and take $\Delta x = 0.1$.

Solution   The slope at the point $P_0 = (0, 1)$ is $dy/dx = 1$. (See Figure 11.25.) As we move from P_0 to P_1, y increases by Δy, where

$$\Delta y = (\text{slope at } P_0)\Delta x = 1(0.1) = 0.1.$$

So we have

$$y\text{-value at } P_1 = (y \text{ value at } P_0) + \Delta y = 1 + 0.1 = 1.1.$$

Table 11.2 *Euler's method for $dy/dx = y$, starting at $(0, 1)$*

	x	y	$\Delta y = (\text{Slope})\Delta x$
P_0	0	1	$0.1 = (1)(0.1)$
P_1	0.1	1.1	$0.11 = (1.1)(0.1)$
P_2	0.2	1.21	$0.121 = (1.21)(0.1)$
P_3	0.3	1.331	$0.1331 = (1.331)(0.1)$
P_4	0.4	1.4641	$0.14641 = (1.4641)(0.1)$
P_5	0.5	1.61051	$0.161051 = (1.61051)(0.1)$

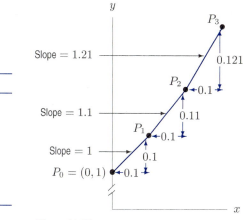

Figure 11.25: Euler's approximate solution to $dy/dx = y$

Thus the point P_1 is $(0.1, 1.1)$. Now, using the differential equation again, we see that

$$\text{Slope at } P_1 = 1.1,$$

so if we move to P_2, then y changes by

$$\Delta y = (\text{slope at } P_1)\Delta x = (1.1)(0.1) = 0.11.$$

This means

$$y\text{-value at } P_2 = (y \text{ value at } P_1) + \Delta y = 1.1 + 0.11 = 1.21.$$

Thus P_2 is $(0.2, 1.21)$. Continuing gives the results in Table 11.2.

Since the solution curves of $dy/dx = y$ are exponentials, they are concave up and bend upward away from the line segments of the slope field. Therefore, in this case, Euler's method produces y-values which are too small.

Notice that Euler's method calculates approximate y-values for points on a solution curve; it does not give a formula for y in terms of x.

Example 2 Show that Euler's method for $dy/dx = y$ starting at $(0, 1)$ and using two steps with $\Delta x = 0.05$ gives $y \approx 1.1025$ when $x = 0.1$.

Solution At $(0, 1)$, the slope is 1 and $\Delta y = (1)(0.05) = 0.05$, so new $y = 1 + 0.05 = 1.05$. At $(0.05, 1.05)$, the slope is 1.05 and $\Delta y = (1.05)(0.05) = 0.0525$, so new $y = 1.05 + 0.0525 = 1.1025$ at $x = 0.1$.

In general, dy/dx may be a function of both x and y. Euler's method still works, as the next example shows.

Example 3 Approximate four points on the solution curve to $dy/dx = -x/y$ starting at $(0, 1)$; use $\Delta x = 0.1$. Are the approximate values overestimates or underestimates?

Solution The results from Euler's method are in Table 11.3, along with the exact y-values (to two decimals) calculated from the equation of the circle $x^2 + y^2 = 1$, which is the solution curve through $(0, 1)$.

Since the curve is concave down, the approximate y-values are above the exact ones. (See Figure 11.26.)

Table 11.3 *Euler's method for $dy/dx = -x/y$, starting at $(0, 1)$*

x	Approximate y-value	$\Delta y = (\text{Slope})\Delta x$	True y-value
0	1	$0 = (0)(0.1)$	1
0.1	1	$-0.01 = (-0.1/1)(0.1)$	0.99
0.2	0.99	$-0.02 = (-0.2/0.99)(0.1)$	0.98
0.3	0.97		0.95

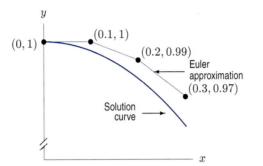

Figure 11.26: Euler's approximate solution to $dy/dx = -x/y$

The Accuracy of Euler's Method

To improve the accuracy of Euler's method, we choose Δx smaller. Let's go back to the differential equation $dy/dx = y$ and compare the exact and approximate values for different Δx's. The exact solution going through the point $(0, 1)$ is $y = e^x$, so the exact values are calculated using this function. (See Figure 11.27.) Where $x = 0.1$,

$$\text{Exact } y\text{-value} = e^{0.1} \approx 1.1051709.$$

In Example 1 we had $\Delta x = 0.1$, and where $x = 0.1$,

$$\text{Approximate } y\text{-value} = 1.1, \quad \text{so the error} \approx 0.005.$$

In Example 2 we decreased Δx to 0.05. After two steps, $x = 0.1$, and we had

$$\text{Approximate } y\text{-value} = 1.1025, \quad \text{so error} \approx 0.00267.$$

Thus, it appears that halving the step size has approximately halved the error.

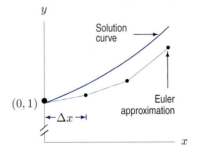

Figure 11.27: Euler's approximate solution to $dy/dx = y$

The *error* in using Euler's method is the difference between the approximate value and the exact value. If the number of steps used is n, the error is approximately proportional to $1/n$.

Just as there are more accurate numerical integration methods than left and right Riemann sums, there are more accurate methods than Euler's for approximating solution curves. However, Euler's method will be all we need.

Exercises and Problems for Section 11.3

Exercises

1. Consider the differential equation $y' = x + y$ whose slope field is in Figure 11.17 on page 486. Use Euler's method with $\Delta x = 0.1$ to estimate y when $x = 0.4$ for the solution curves satisfying

   (a) $y(0) = 1$ (b) $y(-1) = 0$.

2. Consider the solution of the differential equation $y' = y$ passing through $y(0) = 1$.

   (a) Sketch the slope field for this differential equation, and sketch the solution passing through the point $(0,1)$.
   (b) Use Euler's method with step size $\Delta x = 0.1$ to estimate the solution at $x = 0.1, 0.2, \ldots, 1$.
   (c) Plot the estimated solution on the slope field; compare the solution and the slope field.
   (d) Check that $y = e^x$ is the solution of $y' = y$ with $y(0) = 1$.

3. (a) Use ten steps of Euler's method to determine an approximate solution for the differential equation $y' = x^3$, $y(0) = 0$, using a step size $\Delta x = 0.1$.
   (b) What is the exact solution? Compare it to the computed approximation.
   (c) Use a sketch of the slope field for this equation to explain the results of part (b).

4. (a) Use Euler's method to approximate the value of y at $x = 1$ on the solution curve to the differential equation

   $$\frac{dy}{dx} = x^3 - y^3$$

   that passes through $(0,0)$. Use $\Delta x = 1/5$ (i.e., 5 steps).
   (b) Using Figure 11.28, sketch the solution that passes through $(0,0)$. Show the approximation you made in part (a).
   (c) Using the the slope field, say whether your answer to part (a) is an overestimate or an underestimate.

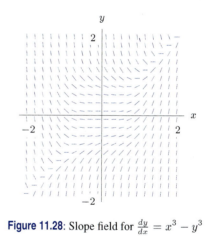

Figure 11.28: Slope field for $\frac{dy}{dx} = x^3 - y^3$

Problems

5. Consider the differential equation $y' = (\sin x)(\sin y)$.

   (a) Calculate approximate y-values using Euler's method with three steps and $\Delta x = 0.1$, starting at each of the following points:
   (i) $(0, 2)$ (ii) $(0, \pi)$.
   (b) Use the slope field in Figure 11.19 on page 486 to explain your solution to part (a)(ii).

6. (a) Use ten steps of Euler's method to approximate y-values for $dy/dt = 1/t$, starting at $(1, 0)$ and using $\Delta t = 0.1$.
   (b) Using integration, solve the differential equation to find the exact value of y at the end of these ten steps.

   (c) Is your approximate value of y at the end of ten steps bigger or smaller than the exact value? Use a slope field to explain your answer.

7. Consider the differential equation

   $$\frac{dy}{dx} = 2x, \quad \text{with initial condition } y(0) = 1.$$

   (a) Use Euler's method with two steps to estimate y when $x = 1$. Now use four steps.
   (b) What is the formula for the exact value of y?
   (c) Does the error in Euler's approximation behave as predicted in the box on page 491?

8. Consider the differential equation

$$\frac{dy}{dx} = \sin(xy), \qquad \text{with initial condition } y(1) = 1.$$

Estimate $y(2)$, using Euler's method with step sizes $\Delta x = 0.2, 0.1, 0.05$.

Plot the computed approximations for $y(2)$ against Δx. What do you conclude? Use your observations to estimate the exact value of $y(2)$.

9. Use Euler's method to solve

$$\frac{dB}{dt} = 0.05B$$

with initial value $B = 1000$ when $t = 0$. Take:

(a) $\Delta t = 1$ and 1 step. (b) $\Delta t = 0.5$ and 2 steps.

(c) $\Delta t = 0.25$ and 4 steps.

(d) Suppose B is the balance in a bank account earning interest. Explain why the result of your calculation in part (a) is equivalent to compounding the interest once a year instead of continuously.

(e) Interpret the result of your calculations in parts (b) and (c) in terms of compound interest.

10. Consider the differential equation $dy/dx = f(x)$ with initial value $y(0) = 0$. Explain why using Euler's method to approximate the solution curve gives the same results as using left Riemann sums to approximate $\int_0^x f(t)\,dt$.

11.4 SEPARATION OF VARIABLES

We have seen how to sketch solution curves of a differential equation using a slope field and how to calculate approximate numerical solutions. Now we see how solve certain differential equations analytically, finding an equation for the solution curve.

First, we look at a familiar example, the differential equation

$$\frac{dy}{dx} = -\frac{x}{y},$$

whose solution curves are the circles

$$x^2 + y^2 = C.$$

We can check that these circles are solutions by differentiation; the question now is how they were obtained. The method of *separation of variables* works by putting all the x's on one side of the equation and all the y's on the other, giving

$$y\,dy = -x\,dx.$$

We then integrate each side separately:

$$\int y\,dy = -\int x\,dx,$$

$$\frac{y^2}{2} = \frac{-x^2}{2} + k.$$

This gives the circles we were expecting:

$$x^2 + y^2 = C \qquad \text{where } C = 2k.$$

You might worry about whether it is legitimate to separate the dx and the dy. The reason it can be done is explained at the end of this section.

The Exponential Growth and Decay Equations

Let's use separation of variables on an equation that occurs frequently in practice:

$$\frac{dy}{dx} = ky.$$

Separating variables,

$$\frac{1}{y}\,dy = k\,dx,$$

and integrating,

$$\int \frac{1}{y}\,dy = \int k\,dx,$$

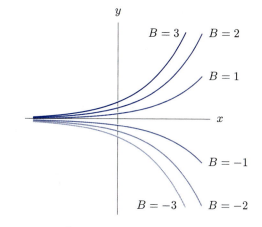

Figure 11.29: Graphs of $y = Be^{kx}$, which are solutions to $dy/dx = ky$, for some fixed $k > 0$

gives

$$\ln|y| = kx + C \quad \text{for some constant } C.$$

Solving for $|y|$ leads to

$$|y| = e^{kx+C} = e^{kx}e^C = Ae^{kx}$$

where $A = e^C$, so A is positive. Thus

$$y = (\pm A)e^{kx} = Be^{kx}$$

where $B = \pm A$, so B is any nonzero constant. Even though there's no C leading to $B = 0$, B can be 0 because $y = 0$ is a solution to the differential equation. We lost this solution when we divided through by y at the first step.

> The general solution to $\dfrac{dy}{dx} = ky$ is $y = Be^{kx}$ for any B.

The differential equation $dy/dx = ky$ always leads to exponential growth (if $k > 0$) or exponential decay (if $k < 0$). Graphs of solution curves for some fixed $k > 0$ are in Figure 11.29. For $k < 0$, the graphs are reflected in the y-axis.

Example 1   For $k > 0$, find and graph solutions of

$$\frac{dH}{dt} = -k(H - 20).$$

Solution   The slope field in Figure 11.30 shows the qualitative behavior of the solutions. To find the equation of the solution curves, we separate variables and integrate:

$$\int \frac{1}{H - 20}\, dH = -\int k\, dt.$$

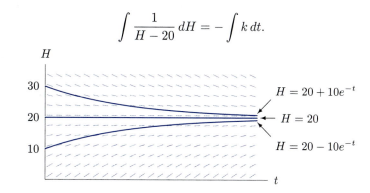

Figure 11.30: Slope field and some solution curves for $dH/dt = -k(H - 20)$, with $k = 1$

This gives
$$\ln |H - 20| = -kt + C.$$

Solving for H leads to:
$$|H - 20| = e^{-kt+C} = e^{-kt}e^C = Ae^{-kt}$$

or
$$H - 20 = (\pm A)e^{-kt} = Be^{-kt}$$
$$H = 20 + Be^{-kt}.$$

Again, $B = 0$ also gives a solution. Graphs for $k = 1$ and i $B = -10, 0, 10$, with $t \geq 0$, are in Figure 11.30.

Example 2   Find and sketch the solution to
$$\frac{dP}{dt} = 2P - 2Pt$$
satisfying $P = 5$ when $t = 0$.

Solution   Factoring the right-hand side gives
$$\frac{dP}{dt} = P(2 - 2t).$$

Separating variables, we get
$$\int \frac{dP}{P} = \int (2 - 2t)\, dt,$$

so
$$\ln |P| = 2t - t^2 + C.$$

Solving for P leads to
$$|P| = e^{2t-t^2+C} = e^C e^{2t-t^2} = Ae^{2t-t^2}$$

with $A = e^C$, so $A > 0$. In addition, $A = 0$ gives a solution. Thus the general solution to the differential equation is
$$P = Be^{2t-t^2} \quad \text{for any } B.$$

To find the value of B, substitute $P = 5$ and $t = 0$ into the general solution, giving
$$5 = Be^{2(0)-0^2} = B$$

so
$$P = 5e^{2t-t^2}.$$

The graph of this function is in Figure 11.31. Since the solution can be rewritten as
$$P = 5e^{1-1+2t-t^2} = 5e^1 e^{-1+2t-t^2} = (5e)e^{(t-1)^2},$$

the graph has the same shape as the graph of $y = e^{-t^2}$, the bell-shaped curve of statistics. Here the maximum, normally at $t = 0$, is shifted one unit to the right and occurs at $t = 1$.

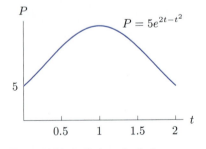

Figure 11.31: Bell-shaped solution curve

Justification For Separation of Variables

Suppose a differential equation can be written in the form

$$\frac{dy}{dx} = g(x)f(y).$$

Provided $f(y) \neq 0$, we write $f(y) = 1/h(y)$ so the right-hand side can be thought of as a fraction,

$$\frac{dy}{dx} = \frac{g(x)}{h(y)}.$$

If we multiply through by $h(y)$ we get

$$h(y)\frac{dy}{dx} = g(x).$$

Thinking of y as a function of x, so $y = y(x)$, and $dy/dx = y'(x)$, we can rewrite the equation as

$$h(y(x)) \cdot y'(x) = g(x).$$

Now integrate both sides with respect to x:

$$\int h(y(x)) \cdot y'(x)\, dx = \int g(x)\, dx.$$

The form of the integral on the left suggests that we use the substitution $y = y(x)$. Since $dy = y'(x)\, dx$, we get

$$\int h(y)\, dy = \int g(x)\, dx.$$

If we can find antiderivatives of h and g, then this gives the equation of the solution curve.
Note that transforming the original differential equation,

$$\frac{dy}{dx} = \frac{g(x)}{h(y)},$$

into

$$\int h(y)\, dy = \int g(x)\, dx$$

looks as though we have treated dy/dx as a fraction, cross-multiplied and then integrated. Although that's not exactly what we've done, you may find this a helpful way of remembering the method. In fact, the dy/dx notation was introduced by Leibniz to allow shortcuts like this (more specifically, to make the chain rule look like cancellation).

Exercises and Problems for Section 11.4

Exercises

Find the solutions to the differential equations in Exercises 1–27, subject to the given initial conditions.

1. $\dfrac{dP}{dt} = 0.02P, \quad P(0) = 20$

2. $\dfrac{dP}{dt} = -2P, \quad P(0) = 1$

3. $P\dfrac{dP}{dt} = 1, \quad P(0) = 1$

4. $\dfrac{dQ}{dt} = \dfrac{Q}{5}, \quad Q = 50$ when $t = 0$

5. $\dfrac{dL}{dp} = \dfrac{L}{2}, \quad L(0) = 100$

6. $\dfrac{dy}{dx} + \dfrac{y}{3} = 0, \quad y(0) = 10$

7. $\dfrac{dm}{dt} = 3m, \quad m = 5$ when $t = 1$

8. $\dfrac{dI}{dx} = 0.2I$, $I = 6$ where $x = -1$

9. $\dfrac{1}{z}\dfrac{dz}{dt} = 5$, $z(1) = 5$

10. $\dfrac{dm}{ds} = m$, $m(1) = 2$

11. $\dfrac{dz}{dy} = zy$, $z = 1$ when $y = 0$

12. $2\dfrac{du}{dt} = u^2$, $u(0) = 1$

13. $\dfrac{dP}{dt} = P + 4$, $P = 100$ when $t = 0$

14. $\dfrac{dy}{dx} = 2y - 4$,   through $(2, 5)$

15. $\dfrac{dm}{dt} = 0.1m + 200$, $m(0) = 1000$

16. $\dfrac{dB}{dt} + 2B = 50$, $B(1) = 100$

17. $\dfrac{dy}{dt} = 0.5(y - 200)$, $y = 50$ when $t = 0$

18. $\dfrac{dQ}{dt} = 0.3Q - 120$, $Q = 50$ when $t = 0$

19. $\dfrac{dR}{dr} + R = 1$, $R(1) = 0.1$

20. $\dfrac{dy}{dt} = \dfrac{y}{3 + t}$, $y(0) = 1$

21. $\dfrac{dz}{dt} = te^z$,   through the origin

22. $\dfrac{dy}{dx} = \dfrac{5y}{x}$, $y = 3$ where $x = 1$

23. $\dfrac{dy}{dt} = y^2(1 + t)$, $y = 2$ when $t = 1$

24. $\dfrac{dz}{dt} = z + zt^2$, $z = 5$ when $t = 0$

25. $\dfrac{dw}{d\theta} = \theta w^2 \sin \theta^2$, $w(0) = 1$

26. $x(x + 1)\dfrac{du}{dx} = u^2$, $u(1) = 1$

27. $\dfrac{dw}{d\psi} = \psi w^2 \cos(\psi^2)$, $w(0) = 1$

Problems

Solve the differential equations in Problems 28–37. Assume a, b, and k are constants.

28. $\dfrac{dR}{dt} = kR$

29. $\dfrac{dQ}{dt} - \dfrac{Q}{k} = 0$

30. $\dfrac{dP}{dt} = P - a$

31. $\dfrac{dQ}{dt} = b - Q$

32. $\dfrac{dP}{dt} = k(P - a)$

33. $\dfrac{dP}{dt} - aP = b$

34. $\dfrac{dR}{dt} = aR + b$

35. $\dfrac{dy}{dt} = ky^2(1 + t^2)$

36. $\dfrac{dR}{dx} = a(R^2 + 1)$

37. $\dfrac{dL}{dx} = k(x + a)(L - b)$

Solve the differential equations in Problems 38–41. Assume $x \geq 0$, $y \geq 0$.

38. $\dfrac{dy}{dt} = y(2 - y)$, $y(0) = 1$.

39. $t\dfrac{dx}{dt} = (1 + 2\ln t)\tan x$.

40. $\dfrac{dx}{dt} = \dfrac{x \ln x}{t}$.

41. $\dfrac{dy}{dt} = -y\ln\left(\dfrac{y}{2}\right)$, $y(0) = 1$.

42. (a) Find the general solution to the differential equation modeling how a person learns:
$$\frac{dy}{dt} = 100 - y.$$

(b) Plot the slope field of this differential equation and sketch solutions with $y(0) = 25$ and $y(0) = 110$.

(c) For each of the initial conditions in part (b), find the particular solution and add to your sketch.

(d) Which of these two particular solutions could represent how a person learns?

43. (a) Sketch the slope field for the differential equation $dy/dx = xy$.
(b) Sketch several solution curves.
(c) Solve the differential equation analytically.

44. (a) Sketch the slope field for the differential equation $dy/dx = y/x$.
(b) Sketch several solution curves.
(c) Solve the differential equation analytically and compare the result with the sketches in part (b).

45. (a) Sketch the slope field for $y' = x/y$.
(b) Sketch several solution curves.
(c) Solve the differential equation analytically.

46. (a) Sketch the slope field for $y' = -y/x$.
(b) Sketch several solution curves.
(c) Solve the differential equation analytically.

47. Compare the slope field for $y' = x/y$, Problem 45, with that for $y' = -y/x$, Problem 46. Show that the solution curves of Problem 45 intersect the solution curves of Problem 46 at right angles.

11.5 GROWTH AND DECAY

In this section we look at exponential growth and decay equations. Consider the population of a region. If there is no immigration or emigration, the rate at which the population is changing is often proportional to the population. In other words, the larger the population, the faster it is growing, because there are more people to have babies. If the population at time t is P, and its continuous growth rate is 2% per unit time, then we know

$$\text{Rate of growth of population} = 2\%(\text{Current population})$$

and we can write this as

$$\frac{dP}{dt} = 0.02P.$$

The 2% growth rate is called the *relative growth rate* to distinguish it from the *absolute growth rate*, dP/dt. Notice they measure different quantities. Since

$$\text{Relative growth rate} = 2\% = \frac{1}{P}\frac{dP}{dt},$$

the relative growth rate is a percent change per unit time, while

$$\text{Absolute growth rate} = \text{Rate of change of population} = \frac{dP}{dt}$$

is a change in population per unit time.

Since the equation $dP/dt = 0.02P$ is of the form $dP/dt = kP$ for $k = 0.02$, it has solution

$$P = P_0 e^{0.02t}.$$

Other processes are described by differential equations similar to that for population growth, but with negative values for k. In summary, we have the following result from the preceding section:

Every solution to the equation

$$\frac{dP}{dt} = kP$$

can be written in the form

$$P = P_0 e^{kt},$$

where P_0 is the value of P at $t = 0$, and $k > 0$ represents growth, whereas $k < 0$ represents decay.

Recall that the *doubling time* of an exponentially growing quantity is the time required for it to double. The *half-life* of an exponentially decaying quantity is the time for half of it to decay.

Continuously Compounded Interest

At a bank, continuous compounding means that interest is accrued at a rate that is a fixed percentage of the balance at that moment. Thus, the larger the balance, the faster interest is earned and the faster the balance grows.

Example 1   A bank account earns interest continuously at a rate of 5% of the current balance per year. Assume that the initial deposit is $\$1000$, and that no other deposits or withdrawals are made.

(a) Write the differential equation satisfied by the balance in the account.

(b) Solve the differential equation and graph the solution.

Solution   (a) We are looking for B, the balance in the account in dollars, as a function of t, time in years. Interest is being added to the account continuously at a rate of 5% of the balance at that moment. Since no deposits or withdrawals are made, at any instant,

$$\text{Rate balance increasing} = \text{Rate interest earned} = 5\%(\text{Current balance}),$$

which we write as

$$\frac{dB}{dt} = 0.05B.$$

This is the differential equation that describes the process. It does not involve the initial condition $1000, because the initial deposit does not affect the process by which interest is earned.

(b) Solving the differential equation by separation of variables gives

$$B = B_0 e^{0.05t},$$

where B_0 is the value of B at $t = 0$, so $B_0 = 1000$. Thus

$$B = 1000 e^{0.05t}$$

and this function is graphed in Figure 11.32.

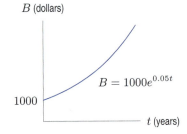

Figure 11.32: Bank balance against time

The Difference Between Continuous and Annual Percentage Growth Rates

If $P = P_0(1 + r)^t$ with t in years, we say that r is the *annual* growth rate, while if $P = P_0 e^{kt}$, we say that k is the *continuous* growth rate.

The constant k in the differential equation $dP/dt = kP$ is not the annual growth rate, but the continuous growth rate. In Example 1, with a continuous interest rate of 5%, we obtain a balance of $B = B_0 e^{0.05t}$, where time, t, is in years. At the end of one year the balance is $B_0 e^{0.05}$. In that one year, our balance has changed from B_0 to $B_0 e^{0.05}$; that is, by a factor of $e^{0.05} = 1.0513$. Thus the annual growth rate is 5.13%. This is what the bank means when it says "5% compounded continuously for an effective annual yield of 5.13%." Since $P_0 e^{0.05t} = P_0(1.0513)^t$, we have two different ways to represent the same function.

Since most growth is measured over discrete time intervals, a continuous growth rate is an idealized concept. A demographer who says a population is growing at the rate of 2% per year usually means that after t years the population is $P = P_0(1.02)^t$. To find the continuous growth rate, k, we express the population i as $P = P_0 e^{kt}$. At the end of one year $P = P_0 e^k$, so $e^k = 1.02$. Thus $k = \ln 1.02 \approx 0.0198$. The continuous growth rate, $k = 1.98\%$, is close to the annual growth rate of 2%, but it is not the same. Again, we have two different representations of the same function, since $P_0(1.02)^t = P_0 e^{0.0198t}$.

Pollution in the Great Lakes

In the 1960s pollution in the Great Lakes became an issue of public concern. We set up a model for how long it would take for the lakes to flush themselves clean, assuming no further pollutants are being dumped in the lake.

Suppose Q is the total quantity of pollutant in a lake of volume V at time t. Suppose that clean water is flowing into the lake at a constant rate r and that water flows out at the same rate. Assume that the pollutant is evenly spread throughout the lake, and that the clean water coming into the lake immediately mixes with the rest of the water.

We investigate how Q varies with time. Since pollutants are being taken out of the lake but not added, Q decreases, and the water leaving the lake becomes less polluted, so the rate at which the pollutants leave decreases. This tells us that Q is decreasing and concave up. In addition, the pollutants are never completely removed from the lake though the quantity remaining becomes arbitrarily small: in other words, Q is asymptotic to the t-axis. (See Figure 11.33.)

Setting Up a Differential Equation for the Pollution

To model exactly how Q changes with time, we write an equation for the rate at which Q changes. We know that

$$\begin{pmatrix} \text{Rate } Q \\ \text{changes} \end{pmatrix} = - \begin{pmatrix} \text{Rate pollutants} \\ \text{leave in outflow} \end{pmatrix}$$

where the negative sign represents the fact that Q is decreasing. At time t, the concentration of pollutants is Q/V and water containing this concentration is leaving at rate r. Thus

$$\begin{pmatrix} \text{Rate pollutants} \\ \text{leave in outflow} \end{pmatrix} = \begin{pmatrix} \text{Rate of} \\ \text{outflow} \end{pmatrix} \times \text{Concentration} = r \cdot \frac{Q}{V}.$$

So the differential equation is

$$\frac{dQ}{dt} = -\frac{r}{V}Q$$

and its solution is

$$Q = Q_0 e^{-rt/V}.$$

Table 11.4 contains values of r and V for four of the Great Lakes.[2] We use this data to calculate how long it would take for certain fractions of the pollution to be removed.

Table 11.4   *Volume and outflow in Great Lakes*

	V (thousands of km^3)	r (km^3/year)
Superior	12.2	65.2
Michigan	4.9	158
Erie	0.46	175
Ontario	1.6	209

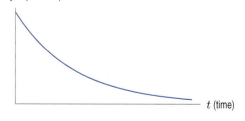

Q (quantity of pollutant)

t (time)

Figure 11.33: Pollutant in lake versus time

[2]Data from William E. Boyce and Richard C. DiPrima, *Elementary Differential Equations* (New York: Wiley, 1977).

Example 2   According to this model, how long will it take for 90% of the pollution to be removed from Lake Erie? For 99% to be removed?

Solution   Substituting r and V for Lake Erie into the differential equation for Q gives

$$\frac{dQ}{dt} = -\frac{r}{V}Q = \frac{-175}{0.46 \times 10^3}Q = -0.38Q$$

where t is measured in years. Thus Q is given by

$$Q = Q_0 e^{-0.38t}.$$

When 90% of the pollution has been removed, 10% remains, so $Q = 0.1Q_0$. Substituting gives

$$0.1Q_0 = Q_0 e^{-0.38t}.$$

Canceling Q_0 and solving for t, we get

$$t = \frac{-\ln(0.1)}{0.38} \approx 6 \text{ years}.$$

When 99% of the pollution has been removed, $Q = 0.01Q_0$, so t satisfies

$$0.01Q_0 = Q_0 e^{-0.38t}.$$

Solving for t gives

$$t = \frac{-\ln(0.01)}{0.38} \approx 12 \text{ years}.$$

Newton's Law of Heating and Cooling

Newton proposed that the temperature of a hot object decreases at a rate proportional to the difference between its temperature and that of its surroundings. Similarly, a cold object heats up at a rate proportional to the temperature difference between the object and its surroundings.

For example, a hot cup of coffee standing on the kitchen table cools at a rate proportional to the temperature difference between the coffee and the surrounding air. As the coffee cools, the rate at which it cools decreases, because the temperature difference between the coffee and the air decreases. In the long run, the rate of cooling tends to zero, and the temperature of the coffee approaches room temperature. See Figure 11.34.

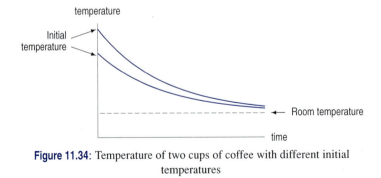

Figure 11.34: Temperature of two cups of coffee with different initial temperatures

Example 3   When a murder is committed, the body, originally at 37°C, cools according to Newton's Law of Cooling. Suppose that after two hours the temperature is 35°C, and that the temperature of the surrounding air is a constant 20°C.

(a) Find the temperature, H, of the body as a function of t, the time in hours since the murder was committed.

(b) Sketch a graph of temperature against time.

(c) What happens to the temperature in the long run? Show this on the graph and algebraically.

(d) If the body is found at 4 pm at a temperature of 30°C, when was the murder committed?

Solution   (a) We first find a differential equation for the temperature of the body as a function of time. Newton's Law of Cooling says that for some constant α,

$$\text{Rate of change of temperature} = \alpha(\text{Temperature difference}).$$

If H is the temperature of the body, then

$$\text{Temperature difference} = H - 20$$

so

$$\frac{dH}{dt} = \alpha(H - 20).$$

What about the sign of α? If the temperature difference is positive (i.e., $H > 20$), then H is falling, so the rate of change must be negative. Thus α should be negative, so we write:

$$\frac{dH}{dt} = -k(H - 20), \qquad \text{for some } k > 0.$$

Separating variables and solving, as in Example 1 on page 493, gives:

$$H - 20 = Be^{-kt}.$$

To find B, substitute the initial condition that $H = 37$ when $t = 0$:

$$37 - 20 = Be^{-k(0)} = B,$$

so $B = 17$. Thus,

$$H - 20 = 17e^{-kt}.$$

To find k, we use the fact that after 2 hours, the temperature is 35°C, so

$$35 - 20 = 17e^{-k(2)}.$$

Dividing by 17 and taking natural logs, we get:

$$\ln\left(\frac{15}{17}\right) = \ln(e^{-2k})$$
$$-0.125 = -2k$$
$$k \approx 0.063.$$

Therefore, the temperature is given by

$$H - 20 = 17e^{-0.063t}$$

or

$$H = 20 + 17e^{-0.063t}.$$

(b) The graph of $H = 20 + 17e^{-0.063t}$ has a vertical intercept of $H = 37$, because the temperature of the body starts at $37°C$. The temperature decays exponentially with $H = 20$ as the horizontal asymptote. (See Figure 11.35.)

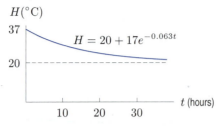

Figure 11.35: Temperature of dead body

(c) "In the long run" means as $t \to \infty$. The graph shows that as $t \to \infty$, $H \to 20$. Algebraically, since $e^{-0.063t} \to 0$ as $t \to \infty$, we have

$$H = 20 + \underbrace{17e^{-0.063t}}_{\text{goes to 0 as } t \to \infty} \longrightarrow 20 \quad \text{as } t \to \infty.$$

(d) We want to know when the temperature reaches $30°C$. Substitute $H = 30$ and solve for t:

$$30 = 20 + 17e^{-0.063t}$$
$$\frac{10}{17} = e^{-0.063t}.$$

Taking natural logs:

$$-0.531 = -0.063t,$$

which gives

$$t \approx 8.4 \, \text{hours}.$$

Thus the murder must have been committed about 8.4 hours before 4 pm. Since 8.4 hours $= 8$ hours 24 minutes, the murder was committed at about 7:30 am.

Equilibrium Solutions

Figure 11.36 shows the temperature of several objects in a $20°C$ room. One is initially hotter than $20°C$ and cools down toward $20°C$; another is initially cooler and warms up toward $20°C$. All these curves are solutions to the differential equation

$$\frac{dH}{dt} = -k(H - 20)$$

for some fixed $k > 0$, and all the solutions have the form

$$H = 20 + Ae^{-kt}$$

for some A. Notice that $H \to 20$ as $t \to \infty$ because $e^{-kt} \to 0$ as $t \to \infty$. In other words, in the long run, the temperature of the object always tends toward $20°C$, the temperature of the room. This means that what happens in the long run is independent of the initial condition.

In the special case when $A = 0$, we have the *equilibrium solution*

$$H = 20$$

for all t. This means that if the object starts at $20°C$, it remains at $20°C$ for all time. Notice that such a solution can be found directly from the differential equation by solving $dH/dt = 0$:

$$\frac{dH}{dt} = -k(H - 20) = 0$$

giving

$$H = 20.$$

Regardless of the initial temperature, H always gets closer and closer to 20 as $t \to \infty$. As a result, $H = 20$ is called a *stable* equilibrium[3] for H.

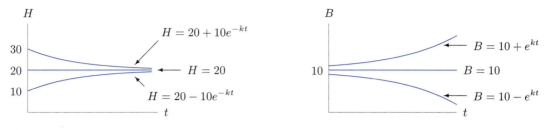

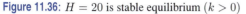

Figure 11.36: $H = 20$ is stable equilibrium $(k > 0)$ **Figure 11.37**: $B = 10$ is unstable equilibrium $(k > 0)$

A different situation is displayed in Figure 11.37, which shows solutions to the differential equation

$$\frac{dB}{dt} = k(B - 10)$$

for some fixed $k > 0$. Solving $dB/dt = 0$ gives the equilibrium $B = 10$, which is *unstable* because if B starts near 10, it moves away as $t \to \infty$.

In general, we have the following definitions.

- An **equilibrium solution** is constant for all values of the independent variable. The graph is a horizontal line.

- An equilibrium is **stable** if a small change in the initial conditions gives a solution which tends toward the equilibrium as the independent variable tends to positive infinity.

- An equilibrium is **unstable** if a small change in the initial conditions gives a solution curve which veers away from the equilibrium as the independent variable tends to positive infinity.

Solutions which do not veer away from an equilibrium solution are also called stable. If the differential equation is of the form $y' = f(y)$, equilibrium solutions can be found by setting y' to zero.

Exercises and Problems for Section 11.5

Exercises

1. The graphs in Figure 11.38 represent the temperature, $H(°C)$, of four eggs as a function of time, t, in minutes. Match three of the graphs with the descriptions below. Write a similar description for the fourth graph, including an interpretation of any intercepts and asymptotes.

(a) An egg is taken out of the refrigerator (just above $0°C$) and put into boiling water.

(b) Twenty minutes after the egg in part (a) is taken out of the fridge and put into boiling water, the same thing is done with another egg.

(c) An egg is taken out of the refrigerator at the same time as the egg in part (a) and left to sit on the kitchen table.

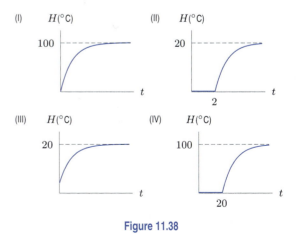

Figure 11.38

---

[3]In more advanced work, this behavior is described as asymptotic stability.

2. Each of the curves in Figure 11.39 represents the balance in a bank account into which a single deposit was made at time zero. Assuming continuously compounded interest, find:

(a) The curve representing the largest initial deposit.
(b) The curve representing the largest interest rate.
(c) Two curves representing the same initial deposit.
(d) Two curves representing the same interest rate.

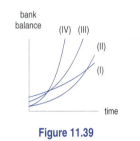

Figure 11.39

3. The slope field for $y' = 0.5(1 + y)(2 - y)$ is given in Figure 11.40.

(a) List equilibrium solutions and state whether each is stable or unstable.
(b) Draw solution curves on the slope field through each of the three marked points.

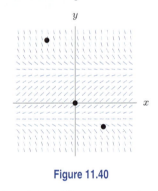

Figure 11.40

Problems

7. The rate of growth of a tumor is proportional to the size of the tumor.

(a) Write a differential equation satisfied by S, the size of the tumor, in mm, as a function of time, t.
(b) Find the general solution to the differential equation.
(c) If the tumor is 5 mm across at time $t = 0$, what does that tell you about the solution?
(d) If, in addition, the tumor is 8 mm across at time $t = 3$, what does that tell you about the solution?

8. Hydrocodone bitartrate is used as a cough suppressant. After the drug is fully absorbed, the quantity of drug in the body decreases at a rate proportional to the amount left in the body. The half-life of hydrocodone bitartrate in the body is 3.8 hours, and the usual oral dose is 10 mg.

(a) Write a differential equation for the quantity, Q,

4. (a) What are the equilibrium solutions for the differential equation

$$\frac{dy}{dt} = 0.2(y - 3)(y + 2)?$$

(b) Use a slope field to determine whether each equilibrium solution is stable or unstable.

5. The slope field for a differential equation is given in Figure 11.41. Estimate all equilibrium solutions for this differential equation, and indicate whether each is stable or unstable.

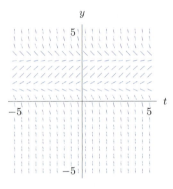

Figure 11.41

6. A yam is put in a 200°C oven and heats up according to the differential equation

$$\frac{dH}{dt} = -k(H - 200), \quad \text{for } k \text{ a positive constant.}$$

(a) If the yam is at 20°C when it is put in the oven, solve the differential equation.
(b) Find k using the fact that after 30 minutes the temperature of the yam is 120°C.

of hydrocodone bitartrate in the body at time t, in hours, since the drug was fully absorbed.

(b) Solve the differential equation given in part (a).
(c) Use the half-life to find the constant of proportionality, k.
(d) How much of the 10 mg dose is still in the body after 12 hours?

9. Oil is pumped continuously from a well at a rate proportional to the amount of oil left in the well. Initially there were 1 million barrels of oil in the well; six years later 500,000 barrels remain.

(a) At what rate was the amount of oil in the well decreasing when there were 600,000 barrels remaining?
(b) When will there be 50,000 barrels remaining?

10. The amount of land in use for growing crops increases as the world's population increases. Suppose $A(t)$ represents the total number of hectares of land in use in year t. (A hectare is about $2\frac{1}{2}$ acres.)

   (a) Explain why it is plausible that $A(t)$ satisfies the equation $A'(t) = kA(t)$. What assumptions are you making about the world's population and its relation to the amount of land used?

   (b) In 1950 about $1 \cdot 10^9$ hectares of land were in use; in 1980 the figure was $2 \cdot 10^9$. If the total amount of land available for growing crops is thought to be $3.2 \cdot 10^9$ hectares, when does this model predict it is exhausted? (Let $t = 0$ in 1950.)

11. (a) An object is placed in a 68°F room. Write a differential equation for H, the temperature of the object at time t.

   (b) Give the general solution for the differential equation.

   (c) The temperature of the object is 40°F initially, and 48°F one hour later. Find the temperature of the object after 3 hours.

12. Money in a bank account grows continuously at an annual rate of r (when the interest rate is 5%, $r = 0.05$, and so on). Suppose $1000 is put into the account in 2000.

   (a) Write a differential equation satisfied by M, the amount of money in the account at time t, measured in years since 2000.

   (b) Solve the equation.

   (c) Sketch the solution until the year 2030 for interest rates of 5% and 10%.

13. (a) If $B = f(t)$ is the balance at time t of a bank account that earns interest at a rate of $r\%$, compounded continuously, what is the differential equation describing the rate at which the balance changes? What is the constant of proportionality, in terms of r?

   (b) What is the solution to this differential equation?

   (c) Sketch the graph of $B = f(t)$ for an account that starts with $1000 and earns interest at the following rates:

   (i)   4%   (ii)   10%   (iii)   15%

14. In Example 2 on page 500 we saw that it would take about 6 years for 90% of the pollution in Lake Erie to be removed, and about 12 years for 99% to be removed. Explain why one time is double the other.

15. Using the model in the text and the data in Table 11.4 on page 499, find how long it would take for 90% of the pollution to be removed from Lake Michigan and from Lake Ontario, assuming no new pollutants are added. Explain how you can tell which lake will take longer to be purified just by looking at the data in the table.

16. Use the model in the text and the data in Table 11.4 on page 499 to determine which of the Great Lakes would require the longest time and which would require the shortest time for 80% of the pollution to be removed, assuming no new pollutants are being added. Find the ratio of these two times.

17. Warfarin is a drug used as an anticoagulant. After administration of the drug is stopped, the quantity remaining in a patient's body decreases at a rate proportional to the quantity remaining. The half-life of warfarin in the body is 37 hours.

   (a) Sketch a rough graph of the quantity, Q, of warfarin in a patient's body as function of the time, t, since stopping administration of the drug. Mark the 37 hours on your graph.

   (b) Write a differential equation satisfied by Q.

   (c) How many days does it take for the drug level in the body to be reduced to 25% of the original level?

18. Morphine is often used as a pain-relieving drug. The half-life of morphine in the body is 2 hours. Suppose morphine is administered to a patient intravenously at a rate of 2.5 mg per hour, and the rate at which the morphine is eliminated is proportional to the amount present.

   (a) Show that the constant of proportionality for the rate at which morphine leaves the body (in mg/hour) is $k = -0.347$.

   (b) Write a differential equation for the quantity, Q, of morphine in the blood after t hours.

   (c) Use the differential equation to find the equilibrium solution. (This is the long-term amount of morphine in the body, once the system has stabilized.)

19. A detective finds a murder victim at 9 am. The temperature of the body is measured at 90.3°F. One hour later, the temperature of the body is 89.0°F. The temperature of the room has been maintained at a constant 68°F.

   (a) Assuming the temperature, T, of the body obeys Newton's Law of Cooling, write a differential equation for T.

   (b) Solve the differential equation to estimate the time the murder occurred.

20. At 1:00 pm one winter afternoon, there is a power failure at your house in Wisconsin, and your heat does not work without electricity. When the power goes out, it is 68°F in your house. At 10:00 pm, it is 57°F in the house, and you notice that it is 10°F outside.

   (a) Assuming that the temperature, T, in your home obeys Newton's Law of Cooling, write the differential equation satisfied by T.

   (b) Solve the differential equation to estimate the temperature in the house when you get up at 7:00 am the next morning. Should you worry about your water pipes freezing?

   (c) What assumption did you make in part (a) about the temperature outside? Given this (probably incorrect) assumption, would you revise your estimate up or down? Why?

21. The amount of radioactive carbon-14 in a sample is measured using a Geiger counter, which records each disintegration of an atom. Living tissue disintegrates at a rate of about 13.5 atoms per minute per gram of carbon. In 1977 a charcoal fragment found at Stonehenge, England, recorded 8.2 disintegrations per minute per gram of carbon. Assuming that the half-life of carbon-14 is 5730 years and that the charcoal was formed during the building of the site, estimate the date at which Stonehenge was built.

22. The radioactive isotope carbon-14 is present in small quantities in all life forms, and it is constantly replenished until the organism dies, after which it decays to stable carbon-12 at a rate proportional to the amount of carbon-14 present, with a half-life of 5730 years. Suppose $C(t)$ is the amount of carbon-14 present at time t.

   (a) Find the value of the constant k in the differential equation $C' = -kC$.

(b) In 1988 three teams of scientists found that the Shroud of Turin, which was reputed to be the burial cloth of Jesus, contained 91% of the amount of carbon-14 contained in freshly made cloth of the same material.[4] How old is the Shroud of Turin, according to these data?

23. Before Galileo discovered that the speed of a falling body with no air resistance is proportional to the time since it was dropped, he mistakenly conjectured that the speed was proportional to the distance it had fallen.

   (a) Assume the mistaken conjecture to be true and write an equation relating the distance fallen, $D(t)$, at time t, and its derivative.

   (b) Using your answer to part (a) and the correct initial conditions, show that D would have to be equal to 0 for all t, and therefore the conjecture must be wrong.

11.6   APPLICATIONS AND MODELING

Much of this book involves functions that represent real processes, such as how the temperature of a yam or the population of Mexico is changing with time. You may wonder where such functions come from. In some cases, we fit functions to experimental data by trial and error. In other cases, we take a more theoretical approach, leading to a differential equation whose solution is the function we want. In this section we give examples of the more theoretical approach.

How a Layer of Ice Forms

When ice forms on a lake, the water on the surface freezes first. As heat from the water travels up through the ice and is lost to the air, more ice is formed. The question we will consider is: How thick is the layer of ice as a function of time? Since the thickness of the ice increases with time, the thickness function is increasing. In addition, as the ice gets thicker, it insulates better, therefore we expect the layer of ice to form more slowly as time goes on. Hence the thickness function is increasing at a decreasing rate, so its graph is concave down.

A Differential Equation for the Thickness of the Ice

To get more detailed information about the thickness function, we have to make some assumptions. Suppose y represents the thickness of the ice as a function of time, t. Since the thicker the ice, the longer it takes the heat to get through it, we will assume that the rate at which ice is formed is inversely proportional to the thickness. In other words, we assume that for some constant k,

$$\frac{\text{Rate thickness}}{\text{is increasing}} = \frac{k}{\text{Thickness}},$$

so

$$\frac{dy}{dt} = \frac{k}{y} \quad \text{where} \quad k > 0.$$

This differential equation enables us to find a formula for y. Using separation of variables:

$$\int y \, dy = \int k \, dt$$

$$\frac{y^2}{2} = kt + C.$$

---
[4]*The New York Times*, October 18, 1988

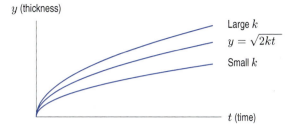

Figure 11.42: Thickness of ice as a function of time

If we measure time so that $y = 0$ when $t = 0$, then $C = 0$. Since y must be non-negative, we have

$$y = \sqrt{2kt}.$$

Graphs of y against t are in Figure 11.42. Notice that the larger y is, the more slowly y increases. In addition, this model suggests that y increases indefinitely as time passes. (Of course, the value of y cannot increase beyond the depth of the lake.)

The Net Worth of a Company

In the preceding section, we saw an example in which money in a bank account was earning interest (Example 1, page 497). Consider a company whose revenues are proportional to its net worth (like interest on a bank account) but which must also make payroll payments. The question is: under what circumstances does the company make money and under what circumstances does it go bankrupt?

Common sense says that if the payroll exceeds the rate at which revenue is earned, the company will eventually be in trouble, whereas if revenue exceeds payroll, the company should do well. We assume that revenue is earned continuously and that payments are made continuously. (For a large company, this is a good approximation.) We also assume that the only factors affecting net worth are revenue and payroll.

Example 1   A company's revenue is earned at a continuous annual rate of 5% of its net worth. At the same time, the company's payroll obligations amount to $200 million a year, paid out continuously.
(a) Write a differential equation that governs the net worth of the company, W million dollars.
(b) Solve the differential equation, assuming an initial net worth of W_0 million dollars.
(c) Sketch the solution for $W_0 = 3000$, 4000, and 5000.

Solution   First, let's see what we can learn without writing a differential equation. For example, we can ask if there is any initial net worth W_0 which will exactly keep the net worth constant. If there's such an equilibrium, the rate at which revenue is earned must exactly balance the payments made, so

Rate revenue is earned $=$ Rate payments are made.

If net worth is a constant W_0, revenue is earned at a constant rate of $0.05W_0$ per year, so we have

$$0.05W_0 = 200 \quad \text{giving} \quad W_0 = 4000.$$

Therefore, if the net worth starts at $4000 million, the revenue and payments are equal, and the net worth remains constant. Therefore $4000 million is an equilibrium solution.

Suppose, however, the initial net worth is above $4000 million. Then, the revenue earned is more than the payroll expenses, and the net worth of the company increases, thereby increasing the revenue still further. Thus the net worth increases faster and faster. On the other hand, if the initial net worth is below $4000 million, the revenue is not enough to meet the payments, and the net worth of the company declines. This decreases the revenue, making the net worth decrease still faster. The net worth will eventually go to zero, and the company goes bankrupt.

(a) Now we set up a differential equation for the net worth, using the fact that

$$\begin{array}{ccc} \text{Rate net worth} \\ \text{is increasing} \end{array} = \begin{array}{ccc} \text{Rate revenue} \\ \text{is earned} \end{array} - \begin{array}{ccc} \text{Rate payroll payments} \\ \text{are made} \end{array}.$$

In millions of dollars per year, revenue is earned at a rate of $0.05W$, and payments are made at a rate of 200 per year, so, for t in years,

$$\frac{dW}{dt} = 0.05W - 200.$$

The equilibrium solution, $W = 4000$, is obtained by setting $dW/dt = 0$.

(b) We solve this equation by separation of variables. It is helpful to factor out 0.05 before separating, so that the W moves over to the left-hand side without a coefficient:

$$\frac{dW}{dt} = 0.05(W - 4000).$$

Separating and integrating gives

$$\int \frac{dW}{W - 4000} = \int 0.05 \, dt,$$

so

$$\ln |W - 4000| = 0.05t + C,$$

or

$$|W - 4000| = e^{0.05t+C} = e^C e^{0.05t}.$$

This means

$$W - 4000 = Ae^{0.05t} \qquad \text{where } A = \pm e^C.$$

To find A we use the initial condition that $W = W_0$ when $t = 0$.

$$W_0 - 4000 = Ae^0 = A.$$

Substituting this value for A into $W = 4000 + Ae^{0.05t}$ gives

$$W = 4000 + (W_0 - 4000)e^{0.05t}.$$

(c) If $W_0 = 4000$, then $W = 4000$, the equilibrium solution.
If $W_0 = 5000$, then $W = 4000 + 1000e^{0.05t}$.
If $W_0 = 3000$, then $W = 4000 - 1000e^{0.05t}$. Substituting $t \approx 27.7$ gives $W = 0$, so the company goes bankrupt in its twenty-eighth year.

In Figure 11.43, notice that if the net worth starts with W_0 near, but not equal to, $4000 million, then W moves further away. Thus, $W = 4000$ is an unstable equilibrium.

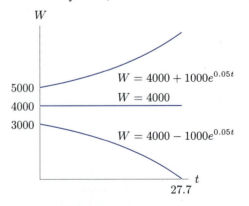

Figure 11.43: Solutions to $dW/dt = 0.05W - 200$

The Velocity of a Falling Body: Terminal Velocity

Think about the velocity of a sky-diver jumping out of a plane. When the sky-diver first jumps, his velocity is zero. The pull of gravity then makes his velocity increase. As the sky-diver speeds up, the air resistance also increases. Since the air resistance partly balances the pull of gravity, the force causing him to accelerate decreases. Thus, the velocity is an increasing function of time, but it is increasing at a decreasing rate. The air resistance increases until it balances gravity, when the sky-diver's velocity levels off. Thus, we expect the the graph of velocity against time to be concave down with a horizontal asymptote.

A Differential Equation: Air Resistance Proportional to Velocity

In order to compute the velocity function, we need to know exactly how air resistance depends on velocity. To decide whether air resistance is proportional to the velocity, say, or is some other function of velocity, requires either lab experiments or a theoretical idea of how the air resistance is created. We consider a very small object, such as a dust particle settling on a computer component during manufacturing,[5] and assume that air resistance is proportional to velocity. Thus, the net force on the object is $F = mg - kv$, where mg is the gravitational force, which acts downward, and kv is the air resistance, which acts upward, so $k > 0$. (See Figure 11.44.) Then, by Newton's Second Law of Motion,

$$\text{Force} = \text{Mass} \cdot \text{Acceleration},$$

we have

$$mg - kv = m\frac{dv}{dt}.$$

This differential equation can be solved by separation of variables. It is easier if we factor out $-k/m$ before separating, giving

$$\frac{dv}{dt} = -\frac{k}{m}\left(v - \frac{mg}{k}\right).$$

Separating and integrating gives

$$\int \frac{dv}{v - mg/k} = -\frac{k}{m}\int dt$$

$$\ln\left|v - \frac{mg}{k}\right| = -\frac{k}{m}t + C.$$

Solving for v:

$$\left|v - \frac{mg}{k}\right| = e^{-kt/m+C} = e^C e^{-kt/m}$$

$$v - \frac{mg}{k} = Ae^{-kt/m},$$

where A is an arbitrary constant. We find A from the initial condition that the object starts from rest, so $v = 0$ when $t = 0$. Substituting

$$0 - \frac{mg}{k} = Ae^0$$

gives

$$A = -\frac{mg}{k}.$$

Thus

$$v = \frac{mg}{k} - \frac{mg}{k}e^{-kt/m} = \frac{mg}{k}(1 - e^{-kt/m}).$$

The graph of this function is in Figure 11.45. The horizontal asymptote represents the *terminal velocity*, mg/k.

---

[5]Example suggested by Howard Stone.

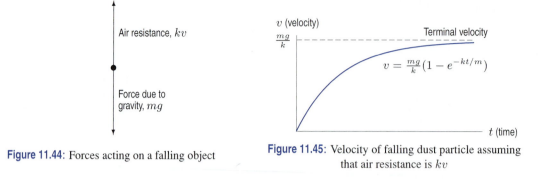

Figure 11.44: Forces acting on a falling object

Figure 11.45: Velocity of falling dust particle assuming that air resistance is kv

Notice that the terminal velocity can also be obtained from the differential equation by setting $dv/dt = 0$ and solving for v:

$$m\frac{dv}{dt} = mg - kv = 0 \quad \text{so} \quad v = \frac{mg}{k}.$$

Compartmental Analysis: A Reservoir

Many processes can be modeled as a container with various solutions flowing in and out—for example, drugs given intravenously or the discharge of pollutants into a lake. We consider a city's water reservoir, fed partly by clean water from a spring and partly by run-off from the surrounding land. In New England, and many other areas with much snow in the winter, the run-off contains salt which has been put on the roads to make them safe for driving. We consider the concentration of salt in the reservoir. If there is no salt in the reservoir initially, the concentration builds up until the rate at which the salt is entering into the reservoir balances the rate at which salt flows out. If, on the other hand, the reservoir starts with a great deal of salt in it, then initially the rate at which the salt is entering is less than the rate at which it is flowing out, and the quantity of salt in the lake decreases. In either case, the salt concentration levels off at an equilibrium value.

A Differential Equation for Salt Concentration

A water reservoir holds 100 million gallons of water and supplies a city with 1 million gallons a day. The reservoir is partly refilled by a spring which provides 0.9 million gallons a day, and the rest of the water, 0.1 million gallons a day, comes from run-off from the surrounding land. The spring is clean, but the run-off contains salt with a concentration of 0.0001 pound per gallon. There was no salt in the reservoir initially and the water is well mixed (that is, the out-flow contains the concentration of salt in the tank at that instant). We find the concentration of salt in the reservoir as a function of time.

It is important to distinguish between the total quantity, Q, of salt in pounds, and the concentration, C, of salt, in pounds/gallon, where

$$\text{Concentration} = C = \frac{\text{Quantity of salt}}{\text{Volume of water}} = \frac{Q}{100 \, \text{million}} \left(\frac{\text{lb}}{\text{gal}} \right).$$

(The volume of the reservoir is 100 million pounds.) We will find Q first, and then C. We know that

$$\begin{array}{c} \text{Rate of change of} \\ \text{quantity of salt} \end{array} = \text{Rate salt entering} - \text{Rate salt leaving}.$$

Salt is entering through the run-off of 0.1 million gallons per day, with each gallon containing 0.0001 pound of salt. Therefore

$$\text{Rate salt entering} = \text{Concentration} \cdot \text{Volume per day}$$

$$= 0.0001 \left(\frac{\text{lb}}{\text{gal}} \right) \cdot 0.1 \left(\frac{\text{million gal}}{\text{day}} \right)$$

$$= 0.00001 \left(\frac{\text{million lb}}{\text{day}} \right) = 10 \text{ lb/day}.$$

Salt is leaving in the million gallons of water used by the city each day . Thus

$$\text{Rate salt leaving} = \text{Concentration} \cdot \text{Volume per day}$$

$$= \frac{Q}{100 \text{ million}} \left(\frac{\text{lb}}{\text{gal}} \right) \cdot 1 \left(\frac{\text{million gal}}{\text{day}} \right) = \frac{Q}{100} \text{ lb/day}.$$

Therefore Q satisfies the differential equation

$$\frac{dQ}{dt} = 10 - \frac{Q}{100}.$$

We factor out $-1/100 = -0.1$ and separate variables, giving

$$\frac{dQ}{dt} = -0.01(Q - 1000),$$

$$\int \frac{dQ}{Q - 1000} = -\int 0.01 \, dt$$

$$\ln |Q - 1000| = -0.01t + k,$$

$$Q - 1000 = Ae^{-0.01t}.$$

There is no salt initially, so we substitute $Q = 0$ when $t = 0$:

$$0 - 1000 = Ae^0 \qquad \text{giving} \qquad A = -1000.$$

Thus

$$Q - 1000 = -1000e^{-0.01t}$$

so

$$Q = 1000(1 - e^{-0.01t}) \quad \text{pounds}.$$

Therefore

$$\text{Concentration } = C = \frac{Q}{100 \text{ million}} = \frac{1000}{10^8}(1 - e^{-0.01t}) = 10^{-5}(1 - e^{-0.01t}) \text{ lbs/gal}.$$

A sketch of concentration against time is in Figure 11.46.

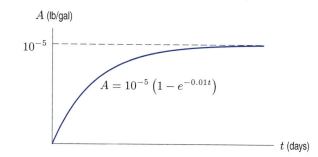

Figure 11.46: Concentration of salt in reservoir

Exercises and Problems for Section 11.6

Exercises

1. The velocity, v, of a dust particle of mass m and acceleration a satisfies the equation

$$ma = m\frac{dv}{dt} = mg - kv, \quad \text{where } g, k \text{ are constant.}$$

By differentiating this equation, find a differential equation satisfied by a. (Your answer may contain m, g, k, but not v.) Solve for a, given that $a(0) = g$.

2. A bank account that earns 10% interest compounded continuously has an initial balance of zero. Money is deposited into the account at a constant rate of $1000 per year.

  (a) Write a differential equation that describes the rate of change of the balance $B = f(t)$.
  (b) Solve the differential equation to find the balance as a function of time.

3. A deposit is made to a bank account paying 8% interest compounded continuously. Payments totalling $2000 per year are made from this account.

  (a) Write a differential equation for the balance, B, in the account after t years.
  (b) Write the solution to the differential equation.
  (c) How much is in the account after 5 years if the initial deposit is   (i) $20,000?   (ii) $30,000?

4. At time $t = 0$, a bottle of juice at $90°F$ is stood in a mountain stream whose temperature is $50°F$. After 5 minutes, its temperature is $80°F$. Let $H(t)$ denote the temperature of the juice at time t, in minutes.

  (a) Write a differential equation for $H(t)$ using Newton's Law of Cooling.
  (b) Solve the differential equation.
  (c) When will the temperature of the juice have dropped to $60°F$?

Problems

5. Dead leaves accumulate on the ground in a forest at a rate of 3 grams per square centimeter per year. At the same time, these leaves decompose at a continuous rate of 75% per year. Write a differential equation for the total quantity of dead leaves (per square centimeter) at time t. Sketch a solution showing that the quantity of dead leaves tends towards an equilibrium level. What is that equilibrium level?

6. According to a simple physiological model, an athletic adult needs 20 calories per day per pound of body weight to maintain his weight. If he consumes more or fewer calories than those required to maintain his weight, his weight changes at a rate proportional to the difference between the number of calories consumed and the number needed to maintain his current weight; the constant of proportionality is $1/3500$ pounds per calorie. Suppose that a particular person has a constant caloric intake of I calories per day. Let $W(t)$ be the person's weight in pounds at time t (measured in days).

  (a) What differential equation has solution $W(t)$?
  (b) Solve this differential equation.
  (c) Graph $W(t)$ if the person starts out weighing 160 pounds and consumes 3000 calories a day.

7. Water leaks out of a barrel at a rate proportional to the square root of the depth of the water at that time. If the water level starts at 36 inches and drops to 35 inches in 1 hour, how long will it take for all of the water to leak out of the barrel?

8. When a gas expands without gain or loss of heat, the rate of change of pressure with respect to volume is proportional to pressure divided by volume. Find a law connecting pressure and volume in this case.

9. In 1692, Johann Bernoulli was teaching the Marquis de l'Hopital calculus in Paris. Solve the following problem, which is similar to the one that they did. What is the equation of the curve which has subtangent (distance BC in Figure 11.47) equal to twice its abscissa (distance OC)?

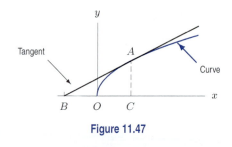

Figure 11.47

10. When the electromotive force (emf) is removed from a circuit containing inductance and resistance but no capacitors, the rate of decrease of current is proportional to the current. If the initial current is 30 amps but decays to 11 amps after 0.01 seconds, find an expression for the current as a function of time.

11. In some chemical reactions, the rate at which the amount of a substance changes with time is proportional to the amount present. For example, this is the case as δ-glucono-lactone changes into gluconic acid.

  (a) Write a differential equation satisfied by y, the quantity of δ-glucono-lactone present at time t.
  (b) If 100 grams of δ-glucono-lactone is reduced to 54.9 grams in one hour, how many grams will remain after 10 hours?

12. The rate at which barometric pressure decreases with altitude is proportional to the barometric pressure at that altitude. If the barometric pressure is measured in inches of mercury, and the altitude in feet, then the constant of proportionality is $3.7 \cdot 10^{-5}$. The barometric pressure at sea level is 29.92 inches of mercury.

  (a) Calculate the barometric pressure at the top of Mount Whitney, 14,500 feet (the highest mountain in the US outside Alaska), and at the top of Mount Everest, 29,000 feet (the highest mountain in the world).

  (b) People cannot easily survive at a pressure below 15 inches of mercury. What is the highest altitude to which people can safely go?

13. The rate (per foot) at which light is absorbed as it passes through water is proportional to the intensity, or brightness, at that point.

  (a) Find the intensity as a function of the distance the light has traveled through the water.

  (b) If 50% of the light is absorbed in 10 feet, how much is absorbed in 20 feet? 25 feet?

14. A spherical snowball melts at a rate proportional to its surface area.

  (a) Write a differential equation for its volume, V.

  (b) If the initial volume is V_0, solve the differential equation and graph the solution.

  (c) When does the snowball disappear?

15. A chemical reaction involves one molecule of a substance A combining with one molecule of substance B to form one molecule of substance C, written $A + B \to C$. The Law of Mass Action states that the rate at which C is formed is proportional to the product of the quantities of A and B present. Assume a and b are the initial quantities of A and B, and x is the quantity of C present at time t.

  (a) Write a differential equation for x.

  (b) Solve the equation with $x(0) = 0$.

16. If the initial quantities, a and b, in Problem 15 are the same, write and solve a differential equation for x, with $x(0) = 0$.

17. A patient is given the drug theophylline intravenously at a rate of 43.2 mg/hour to relieve acute asthma. You can imagine the drug as entering a compartment of volume 35,000 ml. (This is the volume of the part of the body through which the drug circulates.) The rate at which the drug leaves the patient is proportional to the quantity there, with proportionality constant 0.082. Assume the patient's body contains none of the drug initially. [Note: Concentration = Quantity/Volume.]

  (a) Describe in words how you would expect the concentration of theophylline in the patient to vary with time and sketch an approximate graph.

  (b) Write a differential equation satisfied by the concentration of the drug, $c(t)$.

  (c) Solve the differential equation and graph the solution. What happens to the concentration in the long run?

18. As you know, when a course ends, students start to forget the material they have learned. One model (called the Ebbinghaus model) assumes that the rate at which a student forgets material is proportional to the difference between the material currently remembered and some positive constant, a.

  (a) Let $y = f(t)$ be the fraction of the original material remembered t weeks after the course has ended. Set up a differential equation for y. Your equation will contain two constants; the constant a is less than y for all t.

  (b) Solve the differential equation.

  (c) Describe the practical meaning (in terms of the amount remembered) of the constants in the solution $y = f(t)$.

19. An item is initially sold at a price of $\$p$ per unit. Over time, market forces push the price toward the equilibrium price, $\$p^*$, at which supply balances demand. The Evans Price Adjustment model says that the rate of change in the market price, $\$p$, is proportional to the difference between the market price and the equilibrium price.

  (a) Write a differential equation for p as a function of t.

  (b) Solve for p.

  (c) Sketch solutions for various different initial prices, both above and below the equilibrium price.

  (d) What happens to p as $t \to \infty$?

20. A drug is administered intravenously at a constant rate of r mg/hour and is excreted at a rate proportional to the quantity present, with constant of proportionality $\alpha > 0$.

  (a) Solve a differential equation for the quantity, Q, in milligrams, of the drug in the body at time t hours. Your answer will contain r and α. Sketch a graph of Q against t. What is Q_∞, the limiting long-run value of Q?

  (b) What effect does doubling r have on Q_∞? What effect does doubling r have on the time to reach half the limiting value, $\frac{1}{2}Q_\infty$?

  (c) What effect does doubling α have on Q_∞? On the time to reach $\frac{1}{2}Q_\infty$?

21. When people smoke, carbon monoxide is released into the air. In a room of volume 60 m^3, air containing 5% carbon monoxide is introduced at a rate of 0.002 m^3/min. (This means that 5% of the volume of the incoming air is carbon monoxide.) The carbon monoxide mixes immediately with the rest of the air and the mixture leaves the room at the same rate as it enters.

  (a) Write a differential equation for $c(t)$, the concentration of carbon monoxide at time t, in minutes.

  (b) Solve the differential equation, assuming there is no carbon monoxide in the room initially.

  (c) What happens to the value of $c(t)$ in the long run?

The agreement between the predicted and actual values is remarkable. Of course, since we used the data from the entire 70 year period to estimate k, we should expect good agreement throughout that period. What is surprising is that if we had used only the populations in 1790 and 1800 to estimate k, the predictions are still quite good. Let's find a new value of k using only the 1790 and 1800 data and compare predictions. The 10-year growth from 1790 to 1800 is 35.9% so $k = \ln(1.359)/10 = 0.0307$. We would then predict the population in 1860 to be

$$3.9e^{0.0307(70)} = 33.4,$$

which is within about 6% of the actual population of 31.4. It is remarkable that a person in 1800 could accurately predict what the population of the US would be 60 years later, especially considering all the wars, recessions, epidemics, additions of new territory, and immigration that took place from 1800 to 1860.

Predictions From The Exponential Model

The exponential model led to the pessimistic predictions of Thomas Malthus, an early nineteenth-century clergyman and political philosopher, who believed that, if unchecked, a population would grow exponentially, whereas the food supply would grow linearly, and therefore the population would eventually outstrip the food supply. Interestingly enough, an exponential model has fit the growth of world population and the population of many regions remarkably well for decades, even centuries. However, the model must break down at some point because it predicts that the population will continue to grow without bound as time goes on—and this cannot be true forever. Eventually the effects of crowding, emigration, disease, war, and lack of food will have to curb growth. In searching for an improvement, then, we should look for a model whose solution is approximately an exponential function for small values of the population, but which levels off later.

The US Population: 1790–1940

If we try to predict the population of the US in 2000 using our exponential model with $k = 0.0298$, we get

$$\text{US population in } 2000 = 3.9e^{(0.0298)(210)} = 2{,}036 \text{ million,}$$

which is far from the actual figure of around 280 million.

The problem is that growth rate in the US population between 1790 and 1860 did not stay constant in later decades. The 10-year percentage growths[7] from 1860 to 1930 are listed in Table 11.8:

Table 11.8   *Estimated 10-year growth rate of US population, 1860–1930*

Year	1860	1870	1880	1890	1900	1910	1920	1930
Relative growth rate	22.9%	30.1%	25.3%	20.8%	21.1%	14.9%	16.2%	7.2%

These figures are nothing like those during the period 1790 to 1860, where they hovered around 34%. The dramatic drop to 22.9% for the decade 1860–1870 is explainable by the Civil War (but don't ascribe the entire drop to deaths of the war—see Problem 4). The growth rate goes back up during the decade 1870–1880, but by 1890–1900 it has dropped below the rate during the Civil War. There is a slight increase in the rate during the immigrations of 1900 to 1910, another drop during World War I, a small bounce back, and finally the rate plummets during the recession of the 1930s. Notice that the effect of the recession is more dramatic than the effect of wars, which suggests that it is a decrease in birth rate rather than an increase in death rate that is a major factor in declines in the relative population growth rate.

---

[7] Calculated using estimates of the form $\frac{1}{P}\frac{dP}{dt}$ at $1860 \approx \frac{\text{Population in 1870} - \text{Population in 1860}}{10(\text{Population in 1860})}$.

The exponential model gives accurate predictions up to 1860. But it is inadequate for the following years. We look for a new model that takes into account the effects of overcrowding by having a relative growth rate which decreases as the population increases. Using the data in Table 11.8, it can be shown that, approximately,

$$\frac{1}{P}\frac{dP}{dt} = 0.0318 - 0.000170P.$$

Therefore, in our new model, P satisfies the *logistic* differential equation

$$\frac{dP}{dt} = 0.0318P - 0.000170P^2.$$

Its slope field is shown in Figure 11.48, together with the solution with $P(0) = 3.9$.

The most striking difference between this model and the exponential model is that the logistic model predicts that the US population will level off. The logistic model predicts that the population will continue to grow until

$$\frac{dP}{dt} = 0.0318P - 0.000170P^2 = 0,$$

that is, when $P = 0$ or $P \approx 187$ million.

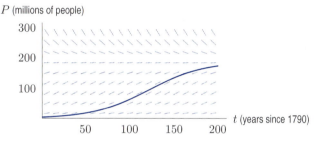

Figure 11.48: Solution to $dP/dt = 0.0318P - 0.000170P^2$ with $P(0) = 3.9$

The shape of the solution curve in Figure 11.48 suggests that initially the population grows faster and faster and then slows down as the limiting value of 187 is approached; the fastest growth rate appears to be about half-way to the limiting value.

Later in this section we derive the formula for the solution to the logistic equation. For the US population, it turns out to be

$$P = \frac{187}{1 + 47e^{-0.0318t}}.$$

(The numbers 187 and the 47 are not exact values, but have been rounded.) The values predicted by this formula for P agree with the actual populations up to 1940. See Table 11.9. During the period from 1700 to 1940, the largest deviation is about 3% in 1840 and 1870 (the Civil War accounts for the second one). All other errors are less than 2%.

Table 11.9 *Predicted versus actual US population in millions, 1790–1980 (logistic model)*

Year	Actual	Predicted	Year	Actual	Predicted	Year	Actual	Predicted
1790	3.9	3.9	1870	38.6	39.9	1950	150.7	145.0
1800	5.3	5.3	1880	50.2	50.7	1960	179.3	154.4
1810	7.2	7.2	1890	62.9	63.3	1970	203.3	162.1
1820	9.6	9.8	1900	76.0	77.2	1980	226.5	168.2
1830	12.9	13.2	1910	92.0	91.9	1990	248.7	172.9
1840	17.1	17.7	1920	105.7	106.7	2000	281.4	176.6
1850	23.1	23.5	1930	122.8	120.8			
1860	31.4	30.8	1940	131.7	133.7			

Of course, the real test is how well our model, based on data from 1790 to 1940, predicts the population in the "future," 1950 to 2000. The fit between predicted and actual population in Table 11.9 is not good from 1950 on. Despite World War II, which undoubtedly depressed population growth between 1942 and 1945, in the last half of the 1940s the US population surged, wiping out in five years a deficit caused by 15 years of depression and war. The 1950s saw a population growth of 28 million, leaving our logistic model in the dust. This surge in population is referred to as the baby boom.

Once again we have reached a point where our model is no longer useful. This should not lead you to believe that a reasonable mathematical model cannot be found. No model is perfect and when one model fails, we seek a better one. Just as we abandoned the exponential model in favor of the logistic model for the US population, we could look further.

The Logistic Model

The logistic model we used to model the US population from 1790 to 1940 assumed that the relative growth rate of the population was a linearly decreasing function of P:

$$\frac{1}{P}\frac{dP}{dt} = k - aP.$$

(We took $k = 0.0318$ and $a = 0.000170$). For small P, we have approximately

$$\frac{1}{P}\frac{dP}{dt} \approx k.$$

The solution to this equation is an exponential function. This is why an exponential model fit the US population well during the years 1790-1860 when the population was relatively small. In the logistic model, as P increases, the relative growth rate decreases to zero; it reaches zero when P is given by

$$\frac{1}{P}\frac{dP}{dt} = k - aP = 0.$$

Solving gives $P = k/a$, the limiting value of the population, which we call L. The value L is the *carrying capacity* of the environment, and represents the largest population the environment can support. Writing $a = k/L$, the logistic equation becomes

$$\frac{1}{P}\frac{dP}{dt} = k - \frac{k}{L}P$$
$$\frac{dP}{dt} = kP\left(1 - \frac{P}{L}\right).$$

This is the general logistic differential equation, first proposed as a model for population growth by the Belgian mathematician P. F. Verhulst in the 1830s.

Qualitative Solution to the Logistic Equation

Figure 11.49 shows the slope field and characteristic *sigmoid*, or S-shaped, solution curve for the logistic model. Notice that for each fixed value of P, that is, along each horizontal line, the slopes are constant because dP/dt depends only on P and not on t. The slopes are small near $P = 0$ and near $P = L$; they are steepest around $P = L/2$. For $P > L$, the slopes are negative, so if the population is above the carrying capacity, the population decreases.

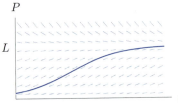

Figure 11.49: Slope field for
$dP/dt = kP(1 - P/L)$

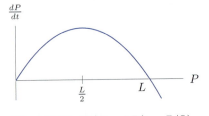

Figure 11.50: $dP/dt = kP(1 - P/L)$

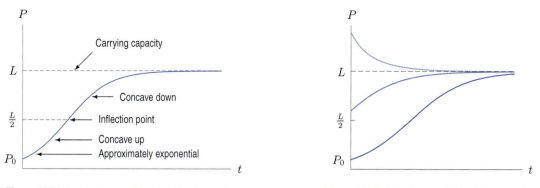

Figure 11.51: Logistic growth with inflection point

Figure 11.52: Solutions to the logistic equation

We can locate precisely the inflection point where the slopes are greatest using the graph of dP/dt against P in Figure 11.50. The graph is a parabola because dP/dt is a quadratic function of P. The horizontal intercepts are at $P = 0$ and $P = L$, so the maximum, where the slope is greatest, is at $P = L/2$. The graph in Figure 11.50 also tells us that for $0 < P < L/2$, the slope dP/dt is positive and increasing, so the graph of P against t is concave up. (See Figure 11.51.) For $L/2 < P < L$, the slope dP/dt is positive and decreasing, so the graph of P against t is concave down. For $P > L$, the slope dP/dt is negative, so the graph of P against t is decreasing.

If $P = 0$ or $P = L$, there is an equilibrium solution (not a very interesting one if $P = 0$). Figure 11.52 shows that $P = 0$ is an unstable equilibrium because solutions which start near 0 move away. However, $P = L$ is a stable equilibrium.

The Analytic Solution to the Logistic Equation

We have already obtained a lot of information about logistic growth without finding a formula for the solution. However, the equation can be solved analytically by separating variables:

$$\frac{dP}{dt} = kP\left(1 - \frac{P}{L}\right) = kP\left(\frac{L - P}{L}\right)$$

giving

$$\int \frac{dP}{P(L - P)} = \int \frac{k}{L}\,dt.$$

We can integrate the left side using the integral tables (Formula 26), or by partial fractions:

$$\int \frac{1}{L}\left(\frac{1}{P} + \frac{1}{L - P}\right) dP = \int \frac{k}{L}\,dt.$$

Canceling the constant L, we integrate to get

$$\ln|P| - \ln|L - P| = kt + C.$$

Multiplying through by (-1) and using properties of logarithms, we have

$$\ln\left|\frac{L - P}{P}\right| = -kt - C.$$

Exponentiating both sides gives

$$\left|\frac{L - P}{P}\right| = e^{-kt - C} = e^{-C}e^{-kt},$$

so, writing $A = \pm e^{-C}$, we have

$$\frac{L - P}{P} = Ae^{-kt}.$$

We find A by substituting $P = P_0$ when $t = 0$, which gives

$$\frac{L - P_0}{P_0} = Ae^0 = A.$$

Since $(L - P)/P = (L/P) - 1$, we have

$$\frac{L}{P} = 1 + Ae^{-kt}$$

giving the formula for the logistic curve:

$$P = \frac{L}{1 + Ae^{-kt}} \quad \text{where} \quad A = \frac{L - P_0}{P_0}.$$

Exercises and Problems for Section 11.7

Exercises

1. Assuming that Switzerland's population is growing exponentially at a continuous rate of 0.2% a year and that its 1988 population was 6.6 million, write an expression for the population as a function of time in years. (Let $t = 0$ in 1988.)

2. (a) On the slope field for $dP/dt = 3P - 3P^2$ in Figure 11.53, sketch three solution curves showing different types of behavior for the population, P.

   (b) Is there a stable value of the population? If so, what is it?

   (c) Describe the meaning of the shape of the solution curves for the population: Where is P increasing? Decreasing? What happens in the long run? Are there any inflection points? Where? What do they mean for the population?

   (d) Sketch a graph of dP/dt against P. Where is dP/dt positive? Negative? Zero? Maximum? How do your observations about dP/dt explain the shapes of your solution curves?

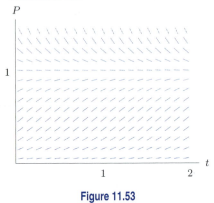

P

1

2

t

Figure 11.53

3. The Tojolobal Mayan Indian community in Southern Mexico has available a fixed amount of land. The proportion, P, of land in use for farming t years after 1935 is modeled with the logistic function in Figure 11.54:[8]

$$P = \frac{1}{1 + 2.968e^{-0.0275t}}.$$

   (a) What proportion of the land was in use for farming in 1935?

   (b) What is the long run prediction of this model?

   (c) When was half the land in use for farming?

   (d) When is the proportion of land used for farming increasing most rapidly?

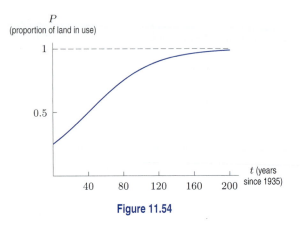

P
(proportion of land in use)

1

0.5

40   80   120   160   200

t (years since 1935)

Figure 11.54

---

[8]Adapted from J. S. Thomas and M. C. Robbins, "The Limits to Growth in a Tojolobal Maya Ejido", *Geoscience and Man 26* (Baton Rouge: Geoscience Publications, 1988), pp. 9–16.

Problems

4. Estimate the US population in 1870 from the 1860 population of 31.4 million, assuming that the population increased at the same percentage rate during the 1860s as it did in the decades previous to 1860 (about 34.7% each decade). Compare your estimate with the actual 1870 population of 38.6 million. Find some estimate of the number of people who died in the Civil War. Does the number of deaths in the Civil War explain the shortfall in the actual population in 1870? What else might influence the shortfall?

5. Another way to estimate the limiting population L for the logistic differential equation is to note that the derivative dP/dt is largest when $P = L/2$. Estimate dP/dt for the US census data for 1790–1940 using the two-sided difference quotient $(P(t+10) - P(t-10))/20$. Estimate L by doubling the population when dP/dt is largest. Compare this estimate with the text's estimate of $L = 187$.

6. The growth of an animal population is governed by the equation

$$\frac{1000}{P}\frac{dP}{dt} = 100 - P,$$

where $P(t)$ is the number of individuals in the colony at time t. The initial population is known to be 200 individuals. Sketch a graph of $P(t)$. Will there ever be more than 200 individuals in the colony? Will there ever be fewer than 100 individuals? Explain.

7. The total number of people infected with a virus often grows like a logistic curve. Suppose that 10 people originally have the virus, and that in the early stages of the virus (with time, t, measured in weeks), the number of people infected is increasing exponentially with $k = 1.78$. It is estimated that, in the long run, approximately 5000 people become infected.

 (a) Use this information to find a logistic function to model this situation.
 (b) Sketch a graph of your answer to part (a).
 (c) Use your graph to estimate the length of time until the rate at which people are becoming infected starts to decrease. What is the vertical coordinate at this point?

8. Table 11.10 gives the percentage, P, of households with a VCR, as a function of year.

 (a) Explain why a logistic model is a reasonable one to use for this data.
 (b) Use the data to estimate the point of inflection of P. What limiting value L does this point of inflection predict? Does this limiting value appear to be accurate given the percentages for 1990 and 1991?
 (c) The best logistic equation for this data turns out to be the following. What limiting value does this model predict?

$$P = \frac{75}{1 + 316.75e^{-0.699t}}.$$

Table 11.10 *Percentage of households with a VCR*

Year	1978	1979	1980	1981	1982	1983	1984
$P(\%)$	0.3	0.5	1.1	1.8	3.1	5.5	10.6
Year	1985	1986	1987	1988	1989	1990	1991
$P(\%)$	20.8	36.0	48.7	58.0	64.6	71.9	71.9

9. It is of considerable interest to policy makers to model the spread of information through a population. For example, various agricultural ministries use models to help them understand the spread of technical innovations or new seed types through their countries. Two models, based on how the information is spread, are given below. Assume the population is of a constant size M.

 (a) If the information is spread by mass media (TV, radio, newspapers), the rate at which information is spread is believed to be proportional to the number of people not having the information at that time. Write a differential equation for the number of people having the information by time t. Sketch a solution assuming that no one (except the mass media) has the information initially.
 (b) If the information is spread by word of mouth, the rate of spread of information is believed to be proportional to the product of the number of people who know and the number who don't. Write a differential equation for the number of people having the information by time t. Sketch the solution for the cases in which

 (i) no one

 (ii) 5% of the population

 (iii) 75% of the population

 knows initially. In each case, when is the information spreading fastest?

10. The population of a species of elk on Reading Island in Canada has been monitored for some years. When the population was 600, the relative birth rate was 35% and the relative death rate was 15%. As the population grew to 800, the corresponding figures were 30% and 20%. The island is isolated so there is no hunting or migration.

 (a) Write a differential equation to model the population as a function of time. Assume that relative growth rate is a linear function of population.
 (b) Find the equilibrium size of the population. Today there are 900 elk on Reading Island. How do you expect the population to change in the future?
 (c) Oil has been discovered on a neighboring island and the oil companies want to move 450 elk of the same species to Reading Island. What effect would this move have on the elk population on Reading Island in the future?

(d) Assuming the elk are moved to Reading Island, sketch the population on Reading Island as a function of time. Start before the elk are transferred and continue for some time into the future. Comment on the significance of your results.

11. Many organ pipes in old European churches are made of tin. In cold climates such pipes can be affected with *tin pest*, when the tin becomes brittle and crumbles into a grey powder. This transformation can appear to take place very suddenly because the presence of the grey powder encourages the reaction to proceed. At the start, when there is little grey powder, the reaction proceeds slowly. Similarly, toward the end, when there is little metallic tin left, the reaction is also slow. In between, however, when there is plenty of both metallic tin and powder, the reaction can be alarmingly fast.

Suppose that the rate of the reaction is proportional to the product of the amount of tin left and the quantity of grey powder, p, present at time t. Assume also that when metallic tin is converted to grey powder, its mass does not change.

(a) Write a differential equation for p. Let the total quantity of metallic tin present originally be B.

(b) Sketch a graph of the solution $p = f(t)$ if there is a small quantity of powder initially. How much metallic tin has crumbled when it is crumbling fastest?

(c) Suppose there is no grey powder initially. (For example, suppose the tin is completely new.) What does this model predict will happen? How do you reconcile this with the fact that many organ pipes do get tin pest?

12. An alternative method of finding the analytic solution to the logistic equation

$$\frac{dP}{dt} = kP\left(1 - \frac{P}{L}\right)$$

uses the substitution $P = 1/u$.

(a) Show that

$$\frac{dP}{dt} = -\frac{1}{u^2}\frac{du}{dt}.$$

(b) Rewrite the logistic equation in terms of u and t, and solve for u in terms of t.

(c) Using your answer to part (b), find P as a function of t.

Any population, P, for which we can ignore immigration, satisfies

$$\frac{dP}{dt} = \text{Birth rate} - \text{Death rate}.$$

For organisms which need a partner for reproduction but rely on a chance encounter for meeting a mate, the birth rate is proportional to the square of the population. Thus, the population

of such a type of organism satisfies a differential equation of the form

$$\frac{dP}{dt} = aP^2 - bP \quad \text{with } a, b > 0.$$

Problems 13–15 investigate the solutions to such an equation.

13. Consider the equation

$$\frac{dP}{dt} = 0.02P^2 - 0.08P.$$

(a) Sketch the slope field for this differential equation for $0 \le t \le 50$, $0 \le P \le 8$.

(b) Use your slope field to sketch the general shape of the solutions to the differential equation satisfying the following initial conditions:

   (i) $P(0) = 1$ (ii) $P(0) = 3$
   (iii) $P(0) = 4$ (iv) $P(0) = 5$

(c) Are there any equilibrium values of the population? If so, are they stable?

14. Consider the equation

$$\frac{dP}{dt} = P^2 - 6P.$$

(a) Sketch a graph of dP/dt against P for positive P.

(b) Use the graph you drew in part (a) to sketch the approximate shape of the solution curve with $P(0) = 5$. To do this, consider the following question. For $0 < P < 6$, is dP/dt positive or negative? What does this tell you about the graph of P against t? As you move along the solution curve with $P(0) = 5$, how does the value of dP/dt change? What does this tell you about the concavity of the graph of P against t?

(c) Use the graph you drew in part (a) to sketch the solution curve with $P(0) = 8$.

(d) Describe the qualitative differences in the behavior of populations with initial value less than 6 and initial value more than 6. Why do you think $P = 6$ is called the *threshold population*?

15. Consider a population satisfying

$$\frac{dP}{dt} = aP^2 - bP \quad \text{with constants } a, b > 0.$$

(a) Sketch a graph of dP/dt against P.

(b) Use this graph to sketch the shape of solution curves with various initial values. Use your graph from part (a) to decide where dP/dt is positive or negative, and where it is increasing or decreasing. What does this tell you about the graph of P against t?

(c) Why is $P = b/a$ called the threshold population? What happens if $P(0) = b/a$? What happens in the long-run if $P(0) > b/a$? What if $P(0) < b/a$?

11.8   SYSTEMS OF DIFFERENTIAL EQUATIONS

In Section 11.7 we modeled the growth of a single population over time. We now consider the growth of two populations which interact, such as a population of sick people infecting the healthy people around them. This involves not just one differential equation, but a system of two.

Diseases and Epidemics

Differential equations can be used to predict when an outbreak of a disease will become so severe that it is called an *epidemic*,[9] and to decide what level of vaccination is necessary to prevent an epidemic. Let's consider a specific example.

Flu in a British Boarding School

In January 1978, 763 students returned to a boys' boarding school after their winter vacation. A week later, one boy developed the flu, followed by two others the next day. By the end of the month, nearly half the boys were sick. Most of the school had been affected by the time the epidemic was over in mid-February.[10]

Being able to predict how many people will get sick, and when, is an important step toward controlling an epidemic. This is one of the responsibilities of Britain's Communicable Disease Surveillance Centre and the US's Center for Disease Control and Prevention.

The S-I-R model

We apply one of the most commonly used models for an epidemic, called the S-I-R model, to the boarding school flu example. The population of the school is divided into three groups:

S = the number of *susceptibles*, the people who are not yet sick
   but who could become sick

I = the number of *infecteds*, the people who are currently sick

R = the number of *recovered*, or *removed*, the people who have
   been sick and can no longer infect others or be reinfected.

The number of susceptibles decreases with time, as people become infected. We assume that the rate people become infected is proportional to the number of contacts between susceptible and infected people. We expect the number of contacts between the two groups to be proportional to both S and I. (If S doubles, we expect the number of contacts to double; similarly, if I doubles, we expect the number of contacts to double.) Thus we assume that the number of contacts is proportional to the product, SI. In other words, we assume that for some constant $a > 0$,

$$\frac{dS}{dt} = -\left(\begin{array}{c} \text{Rate susceptibles} \\ \text{get sick} \end{array} \right) = -aSI.$$

(The negative sign is used because S is decreasing.)

The number of infecteds is changing in two ways: newly sick people are added to the infected group, and others are removed. The newly sick people are exactly those people leaving the susceptible group and so accrue at a rate of aSI (with a positive sign this time). People leave the infected group either because they recover (or die), or because they are physically removed from the rest of

---

[9]Exactly when a disease should be called an epidemic is not always clear. The medical profession generally classifies a disease an epidemic when the frequency is higher than usually expected—leaving open the question of what is usually expected. See, for example, *Epidemiology in Medicine* by C. H. Hennekens and J. Buring (Boston: Little, Brown, 1987).

[10]Data from the Communicable Disease Surveillance Centre (UK); reported in "Influenza in a Boarding School," *British Medical Journal* March 4, 1978, and by J. D. Murray in *Mathematical Biology* (New York: Springer Verlag, 1990).

the group and can no longer infect others. We assume that people are removed at a rate proportional to the number sick, or bI, where b is a positive constant. Thus,

$$\frac{dI}{dt} = \begin{matrix} \text{Rate susceptibles} \\ \text{get sick} \end{matrix} - \begin{matrix} \text{Rate infecteds} \\ \text{get removed} \end{matrix} = aSI - bI.$$

Assuming that those who have recovered from the disease are no longer susceptible, the recovered group increases at the rate of bI, so

$$\frac{dR}{dt} = bI.$$

We are assuming that having the flu confers immunity on a person, that is, that the person cannot get the flu again. (This is true for a given strain of flu, at least in the short run.)

In analyzing the flu, we can use the fact that the total population $S + I + R$ is not changing. (The total population, the total number of boys in the school, did not change during the epidemic.) Thus, once we know S and I, we can calculate R. So we restrict our attention to the two equations

$$\frac{dS}{dt} = -aSI$$

$$\frac{dI}{dt} = aSI - bI.$$

The Constants a and b

The constant a measures how infectious the disease is—that is, how quickly it is transmitted from the infecteds to the susceptibles. In the case of the flu, we know from medical accounts that the epidemic started with one sick boy, with two more becoming sick a day later. Thus, when $I = 1$ and $S = 762$, we have $dS/dt \approx -2$, enabling us to roughly[11] approximate a:

$$a = -\frac{dS/dt}{SI} = \frac{2}{(762)(1)} = 0.0026.$$

The constant b represents the rate at which infected people are removed from the infected population. In this case of the flu, boys were generally taken to the infirmary within one or two days of becoming sick. About half the infected population was removed each day, so we take $b \approx 0.5$. Thus, our equations are:

$$\frac{dS}{dt} = -0.0026SI$$

$$\frac{dI}{dt} = 0.0026SI - 0.5I.$$

The Phase Plane

We can get a good idea of the progress of the disease from graphs. You might expect that we would look for graphs of S and I against t, and eventually we will. However, we first look at a graph of I against S. If we plot a point (S, I) representing the number of susceptibles and the number of infecteds at any moment in time, then, as the numbers of susceptibles and infecteds change, the point moves. The SI-plane on which the point moves is called the *phase plane*. The path along which the point moves is called the *phase trajectory*, or *orbit*, of the point. To find the phase trajectory, we need a differential equation relating S and I directly. Thinking of I as a function of S, and S as a function of t, we use the chain rule to get

$$\frac{dI}{dt} = \frac{dI}{dS} \cdot \frac{dS}{dt},$$

---

[11]The values of a and b are close to those obtained by J. D. Murray in *Mathematical Biology* (New York: Springer Verlag, 1990).

giving

$$\frac{dI}{dS} = \frac{dI/dt}{dS/dt}.$$

Substituting for dI/dt and dS/dt, we get

$$\frac{dI}{dS} = \frac{0.0026SI - 0.5I}{-0.0026SI}.$$

Assuming I is not zero, this equation simplifies to approximately

$$\frac{dI}{dS} = -1 + \frac{192}{S}.$$

The slope field of this differential equation is shown in Figure 11.55. The trajectory with initial condition $S_0 = 762$, $I_0 = 1$ is shown in Figure 11.56. Time is represented by the arrow showing the direction that a point moves on the trajectory. The disease starts at the point $S_0 = 762$, $I_0 = 1$. At first, more people become infected and fewer are susceptible. In other words, S decreases and I increases. Later, I decreases as S continues to decrease.

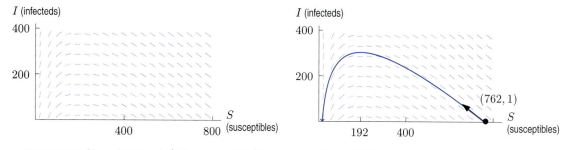

Figure 11.55: Slope field for $dI/dS = -1 + 192/S$

Figure 11.56: Trajectory for $S_0 = 762$, $I_0 = 1$

What does the SI-Phase Plane Tell Us?

To learn how the disease progresses, look at the shape of the curve in Figure 11.56. The value of I increases to about 300 (the maximum number infected and infectious at any one time); then I decreases to zero. This peak value of I occurs when $S \approx 200$. We can determine exactly when the peak value occurs by solving

$$\frac{dI}{dS} = -1 + \frac{192}{S} = 0,$$

which gives

$$S = 192.$$

Notice that the peak value for I always occurs at the same value of S, namely $S = 192$. The graph shows that if a trajectory starts with $S_0 > 192$, then I first increases and then decreases to zero. On the other hand, if $S_0 < 192$, there is no peak as I decreases right away.

> For this example, the value $S_0 = 192$ is called a *threshold value*. If S_0 is around or below 192, there is no epidemic. If S_0 is significantly greater than 192, an epidemic occurs.[12]

The phase diagram makes clear that the maximum value of I is about 300. Another question answered by the phase plane diagram is the total number of students who are expected to get sick during the epidemic. (This is not the maximum value reached by I, which gives the maximum number infected at any one time.) The point at which the trajectory crosses the S-axis represents the time when the epidemic has passed (since $I = 0$). The S-intercept shows how many boys never get the flu and thus, how many do get it.

---

[12]Here we are using J. D. Murray's definition of an epidemic as an outbreak in which the number of infecteds increases from the initial value, I_0. See *Mathematical Biology* (New York: Springer Verlag, 1990).

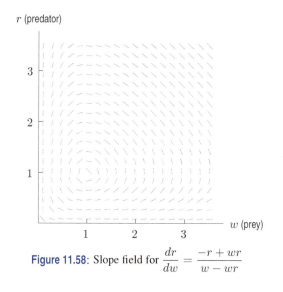

Figure 11.58: Slope field for $\dfrac{dr}{dw} = \dfrac{-r + wr}{w - wr}$

The rate of change of the robin population with respect to time is also zero:

$$\frac{dr}{dt} = -1 + (1)(1) = 0.$$

Thus dr/dw is undefined. In terms of worms and robins, this means that if at some moment $w = 1$ and $r = 1$ (that is, there are 1 million worms and 1 thousand robins), then w and r remain constant forever. The point $w = 1$, $r = 1$ is therefore an equilibrium solution. The slope field suggests that there are no other equilibrium points except the origin.

At an **equilibrium point**, both w and r are constant, so

$$\frac{dw}{dt} = 0 \qquad \text{and} \qquad \frac{dr}{dt} = 0.$$

Therefore, we look for equilibrium points by solving

$$\frac{dw}{dt} = w - wr = 0 \quad \text{and} \quad \frac{dr}{dt} = -r + rw = 0,$$

which has $w = 0, r = 0$ and $w = 1, r = 1$ as the only solutions.

Trajectories in the wr-Phase Plane

Let's look at the trajectories in the phase plane. Remember that a point on a curve represents a pair of populations (w, r) existing at the same time t (though t is not shown on the graph). A short time later, the pair of populations is represented by a nearby point. As time passes, the point traces out a trajectory. The direction is marked on the curve by an arrow. (See Figure 11.59.)

How do we figure out which way to move on the trajectory? Approximating the solution numerically shows that the trajectory is traversed counterclockwise. Alternatively, look at the original pair of differential equations. At the point P_0 in Figure 11.60, where $w > 1$ and $r = 1$,

$$\frac{dr}{dt} = -r + wr = -1 + w > 0.$$

Therefore, r is increasing, so the point is moving counterclockwise around the closed curve.

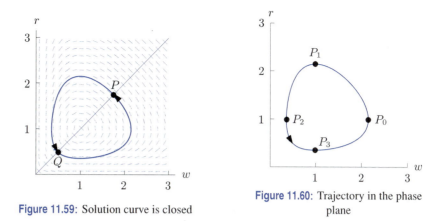

Figure 11.59: Solution curve is closed

Figure 11.60: Trajectory in the phase plane

Now let's think about why the solution curves are closed curves (that is, why they come back and meet themselves). Notice that the slope field is symmetric about the line $w = r$. We can confirm this by observing that interchanging w and r does not alter the differential equation for dr/dw. This means that if we start at point P on the line $w = r$ and travel once around the point $(1, 1)$, we arrive back at the same point P. The reason is that the second half of the path, from Q to P, is the reflection of the first half, from P to Q, in the line $w = r$. (See Figure 11.59.) If we did not end up at P again, the second half of our path would have a different shape from the first half.

The Populations as Functions of Time

The shape of the trajectories tells us how the populations vary with time. We start at $t = 0$ at the point P_0 in Figure 11.60. Then we move to P_1 at time t_1, to P_2 at time t_2, to P_3 at time t_3, and so on. At time t_4 we are back at P_0, and the whole cycle repeats. Since the trajectory is a closed curve, both populations oscillate periodically with the same period. The worms (the prey) are at their maximum a quarter of a cycle before the robins. (See Figure 11.61.)

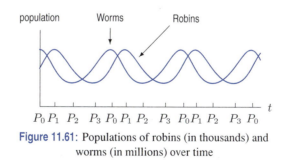

Figure 11.61: Populations of robins (in thousands) and worms (in millions) over time

Lynxes and Hares

A predator-prey system for which there are long-term data is the Canadian lynx and the hare. Since both animals were of interest to fur trappers, records of the Hudson Bay Company shed some light on their populations through much of the last century. These records show that both populations oscillated quite regularly, with a period of about ten years.

Exercises and Problems for Section 11.8

Exercises

1. Show that if S, I, and R satisfy the differential equations on page 523, the total population, $S+I+R$, is a constant.

For Exercises 2–10, let w be the number of worms (in millions) and r the number of robins (in thousands) living on an island. Suppose w and r satisfy the following differential equations, which correspond to the slope field in Figure 11.62.

$$\frac{dw}{dt} = w - wr$$
$$\frac{dr}{dt} = -r + wr.$$

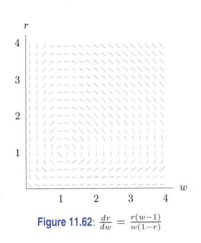

Figure 11.62: $\frac{dr}{dw} = \frac{r(w-1)}{w(1-r)}$

2. Explain why these differential equations are a reasonable model for interaction between the two populations. Why have the signs been chosen this way?

3. Solve these differential equations in the two special cases when there are no robins and when there are no worms living on the island.

4. Describe and explain the symmetry you observe in the slope field. What consequences does this symmetry have for the solution curves?

5. Assume $w = 2$ and $r = 2$ when $t = 0$. Do the numbers of robins and worms increase or decrease at first? What happens in the long run?

6. For the case discussed in Problem 5, estimate the maximum and the minimum values of the robin population. How many worms are there at the time when the robin population reaches its maximum?

7. On the same axes, graph w and r (the worm and the robin populations) against time. Use initial values of 1.5 for w and 1 for r. You may do this without units for t.

8. People on the island like robins so much that they decide to import 200 robins all the way from England, to increase the initial population to $r = 2.2$ when $t = 0$. Does this make sense? Why or why not?

9. Assume that $w = 3$ and $r = 1$ when $t = 0$. Do the numbers of robins and worms increase or decrease initially? What happens in the long run?

10. For the case discussed in Problem 9, estimate the maximum and minimum values of the robin population. Estimate the number of worms when the robin population reaches its minimum.

For Exercises 11–14, suppose x and y are the populations of two different species. Describe in words how each population changes with time.

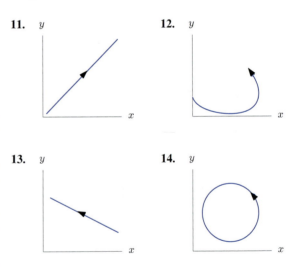

Problems

The systems of differential equations in Problems 15–17 model the interaction of two populations x and y. In each case, answer the following two questions:

(a) What kinds of interaction (symbiosis[15], competition,

predator-prey) do the equations describe?

(b) What happens in the long run? (For one of the systems, your answer will depend on the initial populations.) Use a calculator or computer to draw slope fields.

---

[15]Symbiosis takes place when the interaction of two species benefits both. An example is the pollination of plants by insects.

15. $\dfrac{1}{x}\dfrac{dx}{dt} = y - 1$

$\dfrac{1}{y}\dfrac{dy}{dt} = x - 1$

16. $\dfrac{1}{x}\dfrac{dx}{dt} = 1 - \dfrac{x}{2} - \dfrac{y}{2}$

$\dfrac{1}{y}\dfrac{dy}{dt} = 1 - x - y$

17. $\dfrac{1}{x}\dfrac{dx}{dt} = y - 1 - 0.05x$

$\dfrac{1}{y}\dfrac{dy}{dt} = 1 - x - 0.05y$

For Problems 18–20, consider a conflict between two armies of x and y soldiers, respectively. During World War I, F. W. Lanchester assumed that if both armies are fighting a conventional battle within sight of one another, the rate at which soldiers in one army are put out of action (killed or wounded) is proportional to the amount of fire the other army can concentrate on them, which is in turn proportional to the number of soldiers in the opposing army. Thus Lanchester assumed that if there are no reinforcements and t represents time since the start of the battle, then x and y obey the differential equations

$$\frac{dx}{dt} = -ay$$
$$\frac{dy}{dt} = -bx \quad a, b > 0.$$

18. Near the end of World War II a fierce battle took place between US and Japanese troops over the island of Iwo Jima, off the coast of Japan. Applying Lanchester's analysis to this battle, with x representing the number of US troops and y the number of Japanese troops, it has been estimated[16] that $a = 0.05$ and $b = 0.01$.

(a) Using these values for a and b, and ignoring reinforcements, write a differential equation involving dy/dx and sketch its slope field.

(b) Assuming that the initial strength of the US forces was 54,000 and that of the Japanese was 21,500, draw the trajectory which describes the battle. What outcome is predicted? (That is, which side do the differential equations predict will win?)

(c) Would knowing that the US in fact had 19,000 reinforcements, while the Japanese had none, alter the outcome predicted?

19. (a) For two armies of strengths x and y fighting a conventional battle governed by Lanchester's differential equations, write a differential equation involving dy/dx and the constants of attrition a and b.

(b) Solve the differential equation and hence show that the equation of the phase trajectory is

$$ay^2 - bx^2 = C$$

for some constant C. This equation is called *Lanchester's square law*. The value of C depends on the initial sizes of the two armies.

20. Consider the battle of Iwo Jima, described in Problem 18. Take $a = 0.05$, $b = 0.01$ and assume the initial strength of the US troops to be 54,000 and that of the Japanese troops to be 21,500. (Again, ignore reinforcements.)

(a) Using Lanchester's square law derived in Problem 19, find the equation of the trajectory describing the battle.

(b) Assuming that the Japanese fought without surrendering until they had all been killed, as was the case, how many US troops does this model predict would be left when the battle ended?

21. In this problem we adapt Lanchester's model for a conventional battle (described before Problem 18), to the case in which one or both of the armies is a guerrilla force. We assume that the rate at which a guerrilla force is put out of action is proportional to the product of the strengths of the two armies.

(a) Give a justification for the assumption that the rate at which a guerrilla force is put out of action is proportional to the product of the strengths of the two armies.

(b) Write the differential equations which describe a conflict between a guerrilla army of strength x and a conventional army of strength y, assuming all the constants of proportionality are 1.

(c) Find a differential equation involving dy/dx and solve it to find equations of phase trajectories.

(d) If $C > 0$, which side wins? If $C < 0$, which side wins? What if $C = 0$?

(e) Use your solution to part (d) to divide the phase plane into regions according to which side wins.

22. To model a conflict between two guerrilla armies, we assume that the rate that each one is put out of action is proportional to the product of the strengths of the two armies.

(a) Write the differential equations which describe a conflict between two guerrilla armies, of strengths x and y, respectively.

(b) Find a differential equation involving dy/dx and solve to find equations of phase trajectories.

(c) If $C > 0$, which side wins? If $C < 0$, which side wins? What if $C = 0$?

(d) Use your solution to part (b) to divide the phase plane into regions according to which side wins.

23. The concentrations of two chemicals A and B as functions of time are denoted by x and y respectively. Each alone decays at a rate proportional to its concentration. Put together, they also interact to form a third substance, at a rate proportional to the product of their concentrations. All this is expressed in the equations:

$$\frac{dx}{dt} = -2x - xy, \qquad \frac{dy}{dt} = -3y - xy.$$

[16]See Martin Braun, *Differential Equations and their Applications*, 2nd ed. (New York: Springer and Verlag, 1975).

(a) Find a differential equation describing the relationship between x and y, and solve it.

(b) Show that the only equilibrium state is $x = y = 0$. (Note that the concentrations are nonnegative.)

(c) Show that when x and y are positive and very small, y^2/x^3 is roughly constant. [Hint: When x is small, x is negligible compared to $\ln x$.]

If now the initial concentrations are $x(0) = 4$, $y(0) = 8$:

(d) Find the equation of the phase trajectory.

(e) What would be the concentrations of each substance if they become equal?

(f) If $x = e^{-10}$, find an approximate value for y.

11.9 ANALYZING THE PHASE PLANE

In the previous section we analyzed a system of differential equations using a slope field. In this section we analyze a system of differential equations using *nullclines*. We will consider two species having similar *niches*, or ways of living, and in competition for food and space. In such cases, one species often becomes extinct. This phenomenon is called the *Principle of Competitive Exclusion*. We see how differential equations predict this in a particular case.

Competitive Exclusion: Citrus Tree Parasites

The citrus farmers of Southern California are interested in controlling the insects that live on their trees. Some of these insects can be controlled by parasites which live on the trees too. Scientists are therefore interested in understanding under what circumstances these parasites flourish or die out. One such parasite was introduced accidentally from the Mediterranean; later, other parasites were introduced from China and India; in each case the previous parasite became extinct over part of its habitat. In 1963 a lab experiment was carried out to determine which one of a pair of species became extinct when they were in competition with each other. The data on one pair of species, called *A. fisheri* and *A. melinus*, with populations P_1 and P_2 respectively, is given in Table 11.11 and shows that *A. melinus* (P_2) became extinct after 8 generations.[17]

Table 11.11 *Population (in thousands) of two species of parasite as a function of time*

Generation number	1	2	3	4	5	6	7	8
Population P_1 (thousands)	0.193	1.093	1.834	5.819	13.705	16.965	18.381	16.234
Population P_2 (thousands)	0.083	0.229	0.282	0.378	0.737	0.507	0.13	0

Data from the same experimenters indicates that, when alone, each population grows logistically. In fact, their data suggests that, when alone, the population of P_1 might grow according to the equation

$$\frac{dP_1}{dt} = 0.05P_1\left(1 - \frac{P_1}{20}\right),$$

and when alone, the population of P_2 might grow according to the equation

$$\frac{dP_2}{dt} = 0.09P_2\left(1 - \frac{P_2}{15}\right).$$

Now suppose both parasites are present. Each tends to reduce the growth rate of the other, so each differential equation is modified by subtracting a term on the right. The experimental data shows that together P_1 and P_2 can be well described by the equations

$$\frac{dP_1}{dt} = 0.05P_1\left(1 - \frac{P_1}{20}\right) - 0.002P_1P_2$$

---

[17]Data adapted from Paul DeBach and Ragnhild Sundby, "Competitive Displacement between Ecological Homologues," *Hilgardia 34*:17 (1963).

$$\frac{dP_2}{dt} = 0.09P_2\left(1 - \frac{P_2}{15}\right) - 0.15P_1P_2.$$

The fact that P_2 dies out with time is reflected in these equations: the coefficient of P_1P_2 is so much larger in the equation for P_2 than in the equation for P_1. This indicates that the interaction has a much more devastating effect upon the growth of P_2 than on the growth of P_1.

The Phase Plane and Nullclines

We consider the phase plane with the P_1 axis horizontal and the P_2 axis vertical. To find the trajectories in the P_1P_2 phase plane, we could draw a slope field as in the previous section. Instead, we use a method which gives a good qualitative picture of the behavior of the trajectories even without a calculator or computer. We find the *nullclines* or curves along which $dP_1/dt = 0$ or $dP_2/dt = 0$. At points where $dP_2/dt = 0$, the population P_2 is momentarily constant, so only population P_1 is changing with time. Therefore, at this point the trajectory is horizontal. (See Figure 11.63.) Similarly, at points where $dP_1/dt = 0$, the population P_1 is momentarily constant and population P_2 is the only one changing, so the trajectory is vertical there. A point where both $dP_1/dt = 0$ and $dP_2/dt = 0$ is called an *equilibrium point* because P_1 and P_2 both remain constant if they reach these values.

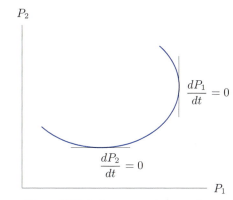

Figure 11.63: Points on a trajectory where $dP_1/dt = 0$ or $dP_2/dt = 0$

On the P_1P_2 phase plane:

- If $\dfrac{dP_1}{dt} = 0$, the trajectory is vertical.

- If $\dfrac{dP_2}{dt} = 0$, the trajectory is horizontal.

- If $\dfrac{dP_1}{dt} = \dfrac{dP_2}{dt} = 0$, there is an equilibrium point.

Using Nullclines to Analyze the Parasite Populations

In order to see where $dP_1/dt = 0$ or $dP_2/dt = 0$, we factor the right side of our differential equations:

$$\frac{dP_1}{dt} = 0.05P_1\left(1 - \frac{P_1}{20}\right) - 0.002P_1P_2 = 0.001P_1(50 - 2.5P_1 - 2P_2)$$

$$\frac{dP_2}{dt} = 0.09P_2\left(1 - \frac{P_2}{15}\right) - 0.15P_1P_2 = 0.001P_2(90 - 150P_1 - 6P_2).$$

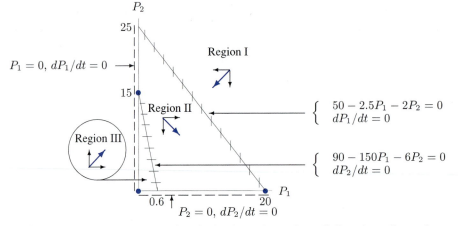

Figure 11.64: Analyzing three regions in the phase plane using nullclines (axes distorted) with equilibrium points represented by dots

Thus $dP_1/dt = 0$ where $P_1 = 0$ or where $50 - 2.5P_1 - 2P_2 = 0$. Graphing these equations in the phase plane gives two lines, which are nullclines. Since the trajectory is vertical where $dP_1/dt = 0$, in Figure 11.64 we draw small vertical line segments on these nullclines to represent the direction of the trajectories as they cross the nullcline. Similarly $dP_2/dt = 0$ where $P_2 = 0$ or where $90 - 150P_1 - 6P_2 = 0$. These equations are graphed in Figure 11.64 with small horizontal line segments on them.

The equilibrium points are where both $dP_1/dt = 0$ and $dP_2/dt = 0$, namely the points $P_1 = 0, P_2 = 0$ (meaning that both species die out); $P_1 = 0, P_2 = 15$ (where P_1 is extinct); and $P_1 = 20, P_2 = 0$ (where P_2 is extinct).

What Happens in the Regions Between the Nullclines?

Nullclines are useful because they divide the plane into regions in which the signs of dP_1/dt and dP_2/dt are constant. In each region, the direction of every trajectory remains roughly the same.

In region I, for example, we might try the point $P_1 = 20, P_2 = 25$. Then

$$\frac{dP_1}{dt} = 0.001(20)(50 - 2.5(20) - 2(25)) < 0$$

$$\frac{dP_2}{dt} = 0.001(25)(90 - 150(20) - 6(25)) < 0$$

Now $dP_1/dt < 0$, so P_1 is decreasing, which can be represented by an arrow in the direction $\leftarrow$. Also, $dP_2/dt < 0$, so P_2 is decreasing, as represented by the arrow $\downarrow$. Combining these directions, we know that the trajectories in this region go approximately in the diagonal direction ↙. (See region I in Figure 11.64.)

In region II, try, for example, $P_1 = 1, P_2 = 1$. Then we have

$$\frac{dP_1}{dt} = 0.001(1)(50 - 2.5 - 2) > 0$$

$$\frac{dP_2}{dt} = 0.001(1)(90 - 150 - 6) < 0$$

So here, P_1 is increasing while P_2 is decreasing. (See region II in Figure 11.64.)

In region III, try $P_1 = 0.1, P_2 = 0.1$:

$$\frac{dP_1}{dt} = 0.001(0.1)(50 - 2.5(0.1) - 2(0.1)) > 0$$

$$\frac{dP_2}{dt} = 0.001(0.1)(90 - 150(0.1) - 6(0.1)) > 0$$

So here, both P_1 and P_2 are increasing. (See region III in Figure 11.64.)

Notice that the behavior of the populations in each region makes biological sense. In region I both populations are so large that overpopulation is a problem, so both populations decrease. In region III both populations are so small that they are effectively not in competition, so both grow. In region II competition between the species comes into play. The fact that P_1 increases while P_2 decreases in region II means that P_1 wins.

Solution Trajectories

Suppose the system starts with some of each population. This means that the initial point of the trajectory is not on one of the axes, and so it is in region I, II, or III. Then the point moves on a trajectory like one of those computed numerically and shown in Figure 11.65. Notice that *all* these trajectories tend toward the point $P_1 = 20$, $P_2 = 0$, corresponding to a population of 20,000 for P_1 and extinction for P_2. Consequently, this model predicts that no matter what the initial populations are, provided $P_1 \neq 0$, the population of P_2 is excluded by P_1, and P_1 tends to a constant value. This makes biological sense: in the absence of P_2, we would expect P_1 to settle down to the carrying capacity of the niche, which is 20,000.

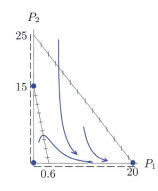

Figure 11.65: Trajectories showing exclusion of population P_2 (not to scale)

Exercises and Problems for Section 11.9

Exercises

1. The equations describing the flu epidemic in a boarding school are

$$\frac{dS}{dt} = -0.0026SI$$

$$\frac{dI}{dt} = 0.0026SI - 0.5I.$$

(a) Find the nullclines and equilibrium points in the SI phase plane.
(b) Find the direction of the trajectories in each region.
(c) Sketch some typical trajectories and describe their behavior in words.

2. Use the idea of nullclines dividing the plane into sectors to analyze the equations describing the interactions

of robins and worms.

$$\frac{dw}{dt} = w - wr$$

$$\frac{dr}{dt} = -r + rw.$$

3. (a) Find the equilibrium points for the following system of equations

$$\frac{dx}{dt} = 20x - 10xy$$

$$\frac{dy}{dt} = 25y - 5xy.$$

(b) Explain why $x = 2$, $y = 4$ is not an equilibrium point for this system.

Problems

For Problems 4–9, analyze the phase plane of the differential equations for x, $y \geq 0$. Show the nullclines and equilibrium points, and sketch the direction of the trajectories in each region.

4. $\dfrac{dx}{dt} = x(2 - x - y)$

$\dfrac{dy}{dt} = y(1 - x - y)$

5. $\dfrac{dx}{dt} = x(2 - x - 3y)$

$\dfrac{dy}{dt} = y(1 - 2x)$

6. $\dfrac{dx}{dt} = x(2 - x - 2y)$

$\dfrac{dy}{dt} = y(1 - 2x - y)$

7. $\dfrac{dx}{dt} = x\left(1 - y - \dfrac{x}{3}\right)$

$\dfrac{dy}{dt} = y\left(1 - \dfrac{y}{2} - x\right)$

8. $\dfrac{dx}{dt} = x\left(1 - x - \dfrac{y}{3}\right)$

$\dfrac{dy}{dt} = y\left(1 - y - \dfrac{x}{2}\right)$

9. $\dfrac{dx}{dt} = x\left(1 - \dfrac{x}{2} - y\right)$

$\dfrac{dy}{dt} = y\left(1 - \dfrac{y}{3} - x\right)$

10. Two companies share the market for a new technology. They have no competition except each other. Let $A(t)$ be the net worth of one company and $B(t)$ the net worth of the other at time t. Assume that net worth cannot be negative, because a company goes out of business if its net worth is zero. Suppose A and B satisfy the differential equations

$$A' = 2A - AB$$
$$B' = B - AB.$$

(a) What do these equations predict about the net worth of each company if the other were not present? What effect do the companies have on each other?

(b) Are there any equilibrium points? If so, what are they?

(c) Sketch a slope field for these equations (using a computer or calculator), and hence describe the different possible long-run behaviors.

11. In the 1930s, the Soviet ecologist G. F. Gause performed a series of experiments on competition among two yeasts with populations P_1 and P_2, respectively. By performing population studies at low density in large volumes, he determined what he called the *coefficients of geometric increase* (and we would call continuous exponential growth rates). These coefficients described the growth of each yeast alone:

$$\frac{1}{P_1}\frac{dP_1}{dt} = 0.2$$

$$\frac{1}{P_2}\frac{dP_2}{dt} = 0.06$$

where P_1 and P_2 are measured in units that Gause established.

He also determined that, in his units, the carrying capacity of P_1 was 13 and the carrying capacity of P_2 was 6. He then observed that one P_2 occupies the niche space of 3 P_1 and that one P_1 occupied the niche space of 0.4 P_2. This led him to the following differential equations to describe the interaction of P_1 and P_2:

$$\frac{dP_1}{dt} = 0.2P_1\left(\frac{13 - (P_1 + 3P_2)}{13}\right),$$

$$\frac{dP_2}{dt} = 0.06P_2\left(\frac{6 - (P_2 + 0.4P_1)}{6}\right).$$

When both yeasts were growing together, Gause recorded the data in Table 11.12.

Table 11.12 *Gause's yeast populations*

Time (hours)	6	16	24	29	48	53
P_1	0.375	3.99	4.69	6.15	7.27	8.30
P_2	0.29	0.98	1.47	1.46	1.71	1.84

(a) Carry out a phase plane analysis of Gause's equations.

(b) Mark the data points on the phase plane and describe what would have happened had Gause continued the experiment.

12. When two countries are in an arms race, the rate at which each one spends money on arms is determined by its own current level of spending and by its opponent's level of spending. We expect that the more a country is already spending on armaments, the less willing it will be to increase its military expenditures. On the other hand, the more a country's opponent spends on armaments, the more rapidly the country will arm. If $\$x$ billion is the country's yearly expenditure on arms, and $\$y$ billion is its opponent's, then the *Richardson arms race model* proposes that x and y are determined by differential equations. Suppose that for some particular arms race the equations are

$$\frac{dx}{dt} = -0.2x + 0.15y + 20$$

$$\frac{dy}{dt} = 0.1x - 0.2y + 40.$$

(a) Explain the signs of the three terms on the right side of the equations for dx/dt.

(b) Find the equilibrium points for this system.

(c) Analyze the direction of the trajectories in each region.

(d) Are the equilibrium points stable or unstable?

(e) What does this model predict will happen if both countries disarm?

(f) What does this model predict will happen in the case of unilateral disarmament (one country disarms, and the other country does not)?

(g) What does the model predict will happen in the long run?

13. In the 1930's L. F. Richardson proposed that an arms race between two countries could be modeled by a system of differential equations. One arms race which can be reasonably well described by differential equations is the US-Soviet Union arms race between 1945 and 1960. If x represents the annual Soviet expenditures on armaments (in billions of dollars) and y represents the corresponding US expenditures, it has been suggested[18] that x and y obey the following differential equations

$$\frac{dx}{dt} = -0.45x + 10.5,$$

$$\frac{dy}{dt} = 8.2x - 0.8y - 142.$$

(a) Find the nullclines and equilibrium points for these differential equations. Which direction do the trajectories go in each region?

(b) Sketch some typical trajectories in the phase plane.

(c) What do these differential equations predict will be the long term outcome of the US-Soviet arms race?

(d) Discuss these predictions in the light of the actual expenditures in Table 11.13.

Table 11.13 *Arms budgets of the United States and the Soviet Union for the years 1945–1960 (billions of dollars)*

	USSR	USA		USSR	USA
1945	14	97	1953	25.7	71.4
1946	14	80	1954	23.9	61.6
1947	15	29	1955	25.5	58.3
1948	20	20	1956	23.2	59.4
1949	20	22	1957	23.0	61.4
1950	21	23	1958	22.3	61.4
1951	22.7	49.6	1959	22.3	61.7
1952	26.0	69.6	1960	22.1	59.6

11.10   SECOND-ORDER DIFFERENTIAL EQUATIONS: OSCILLATIONS

A Second-Order Differential Equation

When a body moves freely under gravity, we know that

$$\frac{d^2 s}{dt^2} = -g,$$

where s is the height of the body above ground at time t and g is the acceleration due to gravity. To solve this equation, we first integrate to get the velocity, $v = ds/dt$:

$$\frac{ds}{dt} = -gt + v_0,$$

where v_0 is the initial velocity. Then we integrate again, giving

$$s = -\frac{1}{2}gt^2 + v_0 t + s_0,$$

where s_0 is the initial height.

The differential equation $d^2 s/dt^2 = -g$ is called *second order* because the equation contains a second derivative but no higher derivatives. The general solution to a second-order differential equation is a family of functions with two parameters, here v_0 and s_0. Finding values for the two constants corresponds to picking a particular function out of this family.

A Mass on a Spring

Not every second-order differential equation can be solved simply by integrating twice. Consider a mass m attached to the end of a spring hanging from the ceiling. We assume that the mass of the spring itself is negligible in comparison with the mass m. (See Figure 11.66.)

---

[18]R. Taagepera, G. M. Schiffler, R. T. Perkins and D. L. Wagner, *Soviet-American and Israeli-Arab Arms Races and the Richardson Model* (General Systems, XX, 1975).

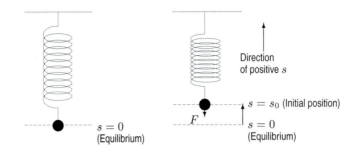

Figure 11.66: Spring and mass in equilibrium position and after upward displacement

When the system is left undisturbed, no net force acts on the mass. The force of gravity is balanced by the force the spring exerts on the mass, and the spring is in the *equilibrium position.* If we pull down on the mass, we feel a force pulling upwards. If instead, we push upward on the mass, the opposite happens: a force pushes the mass down.[19]

What happens if we push the mass upward and then release it? We expect the mass to oscillate up and down around the equilibrium position.

Springs and Hooke's Law

In order to figure out how the mass moves, we need to know the exact relationship between its displacement, s, from the equilibrium position and the net force, F, exerted on the mass. (See Figure 11.66.) We would expect that the further the mass is from the equilibrium position, and the more the spring is stretched or compressed, the larger the force. In fact, provided that the displacement is not large enough to deform the spring permanently, experiments show that the net force, F, is approximately proportional to the displacement, s:

$$F = -ks,$$

where k is the *spring constant* $(k > 0)$ and the negative sign means that the net force is in the opposite direction to the displacement. The value of k depends on the physical properties of the particular spring. This relationship is known as *Hooke's Law.* Suppose we push the mass upward some distance and then release it. After we let go the net force causes the mass to accelerate toward the equilibrium position. By Newton's Second Law of Motion we have

$$\text{Force} = \text{Mass} \times \text{Acceleration}.$$

Since acceleration is d^2s/dt^2 and force is $F = -ks$ by Hooke's law, we have

$$-ks = m\frac{d^2 s}{dt^2},$$

which is equivalent to the following differential equation.

Equation for Oscillations of a Mass on a Spring

$$\frac{d^2 s}{dt^2} = -\frac{k}{m}s$$

Thus, the motion of the mass is described by a second-order differential equation. Since we expect the mass to oscillate, we guess that the solution to this equation involves trigonometric functions.

---

[19]Pulling down on the mass stretches the spring, increasing the tension, so the combination of gravity and spring force is upward. Pushing up the spring decreases tension in the spring, so the combination of gravity and spring force is downward.

Solving the Differential Equation by Guess-and-Check

Guessing is the tried-and-true method for solving differential equations. It may surprise you to learn that there is no systematic method for solving most differential equations analytically, so guesswork is often extremely important.

Example 1   By guessing, find the general solution to the equation

$$\frac{d^2 s}{dt^2} = -s.$$

Solution   We want to find functions whose second derivative is the negative of the original function. We are already familiar with two functions that have this property: $s(t) = \cos t$ and $s(t) = \sin t$. We check that they are solutions by substituting:

$$\frac{d^2}{dt^2}(\cos t) = \frac{d}{dt}(-\sin t) = -\cos t,$$

and

$$\frac{d^2}{dt^2}(\sin t) = \frac{d}{dt}(\cos t) = -\sin t.$$

The remarkable thing is that starting from these two particular solutions, we can build up *all* the solutions to our equation. Given two constants C_1 and C_2, the function

$$s(t) = C_1 \cos t + C_2 \sin t$$

satisfies the differential equation, since

$$\frac{d^2}{dt^2}(C_1 \cos t + C_2 \sin t) = \frac{d}{dt}(-C_1 \sin t + C_2 \cos t)$$

$$= -C_1 \cos t - C_2 \sin t$$

$$= -(C_1 \cos t + C_2 \sin t).$$

It can be shown (though we will not do so) that $s(t) = C_1 \cos t + C_2 \sin t$ is the most general form of the solution. As expected, it contains two constants, C_1 and C_2.

If the differential equation represents a physical problem, then C_1 and C_2 can often be computed, as shown in the next example.

Example 2   If the mass is displaced by a distance of s_0 and then released, find the solution to

$$\frac{d^2 s}{dt^2} = -s$$

Solution   The position of the mass is given by the equation

$$s(t) = C_1 \cos t + C_2 \sin t.$$

We also know that the initial position is s_0; thus,

$$s(0) = C_1 \cos 0 + C_2 \sin 0 = C_1 \cdot 1 + C_2 \cdot 0 = s_0,$$

so $C_1 = s_0$, the initial displacement. What is C_2? To find it, we use the fact that at $t = 0$, when the mass has just been released, its velocity is 0. Velocity is the derivative of the displacement, so

$$\left.\frac{ds}{dt}\right|_{t=0} = (-C_1 \sin t + C_2 \cos t)\Big|_{t=0} = -C_1 \cdot 0 + C_2 \cdot 1 = 0,$$

so $C_2 = 0$. Therefore the solution is $s = s_0 \cos t$.

Solution to the General Spring Equation

Having found the general solution to the equation $d^2 s/dt^2 = -s$, let us return to the equation

$$\frac{d^2 s}{dt^2} = -\frac{k}{m}s.$$

We write $\omega = \sqrt{k/m}$ (why will be clear in a moment), so the differential equation becomes

$$\frac{d^2 s}{dt^2} = -\omega^2 s.$$

This equation no longer has $\sin t$ and $\cos t$ as solutions, since

$$\frac{d^2}{dt^2}(\sin t) = -\sin t \neq -\omega^2 \sin t.$$

However, since we want a factor of ω^2 after differentiating twice, the chain rule leads us to guess that $\sin \omega t$ may be a solution. Checking this:

$$\frac{d^2}{dt^2}(\sin \omega t) = \frac{d}{dt}(\omega \cos \omega t) = -\omega^2 \sin \omega t,$$

we see that $\sin \omega t$ is a solution, and you can check that $\cos \omega t$ is a solution, too.

The general solution to the equation

$$\frac{d^2 s}{dt^2} + \omega^2 s = 0$$

is of the form

$$s(t) = C_1 \cos \omega t + C_2 \sin \omega t,$$

where C_1 and C_2 are arbitrary constants. (We assume $\omega > 0$.) The period of this oscillation is

$$T = \frac{2\pi}{\omega}.$$

Such oscillations are called **simple harmonic motion**.

The solution to our original equation, $\dfrac{d^2 s}{dt^2} + \dfrac{k}{m}s = 0$, is thus $s = C_1 \cos \sqrt{\frac{k}{m}}t + C_2 \sin \sqrt{\frac{k}{m}}t$.

Initial-Value and Boundary-Value Problems

A problem in which the initial position and the initial velocity are used to determine the particular solution is called an *initial-value problem*. (See Example 2.) Alternatively, we may be given the position at two known times. Such a problem is known as a *boundary-value problem*.

Example 3   Find a solution to the differential equation satisfying each set of conditions below

$$\frac{d^2 s}{dt^2} + 4s = 0.$$

(a) The boundary conditions $s(0) = 0$, $s(\pi/4) = 20$.
(b) The initial conditions $s(0) = 1$, $s'(0) = -6$.

Solution   Since $\omega^2 = 4$, $\omega = 2$, the general solution to the differential equation is

$$s(t) = C_1 \cos 2t + C_2 \sin 2t.$$

(a) Substituting the boundary condition $s(0) = 0$ into the general solution gives

$$s(0) = C_1 \cos(2 \cdot 0) + C_2 \sin(2 \cdot 0) = C_1 \cdot 1 + C_2 \cdot 0 = C_1 = 0.$$

Thus $s(t)$ must have the form $s(t) = C_2 \sin 2t$. The second condition yields the value of C_2:

$$s\left(\frac{\pi}{4}\right) = C_2 \sin\left(2 \cdot \frac{\pi}{4}\right) = C_2 = 20.$$

Therefore, the solution satisfying the boundary conditions is

$$s(t) = 20 \sin 2t.$$

(b) For the initial-value problem, we start from the same general solution: $s(t) = C_1 \cos 2t + C_2 \sin 2t$. Substituting 0 for t once again, we find

$$s(0) = C_1 \cos(2 \cdot 0) + C_2 \sin(2 \cdot 0) = C_1 = 1.$$

Differentiating $s(t) = \cos 2t + C_2 \sin 2t$ gives

$$s'(t) = -2 \sin 2t + 2C_2 \cos 2t,$$

and applying the second initial condition gives us C_2:

$$s'(0) = -2 \sin(2 \cdot 0) + 2C_2 \cos(2 \cdot 0) = 2C_2 = -6,$$

so $C_2 = -3$ and our solution is

$$s(t) = \cos 2t - 3 \sin 2t.$$

What do the Graphs of Our Solutions Look Like?

Since the general solution of the equation $d^2 s/dt^2 + \omega^2 s = 0$ is of the form

$$s(t) = C_1 \cos \omega t + C_2 \sin \omega t$$

it would be useful to know what the graph of such a sum of sines and cosines looks like. We start with the example $s(t) = \cos t + \sin t$, which is graphed in Figure 11.67.

Interestingly, the graph in Figure 11.67 looks like another sine function, and in fact it is one. If we measure the graph carefully, we find that its period is 2π, the same as $\sin t$ and $\cos t$, but the amplitude is approximately 1.414 (in fact, it's $\sqrt{2}$), and the graph is shifted along the t-axis (in fact by $\pi/4$). If we plot $C_1 \cos t + C_2 \sin t$ for any C_1 and C_2, the resulting graph is always a sine function with period 2π, though the amplitude and shift can vary. For example, the graph of $s = 6 \cos t - 8 \sin t$ is in Figure 11.68; it has period 2π.

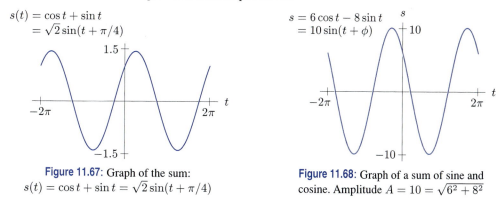

Figure 11.67: Graph of the sum:
$s(t) = \cos t + \sin t = \sqrt{2} \sin(t + \pi/4)$

Figure 11.68: Graph of a sum of sine and cosine. Amplitude $A = 10 = \sqrt{6^2 + 8^2}$

These graphs suggest that we can write the sum of a sine and a cosine of the same argument as one single sine function.[20] Problem 23 shows that this can always be done using the following relations.

---

[20]The sum of a sine and a cosine can also be written as a single cosine function.

If $C_1 \cos \omega t + C_2 \sin \omega t = A \sin(\omega t + \phi)$, the *amplitude*, A, is given by

$$A = \sqrt{C_1^2 + C_2^2}.$$

The angle ϕ is called the *phase shift* and satisfies

$$\tan \phi = \frac{C_1}{C_2}.$$

We choose ϕ in $(-\pi, \pi]$ such that if $C_1 > 0$, ϕ is positive, and if $C_1 < 0$, ϕ is negative.

The reason that it is often useful to write the solution to the differential equation as a single sine function $A \sin(\omega t + \phi)$, as opposed to the sum of a sine and a cosine, is that the amplitude A and the phase shift ϕ are easier to recognize in this form.

Warning: Phase Shift Is Not the Same as Horizontal Translation

Let's look at $s = \sin(3t + \pi/2)$. If we rewrite this as $s = \sin(3(t + \pi/6))$, we can see from Figure 11.69 that the graph of this function is the graph of $s = \sin 3t$ shifted to the left a distance of $\pi/6$. (Remember that replacing x by $x - 2$ shifts a graph by 2 to the right.) But $\pi/6$ is *not* the phase shift; the phase shift[21] is $\pi/2$. From the point of view of a scientist, the important question is often not the distance the curve has shifted, but the relation between the distance shifted and the period.

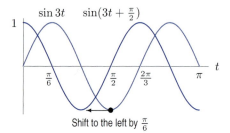

Figure 11.69: Phase shift is $\frac{\pi}{2}$; horizontal translation is $\frac{\pi}{6}$

Exercises and Problems for Section 11.10

Exercises

1. Check by differentiation that $y = 2 \cos t + 3 \sin t$ is a solution to $y'' + y = 0$.

2. Check by differentiation that $y(t) = 3 \sin 2t + 2 \cos 2t$ is a solution of $y'' + 4y = 0$.

3. Check by differentiation that $y = A \cos t + B \sin t$ is a solution to $y'' + y = 0$ for any constants A and B.

4. Check by differentiation that $y(t) = A \sin 2t + B \cos 2t$ is a solution of $y'' + 4y = 0$ for all values of A and B.

5. Check by differentiation that $y(t) = A \sin \omega t + B \cos \omega t$ is a solution of $y'' + \omega^2 y = 0$ for all values of A and B.

6. What values of α and A make $y = A \cos \alpha t$ a solution to $y'' + 5y = 0$ such that $y'(1) = 3$?

7. (a) Using a calculator or computer, graph $s = 4 \cos t + 3 \sin t$ for $-2\pi \leq t \leq 2\pi$.
   (b) If $4 \cos t + 3 \sin t = A \sin(t + \phi)$, use your graph to estimate the values of A and ϕ.
   (c) Calculate A and ϕ analytically.

8. Write the function $s(t) = \cos t - \sin t$ as a single sine function. Draw its graph.

9. Find the amplitude of $3 \sin 2t + 7 \cos 2t$.

10. Find the amplitude of the oscillation $y = 3 \sin 2t + 4 \cos x2t$.

11. Write $y(t) = 5 \sin(2t) + 12 \cos(2t)$ in the form $y(t) = A \sin(\omega t + \psi)$.

12. Write $7 \sin \omega t + 24 \cos \omega t$ in the form $A \sin(\omega t + \phi)$.

---

[21]This definition of phase shift is the one usually used in the sciences.

Problems

The functions in Problems 13–15 describe the motion of a mass on a spring satisfying the differential equation $y'' = -9y$, where y is the displacement of the mass from the equilibrium position at time t, with upwards as positive. In each case, describe in words how the motion starts when $t = 0$. For example, is the mass at the highest point, the lowest point, or in the middle? Is it moving up or down or is it at rest?

13. $y = 2\cos 3t$ **14.** $y = -0.5\sin 3t$

15. $y = -\cos 3t$

16. **(a)** Find the general solution of the differential equation

$$y'' + 9y = 0.$$

   (b) For each of the following initial conditions, find a particular solution.
   (i) $y(0) = 0$, $y'(0) = 1$
   (ii) $y(0) = 1$, $y'(0) = 0$
   (iii) $y(0) = 1$, $y(1) = 0$
   (iv) $y(0) = 0$, $y(1) = 1$

   (c) Sketch a graph of the solutions found in part b).

17. Each graph in Figure 11.70 represents a solution to one of the differential equations:
   (a) $x'' + x = 0$,   **(b)** $x'' + 4x = 0$,
   (c) $x'' + 16x = 0$.

Assuming the t-scales on the four graphs are the same, which graph represents a solution to which equation? Find an equation for each graph.

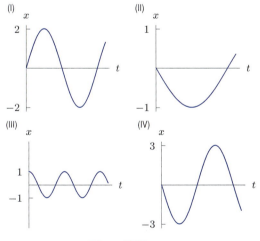

Figure 11.70

18. The following differential equations represent oscillating springs.

   (i) $s'' + 4s = 0$ $s(0) = 5$, $s'(0) = 0$
   (ii) $4s'' + s = 0$ $s(0) = 10$, $s'(0) = 0$
   (iii) $s'' + 6s = 0$ $s(0) = 4$, $s'(0) = 0$
   (iv) $6s'' + s = 0$ $s(0) = 20$, $s'(0) = 0$

Which differential equation represents:

   (a) The spring oscillating most quickly (with the shortest period)?
   (b) The spring oscillating with the largest amplitude?
   (c) The spring oscillating most slowly (with the longest period)?
   (d) The spring with largest maximum velocity?

19. A pendulum of length l makes an angle of x (radians) with the vertical (see Figure 11.71). When x is small, it can be shown that, approximately:

$$\frac{d^2 x}{dt^2} = -\frac{g}{l}x,$$

where g is the acceleration due to gravity.

Figure 11.71

   (a) Solve this equation assuming that $x(0) = 0$ and $x'(0) = v_0$.
   (b) Solve this equation assuming that the pendulum is let go from the position where $x = x_0$. ("Let go" means that the velocity of the pendulum is zero when $x = x_0$. Measure t from the moment when the pendulum is let go.)

20. Look at the pendulum motion in Problem 19. What effect does it have on x as a function of time if:

   (a) x_0 is increased?   **(b)** l is increased?

21. A brick of mass 3 kg hangs from the end of a spring. When the brick is at rest, the spring is stretched by 2 cm. The spring is then stretched an additional 5 cm and released. Assume there is no air resistance.

   (a) Set up a differential equation with initial conditions describing the motion.
   (b) Solve the differential equation.

22. Consider the motion described by the differential equations:

   (a) $x'' + 16x = 0$, $x(0) = 5$, $x'(0) = 0$,
   (b) $25x'' + x = 0$, $x(0) = -1$, $x'(0) = 2$.

In each case, find a formula for $x(t)$ and calculate the amplitude and period of the motion.

23. **(a)** Expand $A\sin(\omega t + \phi)$ using the trigonometric identity $\sin(x + y) = \sin x \cos y + \cos x \sin y$.
   (b) Assume $A > 0$. If $A\sin(\omega t + \phi) = C_1 \cos \omega t + C_2 \sin \omega t$, show that we must have

$$A = \sqrt{C_1^2 + C_2^2} \quad \text{and} \quad \tan \phi = C_1/C_2.$$

Problems 24–26 concern the electric circuit in Figure 11.72. A charged capacitor connected to an inductor causes a current to flow through the inductor until the capacitor is fully discharged. The current in the inductor, in turn, charges up the capacitor until the capacitor is fully charged again. If $Q(t)$ is the charge on the capacitor at time t, and I is the current, then

$$I = \frac{dQ}{dt}.$$

If the circuit resistance is zero, then the charge Q and the current I in the circuit satisfy the differential equation

$$L\frac{dI}{dt} + \frac{Q}{C} = 0,$$

where C is the capacitance, and L is the inductance, so

$$L\frac{d^2Q}{dt^2} + \frac{Q}{C} = 0.$$

The unit of charge is the coulomb, the unit of capacitance the farad, the unit of inductance the henry, the unit of current is the ampere, and time is measured in seconds.

Capacitor

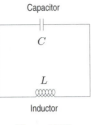

Inductor

Figure 11.72

24. If $L = 36$ henry and $C = 9$ farad, find a formula for $Q(t)$ if
    (a) $Q(0) = 0$ $I(0) = 2$
    (b) $Q(0) = 6$ $I(0) = 0$

25. Suppose $Q(0) = 0$, $Q'(0) = I(0) = 4$, and the maximum possible charge is $2\sqrt{2}$ coulombs. What is the capacitance if the inductance is 10 henry?

26. What happens to the charge and current as t goes to infinity? What does it mean that the charge and current are sometimes positive and sometimes negative?

11.11 LINEAR SECOND-ORDER DIFFERENTIAL EQUATIONS

A Spring with Friction: Damped Oscillations

The differential equation $d^2s/dt^2 = -(k/m)s$, which we used to describe the motion of a spring, disregards friction. But there is friction in every real system. For a mass on a spring, the frictional force from air resistance increases with the velocity of the mass. The frictional force is often approximately proportional to velocity, and so we introduce a *damping term* of the form $a(ds/dt)$, where a is a constant called the *damping coefficient* and ds/dt is the velocity of the mass.

Remember that without damping, the differential equation was obtained from

$$\text{Force} = \text{Mass} \cdot \text{Acceleration}.$$

With damping, the spring force $-ks$ is replaced by $-ks - a(ds/dt)$, where a is positive and the $a(ds/dt)$ term is subtracted because the frictional force is in the direction opposite to the motion. The new differential equation is therefore

$$-ks - a\frac{ds}{dt} = m\frac{d^2s}{dt^2}$$

which is equivalent to the following differential equation:

Equation for Damped Oscillations of a Spring

$$\frac{d^2s}{dt^2} + \frac{a}{m}\frac{ds}{dt} + \frac{k}{m}s = 0$$

We expect the solution to this equation to die away with time, as friction brings the motion to a stop.

The General Solution to a Linear Differential Equation

The equation for damped oscillations is an example of a *linear second-order differential equation with constant coefficients*. This section gives an analytic method of solving the equation,

$$\frac{d^2y}{dt^2} + b\frac{dy}{dt} + cy = 0,$$

for constant b and c. As we have seen for the spring equation, if $f_1(t)$ and $f_2(t)$ satisfy the differential equation, then the *principle of superposition* says that, for any constants C_1 and C_2, the function

$$y(t) = C_1 f_1(t) + C_2 f_2(t)$$

is also a solution. It can be shown that the general solution is of this form, provided $f_1(t)$ is not a multiple of $f_2(t)$.

Finding Solutions: The Characteristic Equation

We now use complex numbers to solve the differential equation

$$\frac{d^2y}{dt^2} + b\frac{dy}{dt} + cy = 0.$$

The method is a form of guess-and-check. We ask what kind of function might satisfy a differential equation in which the second derivative d^2y/dt^2 is a sum of multiples of dy/dt and y. One possibility is an exponential function, so we try to find a solution of the form:

$$y = Ce^{rt},$$

where r may be a complex number.[22] To find r, we substitute into the differential equation:

$$\frac{d^2y}{dt^2} + b\frac{dy}{dt} + cy = r^2Ce^{rt} + b \cdot rCe^{rt} + c \cdot Ce^{rt} = Ce^{rt}(r^2 + br + c) = 0.$$

We can divide by $y = Ce^{rt}$ provided $C \neq 0$, because the exponential function is never zero. If $C = 0$, then $y = 0$, which is not a very interesting solution (though it is a solution). So we assume $C \neq 0$. Then $y = Ce^{rt}$ is a solution to the differential equation if

$$r^2 + br + c = 0.$$

This quadratic is called the *characteristic equation* of the differential equation. Its solutions are

$$r = -\frac{1}{2}b \pm \frac{1}{2}\sqrt{b^2 - 4c}.$$

There are three different types of solutions to the differential equation, depending on whether the solutions to the characteristic equation are real and distinct, complex, or repeated. The sign of $b^2 - 4c$ determines the type of solutions.

The Case with $b^2 - 4c > 0$

There are two real solutions r_1 and r_2 to the characteristic equation, and the following two functions satisfy the differential equation:

$$C_1e^{r_1t} \quad \text{and} \quad C_2e^{r_2t}.$$

The sum of these two solutions is the general solution to the differential equation:

---

[22]See Appendix C on complex numbers.

If $b^2 - 4c > 0$, the general solution to

$$\frac{d^2y}{dt^2} + b\frac{dy}{dt} + cy = 0$$

is

$$y(t) = C_1 e^{r_1 t} + C_2 e^{r_2 t}$$

where r_1 and r_2 are the solutions to the characteristic equation.
If $r_1 < 0$ and $r_2 < 0$, the motion is called **overdamped**.

A physical system satisfying a differential equation of this type is said to be overdamped because it occurs when there is a lot of friction in the system. For example, a spring moving in a thick fluid such as oil or molasses is overdamped: it will not oscillate.

Example 1   A spring is placed in oil, where it satisfies the differential equation

$$\frac{d^2s}{dt^2} + 3\frac{ds}{dt} + 2s = 0.$$

Solve this equation with the initial conditions $s = -0.5$ and $ds/dt = 3$ when $t = 0$.

Solution   The characteristic equation is

$$r^2 + 3r + 2 = 0,$$

with solutions $r = -1$ and $r = -2$, so the general solution to the differential equation is

$$s(t) = C_1 e^{-t} + C_2 e^{-2t}.$$

We use the initial conditions to find C_1 and C_2. At $t = 0$, we have

$$s = C_1 e^{-0} + C_2 e^{-2(0)} = C_1 + C_2 = -0.5.$$

Furthermore, since $ds/dt = -C_1 e^{-t} - 2C_2 e^{-2t}$, we have

$$\left.\frac{ds}{dt}\right|_{t=0} = -C_1 e^{-0} - 2C_2 e^{-2(0)} = -C_1 - 2C_2 = 3.$$

Solving these equations simultaneously, we find $C_1 = 2$ and $C_2 = -2.5$, so that the solution is

$$s(t) = 2e^{-t} - 2.5e^{-2t}.$$

The graph of this function is in Figure 11.73. The mass is so slowed by the oil that it passes through the equilibrium point only once (when $t \approx 1/4$) and for all practical purposes, it comes to rest after a short time. The motion has been "damped out" by the oil.

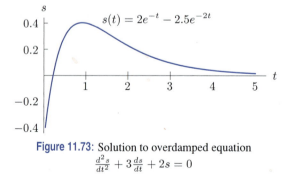

Figure 11.73: Solution to overdamped equation
$$\frac{d^2s}{dt^2} + 3\frac{ds}{dt} + 2s = 0$$

The Case with $b^2 - 4c = 0$

In this case, the characteristic equation has only one solution, $r = -b/2$. By substitution, we can check that both $y = e^{-bt/2}$ and $y = te^{-bt/2}$ are solutions.

If $b^2 - 4c = 0$,

$$\frac{d^2y}{dt^2} + b\frac{dy}{dt} + cy = 0$$

has general solution

$$y(t) = (C_1 t + C_2)e^{-bt/2}.$$

If $b > 0$, the system is said to be **critically damped**.

The Case with $b^2 - 4c < 0$

In this case, the characteristic equation has complex roots. Using Euler's formula,[23] these complex roots lead to trigonometric functions which represent oscillations.

Example 2   An object of mass $m = 10$ kg is attached to a spring with spring constant $k = 20$ kg/sec^2, and the object experiences a frictional force proportional to the velocity, with constant of proportionality $a = 20$ kg/sec. At time $t = 0$, the object is released from rest 2 meters above the equilibrium position. Write the differential equation that describes the motion.

Solution   The differential equation that describes the motion can be obtained from the following general expression for the damped motion of a spring:

$$\underbrace{m\frac{d^2s}{dt^2}}_{\text{Mass·Acceleration}} = \underbrace{F}_{\text{Net force}} = -\underbrace{ks}_{\text{Spring force}} - \underbrace{a\frac{ds}{dt}}_{\text{Frictional force}}.$$

Substituting $m = 10$ kg, $k = 20$ kg/sec^2, and $a = 20$ kg/sec, we obtain the differential equation:

$$\frac{d^2s}{dt^2} + 2\frac{ds}{dt} + 2s = 0.$$

At $t = 0$, the object is at rest 2 meters above equilibrium, so the initial conditions are $s(0) = 2$ and $s'(0) = 0$, where s is in meters and t in seconds.

Notice that this is the same differential equation as in Example 1 except that the coefficient of ds/dt has decreased from 3 to 2, which means that the frictional force has been reduced. This time, the roots of the characteristic equation have imaginary parts which lead to oscillations.

Example 3   Solve the differential equation

$$\frac{d^2s}{dt^2} + 2\frac{ds}{dt} + 2s = 0,$$

subject to $s(0) = 2$, $s'(0) = 0$.

Solution   The characteristic equation is

$$r^2 + 2r + 2 = 0 \quad \text{giving} \quad r = -1 \pm i.$$

The solution to the differential equation is

$$s(t) = A_1 e^{(-1+i)t} + A_2 e^{(-1-i)t},$$

where A_1 and A_2 are arbitrary complex numbers. The initial condition $s(0) = 2$ gives

$$2 = A_1 e^{(-1+i)\cdot 0} + A_2 e^{(-1-i)\cdot 0} = A_1 + A_2.$$

---

[23]See Appendix C on complex numbers.

Also,

$$s'(t) = A_1(-1+i)e^{(-1+i)t} + A_2(-1-i)e^{(-1-i)t},$$

so $s'(0) = 0$ gives

$$0 = A_1(-1+i) + A_2(-1-i).$$

Solving the simultaneous equations for A_1 and A_2 gives (after some algebra)

$$A_1 = 1 - i \quad \text{and} \quad A_2 = 1 + i.$$

The solution is therefore

$$s(t) = (1-i)e^{(-1+i)t} + (1+i)e^{(-1-i)t} = (1-i)e^{-t}e^{it} + (1+i)e^{-t}e^{-it}.$$

Using Euler's formula, $e^{it} = \cos t + i \sin t$ and $e^{-it} = \cos t - i \sin t$, we get

$$s(t) = (1-i)e^{-t}(\cos t + i \sin t) + (1+i)e^{-t}(\cos t - i \sin t).$$

Multiplying out and simplifying, all the complex terms drop out, giving

$$\begin{aligned} s(t) &= e^{-t}\cos t + ie^{-t}\sin t - ie^{-t}\cos t + e^{-t}\sin t \\ &\quad + e^{-t}\cos t - ie^{-t}\sin t + ie^{-t}\cos t + e^{-t}\sin t \\ &= 2e^{-t}\cos t + 2e^{-t}\sin t. \end{aligned}$$

The $\cos t$ and $\sin t$ terms tell us that the solution oscillates; the factor of e^{-t} tells us that the oscillations are damped. See Figure 11.74. However, the period of the oscillations does not change as the amplitude decreases. This is why a spring-driven clock can keep accurate time even as it is running down.

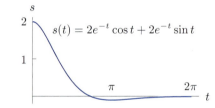

Figure 11.74: Solution to underdamped equation $\frac{d^2s}{dt^2} + 2\frac{ds}{dt} + 2s = 0$

In Example 3, the coefficients A_1 and A_2 are complex, but the solution, $s(t)$, is real. (We expect this, since $s(t)$ represents a real displacement.) In general, provided the coefficients b and c in the original differential equation and the initial values are real, the solution is always real. The coefficients A_1 and A_2 are always complex conjugates (that is, of the form $\alpha \pm i\beta$).

If $b^2 - 4c < 0$, to solve

$$\frac{d^2y}{dt^2} + b\frac{dy}{dt} + cy = 0,$$

- Find the solutions $r = \alpha \pm i\beta$ to the characteristic equation $r^2 + br + c = 0$.
- The general solution to the differential equation is, for some real C_1 and C_2,

$$y = C_1 e^{\alpha t} \cos \beta t + C_2 e^{\alpha t} \sin \beta t.$$

If $\alpha < 0$, such oscillations are called **underdamped**. If $\alpha = 0$, the oscillations are **undamped**.

Example 4   Find the general solution of the equations   (a) $y'' = 9y$ (b) $y'' = -9y$.

Solution   (a) The characteristic equation is $r^2 - 9 = 0$, so $r = \pm 3$. Thus the general solution is

$$y = C_1 e^{3t} + C_2 e^{-3t}.$$

(b) The characteristic equation is $r^2 + 9 = 0$, so $r = 0 \pm 3i$. The general solution is

$$y = C_1 e^{0t} \cos 3t + C_2 e^{0t} \sin 3t = C_1 \cos 3t + C_2 \sin 3t.$$

Notice that we have seen the solution to this equation in Section 11.10; it's the equation of undamped simple harmonic motion.

Example 5   Solve the initial value problem

$$y'' + 4y' + 13y = 0, \quad y(0) = 0, \, y'(0) = 30.$$

Solution   We solve the characteristic equation

$$r^2 + 4r + 13 = 0, \quad \text{getting} \quad r = -2 \pm 3i.$$

The general solution to the differential equation is

$$y = C_1 e^{-2t} \cos 3t + C_2 e^{-2t} \sin 3t.$$

Substituting $t = 0$ gives

$$y(0) = C_1 \cdot 1 \cdot 1 + C_2 \cdot 1 \cdot 0 = 0, \quad \text{so} \quad C_1 = 0.$$

Differentiating $y(t) = C_2 e^{-2t} \sin 3t$ gives

$$y'(t) = C_2(-2e^{-2t} \sin 3t + 3e^{-2t} \cos 3t).$$

Substituting $t = 0$ gives

$$y'(0) = C_2(-2 \cdot 1 \cdot 0 + 3 \cdot 1 \cdot 1) = 3C_2 = 30 \quad \text{so} \quad C_2 = 10.$$

The solution is therefore

$$y(t) = 10e^{-2t} \sin 3t.$$

Summary of Solutions to $y'' + by' + cy = 0$

If $b^2 - 4c > 0$, then $y = C_1 e^{r_1 t} + C_2 e^{r_2 t}$

If $b^2 - 4c = 0$, then $y = (C_1 t + C_2)e^{-bt/2}$

If $b^2 - 4c < 0$, then $y = C_1 e^{\alpha t} \cos \beta t + C_2 e^{\alpha t} \sin \beta t$

Exercises and Problems for Section 11.11

Exercises

For Exercises 1–12, find the general solution to the given differential equation.

1. $y'' + 4y' + 3y = 0$

2. $y'' + 4y' + 4y = 0$

3. $y'' + 4y' + 5y = 0$

4. $s'' - 7s = 0$

5. $s'' + 7s = 0$

6. $y'' - 3y' + 2y = 0$

7. $4z'' + 8z' + 3z = 0$

8. $\dfrac{d^2 x}{dt^2} + 4\dfrac{dx}{dt} + 8x = 0$

9. $\dfrac{d^2 p}{dt^2} + \dfrac{dp}{dt} + p = 0$

10. $z'' + 2z = 0$

11. $z'' + 2z' = 0$

12. $P'' + 2P' + P = 0$

For Exercises 13–20, solve the initial value problem.

13. $y'' + 5y' + 6y = 0$, $y(0) = 1, y'(0) = 0$.

14. $y'' + 5y' + 6y = 0$, $y(0) = 5, y'(0) = 1$.

15. $y'' - 3y' - 4y = 0$, $y(0) = 1, y'(0) = 0$.

16. $y'' - 3y' - 4y = 0$, $y(0) = 0, y'(0) = 0.5$.

17. $y'' + 6y' + 5y = 0$, $y(0) = 1$, $y'(0) = 0$

18. $y'' + 6y' + 5y = 0$, $y(0) = 5$, $y'(0) = 5$

19. $y'' + 6y' + 10y = 0$, $y(0) = 0$, $y'(0) = 2$

20. $y'' + 6y' + 10y = 0$, $y(0) = 0$, $y'(0) = 0$

For Exercises 21–24, solve the boundary value problem.

21. $y'' + 5y' + 6y = 0$, $y(0) = 1, y(1) = 0$.

22. $y'' + 5y' + 6y = 0$, $y(-2) = 0, y(2) = 3$.

23. $p'' + 2p' + 2p = 0$, $p(0) = 0$, $p(\pi/2) = 20$

24. $p'' + 4p' + 5p = 0$, $p(0) = 1$, $p(\pi/2) = 5$

Problems

25. Match the graphs of solutions in Figure 11.75 with the differential equations below.

(a) $x'' + 4x = 0$
(b) $x'' - 4x = 0$
(c) $x'' - 0.2x' + 1.01x = 0$
(d) $x'' + 0.2x' + 1.01x = 0$

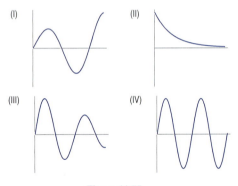

Figure 11.75

26. Match the differential equations to the solution graphs (I)-(IV). Use each graph only once.

(a) $y'' + 5y' + 6y = 0$ (b) $y'' + y' - 6y = 0$
(c) $y'' + 4y' + 9y = 0$ (d) $y'' = -9y$

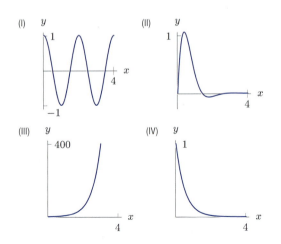

27. If $y = e^{2t}$ is a solution to the differential equation

$$\frac{d^2y}{dt^2} - 5\frac{dy}{dt} + ky = 0,$$

find the value of the constant k and the general solution to this equation.

28. Assuming $b, c > 0$, explain how you know that the solutions of an underdamped differential equation must go to 0 as $t \to \infty$.

Each of the differential equations below represents the position of a 1 gram mass oscillating on the end of a damped spring. For Problems 29–33 below, pick the differential equation representing the system which answers the question.

(i) $s'' + s' + 4s = 0$ (ii) $s'' + 2s' + 5s = 0$
(iii) $s'' + 3s' + 3s = 0$ (iv) $s'' + 0.5s' + 2s = 0$

29. Which spring has the largest coefficient of damping?

30. Which spring exerts the smallest restoring force for a given displacement?

31. In which system does the mass experience the frictional force of smallest magnitude for a given velocity?

32. Which oscillation has the longest period?

33. Which spring is the stiffest? [Hint: You need to determine what it means for a spring to be stiff. Think of an industrial strength spring and a slinky.]

For each of the differential equations in Problems 34–36, find the values of c that make the general solution:

(a) overdamped, (b) underdamped, (c) critically damped.

34. $s'' + 4s' + cs = 0$ **35.** $s'' + 2\sqrt{2}s' + cs = 0$

36. $s'' + 6s' + cs = 0$

37. Find a solution to the following equation which satisfies $z(0) = 3$ and does not tend to infinity as $t \to \infty$:

$$\frac{d^2z}{dt^2} + \frac{dz}{dt} - 2z = 0.$$

38. Consider an overdamped differential equation with $b, c > 0$.

(a) Show that both roots of the characteristic equation are negative.

(b) Show that any solution to the differential equation goes to 0 as $t \to \infty$.

Recall the discussion of electric circuits on page 544. Just as a spring can have a damping force which affects its motion, so can a circuit. Problems 39–42 involve a damping force caused by the resistor in Figure 11.76. The charge Q on a capacitor in a circuit with inductance L, capacitance C, and resistance R, in ohms, satisfies the differential equation

$$L\frac{d^2Q}{dt^2} + R\frac{dQ}{dt} + \frac{1}{C}Q = 0.$$

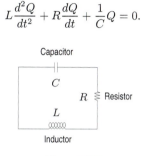

Figure 11.76

39. If $L = 1$ henry, $R = 2$ ohms, and $C = 4$ farads, find a formula for the charge when

(a) $Q(0) = 0, Q'(0) = 2$.
(b) $Q(0) = 2, Q'(0) = 0$.

40. If $L = 1$ henry, $R = 1$ ohm, and $C = 4$ farads, find a formula for the charge when

(a) $Q(0) = 0, Q'(0) = 2$.
(b) $Q(0) = 2, Q'(0) = 0$.
(c) How did reducing the resistance affect the charge? Compare with your solution to Problem 39.

41. If $L = 8$ henry, $R = 2$ ohm, and $C = 4$ farads, find a formula for the charge when

(a) $Q(0) = 0, Q'(0) = 2$.
(b) $Q(0) = 2, Q'(0) = 0$.
(c) How did increasing the inductance affect the charge? Compare with your solution to Problem 39.

42. Given any positive values for R, L and C, what happens to the charge as t goes to infinity?

43. Could the graph in Figure 11.77 show the position of a mass oscillating at the end of an overdamped spring? Why or why not?

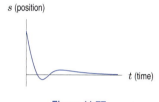

Figure 11.77

44. Consider the system of differential equations

$$\frac{dx}{dt} = -y \qquad \frac{dy}{dt} = -x.$$

(a) Convert this system to a second order differential equation in y by differentiating the second equation with respect to t and substituting for x from the first equation.

(b) Solve the equation you obtained for y as a function of t; hence find x as a function of t.

CHAPTER SUMMARY

- **Differential equations terminology**
  Order, initial conditions, families of solutions, stable/unstable equilibrium solutions.

- **Solving first-order differential equations**
  Slope fields (graphical), Euler's method (numerical), separation of variables (analytical).

- **Modeling with differential equations**
  Growth and decay, Newton's law of heating and cooling,

compartment models, logistic model.

- **Second-order differential equations**
  Initial and boundary conditions, oscillations (with and without damping), characteristic equation, solutions using complex numbers.

- **Systems of differential equations**
  S-I-R model, predator-prey model, phase plane.

REVIEW EXERCISES AND PROBLEMS FOR CHAPTER ELEVEN

Exercises

1. Determine which of the following differential equations is separable. Do not solve the equations.

   (a) $y' = y$
   (b) $y' = x + y$
   (c) $y' = xy$
   (d) $y' = \sin(x + y)$
   (e) $y' - xy = 0$
   (f) $y' = y/x$
   (g) $y' = \ln(xy)$
   (h) $y' = (\sin x)(\cos y)$
   (i) $y' = (\sin x)(\cos xy)$
   (j) $y' = x/y$
   (k) $y' = 2x$
   (l) $y' = (x+y)/(x+2y)$

Find a general solution to the differential equations in Exercises 2–7.

2. $\dfrac{dP}{dt} = t$

3. $\dfrac{dy}{dx} = 0.2y - 8$

4. $\dfrac{dP}{dt} = 10 - 2P$

5. $\dfrac{dH}{dt} = 0.5H + 10$

6. $\dfrac{dR}{dt} = 2R - 6R^2$

7. $\dfrac{dP}{dt} = 0.4P(1 - 0.01P)$

For the differential equations in Exercises 8–23, find a solution which passes through the given point.

8. $\dfrac{dy}{dx} + xy^2 = 0$, $\quad y(1) = 1$

9. $\dfrac{dP}{dt} = 0.03P + 400$, $\quad P(0) = 0$

10. $1 + y^2 - \dfrac{dy}{dx} = 0$, $\quad y(0) = 0$

11. $2\sin x - y^2\dfrac{dy}{dx} = 0$, $\quad y(0) = 3$

12. $\dfrac{dk}{dt} = (1 + \ln t)k$, $\quad k(1) = 1$

13. $\dfrac{dy}{dx} = \dfrac{y(3 - x)}{x(0.5y - 4)}$, $\quad y(1) = 5$

14. $\dfrac{dy}{dx} = \dfrac{0.2y(18 + 0.1x)}{x(100 + 0.5y)}$, $\quad (10, 10)$

15. $\dfrac{dz}{dt} = z(z - 1)$, $\quad z(0) = 10$

16. $\dfrac{dy}{dt} = y(10 - y)$, $\quad y(0) = 1$

17. $\dfrac{dy}{dx} = \dfrac{y(100 - x)}{x(20 - y)}$, $\quad (1, 20)$

18. $\dfrac{df}{dx} = \sqrt{xf(x)}$, $\quad f(1) = 1$

19. $\dfrac{dy}{dx} = e^{x-y}$, $\quad y(0) = 1$

20. $\dfrac{dy}{dx} = e^{x+y}$, $\quad y = 0$ where $x = 1$

21. $e^{-\cos\theta}\dfrac{dz}{d\theta} = \sqrt{1 - z^2}\sin\theta$, $\quad z(0) = \frac{1}{2}$

22. $(1 + t^2)y\dfrac{dy}{dt} = 1 - y$, $\quad y(1) = 0$

23. $\dfrac{dy}{dt} = 2^y \sin^3 t$, $\quad y(0) = 0$

Find a general solution to the differential equations in Exercises 24–29.

24. $\dfrac{d^2z}{dt^2} + \pi^2 z = 0$

25. $9z'' - z = 0$

26. $9z'' + z = 0$

27. $y'' + 6y' + 8y = 0$

28. $y'' + 2y' + 3y = 0$

29. $x'' + 2x' + 10x = 0$

Problems

30. (a) Find all equilibrium solutions for the differential equation
    $$\frac{dy}{dx} = 0.5y(y - 4)(2 + y).$$

    (b) Draw a slope field and use it to determine whether each equilibrium solution is stable or unstable.

31. Consider the initial value problem
    $$y' = 5 - y, \quad y(0) = 1.$$

    (a) Use Euler's method with five steps to estimate $y(1)$.
    (b) Sketch the slope field for this differential equation in the first quadrant, and use it to decide if your estimate is an over or underestimate.

    (c) Find the exact solution to the differential equation and hence find $y(1)$ exactly.
    (d) Without doing the calculation, roughly what would you expect the approximation for $y(1)$ to be with ten steps?

For each of the differential equations in Problems 32–33, find the values of b that make the general solution:

   (a) overdamped,
   (b) underdamped,
   (c) critically damped.

32. $s'' + bs' + 5s = 0$

33. $s'' + bs' - 16s = 0$

34. (a) A cup of coffee is made with boiling water and stands in a room where the temperature is $20°$ C. If $T(t)$ is the temperature of the coffee at time t, explain what the differential equation

$$\frac{dT}{dt} = -k(T - 20)$$

says in everyday terms. What is the sign of k?

(b) Solve this differential equation. If the coffee cools to $90°$ C in 2 minutes, how long will it take to cool to $60°$ C degrees?

35. A roast is taken from the refrigerator (where the temperature is $40°$F) and put in a $350°$F oven. One hour later, the meat thermometer shows a temperature of $90°$F. If the roast is done when its temperature reaches $140°$F, what is the total time the roast should be in the oven? Newton's Law of Cooling says that the rate of change of temperature is proportional to the difference between the temperature of the object and the temperature of the surrounding air.

36. The rate at which a drug leaves the bloodstream and passes into the urine is proportional to the quantity of the drug in the blood at that time. If an initial dose of Q_0 is injected directly into the blood, 20% is left in the blood after 3 hours.

(a) Write and solve a differential equation for the quantity, Q, of the drug in the blood at time, t, in hours.

(b) How much of this drug is in a patient's body after 6 hours if the patient is given 100 mg initially?

37. Water leaks from a vertical cylindrical tank through a small hole in its base at a rate proportional to the square root of the volume of water remaining. If the tank initially contains 200 liters and 20 liters leak out during the first day, when will the tank be half empty? How much water will there be after 4 days?

38. A mothball is in the shape of a sphere and starts with radius 1 cm. The material in the mothball evaporates at a rate proportional to the surface area. After one month, the radius is 0.5 cm. How many months (from the start) is it before the radius is 0.2 cm?

39. A bank account earns 10% annual interest, compounded continuously. Money is deposited in a continuous cash flow at a rate of $1200 per year into the account.

(a) Write a differential equation that describes the rate at which the balance $B = f(t)$ is changing.

(b) Solve the differential equation given an initial balance $B_0 = 0$.

(c) Find the balance after 5 years.

40. (Continuation of Problem 39.) Now suppose the money is deposited once a month (instead of continuously) but still at a rate of $1200 per year.

(a) Write down the sum that gives the balance after 5 years, assuming the first deposit is made one month from today, and today is $t = 0$.

(b) The sum you wrote in part (a) is a Riemann sum approximation to the integral

$$\int_0^5 1200e^{0.1t}\,dt.$$

Determine whether it is a left sum or right sum, and determine what Δt and N are. Then use your calculator to evaluate the sum.

(c) Compare your answer in part (b) to your answer to Problem 39(c).

41. The spread of a non-fatal disease through a population of fixed size M can be modeled as follows. The rate that healthy people are infected, in people per day, is proportional to the product of the numbers of healthy and infected people. The constant of proportionality is $0.01/M$. The rate of recovery, in people per day, is 0.009 times the number of people infected. Construct a differential equation which models the spread of the disease. Assuming that initially only a small number of people are infected, plot a graph of the number of infected people against time. What fraction of the population is infected in the long run?

42. A model for the population, $P(t)$, of carp in a landlocked lake is given by the differential equation

$$\frac{dP}{dt} = 0.25P(1 - 0.0004P).$$

(a) What is the long-term equilibrium population of carp in the lake?

(b) Ten years ago a census was taken and there were found to be 1000 carp in the lake. Estimate the current size of the population.

(c) It is planned to join the lake to a nearby river so that the fish will be able to leave the lake. It is estimated that there will be a net loss of 10% of the carp each year but that the patterns of birth and death are not expected to change. Revise the differential equation to take this into account. Use the revised differential equation to predict the future development of the carp population.

43. Juliet is in love with Romeo, who happens (in our version of this story) to be a fickle lover. The more Juliet loves him, the more he begins to dislike her. When she hates him, his feelings for her warm up. On the other hand, her love for him grows when he loves her and withers when he hates her. A model for their ill-fated romance is

$$\frac{dj}{dt} = Ar, \qquad \frac{dr}{dt} = -Bj,$$

where A and B are positive constants, $r(t)$ represents Romeo's love for Juliet at time t, and $j(t)$ represents Juliet's love for Romeo at time t. (Negative love is hate.)

(a) The constant on the right-hand side of Juliet's equation (the one including dj/dt) has a positive sign, whereas the constant in Romeo's equation is negative. Explain why these signs follow from the story.

(b) Derive a second-order differential equation for $r(t)$ and solve it. (Your equation should involve r and its derivatives, but not j and its derivatives.)

(c) Express $r(t)$ and $j(t)$ as functions of t, given $r(0) = 1$ and $j(0) = 0$. Your answer will contain A and B.

(d) As you may have discovered, the outcome of the relationship is a never-ending cycle of love and hate. Find what fraction of the time they both love one another.

44. The behavior of an electron is described by Schrödinger's equation. Under certain conditions, this equation becomes

$$\frac{-h^2}{8\pi^2 m} \frac{d^2 \Psi}{dx^2} = E\Psi,$$

where Ψ, called a *wave function*, describes the electron's behavior as a function of x, the distance from a fixed point; the constants h, m, E are positive. In addition, Ψ satisfies the boundary conditions

$$\Psi(0) = \Psi(l) = 0$$

where l is another positive constant.

(a) Solutions to Schrödinger's equation are of the form $\Psi = C_1 \cos(\omega x) + C_2 \sin(\omega x)$ for some ω. Show that if Ψ satisfies the boundary conditions, then Ψ must be of the form

$$\Psi = C_2 \sin\left(\frac{n\pi x}{l}\right) \quad \text{for some positive integer } n.$$

(b) Using part (a), find E in terms of other constants.

(c) The constant E represents the energy level of the electron. Assuming that only positive integral values of n are possible, explain why not all values of E are possible. (This is described by saying that energy is quantized.) In addition, show that the possible energy levels are one of the following multiples of the lowest energy level: $1, 4, 9, 16 \ldots$.

CAS Challenge Problems

45. Consider the differential equation

$$\frac{dP}{dt} = P(P-1)(2-P).$$

(a) Find all the equilibrium solutions.

(b) Show that the following two functions are solutions:

$$P_1(t) = 1 - \frac{e^t}{\sqrt{3 + e^{2t}}} \quad \text{and} \quad P_2(t) = 1 + \frac{e^t}{\sqrt{3 + e^{2t}}}.$$

[Hint: Use the computer algebra system to simplify the difference between the right and left-hand sides.]

(c) Find $P_1(0)$, $P_2(0)$, $\lim_{t \to \infty} P_1(t)$, and $\lim_{t \to \infty} P_2(t)$. Explain how you could have predicted the limits as $t \to \infty$ from the values at $t = 0$ without knowing the solutions explicitly.

46. In this problem we investigate *Picard's method* for approximating the solutions of differential equations. Consider the differential equation

$$y'(t) = y(t)^2 + t^2, \quad y(a) = b.$$

Integrating both sides with respect to t gives

$$y(s) - y(a) = \int_a^s y'(t)\, dt = \int_a^s (y(t)^2 + t^2)\, dt.$$

Since $y(a) = b$, we have

$$y(s) = b + \int_a^s (y(t)^2 + t^2)\, dt.$$

We have put the differential equation into the form of an *integral equation*. If we have an approximate solution

$y_0(s)$, we can use the integral form to make a new approximation

$$y_1(s) = b + \int_a^s (y_0(t)^2 + t^2)\, dt.$$

Continuing this process, we get a sequence of approximations $y_0(s)$, $y_1(s)$, ..., $y_n(s)$, ... where each term in the sequence is defined in terms of the previous one by the equation

$$y_{n+1}(s) = b + \int_a^s (y_n(t)^2 + t^2)\, dt.$$

(a) Show that y_n satisfies the initial condition $y_n(a) = b$ for all n.

(b) Using the initial condition $y(1) = 0$, start with the approximation $y_0(s) = 0$ and use a computer algebra system to find the next three approximations y_1, y_2, and y_3.

(c) Use a computer algebra system to find the solution y satisfying $y(1) = 0$, and sketch y, y_1, y_2, y_3 on the same axes. On what domain do the approximations appear to be accurate? [The solution y cannot be expressed in terms of elementary functions. If your computer algebra system cannot solve the equation exactly, use a numerical method such as Euler's method.]

47. Figure 11.78 shows the slope field for the differential equation[24]

$$\frac{dy}{dx} = \sqrt{1 - y^2}.$$

---

[24]Adapted from David Lomen & David Lovelock, *Differential Equations* (New York: John Wiley & Sons, Inc., 1999).

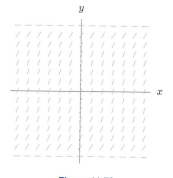

Figure 11.78

(a) Sketch the solution with $y(0) = 0$ for $-3 \leq x \leq 3$.
(b) Use a computer algebra system to find the solution in part (a).
(c) Compare the computer algebra system's solution with your sketch. Do they agree? If not, which one is right?

CHECK YOUR UNDERSTANDING

In Problems 1–16, decide whether the statements are true or false. Give an explanation for your answer.

1. For all constants k, the equation $y'' + ky = 0$ has trigonometric functions as solutions.

2. For all constants k, the equation $y' + ky = 0$ has exponential functions as solutions.

3. Polynomials are never solutions to differential equations.

4. The system of differential equations $x' = -x + xy^2$ and $y' = y - x^2y$ requires initial conditions for both $x(0)$ and $y(0)$ to determine a unique solution.

5. There is a solution curve for the logistic differential equation $P' = P(2-P)$ which goes through the points $(0, 1)$ and $(1, 3)$.

6. For any positive values of the constant k and any positive values of the initial value $P(0)$, the solution to the differential equation $P' = kP(L - P)$ has limiting value L as $t \to \infty$.

7. Two populations modeled by a system of differential equations never die out.

8. All solutions to the differential equation whose slope field is in Figure 11.80 have $\lim_{x \to \infty} y = \infty$.

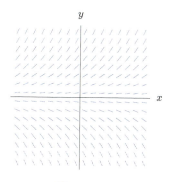

Figure 11.79

9. All solutions to the differential equation whose slope field is in Figure 11.79 have $\lim_{x \to \infty} y = 0$.

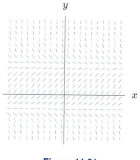

Figure 11.80

10. All solutions to the differential equation whose slope field is in Figure 11.81 have the same limiting value as $x \to \infty$.

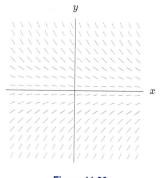

Figure 11.81

11. Euler's method gives the arc length of a solution curve.

12. The differential equation $dy/dx = x + y$ can be solved by separation of variables.

13. The differential equation $dy/dx - xy = x$ can be solved by separation of variables.

14. It would be difficult to find a differential equation having $x^3 + xy + y^3 = 1$ as a solution curve.

15. It could be difficult or impossible to find an equation for a solution curve to the differential equation $dy/dx = x^3 + xy + y^3$.

16. The only solution to the differential equation $dy/dx = 3y^{2/3}$ passing through the point $(0,0)$ is $y = x^3$.

Suppose that $y = f(x)$ is a solution of the differential equation $dy/dx = g(x)$. Are the statements in Problems 17–24 true or false? If a statement is true, explain how you know. If a statement is false, give a counterexample.

17. If $g(x)$ is increasing for all x, then the graph of f is concave up for all x.

18. If $g(x)$ is increasing for $x > 0$, then so is $f(x)$.

19. If $g(0) = 1$ and $g(x)$ is increasing for $x \geq 0$, then $f(x)$ is also increasing for $x \geq 0$.

20. If $g(x)$ is periodic, then $f(x)$ is also periodic.

21. If $\lim_{x \to \infty} g(x) = 0$, then $\lim_{x \to \infty} f(x) = 0$.

22. If $\lim_{x \to \infty} g(x) = \infty$, then $\lim_{x \to \infty} f(x) = \infty$.

23. If $g(x)$ is even, then so is $f(x)$.

24. If $g(x)$ is even, then $f(x)$ is odd.

Suppose that $y = f(x)$ is a solution of the differential equation $dy/dx = 2x - y$. In Problems 25–33, decide whether the statements are true or false. Justify your answer.

25. If $f(a) = b$, the slope of the graph of f at (a, b) is $2a - b$.

26. $f'(x) = 2x - f(x)$.

27. There could be more than one value of x such that $f'(x) = 1$ and $f(x) = 5$.

28. If $y = f(x)$, then $d^2y/dx^2 = 2 - (2x - y)$.

29. If $f(1) = 5$, then $(1, 5)$ could be a critical point of f.

30. The graph of f is decreasing whenever it lies above the line $y = 2x$ and is increasing whenever it lies below the line $y = 2x$.

31. All the inflection points of f lie on the line $y = 2x - 2$.

$$\frac{d^y}{dx^2} = 2 - \frac{dy}{dx} = 2 - (2x - y).$$

Thus at any inflection point of $y = f(x)$, we have $d^2y/dx^2 = 2 - (2x - y) = 0$. That is, any inflection point of f must satisfy $y = 2x - 2$.

32. If $g(x)$ is another solution to the differential equation $dy/dx = 2x - y$, then $g(x) = f(x) + C$.

33. If $g(x)$ is a different solution to the differential equation $dy/dx = 2x - y$, then $\lim_{x \to \infty} (g(x) - f(x)) = 0$. [Hint: Show that $w = g(x) - f(x)$ satisfies the differential equation $dw/dx = -w$.]

In Problems 34–37, give an example of a differential equation with the specified properties.

34. All of whose solutions are increasing and concave up.

35. All of whose solutions have their critical points on the parabola $y = x^2$.

36. All of whose solutions are the family of functions $f(x) = x^2 + C$.

37. All of whose solutions are the family of hyperbolas $x^2 - y^2 = C$.

PROJECTS FOR CHAPTER ELEVEN

1. US Electricity Consumption

The following data shows US electricity consumption, measured in billions of kilowatt hours per year, for much of the 20th century.

Year	1912	1917	1920	1929	1936	1945
Electricity consumption (E)	12	25	39	92	109	222
Year	1955	1960	1965	1970	1980	1987
Electricity consumption (E)	547	755	1055	1531	2286	2455

(a) Decide how well each of the following differential equations model the growth of electricity consumption, E, over time, t, measured in years. Do this by using the data to approximate dE/dt. For each differential equation, make a plot to which you can fit a line and use it to estimate the parameters a, b, or c.

(i) $\dfrac{1}{E}\dfrac{dE}{dt} = c$

(ii) $\dfrac{dE}{dt} = a - bE$

(iii) $\dfrac{1}{E}\dfrac{dE}{dt} = a - bE$

(iv) $\dfrac{1}{E}\dfrac{dE}{dt} = a - bt$

(b) Use each model to predict energy consumption for the year 2020. Which prediction do you think is most reliable? Why? Discuss the long-run predictions from each model. Which differential equation do you think fits best?

2. Pareto's Law

In analyzing a society, sociologists are often interested in how incomes are distributed through the society. Pareto's law asserts that each society has a constant k such that the average income of all people wealthier than you is k times your income ($k > 1$). If $p(x)$ is the number of people in the society with an income of x or above, we define $\Delta p = p(x + \Delta x) - p(x)$, for small Δx.

(a) Explain why the number of people with incomes between x and $x + \Delta x$ is represented by $-\Delta p$. Then show that the total amount of money earned by people with incomes between x and $x + \Delta x$ is approximated by $-x\Delta p$.

(b) Use Pareto's law to show that the total amount of money earned by people with incomes of x or above is $kxp(x)$. Then show that the total amount of money earned by people with incomes between x and $x + \Delta x$ is approximated by $-kp\Delta x - kx\Delta p$.

(c) Using your answers to parts (a) and (b), show that p satisfies the differential equation

$$(1 - k)xp' = kp.$$

(d) Solve this differential equation for $p(x)$.

(e) Sketch a graph of $p(x)$ for various values of k, some large and some near 1. How does the value of k alter the shape of the graph?

3. Vibrations in a Molecule

Suppose the force, F, between two atoms a distance r apart in a molecule is given by

$$F(r) = b\left(\frac{a^2}{r^3} - \frac{a}{r^2}\right)$$

where a and b are positive constants.

(a) Find the equilibrium distance, r, at which $F = 0$.

(b) Expand F in a series about the equilibrium position. Give the first two nonzero terms.

(c) Suppose one atom is fixed. Let x be the displacement of the other atom from the equilibrium position. Give the first two nonzero terms of the series for F in terms of x.

(d) Using Newton's Second Law, show that if x is small, the second atom oscillates about the equilibrium position. Find the period of the oscillation.

4. Harvesting and Logistic Growth

In this project, we look at the effects of *harvesting* a population which is growing logistically. Harvesting could be, for example, fishing or logging. An important question is what level of harvesting leads to a *sustainable yield*. In other words, how much can be harvested without having the population depleted in the long run?

(a) When there is no fishing, suppose a population of fish is governed by the differential equation

$$\frac{dN}{dt} = 2N - 0.01N^2,$$

where N is the number of fish at time t in years. Sketch a graph of dN/dt against N. Mark on your graph the equilibrium values of N. Now sketch N against t for various different initial values.

(b) Suppose now that fish are removed by fishermen at a continuous rate of 75 fish/year. Let P be the number of fish at time t with harvesting. Explain why P satisfies the differential equation

$$\frac{dP}{dt} = 2P - 0.01P^2 - 75.$$

(c) Sketch dP/dt against P. Find and label the intercepts.

(d) Sketch the slope field for the differential equation for P.

(e) Use the graph you drew in part (c) to sketch graphs of P against t, with the initial values:

    (i) $P(0) = 40$ (ii) $P(0) = 60$ (iii) $P(0) = 160$ (iv) $P(0) = 140$

(f) Use the slope field in part (d) to sketch graphs of P against t, with the initial values:

    (i) $P(0) = 40$ (ii) $P(0) = 50$ (iii) $P(0) = 60$

    (iv) $P(0) = 150$ (v) $P(0) = 170$

(g) Using the graphs you drew, decide what the equilibrium values of the populations are and whether or not they are stable.

(h) We now look at the effect of different levels of fishing on a fish population. If fishing takes place at a continuous rate of H fish/year, the fish population P satisfies the differential equation

$$\frac{dP}{dt} = 2P - 0.01P^2 - H.$$

    (i) For each of the values $H = 75, 100, 200$, plot a graph of dP/dt against P.

    (ii) For which of the three values of H that you considered in part (i) is there an initial condition such that the fish population does not die out eventually?

    (iii) Looking at your answer to part (ii), decide for what values of H there is an initial value for P such that the population does not die out eventually.

    (iv) Recommend a policy to ensure long-term survival of the fish population.

Chapter Twelve

FUNCTIONS OF SEVERAL VARIABLES

Many quantities depend on more than one variable: the amount of food grown depends on the amount of rain and the amount of fertilizer used; the rate of a chemical reaction depends on the temperature and the pressure of the environment in which it proceeds; the strength of the gravitational attraction between two bodies depends on their masses and their distance apart; and the rate of fallout from a volcanic explosion depends on the distance from the volcano and the time since the explosion. Each example involves a function of two or more variables. In this chapter, we will see many different ways of looking at functions of several variables.

12.1 FUNCTIONS OF TWO VARIABLES

Function Notation

Suppose you want to calculate your monthly payment on a five-year car loan; this depends on both the amount of money you borrow and the interest rate. These quantities can vary separately: the loan amount can change while the interest rate remains the same, or the interest rate can change while the loan amount remains the same. To calculate your monthly payment you need to know both. If the monthly payment is $\$m$, the loan amount is $\$L$, and the interest rate is $r\%$, then we express the fact that m is a function of L and r by writing:

$$m = f(L, r).$$

This is just like the function notation of one-variable calculus. The variable m is called the dependent variable, and the variables L and r are called the independent variables. The letter f stands for the *function* or rule that gives the value of m corresponding to given values of L and r.

A function of two variables can be represented graphically, numerically by a table of values, or algebraically by a formula. In this section, we give examples of each.

Graphical Example: A Weather Map

Figure 12.1 shows a weather map from a newspaper. What information does it convey? It displays the predicted high temperature, T, in degrees Fahrenheit (°F), throughout the US on that day. The curves on the map, called *isotherms*, separate the country into zones, according to whether T is in the 60s, 70s, 80s, 90s, or 100s. (*Iso* means same and *therm* means heat.) Notice that the isotherm separating the 80s and 90s zones connects all the points where the temperature is exactly 90°F.

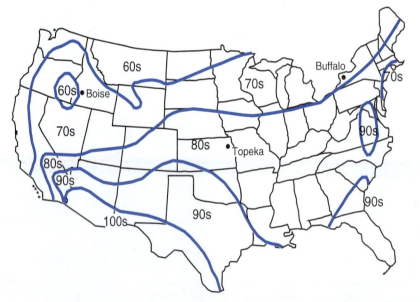

Figure 12.1: Weather map showing predicted high temperatures, T, on a summer day

Example 1   Estimate the predicted value of T in Boise, Idaho; Topeka, Kansas; and Buffalo, New York.

Solution   Boise and Buffalo are in the 70s region, and Topeka is in the 80s region. Thus, the predicted temperature in Boise and Buffalo is between 70 and 80 while the predicted temperature in Topeka is between 80 and 90. In fact, we can say more. Although both Boise and Buffalo are in the 70s, Boise is quite close to the $T = 70$ isotherm, whereas Buffalo is quite close to the $T = 80$ isotherm. So we estimate the temperature to be in the low 70s in Boise and in the high 70s in Buffalo. Topeka is about halfway between the $T = 80$ isotherm and the $T = 90$ isotherm. Thus, we guess the temperature in Topeka to be in the mid 80s. In fact, the actual high temperatures for that day were 71°F for Boise, 79°F for Buffalo, and 86°F for Topeka.

The predicted high temperature, T, illustrated by the weather map is a function of (that is, depends on) two variables, often longitude and latitude, or miles east-west and miles north-south of a fixed point, say, Topeka. The weather map in Figure 12.1 is called a *contour map* or *contour diagram* of that function. Section 12.2 shows another way of visualizing functions of two variables using surfaces; Section 12.3 looks at contour maps in detail.

Numerical Example: Beef Consumption

Suppose you are a beef producer and you want to know how much beef people will buy. This depends on how much money people have and on the price of beef. The consumption of beef, C (in pounds per week per household) is a function of household income, I (in thousands of dollars per year), and the price of beef, p (in dollars per pound). In function notation, we write:

$$C = f(I, p).$$

Table 12.1 contains values of this function. Values of p are shown across the top, values of I are down the left side, and corresponding values of $f(I, p)$ are given in the table.[1] For example, to find the value of $f(40, 3.50)$, we look in the row corresponding to $I = 40$ under $p = 3.50$, where we find the number 4.05. Thus,

$$f(40, 3.50) = 4.05.$$

This means that, on average, if a household's income is \$40,000 a year and the price of beef is \$3.50/lb, the family will buy 4.05 lbs of beef per week.

Table 12.1   *Quantity of beef bought (pounds/household/week)*

		Price of beef, (\$/lb)			
		3.00	3.50	4.00	4.50
Household	20	2.65	2.59	2.51	2.43
income	40	4.14	4.05	3.94	3.88
per year,	60	5.11	5.00	4.97	4.84
I	80	5.35	5.29	5.19	5.07
(1000)	100	5.79	5.77	5.60	5.53

Notice how this differs from the table of values of a one-variable function, where one row or one column is enough to list the values of the function. Here many rows and columns are needed because the function has a value for every *pair* of values of the independent variables.

Algebraic Examples: Formulas

In both the weather map and beef consumption examples, there is no formula for the underlying function. That is usually the case for functions representing real-life data. On the other hand, for many idealized models in physics, engineering, or economics, there are exact formulas.

[1] Adapted from Richard G. Lipsey, *An Introduction to Positive Economics*, 3rd ed., (London: Weidenfeld and Nicolson, 1971).

Example 2   Give a formula for the function $M = f(B, t)$ where M is the amount of money in a bank account t years after an initial investment of B dollars, if interest is accrued at a rate of 5% per year compounded annually.

Solution   Annual compounding means that M increases by a factor of 1.05 every year, so
$$M = f(B, t) = B(1.05)^t.$$

Example 3   A cylinder with closed ends has radius r and height h. If its volume is V and its surface area is A, find formulas for the functions $V = f(r, h)$ and $A = g(r, h)$.

Solution   Since the area of the circular base is πr^2, we have
$$V = f(r, h) = \text{Area of base} \cdot \text{Height} = \pi r^2 h.$$

The surface area of the side is the circumference of the bottom, $2\pi r$, times the height h, giving $2\pi rh$. Thus,
$$A = g(r, h) = 2 \cdot \text{Area of base} + \text{Area of side} = 2\pi r^2 + 2\pi rh.$$

A Tour of 3-Space

In Section 12.2 we see how to visualize a function of two variables as a surface in space. Now we see how to locate points in three-dimensional space (3-space).

Imagine three coordinate axes meeting at the *origin*: a vertical axis, and two horizontal axes at right angles to each other. (See Figure 12.2.) Think of the xy-plane as being horizontal, while the z-axis extends vertically above and below the plane. The labels x, y, and z show which part of each axis is positive; the other side is negative. We generally use *right-handed axes* in which looking down the positive z-axis gives the usual view of the xy-plane. We specify a point in 3-space by giving its coordinates (x, y, z) with respect to these axes. Think of the coordinates as instructions telling you how to get to the point; start at the origin, go x units along the x-axis, then y units in the direction parallel to the y-axis and finally z units in the direction parallel to the z-axis. The coordinates can be positive, zero or negative; a zero coordinate means "don't move in this direction," and a negative coordinate means "go in the negative direction parallel to this axis." For example, the origin has coordinates $(0, 0, 0)$, since we get there from the origin by doing nothing at all.

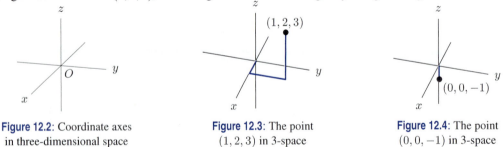

Figure 12.2: Coordinate axes in three-dimensional space

Figure 12.3: The point $(1, 2, 3)$ in 3-space

Figure 12.4: The point $(0, 0, -1)$ in 3-space

Example 4   Describe the position of the points with coordinates $(1, 2, 3)$ and $(0, 0, -1)$.

Solution   We get to the point $(1, 2, 3)$ by starting at the origin, going 1 unit along the x-axis, 2 units in the direction parallel to the y-axis, and 3 units up in the direction parallel to the z-axis. (See Figure 12.3.)

To get to $(0, 0, -1)$, we don't move at all in the x and y directions, but move 1 unit in the negative z direction. So the point is on the negative z-axis. (See Figure 12.4.) You can check that the position of the point is independent of the order of the x, y, and z displacements.

Example 5   You start at the origin, go along the y-axis a distance of 2 units in the positive direction, and then move vertically upward a distance of 1 unit. What are the coordinates of your final position?

Solution   You started at the point $(0, 0, 0)$. When you went along the y-axis, your y-coordinate increased to 2. Moving vertically increased your z-coordinate to 1; your x-coordinate didn't change because you did not move in the x direction. So your final coordinates are $(0, 2, 1)$. (See Figure 12.5.)

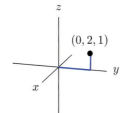

Figure 12.5: The point $(0, 2, 1)$ is reached by moving 2 along the y-axis and 1 upward

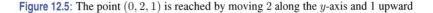

It is often helpful to picture a three dimensional coordinate system in terms of a room. The origin is a corner at floor level where two walls meet the floor. The z-axis is the vertical intersection of the two walls; the x- and y-axes are the intersections of each wall with the floor. Points with negative coordinates lie behind a wall in the next room or below the floor.

Graphing Equations in 3-Space

We can graph an equation involving the variables x, y, and z in 3-space; such a graph is a picture of all points (x, y, z) that satisfy the equation.

Example 6   What do the graphs of the equations $z = 0$, $z = 3$, and $z = -1$ look like?

Solution   To graph $z = 0$, we visualize the set of points whose z-coordinate is zero. If the z-coordinate is 0, then we must be at the same vertical level as the origin, that is, we are in the horizontal plane containing the origin. So the graph of $z = 0$ is the middle plane in Figure 12.6. The graph of $z = 3$ is a plane parallel to the graph of $z = 0$, but three units above it. The graph of $z = -1$ is a plane parallel to the graph of $z = 0$, but one unit below it.

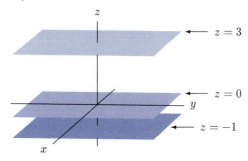

Figure 12.6: The planes $z = -1$, $z = 0$, and $z = 3$

The plane $z = 0$ contains the x- and y-coordinate axes, and is called the xy-plane. There are two other coordinate planes. The yz-plane contains both the y- and the z-axes, and the xz-plane contains the x- and z-axes. (See Figure 12.7.)

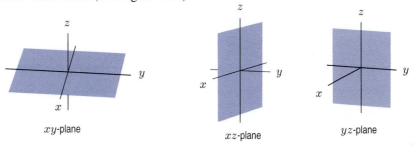

Figure 12.7: The three coordinate planes

Example 7   Which of the points $A = (1, -1, 0)$, $B = (0, 3, 4)$, $C = (2, 2, 1)$, and $D = (0, -4, 0)$ lies closest to the xz-plane? Which point lies on the y-axis?

Solution   The magnitude of the y-coordinate gives the distance to the xz-plane. The point A lies closest to that plane, because it has the smallest y-coordinate in magnitude. To get to a point on the y-axis, we move along the y-axis, but we don't move at all in the x or z directions. Thus, a point on the y-axis has both its x- and z-coordinates equal to zero. The only point of the four that satisfies this is D. (See Figure 12.8.)

In general, if a point has one of its coordinates equal to zero, it lies in one of the coordinate planes. If a point has two of its coordinates equal to zero, it lies on one of the coordinate axes.

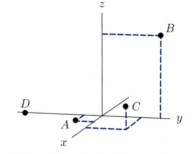

Figure 12.8: Which point lies closest to the xz-plane? Which point lies on the y-axis?

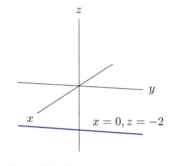

Figure 12.9: The line $x = 0$, $z = -2$

Example 8   You are 2 units below the xy-plane and in the yz-plane. What are your coordinates?

Solution   Since you are 2 units below the xy-plane, your z-coordinate is -2. Since you are in the yz-plane, your x-coordinate is 0; your y-coordinate can be anything. Thus, you are at the point $(0, y, -2)$. The set of all such points forms a line parallel to the y-axis, 2 units below the xy-plane, and in the yz-plane. (See Figure 12.9.)

Example 9   You are standing at the point $(4, 5, 2)$, looking at the point $(0.5, 0, 3)$. Are you looking up or down?

Solution   The point you are standing at has z-coordinate 2, whereas the point you are looking at has z-coordinate 3; hence you are looking up.

Example 10   Imagine that the yz-plane in Figure 12.7 is a page of this book. Describe the region behind the page algebraically.

Solution   The positive part of the x-axis pokes out of the page; moving in the positive x direction brings you out in front of the page. The region behind the page corresponds to negative values of x, so it is the set of all points in 3-space satisfying the inequality $x < 0$.

Distance Between Two Points

In 2-space, the formula for the distance between two points (x, y) and (a, b) is given by

$$\text{Distance} = \sqrt{(x - a)^2 + (y - b)^2}.$$

The distance between two points (x, y, z) and (a, b, c) in 3-space is represented by PG in Figure 12.10. The side PE is parallel to the x-axis, EF is parallel to the y-axis, and FG is parallel to the z-axis.

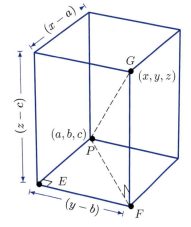

Figure 12.10: The diagonal PG gives the distance between the points (x, y, z) and (a, b, c)

Using Pythagoras' theorem twice gives

$$(PG)^2 = (PF)^2 + (FG)^2 = (PE)^2 + (EF)^2 + (FG)^2 = (x - a)^2 + (y - b)^2 + (z - c)^2.$$

Thus, a formula for the distance between the points (x, y, z) and (a, b, c) in 3-space is

$$\text{Distance} = \sqrt{(x - a)^2 + (y - b)^2 + (z - c)^2}.$$

Example 11   Find the distance between $(1, 2, 1)$ and $(-3, 1, 2)$.

Solution $\text{Distance} = \sqrt{(-3 - 1)^2 + (1 - 2)^2 + (2 - 1)^2} = \sqrt{18} = 4.24$.

Example 12   Find an expression for the distance from the origin to the point (x, y, z).

Solution   The origin has coordinates $(0, 0, 0)$, so the distance from the origin to (x, y, z) is given by

$$\text{Distance} = \sqrt{(x - 0)^2 + (y - 0)^2 + (z - 0)^2} = \sqrt{x^2 + y^2 + z^2}.$$

Example 13   Find an equation for a sphere of radius 1 with center at the origin.

Solution   The sphere consists of all points (x, y, z) whose distance from the origin is 1, that is, which satisfy the equation

$$\sqrt{x^2 + y^2 + z^2} = 1.$$

This is an equation for the sphere. If we square both sides we get the equation in the form

$$x^2 + y^2 + z^2 = 1.$$

Note that this equation represents the *surface* of the sphere. The solid ball enclosed by the sphere is represented by the inequality $x^2 + y^2 + z^2 \leq 1$.

Exercises and Problems for Section 12.1

Exercises

1. The gravitational force, F newtons, exerted on an object by the earth depends on its mass, m kilograms, and its distance, r meters, from the center of the earth, so $F = f(m, r)$. Interpret the following statement in terms of gravitation: $f(100, 7000000) \approx 820$.

Exercises 2–4 refer to the map in Figure 12.1 on page 560.

2. Give the range of daily high temperatures for

   (a) Pennsylvania   (b) North Dakota

   (c) California.

3. Sketch a possible graph of the predicted high temperature T on a line north-south through Topeka.

4. Sketch possible graphs of the predicted high temperature on a north-south line and an east-west line through Boise.

For Exercises 5–7, refer to Table 12.1 on page 561, where p is the price of beef and I is annual household income.

5. Make a table showing the amount of money, M, that the average household spends on beef (in dollars per household per week) as a function of the price of beef and household income.

6. Give tables for beef consumption as a function of p, with I fixed at $I = 20$ and $I = 100$. Give tables for beef consumption as a function of I, with p fixed at $p = 3.00$ and $p = 4.00$. Comment on what you see in the tables.

7. How does beef consumption vary as a function of household income if the price of beef is held constant?

8. Which of the points $A = (1.3, -2.7, 0)$, $B = (0.9, 0, 3.2)$, $C = (2.5, 0.1, -0.3)$ is closest to the yz-plane? Which one lies on the xz-plane? Which one is farthest from the xy-plane?

9. Which of the points $A = (23, 92, 48)$, $B = (-60, 0, 0)$, $C = (60, 1, -92)$ is closest to the yz-plane? Which one lies on the xz-plane? Which one is farthest from the xy-plane?

10. You are at the point $(-1, -3, -3)$, standing upright and facing the yz-plane. You walk 2 units forward, turn left, and walk for another 2 units. What is your final position? From the point of view of an observer looking at the coordinate system in Figure 12.2 on page 562, are you in front of or behind the yz-plane? To the left or to the right of the xz-plane? Above or below the xy-plane?

11. You are at the point $(3, 1, 1)$, standing upright and facing the yz-plane. You walk 2 units forward, turn left, and walk another 2 units. What is your final position? From the point of view of an observer looking at the coordinate system in Figure 12.2 on page 562, are you in front of or behind the yz-plane? To the left or to the right of the xz-plane? Above or below the xy-plane?

12. Which of the points $P = (1, 2, 1)$ and $Q = (2, 0, 0)$ is closest to the origin?

13. Which two of the three points $P_1 = (1, 2, 3)$, $P_2 = (3, 2, 1)$ and $P_3 = (1, 1, 0)$ are closest to each other?

14. Find the equation of the sphere of radius 5 centered at the origin.

15. Find the equation of the sphere of radius 5 centered at $(1, 2, 3)$.

Sketch graphs of the equations in Exercises 16–18 in 3-space.

16. $x = -3$ 17. $y = 1$

18. $z = 2$ and $y = 4$

Problems

19. A car rental company charges $40 a day and 15 cents a mile for its cars.

    (a) Write a formula for the cost, C, of renting a car as a function, f, of the number of days, d, and the number of miles driven, m.
    (b) If $C = f(d, m)$, find $f(5, 300)$ and interpret it.

20. Consider the acceleration due to gravity, g, at a height h above the surface of a planet of mass m.

    (a) If m is held constant, is g an increasing or decreasing function of h? Why?
    (b) If h is held constant, is g an increasing or decreasing function of m? Why?

21. Find a possible formula for the function $f(x, y)$ whose values are in Table 12.2.

Table 12.2

		y				
		-2	-1	0	1	2
x	-2	4	1	0	1	4
	-1	4	1	0	1	4
	0	4	1	0	1	4
	1	4	1	0	1	4
	2	4	1	0	1	4

22. The temperature adjusted for wind-chill is a temperature which tells you how cold it feels, as a result of the combination of wind and temperature. See Table 12.3.

    (a) If the temperature is $0°$F and the wind speed is 15 mph, how cold does it feel?

(b) If the temperature is $35°$F, what wind speed makes it feel like $22°$F?

(c) If the temperature is $25°$F, what wind speed makes it feel like $20°$F?

(d) If the wind is blowing at 15 mph, what temperature feels like $0°$F?

Table 12.3 *Temperature adjusted for wind-chill ($°F$) as a function of wind speed and temperature*

		Temperature (°F)						
		35	30	25	20	15	10	5	0
Wind Speed (mph)	5	33	27	21	16	12	7	0	−5
	10	22	16	10	3	−3	−9	−15	−22
	15	16	9	2	−5	−11	−18	−25	−31
	20	12	4	−3	−10	−17	−24	−31	−39
	25	8	1	−7	−15	−22	−29	−36	−44

23. Using Table 12.3, make tables of the temperature adjusted for wind-chill as a function of wind speed for temperatures of $20°$F and $0°$F.

Problems 24–26 concern the concentration, C, in mg per liter, of a drug in the blood as a function of x, the amount, in mg, of the drug given and t, the time in hours since the injection. For $0 \le x \le 4$ and $t \ge 0$, we have $C = f(x, t) = te^{-t(5-x)}$.

24. Find $f(3, 2)$. Give units and interpret in terms of drug concentration.

25. Graph the following two single variable functions and explain their significance in terms of drug concentration.

(a) $f(4, t)$ **(b)** $f(x, 1)$

26. Graph $f(a, t)$ for $a = 1, 2, 3, 4$ on the same axes. Describe how the graph changes as a increases and explain what this means in terms of drug concentration.

27. Describe the set of points whose distance from the x-axis is 2.

28. Describe the set of points whose distance from the x-axis equals the distance from the yz-plane.

29. A cube is located such that its top four corners have the coordinates $(-1, -2, 2)$, $(-1, 3, 2)$, $(4, -2, 2)$ and $(4, 3, 2)$. Give the coordinates of the center of the cube.

30. Find a formula for the shortest distance between a point (a, b, c) and the y-axis.

31. Find the equations of planes that just touch the sphere $(x-2)^2 + (y-3)^2 + (z-3)^2 = 16$ and are parallel to

(a) The xy-plane   **(b)** The yz-plane

(c) The xz-plane

32. Find an equation of the largest sphere contained in the cube determined by the planes $x = 2, x = 6; y = 5, y = 9$; and $z = -1, z = 3$.

33. Which of the following functions can be expressed in terms of the distance from the y-axis?

(a) $f(x, y, z) = 10e^y$

(b) $g(x, y, z) = x^2 z^2$

(c) $h(x, y, z) = 1/\sqrt{x^2 + b^2 + z^2}$, for constant b.

34. Let $f(x, y, z) = x^2 + y^2 + z^2 - 2x + 3y + z$.

(a) Find a point (a, b, c) so that f can be expressed in terms of distance from (a, b, c). [Hint: Complete the square.]

(b) Express f in terms of the distance d from the point you found in part (a).

35. Values of $f(x, y) = \frac{1}{2}(x + y - 2)(x + y - 1) + y$ are in Table 12.4.

(a) Find a pattern in the table. Make a conjecture and use it to complete Table 12.4 without computation. Check by using the formula for f.

(b) Using the formula, check that the pattern holds for all $x \ge 1$ and $y \ge 1$.

Table 12.4

				y			
		1	2	3	4	5	6
	1	1	3	6	10	15	21
	2	2	5	9	14	20	
x	3	4	8	13	19		
	4	7	12	18			
	5	11	17				
	6	16					

12.2   GRAPHS OF FUNCTIONS OF TWO VARIABLES

The weather map on page 560 is one way of visualizing a function of two variables. In this section we see how to visualize a function of two variables in another way, using a surface in 3-space.

Visualizing a Function of Two Variables Using a Graph

For a function of one variable, $y = f(x)$, the graph of f is the set of all points (x, y) in 2-space such that $y = f(x)$. In general, these points lie on a curve in the plane. When a computer or calculator graphs f, it approximates by plotting points in the xy-plane and joining consecutive points by line segments. The more points, the better the approximation.

Now consider a function of two variables.

> The **graph** of a function of two variables, f, is the set of all points (x, y, z) such that $z = f(x, y)$. In general, the graph of a function of two variables is a surface in 3-space.

Plotting the Graph of the Function $f(x, y) = x^2 + y^2$

To sketch the graph of f we connect points as for a function of one variable. We first make a table of values of f, such as in Table 12.5.

Table 12.5   *Table of values of $f(x, y) = x^2 + y^2$*

		\multicolumn{7}{c}{y}
		-3	-2	-1	0	1	2	3
	-3	18	13	10	9	10	13	18
	-2	13	8	5	4	5	8	13
	-1	10	5	2	1	2	5	10
x	0	9	4	1	0	1	4	9
	1	10	5	2	1	2	5	10
	2	13	8	5	4	5	8	13
	3	18	13	10	9	10	13	18

Now we plot points. For example, we plot $(1, 2, 5)$ because $f(1, 2) = 5$ and we plot $(0, 2, 4)$ because $f(0, 2) = 4$. Then, we connect the points corresponding to the rows and columns in the table. The result is called a *wire-frame* picture of the graph. Filling in between the wires gives a surface. That is the way a computer drew the graphs in Figure 12.11 and 12.12. As more points are plotted, we get the surface in Figure 12.13, called a *paraboloid*.

You should check to see if the sketches make sense. Notice that the graph goes through the origin since $(x, y, z) = (0, 0, 0)$ satisfies $z = x^2 + y^2$. Observe that if x is held fixed and y is allowed to vary, the graph dips down and then goes back up, just like the entries in the rows of Table 12.5. Similarly, if y is held fixed and x is allowed to vary, the graph dips down and then goes back up, just like the columns of Table 12.5.

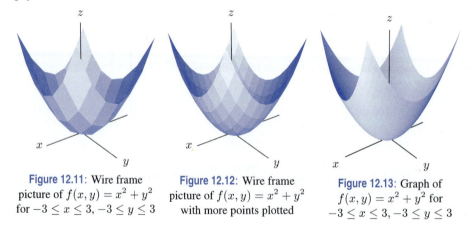

Figure 12.11: Wire frame picture of $f(x, y) = x^2 + y^2$ for $-3 \le x \le 3, -3 \le y \le 3$

Figure 12.12: Wire frame picture of $f(x, y) = x^2 + y^2$ with more points plotted

Figure 12.13: Graph of $f(x, y) = x^2 + y^2$ for $-3 \le x \le 3, -3 \le y \le 3$

New Graphs from Old

We can use the graph of a function to visualize the graphs of related functions.

Example 1 Let $f(x, y) = x^2 + y^2$. Describe in words the graphs of the following functions:
(a) $g(x, y) = x^2 + y^2 + 3$, (b) $h(x, y) = 5 - x^2 - y^2$, (c) $k(x, y) = x^2 + (y - 1)^2$.

Solution   We know from Figure 12.13 that the graph of f is a paraboloid, or bowl with its vertex at the origin. From this we can work out what the graphs of g, h, and k will look like.

(a) The function $g(x, y) = x^2 + y^2 + 3 = f(x, y) + 3$, so the graph of g is the graph of f, but raised by 3 units. See Figure 12.14.

(b) Since $-x^2 - y^2$ is the negative of $x^2 + y^2$, the graph of $-x^2 - y^2$ is an upside down paraboloid. Thus, the graph of $h(x, y) = 5 - x^2 - y^2 = 5 - f(x, y)$ looks like an upside down paraboloid with vertex at $(0, 0, 5)$, as in Figure 12.15.

(c) The graph of $k(x, y) = x^2 + (y - 1)^2 = f(x, y - 1)$ is a paraboloid with vertex at $x = 0$, $y = 1$, since that is where $k(x, y) = 0$, as in Figure 12.16.

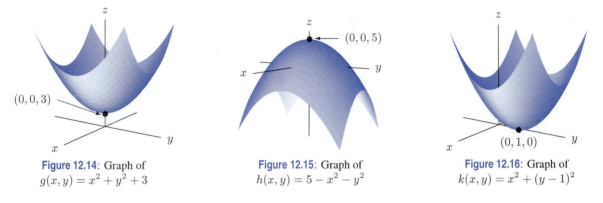

Figure 12.14: Graph of
$g(x, y) = x^2 + y^2 + 3$

Figure 12.15: Graph of
$h(x, y) = 5 - x^2 - y^2$

Figure 12.16: Graph of
$k(x, y) = x^2 + (y - 1)^2$

Example 2   Describe the graph of $G(x, y) = e^{-(x^2 + y^2)}$. What symmetry does it have?

Solution   Since the exponential function is always positive, the graph lies entirely above the xy-plane. From the graph of $x^2 + y^2$ we see that $x^2 + y^2$ is zero at the origin and gets larger as we move farther from the origin in any direction. Thus, $e^{-(x^2 + y^2)}$ is 1 at the origin, and gets smaller as we move away from the origin in any direction. It can't go below the xy-plane; instead it flattens out, getting closer and closer to the plane. We say the surface is *asymptotic* to the xy-plane. (See Figure 12.17.)

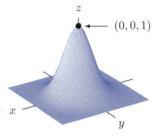

Figure 12.17: Graph of $G(x, y) = e^{-(x^2 + y^2)}$

Now consider a point (x, y) on the circle $x^2 + y^2 = r^2$. Since

$$G(x, y) = e^{-(x^2 + y^2)} = e^{-r^2},$$

the value of the function G is the same at all points on this circle. Thus, we say the graph of G has *circular symmetry*.

Cross-Sections and the Graph of a Function

We have seen that a good way to analyze a function of two variables is to let one variable vary at a time while the other is kept fixed.

For a function $f(x, y)$, the function we get by holding x fixed and letting y vary is called a **cross-section** of f with x fixed. The graph of the cross-section of $f(x, y)$ with $x = c$ is the curve, or cross-section, we get by intersecting the graph of f with the plane $x = c$. We define a cross-section of f with y fixed similarly.

For example, the cross-section of $f(x, y) = x^2 + y^2$ with $x = 2$ is $f(2, y) = 4 + y^2$. The graph of this cross-section is the curve we get by intersecting the graph of f with the plane perpendicular to the x-axis at $x = 2$. (See Figure 12.18.)

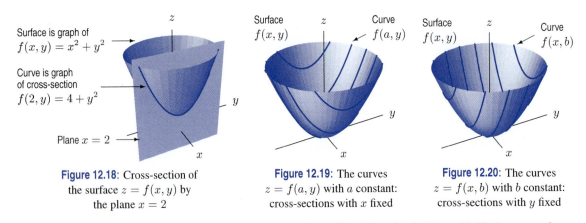

Figure 12.18: Cross-section of the surface $z = f(x, y)$ by the plane $x = 2$

Figure 12.19: The curves $z = f(a, y)$ with a constant: cross-sections with x fixed

Figure 12.20: The curves $z = f(x, b)$ with b constant: cross-sections with y fixed

Figure 12.19 shows graphs of other cross-sections of f with x fixed; Figure 12.20 shows graphs of cross-sections with y fixed.

Example 3   Describe the cross-sections of the function $g(x, y) = x^2 - y^2$ with y fixed and then with x fixed. Use these cross-sections to describe the shape of the graph of g.

Solution   The cross-sections with y fixed at $y = b$ are given by

$$z = g(x, b) = x^2 - b^2.$$

Thus, each cross-section with y fixed gives a parabola opening upwards, with minimum $z = -b^2$. The cross-sections with x fixed are of the form

$$z = g(a, y) = a^2 - y^2,$$

which are parabolas opening downwards with a maximum of $z = a^2$. (See Figures 12.21 and 12.22.) The graph of g is shown in Figure 12.23. Notice the upward opening parabolas in the x-direction and the downward opening parabolas in the y-direction. We say that the surface is *saddle-shaped*.

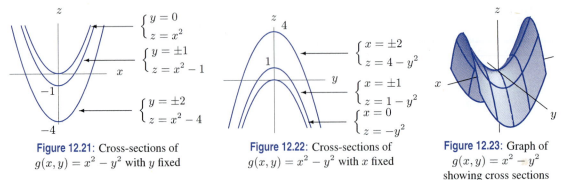

Figure 12.21: Cross-sections of $g(x, y) = x^2 - y^2$ with y fixed

Figure 12.22: Cross-sections of $g(x, y) = x^2 - y^2$ with x fixed

Figure 12.23: Graph of $g(x, y) = x^2 - y^2$ showing cross sections

Linear Functions

Linear functions are central to single variable calculus; they are equally important in multivariable calculus. You may be able to guess the shape of the graph of a linear function of two variables. (It's a plane.) Let's look at an example.

Example 4   Describe the graph of $f(x, y) = 1 + x - y$.

Solution   The plane $x = a$ is vertical and parallel to the yz-plane. Thus, the cross-section with $x = a$ is the line $z = 1 + a - y$ which slopes downward in the y-direction. Similarly, the plane $y = b$ is parallel to the xz-plane. Thus, the cross-section with $y = b$ is the line $z = 1 + x - b$ which slopes upward in the x-direction. Since all the cross-sections are lines, you might expect the graph to be a flat plane, sloping down in the y-direction and up in the x-direction. This is indeed the case. (See Figure 12.24.)

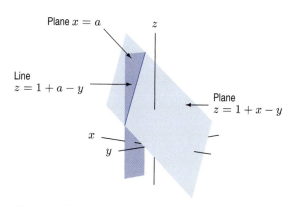

Figure 12.24: Graph of the plane $z = 1 + x - y$ showing cross-section with $x = a$

When One Variable is Missing: Cylinders

Suppose we graph an equation like $z = x^2$ which has one variable missing. What does the surface look like? Since y is missing from the equation, the cross-sections with y fixed are all the same parabola, $z = x^2$. Letting y vary up and down the y-axis, this parabola sweeps out the trough-shaped surface shown in Figure 12.25. The cross-sections with x fixed are horizontal lines obtained by cutting the surface by a plane perpendicular to the x-axis. This surface is called a *parabolic cylinder*, because it is formed from a parabola in the same way that an ordinary cylinder is formed from a circle; it has a parabolic cross-section instead of a circular one.

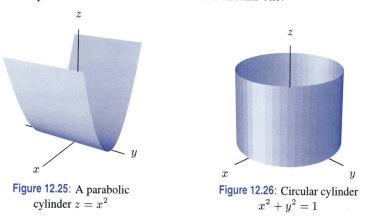

Figure 12.25: A parabolic cylinder $z = x^2$

Figure 12.26: Circular cylinder $x^2 + y^2 = 1$

Example 5   Graph the equation $x^2 + y^2 = 1$ in 3-space.

Solution   Although the equation $x^2 + y^2 = 1$ does not represent a function, the surface representing it can be graphed by the method used for $z = x^2$. The graph of $x^2 + y^2 = 1$ in the xy-plane is a circle. Since z does not appear in the equation, the intersection of the surface with any horizontal plane will be the same circle $x^2 + y^2 = 1$. Thus, the surface is the cylinder shown in Figure 12.26.

Exercises and Problems for Section 12.2

Exercises

In Exercises 1–7, sketch a graph of the surface and briefly describe it in words.

1. $z = 3$ **2.** $z = y^2$

3. $x^2 + y^2 + z^2 = 9$ **4.** $z = x^2 + y^2 + 4$

5. $2x + 4y + 3z = 12$ **6.** $x^2 + z^2 = 4$

7. $z = 5 - x^2 - y^2$

8. Figure 12.27 shows the graph of $z = f(x, y)$.

   (a) Suppose y is fixed and positive. Does z increase or decrease as x increases? Graph z against x.
   (b) Suppose x is fixed and positive. Does z increase or decrease as y increases? Graph z against y.

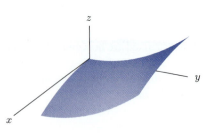

Figure 12.27

9. Without a calculator or computer, match the functions with their graphs in Figure 12.28.

   (a) $z = \dfrac{1}{x^2 + y^2}$ **(b)** $z = -e^{-x^2 - y^2}$

   (c) $z = x + 2y + 3$ **(d)** $z = -y^2$

   (e) $z = x^3 - \sin y$.

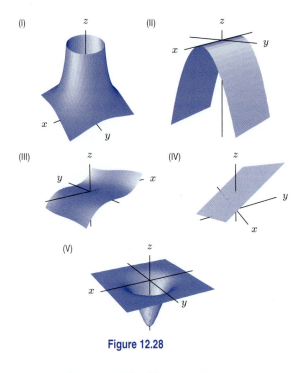

Figure 12.28

10. Match the following descriptions of a company's success with the graphs in Figure 12.29.

   (a) Our success is measured in dollars, plain and simple. More hard work won't hurt, but it also won't help.
   (b) No matter how much money or hard work we put into the company, we just couldn't make a go of it.
   (c) Although we aren't always totally successful, it seems that the amount of money invested doesn't matter. As long as we put hard work into the company our success will increase.
   (d) The company's success is based on both hard work and investment.

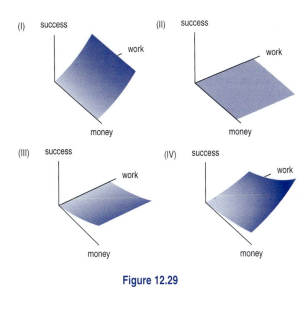

Figure 12.29

(c) There is such a thing as too much cola but no such thing as too many pizzas?

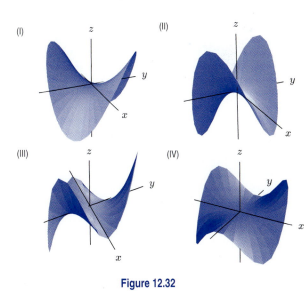

Figure 12.30

11. You like pizza and you like cola. Which of the graphs in Figure 12.30 represents your happiness as a function of how many pizzas and how much cola you have if

    (a) There is no such thing as too many pizzas and too much cola?

    (b) There is such a thing as too many pizzas or too much cola?

12. For each of the graphs I–IV in Problem 11 draw:

    (a) Two cross-sections with pizza fixed

    (b) Two cross-sections with cola fixed.

Problems

13. For each of the following functions, decide whether it could be a bowl, a plate, or neither. Consider a plate to be any fairly flat surface and a bowl to be anything that could hold water, assuming the positive z-axis is up.

    (a) $z = x^2 + y^2$ **(b)** $z = 1 - x^2 - y^2$

    (c) $x + y + z = 1$

    (d) $z = -\sqrt{5 - x^2 - y^2}$ **(e)** $z = 3$

14. For each function in Problem 13 sketch cross-sections.

15. Figure 12.31 contains graphs of the parabolas $z = f(x, b)$ for $b = -2, -1, 0, 1, 2$. Which of the graphs of $z = f(x, y)$ in Figure 12.32 best fits this information?

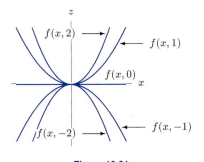

Figure 12.31

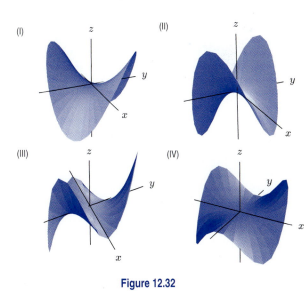

Figure 12.32

16. Without a computer or calculator, match the equations (a)–(i) with the graphs (I)–(IX).

    (a) $z = xye^{-(x^2+y^2)}$ **(b)** $z = \cos(\sqrt{x^2 + y^2})$

    (c) $z = \sin y$ **(d)** $z = -\dfrac{1}{x^2 + y^2}$

(e) $z = \cos^2 x \cos^2 y$ **(f)** $z = \dfrac{\sin(x^2 + y^2)}{x^2 + y^2}$

(g) $z = \cos(xy)$ **(h)** $z = |x||y|$

(i) $z = (2x^2 + y^2)e^{1-x^2-y^2}$

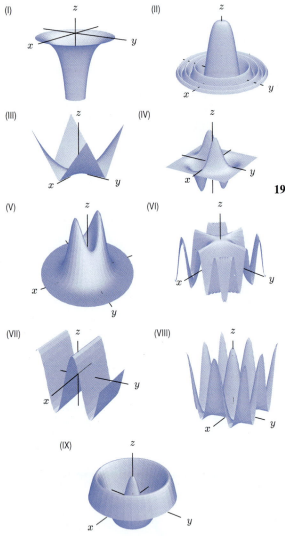

(b) Is the wave traveling in the direction of increasing or decreasing x?

(c) Sketch a surface representing a wave traveling in the opposite direction.

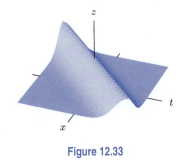

Figure 12.33

19. A swinging pendulum consists of a mass at the end of a string. At one moment the string makes an angle x with the vertical and the mass has speed y. At that time, the energy, E, of the pendulum is given by the expression[2]

$$E = 1 - \cos x + \frac{y^2}{2}.$$

(a) Consider the surface representing the energy. Sketch a cross-section of the surface:

   (i) Perpendicular to the x-axis at $x = c$.

   (ii) Perpendicular to the y-axis at $y = c$.

(b) For each of the graphs in Figures 12.34 and 12.35 use your answer to part (a) to decide which is the x-axis and which is the y-axis and to put reasonable units on each one.

17. Consider the function f given by $f(x, y) = y^3 + xy$. Draw graphs of cross-sections with:

(a) x fixed at $x = -1$, $x = 0$, and $x = 1$.

(b) y fixed at $y = -1$, $y = 0$, and $y = 1$.

18. A wave travels along a canal. Let x be the distance along the canal from the middle, t be the time, and z be the height of the water above the equilibrium level. The graph of z as a function of x and t is in Figure 12.33.

(a) Draw the profile of the wave for $t = -1, 0, 1, 2$. (Put the x-axis to the right and the z-axis vertical.)

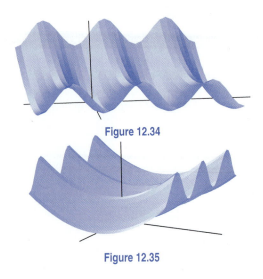

Figure 12.34

Figure 12.35

[2]Adapted from *Calculus in Context*, by James Callahan, Kenneth Hoffman, (New York: W.H. Freeman, 1995)

20. By setting one variable constant, find a plane that intersects the graph of $z = 4x^2 - y^2 + 1$ in a:

   (a) Parabola opening upward
   (b) Parabola opening downward
   (c) Pair of intersecting straight lines

21. By setting one variable constant, find a plane that intersects the graph of $z = (x^2 + 1)\sin y + xy^2$ in a:

   (a) Parabola
   (b) Straight line
   (c) Sine curve

12.3 CONTOUR DIAGRAMS

The surface which represents a function of two variables often gives a good idea of the function's general behavior—for example, whether it is increasing or decreasing as one of the variables increases. However it is difficult to read numerical values off a surface and it can be hard to see all of the function's behavior from a surface. Thus, functions of two variables are often represented by contour diagrams like the weather map on page 560. Contour diagrams have the additional advantage that they can be extended to functions of three variables.

Topographical Maps

One of the most common examples of a contour diagram is a topographical map like that shown in Figure 12.36. It gives the elevation in the region and is a good way of getting an overall picture of the terrain: where the mountains are, where the flat areas are. Such topographical maps are frequently colored green at the lower elevations and brown, red, or white at the higher elevations.

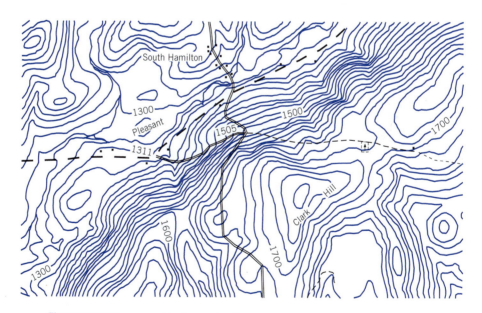

Figure 12.36: A topographical map showing the region around South Hamilton, NY

The curves on a topographical map that separate lower elevations from higher elevations are called *contour lines* because they outline the contour or shape of the land.[3] Because every point along the same contour has the same elevation, contour lines are also called *level curves* or *level sets*. The more closely spaced the contours, the steeper the terrain; the more widely spaced the contours, the flatter the terrain. (Provided, of course, that the elevation between contours varies by a constant amount.) Certain features have distinctive characteristics. A mountain peak is typically surrounded by contour lines like those in Figure 12.37. A pass in a range of mountains may have

---

[3] In fact they are usually not straight lines, but curves. They may also be in disconnected pieces.

contours that look like Figure 12.38. A long valley has parallel contour lines indicating the rising elevations on both sides of the valley (see Figure 12.39); a long ridge of mountains has the same type of contour lines, only the elevations decrease on both sides of the ridge. Notice that the elevation numbers on the contour lines are as important as the curves themselves. We usually draw contours for equally spaced values of z.

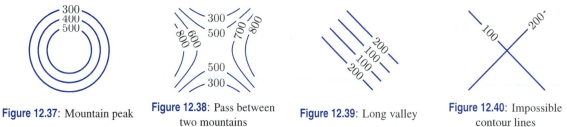

Figure 12.37: Mountain peak

Figure 12.38: Pass between two mountains

Figure 12.39: Long valley

Figure 12.40: Impossible contour lines

Notice that two contours corresponding to different elevations cannot cross each other as shown in Figure 12.40. If they did, the point of intersection of the two curves would have two different elevations, which is impossible (assuming the terrain has no overhangs).

Corn Production

Contour maps can display information about a function of two variables without reference to a surface. Consider the effect of weather conditions on US corn production. Figure 12.41 gives corn production $C = f(R, T)$ as a function of the total rainfall, R, in inches, and average temperature, T, in degrees Fahrenheit, during the growing season.[4] At the present time, $R = 15$ inches and $T = 76°$F. Production is measured as a percentage of the present production; thus, the contour through $R = 15, T = 76$, has value 100, that is, $C = f(15, 76) = 100$.

Example 1 Use Figure 12.41 to estimate $f(18, 78)$ and $f(12, 76)$ and interpret in terms of corn production.

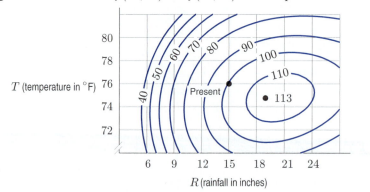

Figure 12.41: Corn production, C, as a function of rainfall and temperature

Solution The point with R-coordinate 18 and T-coordinate 78 is on the contour $C = 100$, so $f(18, 78) = 100$. This means that if the annual rainfall were 18 inches and the temperature were 78°F, the country would produce about the same amount of corn as at present, although it would be wetter and warmer than it is now.

The point with R-coordinate 12 and T-coordinate 76 is about halfway between the $C = 80$ and the $C = 90$ contours, so $f(12, 76) \approx 85$. This means that if the rainfall fell to 12 inches and the temperature stayed at 76°, then corn production would drop to about 85% of what it is now.

Example 2 Use Figure 12.41 to describe in words the cross-sections with T and R constant through the point representing present conditions. Give a common sense explanation of your answer.

[4]Adapted from S. Beaty and R. Healy, "The Future of American Agriculture," *Scientific American* 248, No.2, February 1983.

Solution   To see what happens to corn production if the temperature stays fixed at 76°F but the rainfall changes, look along the horizontal line $T = 76$. Starting from the present and moving left along the line $T = 76$, the values on the contours decrease. In other words, if there is a drought, corn production decreases. Conversely, as rainfall increases, that is, as we move from the present to the right along the line $T = 76$, corn production increases, reaching a maximum of more than 110% when $R = 21$, and then decreases (too much rainfall floods the fields).

  If, instead, rainfall remains at the present value and temperature increases, we move up the vertical line $R = 15$. Under these circumstances corn production decreases; a 2° increase causes a 10% drop in production. This makes sense since hotter temperatures lead to greater evaporation and hence drier conditions, even with rainfall constant at 15 inches. Similarly, a decrease in temperature leads to a very slight increase in production, reaching a maximum of around 102% when $T = 74$, followed by a decrease (the corn won't grow if it is too cold).

Contour Diagrams and Graphs

Contour diagrams and graphs are two different ways of representing a function of two variables. How do we go from one to the other? In the case of the topographical map, the contour diagram was created by joining all the points at the same height on the surface and dropping the curve into the xy-plane.

  How do we go the other way? Suppose we wanted to plot the surface representing the corn production function $C = f(R, T)$ given by the contour diagram in Figure 12.41. Along each contour the function has a constant value; if we take each contour and lift it above the plane to a height equal to this value, we get the surface in Figure 12.42.

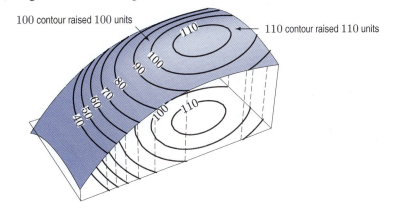

Figure 12.42: Getting the graph of the corn yield function from the contour diagram

Notice that the raised contours are the curves we get by slicing the surface horizontally. In general, we have the following result:

> Contour lines, or level curves, are obtained from a surface by slicing it with horizontal planes.

Finding Contours Algebraically

Algebraic equations for the contours of a function f are easy to find if we have a formula for $f(x, y)$. Suppose the surface has equation

$$z = f(x, y).$$

A contour is obtained by slicing the surface with a horizontal plane with equation $z = c$. Thus, the equation for the contour at height c is given by:

$$f(x, y) = c.$$

Example 3   Find equations for the contours of $f(x, y) = x^2 + y^2$ and draw a contour diagram for f. Relate the contour diagram to the graph of f.

Solution   The contour at height c is given by

$$f(x, y) = x^2 + y^2 = c.$$

This is a contour only for $c \geq 0$, For $c > 0$ it is a circle of radius $\sqrt{c}$. For $c = 0$, it is a single point (the origin). Thus, the contours at an elevation of $c = 1, 2, 3, 4, \dots$ are all circles centered at the origin of radius $1, \sqrt{2}, \sqrt{3}, 2, \dots$. The contour diagram is shown in Figure 12.43. The bowl–shaped graph of f is shown in Figure 12.44. Notice that the graph of f gets steeper as we move further away from the origin. This is reflected in the fact that the contours become more closely packed as we move further from the origin; for example, the contours for $c = 6$ and $c = 8$ are closer together than the contours for $c = 2$ and $c = 4$.

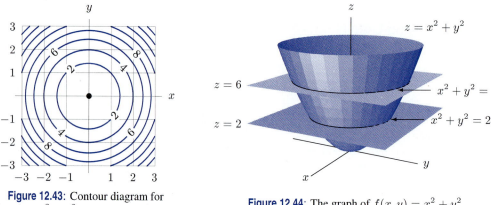

Figure 12.43: Contour diagram for $f(x, y) = x^2 + y^2$ (even values of c only)

Figure 12.44: The graph of $f(x, y) = x^2 + y^2$

Example 4   Draw a contour diagram for $f(x, y) = \sqrt{x^2 + y^2}$ and relate it to the graph of f.

Solution   The contour at level c is given by

$$f(x, y) = \sqrt{x^2 + y^2} = c.$$

For $c > 0$ this is a circle, just as in the previous example, but here the radius is c instead of $\sqrt{c}$. For $c = 0$, it is the origin. Thus, if the level c increases by 1, the radius of the contour increases by 1. This means the contours are equally spaced concentric circles (see Figure 12.45) which do not become more closely packed further from the origin. Thus, the graph of f has the same constant slope as we move away from the origin (see Figure 12.46), making it a cone rather than a bowl.

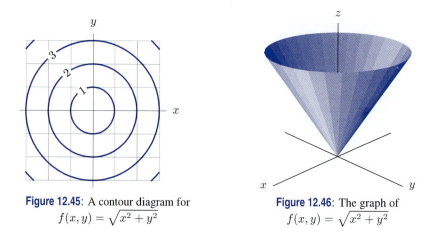

Figure 12.45: A contour diagram for $f(x, y) = \sqrt{x^2 + y^2}$

Figure 12.46: The graph of $f(x, y) = \sqrt{x^2 + y^2}$

In both of the previous examples the level curves are concentric circles because the surfaces have circular symmetry. Any function of two variables which depends only on the quantity (x^2+y^2) has such symmetry: for example, $G(x,y) = e^{-(x^2+y^2)}$ or $H(x,y) = \sin(\sqrt{x^2+y^2})$.

Example 5   Draw a contour diagram for $f(x,y) = 2x + 3y + 1$.

Solution   The contour at level c has equation $2x + 3y + 1 = c$. Rewriting this as $y = -(2/3)x + (c-1)/3$, we see that the contours are parallel lines with slope $-2/3$. The y-intercept for the contour at level c is $(c-1)/3$; each time c increases by 3, the y-intercept moves up by 1. The contour diagram is shown in Figure 12.47.

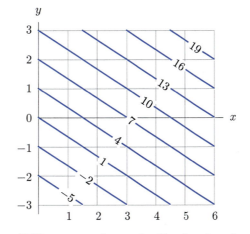

Figure 12.47: A contour diagram for $f(x,y) = 2x + 3y + 1$

Contour Diagrams and Tables

Sometimes we can get an idea of what the contour diagram of a function looks like from its table.

Example 6   Relate the values of $f(x,y) = x^2 - y^2$ in Table 12.6 to its contour diagram in Figure 12.48.

Table 12.6   *Table of values of $f(x,y) = x^2 - y^2$*

y	-3	-2	-1	0	1	2	3
3	0	-5	-8	-9	-8	-5	0
2	5	0	-3	-4	-3	0	5
1	8	3	0	-1	0	3	8
0	9	4	1	0	1	4	9
-1	8	3	0	-1	0	3	8
-2	5	0	-3	-4	-3	0	5
-3	0	-5	-8	-9	-8	-5	0

x

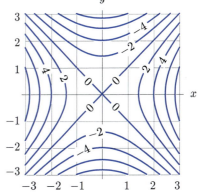

Figure 12.48: Contour map of $f(x,y) = x^2 - y^2$

Solution   One striking feature of the values in Table 12.6 is the zeros along the diagonals. This occurs because $x^2 - y^2 = 0$ along the lines $y = x$ and $y = -x$. So the $z = 0$ contour consists of these two lines. In the triangular region of the table that lies to the right of both diagonals, the entries are positive. To the left of both diagonals, the entries are also positive. Thus, in the contour diagram, the positive contours lie in the triangular regions to the right and left of the lines $y = x$ and $y = -x$. Further, the table shows that the numbers on the left are the same as the numbers on the right; thus, each

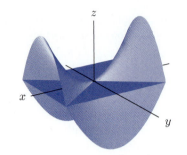

Figure 12.49: Graph of $f(x, y) = x^2 - y^2$ showing plane $z = 0$

contour has two pieces, one on the left and one on the right. See Figure 12.48. As we move away from the origin along the x-axis, we cross contours corresponding to successively larger values. On the saddle-shaped graph of $f(x, y) = x^2 - y^2$ shown in Figure 12.49, this corresponds to climbing out of the saddle along one of the ridges. Similarly, the negative contours occur in pairs in the top and bottom triangular regions; the values get more and more negative as we go out along the y-axis. This corresponds to descending from the saddle along the valleys that are submerged below the xy-plane in Figure 12.49. Notice that we could also get the contour diagram by graphing the family of hyperbolas $x^2 - y^2 = 0, \pm 2, \pm 4, \ldots$.

Using Contour Diagrams: The Cobb-Douglas Production Function

Suppose you decide to expand your small printing business. How should you expand? Should you start a night-shift and hire more workers? Should you buy more expensive but faster computers which will enable the current staff to keep up with the work? Or should you do some combination of the two?

Obviously, the way such a decision is made in practice involves many other considerations—such as whether you could get a suitably trained night shift, or whether there are any faster computers available. Nevertheless, you might model the quantity, P, of work produced by your business as a function of two variables: your total number, N, of workers, and the total value, V, of your equipment.

How would you expect such a production function to behave? In general, having more equipment and more workers enables you to produce more. However, increasing equipment without increasing the number of workers will increase production a bit, but not beyond a point. (If equipment is already lying idle, having more of it won't help.) Similarly, increasing the number of workers without increasing equipment will increase production, but not past the point where the equipment is fully utilized, as any new workers would have no equipment available to them.

Example 7   Explain why the contour diagram in Figure 12.50 does not model the behavior expected of the production function, whereas the contour diagram in Figure 12.51 does.

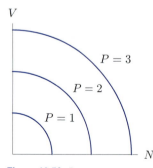

Figure 12.50: Incorrect contours for printing production

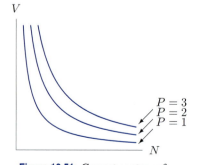

Figure 12.51: Correct contours for printing production

Solution   Look at Figure 12.50. Fixing V and letting N increase corresponds to moving to the right on the contour diagram. As you do so, you cross contours with larger and larger P values, meaning that production increases indefinitely. On the other hand, in Figure 12.51, as you move in the same direction you move nearly parallel to the contours, crossing them less and less frequently. Therefore, production increases more and more slowly as N increases with V fixed. Similarly, if you fix N and let V increase, the contour diagram in Figure 12.50 shows production increasing at a steady rate, whereas Figure 12.51 shows production increasing, but at a decreasing rate. Thus, Figure 12.51 fits the expected behavior of the production function best.

Formula for a Production Function

Production functions are often approximated by formulas of the form

$$P = f(N, V) = cN^{\alpha}V^{\beta}$$

where P is the quantity produced and c, α, and β are positive constants, $0 < \alpha < 1$ and $0 < \beta < 1$.

Example 8   Show that the contours of the function $P = cN^{\alpha}V^{\beta}$ have approximately the shape of the contours in Figure 12.51.

Solution   The contours are the curves where P is equal to a constant value, say P_0, that is, where

$$cN^{\alpha}V^{\beta} = P_0.$$

Solving for V we get

$$V = \left(\frac{P_0}{c}\right)^{1/\beta} N^{-\alpha/\beta}.$$

Thus, V is a power function of N with a negative exponent, so its graph has the shape shown in Figure 12.51.

The Cobb-Douglas Production Model

In 1928, Cobb and Douglas used a similar function to model the production of the entire US economy in the first quarter of this century. Using government estimates of P, the total yearly production between 1899 and 1922, of K, the total capital investment over the same period, and of L, the total labor force, they found that P was well approximated by the *Cobb-Douglas production function*

$$P = 1.01L^{0.75}K^{0.25}.$$

This function turned out to model the US economy surprisingly well, both for the period on which it was based, and for some time afterwards.

Exercises and Problems for Section 12.3

Exercises

In Exercises 1–9, sketch a contour diagram for the function with at least four labeled contours. Describe in words the contours and how they are spaced.

1. $f(x, y) = x + y$
2. $f(x, y) = xy$
3. $f(x, y) = x^2 + y^2$
4. $f(x, y) = 3x + 3y$
5. $f(x, y) = -x^2 - y^2 + 1$
6. $f(x, y) = x^2 + 2y^2$
7. $f(x, y) = \sqrt{x^2 + 2y^2}$
8. $f(x, y) = y - x^2$
9. $f(x, y) = \cos\sqrt{x^2 + y^2}$

10. Match the surfaces (a)–(e) in Figure 12.52 with the contour diagrams (I)–(V) in Figure 12.53.

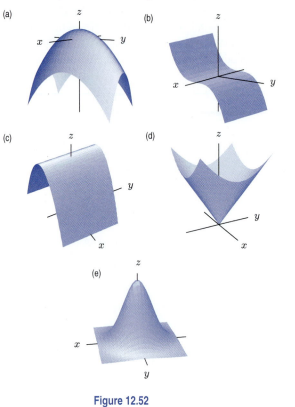

Figure 12.52

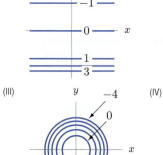

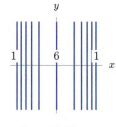

Figure 12.53

11. Match Tables 12.7–12.10 with the contour diagrams (I)–(IV) in Figure 12.54.

Table 12.7

$y \backslash x$	−1	0	1
−1	2	1	2
0	1	0	1
1	2	1	2

Table 12.8

$y \backslash x$	−1	0	1
−1	0	1	0
0	1	2	1
1	0	1	0

Table 12.9

$y \backslash x$	−1	0	1
−1	2	0	2
0	2	0	2
1	2	0	2

Table 12.10

$y \backslash x$	−1	0	1
−1	2	2	2
0	0	0	0
1	2	2	2

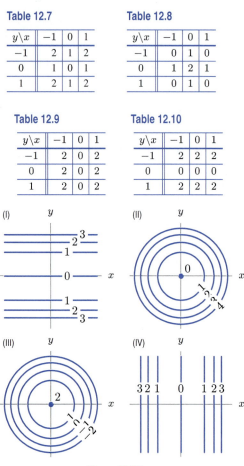

Figure 12.54

For each of the surfaces in Problems 12–15, sketch a possible contour diagram, marked with reasonable z-values. (Note: There are many possible answers.)

12.

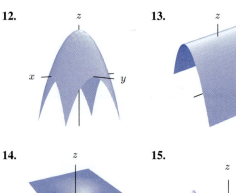

13.

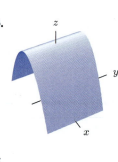

14.

15.

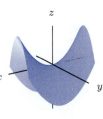

Problems

16. (a) For $z = f(x, y) = xy$, sketch and label the level curves $z = \pm 1$, $z = \pm 2$.
(b) Sketch and label cross-sections of f with $x = \pm 1$, $x = \pm 2$.
(c) The surface $z = xy$ is cut by a vertical plane containing the line $y = x$. Sketch the cross-section.

17. Figure 12.55 is a contour diagram of the monthly payment on a 5-year car loan as a function of the interest rate and the amount you borrow. The interest rate is 13% and you borrow $6000.

(a) What is your monthly payment?
(b) If interest rates drop to 11%, how much more can you borrow without increasing your monthly payment?
(c) Make a table of how much you can borrow, without increasing your monthly payment, as a function of the interest rate.

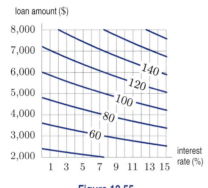

Figure 12.55

18. Total sales, Q, of a product is a function of its price and the amount spent on advertising. Figure 12.56 shows a contour diagram for total sales. Which axis corresponds to the price of the product and which to the amount spent on advertising? Explain.

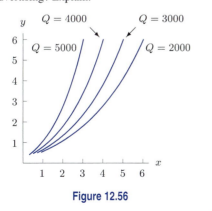

Figure 12.56

19. Figure 12.57 shows a contour map of a hill with two paths, A and B.

(a) On which path, A or B, will you have to climb more steeply?

(b) On which path, A or B, will you probably have a better view of the surrounding countryside? (Assuming trees do not block your view.)
(c) Alongside which path is there more likely to be a stream?

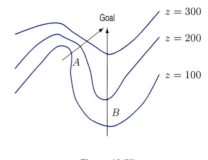

Figure 12.57

20. Match the contour diagrams (a)–(d) with the surfaces (I)–(IV). Give reasons for your choice.

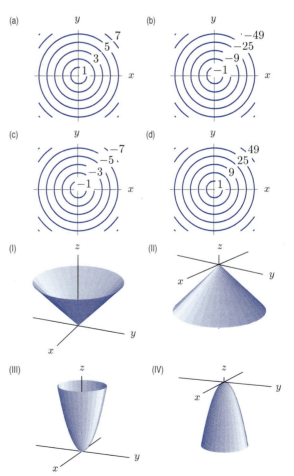

21. A manufacturer sells two goods, one at a price of $3000 a unit and the other at a price of $12,000 a unit. A quantity q_1 of the first good and q_2 of the second good are sold at a total cost of $4000 to the manufacturer.

   (a) Express the manufacturer's profit, π, as a function of q_1 and q_2.

   (b) Sketch curves of constant profit in the q_1q_2-plane for $\pi = 10,000$, $\pi = 20,000$, and $\pi = 30,000$ and the break-even curve $\pi = 0$.

22. Figure 12.58 shows the density of the fox population P (in foxes per square kilometer) for southern England. Draw two different cross-sections along a north-south line and two different cross-sections along an east-west line of the population density P.

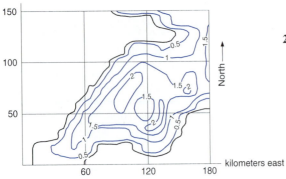

kilometers north

Figure 12.58: Population density of foxes in southwestern England

23. Match each Cobb-Douglas production function (a)-(c) with a graph in Figure 12.59 and a statement (D)-(G).

   (a) $F(L, K) = L^{0.25} K^{0.25}$
   (b) $F(L, K) = L^{0.5} K^{0.5}$
   (c) $F(L, K) = L^{0.75} K^{0.75}$

   (D) Tripling each input triples output.
   (E) Quadrupling each input doubles output.
   (G) Doubling each input almost triples output.

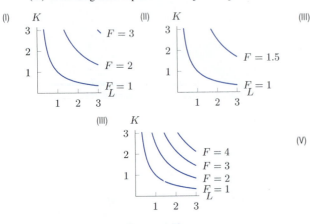

Figure 12.59

24. Consider the Cobb-Douglas production function $P = f(L, K) = 1.01L^{0.75}K^{0.25}$. What is the effect on production of doubling both labor and capital?

25. A general Cobb-Douglas production function has the form

$$P = cL^{\alpha}K^{\beta}.$$

What happens to production if labor and capital are both scaled up? For example, does production double if both labor and capital are doubled? Economists talk about

   - *increasing returns to scale* if doubling L and K more than doubles P,
   - *constant returns to scale* if doubling L and K exactly doubles P,
   - *decreasing returns to scale* if doubling L and K less than doubles P.

What conditions on α and β lead to increasing, constant, or decreasing returns to scale?

26. Figure 12.60 is the contour diagram of $f(x, y)$. Match the functions (a)–(f) with their contour diagrams (I)–(VI). All diagrams are drawn on the same region in the xy-plane.

   (a) $2f(x, y)$ **(b)** $f(x, y) + 2$ **(c)** $f(x - 1, y)$
   (d) $f(x, -y)$ **(e)** $f(2x, y)$ **(f)** $f(y, x)$

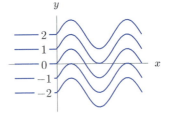

Figure 12.60

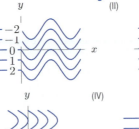

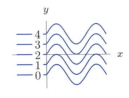

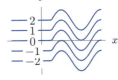

27. Match the functions (a)–(d) with the shapes of their level curves (I)–(IV). Sketch each contour diagram.

   (a) $f(x, y) = x^2$ **(b)** $f(x, y) = x^2 + 2y^2$

   (c) $f(x, y) = y - x^2$ **(d)** $f(x, y) = x^2 - y^2$

   I.   Lines   II.   Parabolas

   III. Hyperbolas   IV. Ellipses

28. Figure 12.61 is the contour diagram of $f(x, y)$. Sketch the contour diagram of each of the following functions.

   (a) $3f(x, y)$ **(b)** $f(x, y) - 10$

   (c) $f(x - 2, y - 2)$ **(d)** $f(-x, y)$

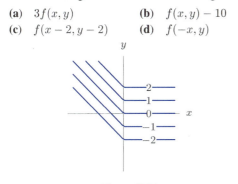

Figure 12.61

29. Figure 12.62 shows part of the contour diagram of $f(x, y)$. Complete the diagram for $x < 0$ if

   (a) $f(-x, y) = f(x, y)$ **(b)** $f(-x, y) = -f(x, y)$

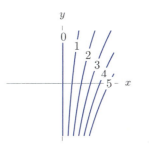

Figure 12.62

30. Figure 12.63 is the contour diagram of a function $f(x, y)$. Sketch a contour diagram of

   (a) $g(x, y) = (f(x, y))^2$ **(b)** $h(x, y) = \sin(f(x, y))$.

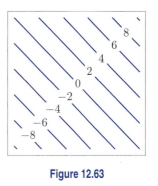

Figure 12.63

31. Use the factored form of $f(x, y) = x^2 - y^2 = (x - y)(x + y)$ to sketch the contour $f(x, y) = 0$ and to find the regions in the xy-plane where $f(x, y) > 0$ and the regions where $f(x, y) < 0$. Explain how this sketch shows that the graph of $f(x, y)$ is saddle-shaped at the origin.

32. Use Problem 31 to find a formula for a "monkey-saddle" surface $z = g(x, y)$ which has three regions with $g(x, y) > 0$ and three with $g(x, y) < 0$.

33. **(a)** Draw the contour diagram of $f(x, y) = g(y - x)$ if

      (i) $g(t) = t^2$ **(ii)** $g(t) = \sin t$

   (b) What can you say about the level curves of a function of the form $f(x, y) = g(y - x)$ where $g(t)$ is a one-variable function?

12.4   LINEAR FUNCTIONS

What is a Linear Function of Two Variables?

Linear functions played a central role in one-variable calculus because many one-variable functions have graphs that look like a line when we zoom in. In two-variable calculus, *linear functions* are those whose graph is a plane. In Chapter 14, we see that many two-variable functions have graphs which look like planes when we zoom in.

What Makes a Plane Flat?

What makes the graph of the function $z = f(x, y)$ a plane? Linear functions of *one* variable have straight line graphs because they have constant slope. On a plane, the situation is a bit more complicated. If we walk around on a tilted plane, the slope is not always the same: it depends on the direction in which we walk. However, at every point on the plane, the slope is the same as long as we choose the same direction. If we walk parallel to the x-axis, we always find ourselves walking up or down with the same slope; the same is true if we walk parallel to the y-axis. In other words, the slope ratios $\Delta z/\Delta x$ (with y fixed) and $\Delta z/\Delta y$ (with x fixed) are each constant.

Example 1   A plane cuts the z-axis at $z = 5$, has slope 2 in the x direction and slope -1 in the y direction. What is the equation of the plane?

Solution   Finding the equation of the plane means constructing a formula for the z-coordinate of the point on the plane directly above the point (x, y) in the xy-plane. To get to that point start from the point above the origin, where $z = 5$. Then walk x units in the x direction. Since the slope in the x direction is 2, the height increases by $2x$. Then walk y units in the y direction; since the slope in the y direction is -1, the height decreases by y units. Since the height has changed by $2x - y$ units, the z-coordinate is $5 + 2x - y$. Thus, the equation for the plane is

$$z = 5 + 2x - y.$$

For any linear function, if we know its value at a point (x_0, y_0), its slope in the x direction, and its slope in the y direction, then we can write the equation of the function. This is just like the equation of a line in the one-variable case, except that there are two slopes instead of one.

If a **plane** has slope m in the x direction, slope n in the y direction, and passes through the point (x_0, y_0, z_0), then its equation is

$$z = z_0 + m(x - x_0) + n(y - y_0).$$

This plane is the graph of the **linear function**

$$f(x, y) = z_0 + m(x - x_0) + n(y - y_0).$$

If we write $c = z_0 - mx_0 - ny_0$, then we can write $f(x, y)$ in the equivalent form

$$f(x, y) = c + mx + ny.$$

Just as in 2-space a line is determined by two points, so in 3-space a plane is determined by three points, provided they do not lie on a line.

Example 2   Find the equation of the plane passing through the points $(1, 0, 1)$, $(1, -1, 3)$, and $(3, 0, -1)$.

Solution   The first two points have the same x-coordinate, so we use them to find the slope of the plane in the y-direction. As the y-coordinate changes from 0 to -1, the z-coordinate changes from 1 to 3, so the slope in the y-direction is $n = \Delta z / \Delta y = (3 - 1)/(-1 - 0) = -2$. The first and third points have the same y-coordinate, so we use them to find the slope in the x-direction; it is $m = \Delta z / \Delta x = (-1 - 1)/(3 - 1) = -1$. Because the plane passes through $(1, 0, 1)$, its equation is

$$z = 1 - (x - 1) - 2(y - 0) \quad \text{or} \quad z = 2 - x - 2y.$$

You should check that this equation is also satisfied by the points $(1, -1, 3)$ and $(3, 0, -1)$.

Example 2 was made easier by the fact that two of the points had the same x-coordinate and two had the same y-coordinate. An alternative method, which works for any three points, is to substitute the x, y, and z-values of each of the three points into the equation $z = c + mx + ny$. The resulting three equations in c, m, n are then solved simultaneously.

Linear Functions from a Numerical Point of View

To avoid flying planes with empty seats, airlines sell some tickets at full price and some at a discount. Table 12.11 shows an airline's revenue in dollars from tickets sold on a particular route, as a function of the number of full-price tickets sold, f, and the number of discount tickets sold, d.

Table 12.11   *Revenue from ticket sales (dollars)*

		Full-price tickets (f)			
		100	200	300	400
	200	39,700	63,600	87,500	111,400
	400	55,500	79,400	103,300	127,200
Discount tickets (d)	600	71,300	95,200	119,100	143,000
	800	87,100	111,000	134,900	158,800
	1000	102,900	126,800	150,700	174,600

In every column, the revenue jumps by $15,800 for each extra 200 discount tickets. Thus, each column is a linear function of the number of discount tickets sold. In addition, every column has the same slope, $15,800/200 = 79$ dollars/ticket. This is the price of a discount ticket. Similarly, each row is a linear function and all the rows have the same slope, 239, which is the price in dollars of a full-fare ticket. Thus, R is a linear function of f and d, given by:

$$R = 239f + 79d.$$

We have the following general result:

A **linear function** can be recognized from its table by the following features:
- Each row and each column is linear.
- All the rows have the same slope.
- All the columns have the same slope (although the slope of the rows and the slope of the columns are generally different).

Example 3   The table contains values of a linear function. Fill in the blank and give a formula for the function.

$x \backslash y$	1.5	2.0
2	0.5	1.5
3	-0.5	?

Solution   In the first column the function decreases by 1 (from 0.5 to -0.5) as x goes from 2 to 3. Since the function is linear, it must decrease by the same amount in the second column. So the missing entry must be $1.5 - 1 = 0.5$. The slope of the function in the x-direction is -1. The slope in the y-direction is 2, since in each row the function increases by 1 when y increases by 0.5. From the table we get $f(2, 1.5) = 0.5$. Therefore, the formula is

$$f(x, y) = 0.5 - (x - 2) + 2(y - 1.5) = -0.5 - x + 2y.$$

What Does the Contour Diagram of a Linear Function Look Like?

The formula for the airline revenue function in Table 12.11 is $R = 239f + 79d$, where f is the number of full-fares and d is the number of discount fares sold.

Notice that the contours of this function in Figure 12.64 are parallel straight lines. What is the practical significance of the slope of these contour lines? Consider the contour $R = 100,000$; that means we are looking at combinations of ticket sales that yield $100,000 in revenue. If we move down and to the right on the contour, the f-coordinate increases and the d-coordinate decreases, so we sell more full-fares and fewer discount fares. This is because to receive a fixed revenue of $100,000, we must sell more full-fares if we sell fewer discount fares. The exact trade-off depends on the slope of the contour; the diagram shows that each contour has a slope of about -3. This means that for a fixed revenue, we must sell three discount fares to replace one full-fare. This can also be seen by comparing prices. Each full fare brings in \$239; to earn the same amount in discount fares we need to sell $239/79 \approx 3.03 \approx 3$ fares. Since the price ratio is independent of how many of each type of fare we sell, this slope remains constant over the whole contour map; thus, the contours are all parallel straight lines.

Notice also that the contours are evenly spaced. Thus, no matter which contour we are on, a fixed increase in one of the variables causes the same increase in the value of the function. In terms of revenue, no matter how many fares we have sold, an extra fare, whether full or discount, brings the same revenue as before.

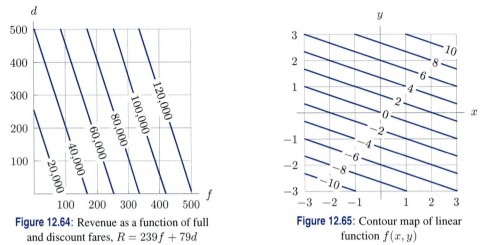

Figure 12.64: Revenue as a function of full and discount fares, $R = 239f + 79d$

Figure 12.65: Contour map of linear function $f(x, y)$

Example 4 Find the equation of the linear function whose contour diagram is in Figure 12.65.

Solution Suppose we start at the origin on the $z = 0$ contour. Moving 2 units in the y direction takes us to the $z = 6$ contour; so the slope in the y direction is $\Delta z/\Delta y = 6/2 = 3$. Similarly, a move of 2 in the x-direction from the origin takes us to the $z = 2$ contour, so the slope in the x direction is $\Delta z/\Delta x = 2/2 = 1$. Since $f(0, 0) = 0$, we have $f(x, y) = x + 3y$.

Exercises and Problems for Section 12.4

Exercises

Which of the tables of values in Exercises 1–4 could represent linear functions?

1.

		y	
	0	1	2
x 0	10	13	16
1	6	9	12
2	2	5	8

2.

		y	
	0	1	2
x 0	0	1	4
1	1	0	1
2	4	1	0

3.

		y	
	0	1	2
x 0	0	5	10
1	2	7	12
2	4	9	14

4.

		y	
	0	1	2
x 0	5	7	9
1	6	9	12
2	7	11	15

5. The total charge, C, in dollars, to use an internet service is a function of m, the number of months of use, and t, the total number of minutes on-line:

$$C = f(m, t) = 35 + 15m + 0.05t.$$

(a) Is f a linear function?
(b) Give units for the coefficients of m and t, and interpret them in terms of charges.
(c) Interpret the intercept 35 as a charge.
(d) Find $f(3, 800)$ and interpret your answer.

6. Suppose that z is a linear function of x and y with slope 2 in the x direction and slope 3 in the y direction.

(a) A change of 0.5 in x and -0.2 in y produces what change in z?
(b) If $z = 2$ when $x = 5$ and $y = 7$, what is the value of z when $x = 4.9$ and $y = 7.2$?

Which of the contour diagrams in Exercises 7–10 could represent linear functions?

7.

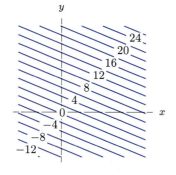

8.

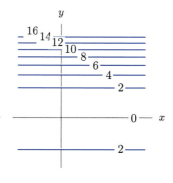

9.

10.

11. **(a)** Find a formula for the linear function whose graph is a plane passing through point $(4, 3, -2)$ with slope 5 in the x-direction and slope -3 in the y-direction.

(b) Sketch the contour diagram for this function.

For Exercises 12–13, find possible equations for linear functions with the given contour diagrams.

12.

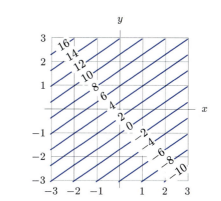

13.

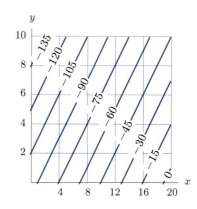

For Exercises 14–15, find equations for linear functions with the given values.

14.

$x \backslash y$	−1	0	1	2
0	1.5	1	0.5	0
1	3.5	3	2.5	2
2	5.5	5	4.5	4
3	7.5	7	6.5	6

15.

$x \backslash y$	10	20	30	40
100	3	6	9	12
200	2	5	8	11
300	1	4	7	10
400	0	3	6	9

Problems

16. Find the equation of the linear function $z = c + mx + ny$ whose graph contains the points $(0, 0, 0)$, $(0, 2, -1)$, and $(-3, 0, -4)$.

17. Find the linear function whose graph is the plane through the points $(4, 0, 0)$, $(0, 3, 0)$ and $(0, 0, 2)$.

18. Find an equation for the plane containing the line in the xy-plane where $y = 1$, and the line in the xz-plane where $z = 2$.

19. Find the equation of the linear function $z = c + mx + ny$ whose graph intersects the xz-plane in the line $z = 3x + 4$ and intersects the yz-plane in the line $z = y + 4$.

Problems 20–22 concern Table 12.12, which gives the number of calories burned per minute for someone roller-blading, as a function of the person's weight and speed.[5]

Table 12.12

	Calories burned per minute			
Weight	8 mph	9 mph	10 mph	11 mph
120 lbs	4.2	5.8	7.4	8.9
140 lbs	5.1	6.7	8.3	9.9
160 lbs	6.1	7.7	9.2	10.8
180 lbs	7.0	8.6	10.2	11.7
200 lbs	7.9	9.5	11.1	12.6

20. Does the data in Table 12.12 look approximately linear? Give a formula for B, the number of calories burned per minute in terms of the weight, w, and the speed, s. Does the formula make sense for all weights or speeds?

21. Who burns more total calories to go 10 miles: A 120 lb person going 10 mph or a 180 lb person going 8 mph? Which of these two people burns more calories per pound for the 10-mile trip?

22. Use Problem 20 to give a formula for P, the number of calories burned per pound, in terms of w and s, for a person weighing w lbs roller-blading 10 miles at s mph.

Problems 23–24 refer to the linear function $z = f(x, y)$ whose values are in Table 12.13.

Table 12.13

x \ y	4	6	8	10	12
5	3	6	9	12	15
10	7	10	13	16	19
15	11	14	17	20	23
20	15	18	21	24	27
25	19	22	25	28	31

23. Each column of Table 12.13 is linear with the same slope, $m = \Delta z/\Delta x = 4/5$. Each row is linear with the same slope, $n = \Delta z/\Delta y = 3/2$. We now investigate the slope obtained by moving through the table along lines that are neither rows nor columns.

(a) Move down the diagonal of the table from the upper left corner ($z = 3$) to the lower right corner ($z = 31$). What do you notice about the changes in z? Now move diagonally from $z = 6$ to $z = 27$. What do you notice about the changes in z now?

(b) Move in the table along a line right one step, up two steps from $z = 19$ to $z = 9$. Then move in the same direction from $z = 22$ to $z = 12$. What do you notice about the changes in z?

(c) Show that $\Delta z = m\Delta x + n\Delta y$. Use this to explain what you observed in parts (a) and (b).

24. If we hold y fixed, that is we keep $\Delta y = 0$, and step in the positive x-direction, we get the x-slope, m. If instead we keep $\Delta x = 0$ and step in the positive y-direction, we get the y-slope, n. Fix a step in which neither $\Delta x = 0$ nor $\Delta y = 0$. The slope in the $\Delta x, \Delta y$ direction is

$$\text{Slope} = \frac{\text{Rise}}{\text{Run}} = \frac{\Delta z}{\text{Length of step}} = \frac{\Delta z}{\sqrt{(\Delta x)^2 + (\Delta y)^2}}.$$

(a) Compute the slopes for the linear function in Table 12.13 in the direction of $\Delta x = 5, \Delta y = 2$.

(b) Compute the slopes for the linear function in Table 12.13 in the direction of $\Delta x = -10, \Delta y = 2$.

It is difficult to graph a linear function by hand. One method that works if the x, y, and z-intercepts are positive is to plot the intercepts and join them by a triangle as shown in Figure 12.66; this shows the part of the plane in the octant where $x \geq 0$, $y \geq 0$, $z \geq 0$. If the intercepts are not all positive, the same method works if the x, y, and z-axes are drawn from a different perspective. Use this method to graph the linear functions in Problems 25–28.

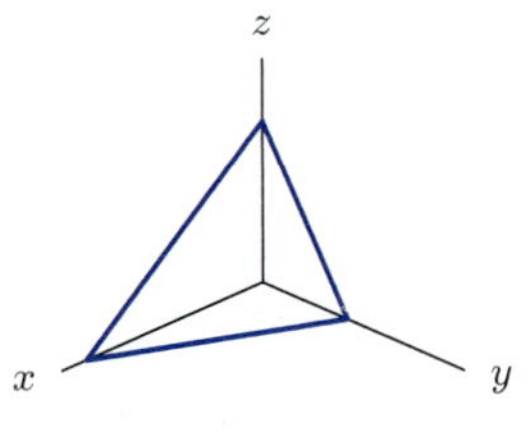

Figure 12.66

25. $z = 2 - x - 2y$

26. $z = 6 - 2x - 3y$

27. $z = 4 + x - 2y$

28. $z = 2 - 2x + y$

[5]From the August 28,1994, issue of *Parade Magazine*.

29. A manufacturer makes two products out of two raw materials. Let q_1, q_2 be the quantities sold of the two products, p_1, p_2 their prices, and m_1, m_2 the quantities purchased of the two raw materials. Which of the following functions do you expect to be linear, and why? In each case, assume that all variables except the ones mentioned are held fixed.

 (a) Expenditure on raw materials as a function of m_1 and m_2.

 (b) Revenue as a function of q_1 and q_2.

 (c) Revenue as a function of p_1 and q_1.

30. Let f be the linear function $f(x, y) = c + mx + ny$, where c, m, n are constants and $n \neq 0$.

 (a) Show that all the contours of f are lines of slope $-m/n$.

 (b) For all x and y, show $f(x + n, y - m) = f(x, y)$.

 (c) Explain the relation between parts (a) and (b).

12.5   FUNCTIONS OF THREE VARIABLES

In applications of calculus, functions of any number of variables can arise. The density of matter in the universe is a function of three variables, since it takes three numbers to specify a point in space. Models of the US economy often use functions of ten or more variables. We need to be able to apply calculus to functions of arbitrarily many variables.

One difficulty with functions of more than two variables is that it is hard to visualize them. The graph of a function of one variable is a curve in 2-space, the graph of a function of two variables is a surface in 3-space, so the graph of a function of three variables would be a solid in 4-space. Since we can't easily visualize 4-space, we won't use the graphs of functions of three variables. On the other hand, it is possible to draw contour diagrams for functions of three variables, only now the contours are surfaces in 3-space.

Representing a Function of Three Variables Using a Family of Level Surfaces

A function of two variables, $f(x, y)$, can be represented by a family of level curves of the form $f(x, y) = c$ for various values of the constant, c.

> A **level surface**, or **level set** of a function of three variables, $f(x, y, z)$, is a surface of the form $f(x, y, z) = c$, where c is a constant. The function f can be represented by the family of level surfaces obtained by allowing c to vary.

The value of the function, f, is constant on each level surface.

Example 1   The temperature, in °C, at a point (x, y, z) is given by $T = f(x, y, z) = x^2 + y^2 + z^2$. What do the level surfaces of the function f look like and what do they mean in terms of temperature?

Solution   The level surface corresponding to $T = 100$ is the set of all points where the temperature is 100°C. That is, where $f(x, y, z) = 100$, so

$$x^2 + y^2 + z^2 = 100.$$

This is the equation of a sphere of radius 10, with center at the origin. Similarly, the level surface corresponding to $T = 200$ is the sphere with radius $\sqrt{200}$. The other level surfaces are concentric spheres. The temperature is constant on each sphere. We may view the temperature distribution as a set of nested spheres, like concentric layers of an onion, each one labeled with a different temperature, starting from low temperatures in the middle and getting hotter as we go out from the center. (See Figure 12.67.) The level surfaces become more closely spaced as we move farther from the origin because the temperature increases more rapidly the farther we get from the origin.

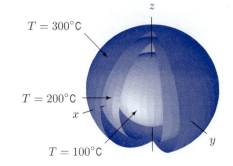

Figure 12.67: Level surfaces of $T = f(x, y, z) = x^2 + y^2 + z^2$, each one having a constant temperature

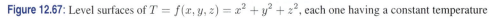

Example 2   What do the level surfaces of $f(x, y, z) = x^2 + y^2$ and $g(x, y, z) = z - y$ look like?

Solution   The level surface of f corresponding to the constant c is the surface consisting of all points satisfying the equation

$$x^2 + y^2 = c.$$

Since there is no z-coordinate in the equation, z can take any value. For $c > 0$, this is a circular cylinder of radius $\sqrt{c}$ around the z-axis. The level surfaces are concentric cylinders; on the narrow ones near the z-axis, f has small values; on the wider ones, f has larger values. See Figure 12.68.

The level surface of g corresponding to the constant c is the surface

$$z - y = c.$$

This time there is no x variable, so this surface is the one we get by taking each point on the straight line $z - y = c$ in the yz-plane and letting x vary. We get a plane which cuts the yz-plane diagonally; the x-axis is parallel to this plane. See Figure 12.69.

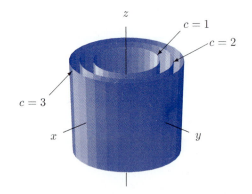

Figure 12.68: Level surfaces of $f(x, y, z) = x^2 + y^2$

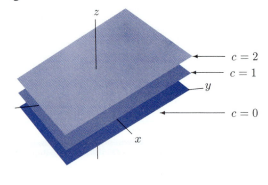

Figure 12.69: Level surfaces of $g(x, y, z) = z - y$

Example 3   What do the level surfaces of $f(x, y, z) = x^2 + y^2 - z^2$ look like?

Solution   In Section 12.3, we saw that the two-variable quadratic function $g(x, y) = x^2 - y^2$ has a saddle-shaped graph and three types of contours. The contour equation $x^2 - y^2 = c$ gives a hyperbola opening right-left when $c > 0$, a hyperbola opening up-down when $c < 0$, and a pair of intersecting lines when $c = 0$. Similarly, the three-variable quadratic function $f(x, y, z) = x^2 + y^2 - z^2$ has three types of level surfaces depending on the value of c in the equation $x^2 + y^2 - z^2 = c$.

Suppose that $c > 0$, say $c = 1$. Rewrite the equation as $x^2 + y^2 = z^2 + 1$ and think of what happens as we cut the surface perpendicular to the z-axis by holding z fixed. The result is a circle, $x^2 + y^2 = $ constant, of radius at least 1 (since the constant $z^2 + 1 \geq 1$). The circles get larger as z gets larger. If we take the $x = 0$ cross-section instead, we get the hyperbola $y^2 - z^2 = 1$. The result is shown in Figure 12.73, with $a = b = c = 1$.

Suppose instead $c < 0$, say $c = -1$. Then the horizontal cross-sections of $x^2 + y^2 = z^2 - 1$ are again circles except that the radii shrink to 0 at $z = \pm 1$ and between $z = -1$ and $z = 1$ there are no cross-sections at all. The result is shown in Figure 12.74 with $a = b = c = 1$.

When $c = 0$, we get the equation $x^2 + y^2 = z^2$. Again the horizontal cross-sections are circles, this time with the radius shrinking down to exactly 0 when $z = 0$. The resulting surface, shown in Figure 12.75 with $a = b = c = 1$, is the cone $z = \sqrt{x^2 + y^2}$ studied in Section 12.3, together with the lower cone $z = -\sqrt{x^2 + y^2}$.

A Catalog of Surfaces

For later reference, here is a small catalog of the surfaces we have encountered.

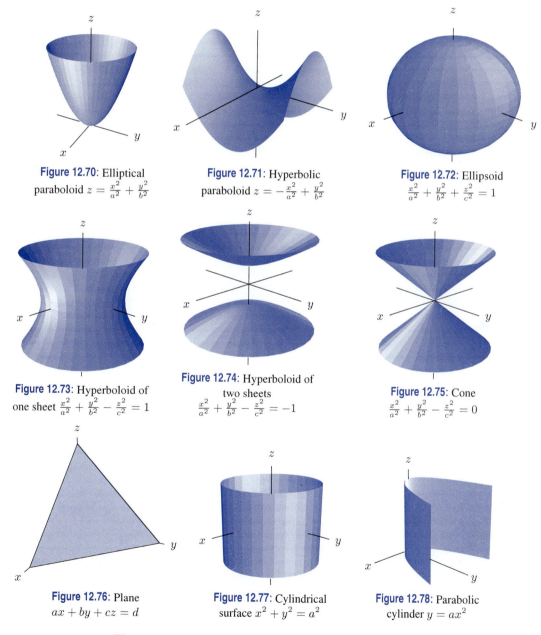

Figure 12.70: Elliptical paraboloid $z = \frac{x^2}{a^2} + \frac{y^2}{b^2}$

Figure 12.71: Hyperbolic paraboloid $z = -\frac{x^2}{a^2} + \frac{y^2}{b^2}$

Figure 12.72: Ellipsoid $\frac{x^2}{a^2} + \frac{y^2}{b^2} + \frac{z^2}{c^2} = 1$

Figure 12.73: Hyperboloid of one sheet $\frac{x^2}{a^2} + \frac{y^2}{b^2} - \frac{z^2}{c^2} = 1$

Figure 12.74: Hyperboloid of two sheets $\frac{x^2}{a^2} + \frac{y^2}{b^2} - \frac{z^2}{c^2} = -1$

Figure 12.75: Cone $\frac{x^2}{a^2} + \frac{y^2}{b^2} - \frac{z^2}{c^2} = 0$

Figure 12.76: Plane $ax + by + cz = d$

Figure 12.77: Cylindrical surface $x^2 + y^2 = a^2$

Figure 12.78: Parabolic cylinder $y = ax^2$

(These are viewed as equations in three variables $x, y,$ and z)

How Surfaces Can Represent Functions of Two Variables and Functions of Three Variables

You may have noticed that we have used surfaces to represent functions in two different ways. First, we used a *single* surface to represent a two-variable function $f(x, y)$. Second, we used a *family* of level surfaces to represent a three-variable function $g(x, y, z)$. These level surfaces have equation $g(x, y, z) = c$.

What is the relation between these two uses of surfaces? For example, consider the function

$$f(x, y) = x^2 + y^2 + 3.$$

Define

$$g(x, y, z) = x^2 + y^2 + 3 - z$$

The points on the graph of f satisfy $z = x^2 + y^2 + 3$, so they also satisfy $x^2 + y^2 + 3 - z = 0$. Thus the graph of f is the same as the level surface

$$g(x, y, z) = x^2 + y^2 + 3 - z = 0.$$

In general, we have the following result:

A single surface that is the graph of a two-variable function $f(x, y)$ can be thought of as one member of the family of level surfaces representing the three-variable function

$$g(x, y, z) = f(x, y) - z.$$

The graph of f is the level surface $g = 0$.

Conversely, a single level surface $g(x, y, z) = c$ can be regarded as the graph of a function $f(x, y)$ if it is possible to solve for z. Sometimes the level surface is pieced together from the graphs of two or more two-variable functions. For example, if $g(x, y, z) = x^2 + y^2 + z^2$, then one member of the family of level surfaces is the sphere

$$x^2 + y^2 + z^2 = 1.$$

This equation defines z implicitly as a function of x and y. Solving it gives two functions

$$z = \sqrt{1 - x^2 - y^2} \qquad \text{and} \qquad z = -\sqrt{1 - x^2 - y^2}.$$

The graph of the first function is the top half of the sphere and the graph of the second function is the bottom half.

Exercises and Problems for Section 12.5

Exercises

1. Match the following functions with the level surfaces in Figure 12.79.

   (a) $f(x, y, z) = y^2 + z^2$ **(b)** $h(x, y, z) = x^2 + z^2$.

2. Find a formula for a function $f(x, y, z)$ whose level surfaces look like those in Figure 12.80.

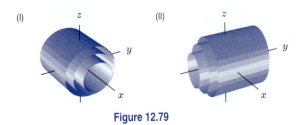

Figure 12.79

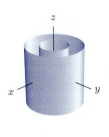

Figure 12.80

In Exercises 3–5, represent the surface whose equation is given as the graph of a two-variable function, $f(x, y)$, and as the level surface of a three-variable function, $g(x, y, z) = c$. There are many possible answers.

3. The plane $4x - y - 2z = 6$

4. The top half of the sphere $x^2 + y^2 + z^2 - 10 = 0$

5. The bottom half of the ellipsoid $x^2 + y^2 + z^2/2 = 1$

In Exercises 6–9, decide if the given level surface can be expressed as the graph of a function, $f(x, y)$.

6. $2x + 3y - 5z - 10 = 0$ **7.** $x^2 + y^2 + z^2 - 1 = 0$

8. $z - x^2 - 3y^2 = 0$ **9.** $z^2 = x^2 + 3y^2$

10. Find a formula for a function $f(x, y, z)$ whose level surface $f = 4$ is a sphere of radius 2, centered at the origin.

11. Find a formula for a function $f(x, y, z)$ whose level surfaces are spheres centered at the point (a, b, c).

12. Write the level surface $x + 2y + 3z = 5$ as the graph of a function $f(x, y)$.

13. Write the level surface $x^2 + y + \sqrt{z} = 1$ as the graph of a function $f(x, y)$.

14. Which of the graphs in the catalog of surfaces on page 593 is the graph of a function of x and y?

Use the catalog on page 593 to identify the surfaces in Exercises 15–18.

15. $-x^2 - y^2 + z^2 = 1$ **16.** $-x^2 + y^2 - z^2 = 0$

17. $x^2 + y^2 - z = 0$ **18.** $x^2 + z^2 = 1$

Problems

19. Find a function $f(x, y, z)$ whose level surface $f = 1$ is the graph of the function $g(x, y) = x + 2y$.

20. Find two functions $f(x, y)$ and $g(x, y)$ so that the graphs of both together form the ellipsoid $x^2 + y^2/4 + z^2/9 = 1$.

21. Describe, in words, the level surface $f(x, y, z) = x^2/4 + z^2 = 1$.

22. Describe, in words, the level surface $g(x, y, z) = x^2 + y^2/4 + z^2 = 1$. [Hint: Look at cross-sections with constant x, y and z values.]

23. Describe in words the level surfaces of the function $g(x, y, z) = x + y + z$.

24. Find a formula for a function $g(x, y, z)$ whose level surfaces are planes parallel to the plane $z = 2x + 3y - 5$.

25. Describe in words the level surfaces of $f(x, y, z) = \sin(x + y + z)$.

26. Describe in words the level surfaces of $g(x, y, z) = e^{-(x^2 + y^2 + z^2)}$.

27. The surface S is the graph of $f(x, y) = \sqrt{1 - x^2 - y^2}$.

  (a) Explain why S is the upper hemisphere of radius 1, with equator in the xy-plane, centered at the origin.

  (b) Find a level surface $g(x, y, z) = c$ representing S.

28. The surface S is the graph of $f(x, y) = \sqrt{1 - y^2}$.

  (a) Explain why S is the upper half of a circular cylinder of radius 1, centered along the x-axis.

  (b) Find a level surface $g(x, y, z) = c$ representing S.

29. A cone C, with height 1 and radius 1, has its base in the xz-plane and its vertex on the positive y-axis. Find a function $g(x, y, z)$ such that C is part of the level surface $g(x, y, z) = 0$. [Hint: The graph of $f(x, y) = \sqrt{x^2 + y^2}$ is a cone which opens up and has vertex at the origin.]

30. What do the level surfaces of $f(x, y, z) = x^2 - y^2 + z^2$ look like? [Hint: Use cross-sections with y constant instead of cross-sections with z constant.]

31. Describe the surface $x^2 + y^2 = (2 + \sin z)^2$. In general, if $f(z) \geq 0$ for all z, describe the surface $x^2 + y^2 = (f(z))^2$.

12.6 LIMITS AND CONTINUITY

The sheer vertical face of Half Dome, in Yosemite National Park in California, was caused by glacial activity during the Ice Age. (See Figure 12.81.) The height of the terrain rises abruptly by nearly 1000 feet as we scale the rock from the west, whereas it is possible to make a gradual climb to the top from the east.

If we consider the function h giving the height of the terrain above sea level in terms of longitude and latitude, then h has a *discontinuity* along the path at the base of the cliff of Half Dome. Looking at the contour map of the region in Figure 12.82, we see that in most places a small change in position results in a small change in height, except near the cliff. There, no matter how small a step we take, we get a large change in height. (You can see how crowded the contours get near the cliff; some end abruptly along the discontinuity.)

This geological feature illustrates the ideas of continuity and discontinuity. Roughly speaking, a function is said to be *continuous* at a point if its values at places near the point are close to the value at the point. If this is not the case, the function is said to be *discontinuous*.

Figure 12.81: Half Dome in Yosemite National Park

Figure 12.82: A contour map of Half Dome

The property of continuity is one that, practically speaking, we usually assume of the functions we are studying. Informally, we expect (except under special circumstances) that values of a function do not change drastically when making small changes to the input variables. Whenever we model a one-variable function by an unbroken curve, we are making this assumption. Even when functions come to us as tables of data, we usually make the assumption that the missing function values between data points are close to the measured ones.

In this section we study limits and continuity a bit more formally in the context of functions of several variables. For simplicity we study these concepts for functions of two variables, but our discussion can be adapted to functions of three or more variables.

One can show that sums, products, and compositions of continuous functions are continuous, while the quotient of two continuous functions is continuous everywhere the denominator function is nonzero. Thus, each of the functions

$$\cos(x^2 y), \qquad \ln(x^2 + y^2), \qquad \frac{e^{x+y}}{x + y}, \qquad \ln(\sin(x^2 + y^2))$$

is continuous at all points (x, y) where it is defined. As for functions of one variable, the graph of a continuous function over an unbroken domain is unbroken—that is, the surface has no holes or rips in it.

Example 1   From Figures 12.83–12.86, which of the following functions appear to be continuous at $(0, 0)$?

(a) $f(x, y) = \begin{cases} \dfrac{x^2 y}{x^2 + y^2}, & (x, y) \neq (0, 0), \\ 0, & (x, y) = (0, 0). \end{cases}$ (b) $g(x, y) = \begin{cases} \dfrac{x^2}{x^2 + y^2}, & (x, y) \neq (0, 0), \\ 0, & (x, y) = (0, 0). \end{cases}$

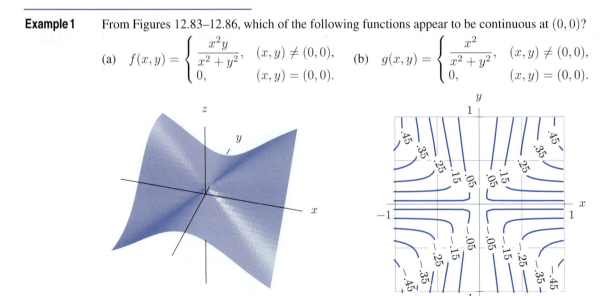

Figure 12.83: Graph of $z = x^2 y/(x^2 + y^2)$

Figure 12.84: Contour diagram of $z = x^2 y/(x^2 + y^2)$

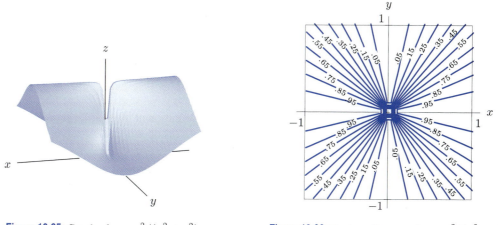

Figure 12.85: Graph of $z = x^2/(x^2 + y^2)$ **Figure 12.86**: Contour diagram of $z = x^2/(x^2 + y^2)$

Solution

(a) The graph and contour diagram of f in Figures 12.83 and 12.84 suggest that f is close to 0 when (x, y) is close to $(0, 0)$. That is, the figures suggest that f is continuous at the point $(0, 0)$; the graph appears to have no rips or holes there.

However, the figures cannot tell us for sure whether f is continuous. To be certain we must investigate the limit analytically, as is done in Example 2(a) on page 598.

(b) The graph of g and its contours near $(0, 0)$ in Figure 12.85 and 12.86 suggest that g behaves differently from f: The contours of g seem to "crash" at the origin and the graph rises rapidly from 0 to 1 near $(0, 0)$. Small changes in (x, y) near $(0, 0)$ can yield large changes in g, so we expect that g is not continuous at the point $(0, 0)$. Again, a more precise analysis is given in Example 2(b) on page 598.

The previous example suggests that continuity *at* a point depends on a function's behavior *near* the point. To study behavior near a point more carefully we need the idea of a limit of a function of two variables. Suppose that $f(x, y)$ is a function defined on a set in 2-space, not necessarily containing the point (a, b), but containing points (x, y) arbitrarily close to (a, b); suppose that L is a number.

The function f has a **limit** L at the point (a, b), written

$$\lim_{(x,y) \to (a,b)} f(x, y) = L,$$

if $f(x, y)$ is as close to L as we please whenever the distance from the point (x, y) to the point (a, b) is sufficiently small, but not zero.

We define continuity for functions of two variables in the same way as for functions of one variable:

A function f is **continuous at the point** (a, b) if

$$\lim_{(x,y) \to (a,b)} f(x, y) = f(a, b).$$

A function is **continuous on a region** R in the xy-plane if it is continuous at each point in R.

Thus, if f is continuous at the point (a, b), then f must be defined at (a, b) and the limit, $\lim_{(x,y)\to(a,b)} f(x, y)$, must exist and be equal to the value $f(a, b)$. If a function is defined at a point (a, b) but is not continuous there, then we say that f is *discontinuous* at (a, b).

We now apply the definition of continuity to the functions in Example 1, showing that f is continuous at $(0, 0)$ and that g is discontinuous at $(0, 0)$.

Example 2 Let f and g be the functions in Example 1. Use the definition of the limit to show that:

(a) $\displaystyle\lim_{(x,y)\to(0,0)} f(x, y) = 0$ (b) $\displaystyle\lim_{(x,y)\to(0,0)} g(x, y)$ does not exist.

Solution To investigate these limits of f and g, we consider values of these functions near, but not at, the origin, where they are given by the formulas

$$f(x, y) = \frac{x^2 y}{x^2 + y^2} \qquad g(x, y) = \frac{x^2}{x^2 + y^2}.$$

(a) The graph and contour diagram of f both suggest that $\lim_{(x,y)\to(0,0)} f(x, y) = 0$. To use the definition of the limit, we estimate $|f(x, y) - L|$ with $L = 0$:

$$|f(x, y) - L| = \left|\frac{x^2 y}{x^2 + y^2} - 0\right| = \left|\frac{x^2}{x^2 + y^2}\right| |y| \leq |y| \leq \sqrt{x^2 + y^2},$$

Now $\sqrt{x^2 + y^2}$ is the distance from (x, y) to $(0, 0)$. Thus, to make $|f(x, y) - 0| < 0.001$, for example, we need only require (x, y) be within 0.001 of $(0, 0)$. More generally, for any positive number u, no matter how small, we are sure that $|f(x, y) - 0| < u$ whenever (x, y) is no farther than u from $(0, 0)$. This is what we mean by saying that the difference $|f(x, y) - 0|$ can be made as small as we wish by choosing the distance to be sufficiently small. Thus, we conclude that

$$\lim_{(x,y)\to(0,0)} f(x, y) = \lim_{(x,y)\to(0,0)} \frac{x^2 y}{x^2 + y^2} = 0.$$

Notice that since this limit equals $f(0, 0)$, the function f is continuous at $(0, 0)$.

(b) Although the formula defining the function g looks similar to that of f, we saw in Example 1 that g's behavior near the origin is quite different. If we consider points $(x, 0)$ lying along the x-axis near $(0, 0)$, then the values $g(x, 0)$ are equal to 1, while if we consider points $(0, y)$ lying along the y-axis near $(0, 0)$, then the values $g(0, y)$ are equal to 0. Thus, within any distance (no matter how small) from the origin, there are points where $g = 0$ and points where $g = 1$. Therefore the limit $\lim_{(x,y)\to(0,0)} g(x, y)$ does not exist, and thus g is not continuous at $(0, 0)$.

While the notions of limit and continuity look formally the same for one- and two-variable functions, they are somewhat more subtle in the multivariable case. The reason for this is that on the line (1-space), we can approach a point from just two directions (left or right) but in 2-space there are an infinite number of ways to approach a given point.

Exercises and Problems for Section 12.6

Exercises

Are the functions in Exercises 1-6 continuous at all points in the given regions?

1. $\dfrac{1}{x^2 + y^2}$ on the square $-1 \leq x \leq 1, -1 \leq y \leq 1$

2. $\dfrac{1}{x^2 + y^2}$ on the square $1 \leq x \leq 2, 1 \leq y \leq 2$

3. $\dfrac{y}{x^2 + 2}$ on the disk $x^2 + y^2 \leq 1$

4. $\dfrac{e^{\sin x}}{\cos y}$ on the rectangle $-\frac{\pi}{2} \leq x \leq \frac{\pi}{2}, 0 \leq y \leq \frac{\pi}{4}$

5. $\tan(xy)$ on the square $-2 \leq x \leq 2, -2 \leq y \leq 2$

6. $\sqrt{2x - y}$ on the disk $x^2 + y^2 \leq 4$

In Exercises 7–11, find the limits of the functions $f(x, y)$ as $(x, y) \to (0, 0)$. Assume that polynomials, exponentials, logarithmic, and trigonometric functions are continuous.

7. $f(x, y) = e^{-x-y}$

8. $f(x, y) = x^2 + y^2$

9. $f(x, y) = \dfrac{x}{x^2 + 1}$

10. $f(x, y) = \dfrac{x + y}{(\sin y) + 2}$

11. $f(x, y) = \dfrac{\sin(x^2 + y^2)}{x^2 + y^2}$ [Hint: $\lim\limits_{t \to 0} \dfrac{\sin t}{t} = 1$.]

Problems

12. By approaching the origin along the positive x-axis and the positive y-axis, show that the following limit does not exist:
$$\lim_{(x,y)\to(0,0)} \frac{2x - y^2}{2x + y^2}.$$

13. By approaching the origin along the positive x-axis and the positive y-axis, show that the following limit does not exist:
$$\lim_{(x,y)\to(0,0)} \frac{x + y^2}{2x + y}.$$

14. Is the following function continuous at $(0, 0)$?
$$f(x, y) = \begin{cases} x^2 + y^2 & \text{if } (x, y) \neq (0, 0) \\ 2 & \text{if } (x, y) = (0, 0) \end{cases}$$

15. What value of c makes the following function continuous at $(0, 0)$?
$$f(x, y) = \begin{cases} x^2 + y^2 + 1 & \text{if } (x, y) \neq (0, 0) \\ c & \text{if } (x, y) = (0, 0) \end{cases}$$

16. (a) Use a computer to draw the graph and the contour diagram of the following function:
$$f(x, y) = \begin{cases} \dfrac{xy(x^2 - y^2)}{x^2 + y^2}, & (x, y) \neq (0, 0), \\ 0, & (x, y) = (0, 0). \end{cases}$$

(b) Do your answers to part (a) suggest that f is continuous at $(0, 0)$? Explain your answer.

17. The function f, whose graph and contour diagram are in Figures 12.87 and 12.88, is given by
$$f(x, y) = \begin{cases} \dfrac{xy}{x^2 + y^2}, & (x, y) \neq (0, 0), \\ 0, & (x, y) = (0, 0). \end{cases}$$

(a) Show that $f(0, y)$ and $f(x, 0)$ are each continuous functions of one variable.
(b) Show that rays emanating from the origin are contained in contours of f.
(c) Is f continuous at $(0, 0)$?

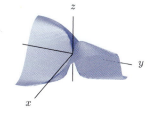

Figure 12.87: Graph of $z = xy/(x^2 + y^2)$

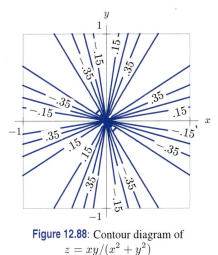

Figure 12.88: Contour diagram of $z = xy/(x^2 + y^2)$

18. Show that the function f does not have a limit at $(0, 0)$ by examining the limits of f as $(x, y) \to (0, 0)$ along the line $y = x$ and along the parabola $y = x^2$:
$$f(x, y) = \frac{x^2 y}{x^4 + y^2}, \qquad (x, y) \neq (0, 0).$$

19. Show that the function f does not have a limit at $(0, 0)$ by examining the limits of f as $(x, y) \to (0, 0)$ along the curve $y = kx^2$ for different values of k:
$$f(x, y) = \frac{x^2}{x^2 + y}, \qquad x^2 + y \neq 0.$$

20. Explain why the following function is not continuous along the line $y = 0$:
$$f(x, y) = \begin{cases} 1 - x, & y \geq 0, \\ -2, & y < 0, \end{cases}$$

In Problems 21–22, determine whether there is a value for c making the function continuous everywhere. If so, find it. If not, explain why not.

21. $f(x, y) = \begin{cases} c + y, & x \leq 3, \\ 5 - x, & x > 3. \end{cases}$

22. $f(x, y) = \begin{cases} c + y, & x \leq 3, \\ 5 - y, & x > 3. \end{cases}$

CHAPTER SUMMARY

- **3-Space**
   Cartesian coordinates, x-, y- and z-axes, xy-, xz- and yz-planes, distance formula.
- **Functions of Two Variables**
   Represented by: tables, graphs, formulas, cross-sections (one variable fixed), contours (function value fixed); cylinders (one variable missing).
- **Linear Functions**

   Recognizing linear functions from tables, graphs, contour diagrams, formulas. Converting from one representation to another.
- **Functions of Three Variables**
   Sketching level surfaces (function value fixed) in 3-space; graph of $z = f(x, y)$ is same as level surface $g(x, y, z) = f(x, y) - z = 0$.
- **Continuity**

REVIEW EXERCISES AND PROBLEMS FOR CHAPTER TWELVE

Exercises

1. Describe the set of points whose x coordinate is 2 and whose y coordinate is 1.

2. Find the equation of the plane through the points $(0, 0, 2), (0, 3, 0), (5, 0, 0)$.

3. Sketch cross-sections of $f(r, h) = \pi r^2 h$, first keeping h fixed, then keeping r fixed.

For each of the functions in Exercises 4–7, make a contour plot in the region $-2 < x < 2$ and $-2 < y < 2$. In each case, what is the equation and the shape of the contour lines?

4. $z = \sin y$

5. $z = 3x - 5y + 1$

6. $z = 2x^2 + y^2$

7. $z = e^{-2x^2 - y^2}$

Find equations for the linear functions with the contour diagrams in Exercises 8–9.

Represent the surfaces in Exercises 10–13 as graphs of functions, $f(x, y)$, and as level surfaces of the form $g(x, y, z) = c$. (There are many possible answers.)

10. Upper half of unit sphere centered at the origin

11. Lower half of sphere of radius 2 centered at $(3, 0, 0)$.

12. Plane with intercepts $x = 2$, $y = 3$, $z = 4$.

13. Paraboloid obtained by shifting $z = x^2 + y^2$ vertically 5 units

14. Match the functions with the level surfaces in Figure 12.89.

   (a) $f(x, y, z) = x^2 + y^2 + z^2$
   (b) $g(x, y, z) = x^2 + z^2$.

8.

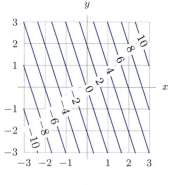

9.

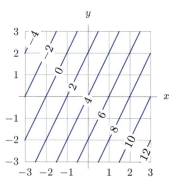

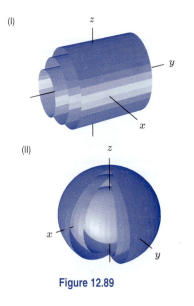

Figure 12.89

Problems

Decide if the statements in Problems 15–19 must be true, might be true, or could not be true. The function $z = f(x, y)$ is defined everywhere.

15. The level curves corresponding to $z = 1$ and $z = -1$ cross at the origin.

16. The level curve $z = 1$ consists of the circle $x^2 + y^2 = 2$ and the circle $x^2 + y^2 = 3$, but no other points.

17. The level curve $z = 1$ consists of two lines which intersect at the origin.

18. If $z = e^{-(x^2+y^2)}$, there is a level curve for every value of z.

19. If $z = e^{-(x^2+y^2)}$, there is a level curve through every point (x, y).

20. Find a linear function whose graph is the plane that intersects the xy-plane along the line $y = 2x+2$ and contains the point $(1, 2, 2)$.

21. Find the center and radius of the sphere with equation $x^2 + 4x + y^2 - 6y + z^2 + 12z = 0$.

22. You are in a room 30 feet long with a heater at one end. In the morning the room is 65°F. You turn on the heater, which quickly warms up to 85°F. Let $H(x, t)$ be the temperature x feet from the heater, t minutes after the heater is turned on. Figure 12.90 shows the contour diagram for H. How warm is it 10 feet from the heater 5 minutes after it was turned on? 10 minutes after it was turned on?

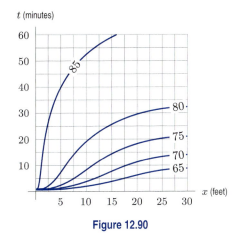

Figure 12.90

23. Using the contour diagram in Figure 12.90, sketch the graphs of the one-variable functions $H(x, 5)$ and $H(x, 20)$. Interpret the two graphs in practical terms, and explain the difference between them.

24. **(a)** Sketch the level curves of $z = \cos \sqrt{x^2 + y^2}$.
(b) Sketch a cross-section through the surface $z = \cos \sqrt{x^2 + y^2}$ in the plane containing the x- and z-axes. Put units on your axes.
(c) Sketch the cross-section through the surface $z = \cos \sqrt{x^2 + y^2}$ in the plane containing the z-axis and the line $y = x$ in the xy-plane.

25. Describe in words the level surfaces of the function $g(x, y, z) = \cos(x + y + z)$.

For Problems 26–29, use a computer or calculator to sketch the graph of a function with the given shapes. Include the axes and the equation used to generate it in your sketch.

26. A cone of circular cross-section opening downward and with its vertex at the origin.

27. A bowl which opens upward and has its vertex at 5 on the z-axis.

28. A plane which has its x, y, and z intercepts all positive.

29. A parabolic cylinder opening upward from along the line $y = x$ in the xy-plane.

Problems 30–33 concern a vibrating guitar string. Snapshots of the guitar string at millisecond intervals are shown in Figure 12.91.

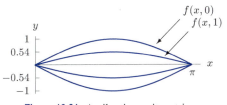

Figure 12.91: A vibrating guitar string: $f(x, t) = \cos t \sin x$ for four t values.

The guitar string is stretched tight along the x-axis from $x = 0$ to $x = \pi$. Each point on the string has an x-value, $0 \le x \le \pi$. As the string vibrates, each point on the string moves back and forth on either side of the x-axis. Let $y = f(x, t)$ be the displacement at time t of the point on the string located x units from the left end. A possible formula is

$$y = f(x, t) = \cos t \sin x, \quad 0 \le x \le \pi, \quad t \text{ in milliseconds.}$$

30. **(a)** Sketch graphs of y versus x for fixed t values, $t = 0$, $\pi/4$, $\pi/2$, $3\pi/4$, π.
(b) Use your graphs to explain why this function could represent a vibrating guitar string.

31. Explain what the functions $f(x, 0)$ and $f(x, 1)$ represent in terms of the vibrating string.

32. Explain what the functions $f(0, t)$ and $f(1, t)$ represent in terms of the vibrating string.

33. Describe the motion of the guitar strings whose displacements are given by the following:
(a) $y = g(x, t) = \cos 2t \sin x$
(b) $y = h(x, t) = \cos t \sin 2x$

CAS Challenge Problems

34. Let $A = (0, 0, 0)$ and $B = (2, 0, 0)$.

(a) Find a point C in the xy-plane that is a distance 2 from both A and B.

(b) Find a point D in 3-space that is a distance 2 from each of A, B, and C.

(c) Describe the figure obtained by joining A, B, C, and D with straight lines.

35. Let $f(x, y) = 3 + x + 2y$.

(a) Find formulas for $f(x, f(x, y))$, $f(x, f(x, f(x, y)))$ by hand.

(b) Consider $f(x, f(x, f(x, f(x, f(x, f(x, y))))))$. Conjecture a formula for this function and check your answer with a computer algebra system.

36. A function $f(x, y, z)$ has the property that $f(1, 0, 1) = 20$, $f(1, 1, 1) = 16$, and $f(1, 1, 2) = 21$.

(a) Estimate $f(1, 1, 3)$ and $f(1, 2, 1)$, assuming f is a linear function of each variable with the other variables held fixed.

(b) Suppose in fact that $f(x, y, z) = ax^2 + byz + czx^3 + d2^{x-y}$, for constants a, b, c and d. Which of your estimates in part (a) do you expect to be exact?

(c) Suppose in addition that $f(0, 0, 1) = 6$. Find an exact formula for f by solving for a, b, c, and d.

(d) Use the formula in part (c) to evaluate $f(1, 1, 3)$ and $f(1, 2, 1)$ exactly. Do the values confirm your answer to part (b)?

CHECK YOUR UNDERSTANDING

Are the statements in Problems 1–60 true or false? Give reasons for your answer.

1. If $f(x, y)$ is a function of two variables defined for all x and y, then $f(10, y)$ is a function of one variable.

2. The volume V of a box of height h and square base of side length s is a function of h and s.

3. If $H = f(t, d)$ is the function giving the water temperature $H°C$ of a lake at time t hours after midnight and depth d meters, then t is a function of d and H.

4. A table for a function $f(x, y)$ cannot have any values of f appearing twice.

5. The function given by the formula $f(v, w) = e^v/w$ is an increasing function of v when w is a nonzero constant.

6. If $f(x)$ and $g(y)$ are both functions of a single variable, then the product $f(x) \cdot g(y)$ is a function of two variables.

7. Two isotherms representing distinct temperatures on a weather map cannot intersect.

8. On a weather map, there can be two isotherms which represent the same temperature but that do not intersect.

9. A function $f(x, y)$ can be an increasing function of x with y held fixed, and be a decreasing function of y with x held fixed.

10. A function $f(x, y)$ can have the property that $g(x) = f(x, 5)$ is increasing, whereas $h(x) = f(x, 10)$ is decreasing.

11. The plane $x + 2y - 3z = 1$ passes through the origin.

12. The point $(1, 2, 3)$ lies above the plane $z = 2$.

13. The graph of the equation $z = 2$ is a plane parallel to the xz-plane.

14. The points $(1, 0, 1)$ and $(0, -1, 1)$ are the same distance from the origin.

15. The plane $x + y + z = 3$ intersects the x-axis when $x = 3$.

16. The point $(2, -1, 3)$ lies on the graph of the sphere $(x - 2)^2 + (y + 1)^2 + (z - 3)^2 = 25$.

17. There is only one point in the yz-plane that is a distance 3 from the point $(3, 0, 0)$.

18. There is only one point in the yz-plane that is distance 5 from the point $(3, 0, 0)$.

19. The sphere $x^2 + y^2 + z^2 = 10$ intersects the plane $x = 10$.

20. If the point $(0, b, 0)$ has distance 4 from the plane $y = 0$, then b must be 4.

21. The cross-section of the function $f(x, y) = x + y^2$ with $y = 1$ is a line.

22. The function $g(x, y) = 1 - y^2$ has identical parabolas for all cross-sections with $x = c$.

23. The function $g(x, y) = 1 - y^2$ has lines for all cross-sections with $y = c$.

24. The graphs of $f(x, y) = \sin(xy)$ and $g(x, y) = \sin(xy) + 2$ never intersect.

25. The graphs of $f(x, y) = x^2 + y^2$ and $g(x, y) = 1 - x^2 - y^2$ intersect in a circle.

26. If all of the $x = c$ cross-sections of the graph of $f(x, y)$ are lines, then the graph of f is a plane.

27. The only point of intersection of the graphs of $f(x, y)$ and $-f(x, y)$ is the origin.

28. A line parallel to the z-axis can intersect the graph of $f(x, y)$ at most once.

29. A line parallel to the y-axis can intersect the graph of $f(x, y)$ at most once.

30. The point $(0, 0, 10)$ is the highest point on the graph of the function $f(x, y) = 10 - x^2 - y^2$.

31. The contours of the function $f(x, y) = y^2 + (x - 2)^2$ are either circles or a single point.

32. Two contours of $f(x, y)$ with different heights never intersect.

33. If the contours of $g(x, y)$ are concentric circles, then the graph of g is a cone.

34. If the contours for $f(x, y)$ get closer together in a certain direction, then f is increasing in that direction.

35. If all of the contours of $f(x, y)$ are parallel lines, then the graph of f is a plane.

36. The $h = 1$ contour of the function $h(x, y) = xy$ is a hyperbola.

37. If the $f = 10$ contour of the function $f(x, y)$ is identical to the $g = 10$ contour of the function $g(x, y)$, then $f(x, y) = g(x, y)$ for all (x, y).

38. A function $f(x, y)$ has a contour $f = c$ for every value of c.

39. The $f = 5$ contour of the function $f(x, y)$ is identical to the $g = 0$ contour of the function $g(x, y) = f(x, y) - 5$.

40. Contours of $f(x, y) = 3x + 2y$ are lines with slope 3.

41. If the contours of f are all parallel lines, then f is linear.

42. If f is linear, then the contours of f are parallel lines.

43. The function f satisfying $f(0, 0) = 1, f(0, 1) = 4, f(0, 3) = 5$ cannot be linear.

44. The graph of a linear function is always a plane.

45. The cross-section $x = c$ of a linear function $f(x, y)$ is always a line.

46. There is no linear function $f(x, y)$ with a graph parallel to the xy-plane.

47. There is no linear function $f(x, y)$ with a graph parallel to the xz-plane.

48. A linear function $f(x, y) = 2x + 3y - 5$, has exactly one point (a, b) satisfying $f(a, b) = 0$.

49. In a table of values of a linear function, the columns have the same slope as the rows.

50. There is exactly one linear function $f(x, y)$ whose $f = 0$ contour is $y = 2x + 1$.

51. The graph of the function $f(x, y) = x^2 + y^2$ is the same as the level surface $g(x, y, z) = x^2 + y^2 - z = 0$.

52. The graph of $f(x, y) = \sqrt{1 - x^2 - y^2}$ is the same as the level surface $g(x, y, z) = x^2 + y^2 + z^2 = 1$.

53. Any surface which is the graph of a two-variable function $f(x, y)$ can also be represented as the level surface of a three-variable function $g(x, y, z)$.

54. Any surface which is the level surface of a three-variable function $g(x, y, z)$ can also be represented as the graph of a two-variable function $f(x, y)$.

55. The level surfaces of the function $g(x, y, z) = x + 2y + z$ are parallel planes.

56. The level surfaces of $g(x, y, z) = x^2 + y + z^2$ are cylinders with axis along the y-axis.

57. A level surface of a function $g(x, y, z)$ cannot be a single point.

58. If $g(x, y, z) = ax + by + cz + d$, where a, b, c, d are nonzero constants, then the level surfaces of g are planes.

59. If the level surfaces of g are planes, then $g(x, y, z) = ax + by + cz + d$, where a, b, c, d are constants.

60. If the level surfaces $g(x, y, z) = k_1$ and $g(x, y, z) = k_2$ are the same surface, then $k_1 = k_2$.

PROJECTS FOR CHAPTER TWELVE

1. A Heater in a Room

Figure 12.92 shows the contours of the temperature along one wall of a heated room through one winter day, with time indicated as on a 24-hour clock. The room has a heater located at the left-most corner of the wall and one window in the wall. The heater is controlled by a thermostat about 2 feet from the window.

(a) Where is the window?

(b) When is the window open?

(c) When is the heat on?

(d) Draw graphs of the temperature along the wall of the room at 6 am, at 11 am, at 3 pm (15 hours) and at 5 pm (17 hours).

(e) Draw a graph of the temperature as a function of time at the heater, at the window and midway between them.

(f) The temperature at the window at 5 pm (17 hours) is less than at 11 am. Why do you think this might be?

(g) To what temperature do you think the thermostat is set? How do you know?

(h) Where is the thermostat?

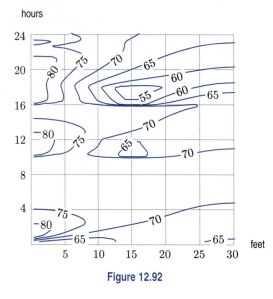

Figure 12.92

2. Light in a Wave-guide

Figure 12.93 shows the contours of light intensity as a function of location and time in a microscopic wave-guide.

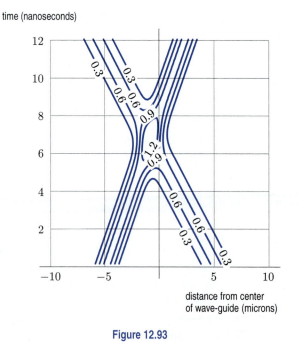

Figure 12.93

(a) Draw graphs showing intensity as a function of location at times 0, 2, 4, 6, 8, and 10 nanoseconds.

(b) If you could create an animation showing how the graph of intensity as a function of location varies as time passes, what would it look like?

(c) Draw a graph of intensity as a function of time at locations −5, 0, and 5 microns from center of wave-guide.

(d) Describe what the light beams are doing in this wave-guide.

Chapter Thirteen

A FUNDAMENTAL TOOL: VECTORS

In one-variable calculus we represented quantities such as velocity by numbers. However, to specify the velocity of a moving object in space, we need to say how fast it is moving and in what direction it is moving. In this chapter *vectors* are used to represent quantities that have direction as well as magnitude.

13.1 DISPLACEMENT VECTORS

Suppose you are a pilot planning a flight from Dallas to Pittsburgh. There are two things you must know: the distance to be traveled (so you have enough fuel to make it) and in what direction to go (so you don't miss Pittsburgh). Both these quantities together specify the displacement or *displacement vector* between the two cities.

> The **displacement vector** from one point to another is an arrow with its tail at the first point and its tip at the second. The **magnitude** (or length) of the displacement vector is the distance between the points, and is represented by the length of the arrow. The **direction** of the displacement vector is the direction of the arrow.

Figure 13.1 shows the displacement vectors from Dallas to Pittsburgh, from Albuquerque to Oshkosh, and from Los Angeles to Buffalo, SD. These displacement vectors have the same length and the same direction. We say that the displacement vectors between the corresponding cities are the same, even though they do not coincide. In other words

> Displacement vectors which point in the same direction and have the same magnitude are considered to be the same, even if they do not coincide.

Figure 13.1: Displacement vectors between cities

Notation and Terminology

The displacement vector is our first example of a vector. Vectors have both magnitude and direction; in comparison, a quantity specified only by a number, but no direction, is called a *scalar*.[1] For instance, the time taken by the flight from Dallas to Pittsburgh is a scalar quantity. Displacement is a vector since it requires both distance and direction to specify it.

In this book, vectors are written with an arrow over them, $\vec{v}$, to distinguish them from scalars. Other books use a bold **v** to denote a vector. We use the notation $\overrightarrow{PQ}$ to denote the displacement vector from a point P to a point Q. The magnitude, or length, of a vector $\vec{v}$ is written $\|\vec{v}\|$.

Addition and Subtraction of Displacement Vectors

Suppose NASA commands a robot on Mars to move 75 meters in one direction and then 50 meters in another direction. (See Figure 13.2.) Where does the robot end up? Suppose the displacements are represented by the vectors $\vec{v}$ and $\vec{w}$, respectively. Then the sum $\vec{v} + \vec{w}$ gives the final position.

---

[1] So named by W. R. Hamilton because they are merely numbers on the *scale* from $-\infty$ to ∞.

The **sum**, $\vec{v} + \vec{w}$, of two vectors $\vec{v}$ and $\vec{w}$ is the combined displacement resulting from first applying $\vec{v}$ and then $\vec{w}$. (See Figure 13.3.) The sum $\vec{w} + \vec{v}$ gives the same displacement.

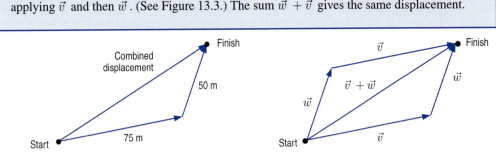

Figure 13.2: Sum of displacements of robots on Mars   Figure 13.3: The sum $\vec{v} + \vec{w} = \vec{w} + \vec{v}$

Suppose two different robots start from the same location. One moves along a displacement vector $\vec{v}$ and the second along a displacement vector $\vec{w}$. What is the displacement vector, $\vec{x}$, from the first robot to the second? (See Figure 13.4.) Since $\vec{v} + \vec{x} = \vec{w}$, we define $\vec{x}$ to be the difference $\vec{x} = \vec{w} - \vec{v}$. In other words, $\vec{w} - \vec{v}$ gets you from $\vec{v}$ to $\vec{w}$.

The **difference**, $\vec{w} - \vec{v}$, is the displacement vector which, when added to $\vec{v}$, gives $\vec{w}$. That is, $\vec{w} = \vec{v} + (\vec{w} - \vec{v})$. (See Figure 13.4.)

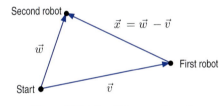

Figure 13.4: The difference $\vec{w} - \vec{v}$

If the robot ends up where it started, then its total displacement vector is the *zero vector*, $\vec{0}$. The zero vector has no direction.

The **zero vector**, $\vec{0}$, is a displacement vector with zero length.

Scalar Multiplication of Displacement Vectors

If $\vec{v}$ represents a displacement vector, the vector $2\vec{v}$ represents a displacement of twice the magnitude in the same direction as $\vec{v}$. Similarly, $-2\vec{v}$ represents a displacement of twice the magnitude in the opposite direction. (See Figure 13.5.)

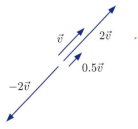

Figure 13.5: Scalar multiples of the vector $\vec{v}$

If λ is a scalar and $\vec{v}$ is a displacement vector, the **scalar multiple of $\vec{v}$ by** λ, written $\lambda\vec{v}$, is the displacement vector with the following properties:

- The displacement vector $\lambda\vec{v}$ is parallel to $\vec{v}$, pointing in the same direction if $\lambda > 0$, and in the opposite direction if $\lambda < 0$.
- The magnitude of $\lambda\vec{v}$ is $|\lambda|$ times the magnitude of $\vec{v}$, that is, $\|\lambda\vec{v}\| = |\lambda|\,\|\vec{v}\|$.

Note that $|\lambda|$ represents the absolute value of the scalar λ while $\|\lambda\vec{v}\|$ represents the magnitude of the vector $\lambda\vec{v}$.

Example 1   Explain why $\vec{w} - \vec{v} = \vec{w} + (-1)\vec{v}$.

Solution   The vector $(-1)\vec{v}$ has the same magnitude as $\vec{v}$, but points in the opposite direction. Figure 13.6 shows that the combined displacement $\vec{w} + (-1)\vec{v}$ is the same as the displacement $\vec{w} - \vec{v}$.

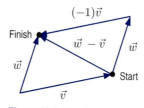

Figure 13.6: Explanation for why $\vec{w} - \vec{v} = \vec{w} + (-1)\vec{v}$

Parallel Vectors

Two vectors $\vec{v}$ and $\vec{w}$ are *parallel* if one is a scalar multiple of the other, that is, if $\vec{w} = \lambda\vec{v}$, for some scalar λ.

Components of Displacement Vectors: The Vectors $\vec{i}$, $\vec{j}$, and $\vec{k}$

Suppose that you live in a city with equally spaced streets running east-west and north-south and that you want to tell someone how to get from one place to another. You'd be likely to tell them how many blocks east-west and how many blocks north-south to go. For example, to get from P to Q in Figure 13.7, we go 4 blocks east and 1 block south. If $\vec{i}$ and $\vec{j}$ are as shown in Figure 13.7, then the displacement vector from P to Q is $4\vec{i} - \vec{j}$.

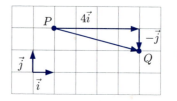

Figure 13.7: The displacement vector from P to Q is $4\vec{i} - \vec{j}$

We extend the same idea to 3-dimensions. First we choose a Cartesian system of coordinate axes. The three vectors of length 1 shown in Figure 13.8 are the vector $\vec{i}$, which points along the positive x-axis, the vector $\vec{j}$, along the positive y-axis, and the vector $\vec{k}$, along the positive z-axis.

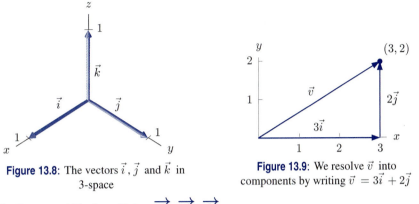

Figure 13.8: The vectors $\vec{i}$, $\vec{j}$ and $\vec{k}$ in 3-space

Figure 13.9: We resolve $\vec{v}$ into components by writing $\vec{v} = 3\vec{i} + 2\vec{j}$

Writing Displacement Vectors Using $\vec{i}$, $\vec{j}$, $\vec{k}$

Any displacement in 3-space or the plane can be expressed as a combination of displacements in the coordinate directions. For example, Figure 13.9 shows that the displacement vector $\vec{v}$ from the origin to the point $(3, 2)$ can be written as a sum of displacement vectors along the x and y-axes:

$$\vec{v} = 3\vec{i} + 2\vec{j}.$$

This is called *resolving $\vec{v}$ into components*. In general:

We **resolve** $\vec{v}$ into components by writing $\vec{v}$ in the form

$$\vec{v} = v_1\vec{i} + v_2\vec{j} + v_3\vec{k}.$$

We call $v_1\vec{i}$, $v_2\vec{j}$, and $v_3\vec{k}$ the **components** of $\vec{v}$.

An Alternative Notation for Vectors

Many people write a vector in 3-dimensions as a string of three numbers, that is, as

$$\vec{v} = (v_1, v_2, v_3) \quad \text{instead of} \quad \vec{v} = v_1\vec{i} + v_2\vec{j} + v_3\vec{k}.$$

Since the first notation can be confused with a point and the second cannot, we usually use the second form.

Example 2   Resolve the displacement vector, $\vec{v}$, from the point $P_1 = (2, 4, 10)$ to the point $P_2 = (3, 7, 6)$ into components.

Solution   To get from P_1 to P_2, we move 1 unit in the positive x-direction, 3 units in the positive y-direction, and 4 units in the negative z-direction. Hence $\vec{v} = \vec{i} + 3\vec{j} - 4\vec{k}$.

Example 3   Decide whether the vector $\vec{v} = 2\vec{i} + 3\vec{j} + 5\vec{k}$ is parallel to each of the following vectors:
$$\vec{w} = 4\vec{i} + 6\vec{j} + 10\vec{k}, \quad \vec{a} = -\vec{i} - 1.5\vec{j} - 2.5\vec{k}, \quad \vec{b} = 4\vec{i} + 6\vec{j} + 9\vec{k}$$

Solution   Since $\vec{w} = 2\vec{v}$ and $\vec{a} = -0.5\vec{v}$, the vectors $\vec{v}$, $\vec{w}$, and $\vec{a}$ are parallel. However, $\vec{b}$ is not a multiple of $\vec{v}$ (since, for example, $4/2 \neq 9/5$), so $\vec{v}$ and $\vec{b}$ are not parallel.

In general, Figure 13.10 shows us how to express the displacement vector between two points in components:

Components of Displacement Vectors

The displacement vector from the point $P_1 = (x_1, y_1, z_1)$ to the point $P_2 = (x_2, y_2, z_2)$ is given in components by

$$\overrightarrow{P_1P_2} = (x_2 - x_1)\vec{i} + (y_2 - y_1)\vec{j} + (z_2 - z_1)\vec{k}.$$

Position Vectors: Displacement of a Point from the Origin

A displacement vector whose tail is at the origin is called a *position vector*. Thus, any point (x_0, y_0, z_0) in space has associated with it the position vector $\vec{r}_0 = x_0\vec{i} + y_0\vec{j} + z_0\vec{k}$. (See Figure 13.11.) In general, a position vector gives the displacement of a point from the origin.

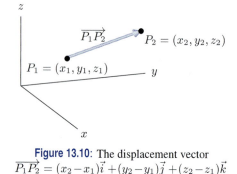

Figure 13.10: The displacement vector
$\overrightarrow{P_1P_2} = (x_2-x_1)\vec{i} + (y_2-y_1)\vec{j} + (z_2-z_1)\vec{k}$

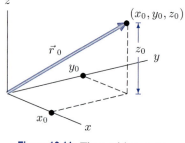

Figure 13.11: The position vector
$\vec{r}_0 = x_0\vec{i} + y_0\vec{j} + z_0\vec{k}$

The Components of the Zero Vector

The zero displacement vector has magnitude equal to zero and is written $\vec{0}$. So $\vec{0} = 0\vec{i} + 0\vec{j} + 0\vec{k}$.

The Magnitude of a Vector in Components

For a vector, $\vec{v} = v_1\vec{i} + v_2\vec{j}$, the Pythagorean theorem is used to find its magnitude, $\|\vec{v}\|$. (See Figure 13.12.)

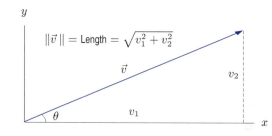

Figure 13.12: Magnitude, $\|\vec{v}\|$, of a 2-dimensional vector, $\vec{v}$

In 3-dimensions, for a vector $\vec{v} = v_1\vec{i} + v_2\vec{j} + v_3\vec{k}$, we have

---

Magnitude of $\vec{v}$ $= \|\vec{v}\| = $ Length of the arrow $= \sqrt{v_1^2 + v_2^2 + v_3^2}.$

For instance, if $\vec{v} = 3\vec{i} - 4\vec{j} + 5\vec{k}$, then $\|\vec{v}\| = \sqrt{3^2 + (-4)^2 + 5^2} = \sqrt{50}$.

Addition and Scalar Multiplication of Vectors in Components

Suppose the vectors $\vec{v}$ and $\vec{w}$ are given in components:

$$\vec{v} = v_1\vec{i} + v_2\vec{j} + v_3\vec{k} \quad \text{and} \quad \vec{w} = w_1\vec{i} + w_2\vec{j} + w_3\vec{k}.$$

Then

$$\vec{v} + \vec{w} = (v_1 + w_1)\vec{i} + (v_2 + w_2)\vec{j} + (v_3 + w_3)\vec{k},$$

and

$$\lambda\vec{v} = \lambda v_1\vec{i} + \lambda v_2\vec{j} + \lambda v_3\vec{k}.$$

Figures 13.13 and 13.14 illustrate these properties in two dimensions. Finally, $\vec{v} - \vec{w} = \vec{v} + (-1)\vec{w}$, so we can write $\vec{v} - \vec{w} = (v_1 - w_1)\vec{i} + (v_2 - w_2)\vec{j} + (v_3 - w_3)\vec{k}$.

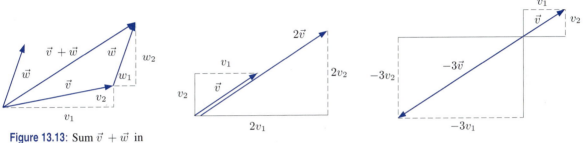

Figure 13.13: Sum $\vec{v} + \vec{w}$ in components

Figure 13.14: Scalar multiples of vectors showing $\vec{v}$, $2\vec{v}$, and $-3\vec{v}$

How to Resolve a Vector into Components

You may wonder how we find the components of a 2-dimensional vector, given its length and direction. Suppose the vector $\vec{v}$ has length v and makes an angle of θ with the x-axis, measured counterclockwise, as in Figure 13.15. If $\vec{v} = v_1\vec{i} + v_2\vec{j}$, Figure 13.15 shows that

$$v_1 = v\cos\theta \quad \text{and} \quad v_2 = v\sin\theta.$$

Thus, we resolve $\vec{v}$ into components by writing

$$\vec{v} = (v\cos\theta)\vec{i} + (v\sin\theta)\vec{j}.$$

Vectors in 3-space are resolved using direction cosines; see Problem 31 on page 636.

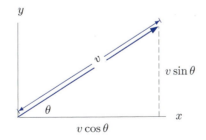

Figure 13.15: Resolving a vector: $\vec{v} = (v\cos\theta)\vec{i} + (v\sin\theta)\vec{j}$

Example 4   Resolve $\vec{v}$ into components if $\|\vec{v} = 2\|$ and $\theta = \pi/6$.

Solution   We have $\vec{v} = 2\cos(\pi/6)\vec{i} + 2\sin(\pi/6)\vec{j} = 2(\sqrt{3}/2)\vec{i} + 2(1/2)\vec{j} = \sqrt{3}\vec{i} + \vec{j}$.

Unit Vectors

A *unit vector* is a vector whose magnitude is 1. The vectors $\vec{i}$, $\vec{j}$, and $\vec{k}$ are unit vectors in the directions of the coordinate axes. It is often helpful to find a unit vector in the same direction as a given vector $\vec{v}$. Suppose that $\|\vec{v}\| = 10$; a unit vector in the same direction as $\vec{v}$ is $\vec{v}/10$. In general, a unit vector in the direction of any nonzero vector $\vec{v}$ is

$$\vec{u} = \frac{\vec{v}}{\|\vec{v}\|}.$$

Example 5   Find a unit vector, $\vec{u}$, in the direction of the vector $\vec{v} = \vec{i} + 3\vec{j}$.

Solution   If $\vec{v} = \vec{i} + 3\vec{j}$, then $\|\vec{v}\| = \sqrt{1^2 + 3^2} = \sqrt{10}$. Thus, a unit vector in the same direction is given by

$$\vec{u} = \frac{\vec{v}}{\sqrt{10}} = \frac{1}{\sqrt{10}}(\vec{i} + 3\vec{j}) = \frac{1}{\sqrt{10}}\vec{i} + \frac{3}{\sqrt{10}}\vec{j} \approx 0.32\vec{i} + 0.95\vec{j}.$$

Example 6   Find a unit vector at the point (x, y, z) that points radially outward away from the origin.

Solution   The vector from the origin to (x, y, z) is the position vector

$$\vec{r} = x\vec{i} + y\vec{j} + z\vec{k}.$$

Thus, if we put its tail at (x, y, z) it will point away from the origin. Its magnitude is

$$\|\vec{r}\| = \sqrt{x^2 + y^2 + z^2},$$

so a unit vector pointing in the same direction is

$$\frac{\vec{r}}{\|\vec{r}\|} = \frac{x\vec{i} + y\vec{j} + z\vec{k}}{\sqrt{x^2 + y^2 + z^2}} = \frac{x}{\sqrt{x^2 + y^2 + z^2}}\vec{i} + \frac{y}{\sqrt{x^2 + y^2 + z^2}}\vec{j} + \frac{z}{\sqrt{x^2 + y^2 + z^2}}\vec{k}.$$

Exercises and Problems for Section 13.1

Exercises

For Exercises 1–6, perform the indicated operations on the following vectors:

$$\vec{a} = 2\vec{j} + \vec{k}, \quad \vec{b} = -3\vec{i} + 5\vec{j} + 4\vec{k}, \quad \vec{c} = \vec{i} + 6\vec{j},$$
$$\vec{x} = -2\vec{i} + 9\vec{j}, \quad \vec{y} = 4\vec{i} - 7\vec{j}, \quad \vec{z} = \vec{i} - 3\vec{j} - \vec{k}.$$

1. $\|\vec{z}\|$

2. $\vec{a} + \vec{z}$

3. $5\vec{b}$

4. $2\vec{c} + \vec{x}$

5. $\|\vec{y}\|$

6. $2\vec{a} + 7\vec{b} - 5\vec{z}$

For Exercises 7–10, perform the indicated computation.

7. $(4\vec{i} + 2\vec{j}) - (3\vec{i} - \vec{j})$

8. $(\vec{i} + 2\vec{j}) + (-3)(2\vec{i} + \vec{j})$

9. $-4(\vec{i} - 2\vec{j}) - 0.5(\vec{i} - \vec{k})$

10. $2(0.45\vec{i} - 0.9\vec{j} - 0.01\vec{k}) - 0.5(1.2\vec{i} - 0.1\vec{k})$

Find the length of the vectors in Exercises 11–14.

11. $\vec{v} = \vec{i} - \vec{j} + 3\vec{k}$

12. $\vec{v} = \vec{i} - \vec{j} + 2\vec{k}$

13. $\vec{v} = 1.2\vec{i} - 3.6\vec{j} + 4.1\vec{k}$

14. $\vec{v} = 7.2\vec{i} - 1.5\vec{j} + 2.1\vec{k}$

15. Resolve the vectors in Figure 13.16 into components.

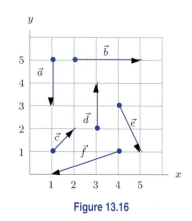

Figure 13.16

16. Resolve vector $\vec{v}$ into components if $\|\vec{v}\| = 8$ and the direction of $\vec{v}$ is shown in Figure 13.17.

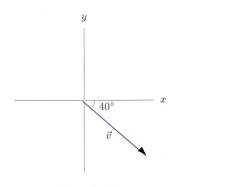

Figure 13.17

17. (a) Draw the position vector for $\vec{v} = 5\vec{i} - 7\vec{j}$.
(b) What is $\|\vec{v}\|$?
(c) Find the angle between $\vec{v}$ and the positive x-axis.

Resolve the vectors in Exercises 18–20 into components.

18. A vector starting at the point $Q = (4,6)$ and ending at the point $P = (1,2)$.

Problems

21. Figure 13.18 shows a rectangular box containing several vectors. Are the following statements true or false? Explain.
(a) $\vec{c} = \vec{f}$ **(b)** $\vec{a} = \vec{d}$ **(c)** $\vec{a} = -\vec{b}$
(d) $\vec{g} = \vec{f} + \vec{a}$ **(e)** $\vec{e} = \vec{a} - \vec{b}$ **(f)** $\vec{d} = \vec{g} - \vec{c}$

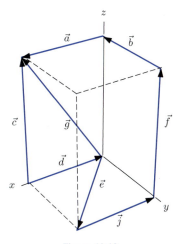

Figure 13.18

22. In Figure 13.19, a ship is at point B, a submarine is 6 units below point C, and a helicopter is 10 units above point A.

(a) Find each of the following displacement vectors: from the submarine to the ship, from the helicopter to the ship, and from the submarine to the helicopter.

19.

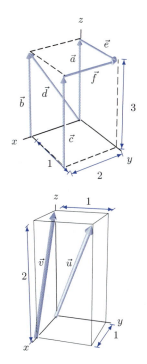

20.

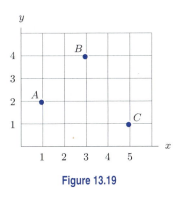

(b) Which two objects are closest together?

Figure 13.19

A cat on the ground at the point $(1,4,0)$ watches a squirrel at the top of a tree. The tree is one unit high with its base at $(2,4,0)$. Find the displacement vectors in Problems 23–26.

23. From the origin to the cat.

24. From the bottom of the tree to the squirrel.

25. From the bottom of the tree to the cat.

26. From the cat to the squirrel.

27. Find a vector with length 2 that points in the same direction as $\vec{i} - \vec{j} + 2\vec{k}$.

28. (a) Find a unit vector from the point $P = (1,2)$ and toward the point $Q = (4,6)$.
(b) Find a vector of length 10 pointing in the same direction.

29. Find all vectors $\vec{v}$ in 2 dimensions having $\|\vec{v}\| = 5$ and the $\vec{i}$-component of $\vec{v}$ is $3\vec{i}$.

30. Which of the following vectors are parallel?

$$\vec{u} = 2\vec{i} + 4\vec{j} - 2\vec{k}, \quad \vec{p} = \vec{i} + \vec{j} + \vec{k},$$
$$\vec{v} = \vec{i} - \vec{j} + 3\vec{k}, \quad \vec{q} = 4\vec{i} - 4\vec{j} + 12\vec{k},$$
$$\vec{w} = -\vec{i} - 2\vec{j} + \vec{k}, \quad \vec{r} = \vec{i} - \vec{j} + \vec{k}.$$

31. A truck is traveling due north at 30 km/hr approaching a crossroad. On a perpendicular road a police car is traveling west toward the intersection at 40 km/hr. Both vehicles will reach the crossroad in exactly one hour. Find the vector currently representing the displacement of the truck with respect to the police car.

32. Figure 13.20 shows a molecule with four atoms at O, A, B and C. Verify that every atom in the molecule is 2 units away from every other atom.

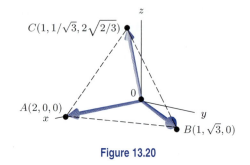

Figure 13.20

33. Show that the medians of a triangle intersect at a point $\frac{1}{3}$ of the way along each median from the side it bisects.

34. Show that the lines joining the centroid (the intersection point of the medians) of a face of the tetrahedron and the opposite vertex meet at a point $\frac{1}{4}$ of the way from each centroid to its opposite vertex.

13.2 VECTORS IN GENERAL

Besides displacement, there are many quantities that have both magnitude and direction and are added and multiplied by scalars in the same way as displacements. Any such quantity is called a *vector* and is represented by an arrow in the same manner we represent displacements. The length of the arrow is the *magnitude* of the vector, and the direction of the arrow is the direction of the vector.

Velocity Versus Speed

The speed of a moving body tells us how fast it is moving, say 80 km/hr. The speed is just a number; it is therefore a scalar. The velocity, on the other hand, tells us both how fast the body is moving and the direction of motion; it is a vector. For instance, if a car is heading northeast at 80 km/hr, then its velocity is a vector of length 80 pointing northeast.

> The **velocity vector** of a moving object is a vector whose magnitude is the speed of the object and whose direction is the direction of its motion.

Example 1   A car is traveling north at a speed of 100 km/hr, while a plane above is flying horizontally south-west at a speed of 500 km/hr. Draw the velocity vectors of the car and the plane.

Solution   Figure 13.21 shows the velocity vectors. The plane's velocity vector is five times as long as the car's, because its speed is five times as great.

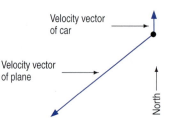

Velocity vector of car

Velocity vector of plane

North

Figure 13.21: Velocity vector of the car is 100 km/hr north and of the plane is 500 km/hr south-west

The next example illustrates that the velocity vectors for two motions add to give the velocity vector for the combined motion, just as displacements do.

Example 2   A riverboat is moving with velocity $\vec{v}$ and a speed of 8 km/hr relative to the water. In addition, the river has a current $\vec{c}$ and a speed of 1 km/hr. (See Figure 13.22.) What is the physical significance of the vector $\vec{v} + \vec{c}$?

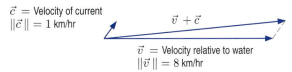

$\vec{c}$ = Velocity of current
$\|\vec{c}\| = 1$ km/hr

$\vec{v} + \vec{c}$

$\vec{v}$ = Velocity relative to water
$\|\vec{v}\| = 8$ km/hr

Figure 13.22: Boat's velocity relative to the river bed is the sum, $\vec{v} + \vec{c}$

Solution   The vector $\vec{v}$ shows how the boat is moving relative to the water, while $\vec{c}$ shows how the water is moving relative to the riverbed. During an hour, imagine that the boat first moves 8 km relative to the water, which remains still; this displacement is represented by $\vec{v}$. Then imagine the water moving 1 km while the boat remains stationary relative to the water; this displacement is represented by $\vec{c}$. The combined displacement is represented by $\vec{v} + \vec{c}$. Thus, the vector $\vec{v} + \vec{c}$ is the velocity of the boat relative to the riverbed.

Note that the effective speed of the boat is not necessarily 9 km/hr unless the boat is moving in the direction of the current. Although we add the velocity vectors, we do not necessarily add their lengths.

Scalar multiplication also makes sense for velocity vectors. For example, if $\vec{v}$ is a velocity vector, then $-2\vec{v}$ represents a velocity of twice the magnitude in the opposite direction.

Example 3   A ball is moving with velocity $\vec{v}$ when it hits a wall at a right angle and bounces straight back, with its speed reduced by 20%. Express its new velocity in terms of the old one.

Solution   The new velocity is $-0.8\vec{v}$, where the negative sign expresses the fact that the new velocity is in the direction opposite to the old.

We can represent velocity vectors in components in the same way we did on page 611.

Example 4   Represent the velocity vectors of the car and the plane in Example 1 using components. Take north to be the positive y-axis, east to be the positive x-axis, and upward to be the positive z-axis.

Solution   The car is traveling north at 100 km/hr, so the y-component of its velocity is $100\vec{j}$ and the x-component is $0\vec{i}$. Since it is traveling horizontally, the z-component is $0\vec{k}$. So we have

$$\text{Velocity of car } = 0\vec{i} + 100\vec{j} + 0\vec{k} = 100\vec{j}.$$

The plane's velocity vector also has $\vec{k}$ component equal to zero. Since it is traveling southwest, its $\vec{i}$ and $\vec{j}$ components have negative coefficients (north and east are positive). Since the plane is traveling at 500 km/hr, in one hour it is displaced $500/\sqrt{2} \approx 354$ km to the west and 354 km to the south. (See Figure 13.23.) Thus,

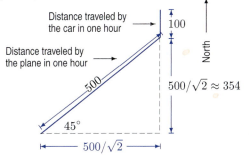

Distance traveled by the car in one hour → 100

Distance traveled by the plane in one hour →

North

500

$500/\sqrt{2} \approx 354$

$45°$

$500/\sqrt{2}$

Figure 13.23: Distance traveled by the plane and car in one hour

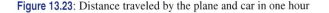

$$\text{Velocity of plane} = -(500\cos 45°)\vec{i} - (500\sin 45°)\vec{j} \approx -354\vec{i} - 354\vec{j}.$$

Of course, if the car were climbing a hill or if the plane were descending for a landing, then the $\vec{k}$ component would not be zero.

Acceleration

Another example of a vector quantity is acceleration. Acceleration, like velocity, is specified by both a magnitude and a direction — for example, the acceleration due to gravity is 9.81 m/sec^2 vertically downward.

Force

Force is another example of a vector quantity. Suppose you push on an open door. The result depends both on how hard you push and in what direction. Thus, to specify a force we must give its magnitude (or strength) and the direction in which it is acting. For example, the gravitational force exerted on an object by the earth is a vector pointing from the object toward the center of the earth; its magnitude is the strength of the gravitational force.

Example 5   The earth travels around the sun in an ellipse. The gravitational force on the earth and the velocity of the earth are governed by the following laws:
Newton's Law of Gravitation: The magnitude of the gravitational attraction, F, between two masses m_1 and m_2 at a distance r apart is given by $F = Gm_1m_2/r^2$, where G is a constant. The force vector lies along the line between the masses.
Kepler's Second Law: The line joining a planet to the sun sweeps out equal areas in equal times.
   (a) Sketch vectors representing the gravitational force of the sun on the earth at two different positions in the earth's orbit.
   (b) Sketch the velocity vector of the earth at two points in its orbit.

Solution   (a) Figure 13.24 shows the earth orbiting the sun. Note that the gravitational force vector always points toward the sun and is larger when the earth is closer to the sun because of the r^2 term in the denominator. (In fact, the real orbit looks much more like a circle than we have shown here.)
   (b) The velocity vector points in the direction of motion of the earth. Thus, the velocity vector is tangent to the ellipse. See Figure 13.25. Furthermore, the velocity vector is longer at points of the orbit where the planet is moving quickly, because the magnitude of the velocity vector is the speed. Kepler's Second Law enables us to determine when the earth is moving quickly and when it is moving slowly. Over a fixed period of time, say one month, the line joining the earth to the sun sweeps out a sector having a certain area. Figure 13.25 shows two sectors swept out in two different one-month time-intervals. Kepler's law says that the areas of the two sectors are the same. Thus, the earth must move farther in a month when it is close to the sun than when it is far from the sun. Therefore, the earth moves faster when it is closer to the sun and slower when it is farther away.

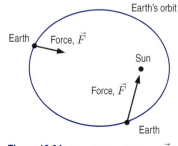

Figure 13.24: Gravitational force, $\vec{F}$, exerted by the sun on the earth: Greater magnitude closer to sun

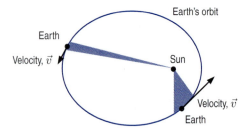

Figure 13.25: The velocity vector, $\vec{v}$, of the earth: Greater magnitude closer to the sun

Properties of Addition and Scalar Multiplication

In general, vectors add, subtract, and are multiplied by scalars in the same way as displacement vectors. Thus, for any vectors $\vec{u}$, $\vec{v}$, and $\vec{w}$ and any scalars α and β, we have the following properties:

Commutativity
1. $\vec{v} + \vec{w} = \vec{w} + \vec{v}$

Distributivity
4. $(\alpha + \beta)\vec{v} = \alpha\vec{v} + \beta\vec{v}$
5. $\alpha(\vec{v} + \vec{w}) = \alpha\vec{v} + \alpha\vec{w}$

Associativity
2. $(\vec{u} + \vec{v}) + \vec{w} = \vec{u} + (\vec{v} + \vec{w})$
3. $\alpha(\beta\vec{v}) = (\alpha\beta)\vec{v}$

Identity
6. $1 \cdot \vec{v} = \vec{v}$
7. $0 \cdot \vec{v} = \vec{0}$
8. $\vec{v} + \vec{0} = \vec{v}$
9. $\vec{w} + (-1) \cdot \vec{v} = \vec{w} - \vec{v}$

Problems 23–30 at the end of this section ask for a justification of these results in terms of displacement vectors.

Using Components

Example 6   A plane, heading due east at an airspeed of 600 km/hr, experiences a wind of 50 km/hr blowing toward the northeast. Find the plane's direction and ground speed.

Solution   We choose a coordinate system with the x-axis pointing east and the y-axis pointing north. See Figure 13.26.

The airspeed tells us the speed of the plane relative to still air. Thus, the plane is moving due east with velocity $\vec{v} = 600\vec{i}$ relative to still air. In addition, the air is moving with a velocity $\vec{w}$. Writing $\vec{w}$ in components, we have

$$\vec{w} = (50\cos 45°)\vec{i} + (50\sin 45°)\vec{j} = 35.4\vec{i} + 35.4\vec{j}.$$

The vector $\vec{v} + \vec{w}$ represents the displacement of the plane in one hour relative to the ground. Therefore, $\vec{v} + \vec{w}$ is the velocity of the plane relative to the ground. In components, we have

$$\vec{v} + \vec{w} = 600\vec{i} + \left(35.4\vec{i} + 35.4\vec{j}\right) = 635.4\vec{i} + 35.4\vec{j}.$$

The direction of the plane's motion relative to the ground is given by the angle θ in Figure 13.26, where

$$\tan\theta = \frac{35.4}{635.4}$$

so

$$\theta = \arctan\left(\frac{35.4}{635.4}\right) = 3.2°.$$

The ground speed is the speed of the plane relative to the ground, so

$$\text{Groundspeed} = \sqrt{635.4^2 + 35.4^2} = 636.4 \text{ km/hr}.$$

Thus, the speed of the plane relative to the ground has been increased slightly by the wind. (This is as we would expect, as the wind has a positive component in the direction in which the plane is traveling.) The angle θ shows how far the plane is blown off course by the wind.

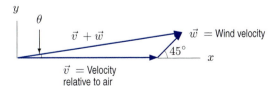

Figure 13.26: Plane's velocity relative to the ground is the sum $\vec{v} + \vec{w}$

Vectors in n Dimensions

Using the alternative notation $\vec{v} = (v_1, v_2, v_3)$ for a vector in 3-space, we can define a vector in n dimensions as a string of n numbers. Thus, a vector in n dimensions can be written as

$$\vec{c} = (c_1, c_2, \ldots, c_n).$$

Addition and scalar multiplication are defined by the formulas

$$\vec{v} + \vec{w} = (v_1, v_2, \ldots, v_n) + (w_1, w_2, \ldots, w_n) = (v_1 + w_1, v_2 + w_2, \ldots, v_n + w_n)$$

and

$$\lambda \vec{v} = \lambda(v_1, v_2, \ldots, v_n) = (\lambda v_1, \lambda v_2, \ldots, \lambda v_n).$$

Why Do We Want Vectors in n Dimensions?

Vectors in two and three dimensions can be used to model displacement, velocities, or forces. But what about vectors in n dimensions? There is another interpretation of 3-dimensional vectors (or 3-vectors) which is useful: they can be thought of as listing 3 different quantities — for example, the displacements parallel to the x, y, and z axes. Similarly, the n-vector

$$\vec{c} = (c_1, c_2, \ldots, c_n)$$

can be thought of as a way of keeping n different quantities organized. For example, a *population* vector $\vec{N}$ shows the number of children and adults in a population:

$$\vec{N} = (\text{Number of children, Number of adults}),$$

or, if we are interested in a more detailed breakdown of ages, we might give the number in each ten-year age bracket in the population (up to age 110) in the form

$$\vec{N} = (N_1, N_2, N_3, N_4, \ldots, N_{10}, N_{11}),$$

where N_1 is the population aged 0–9, and N_2 is the population aged 10–19, and so on.

A *consumption* vector ,

$$\vec{q} = (q_1, q_2, \ldots, q_n)$$

shows the quantities $q_1, q_2, \ldots, q_n$ consumed of each of n different goods. A *price* vector

$$\vec{p} = (p_1, p_2, \ldots, p_n)$$

contains the prices of n different items.

In 1907, Hermann Minkowski used vectors with four components when he introduced *space-time coordinates*, whereby each event is assigned a vector position $\vec{v}$ with four coordinates, three for its position in space and one for time:

$$\vec{v} = (x, y, z, t).$$

Example 7   Suppose the vector $\vec{I}$ represents the number of copies, in thousands, made by each of four copy centers in the month of December and $\vec{J}$ represents the number of copies made at the same four copy centers during the previous eleven months (the "year-to-date"). If $\vec{I} = (25, 211, 818, 642)$, and $\vec{J} = (331, 3227, 1377, 2570)$, compute $\vec{I} + \vec{J}$. What does this sum represent?

Solution   The sum is

$$\vec{I} + \vec{J} = (25 + 331, 211 + 3227, 818 + 1377, 642 + 2570) = (356, 3438, 2195, 3212).$$

Each term in $\vec{I} + \vec{J}$ represents the sum of the number of copies made in December plus those in the previous eleven months, that is, the total number of copies made during the entire year at that particular copy center.

___

Example 8   The price vector $\vec{p} = (p_1, p_2, p_3)$ represents the prices in dollars of three goods. Write a vector which gives the prices of the same goods in cents.

Solution   The prices in cents are $100p_1$, $100p_2$, and $100p_3$ respectively, so the new price vector is

$$(100p_1, 100p_2, 100p_3) = 100\vec{p}.$$

Exercises and Problems for Section 13.2

Exercises

In Exercises 1–5, say whether the given quantity is a vector or a scalar.

1. The distance from Seattle to St. Louis.

2. The population of the US.

3. The magnetic field at a point on the earth's surface.

4. The temperature at a point on the earth's surface.

5. The populations of each of the 50 states.

6. Give the components of the velocity vector for wind blowing at 10 km/hr toward the southeast. (Assume north is the positive y-direction.)

7. Give the components of the velocity vector of a boat which is moving at 40 km/hr in a direction $20°$ south of west. (Assume north is in the positive y-direction.)

8. A car is traveling at a speed of 50 km/hr. The positive y-axis is north and the positive x-axis is east. Resolve the car's velocity vector (in 2-space) into components if the car is traveling in each of the following directions:

   (a) East   (b) South
   (c) Southeast   (d) Northwest.

9. Which is traveling faster, a car whose velocity vector is $21\vec{i} + 35\vec{j}$, or a car whose velocity vector is $40\vec{i}$, assuming that the units are the same for both directions?

Problems

10. A model rocket is shot into the air at an angle with the earth of about $60°$. The rocket is going fast initially but slows down as it reaches its highest point. It picks up speed again as it falls to earth.

    (a) Sketch a graph showing the path of the rocket. Draw several velocity vectors on your graph.
    (b) A second rocket has a parachute that deploys as it begins its descent. How do the velocity vectors from part (a) change for this rocket?

11. A car drives clockwise around the track in Figure 13.27, slowing down at the curves and speeding up along the straight portions. Sketch velocity vectors at the points P, Q, and R.

12. A racing car drives clockwise around the track shown in Figure 13.27 at a constant speed. At what point on the track does the car have the longest acceleration vector, and in roughly what direction is it pointing? (Recall that acceleration is the rate of change of velocity.)

13. A boat is heading due east at 25 km/hr (relative to the water). The current is moving toward the southwest at 10 km/hr.

    (a) Give the vector representing the actual movement of the boat.
    (b) How fast is the boat going, relative to the ground?
    (c) By what angle does the current push the boat off of its due east course?

14. An airplane is flying at an airspeed of 500 km/hr in a wind blowing at 60 km/hr toward the southeast. In what direction should the plane head to end up going due east? What is the airplane's speed relative to the ground?

Figure 13.27

15. Shortly after takeoff, a plane is climbing northwest through still air at an airspeed of 200 km/hr, and rising at a rate of 300 m/min. Resolve its velocity vector into components. The x-axis points east, the y-axis points north, and the z-axis points up.

16. An airplane heads northeast at an airspeed of 700 km/hr, but there is a wind blowing from the west at 60 km/hr. In what direction does the plane end up flying? What is its speed relative to the ground?

17. An airplane is flying at an airspeed of 600 km/hr in a cross-wind that is blowing from the northeast at a speed of 50 km/hr. In what direction should the plane head to end up going due east?

18. A plane is heading due east and climbing at the rate of 80 km/hr. If its airspeed is 480 km/hr and there is a wind blowing 100 km/hr to the northeast, what is the ground speed of the plane?

19. There are five students in a class. Their scores on the midterm (out of 100) are given by the vector $\vec{v} = (73, 80, 91, 65, 84)$. Their scores on the final (out of 100) are given by $\vec{w} = (82, 79, 88, 70, 92)$. If the final counts twice as much as the midterm, find a vector giving the total scores (as a percentage) of the students.

20. The price vector of beans, rice, and tofu is $(0.30, 0.20, 0.50)$ in dollars per pound. Express it in dollars per ounce.

21. Two forces, represented by the vectors $\vec{F}_1 = 8\vec{i} - 6\vec{j}$ and $\vec{F}_2 = 3\vec{i} + 2\vec{j}$, are acting on an object. Give a vector representing the force that must be applied to the object if it is to remain stationary.

22. An object is to be moved vertically upwards by a crane. As the crane cannot get directly above the object, three ropes are attached to guide the object. One rope is pulled parallel to the ground with a force of 100 newtons in a direction 30° north of east. The second rope is pulled parallel to the ground with a force of 70 newtons in a direction 80° south of east. If the crane is attached to the third rope and can pull with a total force of 3000 newtons, find the force vector for the crane. What is the resulting (total) force on the object? (Assume vector $\vec{i}$ points east, vector $\vec{j}$ points north, and vector $\vec{k}$ points vertically up.)

Use the geometric definition of addition and scalar multiplication to explain each of the properties in Problems 23–30.

23. $\vec{w} + \vec{v} = \vec{v} + \vec{w}$

24. $(\alpha + \beta)\vec{v} = \alpha\vec{v} + \beta\vec{v}$

25. $\alpha(\vec{v} + \vec{w}) = \alpha\vec{v} + \alpha\vec{w}$

26. $(\vec{u} + \vec{v}) + \vec{w} = \vec{u} + (\vec{v} + \vec{w})$

27. $\alpha(\beta\vec{v}) = (\alpha\beta)\vec{v}$

28. $\vec{v} + \vec{0} = \vec{v}$

29. $1\vec{v} = \vec{v}$

30. $\vec{v} + (-1)\vec{w} = \vec{v} - \vec{w}$

13.3 THE DOT PRODUCT

We have seen how to add vectors; can we multiply two vectors together? In the next two sections we will see two different ways of doing so: the *scalar product* (or *dot product*) which produces a scalar, and the *vector product* (or *cross product*), which produces a vector.

Definition of the Dot Product

The dot product links geometry and algebra. We already know how to calculate the length of a vector from its components; the dot product gives us a way of computing the angle between two vectors. For any two vectors $\vec{v} = v_1\vec{i} + v_2\vec{j} + v_3\vec{k}$ and $\vec{w} = w_1\vec{i} + w_2\vec{j} + w_3\vec{k}$, shown in Figure 13.28, we define a scalar as follows:

The following two definitions of the **dot product**, or **scalar product**, $\vec{v} \cdot \vec{w}$, are equivalent:
- **Geometric definition**
  $\vec{v} \cdot \vec{w} = \|\vec{v}\|\|\vec{w}\|\cos\theta$ where θ is the angle between $\vec{v}$ and $\vec{w}$ and $0 \leq \theta \leq \pi$.
- **Algebraic definition**
  $\vec{v} \cdot \vec{w} = v_1w_1 + v_2w_2 + v_3w_3$.

Notice that the dot product of two vectors is a *number*.

Why don't we give just one definition of $\vec{v} \cdot \vec{w}$? The reason is that both definitions are equally important; the geometric definition gives us a picture of what the dot product means and the algebraic definition gives us a way of calculating it.

How do we know the two definitions are equivalent — that is, they really do define the same thing? First, we observe that the two definitions give the same result in a particular example. Then we show why they are equivalent in general.

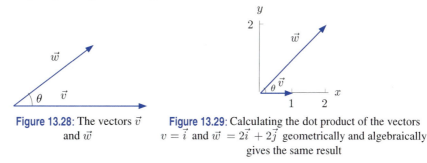

Figure 13.28: The vectors $\vec{v}$ and $\vec{w}$

Figure 13.29: Calculating the dot product of the vectors $v = \vec{i}$ and $\vec{w} = 2\vec{i} + 2\vec{j}$ geometrically and algebraically gives the same result

Example 1   Suppose $\vec{v} = \vec{i}$ and $\vec{w} = 2\vec{i} + 2\vec{j}$. Compute $\vec{v} \cdot \vec{w}$ both geometrically and algebraically.

Solution   To use the geometric definition, see Figure 13.29. The angle between the vectors is $\pi/4$, or $45°$, and the lengths of the vectors are given by

$$\|\vec{v}\| = 1 \quad \text{and} \quad \|\vec{w}\| = 2\sqrt{2}.$$

Thus,

$$\vec{v} \cdot \vec{w} = \|\vec{v}\|\|\vec{w}\|\cos\theta = 1 \cdot 2\sqrt{2}\cos\left(\frac{\pi}{4}\right) = 2.$$

Using the algebraic definition, we get the same result:

$$\vec{v} \cdot \vec{w} = 1 \cdot 2 + 0 \cdot 2 = 2.$$

Why the Two Definitions of the Dot Product Give the Same Result

In the previous example, the two definitions give the same value for the dot product. To show that the geometric and algebraic definitions of the dot product always give the same result, we must show that, for any vectors $\vec{v} = v_1\vec{i} + v_2\vec{j} + v_3\vec{k}$ and $\vec{w} = w_1\vec{i} + w_2\vec{j} + w_3\vec{k}$ with an angle θ between them:

$$\|\vec{v}\|\|\vec{w}\|\cos\theta = v_1w_1 + v_2w_2 + v_3w_3.$$

One method follows; a method which does not use trigonometry is given in Problem 41 on page 628.

Using the Law of Cosines. Suppose that $0 < \theta < \pi$, so that the vectors $\vec{v}$ and $\vec{w}$ form a triangle. (See Figure 13.30.) By the Law of Cosines, we have

$$\|\vec{v} - \vec{w}\|^2 = \|\vec{v}\|^2 + \|\vec{w}\|^2 - 2\|\vec{v}\|\|\vec{w}\|\cos\theta.$$

This result is also true for $\theta = 0$ and $\theta = \pi$. We calculate the lengths using components:

$$\|\vec{v}\|^2 = v_1^2 + v_2^2 + v_3^2$$
$$\|\vec{w}\|^2 = w_1^2 + w_2^2 + w_3^2$$
$$\|\vec{v} - \vec{w}\|^2 = (v_1 - w_1)^2 + (v_2 - w_2)^2 + (v_3 - w_3)^2$$
$$= v_1^2 - 2v_1w_1 + w_1^2 + v_2^2 - 2v_2w_2 + w_2^2 + v_3^2 - 2v_3w_3 + w_3^2.$$

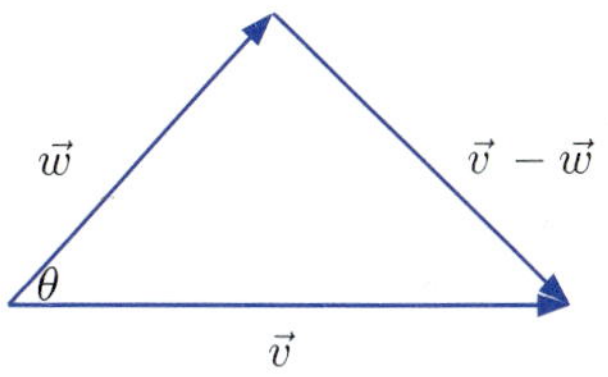

Figure 13.30: Triangle used in the justification of $\|\vec{v}\|\|\vec{w}\|\cos\theta = v_1w_1 + v_2w_2 + v_3w_3$

Substituting into the Law of Cosines and canceling, we see that

$$-2v_1w_1 - 2v_2w_2 - 2v_3w_3 = -2\|\vec{v}\|\|\vec{w}\|\cos\theta.$$

Therefore we have the result we wanted, namely that:

$$v_1w_1 + v_2w_2 + v_3w_3 = \|\vec{v}\|\|\vec{w}\|\cos\theta.$$

Properties of the Dot Product

The following properties of the dot product can be justified using the algebraic definition; see Problem 34 on page 628. For a geometric interpretation of Property 3, see Problem 38.

Properties of the Dot Product. For any vectors $\vec{u}$, $\vec{v}$, and $\vec{w}$ and any scalar λ,
1. $\vec{v} \cdot \vec{w} = \vec{w} \cdot \vec{v}$
2. $\vec{v} \cdot (\lambda\vec{w}) = \lambda(\vec{v} \cdot \vec{w}) = (\lambda\vec{v}) \cdot \vec{w}$
3. $(\vec{v} + \vec{w}) \cdot \vec{u} = \vec{v} \cdot \vec{u} + \vec{w} \cdot \vec{u}$

Perpendicularity, Magnitude, and Dot Products

Two vectors are perpendicular if the angle between them is $\pi/2$ or $90°$. Since $\cos(\pi/2) = 0$, if $\vec{v}$ and $\vec{w}$ are perpendicular, then $\vec{v} \cdot \vec{w} = 0$. Conversely, provided that $\vec{v} \cdot \vec{w} = 0$, then $\cos\theta = 0$, so $\theta = \pi/2$ and the vectors are perpendicular. Thus, we have the following result:

Two nonzero vectors $\vec{v}$ and $\vec{w}$ are **perpendicular**, or **orthogonal**, if and only if

$$\vec{v} \cdot \vec{w} = 0.$$

For example: $\vec{i} \cdot \vec{j} = 0, \vec{j} \cdot \vec{k} = 0, \vec{i} \cdot \vec{k} = 0$.

If we take the dot product of a vector with itself, then $\theta = 0$ and $\cos\theta = 1$. For any vector $\vec{v}$:

Magnitude and dot product are related as follows:

$$\vec{v} \cdot \vec{v} = \|\vec{v}\|^2.$$

For example: $\vec{i} \cdot \vec{i} = 1, \vec{j} \cdot \vec{j} = 1, \vec{k} \cdot \vec{k} = 1$.

Using the Dot Product

Depending on the situation, one definition of the dot product may be more convenient to use than the other. In Example 2 which follows, the geometric definition is the only one which can be used because we are not given components. In Example 3, the algebraic definition is used.

Example 2   Suppose the vector $\vec{b}$ is fixed and has length 2; the vector $\vec{a}$ is free to rotate and has length 3. What are the maximum and minimum values of the dot product $\vec{a} \cdot \vec{b}$ as the vector $\vec{a}$ rotates through all possible positions? What positions of $\vec{a}$ and $\vec{b}$ lead to these values?

Solution   The geometric definition gives $\vec{a} \cdot \vec{b} = \|\vec{a}\|\|\vec{b}\|\cos\theta = 3 \cdot 2\cos\theta = 6\cos\theta$. Thus, the maximum value of $\vec{a} \cdot \vec{b}$ is 6, and it occurs when $\cos\theta = 1$ so $\theta = 0$, that is, when $\vec{a}$ and $\vec{b}$ point in the same direction. The minimum value of $\vec{a} \cdot \vec{b}$ is -6, and it occurs when $\cos\theta = -1$ so $\theta = \pi$, that is, when $\vec{a}$ and $\vec{b}$ point in opposite directions. (See Figure 13.31.)

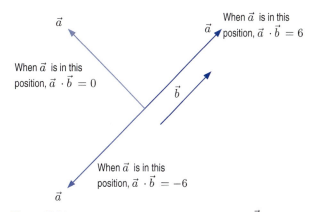

Figure 13.31: Maximum and minimum values of $\vec{a} \cdot \vec{b}$ obtained from a fixed vector $\vec{b}$ of length 2 and rotating vector $\vec{a}$ of length 3

Example 3

Which pairs from the following list of 3-dimensional vectors are perpendicular to one another?

$$\vec{u} = \vec{i} + \sqrt{3}\,\vec{k}, \quad \vec{v} = \vec{i} + \sqrt{3}\,\vec{j}, \quad \vec{w} = \sqrt{3}\,\vec{i} + \vec{j} - \vec{k}.$$

Solution

The geometric definition tells us that two vectors are perpendicular if and only if their dot product is zero. Since the vectors are given in components, we calculate dot products using the algebraic definition:

$$\vec{v} \cdot \vec{u} = (\vec{i} + \sqrt{3}\,\vec{j} + 0\vec{k}) \cdot (\vec{i} + 0\vec{j} + \sqrt{3}\,\vec{k}) = 1 \cdot 1 + \sqrt{3} \cdot 0 + 0 \cdot \sqrt{3} = 1,$$
$$\vec{v} \cdot \vec{w} = (\vec{i} + \sqrt{3}\,\vec{j} + 0\vec{k}) \cdot (\sqrt{3}\,\vec{i} + \vec{j} - \vec{k}) = 1 \cdot \sqrt{3} + \sqrt{3} \cdot 1 + 0(-1) = 2\sqrt{3},$$
$$\vec{w} \cdot \vec{u} = (\sqrt{3}\,\vec{i} + \vec{j} - \vec{k}) \cdot (\vec{i} + 0\vec{j} + \sqrt{3}\,\vec{k}) = \sqrt{3} \cdot 1 + 1 \cdot 0 + (-1) \cdot \sqrt{3} = 0.$$

So the only two vectors which are perpendicular are $\vec{w}$ and $\vec{u}$.

Normal Vectors and the Equation of a Plane

In Section 12.4 we wrote the equation of a plane given its x-slope, y-slope and z-intercept. Now we write the equation of a plane using a vector and a point lying on the plane. A *normal vector* to a plane is a vector that is perpendicular to the plane, that is, it is perpendicular to every displacement vector between any two points in the plane. Let $\vec{n} = a\vec{i} + b\vec{j} + c\vec{k}$ be a normal vector to the plane, let $P_0 = (x_0, y_0, z_0)$ be a fixed point in the plane, and let $P = (x, y, z)$ be any other point in the plane. Then $\overrightarrow{P_0P} = (x - x_0)\vec{i} + (y - y_0)\vec{j} + (z - z_0)\vec{k}$ is a vector whose head and tail both lie in the plane. (See Figure 13.32.) Thus, the vectors $\vec{n}$ and $\overrightarrow{P_0P}$ are perpendicular, so $\vec{n} \cdot \overrightarrow{P_0P} = 0$. The algebraic definition of the dot product gives $\vec{n} \cdot \overrightarrow{P_0P} = a(x - x_0) + b(y - y_0) + c(z - z_0)$, so we obtain the following result:

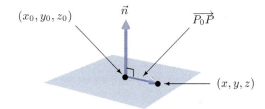

Figure 13.32: Plane with normal $\vec{n}$ and containing a fixed point (x_0, y_0, z_0)

The **equation of the plane** with normal vector $\vec{n} = a\vec{i} + b\vec{j} + c\vec{k}$ and containing the point $P_0 = (x_0, y_0, z_0)$ is

$$a(x - x_0) + b(y - y_0) + c(z - z_0) = 0.$$

Letting $d = ax_0 + by_0 + cz_0$ (a constant), we can write the equation of the plane in the form

$$ax + by + cz = d.$$

Example 4   Find the equation of the plane perpendicular to $-\vec{i} + 3\vec{j} + 2\vec{k}$ and passing through the point $(1, 0, 4)$.

Solution   The equation of the plane is

$$-(x - 1) + 3(y - 0) + 2(z - 4) = 0,$$

which simplifies to

$$-x + 3y + 2z = 7.$$

Example 5   Find a normal vector to the plane with equation   (a) $x - y + 2z = 5$ (b) $z = 0.5x + 1.2y$.

Solution   (a) Since the coefficients of $\vec{i}$, $\vec{j}$, and $\vec{k}$ in a normal vector are the coefficients of x, y, z in the equation of the plane, a normal vector is $\vec{n} = \vec{i} - \vec{j} + 2\vec{k}$.

(b) Before we can find a normal vector, we rewrite the equation of the plane in the form

$$0.5x + 1.2y - z = 0.$$

Thus, a normal vector is $\vec{n} = 0.5\vec{i} + 1.2\vec{j} - \vec{k}$.

The Dot Product in n Dimensions

The algebraic definition of the dot product can be extended to vectors in higher dimensions.

If $\vec{u} = (u_1, \ldots, u_n)$ and $\vec{v} = (v_1, \ldots, v_n)$ then the dot product of $\vec{u}$ and $\vec{v}$ is the **scalar**

$$\vec{u} \cdot \vec{v} = u_1 v_1 + \ldots + u_n v_n.$$

Example 6   A video store sells videos, tapes, CDs, and computer games. We define the quantity vector $\vec{q} = (q_1, q_2, q_3, q_4)$, where q_1, q_2, q_3, q_4 denote the quantities sold of each of the items, and the price vector $\vec{p} = (p_1, p_2, p_3, p_4)$, where p_1, p_2, p_3, p_4 denote the price per unit of each item. What does the dot product $\vec{p} \cdot \vec{q}$ represent?

Solution   The dot product is $\vec{p} \cdot \vec{q} = p_1 q_1 + p_2 q_2 + p_3 q_3 + p_4 q_4$. The quantity $p_1 q_1$ represents the revenue received by the store for the videos, $p_2 q_2$ represents the revenue for the tapes, and so on. The dot product represents the total revenue received by the store for the sale of these four items.

Resolving a Vector into Components: Projections

In Section 13.1, we resolved a vector into components parallel to the axes. Now we see how to resolve a vector, $\vec{v}$, into components, called $\vec{v}$ parallel and $\vec{v}$ perp, which are parallel and perpendicular, respectively, to a given nonzero vector, $\vec{u}$. (See Figure 13.33.)

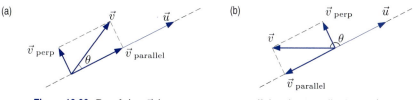

Figure 13.33: Resolving $\vec{v}$ into components parallel and perpendicular to $\vec{u}$
(a) $0 < \theta < \pi/2$ (b) $\pi/2 < \theta < \pi$

The projection of $\vec{v}$ on $\vec{u}$, written $\vec{v}_{\text{parallel}}$, measures (in some sense) how much the vector $\vec{v}$ is aligned with the vector $\vec{u}$. The length of $\vec{v}_{\text{parallel}}$ is the length of the shadow cast by $\vec{v}$ on a line in the direction of $\vec{u}$.

To compute $\vec{v}_{\text{parallel}}$, we assume $\vec{u}$ is a unit vector. (If not, create one by dividing by its length.) Then Figure 13.33(a) shows that, if $0 \leq \theta \leq \pi/2$:

$$\|\vec{v}_{\text{parallel}}\| = \|\vec{v}\| \cos\theta = \vec{v} \cdot \vec{u} \qquad (\text{since } \|\vec{u}\| = 1).$$

Now $\vec{v}_{\text{parallel}}$ is a scalar multiple of $\vec{u}$, and since $\vec{u}$ is a unit vector,

$$\vec{v}_{\text{parallel}} = (\|\vec{v}\| \cos\theta)\vec{u} = (\vec{v} \cdot \vec{u})\vec{u}.$$

A similar argument shows that if $\pi/2 < \theta \leq \pi$, as in Figure 13.33(b), this formula for $\vec{v}_{\text{parallel}}$ still holds. The vector $\vec{v}_{\text{perp}}$ is specified by

$$\vec{v}_{\text{perp}} = \vec{v} - \vec{v}_{\text{parallel}}.$$

Thus, we have the following results:

Projection of $\vec{v}$ on the Line in the Direction of the Unit Vector $\vec{u}$

If $\vec{v}_{\text{parallel}}$ and $\vec{v}_{\text{perp}}$ are components of $\vec{v}$ which are parallel and perpendicular, respectively, to $\vec{u}$, then

$$\text{Projection of } \vec{v} \text{ on to } \vec{u} = \vec{v}_{\text{parallel}} = (\vec{v} \cdot \vec{u})\vec{u} \qquad \text{provided } \|\vec{u}\| = 1$$

and $\qquad \vec{v} = \vec{v}_{\text{parallel}} + \vec{v}_{\text{perp}} \qquad$ so $\qquad \vec{v}_{\text{perp}} = \vec{v} - \vec{v}_{\text{parallel}}.$

Example 7   Figure 13.34 shows the force the wind exerts on the sail of a sailboat. Find the component of the force in the direction in which the sailboat is traveling.

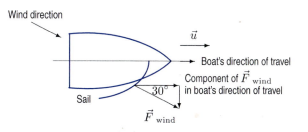

Figure 13.34: Wind moving a sailboat

Solution

Let $\vec{u}$ be a unit vector in the direction of travel. The force of the wind on the sail makes an angle of $30°$ with $\vec{u}$. Thus, the component of this force in the direction of $\vec{u}$ is

$$\vec{F}_{\text{parallel}} = (\vec{F} \cdot \vec{u})\vec{u} = \|\vec{F}\|(\cos 30°)\vec{u} = 0.87\|\vec{F}\|\vec{u}.$$

Thus, the boat is being pushed forward with about 87% of the total force due to the wind. (In fact, the interaction of wind and sail is much more complex than this model suggests.)

A Physical Interpretation of the Dot Product: Work

In physics, the word "work" has a slightly different meaning from its everyday meaning. In physics, when a force of magnitude F acts on an object through a distance d, we say the *work*, W, done by the force is

$$W = Fd,$$

provided the force and the displacement are in the same direction. For example, if a 1 kg body falls 10 meters under the force of gravity, which is 9.8 newtons, then the work done by gravity is

$$W = (9.8 \text{ newtons}) \cdot (10 \text{ meters}) = 98 \text{ joules}.$$

What if the force and the displacement are not in the same direction? Suppose a force $\vec{F}$ acts on an object as it moves along a displacement vector $\vec{d}$. Let θ be the angle between $\vec{F}$ and $\vec{d}$. First, we assume $0 \leq \theta \leq \pi/2$. Figure 13.35 shows how we can resolve $\vec{F}$ into components that are parallel and perpendicular to $\vec{d}$:

$$\vec{F} = \vec{F}_{\text{parallel}} + \vec{F}_{\text{perp}},$$

Then the work done by $\vec{F}$ is defined to be

$$W = \|\vec{F}_{\text{parallel}}\| \|\vec{d}\|.$$

We see from Figure 13.35 that $\vec{F}_{\text{parallel}}$ has magnitude $\|\vec{F}\| \cos \theta$. So the work is given by the dot product:

$$W = (\|\vec{F}\| \cos \theta)\|\vec{d}\| = \|\vec{F}\|\|\vec{d}\| \cos \theta = \vec{F} \cdot \vec{d}.$$

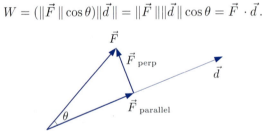

Figure 13.35: Resolving the force $\vec{F}$ into two forces, one parallel to $\vec{d}$, one perpendicular to $\vec{d}$

The formula $W = \vec{F} \cdot \vec{d}$ holds when $\pi/2 < \theta \leq \pi$ also. In that case, the work done by the force is negative and the object is moving against the force. Thus, we have the following definition:

> The **work**, W, done by a force $\vec{F}$ acting on an object through a displacement $\vec{d}$ is given by
>
> $$W = \vec{F} \cdot \vec{d}.$$

Notice that if the vectors $\vec{F}$ and $\vec{d}$ are parallel and in the same direction, with magnitudes F and d, then $\cos \theta = \cos 0 = 1$, so $W = \|\vec{F}\|\|\vec{d}\| = Fd$, which is the original definition. When the vectors are perpendicular, $\cos \theta = \cos \frac{\pi}{2} = 0$, so $W = 0$ and no work is done in the technical definition of the word. For example, if you carry a heavy box across the room at the same horizontal height, no work is done by gravity because the force of gravity is vertical but the motion is horizontal.

Exercises and Problems for Section 13.3

Exercises

For Exercises 1–6, perform the following operations on the given 3-dimensional vectors.

$$\vec{a} = 2\vec{j} + \vec{k} \qquad \vec{b} = -3\vec{i} + 5\vec{j} + 4\vec{k} \qquad \vec{c} = \vec{i} + 6\vec{j}$$

$$\vec{y} = 4\vec{i} - 7\vec{j} \qquad \vec{z} = \vec{i} - 3\vec{j} - \vec{k}$$

1. $\vec{c} \cdot \vec{y}$

2. $\vec{a} \cdot \vec{z}$

3. $\vec{a} \cdot \vec{b}$

4. $(\vec{a} \cdot \vec{b})\vec{a}$

5. $(\vec{a} \cdot \vec{y})(\vec{c} \cdot \vec{z})$

6. $((\vec{c} \cdot \vec{c})\vec{a}) \cdot \vec{a}$

7. **(a)** Give a vector that is parallel to, but not equal to, $\vec{v} = 4\vec{i} + 3\vec{j}$.
   (b) Give a vector that is perpendicular to $\vec{v}$.

In Exercises 8–11, find a normal vector to the plane.

8. $2x + y - z = 5$

9. $z = 3x + 4y - 7$

10. $z - 5(x - 2) = 3(5 - y)$

11. $2(x - z) = 3(x + y)$

Problems

12. Compute the angle between the vectors $\vec{i} + \vec{j} + \vec{k}$ and $\vec{i} - \vec{j} - \vec{k}$.

13. Which pairs of the vectors $\sqrt{3}\vec{i} + \vec{j}, 3\vec{i} + \sqrt{3}\vec{j}, \vec{i} - \sqrt{3}\vec{j}$ are parallel and which are perpendicular?

14. For what values of t are $\vec{u} = t\vec{i} - \vec{j} + \vec{k}$ and $\vec{v} = t\vec{i} + t\vec{j} - 2\vec{k}$ perpendicular? Are there values of t for which $\vec{u}$ and $\vec{v}$ are parallel?

15. List any vectors which are parallel to each other and any vectors which are perpendicular to each other:

$$\vec{v}_1 = \vec{i} - 2\vec{j}, \qquad \vec{v}_2 = 2\vec{i} + 4\vec{j}, \qquad \vec{v}_3 = 3\vec{i} + 1.5\vec{j},$$
$$\vec{v}_4 = -1.2\vec{i} + 2.4\vec{j}, \vec{v}_5 = -5\vec{i} - 2.5\vec{j}, \vec{v}_6 = 12\vec{i} - 12\vec{j},$$
$$\vec{v}_7 = 4\vec{i} + 2\vec{j}, \qquad \vec{v}_8 = 3\vec{i} - 6\vec{j}, \qquad \vec{v}_9 = 0.70\vec{i} - 0.35\vec{j}$$

In Problems 16–20, find an equation of a plane that satisfies the given conditions.

16. Perpendicular to the vector $-\vec{i} + 2\vec{j} + \vec{k}$ and passing through the point $(1, 0, 2)$.

17. Perpendicular to the vector $5\vec{i} + \vec{j} - 2\vec{k}$ and passing through the point $(0, 1, -1)$.

18. Perpendicular to the vector $2\vec{i} - 3\vec{j} + 7\vec{k}$ and passing through the point $(1, -1, 2)$.

19. Parallel to the plane $2x + 4y - 3z = 1$ and through the point $(1, 0, -1)$.

20. Through $(-2, 3, 2)$ and parallel to $3x + y + z = 4$.

21. A plane has equation $z = 5x - 2y + 7$.
   (a) Find a value of λ making the vector $\lambda\vec{i} + \vec{j} + 0.5\vec{k}$ normal to the plane.
   (b) Find a value of a so that the point $(a + 1, a, a - 1)$ lies on the plane.

22. Find angle BAC if $A = (2, 2, 2)$, $B = (4, 2, 1)$, and $C = (2, 3, 1)$.

23. Let $A = (0, 4)$, $B = (-1, -3)$, and $C = (-5, 1)$. Draw triangle ABC and find each of its interior angles.

24. Let S be the triangle with vertices $A = (2, 2, 2), B = (4, 2, 1)$, and $C = (2, 3, 1)$.
   (a) Find the length of the shortest side of S.
   (b) Find the cosine of the angle BAC at vertex A.

25. Write $\vec{a} = 3\vec{i} + 2\vec{j} - 6\vec{k}$ as the sum of two vectors, one parallel, and one perpendicular, to $\vec{d} = 2\vec{i} - 4\vec{j} + \vec{k}$.

26. Find the points where the plane $z = 5x - 4y + 3$ intersects each of the coordinate axes. Find the lengths of the sides and the angles of the triangle formed by these points.

27. Find the angle between the planes $5(x - 1) + 3(y + 2) + 2z = 0$ and $x + 3(y - 1) + 2(z + 4) = 0$.

28. A basketball gymnasium is 25 meters high, 80 meters wide and 200 meters long. For a half time stunt, the cheerleaders want to run two strings, one from each of the two corners above one basket to the diagonally opposite corners of the gym floor. What is the cosine of the angle made by the strings as they cross?

29. A street vendor sells six items, with prices p_1 dollars per unit, p_2 dollars per unit, and so on. The vendor's price vector is $\vec{p} = (p_1, p_2, p_3, p_4, p_5, p_6) = (1.00, 3.50, 4.00, 2.75, 5.00, 3.00)$. The vendor sells q_1 units of the first item, q_2 units of the second item, and so on. The vendor's quantity vector is $\vec{q} = (q_1, q_2, q_3, q_4, q_5, q_6) = (43, 57, 12, 78, 20, 35)$. Find $\vec{p} \cdot \vec{q}$, give its units, and explain its significance to the vendor.

30. A course has four exams, weighted 10%, 15%, 25%, 50%, respectively. The class average on each of these exams is 75%, 91%, 84%, 87%, respectively. What do the vectors $\vec{a} = (0.75, 0.91, 0.84, 0.87)$ and $\vec{w} = (0.1, 0.15, 0.25, 0.5)$ represent, in terms of the course? Calculate the dot product $\vec{w} \cdot \vec{a}$. What does it represent, in terms of the course?

31. A consumption vector of three goods is defined by $\vec{x} = (x_1, x_2, x_3)$, where x_1, x_2 and x_3 are the quantities consumed of the three goods. A budget constraint is represented by the equation $\vec{p} \cdot \vec{x} = k$, where $\vec{p}$ is the price vector of the three goods and k is a constant. Show that the difference between two consumption vectors corresponding to points satisfying the same budget constraint is perpendicular to the price vector $\vec{p}$.

32. A 100-meter dash is run on a track in the direction of the vector $\vec{v} = 2\vec{i} + 6\vec{j}$. The wind velocity $\vec{w}$ is $5\vec{i} + \vec{j}$ km/hr. The rules say that a legal wind speed measured in the direction of the dash must not exceed 5 km/hr. Will the race results be disqualified due to an illegal wind? Justify your answer.

33. Consider a point P and plane through the point P_0 with normal vector $\vec{n}$.

    (a) What do all the points satisfying the following condition have in common:
    $$\vec{n} \cdot \overrightarrow{P_0P} > 0?$$
    What about all the points satisfying $\vec{n} \cdot \overrightarrow{P_0P} < 0$?

    (b) Express the conditions $\vec{n} \cdot \overrightarrow{P_0P} > 0$ and $\vec{n} \cdot \overrightarrow{P_0P} < 0$ in terms of coordinates.

    (c) Use parts (a) and (b) to decide which of the points
    $$P = (-1, -1, 1), \quad Q = (-1, -1, -1)$$
    $$R = (1, 1, 1)$$
    are on the same side of the plane $2x - 3y + 4z = 4$.

34. Show why each of the properties of the dot product in the box on page 622 follows from the algebraic definition of the dot product:
    $$\vec{v} \cdot \vec{w} = v_1 w_1 + v_2 w_2 + v_3 w_3$$

35. What does Property 2 of the dot product in the box on page 622 say geometrically?

36. Show that the vectors $(\vec{b} \cdot \vec{c})\vec{a} - (\vec{a} \cdot \vec{c})\vec{b}$ and $\vec{c}$ are perpendicular.

37. Show that if $\vec{u}$ and $\vec{v}$ are two vectors such that
    $$\vec{u} \cdot \vec{w} = \vec{v} \cdot \vec{w}$$
    for every vector $\vec{w}$, then
    $$\vec{u} = \vec{v}.$$

38. Figure 13.36 shows that, given three vectors $\vec{u}$, $\vec{v}$, and $\vec{w}$, the sum of the components of $\vec{v}$ and $\vec{w}$ in the direction of $\vec{u}$ is the component of $\vec{v} + \vec{w}$ in the direction of $\vec{u}$. (Although the figure is drawn in two dimensions, this result is also true in three dimensions.) Use this figure to explain why the geometric definition of the dot product satisfies $(\vec{v} + \vec{w}) \cdot \vec{u} = \vec{v} \cdot \vec{u} + \vec{w} \cdot \vec{u}$.

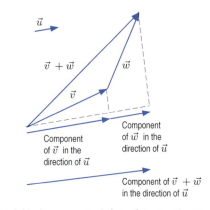

Figure 13.36: Component of $\vec{v} + \vec{w}$ in the direction of $\vec{u}$ is the sum of the components of $\vec{v}$ and $\vec{w}$ in that direction

39. (a) Using the geometric definition of the dot product, show that
    $$\vec{u} \cdot (-\vec{v}) = -(\vec{u} \cdot \vec{v}).$$
    [Hint: What happens to the angle when you multiply $\vec{v}$ by -1?]

    (b) Using the geometric definition of the dot product, show that for any negative scalar λ
    $$\vec{u} \cdot (\lambda \vec{v}) = \lambda(\vec{u} \cdot \vec{v})$$
    $$(\lambda \vec{u}) \cdot \vec{v} = \lambda(\vec{u} \cdot \vec{v}).$$

40. The Law of Cosines for a triangle with side lengths a, b, and c, and with angle C opposite side c, says
    $$c^2 = a^2 + b^2 - 2ab \cos C.$$
    On page 621, we used the Law of Cosines to show that the two definitions of the dot product are equivalent. In this problem, use the geometric definition of the dot product and its properties in the box on page 622 to prove the Law of Cosines. [Hint: Let $\vec{u}$ and $\vec{v}$ be the displacement vectors from C to the other two vertices, and express c^2 in terms of $\vec{u}$ and $\vec{v}$.]

41. Use the following steps and the results of Problems 38–39 to show (without trigonometry) that the geometric and algebraic definitions of the dot product are equivalent.

    Let $\vec{u} = u_1\vec{i} + u_2\vec{j} + u_3\vec{k}$ and $\vec{v} = v_1\vec{i} + v_2\vec{j} + v_3\vec{k}$ be any vectors. Write $(\vec{u} \cdot \vec{v})_{\text{geom}}$ for the result of the dot product computed geometrically. Substitute $\vec{u} = u_1\vec{i} + u_2\vec{j} + u_3\vec{k}$ and use Problems 38–39 to expand $(\vec{u} \cdot \vec{v})_{\text{geom}}$. Substitute for $\vec{v}$ and expand. Then calculate the dot products $\vec{i} \cdot \vec{i}$, $\vec{i} \cdot \vec{j}$, etc. geometrically.

42. For any vectors $\vec{v}$ and $\vec{w}$, consider the following function of t:
    $$q(t) = (\vec{v} + t\vec{w}) \cdot (\vec{v} + t\vec{w}).$$

    (a) Explain why $q(t) \geq 0$ for all real t.

    (b) Expand $q(t)$ as a quadratic polynomial in t using the properties on page 622.

    (c) Using the discriminant of the quadratic, show that,
    $$|\vec{v} \cdot \vec{w}| \leq \|\vec{v}\|\|\vec{w}\|.$$

13.4 THE CROSS PRODUCT

In the previous section we combined two vectors to get a number, the dot product. In this section we see another way of combining two vectors, this time to get a vector, the *cross product*. Any two vectors in 3-space form a parallelogram. We define the cross product using this parallelogram.

The Area of a Parallelogram

Consider the parallelogram formed by the vectors $\vec{v}$ and $\vec{w}$ with an angle of θ between them. Then Figure 13.37 shows

$$\text{Area of parallelogram} = \text{Base} \cdot \text{Height} = \|\vec{v}\|\|\vec{w}\|\sin\theta.$$

How would we compute the area of the parallelogram if we were given $\vec{v}$ and $\vec{w}$ in components, $\vec{v} = v_1\vec{i} + v_2\vec{j} + v_3\vec{k}$ and $\vec{w} = w_1\vec{i} + w_2\vec{j} + w_3\vec{k}$? Project 1 on page 638 shows that if $\vec{v}$ and $\vec{w}$ are in the xy-plane, so $v_3 = w_3 = 0$, then

$$\text{Area of parallelogram} = |v_1 w_2 - v_2 w_1|.$$

What if $\vec{v}$ and $\vec{w}$ do not lie in the xy-plane? The cross product will enable us to compute the area of the parallelogram formed by any two vectors.

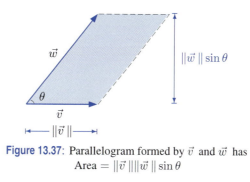

Figure 13.37: Parallelogram formed by $\vec{v}$ and $\vec{w}$ has
Area $= \|\vec{v}\|\|\vec{w}\|\sin\theta$

Definition of the Cross Product

We define the cross product of the vectors $\vec{v}$ and $\vec{w}$, written $\vec{v} \times \vec{w}$, to be a vector perpendicular to both $\vec{v}$ and $\vec{w}$. The magnitude of this vector is the area of the parallelogram formed by the two vectors. The direction of $\vec{v} \times \vec{w}$ is given by the normal vector, $\vec{n}$, to the plane defined by $\vec{v}$ and $\vec{w}$. If we require that $\vec{n}$ be a unit vector, there are two choices for $\vec{n}$, pointing out of the plane in opposite directions. We pick one by the following rule (see Figure 13.38):

> **The right-hand rule:** Place $\vec{v}$ and $\vec{w}$ so that their tails coincide and curl the fingers of your right hand through the smaller of the two angles from $\vec{v}$ to $\vec{w}$; your thumb points in the direction of the normal vector, $\vec{n}$.

Like the dot product, there are two equivalent definitions of the cross product:

The following two definitions of the **cross product** or **vector product** $\vec{v} \times \vec{w}$ are equivalent:

- **Geometric definition**

  If $\vec{v}$ and $\vec{w}$ are not parallel, then

  $$\vec{v} \times \vec{w} = \begin{pmatrix} \text{Area of parallelogram} \\ \text{with edges } \vec{v} \text{ and } \vec{w} \end{pmatrix} \vec{n} = (\|\vec{v}\|\|\vec{w}\| \sin \theta)\vec{n},$$

  where $0 \leq \theta \leq \pi$ is the angle between $\vec{v}$ and $\vec{w}$ and $\vec{n}$ is the unit vector perpendicular to $\vec{v}$ and $\vec{w}$ pointing in the direction given by the right-hand rule. If $\vec{v}$ and $\vec{w}$ are parallel, then $\vec{v} \times \vec{w} = \vec{0}$.

- **Algebraic definition**

  $$\vec{v} \times \vec{w} = (v_2 w_3 - v_3 w_2)\vec{i} + (v_3 w_1 - v_1 w_3)\vec{j} + (v_1 w_2 - v_2 w_1)\vec{k}$$

  where $\vec{v} = v_1\vec{i} + v_2\vec{j} + v_3\vec{k}$ and $\vec{w} = w_1\vec{i} + w_2\vec{j} + w_3\vec{k}$.

Notice that the magnitude of the $\vec{k}$ component is the area of a 2-dimensional parallelogram and the other components have a similar form. Problems 29 and 30 at the end of this section show that the geometric and algebraic definitions of the cross product give the same result.

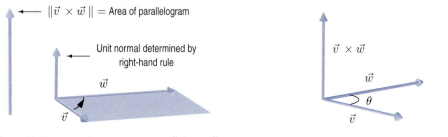

Figure 13.38: Area of parallelogram $= \|\vec{v} \times \vec{w}\|$ **Figure 13.39**: The cross product $\vec{v} \times \vec{w}$

Unlike the dot product, the cross product is only defined for three-dimensional vectors. The geometric definition shows us that the cross product is *rotation invariant*. Imagine the two vectors $\vec{v}$ and $\vec{w}$ as two metal rods welded together. Attach a third rod whose direction and length correspond to $\vec{v} \times \vec{w}$. (See Figure 13.39.) Then, no matter how we turn this set of rods, the third will still be the cross product of the first two.

The algebraic definition is more easily remembered by writing it as a 3×3 determinant. (See Appendix E.)

$$\vec{v} \times \vec{w} = \begin{vmatrix} \vec{i} & \vec{j} & \vec{k} \\ v_1 & v_2 & v_3 \\ w_1 & w_2 & w_3 \end{vmatrix} = (v_2 w_3 - v_3 w_2)\vec{i} + (v_3 w_1 - v_1 w_3)\vec{j} + (v_1 w_2 - v_2 w_1)\vec{k}.$$

Example 1   Find $\vec{i} \times \vec{j}$ and $\vec{j} \times \vec{i}$.

Solution   The vectors $\vec{i}$ and $\vec{j}$ both have magnitude 1 and the angle between them is $\pi/2$. By the right-hand rule, the vector $\vec{i} \times \vec{j}$ is in the direction of $\vec{k}$, so $\vec{n} = \vec{k}$ and we have

$$\vec{i} \times \vec{j} = \left(\|\vec{i}\|\|\vec{j}\| \sin \frac{\pi}{2}\right) \vec{k} = \vec{k}.$$

Similarly, the right-hand rule says that the direction of $\vec{j} \times \vec{i}$ is $-\vec{k}$, so

$$\vec{j} \times \vec{i} = (\|\vec{j}\|\|\vec{i}\| \sin \frac{\pi}{2})(-\vec{k}) = -\vec{k}.$$

Similar calculations show that $\vec{j} \times \vec{k} = \vec{i}$ and $\vec{k} \times \vec{i} = \vec{j}$.

Example 2   For any vector $\vec{v}$, find $\vec{v} \times \vec{v}$.

Solution   Since $\vec{v}$ is parallel to itself, $\vec{v} \times \vec{v} = \vec{0}$.

Example 3   Find the cross product of $\vec{v} = 2\vec{i} + \vec{j} - 2\vec{k}$ and $\vec{w} = 3\vec{i} + \vec{k}$ and check that the cross product is perpendicular to both $\vec{v}$ and $\vec{w}$.

Solution   Writing $\vec{v} \times \vec{w}$ as a determinant and expanding it into three two-by-two determinants, we have

$$\vec{v} \times \vec{w} = \begin{vmatrix} \vec{i} & \vec{j} & \vec{k} \\ 2 & 1 & -2 \\ 3 & 0 & 1 \end{vmatrix} = \vec{i} \begin{vmatrix} 1 & -2 \\ 0 & 1 \end{vmatrix} - \vec{j} \begin{vmatrix} 2 & -2 \\ 3 & 1 \end{vmatrix} + \vec{k} \begin{vmatrix} 2 & 1 \\ 3 & 0 \end{vmatrix}$$

$$= \vec{i} \left(1(1) - 0(-2) \right) - \vec{j} \left(2(1) - 3(-2) \right) + \vec{k} \left(2(0) - 3(1) \right)$$

$$= \vec{i} - 8\vec{j} - 3\vec{k}.$$

To check that $\vec{v} \times \vec{w}$ is perpendicular to $\vec{v}$, we compute the dot product:

$$\vec{v} \cdot (\vec{v} \times \vec{w}) = (2\vec{i} + \vec{j} - 2\vec{k}) \cdot (\vec{i} - 8\vec{j} - 3\vec{k}) = 2 - 8 + 6 = 0.$$

Similarly,

$$\vec{w} \cdot (\vec{v} \times \vec{w}) = (3\vec{i} + 0\vec{j} + \vec{k}) \cdot (\vec{i} - 8\vec{j} - 3\vec{k}) = 3 + 0 - 3 = 0.$$

Thus, $\vec{v} \times \vec{w}$ is perpendicular to both $\vec{v}$ and $\vec{w}$.

Properties of the Cross Product

The right-hand rule tells us that $\vec{v} \times \vec{w}$ and $\vec{w} \times \vec{v}$ point in opposite directions. The magnitudes of $\vec{v} \times \vec{w}$ and $\vec{w} \times \vec{v}$ are the same, so $\vec{w} \times \vec{v} = -(\vec{v} \times \vec{w})$. (See Figure 13.40.)

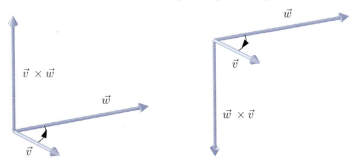

Figure 13.40: Diagram showing $\vec{v} \times \vec{w} = -(\vec{w} \times \vec{v})$

This explains the first of the following properties. The other two are derived in Problems 22, 25, and 29 at the end of this section.

Properties of the Cross Product

For vectors $\vec{u}, \vec{v}, \vec{w}$ and scalar λ

1. $\vec{w} \times \vec{v} = -(\vec{v} \times \vec{w})$
2. $(\lambda \vec{v}) \times \vec{w} = \lambda(\vec{v} \times \vec{w}) = \vec{v} \times (\lambda \vec{w})$
3. $\vec{u} \times (\vec{v} + \vec{w}) = \vec{u} \times \vec{v} + \vec{u} \times \vec{w}.$

The Equivalence of the Two Definitions of the Cross Product

Problem 29 on page 634 uses geometric arguments to show that the cross product distributes over addition. Problem 30 then shows how the formula in the algebraic definition of the cross product can be derived from the geometric definition.

The Equation of a Plane Through Three Points

The equation of a plane is determined by a point $P_0 = (x_0, y_0, z_0)$ on the plane, and a normal vector, $\vec{n} = a\vec{i} + b\vec{j} + c\vec{k}$:

$$a(x - x_0) + b(y - y_0) + c(z - z_0) = 0.$$

However, a plane can also be determined by three points on it (provided they do not lie on a line). In that case we can find an equation of the plane by first determining two vectors in the plane and then finding a normal vector using the cross product, as in the following example.

Example 4   Find an equation of the plane containing the points $P = (1, 3, 0)$, $Q = (3, 4, -3)$, and $R = (3, 6, 2)$.

Solution   Since the points P and Q are in the plane, the displacement vector between them, $\overrightarrow{PQ}$, is in the plane, where

$$\overrightarrow{PQ} = (3 - 1)\vec{i} + (4 - 3)\vec{j} + (-3 - 0)\vec{k} = 2\vec{i} + \vec{j} - 3\vec{k}.$$

The displacement vector $\overrightarrow{PR}$ is also in the plane, where

$$\overrightarrow{PR} = (3 - 1)\vec{i} + (6 - 3)\vec{j} + (2 - 0)\vec{k} = 2\vec{i} + 3\vec{j} + 2\vec{k}.$$

Thus, a normal vector, $\vec{n}$, to the plane is given by

$$\vec{n} = \overrightarrow{PQ} \times \overrightarrow{PR} = \begin{vmatrix} \vec{i} & \vec{j} & \vec{k} \\ 2 & 1 & -3 \\ 2 & 3 & 2 \end{vmatrix} = 11\vec{i} - 10\vec{j} + 4\vec{k}.$$

Since the point $(1, 3, 0)$ is on the plane, the equation of the plane is

$$11(x - 1) - 10(y - 3) + 4(z - 0) = 0,$$

which simplifies to

$$11x - 10y + 4z = -19.$$

You should check that P, Q, and R satisfy the equation of the plane.

Areas and Volumes Using the Cross Product and Determinants

We can use the cross product to calculate the area of the parallelogram with sides $\vec{v}$ and $\vec{w}$. We say that $\vec{v} \times \vec{w}$ is the *area vector* of the parallelogram. The geometric definition of the cross product tells us that $\vec{v} \times \vec{w}$ is normal to the parallelogram and gives us the following result:

> **Area of a parallelogram** with edges $\vec{v} = v_1\vec{i} + v_2\vec{j} + v_3\vec{k}$ and $\vec{w} = w_1\vec{i} + w_2\vec{j} + w_3\vec{k}$ is given by
>
> $$\text{Area} = \|\vec{v} \times \vec{w}\|, \quad \text{where} \quad \vec{v} \times \vec{w} = \begin{vmatrix} \vec{i} & \vec{j} & \vec{k} \\ v_1 & v_2 & v_3 \\ w_1 & w_2 & w_3 \end{vmatrix}.$$

Example 5   Find the area of the parallelogram with edges $\vec{v} = 2\vec{i} + \vec{j} - 3\vec{k}$ and $\vec{w} = \vec{i} + 3\vec{j} + 2\vec{k}$.

Solution   We calculate the cross product:

$$\vec{v} \times \vec{w} = \begin{vmatrix} \vec{i} & \vec{j} & \vec{k} \\ 2 & 1 & -3 \\ 1 & 3 & 2 \end{vmatrix} = (2+9)\vec{i} - (4+3)\vec{j} + (6-1)\vec{k} = 11\vec{i} - 7\vec{j} + 5\vec{k}.$$

The area of the parallelogram with edges $\vec{v}$ and $\vec{w}$ is the magnitude of the vector $\vec{v} \times \vec{w}$:

$$\text{Area} = \|\vec{v} \times \vec{w}\| = \sqrt{11^2 + (-7)^2 + 5^2} = \sqrt{195}.$$

Volume of a Parallelepiped

Consider the parallelepiped with sides formed by $\vec{a}$, $\vec{b}$, and $\vec{c}$. (See Figure 13.41.) Since the base is formed by the vectors $\vec{b}$ and $\vec{c}$, we have

$$\text{Area of base of parallelepiped} = \|\vec{b} \times \vec{c}\|.$$

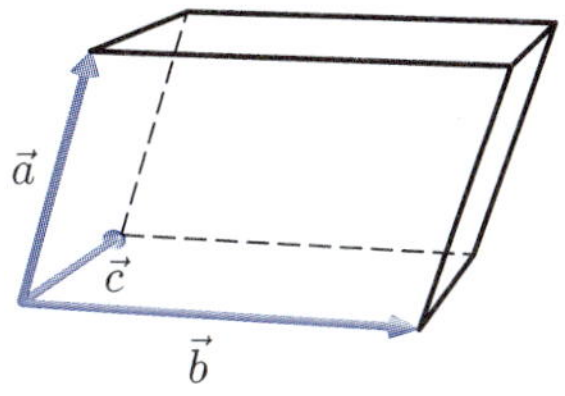

Figure 13.41: Volume of a Parallelepiped 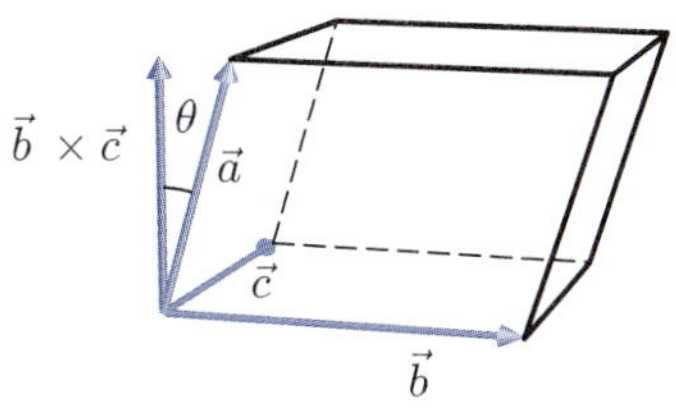   Figure 13.42: The vectors $\vec{a}$, $\vec{b}$, $\vec{c}$ are called a right-handed set   Figure 13.43: The vectors $\vec{a}$, $\vec{b}$, $\vec{c}$ are called a left-handed set

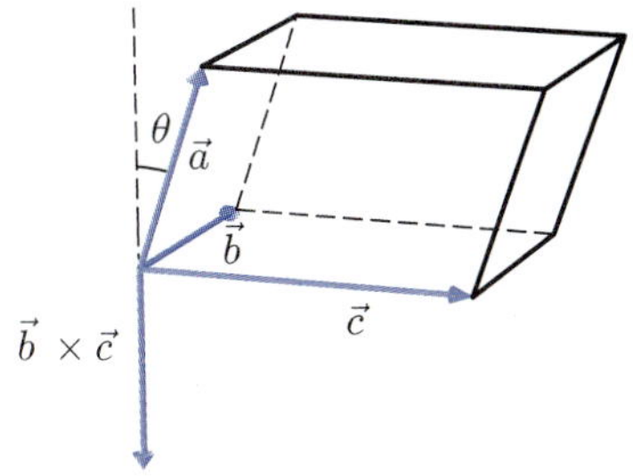

The vectors $\vec{a}$, $\vec{b}$, and $\vec{c}$ can be arranged either as in Figure 13.42 or as in Figure 13.43. In either case,

$$\text{Height of parallelepiped} = \|\vec{a}\| \cos\theta,$$

where θ is the angle shown in the figures. In Figure 13.42 the angle θ is less than $\pi/2$, so the product, $(\vec{b} \times \vec{c}) \cdot \vec{a}$, called the *triple product*, is positive. Thus, in this case

$$\text{Volume of parallelepiped} = \text{Base} \cdot \text{Height} = \|\vec{b} \times \vec{c}\| \cdot \|\vec{a}\| \cos\theta = (\vec{b} \times \vec{c}) \cdot \vec{a}.$$

In Figure 13.43, the angle, $\pi - \theta$, between $\vec{a}$ and $\vec{b} \times \vec{c}$ is more than $\pi/2$, so the product $(\vec{b} \times \vec{c}) \cdot \vec{a}$ is negative. Thus, in this case we have

$$\text{Volume} = \text{Base} \cdot \text{Height} = \|\vec{b} \times \vec{c}\| \cdot \|\vec{a}\| \cos\theta = -\|\vec{b} \times \vec{c}\| \cdot \|\vec{a}\| \cos(\pi - \theta)$$
$$= -(\vec{b} \times \vec{c}) \cdot \vec{a} = \left| (\vec{b} \times \vec{c}) \cdot \vec{a} \right|.$$

Therefore, in both cases the volume is given by $\left| (\vec{b} \times \vec{c}) \cdot \vec{a} \right|$. Using determinants, we can write

Volume of a parallelepiped with edges $\vec{a}$, $\vec{b}$, $\vec{c}$ is given by

$$\text{Volume} = \left| (\vec{b} \times \vec{c}) \cdot \vec{a} \right| = \text{Absolute value of the determinant } \begin{vmatrix} a_1 & a_2 & a_3 \\ b_1 & b_2 & b_3 \\ c_1 & c_2 & c_3 \end{vmatrix}.$$

Exercises and Problems for Section 13.4

Exercises

In Exercises 1–5, use the algebraic definition to find $\vec{v} \times \vec{w}$.

1. $\vec{v} = \vec{k}, \vec{w} = \vec{j}$

2. $\vec{v} = \vec{i} + \vec{k}, \vec{w} = \vec{i} + \vec{j}$

3. $\vec{v} = -\vec{i}, \vec{w} = \vec{j} + \vec{k}$

4. $\vec{v} = \vec{i} + \vec{j} + \vec{k}, \vec{w} = \vec{i} + \vec{j} + -\vec{k}$

5. $\vec{v} = 2\vec{i} - 3\vec{j} + \vec{k}, \vec{w} = \vec{i} + 2\vec{j} - \vec{k}$

6. If $\vec{v} = 3\vec{i} - 2\vec{j} + 4\vec{k}$ and $\vec{w} = \vec{i} + 2\vec{j} - \vec{k}$, find $\vec{v} \times \vec{w}$ and $\vec{w} \times \vec{v}$. What is the relation between the two answers?

7. For $\vec{a} = 3\vec{i} + \vec{j} - \vec{k}$ and $\vec{b} = \vec{i} - 4\vec{j} + 2\vec{k}$, find $\vec{a} \times \vec{b}$ and check that it is perpendicular to both $\vec{a}$ and $\vec{b}$.

In Exercises 8–9, use the properties on page 631.

8. $\left((\vec{i} + \vec{j}) \times \vec{i}\right) \times \vec{j}$

9. $(\vec{i} + \vec{j}) \times (\vec{i} \times \vec{j})$

Use the geometric definition in Exercises 10–11.

10. $2\vec{i} \times (\vec{i} + \vec{j})$

11. $(\vec{i} + \vec{j}) \times (\vec{i} - \vec{j})$

Problems

12. If $\vec{v} \times \vec{w} = 2\vec{i} - 3\vec{j} + 5\vec{k}$, and $\vec{v} \cdot \vec{w} = 3$, find $\tan\theta$ where θ is the angle between $\vec{v}$ and $\vec{w}$.

13. If $\vec{v}$ and $\vec{w}$ are both parallel to the xy-plane, what can you conclude about $\vec{v} \times \vec{w}$? Explain.

Find an equation for the plane through the points in Problems 14–15.

14. $(1, 0, 0), (0, 1, 0), (0, 0, 1)$.

15. $(3, 4, 2), (-2, 1, 0), (0, 2, 1)$.

16. Let $A = (-1, 3, 0)$, $B = (3, 2, 4)$, and $C = (1, -1, 5)$.

    (a) Find an equation for the plane that passes through these three points.
    (b) Find the area of the triangle determined by these three points.

17. Let $P = (0, 1, 0)$, $Q = (-1, 1, 2)$, $R = (2, 1, -1)$. Find

    (a) The area of the triangle PQR.
    (b) The equation for a plane that contains $P, Q,$ and R.

18. Find a vector parallel to the line of intersection of the two planes $4x - 3y + 2z = 12$ and $x + 5y - z = 25$.

19. Find a vector parallel to the intersection of the planes $2x - 3y + 5z = 2$ and $4x + y - 3z = 7$.

20. Find the equation of the plane through the origin which is perpendicular to the line of intersection of the planes in Problem 19.

21. Find the equation of the plane through the point $(4, 5, 6)$ which is perpendicular to the line of intersection of the planes in Problem 19.

22. Use the algebraic definition to check that

$$\vec{a} \times (\vec{b} + \vec{c}) = (\vec{a} \times \vec{b}) + (\vec{a} \times \vec{c}).$$

23. In this problem, we arrive at the algebraic definition for the cross product by a different route. Let $\vec{a} = a_1\vec{i} + a_2\vec{j} + a_3\vec{k}$ and $\vec{b} = b_1\vec{i} + b_2\vec{j} + b_3\vec{k}$. We seek a vector $\vec{v} = x\vec{i} + y\vec{j} + z\vec{k}$ which is perpendicular to both $\vec{a}$ and $\vec{b}$. Use this requirement to construct two equations for x, y, and z. Eliminate x and solve for y in terms of z. Then eliminate y and solve for x in terms of z. Since z can be any value whatsoever (the direction of $\vec{v}$ is unaffected), select the value for z which eliminates the denominator in the equation you obtained. How does the resulting expression for $\vec{v}$ compare to the formula we derived on page 630?

24. Suppose $\vec{a}$ and $\vec{b}$ are vectors in the xy-plane, such that $\vec{a} = a_1\vec{i} + a_2\vec{j}$ and $\vec{b} = b_1\vec{i} + b_2\vec{j}$ with $0 < a_2 < a_1$ and $0 < b_1 < b_2$.

    (a) Sketch $\vec{a}$ and $\vec{b}$ and the vector $\vec{c} = -a_2\vec{i} + a_1\vec{j}$. Shade the parallelogram formed by $\vec{a}$ and $\vec{b}$.
    (b) What is the relation between $\vec{a}$ and $\vec{c}$? [Hint: Find $\vec{c} \cdot \vec{a}$ and $\vec{c} \cdot \vec{c}$.]
    (c) Find $\vec{c} \cdot \vec{b}$.
    (d) Explain why $\vec{c} \cdot \vec{b}$ gives the area of the parallelogram formed by $\vec{a}$ and $\vec{b}$.
    (e) Check that in this case $\vec{a} \times \vec{b} = (a_1 b_2 - a_2 b_1)\vec{k}$.

25. If $\vec{v}$ and $\vec{w}$ are nonzero vectors, use the geometric definition of the cross product to explain why

$$(\lambda\vec{v}) \times \vec{w} = \lambda(\vec{v} \times \vec{w}) = \vec{v} \times (\lambda\vec{w}).$$

Consider the cases $\lambda > 0$, and $\lambda = 0$, and $\lambda < 0$ separately.

26. Use a parallelepiped to show that $\vec{a} \cdot (\vec{b} \times \vec{c}) = (\vec{a} \times \vec{b}) \cdot \vec{c}$ for any vectors $\vec{a}$, $\vec{b}$, and $\vec{c}$.

27. Show that $\|\vec{a} \times \vec{b}\|^2 = \|\vec{a}\|^2\|\vec{b}\|^2 - (\vec{a} \cdot \vec{b})^2$.

28. For vectors $\vec{a}$ and $\vec{b}$, let $\vec{c} = \vec{a} \times (\vec{b} \times \vec{a})$.

    (a) Show that $\vec{c}$ lies in the plane containing $\vec{a}$ and $\vec{b}$.
    (b) Use Problems 26 and 27 to show that $\vec{a} \cdot \vec{c} = 0$ and $\vec{b} \cdot \vec{c} = \|\vec{a}\|^2\|\vec{b}\|^2 - (\vec{a} \cdot \vec{b})^2$.
    (c) Show that

$$\vec{a} \times (\vec{b} \times \vec{a}) = \|\vec{a}\|^2\vec{b} - (\vec{a} \cdot \vec{b})\vec{a}.$$

29. Use the result of Problem 26 to show that the cross product distributes over addition. First, use distributivity for the dot product to show that for any vector $\vec{d}$,

$$[(\vec{a} + \vec{b}) \times \vec{c}] \cdot \vec{d} = [(\vec{a} \times \vec{c}) + (\vec{b} \times \vec{c})] \cdot \vec{d}.$$

Next, show that for any vector $\vec{d}$,

$$[((\vec{a} + \vec{b}) \times \vec{c}) - (\vec{a} \times \vec{c}) - (\vec{b} \times \vec{c})] \cdot \vec{d} = 0.$$

Finally, explain why you can conclude that

$$(\vec{a} + \vec{b}) \times \vec{c} = (\vec{a} \times \vec{c}) + (\vec{b} \times \vec{c}).$$

30. Use the fact that $\vec{i} \times \vec{i} = \vec{0}, \vec{i} \times \vec{j} = \vec{k}, \vec{i} \times \vec{k} = -\vec{j}$, and so on, together with the properties on page 631 to derive the algebraic definition for the cross product.

31. Figure 13.44 shows the tetrahedron determined by three vectors $\vec{a}, \vec{b}, \vec{c}$. The *area vector* of a face is a vector perpendicular to the face, pointing outward, whose magnitude is the area of the face. Show that the sum of the four outward pointing area vectors of the faces equals the zero vector.

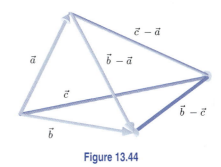

Figure 13.44

CHAPTER SUMMARY

- **Vectors**

    Geometric definition of vector addition, subtraction and scalar multiplication, resolving into $\vec{i}, \vec{j}$, and $\vec{k}$ components, magnitude of a vector, algebraic properties of addition and scalar multiplication.

- **Dot Product**

    Geometric and algebraic definition, algebraic properties, using dot products to find angles and determine perpendicularity, the equation of a plane with given normal vector passing through a given point, projection of a vector in a direction given by a unit vector.

- **Cross Product**

    Geometric and algebraic definition, algebraic properties, cross product and volume, finding the equation of a plane through three points.

REVIEW EXERCISES AND PROBLEMS FOR CHAPTER THIRTEEN

Exercises

In Exercises 1–8, use $\vec{v} = 2\vec{i} + 3\vec{j} - \vec{k}$ and $\vec{w} = \vec{i} - \vec{j} + 2\vec{k}$ to calculate the given quantities.

1. $\vec{v} + 2\vec{w}$

2. $3\vec{v} - \vec{w} - \vec{v}$

3. $\vec{v} \cdot \vec{w}$

4. $\vec{v} \times \vec{w}$

5. $(\vec{v} \cdot \vec{w})\vec{v}$

6. $(\vec{v} \times \vec{w}) \times \vec{w}$

7. $(\vec{v} \times \vec{w}) \cdot \vec{w}$

8. $(\vec{v} \times \vec{w}) \times (\vec{v} \times \vec{w})$

9. Let $\vec{v} = 3\vec{i} + 2\vec{j} - 2\vec{k}$ and $\vec{w} = 4\vec{i} - 3\vec{j} + \vec{k}$. Find each of the following:

(a) $\vec{v} \cdot \vec{w}$
(b) $\vec{v} \times \vec{w}$

(c) A vector of length 5 parallel to vector $\vec{v}$
(d) The angle between vectors $\vec{v}$ and $\vec{w}$
(e) The component of vector $\vec{v}$ in the direction of vector $\vec{w}$
(f) A vector perpendicular to vector $\vec{v}$
(g) A vector perpendicular to both vectors $\vec{v}$ and $\vec{w}$

In Exercises 10–13, find a vector with the given property.

10. Length 10, parallel to $2\vec{i} + 3\vec{j} - \vec{k}$.

11. Unit vector perpendicular to $\vec{i} + \vec{j}$ and $\vec{i} - \vec{j} - \vec{k}$

12. Unit vector in the xy-plane perpendicular to $3\vec{i} - 2\vec{j}$.

13. The vector obtained from $4\vec{i} + 3\vec{j}$ by rotating it $90°$ counterclockwise.

Problems

14. For what values of t are the following pairs of vectors parallel?

(a) $2\vec{i} + (t^2 + \frac{2}{3}t + 1)\vec{j} + t\vec{k}$, $6\vec{i} + 8\vec{j} + 3\vec{k}$
(b) $t\vec{i} + \vec{j} + (t - 1)\vec{k}$, $2\vec{i} - 4\vec{j} + \vec{k}$
(c) $2t\vec{i} + t\vec{j} + t\vec{k}$, $6\vec{i} + 3\vec{j} + 3\vec{k}$.

15. Find a vector normal to $4(x - 1) + 6(z + 3) = 12$.

16. Find the equation of the plane through the origin which is parallel to $z = 4x - 3y + 8$.

17. Consider the plane $5x - y + 7z = 21$.

  (a) Find a point on the x-axis on this plane.
  (b) Find two other points on the plane.
  (c) Find a vector perpendicular to the plane.
  (d) Find a vector parallel to the plane.

18. Find the area of the triangle with vertices
  $P = (-2, 2, 0)$, $Q = (1, 3, -1)$, and $R = (-4, 2, 1)$.

19. A plane is drawn through the points $A = (2, 1, 0)$,
  $B = (0, 1, 3)$ and $C = (1, 0, 1)$. Find

  (a) Two vectors lying in the plane.
  (b) A vector perpendicular to the plane.
  (c) The equation of the plane.

20. A particle moving with speed v hits a barrier at an angle of $60°$ and bounces off at an angle of $60°$ in the opposite direction with speed reduced by 20 percent. See Figure 13.45. Find the velocity vector of the object after impact.

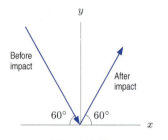

Figure 13.45

21. An object is attached by a string to a fixed point and rotates 30 times per minute in a horizontal plane. Show that the speed of the object is constant but the velocity is not. What does this imply about the acceleration?

22. An object is moving counterclockwise at a constant speed around the circle $x^2 + y^2 = 1$, where x and y are measured in meters. It completes one revolution every minute.

  (a) What is its speed?
  (b) What is its velocity vector 30 seconds after it passes the point $(1, 0)$? Does your answer change if the object is moving clockwise? Explain.

23. An airport is at the point $(200, 10, 0)$ and an approaching plane is at the point $(550, 60, 4)$. Assume that the xy-plane is horizontal, with the x-axis pointing eastward and the y-axis pointing northward. Also assume that the z-axis is upward and that all distances are measured in kilometers. The plane flies due west at a constant altitude at a speed of 500 km/hr for half an hour. It then descends at 200 km/hr, heading straight for the airport.

  (a) Find the velocity vector of the plane while it is flying at constant altitude.
  (b) Find the coordinates of the point at which the plane starts to descend.
  (c) Find a vector representing the velocity of the plane when it is descending.

24. A large ship is being towed by two tugs. The larger tug exerts a force which is 25% greater than the smaller tug and at an angle of 30 degrees north of east. Which direction must the smaller tug pull to ensure that the ship travels due east?

25. Three people are trying to hold a ferocious lion still for the veterinarian. The lion, in the center, is wearing a collar with three ropes attached to it and each person has hold of a rope. Charlie is pulling in the direction $62°$ west of north with a force of 175 newtons and Sam is pulling in the direction $43°$ east of north with a force of 200 newtons. What is the direction and magnitude of the force which must be exerted by Alice on the third rope to counterbalance Sam and Charlie?

26. A man wishes to row the shortest possible distance from north to south across a river which is flowing at 4 km/hr from the east. He can row at 5 km/hr.

  (a) In which direction should he steer?
  (b) If there is a wind of 10 km/hr from the southwest, in which direction should he steer to try and go directly across the river? What happens?

27. Given the points $P = (1, 2, 3)$, $Q = (3, 5, 7)$, and $R = (2, 5, 3)$, find:

  (a) A unit vector perpendicular to a plane containing P, Q, R.
  (b) The angle between PQ and PR.
  (c) The area of the triangle PQR.
  (d) The distance from R to the line through P and Q.

28. Find the distance from the point $P = (2, -1, 3)$ to the plane $2x + 4y - z = -1$.

29. Find an equation of the plane passing through the three points $(1, 1, 1)$, $(1, 4, 5)$, $(-3, -2, 0)$. Find the distance from the origin to the plane.

30. Two lines in space are skew if they are not parallel and do not intersect. Determine the minimum distance between two such lines.

31. **(a)** A vector $\vec{v}$ of magnitude v makes an angle α with the positive x-axis, angle β with the positive y-axis, and angle γ with the positive z-axis. Show that

$$\vec{v} = v\cos\alpha\vec{i} + v\cos\beta\vec{j} + v\cos\gamma\vec{k}.$$

  (b) Cos α, cos β, and cos γ are called *direction cosines*. Show that

$$\cos^2\alpha + \cos^2\beta + \cos^2\gamma = 1.$$

32. Find the vector $\vec{v}$ with all of the following properties:

  - Magnitude 10
  - Angle of $45°$ with positive x-axis
  - Angle of $75°$ with positive y-axis
  - Positive $\vec{k}$ -component.

CAS Challenge Problems

33. Let $\vec{a} = x\vec{i} + y\vec{j} + z\vec{k}$, $\vec{b} = u\vec{i} + v\vec{j} + w\vec{k}$, and $\vec{c} = m\vec{a} + n\vec{b}$. Compute $(\vec{a} \times \vec{b}) \cdot \vec{c}$ and $(\vec{a} \times \vec{b}) \times (\vec{a} \times \vec{c})$, and explain the geometric meaning of your answers.

34. Let $\vec{a} = x\vec{i} + y\vec{j} + z\vec{k}$, $\vec{b} = u\vec{i} + v\vec{j} + w\vec{k}$ and $\vec{c} = r\vec{i} + s\vec{j} + t\vec{k}$. Show that the parallelepiped with edges $\vec{a}, \vec{b}, \vec{c}$ has the same volume as the parallelepiped with edges $\vec{a}, \vec{b}, 2\vec{a} - \vec{b} + \vec{c}$. Explain this result geometrically.

35. Let $\vec{a} = \vec{i} + 2\vec{j} + 3\vec{k}$, $\vec{b} = 2\vec{i} + \vec{j} + 2\vec{k}$ and let θ be the angle between $\vec{a}$ and $\vec{b}$.

(a) For $\vec{c} = x\vec{i} + y\vec{j} + z\vec{k}$, write the following conditions as equations in x, y, z and solve them:
$$\vec{a} \cdot \vec{c} = 0, \quad \vec{b} \cdot \vec{c} = 0, \quad \|\vec{c}\|^2 = \|\vec{a}\|^2\|\vec{b}\|^2 \sin^2 \theta.$$
[Hint: Use the dot product to find $\sin^2 \theta$.]

(b) Compute the cross product $\vec{a} \times \vec{b}$ and compare with your answer in part (a). What do you notice? Explain.

36. Let $A = (0, 0, 0)$, $B = (2, 0, 0)$, $C = (1, \sqrt{3}, 0)$ and $D = (1, 1/\sqrt{3}, 2\sqrt{2/3})$.

(a) Show that A, B, C, D are all the same distance from each other.

(b) Find the point $P = (x, y, z)$ which is equidistant from A, B, C and D by setting up and solving three equations in x, y, and z.

(c) Use the dot product to find the angle APB. (In chemistry, this angle is often approximated by $109.5°$. A methane molecule can be represented by four hydrogen atoms at points A, B, C and D, and a carbon atom at P.)

37. Let $P = (x, y, z)$, $Q = (u, v, w)$ and $R = (r, s, t)$ be points on the plane $ax + by + cz = d$.

(a) What is the relation between $\overrightarrow{PQ} \times \overrightarrow{PR}$ and the normal vector to the plane, $a\vec{i} + b\vec{j} + c\vec{k}$?

(b) Express $\overrightarrow{PQ} \times \overrightarrow{PR}$ in terms of $x, y, z, u, v, w, r, s, t$.

(c) Use the equation for the plane to eliminate z, w, and t from the expression you obtained in part (b), and simplify. Does your answer agree with what you said in part (a)?

CHECK YOUR UNDERSTANDING

Are the statements in Problems 1–30 true or false? Give reasons for your answer.

1. There is exactly one unit vector parallel to a given nonzero vector $\vec{v}$.

2. The vector $\frac{1}{\sqrt{3}}\vec{i} + \frac{-1}{\sqrt{3}}\vec{j} + \frac{2}{\sqrt{3}}\vec{k}$ is a unit vector.

3. The length of the vector $2\vec{v}$ is twice the length of the vector $\vec{v}$.

4. If $\vec{v}$ and $\vec{w}$ are any two vectors, then $\|\vec{v} + \vec{w}\| = \|\vec{v}\| + \|\vec{w}\|$.

5. If $\vec{v}$ and $\vec{w}$ are any two vectors, then $\|\vec{v} - \vec{w}\| = \|\vec{v}\| - \|\vec{w}\|$.

6. The vectors $2\vec{i} - \vec{j} + \vec{k}$ and $\vec{i} - 2\vec{j} + \vec{k}$ are parallel.

7. The vector $\vec{u} + \vec{v}$ is always larger in magnitude than both $\vec{u}$ and $\vec{v}$.

8. For any scalar c and vector $\vec{v}$ we have $\|c\vec{v}\| = c\|\vec{v}\|$.

9. The displacement vector from $(1, 1, 1)$ to $(1, 2, 3)$ is $-\vec{j} - 2\vec{k}$.

10. The displacement vector from (a, b) to (c, d) is the same as the displacement vector from (c, d) to (a, b).

11. The quantity $\vec{u} \cdot \vec{v}$ is a vector.

12. The plane $x + 2y - 3z = 5$ has normal vector $\vec{i} + 2\vec{j} - 3\vec{k}$.

13. If $\vec{u} \cdot \vec{v} < 0$ then the angle between $\vec{u}$ and $\vec{v}$ is greater than $\pi/2$.

14. An equation of the plane with normal vector $\vec{i} + \vec{j} + \vec{k}$ containing the point $(1, 2, 3)$ is $z = x + y$.

15. The triangle in 3-space with vertices $(1, 1, 0), (0, 1, 0)$ and $(0, 1, 1)$ has a right angle.

16. The dot product $\vec{v} \cdot \vec{v}$ is never negative.

17. If $\vec{u} \cdot \vec{v} = 0$ then either $\vec{u} = 0$ or $\vec{v} = 0$.

18. If $\vec{u}, \vec{v}$ and $\vec{w}$ are all nonzero, and $\vec{u} \cdot \vec{v} = \vec{u} \cdot \vec{w}$, then $\vec{v} = \vec{w}$.

19. For any vectors $\vec{u}$ and $\vec{v}$: $(\vec{u} + \vec{v}) \cdot (\vec{u} - \vec{v}) = \|\vec{u}\|^2 - \|\vec{v}\|^2$.

20. If $\|\vec{u}\| = 1$, then the vector $\vec{v} - (\vec{v} \cdot \vec{u})\vec{u}$ is perpendicular to $\vec{u}$.

21. $\vec{u} \times \vec{v}$ is a vector.

22. $\vec{u} \times \vec{v}$ has direction parallel to both $\vec{u}$ and $\vec{v}$.

23. $\|\vec{u} \times \vec{v}\| = \|\vec{u}\|\|\vec{v}\|$.

24. $(\vec{i} \times \vec{j}) \cdot \vec{k} = \vec{i} \cdot (\vec{j} \times \vec{k})$.

25. If $\vec{v}$ is a nonzero vector and $\vec{v} \times \vec{u} = \vec{v} \times \vec{w}$, then $\vec{u} = \vec{w}$.

26. The value of $\vec{v} \cdot (\vec{v} \times \vec{w})$ is always 0.

27. The value of $\vec{v} \times \vec{w}$ is never the same as $\vec{v} \cdot \vec{w}$.

28. The area of the triangle with two sides given by $\vec{i} + \vec{j}$ and $\vec{j} + 2\vec{k}$ is $3/2$.

29. Given a nonzero vector $\vec{v}$ in 3-space, there is a nonzero vector $\vec{w}$ such that $\vec{v} \times \vec{w} = \vec{0}$.

30. It is never true that $\vec{v} \times \vec{w} = \vec{w} \times \vec{v}$.

PROJECTS FOR CHAPTER THIRTEEN

1. **Cross Product of Vectors in the Plane**

   Let $\vec{a} = a_1\vec{i} + a_2\vec{j}$ and $\vec{b} = b_1\vec{i} + b_2\vec{j}$ be two nonparallel vectors in 2-space, as in Figure 13.46.

   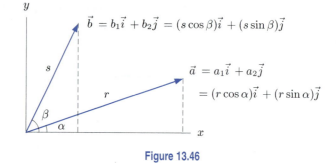

   Figure 13.46

   (a) Use the identity $\sin(\beta - \alpha) = (\sin\beta\cos\alpha - \cos\beta\sin\alpha)$ to derive the formula for the area of the parallelogram formed by $\vec{a}$ and $\vec{b}$:

   $$\text{Area of parallelogram} = |a_1b_2 - a_2b_1|.$$

   (b) Show that $a_1b_2 - a_2b_1$ is positive when the rotation from $\vec{a}$ to $\vec{b}$ is counterclockwise, and negative when it is clockwise.

   (c) Use parts (a) and (b) to show that the geometric and algebraic definitions of $\vec{a} \times \vec{b}$ give the same result.

2. **The Dot Product in Genetics**[2]

   We define[3] the angle between two n-dimensional vectors, $\vec{v}$ and $\vec{w}$, using the dot product:

   $$\cos\theta = \frac{\vec{v} \cdot \vec{w}}{\|\vec{v}\|\|\vec{w}\|} = \frac{v_1w_1 + v_2w_2 + \cdots + v_nw_n}{\|\vec{v}\|\|\vec{w}\|}, \qquad \text{provided } \|\vec{v}\|, \|\vec{w}\| \neq 0.$$

   We use this idea of angle to measure how close two populations are to one another genetically. The table shows the relative frequencies of four alleles (variants of a gene) in four populations.

Allele	Eskimo	Bantu	English	Korean
A_1	0.29	0.10	0.20	0.22
A_2	0.00	0.08	0.06	0.00
B	0.03	0.12	0.06	0.20
O	0.67	0.69	0.66	0.57

   Let $\vec{a}_1$ be the 4-vector showing the relative frequencies of the alleles in the Eskimo population. Let $\vec{a}_2, \vec{a}_3, \vec{a}_4$ be the corresponding vectors for the Bantu, English, and Korean populations, respectively. The genetic distance between two populations is defined as the angle between the corresponding vectors.

   (a) Using this definition, is the English population closer genetically to the Bantus or to the Koreans? Explain.

   (b) Is the English population closer to a half Eskimo, half Bantu population than to the Bantu population alone?

   (c) Among all possible populations that are a mix of Eskimo and Bantu, find the mix that is closest to the English population.

   ---

   [2]Adapted from Cavalli-Sforza and Edwards, "Models and Estimation Procedures," Am J. Hum. Genet., Vol. 19 (1967), pp. 223-57.

   [3]The result of Problem 42 on page 628 shows that the quantity on the right-hand side of this equation is between -1 and 1, so this definition makes sense.

Chapter Fourteen

DIFFERENTIATING FUNCTIONS OF SEVERAL VARIABLES

For a function of one variable, $y = f(x)$, the derivative $dy/dx = f'(x)$ gives the rate of change of y with respect to x. For a function of two variables, $z = f(x, y)$, there is no such thing as *the* rate of change, since x and y can each vary while the other is held fixed or both can vary at once. However, we can consider the rate of change with respect to each one of the independent variables. This chapter introduces these *partial derivatives* and several ways they can be used to get a complete picture of the way the function varies.

14.1 THE PARTIAL DERIVATIVE

The derivative of a one-variable function measures its rate of change. In this section we see how a two variable function has two rates of change: one as x changes (with y held constant) and one as y changes (with x held constant).

Rate of Change of Temperature in a Metal Rod: a One-Variable Problem

Imagine an unevenly heated metal rod lying along the x-axis, with its left end at the origin and x measured in meters. (See Figure 14.1.) Let $u(x)$ be the temperature (in °C) of the rod at the point x. Table 14.1 gives values of $u(x)$. We see that the temperature increases as we move along the rod, reaching its maximum at $x = 4$, after which it starts to decrease.

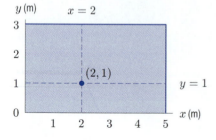

Figure 14.1: Unevenly heated metal rod

Table 14.1 *Temperature $u(x)$ of the rod*

x (m)	0	1	2	3	4	5
$u(x)$ (°C)	125	128	135	160	175	160

Example 1 Estimate the derivative $u'(2)$ using Table 14.1 and explain what the answer means in terms of temperature.

Solution The derivative $u'(2)$ is defined as a limit of difference quotients:

$$u'(2) = \lim_{h \to 0} \frac{u(2+h) - u(2)}{h}.$$

Choosing $h = 1$ so that we can use the data in Table 14.1, we get

$$u'(2) \approx \frac{u(2+1) - u(2)}{1} = \frac{160 - 135}{1} = 25.$$

This means that the temperature increases at a rate of approximately 25°C per meter as we go from left to right, past $x = 2$.

Rate of Change of Temperature in a Metal Plate

Imagine an unevenly heated thin rectangular metal plate lying in the xy-plane with its lower left corner at the origin and x and y measured in meters. The temperature (in °C) at the point (x, y) is $T(x, y)$. See Figure 14.2 and Table 14.2. How does T vary near the point $(2, 1)$? We consider the horizontal line $y = 1$ containing the point $(2, 1)$. The temperature along this line is the cross-section, $T(x, 1)$, of the function $T(x, y)$ with $y = 1$. Suppose we write $u(x) = T(x, 1)$.

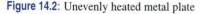

Figure 14.2: Unevenly heated metal plate

Table 14.2 *Temperature (°C) of a metal plate*

y (m)						
3	85	90	**110**	135	155	180
2	100	110	**120**	145	190	170
1	**125**	**128**	**135**	**160**	**175**	**160**
0	120	135	**155**	160	160	150
	0	1	2	3	4	5

x (m)

What is the meaning of the derivative $u'(2)$? It is the rate of change of temperature T *in the x-direction* at the point $(2, 1)$, keeping y fixed. Denote this rate of change by $T_x(2, 1)$, so that

$$T_x(2, 1) = u'(2) = \lim_{h \to 0} \frac{u(2+h) - u(2)}{h} = \lim_{h \to 0} \frac{T(2+h, 1) - T(2, 1)}{h}.$$

We call $T_x(2,1)$ the *partial derivative of T with respect to x at the point* $(2,1)$. Taking $h = 1$, we can read values of T from the row with $y = 1$ in Table 13.2, giving

$$T_x(2,1) \approx \frac{T(3,1) - T(2,1)}{1} = \frac{160 - 135}{1} = 25°\text{C/m}.$$

The fact that $T_x(2,1)$ is positive means that the temperature of the plate is increasing as we move past the point $(2,1)$ in the direction of increasing x (that is, horizontally from left to right in Figure 14.2).

Example 2   Estimate the rate of change of T in the y-direction at the point $(2,1)$.

Solution   The temperature along the line $x = 2$ is the cross-section of T with $x = 2$, that is, the function $v(y) = T(2,y)$. If we denote the rate of change of T in the y-direction at $(2,1)$ by $T_y(2,1)$, then

$$T_y(2,1) = v'(1) = \lim_{h \to 0} \frac{v(1+h) - v(1)}{h} = \lim_{h \to 0} \frac{T(2,1+h) - T(2,1)}{h}.$$

We call $T_y(2,1)$ the *partial derivative of T with respect to y at the point* $(2,1)$. Taking $h = 1$ so that we can use the column with $x = 2$ in Table 13.2, we get

$$T_y(2,1) \approx \frac{T(2,1+1) - T(2,1)}{1} = \frac{120 - 135}{1} = -15°\text{C/m}.$$

The fact that $T_y(2,1)$ is negative means that the temperature decreases as y increases.

Definition of the Partial Derivative

We study the influence of x and y separately on the value of the function $f(x,y)$ by holding one fixed and letting the other vary. This leads to the following definitions.

> **Partial Derivatives of f With Respect to x and y**
>
> For all points at which the limits exist, we define the **partial derivatives at the point** (a, b) by
>
> $$f_x(a,b) = \begin{matrix} \text{Rate of change of } f \text{ with respect to } x \\ \text{at the point } (a,b) \end{matrix} = \lim_{h \to 0} \frac{f(a+h,b) - f(a,b)}{h},$$
>
> $$f_y(a,b) = \begin{matrix} \text{Rate of change of } f \text{ with respect to } y \\ \text{at the point } (a,b) \end{matrix} = \lim_{h \to 0} \frac{f(a,b+h) - f(a,b)}{h}.$$
>
> If we let a and b vary, we have the **partial derivative functions** $f_x(x,y)$ and $f_y(x,y)$.

Just as with ordinary derivatives, there is an alternative notation:

> **Alternative Notation for Partial Derivatives**
>
> If $z = f(x,y)$, we can write
>
> $$f_x(x,y) = \frac{\partial z}{\partial x} \quad \text{and} \quad f_y(x,y) = \frac{\partial z}{\partial y},$$
>
> $$f_x(a,b) = \frac{\partial z}{\partial x}\bigg|_{(a,b)} \quad \text{and} \quad f_y(a,b) = \frac{\partial z}{\partial y}\bigg|_{(a,b)}.$$

We use the symbol ∂ to distinguish partial derivatives from ordinary derivatives. In cases where the independent variables have names different from x and y, we adjust the notation accordingly. For example, the partial derivatives of $f(u, v)$ are denoted by f_u and f_v.

Visualizing Partial Derivatives on a Graph

The ordinary derivative of a one-variable function is the slope of its graph. How do we visualize the partial derivative $f_x(a, b)$? The graph of the one-variable function $f(x, b)$ is the curve where the vertical plane $y = b$ cuts the graph of $f(x, y)$. (See Figure 14.3.) Thus, $f_x(a, b)$ is the slope of the tangent line to this curve at $x = a$.

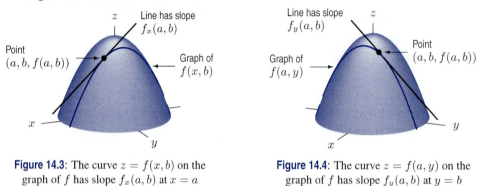

Figure 14.3: The curve $z = f(x, b)$ on the graph of f has slope $f_x(a, b)$ at $x = a$

Figure 14.4: The curve $z = f(a, y)$ on the graph of f has slope $f_y(a, b)$ at $y = b$

Similarly, the graph of the function $f(a, y)$ is the curve where the vertical plane $x = a$ cuts the graph of f, and the partial derivative $f_y(a, b)$ is the slope of this curve at $y = b$. (See Figure 14.4.)

Example 3   At each point labeled on the graph of the surface $z = f(x, y)$ in Figure 14.5, say whether each partial derivative is positive or negative.

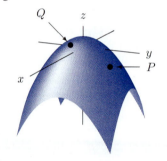

Figure 14.5: Decide the signs of f_x and f_y at P and Q

Solution   The positive x-axis points out of the page. Imagine heading off in this direction from the point marked P; we descend steeply. So the partial derivative with respect to x is negative at P, with quite a large absolute value. The same is true for the partial derivative with respect to y at P, since there is also a steep descent in the positive y-direction.

At the point marked Q, heading in the positive x-direction results in a gentle descent, whereas heading in the positive y-direction results in a gentle ascent. Thus, the partial derivative f_x at Q is negative but small (that is, near zero), and the partial derivative f_y is positive but small.

Estimating Partial Derivatives From a Contour Diagram

The graph of a function $f(x, y)$ often makes clear the sign of the partial derivatives. However, numerical estimates of these derivatives are more easily made from a contour diagram than a surface graph. If we move parallel to one of the axes on a contour diagram, the partial derivative is the rate of change of the value of the function on the contours. For example, if the values on the contours are increasing as we move in the positive direction, then the partial derivative must be positive.

Example 4   Figure 14.6 shows the contour diagram for the temperature $H(x, t)$ (in °C) in a room as a function of distance x (in meters) from a heater and time t (in minutes) after the heater has been turned on. What are the signs of $H_x(10, 20)$ and $H_t(10, 20)$? Estimate these partial derivatives and explain the answers in practical terms.

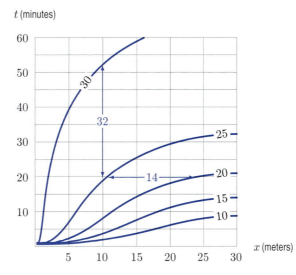

Figure 14.6: Temperature in a heated room: Heater at $x = 0$ is turned on at $t = 0$

Solution   The point $(10, 20)$ is nearly on the $H = 25$ contour. As x increases, we move towards the $H = 20$ contour, so H is decreasing and $H_x(10, 20)$ is negative. This makes sense because the $H = 30$ contour is to the left: As we move further from the heater, the temperature drops. On the other hand, as t increases, we move towards the $H = 30$ contour, so H is increasing; as t decreases H decreases. Thus, $H_t(10, 20)$ is positive. This says that as time passes, the room warms up.

To estimate the partial derivatives, use a difference quotient. Looking at the contour diagram, we see there is a point on the $H = 20$ contour about 14 units to the right of the point $(10, 20)$. Hence, H decreases by 5 when x increases by 14, so we find

$$\text{Rate of change of } H \text{ with respect to } x = H_x(10, 20) \approx \frac{-5}{14} \approx -0.36°\text{C/meter.}$$

This means that near the point 10 m from the heater, after 20 minutes the temperature drops about 0.36, or one third, of a degree, for each meter we move away from the heater.

To estimate $H_t(10, 20)$, we notice that the $H = 30$ contour is about 32 units directly above the point $(10, 20)$. So H increases by 5 when t increases by 32. Hence,

$$\text{Rate of change of } H \text{ with respect to } t = H_t(10, 20) \approx \frac{5}{32} = 0.16°\text{C/meter.}$$

This means that after 20 minutes the temperature is going up about 0.16, or 1/6, of a degree each minute at the point 10 m from the heater.

Using Units to Interpret Partial Derivatives

The meaning of a partial derivative can often be explained using units.

Example 5   Suppose that your weight w in pounds is a function $f(c, n)$ of the number c of calories you consume daily and the number n of minutes you exercise daily. Using the units for w, c and n, interpret in everyday terms the statements

$$\frac{\partial w}{\partial c}(2000, 15) = 0.02 \quad \text{and} \quad \frac{\partial w}{\partial n}(2000, 15) = -0.025.$$

Solution   The units of $\partial w/\partial c$ are pounds per calorie. The statement

$$\frac{\partial w}{\partial c}(2000, 15) = 0.02$$

means that if you are presently consuming 2000 calories daily and exercising 15 minutes daily, you will weigh 0.02 pounds more for each extra calorie you consume daily, or about 2 pounds for each extra 100 calories per day. The units of $\partial w/\partial n$ are pounds per minute. The statement

$$\frac{\partial w}{\partial n}(2000, 15) = -0.025$$

means that for the same calorie consumption and number of minutes of exercise, you will weigh 0.025 pounds less for each extra minute you exercise daily, or about 1 pound less for each extra 40 minutes per day. So if you eat an extra 100 calories each day and exercise about 80 minutes more each day, your weight should remain roughly steady.

Exercises and Problems for Section 14.1

Exercises

1. Using difference quotients, estimate $f_x(3, 2)$ and $f_y(3, 2)$ for the function given by

$$f(x, y) = \frac{x^2}{y + 1}.$$

[Recall: A difference quotient is an expression of the form $(f(a + h, b) - f(a, b))/h$.]

2. Use difference quotients with $\Delta x = 0.1$ and $\Delta y = 0.1$ to estimate $f_x(1, 3)$ and $f_y(1, 3)$ where

$$f(x, y) = e^{-x} \sin y.$$

Then give better estimates by using $\Delta x = 0.01$ and $\Delta y = 0.01$.

3. A contour diagram for $z = f(x, y)$ is given in Figure 14.7. Is f_x positive or negative? Is f_y positive or negative? Estimate $f(2, 1)$, $f_x(2, 1)$, and $f_y(2, 1)$.

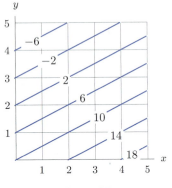

Figure 14.7

4. The quantity, Q, of beef purchased at a store, in kilograms per week, is a function of the price of beef, b, and the price of chicken, c, both in dollars per kilogram.

   (a) Do you expect $\partial Q/\partial b$ to be positive or negative? Explain.

   (b) Do you expect $\partial Q/\partial c$ to be positive or negative? Explain.

   (c) Interpret the statement $\partial Q/\partial b = -213$ in terms of quantity of beef purchased.

5. The sales of a product, $S = f(p, a)$, is a function of the price, p, of the product (in dollars per unit) and the amount, a, spent on advertising (in thousands of dollars).

   (a) Do you expect f_p to be positive or negative? Why?

   (b) Explain the meaning of the statement $f_a(8, 12) = 150$ in terms of sales.

6. The monthly mortgage payment in dollars, P, for a house is a function of three variables

$$P = f(A, r, N),$$

where A is the amount borrowed in dollars, r is the interest rate, and N is the number of years before the mortgage is paid off.

   (a) $f(92000, 14, 30) = 1090.08$. What does this tell you, in financial terms?

   (b) $\left.\dfrac{\partial P}{\partial r}\right|_{(92000, 14, 30)} = 72.82$. What is the financial significance of the number 72.82?

   (c) Would you expect $\partial P/\partial A$ to be positive or negative? Why?

   (d) Would you expect $\partial P/\partial N$ to be positive or negative? Why?

7. A drug is injected into a patient's blood vessel. The function $c = f(x, t)$ represents the concentration of the drug at a distance x mm in the direction of the blood flow measured from the point of injection and at time t seconds since the injection. What are the units of the following partial derivatives? What are their practical interpretations? What do you expect their signs to be?

   (a) $\partial c/\partial x$ (b) $\partial c/\partial t$

8. Suppose that x is the average price of a new car and that y is the average price of a gallon of gasoline. Then q_1, the number of new cars bought in a year, depends on both x and y, so $q_1 = f(x, y)$. Similarly, if q_2 is the quantity of gas bought in a year, then $q_2 = g(x, y)$.

  (a) What do you expect the signs of $\partial q_1 / \partial x$ and $\partial q_2 / \partial y$ to be? Explain.
  (b) What do you expect the signs of $\partial q_1 / \partial y$ and $\partial q_2 / \partial x$ to be? Explain.

For Exercises 9–11, refer to Table 12.3 on page 567 giving the temperature adjusted for wind-chill, C, in °F, as a function $f(w, T)$ of the wind speed, w, in mph, and the temperature, T, in °F. The temperature adjusted for wind-chill tells you how cold it feels, as a result of the combination of wind and temperature.

9. Estimate $f_w(10, 25)$. What does your answer mean in practical terms?

10. Estimate $f_T(5, 20)$. What does your answer mean in practical terms?

11. From Table 12.3 you can see that when the temperature is 20°F, the temperature adjusted for wind-chill drops by an average of about 2.6°F with every 1 mph increase in wind speed from 5 mph to 10 mph. Which partial derivative is this telling you about?

Problems

12. The surface $z = f(x, y)$ is shown in Figure 14.8. The points A and B are in the xy-plane.

  (a) What is the sign of (i) $f_x(A)$?
  (ii) $f_y(A)$?
  (b) The point P in the xy-plane moves along a straight line from A to B. How does the sign of $f_x(P)$ change? How does the sign of $f_y(P)$ change?

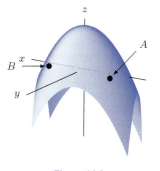

Figure 14.8

13. Figure 14.9 shows the saddle-shaped surface $z = f(x, y)$.

  (a) What is the sign of $f_x(0, 5)$?
  (b) What is the sign of $f_y(0, 5)$?

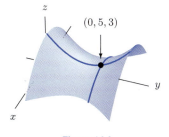

Figure 14.9

14. In each case, give a possible contour diagram for the function $f(x, y)$ if

$f_x > 0$ and $f_y > 0$ $f_x > 0$ and $f_y < 0$
$f_x < 0$ and $f_y > 0$ $f_x < 0$ and $f_y < 0$

15. Figure 14.10 shows contours of $f(x, y)$ with values of f on the contours omitted. If $f_x(P) > 0$, find the sign of

  (a) $f_y(P)$ **(b)** $f_y(Q)$ **(c)** $f_x(Q)$

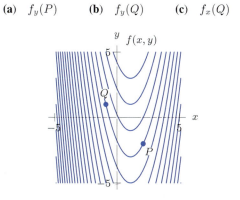

Figure 14.10

16. Figure 14.11 gives a contour diagram for the number n of foxes per square kilometer in southwestern England. Estimate $\partial n / \partial x$ and $\partial n / \partial y$ at the points A, B, and C, where x is kilometers east and y is kilometers north.

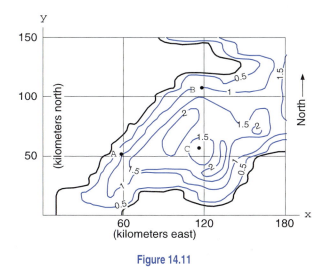

Figure 14.11

17. Figure 14.12 shows a contour diagram for the monthly payment P as a function of the interest rate, $r\%$, and the amount, L, of a 5-year loan. Estimate $\partial P/\partial r$ and $\partial P/\partial L$ at the following points. In each case, give the units and the everyday meaning of your answer.

(a) $r = 8, L = 4000$ **(b)** $r = 8, L = 6000$

(c) $r = 13, L = 7000$

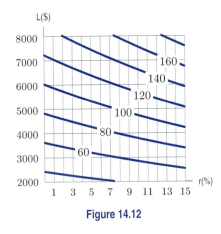

Figure 14.12

18. An experiment to measure the toxicity of formaldehyde yielded the data in Table 14.3. The values show the percent, $P = f(t, c)$, of rats surviving an exposure to formaldehyde at a concentration of c (in parts per million, ppm) after t months. Estimate $f_t(18, 6)$ and $f_c(18, 6)$. Interpret your answers in terms of formaldehyde toxicity.

Table 14.3

		Time t (months)					
		14	16	18	20	22	24
	0	100	100	100	99	97	95
Conc. c	2	100	99	98	97	95	92
(ppm)	6	96	95	93	90	86	80
	15	96	93	82	70	58	36

An airport can be cleared of fog by heating the air. The amount of heat required depends on the air temperature and the wetness of the fog. Problems 19–21 involve Figure 14.13 which shows the heat $H(T, w)$ required (in calories per cubic meter of fog) as a function of the temperature T (in degrees Celsius) and the water content w (in grams per cubic meter of fog). Note that Figure 14.13 is not a contour diagram, but shows cross-sections of H with w fixed at 0.1, 0.2, 0.3, and 0.4.

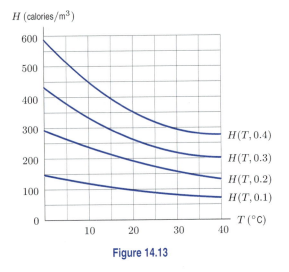

Figure 14.13

19. Use Figure 14.13 to estimate $H_T(10, 0.1)$. Interpret the partial derivative in practical terms.

20. Make a table of values for $H(T, w)$ from Figure 14.13, and use it to estimate $H_T(T, w)$ for $T = 10, 20$, and 30 and $w = 0.1, 0.2$, and 0.3.

21. Repeat Problem 20 for $H_w(T, w)$ at $T = 10, 20$, and 30 and $w = 0.1, 0.2$, and 0.3. What is the practical meaning of these partial derivatives?

22. The cardiac output, represented by c, is the volume of blood flowing through a person's heart, per unit time. The systemic vascular resistance (SVR), represented by s, is the resistance to blood flowing through veins and arteries. Let p be a person's blood pressure. Then p is a function of c and s, so $p = f(c, s)$.

(a) What does $\partial p/\partial c$ represent?

Suppose now that $p = kcs$, where k is a constant.

(b) Sketch the level curves of p. What do they represent? Label your axes.

(c) For a person with a weak heart, it is desirable to have the heart pumping against less resistance, while maintaining the same blood pressure. Such a person may be given the drug Nitroglycerine to decrease the SVR and the drug Dopamine to increase the cardiac output. Represent this on a graph showing level curves. Put a point A on the graph representing the person's state before drugs are given and a point B for after.

(d) Right after a heart attack, a patient's cardiac output drops, thereby causing the blood pressure to drop. A common mistake made by medical residents is to get the patient's blood pressure back to normal by using drugs to increase the SVR, rather than by increasing the cardiac output. On a graph of the level curves of p, put a point D representing the patient before the heart attack, a point E representing the patient right after the heart attack, and a third point F representing the patient after the resident has given the drugs to increase the SVR.

14.2 COMPUTING PARTIAL DERIVATIVES ALGEBRAICALLY

Since the partial derivative $f_x(x, y)$ is the ordinary derivative of the function $f(x, y)$ with y held constant and $f_y(x, y)$ is the ordinary derivative of $f(x, y)$ with x held constant, we can use all the differentiation formulas from one-variable calculus to find partial derivatives.

Example 1   Let $f(x, y) = \dfrac{x^2}{y + 1}$. Find $f_x(3, 2)$ algebraically.

Solution   We use the fact that $f_x(3, 2)$ equals the derivative of $f(x, 2)$ at $x = 3$. Since

$$f(x, 2) = \frac{x^2}{2 + 1} = \frac{x^2}{3},$$

differentiating with respect to x, we have

$$f_x(x, 2) = \frac{\partial}{\partial x}\left(\frac{x^2}{3}\right) = \frac{2x}{3}, \qquad \text{and so} \qquad f_x(3, 2) = 2.$$

Example 2   Compute the partial derivatives with respect to x and with respect to y for the following functions.
   (a) $f(x, y) = y^2 e^{3x}$ (b) $z = (3xy + 2x)^5$ (c) $g(x, y) = e^{x+3y} \sin(xy)$

Solution   (a) This is the product of a function of x (namely e^{3x}) and a function of y (namely y^2). When we differentiate with respect to x, we think of the function of y as a constant, and vice versa. Thus,

$$f_x(x, y) = y^2 \frac{\partial}{\partial x}\left(e^{3x}\right) = 3y^2 e^{3x},$$

$$f_y(x, y) = e^{3x} \frac{\partial}{\partial y}(y^2) = 2y e^{3x}.$$

(b) Here we use the chain rule:

$$\frac{\partial z}{\partial x} = 5(3xy + 2x)^4 \frac{\partial}{\partial x}(3xy + 2x) = 5(3xy + 2x)^4(3y + 2),$$

$$\frac{\partial z}{\partial y} = 5(3xy + 2x)^4 \frac{\partial}{\partial y}(3xy + 2x) = 5(3xy + 2x)^4 3x = 15x(3xy + 2x)^4.$$

(c) Since each function in the product is a function of both x and y, we need to use the product rule for each partial derivative:

$$g_x(x, y) = \left(\frac{\partial}{\partial x}(e^{x+3y})\right)\sin(xy) + e^{x+3y}\frac{\partial}{\partial x}(\sin(xy)) = e^{x+3y}\sin(xy) + e^{x+3y}y\cos(xy),$$

$$g_y(x, y) = \left(\frac{\partial}{\partial y}(e^{x+3y})\right)\sin(xy) + e^{x+3y}\frac{\partial}{\partial y}(\sin(xy)) = 3e^{x+3y}\sin(xy) + e^{x+3y}x\cos(xy).$$

For functions of three or more variables, we find partial derivatives by the same method: Differentiate with respect to one variable, regarding the other variables as constants. For a function $f(x, y, z)$, the partial derivative $f_x(a, b, c)$ gives the rate of change of f with respect to x along the line $y = b$, $z = c$.

Example 3   Find all the partial derivatives of $f(x, y, z) = \dfrac{x^2 y^3}{z}$.

Solution   To find $f_x(x, y, z)$, for example, we consider y and z as fixed, giving

$$f_x(x, y, z) = \frac{2xy^3}{z}, \quad \text{and} \quad f_y(x, y, z) = \frac{3x^2 y^2}{z}, \quad \text{and} \quad f_z(x, y, z) = -\frac{x^2 y^3}{z^2}.$$

Interpretation of Partial Derivatives

Example 4   A vibrating guitar string, originally at rest along the x-axis, is shown in Figure 14.14. Let x be the distance in meters from the left end of the string. At time t seconds the point x has been displaced $y = f(x, t)$ meters vertically from its rest position, where

$$y = f(x, t) = 0.003 \sin(\pi x) \sin(2765t).$$

Evaluate $f_x(0.3, 1)$ and $f_t(0.3, 1)$ and explain what each means in practical terms.

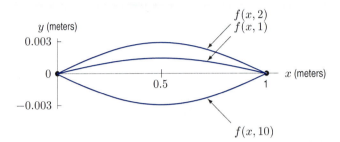

Figure 14.14: The position of a vibrating guitar string at several different times: Graph of $f(x, t)$ for $t = 1, 2, 10$.

Solution   Differentiating $f(x, t) = 0.003 \sin(\pi x) \sin(2765t)$ with respect to x, we have

$$f_x(x, t) = 0.003\pi \cos(\pi x) \sin(2765t).$$

In particular, substituting $x = 0.3$ and $t = 1$ gives

$$f_x(0.3, 1) = 0.003\pi \cos(\pi(0.3)) \sin(2765) \approx 0.002.$$

To see what $f_x(0.3, 1)$ means, think about the function $f(x, 1)$. The graph of $f(x, 1)$ in Figure 14.15 is a snapshot of the string at the time $t = 1$. Thus, the derivative $f_x(0.3, 1)$ is the slope of the string at the point $x = 0.3$ at the instant when $t = 1$.

Similarly, taking the derivative of $f(x, t) = 0.003 \sin(\pi x) \sin(2765t)$ with respect to t, we get

$$f_t(x, t) = (0.003)(2765) \sin(\pi x) \cos(2765t) = 8.3 \sin(\pi x) \cos(2765t).$$

Since $f(x, t)$ is in meters and t is in seconds, the derivative $f_t(0.3, 1)$ is in m/sec. Thus, substituting $x = 0.3$ and $t = 1$,

$$f_t(0.3, 1) = 8.3 \sin(\pi(0.3)) \cos(2765(1)) \approx 6 \text{ m/sec.}$$

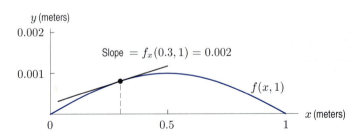

Figure 14.15: Graph of $f(x, 1)$: Snapshot of the shape of the string at $t = 1$ sec

To see what $f_t(0.3, 1)$ means, think about the function $f(0.3, t)$. The graph of $f(0.3, t)$ is a position versus time graph that tracks the up and down movement of the point on the string where $x = 0.3$.

(See Figure 14.16.) The derivative $f_t(0.3, 1) = 6$ m/sec is the velocity of that point on the string at time $t = 1$. The fact that $f_t(0.3, 1)$ is positive indicates that the point is moving upward when $t = 1$.

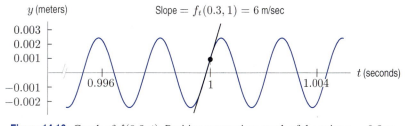

Figure 14.16: Graph of $f(0.3, t)$: Position versus time graph of the point $x = 0.3$ m from the end of the guitar string

Exercises and Problems for Section 14.2

Exercises

Find the partial derivatives in Exercises 1–34. Assume the variables are restricted to a domain on which the function is defined.

1. f_x and f_y if $f(x, y) = 5x^2 y^3 + 8xy^2 - 3x^2$

2. z_x if $z = x^2 y + 2x^5 y$

3. $\dfrac{\partial}{\partial y}(3x^5 y^7 - 32x^4 y^3 + 5xy)$

4. z_y if $z = \dfrac{3x^2 y^7 - y^2}{15xy - 8}$

5. V_r if $V = \frac{1}{3}\pi r^2 h$

6. f_a if $f(a, b) = e^a \sin(a + b)$

7. z_x if $z = \sin(5x^3 y - 3xy^2)$

8. $\dfrac{\partial}{\partial x}(a\sqrt{x})$

9. g_x if $g(x, y) = \ln(ye^{xy})$

10. $\dfrac{\partial}{\partial x}(xe^{\sqrt{xy}})$

11. F_m if $F = mg$

12. $\dfrac{\partial A}{\partial h}$ if $A = \frac{1}{2}(a + b)h$

13. $\dfrac{\partial}{\partial m}\left(\frac{1}{2}mv^2\right)$

14. $\dfrac{\partial}{\partial B}\left(\frac{1}{u_0}B^2\right)$

15. $\dfrac{\partial}{\partial r}\left(\dfrac{2\pi r}{v}\right)$

16. F_v if $F = \dfrac{mv^2}{r}$

17. $\dfrac{\partial V}{\partial r}$ and $\dfrac{\partial V}{\partial h}$ if $V = \frac{4}{3}\pi r^2 h$

18. $\dfrac{\partial}{\partial v_0}(v_0 + at)$

19. $\dfrac{\partial F}{\partial m_2}$ if $F = \dfrac{Gm_1 m_2}{r^2}$

20. a_v if $a = \dfrac{v^2}{r}$

21. $\dfrac{\partial}{\partial T}\left(\dfrac{2\pi r}{T}\right)$

22. $\dfrac{\partial}{\partial t}\left(v_0 t + \frac{1}{2}at^2\right)$

23. u_E if $u = \frac{1}{2}\epsilon_0 E^2 + \dfrac{1}{2\mu_0}B^2$

24. $\dfrac{\partial f_0}{\partial L}$ if $f_0 = \dfrac{1}{2\pi\sqrt{LC}}$

25. $\dfrac{\partial T}{\partial l}$ if $T = 2\pi\sqrt{\dfrac{l}{g}}$

26. $\dfrac{\partial}{\partial x}\left(\dfrac{1}{\sqrt{2\pi}\sigma}e^{-(x-\mu)^2/(2\sigma^2)}\right)$

27. $\dfrac{\partial Q}{\partial K}$ if $Q = c(a_1 K^{b_1} + a_2 L^{b_2})^\gamma$

28. z_x and z_y for $z = x^7 + 2^y + x^y$

29. $\dfrac{\partial}{\partial M}\left(\dfrac{2\pi r^{3/2}}{\sqrt{GM}}\right)$

30. $\dfrac{\partial m}{\partial v}$ if $m = \dfrac{m_0}{\sqrt{1 - v^2/c^2}}$

31. $f_x(1, 2)$ and $f_y(1, 2)$ if $f(x, y) = x^3 + 3x^2 y - 2y^2$

32. $z_x(2, 3)$ if $z = (\cos x) + y$

33. $\left.\dfrac{\partial z}{\partial y}\right|_{(1, 0.5)}$ if $z = e^{x+2y}\sin y$

34. $\left.\dfrac{\partial f}{\partial x}\right|_{(\pi/3, 1)}$ if $f(x, y) = x\ln(y\cos x)$

Problems

35. (a) Let $f(u,v) = u(u^2 + v^2)^{3/2}$. Use a difference quotient to approximate $f_u(1,3)$ with $h = 0.001$.
   (b) Now evaluate $f_u(1,3)$ exactly. Was the approximation in part (a) reasonable?

36. (a) Let $f(x,y) = x^2 + y^2$. Estimate $f_x(2,1)$ and $f_y(2,1)$ using the contour diagram for f in Figure 14.17.
   (b) Estimate $f_x(2,1)$ and $f_y(2,1)$ from a table of values for f with $x = 1.9, 2, 2.1$ and $y = 0.9, 1, 1.1$.
   (c) Compare your estimates in parts (a) and (b) with the exact values of $f_x(2,1)$ and $f_y(2,1)$ found algebraically.

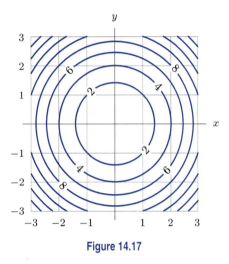

Figure 14.17

37. The Dubois formula relates a person's surface area, s, in m^2, to weight, w, in kg, and height, h, in cm, by

$$s = f(w,h) = 0.01w^{0.25}h^{0.75}.$$

Find $f(65, 160)$, $f_w(65, 160)$, and $f_h(65, 160)$. Interpret your answers in terms of surface area, height, and weight.

38. Money in a bank account earns interest at a continuous rate, r. The amount of money, $\$B$, in the account depends on the amount deposited, $\$P$, and the time, t, it has been in the bank according to the formula

$$B = Pe^{rt}.$$

Find $\partial B/\partial t$ and $\partial B/\partial P$ and interpret each in financial terms.

39. The gravitational force, F newtons, exerted on a mass of m kg at a distance of r meters from the center of the earth is given by

$$F = \frac{GMm}{r^2}$$

where the mass of the earth $M = 6 \cdot 10^{24}$ kilograms, and $G = 6.67 \cdot 10^{-11}$. Find the gravitational force on a person with mass 70 kg at the surface of the earth ($r = 6.4 \cdot 10^6$). Calculate $\partial F/\partial m$ and $\partial F/\partial r$ for these values of m and r. Interpret these partial derivatives in terms of gravitational force.

40. The energy, E, of a body of mass m moving with speed v is given by the formula

$$E = mc^2 \left(\frac{1}{\sqrt{1 - v^2/c^2}} - 1 \right).$$

The speed, v, is nonnegative and less than the speed of light, c, which is a constant.

   (a) Find $\partial E/\partial m$. What would you expect the sign of $\partial E/\partial m$ to be? Explain.
   (b) Find $\partial E/\partial v$. Explain what you would expect the sign of $\partial E/\partial v$ to be and why.

41. Show that the Cobb-Douglas function

$$Q = bK^\alpha L^{1-\alpha} \quad \text{where} \quad 0 < \alpha < 1$$

satisfies the equation

$$K\frac{\partial Q}{\partial K} + L\frac{\partial Q}{\partial L} = Q.$$

42. A one-meter long bar is heated unevenly, with temperature in °C at a distance x meters from one end at time t given by

$$H(x,t) = 100e^{-0.1t}\sin(\pi x) \qquad 0 \le x \le 1.$$

   (a) Sketch a graph of H against x for $t = 0$ and $t = 1$.
   (b) Calculate $H_x(0.2, t)$ and $H_x(0.8, t)$. What is the practical interpretation (in terms of temperature) of these two partial derivatives? Explain why each one has the sign it does.
   (c) Calculate $H_t(x, t)$. What is its sign? What is its interpretation in terms of temperature?

43. Let $h(x, t) = 5 + \cos(0.5x - t)$ describe a wave. The value of $h(x, t)$ gives the depth of the water in cm at a distance x meters from a fixed point and at time t seconds. Evaluate $h_x(2, 5)$ and $h_t(2, 5)$ and interpret each in terms of the wave.

44. Is there a function f which has the following partial derivatives? If so what is it? Are there any others?

$$f_x(x,y) = 4x^3y^2 - 3y^4,$$
$$f_y(x,y) = 2x^4y - 12xy^3.$$

14.3   LOCAL LINEARITY AND THE DIFFERENTIAL

In Sections 14.1 and 14.2 we studied a function of two variables by allowing one variable at a time to change. We now let both variables change at once to develop a linear approximation for functions of two variables.

Zooming In to See Local Linearity

For a function of one variable, local linearity means that as we zoom in on the graph, it looks like a straight line. As we zoom in on the graph of a two-variable function, the graph usually looks like a plane, which is the graph of a linear function of two variables. (See Figure 14.18.)

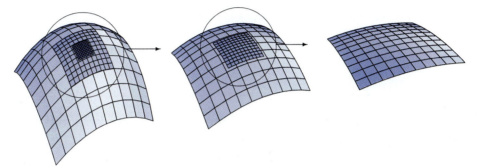

Figure 14.18: Zooming in on the graph of a function of two variables until the graph looks like a plane

Similarly, Figure 14.19 shows three successive views of the contours near a point. As we zoom in, the contours look more like equally spaced parallel lines, which are the contours of a linear function. (As we zoom in, we have to add more contours.)

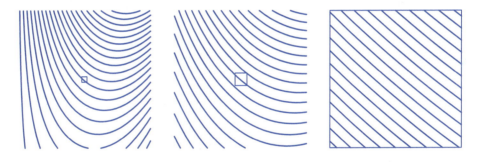

Figure 14.19: Zooming in on a contour diagram until the lines look parallel and equally spaced

This effect can also be seen numerically by zooming in with tables of values. Table 14.4 shows three tables of values for $f(x, y) = x^2 + y^3$ near $x = 2$, $y = 1$, each one a closer view than the previous one. Notice how each table looks more like the table of a linear function.

Table 14.4   *Zooming in on values of $f(x, y) = x^2 + y^3$ near $(2, 1)$ until the table looks linear*

		y	
	0	1	2
1	1	2	9
x 2	4	5	12
3	9	10	17

		y	
	0.9	1.0	1.1
1.9	4.34	4.61	4.94
x 2.0	4.73	5.00	5.33
2.1	5.14	5.41	5.74

		y	
	0.99	1.00	1.01
1.99	4.93	4.96	4.99
x 2.00	4.97	5.00	5.03
2.01	5.01	5.04	5.07

Zooming in Algebraically: Differentiability

Seeing a plane when we zoom in at a point tells us (provided the plane is not vertical) that $f(x, y)$ is closely approximated near that point by a linear function, $L(x, y)$:

$$f(x, y) \approx L(x, y).$$

The graph of the function $z = L(x, y)$ is the tangent plane at that point. Provided the approximation is sufficiently good, we say that $f(x, y)$ is *differentiable* at the point. Section 14.8 on page 689 defines precisely what is meant by the approximation being sufficiently good. The functions we encounter are differentiable at most points in their domain. We usually only consider the question of differentiability at points that are in the interior of the domain of f, not on the boundary[1].

The Tangent Plane

The plane that we see when we zoom in on a surface is called the *tangent plane* to the surface at the point. Figure 14.20 shows the graph of a function with a tangent plane.

What is the equation of the tangent plane? At the point (a, b), the x-slope of the graph of f is the partial derivative $f_x(a, b)$ and the y-slope is $f_y(a, b)$. Thus, using the equation for a plane on page 586 of Chapter 12, we have the following result:

Tangent Plane to the Surface $z = f(x, y)$ at the Point (a, b)

Assuming f is differentiable at (a, b), the equation of the tangent plane is

$$z = f(a, b) + f_x(a, b)(x - a) + f_y(a, b)(y - b).$$

Here we are thinking of a and b as fixed, so $f(a, b)$, and $f_x(a, b)$, and $f_y(a, b)$ are constants. Thus, the right side of the equation is a linear function of x and y.

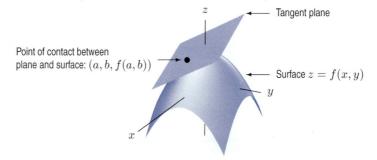

Point of contact between plane and surface: $(a, b, f(a, b))$

Tangent plane

Surface $z = f(x, y)$

Figure 14.20: The tangent plane to the surface $z = f(x, y)$ at the point (a, b)

Example 1   Find the equation for the tangent plane to the surface $z = x^2 + y^2$ at the point $(3, 4)$.

Solution   We have $f_x(x, y) = 2x$, so $f_x(3, 4) = 6$, and $f_y(x, y) = 2y$, so $f_y(3, 4) = 8$. Also, $f(3, 4) = 3^2 + 4^2 = 25$. Thus, the equation for the tangent plane at $(3, 4)$ is

$$z = 25 + 6(x - 3) + 8(y - 4) = -25 + 6x + 8y.$$

Local Linearization

Since the tangent plane lies close to the surface near the point at which they meet, z-values on the tangent plane are close to values of $f(x, y)$ for points near (a, b). Thus, replacing z by $f(x, y)$ in the equation of the tangent plane, we get the following approximation:

---

[1]Interior points and boundary points are defined precisely on page 715

> ### Tangent Plane Approximation to $f(x, y)$ for (x, y) Near the Point (a, b)
>
> Provided f is differentiable at (a, b), we can approximate $f(x, y)$:
>
> $$f(x, y) \approx f(a, b) + f_x(a, b)(x - a) + f_y(a, b)(y - b).$$
>
> We are thinking of a and b as fixed, so the expression on the right side is linear in x and y. The right side of this approximation is called the **local linearization** of f near $x = a$, $y = b$.

Figure 14.21 shows the tangent plane approximation graphically.

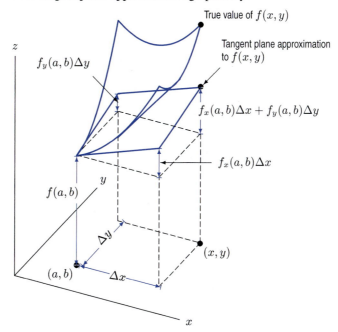

Figure 14.21: Local linearization: Approximating $f(x, y)$ by the z-value from the tangent plane

---

Example 2   Find the local linearization of $f(x, y) = x^2 + y^2$ at the point $(3, 4)$. Estimate $f(2.9, 4.2)$ and $f(2, 2)$ using the linearization and compare your answers to the true values.

Solution   Let $z = f(x, y) = x^2 + y^2$. In Example 1 on page 652, we found the equation of the tangent plane at $(3, 4)$ to be

$$z = 25 + 6(x - 3) + 8(y - 4).$$

Therefore, for (x, y) near $(3, 4)$, we have the local linearization

$$f(x, y) \approx 25 + 6(x - 3) + 8(y - 4).$$

Substituting $x = 2.9$, $y = 4.2$ gives

$$f(2.9, 4.2) \approx 25 + 6(-0.1) + 8(0.2) = 26.$$

This compares favorably with the true value $f(2.9, 4.2) = (2.9)^2 + (4.2)^2 = 26.05$.

However, the local linearization does not give a good approximation at points far away from $(3, 4)$. For example, if $x = 2$, $y = 2$, the local linearization gives

$$f(2, 2) \approx 25 + 6(-1) + 8(-2) = 3,$$

whereas the true value of the function is $f(2, 2) = 2^2 + 2^2 = 8$.

Example 3   Designing safe boilers depends on knowing how steam behaves under changes in temperature and pressure. Steam tables, such as Table 14.5, are published giving values of the function $V = f(T, P)$ where V is the volume (in ft^3) of one pound of steam at a temperature T (in °F) and pressure P (in lb/in^2).

(a) Give a linear function approximating $V = f(T, P)$ for T near $500°$F and P near 24 lb/in^2.
(b) Estimate the volume of a pound of steam at a temperature of $505°$F and a pressure of 24.3 lb/in^2.

Table 14.5   *Volume (in cubic feet) of one pound of steam at various temperatures and pressures*

		Pressure P (lb/in^2)			
		20	22	24	26
Temperature T (°F)	480	27.85	25.31	23.19	21.39
	500	28.46	25.86	23.69	21.86
	520	29.06	26.41	24.20	22.33
	540	29.66	26.95	24.70	22.79

Solution   (a) We want the local linearization around the point $T = 500$, $P = 24$, which is

$$f(T, P) \approx f(500, 24) + f_T(500, 24)(T - 500) + f_P(500, 24)(P - 24).$$

We read the value $f(500, 24) = 23.69$ from the table.

Next we approximate $f_T(500, 24)$ by a difference quotient. From the $P = 24$ column, we compute the average rate of change between $T = 500$ and $T = 520$:

$$f_T(500, 24) \approx \frac{f(520, 24) - f(500, 24)}{520 - 500} = \frac{24.20 - 23.69}{20} = 0.0255.$$

Note that $f_T(500, 24)$ is positive, because steam expands when heated.

Next we approximate $f_P(500, 24)$ by looking at the $T = 500$ row and computing the average rate of change between $P = 24$ and $P = 26$:

$$f_P(500, 24) \approx \frac{f(500, 26) - f(500, 24)}{26 - 24} = \frac{21.86 - 23.69}{2} = -0.915.$$

Note that $f_P(500, 24)$ is negative, because increasing the pressure on steam decreases its volume. Using these approximations for the partial derivatives, we obtain the local linearization:

$$V = f(T, P) \approx 23.69 + 0.0255(T - 500) - 0.915(P - 24) \text{ ft}^3 \quad \begin{matrix} \text{for } T \text{ near } 500 \text{ °F} \\ \text{and } P \text{ near } 24 \text{ lb/in}^2. \end{matrix}$$

(b) We are interested in the volume at $T = 505°$F and $P = 24.3$ lb/in^2. Since these values are close to $T = 500°$F and $P = 24$ lb/in^2, we use the linear relation obtained in part (a).

$$V \approx 23.69 + 0.0255(505 - 500) - 0.915(24.3 - 24) = 23.54 \text{ ft}^3.$$

Local Linearity With Three or More Variables

Local linear approximations for functions of three or more variables follow the same pattern as for functions of two variables. The local linearization of $f(x, y, z)$ at (a, b, c) is given by

$$f(x, y, z) \approx f(a, b, c) + f_x(a, b, c)(x - a) + f_y(a, b, c)(y - b) + f_z(a, b, c)(z - c).$$

The Differential

We are often interested in the change in the value of the function as we move from the point (a, b) to a nearby point (x, y). Then we use the notation

$$\Delta f = f(x, y) - f(a, b) \quad \text{and} \quad \Delta x = x - a \quad \text{and} \quad \Delta y = y - b$$

to rewrite the tangent plane approximation

$$f(x, y) \approx f(a, b) + f_x(a, b)(x - a) + f_y(a, b)(y - b)$$

in the form

$$\Delta f \approx f_x(a, b)\Delta x + f_y(a, b)\Delta y.$$

For fixed a and b, the right side of this is a linear function of Δx and Δy that can be used to estimate Δf. We call this linear function the *differential*. To define the differential in general, we introduce new variables dx and dy to represent changes in x and y.

The Differential of a Function $z = f(x, y)$

The **differential**, df (or dz), at a point (a, b) is the linear function of dx and dy given by the formula

$$df = f_x(a, b)\, dx + f_y(a, b)\, dy.$$

The differential at a general point is often written $df = f_x\, dx + f_y\, dy$.

Note that the differential, df, is a function of four variables a, b, and dx, dy.

Example 4   Compute the differentials of the following functions.

(a) $f(x, y) = x^2 e^{5y}$ (b) $z = x\sin(xy)$ (c) $f(x, y) = x\cos(2x)$

Solution   (a) Since $f_x(x, y) = 2xe^{5y}$ and $f_y(x, y) = 5x^2 e^{5y}$, we have

$$df = 2xe^{5y}\, dx + 5x^2 e^{5y}\, dy.$$

(b) Since $\partial z/\partial x = \sin(xy) + xy\cos(xy)$ and $\partial z/\partial y = x^2\cos(xy)$, we have

$$dz = (\sin(xy) + xy\cos(xy))\, dx + x^2\cos(xy)\, dy.$$

(c) Since $f_x(x, y) = \cos(2x) - 2x\sin(2x)$ and $f_y(x, y) = 0$, we have

$$df = (\cos(2x) - 2x\sin(2x))\, dx + 0\, dy = (\cos(2x) - 2x\sin(2x))\, dx.$$

Example 5   The density ρ (in g/cm^3) of carbon dioxide gas CO_2 depends upon its temperature T (in °C) and pressure P (in atmospheres). The ideal gas model for CO_2 gives what is called the state equation

$$\rho = \frac{0.5363P}{T + 273.15}.$$

Compute the differential $d\rho$. Explain the signs of the coefficients of dT and dP.

Solution   The differential for $\rho = f(T, P)$ is

$$d\rho = f_T(T, P)\, dT + f_P(T, P)dP = \frac{-0.5363P}{(T + 273.15)^2}\, dT + \frac{0.5363}{T + 273.15}\, dP.$$

The coefficient of dT is negative because increasing the temperature expands the gas (if the pressure is kept constant) and therefore decreases its density. The coefficient of dP is positive because increasing the pressure compresses the gas (if the temperature is kept constant) and therefore increases its density.

Where Does the Notation for the Differential Come From?

We write the differential as a linear function of the new variables dx and dy. You may wonder why we chose these names for our variables. The reason is historical: The people who invented calculus thought of dx and dy as "infinitesimal" changes in x and y. The equation

$$df = f_x dx + f_y dy$$

was regarded as an infinitesimal version of the local linear approximation

$$\Delta f \approx f_x \Delta x + f_y \Delta y.$$

In spite of the problems with defining exactly what "infinitesimal" means, some mathematicians, scientists, and engineers think of the differential in terms of infinitesimals.

Figure 14.22 illustrates a way of thinking about differentials that combines the definition with this informal point of view. It shows the graph of f along with a view of the graph around the point $(a, b, f(a, b))$ under a microscope. Since f is locally linear at the point, the magnified view looks like the tangent plane. Under the microscope, we use a magnified coordinate system with its origin at the point $(a, b, f(a, b))$ and with coordinates dx, dy, and dz along the three axes. The graph of the differential df is the tangent plane, which has equation $df = f_x(a, b)\, dx + f_y(a, b)\, dy$ in the magnified coordinates.

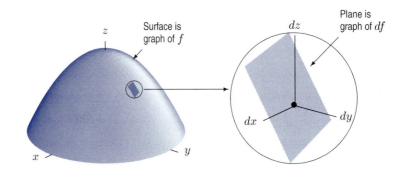

Figure 14.22: The graph of f, with a view through a microscope showing the tangent plane in the magnified coordinate system

Exercises and Problems for Section 14.3

Exercises

For the functions in Exercises 1–3, find the equation of the tangent plane at the given point.

1. $z = e^y + x + x^2 + 6$ at the point $(1, 0, 9)$.

2. $z = ye^{x/y}$ at the point $(1, 1, e)$.

3. $z = \frac{1}{2}(x^2 + 4y^2)$ at the point $(2, 1, 4)$.

Find the differentials of the functions in Exercises 4–7.

4. $f(x, y) = \sin(xy)$ **5.** $z = e^{-x} \cos y$

6. $g(u, v) = u^2 + uv$

7. $h(x, t) = e^{-3t} \sin(x + 5t)$

Find differentials of the functions in Exercises 8–11 at the given point.

8. $f(x, y) = xe^{-y}$ at $(1, 0)$

9. $g(x, t) = x^2 \sin(2t)$ at $(2, \pi/4)$

10. $P(L, K) = 1.01L^{0.25}K^{0.75}$ at $(100, 1)$

11. $F(m, r) = Gm/r^2$ at $(100, 10)$

Problems

12. (a) Check the local linearity of $f(x,y) = e^{-x}\sin y$ near $x = 1$, $y = 2$ by making a table of values of f for $x = 0.9, 1.0, 1.1$ and $y = 1.9, 2.0, 2.1$. Express values of f with 4 digits after the decimal point. Then make a table of values for $x = 0.99, 1.00, 1.01$ and $y = 1.99, 2.00, 2.01$, again showing 4 digits after the decimal point. Do both tables look nearly linear? Does the second table look more linear than the first?

(b) Give the local linearization of $f(x,y) = e^{-x}\sin y$ at $(1,2)$, first using your tables, and second using the fact that $f_x(x,y) = -e^{-x}\sin y$ and $f_y(x,y) = e^{-x}\cos y$.

13. A student was asked to find the equation of the tangent plane to the surface $z = x^3 - y^2$ at the point $(x,y) = (2,3)$. The student's answer was

$$z = 3x^2(x-2) - 2y(y-3) - 1.$$

(a) At a glance, how do you know this is wrong?

(b) What mistake did the student make?

(c) Answer the question correctly.

14. Find the local linearization of the function $f(x,y) = x^2y$ at the point $(3,1)$.

15. Find the differential of $f(x,y) = \sqrt{x^2 + y^3}$ at the point $(1,2)$. Use it to estimate $f(1.04, 1.98)$.

16. An unevenly heated plate has temperature $T(x,y)$ in $°C$ at the point (x,y). If $T(2,1) = 135$, and $T_x(2,1) = 16$, and $T_y(2,1) = -15$, estimate the temperature at the point $(2.04, 0.97)$.

17. Give the local linearization for the monthly car-loan payment function at each of the points investigated in Problem 17 on page 646.

18. In Example 3 on page 654 we found a linear approximation for $V = f(T,P)$ near $(500, 24)$. Now find a linear approximation near $(480, 20)$.

19. In Example 3 on page 654 we found a linear approximation for $V = f(T,p)$ near $(500, 24)$.

(a) Test the accuracy of this approximation by comparing its predicted value with the four neighboring values in the table. What do you notice? Which predicted values are accurate? Which are not? Explain your answer.

(b) Suggest a linear approximation for $f(T,p)$ near $(500, 24)$ that does not have the property you noticed in part (a). [Hint: Estimate the partial derivatives in a different way.]

20. The gas equation for one mole of oxygen relates its pressure, P (in atmospheres), its temperature, T (in K), and its volume, V (in cubic decimeters, dm^3):

$$T = 16.574\frac{1}{V} - 0.52754\frac{1}{V^2} - 0.3879P + 12.187VP.$$

(a) Find the temperature T and differential dT if the volume is 25 dm^3 and the pressure is 1 atmosphere.

(b) Use your answer to part (a) to estimate how much the volume would have to change if the pressure increased by 0.1 atmosphere and the temperature remained constant.

21. A fluid moves through a tube of radius $r = 0.005 \pm 0.00025$ m, and length 1 m under a pressure $p = 10^5 \pm 1000$ pascals, at a rate $v = 0.625 \cdot 10^{-9}$ m^3 per unit time. Find the maximum error in the viscosity η given by

$$\eta = \frac{\pi}{8}\frac{pr^4}{v}.$$

22. One mole of ammonia gas is contained in a vessel which is capable of changing its volume (a compartment sealed by a piston, for example). The total energy U (in joules) of the ammonia, is a function of the volume V (in m^3) of the container, and the temperature T (in K) of the gas. The differential dU is given by

$$dU = 840\,dV + 27.32\,dT.$$

(a) How does the energy change if the volume is held constant and the temperature is increased slightly?

(b) How does the energy change if the temperature is held constant and the volume is increased slightly?

(c) Find the approximate change in energy if the gas is compressed by 100 cm^3 and heated by 2 K.

23. The coefficient, β, of thermal expansion of a liquid relates the change in the volume V (in m^3) of a fixed quantity of a liquid to an increase in its temperature T (in $°C$):

$$dV = \beta V\,dT.$$

(a) Let ρ be the density (in kg/m^3) of water as a function of temperature. Write an expression for $d\rho$ in terms of ρ and dT.

(b) The graph in Figure 14.23 shows density of water as a function of temperature. Use it to estimate β when $T = 20°C$ and when $T = 80°C$.

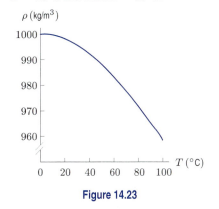

Figure 14.23

24. The period, T, of oscillation in seconds of a pendulum clock is given by $T = 2\pi\sqrt{l/g}$, where g is the acceleration due to gravity. The length of the pendulum, l, depends on the temperature, t, according to the formula $l = l_0(1+\alpha t)$ where l_0 is the length of the pendulum at temperature t_0 and α is a constant which characterizes the clock. The clock is set to the correct period at the temperature t_0. How many seconds a day does the clock lose or gain when the temperature is $t_0 + \Delta t$? Show that this loss or gain is independent of l_0.

25. (a) Write a formula for the number π using only the perimeter L and the area A of a circle.

(b) Suppose that L and A are determined experimentally. Show that if the relative, or percent, errors in the measured values of L and A are λ and μ, respectively, then the resulting relative, or percent, error in π is $2\lambda - \mu$.

14.4 GRADIENTS AND DIRECTIONAL DERIVATIVES IN THE PLANE

The Rate of Change in an Arbitrary Direction: The Directional Derivative

The partial derivatives of a function f tell us the rate of change of f in the directions parallel to the coordinate axes. In this section we see how to compute the rate of change of f in an arbitrary direction.

Example 1   Figure 14.24 shows the temperature, in °C, at the point (x, y). Estimate the average rate of change of temperature as we walk from point A to point B.

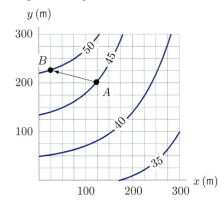

Figure 14.24: Estimating rate of change on a temperature map

Solution   At the point A we are on the $H = 45°C$ contour. At B we are on the $H = 50°C$ contour. The displacement vector from A to B has x component approximately $-100\vec{i}$ and y component approximately $25\vec{j}$, so its length is $\sqrt{(-100)^2 + 25^2} \approx 103$. Thus the temperature rises by 5°C as we move 103 meters, so the average rate of change of the temperature in that direction is about $5/103 \approx 0.05°C/m$.

Suppose we want to compute the rate of change of a function $f(x, y)$ at the point $P = (a, b)$ in the direction of the unit vector $\vec{u} = u_1\vec{i} + u_2\vec{j}$. For $h > 0$, consider the point $Q = (a+hu_1, b+hu_2)$ whose displacement from P is $h\vec{u}$. (See Figure 14.25.) Since $\|\vec{u}\| = 1$, the distance from P to Q is h. Thus,

$$\begin{array}{l}\text{Average rate of change} \\ \text{in } f \text{ from } P \text{ to } Q\end{array} = \frac{\text{Change in } f}{\text{Distance from } P \text{ to } Q} = \frac{f(a + hu_1, b + hu_2) - f(a, b)}{h}.$$

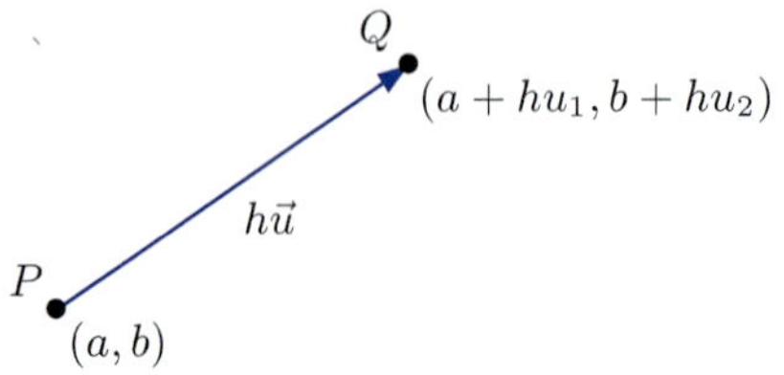

Figure 14.25: Displacement of $h\vec{u}$ from the point (a, b)

Taking the limit as $h \to 0$ gives the instantaneous rate of change and the following definition:

Directional Derivative of f at (a, b) in the Direction of a Unit Vector $\vec{u}$

If $\vec{u} = u_1\vec{i} + u_2\vec{j}$ is a unit vector, we define the directional derivative, $f_{\vec{u}}$, by

$$
f_{\vec{u}}(a, b) = \begin{array}{c} \text{Rate of change} \\ \text{of } f \text{ in direction} \\ \text{of } \vec{u} \text{ at } (a, b) \end{array} = \lim_{h \to 0} \frac{f(a + hu_1, b + hu_2) - f(a, b)}{h},
$$

provided the limit exists.

Notice that if $\vec{u} = \vec{i}$, so $u_1 = 1, u_2 = 0$, then the directional derivative is f_x, since

$$
f_{\vec{i}}(a, b) = \lim_{h \to 0} \frac{f(a + h, b) - f(a, b)}{h} = f_x(a, b).
$$

Similarly, if $\vec{u} = \vec{j}$ then the directional derivative $f_{\vec{j}} = f_y$.

What If We Do Not Have a Unit Vector?

We defined $f_{\vec{u}}$ for $\vec{u}$ a unit vector. If $\vec{v}$ is not a unit vector, $\vec{v} \neq \vec{0}$, we construct a unit vector $\vec{u} = \vec{v}/\|\vec{v}\|$ in the same direction as $\vec{v}$ and define the rate of change of f in the direction of $\vec{v}$ as $f_{\vec{u}}$.

Example 2   For each of the functions f, g, and h in Figure 14.26, decide whether the directional derivative at the indicated point is positive, negative, or zero, in the direction of the vector $\vec{v} = \vec{i} + 2\vec{j}$, and in the direction of the vector $\vec{w} = 2\vec{i} + \vec{j}$.

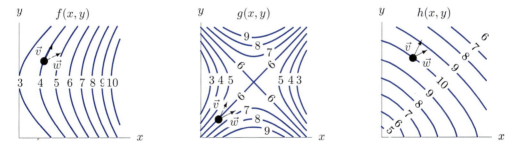

Figure 14.26: Contour diagrams of three functions with direction vectors $\vec{v} = \vec{i} + 2\vec{j}$ and $\vec{w} = 2\vec{i} + \vec{j}$ marked on each

Solution   On the contour diagram for f, the vector $\vec{v} = \vec{i} + 2\vec{j}$ appears to be tangent to the contour. Thus, in this direction, the value of the function is not changing, so the directional derivative in the direction of $\vec{v}$ is zero. The vector $\vec{w} = 2\vec{i} + \vec{j}$ points from the contour marked 4 toward the contour marked 5. Thus, the values of the function are increasing and the directional derivative in the direction of $\vec{w}$ is positive.

On the contour diagram for g, the vector $\vec{v} = \vec{i} + 2\vec{j}$ points from the contour marked 6 toward the contour marked 5, so the function is decreasing in that direction. Thus, the rate of change is negative. On the other hand, the vector $\vec{w} = 2\vec{i} + \vec{j}$ points from the contour marked 6 toward the contour marked 7, and hence the directional derivative in the direction of $\vec{w}$ is positive.

Finally, on the contour diagram for h, both vectors point from the $h = 10$ contour to the $h = 9$ contour, so both directional derivatives are negative.

Example 3   Calculate the directional derivative of $f(x, y) = x^2 + y^2$ at $(1, 0)$ in the direction of the vector $\vec{i} + \vec{j}$.

Solution   First we have to find the unit vector in the same direction as the vector $\vec{i} + \vec{j}$. Since this vector has magnitude $\sqrt{2}$, the unit vector is

$$\vec{u} = \frac{1}{\sqrt{2}}(\vec{i} + \vec{j}) = \frac{1}{\sqrt{2}}\vec{i} + \frac{1}{\sqrt{2}}\vec{j}.$$

Thus,

$$f_{\vec{u}}(1, 0) = \lim_{h \to 0} \frac{f(1 + h/\sqrt{2}, h/\sqrt{2}) - f(1, 0)}{h} = \lim_{h \to 0} \frac{(1 + h/\sqrt{2})^2 + (h/\sqrt{2})^2 - 1}{h}$$

$$= \lim_{h \to 0} \frac{\sqrt{2}h + h^2}{h} = \lim_{h \to 0} (\sqrt{2} + h) = \sqrt{2}.$$

Computing Directional Derivatives From Partial Derivatives

If f is differentiable, we will now see how to use local linearity to find a formula for the directional derivative which does not involve a limit. If $\vec{u}$ is a unit vector, the definition of $f_{\vec{u}}$ says

$$f_{\vec{u}}(a, b) = \lim_{h \to 0} \frac{f(a + hu_1, b + hu_2) - f(a, b)}{h} = \lim_{h \to 0} \frac{\Delta f}{h},$$

where $\Delta f = f(a + hu_1, b + hu_2) - f(a, b)$ is the change in f. We write Δx for the change in x, so $\Delta x = (a + hu_1) - a = hu_1$; similarly $\Delta y = hu_2$. Using local linearity, we have

$$\Delta f \approx f_x(a, b)\Delta x + f_y(a, b)\Delta y = f_x(a, b)hu_1 + f_y(a, b)hu_2.$$

Thus, dividing by h gives

$$\frac{\Delta f}{h} \approx \frac{f_x(a, b)hu_1 + f_y(a, b)hu_2}{h} = f_x(a, b)u_1 + f_y(a, b)u_2.$$

This approximation becomes exact as $h \to 0$, so we have the following formula:

$$f_{\vec{u}}(a, b) = f_x(a, b)u_1 + f_y(a, b)u_2.$$

Example 4   Use the preceding formula to compute the directional derivative in Example 3. Check that we get the same answer as before.

Solution   We calculate $f_{\vec{u}}(1, 0)$, where $f(x, y) = x^2 + y^2$ and $\vec{u} = \frac{1}{\sqrt{2}}\vec{i} + \frac{1}{\sqrt{2}}\vec{j}$.

The partial derivatives are $f_x(x, y) = 2x$ and $f_y(x, y) = 2y$. So, as before

$$f_{\vec{u}}(1, 0) = f_x(1, 0)u_1 + f_y(1, 0)u_2 = (2)\left(\frac{1}{\sqrt{2}}\right) + (0)\left(\frac{1}{\sqrt{2}}\right) = \sqrt{2}.$$

The Gradient Vector

Notice that the expression for $f_{\vec{u}}(a, b)$ can be written as a dot product of $\vec{u}$ and a new vector:

$$f_{\vec{u}}(a, b) = f_x(a, b)u_1 + f_y(a, b)u_2 = (f_x(a, b)\vec{i} + f_y(a, b)\vec{j}) \cdot (u_1\vec{i} + u_2\vec{j}).$$

The new vector, $f_x(a, b)\vec{i} + f_y(a, b)\vec{j}$, turns out to be important. Thus, we make the following definition:

> **The Gradient Vector** of a differentiable function f at the point (a, b) is
>
> $$\text{grad } f(a, b) = f_x(a, b)\vec{i} + f_y(a, b)\vec{j}$$

The formula for the directional derivative can be written in terms of the gradient as follows:

The Directional Derivative and the Gradient

If f is differentiable at (a, b) and $\vec{u} = u_1\vec{i} + u_2\vec{j}$ is a unit vector, then

$$f_{\vec{u}}(a, b) = f_x(a, b)u_1 + f_y(a, b)u_2 = \text{grad } f(a, b) \cdot \vec{u},$$

Example 5   Find the gradient vector of $f(x, y) = x + e^y$ at the point $(1, 1)$.

Solution   Using the definition we have

$$\text{grad } f = f_x\vec{i} + f_y\vec{j} = \vec{i} + e^y\vec{j},$$

so at the point $(1, 1)$

$$\text{grad } f(1, 1) = \vec{i} + e\vec{j}.$$

Alternative Notation for the Gradient

You can think of $\dfrac{\partial f}{\partial x}\vec{i} + \dfrac{\partial f}{\partial y}\vec{j}$ as the result of applying the vector operator (pronounced "del")

$$\nabla = \frac{\partial}{\partial x}\vec{i} + \frac{\partial}{\partial y}\vec{j}$$

to the function f. Thus, we get the alternative notation

$$\text{grad } f = \nabla f.$$

If $z = f(x, y)$, we can write grad z or ∇z for grad for ∇f.

What Does the Gradient Tell Us?

The fact that $f_{\vec{u}} = \text{grad } f \cdot \vec{u}$ enables us to see what the gradient vector represents. Suppose θ is the angle between the vectors grad$f$ and $\vec{u}$. At the point (a, b), we have

$$f_{\vec{u}} = \text{grad } f \cdot \vec{u} = \| \text{grad } f\| \underbrace{\|\vec{u}\|}_{1} \cos\theta = \| \text{grad } f\| \cos\theta.$$

Imagine that grad f is fixed and that $\vec{u}$ can rotate. (See Figure 14.27.) The maximum value of $f_{\vec{u}}$ occurs when $\cos\theta = 1$, so $\theta = 0$ and $\vec{u}$ is pointing in the direction of grad f. Then

$$\text{Maximum } f_{\vec{u}} = \| \text{grad } f\| \cos 0 = \| \text{grad } f\|.$$

The minimum value of $f_{\vec{u}}$ occurs when $\cos\theta = -1$, so $\theta = \pi$ and $\vec{u}$ is pointing in the direction opposite to grad f. Then

$$\text{Minimum } f_{\vec{u}} = \| \text{grad } f\| \cos \pi = -\| \text{grad } f\|.$$

When $\theta = \pi/2$ or $3\pi/2$, so $\cos\theta = 0$, the directional derivative is zero.

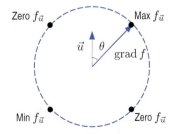

Figure 14.27: Values of the directional derivative at different angles to the gradient

Properties of The Gradient Vector

We have seen that the gradient vector points in the direction of the greatest rate of change at a point and the magnitude of the gradient vector is that rate of change.

Figure 14.28 shows that the gradient vector at any point is perpendicular to the contour through that point. Assuming f is differentiable at the point (a, b), local linearity tells us that the contours of f around the point (a, b) appear straight, parallel, and equally spaced. The greatest rate of change is obtained by moving in the direction that takes us to the next contour in the shortest possible distance, which is the direction perpendicular to the contour. Thus, we have the following:

Geometric Properties of the Gradient Vector in the Plane

If f is a differentiable function at the point (a, b) and grad $f(a, b) \neq \vec{0}$, then:
- The direction of grad $f(a, b)$ is
  - Perpendicular to the contour of f through (a, b)
  - In the direction of increasing f
- The magnitude of the gradient vector, $\| \text{grad } f \|$, is
  - The maximum rate of change of f at that point
  - Large when the contours are close together and small when they are far apart.

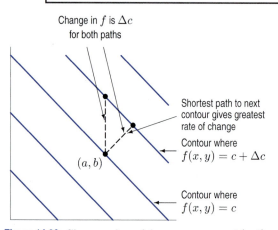

Figure 14.28: Close-up view of the contours around (a, b), showing the gradient is perpendicular to the contours

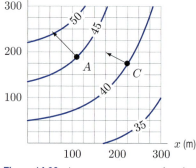
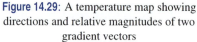

Figure 14.29: A temperature map showing directions and relative magnitudes of two gradient vectors

Examples of Directional Derivatives and Gradient Vectors

Example 6   Explain why the gradient vectors at points A and C in Figure 14.29 have the direction and the relative magnitudes they do.

Solution   The gradient vector points in the direction of greatest increase of the function. This means that in Figure 14.29, the gradient points directly towards warmer temperatures. The magnitude of the gradient vector measures the rate of change. The gradient vector at A is longer than the gradient vector at C because the contours are closer together at A, so the rate of change is larger.

Example 2 on page 659 shows how the contour diagram can tell us the sign of the directional derivative. In the next example we compute the directional derivative in three directions, two close to that of the gradient vector and one that is not.

Example 7   Use the gradient to find the directional derivative of $f(x, y) = x + e^y$ at the point $(1, 1)$ in the direction of the vectors $\vec{i} - \vec{j}, \vec{i} + 2\vec{j}, \vec{i} + 3\vec{j}$.

Solution   In Example 5 we found

$$\operatorname{grad} f(1, 1) = \vec{i} + e\vec{j}.$$

A unit vector in the direction of $\vec{i} - \vec{j}$ is $\vec{s} = (\vec{i} - \vec{j})/\sqrt{2}$, so

$$f_{\vec{s}}(1, 1) = \operatorname{grad} f(1, 1) \cdot \vec{s} = (\vec{i} + e\vec{j}) \cdot \left(\frac{\vec{i} - \vec{j}}{\sqrt{2}} \right) = \frac{1 - e}{\sqrt{2}} \approx -1.215.$$

A unit vector in the direction of $\vec{i} + 2\vec{j}$ is $\vec{v} = (\vec{i} + 2\vec{j})/\sqrt{5}$, so

$$f_{\vec{v}}(1, 1) = \operatorname{grad} f(1, 1) \cdot \vec{v} = (\vec{i} + e\vec{j}) \cdot \left(\frac{\vec{i} + 2\vec{j}}{\sqrt{5}} \right) = \frac{1 + 2e}{\sqrt{5}} \approx 2.879.$$

A unit vector in the direction of $\vec{i} + 3\vec{j}$ is $\vec{w} = (\vec{i} + 3\vec{j})/\sqrt{10}$, so

$$f_{\vec{w}}(1, 1) = \operatorname{grad} f(1, 1) \cdot \vec{w} = (\vec{i} + e\vec{j}) \cdot \left(\frac{\vec{i} + 3\vec{j}}{\sqrt{10}} \right) = \frac{1 + 3e}{\sqrt{10}} \approx 2.895.$$

Now look back at the answers and compare with the value of $\| \operatorname{grad} f \| = \sqrt{1 + e^2} \approx 2.896$. One answer is not close to this value; the other two, $f_{\vec{v}} = 2.879$ and $f_{\vec{w}} = 2.895$, are close but slightly smaller than $\| \operatorname{grad} f \|$. Since $\| \operatorname{grad} f \|$ is the maximum rate of change of f at the point, we would expect for *any* unit vector $\vec{u}$:

$$f_{\vec{u}}(1, 1) \leq \| \operatorname{grad} f \|.$$

with equality when $\vec{u}$ is in the direction of $\operatorname{grad} f$. Since $e \approx 2.718$, the vectors $\vec{i} + 2\vec{j}$ and $\vec{i} + 3\vec{j}$ both point roughly, but not exactly, in the direction of the gradient vector $\operatorname{grad} f(1, 1) = \vec{i} + e\vec{j}$. Thus, the values of $f_{\vec{v}}$ and $f_{\vec{w}}$ are both close to the value of $\| \operatorname{grad} f \|$. The direction of the vector $\vec{i} - \vec{j}$ is not close to the direction of $\operatorname{grad} f$ and the value of $f_{\vec{s}}$ is not close to the value of $\| \operatorname{grad} f \|$.

Exercises and Problems for Section 14.4

Exercises

In Exercises 1–14 find the gradient of the given function. Assume the variables are restricted to a domain on which the function is defined.

1. $f(x, y) = \frac{3}{2}x^5 - \frac{4}{7}y^6$

2. $z = xe^y$

3. $z = (x + y)e^y$

4. $f(m, n) = m^2 + n^2$

5. $f(x, y) = \sqrt{x^2 + y^2}$

6. $f(K, L) = K^{0.3}L^{0.7}$

7. $f(r, \theta) = r\sin\theta$

8. $f(r, h) = \pi r^2 h$

9. $z = \sin(x^2 + y^2)$

10. $z = \sin(x/y)$

11. $z = \tan^{-1}(x/y)$

12. $f(\alpha, \beta) = \dfrac{2\alpha + 3\beta}{2\alpha - 3\beta}$

13. $z = x\dfrac{e^y}{x + y}$

14. $f(K, L) = 50K + 100L$

In Exercises 15–20, find the gradient at the given point.

15. $f(m, n) = 5m^2 + 3n^4$, at $(5, 2)$

16. $f(x, y) = x^2 y + 7xy^3$, at $(1, 2)$

17. $f(r, h) = 2\pi rh + \pi r^2$, at $(2, 3)$

18. $f(x, y) = 1/(x^2 + y^2)$, at $(-1, 3)$

19. $f(x, y) = \sqrt{\tan x + y}$, at $(0, 1)$

20. $f(x, y) = \sin(x^2) + \cos y$, at $(\frac{\sqrt{\pi}}{2}, 0)$

In Exercises 21–26, use the contour diagram of f in Figure 14.30 to decide if the specified directional derivative is positive, negative, or approximately zero.

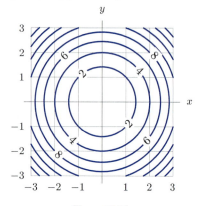

Figure 14.30

21. At point $(-2, 2)$, in direction $\vec{i}$.

22. At point $(-1, 1)$, in direction $-\vec{i} + \vec{j}$.

23. At point $(-1, 1)$, in direction $\vec{i} + \vec{j}$.

24. At point $(0, -2)$, in direction $\vec{j}$.

25. At point $(0, -2)$, in direction $\vec{i} + 2\vec{j}$.

26. At point $(0, -2)$, in direction $\vec{i} - 2\vec{j}$.

In Exercises 27–34, use the contour diagram of f in Figure 14.30 to find the approximate direction of the gradient vector at the given point.

27. $(2, 0)$ **28.** $(0, 2)$ **29.** $(-2, 0)$ **30.** $(0, -2)$

31. $(2, 2)$ **32.** $(2, -2)$ **33.** $(-2, 2)$ **34.** $(-2, -2)$

In Exercises 35–38, find the directional derivative $f_{\vec{u}}(1, 2)$ for the function f with $\vec{u} = (3\vec{i} - 4\vec{j})/5$.

35. $f(x, y) = 3x - 4y$ **36.** $f(x, y) = xy + y^3$

37. $f(x, y) = \sin(2x - y)$ **38.** $f(x, y) = x^2 - y^2$

Problems

For Problems 39–42 use Figure 14.31, showing level curves of $f(x, y)$, to estimate the directional derivatives.

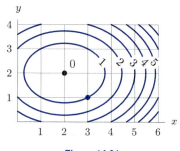

Figure 14.31

39. $f_{\vec{i}}(3, 1)$ **40.** $f_{\vec{j}}(3, 1)$

41. $f_{\vec{u}}(3, 1)$ where $\vec{u} = (\vec{i} - \vec{j})/\sqrt{2}$

42. $f_{\vec{u}}(3, 1)$ where $\vec{u} = (-\vec{i} + \vec{j})/\sqrt{2}$

In Problems 43–47, use the contour diagram for $f(x, y)$ in Figure 14.32 to estimate the directional derivative of $f(x, y)$ in the direction $\vec{v}$ at the point given.

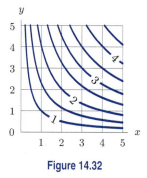

Figure 14.32

43. $\vec{v} = \vec{i}$ at $(1, 1)$ **44.** $\vec{v} = \vec{j}$ at $(1, 1)$

45. $\vec{v} = \vec{i} + \vec{j}$ at $(1, 1)$ **46.** $\vec{v} = -2\vec{i} + \vec{j}$ at $(3, 3)$

47. $\vec{v} = -2\vec{i} + \vec{j}$ at $(4, 1)$

48. Figure 14.32 shows the level curves of $f(x, y)$. At the points $(1, 1)$ and $(1, 4)$, draw a vector representing grad f. Explain how you know the direction and length of each vector.

49. The surface $z = g(x, y)$ is in Figure 14.33. What is the sign of each of the following directional derivatives?

(a) $g_{\vec{u}}(2, 5)$ where $\vec{u} = (\vec{i} - \vec{j})/\sqrt{2}$.
(b) $g_{\vec{u}}(2, 5)$ where $\vec{u} = (\vec{i} + \vec{j})/\sqrt{2}$.

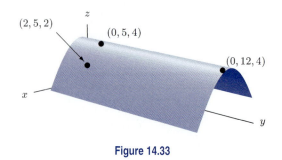

Figure 14.33

50. Let $f(x, y) = x^2 y^3$. At the point $(-1, 2)$, find a vector

(a) In the direction of maximum rate of change.
(b) In the direction of minimum rate of change.
(c) In a direction in which the rate of change is zero.

51. (a) What is the rate of change of $f(x,y) = 3xy + y^2$ at the point $(2,3)$ in the direction $\vec{v} = 3\vec{i} - \vec{j}$?

(b) What is the direction of maximum rate of change of f at $(2,3)$?

(c) What is the maximum rate of change?

52. (a) Let $f(x,y) = x^2 + \ln y$. Find the average rate of change of f as you go from $(3,1)$ to $(1,2)$.

(b) Find the instantaneous rate of change of f as you leave the point $(3,1)$ heading toward $(1,2)$.

53. (a) Let $f(x,y) = (x+y)/(1+x^2)$. Find the directional derivative of f at $P = (1,-2)$ in the direction of:

   (i) $\vec{v} = 3\vec{i} - 2\vec{j}$ (ii) $\vec{v} = -\vec{i} + 4\vec{j}$

(b) What is the direction of greatest increase of f at P ?

54. Find the directional derivative of $z = x^2 y$ at $(1,2)$ in the direction making an angle of $5\pi/4$ with the x-axis. In which direction is the directional derivative the largest?

55. The vector ∇f at point P and four unit vectors $\vec{u}_1, \vec{u}_2, \vec{u}_3, \vec{u}_4$ are shown in Figure 14.34. Arrange the following quantities in ascending order

$$f_{\vec{u}_1}, \quad f_{\vec{u}_2}, \quad f_{\vec{u}_3}, \quad f_{\vec{u}_4}, \quad \text{the number } 0.$$

The directional derivatives are all evaluated at the point P and the function $f(x,y)$ is differentiable at P.

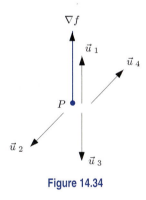

Figure 14.34

56. At a certain point on a heated plate, the greatest rate of temperature increase, $5°$ C per meter, is toward the northeast. If an object at this point moves directly north, at what rate is the temperature increasing?

57. You are climbing a mountain by the steepest route at a slope of $20°$ when you come upon a trail branching off at a $30°$ angle from yours. What is the angle of ascent of the branch trail?

58. The directional derivative of $z = f(x,y)$ at $(2,1)$ in the direction toward the point $(1,3)$ is $-2/\sqrt{5}$, and the directional derivative in the direction toward the point $(5,5)$ is 1. Compute $\partial z/\partial x$ and $\partial z/\partial y$ at $(2,1)$.

59. Consider the function $f(x,y)$. If you start at the point $(4,5)$ and move toward the point $(5,6)$, the directional derivative is 2. Starting at the point $(4,5)$ and moving toward the point $(6,6)$ gives a directional derivative of 3. Find ∇f at the point $(4,5)$.

In Problems 60–63, check that the point $(2,3)$ lies on the given curve. Then, viewing the curve as a contour for a function $f(x,y)$, use grad $f(2,3)$ to find a vector normal to the curve at $(2,3)$ and an equation for the tangent line to the curve at $(2,3)$.

60. $x^2 + y^2 = 13$

61. $xy = 6$

62. $y = x^2 - 1$

63. $(y-x)^2 + 2 = xy - 3$

64. A differentiable function $f(x,y)$ has $f_x(4,1) = 2$ and $f_y(4,1) = -1$. Find the equation of the tangent line to the contour of f through the point $(4,1)$.

65. The table gives values of a differentiable function $f(x,y)$. At the point $(1.2, 0)$, into which quadrant does the gradient vector of f point? Justify your answer.

		y		
		-1	0	1
	1.0	0.7	0.1	-0.5
x	1.2	4.8	4.2	3.6
	1.4	8.9	8.3	7.7

66. The temperature at any point in the plane is given by the function

$$T(x,y) = \frac{100}{x^2 + y^2 + 1}.$$

(a) What shape are the level curves of T?

(b) Where on the plane is it hottest? What is the temperature at that point?

(c) Find the direction of the greatest increase in temperature at the point $(3,2)$. What is the magnitude of that greatest increase?

(d) Find the direction of the greatest decrease in temperature at the point $(3,2)$.

(e) Find a direction at the point $(3,2)$ in which the temperature does not increase or decrease.

67. Figure 14.35 represents the level curves $f(x,y) = c$; the values of f on each curve are marked. In each of the following parts, decide whether the given quantity is positive, negative or zero. Explain your answer.

(a) The value of $\nabla f \cdot \vec{i}$ at P.

(b) The value of $\nabla f \cdot \vec{j}$ at P.

(c) $\partial f/\partial x$ at Q.

(d) $\partial f/\partial y$ at Q.

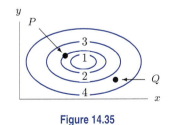

Figure 14.35

68. In Figure 14.35, which is larger: $\|\nabla f\|$ at P or $\|\nabla f\|$ at Q? Explain how you know.

69. Figure 14.37 is a graph of the directional derivative, $f_{\vec{u}}$, at the point (a, b) versus θ, the angle in Figure 14.36.

  (a) Which points on the graph in Figure 14.37 correspond to the greatest rate of increase of f? The greatest rate of decrease?

  (b) Mark points on the circle in Figure 14.36 corresponding to the points P, Q, R, S.

  (c) What is the amplitude of the function graphed in Figure 14.37? What is its formula?

approximated by the difference quotient

$$\frac{\text{Change in } f \text{ between } P \text{ and } Q}{h}.$$

  (a) Use the gradient to show that

$$\text{Change in } f \approx \| \operatorname{grad} f \| (h \cos \theta).$$

  (b) Use part (a) to obtain $f_{\vec{u}}(a, b) = \operatorname{grad} f(a, b) \cdot \vec{u}$.

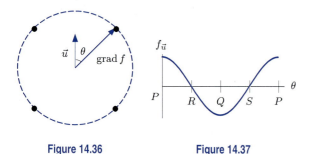

Figure 14.36   **Figure 14.37**

70. In this problem we see another way of obtaining the formula $f_{\vec{u}}(a, b) = \operatorname{grad} f(a, b) \cdot \vec{u}$. Imagine zooming in on a function $f(x, y)$ at a point (a, b). By local linearity, the contours around (a, b) look like the contours of a linear function. See Figure 14.38. Suppose you want to find the directional derivative $f_{\vec{u}}(a, b)$ in the direction of a unit vector $\vec{u}$. If you move from P to Q, a small distance h in the direction of $\vec{u}$, then the directional derivative is

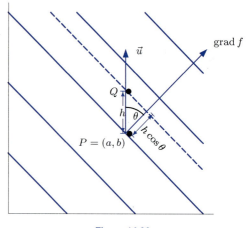

Figure 14.38

14.5 GRADIENTS AND DIRECTIONAL DERIVATIVES IN SPACE

Directional Derivatives of Functions of Three Variables

We calculate directional derivatives of a function of three variables in the same way as for a function of two variables. If the function f is differentiable at the point (a, b, c), then the rate of change of $f(x, y, z)$ at the point (a, b, c) in the direction of a unit vector $\vec{u} = u_1 \vec{i} + u_2 \vec{j} + u_3 \vec{k}$ is

$$f_{\vec{u}}(a, b, c) = f_x(a, b, c)u_1 + f_y(a, b, c)u_2 + f_z(a, b, c)u_3.$$

This can be justified using local linearity in the same way as for functions of two variables.

Example 1   Find the directional derivative of $f(x, y, z) = xy + z$ at the point $(-1, 0, 1)$ in the direction of the vector $\vec{v} = 2\vec{i} + \vec{k}$.

Solution   The magnitude of $\vec{v}$ is $\| \vec{v} \| = \sqrt{2^2 + 1} = \sqrt{5}$, so a unit vector in the same direction as $\vec{v}$ is

$$\vec{u} = \frac{\vec{v}}{\| \vec{v} \|} = \frac{2}{\sqrt{5}} \vec{i} + 0\vec{j} + \frac{1}{\sqrt{5}} \vec{k}.$$

The partial derivatives of f are $f_x(x, y, z) = y$ and $f_y(x, y, z) = x$ and $f_z(x, y, z) = 1$. Thus,

$$f_{\vec{u}}(-1, 0, 1) = f_x(-1, 0, 1)u_1 + f_y(-1, 0, 1)u_2 + f_z(-1, 0, 1)u_3$$

$$= (0)\left(\frac{2}{\sqrt{5}}\right) + (-1)(0) + (1)\left(\frac{1}{\sqrt{5}}\right) = \frac{1}{\sqrt{5}}.$$

The Gradient Vector of a Function of Three Variables

The gradient of a function of three variables is defined in the same way as for two variables:

$$\text{grad } f(a, b, c) = f_x(a, b, c)\vec{i} + f_y(a, b, c)\vec{j} + f_z(a, b, c)\vec{k}.$$

Directional derivatives are related to gradients in the same way as for functions of two variables:

$$f_{\vec{u}}(a, b, c) = f_x(a, b, c)u_1 + f_y(a, b, c)u_2 + f_z(a, b, c)u_3 = \text{grad } f(a, b, c) \cdot \vec{u}.$$

Since $\text{grad } f(a, b, c) \cdot \vec{u} = \| \text{grad } f(a, b, c) \| \cos\theta$, where θ is the angle between $\text{grad } f(a, b, c)$ and $\vec{u}$, the value of $f_{\vec{u}}(a, b, c)$ is largest when $\theta = 0$, that is, when $\vec{u}$ is in the same direction as $\text{grad } f(a, b, c)$. In addition, $f_{\vec{u}}(a, b, c) = 0$ when $\theta = \pi/2$, so $\text{grad } f(a, b, c)$ is perpendicular to the level surface of f. The properties of gradients in space are similar to those in the plane:

Properties of the Gradient Vector in Space

If f is differentiable at (a, b, c) and $\vec{u}$ is a unit vector, then

$$f_{\vec{u}}(a, b, c) = \text{grad } f(a, b, c) \cdot \vec{u}.$$

If, in addition, $\text{grad } f(a, b, c) \neq \vec{0}$, then
- $\text{grad } f(a, b, c)$ is in the direction of the gratest rate of increase of f
- $\text{grad } f(a, b, c)$ is perpendicular to the level surface of f at (a, b, c)
- $\| \text{grad } f(a, b, c) \|$ is the maximum rate of change of f at (a, b, c).

Example 2   Let $f(x, y, z) = x^2 + y^2$ and $g(x, y, z) = -x^2 - y^2 - z^2$. What can we say about the direction of the following vectors?

(a) $\text{grad } f(0, 1, 1)$ (b) $\text{grad } f(1, 0, 1)$ (c) $\text{grad } g(0, 1, 1)$ (d) $\text{grad } g(1, 0, 1)$.

Solution   The cylinder $x^2 + y^2 = 1$ in Figure 14.39 is a level surface of f and contains both the points $(0, 1, 1)$ and $(1, 0, 1)$. Since the value of f does not change at all in the z-direction, all the gradient vectors are horizontal. They are perpendicular to the cylinder and point outward because the value of f increases as we move out.

Similarly, the points $(0, 1, 1)$ and $(1, 0, 1)$ also lie on the same level surface of g, namely $g(x, y, z) = -x^2 - y^2 - z^2 = -2$, which is the sphere $x^2 + y^2 + z^2 = 2$. Part of this level surface is shown in Figure 14.40. This time the gradient vectors point inward, since the negative signs mean that the function increases (from large negative values to small negative values) as we move inward.

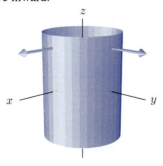

Figure 14.39: The level surface $f(x, y, z) = x^2 + y^2 = 1$ with two gradient vectors

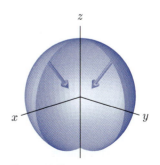

Figure 14.40: The level surface $g(x, y, z) = -x^2 - y^2 - z^2 = -2$ with two gradient vectors

Example 3   Consider the functions $f(x, y) = 4 - x^2 - 2y^2$ and $g(x, y) = 4 - x^2$. Calculate a vector perpendicular to each of the following:

(a)   The level curve of f at the point $(1, 1)$ (b)   The surface $z = f(x, y)$ at the point $(1, 1, 1)$
(c)   The level curve of g at the point $(1, 1)$ (d)   The surface $z = g(x, y)$ at the point $(1, 1, 3)$

Solution   (a) The vector we want is a 2-vector in the plane. Since grad $f = -2x\vec{i} - 4y\vec{j}$, we have

$$\text{grad } f(1, 1) = -2\vec{i} - 4\vec{j}.$$

Any nonzero multiple of this vector is perpendicular to the level curve at the point $(1, 1)$.

(b) In this case we want a 3-vector in space. To find it we rewrite $z = 4 - x^2 - 2y^2$ as the level surface of the function F, where

$$F(x, y, z) = 4 - x^2 - 2y^2 - z = 0.$$

Then

$$\text{grad } F = -2x\vec{i} - 4y\vec{j} - \vec{k},$$

so

$$\text{grad } F(1, 1, 1) = -2\vec{i} - 4\vec{j} - \vec{k},$$

and grad $F(1, 1, 1)$ is perpendicular to the surface $z = 4 - x^2 - 2y^2$ at the point $(1, 1, 1)$. Notice that $-2\vec{i} - 4\vec{j} - \vec{k}$ is not the only possible answer: any multiple of this vector will do.

(c) We are looking for a 2-vector. Since grad $g = -2x\vec{i} + 0\vec{j}$, we have

$$\text{grad } g(1, 1) = -2\vec{i}.$$

Any multiple of this vector is perpendicular to the level curve also.

(d) We are looking for a 3-vector. We rewrite $z = 4 - x^2$ as the level surface of the function G, where

$$G(x, y, z) = 4 - x^2 - z = 0.$$

Then

$$\text{grad } G = -2x\vec{i} - \vec{k}$$

So

$$\text{grad } G(1, 1, 3) = -2\vec{i} - \vec{k},$$

and any multiple of grad $G(1, 1, 3)$ is perpendicular to the surface $z = 4 - x^2$ at this point.

Example 4   (a) A hiker on the surface $f(x, y) = 4 - x^2 - 2y^2$ at the point $(1, -1, 1)$ starts to climb along the path of steepest ascent. What is the relation between the vector grad $f(1, -1)$ and a vector tangent to the path at the point $(1, -1, 1)$ and pointing uphill?

(b) Consider the surface $g(x, y) = 4 - x^2$. What is the relation between grad $g(-1, -1)$ and a vector tangent to the path of steepest ascent at $(-1, -1, 3)$?

(c) At the point $(1, -1, 1)$ on the surface $f(x, y) = 4 - x^2 - 2y^2$, calculate a vector perpendicular to the surface and a vector, $\vec{T}$, tangent to the curve of steepest ascent.

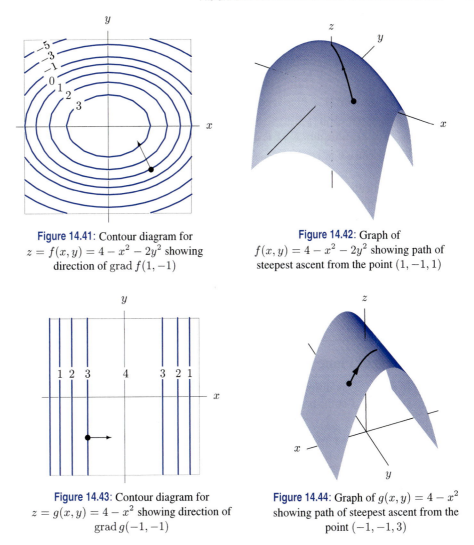

Figure 14.41: Contour diagram for $z = f(x, y) = 4 - x^2 - 2y^2$ showing direction of grad $f(1, -1)$

Figure 14.42: Graph of $f(x, y) = 4 - x^2 - 2y^2$ showing path of steepest ascent from the point $(1, -1, 1)$

Figure 14.43: Contour diagram for $z = g(x, y) = 4 - x^2$ showing direction of grad $g(-1, -1)$

Figure 14.44: Graph of $g(x, y) = 4 - x^2$ showing path of steepest ascent from the point $(-1, -1, 3)$

Solution

(a) The hiker at the point $(1, -1, 1)$ lies directly above the point $(1, -1)$ in the xy-plane. The vector grad $f(1, -1)$ lies in 2-space, pointing like a compass in the direction in which f increases most rapidly. Therefore, grad $f(1, -1)$ lies directly under a vector tangent to the hiker's path at $(1, -1, 1)$ and pointing uphill. (See Figures 14.41 and 14.42.)

(b) The point $(-1, -1, 3)$ lies above the point $(-1, -1)$. The vector grad $g(-1, -1)$ points in the direction in which g increases most rapidly and lies directly under the path of steepest ascent. (See Figures 14.43 and 14.44.)

(c) The surface is represented by $F(x, y, z) = 4 - x^2 - 2y^2 - z = 0$. Since grad $F = -2x\vec{i} - 4y\vec{j} - \vec{k}$, the normal, $\vec{N}$, to the surface is given by

$$\vec{N} = \text{grad } F(1, -1, 1) = -2(1)\vec{i} - 4(-1)\vec{j} - \vec{k} = -2\vec{i} + 4\vec{j} - \vec{k}.$$

We take the $\vec{i}$ and $\vec{j}$ components of $\vec{T}$ to be the vector grad $f(1, -1) = -2\vec{i} + 4\vec{j}$. Thus we have that, for some $a > 0$,

$$\vec{T} = -2\vec{i} + 4\vec{j} + a\vec{k}$$

We want $\vec{N} \cdot \vec{T} = 0$, so

$$\vec{N} \cdot \vec{T} = (-2\vec{i} + 4\vec{j} - \vec{k}) \cdot (-2\vec{i} + 4\vec{j} + a\vec{k}) = 4 + 16 - a = 0$$

So $a = 20$ and hence

$$\vec{T} = -2\vec{i} + 4\vec{j} + 20\vec{k}.$$

Comparing with the formula for dU obtained from U_1:

$$dU = \left(\frac{\partial U}{\partial T}\right)_P dT + \left(\frac{\partial U}{\partial P}\right)_T dP,$$

we have

$$\left(\frac{\partial U}{\partial T}\right)_P = \frac{16}{3} \quad \text{and} \quad \left(\frac{\partial U}{\partial P}\right)_T = -\frac{11}{3}.$$

In Example 6, we could have substituted for dP instead of dV, leading to values of $\left(\frac{\partial U}{\partial T}\right)_V$ and $\left(\frac{\partial U}{\partial V}\right)_T$. See Problem 27.

In general, if for some particular P, V, and T, we can measure two of the six quantities $\left(\frac{\partial U}{\partial P}\right)_V$, $\left(\frac{\partial U}{\partial V}\right)_P$, $\left(\frac{\partial U}{\partial P}\right)_T$, $\left(\frac{\partial U}{\partial T}\right)_P$, $\left(\frac{\partial U}{\partial V}\right)_T$, $\left(\frac{\partial U}{\partial T}\right)_V$, then we can compute the other four using the relationship between dP, dV, and dT given by the gas law.

General formulas for each partial derivative in terms of others can be obtained in the same way. See the following example and Problem 27.

Example 7   Express $\left(\frac{\partial U}{\partial T}\right)_P$ in terms of $\left(\frac{\partial U}{\partial T}\right)_V$ and $\left(\frac{\partial U}{\partial V}\right)_T$ and $\left(\frac{\partial V}{\partial T}\right)_P$

Solution   Since we are interested in the derivatives $\left(\frac{\partial U}{\partial T}\right)_V$ and $\left(\frac{\partial U}{\partial V}\right)_T$, we think of U as a function of T and V and use the formula

$$dU = \left(\frac{\partial U}{\partial T}\right)_V dT + \left(\frac{\partial U}{\partial V}\right)_T dV \qquad \text{corresponding to } U_2.$$

We want to find a formula for $\left(\frac{\partial U}{\partial T}\right)_P$, which means thinking of U as a function of T and P. Thus, we want to substitute for dV. Since V is a function of T and P, we have

$$dV = \left(\frac{\partial V}{\partial T}\right)_P dT + \left(\frac{\partial V}{\partial P}\right)_T dP.$$

Substituting for dV into the formula for dU corresponding to U_2 gives

$$dU = \left(\frac{\partial U}{\partial T}\right)_V dT + \left(\frac{\partial U}{\partial V}\right)_T \left(\left(\frac{\partial V}{\partial T}\right)_P dT + \left(\frac{\partial V}{\partial P}\right)_T dP\right).$$

Collecting the terms containing dT and the terms containing dP gives

$$dU = \left(\left(\frac{\partial U}{\partial T}\right)_V + \left(\frac{\partial U}{\partial V}\right)_T \left(\frac{\partial V}{\partial T}\right)_P\right) dT + \left(\frac{\partial U}{\partial V}\right)_T \left(\frac{\partial V}{\partial P}\right)_T dP.$$

But we also have the formula

$$dU = \left(\frac{\partial U}{\partial T}\right)_P dT + \left(\frac{\partial U}{\partial P}\right)_T dP \qquad \text{corresponding to } U_1.$$

We now have two formulas for dU in terms of dT and dP. The coefficients of dT must be identical, so we conclude

$$\left(\frac{\partial U}{\partial T}\right)_P = \left(\frac{\partial U}{\partial T}\right)_V + \left(\frac{\partial U}{\partial V}\right)_T \left(\frac{\partial V}{\partial T}\right)_P.$$

Example 7 expresses $\left(\frac{\partial U}{\partial T}\right)_P$ in terms of three other partial derivatives. Two of them, namely $\left(\frac{\partial U}{\partial T}\right)_V$, the constant-volume heat capacity, and $\left(\frac{\partial V}{\partial T}\right)_P$, the expansion coefficient, can be easily measured experimentally. The third, the internal pressure, $\left(\frac{\partial U}{\partial V}\right)_T$, cannot be measured directly but can be related to $\left(\frac{\partial P}{\partial T}\right)_V$, which is measurable. Thus, $\left(\frac{\partial U}{\partial T}\right)_P$ can be determined indirectly using this identity.

Exercises and Problems for Section 14.6

Exercises

For Exercises 1–6, find dz/dt using the chain rule. Assume the variables are restricted to domains on which the functions are defined.

1. $z = xy^2$, $x = e^{-t}$, $y = \sin t$

2. $z = x \sin y + y \sin x$, $x = t^2$, $y = \ln t$

3. $z = \ln(x^2 + y^2)$, $x = 1/t$, $y = \sqrt{t}$

4. $z = \sin(x/y)$, $x = 2t$, $y = 1 - t^2$

5. $z = xe^y$, $x = 2t$, $y = 1 - t^2$

6. $z = (x + y)e^y$, $x = 2t$, $y = 1 - t^2$

For Exercises 7–14, find $\partial z/\partial u$ and $\partial z/\partial v$. The variables are restricted to domains on which the functions are defined.

7. $z = xe^{-y} + ye^{-x}$, $x = u \sin v$, $y = v \cos u$

8. $z = \cos(x^2 + y^2)$, $x = u \cos v$, $y = u \sin v$

9. $z = xe^y$, $x = \ln u$, $y = v$

10. $z = (x + y)e^y$, $x = \ln u$, $y = v$

11. $z = xe^y$, $x = u^2 + v^2$, $y = u^2 - v^2$

12. $z = (x + y)e^y$, $x = u^2 + v^2$, $y = u^2 - v^2$

13. $z = \sin(x/y)$, $x = \ln u$, $y = v$

14. $z = \tan^{-1}(x/y)$, $x = u^2 + v^2$, $y = u^2 - v^2$

Problems

15. Suppose $w = f(x, y, z)$ and that x, y, z are functions of u and v. Use a tree diagram to write down the chain rule formula for $\partial w/\partial u$ and $\partial w/\partial v$.

16. Suppose $w = f(x, y, z)$ and that x, y, z are all functions of t. Use a tree diagram to write down the chain rule for dw/dt.

17. Corn production, C, is a function of rainfall, R, and temperature, T. Figures 14.49 and 14.50 show how rainfall and temperature are predicted to vary with time because of global warming. Suppose we know that $\Delta C \approx 3.3\Delta R - 5\Delta T$. Use this to estimate the change in corn production between the year 2020 and the year 2021. Hence, estimate dC/dt when $t = 2020$.

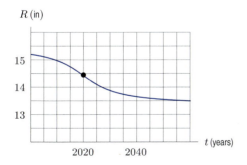

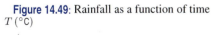

Figure 14.49: Rainfall as a function of time

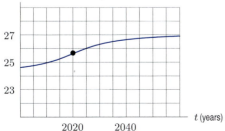

Figure 14.50: Temperature as a function of time

18. The voltage, V, (in volts) across a circuit is given by Ohm's law: $V = IR$, where I is the current (in amps) flowing through the circuit and R is the resistance (in ohms). If we place two circuits, with resistance R_1 and R_2, in parallel, then their combined resistance, R, is given by

$$\frac{1}{R} = \frac{1}{R_1} + \frac{1}{R_2}.$$

Suppose the current is 2 amps and increasing at 10^{-2} amp/sec and R_1 is 3 ohms and increasing at 0.5 ohm/sec, while R_2 is 5 ohms and decreasing at 0.1 ohm/sec. Calculate the rate at which the voltage is changing.

19. Air pressure decreases at a rate of 2 pascals per kilometer in the eastward direction. In addition, the air pressure is dropping at a constant rate with respect to time everywhere. A ship sailing eastward at 10 km/hour past an island takes barometer readings and records a pressure drop of 50 pascals in 2 hours. Estimate the time rate of change of air pressure on the island. (A pascal is a unit of air pressure.)

20. Given $z = f(x, y)$, $x = x(u, v)$, $y = y(u, v)$ and $x(1, 2) = 5$, $y(1, 2) = 3$, calculate $z_u(1, 2)$ in terms of some of the numbers a, b, c, d, e, k, p, q, where

$$\begin{array}{llll} f_x(1,2) = a & f_y(1,2) = c & x_u(1,2) = e & y_u(1,2) = p \\ f_x(5,3) = b & f_y(5,3) = d & x_v(1,2) = k & y_v(1,2) = q \end{array}$$

21. Let $F(u, v)$ be a function of two variables. Find $f'(x)$ if

    (a) $f(x) = F(x, 3)$ (b) $f(x) = F(3, x)$

    (c) $f(x) = F(x, x)$ (d) $f(x) = F(5x, x^2)$

22. Let $F(u, v, w)$ be a function of three variables. Find $G_x(x, y)$ if

    (a) $G(x, y) = F(x, y, 3)$ (b) $G(x, y) = F(3, y, x)$

    (c) $G(x, y) = F(x, y, x)$ (d) $G(x, y) = F(x, y, xy)$

For Problems 23–24, suppose the quantity z can be expressed either as a function of Cartesian coordinates (x, y) or as a function of polar coordinates (r, θ), so that $z = f(x, y) = g(r, \theta)$. [Recall that $x = r\cos\theta, y = r\sin\theta$ and $r = \sqrt{x^2 + y^2}, \theta = \arctan(y/x)$]

23. (a) Use the chain rule to find $\partial z/\partial r$ and $\partial z/\partial\theta$ in terms of $\partial z/\partial x$ and $\partial z/\partial y$.
   (b) Solve the equations you have just written down for $\partial z/\partial x$ and $\partial z/\partial y$ in terms of $\partial z/\partial r$ and $\partial z/\partial\theta$.
   (c) Show that the expressions you get in part (b) are the same as you would get by using the chain rule to find $\partial z/\partial x$ and $\partial z/\partial y$ in terms of $\partial z/\partial r$ and $\partial z/\partial\theta$.

24. Show that

$$\left(\frac{\partial z}{\partial x}\right)^2 + \left(\frac{\partial z}{\partial y}\right)^2 = \left(\frac{\partial z}{\partial r}\right)^2 + \frac{1}{r^2}\left(\frac{\partial z}{\partial\theta}\right)^2.$$

Problems 25–31 are continuations of the physical chemistry example on page 678.

25. Write $\left(\frac{\partial U}{\partial P}\right)_V$ as a partial derivative of one of the functions $U_1, U_2,$ or U_3.

26. Write $\left(\frac{\partial U}{\partial P}\right)_T$ as a partial derivative of one of the functions U_1, U_2, U_3.

27. For the gas in Example 6, find $\left(\frac{\partial U}{\partial T}\right)_V$ and $\left(\frac{\partial U}{\partial V}\right)_T$. [Hint: Use the same method as the example, but substitute for dP instead of dV.]

28. Show that $\left(\frac{\partial T}{\partial V}\right)_P = 1 \Big/ \left(\frac{\partial V}{\partial T}\right)_P$.

29. Express $\left(\frac{\partial U}{\partial P}\right)_T$ in terms of $\left(\frac{\partial U}{\partial V}\right)_T$ and $\left(\frac{\partial V}{\partial P}\right)_T$.

30. Use Example 7 and Problem 28 to show that

$$\left(\frac{\partial U}{\partial V}\right)_P = \left(\frac{\partial U}{\partial V}\right)_T + \frac{\left(\frac{\partial U}{\partial T}\right)_V}{\left(\frac{\partial V}{\partial T}\right)_P}.$$

31. In Example 6, we calculated values of $(\partial U/\partial T)_P$ and $(\partial U/\partial P)_T$ using the relationship $PV = 2T$ for a specific gas. In this problem, you will derive general relationships for these two partial derivatives.

   (a) Think of V as a function of P and T and write an expression for dV.
   (b) Substitute for dV into the following formula for dU (obtained by thinking of U as a function of P and V):

$$dU = \left(\frac{\partial U}{\partial P}\right)_V dP + \left(\frac{\partial U}{\partial V}\right)_P dV.$$

   (c) Thinking of U as a function of P and T, write an expression for dU.
   (d) By comparing coefficients of dP and dT in your answers to parts (b) and (c), show that

$$\left(\frac{\partial U}{\partial T}\right)_P = \left(\frac{\partial U}{\partial V}\right)_P \cdot \left(\frac{\partial V}{\partial T}\right)_P$$
$$\left(\frac{\partial U}{\partial P}\right)_T = \left(\frac{\partial U}{\partial P}\right)_V + \left(\frac{\partial U}{\partial V}\right)_P \cdot \left(\frac{\partial V}{\partial P}\right)_T.$$

32. A function $f(x, y)$ is *homogeneous of degree* p if $f(tx, ty) = t^p f(x, y)$ for all t. Show that any differentiable, homogeneous function of degree p satisfies Euler's Theorem:

$$x f_x(x, y) + y f_y(x, y) = p f(x, y).$$

[Hint: Define $g(t) = f(tx, ty)$ and compute $g'(1)$.]

33. Let $F(x, y, z)$ be a function and define a function $z = f(x, y)$ implicitly by letting $F(x, y, f(x, y)) = 0$. Use the chain rule to show that

$$\frac{\partial z}{\partial x} = -\frac{\partial F/\partial x}{\partial F/\partial z} \quad \text{and} \quad \frac{\partial z}{\partial y} = -\frac{\partial F/\partial y}{\partial F/\partial z}.$$

Problems 34–36 concern differentiating an integral in one variable, y, which also involves another variable x, either in the integrand, or in the limits, or both:

$$\int_0^5 (x^2y+4)\,dy \quad \text{or} \quad \int_0^x (y+4)\,dy \quad \text{or} \quad \int_0^x (xy+4)\,dy.$$

To differentiate the first integral with respect to x, it can be shown that in most cases we can differentiate with respect to x inside the integral:

$$\frac{d}{dx}\left(\int_0^5 (x^2y + 4)\,dy\right) = \int_0^5 2xy\,dy.$$

Differentiating the second integral with respect to x uses the Fundamental Theorem of Calculus:

$$\frac{d}{dx}\int_0^x (y + 4)\,dy = x + 4.$$

Differentiating the third integral involves the chain rule, as shown in Problem 36. Assume that the function F is continuously differentiable and b is constant throughout.

34. Let $f(x) = \int_0^b F(x, y)\,dy$. Find $f'(x)$.

35. Let $f(x) = \int_0^x F(b, y)\,dy$. Find $f'(x)$.

36. Let $f(x) = \int_0^x F(x, y)\,dy$. Use Problem 34 and Problem 35 to find $f'(x)$ by the following steps:

   (a) Let $G(u, w) = \int_0^w F(u, y)\,dy$. Find $G_u(u, w)$ and $G_w(u, w)$.
   (b) Use part (a) and the chain rule applied to $G(x, x) = f(x)$ to show:

$$f'(x) = \int_0^x F_u(x, y)\,dy + F(x, x).$$

14.7   SECOND-ORDER PARTIAL DERIVATIVES

Since the partial derivatives of a function are themselves functions, we can differentiate them, giving *second-order partial derivatives*. A function $z = f(x, y)$ has two first-order partial derivatives, f_x and f_y, and four second-order partial derivatives.

The Second-Order Partial Derivatives of $z = f(x, y)$

$$\frac{\partial^2 z}{\partial x^2} = f_{xx} = (f_x)_x, \qquad \frac{\partial^2 z}{\partial x \partial y} = f_{yx} = (f_y)_x,$$

$$\frac{\partial^2 z}{\partial y \partial x} = f_{xy} = (f_x)_y, \qquad \frac{\partial^2 z}{\partial y^2} = f_{yy} = (f_y)_y.$$

It is usual to omit the parentheses, writing f_{xy} instead of $(f_x)_y$ and $\dfrac{\partial^2 z}{\partial y\, \partial x}$ instead of $\dfrac{\partial}{\partial y}\left(\dfrac{\partial z}{\partial x}\right)$.

Example 1   Compute the four second-order partial derivatives of $f(x, y) = xy^2 + 3x^2 e^y$.

Solution   From $f_x(x, y) = y^2 + 6xe^y$ we get

$$f_{xx}(x, y) = \frac{\partial}{\partial x}(y^2 + 6xe^y) = 6e^y \quad \text{and} \quad f_{xy}(x, y) = \frac{\partial}{\partial y}(y^2 + 6xe^y) = 2y + 6xe^y.$$

From $f_y(x, y) = 2xy + 3x^2 e^y$ we get

$$f_{yx}(x, y) = \frac{\partial}{\partial x}(2xy + 3x^2 e^y) = 2y + 6xe^y \quad \text{and} \quad f_{yy}(x, y) = \frac{\partial}{\partial y}(2xy + 3x^2 e^y) = 2x + 3x^2 e^y.$$

Observe that $f_{xy} = f_{yx}$ in this example.

Example 2   Use the values of the function $f(x, y)$ in Table 14.6 to estimate $f_{xy}(1, 2)$ and $f_{yx}(1, 2)$.

Table 14.6   *Values of $f(x, y)$*

$y \backslash x$	0.9	1.0	1.1
1.8	4.72	5.83	7.06
2.0	6.48	8.00	9.60
2.2	8.62	10.65	12.88

Solution   Since $f_{xy} = (f_x)_y$, we first estimate f_x

$$f_x(1, 2) \approx \frac{f(1.1, 2) - f(1, 2)}{0.1} = \frac{9.60 - 8.00}{0.1} = 16.0,$$

$$f_x(1, 2.2) \approx \frac{f(1.1, 2.2) - f(1, 2.2)}{0.1} = \frac{12.88 - 10.65}{0.1} = 22.3.$$

Thus,

$$f_{xy}(1, 2) \approx \frac{f_x(1, 2.2) - f_x(1, 2)}{0.2} = \frac{22.3 - 16.0}{0.2} = 31.5.$$

Similarly,

$$f_{yx}(1,2) \approx \frac{f_y(1.1,2) - f_y(1,2)}{0.1} \approx \frac{1}{0.1}\left(\frac{f(1.1,2.2) - f(1.1,2)}{0.2} - \frac{f(1,2.2) - f(1,2)}{0.2}\right)$$

$$= \frac{1}{0.1}\left(\frac{12.88 - 9.60}{0.2} - \frac{10.65 - 8.00}{0.2}\right) = 31.5.$$

Observe that in this example also, $f_{xy} = f_{yx}$.

The Mixed Partial Derivatives Are Equal

It is not an accident that the estimates for $f_{xy}(1,2)$ and $f_{yx}(1,2)$ are equal in Example 2, because the same values of the function are used to calculate each one. The fact that $f_{xy} = f_{yx}$ in Examples 1 and 2 corroborates the following general result; Problem 43 suggests why you might expect it to be true[2].

Theorem 14.1: Equality of Mixed Partial Derivatives

If f_{xy} and f_{yx} are continuous at (a,b), an interior point of their domain, then

$$f_{xy}(a,b) = f_{yx}(a,b).$$

For most functions f we encounter and most points (a,b) in their domains, not only are f_{xy} and f_{yx} continuous at (a,b), but all their higher order partial derivatives (such as f_{xxy} or f_{xyyy}) exist and are continuous at (a,b). In that case we say f is *smooth* at (a,b). We say f is smooth on a region R if it is smooth at every point of R.

What Do the Second-Order Partial Derivatives Tell Us?

Example 3   Let us return to the guitar string of Example 14.2, page 648. The string is 1 meter long and at time t seconds, the point x meters from one end is displaced $f(x,t)$ meters from its rest position, where

$$f(x,t) = 0.003\sin(\pi x)\sin(2765t).$$

Compute the four second-order partial derivatives of f at the point $(x,t) = (0.3,1)$ and describe the meaning of their signs in practical terms.

Solution   First we compute $f_x(x,t) = 0.003\pi\cos(\pi x)\sin(2765t)$, from which we get

$$f_{xx}(x,t) = \frac{\partial}{\partial x}(f_x(x,t)) = -0.003\pi^2\sin(\pi x)\sin(2765t), \qquad \text{so} \qquad f_{xx}(0.3,1) \approx -0.01;$$

and

$$f_{xt}(x,t) = \frac{\partial}{\partial t}(f_x(x,t)) = (0.003)(2765)\pi\cos(\pi x)\cos(2765t), \qquad \text{so} \qquad f_{xt}(0.3,1) \approx 14.$$

On page 648 we saw that $f_x(x,t)$ gives the slope of the string at any point and time. Therefore, $f_{xx}(x,t)$ measures the concavity of the string. The fact that $f_{xx}(0.3,1) < 0$ means the string is concave down at the point $x = 0.3$ when $t = 1$. (See Figure 14.51.)

On the other hand, $f_{xt}(x,t)$ is the rate of change of the slope of the string with respect to time. Thus $f_{xt}(0.3,1) > 0$ means that at time $t = 1$ the slope at the point $x = 0.3$ is increasing. (See Figure 14.52.)

---

[2]For a proof, see M. Spivak, *Calculus on Manifolds*, p. 26 (New York: Benjamin, 1965)

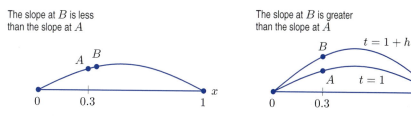

The slope at B is less
than the slope at A

Figure 14.51: Interpretation of $f_{xx}(0.3, 1) < 0$:
The concavity of the string at $t = 1$

The slope at B is greater
than the slope at A

$t = 1 + h$

$t = 1$

Figure 14.52: Interpretation of
$f_{xt}(0.3, 1) > 0$: The slope of one point on
the string at two different times

Now we compute $f_t(x, t) = (0.003)(2765) \sin(\pi x) \cos(2765t)$, from which we get

$$f_{tx}(x, t) = \frac{\partial}{\partial x}(f_t(x, t)) = (0.003)(2765)\pi \cos(\pi x) \cos(2765t), \quad \text{so} \quad f_{tx}(0.3, 1) \approx 14$$

and

$$f_{tt}(x, t) = \frac{\partial}{\partial t}(f_t(x, t)) = -(0.003)(2765)^2 \sin(\pi x) \sin(2765t), \quad \text{so} \quad f_{tt}(0.3, 1) \approx -7200.$$

On page 648 we saw that $f_t(x, t)$ gives the velocity of the string at any point and time. Therefore, $f_{tx}(x, t)$ and $f_{tt}(x, t)$ will both be rates of change of velocity. That $f_{tx}(0.3, 1) > 0$ means that at time $t = 1$ the velocities of points just to the right of $x = 0.3$ are greater than the velocity at $x = 0.3$. (See Figure 14.53.) That $f_{tt}(0.3, 1) < 0$ means that the velocity of the point $x = 0.3$ is decreasing at time $t = 1$. Thus, $f_{tt}(0.3, 1) = -7200$ m/sec^2 is the acceleration of this point. (See Figure 14.54.)

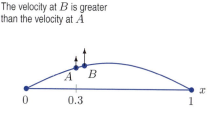

The velocity at B is greater
than the velocity at A

Figure 14.53: Interpretation of
$f_{tx}(0.3, 1) > 0$: The velocity of different
points on the string at $t = 1$

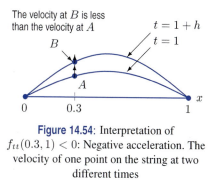

The velocity at B is less
than the velocity at A

$t = 1 + h$

$t = 1$

Figure 14.54: Interpretation of
$f_{tt}(0.3, 1) < 0$: Negative acceleration. The
velocity of one point on the string at two
different times

Taylor Approximations

We use second derivatives to construct quadratic Taylor approximations. In Section 14.3, we saw how to approximate $f(x, y)$ by a linear function (its local linearization). We now see how to improve this approximation of $f(x, y)$ using a quadratic function.

Linear and Quadratic Approximations Near (0,0)

For a function of one variable, local linearity tells us that the best *linear* approximation is the degree 1 Taylor polynomial

$$f(x) \approx f(a) + f'(a)(x - a) \quad \text{for } x \text{ near } a.$$

A better approximation to $f(x)$ is given by the degree 2 Taylor polynomial:

$$f(x) \approx f(a) + f'(a)(x - a) + \frac{f''(a)}{2}(x - a)^2 \quad \text{for } x \text{ near } a.$$

For a function of two variables the local linearization for (x, y) near (a, b) is

$$f(x, y) \approx L(x, y) = f(a, b) + f_x(a, b)(x - a) + f_y(a, b)(y - b).$$

In the case $(a, b) = (0, 0)$, we have:

> **Taylor Polynomial of Degree 1 Approximating $f(x, y)$ for (x, y) near (0,0)**
> If f has continuous first-order partial derivatives, then
>
> $$f(x, y) \approx L(x, y) = f(0, 0) + f_x(0, 0)x + f_y(0, 0)y.$$

We get a better approximation to f by using a quadratic polynomial. We choose a quadratic polynomial $Q(x, y)$, with the same partial derivatives as the original function f. You can check that the following Taylor polynomial of degree 2 has this property.

> **Taylor Polynomial of Degree 2 Approximating $f(x, y)$ for (x, y) near (0,0)** If f has continuous second-order partial derivatives, then
>
> $$f(x, y) \approx Q(x, y)$$
> $$= f(0, 0) + f_x(0, 0)x + f_y(0, 0)y + \frac{f_{xx}(0, 0)}{2}x^2 + f_{xy}(0, 0)xy + \frac{f_{yy}(0, 0)}{2}y^2.$$

Example 4   Let $f(x, y) = \cos(2x + y) + 3\sin(x + y)$
(a) Compute the linear and quadratic Taylor polynomials, L and Q, approximating f near $(0, 0)$.
(b) Explain why the contour plots of L and Q for $-1 \leq x \leq 1$, $-1 \leq y \leq 1$ look the way they do.

Solution   (a) We have $f(0, 0) = 1$. The derivatives we need are as follows:

$$
\begin{aligned}
f_x(x, y) &= -2\sin(2x + y) + 3\cos(x + y) &\quad\text{so}\quad& f_x(0, 0) = 3, \\
f_y(x, y) &= -\sin(2x + y) + 3\cos(x + y) &\quad\text{so}\quad& f_y(0, 0) = 3, \\
f_{xx}(x, y) &= -4\cos(2x + y) - 3\sin(x + y) &\quad\text{so}\quad& f_{xx}(0, 0) = -4, \\
f_{xy}(x, y) &= -2\cos(2x + y) - 3\sin(x + y) &\quad\text{so}\quad& f_{xy}(0, 0) = -2, \\
f_{yy}(x, y) &= -\cos(2x + y) - 3\sin(x + y) &\quad\text{so}\quad& f_{yy}(0, 0) = -1.
\end{aligned}
$$

Thus, the linear approximation, $L(x, y)$, to $f(x, y)$ at $(0, 0)$ is given by

$$f(x, y) \approx L(x, y) = f(0, 0) + f_x(0, 0)x + f_y(0, 0)y = 1 + 3x + 3y.$$

The quadratic approximation, $Q(x, y)$, to $f(x, y)$ near $(0, 0)$ is given by

$$f(x, y) \approx Q(x, y)$$
$$= f(0, 0) + f_x(0, 0)x + f_y(0, 0)y + \frac{f_{xx}(0, 0)}{2}x^2 + f_{xy}(0, 0)xy + \frac{f_{yy}(0, 0)}{2}y^2$$
$$= 1 + 3x + 3y - 2x^2 - 2xy - \frac{1}{2}y^2.$$

Notice that the linear terms in $Q(x, y)$ are the same as the linear terms in $L(x, y)$. The quadratic terms in $Q(x, y)$ can be thought of as "correction terms" to the linear approximation.
(b) The contour plots of $f(x, y)$, $L(x, y)$, and $Q(x, y)$ are in Figures 14.55–14.57.

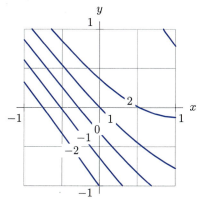

Figure 14.55: Original function, $f(x, y)$

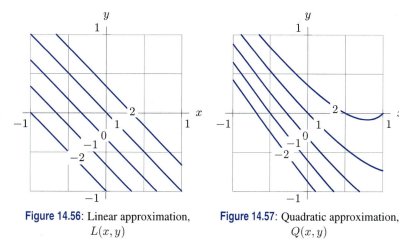

Figure 14.56: Linear approximation, $L(x, y)$

Figure 14.57: Quadratic approximation, $Q(x, y)$

Notice that the contour plot of Q is more similar to the contour plot of f than is the contour plot of L. Since L is linear, the contour plot of L consists of parallel, equally spaced lines.

An alternative, and much quicker, way to find the Taylor polynomial in the previous example is to use the single-variable approximations. For example, since

$$\cos u = 1 - \frac{u^2}{2!} + \frac{u^4}{4!} + \cdots \quad \text{and} \quad \sin v = v - \frac{v^3}{3!} + \cdots,$$

we can substitute $u = 2x + y$ and $v = x + y$ and expand. We discard terms beyond the second (since we want the quadratic polynomial) getting

$$\cos(2x+y) = 1 - \frac{(2x+y)^2}{2!} + \frac{(2x+y)^4}{4!} + \cdots \approx 1 - \frac{1}{2}(4x^2 + 4xy + y^2) = 1 - 2x^2 - 2xy - \frac{1}{2}y^2$$

and

$$\sin(x+y) = (x+y) - \frac{(x+y)^3}{3!} + \cdots \approx x + y.$$

Combining these results, we get

$$\cos(2x+y) + 3\sin(x+y) \approx 1 - 2x^2 - 2xy - \frac{1}{2}y^2 + 3(x+y) = 1 + 3x + 3y - 2x^2 - 2xy - \frac{1}{2}y^2.$$

Linear and Quadratic Approximations Near (a, b)

The local linearization for a function $f(x, y)$ at a point (a, b) is

> **Taylor Polynomial of Degree 1 Approximating $f(x, y)$ for (x, y) near (a, b)**
> If f has continuous first-order partial derivatives, then
>
> $$f(x, y) \approx L(x, y) = f(a, b) + f_x(a, b)(x - a) + f_y(a, b)(y - b).$$

This suggests that a quadratic polynomial approximation $Q(x, y)$ for $f(x, y)$ near a point (a, b) should be written in terms of $(x - a)$ and $(y - b)$ instead of x and y. If we require that $Q(a, b) = f(a, b)$ and that the first- and second-order partial derivatives of Q and f at (a, b) be equal, then we get the following polynomial:

> **Taylor Polynomial of Degree** 2 **Approximating** $f(x, y)$ **for** (x, y) **near** (a, b)
> If f has continuous second-order partial derivatives, then
>
> $$f(x, y) \approx Q(x, y)$$
> $$= f(a, b) + f_x(a, b)(x - a) + f_y(a, b)(y - b)$$
> $$+ \frac{f_{xx}(a, b)}{2}(x - a)^2 + f_{xy}(a, b)(x - a)(y - b) + \frac{f_{yy}(a, b)}{2}(y - b)^2.$$

These coefficients are derived in exactly the same way as for $(a, b) = (0, 0)$.

Example 5   Find the Taylor polynomial of degree 2 at the point $(1, 2)$ for the function $f(x, y) = \dfrac{1}{xy}$.

Solution   Table 14.7 contains the partial derivatives and their values at the point $(1, 2)$.

Table 14.7   *Partial derivatives of* $f(x, y) = 1/(xy)$

Derivative	Formula	Value at $(1, 2)$	Derivative	Formula	Value at $(1, 2)$
$f(x, y)$	$1/(xy)$	$1/2$	$f_{xx}(x, y)$	$2/(x^3 y)$	1
$f_x(x, y)$	$-1/(x^2 y)$	$-1/2$	$f_{xy}(x, y)$	$1/(x^2 y^2)$	$1/4$
$f_y(x, y)$	$-1/(xy^2)$	$-1/4$	$f_{yy}(x, y)$	$2/(xy^3)$	$1/4$

So, the quadratic Taylor polynomial for f near $(1, 2)$ is

$$\frac{1}{xy} \approx Q(x, y)$$

$$= \frac{1}{2} - \frac{1}{2}(x - 1) - \frac{1}{4}(y - 2) + \frac{1}{2}(1)(x - 1)^2 + \frac{1}{4}(x - 1)(y - 2) + \left(\frac{1}{2}\right)\left(\frac{1}{4}\right)(y - 2)^2$$

$$= \frac{1}{2} - \frac{x - 1}{2} - \frac{y - 2}{4} + \frac{(x - 1)^2}{2} + \frac{(x - 1)(y - 2)}{4} + \frac{(y - 2)^2}{8}.$$

Exercises and Problems for Section 14.7

Exercises

In Exercises 1–10, calculate all four second-order partial derivatives and check that $f_{xy} = f_{yx}$. Assume the variables are restricted to a domain on which the function is defined.

1. $f(x, y) = 3x^2 y + 5xy^3$

2. $f(x, y) = (x + y)^2$

3. $f(x, y) = xe^y$

4. $f(x, y) = (x + y)^3$

5. $f(x, y) = (x + y)e^y$

6. $f(x, y) = e^{2xy}$

7. $f(x, y) = \sqrt{x^2 + y^2}$

8. $f(x, y) = \sin(x/y)$

9. $f(x, y) = 3\sin 2x \cos 5y$

10. $f(x, y) = \sin(x^2 + y^2)$

Find the quadratic Taylor polynomials about $(0, 0)$ for the functions in Exercises 11–18.

11. $(x - y + 1)^2$

12. $(y - 1)(x + 1)^2$

13. $\cos(x + 3y)$

14. $e^x \cos y$

15. $1/(1 + 2x - y)$

16. $e^{-2x^2 - y^2}$

17. $\sin 2x + \cos y$

18. $\ln(1 + x^2 - y)$

In Exercises 19–27, use the level curves of the function $z = f(x, y)$ to decide the sign (positive, negative, or zero) of each of the following partial derivatives at the point P. Assume the x- and y-axes are in the usual positions.

(a) $f_x(P)$ **(b)** $f_y(P)$ **(c)** $f_{xx}(P)$

(d) $f_{yy}(P)$ **(e)** $f_{xy}(P)$

19.

20.

21.

22.

23.

24.

25.

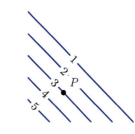

26.

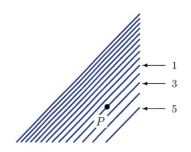

27.

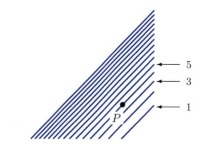

In Exercises 28–31, find the linear, $L(x, y)$, and quadratic, $Q(x, y)$, Taylor polynomials valid near $(1, 0)$. Compare the values of the approximations $L(0.9, 0.2)$ and $Q(0.9, 0.2)$ with the exact value of the function $f(0.9, 0.2)$.

28. $f(x, y) = x^2 y$ **29.** $f(x, y) = \sqrt{x + 2y}$

30. $f(x, y) = xe^{-y}$

31. $f(x, y) = \sin(x - 1) \cos y$

Problems

32. If $z = f(x) + yg(x)$, what can you say about z_{yy}? Explain your answer.

33. If $z_{xy} = 4y$, what can you say about the value of
(a) z_{yx}?   (b) z_{xyx}?   (c) z_{xyy}?

34. Figure 14.58 shows a graph of $z = f(x, y)$. Is $f_{xx}(0, 0)$ positive or negative? Is $f_{yy}(0, 0)$ positive or negative? Give reasons for your answers.

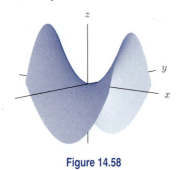

Figure 14.58

35. A contour diagram for the smooth function $z = f(x, y)$ is in Figure 14.59.

(a) Is z an increasing or decreasing function of x? Of y? Explain.
(b) Is f_x positive or negative? How about f_y? Explain.
(c) Is f_{xx} positive or negative? How about f_{yy}? Explain.
(d) Sketch the direction of grad f at points P and Q.
(e) Is grad f longer at P or at Q? How do you know?

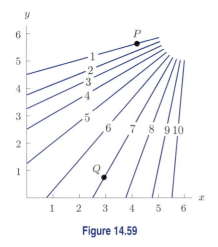

Figure 14.59

36. Figure 14.60 shows the level curves of a function $f(x, y)$ around a maximum or minimum, M. One of the points P and Q has coordinates (x_1, y_1) and the other has coordinates (x_2, y_2). Suppose $b > 0$ and $c > 0$. Consider the two linear approximations to f given by

$$f(x, y) \approx a + b(x - x_1) + c(y - y_1)$$
$$f(x, y) \approx k + m(x - x_2) + n(y - y_2).$$

(a) What is the relationship between the values of a and k?
(b) What are the coordinates of P?
(c) Is M a maximum or a minimum?
(d) What can you say about the sign of the constants m and n?

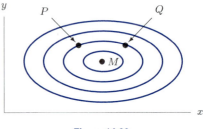

Figure 14.60

37. Consider the function $f(x, y) = (\sin x)(\sin y)$.

(a) Find the Taylor polynomials of degree 2 for f about the points $(0, 0)$ and $(\pi/2, \pi/2)$.
(b) Use the Taylor polynomials to sketch the contours of f close to each of the points $(0, 0)$ and $(\pi/2, \pi/2)$.

38. Give contour diagrams of two different functions $f(x, y)$ and $g(x, y)$ that have the same quadratic approximations near $(0, 0)$.

39. Let $f(x, y) = \sqrt{x + 2y + 1}$.

(a) Compute the local linearization of f at $(0, 0)$.
(b) Compute the quadratic Taylor polynomial for f at $(0, 0)$.
(c) Compare the values of the linear and quadratic approximations in part (a) and part (b) with the true values for $f(x, y)$ at the points $(0.1, 0.1)$, $(-0.1, 0.1)$, $(0.1, -0.1)$, $(-0.1, -0.1)$. Which approximation gives the closest values?

40. Using a computer and your answer to Problem 39, draw the six contour diagrams of $f(x, y) = \sqrt{x + 2y + 1}$ and its linear and quadratic approximations, $L(x, y)$ and $Q(x, y)$, in the two windows $[-0.6, 0.6] \times [-0.6, 0.6]$ and $[-2, 2] \times [-2, 2]$. Explain the shape of the contours, their spacing, and the relationship between the contours of f, L, and Q.

41. Suppose that $f(x, y)$ has continuous partial derivatives f_x and f_y. Using the Fundamental Theorem of Calculus to evaluate the integrals, show that

$$f(a, b) = f(0, 0) + \int_{t=0}^{a} f_x(t, 0)dt + \int_{t=0}^{b} f_y(a, t)dt.$$

42. Suppose that $f(x, y)$ has continuous partial derivatives and that $f(0, 0) = 0$ and $|f_x(x, y)| \leq A$ and $|f_y(x, y)| \leq B$ for all points (x, y) in the plane. Use Problem 41 to show that $|f(x, y)| \leq A|x| + B|y|$.

This inequality shows how bounds on the partial derivatives of f limit the growth of f.

43. Give an explanation of why you might expect $f_{xy}(a, b) = f_{yx}(a, b)$ using the following steps.

(a) Write the definition of $f_x(a, b)$.

(b) Write a definition of $f_{xy}(a, b)$ as $(f_x)_y$.

(c) Substitute for f_x in the definition of f_{xy}.

(d) Write an expression for f_{yx} similar to the one for f_{xy} you obtained in part (c).

(e) Compare your answers to parts (c) and (d). What do you have to assume to conclude that f_{xy} and f_{yx} are equal?

14.8 DIFFERENTIABILITY

In Section 14.3 we gave an informal introduction to the concept of differentiability. We called a function $f(x, y)$ *differentiable* at a point (a, b) if it is well-approximated by a linear function near (a, b). This section focuses on the precise meaning of the phrase "well-approximated." By looking at examples, we shall see that local linearity requires the existence of partial derivatives, but they do not tell the whole story. In particular, existence of partial derivatives at a point is not sufficient to guarantee local linearity at that point.

We begin by discussing the relation between continuity and differentiability. As an illustration, take a sheet of paper, crumple it into a ball and smooth it out again. Wherever there is a crease it would be difficult to approximate the surface by a plane—these are points of nondifferentiability of the function giving the height of the paper above the floor. Yet the sheet of paper models a graph which is continuous—there are no breaks. As in the case of one-variable calculus, continuity does not imply differentiability. But differentiability does *require* continuity: there cannot be linear approximations to a surface at points where there are abrupt changes in height.

Differentiability For Functions Of Two Variables

For a function of two variables, as for a function of one variable, we define differentiability at a point in terms of the error and the distance from the point. If the point is (a, b) and a nearby point is $(a + h, b + k)$, the distance between them is $\sqrt{h^2 + k^2}$. (See Figure 14.61.)

A function $f(x, y)$ is **differentiable at the point** (a, b) if there is a linear function $L(x, y) = f(a, b) + m(x - a) + n(y - b)$ such that if the *error* $E(x, y)$ is defined by

$$f(x, y) = L(x, y) + E(x, y),$$

and if $h = x - a, k = y - b$, then the *relative error* $E(a + h, b + k)/\sqrt{h^2 + k^2}$ satisfies

$$\lim_{\substack{h \to 0 \\ k \to 0}} \frac{E(a + h, b + k)}{\sqrt{h^2 + k^2}} = 0.$$

The function f is **differentiable on a region** R if it is differentiable at each point of R. The function $L(x, y)$ is called the *local linearization* of $f(x, y)$ near (a, b).

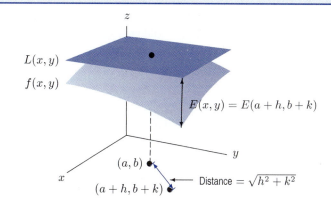

Figure 14.61: Graph of function $z = f(x, y)$ and its local linearization $z = L(x, y)$ near the point (a, b)

Partial Derivatives and Differentiability

In the next example, we show that this definition of differentiability is consistent with our previous notion — that is, that $m = f_x$ and $n = f_y$ and that the graph of $L(x, y)$ is the tangent plane.

Example 1   Show that if f is a differentiable function with local linearization $L(x, y) = f(a, b) + m(x - a) + n(y - b)$, then $m = f_x(a, b)$ and $n = f_y(a, b)$.

Solution   Since f is differentiable, we know that the relative error in $L(x, y)$ tends to 0 as we get close to (a, b). Suppose $h > 0$ and $k = 0$. Then we know that

$$0 = \lim_{h \to 0} \frac{E(a + h, b)}{\sqrt{h^2 + k^2}} = \lim_{h \to 0} \frac{E(a + h, b)}{h} = \lim_{h \to 0} \frac{f(a + h, b) - L(a + h, b)}{h}$$

$$= \lim_{h \to 0} \frac{f(a + h, b) - f(a, b) - mh}{h}$$

$$= \lim_{h \to 0} \left(\frac{f(a + h, b) - f(a, b)}{h} \right) - m = f_x(a, b) - m.$$

A similar result holds if $h < 0$, so we have $m = f_x(a, b)$. The result $n = f_y(a, b)$ is found in a similar manner.

The previous example shows that if a function is differentiable at a point, it has partial derivatives there. Therefore, if any of the partial derivatives fail to exist, then the function cannot be differentiable. This is what happens in the following example of a cone.

Example 2   Consider the function $f(x, y) = \sqrt{x^2 + y^2}$. Is f differentiable at the origin?

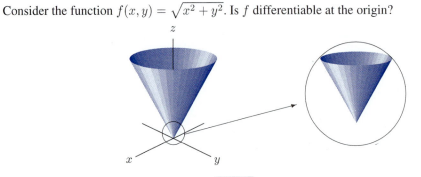

Figure 14.62: The function $f(x, y) = \sqrt{x^2 + y^2}$ is not locally linear at $(0, 0)$: Zooming in around $(0, 0)$ does not make the graph look like a plane

Solution   If we zoom in on the graph of the function $f(x, y) = \sqrt{x^2 + y^2}$ at the origin, as shown in Figure 14.62, the sharp point remains; the graph never flattens out to look like a plane. Near its vertex, the graph does not look like it is well approximated (in any reasonable sense) by any plane.

Judging from the graph of f, we would not expect f to be differentiable at $(0, 0)$. Let us check this by trying to compute the partial derivatives of f at $(0, 0)$:

$$f_x(0, 0) = \lim_{h \to 0} \frac{f(h, 0) - f(0, 0)}{h} = \lim_{h \to 0} \frac{\sqrt{h^2 + 0} - 0}{h} = \lim_{h \to 0} \frac{|h|}{h}.$$

Since $|h|/h = \pm 1$, depending on whether h approaches 0 from the left or right, this limit does not exist and so neither does the partial derivative $f_x(0, 0)$. Thus, f cannot be differentiable at the origin. If it were, both of the partial derivatives, $f_x(0, 0)$ and $f_y(0, 0)$, would exist.

Alternatively, we could show directly that there is no linear approximation near $(0, 0)$ that satisfies the small relative error criterion for differentiability. Any plane passing through the point $(0, 0, 0)$ has the form $L(x, y) = mx + ny$ for some constants m and n. If $E(x, y) = f(x, y) - L(x, y)$, then

$$E(x, y) = \sqrt{x^2 + y^2} - mx - ny.$$

Then for f to be differentiable at the origin, we would need to show that

$$\lim_{\substack{h \to 0 \\ k \to 0}} \frac{\sqrt{h^2 + k^2} - mh - nk}{\sqrt{h^2 + k^2}} = 0.$$

Taking $k = 0$ gives

$$\lim_{h \to 0} \frac{|h| - mh}{|h|} = 1 - m \lim_{h \to 0} \frac{h}{|h|}.$$

This limit exists only if $m = 0$ for the same reason as before. But then the value of the limit is 1 and not 0 as required. Thus, we again conclude f is not differentiable.

In Example 2 the partial derivatives f_x and f_y did not exist at the origin and this was sufficient to establish nondifferentiability there. We might expect that if both partial derivatives do exist, then f *is* differentiable. But the next example shows that this not necessarily true: the existence of both partial derivatives at a point is *not* sufficient to guarantee differentiability.

Example 3 Consider the function $f(x, y) = x^{1/3}y^{1/3}$. Show that the partial derivatives $f_x(0, 0)$ and $f_y(0, 0)$ exist, but that f is not differentiable at $(0, 0)$.

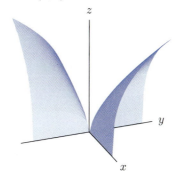

Figure 14.63: Graph of $z = x^{1/3}y^{1/3}$ for $z \geq 0$

Solution See Figure 14.63 for the part of the graph of $z = x^{1/3}y^{1/3}$ when $z \geq 0$. We have $f(0, 0) = 0$ and we compute the partial derivatives using the definition:

$$f_x(0, 0) = \lim_{h \to 0} \frac{f(h, 0) - f(0, 0)}{h} = \lim_{h \to 0} \frac{0 - 0}{h} = 0,$$

and similarly

$$f_y(0, 0) = 0.$$

So, if there did exist a linear approximation near the origin, it would have to be $L(x, y) = 0$. But we can show that this choice of $L(x, y)$ doesn't result in the small relative error that is required for differentiability. In fact, since $E(x, y) = f(x, y) - L(x, y) = f(x, y)$, we need to look at the limit

$$\lim_{\substack{h \to 0 \\ k \to 0}} \frac{h^{1/3}k^{1/3}}{\sqrt{h^2 + k^2}}.$$

If this limit exists, we get the same value no matter how h and k approach 0. Suppose we take $k = h > 0$. Then the limit becomes

$$\lim_{h \to 0} \frac{h^{1/3}h^{1/3}}{\sqrt{h^2 + h^2}} = \lim_{h \to 0} \frac{h^{2/3}}{h\sqrt{2}} = \lim_{h \to 0} \frac{1}{h^{1/3}\sqrt{2}}.$$

But this limit does not exist, since small values for h will make the fraction arbitrarily large. So the only possible candidate for a linear approximation at the origin does not have a sufficiently small relative error. Thus, this function is *not* differentiable at the origin, even though the partial derivatives $f_x(0, 0)$ and $f_y(0, 0)$ exist. Figure 14.63 confirms that near the origin the graph of $z = f(x, y)$ is not well approximated by any plane.

In summary,

- If a function is differentiable at a point, then both partial derivatives exist there.
- Having both partial derivatives at a point does not guarantee that a function is differentiable there.

Continuity and Differentiability

We know that differentiable functions of one variable are continuous. Similarly, it can be shown that if a function of two variables is differentiable at a point, then the function is continuous there.

In Example 3 the function f was continuous at the point where it was not differentiable. Example 4 shows that even if the partial derivatives of a function exist at a point, the function is not necessarily continuous at that point if it is not differentiable there.

Example 4   Suppose that f is the function of two variables defined by

$$f(x, y) = \begin{cases} \dfrac{xy}{x^2 + y^2}, & (x, y) \neq (0, 0), \\ 0, & (x, y) = (0, 0). \end{cases}$$

Problem 17 on page 599 showed that $f(x, y)$ is not continuous at the origin. Show that the partial derivatives $f_x(0, 0)$ and $f_y(0, 0)$ exist. Could f be differentiable at $(0, 0)$?

Solution   From the definition of the partial derivative we see that

$$f_x(0, 0) = \lim_{h \to 0} \frac{f(h, 0) - f(0, 0)}{h} = \lim_{h \to 0} \left(\frac{1}{h} \cdot \frac{0}{h^2 + 0^2} \right) = \lim_{h \to 0} \frac{0}{h} = 0,$$

and similarly

$$f_y(0, 0) = 0.$$

So, the partial derivatives $f_x(0, 0)$ and $f_y(0, 0)$ exist. However, f cannot be differentiable at the origin since it is not continuous there.

In summary,

- If a function is differentiable at a point, then it is continuous there.
- Having both partial derivatives at a point does not guarantee that a function is continuous there.

How Do We Know If a Function Is Differentiable?

Can we use partial derivatives to tell us if a function is differentiable? As we see from Examples 3 and 4, it is not enough that the partial derivatives exist. However, the following theorem gives conditions that *do* guarantee differentiability[3]:

Theorem 14.2: Continuity of Partial Derivatives Implies Differentiability

If the partial derivatives, f_x and f_y, of a function f exist and are continuous on a small disk centered at the point (a, b), then f is differentiable at (a, b).

We will not prove this theorem, although it provides a criterion for differentiability which is often simpler to use than the definition. It turns out that the requirement of continuous partial derivatives is more stringent than that of differentiability, so there exist differentiable functions which do

---

[3]For a proof, see M. Spivak, *Calculus on Manifolds*, p. 31, (New York: Benjamin, 1965)

not have continuous partial derivatives. However, most functions we encounter will have continuous partial derivatives. The class of functions with continuous partial derivatives is given the name C^1.

Example 5   Show that the function $f(x, y) = \ln(x^2 + y^2)$ is differentiable everywhere in its domain.

Solution   The domain of f is all of 2-space except for the origin. We shall show that f has continuous partial derivatives everywhere in its domain (that is, the function f is in C^1). The partial derivatives are

$$f_x = \frac{2x}{x^2 + y^2} \quad \text{and} \quad f_y = \frac{2y}{x^2 + y^2}.$$

Since each of f_x and f_y is the quotient of continuous functions, the partial derivatives are continuous everywhere except the origin (where the denominators are zero). Thus, f is differentiable everywhere in its domain.

Most functions built up from elementary functions have continuous partial derivatives, except perhaps at a few obvious points. Thus, in practice, we can often identify functions as being C^1 without explicitly computing the partial derivatives.

Exercises and Problems for Section 14.8

Exercises

In Exercises 1–6, list the points in the xy-plane, if any, at which the function $z = f(x, y)$ is not differentiable.

1. $z = -\sqrt{x^2 + y^2}$

2. $z = 4 + \sqrt{(x-1)^2 + (y-2)^2}$

3. $z = 1 + \left((x-1)^2 + (y-2)^2\right)^2$

4. $z = e^{-(x^2+y^2)}$

5. $z = |x| + |y|$

6. $z = (\sin x)(\cos |y|)$

Problems

For the functions f in Problems 7–10 answer the following questions. Justify your answers.

(a) Use a computer to draw a contour diagram for f.

(b) Is f differentiable at all points $(x, y) \neq (0, 0)$?

(c) Do the partial derivatives f_x and f_y exist and are they continuous at all points $(x, y) \neq (0, 0)$?

(d) Is f differentiable at $(0, 0)$?

(e) Do the partial derivatives f_x and f_y exist and are they continuous at $(0, 0)$?

7. $f(x, y) = \begin{cases} \dfrac{x}{y} + \dfrac{y}{x}, & x \neq 0 \text{ and } y \neq 0, \\ 0, & x = 0 \text{ or } y = 0. \end{cases}$

8. $f(x, y) = \begin{cases} \dfrac{2xy}{(x^2 + y^2)^2}, & (x, y) \neq (0, 0), \\ 0, & (x, y) = (0, 0). \end{cases}$

9. $f(x, y) = \begin{cases} \dfrac{xy}{\sqrt{x^2 + y^2}}, & (x, y) \neq (0, 0), \\ 0, & (x, y) = (0, 0). \end{cases}$

10. $f(x, y) = \begin{cases} \dfrac{x^2 y}{x^4 + y^2}, & (x, y) \neq (0, 0), \\ 0, & (x, y) = (0, 0). \end{cases}$

11. Consider the function

$$f(x, y) = \begin{cases} \dfrac{xy^2}{x^2 + y^2}, & (x, y) \neq (0, 0), \\ 0, & (x, y) = (0, 0). \end{cases}$$

(a) Use a computer to draw the contour diagram for f.

(b) Is f differentiable for $(x, y) \neq (0, 0)$?

(c) Show that $f_x(0, 0)$ and $f_y(0, 0)$ exist.

(d) Is f differentiable at $(0, 0)$?

(e) Suppose $x(t) = at$ and $y(t) = bt$, where a and b are constants, not both zero. If $g(t) = f(x(t), y(t))$, show that

$$g'(0) = \frac{ab^2}{a^2 + b^2}.$$

(f) Show that

$$f_x(0, 0)x'(0) + f_y(0, 0)y'(0) = 0.$$

Does the chain rule hold for the composite function $g(t)$ at $t = 0$? Explain.

(g) Show that the directional derivative $f_{\vec{u}}(0, 0)$ exists for each unit vector $\vec{u}$. Does this imply that f is differentiable at $(0, 0)$?

12. Consider the function

$$f(x, y) = \begin{cases} \dfrac{xy^2}{x^2 + y^4}, & (x, y) \neq (0, 0), \\ 0, & (x, y) = (0, 0). \end{cases}$$

(a) Use a computer to draw the contour diagram for f.

(b) Show that the directional derivative $f_{\vec{u}}(0, 0)$ exists for each unit vector $\vec{u}$.

(c) Is f continuous at $(0, 0)$? Is f differentiable at $(0, 0)$? Explain.

13. Consider the function $f(x, y) = \sqrt{|xy|}$.

(a) Use a computer to draw the contour diagram for f. Does the contour diagram look like that of a plane when we zoom in on the origin?

(b) Use a computer to draw the graph of f. Does the graph look like a plane when we zoom in on the origin?

(c) Is f differentiable for $(x, y) \neq (0, 0)$?

(d) Show that $f_x(0, 0)$ and $f_y(0, 0)$ exist.

(e) Is f differentiable at $(0, 0)$? [Hint: Consider the directional derivative $f_{\vec{u}}(0, 0)$ for $\vec{u} = (\vec{i} + \vec{j})/\sqrt{2}$.]

14. Suppose a function f is differentiable at the point (a, b). Show that f is continuous at (a, b).

15. Suppose $f(x, y)$ is a function such that $f_x(0, 0) = 0$ and $f_y(0, 0) = 0$, and $f_{\vec{u}}(0, 0) = 3$ for $\vec{u} = (\vec{i} + \vec{j})/\sqrt{2}$.

(a) Is f differentiable at $(0, 0)$? Explain.

(b) Give an example of a function f defined on 2-space which satisfies these conditions. [Hint: The function f does not have to be defined by a single formula valid over all of 2-space.]

16. Consider the following function:

$$f(x, y) = \begin{cases} \dfrac{xy(x^2 - y^2)}{x^2 + y^2}, & (x, y) \neq (0, 0), \\ 0, & (x, y) = (0, 0). \end{cases}$$

The graph of f is shown in Figure 14.64, and the contour diagram of f is shown in Figure 14.65.

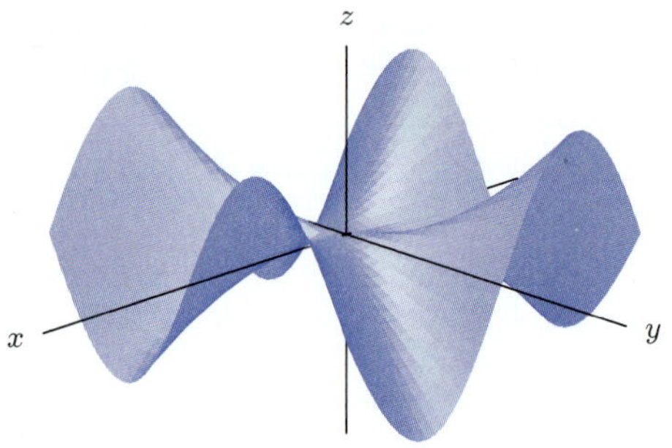

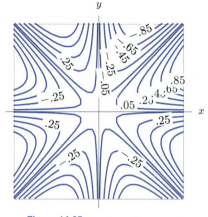

Figure 14.64: Graph of $\dfrac{xy(x^2 - y^2)}{x^2 + y^2}$

Figure 14.65: Contour diagram of $\dfrac{xy(x^2 - y^2)}{x^2 + y^2}$

(a) Find $f_x(x, y)$ and $f_y(x, y)$ for $(x, y) \neq (0, 0)$.

(b) Show that $f_x(0, 0) = 0$ and $f_y(0, 0) = 0$.

(c) Are the functions f_x and f_y continuous at $(0, 0)$?

(d) Is f differentiable at $(0, 0)$?

CHAPTER SUMMARY

- **Partial Derivatives**
  Definition as a difference quotient, visualizing on a graph, estimating from a contour diagram, computing from a formula, interpreting units, alternative notation.

- **Local Linearity**
  Zooming on a surface, contour diagram or table to see local linearity, the idea of tangent plane, formula for a tangent plane in terms of partials, the differential.

- **Directional Derivatives**
  Definition as a difference quotient, interpretation as a rate of change, computation using partial derivatives.

- **Gradient Vector**
  Definition in terms of partial derivatives, geometric properties of gradient's length and direction, relation to directional derivative, relation to contours and level surfaces.

- **Chain Rule**
  Local linearity and differentials for composition of functions, tree diagrams, chain rule in general, application to physical chemistry.

- **Second and Higher Order Partial Derivatives**
  Interpretations, mixed partials are equal.

- **Taylor Approximations**
  Linear and quadratic polynomial approximations to functions near a point.

REVIEW EXERCISES AND PROBLEMS FOR CHAPTER FOURTEEN

Exercises

For Exercises 1–8, find the indicated partial derivatives. Assume the variables are restricted to a domain on which the function is defined.

1. $\dfrac{\partial z}{\partial x}$ and $\dfrac{\partial z}{\partial y}$ if $z = (x^2 + x - y)^7$

2. $\dfrac{\partial F}{\partial L}$ if $F(L, K) = 3\sqrt{LK}$

3. $\dfrac{\partial f}{\partial p}$ and $\dfrac{\partial f}{\partial q}$ if $f(p, q) = e^{p/q}$

4. $\dfrac{\partial f}{\partial x}$ if $f(x, y) = e^{xy}(\ln y)$

5. f_{xx} and f_{xy} if $f(x, y) = 1/\sqrt{x^2 + y^2}$

6. u_{xx} and u_{yy} if $u = e^x \sin y$

7. V_{rr} and V_{rh} if $V = \pi r^2 h$

8. f_{xxy} and f_{yxx} if $f(x, y) = \sin(x - 2y)$

In Exercises 9–12, find the gradient for the given function.

9. $f(x, y) = \ln(x^2 + y^2)$

10. $f(x, y) = \sin(y^2 - xy)$

11. $f(x, y, z) = x^3 + z^3 - xyz$

12. $f(\rho, \phi, \theta) = \rho \sin \phi \cos \theta$

In Exercises 13–18, find the directional derivative for the given function.

13. $f(x, y) = x^3 - y^3$ at $(2, -1)$ in the direction of $\vec{i} - \vec{j}$

14. $f(x, y) = xe^y$ at $(3, 0)$ in the direction of $4\vec{i} - 3\vec{j}$

15. $f(x, y, z) = x^2 + y^2 - z^2$ at $(2, 3, 4)$ in the direction of $2\vec{i} - 2\vec{j} + \vec{k}$

16. $f(x, y, z) = e^{x+z} \cos y$ at $(1, 0, -1)$ in the direction of $\vec{i} + \vec{j} + \vec{k}$

17. $f(x, y, z) = 3x^2y^2 + 2yz$ at $(-1, 0, 4)$ in the direction of $\vec{i} - \vec{k}$

18. $f(x, y, z) = 3x^2y^2 + 2yz$ at $(-1, 0, 4)$ in the direction of $-\vec{i} + 3\vec{j} + 3\vec{k}$

In Exercises 19–21, find a vector normal to the given curve or surface at the given point.

19. $x^2 - y^2 = 3$ at $(2, 1)$

20. $xy + xz + yz = 11$ at $(1, 2, 3)$

21. $z^2 - 2xyz = x^2 + y^2$ at $(1, 2, -1)$

In Exercises 22–24, find the quadratic Taylor polynomial for the given function.

22. $f(x, y) = \cos x \cos 3y$ about $(0, 0)$

23. $f(x, y) = (x + 1)^3(y + 2)$ about $(0, 0)$

24. $f(x, y) = \sqrt{2x - y}$ about $(3, 5)$

Problems

25. (a) Let $f(w, z) = e^{w \ln z}$. Use difference quotients with $h = 0.01$ to approximate $f_w(2, 2)$ and $f_z(2, 2)$.
    (b) Now evaluate $f_w(2, 2)$ and $f_z(2, 2)$ exactly.

In Problems 26–33, the function f is differentiable and $f_x(2, 1) = -3$, $f_y(2, 1) = 4$, and $f(2, 1) = 7$.

26. (a) Give an equation for the tangent plane to the graph of f at $x = 2, y = 1$.
    (b) Give an equation for the tangent line to the contour for f at $x = 2, y = 1$.

27. (a) Find a vector perpendicular to the tangent plane to the graph of f at $x = 2, y = 1$.
    (b) Find a vector perpendicular to the tangent line to the contour for f at $x = 2, y = 1$.

28. Near $x = 2$ and $y = 1$, how far apart are the contours $f(x, y) = 7$ and $f(x, y) = 7.3$?

29. Give an approximate table of values of f for $x = 1.8, 2.0, 2.2$ and $y = 0.9, 1.0, 1.1$.

30. Give an approximate contour diagram for f for $1 \le x \le 3, 0 \le y \le 2$, using contour values ... $5, 6, 7, 8, 9$ ....

31. The function f gives temperature in $^\circ$C and x and y are in centimeters. A bug leaves $(2, 1)$ at 3 cm/min, trying to cool off as fast as possible. In which direction should the bug head? At what rate does the bug cool off, in $^\circ$C/min?

32. Find $f_r(2, 1)$ and $f_\theta(2, 1)$, where r and θ are polar coordinates, $x = r \cos \theta$ and $y = r \sin \theta$. If $\vec{u}$ is the unit vector in the direction $2\vec{i} + \vec{j}$, show that $f_{\vec{u}}(2, 1) = f_r(2, 1)$ and explain why this should be the case.

33. Find approximately the largest value of f on or inside the circle of radius 0.1 about the point $(2, 1)$. At what point does f achieve this value?

34. Figure 14.66 shows a contour diagram for the temperature T (in $^\circ$C) along a wall in a heated room as a function of distance x along the wall and time t in minutes. Estimate $\partial T/\partial x$ and $\partial T/\partial t$ at the given points. Give units for your answers and say what the answers mean.

    (a) $x = 15, t = 20$ (b) $x = 5, t = 12$

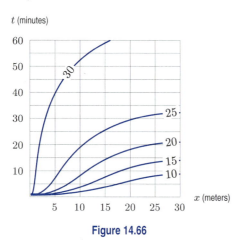

Figure 14.66

35. The quantity Q (in pounds) of beef that a certain community buys during a week is a function $Q = f(b, c)$ of the prices of beef, b, and chicken, c, during the week. Do you expect $\partial Q / \partial b$ to be positive or negative? What about $\partial Q / \partial c$?

36. The cost of producing one unit of a product is given by

$$c = a + bx + ky,$$

where x is the amount of labor used (in man hours) and y is the amount of raw material used (by weight) and a and b and k are constants. What does $\partial c / \partial x = b$ mean? What is the practical interpretation of b?

37. People commuting to a city can choose to go either by bus or by train. The number of people who choose either method depends in part upon the price of each. Let $f(P_1, P_2)$ be the number of people who take the bus when P_1 is the price of a bus ride and P_2 is the price of a train ride. What can you say about the signs of $\partial f / \partial P_1$ and $\partial f / \partial P_2$? Explain your answers.

38. The acceleration g due to gravity, at a distance r from the center of a planet of mass m, is given by

$$g = \frac{Gm}{r^2},$$

where G is the universal gravitational constant.

(a) Find $\partial g / \partial m$ and $\partial g / \partial r$.
(b) Interpret each of the partial derivatives you found in part (a) as the slope of a graph in the plane and sketch the graph.

39. In analyzing a factory and deciding whether or not to hire more workers, it is useful to know under what circumstances productivity increases. Suppose $P = f(x_1, x_2, x_3)$ is the total quantity produced as a function of x_1, the number of workers, and any other variables x_2, x_3. We define the average productivity of a worker as P / x_1. Show that the average productivity increases as x_1 increases when marginal production, $\partial P / \partial x_1$, is greater than the average productivity, P / x_1.

40. Figure 14.67 shows a contour diagram for a vibrating string function, $f(x, t)$.

(a) Is $f_t(\pi/2, \pi/2)$ positive or negative? How about $f_t(\pi/2, \pi)$? What does the sign of $f_t(\pi/2, b)$ tell you about the motion of the point on the string at $x = \pi/2$ when $t = b$?
(b) Find all t for which f_t is positive, for $0 \le t \le 5\pi/2$.
(c) Find all x and t such that f_x is positive.

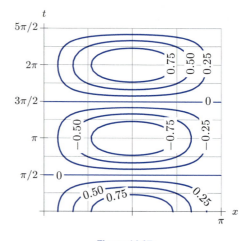

Figure 14.67

41. For the Cobb-Douglas production function $P = 40L^{0.25}K^{0.75}$, find the differential dP when $L = 2$ and $K = 16$.

42. The period, T, of a pendulum is $T = 2\pi \sqrt{l/g}$. If the approximate length is $l = 2$ meters, find the approximate error in T if the true length is $l = 1.99$ and we take $g = 9.8$ as an approximation for $g = 9.81$ m/s^2.

43. Figure 14.68 shows the monthly payment, m, on a 5-year car loan if you borrow P dollars at r percent interest. Find a formula for a linear function which approximates m. What is the practical significance of the constants in your formula?

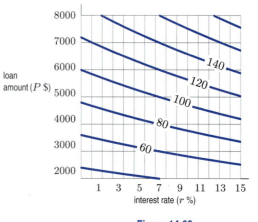

Figure 14.68

44. If $f(x, y) = x^2 y$ and $\vec{v} = 4\vec{i} - 3\vec{j}$, find the directional derivative at the point $(2, 6)$ in the direction of $\vec{v}$.

45. Find the directional derivative of $z = x^2 - y^2$ at the point $(3, -1)$ in the direction making an angle $\theta = \pi/4$ with the x-axis. In which direction is the directional derivative the largest?

46. Figure 14.69 shows the level curves of a function $f(x, y)$. Give the approximate value of $f_{\vec{u}}(3, 1)$ with $\vec{u} = (-2\vec{i} + \vec{j})/\sqrt{5}$. Explain your answer.

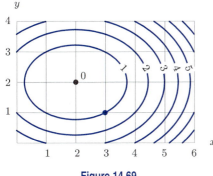

Figure 14.69

47. The temperature at (x, y) is $T(x, y) = 100 - x^2 - y^2$. In which direction should a heat-seeking bug move from the point (x, y) to increase its temperature fastest?

48. Find the point(s) on $x^2 + y^2 + z^2 = 8$ where the tangent plane is parallel to the plane $x - y + 3z = 0$.

49. Values of the function $f(x, y)$ near the point $x = 2$, $y = 3$ are given in Table 14.8. Estimate the following.

(a) $\left.\dfrac{\partial f}{\partial x}\right|_{(2,3)}$ and $\left.\dfrac{\partial f}{\partial y}\right|_{(2,3)}$.

(b) The rate of change of f at $(2, 3)$ in the direction of the vector $\vec{i} + 3\vec{j}$.

(c) The maximum possible rate of change of f as you move away from the point $(2, 3)$. In which direction should you move to obtain this rate of change?

(d) Write an equation for the level curve through the point $(2, 3)$.

(e) Find a vector tangent to the level curve of f through the point $(2, 3)$.

(f) Find the differential of f at the point $(2, 3)$. If $dx = 0.03$ and $dy = 0.04$, find df. What does df represent in this case?

Table 14.8

		x	
		2.00	2.01
y	3.00	7.56	7.42
	3.02	7.61	7.47

50. Find the quadratic Taylor polynomial about $(0, 0)$ for $f(x, y) = \cos(x + 2y)\sin(x - y)$.

51. Suppose $f(x, y) = e^{(x-1)^2 + (y-3)^2}$.

(a) Find the first-order Taylor polynomial about $(0, 0)$.

(b) Find the second-order (quadratic) Taylor polynomial about the point $(1, 3)$.

(c) Find a 2-vector perpendicular to the level curve through $(0, 0)$.

(d) Find a 3-vector perpendicular to the surface $z = f(x, y)$ at the point $(0, 0)$.

52. Show that if F is any differentiable function of one variable, then $V(x, y) = xF(2x + y)$ satisfies the equation

$$x\frac{\partial V}{\partial x} - 2x\frac{\partial V}{\partial y} = V.$$

53. Find a particular solution to the differential equation in Problem 52 satisfying

$$V(1, y) = y^2.$$

54. Each diagram (I) – (IV) in Figure 14.70 represents the level curves of a function $f(x, y)$. For each function f, consider the point above P on the surface $z = f(x, y)$ and choose from the lists which follow:

(a) A vector which could be the normal to the surface at that point;

(b) An equation which could be the equation of the tangent plane to the surface at that point.

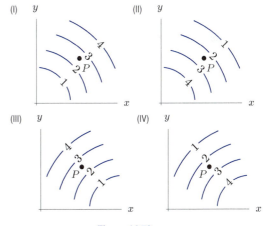

Figure 14.70

Vectors	Equations
(E) $2\vec{i} + 2\vec{j} - 2\vec{k}$	(J) $x + y + z = 4$
(F) $2\vec{i} + 2\vec{j} + 2\vec{k}$	(K) $2x - 2y - 2z = 2$
(G) $2\vec{i} - 2\vec{j} + 2\vec{k}$	(L) $-3x - 3y + 3z = 6$
(H) $-2\vec{i} + 2\vec{j} + 2\vec{k}$	(M) $-\dfrac{x}{2} + \dfrac{y}{2} - \dfrac{z}{2} = -7$

CAS Challenge Problems

55. (a) Find the quadratic Taylor polynomial about $(0, 0)$ of
$$f(x, y) = \frac{e^x (1 + \sin(3y))^2}{5 + e^{2x}}.$$
   (b) Find the quadratic Taylor polynomial about 0 of the one-variable functions $g(x) = e^x/(5 + e^{2x})$ and $h(y) = (1 + \sin(3y))^2$. Multiply these polynomials together and compare with your answer to part (a).
   (c) Show that if $f(x, y) = g(x)h(y)$, then the quadratic Taylor polynomial of f about $(0, 0)$ is the product of the quadratic Taylor polynomials of $g(x)$ and $h(y)$ about 0. [If you use a computer algebra system, make sure that f, g and h do not have any previously assigned formula.]

56. Let
$$f(x, y) = A_0 + A_1 x + A_2 y + A_3 x^2 + A_4 xy + A_5 y^2,$$
$$g(t) = 1 + B_1 t + B_2 t^2,$$
$$h(t) = 2 + C_1 t + C_2 t^2.$$

   (a) Find $L(x, y)$, the linear approximation to $f(x, y)$ at the point $(1, 2)$. Also find $m(t)$ and $n(t)$, the linear approximations to $g(t)$ and $h(t)$ at $t = 0$.
   (b) Calculate $(d/dt)f(g(t), h(t))|_{t=0}$ and $(d/dt)L(m(t), n(t))|_{t=0}$. Describe what you notice and explain it in terms of the chain rule.

57. Let $f(x, y) = A_0 + A_1 x + A_2 y + A_3 x^2 + A_4 xy + A_5 y^2$.

   (a) Find the quadratic Taylor approximation for $f(x, y)$ near the point $(1, 2)$, and expand the result in powers of x and y.
   (b) Explain what you notice in part (a) and formulate a generalization to points other than $(1, 2)$.
   (c) Repeat part (a) for the linear approximation. How does it differ from the quadratic?

58. Suppose that $w = f(x, y, z)$, that x and y are functions of u and v, and that z, u, and v are functions of t. Use a computer algebra system to find the derivative
$$\frac{d}{dt} f(x(u(t), v(t)), y(u(t), v(t)), z(t))$$
and explain the answer using a tree diagram.

CHECK YOUR UNDERSTANDING

Are the statements in Problems 1–40 true or false? Give reasons for your answer.

1. If $f(x, y)$ is a function of two variables and $f_x(10, 20)$ is defined, then $f_x(10, 20)$ is a scalar.

2. If $f(x, y) = x^2 + y^2$, then $f_y(1, 1) < 0$.

3. If the graph of $f(x, y)$ is a hemisphere centered at the origin, then $f_x(0, 0) = f_y(0, 0) = 0$.

4. If $P = f(T, V)$ is a function expressing the pressure P (in grams/cm³) of gas in a piston in terms of the temperature T (in degrees °C) and volume V (in cm³), then $\partial P/\partial V$ has units of grams.

5. If $f_x(a, b) > 0$, then the values of f decrease as we move in the negative x-direction near (a, b).

6. If $g(r, s) = r^2 + s$, then for fixed s, the partial derivative g_r increases as r increases.

7. If $g(u, v) = (u + v)^u$, then $2.3 \leq g_u(1, 1) \leq 2.4$.

8. Let $P = f(m, d)$ be the purchase price (in dollars) of a used car that has m miles on its engine and originally cost d dollars when new. Then $\partial P/\partial m$ and $\partial P/\partial d$ have the same sign.

9. If $f(x, y)$ is a function with the property that $f_x(x, y)$ and $f_y(x, y)$ are both constant, then f is linear.

10. If the point (a, b) is on the contour $f(x, y) = k$, then the slope of the line tangent to this contour at (a, b) is $f_y(a, b)/f_x(a, b)$.

11. There is a function $f(x, y)$ with $f_x(x, y) = y$ and $f_y(x, y) = x$.

12. The function $z(u, v) = u \cos v$ satisfies the equation
$$\cos v \frac{\partial z}{\partial u} - \frac{\sin v}{u} \frac{\partial z}{\partial v} = 1.$$

13. If $f(x, y)$ is a function of two variables and $g(x)$ is a function of a single variable, then
$$\frac{\partial}{\partial y} (g(x)f(x, y)) = g(x)f_y(x, y).$$

14. The function $k(r, s) = rse^s$ is increasing in the s-direction at the point $(r, s) = (-1, 2)$.

15. There is a function $f(x, y)$ with $f_x(x, y) = y^2$ and $f_y(x, y) = x^2$.

16. If f is a symmetric two-variable function, that is $f(x, y) = f(y, x)$, then $f_x(x, y) = f_y(x, y)$.

17. If f is a symmetric two-variable function, that is $f(x, y) = f(y, x)$, then $f_x(x, y) = f_y(y, x)$.

18. If $f(x, y)$ has $f_y(x, y) = 0$ then f must be a constant.

19. If $f(x, y) = ye^{g(x)}$ then $f_x = f$.

20. If $f(x, y)$ has $f_x(a, b) = f_y(a, b) = 0$ at the point (a, b), then f is constant everywhere.

21. An equation for the tangent plane to the surface $z = x^2 + y^3$ at $(1, 1)$ is $z = 2 + 2x(x - 1) + 3y^2(y - 1)$.

22. There is a function $f(x, y)$ which has a tangent plane with equation $z = 0$ at a point (a, b).

23. The tangent plane approximation of $f(x,y) = ye^{x^2}$ at the point $(0,1)$ is $f(x,y) \approx y$.

24. If f is a function with differential $df = 2y\,dx + \sin(xy)\,dy$, then f changes by about -0.4 between the points $(1,2)$ and $(0.9, 2.0002)$.

25. The local linearization of $f(x,y) = x^2 + y^2$ at $(1,1)$ gives an overestimate of the value of $f(x,y)$ at the point $(1.04, 0.95)$.

26. If two functions f and g have the same differential at the point $(1,1)$, then $f = g$.

27. If two functions f and g have the same tangent plane at a point $(1,1)$, then $f = g$.

28. If $f(x,y)$ is a constant function, then $df = 0$.

29. If $f(x,y)$ is a linear function, then df is a linear function of dx and dy.

30. If you zoom close enough near a point (a,b) on the contour diagram of any differentiable function, the contours will be *precisely* parallel and *exactly* equally spaced.

31. The gradient vector grad $f(a,b)$ is a vector in 3-space.

32. The gradient vector grad $f(a,b)$ is tangent to the contour of f at (a,b).

33. If you know the gradient vector of f at (a,b) then you can find the directional derivative $f_{\vec{u}}(a,b)$ for any unit vector $\vec{u}$.

34. If you know the directional derivative $f_{\vec{u}}(a,b)$ for all unit vectors $\vec{u}$ then you can find the gradient vector of f at (a,b).

35. The directional derivative $f_{\vec{u}}(a,b)$ is parallel to $\vec{u}$.

36. The gradient grad $f(3,4)$ is perpendicular to the vector $3\vec{i} + 4\vec{j}$.

37. If grad $f(1,2) = \vec{i}$, then f decreases in the $-\vec{i}$ direction at $(1,2)$.

38. If grad $f(1,2) = \vec{i}$, then $f(10,2) > f(1,2)$.

39. At the point $(3,0)$, the function $g(x,y) = x^2 + y^2$ has the same maximal rate of increase as that of the function $h(x,y) = 2xy$.

40. If $f(x,y) = e^{x+y}$, then the directional derivative in any direction $\vec{u}$ (with $\|\vec{u}\| = 1$) at the point $(0,0)$ is always less than or equal to $\sqrt{2}$.

Assume that $f(x,y)$ is a differentiable function. Are the statements in Problems 41–47 true or false? Explain your answer.

41. $f_{\vec{u}}(x_0, y_0)$ is a scalar.

42. $f_{\vec{u}}(a,b) = \|\nabla f(a,b)\|$

43. If $\vec{u}$ is tangent to the level curve of f at some point, then grad $f \cdot \vec{u} = 0$ there.

44. Suppose that f is differentiable at (a,b). Then there is always a direction in which the rate of change of f at (a,b) is 0.

45. There is a function with a point in its domain where $\|\text{grad } f\| = 0$ and where there is a nonzero directional derivative.

46. There is a function with $\|\text{grad } f\| = 4$ and $f_{\vec{i}} = 5$ at some point.

47. There is a function with $\|\text{grad } f\| = 4$ and $f_{\vec{j}} = -3$ at some point.

PROJECTS FOR CHAPTER FOURTEEN

1. **Heat Equation**

   The function $T(x,y,z,t)$ is a solution to the *heat equation*

   $$T_t = K(T_{xx} + T_{yy} + T_{zz}),$$

   and gives the temperature at the point (x,y,z) in 3-space and time t. The constant K is the *thermal conductivity* of the medium through which the heat is flowing.

   (a) Show that the function

   $$T(x,y,z,t) = \frac{1}{(4\pi Kt)^{3/2}} e^{-(x^2+y^2+z^2)/4Kt}$$

   is a solution to the heat equation for all (x,y,z) in 3-space and $t > 0$.

   (b) For each fixed time t, what are the level surfaces of the function $T(x,y,z,t)$ in 3-space?

   (c) Regard t as fixed and compute grad $T(x,y,z,t)$. What does grad $T(x,y,z,t)$ tell us about the direction and magnitude of the heat flow?

2. **Matching Birthdays**

   Consider a class of m students and a year with n days. Let $q(m,n)$ denote the probability, expressed as a number between 0 and 1, that at least two students have the same birthday. Surprisingly, $q(23, 365) \approx 0.5073$. (This means that there is slightly better than an even chance that at least two students in a class of 23 have the same birthday.) A general formula for q is complicated, but it can be shown that

$$\frac{\partial q}{\partial m} \approx +\frac{m}{n}(1-q) \qquad \text{and} \qquad \frac{\partial q}{\partial n} \approx -\frac{m^2}{2n^2}(1-q).$$

(These approximations hold when n is a good deal larger than m, and m is a good deal larger than 1.)

(a) Explain why the $+$ and $-$ signs in the approximations for $\partial q/\partial m$ and $\partial q/\partial n$ are to be expected.

(b) Suppose there are 21 students in a class. What is the approximate probability that at least two students in the class have the same birthday? (Assume that a year always has 365 days.)

(c) Suppose there is a class of 24 students and you know that no one was born in the first week of the year. (This has the effect of making $n = 358$.) What is the approximate value of q for this class?

(d) If you want to bet that a certain class of 23 students has at least two matching birthdays, would you prefer to have two more students added to the class or to be told that no one in the class was born in December?

(e) (Optional) Find the actual formula for q. [Hint: It's easier to find $1 - q$. There are $n \cdot n \cdot n \cdots n = n^m$ different choices for the students' birthdays. How many such choices have no matching birthdays?]

Chapter Fifteen

OPTIMIZATION: LOCAL AND GLOBAL EXTREMA

In one-variable calculus we saw how to find the maximum and minimum values of a function of one variable. In practice, there are often several variables in an optimization problem. For example, you may have $10,000 to invest in new equipment and advertising for your business. What combination of equipment and advertising will yield the greatest profit? Or, what combination of drugs will lower a patient's temperature the most? In this chapter we consider optimization problems, where the variables are completely free to vary (unconstrained optimization) and where there is a constraint on the variables (for example, a budget constraint).

Exa

Solu

Exampl

Solution

Exampl

Solution

23. (a) Find the critical point of $f(x, y) = (x^2 - y)(x^2 + y)$.
(b) Show that at the critical point, the discriminant $D = 0$, so the second derivative test gives no information about the nature of the critical point.
(c) Sketch contours near the critical point to determine whether it is a local maximum, a local minimum, a saddle point, or none of these.

24. The behavior of a function can be complicated near a critical point where $D = 0$. Suppose that

$$f(x, y) = x^3 - 3xy^2.$$

Show that there is one critical point at $(0, 0)$ and that $D = 0$ there. Show that the contour for $f(x, y) = 0$ consists of three lines intersecting at the origin and that these lines divide the plane into six regions around the origin where f alternates from positive to negative. Sketch a contour diagram for f near $(0, 0)$. The graph of this function is called a *monkey saddle*.

25. On a computer, draw contour diagrams for functions

$$f(x, y) = k(x^2 + y^2) - 2xy$$

for $k = -2, -1, 0, 1, 2$. Use these figures to classify the critical point at $(0, 0)$ for each value of k. Explain your observations using the discriminant, D.

26. Let $f(x, y) = kx^2 + y^2 - 4xy$. Determine the values of k (if any) for which the critical point at $(0, 0)$ is:

(a) A saddle point
(b) A local maximum
(c) A local minimum

27. Draw a possible contour diagram of f such that $f_x(-1, 0) = 0$, $f_y(-1, 0) < 0$, $f_x(3, 3) > 0$, $f_y(3, 3) > 0$, and f has a local maximum at $(3, -3)$.

28. Draw a possible contour diagram of a function with a saddle point at $(2, 1)$, a local minimum at $(2, 4)$, and no other critical points. Label the contours.

15.2   OPTIMIZATION

Suppose we want to find the highest and the lowest points in Colorado. A contour map is shown in Figure 15.19. The highest point is the top of a mountain peak (point A on the map, Mt. Elbert, 14,431 feet high). What about the lowest point? Colorado does not have large pits without drainage, like Death Valley in California. A drop of rain falling at any point in Colorado will eventually flow out of the state. If there is no local minimum inside the state, where is the lowest point? It must be on the state boundary at a point where a river is flowing out of the state (point B where the Arkansas River leaves the state, 3,400 feet high). The highest point in Colorado is a global maximum for the elevation function in Colorado and the lowest point is the global minimum.

In general, if we are given a function f defined on a region R, we say:

- f has a **global maximum on** R at the point P_0 if $f(P_0) \geq f(P)$ for all points P in R.
- f has a **global minimum on** R at the point P_0 if $f(P_0) \leq f(P)$ for all points P in R.

The process of finding a global maximum or minimum for a function f on a region R is called *optimization*. If the region R is the entire xy-plane, we have *unconstrained optimization*; if the region R is not the entire xy-plane, that is, if x or y is restricted in some way, then we have *constrained optimization*. If the region R is not stated explicitly, it is understood to be the whole xy-plane.

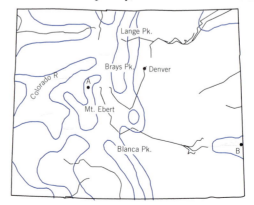

Figure 15.19: The highest and lowest points in the state of Colorado

How Do We Find Global Maxima and Minima?

As the Colorado example illustrates, a global extremum can occur either at a critical point inside the region or at a point on the boundary of the region. This is analogous to single-variable calculus, where a function achieves its global extrema on an interval either at a critical point inside the interval or at an endpoint of the interval. Optimization for functions of more than one variable, however, is more difficult because regions in 2-space can have very complicated boundaries.

For an Unconstrained Optimization Problem

- Find the critical points.
- Investigate whether the critical points give global maxima or minima.

Not all functions have a global maximum or minimum: it depends on the function and the region. First, we consider applications in which global extrema are expected from practical considerations. At the end of this section, we examine the conditions that lead to global extrema. In general, the fact that a function has a single local maximum or minimum does not guarantee that the point is the global maximum or minimum. (See Problem 27.) An exception is if the function is quadratic, in which case the local maximum or minimum is the global maximum or minimum. (See Example 1 on page 703 and Example 5 on page 706.)

Maximizing Profit and Minimizing Cost

In planning production of an item, a company often chooses the combination of price and quantity that maximizes its profit. We use

$$\text{Profit} = \text{Revenue} - \text{Cost},$$

and, provided the price is constant,

$$\text{Revenue} = \text{Price} \cdot \text{Quantity} = pq.$$

In addition, we need to know how the cost and price depend on quantity.

Example 1 A company manufactures two items which are sold in two separate markets. The quantities, q_1 and q_2, demanded by consumers, and the prices, p_1 and p_2 (in dollars), of each item are related by

$$p_1 = 600 - 0.3q_1 \quad \text{and} \quad p_2 = 500 - 0.2q_2.$$

Thus, if the price for either item increases, the demand for it decreases. The company's total production cost is given by

$$C = 16 + 1.2q_1 + 1.5q_2 + 0.2q_1q_2.$$

To maximize its total profit, how much of each product should be produced? What is the maximum profit? [2]

Solution The total revenue, R, is the sum of the revenues, p_1q_1 and p_2q_2, from each market. Substituting for p_1 and p_2, we get

$$R = p_1q_1 + p_2q_2 = (600 - 0.3q_1)q_1 + (500 - 0.2q_2)q_2$$
$$= 600q_1 - 0.3q_1^2 + 500q_2 - 0.2q_2^2.$$

Thus, the total profit P is given by

$$P = R - C = 600q_1 - 0.3q_1^2 + 500q_2 - 0.2q_2^2 - (16 + 1.2q_1 + 1.5q_2 + 0.2q_1q_2)$$
$$= -16 + 598.8q_1 - 0.3q_1^2 + 498.5q_2 - 0.2q_2^2 - 0.2q_1q_2.$$

---

[2] Adapted from M. Rosser, *Basic Mathematics for Economists*, p. 316 (New York: Routledge, 1993).

To maximize P, we compute partial derivatives and set them equal to 0:

$$\frac{\partial P}{\partial q_1} = 598.8 - 0.6q_1 - 0.2q_2 = 0,$$

$$\frac{\partial P}{\partial q_2} = 498.5 - 0.4q_2 - 0.2q_1 = 0.$$

Since grad P is defined everywhere, the only critical points of P are those where grad $P = \vec{0}$. Thus, solving for q_1, q_2, we find that

$$q_1 = 699.1 \quad \text{and} \quad q_2 = 896.7.$$

The corresponding prices are

$$p_1 = 390.27 \quad \text{and} \quad p_2 = 320.66.$$

To see whether or not we have found a maximum, we compute second partial derivatives:

$$\frac{\partial^2 P}{\partial q_1^2} = -0.6, \quad \frac{\partial^2 P}{\partial q_2^2} = -0.4, \quad \frac{\partial^2 P}{\partial q_1 \partial q_2} = -0.2,$$

so,

$$D = \frac{\partial^2 P}{\partial q_1^2} \frac{\partial^2 P}{\partial q_2^2} - \left(\frac{\partial^2 P}{\partial q_1 \partial q_2} \right)^2 = (-0.6)(-0.4) - (-0.2)^2 = 0.2.$$

Therefore we have found a local maximum. The graph of P is an upside-down paraboloid, so $(699.1, 896.7)$ is a global maximum. The company should produce 699.1 units of the first item priced at \$390.27 per unit, and 896.7 units of the second item priced at \$320.66 per unit. The maximum profit $P(699.1, 896.7) \approx \$433,000$.

Example 2   Twenty cubic meters of gravel are to be delivered to a landfill. The trucker plans to purchase an open-top box in which to transport the gravel in numerous trips. The total cost to the trucker is the cost of the box plus \$2 per trip. The box must have height 0.5 m, but the trucker can choose the length and width. The cost of the box is \$20/m² for the ends and \$10/m² for the bottom and sides. Notice the tradeoff: A smaller box is cheaper to buy but requires more trips. What size box should the trucker buy to minimize the total cost? [3]

Solution   We first get an algebraic expression for the trucker's cost. Let the length of the box be x meters and the width be y meters and the height be 0.5 m (See Figure 15.20.)

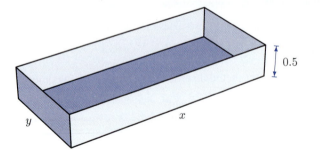

Table 15.2   *Trucker's itemized cost*

20/(0.5xy) at \$2/trip	80/(xy)
2 ends at \$20/m² · 0.5y m²	20y
2 sides at \$10/m² · 0.5x m²	10x
1 bottom at \$10/m² · xy m²	10xy
Total cost	$f(x, y)$

Figure 15.20: The box for transporting gravel

The volume of the box is $0.5xy$ m³, so delivery of 20 m³ of gravel requires $20/(0.5xy)$ trips. The trucker's cost is itemized in Table 15.2. The problem is to minimize

$$\text{Total cost} = f(x, y) = \frac{80}{xy} + 20y + 10x + 10xy.$$

[3] Adapted from Claude McMillan, Jr., *Mathematical Programming*, 2nd ed., p. 156-157 (New York: Wiley, 1978).

The critical points of this function occur where

$$f_x(x, y) = -\frac{80}{x^2 y} + 10 + 10y = 0$$

$$f_y(x, y) = -\frac{80}{xy^2} + 10x + 20 = 0$$

We put the $80/(x^2 y)$ and $80/(xy^2)$ terms on the other side of the the equation, divide, and simplify:

$$\frac{80/(x^2 y)}{80/(xy^2)} = \frac{10 + 10y}{10x + 20} \quad \text{so} \quad \frac{y}{x} = \frac{1 + y}{x + 2} \quad \text{giving} \quad 2y = x.$$

Substituting $x = 2y$ in the equation $f_y(x, y) = 0$ gives

$$-\frac{80}{2y \cdot y^2} + 10(2y) + 20 = 0$$

$$y^4 + y - 2 = 0.$$

The only positive solution to this equation is $y = 1$, so the only critical point is $(2, 1)$.

To check that the critical point is a minimum, we use the second derivative test. Since

$$D(2, 1) = f_{xx} f_{yy} - (f_{xy})^2 = \frac{160}{2^3 \cdot 1} \cdot \frac{160}{2 \cdot 1^3} - \left(\frac{80}{2^2 \cdot 1^2} + 10 \right)^2 = 700 > 0$$

and $f_{xx}(2, 1) > 0$, the point $(2, 1)$ is a local minimum. Since the value of f increases without bound as x or y increases without bound and as $x \to 0^+$ and $y \to 0^+$, it can be shown that $(2, 1)$ is a global minimum. (See Problem 28.) Thus, the optimal box is 2 meters long and 1 meter wide.

Fitting a Line to Data

Suppose we want to fit the "best" line to some data in the plane. We measure the distance from a line to the data points by adding the squares of the vertical distances from each point to the line. The smaller this sum of squares is, the better the line fits the data. The line with the minimum sum of square distances is called the *least squares line*, or the *regression line*. If the data is nearly linear, the least squares line is a good fit; otherwise it may not be. (See Figure 15.21.)

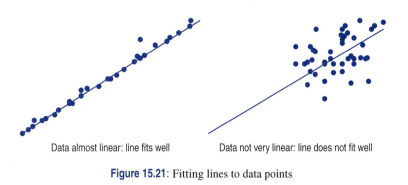

Data almost linear: line fits well   Data not very linear: line does not fit well

Figure 15.21: Fitting lines to data points

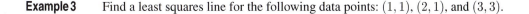

Example 3   Find a least squares line for the following data points: $(1, 1)$, $(2, 1)$, and $(3, 3)$.

Solution   Suppose the line has equation $y = b + mx$. If we find b and m then we have found the line. So, for this problem, b and m are the two variables. We want to minimize the function $f(b, m)$ that gives the sum of the three squared vertical distances from the points to the line in Figure 15.22.

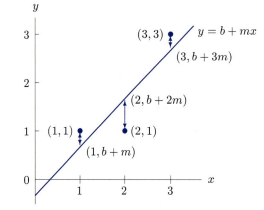

Figure 15.22: The least squares line minimizes the sum of the squares of these vertical distances

The vertical distance from the point $(1, 1)$ to the line is the difference in the y-coordinates $1 - (b + m)$; similarly for the other points. Thus, the sum of squares is

$$f(b, m) = (1 - (b + m))^2 + (1 - (b + 2m))^2 + (3 - (b + 3m))^2.$$

To minimize f we look for critical points. First we differentiate f with respect to b:

$$\frac{\partial f}{\partial b} = -2(1 - (b + m)) - 2(1 - (b + 2m)) - 2(3 - (b + 3m))$$
$$= -2 + 2b + 2m - 2 + 2b + 4m - 6 + 2b + 6m$$
$$= -10 + 6b + 12m.$$

Now we differentiate with respect to m:

$$\frac{\partial f}{\partial m} = 2(1 - (b + m))(-1) + 2(1 - (b + 2m))(-2) + 2(3 - (b + 3m))(-3)$$
$$= -2 + 2b + 2m - 4 + 4b + 8m - 18 + 6b + 18m$$
$$= -24 + 12b + 28m.$$

The equations $\dfrac{\partial f}{\partial b} = 0$ and $\dfrac{\partial f}{\partial m} = 0$ give a system of two linear equations in two unknowns:

$$-10 + 6b + 12m = 0,$$
$$-24 + 12b + 28m = 0.$$

The solution to this pair of equations is the critical point $b = -1/3$ and $m = 1$. Since

$$D = f_{bb}f_{mm} - (f_{mb})^2 = (6)(28) - 12^2 = 24 \quad \text{and} \quad f_{bb} = 6 > 0,$$

we have found a local minimum. The graph of $f(b, m)$ is a parabolic bowl, so the local minimum is the global minimum of f. Thus, the least squares line is

$$y = x - \frac{1}{3}.$$

As a check, notice that the line $y = x$ passes through the points $(1, 1)$ and $(3, 3)$. It is reasonable that introducing the point $(2, 1)$ moves the y-intercept down from 0 to $-1/3$.

The general formulas for the slope and y-intercept of a least squares line are in Project 2 on page 734. Many calculators have built in formulas for b and m, as well as for the *correlation coefficient*, which measures how well the data points fit the least squares line.

How Do We Know Whether a Function Has a Global Maximum or Minimum?

Under what circumstances does a function of two variables have a global maximum or minimum? The next example shows that a function may have both a global maximum and a global minimum on a region, or just one, or neither.

Example 4

Investigate the global maxima and minima of the following functions:

(a) $h(x, y) = 1 + x^2 + y^2$ on the disk $x^2 + y^2 \leq 1$.
(b) $f(x, y) = x^2 - 2x + y^2 - 4y + 5$ on the xy-plane.
(c) $g(x, y) = x^2 - y^2$ on the xy-plane.

Solution

(a) The graph of $h(x, y) = 1 + x^2 + y^2$ is a bowl shaped paraboloid with a global minimum of 1 at $(0, 0)$, and a global maximum of 2 on the edge of the region, $x^2 + y^2 = 1$.
(b) The graph of f in Figure 15.5 on page 703 shows that f has a global minimum at the point $(1, 2)$ and no global maximum (because the value of f increases without bound as $x \to \infty$, $y \to \infty$).
(c) The graph of g in Figure 15.7 on page 704 shows that g has no global maximum because $g(x, y) \to \infty$ as $x \to \infty$ if y is constant. Similarly, g has no global minimum because $g(x, y) \to -\infty$ as $y \to \infty$ if x is constant.

There are, however, conditions that guarantee that a function has a global maximum and minimum. For $h(x)$, a function of one variable, the function must be continuous on a closed interval $a \leq x \leq b$. If h is continuous on a non-closed interval, such as $a \leq x < b$ or $a < x < b$, or on an interval which is not bounded, such as $a < x < \infty$, then h need not have a maximum or minimum value. What is the situation for functions of two variables? As it turns out, a similar result is true for continuous functions defined on regions which are closed and bounded, analogous to the closed and bounded interval $a \leq x \leq b$. In everyday language we say

- A **closed** region is one which contains its boundary;
- A **bounded** region is one which does not stretch to infinity in any direction.

More precise definitions follow. Suppose R is a region in 2-space. A point (x_0, y_0) is a *boundary point* of R if, for every $r > 0$, the disk $(x - x_0)^2 + (y - y_0)^2 < r^2$ with center (x_0, y_0) and radius r contains both points which are in R and points which are not in R. See Figure 15.23. A point (x_0, y_0) can be a boundary point of the region R without belonging to R. A point (x_0, y_0) in R is an *interior point* if it is not a boundary point; thus, for small enough $r > 0$, the disk of radius r centered at (x_0, y_0) lies entirely in the region R. See Figure 15.24. The collection of all the boundary points is the *boundary of R* and the collection of all the interior points is the *interior of R*. The region R is *closed* if it contains its boundary; it is *open* if every point in R is an interior point.

A region R in 2-space is *bounded* if the distance between every point (x, y) in R and the origin is less than some constant K. Closed and bounded regions in 3-space are defined in the same way.

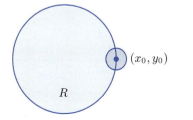

Figure 15.23: Boundary point (x_0, y_0) of R

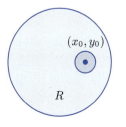

Figure 15.24: Interior point (x_0, y_0) of R

Example 5   (a) The square $-1 \leq x \leq 1$, $-1 \leq y \leq 1$ is closed and bounded.
(b) The first quadrant $x \geq 0$, $y \geq 0$ is closed but is not bounded.
(c) The disk $x^2 + y^2 < 1$ is open and bounded, but is not closed.
(d) The half-plane $y > 0$ is open, but is neither closed nor bounded.

The reason that closed and bounded regions are useful is the following theorem, which is also true for functions of three or more variables:[4]

Theorem 15.1: Extreme Value Theorem for Multivariable Functions

If f is a continuous function on a closed and bounded region R, then f has a global maximum at some point (x_0, y_0) in R and a global minimum at some point (x_1, y_1) in R.

If f is not continuous or the region R is not closed and bounded, there is no guarantee that f achieves a global maximum or global minimum on R. In Example 4, the function g is continuous but does not achieve a global maximum or minimum in 2-space, a region which is closed but not bounded. Example 6 illustrates what can go wrong when the region is bounded but not closed.

Example 6   Does the function f have a global maximum or minimum on the region R given by $0 < x^2 + y^2 \leq 1$?

$$f(x, y) = \frac{1}{x^2 + y^2}$$

Solution   The region R is bounded, but it is not closed since it does not contain the boundary point $(0, 0)$. We see from the graph of $z = f(x, y)$ in Figure 15.25 that f has a global minimum on the circle $x^2 + y^2 = 1$. However, $f(x, y) \to \infty$ as $(x, y) \to (0, 0)$, so f has no global maximum.

Figure 15.25: Graph showing $f(x, y) = \frac{1}{x^2 + y^2}$ has no global maximum on $0 < x^2 + y^2 \leq 1$

Exercises and Problems for Section 15.2

Exercises

1. By looking at the weather map in Figure 12.1 on page 560, find the maximum and minimum daily high temperatures in the states of Mississippi, Alabama, Pennsylvania, New York, California, Arizona, and Massachusetts.

2. Which of the following functions have both a global maximum and global minimum? Which have neither?

$$f(x, y) = x^2 - 2y^2, \ g(x, y) = x^2 y^2, \ h(x, y) = x^3 + y^3$$

3. Does the function $f(x, y) = -2x^2 - 7y^2$ have global maxima and minima? Explain.

4. Does the function $f(x, y) = x^2/2 + 3y^3 + 9y^2 - 3x$ have global maxima and minima? Explain.

In Exercises 5–7, find the global maximum and minimum of the function on $-1 \leq x \leq 1$, $-1 \leq y \leq 1$, and say whether it occurs on the boundary of the square. [Hint: Use graphs.]

5. $z = x^2 + y^2$ **6.** $z = x^2 - y^2$ **7.** $z = -x^2 - y^2$

8. Compute the regression line for the points $(-1, 2)$, $(0, -1)$, $(1, 1)$ using least squares.

---

[4]For a proof, see W. Rudin, *Principles of Mathematical Analysis*, 2nd ed., p. 89, (New York: McGraw-Hill, 1976)

In Exercises 9–11, estimate the position and approximate value of the global maxima and minima on the region shown.

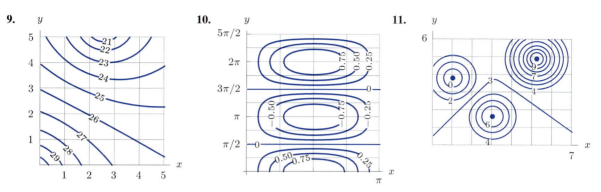

Problems

12. Find the parabola of the form $y = ax^2 + b$ which best fits the points $(1, 0)$, $(2, 2)$, $(3, 4)$ by minimizing the sum of squares, S, given by

$$S = (a + b)^2 + (4a + b - 2)^2 + (9a + b - 4)^2.$$

13. A closed rectangular box has volume $32\,\text{cm}^3$. What are the lengths of the edges giving the minimum surface area?

14. An open rectangular box has volume $32\,\text{cm}^3$. What are the lengths of the edges giving the minimum surface area?

15. A closed rectangular box with faces parallel to the coordinate planes has one bottom corner at the origin and the opposite top corner in the first octant on the plane $3x + 2y + z = 1$. What is the maximum volume of such a box?

16. Find the point on the plane $3x + 2y + z = 1$ that is closest to the origin by minimizing the square of the distance.

17. What is the shortest distance from the surface $xy + 3x + z^2 = 9$ to the origin?

18. The quantity of a product demanded by consumers is a function of its price. The quantity of one product demanded may also depend on the price of other products. For example, the demand for tea is affected by the price of coffee; the demand for cars is affected by the price of gas. The quantities demanded, q_1 and q_2, of two products depend on their prices, p_1 and p_2, as follows

$$q_1 = 150 - 2p_1 - p_2$$

$$q_2 = 200 - p_1 - 3p_2.$$

(a) What does the fact that the coefficients of p_1 and p_2 are negative tell you? Give an example of two products that might be related this way.

(b) If one manufacturer sells both products, how should the prices be set to generate the maximum possible revenue? What is that maximum possible revenue?

19. A company operates two plants which manufacture the same item and whose total cost functions are

$$C_1 = 8.5 + 0.03q_1^2 \quad \text{and} \quad C_2 = 5.2 + 0.04q_2^2,$$

where q_1 and q_2 are the quantities produced by each plant. The total quantity demanded, $q = q_1 + q_2$, is related to the price, p, by

$$p = 60 - 0.04q.$$

How much should each plant produce in order to maximize the company's profit? [5]

20. Two products are manufactured in quantities q_1 and q_2 and sold at prices of p_1 and p_2, respectively. The cost of producing them is given by

$$C = 2q_1^2 + 2q_2^2 + 10.$$

(a) Find the maximum profit that can be made, assuming the prices are fixed.

(b) Find the rate of change of that maximum profit as p_1 increases.

21. A missile has a guidance device which is sensitive to both temperature, $t°C$, and humidity, h. The range in km over which the missile can be controlled is given by

$$\text{Range} = 27{,}800 - 5t^2 - 6ht - 3h^2 + 400t + 300h.$$

What are the optimal atmospheric conditions for controlling the missile?

---

[5] Adapted from M. Rosser, *Basic Mathematics for Economists*, p. 318 (New York: Routledge, 1993).

22. Some items are sold at a discount to senior citizens or children. The reason is that these groups are more sensitive to price, so a discount has greater impact on their purchasing decisions. The seller faces an optimization problem: How large a discount to offer in order to maximize profits? Suppose a theater can sell q_c child tickets and q_a adult tickets at prices p_c and p_a, according to the demand functions:

$$q_c = r p_c^{-4} \quad \text{and} \quad q_a = s p_a^{-2},$$

and has operating costs proportional to the total number of tickets sold. What should be the relative price of children's and adults' tickets?

23. Design a rectangular milk carton box of width w, length l, and height h which holds 512 cm^3 of milk. The sides of the box cost 1 cent/cm^2 and the top and bottom cost 2 cent/cm^2. Find the dimensions of the box that minimize the total cost of materials used.

24. An international airline has a regulation that each passenger can carry a suitcase having the sum of its width, length and height less than or equal to 135 cm. Find the dimensions of the suitcase of maximum volume that a passenger may carry under this regulation.

25. A company manufactures a product which requires capital and labor to produce. The quantity, Q, of the product manufactured is given by the Cobb-Douglas function

$$Q = AK^a L^b,$$

where K is the quantity of capital and L is the quantity of labor used and A, a, and b are positive constants with $0 < a < 1$ and $0 < b < 1$. One unit of capital costs $\$k$ and one unit of labor costs $\$\ell$. The price of the product is fixed at $\$p$ per unit.

(a) If $a + b < 1$, how much capital and labor should the company use to maximize its profit?

(b) Is there a maximum profit in the case $a + b = 1$? What about $a + b \geq 1$? Explain.

26. The population of the US was about 180 million in 1960, grew to 206 million in 1970, and 226 million in 1980.

(a) Assuming that the population was growing exponentially, use logarithms and the method of least squares to estimate the population in 1990.

(b) According to the national census, the 1990 population was 249 million. What does this say about the assumption of exponential growth?

(c) Predict the population in the year 2010.

27. Let $f(x, y) = x^2(y + 1)^3 + y^2$. Show that f has only one critical point, namely $(0, 0)$, and that point is a local minimum but not a global minimum. Contrast this with the case of a function with a single local minimum in one-variable calculus.

28. Let $f(x, y) = 80/(xy) + 20y + 10x + 10xy$ in the region R where $x, y > 0$.

(a) Explain why $f(x, y) > f(2, 1)$ at every point in R where

(i) $x > 20$ (ii) $y > 20$

(iii) $x < 0.01$ and $y \leq 20$

(iv) $y < 0.01$ and $x \leq 20$

(b) Explain why f must have a global minimum at a critical point in R.

(c) Explain why f must have a global minimum in R at the point $(2, 1)$.

29. Let $f(x, y) = 2/x + 3/y + 4x + 5y$ in the region R where $x, y > 0$.

(a) Explain why f must have a global minimum at some point in R.

(b) Find the global minimum.

15.3 CONSTRAINED OPTIMIZATION: LAGRANGE MULTIPLIERS

Many, perhaps most, real optimization problems are constrained by external circumstances. For example, a city wanting to build a public transportation system has only a limited number of tax dollars it can spend on the project. In this section, we see how to find an optimum value under such constraints.

Graphical Approach: Maximizing Production Subject to a Budget Constraint

Suppose we want to maximize the production under a budget constraint. Suppose production, f, is a function of two variables, x and y, which are quantities of two raw materials, and that

$$f(x, y) = x^{2/3} y^{1/3}.$$

If x and y are purchased at prices of p_1 and p_2 thousands of dollars per unit, what is the maximum production f that can be obtained with a budget of c thousand dollars?

To maximize f without regard to the budget, we simply increase x and y. However, the budget constraint prevents us from increasing x and y beyond a certain point. Exactly how does the budget

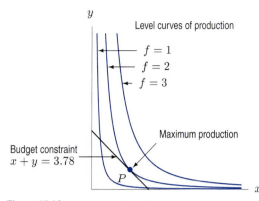

Figure 15.26: Optimal point, P, where budget constraint
is tangent to a level of production function

constrain us? With prices of p_1 and p_2, the amount spent on x is $p_1 x$ and the amount spent on y is $p_2 y$, so we must have

$$g(x, y) = p_1 x + p_2 y \leq c,$$

where $g(x, y)$ is the total cost of the raw materials and c is the budget in thousands of dollars.

Let's look at the case when $p_1 = p_2 = 1$ and $c = 3.78$. Then

$$x + y \leq 3.78.$$

Figure 15.26 shows some contours of f and the budget constraint represented by the line $x + y = 3.78$. Any point on or below the line represents a pair of values of x and y that we can afford. A point on the line completely exhausts the budget, while a point below the line represents values of x and y which can be bought without using up the budget. Any point above the line represents a pair of values that we cannot afford. To maximize f, we find the point which lies on the level curve with the largest possible value of f *and* which lies within the budget. The point must lie on the budget constraint because we should spend all the available money. Unless we are at the point where the budget constraint is tangent to the contour $f = 2$, we can increase f by moving along the line representing the budget constraint in Figure 15.26. For example, if we are on the line to the left of the point of tangency, moving right on the constraint will increase f; if we are on the line to the right of the point of tangency, moving left will increase f. Thus, the maximum value of f on the budget constraint occurs at the point where the budget constraint is tangent to the contour $f = 2$.

Analytical Solution: Lagrange Multipliers

Figure 15.26 suggests that maximum production is achieved at the point where the budget constraint is tangent to a level curve of the production function. The method of Lagrange multipliers uses this fact in algebraic form. Figure 15.27 shows that at the optimum point, P, the gradient of f and the normal to the budget line $g(x, y) = 3.78$ are parallel. Thus, at P, grad f and grad g are parallel, so for some scalar λ, called the *Lagrange multiplier*

$$\text{grad } f = \lambda \, \text{grad } g.$$

Since grad $f = \left(\frac{2}{3} x^{-1/3} y^{1/3}\right) \vec{i} + \left(\frac{1}{3} x^{2/3} y^{-2/3}\right) \vec{j}$ and grad $g = \vec{i} + \vec{j}$, equating components gives

$$\frac{2}{3} x^{-1/3} y^{1/3} = \lambda \quad \text{and} \quad \frac{1}{3} x^{2/3} y^{-2/3} = \lambda.$$

Eliminating λ gives

$$\frac{2}{3} x^{-1/3} y^{1/3} = \frac{1}{3} x^{2/3} y^{-2/3}, \quad \text{which leads to} \quad 2y = x.$$

Since the constraint $x + y = 3.78$ must be satisfied, we have $x = 2.52$ and $y = 1.26$. Then

$$f(2.52, 1.26) = (2.52)^{2/3}(1.26)^{1/3} \approx 2.$$

As before, we see that the maximum value of f is approximately 2. Thus, to maximize production on a budget of \$3780, we should use 2.52 units of one raw material and 1.26 units of the other.

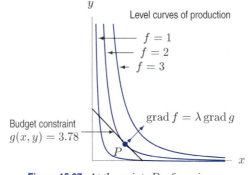

Figure 15.27: At the point, P, of maximum production, the vectors $\operatorname{grad} f$ and $\operatorname{grad} g$ are parallel

Lagrange Multipliers in General

Suppose we want to optimize an *objective function* $f(x, y)$ subject to a *constraint* $g(x, y) = c$. We look for extrema among the points which satisfy the constraint. We make the following definition.

> Suppose P_0 is a point satisfying the constraint $g(x, y) = c$.
> - f has a **local maximum** at P_0 **subject to the constraint** if $f(P_0) \geq f(P)$ for all points P near P_0 satisfying the constraint.
> - f has a **global maximum** at P_0 **subject to the constraint** if $f(P_0) \geq f(P)$ for all points P satisfying the constraint.
>
> Local and global minima are defined similarly.

As we saw in the production example, constrained extrema occur at points where the contours of f are tangent to the contours of g; they can also occur at endpoints of the constraint. In general, at any point, $\operatorname{grad} f$ is perpendicular to the directions in which the directional derivative of f is zero (provided $\operatorname{grad} f \neq \vec{0}$ there). Thus if $\operatorname{grad} f$ is not perpendicular to the constraint, it is possible to increase and decrease f by moving along the constraint. Therefore, at a constrained extremum, $\operatorname{grad} f$ must be perpendicular to the constraint and parallel to $\operatorname{grad} g$. (See Figure 15.28.) Thus, provided $\operatorname{grad} g \neq \vec{0}$ at P_0, there is a constant λ such that $\operatorname{grad} f = \lambda \operatorname{grad} g$ at P_0.

> **Optimizing f Subject to the Constraint $g = c$:**
> If a smooth function, f, has a maximum or minimum subject to a smooth constraint $g = c$ at a point P_0, then either P_0 satifies the equations
>
> $$\operatorname{grad} f = \lambda \operatorname{grad} g \quad \text{and} \quad g = c,$$
>
> or P_0 is an endpoint of the constraint, or $\operatorname{grad} g(P_0) = \vec{0}$. To find P_0, compare values of f at these points. The number λ is called the **Lagrange multiplier**.

If the set of points satisfying the constraint is closed and bounded, such as a circle or line segment, then there must be a global maximum and minimum of f subject to the constraint. If the constraint is not closed and bounded, such as a line or hyperbola, then there may or may not be a global maximum and minimum.

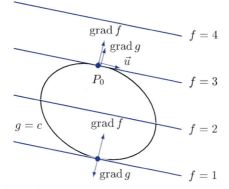

Figure 15.28: Maximum and minimum values
of $f(x, y)$ on $g(x, y) = c$ are at points where
grad f is parallel to grad g

Example 1   Find the maximum and minimum values of $x + y$ on the circle $x^2 + y^2 = 4$.

Solution   The objective function is

$$f(x, y) = x + y,$$

and the constraint is

$$g(x, y) = x^2 + y^2 = 4.$$

Since grad $f = f_x \vec{i} + f_y \vec{j} = \vec{i} + \vec{j}$ and grad $g = g_x \vec{i} + g_y \vec{j} = 2x\vec{i} + 2y\vec{j}$, then grad $f = \lambda$ grad g
gives

$$1 = 2\lambda x \quad \text{and} \quad 1 = 2\lambda y,$$

so

$$x = y.$$

We also know that

$$x^2 + y^2 = 4,$$

giving $x = y = \sqrt{2}$ or $x = y = -\sqrt{2}$. The constraint has no endpoints (it's a circle) and grad $g \neq \vec{0}$
on the circle, so we compare values of f at $(\sqrt{2}, \sqrt{2})$ and $(-\sqrt{2}, -\sqrt{2})$. Since $f(x, y) = x + y$, the
maximum value of f is $f(\sqrt{2}, \sqrt{2}) = 2\sqrt{2}$; the minimum value is $f(-\sqrt{2}, -\sqrt{2}) = -2\sqrt{2}$. (See
Figure 15.29.)

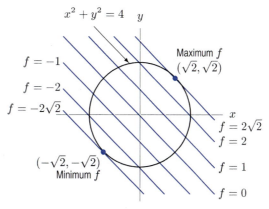

Figure 15.29: Maximum and minimum values of
$f(x, y) = x + y$ on the circle $x^2 + y^2 = 4$ are at points
where contours of f are tangent to the circle

How to Distinguish Maxima from Minima

There is a second derivative test[6] for classifying the critical points of constrained optimization problems, but it is more complicated than the test in Section 15.1. However, a graph of the constraint and some contours usually shows which points are maxima, which points are minima, and which are neither.

Optimization with Inequality Constraints

The production problem that we looked at first was to maximize production $f(x, y)$ subject to a budget constraint

$$g(x, y) = p_1 x + p_2 y \leq c.$$

Since the inputs are nonnegative, $x \geq 0$ and $y \geq 0$, we have three inequality constraints, which restrict (x, y) to a region of the plane rather than to a curve in the plane. In principle, we should first check to see whether or not $f(x, y)$ has any critical points in the interior:

$$p_1 x + p_2 y < c, \qquad x > 0 \quad y > 0.$$

However, in the case of a budget constraint, we can see that the maximum of f must occur when the budget is exhausted, so we look for the maximum value of f on the boundary line:

$$p_1 x + p_2 y = c, \qquad x \geq 0 \quad y \geq 0.$$

Strategy for Optimizing $f(x, y)$ Subject to the Constraint $g(x, y) \leq c$

- Find all points in the interior $g(x, y) < c$ where grad f is zero or undefined.
- Use Lagrange multipliers to find the local extrema of f on the boundary $g(x, y) = c$.
- Evaluate f at the points found in the previous two steps and compare the values.

From Section 15.2 we know that if f is continuous on a closed and bounded region, R, then f is guaranteed to attain its global maximum and minimum values on R.

Example 2   Find the maximum and minimum values of $f(x, y) = (x - 1)^2 + (y - 2)^2$ subject to the constraint $x^2 + y^2 \leq 45$.

Solution   First, we look for all critical points for f in the interior of the region. Setting

$$f_x(x, y) = 2(x - 1) = 0$$
$$f_y(x, y) = 2(y - 2) = 0$$

we find f has exactly one critical point at $x = 1$, $y = 2$. Since $1^2 + 2^2 < 45$, that critical point is in the interior of the region.

Next, we find the local extrema of f on the boundary curve $x^2 + y^2 = 45$. To do this, we use Lagrange multipliers with constraint $g(x, y) = x^2 + y^2 = 45$. Setting grad $f = \lambda$ grad g, we get

$$2(x - 1) = \lambda \cdot 2x,$$
$$2(y - 2) = \lambda \cdot 2y.$$

We can't have $x = 0$ since the first equation would become $-2 = 0$. Similarly, $y \neq 0$. So we can solve each equation for λ by dividing by x and y. Setting the expressions for λ equal gives

$$\frac{x - 1}{x} = \frac{y - 2}{y},$$

---

[6]See J. E. Marsden and A. J. Tromba, *Vector Calculus*, 4th ed., p.218 (New York: W.H. Freeman, 1996).

so
$$y = 2x.$$
Combining this with the constraint $x^2 + y^2 = 45$, we get
$$5x^2 = 45$$
so
$$x = \pm 3.$$
Since $y = 2x$, we have possible local extrema at $x = 3, y = 6$ and $x = -3, y = -6$.

We conclude that the only candidates for the maximum and minimum values of f in the region occur at $(1, 2)$, $(3, 6)$, and $(-3, -6)$. Evaluating f at these three points we find
$$f(1, 2) = 0, \qquad f(3, 6) = 20, \qquad f(-3, -6) = 80.$$
Therefore, the minimum value of f is 0 at $(1, 2)$ and the maximum value is 80 at $(-3, -6)$.

The Meaning of λ

In the uses of Lagrange multipliers so far, we never found (or needed) the value of λ. However, λ does have a practical interpretation. In the production example, we wanted to maximize
$$f(x, y) = x^{2/3} y^{1/3}$$
subject to the constraint
$$g(x, y) = x + y = 3.78.$$
We solved the equations
$$\frac{2}{3} x^{-1/3} y^{1/3} = \lambda,$$
$$\frac{1}{3} x^{2/3} y^{-2/3} = \lambda,$$
$$x + y = 3.78,$$
to get $x = 2.52, y = 1.26$ and $f(2.52, 1.26) \approx 2$. Continuing to find λ gives us
$$\lambda \approx 0.53.$$
Now we do another, apparently unrelated, calculation. Suppose our budget is increased slightly, from 3.78 to 4.78, giving a new budget constraint of $x + y = 4.78$. Then the corresponding solution is at $x = 3.19$ and $y = 1.59$ and the new maximum value (instead of $f = 2$) is
$$f = (3.19)^{2/3} (1.59)^{1/3} \approx 2.53.$$
Notice that the amount by which f has increased is 0.53, the value of λ. Thus, in this example, the value of λ represents the extra production achieved by increasing the budget by one—in other words, the extra "bang" you get for an extra "buck" of budget. In fact, this is true in general:

- The value of λ is approximately the increase in the optimum value of f when the budget is increased by 1 unit.

More precisely:

- The value of λ represents the rate of change of the optimum value of f as the budget increases.

An Expression for λ

To interpret λ, we look at how the optimum value of the objective function f changes as the value c of the constraint function g is varied. In general, the optimum point (x_0, y_0) depends on the constraint value c. So, provided x_0 and y_0 are differentiable functions of c, we can use the chain rule to differentiate the optimum value $f(x_0(c), y_0(c))$ with respect to c:
$$\frac{df}{dc} = \frac{\partial f}{\partial x} \frac{dx_0}{dc} + \frac{\partial f}{\partial y} \frac{dy_0}{dc}.$$

At the optimum point (x_0, y_0), we have $f_x = \lambda g_x$ and $f_y = \lambda g_y$, and therefore

$$\frac{df}{dc} = \lambda \left(\frac{\partial g}{\partial x} \frac{dx_0}{dc} + \frac{\partial g}{\partial y} \frac{dy_0}{dc} \right) = \lambda \frac{dg}{dc}.$$

But, as $g(x_0(c), y_0(c)) = c$, we see that $dg/dc = 1$, so $df/dc = \lambda$. Thus, we have the following interpretation of the Lagrange multiplier λ:

> The value of λ is the rate of change of the optimum value of f as c increases (where $g(x, y) = c$). If the optimum value of f is written as $f(x_0(c), y_0(c))$, then
>
> $$\frac{d}{dc} f(x_0(c), y_0(c)) = \lambda.$$

Example 3 The quantity of goods produced according to the function $f(x, y) = x^{2/3} y^{1/3}$ is maximized subject to the budget constraint $x + y \leq 3.78$. The budget is increased to allow for a small increase in production. What is the price of the product if the sale of the additional goods covers the budget increase?

Solution We know that $\lambda = 0.53$, which tells us that $df/dc = 0.53$. The constraint corresponds to a budget of \$3.78 thousand. Therefore increasing the budget by \$1000 increases production by about 0.53 units. In order to make the increase in budget profitable, the extra goods produced must sell for more than \$1000. Thus, if p is the price of each unit of the good, then $0.53p$ is the revenue from the extra 0.53 units sold. Thus, we need $0.53p \geq 1000$ so $p \geq 1000/0.53 = \$1890$.

The Lagrangian Function

Constrained optimization problems are frequently solved using a *Lagrangian function*, $\mathcal{L}$. For example, to optimize $f(x, y)$ subject to the constraint $g(x, y) = c$, we use the Lagrangian function

$$\mathcal{L}(x, y, \lambda) = f(x, y) - \lambda(g(x, y) - c).$$

To see how the function $\mathcal{L}$ is used, compute the partial derivatives of $\mathcal{L}$:

$$\frac{\partial \mathcal{L}}{\partial x} = \frac{\partial f}{\partial x} - \lambda \frac{\partial g}{\partial x},$$

$$\frac{\partial \mathcal{L}}{\partial y} = \frac{\partial f}{\partial y} - \lambda \frac{\partial g}{\partial y},$$

$$\frac{\partial \mathcal{L}}{\partial \lambda} = -(g(x, y) - c).$$

Notice that if (x_0, y_0) is an extreme point of $f(x, y)$ subject to the constraint $g(x, y) = c$ and λ_0 is the corresponding Lagrange multiplier, then at the point (x_0, y_0, λ_0) we have

$$\frac{\partial \mathcal{L}}{\partial x} = 0 \quad \text{and} \quad \frac{\partial \mathcal{L}}{\partial y} = 0 \quad \text{and} \quad \frac{\partial \mathcal{L}}{\partial \lambda} = 0.$$

In other words, (x_0, y_0, λ_0) is a critical point for the unconstrained Lagrangian function, $\mathcal{L}(x, y, \lambda)$. Thus, the Lagrangian converts a constrained optimization problem to an unconstrained problem.

Example 4 A company has a production function with three inputs x, y, and z given by

$$f(x, y, z) = 50x^{2/5} y^{1/5} z^{1/5}.$$

The total budget is \$24,000 and the company can buy x, y, and z at \$80, \$12, and \$10 per unit, respectively. What combination of inputs will maximize production? [7]

[7] Adapted from M. Rosser, *Basic Mathematics for Economists*, p. 363 (New York: Routledge, 1993).

Solution   We need to maximize the objective function

$$f(x, y, z) = 50x^{2/5}y^{1/5}z^{1/5},$$

subject to the constraint

$$g(x, y, z) = 80x + 12y + 10z = 24{,}000.$$

Therefore, the Lagrangian function is

$$\mathcal{L}(x, y, z) = 50x^{2/5}y^{1/5}z^{1/5} - \lambda(80x + 12y + 10z - 24{,}000),$$

so we solve the system of equations we get from grad $\mathcal{L} = \vec{0}$:

$$\frac{\partial \mathcal{L}}{\partial x} = 20x^{-3/5}y^{1/5}z^{1/5} - 80\lambda = 0,$$

$$\frac{\partial \mathcal{L}}{\partial y} = 10x^{2/5}y^{-4/5}z^{1/5} - 12\lambda = 0,$$

$$\frac{\partial \mathcal{L}}{\partial z} = 10x^{2/5}y^{1/5}z^{-4/5} - 10\lambda = 0,$$

$$\frac{\partial \mathcal{L}}{\partial \lambda} = -(80x + 12y + 10z - 24{,}000) = 0.$$

We simplify this system to give

$$\lambda = \frac{1}{4}x^{-3/5}y^{1/5}z^{1/5},$$

$$\lambda = \frac{5}{6}x^{2/5}y^{-4/5}z^{1/5},$$

$$\lambda = x^{2/5}y^{1/5}z^{-4/5},$$

$$80x + 12y + 10z = 24{,}000.$$

Eliminating z from the first two equations gives $x = 0.3y$. Eliminating x from the second and third equations gives $z = 1.2y$. Substituting for x and z into $80x + 12y + 10z = 24{,}000$ gives

$$80(0.3y) + 12y + 10(1.2y) = 24{,}000,$$

so $y = 500$. Then $x = 150$ and $z = 600$, and $f(150, 500, 600) = 4{,}622$ units.

The graph of the constraint, $80x + 12y + 10z = 24{,}000$, is a plane. Since the inputs x, y, z must be nonnegative, the graph is a triangle in the first quadrant, with edges on the coordinate planes. On the boundary of the triangle, one (or more) of the variables x, y, z is zero, so the function f is zero. Thus production is maximized within the budget using $x = 150$, $y = 500$, and $z = 600$.

Exercises and Problems for Section 15.3

Exercises

In Exercises 1–15, use Lagrange multipliers to find the maximum and minimum values of f subject to the given constraint.

1. $f(x, y) = x + y, \quad x^2 + y^2 = 1$

2. $f(x, y) = x^2 + y^2, \quad 4x - 2y = 15$

3. $f(x, y) = 3x - 2y, \quad x^2 + 2y^2 = 44$

4. $f(x, y) = x^2 + y, \quad x^2 - y^2 = 1$

5. $f(x, y) = xy, \quad 4x^2 + y^2 = 8$

6. $f(x, y) = x^2 + y^2, \quad x^4 + y^4 = 2$

7. $f(x, y) = x^2 - xy + y^2, \quad x^2 - y^2 = 1$

8. $f(x, y, z) = x + 3y + 5z, \quad x^2 + y^2 + z^2 = 1$

9. $f(x, y, z) = 2x + y + 4z, \quad x^2 + y + z^2 = 16$

10. $f(x, y, z) = x^2 - 2y + 2z^2, \quad x^2 + y^2 + z^2 = 1$

11. $f(x, y) = x^2 + 2y^2, \quad x^2 + y^2 \le 4$

12. $f(x, y) = xy, \quad x^2 + 2y^2 \le 1$

13. $f(x, y) = x + 3y, \quad x^2 + y^2 \le 2$

14. $f(x, y) = x^3 + y, \quad x + y \ge 1$

15. $f(x, y) = x^3 - y^2, \quad x^2 + y^2 \le 1$

Problems

16. Figure 15.30 shows contours of f. Does f have a maximum value subject to the constraint $g(x, y) = c$ for $x \geq 0$, $y \geq 0$? If so, approximately where is it and what is its value? Does f have a minimum value subject to the constraint? If so, approximately where and what?

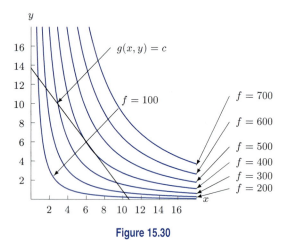

Figure 15.30

17. (a) Draw contours of $f(x, y) = 2x + y$ for $z = -7, -5, -3, -1, 1, 3, 5, 7$.
(b) On the same axes, graph the constraint $x^2 + y^2 = 5$.
(c) Use the graph to approximate the points at which f has a maximum or a minimum value subject to the constraint $x^2 + y^2 = 5$.
(d) Use Lagrange multipliers to find the maximum and minimum values of $f(x, y) = 2x + y$ subject to $x^2 + y^2 = 5$.

18. The function $P(x, y)$ gives the number of units produced and $C(x, y)$ gives the cost of production.

(a) A company wishes to maximize production at a fixed cost of $50,000$. What is the objective function f? What is the constraint equation? What is the meaning of λ in this situation?
(b) A company wishes to minimize costs at a fixed production level of 2000 units. What is the objective function f? What is the constraint equation? What is the meaning of λ in this situation?

19. A chemical manufacturing plant can produce z units of chemical Z given x units of chemical X and y units of chemical Y, where $z = 500x^{0.6}y^{0.3}$. Chemical X costs 10 a unit and chemical Y costs 25 a unit. The company wishes to produce as many units of chemical Z as possible given a fixed budget of 2000.

(a) How many units each of chemical X and chemical Y should the company purchase? How many units of chemical Z can be produced given these purchases?
(b) What is the value of λ? Interpret it in this situation.

20. The quantity, q, of a product manufactured depends on the number of workers, W, and the amount of capital invested, K, and is given by

$$q = 6W^{3/4}K^{1/4}.$$

Labor costs are 10 per worker and capital costs are 20 per unit, and the budget is 3000.

(a) What are the optimum number of workers and the optimum number of units of capital?
(b) Show that at the optimum values of W and K, the ratio of the marginal productivity of labor $(\partial q/\partial W)$ to the marginal productivity of capital $(\partial q/\partial K)$ is the same as the ratio of the cost of a unit of labor to the cost of a unit of capital.
(c) Recompute the optimum values of W and K when the budget is increased by one dollar. Check that increasing the budget by 1 allows the production of λ extra units of the good, where λ is the Lagrange multiplier.

21. The director of a neighborhood health clinic has an annual budget of $600,000$. He wants to allocate his budget so as to maximize the number of patient visits, V, which is given as a function of the number of doctors, D, and the number of nurses, N, by

$$V = 1000D^{0.6}N^{0.3}.$$

A doctor's salary is $40,000$; nurses get $10,000$.

(a) Set up the director's constrained optimization problem.
(b) Describe, in words, the conditions which must be satisfied by $\partial V/\partial D$ and $\partial V/\partial N$ for V to have an optimum value.
(c) Solve the problem formulated in part (a).
(d) Find the value of the Lagrange multiplier and interpret its meaning in this problem.
(e) At the optimum point, what is the marginal cost of a patient visit (that is, the cost of an additional visit)? Will that marginal cost rise or fall with the number of visits? Why?

22. (a) In Problem 20, does the value of λ change if the budget changes from 3000 to 4000?
(b) In Problem 21, does the value of λ change if the budget changes from $600,000$ to $700,000$?
(c) What condition must a Cobb-Douglas production function, $Q = cK^aL^b$, satisfy to ensure that the marginal increase of production (that is, the rate of increase of production with budget) is not affected by the size of the budget?

23. A firm manufactures a commodity at two different factories. The total cost of manufacturing depends on the quantities, q_1 and q_2, supplied by each factory, and is expressed by the *joint cost function*,

$$C = f(q_1, q_2) = 2q_1^2 + q_1q_2 + q_2^2 + 500.$$

The company's objective is to produce 200 units, while minimizing production costs. How many units should be supplied by each factory?

24. Design a closed cylindrical container which holds 100 cm^3 and has the minimal possible surface area. What should its dimensions be?

25. An industry manufactures a product from two raw materials. The quantity produced, Q, can be given by the Cobb-Douglas function:

$$Q = cx^a y^b,$$

where x and y are quantities of each of the two raw materials used and a, b, and c are positive constants. The first raw material costs $\$P_1$ per unit and the second costs $\$P_2$ per unit. Find the maximum production possible if no more than $\$K$ can be spent on raw materials.

26. An electrical current I flows through a circuit containing the resistors R_1 and R_2 in Figure 15.31. The currents i_1 and i_2 through the individual resistors minimize energy loss $i_1^2 R_1 + i_2^2 R_2$ subject to the constraint $i_1 + i_2 = I$ given by Kirchoff's current law.

   (a) Find the currents i_1 and i_2 by the method of Lagrange multipliers.

   (b) If you are familiar with Ohm's law, find the meaning of λ.

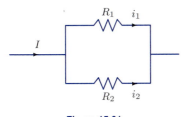

Figure 15.31

27. Each person tries to balance his or her time between leisure and work. The tradeoff is that as you work less your income falls. Therefore each person has *indifference curves* which connect the number of hours of leisure, l, and income, s. If, for example, you are indifferent between 0 hours of leisure and an income of $\$1125$ a week on the one hand, and 10 hours of leisure and an income of $\$750$ a week on the other hand, then the points $l = 0$, $s = 1125$, and $l = 10$, $s = 750$ both lie on the same indifference curve. Table 15.3 gives information on three indifference curves, I, II, and III.

Table 15.3

Weekly income			Weekly leisure hours		
I	II	III	I	II	III
1125	1250	1375	0	20	40
750	875	1000	10	30	50
500	625	750	20	40	60
375	500	625	30	50	70
250	375	500	50	70	90

   (a) Graph the three indifference curves.

   (b) Suppose you have 100 hours a week available for work and leisure combined, and that you earn $\$10$/hour. Write an equation in terms of l and s which represents this constraint.

   (c) On the same axes, graph this constraint.

   (d) Estimate from the graph what combination of leisure hours and income you would choose under these circumstances. Give the corresponding number of hours per week you would work.

28. Figure 15.32 shows ∇f for a function $f(x, y)$ and two curves $g(x, y) = 1$ and $g(x, y) = 2$. Mark the following:

   (a) The point(s) A where f has a local maximum.

   (b) The point(s) B where f has a saddle point.

   (c) The point C where f has a maximum on $g = 1$.

   (d) The point D where f has a minimum on $g = 1$.

   (e) If you used Lagrange multipliers to find C, what would the sign of λ be? Why?

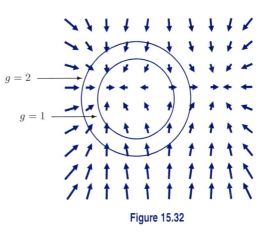

Figure 15.32

29. A mountain climber at the summit of a mountain wants to descend to a lower altitude as fast as possible. The altitude of the mountain is given approximately by

$$h(x, y) = 3000 - \frac{1}{10,000}(5x^2 + 4xy + 2y^2) \text{ meters},$$

where x, y are horizontal coordinates on the earth (in meters), with the mountain summit located above the origin. In thirty minutes, the climber can reach any point (x, y) on a circle of radius 1000 m. In which direction should she travel in order to descend as far as possible?

30. Let $f(x, y, z) = \sqrt{(x - a)^2 + (y - b)^2 + (z - c)^2}$. Minimize f subject to $Ax + By + Cz + D = 0$. What is the geometric meaning of your solution?

31. For constants a, b, c, let $f(x, y) = ax + by + c$ be a linear function, and let R be a region in the xy-plane.

(a) If R is any disk, show that the maximum and minimum values of f on R occur on the boundary of the disk.

(b) If R is any rectangle, show that the maximum and minimum values of f on R occur at the corners of the rectangle. They may occur at other points of the rectangle as well.

(c) Use a graph of the plane $z = f(x, y)$ to explain your answers in parts (a) and (b).

32. The function $f(x, y)$ is defined for all x, y and the origin does not lie on the surface representing $z = f(x, y)$. There is a unique point $P = (a, b, c)$ on the surface which is closest to the origin. Explain why the position vector from the origin to P must be perpendicular to the surface at that point.

33. A company produces one product from two inputs (for example capital and labor). Its production function $g(x, y)$ gives the quantity of the product that can be produced with x units of the first input and y units of the second. The *cost function* (or *expenditure function*) is the three variable function $C(p, q, u)$ where p and q are the unit prices of the two inputs. For fixed p, q, and u, the value $C(p, q, u)$ is the minimum of $f(x, y) = px + qy$ subject to the constraint $g(x, y) = u$.

(a) What is the practical meaning of $C(p, q, u)$?

(b) Find a formula for $C(p, q, u)$ if $g(x, y) = xy$.

34. A *utility function* $U(x, y)$ for two items gives the utility (benefit) to a consumer of x units of item 1 and y units of item 2. The *indirect utility function* is the three variable function $V(p, q, I)$ where p and q are the unit prices of the two items. For fixed p, q, and I, the value $V(p, q, I)$ is the maximum of $U(x, y)$ subject to the constraint $px + qy = I$.

(a) What is the practical meaning of $V(p, q, I)$?

(b) The Lagrange multiplier λ that arises in the maximization defining V is called the *marginal utility of money*. What is its practical meaning?

(c) Find formulas for $V(p, q, I)$ and λ if $U(x, y) = xy$.

CHAPTER SUMMARY

- **Critical Points**

    Definitions of local extrema, critical points, saddle points, finding critical points algebraically, behavior of contours near critical points, discriminant, second derivative test to classify critical points.

- **Unconstrained Optimization**

    Definitions of global extrema, method of least squares, closed and bounded regions, existence of global extrema.

- **Constrained Optimization**

    Objective function and constraint, definitions of extrema subject to a constraint, geometric interpretation of Lagrange multiplier method, solving Lagrange multiplier problems algebraically, inequality constraints, meaning of the Lagrange multiplier λ, the Lagrangian function.

REVIEW EXERCISES AND PROBLEMS FOR CHAPTER FIFTEEN

Exercises

For Exercises 1–4, find the critical points of the given function and classify them as local maxima, local minima, saddle points, or none of these.

1. $f(x, y) = x^2 y + 2y^2 - 2xy + 6$

2. $f(x, y) = 2xy^2 - x^2 - 2y^2 + 1$

3. $f(x, y) = 2x^3 - 3x^2 y + 6x^2 - 6y^2$

4. $f(x, y) = \sin x + \sin y + \sin(x + y)$, $\quad 0 < x < \pi$, $0 < y < \pi$.

For Exercises 5–8, find the local maxima, local minima, and saddle points of the function. Decide if the local maxima or minima are global maxima or minima. Explain.

5. $f(x, y) = 10 + 12x + 6y - 3x^2 - y^2$

6. $f(x, y) = x^2 + y^3 - 3xy$

7. $f(x, y) = xy + \ln x + y^2 - 10$, $\quad x > 0$

8. $f(x, y) = x + y + \dfrac{1}{x} + \dfrac{4}{y}$

For Exercises 9–13, use Lagrange multipliers to find the maximum or minimum values of f subject to the given constraint.

9. $f(x, y) = x^2 + 2y^2$, $\quad 3x + 5y = 200$

10. $f(x, y) = 3x - 4y$, $\quad x^2 + y^2 = 5$

11. $f(x, y) = 2xy$, $\quad 5x + 4y = 100$

12. $f(x, y, z) = x^2 - y^2 - 2z$, $\quad x^2 + y^2 = z$

13. $f(x, y) = x^2 - y^2$, $\quad x^2 \geq y$

14. What are the maximum and minimum values of $f(x, y) = -3x^2 - 2y^2 + 20xy$ on the line $x + y = 100$?

15. Find the minimum and maximum of the function $z = 4x^2 - xy + 4y^2$ over the closed disk $x^2 + y^2 \leq 2$.

16. Find the least squares line for the data points $(0, 4)$, $(1, 3)$, $(2, 1)$.

Problems

17. At the point $(1, 3)$, suppose $f_x = f_y = 0$ and $f_{xx} < 0$, $f_{yy} < 0$, $f_{xy} = 0$. Draw a possible contour diagram.

18. A company sells two products which are partial substitutes for each other, such as coffee and tea. If the price of one product rises, then the demand for the other product rises. The quantities demanded, q_1 and q_2, are given as a function of the prices, p_1 and p_2, by

$$q_1 = 517 - 3.5p_1 + 0.8p_2, \quad q_2 = 770 - 4.4p_2 + 1.4p_1.$$

What prices should the company charge in order to maximize the total sales revenue? [8]

19. The quantity, Q, of a certain product manufactured depends on the quantity of labor, L, and of capital, K, used according to the function

$$Q = 900L^{1/2}K^{2/3}.$$

Labor costs $100 per unit and capital costs $200 per unit. What combination of labor and capital should be used to produce 36,000 units of the goods at minimum cost? What is that minimum cost?

20. An international organization must decide how to spend the $2000 they have been allotted for famine relief in a remote area. They expect to divide the money between buying rice at $5/sack and beans at $10/sack. The number, P, of people who would be fed if they buy x sacks of rice and y sacks of beans is given by

$$P = x + 2y + \frac{x^2 y^2}{2 \cdot 10^8}.$$

What is the maximum number of people that can be fed, and how should the organization allocate its money?

21. The quantity q of goods produced is given by

$$q = 80x^{0.75}y^{0.25},$$

where x is the number of units of labor used and y is the number of units of capital used. Labor costs are $6 per unit, capital costs are $4 per unit, and the budget is $8000. Find the optimal values of x and y in order to maximize the quantity produced.

22. A company manufactures a product using inputs x, y, and z according to the production function

$$Q(x, y, z) = 20x^{1/2}y^{1/4}z^{2/5}.$$

The prices per unit are $20 for x, and $10 for y, and $5 for z. What quantity of each input should be used in order to manufacture 1,200 units at minimum cost? [9]

23. A biological rule of thumb states that as the area A of an island increases tenfold, the number of animal species, N, living on it doubles. The table contains data for islands in the West Indies. Assume that N is a power function of A.

(a) Use the biological rule of thumb to find
    (i) N as a function of A
    (ii) $\ln N$ as a function of $\ln A$

(b) Using the data given, tabulate $\ln N$ against $\ln A$ and find the line of best fit. Does your answer agree with the biological rule of thumb?

Island	Area (sq km)	Number of species
Redonda	3	5
Saba	20	9
Montserrat	192	*15
Puerto Rico	8858	75
Jamaica	10854	70
Hispaniola	75571	130
Cuba	113715	125

24. A company manufactures x units of one item and y units of another. The total cost in dollars, C, of producing these two items is approximated by the function

$$C = 5x^2 + 2xy + 3y^2 + 800.$$

(a) If the production quota for the total number of items (both types combined) is 39, find the minimum production cost.

(b) Estimate the additional production cost or savings if the production quota is raised to 40 or lowered to 38.

25. The quantity, Q, of a product manufactured by a company is given by

$$Q = aK^{0.6}L^{0.4},$$

where a is a positive constant, K is the quantity of capital and L is the quantity of labor used. Capital costs are $20 per unit, labor costs are $10 per unit, and the company wants costs for capital and labor combined to be no higher than $150. Suppose you are asked to consult for the company, and learn that 5 units each of capital and labor are being used.

(a) What do you advise? Should the company use more or less labor? More or less capital? If so, by how much?

(b) Write a one sentence summary that could be used to sell your advice to the board of directors.

---

[8] Adapted from M. Rosser, *Basic Mathematics for Economists*, p. 318 (New York: Routledge, 1993).
[9] Adapted from M. Rosser, *Basic Mathematics for Economists*, p.363 (New York: Routledge, 1993).

26. The Cobb-Douglas function models the quantity, q, of a commodity produced as a function of the number of workers, W, and the amount of capital invested, K:

$$q = cW^{1-a}K^a,$$

where a and c are positive constants. Labor costs are $\$p_1$ per worker, capital costs are $\$p_2$ per unit, and there is a fixed budget of $\$b$. Show that when W and K are at their optimal levels, the ratio of marginal productivity of labor to marginal productivity of capital equals the ratio of the cost of one unit of labor to one unit of capital.

27. Figure 15.33 shows contours labeled with values of $f(x, y)$ and a constraint $g(x, y) = c$. Mark the approximate points at which:

(a) $\operatorname{grad} f = \lambda \operatorname{grad} g$

(b) f has a maximum

(c) f has a maximum on the constraint $g = c$.

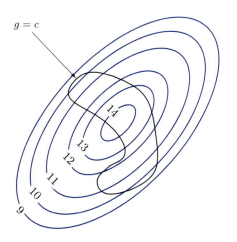

$g = c$

14
13
12
11
10
9

Figure 15.33

28. Let f be differentiable and $\operatorname{grad} f(2, 1) = -3\vec{i} + 4\vec{j}$. You want to see if $(2, 1)$ is a candidate for the maximum and minimum values of f subject to a constraint satisfied by the point $(2, 1)$.

(a) Show $(2, 1)$ is not a candidate if the constraint is $x^2 + y^2 = 5$.

(b) Show $(2, 1)$ is a candidate if the constraint is $(x - 5)^2 + (y + 3)^2 = 25$. From a sketch of the contours for f near $(2, 1)$ and the constraint, decide whether $(2, 1)$ is a candidate for a maximum or minimum.

(c) Do the same as part (b), but using the constraint $(x + 1)^2 + (y - 5)^2 = 25$.

29. Figure 15.34 shows two weightless springs with spring constants k_1 and k_2 attached between a ceiling and floor without tension or compression. A mass m is placed between the springs which settle into equilibrium as in Figure 15.35. The magnitudes f_1 and f_2 of the forces of the springs on the mass minimize the complementary energy

$$\frac{f_1^2}{2k_1} + \frac{f_2^2}{2k_2}$$

subject to the force balance constraint $f_1 + f_2 = mg$.

(a) Determine f_1 and f_2 by the method of Lagrange multipliers.

(b) If you are familiar with Hooke's law, find the meaning of λ.

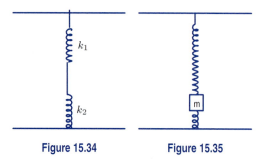

k_1

k_2

m

Figure 15.34   **Figure 15.35**

30. A doctor wants to schedule visits for two patients who have been operated on for tumors so as to minimize the expected delay in detecting a new tumor. Visits for patients 1 and 2 are scheduled at intervals of x_1 and x_2 weeks, respectively. A total of m visits per week is available for both patients combined.

The recurrence rates for tumors for patients 1 and 2 are judged to be v_1 and v_2 tumors per week, respectively. Thus, $v_1/(v_1 + v_2)$ and $v_2/(v_1 + v_2)$ are the probabilities that patient 1 and patient 2, respectively, will have the next tumor. It is known that the expected delay in detecting a tumor for a patient checked every x weeks is $x/2$. Hence, the expected detection delay for both patients combined is given by[10]

$$f(x_1, x_2) = \frac{v_1}{v_1 + v_2} \cdot \frac{x_1}{2} + \frac{v_2}{v_1 + v_2} \cdot \frac{x_2}{2}.$$

Find the values of x_1 and x_2 in terms of v_1 and v_2 that minimize $f(x_1, x_2)$ subject to the fact that m, the number of visits per week, is fixed.

31. What is the value of the Lagrange multiplier in Problem 30? What are the units of λ? What is its practical significance to the doctor?

32. An underground cable is to be laid from point A to point D in Figure 15.36. The cable will go under a field between points A and B, where the cost to lay it is $\$500$/meter. From points B to C, the cable goes under a road, where the cost to lay it is $\$2000$/meter. Between

---

[10]Adapted from Daniel Kent, Ross Shachter, *et al.,* Efficient Scheduling of Cystoscopies in Monitoring for Recurrent Bladder Cancer in *Medical Decision Making* (Philadelphia: Hanley and Belfus, 1989).

points C and D, it goes alongside a stream where the cost is \$300/meter. Find the values of x, y, z in Figure 15.36 minimizing the total cost of laying the cable.

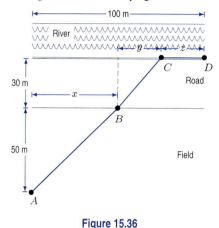

Figure 15.36

33. An irrigation canal has a trapezoidal cross-section of area 50 m^2, as in Figure 15.37. The average flow rate in the canal is inversely proportional to the wetted perimeter, p, of the canal, that is, to the perimeter of the trapezoid in Figure 15.37, excluding the top. Thus, to maximize the flow rate we must minimize p. Find the depth d, base width w, and angle θ that give the maximum flow rate.[11]

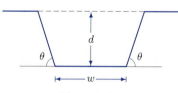

Figure 15.37

34. A light ray crossing the boundary between two different media (for example, vacuum and glass, or air and water) undergoes a change in direction or is *refracted* by an amount which depends on the properties of the media. Suppose that a light ray travels from point A to point B in Figure 15.38, with velocity v_1 in medium 1 and velocity v_2 in medium 2.

(a) Find the time $T(\theta_1, \theta_2)$ taken for the ray to travel from A to B in terms of the angles θ_1, θ_2 and the constants a, b, v_1, and v_2.

CAS Challenge Problems

35. Let $f(x, y) = \dfrac{\sqrt{a + x + y}}{1 + y + \sqrt{a + x}}$, for $x, y > 0$, where a is a positive constant.

(a) Find the critical points of f and classify them as local maximum, local minimum, saddle point, or none of these.

(b) Describe how the position and type of the critical points changes with respect to a, and explain this in terms of the graph of f.

(b) Show that the angles θ_1, θ_2 satisfy the constraint

$$a \tan \theta_1 + b \tan \theta_2 = d.$$

(c) What is the effect on the time, T, of letting $\theta_1 \to -\pi/2$ (that is, moving R far to the left of A') or of letting $\theta_1 \to \pi/2$ (that is, moving R far to the right of B')?

(d) *Fermat's Principle* states that the light ray follows a path such that the time taken T is minimized. Use Lagrange multipliers to show that $T(\theta_1, \theta_2)$ is minimized when *Snell's Law of Refraction* holds:

$$\frac{\sin \theta_1}{\sin \theta_2} = \frac{v_1}{v_2}.$$

(The constant v_1/v_2 is called the *index of refraction* of medium 2 with respect to medium 1. For example, the indices of air, water, and glass with respect to a vacuum are approximately 1.0003, 1.33, and 1.52, respectively. The lenses of modern reading "glasses" are made of plastics with a high index of refraction in order to reduce weight and thickness when the prescription is strong.)

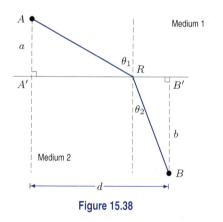

Figure 15.38

36. Students are asked to find the global maximum of $f(x, y) = x^2 + y$ subject to the constraint $g(x, y) = x^2 + 2xy + y^2 - 9 = 0$. Student A uses the method of Lagrange multipliers with the help of a computer algebra system, and says that the global maximum is 11/4. Student B looks at a contour diagram of f and a graph of $g = 0$ and says there is no global maximum. Which student is correct and what mistake is the other one making?

---

[11] Adapted from Robert M. Stark and Robert L. Nichols, *Mathematical Foundations of Design: Civil Engineering Systems*, (New York: McGraw-Hill, 1972).

37. Let $f(x, y) = 3x + 2y + 5$, $g(x, y) = 2x^2 - 4xy + 5y^2$.

(a) Find the maximum of f subject to the constraint $g = 20$.

(b) Using the value of λ in part (a), estimate the maximum of f subject to each of the constraints $g = 20.5$ and $g = 20.2$.

(c) Use Lagrange multipliers to find the two maxima in part (b) exactly. Compare them with the estimates.

CHECK YOUR UNDERSTANDING

Are the statements in Problems 1–25 true or false? Give reasons for your answer.

1. If $f_x(P_0) = f_y(P_0) = 0$, then P_0 is a critical point of f.

2. If $f_x(P_0) = f_y(P_0) = 0$, then P_0 is a local maximum or local minimum of f.

3. If P_0 is a critical point of f, then P_0 is either a local maximum or local minimum of f.

4. If P_0 is a local maximum or local minimum of f, and not on the boundary of the domain of f, then P_0 is a critical point of f.

5. The function $f(x, y) = \sqrt{x^2 + y^2}$ has a local minimum at the origin.

6. The function $f(x, y) = x^2 - y^2$ has a local minimum at the origin.

7. If f has a local minimum at P_0 then so does the function $g(x, y) = f(x, y) + 5$.

8. If f has a local minimum at P_0 then the function $g(x, y) = -f(x, y)$ has a local maximum at P_0.

9. Every function has at least one local maximum.

10. If P_0 is a local maximum for f then $f(a, b) \leq f(P_0)$ for all points (a, b) in 2-space.

11. If P_0 is a local maximum of f, then P_0 is also a global maximum of f.

12. If P_0 is a global maximum of f, where f is defined on all of 2-space, then P_0 is also a local maximum of f.

13. Every function has a global maximum.

14. The region consisting of all points (x, y) satisfying $x^2 + y^2 < 1$ is bounded.

15. The region consisting of all points (x, y) satisfying $x^2 + y^2 < 1$ is closed.

16. The function $f(x, y) = x^2 + y^2$ has a global minimum on the region $x^2 + y^2 < 1$.

17. The function $f(x, y) = x^2 + y^2$ has a global maximum on the region $x^2 + y^2 < 1$.

18. If P and Q are two distinct points in 2-space, and f has a global maximum at P, then f cannot have a global maximum at Q.

19. The function $f(x, y) = \sin(1 + e^{xy})$ must have a global minimum in the square region $0 \leq x \leq 1, 0 \leq y \leq 1$.

20. If P_0 is a global minimum of f on a closed and bounded region, then P_0 need not be a critical point of f.

21. If $f(x, y)$ has a local maximum at (a, b) subject to the constraint $g(x, y) = c$, then $g(a, b) = c$.

22. If $f(x, y)$ has a local maximum at (a, b) subject to the constraint $g(x, y) = c$, then $\operatorname{grad} f(a, b) = \vec{0}$.

23. The function $f(x, y) = x + y$ has no global maximum subject to the constraint $x - y = 0$.

24. The point $(2, -1)$ is a local minimum of $f(x, y) = x^2 + y^2$ subject to the constraint $x + 2y = 0$.

25. If $\operatorname{grad} f(a, b)$ and $\operatorname{grad} g(a, b)$ point in opposite directions, then (a, b) is a local minimum of $f(x, y)$ constrained by $g(x, y) = c$.

In Problems 26–33, suppose that M and m are the maximum and minimum values of $f(x, y)$ subject to the constraint $g(x, y) = c$ and that (a, b) satisfies $g(a, b) = c$. Decide whether the statements are true or false. Give an explanation for your answer.

26. If $f(a, b) = M$, then $f_x(a, b) = f_y(a, b) = 0$.

27. If $f(a, b) = M$, then $f(a, b) = \lambda g(a, b)$ for some value of λ.

28. If $\operatorname{grad} f(a, b) = \lambda \operatorname{grad} g(a, b)$, then $f(a, b) = M$ or $f(x, y) = m$.

29. If $f(a, b) = M$ and $f_x(a, b)/f_y(a, b) = 5$, then $g_x(a, b)/g_y(a, b) = 5$.

30. If $f(a, b) = m$ and $g_x(a, b) = 0$, then $f_x(a, b) = 0$.

31. Increasing the value of c will increase the value of M.

32. Suppose that $f(a, b) = M$ and that $\operatorname{grad} f(a, b) = 3 \operatorname{grad} g(a, b)$. Then increasing the value of c by 0.02 increases the value of M by about 0.06.

33. Suppose that $f(a, b) = m$ and that $\operatorname{grad} f(a, b) = 3 \operatorname{grad} g(a, b)$. Then increasing the value of c by 0.02 decreases the value of m by about 0.06.

PROJECTS FOR CHAPTER FIFTEEN

1. Hockey and Entropy

Thirty teams compete for the Stanley Cup in the National Hockey League (after expansion in 2000). At the beginning of the season an experienced fan estimates that the probability that team i will win is some number p_i, where $0 \le p_i \le 1$ and

$$\sum_{i=1}^{30} p_i = 1.$$

Exactly one team will actually win, so the probabilities have to add to 1. If one of the teams, say team i, is certain to win then p_i is equal to 1 and all the other p_j are zero. Another extreme case occurs if all the teams are equally likely to win, so all the p_i are equal to $1/30$, and the outcome of the hockey season is completely unpredictable. Thus, the *uncertainty* in the outcome of the hockey season depends on the probabilities $p_1, \ldots, p_{30}$. In this problem we measure this uncertainty quantitatively using the following function:

$$S(p_1, \ldots, p_{30}) = -\sum_{i=1}^{30} p_i \frac{\ln p_i}{\ln 2}.$$

Note that as $p_i \le 1$, we have $-\ln p_i \ge 0$ and hence $S \ge 0$.

(a) Show that $\lim_{p \to 0} p \ln p = 0$. (This means that S is a continuous function of the p_i, where $0 \le p_i \le 1$ and $1 \le i \le 30$, if we set $p \ln p|_{p=0}$ equal to zero. Since S is then a continuous function on a closed and bounded region, it attains a maximum and a minimum value on this region.)

(b) Find the maximum value of $S(p_1, \ldots, p_{30})$ subject to the constraint $p_1 + \cdots + p_{30} = 1$. What are the values of p_i in this case? What does your answer mean in terms of the uncertainty in the outcome of the hockey season?

(c) Find the minimum value of $S(p_1, \ldots, p_{30})$, subject to the constraint $p_1 + \cdots + p_{30} = 1$. What are the values of p_i in this case? What does your answer mean in terms of the uncertainty in the outcome of the hockey season?

[Note: The function S is an example of an *entropy* function; the concept of entropy is used in information theory, statistical mechanics, and thermodynamics when measuring the uncertainty in an experiment (the hockey season in this problem) or physical system.]

2. **Fitting a Line to Data Using Least Squares**
   In this problem you will derive the general formulas for the slope and y-intercept of a least squares line. Assume that you have n data points $(x_1, y_1), (x_2, y_2), \cdots, (x_n, y_n)$. Let the equation of the least squares line be $y = b + mx$.

   (a) For each data point (x_i, y_i), show that the corresponding point directly above or below it on the least squares line has y-coordinate $b + mx_i$.

   (b) For each data point (x_i, y_i), show that the square of the vertical distance from it to the point found in part (a) is $(y_i - (b + mx_i))^2$.

   (c) Form the function $f(b, m)$ which is the sum of all of the n squared distances found in part (b). That is,

   $$f(b, m) = \sum_{i=1}^{n} (y_i - (b + mx_i))^2.$$

   Show that the partial derivatives $\dfrac{\partial f}{\partial b}$ and $\dfrac{\partial f}{\partial m}$ are given by

   $$\frac{\partial f}{\partial b} = -2 \sum_{i=1}^{n} (y_i - (b + mx_i))$$

   and

   $$\frac{\partial f}{\partial m} = -2 \sum_{i=1}^{n} (y_i - (b + mx_i)) \cdot x_i.$$

(d) Show that the critical point equations $\dfrac{\partial f}{\partial b} = 0$ and $\dfrac{\partial f}{\partial m} = 0$ lead to a pair of simultaneous linear equations in b and m:

$$nb + \left(\sum x_i\right)m = \sum y_i$$

$$\left(\sum x_i\right)b + \left(\sum x_i^2\right)m = \sum x_i y_i$$

(e) Solve the equations in part (d) for b and m, getting

$$b = \left(\sum_{i=1}^{n} x_i^2 \sum_{i=1}^{n} y_i - \sum_{i=1}^{n} x_i \sum_{i=1}^{n} x_i y_i\right) \Bigg/ \left(n\sum_{i=1}^{n} x_i^2 - \left(\sum_{i=1}^{n} x_i\right)^2\right)$$

$$m = \left(n\sum_{i=1}^{n} x_i y_i - \sum_{i=1}^{n} x_i \sum_{i=1}^{n} y_i\right) \Bigg/ \left(n\sum_{i=1}^{n} x_i^2 - \left(\sum_{i=1}^{n} x_i\right)^2\right)$$

(f) Apply the formulas of part (e) to the data points $(1, 1), (2, 1), (3, 3)$ to check that you get the same result as in Example 3 on page 713.

Chapter Sixteen

INTEGRATING FUNCTIONS OF SEVERAL VARIABLES

A definite integral is the limit of a sum. We use a definite integral to calculate the total population of a region given the population density as a function of position. If the population density is a function of one variable only (for example, the distance from the center of a city), we have an ordinary one-variable definite integral. If the density depends on more than one variable, we need a multivariable integral to calculate the total population. In this chapter we develop double integrals in Cartesian and polar coordinates and triple integrals in Cartesian, cylindrical, and spherical coordinates. We see how to change variables using a Jacobian and apply integration to probability.

16.1 THE DEFINITE INTEGRAL OF A FUNCTION OF TWO VARIABLES

The definite integral of a continuous one-variable function, f, is a limit of Riemann sums:

$$\int_a^b f(x)\,dx = \lim_{\Delta x \to 0} \sum_i f(x_i)\,\Delta x,$$

where x_i is a point in the ith subdivision of the interval $[a, b]$. In this section we extend this definition to functions of two variables. We start by considering how to estimate total population from a two-variable population density.

Population Density of Foxes in England

The fox population in parts of England is important to public health officials concerned about the disease rabies, which is spread by animals. The contour diagram in Figure 16.1 shows the population density $D = f(x, y)$ of foxes in the southwestern corner of England, where x and y are in kilometers from the southwest corner of the map and D is in foxes per square kilometer.[1] The bold contour is the coastline, which may be thought of as the $D = 0$ contour; clearly the density is zero outside it.

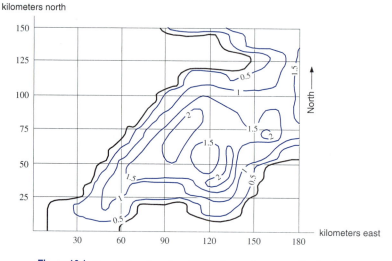

Figure 16.1: Population density of foxes in southwestern England

---

Example 1   Estimate the total fox population in the region represented by the map in Figure 16.1.

Solution   We subdivide the map into the rectangles shown in Figure 16.1 and estimate the population in each rectangle. For simplicity, we use the population density at the northeast corner of each rectangle. For example, in the bottom left rectangle, the density is 0 at the northeast corner, in the next rectangle to the east (right), the density in the northeast corner is 1. Continuing in this way, we get the estimates in Table 16.1. To estimate the population in a rectangle, we multiply the density by the area of the rectangle, $30 \cdot 25 = 750$ km^2. Adding the results, we obtain

$$\begin{aligned}
\text{Estimate of population} = (&0.2 + 0.7 + 1.2 + 1.2 + 0.1 + 1.6 + 0.5 + 1.4 \\
&+ 1.1 + 1.6 + 1.5 + 1.8 + 1.5 + 1.3 + 1.1 + 2.0 \\
&+ 1.4 + 1.0 + 1.0 + 0.6 + 1.2)750 = 18{,}000 \text{ foxes.}
\end{aligned}$$

Taking the upper and lower bounds for the population density on each rectangle enables us to find upper and lower estimates for the population. Using the same rectangles, the upper estimate is approximately 35,000 and the lower estimate is 4,000. There is a wide discrepancy between the upper and lower estimates; we could make them closer by taking finer subdivisions.

[1]Adapted from J. D. Murray, *Mathematical Biology*, Springer-Verlag, 1989.

Table 16.1 *Estimates of population density (northeast corner)*

0.0	0.0	0.2	0.7	1.2	1.2
0.0	0.0	0.0	0.0	0.1	1.6
0.0	0.0	0.5	1.4	1.1	1.6
0.0	0.0	1.5	1.8	1.5	1.3
0.0	1.1	2.0	1.4	1.0	0.0
0.0	1.0	0.6	1.2	0.0	0.0

Definition of the Definite Integral

The sums used to approximate the fox population are Riemann sums. We now define the definite integral for a function f of two variables on a rectangular region. Given a continuous function $f(x, y)$ defined on a region $a \leq x \leq b$ and $c \leq y \leq d$, we subdivide each of the intervals $a \leq x \leq b$ and $c \leq y \leq d$ into n and m equal subintervals respectively, giving nm subrectangles. (See Figure 16.2.)

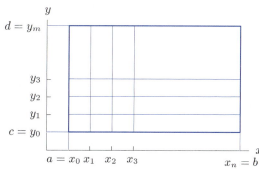

Figure 16.2: Subdivision of a rectangle into nm subrectangles

The area of each subrectangle is $\Delta A = \Delta x\,\Delta y$, where $\Delta x = (b - a)/n$ is the width of each subdivision on the x-axis, and $\Delta y = (d - c)/m$ is the width of each subdivision on the y-axis. To compute the Riemann sum, we multiply the area of each subrectangle by the value of the function at a point in the rectangle and add the resulting numbers. Choosing the point which gives the maximum value, M_{ij}, of the function on each rectangle, we get the *upper sum*, $\sum_{i,j} M_{ij}\Delta x\Delta y$.

The *lower sum*, $\sum_{i,j} L_{ij}\Delta x\Delta y$, is obtained by taking the minimum value on each rectangle. If (u_{ij}, v_{ij}) is any point in the ij-th subrectangle, any other Riemann sum satisfies

$$\sum_{i,j} L_{ij}\Delta x\Delta y \leq \sum_{i,j} f(u_{ij}, v_{ij})\,\Delta x\,\Delta y \leq \sum_{i,j} M_{ij}\Delta x\Delta y.$$

We define the definite integral by taking the limit as the numbers of subdivisions, n and m, tend to infinity. By comparing upper and lower sums, as we did for the fox population, it can be shown that the limit exists when the function, f, is continuous. We get the same limit by letting Δx and Δy tend to 0. Thus, we have the following definition:

Suppose the function f is continuous on R, the rectangle $a \leq x \leq b$, $c \leq y \leq d$. If (u_{ij}, v_{ij}) is any point in the ij-th subrectangle, we define the **definite integral** of f over R

$$\int_R f\,dA = \lim_{\Delta x, \Delta y \to 0} \sum_{i,j} f(u_{ij}, v_{ij})\Delta x\Delta y.$$

Such an integral is called a **double integral**.

The case when R is not rectangular is considered on page 740. Sometimes we think of dA as being the area of an infinitesimal rectangle of length dx and height dy, so that $dA = dx\, dy$. Then we use the notation[2]

$$\int_R f\, dA = \int_R f(x,y)\, dx\, dy.$$

For this definition, we used a particular type of Riemann sum with equal-sized rectangular subdivisions. In a general Riemann sum, the subdivisions do not all have to be the same size.

Interpretation of the Double Integral as Volume

Just as the definite integral of a positive one-variable function can be interpreted as an area, so the double integral of a positive two-variable function can be interpreted as a volume. In the one-variable case we visualize the Riemann sums as the total area of rectangles above the subdivisions. In the two-variable case we get solid bars instead of rectangles. As the number of subdivisions grows, the tops of the bars approximate the surface better, and the volume of the bars gets closer to the volume under the graph of the function. (See Figure 16.3.)

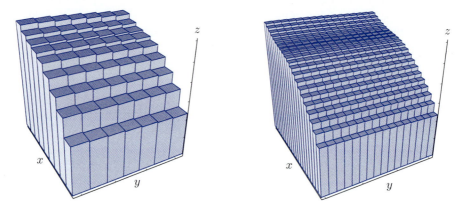

Figure 16.3: Approximating volume under a graph with finer and finer Riemann sums

Thus we have the following result:

> If x, y, z represent length and f is positive, then
>
> $$\begin{array}{c}\text{Volume under graph} \\ \text{of } f \text{ above region } R\end{array} = \int_R f\, dA$$

Example 2   Let R be the rectangle $0 \le x \le 1$ and $0 \le y \le 1$. Use Riemann sums to make upper and lower estimates of the volume of the region above R and under the graph of $z = e^{-(x^2+y^2)}$.

Solution   If R is the rectangle $0 \le x \le 1, 0 \le y \le 1$, the volume we want is given by

$$\text{Volume} = \int_R e^{-(x^2+y^2)}\, dA.$$

We divide R into 16 subrectangles by dividing each edge into 4 parts. Figure 16.4 shows that $f(x,y) = e^{-(x^2+y^2)}$ decreases as we move away from the origin. Thus, to get an upper sum we evaluate f on each subrectangle at the corner nearest the origin. For example, in the rectangle $0 \le x \le 0.25, 0 \le y \le 0.25$, we evaluate f at $(0,0)$. Using Table 16.2 we find that

---

[2]Another common notation for the double integral is $\int \int_R f\, dA$.

Figure 16.4: Graph of $e^{-(x^2+y^2)}$ above the rectangle R

Upper sum $= ($ $1 + 0.9394 + 0.7788 + 0.5698$

$+ 0.9394 + 0.8825 + 0.7316 + 0.5353$

$+ 0.7788 + 0.7316 + 0.6065 + 0.4437$

$+ 0.5698 + 0.5353 + 0.4437 + 0.3247)(0.0625) = 0.68.$

To get a lower sum, we evaluate f at the opposite corner of each rectangle because the surface slopes down in both the x and y directions. This yields a lower sum of 0.44. Thus,

$$0.44 \leq \int_R e^{-(x^2+y^2)} \, dA \leq 0.68.$$

To get a better approximation, we use more subdivisions. See Table 16.3.

Table 16.2   *Values of $f(x,y) = e^{-(x^2+y^2)}$ on the rectangle R*

			y		
		0.0	0.25	0.50	0.75	1.00
	0.0	1	0.9394	0.7788	0.5698	0.3679
	0.25	0.9394	0.8825	0.7316	0.5353	0.3456
x	0.50	0.7788	0.7316	0.6065	0.4437	0.2865
	0.75	0.5698	0.5353	0.4437	0.3247	0.2096
	1.00	0.3679	0.3456	0.2865	0.2096	0.1353

Table 16.3   *Riemann sum approximations to $\int_R e^{-(x^2+y^2)} \, dA$*

	Number of subdivisions in x and y directions			
	8	16	32	64
Upper	0.6168	0.5873	0.5725	0.5651
Lower	0.4989	0.5283	0.5430	0.5504

The true value of the double integral, $0.5577\ldots$, is trapped between the lower and upper sums. Notice that the lower sum increases and the upper sum decreases as the number of subdivisions increases. (Why?) However, even with 64 subdivisions, the lower and upper sums agree with the true value of the integral only in the first decimal place. Hence, we should look for better ways to approximate the integral.

Interpretation of the Double Integral as Area

In the special case that $f(x,y) = 1$ for all points (x,y) in the region R, each term in the Riemann sum is of the form $1 \cdot \Delta A = \Delta A$ and the double integral gives the area of the region R:

$$\text{Area}(R) = \int_R 1 \, dA = \int_R dA.$$

Interpretation of the Double Integral as Average Value

As in the one-variable case, the definite integral can be used to compute the average value of a function:

$$\begin{array}{c} \text{Average value of } f \\ \text{on the region } R \end{array} = \frac{1}{\text{Area of } R} \int_R f \, dA$$

We can rewrite this as

$$\text{Average value} \times \text{Area of } R = \int_R f \, dA.$$

If we interpret the integral as the volume under the graph of f, then we can think of the average value of f as the height of the box with the same volume that is on the same base. (See Figure 16.5.) Imagine that the volume under the graph is made out of wax. If the wax melted within the perimeter of R, then it would end up box-shaped with height equal to the average value of f.

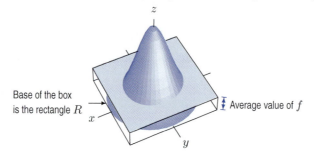

Figure 16.5: Volume and average value

Integral over Regions That are Not Rectangles

We defined the definite integral $\int_R f(x, y) \, dA$, for a rectangular region R. Now we extend the definition to regions of other shapes, including triangles, circles, and regions bounded by the graphs of piecewise continuous functions.

To approximate the definite integral over a region, R, which is not rectangular, we use a grid of rectangles approximating the region. We obtain this grid by enclosing R in a large rectangle and subdividing that rectangle; we consider just the subrectangles which are inside R.

As before, we pick a point (u_{ij}, v_{ij}) in each subrectangle and form a Riemann sum

$$\sum_{i,j} f(u_{ij}, v_{ij}) \Delta x \Delta y.$$

This time, however, the sum is over only those subrectangles within R. For example, in the case of the fox population we can use the rectangles which are entirely on land. As the subdivisions become finer, the grid resembles the region R more closely. For a function, f, which is continuous on R, we define the definite integral as follows:

$$\int_R f \, dA = \lim_{\Delta x, \Delta y \to 0} \sum_{i,j} f(u_{ij}, v_{ij}) \Delta x \Delta y$$

where the Riemann sum is taken over the subrectangles inside R.

You may wonder why we can leave out the rectangles which cover the edge of R—if we included them might we get a different value for the integral? The answer is that for any region that we are likely to meet, the area of the subrectangles covering the edge tends to 0 as the grid becomes finer. Therefore, omitting these rectangles does not affect the limit.

Convergence of Upper and Lower Sums to Same Limit

We have said that if f is continuous on the rectangle R, then the difference between upper and lower sums for f converges to 0 as Δx and Δy approach 0. In the following example, we show this in a particular case. The ideas in this example can be used in a general proof.

Example 3 Let $f(x, y) = x^2 y$ and let R be the rectangle $0 \leq x \leq 1, 0 \leq y \leq 1$. Show that the difference between upper and lower Riemann sums for f on R converges to 0, as Δx and Δy approach 0.

Solution The difference between the sums is

$$\sum M_{ij} \Delta x \Delta y - \sum L_{ij} \Delta x \Delta y = \sum (M_{ij} - L_{ij}) \Delta x \Delta y.$$

where M_{ij} and L_{ij} are the maximum and minimum of f on the ij-th subrectangle. Since f increases in both the x and y directions, M_{ij} occurs at the corner of the subrectangle farthest from the origin and L_{ij} at the closest. Moreover, since the slopes in the x and y directions don't decrease as x and y increase, the difference $M_{ij} - L_{ij}$ is largest in the subrectangle R_{nm}, which is farthest from the origin. Thus,

$$\sum (M_{ij} - L_{ij}) \Delta x \Delta y \leq (M_{nm} - L_{nm}) \sum \Delta x \Delta y = (M_{nm} - L_{nm}) \text{Area}(R).$$

Thus, the difference converges to 0 as long as $(M_{nm} - L_{nm})$ does. The maximum M_{nm} of f on the nm-th subrectangle occurs at $(1, 1)$, the subrectangle's top right corner, and the minimum L_{nm} occurs at the opposite corner, $(1 - 1/n, 1 - 1/m)$. Substituting into $f(x, y) = x^2 y$ gives

$$M_{nm} - L_{nm} = (1)^2(1) - \left(1 - \frac{1}{n}\right)^2 \left(1 - \frac{1}{m}\right) = \frac{2}{n} - \frac{1}{n^2} + \frac{1}{m} - \frac{2}{nm} + \frac{1}{n^2 m}.$$

The right hand side converges to 0 as $n, m \to \infty$, that is, as $\Delta x, \Delta y \to 0$.

Exercises and Problems for Section 16.1

Exercises

1. Figure 16.6 shows contours of $f(x, y)$ on the rectangle R with $0 \leq x \leq 30$ and $0 \leq y \leq 15$. Using $\Delta x = 10$ and $\Delta y = 5$, find an overestimate and an underestimate for $\int_R f(x, y) dA$.

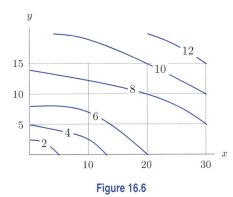

Figure 16.6

2. Values of $f(x, y)$ are in Table 16.4. Let R be the rectangle $1 \leq x \leq 1.2$, $2 \leq y \leq 2.4$. Find Riemann sums which are reasonable over- and under-estimates for $\int_R f(x, y) \, dA$ with $\Delta x = 0.1$ and $\Delta y = 0.2$.

Table 16.4

		x		
		1.0	1.1	1.2
y	2.0	5	7	10
	2.2	4	6	8
	2.4	3	5	4

3. Let R be the rectangle with vertices $(0, 0)$, $(4, 0)$, $(4, 4)$, and $(0, 4)$ and let $f(x, y) = \sqrt{xy}$.

   (a) Find reasonable upper and lower bounds for $\int_R f \, dA$ without subdividing R.

   (b) Estimate $\int_R f \, dA$ by partitioning R into four subrectangles and evaluating f at its maximum and minimum values on each subrectangle.

4. Table 16.5 gives values of $z = f(x, y)$ on the rectangle R with $0 \le x \le 6$ and $0 \le y \le 8$.

(a) Estimate $\int_R f(x, y)dA$ as accurately as possible.
(b) Estimate the average value of $f(x, y)$ on R.

Table 16.5

		x		
		0	3	6
	0	100	90	81
y	4	85	79	68
	8	65	61	55

5. Using Riemann sums with four subdivisions in each direction, find upper and lower bounds for the volume under the graph of $f(x, y) = 2 + xy$ above the rectangle R with $0 \le x \le 2$, $0 \le y \le 4$.

6. If $f(x, y)$ gives pollution density, in micrograms per square meter, and x and y are in meters, give the units and a practical interpretation of $\int_R f(x, y)dA$.

In Exercises 7–8, use a computer or calculator program for two-dimensional Riemann sums to estimate the integral.

7. $\int_R (x^2 + y^2)\, dA$, if R is $1 \le x \le 2, 1 \le y \le 3$.

8. $\int_R \sin(xy)\, dA$, if R is $-\pi \le x \le 0, 0 \le y \le \pi/2$.

Problems

9. Figure 16.7 shows the distribution of temperature, in °C, in a 5 meter by 5 meter heated room. Using Riemann sums, estimate the average temperature in the room.

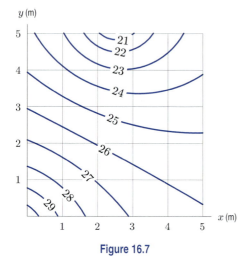

Figure 16.7

10. Figure 16.8 shows contours of $f(x, y)$. Let R be the square $-0.5 \le x \le 1$, $-0.5 \le y \le 1$. Is the integral $\int_R f\, dA$ positive or negative? Explain your reasoning.

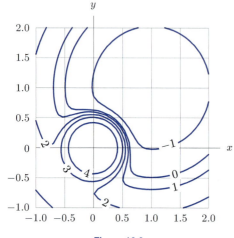

Figure 16.8

11. A biologist studying insect populations measures the population density of flies and mosquitos at various points in a rectangular study region. The graphs of the two population densities for the region are shown in Figures 16.9 and 16.10. Assuming that the units along the corresponding axes are the same in the two graphs, are there more flies or more mosquitos in the region?

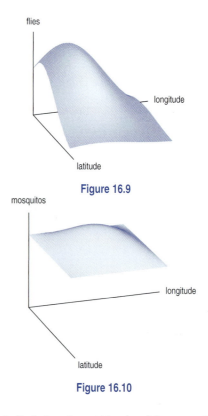

Figure 16.9

Figure 16.10

12. The hull of a boat has width $w(x, y)$ feet at a point x feet from the front and y feet below the water line. Values of w are in Table 16.6. Set up a definite integral that gives the volume of the hull below the waterline, and then estimate the value of the integral.

Table 16.6

	Front of boat → Back of boat						
	0	10	20	30	40	50	60
Depth below waterline (in feet) 0	2	8	13	16	17	16	10
2	1	4	8	10	11	10	8
4	0	3	4	6	7	6	4
6	0	1	2	3	4	3	2
8	0	0	1	1	1	1	1

13. Table 16.7 gives values of $f(x, y)$, the number of milligrams of mosquito larvae per square meter in a swamp. If x and y are in meters and R is the rectangle $0 \le x \le 8$, $0 \le y \le 6$, estimate $\int_R f(x, y)\, dA$. Give units and interpret your answer.

Table 16.7

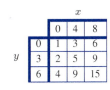

		x		
		0	4	8
	0	1	3	6
y	3	2	5	9
	6	4	9	15

Let D be the region inside the unit circle centered at the origin, let R be the right half of D and let B be the bottom half of D. Decide (without calculation) whether the integrals in Problems 14–29 are positive, negative, or zero.

14. $\int_D dA$

15. $\int_B dA$

16. $\int_R 5x\, dA$

17. $\int_B 5x\, dA$

18. $\int_D 5x\, dA$

19. $\int_D (y^3 + y^5)\, dA$

20. $\int_B (y^3 + y^5)\, dA$

21. $\int_R (y^3 + y^5)\, dA$

22. $\int_B (y - y^3)\, dA$

23. $\int_D (y - y^3)\, dA$

24. $\int_D \sin y\, dA$

25. $\int_D \cos y\, dA$

26. $\int_D e^x\, dA$

27. $\int_D xe^x\, dA$

28. $\int_D xy^2\, dA$

29. $\int_B x\cos y\, dA$

30. Figure 16.11 shows contours for the annual frequency of tornados per 10,000 square miles in the U.S.[3] Each grid square is 100 miles on a side. Use the map to estimate the total number of tornados per year in

  (a) Texas   **(b)** Florida   **(c)** Arizona.

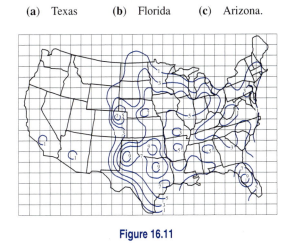

Figure 16.11

31. Let $f(x, y)$ be a function of x and y which is independent of y, that is, $f(x, y) = g(x)$ for some one-variable function g.

  (a) What does the graph of f look like?

  (b) Let R be the rectangle $a \le x \le b, c \le y \le d$. By interpreting the integral as a volume, and using your answer to part (a), express $\int_R f\, dA$ in terms of a one-variable integral.

32. For any region, R, in the plane, does the value of the following integral increase or decrease if the positive constant a is increased? Justify your answer with Riemann sums.

$$\int_R \frac{1}{1 + ax^2 + y^2}\, dA.$$

33. For any numbers a and b, assume that $|a + b| \le |a| + |b|$. Use this to explain why

$$\left| \int_R f\, dA \right| \le \int_R |f|\, dA.$$

16.2 ITERATED INTEGRALS

In Section 16.1 we approximated double integrals using Riemann sums. In this section we see how to compute double integrals exactly using one-variable integrals.

The Fox Population Again: Expressing a Double Integral as an Iterated Integral

To estimate the fox population, we computed a sum of the form

$$\text{Total population} \approx \sum_{i,j} f(u_{ij}, v_{ij})\Delta x\, \Delta y,$$

[3]From *Modern Physical Geography*, Alan H. Strahler and Arthur H. Strahler, Fourth Edition, John Wiley & Sons, New York, 1992, p. 128

where $1 \leq i \leq n$ and $1 \leq j \leq m$ and the values $f(u_{ij}, v_{ij})$ can be arranged as in Table 16.8.

Table 16.8 *Estimates for fox population densities for $n = m = 6$*

0.0	0.0	0.2	0.7	1.2	1.2
0.0	0.0	0.0	0.0	0.1	1.6
0.0	0.0	0.5	1.4	1.1	1.6
0.0	0.0	1.5	1.8	1.5	1.3
0.0	1.1	2.0	1.4	1.0	0.0
0.0	1.0	0.6	1.2	0.0	0.0

For any values of n and m, we can either add across the rows first, or we can add down the columns first. If we add rows first, we can write the sum in the form

$$\text{Total population} \approx \sum_{j=1}^{m} \left(\sum_{i=1}^{n} f(u_{ij}, v_{ij}) \Delta x \right) \Delta y.$$

The inner sum, $\sum_{i=1}^{n} f(u_{ij}, v_{ij}) \Delta x$, approximates the integral $\int_0^{180} f(x, v_{ij}) \, dx$. Thus, we have

$$\text{Total population} \approx \sum_{j=1}^{m} \left(\int_0^{180} f(x, v_{ij}) \, dx \right) \Delta y.$$

The outer Riemann sum approximates another integral, this time with integrand $\int_0^{180} f(x, y) \, dx$, which is a function of y. Thus, we can write the total population in terms of nested, or *iterated*, one-variable integrals:

$$\text{Total population} = \int_0^{150} \left(\int_0^{180} f(x, y) \, dx \right) dy.$$

Since the total population is represented by $\int_R f \, dA$, this suggests the method of computing double integrals in the following theorem:[4]

Theorem 16.1: Writing a Double Integral as an Iterated Integral

If R is the rectangle $a \leq x \leq b$, $c \leq y \leq d$ and f is a continuous function on R, then the integral of f over R exists and is equal to the **iterated integral**

$$\int_R f \, dA = \int_{y=c}^{y=d} \left(\int_{x=a}^{x=b} f(x, y) \, dx \right) dy.$$

The expression $\int_{y=c}^{y=d} \left(\int_{x=a}^{x=b} f(x, y) \, dx \right) dy$ can be written $\int_c^d \int_a^b f(x, y) \, dx \, dy$.

To evaluate the iterated integral, first perform the inside integral with respect to x, holding y constant; then integrate the result with respect to y.

The Parallel Between Repeated Summation and Repeated Integration

It is helpful to notice how the summation and integral notation parallel each other. Viewing a Riemann sum as a sum of sums,

$$\sum_{i,j} f(u_{ij}, v_{ij}) \Delta x \, \Delta y = \sum_j \left(\sum_i f(u_{ij}, v_{ij}) \Delta x \right) \Delta y$$

---

[4]For a proof, see M. Spivak, *Calculus on Manifolds*, pp. 53 and 58 (New York: Benjamin, 1965).

leads us to see a double integral as an integral of integrals:

$$\int_R f(x, y)\, dA = \int_c^d \left(\int_a^b f(x, y)\, dx \right) dy,$$

where R is the rectangle $a \le x \le b$ and $c \le y \le d$.

Example 1   A building is 8 meters wide and 16 meters long. It has a flat roof that is 12 meters high at one corner, and 10 meters high at each of the adjacent corners. What is the volume of the building?

Solution   If we put the high corner on the z-axis, the long side along the y-axis, and the short side along the x-axis, as in Figure 16.12, then the roof is a plane with z-intercept 12, and x slope $(-2)/8 = -1/4$, and y slope $(-2)/16 = -1/8$. Hence, the equation of the roof is

$$z = 12 - \tfrac{1}{4}x - \tfrac{1}{8}y.$$

The volume is given by the double integral

$$\text{Volume} = \int_R \left(12 - \tfrac{1}{4}x - \tfrac{1}{8}y\right) dA,$$

where R is the rectangle $0 \le x \le 8, 0 \le y \le 16$. Setting up an iterated integral, we get

$$\text{Volume} = \int_0^{16} \int_0^8 \left(12 - \tfrac{1}{4}x - \tfrac{1}{8}y\right) dx\, dy.$$

The inside integral is

$$\int_0^8 \left(12 - \tfrac{1}{4}x - \tfrac{1}{8}y\right) dx = \left(12x - \tfrac{1}{8}x^2 - \tfrac{1}{8}xy\right)\Big|_{x=0}^{x=8} = 88 - y.$$

Then the outside integral gives

$$\text{Volume} = \int_0^{16} (88 - y)\, dy = \left(88y - \tfrac{1}{2}y^2\right)\Big|_0^{16} = 1280.$$

The volume of the building is 1280 cubic meters.

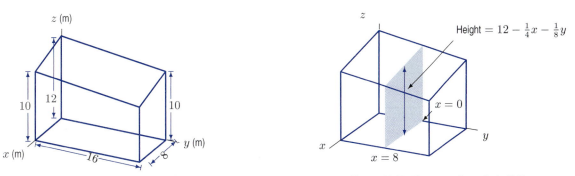

Figure 16.12: A slant-roofed building   **Figure 16.13**: Cross-section of a building

Notice that the inner integral $\int_0^8 (12 - \tfrac{1}{4}x - \tfrac{1}{8}y)\, dx$ in Example 1 gives the area of the cross section of the building perpendicular to the y-axis in Figure 16.13.

The iterated integral $\int_0^{16} \int_0^8 (12 - \tfrac{1}{4}x - \tfrac{1}{8}y)\, dx\, dy$ thus calculates the volume by adding the volumes of thin cross-sectional slabs.

The Order of Integration

In computing the fox population, we could have chosen to add columns (fixed x) first, instead of the rows. This leads to an iterated integral where x is constant in the inner integral instead of y. Thus,

$$\int_R f(x,y) \, dA = \int_a^b \left(\int_c^d f(x,y) \, dy \right) dx$$

where R is the rectangle $a \le x \le b$ and $c \le y \le d$.

For any function we are likely to meet, it doesn't matter in which order we integrate over a rectangular region R; we get the same value for the double integral either way.

$$\int_R f \, dA = \int_c^d \left(\int_a^b f(x,y) \, dx \right) dy = \int_a^b \left(\int_c^d f(x,y) \, dy \right) dx$$

Example 2   Compute the volume of Example 1 as an iterated integral with y fixed in the inner integral.

Solution   Rewriting the integral, we have

$$\text{Volume} = \int_0^8 \left(\int_0^{16} (12 - \tfrac{1}{4}x - \tfrac{1}{8}y) \, dy \right) dx = \int_0^8 \left((12y - \tfrac{1}{4}xy - \tfrac{1}{16}y^2) \Big|_{y=0}^{y=16} \right) dx$$

$$= \int_0^8 (176 - 4x) \, dx = (176x - 2x^2) \Big|_0^8 = 1280 \text{ meter}^3.$$

Iterated Integrals Over Non-Rectangular Regions

Example 3   The density at the point (x,y) of a triangular metal plate, as shown in Figure 16.14, is $\delta(x,y)$. Express its mass as an iterated integral.

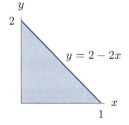

Figure 16.14: A triangular metal plate with density $\delta(x,y)$ at the point (x,y)

Solution   Approximate the triangular region using a grid of small rectangles of sides Δx and Δy. The mass of one rectangle is given by

$$\text{Mass of rectangle} \approx \text{Density} \cdot \text{Area} \approx \delta(x,y)\Delta x \Delta y.$$

Summing over all rectangles gives a Riemann sum which approximates the double integral

$$\text{Mass} = \int_R \delta(x,y) \, dA,$$

where R is the triangle. We want to compute this integral using an iterated integral.

Think about how the iterated integral over the rectangle $a \le x \le b$, $c \le y \le d$ works:

$$\int_a^b \int_c^d f(x,y) \, dy \, dx.$$

The inside integral with respect to y is along vertical strips which begin at the horizontal line $y = c$ and end at the line $y = d$. There is one such strip for each x between $x = a$ and $x = b$. (See Figure 16.15).

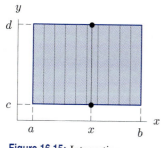

Figure 16.15: Integrating over a rectangle using vertical strips

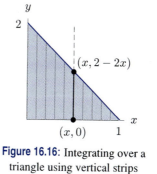

Figure 16.16: Integrating over a triangle using vertical strips

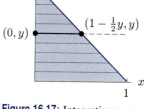

Figure 16.17: Integrating over a triangle using horizontal strips

For the triangular region in Figure 16.14, the idea is the same. The only difference is that the individual vertical strips no longer all go from $y = c$ to $y = d$. The vertical strip that starts at the point $(x, 0)$ ends at the point $(x, 2 - 2x)$, because the top edge of the triangle is the line $y = 2 - 2x$. See Figure 16.16. On this vertical strip, y goes from 0 to $2 - 2x$. Hence, the inside integral is

$$\int_0^{2-2x} \delta(x, y) \, dy.$$

Finally, since there is a vertical strip for each x between 0 and 1, the outside integral goes from $x = 0$ to $x = 1$. Thus, the iterated integral we want is

$$\text{Mass} = \int_0^1 \int_0^{2-2x} \delta(x, y) \, dy \, dx.$$

We could have chosen to integrate in the opposite order, keeping y fixed in the inner integral instead of x. The limits are formed by looking at horizontal strips instead of vertical ones, and expressing the x-values at the end points in terms of y. To find the right endpoint of the strip, we use the equation of the top edge of the triangle in the form $x = 1 - \frac{1}{2}y$. Thus, a horizontal strip goes from $x = 0$ to $x = 1 - \frac{1}{2}y$. Since there is a strip for every y from 0 and 2, the iterated integral is

$$\text{Mass} = \int_0^2 \int_0^{1-\frac{1}{2}y} \delta(x, y) \, dx \, dy.$$

Limits on Iterated Integrals

- The limits on the outer integral must be constants.
- The limits on the inner integral can involve only the variable in the outer integral. For example, if the inner integral is with respect to x, its limits can be functions of y.

Example 4   Find the mass M of a metal plate R bounded by $y = x$ and $y = x^2$, with density given by $\delta(x, y) = 1 + xy$ kg/meter^2. (See Figure 16.18.)

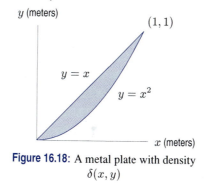

Figure 16.18: A metal plate with density $\delta(x, y)$

Solution   The mass is given by

$$M = \int_R \delta(x,y)\, dA.$$

We integrate along vertical strips first; this means we do the y integral first, which goes from the bottom boundary $y = x^2$ to the top boundary $y = x$. The left edge of the region is at $x = 0$ and the right edge is at the intersection point of $y = x$ and $y = x^2$, which is $(1, 1)$. Thus, the x-coordinate of the vertical strips can vary from $x = 0$ to $x = 1$, and so the mass is given by

$$M = \int_0^1 \int_{x^2}^x \delta(x,y)\, dy\, dx = \int_0^1 \int_{x^2}^x (1 + xy)\, dy\, dx.$$

Calculating the inner integral first gives

$$M = \int_0^1 \int_{x^2}^x (1 + xy)\, dy\, dx = \int_0^1 \left(y + x\frac{y^2}{2} \right) \Bigg|_{y=x^2}^{y=x} dx$$

$$= \int_0^1 \left(x - x^2 + \frac{x^3}{2} - \frac{x^5}{2} \right) dx = \left(\frac{x^2}{2} - \frac{x^3}{3} + \frac{x^4}{8} - \frac{x^6}{12} \right) \Bigg|_0^1 = \frac{5}{24} \approx 0.21 \text{ kg.}$$

Example 5   A city occupies a semicircular region of radius 3 km bordering on the ocean. Find the average distance from points in the city to the ocean.

Solution   Think of the ocean as everything below the x-axis in the xy-plane and think of the city as the upper half of the circular disk of radius 3 bounded by $x^2 + y^2 = 9$. (See Figure 16.19).

The distance from any point (x, y) in the city to the ocean is the vertical distance to the x-axis, namely y. Thus, we want to compute

$$\text{Average distance} = \frac{1}{\text{Area}(R)} \int_R y\, dA,$$

where R is the region between the upper half of the circle $x^2 + y^2 = 9$ and the x-axis. The area of R is $\pi 3^2 / 2 = 9\pi/2$.

To compute the integral, let's take the inner integral with respect to y. Then a vertical strip goes from the x-axis, namely $y = 0$, to the semicircle. The upper limit must be expressed in terms of x so we solve $x^2 + y^2 = 9$ to get $y = \sqrt{9 - x^2}$. Since there is a strip for every x from -3 to 3, the integral is:

$$\int_R y\, dA = \int_{-3}^3 \left(\int_0^{\sqrt{9-x^2}} y\, dy \right) dx = \int_{-3}^3 \left(\frac{y^2}{2} \Bigg|_0^{\sqrt{9-x^2}} \right) dx$$

$$= \int_{-3}^3 \frac{1}{2}(9 - x^2)\, dx = \frac{1}{2} \left(9x - \frac{x^3}{3} \right) \Bigg|_{-3}^3 = \frac{1}{2}(18 - (-18)) = 18.$$

Therefore, the average distance is $18/(9\pi/2) = 4/\pi \approx 1.3$ km.

What if we choose the inner integral with respect to x? Then we get the limits by looking at horizontal strips, not vertical, and we solve $x^2 + y^2 = 9$ for x in terms of y. We get $x = -\sqrt{9 - y^2}$ at the left end of the strip and $x = \sqrt{9 - y^2}$ at the right. There is a strip for every y from 0 to 3, so

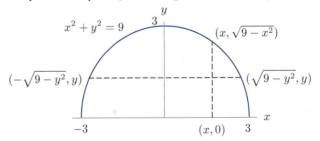

Figure 16.19: The city by the ocean showing a typical vertical strip and a typical horizontal strip

$$\int_R y \, dA = \int_0^3 \left(\int_{-\sqrt{9-y^2}}^{\sqrt{9-y^2}} y \, dx \right) dy = \int_0^3 \left(yx \Big|_{x=-\sqrt{9-y^2}}^{x=\sqrt{9-y^2}} \right) dy = \int_0^3 2y\sqrt{9-y^2} \, dy$$

$$= -\frac{2}{3}(9-y^2)^{3/2} \Big|_0^3 = -\frac{2}{3}(0-27) = 18.$$

We get the same result as before. The average distance to the ocean is $(2/(9\pi))18 = 4/\pi \approx 1.3$ km.

In the examples so far, a region was given and the problem was to determine the limits for an iterated integral. Sometimes the limits are known and we want to determine the region.

Example 6   Sketch the region of integration for the iterated integral $\int_0^6 \int_{x/3}^2 x\sqrt{y^3+1} \, dy \, dx$.

Solution   The inner integral is with respect to y, so we imagine the region built of vertical strips. The bottom of each strip is on the line $y = x/3$, and the top is on the horizontal line $y = 2$. Since the limits of the outer integral are 0 and 6, the whole region is contained between the vertical lines $x = 0$ and $x = 6$. Notice that the lines $y = 2$ and $y = x/3$ meet where $x = 6$. See Figure 16.20.

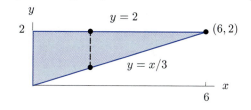

Figure 16.20: The region of integration for Example 6, showing vertical strip

Reversing the Order of Integration

It is sometimes helpful to reverse the order of integration in an iterated integral. An integral which is difficult or impossible with the integration in one order can be quite straightforward in the other. The next example is such a case.

Example 7   Evaluate $\int_0^6 \int_{x/3}^2 x\sqrt{y^3+1} \, dy \, dx$ using the region sketched in Figure 16.20.

Solution   Since $\sqrt{y^3+1}$ has no elementary antiderivative, we cannot calculate the inner integral symbolically. We try reversing the order of integration. From Figure 16.20, we see that horizontal strips go from $x = 0$ to $x = 3y$ and that there is a strip for every y from 0 to 2. Thus, when we change the order of integration we get

$$\int_0^6 \int_{x/3}^2 x\sqrt{y^3+1} \, dy \, dx = \int_0^2 \int_0^{3y} x\sqrt{y^3+1} \, dx \, dy.$$

Now we can at least do the inner integral because we know the antiderivative of x. What about the outer integral?

$$\int_0^2 \int_0^{3y} x\sqrt{y^3+1} \, dx \, dy = \int_0^2 \left(\frac{x^2}{2}\sqrt{y^3+1} \right) \Big|_{x=0}^{x=3y} dy = \int_0^2 \frac{9y^2}{2}(y^3+1)^{1/2} \, dy$$

$$= (y^3+1)^{3/2} \Big|_0^2 = 27 - 1 = 26.$$

Thus, reversing the order of integration made the integral in the previous problem much easier. Notice that to reverse the order it is essential first to sketch the region over which the integration is being performed.

Exercises and Problems for Section 16.2

Exercises

For Exercises 1–4, evaluate the integral.

1. $\int_0^2 \int_0^3 (x^2 + y^2)\,dy\,dx$

2. $\int_0^3 \int_0^2 (6xy)\,dy\,dx$

3. $\int_0^1 \int_0^2 (x^2 y)\,dy\,dx$

4. $\int_0^3 \int_0^4 (4x + 3y)\,dx\,dy$

For each of the regions R in Exercises 5–10, write $\int_R f\,dA$ as an iterated integral.

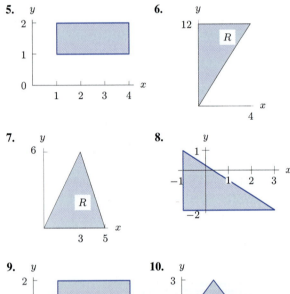

5.

6.

7.

8.

9.

10.

For Exercises 11–16, sketch the region of integration and evaluate the integral.

11. $\int_1^3 \int_0^4 e^{x+y}\,dy\,dx$

12. $\int_0^2 \int_0^x e^{x^2}\,dy\,dx$

13. $\int_1^5 \int_x^{2x} \sin x\,dy\,dx$

14. $\int_1^4 \int_{\sqrt{y}}^y x^2 y^3\,dx\,dy$

15. $\int_0^3 \int_0^{2x} (x^2 + y^2)\,dy\,dx$

16. $\int_{-2}^0 \int_{-\sqrt{9-x^2}}^0 2xy\,dy\,dx$

For Exercises 17–21, evaluate the integral.

17. $\int_R xy\,dA$, where R is the triangle $x + y \leq 1$, $x \geq 0$, $y \geq 0$.

18. $\int_R \sqrt{x+y}\,dA$, where R is the rectangle $0 \leq x \leq 1$, $0 \leq y \leq 2$.

19. Calculate the integral in Exercise 18 using the other order of integration.

20. $\int_R (5x^2 + 1)\sin 3y\,dA$, where R is the rectangle $-1 \leq x \leq 1$, $0 \leq y \leq \pi/3$.

21. $\int_R (2x + 3y)^2\,dA$, where R is the triangle with vertices at $(-1, 0)$, $(0, 1)$, and $(1, 0)$.

22. Find the average value of $f(x, y) = xy^2$ on the rectangle $0 \leq x \leq 4$, $0 \leq y \leq 3$.

23. Find the average value of $f(x, y) = x^2 + 4y$ on the rectangle $0 \leq x \leq 3$ and $0 \leq y \leq 6$.

24. Find the volume under the graph of the function $f(x, y) = 6x^2 y$ over the region shown in Figure 16.21.

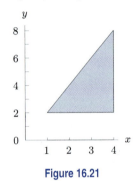

Figure 16.21

25. Consider the integral $\int_0^4 \int_0^{-(y-4)/2} g(x, y)\,dx\,dy$.

   (a) Sketch the region over which the integration is being performed.

   (b) Write the integral with the order of the integration reversed.

Evaluate the integrals in Exercises 26–30 by reversing the order of integration.

26. $\int_0^1 \int_y^1 e^{x^2}\,dx\,dy$

27. $\int_0^1 \int_y^1 \sin(x^2)\,dx\,dy$

28. $\int_0^3 \int_{y^2}^9 y\sin(x^2)\,dx\,dy$

29. $\int_0^1 \int_{\sqrt{y}}^1 \sqrt{2 + x^3}\,dx\,dy$

30. $\int_0^1 \int_{e^y}^e \frac{x}{\ln x}\,dx\,dy$

Problems

31. (a) Use four subrectangles to approximate the volume of the object whose base is the region $0 \leq x \leq 4$ and $0 \leq y \leq 6$, and whose height is given by $f(x, y) = xy$. Find an overestimate and an underestimate and average the two.

(b) Integrate to find the exact volume of the three-dimensional object described in part (a).

32. Reverse the order of integration:
$$\int_{-4}^{0} \int_{0}^{2x+8} f(x, y)\, dy\, dx + \int_{0}^{4} \int_{0}^{-2x+8} f(x, y)\, dy\, dx.$$

For Problems 33–35 set up, but do not evaluate, an iterated integral for the volume of the solid.

33. Under the graph of $f(x, y) = 25 - x^2 - y^2$ and above the xy-plane.

34. Below the graph of $f(x, y) = 25 - x^2 - y^2$ and above the plane $z = 16$.

35. The three-sided pyramid whose base is on the xy-plane and whose three sides are the vertical planes $y = 0$ and $y - x = 4$, and the slanted plane $2x + y + z = 4$.

For Problems 36–41, find the volume of the region.

36. Under the graph of $f(x, y) = xy$ and above the square $0 \leq x \leq 2, 0 \leq y \leq 2$ in the xy-plane.

37. Under the graph of $f(x, y) = x^2 + y^2$ and above the triangle $y \leq x, 0 \leq x \leq 1$.

38. Under the graph of $f(x, y) = x + y$ and above the region $y^2 \leq x, 0 \leq x \leq 9$.

39. Under the graph of $2x + y + z = 4$ in the first octant.

40. The solid between the planes $z = 3x + 2y + 1$ and $z = x + y$, and above the triangle with vertices $(1, 0, 0)$, $(2, 2, 0)$, and $(0, 1, 0)$ in the xy-plane. See Figure 16.22.

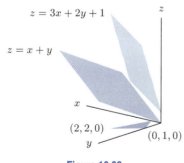

Figure 16.22

41. The region R bounded by the graph of $ax + by + cz = 1$ and the coordinate planes. Assume a, b, and $c > 0$.

42. Find the average distance to the x-axis for points in the region bounded by the x-axis and the graph of $y = x - x^2$.

43. Show that for a right triangle the average distance from any point in the triangle to one of the legs is one-third the length of the other leg. (The legs of a right triangle are the two sides that are not the hypotenuse.)

44. The function $f(x, y) = ax^2 + bxy + cy^2$ has an average value of 20 on the square $0 \leq x \leq 2, 0 \leq y \leq 2$.

(a) What can you say about the constants a, b, and c?

(b) Find two different choices for f that have average value 20 on the square, and give their contour diagrams on the square.

45. The function $f(x, y) = ax + by$ has an average value of 20 on the rectangle $0 \leq x \leq 2, 0 \leq y \leq 3$.

(a) What can you say about the constants a and b?

(b) Find two different choices for f that have average value 20 on the rectangle, and give their contour diagrams on the rectangle.

46. Give the contour diagram of a function f whose average value on the square $0 \leq x \leq 1, 0 \leq 1 \leq y$ is

(a) Greater than the average of the values of f at the four corners of the square.

(b) Less than the average of the values of f at the four corners of the square.

47. A rectangular plate of sides a and b is subjected to a normal force (that is, perpendicular to the plate). The pressure, p, at any point on the plate is proportional to the square of the distance of that point from one corner. Find the total force on the plate. [Note that pressure is force per unit area.]

16.3   TRIPLE INTEGRALS

A continuous function of three variables can be integrated over a solid region W in 3-space in the same way as a function of two variables is integrated over a flat region in 2-space. Again, we start with a Riemann sum. First we subdivide W into smaller regions, then we multiply the volume of each region by a value of the function in that region, and then we add the results. For example, if W is the box $a \leq x \leq b, c \leq y \leq d, p \leq z \leq q$, then we subdivide each side into n, m, and l pieces, thereby chopping W into nml smaller boxes, as shown in Figure 16.23.

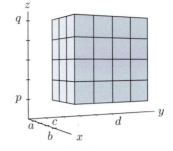

Figure 16.23: Subdividing a
three-dimensional box

The volume of each smaller box is

$$\Delta V = \Delta x \Delta y \Delta z,$$

where $\Delta x = (b - a)/n$, and $\Delta y = (d - c)/m$, and $\Delta z = (q - p)/l$. Using this subdivision, we pick a point $(u_{ijk}, v_{ijk}, w_{ijk})$ in the ijk-th small box and construct a Riemann sum

$$\sum_{i,j,k} f(u_{ijk}, v_{ijk}, w_{ijk})\, \Delta V.$$

If f is continuous, as Δx, Δy, and Δz approach 0, this Riemann sum approaches the definite integral, $\int_W f\, dV$, called a *triple integral*, which is defined as

$$\int_W f\, dV = \lim_{n,m,l \to \infty} \sum_{i,j,k} f(u_{ijk}, v_{ijk}, w_{ijk})\Delta V.$$

As in the case of a double integral, we can evaluate this integral as an iterated integral:

Triple integral as an iterated integral

$$\int_W f\, dV = \int_p^q \left(\int_c^d \left(\int_a^b f(x, y, z)\, dx \right) dy \right) dz,$$

where y and z are treated as constants in the innermost (dx) integral, and z is treated as a constant in the middle (dy) integral. Other orders of integration are possible.

Example 1   A cube C has sides of length 4 cm and is made of a material of variable density. If one corner is at the origin and the adjacent corners are on the positive x, y, and z axes, then the density at the point (x, y, z) is $\delta(x, y, z) = 1 + xyz$ gm/cm^3. Find the mass of the cube.

Solution   Consider a small piece ΔV of the cube, small enough so that the density remains close to constant over the piece. Then

$$\text{Mass of small piece} = \text{Density} \cdot \text{Volume} \approx \delta(x, y, z)\, \Delta V.$$

To get the total mass, we add the masses of the small pieces and take the limit as $\Delta V \to 0$. Thus, the mass is the triple integral

$$M = \int_C \delta\, dV = \int_0^4 \int_0^4 \int_0^4 (1 + xyz)\, dx\, dy\, dz = \int_0^4 \int_0^4 \left(x + \frac{1}{2}x^2 yz \right) \Big|_{x=0}^{x=4} dy\, dz$$

$$= \int_0^4 \int_0^4 (4 + 8yz)\, dy\, dz = \int_0^4 \left(4y + 4y^2 z \right) \Big|_{y=0}^{y=4} dz = \int_0^4 (16 + 64z)\, dz = 576 \text{ gm}.$$

Example 2   Express the volume of the building described in Example 1 on page 745 as a triple integral.

Solution   The building is given by $0 \leq x \leq 8$, $0 \leq y \leq 16$, and $0 \leq z \leq 12 - x/4 - y/8$. (See Figure 16.24 on page 753.) To find its volume, divide it into small cubes of volume $\Delta V = \Delta x \, \Delta y \, \Delta z$ and add. First, make a vertical stack of cubes above the point $(x, y, 0)$. This stack goes from $z = 0$ to $z = 12 - x/4 - y/8$, so

$$\text{Volume of vertical stack} \approx \sum_z \Delta V = \sum_z \Delta x \, \Delta y \, \Delta z = \left(\sum_z \Delta z \right) \Delta x \, \Delta y.$$

Next, line up these stacks parallel to the y-axis to form a slice from $y = 0$ to $y = 16$. So

$$\text{Volume of slice} \approx \left(\sum_y \sum_z \Delta z \, \Delta y \right) \Delta x.$$

Finally, line up the slices along the x-axis from $x = 0$ to $x = 8$ and add up their volumes, to get

$$\text{Volume of building} \approx \sum_x \sum_y \sum_z \Delta z \, \Delta y \, \Delta x.$$

Thus, in the limit,

$$\text{Volume of building} = \int_0^8 \int_0^{16} \int_0^{12 - x/4 - y/8} 1 \, dz \, dy \, dx.$$

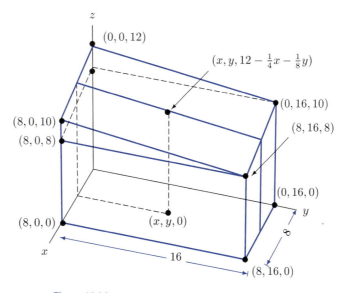

Figure 16.24: Volume of building as triple integral

Example 3   Set up an iterated integral to compute the mass of the solid cone bounded by $z = \sqrt{x^2 + y^2}$ and $z = 3$, if the density is given by $\delta(x, y, z) = z$.

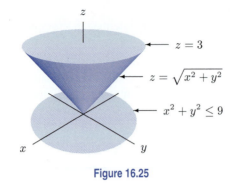

Figure 16.25

Solution   We break the cone in Figure 16.25 into small cubes of volume $\Delta V = \Delta x \, \Delta y \, \Delta z$, on which the density is approximately constant, and approximate the mass of each cube by $\delta(x, y, z) \, \Delta x \, \Delta y \, \Delta z$. Stacking the cubes vertically above the point $(x, y, 0)$, starting on the cone at height $z = \sqrt{x^2 + y^2}$ and going up to $z = 3$, tells us that the inner integral is

$$\int_{\sqrt{x^2+y^2}}^{3} \delta(x, y, z) \, dz = \int_{\sqrt{x^2+y^2}}^{3} z \, dz.$$

There is a stack for every point in the xy-plane in the shadow of the cone. The cone $z = \sqrt{x^2 + y^2}$ intersects the horizontal plane $z = 3$ in the circle $x^2 + y^2 = 9$, so there is a stack for all (x, y) in the region $x^2 + y^2 \le 9$. Lining up the stacks parallel to the y-axis gives a slice from $y = -\sqrt{9 - x^2}$ to $y = \sqrt{9 - x^2}$, for each fixed value of x. Thus, the limits on the middle integral are

$$\int_{-\sqrt{9-x^2}}^{\sqrt{9-x^2}} \int_{\sqrt{x^2+y^2}}^{3} z \, dz \, dy.$$

Finally, there is a slice for each x between -3 and 3, so the integral we want is

$$\text{Mass} = \int_{-3}^{3} \int_{-\sqrt{9-x^2}}^{\sqrt{9-x^2}} \int_{\sqrt{x^2+y^2}}^{3} z \, dz \, dy \, dx.$$

Notice that setting up the limits on the two outer integrals was just like setting up the limits for a double integral over the region $x^2 + y^2 \le 9$.

As the previous example illustrates, for a region W contained between two surfaces, the innermost limits correspond to these surfaces. The middle and outer limits ensure that we integrate over the "shadow" of W in the xy-plane.

Limits on Triple Integrals

- The limits for the outer integral are constants.
- The limits for the middle integral can involve only one variable (that in the outer integral).
- The limits for the inner integral can involve two variables (those on the two outer integrals).

Exercises and Problems for Section 16.3

Exercises

In Exercises 1–4, find the triple integrals of the function over the region W.

1. $f(x, y, z) = x^2 + 5y^2 - z$, W is the rectangular box $0 \leq x \leq 2, -1 \leq y \leq 1, 2 \leq z \leq 3$.

2. $f(x, y, z) = \sin x \cos(y + z)$, W is the cube $0 \leq x \leq \pi$, $0 \leq y \leq \pi, 0 \leq z \leq \pi$.

3. $h(x, y, z) = ax + by + cz$, W is the rectangular box $0 \leq x \leq 1, 0 \leq y \leq 1, 0 \leq z \leq 2$.

4. $f(x, y, z) = e^{-x-y-z}$, W is the rectangular box with corners at $(0, 0, 0), (a, 0, 0), (0, b, 0)$, and $(0, 0, c)$.

Sketch the region of integration in Exercises 5–13.

5. $\int_0^1 \int_{-1}^1 \int_0^{\sqrt{1-z^2}} f(x, y, z)\, dy\, dz\, dx$

6. $\int_0^1 \int_{-1}^1 \int_0^{\sqrt{1-x^2}} f(x, y, z)\, dz\, dx\, dy$

7. $\int_0^1 \int_{-\sqrt{1-z^2}}^{\sqrt{1-z^2}} \int_0^{\sqrt{1-x^2-z^2}} f(x, y, z)\, dy\, dx\, dz$

8. $\int_{-1}^1 \int_0^1 \int_{-\sqrt{1-z^2}}^{\sqrt{1-z^2}} f(x, y, z)\, dy\, dz\, dx$

9. $\int_0^1 \int_{-1}^1 \int_{-\sqrt{1-x^2}}^{\sqrt{1-x^2}} f(x, y, z)\, dz\, dx\, dy$

10. $\int_{-1}^1 \int_{-\sqrt{1-x^2}}^{\sqrt{1-x^2}} \int_0^{\sqrt{1-x^2-z^2}} f(x, y, z)\, dy\, dz\, dx$

11. $\int_0^1 \int_{-\sqrt{1-z^2}}^{\sqrt{1-z^2}} \int_{-\sqrt{1-y^2-z^2}}^{\sqrt{1-y^2-z^2}} f(x, y, z)\, dx\, dy\, dz$

12. $\int_0^1 \int_0^{\sqrt{1-z^2}} \int_{-\sqrt{1-x^2-z^2}}^{\sqrt{1-x^2-z^2}} f(x, y, z)\, dy\, dx\, dz$

13. $\int_0^1 \int_0^{\sqrt{1-y^2}} \int_{-\sqrt{1-x^2-y^2}}^{\sqrt{1-x^2-y^2}} f(x, y, z)\, dz\, dx\, dy$

Problems

14. Find the volume of the region bounded by $z = x + y$, $z = 10$, and the planes $x = 0$, $y = 0$ and $z = 0$.

15. Find the volume of the pyramid with base in the plane $z = -6$ and sides formed by the three planes $y = 0$ and $y - x = 4$ and $2x + y + z = 4$.

16. A solid region D is a half cylinder of radius 1 lying horizontally with its rectangular base in the xy-plane and its axis along the y-axis from $y = 0$ to $y = 10$. (The region is above the xy-plane.)

    (a) What is the equation of the curved surface of this half cylinder?
    (b) Write the limits of integration of the integral $\int_D f(x, y, z)\, dV$ in Cartesian coordinates.

17. Find the mass of the solid bounded by the xy-plane, yz-plane, xz-plane, and the plane $(x/3) + (y/2) + (z/6) = 1$, if the density of the solid is given by $\delta(x, y, z) = x + y$.

18. Find the average value of the sum of the squares of three numbers x, y, z, where each number is between 0 and 2.

Let W be the solid cone bounded by $z = \sqrt{x^2 + y^2}$ and $z = 2$. For Problems 19–27, decide (without calculating its value) whether the integral is positive, negative, or zero.

19. $\int_W \sqrt{x^2 + y^2}\, dV$

20. $\int_W (z - \sqrt{x^2 + y^2})\, dV$

21. $\int_W x\, dV$

22. $\int_W y\, dV$

23. $\int_W z\, dV$

24. $\int_W xy\, dV$

25. $\int_W xyz\, dV$

26. $\int_W (z - 2)\, dV$

27. $\int_W e^{-xyz}\, dV$

Let W be the solid half-cone bounded by $z = \sqrt{x^2 + y^2}$, $z = 2$ and the yz-plane with $x \geq 0$. For Problems 28–36, decide (without calculating its value) whether the integral is positive, negative, or zero.

28. $\int_W \sqrt{x^2 + y^2}\, dV$

29. $\int_W (z - \sqrt{x^2 + y^2})\, dV$

30. $\int_W x\, dV$

31. $\int_W y\, dV$

32. $\int_W z\, dV$

33. $\int_W xy\, dV$

34. $\int_W xyz\, dV$

35. $\int_W (z - 2)\, dV$

36. $\int_W e^{-xyz}\, dV$

37. Figure 16.26 shows part of a spherical ball of radius 5 cm. Write an iterated triple integral which represents the volume of this region.

2 cm

Figure 16.26

38. Set up, but do not evaluate, an iterated integral for the volume of the solid formed by the intersections of the cylinders $x^2 + z^2 = 1$ and $y^2 + z^2 = 1$.

The motion of a solid object can be analyzed by thinking of the mass as concentrated at a single point, the *center of mass*. If the object has density $\rho(x, y, z)$ at the point (x, y, z) and occupies a region W, then the coordinates $(\bar{x}, \bar{y}, \bar{z})$ of the center of mass are given by

$$\bar{x} = \frac{1}{m} \int_W x\rho \, dV \quad \bar{y} = \frac{1}{m} \int_W y\rho \, dV \quad \bar{z} = \frac{1}{m} \int_W z\rho \, dV$$

where $m = \int_W \rho \, dV$ is the total mass of the body. Use these definitions for Problems 39–40.

39. A solid is bounded below by the square $z = 0, 0 \leq x \leq 1, 0 \leq y \leq 1$ and above by the surface $z = x + y + 1$. Find the total mass and the coordinates of the center of mass if the density is 1 gm/cm^3 and x, y, z are measured in centimeters.

40. Find the center of mass of the tetrahedron that is bounded by the xy, yz, xz planes and the plane $x + 2y + 3z = 1$. Assume the density is 1 gm/cm^3 and x, y, z are in centimeters.

The *moment of inertia* of a solid body about an axis in 3-space relates the angular acceleration about this axis to torque (force twisting the body). The moments of inertia about the coordinate axes of a body of constant density and mass m occupying a region W of volume V are defined to be

$$I_x = \frac{m}{V} \int_W (y^2 + z^2) \, dV \qquad I_y = \frac{m}{V} \int_W (x^2 + z^2) \, dV$$

$$I_z = \frac{m}{V} \int_W (x^2 + y^2) \, dV$$

Use these definitions for Problems 41–43.

41. Find the moment of inertia about the z-axis of the rectangular solid of mass m given by $0 \leq x \leq 1, 0 \leq y \leq 2, 0 \leq z \leq 3$.

42. Find the moment of inertia about the x-axis of the rectangular solid $-a \leq x \leq a, -b \leq y \leq b$ and $-c \leq z \leq c$ of mass m.

43. Let a, b, and c denote the moments of inertia of a homogeneous solid object about the x, y and z-axes respectively. Explain why $a + b > c$.

44. A cube of side length 2 m is composed of two materials, separated by a plane slanting down from the top corner to cut the bottom face in a diagonal, as in Figure 16.27. The material below the plane has density δ_1 gm/m^3 and the material above the plane has density δ_2 gm/m^3. Find the mass of the cube.

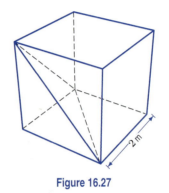

Figure 16.27

16.4   DOUBLE INTEGRALS IN POLAR COORDINATES

Integration in Polar Coordinates

We started this chapter by putting a rectangular grid on the fox population density map, to estimate the total population using a Riemann sum. However, sometimes a polar grid is more appropriate. A review of polar coordinates is in Appendix B.

Example 1   A biologist studying insect populations around a circular lake divides the area into the polar sectors in Figure 16.28. The population density in each sector is shown in millions per square km. Estimate the total insect population around the lake.

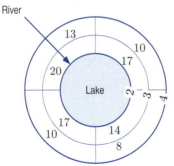

Figure 16.28: An insect infested lake showing the insect population density by sector

Solution   To get the estimate, we multiply the population density in each sector by the area of that sector. Unlike the rectangles in a rectangular grid, the sectors in this grid do not all have the same area. The inner sectors have area

$$\frac{1}{4}(\pi 3^2 - \pi 2^2) = \frac{5\pi}{4} \approx 3.93 \text{ km}^2,$$

and the outer sectors have area

$$\frac{1}{4}(\pi 4^2 - \pi 3^2) = \frac{7\pi}{4} \approx 5.50 \text{ km}^2,$$

so we estimate

$$\text{Population} \approx (20)(3.93) + (17)(3.93) + (14)(3.93) + (17)(3.93) +$$
$$(13)(5.50) + (10)(5.50) + (8)(5.50) + (10)(5.50)$$
$$\approx 493 \text{ million insects.}$$

What is dA in Polar Coordinates?

The previous example used a polar grid rather than a rectangular grid. A rectangular grid is constructed from vertical and horizontal lines of the form $x = k$ (a constant) and $y = l$ (another constant). In polar coordinates, $r = k$ gives a circle of radius k centered at the origin and $\theta = l$ gives a ray emanating from the origin (at angle l with the x-axis). A polar grid is built out of these circles and rays. See Figure 16.29.

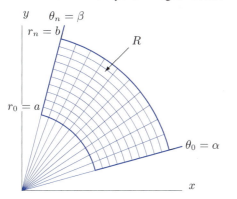

Figure 16.29: Dividing up a region using a polar grid

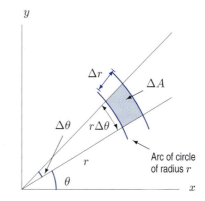

Figure 16.30: Calculating area ΔA in polar coordinates

Choosing (r_{ij}, θ_{ij}) in the ij-th bent rectangle in Figure 16.29 gives a Riemann sum:

$$\sum_{i,j} f(r_{ij}, \theta_{ij})\, \Delta A.$$

To calculate the area ΔA, look at Figure 16.30. If Δr and $\Delta \theta$ are small, the shaded region is approximately a rectangle with sides $r\,\Delta\theta$ and Δr, so

$$\Delta A \approx r\Delta\theta\Delta r.$$

Thus, the Riemann sum is approximately

$$\sum_{i,j} f(r_{ij}, \theta_{ij})\, r_{ij}\, \Delta\theta\, \Delta r.$$

If we take the limit as Δr and $\Delta \theta$ approach 0, we obtain

$$\int_R f\, dA = \int_\alpha^\beta \int_a^b f(r, \theta)\, r\, dr\, d\theta.$$

> When computing integrals in polar coordinates, use $x = r\cos\theta, y = r\sin\theta, x^2 + y^2 = r^2$.
> Put $dA = r\, dr\, d\theta$ or $dA = r\, d\theta\, dr$.

Example 2   Compute the integral of $f(x, y) = 1/(x^2 + y^2)^{3/2}$ over the region R shown in Figure 16.31.

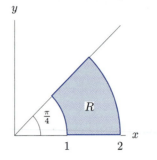

Figure 16.31: Integrate f over the polar region

Solution   The region R is described by the inequalities $1 \le r \le 2$, $0 \le \theta \le \pi/4$. In polar coordinates, $r = \sqrt{x^2 + y^2}$, so we can write f as

$$f(x, y) = \frac{1}{(x^2 + y^2)^{3/2}} = \frac{1}{(r^2)^{3/2}} = \frac{1}{r^3}.$$

Then

$$\int_R f \, dA = \int_0^{\pi/4} \int_1^2 \frac{1}{r^3} r \, dr \, d\theta = \int_0^{\pi/4} \left(\int_1^2 r^{-2} \, dr \right) d\theta$$

$$= \int_0^{\pi/4} -\frac{1}{r} \Big|_{r=1}^{r=2} d\theta = \int_0^{\pi/4} \frac{1}{2} \, d\theta = \frac{\pi}{8}.$$

Example 3   For each region in Figure 16.32, decide whether to integrate using polar or Cartesian coordinates. On the basis of its shape, write an iterated integral of an arbitrary function $f(x, y)$ over the region.

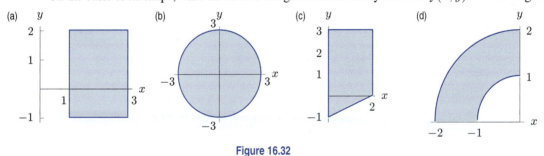

Figure 16.32

Solution   (a) Since this is a rectangular region, Cartesian coordinates are likely to be a better choice. The rectangle is described by the inequalities $1 \le x \le 3$ and $-1 \le y \le 2$, so the integral is

$$\int_{-1}^2 \int_1^3 f(x, y) \, dx \, dy.$$

(b) A circle is best described in polar coordinates. The radius is 3, so r goes from 0 to 3, and to describe the whole circle, θ goes from 0 to 2π. The integral is

$$\int_0^{2\pi} \int_0^3 f(r \cos \theta, r \sin \theta) \, r \, dr \, d\theta.$$

(c) The bottom boundary of this trapezoid is the line $y = (x/2) - 1$ and the top is the line $y = 3$, so we use Cartesian coordinates. If we integrate with respect to y first, the lower limit of the integral is $(x/2) - 1$ and the upper limit is 3. The x limits are $x = 0$ to $x = 2$. So the integral is

$$\int_0^2 \int_{(x/2)-1}^3 f(x, y) \, dy \, dx.$$

(d) This is another polar region: it is a piece of a ring in which r goes from 1 to 2. Since it is in the second quadrant, θ goes from $\pi/2$ to π. The integral is

$$\int_{\pi/2}^{\pi} \int_{1}^{2} f(r\cos\theta, r\sin\theta)\, r\, dr\, d\theta.$$

Exercises and Problems for Section 16.4

Exercises

For the regions R in Exercises 1–4, write $\int_R f\, dA$ as an iterated integral in polar coordinates.

1.

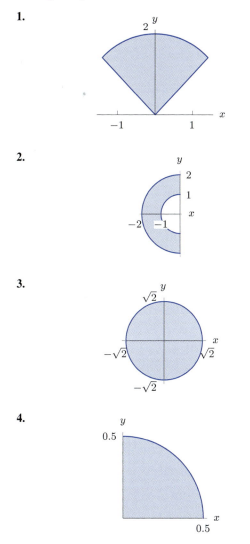

2.

3.

4.

Sketch the region of integration in Exercises 5–11.

5. $\displaystyle\int_{0}^{2\pi} \int_{1}^{2} f(r,\theta)\, r\, dr\, d\theta$

6. $\displaystyle\int_{\pi/2}^{\pi} \int_{0}^{1} f(r,\theta)\, r\, dr\, d\theta$

7. $\displaystyle\int_{\pi/6}^{\pi/3} \int_{0}^{1} f(r,\theta)\, r\, dr\, d\theta$

8. $\displaystyle\int_{3}^{4} \int_{3\pi/4}^{3\pi/2} f(r,\theta)\, r\, d\theta\, dr$

9. $\displaystyle\int_{0}^{\pi/4} \int_{0}^{1/\cos\theta} f(r,\theta)\, r\, dr\, d\theta$

10. $\displaystyle\int_{\pi/4}^{\pi/2} \int_{0}^{2/\sin\theta} f(r,\theta)\, r\, dr\, d\theta$

11. $\displaystyle\int_{0}^{4} \int_{-\pi/2}^{\pi/2} f(r,\theta)\, r\, d\theta\, dr$

In Exercises 12–14, evaluate the integral.

12. $\int_R \sin(x^2 + y^2)\, dA$, where R is the disk of radius 2 centered at the origin.

13. $\int_R (x^2 - y^2)\, dA$, where R is the first quadrant region between the circles of radius 1 and radius 2.

14. $\int_R \sqrt{x^2 + y^2}\, dx\, dy$ where R is $4 \leq x^2 + y^2 \leq 9$.

15. Consider the integral $\int_{0}^{3} \int_{x/3}^{1} f(x,y)\, dy\, dx$.

(a) Sketch the region R over which the integration is being performed.

(b) Rewrite the integral with the order of integration reversed.

(c) Rewrite the integral in polar coordinates.

Convert the integrals in Exercises 16–18 to polar coordinates and evaluate.

16. $\displaystyle\int_{-1}^{0} \int_{-\sqrt{1-x^2}}^{\sqrt{1-x^2}} x\, dy\, dx$

17. $\displaystyle\int_{0}^{\sqrt{6}} \int_{-x}^{x} dy\, dx$

18. $\displaystyle\int_{0}^{\sqrt{2}} \int_{y}^{\sqrt{4-y^2}} xy\, dx\, dy$

Problems

19. A city surrounds a bay as shown in Figure 16.33. The population density of the city (in thousands of people per square km) is $\delta(r, \theta)$, where r and θ are polar coordinates and distances are in km.

(a) Set up an iterated integral in polar coordinates giving the total population of the city.

(b) The population density decreases the farther you live from the shoreline of the bay; it also decreases the farther you live from the ocean. Which of the following functions best describes this situation?

   (i) $\delta(r, \theta) = (4 - r)(2 + \cos \theta)$
   (ii) $\delta(r, \theta) = (4 - r)(2 + \sin \theta)$
   (iii) $\delta(r, \theta) = (r + 4)(2 + \cos \theta)$

(c) Estimate the population using your answers to parts (a) and (b).

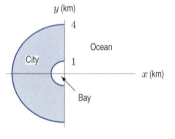

Figure 16.33

20. Find the volume of the region between the graph of $f(x, y) = 25 - x^2 - y^2$ and the xy plane.

21. Find the volume of an ice cream cone bounded by the hemisphere $z = \sqrt{8 - x^2 - y^2}$ and the cone $z = \sqrt{x^2 + y^2}$.

22. A disk of radius 5 cm has density 10 gm/cm^2 at its center, density 0 at its edge, and its density is a linear function of the distance from the center. Find the mass of the disk.

23. (a) For $a > 0$, find the volume under the graph of $z = e^{-(x^2 + y^2)}$ above the disk $x^2 + y^2 \leq a^2$.

(b) What happens to the volume as $a \to \infty$?

24. A watch spring lies flat on the table. It is made of a coiled steel strip standing a height of 0.2 inches above the table. The inner edge is the spiral $r = 0.25 + 0.04\theta$, where $0 \leq \theta \leq 4\pi$ (so the spiral makes two complete turns). The outer edge is given by $r = 0.26 + 0.04\theta$. Find the volume of the spring.

25. Two circular disks, each of radius 1, have centers which are 1 unit apart. Write, but do not evaluate, a double integral, including limits of integration, giving the area of overlap of the disks in

(a) Cartesian coordinates (b) Polar coordinates

26. Electric charge is distributed over the xy-plane, with density inversely proportional to the distance from the origin. Show that the total charge inside a circle of radius R centered at the origin is proportional to R. What is the constant of proportionality?

16.5 INTEGRALS IN CYLINDRICAL AND SPHERICAL COORDINATES

Some double integrals are easier to evaluate in polar, rather than Cartesian, coordinates. Similarly, some triple integrals are easier in non-Cartesian coordinates.

Cylindrical Coordinates

The cylindrical coordinates of a point (x, y, z) in 3-space are obtained by representing the x and y coordinates in polar coordinates and letting the z-coordinate be the z-coordinate of the Cartesian coordinate system. (See Figure 16.34.)

Relation between Cartesian and Cylindrical Coordinates

Each point in 3-space is represented using $0 \leq r < \infty, 0 \leq \theta \leq 2\pi, -\infty < z < \infty$.

$$x = r \cos \theta,$$
$$y = r \sin \theta,$$
$$z = z.$$

As with polar coordinates in the plane, note that $x^2 + y^2 = r^2$.

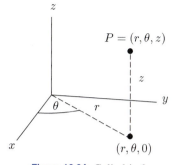

Figure 16.34: Cylindrical
coordinates: (r, θ, z)

A useful way to visualize cylindrical coordinates is to sketch the surfaces obtained by setting one of the coordinates equal to a constant. See Figures 16.35–16.37.

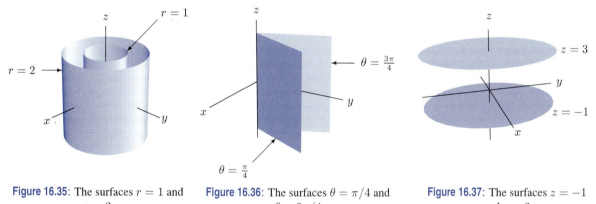

Figure 16.35: The surfaces $r = 1$ and $r = 2$

Figure 16.36: The surfaces $\theta = \pi/4$ and $\theta = 3\pi/4$

Figure 16.37: The surfaces $z = -1$ and $z = 3$

Setting $r = c$ (where c is constant) gives a cylinder around the z-axis whose radius is c. Setting $\theta = c$ gives a half-plane perpendicular to the xy plane, with one edge along the z-axis, making an angle c with the x-axis. Setting $z = c$ gives a horizontal plane $|c|$ units from the xy-plane. We call these *fundamental surfaces*.

The regions that can most easily be described in cylindrical coordinates are those regions whose boundaries are such fundamental surfaces. (For example, vertical cylinders, or wedge-shaped parts of vertical cylinders.)

Example 1   Describe in cylindrical coordinates a wedge of cheese cut from a cylinder 4 cm high and 6 cm in radius; this wedge subtends an angle of $\pi/6$ at the center. (See Figure 16.38.)

Solution   The wedge is described by the inequalities $0 \leq r \leq 6$, and $0 \leq z \leq 4$, and $0 \leq \theta \leq \pi/6$.

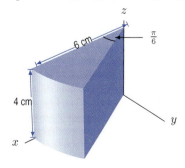

Figure 16.38: A wedge of cheese

Integration in Cylindrical Coordinates

To integrate in polar coordinates, we had to express the area element dA in terms of polar coordinates: $dA = r\,dr\,d\theta$. To evaluate a triple integral $\int_W f\,dV$ in cylindrical coordinates, we need to express the volume element dV in cylindrical coordinates.

Consider the volume element ΔV, shown in Figure 16.39. It is a wedge bounded by fundamental surfaces. The area of the base is $\Delta A \approx r\Delta r\Delta\theta$. Since the height is Δz, the volume element is given approximately by $\Delta V \approx r\,\Delta r\,\Delta\theta\,\Delta z$.

When computing integrals in cylindrical coordinates, put $dV = r\,dr\,d\theta\,dz$. Other orders of integration are also possible.

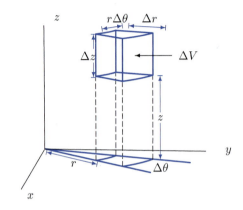

Figure 16.39: Volume element in cylindrical coordinates

Example 2   Find the mass of the wedge of cheese in Example 1, if its density is 1.2 grams/cm^3.

Solution   If the wedge is W, its mass is

$$\int_W 1.2\,dV.$$

In cylindrical coordinates this integral is

$$\int_0^4 \int_0^{\pi/6} \int_0^6 1.2\,r\,dr\,d\theta\,dz = \int_0^4 \int_0^{\pi/6} 0.6r^2 \Big|_0^6 d\theta\,dz = 21.6 \int_0^4 \int_0^{\pi/6} d\theta\,dz$$

$$= 21.6(\frac{\pi}{6})4 \approx 45.24 \text{ grams.}$$

Example 3   A water tank in the shape of a hemisphere has radius a; its base is its plane face. Find the volume, V, of water in the tank as a function of h, the depth of the water.

Solution   In Cartesian coordinates a sphere of radius a has the equation $x^2 + y^2 + z^2 = a^2$. (See Figure 16.40.) In cylindrical coordinates, $r^2 = x^2 + y^2$, so this becomes

$$r^2 + z^2 = a^2.$$

Thus, if we want to describe the amount of water in the tank in cylindrical coordinates, we let r go from 0 to $\sqrt{a^2 - z^2}$, we let θ go from 0 to 2π, and we let z go from 0 to h, giving

$$\text{Volume of water} = \int_W dV = \int_0^{2\pi} \int_0^h \int_0^{\sqrt{a^2-z^2}} r\,dr\,dz\,d\theta = \int_0^{2\pi} \int_0^h \left.\frac{r^2}{2}\right|_{r=0}^{r=\sqrt{a^2-z^2}} dz\,d\theta$$

$$= \int_0^{2\pi} \int_0^h \frac{1}{2}(a^2 - z^2)\,dz\,d\theta = \int_0^{2\pi} \frac{1}{2}\left.\left(a^2 z - \frac{z^3}{3}\right)\right|_{z=0}^{z=h} d\theta$$

$$= \int_0^{2\pi} \frac{1}{2}\left(a^2 h - \frac{h^3}{3}\right) d\theta = \pi\left(a^2 h - \frac{h^3}{3}\right).$$

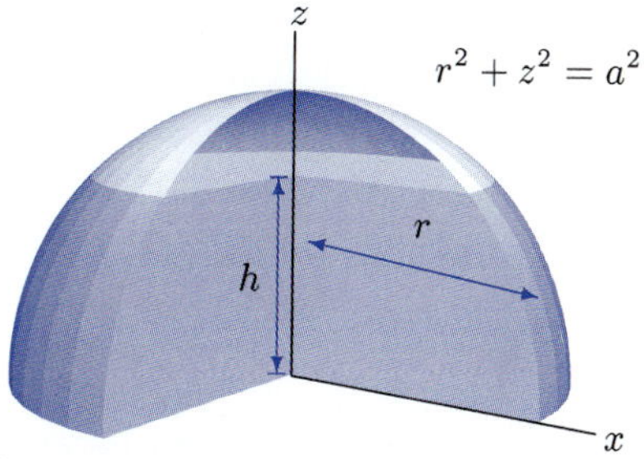

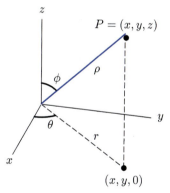

Figure 16.40: Hemispherical water tank with radius a and water of depth h

Spherical Coordinates

In Figure 16.41, the point P has coordinates (x, y, z) in the Cartesian coordinate system. We define spherical coordinates ρ, ϕ, and θ for P as follows: $\rho = \sqrt{x^2 + y^2 + z^2}$ is the distance of P from the origin; ϕ is the angle between the positive z-axis and the line through the origin and the point P; and θ is the same as in cylindrical coordinates.

Figure 16.41: Spherical coordinates

In cylindrical coordinates,

$$x = r\cos\theta, \quad \text{and} \quad y = r\sin\theta, \quad \text{and} \quad z = z.$$

From Figure 16.41 we have $z = \rho\cos\phi$ and $r = \rho\sin\phi$, giving the following relationship:

Relation between Cartesian and Spherical Coordinates

Each point in 3-space is represented using $0 \le \rho < \infty$, $0 \le \phi \le \pi$, and $0 \le \theta \le 2\pi$.

$$x = \rho\sin\phi\cos\theta$$
$$y = \rho\sin\phi\sin\theta$$
$$z = \rho\cos\phi.$$

Also, $\rho^2 = x^2 + y^2 + z^2$.

This system of coordinates is useful when there is spherical symmetry with respect to the origin, either in the region of integration or in the integrand. The fundamental surfaces in spherical coordinates are $\rho = k$ (a constant), which is a sphere of radius k centered at the origin, $\theta = k$ (a constant), which is the half-plane with its edge along the z-axis, and $\phi = k$ (a constant), which is a cone if $k \neq \pi/2$ and the xy-plane if $k = \pi/2$. (See Figures 16.42–16.44.)

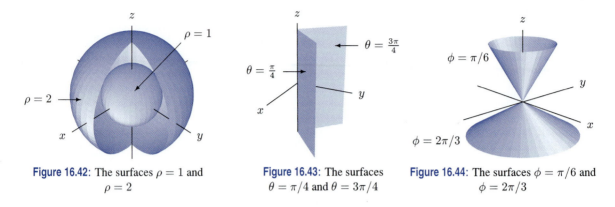

Figure 16.42: The surfaces $\rho = 1$ and $\rho = 2$

Figure 16.43: The surfaces $\theta = \pi/4$ and $\theta = 3\pi/4$

Figure 16.44: The surfaces $\phi = \pi/6$ and $\phi = 2\pi/3$

Integration in Spherical Coordinates

To use spherical coordinates in triple integrals we need to express the volume element, dV, in spherical coordinates. From Figure 16.45, we see that the volume element can be approximated by a box with curved edges. One edge has length $\Delta\rho$. The edge parallel to the xy-plane is an arc of a circle made from rotating the cylindrical radius r ($= \rho \sin \phi$) through an angle $\Delta\theta$, and so has length $\rho \sin \phi \, \Delta\theta$. The remaining side comes from rotating the radius ρ through an angle $\Delta\phi$, and so has length $\rho \, \Delta\phi$. Therefore, $\Delta V \approx \Delta\rho (\rho \, \Delta\phi)(\rho \sin \phi \, \Delta\theta) = \rho^2 \sin \phi \, \Delta\rho \, \Delta\phi \, \Delta\theta$.

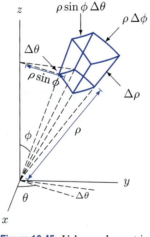

Figure 16.45: Volume element in spherical coordinates

Thus,

When computing integrals in spherical coordinates, put $dV = \rho^2 \sin \phi \, d\rho \, d\phi \, d\theta$. Other orders of integration are also possible.

Example 4   Use spherical coordinates to derive the formula for the volume of a solid sphere of radius a.

Solution   In spherical coordinates, a sphere of radius a is described by the inequalities $0 \leq \rho \leq a$, $0 \leq \theta \leq 2\pi$, and $0 \leq \phi \leq \pi$. Note that θ goes all the way around the circle, whereas ϕ only goes from 0 to π. We find the volume by integrating the constant density function 1 over the sphere:

$$\text{Volume} = \int_R 1 \, dV = \int_0^{2\pi} \int_0^{\pi} \int_0^a \rho^2 \sin\phi \, d\rho \, d\phi \, d\theta = \int_0^{2\pi} \int_0^{\pi} \frac{1}{3} a^3 \sin\phi \, d\phi \, d\theta$$

$$= \frac{1}{3} a^3 \int_0^{2\pi} -\cos\phi \Big|_0^{\pi} \, d\theta = \frac{2}{3} a^3 \int_0^{2\pi} d\theta = \frac{4\pi a^3}{3}.$$

Example 5   Find the magnitude of the gravitational force exerted by a solid hemisphere of radius a and constant density δ on a unit mass located at the center of the base of the hemisphere.

Solution   Assume the base of the hemisphere rests on the xy-plane with center at the origin. (See Figure 16.46.) Newton's law of gravitation says that the force between two masses m_1 and m_2 at a distance r apart is $F = Gm_1m_2/r^2$. In this example, symmetry shows that the net component of the force on the particle at the origin due to the hemisphere is in the z direction only. Any force in the x or y direction from some part of the hemisphere will be canceled by the force from another part of the hemisphere directly opposite the first. To compute the net z component of the gravitational force, we imagine a small piece of the hemisphere with volume ΔV, located at spherical coordinates (ρ, θ, ϕ). This piece has mass $\delta\Delta V$, and exerts a force of magnitude F on the unit mass at the origin. The z-component of this force is given by its projection onto the z-axis, which can be seen from the figure to be $F\cos\phi$. The distance from the mass $\delta\Delta V$ to the unit mass at the origin is the spherical coordinate ρ. Therefore the z-component of the force due to the small piece ΔV is

$$\begin{array}{c}\text{z-component}\\\text{of force}\end{array} = \frac{G(\delta\Delta V)(1)}{\rho^2}\cos\phi.$$

Adding the contributions of the small pieces, we get a vertical force with magnitude

$$F = \int_0^{2\pi} \int_0^{\pi/2} \int_0^a \left(\frac{G\delta}{\rho^2}\right)(\cos\phi)\rho^2 \sin\phi \, d\rho \, d\phi \, d\theta = \int_0^{2\pi} \int_0^{\pi/2} G\delta(\cos\phi\sin\phi)\rho \Big|_{\rho=0}^{\rho=a} d\phi \, d\theta$$

$$= \int_0^{2\pi} \int_0^{\pi/2} G\delta a \cos\phi\sin\phi \, d\phi \, d\theta = \int_0^{2\pi} G\delta a \left(-\frac{(\cos\phi)^2}{2}\right)\Big|_{\phi=0}^{\phi=\pi/2} d\theta$$

$$= \int_0^{2\pi} G\delta a\left(\frac{1}{2}\right) d\theta = G\delta a\pi.$$

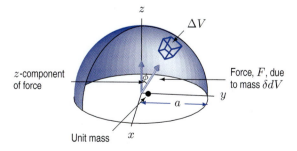

Figure 16.46: Gravitational force of hemisphere on mass at origin

Exercises and Problems for Section 16.5

Exercises

In Exercises 1–2, evaluate the triple integrals in cylindrical coordinates over the region W.

1. $f(x, y, z) = x^2 + y^2 + z^2$, W is the region $0 \leq r \leq 4$, $\pi/4 \leq \theta \leq 3\pi/4$, $-1 \leq z \leq 1$.

2. $f(x, y, z) = \sin(x^2 + y^2)$, W is the solid cylinder with height 4 and with base of radius 1 centered on the z axis at $z = -1$.

In Exercises 3–4, evaluate the triple integrals in spherical coordinates.

3. $f(x, y, z) = 1/(x^2 + y^2 + z^2)^{1/2}$ over the bottom half of the sphere of radius 5 centered at the origin.

4. $f(\rho, \theta, \phi) = \sin \phi$, over the region $0 \leq \theta \leq 2\pi$, $0 \leq \phi \leq \pi/4$, $1 \leq \rho \leq 2$.

For Exercises 5–11, choose coordinates and set up a triple integral, including limits of integration, for a density function f over the given region.

5.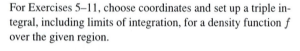

6.

7.

8.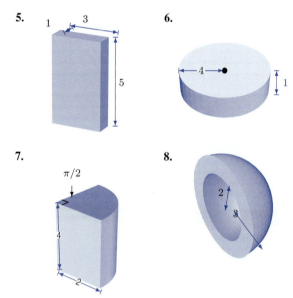

9. A piece of a sphere; angle at the center is $\pi/3$.

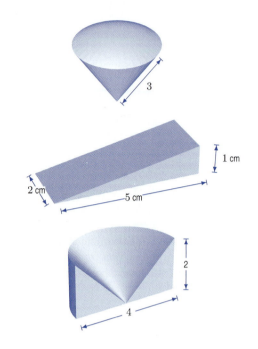

10.

11.

12. Sketch the region R over which the integration is being performed:

$$\int_0^{\pi/2} \int_{\pi/2}^{\pi} \int_0^1 f(\rho, \phi, \theta) \rho^2 \sin \phi \, d\rho \, d\phi \, d\theta.$$

13. **(a)** Convert the following triple integral to spherical coordinates:

$$\int_0^{2\pi} \int_0^3 \int_0^r r \, dz dr d\theta.$$

(b) Evaluate either the original integral or your answer to part (a).

Problems

14. Suppose W is the region outside the cylinder $x^2 + y^2 = 1$ and inside the sphere $x^2 + y^2 + z^2 = 2$. Calculate

$$\int_W (x^2 + y^2) \, dV.$$

Evaluate the integrals in Problems 15–16.

15. $\int_0^1 \int_{-\sqrt{1-x^2}}^{\sqrt{1-x^2}} \int_{-\sqrt{1-x^2-z^2}}^{\sqrt{1-x^2-z^2}} \dfrac{1}{(x^2 + y^2 + z^2)^{1/2}} \, dy \, dz \, dx$

16. $\int_0^1 \int_{-1}^1 \int_{-\sqrt{1-x^2}}^{\sqrt{1-x^2}} \dfrac{1}{(x^2 + y^2)^{1/2}} \, dy \, dx \, dz$

Without performing the integration, decide whether each of the integrals in Problems 17–18 is positive, negative, or zero. Give reasons for your decision.

17. W_1 is the unit ball, $x^2 + y^2 + z^2 \leq 1$.

    (a) $\int_{W_1} \sin \phi \, dV$ **(b)** $\int_{W_1} \cos \phi \, dV$

18. W_2 is $0 \leq z \leq \sqrt{1 - x^2 - y^2}$, the top half of the unit ball.

    (a) $\int_{W_2} (z^2 - z) \, dV$ **(b)** $\int_{W_2} (-xz) \, dV$

19. Write a triple integral representing the volume of the cone in Figure 16.47 and evaluate it.

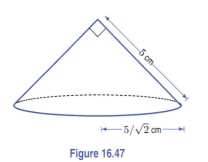

5 cm

5/√2 cm

Figure 16.47

20. Write a triple integral representing the volume of a slice of the cylindrical cake of height 2 and radius 5 between the planes $\theta = \pi/6$ and $\theta = \pi/3$. Evaluate this integral.

21. Write a triple integral representing the volume above the cone $z = \sqrt{x^2 + y^2}$ and below the sphere of radius 2 centered at the origin. Include limits of integration but do not evaluate. Use:

(a) Cylindrical coordinates
(b) Spherical coordinates

22. Write a triple integral representing the volume of the region between spheres of radius 1 and 2, both centered at the origin. Include limits of integration but do not evaluate. Use:

(a) Spherical coordinates.
(b) Cylindrical coordinates. Write your answer as the difference of two integrals.

23. Figure 16.48 shows part of a spherical ball of radius 5 cm. Write an integral in cylindrical coordinates representing the volume of this region and evaluate it.

2 cm

Figure 16.48

24. Figure 16.48 shows part of a spherical ball of radius 5 cm. Write an integral in spherical coordinates representing the volume of this region and evaluate it.

For the regions W shown in Problems 25–27, write the limits of integration for $\int_W dV$ in the following coordinates:

(a) Cartesian   (b) Cylindrical   (c) Spherical

25.

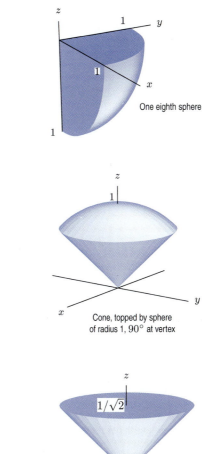

One eighth sphere

26.

Cone, topped by sphere of radius 1, $90°$ at vertex

27.

$1/\sqrt{2}$

Cone, flat on top, $\pi/2$ at vertex

28. Find the mass M of the solid region W given in spherical coordinates by $0 \leq \rho \leq 3, 0 \leq \theta < 2\pi, 0 \leq \phi \leq \pi/4$. The density, $\delta(P)$, at any point P is given by the distance of P from the origin.

29. A spherical cloud of gas of radius 3 km is more dense at the center than toward the edge. At a distance ρ km from the center, the density is $\delta(\rho) = 3 - \rho$. Write an integral representing the total mass of the cloud of gas, and evaluate it.

30. A solid hemisphere of radius 2 cm has density, in gm/cm^3, at each point equal to the distance in centimeters from the point to the center of the base. Write a triple integral representing the total mass of the hemisphere. Evaluate the integral.

31. Use appropriate coordinates to find the average distance to the origin for points in the ice cream cone region bounded by the hemisphere $z = \sqrt{8 - x^2 - y^2}$ and the cone $z = \sqrt{x^2 + y^2}$. [Hint: The volume of this region is computed in Problem 21 on page 760.]

32. (a) Write an iterated integral which represents the mass of a solid ball of radius a. The density at each point in the ball is k times the distance from that point to a fixed plane passing through the center of the ball.

(b) Evaluate the integral.

33. A bead is made by drilling a cylindrical hole of radius 1 mm through a sphere of radius 5 mm. See Figure 16.49.

(a) Set up a triple integral in cylindrical coordinates representing the volume of the bead.

(b) Evaluate the integral.

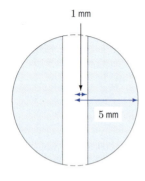

1 mm

5 mm

Figure 16.49

34. A half-melon is approximated by the region between two concentric spheres, one of radius a and the other of radius b, with $0 < a < b$. Write a triple integral, including limits of integration, giving the volume of the half-melon. Evaluate the integral.

For Problems 35–38, use the definition of center of mass given on page 756.

35. Let C be a solid cone with both height and radius 1 and contained between the surfaces $z = \sqrt{x^2 + y^2}$ and $z = 1$. If C has constant mass density of 1 gm/cm^3, find the z-coordinate of C's center of mass.

36. The density of the cone C in Problem 35 is given by $\delta(z) = z^2$ gm/cm^3. Find

(a) The mass of C.

(b) The z-coordinate of C's center of mass.

37. For $a > 0$, consider the family of solids bounded below by the paraboloid $z = a(x^2 + y^2)$ and above by the plane $z = 1$. If the solids all have constant mass density 1 gm/cm^3, show that the z-coordinate of the center of mass is $2/3$ and so independent of the parameter a.

38. Find the location of the center of mass of a hemisphere of radius a and density b gm/cm^3.

For Problems 39–40, use the definition of moment of inertia given on page 756.

39. The moment of inertia of a solid homogeneous ball B of mass 1 and radius a centered at the origin is the same

about any of the coordinate axes (due to the symmetry of the ball). It is easier to evaluate the sum of the three integrals involved in computing the moment of inertia about each of the axes than to compute them individually. Find the sum of the moments of inertia about the x, y and z-axes and thus find the individual moments of inertia.

40. Find the moment of inertia about the z axis of the solid "fat ice cream cone" given in spherical coordinates by $0 \leq \rho \leq a$, $0 \leq \phi \leq \frac{\pi}{3}$ and $0 \leq \theta \leq 2\pi$. Assume that the solid is homogeneous with mass m.

41. Compute the force of gravity exerted by a solid cylinder of radius R, height H, and constant density δ on a unit mass at the center of the base of the cylinder.

42. Figure 16.50 shows an alternative notation for spherical coordinates, used often in electrical engineering. Write the volume element dV in this coordinate system.

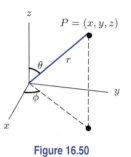

Figure 16.50

43. Electric charge is distributed throughout 3-space, with density proportional to the distance from the xy-plane. Show that the total charge inside a cylinder of radius R and height h, sitting on the xy-plane and centered along the z-axis, is proportional to R^2h^2.

44. Electric charge is distributed throughout 3-space with density inversely proportional to the distance from the origin. Show that the total charge inside a sphere of radius R is proportional to R^2.

Problems 45–46 deal with the energy stored in an electric field. If a region of space W contains an electric field whose magnitude at a point (x, y, z) is $E(x, y, z)$, then

$$\text{Energy stored by field} = \frac{1}{2} \int_W \epsilon E^2 \, dV,$$

where ϵ is a property of the material called the *permittivity*.

45. The region between two concentric spheres, with radii $a < b$, contains an electric field with magnitude $E = q/(4\pi\epsilon\rho^2)$, where ρ is the distance from the center of the spheres and q is the charge on the inner sphere. Assuming the permittivity, ϵ, is constant, find the total energy stored in the region between the two spheres.

46. Figure 16.51 shows a coaxial cable consisting of two cylindrical conductors centered on the same axis, of radii $a < b$. The electric field between the conductors has magnitude $E = q/(2\pi\epsilon r)$, where r is the distance from the axis and q is the charge per unit length on the cable. The permittivity of the material between the conductors is constant.[5] Show that the stored energy per unit length is proportional to $\ln(b/a)$.

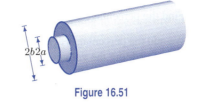

Figure 16.51

16.6 APPLICATIONS OF INTEGRATION TO PROBABILITY

To represent how a quantity such as height or weight is distributed throughout a population, we use a density function. To study two or more quantities at the same time and see how they are related, we use a multivariable density function.

Density Functions

Distribution of Weight and Height in Expectant Mothers

Table 16.9 shows the distribution of weight and height in a survey of expectant mothers. The histogram in Figure 16.52 is constructed in such a way that the volume of each bar represents the percentage in the corresponding weight and height range. For example, the bar representing the mothers who weighed 60–70 kg and were 160–165 cm tall has base of area 10 kg · 5 cm = 50 kg cm. The volume of this bar is 12%, so its height is 12%/50 kg cm = 0.24%/ kg cm. Notice that the units on the vertical axis are percent/ kg cm. Thus, volumes under the histogram are in units of %. The total volume is 100% = 1.

Table 16.9 *Distribution of weight and height in a survey of expectant mothers*

	45-50 kg	50-60 kg	60-70 kg	70-80 kg	80-105 kg	Totals by height
150-155 cm	2	4	4	2	1	13
155-160 cm	0	12	8	2	1	23
160-165 cm	1	7	12	4	3	27
165-170 cm	0	8	12	6	2	28
170-180 cm	0	1	3	4	1	9
Totals by weight	3	32	39	18	8	100

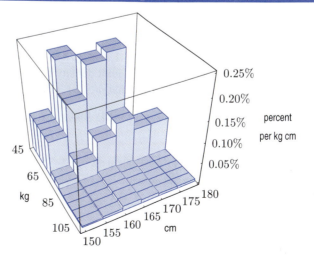

Figure 16.52: Histogram representing the data in Table 16.9.

[5]See C. R. Paul and S. A. Nasar, *Introduction to Electromagnetic Fields*, 2nd ed. (New York: McGraw-Hill, 1987).

16.7   CHANGE OF VARIABLES IN A MULTIPLE INTEGRAL

In the previous sections, we used polar, cylindrical, and spherical coordinates to simplify iterated integrals. In this section, we discuss more general changes of variable. In the process, we will see where the extra factor of r comes from when we change from Cartesian to polar coordinates and the factor $\rho^2 \sin \phi$ when we change from Cartesian to spherical coordinates.

Polar Change of Variables Revisited

Consider the integral $\int_R (x+y)\, dA$ where R is the region in the first quadrant bounded by the circle $x^2 + y^2 = 16$ and the x and y-axes. Writing the integral in Cartesian and polar coordinates we have

$$\int_R (x+y)\, dA = \int_0^4 \int_0^{\sqrt{16-x^2}} (x+y)\, dy\, dx = \int_0^{\pi/2} \int_0^4 (r\cos\theta + r\sin\theta) r\, dr\, d\theta.$$

This is an integral over the rectangle in the $r\theta$-space given by $0 \le r \le 4$, $0 \le \theta \le \pi/2$. The conversion from polar to Cartesian coordinates changes this rectangle into a quarter-disk. Figure 16.56 shows how a typical rectangle (shaded) in the $r\theta$-plane with sides of length Δr and $\Delta\theta$ corresponds to a curved rectangle in the xy-plane with sides of length Δr and $r\Delta\theta$. The extra r is needed because the correspondence between r, θ and x, y not only curves the lines $r = 1, 2, 3 \dots$ into circles, it also stretches those lines around larger and larger circles.

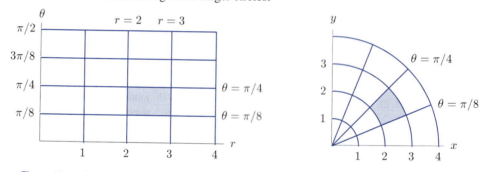

Figure 16.56: A grid in the $r\theta$-plane and the corresponding curved grid in the xy-plane

General Change of Variables

We now consider a general change of variables, where x, y coordinates are related to s, t coordinates by the differentiable functions

$$x = x(s,t) \quad y = y(s,t).$$

Just as a rectangular region in the $r\theta$-plane corresponds to a region in the xy-plane, a rectangular region, T, in the st-plane corresponds to a region, R, in the xy-plane. We assume that the change of coordinates is one-to-one, that is, that each point R corresponds to one point in T.

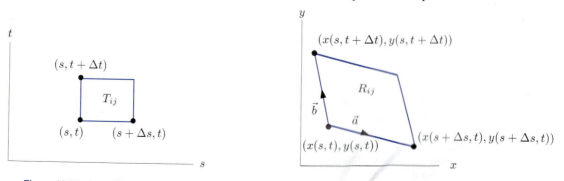

Figure 16.57: A small rectangle T_{ij} in the st-plane and the corresponding region R_{ij} of the xy-plane

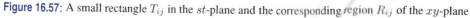

We divide T into small rectangles T_{ij} with sides of length Δs and Δt. (See Figure 16.57.) The corresponding piece R_{ij} of the xy-plane is a quadrilateral with curved sides. If we choose Δs and Δt small, then by local linearity of $x(s,t)$ and $y(s,t)$, we know R_{ij} is approximately a parallelogram.

Recall from Chapter 13 that the area of the parallelogram with sides $\vec{a}$ and $\vec{b}$ is $\|\vec{a} \times \vec{b}\|$. Thus, we need to find the sides of R_{ij} as vectors. The side of R_{ij} corresponding to the bottom side of T_{ij} has endpoints $(x(s,t), y(s,t))$ and $(x(s+\Delta s, t), y(s+\Delta s, t))$, so in vector form that side is

$$\vec{a} = (x(s+\Delta s, t) - x(s,t))\vec{i} + (y(s+\Delta s, t) - y(s,t))\vec{j} \approx \left(\frac{\partial x}{\partial s}\Delta s\right)\vec{i} + \left(\frac{\partial y}{\partial s}\Delta s\right)\vec{j}.$$

Similarly, the side of R_{ij} corresponding to the left edge of T_{ij} is given by

$$\vec{b} \approx \left(\frac{\partial x}{\partial t}\Delta t\right)\vec{i} + \left(\frac{\partial y}{\partial t}\Delta t\right)\vec{j}.$$

Computing the cross product, we get

$$\text{Area } R_{ij} \approx \|\vec{a} \times \vec{b}\| \approx \left|\left(\frac{\partial x}{\partial s}\Delta s\right)\left(\frac{\partial y}{\partial t}\Delta t\right) - \left(\frac{\partial x}{\partial t}\Delta t\right)\left(\frac{\partial y}{\partial s}\Delta s\right)\right|$$

$$= \left|\frac{\partial x}{\partial s}\cdot\frac{\partial y}{\partial t} - \frac{\partial x}{\partial t}\cdot\frac{\partial y}{\partial s}\right|\Delta s\Delta t.$$

Using determinant notation,[6] we define the *Jacobian*, $\dfrac{\partial(x,y)}{\partial(s,t)}$, as follows

$$\frac{\partial(x,y)}{\partial(s,t)} = \frac{\partial x}{\partial s}\cdot\frac{\partial y}{\partial t} - \frac{\partial x}{\partial t}\cdot\frac{\partial y}{\partial s} = \begin{vmatrix} \dfrac{\partial x}{\partial s} & \dfrac{\partial x}{\partial t} \\[2mm] \dfrac{\partial y}{\partial s} & \dfrac{\partial y}{\partial t} \end{vmatrix}.$$

Thus, we can write

$$\text{Area } R_{ij} \approx \left|\frac{\partial(x,y)}{\partial(s,t)}\right|\Delta s\,\Delta t.$$

To compute $\int_R f(x,y)\,dA$, where f is a continuous function, we look at the Riemann sum obtained by dividing the region R into the small curved regions R_{ij}, giving

$$\int_R f(x,y)\,dA \approx \sum_{i,j} f(u_{ij}, v_{ij})\cdot \text{Area of } R_{ij} \approx \sum_{i,j} f(u_{ij}, v_{ij})\left|\frac{\partial(x,y)}{\partial(s,t)}\right|\Delta s\,\Delta t.$$

Each point (u_{ij}, v_{ij}) in R_{ij} corresponds to a point (s_{ij}, t_{ij}) in T_{ij}, so the sum can be written in terms of s and t:

$$\sum_{i,j} f(x(s_{ij}, t_{ij}), y(s_{ij}, t_{ij}))\left|\frac{\partial(x,y)}{\partial(s,t)}\right|\Delta s\,\Delta t.$$

This is a Riemann sum in terms of s and t, so as Δs and Δt approach 0, we get

$$\int_R f(x,y)\,dA = \int_T f(x(s,t), y(s,t))\left|\frac{\partial(x,y)}{\partial(s,t)}\right|\,ds\,dt.$$

To convert an integral from x, y to s, t coordinates we make three changes:

1. Substitute for x and y in the integrand in terms of s and t.

2. Change the xy region R into an st region T.

3. Use the absolute value of the Jacobian to change the area element by making the substitution $dx\,dy = \left|\dfrac{\partial(x,y)}{\partial(s,t)}\right|ds\,dt$.

---

[6]See Appendix E.

Example 1   Check that the Jacobian $\dfrac{\partial(x, y)}{\partial(r, \theta)} = r$ for polar coordinates $x = r\cos\theta$, $y = r\sin\theta$.

Solution   We have $\dfrac{\partial(x, y)}{\partial(r, \theta)} = \begin{vmatrix} \frac{\partial x}{\partial r} & \frac{\partial x}{\partial \theta} \\ \frac{\partial y}{\partial r} & \frac{\partial y}{\partial \theta} \end{vmatrix} = \begin{vmatrix} \cos\theta & \sin\theta \\ -r\sin\theta & r\cos\theta \end{vmatrix} = r\cos^2\theta + r\sin^2\theta = r.$

Example 2   Find the area of the ellipse $\dfrac{x^2}{a^2} + \dfrac{y^2}{b^2} = 1$.

Solution   Let $x = as$, $y = bt$. Then the ellipse $x^2/a^2 + y^2/b^2 = 1$ in the xy-plane corresponds to the circle $s^2 + t^2 = 1$ in the st-plane. The Jacobian is $\begin{vmatrix} a & 0 \\ 0 & b \end{vmatrix} = ab$. Thus, if R is the ellipse in the xy-plane and T is the unit circle in the st-plane, we get

$$\text{Area of } xy\text{-ellipse} = \int_R 1 \, dA = \int_T 1ab \, ds \, dt = ab\int_T ds \, dt = ab \cdot \text{Area of } st\text{-circle} = \pi ab.$$

Change of Variables in Triple Integrals

For triple integrals, there is a similar formula. Suppose the differentiable functions

$$x = x(s, t, u), \quad y = y(s, t, u), \quad z = z(s, t, u)$$

define a change of variables from a region S in stu-space to a region W in xyz-space. Then, the Jacobian of this change of variables is given by the determinant[7]

$$\frac{\partial(x, y, z)}{\partial(s, t, u)} = \begin{vmatrix} \frac{\partial x}{\partial s} & \frac{\partial x}{\partial t} & \frac{\partial x}{\partial u} \\ \frac{\partial y}{\partial s} & \frac{\partial y}{\partial t} & \frac{\partial y}{\partial u} \\ \frac{\partial z}{\partial s} & \frac{\partial z}{\partial t} & \frac{\partial z}{\partial u} \end{vmatrix}.$$

Just as the Jacobian in two dimensions gives us the change in the area element, the Jacobian in three dimensions represents the change in the volume element. Thus, we have

$$\int_W f(x, y, z) \, dx \, dy \, dz = \int_S f(x(s, t, u), y(s, t, u), z(s, t, u)) \left| \frac{\partial(x, y, z)}{\partial(s, t, u)} \right| ds \, dt \, du.$$

Problem 9 at the end of this section asks you to check that the Jacobian for the change of variables to spherical coordinates is $\rho^2 \sin\phi$. The next example generalizes Example 2 to ellipsoids.

Example 3   Find the volume of the ellipsoid $\dfrac{x^2}{a^2} + \dfrac{y^2}{b^2} + \dfrac{z^2}{c^2} = 1$.

Solution   Let $x = as$, $y = bt$, $z = cu$. The Jacobian is computed to be abc. The xyz-ellipsoid corresponds to the stu-sphere $s^2 + t^2 + u^2 = 1$. Thus, as in Example 2,

$$\text{Volume of } xyz\text{-ellipsoid} = abc \cdot \text{Volume of } stu\text{-sphere} = abc\frac{4}{3}\pi = \frac{4}{3}\pi abc.$$

---

[7]See Appendix E.

Exercises and Problems for Section 16.7

Exercises

In Exercises 1–4, find the absolute value of the Jacobian, $\left|\frac{\partial(x,y)}{\partial(s,t)}\right|$, for the given change of variables.

1. $x = 5s + 2t, y = 3s + t$
2. $x = s^2 - t^2, y = 2st$
3. $x = e^s \cos t, y = e^s \sin t$
4. $x = s^3 - 3st^2, y = 3s^2t - t^3$

In Exercises 5–6, find the Jacobian.

5. $\dfrac{\partial(x, y, z)}{\partial(s, t, u)}$, where $x = 3s + t + 2u, y = s + 5t - u, z = 2s - t + u$.

6. $\dfrac{\partial(x, y, z)}{\partial(r, \theta, z)}$, where $x = r \cos \theta, y = r \sin \theta, z = z$.

Problems

7. Find the region R in the xy-plane corresponding to the region $T = \{(s, t) \mid 0 \le s \le 3, 0 \le t \le 2\}$ under the change of variables $x = 2s - 3t, y = s - 2t$. Check that

$$\int_R dx\,dy = \int_T \left|\frac{\partial(x, y)}{\partial(s, t)}\right| ds\,dt.$$

8. Find the region R in the xy-plane corresponding to the region $T = \{(s, t) \mid 0 \le s \le 2, s \le t \le 2\}$ under the change of variables $x = s^2, y = t$. Check that

$$\int_R dx\,dy = \int_T \left|\frac{\partial(x, y)}{\partial(s, t)}\right| ds\,dt.$$

9. Compute the Jacobian for the change of variables into spherical coordinates:

$$x = \rho \sin \phi \cos \theta, \quad y = \rho \sin \phi \sin \theta, \quad z = \rho \cos \phi.$$

10. For the change of variables $x = 3s - 4t$, $y = 5s + 2t$, show that

$$\frac{\partial(x, y)}{\partial(s, t)} \cdot \frac{\partial(s, t)}{\partial(x, y)} = 1$$

11. Use the change of variables $x = 2s + t, y = s - t$ to compute the integral $\int_R (x+y)\,dA$, where R is the parallelogram formed by $(0, 0)$, $(3, -3)$, $(5, -2)$, and $(2, 1)$.

12. Use the change of variables $x = \frac{1}{2}s, y = \frac{1}{3}t$ to compute the integral $\int_R (x^2 + y^2)\,dA$, where R is the region bounded by the curve $4x^2 + 9y^2 = 36$.

13. Use the change of variables $s = xy, t = xy^2$ to compute $\int_R xy^2\,dA$, where R is the region bounded by $xy = 1, xy = 4, xy^2 = 1, xy^2 = 4$.

14. If R is the triangle bounded by $x + y = 1, x = 0$, and $y = 0$, evaluate the integral $\int_R \cos\left(\frac{x-y}{x+y}\right) dx\,dy$.

CHAPTER SUMMARY

- **Double Integral**
    Definition as a limit of Riemann sum, interpretation as volume under graph, as area, as average value, or as total mass from density, estimating from contour diagrams or tables, evaluating using iterated integrals, setting up in polar coordinates.

- **Triple Integral**
    Definition as a limit of Riemann sum, interpretation as volume of solid, as total mass, or as average value,

evaluating using iterated integrals, setting up in cylindrical or spherical coordinates.

- **Change of Variable** Polar coordinates, spherical and cylindrical coordinates, general change of variables and Jacobians

- **Probability**
    Joint density functions, using integrals to calculate probability .

REVIEW EXERCISES AND PROBLEMS FOR CHAPTER SIXTEEN

Exercises

Sketch the regions of integration in Exercises 1–4.

1. $\displaystyle\int_1^4 \int_{-\sqrt{y}}^{\sqrt{y}} f(x, y)\,dx\,dy$

2. $\displaystyle\int_0^1 \int_0^{\sin^{-1} y} f(x, y)\,dx\,dy$

3. $\displaystyle\int_{-1}^1 \int_{-\sqrt{1-x^2}}^{\sqrt{1-x^2}} f(x, y)\,dy\,dx$

4. $\displaystyle\int_0^2 \int_{-\sqrt{4-y^2}}^0 f(x, y)\,dx\,dy$

Choose coordinates and write a triple integral, including limits of integration, for a function over the regions in Exercises 5–8.

5. **6.**

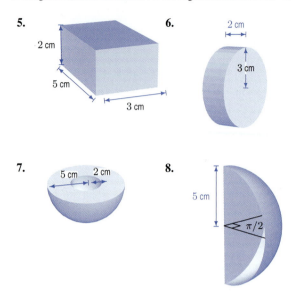

7. **8.**

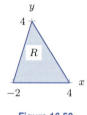

9. Write $\int_R f(x, y)\, dA$ as an iterated integral if R is the region in Figure 16.58.

y
4
R
x
-2 4

Figure 16.58

Calculate exactly the integrals in Exercises 10–15. (Your answer may contain e, π, $\sqrt{2}$, and so on.)

10. $\int_0^1 \int_0^z \int_0^2 (y + z)^7 \, dx \, dy \, dz$

11. $\int_0^1 \int_3^4 (\sin(2 - y)) \cos(3x - 7) \, dx \, dy$

Problems

The finite regions T, B, R, L are in the xy-plane.

    T lies in the region where $y > 0$,
    R lies in the region where $x > 0$,
    B lies in the region where $y < 0$,
    L lies in the region where $x < 0$.

Are the double integrals in Problems 22–26 positive, negative, or is it impossible to tell?

12. $\int_0^{10} \int_0^{0.1} x e^{xy} \, dy \, dx$

13. $\int_0^1 \int_0^y (\sin^3 x)(\cos x)(\cos y) \, dx \, dy$

14. $\int_3^4 \int_0^1 x^2 y \cos(xy) \, dy \, dx$

15. $\int_0^1 \int_{-\sqrt{1-x^2}}^{\sqrt{1-x^2}} e^{-(x^2+y^2)} \, dy \, dx$

16. Evaluate $\int_R \sqrt{x^2 + y^2} \, dA$ where R is the region in Figure 16.59.

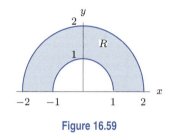

y
2
R
1
x
-2 -1 1   2

Figure 16.59

17. Set up $\int_R f \, dV$ as an iterated integral in all six possible orders of integration, where R is the hemisphere bounded by the upper half of $x^2 + y^2 + z^2 = 1$ and the xy-plane.

Evaluate the integrals in Exercises 18–21 by changing them to cylindrical or spherical coordinates.

18. $\int_{-\sqrt{3}}^{\sqrt{3}} \int_{-\sqrt{3-x^2}}^{\sqrt{3-x^2}} \int_1^{4-x^2-y^2} \frac{1}{z^2} \, dz \, dy \, dx$

19. $\int_0^3 \int_{-\sqrt{9-z^2}}^{\sqrt{9-z^2}} \int_{-\sqrt{9-y^2-z^2}}^{\sqrt{9-y^2-z^2}} x^2 \, dx \, dy \, dz$

20. $\int_0^1 \int_0^{\sqrt{1-x^2}} \int_0^{\sqrt{x^2+y^2}} (z + \sqrt{x^2 + y^2}) \, dz \, dy \, dx$

21. $\int_W \frac{z}{(x^2+y^2)^{3/2}} \, dV$, if W is $1 \le x^2 + y^2 \le 4$, $0 \le z \le 4$

22. $\int_T e^{-x} \, dA$

23. $\int_B y^3 \, dA$

24. $\int_R (x + y^2) \, dA$

25. $\int_L y^3 \, dA$

26. $\int_L (x + y^2) \, dA$

Let W be the solid sphere bounded by $x^2 + y^2 + z^2 = 1$. For Problems 27–35, decide (without calculating its value) whether the integral is positive, negative, or zero.

27. $\int_W z\, dV$

28. $\int_W x\, dV$

29. $\int_W xy\, dV$

30. $\int_W xyz\, dV$

31. $\int_W \sin(\frac{\pi}{2}xy)\, dV$

32. $\int_W \cos(\frac{\pi}{2}xy)\, dV$

33. $\int_W (z^2 - 1)\, dV$

34. $\int_W e^{-xyz}\, dV$

35. $\int_W \sqrt{x^2 + y^2 + z^2}\, dV$

36. Write an integral representing the mass of a sphere of radius 3 if the density of the sphere at any point is twice the distance of that point from the center of the sphere.

37. A forest next to a road has the shape in Figure 16.60. The population density of rabbits is proportional to the distance from the road. It is 0 at the road, and 10 rabbits per square mile at the opposite edge of the forest. Find the total rabbit population in the forest.

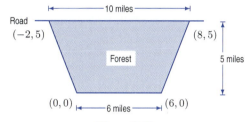

Figure 16.60

38. The insulation surrounding a pipe is the region between two cylinders of length l and with the same axis. The inner cylinder has radius a, the outer radius of the pipe, and the insulation has thickness h. Write a triple integral, including limits of integration, giving the volume of the insulation. Evaluate the integral.

39. Assume p, q, r are positive constants. Find the volume contained between the coordinates planes and the plane

$$\frac{x}{p} + \frac{y}{q} + \frac{z}{r} = 1.$$

40. A cone stands with its flat base on a table. The cone's circular base has radius a; the vertex (tip) is at a height of h above the center of the base. Write a triple integral, including limits of integration, representing the volume of the cone. Evaluate the integral.

41. Find the volume that remains after a cylindrical hole of radius R is bored through a sphere of radius a, where $0 < R < a$, passing through the center of the sphere along the pole.

42. Two spheres, one of radius 1, one of radius $\sqrt{2}$, have centers that are 1 unit apart. Write a triple integral, including limits of integration, giving the volume of the smaller region that is outside one sphere and inside the other. Evaluate the integral.

43. Give the joint density of x and y where x and y are independent, x has a normal distribution with mean 5 and standard deviation $1/10$, and y has a normal distribution with mean 15 and standard deviation $1/6$.

For Problems 44–45, use the definition of moment of inertia on page 756.

44. Consider a rectangular brick with length 5, width 3, and height 1, and of uniform density 1. Compute the moment of inertia about each of the three axes passing through the center of the brick, perpendicular to one of the sides.

45. Compute the moment of inertia of a ball of radius R about an axis passing through its center. Assume that the ball has a constant density of 1.

46. A particle of mass m is placed at the center of one base of a circular cylindrical shell of inner radius r_1, outer radius r_2, height h, and constant density δ. Find the force of gravity exerted by the cylinder on the particle.

47. Find the area of the half-moon shape with circular arcs as edges and the dimensions shown in Figure 16.61.

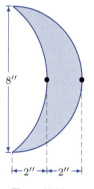

Figure 16.61

48. A river follows the path $y = f(x)$ where x, y are in kilometers. Near the sea, it widens into a lagoon, then narrows again at its mouth. See Figure 16.62. At the point (x, y), the depth, $d(x, y)$, of the lagoon is given by

$$d(x, y) = 40 - 160(y - f(x))^2 - 40x^2 \text{ meters.}$$

The lagoon itself is described by $d(x, y) \geq 0$. What is the volume of the lagoon in cubic meters? [Hint: Use new coordinates $u = x/2$, $v = y - f(x)$ and Jacobians.]

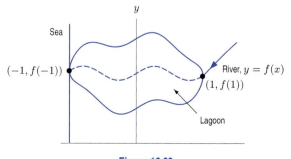

Figure 16.62

CAS Challenge Problems

49. Let D be the region inside the triangle with vertices $(0,0)$, $(1,1)$ and $(0,1)$. Express the double integral $\int_D e^{y^2} \, dA$ as an iterated integral in two different ways. Calculate whichever of the two you can do by hand, and calculate the other with a computer algebra system if possible. Compare the answers.

50. Let D be the region inside the circle $x^2 + y^2 = 1$. Express the integral $\int_D \sqrt[3]{x^2 + y^2} \, dA$ as an iterated integral in both Cartesian and polar coordinates. Calculate whichever of the two you can do by hand, and calculate the other with a computer algebra system if possible. Compare the answers.

51. Compute the iterated integrals $\int_0^1 \int_{-1}^0 \frac{x+y}{(x-y)^3} \, dy\,dx$

and $\int_{-1}^0 \int_0^1 \frac{x+y}{(x-y)^3} \, dx\,dy$. Explain why your answers do not contradict Theorem 16.1 on page 744

52. For each of the following functions, find its average value over the square $-h \le x \le h$, $-h \le y \le h$, calculate the limit of your answer as $h \to 0$, and compare with the value of the function at $(0,0)$. Assume a, b, c, d, e, and k are constants.

$$F(x,y) = a + bx^4 + cy^4 + dx^2y^2 + ex^3y^3$$
$$G(x,y) = a\sin(kx) + b\cos(ky) + c$$
$$H(x,y) = ax^2 e^{x+y} + by^2 e^{x-y}$$

Formulate a conjecture from your results and explain why it makes sense.

CHECK YOUR UNDERSTANDING

Are the statements in Problems 1–30 true or false? Give reasons for your answer.

1. The double integral $\int_R f \, dA$ is always positive.

2. If $f(x,y) = k$ for all points (x,y) in a region R then $\int_R f \, dA = k \cdot \text{Area}(R)$.

3. If R is the rectangle $0 \le x \le 1, 0 \le y \le 1$ then $\int_R e^{xy} \, dA > 3$.

4. If R is the rectangle $0 \le x \le 2, 0 \le y \le 3$ and S is the rectangle $-2 \le x \le 0, -3 \le y \le 0$ then $\int_R f \, dA = -\int_R f \, dA$.

5. Let $\rho(x,y)$ be the population density of a city, in people per km^2. If R is a region in the city, then $\int_R f \, dA$ gives the total number of people in the region R.

6. If $\int_R f \, dA = 0$ then $f(x,y) = 0$ at all points of R.

7. If $g(x,y) = kf(x,y)$, where k is constant, then $\int_R g \, dA = k \int_R f \, dA$.

8. If f and g are two functions continuous on a region R, then $\int_R f \cdot g \, dA = \int_R f \, dA \cdot \int_R g \, dA$.

9. If R is the rectangle $0 \le x \le 1, 0 \le y \le 2$ and S is the square $0 \le x \le 1, 0 \le y \le 1$, then $\int_R f \, dA = 2\int_S f \, dA$.

10. If R is the rectangle $2 \le x \le 4, 5 \le y \le 9$, $f(x,y) = 2x$ and $g(x,y) = x + y$, then the average value of f on R is less than the average value of g on R.

11. The iterated integral $\int_0^1 \int_5^{12} f \, dx\,dy$ is computed over the rectangle $0 \le x \le 1, 5 \le y \le 12$.

12. If R is the region inside the triangle with vertices $(0,0), (1,1)$ and $(0,2)$, then the double integral $\int_R f \, dA$ can be evaluated by an iterated integral of the form $\int_0^2 \int_0^1 f \, dx\,dy$.

13. The region of integration of the iterated integral $\int_1^2 \int_{x^2}^{x^3} f \, dy\,dx$ lies completely in the first quadrant (that is, $x \ge 0, y \ge 0$).

14. If the limits a, b, c and d in the iterated integral $\int_a^b \int_c^d f\,dy\,dx$ are all positive, then the value of $\int_a^b \int_c^d f\,dy\,dx$ is also positive.

15. The iterated integrals $\int_0^1 \int_0^x f \, dy\,dx$ and $\int_0^1 \int_0^y f \, dx\,dy$ give the same answer for any function f.

16. If $f(x,y)$ is a function of y only, then $\int_a^b \int_0^1 f\,dx\,dy = \int_a^b f\,dy$.

17. If R is the region inside a circle of radius a, centered at the origin, then $\int_R f \, dA = \int_{-a}^a \int_0^{\sqrt{a^2-x^2}} f\,dy\,dx$.

18. There is a region R over which $\int_R f \, dA$ cannot be evaluated by a *single* iterated integral in the order $\int \int f\,dx\,dy$ or $\int \int f\,dy\,dx$.

19. If $f(x,y) = g(x) \cdot h(y)$, where g and h are single-variable functions, then
$$\int_a^b \int_c^d f\,dy\,dx = \left(\int_a^b g(x)\,dx\right) \cdot \left(\int_c^d h(y)\,dy\right).$$

20. If $f(x,y) = g(x) + h(y)$, where g and h are single-variable functions, then
$$\int_a^b \int_c^d f\,dx\,dy = \left(\int_a^b g(x)\,dx\right) + \left(\int_c^d h(y)\,dy\right).$$

21. If $\rho(x,y,z)$ is mass density of a material in 3-space, then $\int_W \rho(x,y,z) \, dV$ gives the volume of the solid region W.

22. The region of integration of the triple iterated integral $\int_0^1 \int_0^1 \int_0^x f \, dz\,dy\,dx$ lies above a square in the xy-plane and below a plane.

23. If W is the entire unit ball $x^2 + y^2 + z^2 \leq 1$ then an iterated integral over W has limits $\int_0^1 \int_0^{\sqrt{1-x^2}} \int_0^{\sqrt{1-x^2-y^2}} f \, dz dy dx$.

24. The iterated integrals $\int_0^1 \int_0^{1-x} \int_0^{1-x-y} f \, dz dy dx$ and $\int_0^1 \int_0^{1-z} \int_0^{1-y-z} f \, dx dy dz$ are equal.

25. The iterated integrals $\int_{-1}^1 \int_0^1 \int_0^{1-x^2} f \, dz dy dx$ and $\int_0^1 \int_0^1 \int_{-\sqrt{1-z}}^{\sqrt{1-z}} f \, dx dy dz$ are equal.

26. If W is a rectangular solid in 3-space, then $\int_W f \, dV = \int_a^b \int_c^d \int_e^k f \, dz \, dy \, dx$, where $a, b, c, d, e,$ and k are constants.

27. If W is the unit cube $0 \leq x \leq 1, 0 \leq y \leq 1, 0 \leq z \leq 1$ and $\int_W f \, dV = 0$, then $f = 0$ everywhere in the unit cube.

28. If $f > g$ at all points in the solid region W, then $\int_W f \, dV > \int_W g \, dV$.

29. If W_1 and W_2 are solid regions with volume(W_1) > volume(W_2) then $\int_{W_1} f \, dV > \int_{W_2} f \, dV$.

30. Both double and triple integrals can be used to compute volume.

PROJECTS FOR CHAPTER SIXTEEN

1. **A Connection Between e and π**

   In this problem you will derive one of the remarkable formulas of mathematics, namely that

   $$\int_{-\infty}^{\infty} e^{-x^2} dx = \sqrt{\pi}.$$

   (a) Change the following double integral into polar coordinates and evaluate it:

   $$\int_{-\infty}^{\infty} \int_{-\infty}^{\infty} e^{-(x^2+y^2)} dx dy.$$

   (b) Explain why

   $$\int_{-\infty}^{\infty} \int_{-\infty}^{\infty} e^{-(x^2+y^2)} dx dy = \left(\int_{-\infty}^{\infty} e^{-x^2} dx \right)^2.$$

   (c) Explain why the answers to parts (a) and (b) give the formula we want.

2. **Average Distance Walked to an Airport Gate**

   At airports, departure gates are often lined up in a terminal like points along a line. If you arrive at one gate and proceed to another gate for a connecting flight, what proportion of the length of the terminal will you have to walk, on average?

   (a) One way to model this situation is to randomly choose two numbers, $0 \leq x \leq 1$ and $0 \leq y \leq 1$, and calculate the average value of $|x - y|$. Use a double integral to show that, on average, you have to walk $1/3$ the length of the terminal.

   (b) The terminal gates are not actually located continuously from 0 to 1, as we assumed in part (a). There are only a finite number of gates and they are likely to be equally spaced. Suppose there are $n + 1$ gates located $1/n$ units apart from one end of the terminal ($x_0 = 0$) to the other ($x_n = 1$). Assuming that all pairs (i, j) of arrival and departure gates are equally likely, show that

   $$\text{Average distance between gates} = \frac{1}{(n+1)^2} \cdot \sum_{i=0}^n \sum_{j=0}^n \left| \frac{i}{n} - \frac{j}{n} \right|.$$

   Identify this sum as approximately (but not exactly) a Riemann sum with n subdivisions for the integrand used in part (a). Compute this sum for $n = 5$ and $n = 10$ and compare to the answer of $1/3$ obtained in part (a).

Chapter Seventeen

PARAMETERIZATION AND VECTOR FIELDS

Some physical quantities (such as temperature) are best represented by scalars; others (such as velocity) are best represented by vectors. We have looked at functions whose values are scalars. For example, in single-variable calculus we represented the motion of an object thrown straight up into the air by a function $h(t)$, the height of the object above the ground at time t. In multivariable calculus we considered temperature as a function of position on a weather map. Such functions are called scalar-valued functions.

In this chapter we look at vector-valued functions. We represent the motion of a particle in space by giving its position vector $\vec{r}(t)$ at time t. This enables us to find the velocity and acceleration vectors of the particle. We represent a surface in 3-space by giving the position vector of a point on the surface as a function of two parameters.

Another example of a vector-valued function is a *vector field*, which assigns a vector to each point in space. We have already seen one important example of a vector field, namely the gradient of a scalar-valued function. In this chapter we look at other examples, such as velocity vector fields describing a fluid flow. We also look at the path followed by a particle moving with the flow, which is called a *flow line* of the vector field.

17.1 PARAMETERIZED CURVES

A curve in the plane may be parameterized by a pair of equations of the form $x = f(t)$, $y = g(t)$. As the parameter t changes, the point (x, y) traces out the curve. In this section we find parametric equations for curves in three dimensions, and we see how to write parametric equations using position vectors.

Parametric Equations in Three Dimensions

We describe motion in the plane by giving parametric equations for x and y in terms of t. To describe a motion in 3-dimensional space parametrically, we need a third equation giving z in terms of t.

Example 1   Find parametric equations for the curve $y = x^2$ in the xy-plane.

Solution   A possible parameterization in two dimensions is $x = t$, $y = t^2$. Since the curve is in the xy-plane, the z-coordinate is zero, so a parameterization in three dimensions is

$$x = t, \quad y = t^2, \quad z = 0.$$

Example 2   Find parametric equations for a particle that starts at $(0, 3, 0)$ and moves around a circle as shown in Figure 17.1.

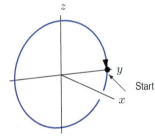

Figure 17.1: Circle of radius 3 in the yz-plane, centered at origin

Solution   Since the motion is in the yz-plane, we have $x = 0$ at all times t. Looking at the yz-plane from the positive x-direction we see motion around a circle of radius 3 in the clockwise direction. Thus

$$x = 0, \quad y = 3\cos t, \quad z = -3\sin t.$$

Example 3   Describe in words the motion given parametrically by

$$x = \cos t, \quad y = \sin t, \quad z = t.$$

Solution   The particle's x- and y-coordinates give circular motion in the xy-plane, while the z-coordinate increases steadily. Thus, the particle traces out a rising spiral, like a coiled spring. (See Figure 17.2.) This curve is called a *helix*.

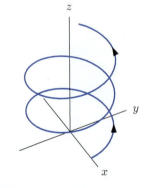

Figure 17.2: The helix $x = \cos t, y = \sin t, z = t$

Example 4   Find parametric equations for the line parallel to the vector $2\vec{i} + 3\vec{j} + 4\vec{k}$ and through the point $(1, 5, 7)$.

Solution   Let's imagine a particle at the point $(1, 5, 7)$ at time $t = 0$ and moving through a displacement of $2\vec{i} + 3\vec{j} + 4\vec{k}$ for each unit of time, t. When $t = 0$, $x = 1$ and x increases by 2 units for every unit of time. Thus, at time t, the x-coordinate of the particle is given by

$$x = 1 + 2t.$$

Similarly, the y-coordinate starts at $y = 5$ and increases at a rate of 3 units for every unit of time. The z-coordinate starts at $y = 7$ and increases by 4 units for every unit of time. Thus, the parametric equations of the line are

$$x = 1 + 2t, \quad y = 5 + 3t, \quad z = 7 + 4t.$$

We can generalize the previous example as follows:

Parametric Equations of a Line through the point (x_0, y_0, z_0) and parallel to the vector $a\vec{i} + b\vec{j} + c\vec{k}$ are

$$x = x_0 + at, \quad y = y_0 + bt, \quad z = z_0 + ct.$$

Notice that the coordinates x, y, and z are linear functions of the parameter t.

Example 5   (a) Describe in words the curve given by the parametric equations $x = 3 + t$, $y = 2t$, $z = 1 - t$.
(b) Find parametric equations for the line through the points $(1, 2, -1)$ and $(3, 3, 4)$.

Solution   (a) The curve is a line through the point $(3, 0, 1)$ and parallel to the vector $\vec{i} + 2\vec{j} - \vec{k}$.
(b) The line is parallel to the vector between the points $P = (1, 2, -1)$ and $Q = (3, 3, 4)$.

$$\overrightarrow{PQ} = (3 - 1)\vec{i} + (3 - 2)\vec{j} + (4 - (-1))\vec{k} = 2\vec{i} + \vec{j} + 5\vec{k}.$$

Thus, using the point P, the parametric equations are

$$x = 1 + 2t, \quad y = 2 + t, \quad z = -1 + 5t.$$

Using the point Q gives the equations $x = 3 + 2t$, $y = 3 + t$, $z = 4 + 5t$, which represent the same line.

Using Position Vectors to Write Parameterized Curves as Vector-Valued Functions

A point in the plane with coordinates (x, y) can be represented by the position vector $\vec{r} = x\vec{i} + y\vec{j}$ in Figure 17.3. Similarly, in 3-space we write $\vec{r} = x\vec{i} + y\vec{j} + z\vec{k}$. (See Figure 17.4.)

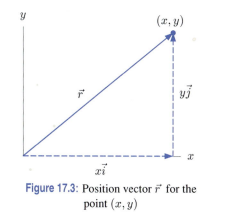

Figure 17.3: Position vector $\vec{r}$ for the point (x, y)

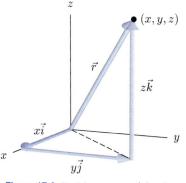

Figure 17.4: Position vector $\vec{r}$ for the point (x, y, z)

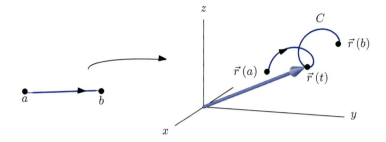

Figure 17.5: The parameterization sends the interval, $a \leq t \leq b$, to the curve, C, in 3-space

We can write the parametric equations $x = f(t)$, $y = g(t)$, $z = h(t)$ as a single vector equation

$$\vec{r}(t) = f(t)\vec{i} + g(t)\vec{j} + h(t)\vec{k}$$

called a *parameterization*. As the parameter t varies, the point with position vector $\vec{r}(t)$ traces out a curve in 3-space. For example, the circular motion in the plane

$$x = \cos t, y = \sin t \quad \text{can be written as} \quad \vec{r} = (\cos t)\vec{i} + (\sin t)\vec{j}$$

and the helix in 3-space

$$x = \cos t, y = \sin t, z = t \quad \text{can be written as} \quad \vec{r} = (\cos t)\vec{i} + (\sin t)\vec{j} + t\vec{k}.$$

See Figure 17.5.

Example 6   Use vectors to give a parameterization for the circle of radius $\frac{1}{2}$ centered at the point $(-1, 2)$.

Solution   The circle of radius 1 centered at the origin is parameterized by the vector-valued function

$$\vec{r}_1(t) = \cos t\vec{i} + \sin t\vec{j}, \quad 0 \leq t \leq 2\pi.$$

The point $(-1, 2)$ has position vector $\vec{r}_0 = -\vec{i} + 2\vec{j}$. The position vector, $\vec{r}(t)$, of a point on the circle of radius $\frac{1}{2}$ centered at $(-1, 2)$ is found by adding $\frac{1}{2}\vec{r}_1$ to $\vec{r}_0$. (See Figures 17.6 and 17.7.) Thus,

$$\vec{r}(t) = \vec{r}_0 + \tfrac{1}{2}\vec{r}_1(t) = -\vec{i} + 2\vec{j} + \tfrac{1}{2}(\cos t\vec{i} + \sin t\vec{j}) = (-1 + \tfrac{1}{2}\cos t)\vec{i} + (2 + \tfrac{1}{2}\sin t)\vec{j},$$

or, equivalently,

$$x = -1 + \tfrac{1}{2}\cos t, \quad y = 2 + \tfrac{1}{2}\sin t, \quad 0 \leq t \leq 2\pi.$$

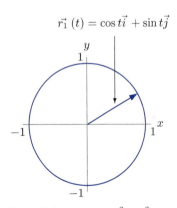

Figure 17.6: The circle $x^2 + y^2 = 1$ parameterized by $\vec{r}_1(t) = \cos t\vec{i} + \sin t\vec{j}$

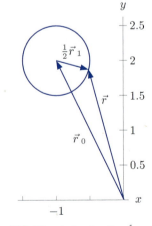

Figure 17.7: The circle of radius $\frac{1}{2}$ and center $(-1, 2)$ parameterized by $\vec{r}(t) = \vec{r}_0 + \frac{1}{2}\vec{r}_1(t)$

Parametric Equation of a Line

Consider a straight line in the direction of a vector $\vec{v}$ passing through the point (x_0, y_0, z_0) with position vector $\vec{r}_0$. We start at $\vec{r}_0$ and move up and down the line, adding different multiples of $\vec{v}$ to $\vec{r}_0$. (See Figure 17.8.)

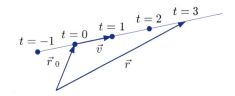

Figure 17.8: The line $\vec{r}(t) = \vec{r}_0 + t\vec{v}$

In this way, every point on the line can be written as $\vec{r}_0 + t\vec{v}$, which yields the following:

Parametric Equation of a Line in Vector Form

The line through the point with position vector $\vec{r_0} = x_0\vec{i} + y_0\vec{j} + z_0\vec{k}$ in the direction of the vector $\vec{v} = a\vec{i} + b\vec{j} + c\vec{k}$ has parametric equation

$$\vec{r}(t) = \vec{r}_0 + t\vec{v}.$$

Example 7 (a) Find parametric equations for the line passing through the points $(2, -1, 3)$ and $(-1, 5, 4)$.
(b) Represent the line segment from $(2, -1, 3)$ to $(-1, 5, 4)$ parametrically.

Solution (a) The line passes through $(2, -1, 3)$ and is parallel to the displacement vector $\vec{v} = -3\vec{i} + 6\vec{j} + \vec{k}$ from $(2, -1, 3)$ to $(-1, 5, 4)$. Thus the parametric equation is

$$\vec{r}(t) = 2\vec{i} - \vec{j} + 3\vec{k} + t(-3\vec{i} + 6\vec{j} + \vec{k}).$$

(b) In the parameterization in part (a), $t = 0$ corresponds to the point $(2, -1, 3)$ and $t = 1$ corresponds to the point $(-1, 5, 4)$. So the parameterization of the segment is

$$\vec{r}(t) = 2\vec{i} - \vec{j} + 3\vec{k} + t(-3\vec{i} + 6\vec{j} + \vec{k}), \qquad 0 \le t \le 1.$$

Intersection of Curves and Surfaces

Parametric equations for a curve enable us to find where a curve intersects a surface.

Example 8 Find the points at which the line $x = t$, $y = 2t$, $z = 1 + t$ pierces the sphere of radius 10 centered at the origin.

Solution The equation for the sphere of radius 10 and center at the origin is

$$x^2 + y^2 + z^2 = 100.$$

To find the intersection points of the line and the sphere, substitute the parametric equations of the line into the equation of the sphere, giving

$$t^2 + 4t^2 + (1 + t)^2 = 100,$$

so

$$6t^2 + 2t - 99 = 0,$$

which has the two solutions at approximately $t = -4.23$ and $t = 3.90$. Using the parametric equation for the line, $(x, y, z) = (t, 2t, 1 + t)$, we see that the line cuts the sphere at the two points

$$(x, y, z) = (-4.23, 2(-4.23), 1 + (-4.23)) = (-4.23, -8.46, -3.23),$$

and

$$(x, y, z) = (3.90, 2(3.90), 1 + 3.90) = (3.90, 7.80, 4.90).$$

We can also use parametric equations to find the intersection of two curves.

Example 9   Two particles move through space, with equations $\vec{r}_1(t) = t\vec{i} + (1 + 2t)\vec{j} + (3 - 2t)\vec{k}$ and $\vec{r}_2(t) = (-2 - 2t)\vec{i} + (1 - 2t)\vec{j} + (1 + t)\vec{k}$. Do the particles ever collide? Do their paths cross?

Solution   To see if the particles collide, we must find out if they pass through the same point at the same time t. So we must find a solution to the vector equation $\vec{r}_1(t) = \vec{r}_2(t)$, which is the same as finding a common solution to the three scalar equations

$$t = -2 - 2t, \qquad 1 + 2t = 1 - 2t, \qquad 3 - 2t = 1 + t.$$

Separately, the solutions are $t = -2/3$, $t = 0$, and $t = 2/3$, so there is no common solution, and the particles don't collide. To see if their paths cross, we find out if they pass through the same point at two possibly different times, t_1 and t_2. So we solve the equations

$$t_1 = -2 - 2t_2, \qquad 1 + 2t_1 = 1 - 2t_2, \qquad 3 - 2t_1 = 1 + t_2.$$

We solve the first two equations simultaneously and get $t_1 = 2$, $t_2 = -2$. Since these values also satisfy the third equation, the paths cross. The position of the first particle at time $t = 2$ is the same as the position of the second particle at time $t = -2$, namely the point $(2, 5, -1)$.

Example 10   Are the lines $x = -1 + t$, $y = 1 + 2t$, $z = 5 - t$ and $x = 2 + 2t$, $y = 4 + t$, $z = 3 + t$ parallel? Do they intersect?

Solution   In vector form the lines are parameterized by

$$\vec{r} = -\vec{i} + \vec{j} + 5\vec{k} + t(\vec{i} + 2\vec{j} - \vec{k})$$
$$\vec{r} = 2\vec{i} + 4\vec{j} + 3\vec{k} + t(2\vec{i} + \vec{j} + \vec{k})$$

Their direction vectors $\vec{i} + 2\vec{j} - \vec{k}$ and $2\vec{i} + \vec{j} + \vec{k}$ are not multiplies of each other, so the lines are not parallel. To find out if they intersect, we see if they pass through same point at two possibly different times, t_1 and t_2:

$$-1 + t_1 = 2 + 2t_2, \qquad 1 + 2t_1 = 4 + t_2, \qquad 5 - t_1 = 3 + t_2.$$

The first two equations give $t_1 = 1$, $t_2 = -1$. Since these values do not satisfy the third equation, the paths do not cross, and so the lines do not intersect.

The next example shows how to tell if two different parameterizations give the same line.

Example 11   Show that the following two lines are the same:

$$\vec{r} = -\vec{i} - \vec{j} + \vec{k} + t(3\vec{i} + 6\vec{j} - 3\vec{k})$$
$$\vec{r} = \vec{i} + 3\vec{j} - \vec{k} + t(-\vec{i} - 2\vec{j} + \vec{k})$$

Solution   The direction vectors of the two lines, $3\vec{i} + 6\vec{j} - 3\vec{k}$ and $-\vec{i} - 2\vec{j} + \vec{k}$, are multiples of each other, so the lines are parallel. To see if they are the same, we pick a point on the first line and see if it is on the second line. For example, the point on the first line with $t = 0$ has position vector $-\vec{i} - \vec{j} + \vec{k}$. Solving

$$\vec{i} + 3\vec{j} - \vec{k} + t(-\vec{i} - 2\vec{j} + \vec{k}) = -\vec{i} - \vec{j} + \vec{k}$$

we get $t = 2$, so the two lines have a point in common. Thus they are the same line, parameterized in two different ways.

Exercises and Problems for Section 17.1

Exercises

1. Explain how you know the following equations parameterize the same line:

$$\vec{r} = (3 - t)\vec{i} + (3 + 4t)\vec{j} - (1 + 2t)\vec{k}$$
$$\vec{r} = (1 + 2t)\vec{i} + (11 - 8t)\vec{j} + (4t - 5)\vec{k}$$

2. A line is parameterized by $\vec{r} = 10\vec{k} + t(\vec{i} + 2\vec{j} + 3\vec{k})$.

   (a) Suppose we restrict ourselves to $t < 0$. What part of the line do we get?

   (b) Suppose we restrict ourselves to $0 \le t \le 1$. What part of the line do we get?

In Exercises 3–15, find a parameterization for the curve.

3. The circle of radius 3 in the xy-plane, centered at the origin, counterclockwise.

4. The circle of radius 2 in the xy-plane, centered at the origin, clockwise.

5. The circle of radius 2 in the xz-plane, centered at the origin.

6. The circle of radius 3 parallel to the xy-plane, centered at the point $(0, 0, 2)$.

7. The circle of radius 3 in the yz-plane, centered at the point $(0, 0, 2)$.

8. The circle of radius 5 parallel to the yz-plane, centered at the point $(-1, 0, -2)$.

9. The curve $y = x^3$ in the xy-plane.

10. The curve $x = y^2$ in the xy-plane.

11. The curve $x = -3z^2$ in the xz-plane.

12. The curve $y = 4 - 5x^4$ through the point $(0, 4, 4)$, parallel to the xy-plane.

13. The ellipse of major diameter 6 along the x-axis and minor diameter 4 along the y-axis, centered at the origin.

14. The ellipse of major diameter 5 parallel to the y-axis and minor diameter 2 parallel to the z-axis, centered at $(0, 1, -2)$.

15. The ellipse of major diameter 3 parallel to the x-axis and minor diameter 2 parallel to the z-axis, centered at $(0, 1, -2)$.

In Exercises 16–26, find parametric equations for the line.

16. The line in the direction of the vector $3\vec{i} - 3\vec{j} + \vec{k}$ and through the point $(1, 2, 3)$.

17. The line in the direction of the vector $\vec{i} + 2\vec{j} - \vec{k}$ and through the point $(3, 0, -4)$.

18. The line in the direction of the vector $5\vec{j} + 2\vec{k}$ and through the point $(5, -1, 1)$.

19. The line in the direction of the vector $2\vec{i} + 2\vec{j} - 3\vec{k}$ and through the point $(-3, 4, -2)$.

20. The line in the direction of the vector $\vec{i} - \vec{k}$ and through the point $(0, 1, 0)$.

21. The line parallel to the z-axis passing through the point $(1, 0, 0)$.

22. The line through the points $(2, 3, -1)$ and $(5, 2, 0)$.

23. The line through the points $(1, 5, 2)$ and $(5, 0, -1)$.

24. The line through the points $(-3, -2, 1)$ and $(-1, -3, -1)$.

25. The line intersecting the x-axis at $x = 3$ and the z-axis at $z = -5$.

26. The line through $(3, -2, 2)$ and intersecting the y-axis at $y = 2$.

27. Do the lines in Exercises 22 and 16 intersect?

In Exercises 28–31, find parametric equations for the curve segment.

28. Line from $P_0 = (1, -3, 2)$ to $P_1 = (4, 1, -3)$.

29. Line from $P_0 = (-1, -3)$ to $P_1 = (5, 2)$.

30. Semicircle from $(1, 0, 0)$ to $(-1, 0, 0)$ in the xy-plane with $y \ge 0$.

31. Semicircle from $(0, 0, 5)$ to $(0, 0, -5)$ in the yz-plane with $y \ge 0$.

Are the lines in Exercises 32–35 parallel? Do they intersect? If so, where?

32. l_1 : $\quad x = 5 + 2t, \quad y = -t, \quad z = -2 + 3t$
    l_2 : $\quad x = 4 - 3t, \quad y = 1 + 5t, \quad z = 4$

33. l_1 : $\quad x = 3 - 2t, \quad y = 5 + t, \quad z = -4 + 2t$
    l_2 : $\quad x = 7 + 6t, \quad y = 1 - 3t, \quad z = 5 - 6t$

34. l_1 : $\quad x = 1 + 4t, \quad y = 1 - 3t, \quad z = 2t$
    l_2 : $\quad x = 3 - 10t, \quad y = 7, \quad z = -6 + 2t$

35. l_1 : $\quad x = 3 + 2t, \quad y = 5 - 2t, \quad z = 1 + t$
    l_2 : $\quad x = -1 - t, \quad y = 4t, \quad z = 4 - 2t$

Problems

In Problems 36–40, parameterize the line through $P = (2, 5)$ and $Q = (12, 9)$ so that the points P and Q correspond to the given parameter values.

36. $t = 0$ and 1 **37.** $t = 0$ and 5

38. $t = 10$ and 11 **39.** $t = 20$ and 30

40. $t = 0$ and -1

In Problems 41–44, find parametric equations for the curve segment.

41. The graph of $y = \sqrt{x}$ from $(1, 1)$ to $(16, 4)$.

42. The quarter-ellipse from $(4, 0, 3)$ to $(0, -3, 3)$ in the plane $z = 3$.

43. A circular arc from $P = (0, 0)$ to $Q = (10, 0)$ if the radius of the circle is 5.

44. A circular arc from $P = (0, 0)$ to $Q = (10, 0)$ if the radius of the circle is 10.

In Problems 45–47, describe in words the motion parameterized by the equations.

45. $x = 3 + t^3$, $y = 5 - t^3$, $z = 7 + 2t^3$

46. $x = 5 \cos t$, $y = 5 \sin t$, $z = t$

47. $x = \cos t$, $y = t^2$, $z = \sin t$

48. **(a)** Explain why the line of intersection of two planes must be parallel to the cross product of a normal vector to the first plane and a normal vector to the second.
  (b) Find a vector parallel to the line of intersection of the two planes $x + 2y - 3z = 7$ and $3x - y + z = 0$.
  (c) Find parametric equations for the line in part (b).

In Problems 49–52, find parametric equations for the line.

49. The line of intersection of the planes $x - y + z = 3$ and $2x + y - z = 5$.

50. The line of intersection of the planes $x + y + z = 3$ and $x - y + 2z = 2$.

51. The line perpendicular to the surface $z = x^2 + y^2$ at the point $(1, 2, 5)$.

52. The line through the point $(-4, 2, 3)$ and parallel to a line in the yz-plane which makes a $45°$ angle with the positive y-axis and the positive z-axis.

53. Show that the equations $x = 3 + t$, $y = 2t$, $z = 1 - t$ satisfy the equations $x + y + 3z = 6$ and $x - y - z = 2$. What does this tell you about the curve parameterized by these equations?

54. Two particles are traveling through space. At time t the first particle is at the point $(-1 + t, 4 - t, -1 + 2t)$ and the second particle is at $(-7 + 2t, -6 + 2t, -1 + t)$.
  (a) Describe the two paths.
  (b) Do the two particles collide? If so, when and where?
  (c) Do the paths of the two particles cross? If so, where?

55. Is the point $(-3, -4, 2)$ visible from the point $(4, 5, 0)$ if there is an opaque ball of radius 1 centered at the origin?

56. A plane from Denver, Colorado, (altitude 1650 meters) flies to Bismark, North Dakota (altitude 550 meters). It travels at 650 km/hour along a horizontal line at 8,000 meters above the line joining Denver and Bismark. Bismark is about 850 km in the direction $60°$ north of east from Denver. Find parametric equations describing the plane's motion. Assume the origin is at sea level beneath Denver and that the x-axis points east and the y-axis points north. Measure distances in kilometers and time in hours.

57. A light shines on the helix of Example 3 on page 784 from far down each axis. Sketch the shadow the helix casts on each of the coordinate planes: xy, xz, and yz.

58. **(a)** Find a parametric equation for the line through the point $(2, 1, 3)$ and in the direction of $a\vec{i} + b\vec{j} + c\vec{k}$.
  (b) Find conditions on a, b, c so that the line you found in part (a) goes through the origin. Give a reason for your answer.

59. Consider the line $x = 5 - 2t, y = 3 + 7t, z = 4t$ and the plane $ax + by + cz = d$. All the following questions have many possible answers. Find values of a, b, c, d such that:
  (a) The plane is perpendicular to the line.
  (b) The plane is perpendicular to the line and through the point $(5, 3, 0)$.
  (c) The line lies in the plane.

60. Consider two points P_0 and P_1 in 3-space.
  (a) Show that the line segment from P_0 to P_1 can be parametrized by
$$\vec{r}(t) = (1 - t)\overrightarrow{OP_0} + t\overrightarrow{OP_1}, \quad 0 \le t \le 1.$$
  (b) What is represented by the parametric equation
$$\vec{r}(t) = t\overrightarrow{OP_0} + (1 - t)\overrightarrow{OP_1}, \quad 0 \le t \le 1?$$

61. **(a)** Find a parametric equation for the line through the point $(1, 5, 2)$ and in the direction of the vector $2\vec{i} + 3\vec{j} - \vec{k}$.
  (b) By minimizing the square of the distance from a point on the line to the origin, find the exact point on the line which is closest to the origin.

62. A line has equation $\vec{r} = \vec{a} + t\vec{b}$ where $\vec{r} = x\vec{i} + y\vec{j} + z\vec{k}$ and $\vec{a}$ and $\vec{b}$ are constant vectors such that $\vec{a} \neq \vec{0}, \vec{b} \neq \vec{0}$, $\vec{a} \neq \vec{b}, \vec{b}$ not perpendicular to $\vec{a}$. For each of the planes (a)–(c), pick the equation (i)–(ix) which represents it. Explain your choice.

   (a) A plane perpendicular to the line and through the origin.

   (b) A plane perpendicular to the line and not through the origin.

   (c) A plane containing the line.

   (i) $\vec{a} \cdot \vec{r} = \|\vec{b}\|$ (ii) $\vec{b} \cdot \vec{r} = \|\vec{a}\|$

   (iii) $\vec{a} \cdot \vec{r} = \vec{b} \cdot \vec{r}$ (iv) $(\vec{a} \times \vec{b}) \cdot (\vec{r} - \vec{a}) = 0$

   (v) $(\vec{a} \times \vec{b}) \cdot (\vec{r} - \vec{b}) = 0$ (vi) $\vec{a} \cdot \vec{r} = 0$

(vii) $\vec{b} \cdot \vec{r} = 0$ (viii) $\vec{a} + \vec{r} = \vec{b}$

(ix) $\vec{r} - \vec{a} = \vec{b}$

63. The plane $x + 3y - 2z = 6$ is colored blue and the plane $2x + y + z = 3$ is colored yellow. The planes intersect in a line, which is colored green. You are at the point $P = (1, -2, -1)$.

   (a) You look in the direction $\vec{v} = \vec{i} + 2\vec{j} + \vec{k}$. Do you see the blue plane or the yellow plane?

   (b) In what direction(s) are you looking directly at the green line?

   (c) In what direction(s) should you look to see the yellow plane? The blue plane?

17.2 MOTION, VELOCITY, AND ACCELERATION

In this section we see how to find the vector quantities of velocity and acceleration from a parametric equation for the motion of an object.

The Velocity Vector

The velocity of a moving particle can be represented by a vector with the following properties:

> The **velocity vector** of a moving object is a vector $\vec{v}$ such that:
> - The magnitude of $\vec{v}$ is the speed of the object.
> - The direction of $\vec{v}$ is the direction of motion.
>
> Thus the speed of the object is $\|\vec{v}\|$ and the velocity vector is tangent to the object's path.

Example 1   A child is sitting on a ferris wheel of diameter 10 meters, making one revolution every 2 minutes. Find the speed of the child and draw velocity vectors at two different times.

Solution   The child moves at a constant speed around a circle of radius 5 meters, completing one revolution every 2 minutes. One revolution around a circle of radius 5 is a distance of 10π, so the child's speed is $10\pi/2 = 5\pi \approx 15.7$ m/min. Hence, the magnitude of the velocity vector is 15.7 m/min. The direction of motion is tangent to the circle, and hence perpendicular to the radius at that point. Figure 17.9 shows the direction of the vector at two different times.

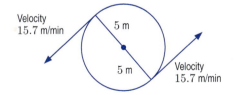

Figure 17.9: Velocity vectors of a child on a ferris wheel (Note that vectors would be in opposite direction if viewed from the other side)

Computing the Velocity

We find the velocity, as in one-variable calculus, by taking a limit. If the position vector of the particle is $\vec{r}(t)$ at time t, then the displacement vector between its positions at times t and $t + \Delta t$ is $\Delta \vec{r} = \vec{r}(t + \Delta t) - \vec{r}(t)$. (See Figure 17.10.) Over this interval,

$$\text{Average velocity} = \frac{\Delta \vec{r}}{\Delta t}.$$

In the limit as Δt goes to zero we have the instantaneous velocity at time t:

> The **velocity vector**, $\vec{v}(t)$, of a moving object with position vector $\vec{r}(t)$ at time t is
>
> $$\vec{v}(t) = \lim_{\Delta t \to 0} \frac{\Delta \vec{r}}{\Delta t} = \lim_{\Delta t \to 0} \frac{\vec{r}(t + \Delta t) - \vec{r}(t)}{\Delta t},$$
>
> whenever the limit exists. We use the notation $\vec{v} = \dfrac{d\vec{r}}{dt} = \vec{r}\,'(t).$

Notice that the direction of the velocity vector $\vec{r}\,'(t)$ in Figure 17.10 is approximated by the direction of the vector $\Delta \vec{r}$ and that the approximation gets better as $\Delta t \to 0$.

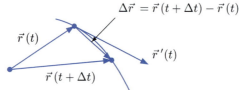

$$\Delta \vec{r} = \vec{r}(t + \Delta t) - \vec{r}(t)$$

$\vec{r}(t)$

$\vec{r}\,'(t)$

$\vec{r}(t + \Delta t)$

Figure 17.10: The change, $\Delta \vec{r}$, in the position vector for a particle moving on a curve and the velocity vector $\vec{v} = \vec{r}\,'(t)$

The Components of the Velocity Vector

If we represent a curve parametrically by $x = f(t), y = g(t), z = h(t)$, then we can write its position vector as: $\vec{r}(t) = f(t)\vec{i} + g(t)\vec{j} + h(t)\vec{k}$. Now we can compute the velocity vector:

$$
\begin{aligned}
\vec{v}(t) &= \lim_{\Delta t \to 0} \frac{\vec{r}(t + \Delta t) - \vec{r}(t)}{\Delta t} \\
&= \lim_{\Delta t \to 0} \frac{(f(t+\Delta t)\vec{i} + g(t+\Delta t)\vec{j} + h(t+\Delta t)\vec{k}) - (f(t)\vec{i} + g(t)\vec{j} + h(t)\vec{k})}{\Delta t} \\
&= \lim_{\Delta t \to 0} \left(\frac{f(t+\Delta t) - f(t)}{\Delta t}\vec{i} + \frac{g(t+\Delta t) - g(t)}{\Delta t}\vec{j} + \frac{h(t+\Delta t) - h(t)}{\Delta t}\vec{k} \right) \\
&= f'(t)\vec{i} + g'(t)\vec{j} + h'(t)\vec{k} \\
&= \frac{dx}{dt}\vec{i} + \frac{dy}{dt}\vec{j} + \frac{dz}{dt}\vec{k}.
\end{aligned}
$$

Thus we have the following result:

> The **components of the velocity vector** of a particle moving in space with position vector $\vec{r}(t) = f(t)\vec{i} + g(t)\vec{j} + h(t)\vec{k}$ at time t are given by
>
> $$\vec{v}(t) = f'(t)\vec{i} + g'(t)\vec{j} + h'(t)\vec{k} = \frac{dx}{dt}\vec{i} + \frac{dy}{dt}\vec{j} + \frac{dz}{dt}\vec{k}.$$

Example 2   Find the components of the velocity vector for the child on the ferris wheel in Example 1 using a coordinate system which has its origin at the center of the ferris wheel and which makes the rotation counterclockwise.

Solution   The ferris wheel has radius 5 meters and completes 1 revolution counterclockwise every 2 minutes. The motion is parameterized by an equation of the form

$$\vec{r}(t) = 5\cos(\omega t)\vec{i} + 5\sin(\omega t)\vec{j},$$

where ω is chosen to make the period 2 minutes. Since the period of $\cos(\omega t)$ and $\sin(\omega t)$ is $2\pi/\omega$, we must have

$$\frac{2\pi}{\omega} = 2, \quad \text{so} \quad \omega = \pi.$$

Thus, the motion is described by the equation

$$\vec{r}(t) = 5\cos(\pi t)\vec{i} + 5\sin(\pi t)\vec{j},$$

where t is in minutes. The velocity is given by

$$\vec{v} = \frac{dx}{dt}\vec{i} + \frac{dy}{dt}\vec{j} = -5\pi\sin(\pi t)\vec{i} + 5\pi\cos(\pi t)\vec{j}.$$

To check, we calculate the magnitude of $\vec{v}$,

$$\|\vec{v}\| = \sqrt{(-5\pi)^2\sin^2(\pi t) + (5\pi)^2\cos^2(\pi t)} = 5\pi\sqrt{\sin^2(\pi t) + \cos^2(\pi t)} = 5\pi \approx 15.7,$$

which agrees with the speed we calculated in Example 1. To see that the direction is correct, we must show that the vector $\vec{v}$ at any time t is perpendicular to the position vector of the particle at time t. To do this, we compute the dot product of $\vec{v}$ and $\vec{r}$:

$$\vec{v} \cdot \vec{r} = (-5\pi\sin(\pi t)\vec{i} + 5\pi\cos(\pi t)\vec{j}) \cdot (5\cos(\pi t)\vec{i} + 5\sin(\pi t)\vec{j})$$
$$= -25\pi\,\sin(\pi t)\,\cos(\pi t) + 25\pi\cos(\pi t)\,\sin(\pi t) = 0.$$

So the velocity vector, $\vec{v}$, is perpendicular to $\vec{r}$ and hence tangent to the circle. (See Figure 17.11.)

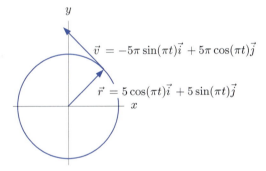

Figure 17.11: Velocity and radius vector of motion around a circle

Velocity Vectors and Tangent Lines

Since the velocity vector is tangent to the path of motion, it can be used to find parametric equations for the tangent line, if there is one.

Example 3   Find the tangent line at the point $(1, 1, 2)$ to the curve defined by the parametric equation

$$\vec{r}(t) = t^2\vec{i} + t^3\vec{j} + 2t\vec{k}.$$

Solution   At time $t = 1$ the particle is at the point $(1, 1, 2)$ with position vector $\vec{r}_0 = \vec{i} + \vec{j} + 2\vec{k}$. The velocity vector at time t is $\vec{r}'(t) = 2t\vec{i} + 3t^2\vec{j} + 2\vec{k}$, so at time $t = 1$ the velocity is $\vec{v} = \vec{r}'(1) = 2\vec{i} + 3\vec{j} + 2\vec{k}$. The tangent line passes through $(1, 1, 2)$ in the direction of $\vec{v}$, so it has the parametric equation

$$\vec{r}(t) = \vec{r}_0 + t\vec{v} = (\vec{i} + \vec{j} + 2\vec{k}) + t(2\vec{i} + 3\vec{j} + 2\vec{k}).$$

The Acceleration Vector

Just as the velocity of a particle moving in 2-space or 3-space is a vector quantity, so is the rate of change of the velocity of the particle, namely its acceleration. Figure 17.12 shows a particle at time t with velocity vector $\vec{v}(t)$ and then a little later at time $t + \Delta t$. The vector $\Delta \vec{v} = \vec{v}(t + \Delta t) - \vec{v}(t)$ is the change in velocity and points approximately in the direction of the acceleration. So,

$$\text{Average acceleration} = \frac{\Delta \vec{v}}{\Delta t}.$$

In the limit as $\Delta t \to 0$, we have the instantaneous acceleration at time t:

> The **acceleration vector** of an object moving with velocity $\vec{v}(t)$ at time t is
>
> $$\vec{a}(t) = \lim_{\Delta t \to 0} \frac{\Delta \vec{v}}{\Delta t} = \lim_{\Delta t \to 0} \frac{\vec{v}(t + \Delta t) - \vec{v}(t)}{\Delta t},$$
>
> if the limit exists. We use the notation $\vec{a} = \dfrac{d\vec{v}}{dt} = \dfrac{d^2 \vec{r}}{dt^2} = \vec{r}''(t)$.

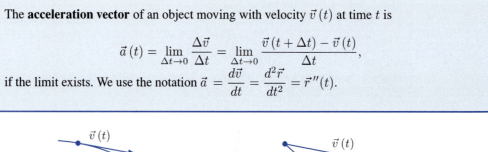

Figure 17.12: Computing the difference between two velocity vectors

Components of the Acceleration Vector

If we represent a curve in space parametrically by $x = f(t)$, $y = g(t)$, $z = h(t)$, we can express the acceleration in components. The velocity vector $\vec{v}(t)$ is given by

$$\vec{v}(t) = f'(t)\vec{i} + g'(t)\vec{j} + h'(t)\vec{k}.$$

From the definition of the acceleration vector, we have

$$\vec{a}(t) = \lim_{\Delta t \to 0} \frac{\vec{v}(t + \Delta t) - \vec{v}(t)}{\Delta t} = \frac{d\vec{v}}{dt}.$$

Using the same method to compute $d\vec{v}/dt$ as we used to compute $d\vec{r}/dt$ on page 792, we obtain

> The **components of the acceleration vector**, $\vec{a}(t)$, at time t of a particle moving in space with position vector $\vec{r}(t) = f(t)\vec{i} + g(t)\vec{j} + h(t)\vec{k}$ at time t are given by
>
> $$\vec{a}(t) = f''(t)\vec{i} + g''(t)\vec{j} + h''(t)\vec{k} = \frac{d^2 x}{dt^2}\vec{i} + \frac{d^2 y}{dt^2}\vec{j} + \frac{d^2 z}{dt^2}\vec{k}.$$

Motion In a Circle and Along a Line

We now consider the velocity and acceleration vectors for two basic motions: uniform motion around a circle, and motion along a straight line.

Example 4   Find the acceleration vector for the child on the ferris wheel in Examples 1 and 2.

Solution   The child's position vector is given by $\vec{r}(t) = 5\cos(\pi t)\vec{i} + 5\sin(\pi t)\vec{j}$. In Example 2 we saw that the velocity vector is

$$\vec{v}(t) = \frac{dx}{dt}\vec{i} + \frac{dy}{dt}\vec{j} = -5\pi\sin(\pi t)\vec{i} + 5\pi\cos(\pi t)\vec{j}.$$

Thus, the acceleration vector is

$$\vec{a}(t) = \frac{d^2x}{dt^2}\vec{i} + \frac{d^2y}{dt^2}\vec{j} = -(5\pi)\cdot\pi\cos(\pi t)\vec{i} - (5\pi)\cdot\pi\sin(\pi t)\vec{j}$$
$$= -5\pi^2\cos(\pi t)\vec{i} - 5\pi^2\sin(\pi t)\vec{j}.$$

Notice that $\vec{a}(t) = -\pi^2\vec{r}(t)$. Thus, the acceleration vector is a multiple of $\vec{r}(t)$ and points toward the origin.

The motion of the child on the ferris wheel is an example of uniform circular motion, whose properties follow. (See Problem 22.)

Uniform Circular Motion: For a particle whose motion is described by

$$\vec{r}(t) = R\cos(\omega t)\vec{i} + R\sin(\omega t)\vec{j}$$

- Motion is in a circle of radius R with period $2\pi/\omega$.
- Velocity, $\vec{v}$, is tangent to the circle and speed is constant $\|\vec{v}\| = \omega R$.
- Acceleration, $\vec{a}$, points toward the center of the circle with $\|\vec{a}\| = \|\vec{v}\|^2/R$.

In uniform circular motion, the acceleration vector is perpendicular to the velocity vector because the velocity vector does not change in magnitude, only in direction. We now look at straight line motion in which the velocity vector always has the same direction but its magnitude changes. In straight line motion, the acceleration vector points in the same direction as the velocity vector if the speed is increasing and in the opposite direction to the velocity vector if the speed is decreasing.

Example 5   Consider the motion given by the vector equation

$$\vec{r}(t) = 2\vec{i} + 6\vec{j} + (t^3 + t)(4\vec{i} + 3\vec{j} + \vec{k}).$$

Show that this is straight line motion in the direction of the vector $4\vec{i} + 3\vec{j} + \vec{k}$ and relate the acceleration vector to the velocity vector.

Solution   The velocity vector is

$$\vec{v} = (3t^2 + 1)(4\vec{i} + 3\vec{j} + \vec{k}).$$

Since $(3t^2 + 1)$ is a positive scalar, the velocity vector $\vec{v}$ always points in the direction of the vector $4\vec{i} + 3\vec{j} + \vec{k}$. In addition,

$$\text{Speed} = \|\vec{v}\| = (3t^2 + 1)\sqrt{4^2 + 3^2 + 1^2} = \sqrt{26}(3t^2 + 1).$$

Notice that the speed is decreasing until $t = 0$, then starts increasing. The acceleration vector is

$$\vec{a} = 6t(4\vec{i} + 3\vec{j} + \vec{k}).$$

For $t > 0$, the acceleration vector points in the same direction as $4\vec{i} + 3\vec{j} + \vec{k}$, which is the same direction as $\vec{v}$. This makes sense because the object is speeding up. For $t < 0$, the acceleration vector $6t(4\vec{i} + 3\vec{j} + \vec{k})$ points in the opposite direction to $\vec{v}$ because the object is slowing down.

Motion in a Straight Line: For a particle whose motion is described by

$$\vec{r}(t) = \vec{r}_0 + f(t)\vec{v}_0$$

- Motion is along a straight line through the point with position vector $\vec{r}_0$ parallel to $\vec{v}_0$.
- Velocity, $\vec{v}$, and acceleration, $\vec{a}$, are parallel to the line.

The Length of a Curve

The speed of a particle is the magnitude of its velocity vector:

$$\text{Speed} = \|\vec{v}\| = \sqrt{\left(\frac{dx}{dt}\right)^2 + \left(\frac{dy}{dt}\right)^2 + \left(\frac{dz}{dt}\right)^2}.$$

As in one dimension, we can find the distance traveled by a particle along a curve by integrating its speed. Thus,

$$\text{Distance traveled} = \int_a^b \|\vec{v}(t)\| \, dt.$$

If the particle never stops or reverses its direction as it moves along the curve, the distance it travels will be the same as the length of the curve. This suggests the following formula, which is justified in Problem 39:

If the curve C is given parametrically for $a \leq t \leq b$ by smooth functions and if the velocity vector $\vec{v}$ is not $\vec{0}$ for $a < t < b$, then

$$\text{Length of } C = \int_a^b \|\vec{v}\| \, dt.$$

Example 6   Find the circumference of the ellipse given by the parametric equations

$$x = 2\cos t, \quad y = \sin t, \quad 0 \leq t \leq 2\pi.$$

Solution   The circumference of this curve is given by an integral which must be calculated numerically:

$$\text{Circumference} = \int_0^{2\pi} \sqrt{\left(\frac{dx}{dt}\right)^2 + \left(\frac{dy}{dt}\right)^2} \, dt = \int_0^{2\pi} \sqrt{(-2\sin t)^2 + (\cos t)^2} \, dt$$

$$= \int_0^{2\pi} \sqrt{4\sin^2 t + \cos^2 t} \, dt = 9.69.$$

Since the ellipse is inscribed in a circle of radius 2 and circumscribes a circle of radius 1, we would expect the length of the ellipse to be between $2\pi(2) \approx 12.57$ and $2\pi(1) \approx 6.28$, so the value of 9.69 is reasonable.

Exercises and Problems for Section 17.2

Exercises

For Exercises 1–4, find the velocity $\vec{v}(t)$ and the speed $\|\vec{v}(t)\|$. Find any times at which the particle stops.

1. $x = 3t^2, \quad y = t^3 + 1$

2. $x = 3\sin(t^2) - 1, \quad y = 3\cos(t^2)$

3. $x = 3\sin^2 t, \quad y = \cos t - 1, \quad z = t^2$

4. $x = (t-1)^2, \, y = 2, \, z = 2t^3 - 3t^2$

In Exercises 5–8, find the velocity and acceleration vectors.

5. $x = 3\cos t, y = 4\sin t$

6. $x = t, y = t^3 - t$

7. $x = 2 + 3t, y = 4 + t, z = 1 - t$

8. $x = 3\cos(t^2), y = 3\sin(t^2), z = t^2$

9. Find the position, velocity vector, and speed, at $t = 3$, of the particle whose motion is given by

$$x = t^3 - 8t + 1, \quad y = t^2 - 4.$$

10. A particle moves with $x = 3\cos 2t, y = 3\sin 2t$. Describe the motion in words and sketch the particle's path. Find the particle's position, velocity, and acceleration vectors at $t = 0$, and add these vectors to your sketch.

In Exercises 11–12, find the velocity and acceleration vectors of the uniform circular motion and check that they are perpendicular. Check that the speed and magnitude of the acceleration are constant.

11. $x = 3\cos(2\pi t), \quad y = 3\sin(2\pi t), \quad z = 0$

12. $x = 2\pi, \quad y = 2\sin(3t), \quad z = 2\cos(3t)$

In Exercises 13–14, find the velocity and acceleration vectors of the straight line motion. Check that the acceleration vector points in the same direction as the velocity vector if the speed is increasing and in the opposite direction if the speed is decreasing.

13. $x = 2 + t^2, y = 3 - 2t^2, z = 5 - t^2$

14. $x = -2t^3 - 3t + 1, y = 4t^3 + 6t - 5, z = 6t^3 + 9t - 2$

Find the length of the curves in Exercises 15–17.

15. $x = 3 + 5t, y = 1 + 4t, z = 3 - t$ for $1 \le t \le 2$. Explain your answer.

16. $x = \cos(e^t), y = \sin(e^t)$ for $0 \le t \le 1$. Explain why your answer is reasonable.

17. $x = \cos 3t, y = \sin 5t$ for $0 \le t \le 2\pi$.

Problems

18. Using time increments of 0.01, give a table of values near $t = 1$ for the position vector of the circular motion

$$\vec{r}(t) = (\cos t)\vec{i} + (\sin t)\vec{j}.$$

Use the table to approximate the velocity vector, $\vec{v}$, at time $t = 1$. Show that $\vec{v}$ is perpendicular to the radius from the origin at $t = 1$.

19. The table gives x and y coordinates of a particle in the plane at time t. Assuming the path is smooth, estimate the following quantities:

(a) The velocity vector and speed at time $t = 2$.
(b) Any times when the particle is moving parallel to the y-axis.
(c) Any times when the particle has come to a stop.

t	0	0.5	1.0	1.5	2.0	2.5	3.0	3.5	4.0
x	1	4	6	7	6	3	2	3	5
y	3	2	3	5	8	10	11	10	9

20. **(a)** Sketch the parameterized curve $x = t\cos t, y = t\sin t$ for $0 \le t \le 4\pi$.
(b) Use difference quotients to approximate the velocity vectors $\vec{v}(t)$ for $t = 2, 4$, and 6.
(c) Compute the velocity vectors $\vec{v}(t)$ for $t = 2, 4$, and 6, exactly and sketch them on the graph of the curve.

21. Find parametric equations for the tangent line at $t = 2$ for Exercise 1.

In Problems 22–23, find all values of t for which the particle is moving parallel to the x-axis and to the y-axis. Determine the end behavior and graph the particle's path.

22. $x = t^3 - 12t, \quad y = t^2 + 10t$

23. $x = t^2 - 6t, \quad y = t^3 - 3t$

24. A particle passes through the point $P = (5, 4, -2)$ at time $t = 4$, moving with constant velocity $\vec{v} = 2\vec{i} - 3\vec{j} + \vec{k}$. Find a parametric equation for its motion.

25. A particle passes through the point $P = (5, 4, 3)$ at time $t = 7$, moving with constant velocity $\vec{v} = 3\vec{i} + \vec{j} + 2\vec{k}$. Find equations for its position at time t.

26. An object moving with constant velocity in 3-space, with coordinates in meters, passes through the point $(1, 1, 1)$, then through $(2, -1, 3)$ five seconds later. What is its velocity vector? What is its acceleration vector?

27. A particle moves at a constant speed along a line from the point $P = (2, -1, 5)$ to the point $Q = (5, 3, -1)$. Find a parametric equation of the line if:

(a) The particle takes 5 seconds to move from P to Q.
(b) The speed of the particle is 5 units per second.

28. A particle starts at the point $P = (3, 2, -5)$ and moves along a straight line toward $Q = (5, 7, -2)$ at a speed of 5 cm/sec. Let x, y, z be measured in centimeters.

(a) Find the particle's velocity vector.
(b) Find parametric equations for the particle's motion.

29. Find parametric equations for a particle moving along the line $y = -2x + 5$ with speed 3.

30. A stone is thrown from a rooftop at time $t = 0$ seconds. Its position at time t is given by

$$\vec{r}(t) = 10t\vec{i} - 5t\vec{j} + (6.4 - 4.9t^2)\vec{k}.$$

The origin is at the base of the building, which is standing on flat ground. Distance is measured in meters. The vector $\vec{i}$ points east, $\vec{j}$ points north, and $\vec{k}$ points up.

(a) How high is the rooftop above the ground?
(b) At what time does the stone hit the ground?
(c) How fast is the stone moving when it hits the ground?
(d) Where does the stone hit the ground?
(e) What is the stone's acceleration when it hits the ground?

31. Determine the position vector $\vec{r}(t)$ for a rocket which is launched from the origin at time $t = 0$ seconds, reaches its highest point of $(x, y, z) = (1000, 3000, 10,000)$, where x, y, z are in meters, and after the launch is subject only to the acceleration due to gravity, 9.8 m/sec^2.

32. A particle moves on a circle of radius 5 cm, centered at the origin, in the xy-plane (x and y measured in centimeters). It starts at the point $(0, 5)$ and moves counterclockwise, going once around the circle in 8 seconds.

(a) Write a parameterization for the particle's motion.
(b) What is the particle's speed? Give units.

33. Suppose $\vec{r}(t) = \cos t\,\vec{i} + \sin t\,\vec{j} + 2t\,\vec{k}$ represents the position of a particle on a helix, where z is the height of the particle above the ground.

(a) Is the particle ever moving downwards? When?
(b) When does the particle reach a point 10 units above the ground?
(c) What is the velocity of the particle when it is 10 units above the ground?
(d) When it is 10 units above the ground, the particle leaves the helix and moves along the tangent. Find parametric equations for this tangent line.

34. A stone is swung around on a string at a constant speed with period 2π seconds in a horizontal circle centered at the point $(0, 0, 8)$. When $t = 0$, the stone is at the point $(0, 5, 8)$; it travels clockwise when viewed from above. When the stone is at the point $(5, 0, 8)$, the string breaks and it moves under gravity.

(a) Parameterize the stone's circular trajectory.
(b) Find the velocity and acceleration of the stone at the moment before the string breaks.
(c) Write, but do not solve, the differential equations (with initial conditions) satisfied by the coordinates x, y, z giving the position of the stone after it has left the circle.

35. Emily is standing on the outer edge of a merry-go-round, 10 meters from the center. The merry-go-round completes one full revolution every 20 seconds. As Emily passes over a point P on the ground, she drops a ball from 3 meters above the ground.

(a) How fast is Emily going?
(b) How far from P does the ball hit the ground? (The acceleration due to gravity is 9.8 m/sec^2.)
(c) How far from Emily does the ball hit the ground?

36. An ant crawls along the radius from the center to the edge of a circular disk of radius 1 meter, moving at a constant rate of 1 cm/sec. Meanwhile, the disk is turning counterclockwise about its center at 1 revolution per second.

(a) Parameterize the motion of the ant.
(b) Find the velocity and speed of the ant.
(c) Determine the acceleration and magnitude of the acceleration of the ant.

37. How do the motions of objects A and B differ, if A has position vector $\vec{r}_A(t)$ and B has position vector $\vec{r}_B(t) = \vec{r}_A(2t)$ for $t \geq 0$. Illustrate your answer with $\vec{r}_A(t) = t\vec{i} + t^2\vec{j}$.

38. How does the motion of objects A and B differ if A has position vector $\vec{r}_A(t)$ and B has position vector $\vec{r}_B(t) = \vec{r}_A(t - 2)$? Illustrate your answer with $\vec{r}_A(t) = t\vec{i} + t^2\vec{j}$.

39. In this problem we justify the formula for the length of a curve given on page 796. Suppose the curve C is given by smooth parametric equations $x = x(t)$, $y = y(t)$, $z = z(t)$ for $a \leq t \leq b$. By dividing the parameter interval $a \leq t \leq b$ at points $t_1, \ldots, t_{n-1}$ into small segments of length $\Delta t = t_{i+1} - t_i$, we get a corresponding division of the curve C into small pieces. See Figure 17.13, where the points $P_i = (x(t_i), y(t_i), z(t_i))$ on the curve C correspond to parameter values $t = t_i$. Let C_i be the portion of the curve C between P_i and P_{i+1}.

(a) Use local linearity to show that

$$\text{Length of } C_i \approx \sqrt{x'(t_i)^2 + y'(t_i)^2 + z'(t_i)^2}\, \Delta t.$$

(b) Use part (a) and a Riemann sum to explain why

$$\text{Length of } C = \int_a^b \sqrt{x'(t)^2 + y'(t)^2 + z'(t)^2}\, dt.$$

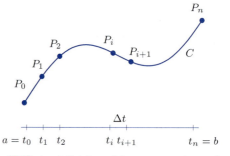

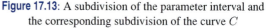

Figure 17.13: A subdivision of the parameter interval and the corresponding subdivision of the curve C

40. At time $t = 0$ an object is moving with velocity vector $\vec{v} = 2\vec{i} + \vec{j}$ and acceleration vector $\vec{a} = \vec{i} + \vec{j}$. Can it be in uniform circular motion about some point in the plane?

41. Figure 17.14 shows the velocity and acceleration vectors of an object in uniform circular motion about a point in the plane at a particular moment. Is it moving round the circle in the clockwise or counterclockwise direction?

Figure 17.14

17.3   VECTOR FIELDS

Introduction to Vector Fields

A *vector field* is a function that assigns a vector to each point in the plane or in 3-space. One example of a vector field is the gradient of a function $f(x, y)$; at each point (x, y) the vector grad $f(x, y)$ points in the direction of maximum rate of increase of f. In this section we will look at other vector fields representing velocities and forces.

Velocity Vector Fields

Figure 17.15 shows the flow of a part of the Gulf stream, a current in the Atlantic Ocean.[1] It is an example of a *velocity vector field*: each vector shows the velocity of the current at that point. The current is fastest where the velocity vectors are longest in the middle of the stream. Beside the stream are eddies where the water flows round and round in circles.

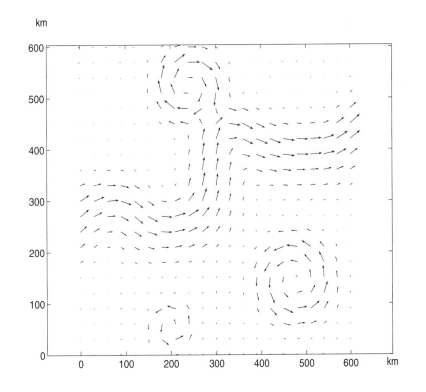

Figure 17.15: The velocity vector field of the Gulf stream

[1] Based on data supplied by Avijit Gangopadhyay of the Jet Propulsion Laboratory

Figure 17.16: The gravitational field of the earth

Force Fields

Another physical quantity represented by a vector is force. When we experience a force, sometimes it results from direct contact with the object that supplies the force (for example, a push). Many forces, however, can be felt at all points in space. For example, the earth exerts a gravitational pull on all other masses. Such forces can be represented by vector fields.

Figure 17.16 shows the gravitational force exerted by the earth on a mass of one kilogram at different points in space. This is a sketch of the vector field in 3-space. You can see that the vectors all point towards the earth (which is not shown in the diagram) and that the vectors further from the earth are smaller in magnitude.

Definition of a Vector Field

Now that you have seen some examples of vector fields we give a more formal definition.

> A **vector field** in 2-space is a function $\vec{F}(x, y)$ whose value at a point (x, y) is a 2-dimensional vector. Similarly, a vector field in 3-space is a function $\vec{F}(x, y, z)$ whose values are 3-dimensional vectors.

Notice the arrow over the function, $\vec{F}$, indicating that its value is a vector, not a scalar. We often represent the point (x, y) or (x, y, z) by its position vector $\vec{r}$ and write the vector field as $\vec{F}(\vec{r})$.

Visualizing a Vector Field Given by a Formula

Since a vector field is a function that assigns a vector to each point, a vector field can often be given by a formula.

Example 1 Sketch the vector field in 2-space given by $\vec{F}(x, y) = -y\vec{i} + x\vec{j}$.

Solution Table 17.1 shows the value of the vector field at a few points. Notice that each value is a vector. To plot the vector field, we plot $\vec{F}(x, y)$ with its tail at (x, y). (See Figure 17.17.)

Table 17.1 *Values of $\vec{F}(x, y) = -y\vec{i} + x\vec{j}$*

			y	
		-1	0	1
	-1	$\vec{i} - \vec{j}$	$-\vec{j}$	$-\vec{i} - \vec{j}$
x	0	$\vec{i}$	$\vec{0}$	$-\vec{i}$
	1	$\vec{i} + \vec{j}$	$\vec{j}$	$-\vec{i} + \vec{j}$

Now we look at the formula. The magnitude of the vector at (x, y) is $\|\vec{F}(x, y)\| = \| - y\vec{i} + x\vec{j}\| = \sqrt{x^2 + y^2}$, which is the distance from (x, y) to the origin. Therefore, all the vectors at a fixed distance from the origin (that is, on a circle centered at the origin) have the same magnitude. The magnitude gets larger as we move further from the origin. What about the direction? Figure 17.17 suggests that at each point (x, y) the vector $\vec{F}(x, y)$ is perpendicular to the position vector $\vec{r} = x\vec{i} + y\vec{j}$. We confirm this using the dot product: $\vec{r} \cdot \vec{F}(x, y) = (x\vec{i} + y\vec{j}) \cdot (-y\vec{i} + x\vec{j}) = 0$. Thus,

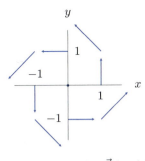

Figure 17.17: The value $\vec{F}(x, y)$ is placed at the point (x, y)

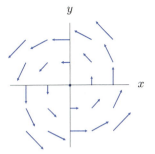

Figure 17.18: The vector field $\vec{F}(x, y) = -y\vec{i} + x\vec{j}$, vectors scaled smaller to fit in diagram

the vectors are tangent to circles centered at the origin and get longer as we go out. In Figure 17.18, the vectors have been scaled so that they do not obscure each other.

Example 2   Sketch the vector fields in 2-space given by (a) $\vec{F}(x, y) = x\vec{j}$ (b) $\vec{G}(x, y) = x\vec{i}$.

Solution   (a) The vector $x\vec{j}$ is parallel to the y-direction, pointing up when x is positive and down when x is negative. Also, the larger $|x|$ is, the longer the vector. The vectors in the field are constant along vertical lines since the vector field does not depend on y. (See Figure 17.19.)

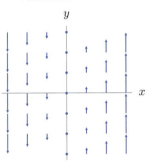

Figure 17.19: The vector field $\vec{F}(x, y) = x\vec{j}$

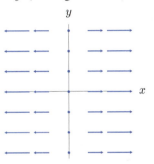

Figure 17.20: The vector field $\vec{F}(x, y) = x\vec{i}$

(b) This is similar to the previous example, except that the vector $x\vec{i}$ is parallel to the x-direction, pointing to the right when x is positive and to the left when x is negative. Again, the larger $|x|$ is the longer the vector, and the vectors are constant along vertical lines, since the vector field does not depend on y. (See Figure 17.20.)

Example 3   Describe the vector field in 3-space given by $\vec{F}(\vec{r}) = \vec{r}$, where $\vec{r} = x\vec{i} + y\vec{j} + z\vec{k}$.

Solution   The notation $\vec{F}(\vec{r}) = \vec{r}$ means that the value of $\vec{F}$ at the point (x, y, z) with position vector $\vec{r}$ is the vector $\vec{r}$ with its tail at (x, y, z). Thus, the vector field points outward. See Figure 17.21. Note that the lengths of the vectors have been scaled down so as to fit into the diagram.

Figure 17.21: The vector field $\vec{F}(\vec{r}) = \vec{r}$

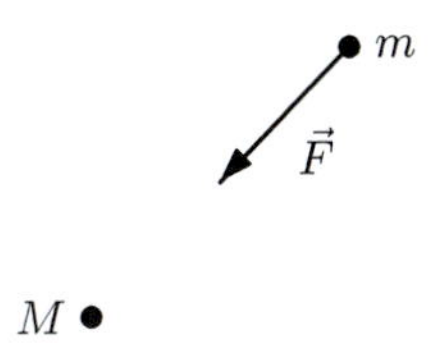

m

$\vec{F}$

M

Figure 17.22: Force exerted on mass m by mass M

Finding a Formula for a Vector Field

Example 4   Newton's Law of Gravitation states that the magnitude of the gravitational force exerted by an object of mass M on an object of mass m is proportional to M and m and inversely proportional to the square of the distance between them. The direction of the force is from m to M along the line connecting them. (See Figure 17.22.) Find a formula for the vector field $\vec{F}(\vec{r})$ that represents the gravitational force, assuming M is located at the origin and m is located at the point with position vector $\vec{r}$.

Solution   Since the mass m is located at $\vec{r}$, Newton's law says that the magnitude of the force is given by

$$\|\vec{F}(\vec{r})\| = \frac{GMm}{\|\vec{r}\|^2},$$

where G is the universal gravitational constant. A unit vector in the direction of the force is $-\vec{r}/\|\vec{r}\|$, where the negative sign indicates that the direction of force is towards the origin (gravity is attractive). By taking the product of the magnitude of the force and a unit vector in the direction of the force, we obtain an expression for the force vector field:

$$\vec{F}(\vec{r}) = \frac{GMm}{\|\vec{r}\|^2}\left(-\frac{\vec{r}}{\|\vec{r}\|}\right) = \frac{-GMm\vec{r}}{\|\vec{r}\|^3}.$$

We have already seen a picture of this vector field in Figure 17.16.

Gradient Vector Fields

The gradient of a scalar function f is a function that assigns a vector to each point, and is therefore a vector field. It is called the *gradient field* of f. Many vector fields in physics are gradient fields.

Example 5   Sketch the gradient field of the functions in Figures 17.23–17.25.

Figure 17.23: The contour map of $f(x, y) = x^2 + 2y^2$

Figure 17.24: The contour map of $g(x, y) = 5 - x^2 - 2y^2$

Figure 17.25: The contour map of $h(x, y) = x + 2y + 3$

Solution   See Figures 17.26–17.28. For a function $f(x, y)$, the gradient vector of f at a point is perpendicular to the contours in the direction of increasing f and its magnitude is the rate of change in that direction. The rate of change is large when the contours are close together and small when they are far apart. Notice that in Figure 17.26 the vectors all point outward, away from the local minimum of f, and in Figure 17.27 the vectors of grad g all point inward, toward the local maximum of g. Since h is a linear function, its gradient is constant, so grad h in Figure 17.28 is a constant vector field.

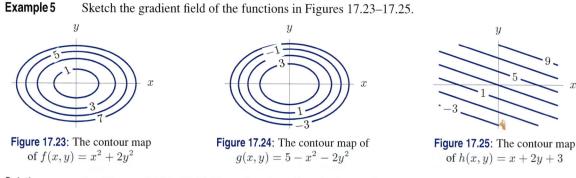

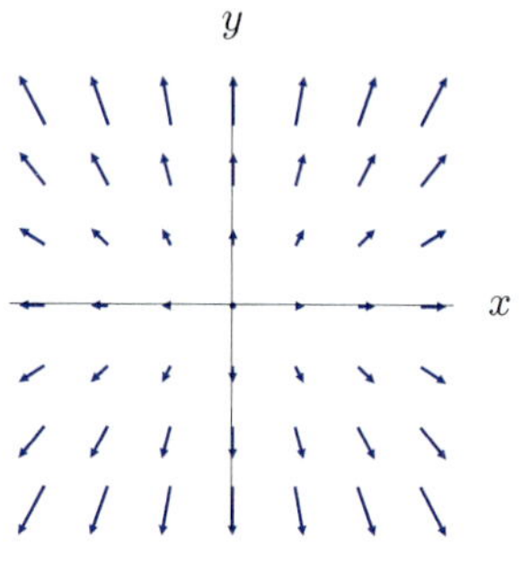

Figure 17.26: grad f

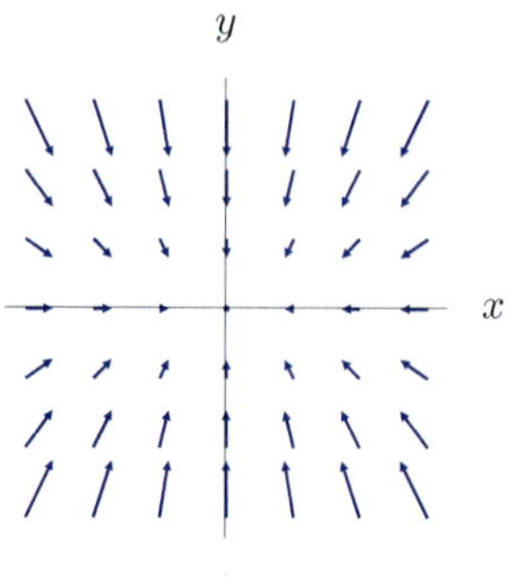

Figure 17.27: grad g

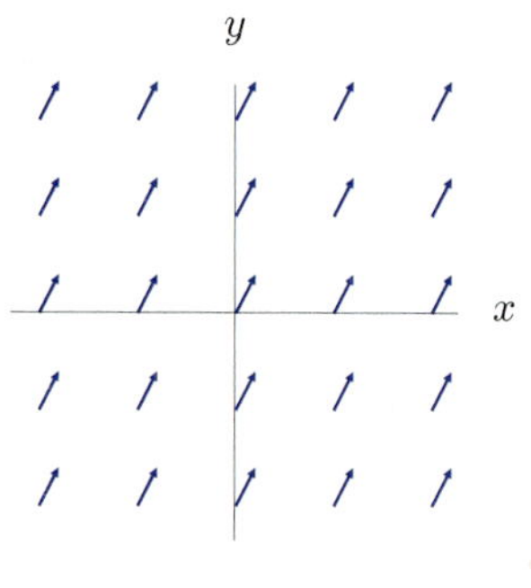

Figure 17.28: grad h

Exercises and Problems for Section 17.3

Exercises

1. Each vector field in Figures (I)–(IV) represents the force on a particle at different points in space as a result of another particle at the origin. Match up the vector fields with the descriptions below.

   (a) A repulsive force whose magnitude decreases as distance increases, such as between electric charges of the same sign.
   (b) A repulsive force whose magnitude increases as distance increases.
   (c) An attractive force whose magnitude decreases as distance increases, such as gravity.
   (d) An attractive force whose magnitude increases as distance increases.

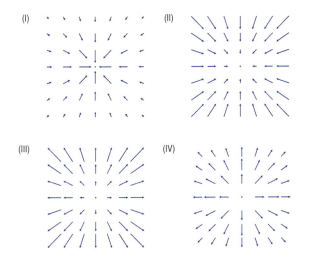

Sketch the vector fields in Exercises 2–10 in the xy-plane.

2. $\vec{F}(x, y) = 2\vec{i} + 3\vec{j}$ 3. $\vec{F}(x, y) = y\vec{i}$

4. $\vec{F}(x, y) = -y\vec{j}$ 5. $\vec{F}(x, y) = -y\vec{i} + x\vec{j}$

6. $\vec{F}(x, y) = 2x\vec{i} + x\vec{j}$ 7. $\vec{F}(\vec{r}) = 2\vec{r}$

8. $\vec{F}(\vec{r}) = \vec{r}/\|\vec{r}\|$ 9. $\vec{F}(\vec{r}) = -\vec{r}/\|\vec{r}\|^3$

10. $\vec{F}(x, y) = (x + y)\vec{i} + (x - y)\vec{j}$

For Exercises 11–16, find formulas for the vector fields. (There are many possible answers.)

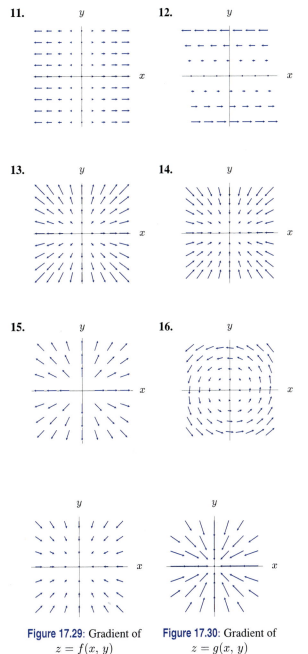

Problems

17. Figures 17.29 and 17.30 show the gradient of the functions $z = f(x, y)$ and $z = g(x, y)$.

    (a) For each function, draw a rough sketch of the level curves, showing possible z-values.
    (b) The xz-plane cuts each of the surfaces $z = f(x, y)$ and $z = g(x, y)$ in a curve. Sketch each of these curves, making clear how they are similar and how they are different from one another.

Figure 17.29: Gradient of $z = f(x, y)$

Figure 17.30: Gradient of $z = g(x, y)$

In Problems 18–20, write formulas for vector fields with the given properties.

18. All vectors are parallel to the x-axis; all vectors on a vertical line have the same magnitude.

19. All vectors point towards the origin and have constant length.

20. All vectors are of unit length and perpendicular to the position vector at that point.

21. In the middle of a wide, steadily flowing river there is a fountain that spouts water horizontally in all directions. The river flows in the $\vec{i}$-direction in the xy-plane and the fountain is at the origin.

    (a) If $A > 0, K > 0$, explain why the following expression could represent the velocity field for the combined flow of the river and the fountain:

$$\vec{v} = A\vec{i} + K(x^2 + y^2)^{-1}(x\vec{i} + y\vec{j}).$$

    (b) What is the significance of the constants A and K?
    (c) Using a computer, sketch the vector field $\vec{v}$ for $K = 1$ and $A = 1$ and $A = 2$, and for $A = 0.2, K = 2$.

Problems 22–23 concern the vector fields $\vec{F} = x\vec{i} + y\vec{j}$, $\vec{G} = -y\vec{i} + x\vec{j}$, and $\vec{H} = x\vec{i} - y\vec{j}$.

22. Match $\vec{F}, \vec{G}, \vec{H}$ with their sketches in (I)–(III).

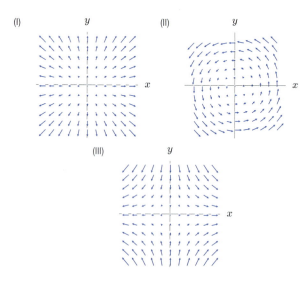

23. Match the vector fields with their sketches, (I)–(IV).

    (a) $\vec{F} + \vec{G}$ **(b)** $\vec{F} + \vec{H}$ **(c)** $\vec{G} + \vec{H}$ **(d)** $-\vec{F} + \vec{G}$

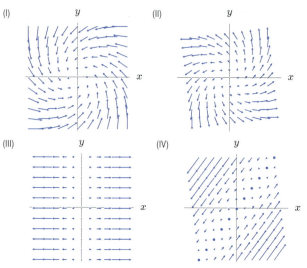

24. Let $\vec{F} = u\vec{i} + v\vec{j}$ be a vector field in 2-space with magnitude $F = \|\vec{F}\|$.

    (a) Let $\vec{T} = (1/F)\vec{F}$. Show that $\vec{T}$ is the unit vector in the direction of $\vec{F}$. See Figure 17.31.
    (b) Let $\vec{N} = (1/F)(\vec{k} \times \vec{F}) = (1/F)(-v\vec{i} + u\vec{j})$. Show that $\vec{N}$ is the unit vector pointing to the left of and at right angles to the direction of $\vec{F}$. See Figure 17.31.

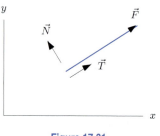

Figure 17.31

Problems 25–26 concern the vector fields $\vec{F} = x\vec{i} + y\vec{j}$ and $\vec{G} = -y\vec{i} + x\vec{j}$.

25. Sketch the vector field $\vec{L} = a\vec{F} + \vec{G}$ if:

    (a) $a = 0$ **(b)** $a > 0$ **(c)** $a < 0$

26. Sketch the vector field $\vec{L} = \vec{F} + b\vec{G}$ if:

    (a) $b = 0$ **(b)** $b > 0$ **(c)** $b < 0$

17.4 THE FLOW OF A VECTOR FIELD

When an iceberg is spotted in the North Atlantic, it is important to be able to predict where the iceberg is likely to be a day or a week later. To do this, one needs to know the velocity vector field of the ocean currents, that is, how fast and in what direction the water is moving at each point.

In this section we use differential equations to find the path of an object in a fluid flow. This path is called a *flow line*. Figure 17.32 shows several flow lines for the Gulf stream velocity vector field in Figure 17.15 on page 799. The arrows on each flow line indicate the direction of flow.

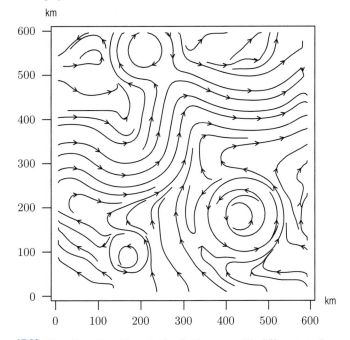

Figure 17.32: Flow lines for objects in the Gulf stream with different starting points

How Do We Find a Flow Line?

Suppose that $\vec{F}$ is the velocity vector field of water on the surface of a creek and imagine a seed being carried along by the current. We want to know the position vector $\vec{r}(t)$ of the seed at time t. We know

$$\text{Velocity of seed at time } t = \text{Velocity of current at seed's position at time } t$$

that is,

$$\vec{r}'(t) = \vec{F}(\vec{r}(t)).$$

We make the following definition:

A **flow line** of a vector field $\vec{v} = \vec{F}(\vec{r})$ is a path $\vec{r}(t)$ whose velocity vector equals $\vec{v}$. Thus

$$\vec{r}'(t) = \vec{v} = \vec{F}(\vec{r}(t)).$$

The **flow** of a vector field is the family of all of its flow lines.

A flow line is also called an *integral curve* or a *streamline*. We define flow lines for any vector field, as it turns out to be useful to study the flow of fields (for example, electric and magnetic) that are not velocity fields.

Resolving $\vec{F}$ and $\vec{r}$ into components, $\vec{F} = F_1\vec{i} + F_2\vec{j}$ and $\vec{r}(t) = x(t)\vec{i} + y(t)\vec{j}$, the definition of a flow line tells us that $x(t)$ and $y(t)$ satisfy the system of differential equations

$$x'(t) = F_1(x(t), y(t)) \quad \text{and} \quad y'(t) = F_2(x(t), y(t)).$$

Solving these differential equations gives a parameterization of the flow line.

Example 1   Find the flow line of the constant velocity field $\vec{v} = 3\vec{i} + 4\vec{j}$ cm/sec that passes through the point $(1, 2)$ at time $t = 0$.

Solution   Let $\vec{r}(t) = x(t)\vec{i} + y(t)\vec{j}$ be the position in cm of a particle at time t, where t is in seconds. We have

$$x'(t) = 3 \quad \text{and} \quad y'(t) = 4.$$

Thus,

$$x(t) = 3t + x_0 \quad \text{and} \quad y(t) = 4t + y_0.$$

Since the path passes the point $(1, 2)$ at $t = 0$, we have $x_0 = 1$ and $y_0 = 2$ and so

$$x(t) = 3t + 1 \quad \text{and} \quad y(t) = 4t + 2.$$

Thus, the path is the line given parametrically by

$$\vec{r}(t) = (3t + 1)\vec{i} + (4t + 2)\vec{j}.$$

(See Figure 17.33.) To find an explicit equation for the path, eliminate t between these expressions to get

$$\frac{x - 1}{3} = \frac{y - 2}{4} \qquad \text{or} \qquad y = \frac{4}{3}x + \frac{2}{3}.$$

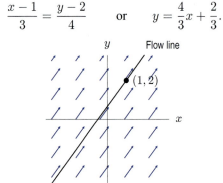

Figure 17.33: Vector field $\vec{F} = 3\vec{i} + 4\vec{j}$ with the flow line through $(1, 2)$

Example 2   The velocity of a flow at the point (x, y) is $\vec{F}(x, y) = \vec{i} + x\vec{j}$. Find the path of motion of an object in the flow that is at the point $(-2, 2)$ at time $t = 0$.

Solution   Figure 17.34 shows a sketch of this field. Since $\vec{r}'(t) = \vec{F}(\vec{r}(t))$, we are looking for the flow line that satisfies the system of differential equations

$$x'(t) = 1, \quad y'(t) = x(t).$$

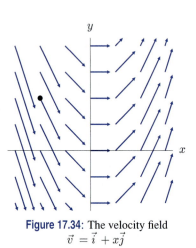

Figure 17.34: The velocity field $\vec{v} = \vec{i} + x\vec{j}$

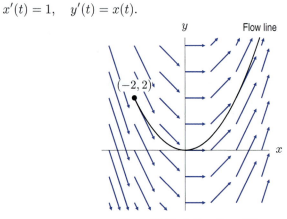

Figure 17.35: A flow line of the velocity field $\vec{v} = \vec{i} + x\vec{j}$

Solving for $x(t)$ first, we get $x(t) = t + x_0$, where x_0 is a constant of integration. Thus, $y'(t) = t + x_0$, so $y(t) = \frac{1}{2}t^2 + x_0 t + y_0$, where y_0 is also a constant of integration. Since $x(0) = x_0 = -2$ and $y(0) = y_0 = 2$, the path of motion is given by

$$x(t) = t - 2, \quad y(t) = \frac{1}{2}t^2 - 2t + 2,$$

or, equivalently,

$$\vec{r}(t) = (t - 2)\vec{i} + (\tfrac{1}{2}t^2 - 2t + 2)\vec{j}.$$

The graph of this flow line in Figure 17.35 looks like a parabola. We check this by seeing that an explicit equation for the path is $y = \frac{1}{2}x^2$.

Example 3   Determine the flow of the vector field $\vec{v} = -y\vec{i} + x\vec{j}$.

Solution   Figure 17.36 suggests that the flow consists of concentric counterclockwise circles, centered at the origin. The system of differential equations for the flow is

$$x'(t) = -y(t) \qquad y'(t) = x(t).$$

The equations $(x(t), y(t)) = (a\cos t, a\sin t)$ parameterize a family of counterclockwise circles of radius a, centered at the origin. We check that this family satisfies the system of differential equations:

$$x'(t) = -a\sin t = -y(t) \quad \text{and} \quad y'(t) = a\cos t = x(t).$$

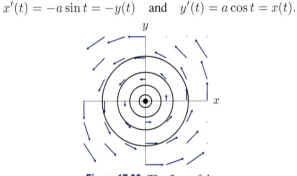

Figure 17.36: The flow of the vector field $\vec{v} = -y\vec{i} + x\vec{j}$

Approximating Flow Lines Numerically

Often it is not possible to find formulas for the flow lines of a vector field. However, we can approximate them numerically by Euler's method for solving differential equations. Since the flow lines $\vec{r}(t) = x(t)\vec{i} + y(t)\vec{j}$ of a vector field $\vec{v} = \vec{F}(x, y)$ satisfy the differential equation $\vec{r}'(t) = \vec{F}(\vec{r}(t))$, we have

$$\vec{r}(t + \Delta t) \approx \vec{r}(t) + (\Delta t)\vec{r}'(t)$$
$$= \vec{r}(t) + (\Delta t)\vec{F}(\vec{r}(t)) \quad \text{for } \Delta t \text{ near } 0.$$

To approximate a flow line, we start at a point $\vec{r}_0 = \vec{r}(0)$ and estimate the position $\vec{r}_1$ of a particle at time Δt later:

$$\vec{r}_1 = \vec{r}(\Delta t) \approx \vec{r}(0) + (\Delta t)\vec{F}(\vec{r}(0))$$
$$= \vec{r}_0 + (\Delta t)\vec{F}(\vec{r}_0).$$

We then repeat the same procedure starting at $\vec{r}_1$, and so on. The general formula for getting from one point to the next is

$$\vec{r}_{n+1} = \vec{r}_n + (\Delta t)\vec{F}(\vec{r}_n).$$

The points with position vectors $\vec{r}_0, \vec{r}_1, \ldots$ trace out the path, as shown in the next example.

Example 4   Use Euler's method to approximate the flow line through $(1, 2)$ for the vector field $\vec{v} = y^2\vec{i} + 2x^2\vec{j}$.

Solution   The flow is determined by the differential equations $\vec{r}\,'(t) = \vec{v}$, or equivalently

$$x'(t) = y^2, \qquad y'(t) = 2x^2.$$

We use Euler's method with $\Delta t = 0.02$, giving

$$\vec{r}_{n+1} = \vec{r}_n + 0.02\,\vec{v}\,(x_n, y_n)$$
$$= x_n\vec{i} + y_n\vec{j} + 0.02(y_n^2\vec{i} + 2x_n^2\vec{j}),$$

or equivalently,

$$x_{n+1} = x_n + 0.02y_n^2, \qquad y_{n+1} = y_n + 0.02 \cdot 2x_n^2.$$

When $t = 0$, we have $(x_0, y_0) = (1, 2)$. Then

$$x_1 = x_0 + 0.02 \cdot y_0^2 = 1 + 0.02 \cdot 2^2 = 1.08,$$
$$y_1 = y_0 + 0.02 \cdot 2x_0^2 = 2 + 0.02 \cdot 2 \cdot 1^2 = 2.04.$$

So after one step $x(0.02) \approx 1.08$ and $y(0.02) \approx 2.04$. Similarly, $x(0.04) = x(2\Delta t) \approx 1.16$, $y(0.04) = y(2\Delta t) \approx 2.08$ and so on. Further values along the flow line are given in Table 17.2 and plotted in Figure 17.37.

Table 17.2   *Approximated flow line starting at $(1, 2)$ for the vector field $\vec{v} = y^2\vec{i} + 2x^2\vec{j}$*

t	0	0.02	0.04	0.06	0.08	0.1	0.12	0.14	0.16	0.18
x	1	1.08	1.16	1.25	1.34	1.44	1.54	1.65	1.77	1.90
y	2	2.04	2.08	2.14	2.20	2.28	2.36	2.45	2.56	2.69

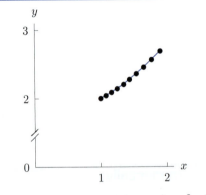

Figure 17.37: Euler's method solution to $x' = y^2$, $y' = 2x^2$

Exercises and Problems for Section 17.4

Exercises

In Exercises 1–3, sketch the vector field and its flow.

1. $\vec{v} = 3\vec{i}$ **2.** $\vec{v} = 2\vec{j}$ **3.** $\vec{v} = 3\vec{i} - 2\vec{j}$

In Exercises 4–7, sketch the vector field and the flow. Then find the system of differential equations associated with the vector field and check that the flow satisfies the system.

4. $\vec{v} = y\vec{i} + x\vec{j}$; $x(t) = a(e^t + e^{-t})$, $y(t) = a(e^t - e^{-t})$.

5. $\vec{v} = y\vec{i} - x\vec{j}$; $x(t) = a\sin t$, $y(t) = a\cos t$.

6. $\vec{v} = x\vec{i} + y\vec{j}$; $x(t) = ae^t$, $y(t) = be^t$.

7. $\vec{v} = x\vec{i} - y\vec{j}$; $x(t) = ae^t$, $y(t) = be^{-t}$.

8. Use a computer or calculator with Euler's method to approximate the flow line through $(1, 2)$ for the vector field $\vec{v} = y^2\vec{i} + 2x^2\vec{j}$ using 5 steps with a time interval $\Delta t = 0.1$.

9. Match the vector fields (a)–(f) with their flow lines (I)–(VI). Put arrows on the flow lines indicating the direction of flow.

(a) $y\vec{i} + x\vec{j}$ (b) $-y\vec{i} + x\vec{j}$

(c) $x\vec{i} + y\vec{j}$ (d) $-y\vec{i} + (x + y/10)\vec{j}$

(e) $-y\vec{i} + (x - y/10)\vec{j}$ (f) $(x - y)\vec{i} + (x - y)\vec{j}$

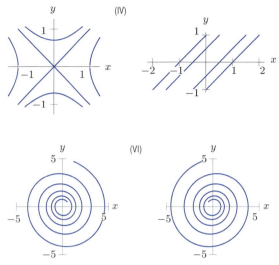

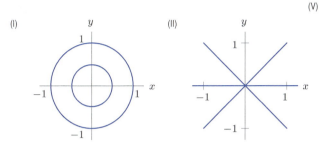

Problems

10. We have a fixed set of axes and a solid metal ball whose center is at the origin. The ball rotates once every 24 hours around the z-axis. The direction of rotation is counterclockwise when viewed from above. Consider the point (x, y, z) in our coordinate system lying within the ball. Let $\vec{v}(x, y, z)$ be the velocity vector of the particle of metal at this point. Assume x, y, z are in meters and time is in hours.

    (a) Find a formula for the vector field $\vec{v}$. Give units for your answer.

    (b) Describe in words the flow lines of $\vec{v}$.

11. $\vec{F}(x, y)$ and $\vec{G}(x, y) = 2\vec{F}(x, y)$ are two vector fields. Illustrating your answer with $\vec{F}(x, y) = -y\vec{i} + x\vec{j}$, describe the graphical difference between:

    (a) The vector fields   (b) Their flows

12. Show that the acceleration $\vec{a}$ of an object flowing in a velocity field $\vec{F}(x, y) = u(x, y)\vec{i} + v(x, y)\vec{j}$ is given by $\vec{a} = (u_x u + u_y v)\vec{i} + (v_x u + v_y v)\vec{j}$

17.5 PARAMETERIZED SURFACES

How Do We Parameterize a Surface?

In Section 17.1 we parameterized a circle in 2-space using the equations

$$x = \cos t, \quad y = \sin t.$$

In 3-space, the same circle in the xy-plane has parametric equations

$$x = \cos t, \quad y = \sin t, \quad z = 0.$$

We add the equation $z = 0$ to specify that the circle is in the xy-plane. If we wanted a circle in the plane $z = 3$, we would use the equations

$$x = \cos t, \quad y = \sin t, \quad z = 3.$$

Suppose now we let z vary freely, as well as t. We get circles in every horizontal plane, forming the cylinder in Figure 17.38. Thus, we need two parameters, t and z, to parameterize the cylinder.

This is true in general. A curve, though it may live in two or three dimensions, is itself one-dimensional; if we move along it we can only move backwards and forwards in one direction. Thus, it only requires one parameter to trace out a curve.

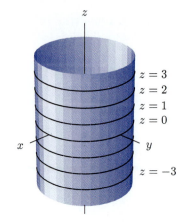

Figure 17.38: The cylinder $x = \cos t,\ y = \sin t,\ z = z$

A surface is 2-dimensional; at any given point there are two independent directions we can move in. For example, on the cylinder we can move vertically, or we can circle around the z-axis horizontally. So we need *two* parameters to describe it. We can think of the parameters as map coordinates, like longitude and latitude on the surface of the earth.

In the case of the cylinder our parameters are t and z, so

$$x = \cos t, \quad y = \sin t, \quad z = z, \qquad 0 \le t < 2\pi, \quad -\infty < z < \infty.$$

The last equation, $z = z$, looks strange, but it reminds us that we are in three dimensions, not two, and that the z-coordinate on our surface is allowed to vary freely.

In general, we express the coordinates, (x, y, z) of a point on a surface S in terms of two parameters, s and t:

$$x = f_1(s, t), \quad y = f_2(s, t), \quad z = f_3(s, t).$$

As the values of s and t vary, the corresponding point (x, y, z) sweeps out the surface, S. (See Figure 17.39.) The function which sends the point (s, t) to the point (x, y, z) is called the *parameterization of the surface.*

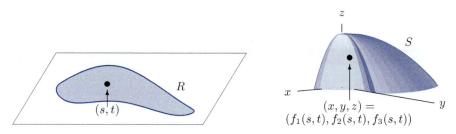

Figure 17.39: The parameterization sends each point (s, t) in the parameter region, R, to a point $(x, y, z) = (f_1(s, t), f_2(s, t), f_3(s, t))$ in the surface, S

Using Position Vectors

We can use the position vector $\vec{r} = x\vec{i} + y\vec{j} + z\vec{k}$ to combine the three parametric equations for a surface into a single vector equation. For example, the parameterization of the cylinder $x = \cos t, y = \sin t, z = z$ can be written as

$$\vec{r}(t, z) = \cos t\,\vec{i} + \sin t\,\vec{j} + z\vec{k} \qquad 0 \le t < 2\pi, \quad -\infty < z < \infty.$$

For a general parameterized surface S, we write

$$\vec{r}(s, t) = f_1(s, t)\vec{i} + f_2(s, t)\vec{j} + f_3(s, t)\vec{k}.$$

Parameterizing a Surface of the Form $z = f(x, y)$

The graph of a function $z = f(x, y)$ can be given parametrically simply by letting the parameters s and t be x and y:

$$x = s, \quad y = t, \quad z = f(s, t).$$

Example 1

Give a parametric description of the lower hemisphere of the sphere $x^2 + y^2 + z^2 = 1$.

Solution

The surface is the graph of the function $z = -\sqrt{1 - x^2 - y^2}$ over the region $x^2 + y^2 \leq 1$ in the plane. Then parametric equations are $x = s$, $y = t$, $z = -\sqrt{1 - s^2 - t^2}$, where the parameters s and t vary inside the unit circle.

In practice we often think of x and y as parameters rather than introduce new variables s and t. Thus, we may write $x = x$, $y = y$, $z = f(x, y)$.

Parameterizing Planes

Consider a plane containing two nonparallel vectors $\vec{v}_1$ and $\vec{v}_2$ and a point P_0 with position vector $\vec{r}_0$. We can get to any point on the plane by starting at P_0 and moving parallel to $\vec{v}_1$ or $\vec{v}_2$, adding multiples of them to $\vec{r}_0$. (See Figure 17.40.)

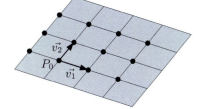

Figure 17.40: The plane $\vec{r}(s, t) = \vec{r}_0 + s\vec{v}_1 + t\vec{v}_2$ and some points corresponding to various choices of s and t

Since $s\vec{v}_1$ is parallel to $\vec{v}_1$ and $t\vec{v}_2$ is parallel to $\vec{v}_2$, we have the following result:

Parametric Equations for a Plane

The plane through the point with position vector $\vec{r}_0$ and containing the two nonparallel vectors $\vec{v}_1$ and $\vec{v}_2$ has parametric equation

$$\vec{r}(s, t) = \vec{r}_0 + s\vec{v}_1 + t\vec{v}_2.$$

If $\vec{r}_0 = x_0\vec{i} + y_0\vec{j} + z_0\vec{k}$, and $\vec{v}_1 = a_1\vec{i} + a_2\vec{j} + a_3\vec{k}$, and $\vec{v}_2 = b_1\vec{i} + b_2\vec{j} + b_3\vec{k}$, then the parametric equations of the plane can be written in the form

$$x = x_0 + sa_1 + tb_1, \quad y = y_0 + sa_2 + tb_2, \quad z = z_0 + sa_3 + tb_3.$$

Notice that the parameterization of the plane expresses the coordinates $x, y,$ and z as linear functions of the parameters s and t.

Example 2

Write a parametric equation for the plane through the point $(2, -1, 3)$ and containing the vectors $\vec{v}_1 = 2\vec{i} + 3\vec{j} - \vec{k}$ and $\vec{v}_2 = \vec{i} - 4\vec{j} + 5\vec{k}$.

Solution

The parametric equation is

$$\vec{r}(s, t) = \vec{r}_0 + s\vec{v}_1 + t\vec{v}_2 = 2\vec{i} - \vec{j} + 3\vec{k} + s(2\vec{i} + 3\vec{j} - \vec{k}) + t(\vec{i} - 4\vec{j} + 5\vec{k})$$
$$= (2 + 2s + t)\vec{i} + (-1 + 3s - 4t)\vec{j} + (3 - s + 5t)\vec{k},$$

or equivalently,

$$x = 2 + 2s + t, \quad y = -1 + 3s - 4t, \quad z = 3 - s + 5t.$$

Parameterizations Using Spherical Coordinates

Recall the spherical coordinates ρ, ϕ, and θ introduced on page 763 of Chapter 16. On a sphere of radius $\rho = a$ we can use ϕ and θ as coordinates, similar to latitude and longitude on the surface of the earth. (See Figure 17.41.) The latitude, however, is measured from the equator, whereas ϕ is measured from the north pole. If the positive x-axis passes through the Greenwich meridian, the longitude and θ are equal for $0 \leq \theta \leq \pi$.

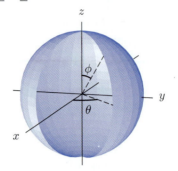

Figure 17.41: Parameterizing the sphere by ϕ and θ

Example 3   You are at a point on a sphere with $\phi = 3\pi/4$. Are you in the northern or southern hemisphere? If ϕ decreases, do you move closer to or farther from the equator?

Solution   The equator has $\phi = \pi/2$. Since $3\pi/4 > \pi/2$, you are in the southern hemisphere. If ϕ decreases, you move closer to the equator.

Example 4   On a sphere, you are standing at a point with coordinates θ_0 and ϕ_0. Your *antipodal* point is the point on the other side of the sphere on a line through you and the center. What are the θ, ϕ coordinates of your antipodal point?

Solution   Figure 17.42 shows that the coordinates are $\theta = \pi + \theta_0$ if $\theta_0 < \pi$ or $\theta = \theta_0 - \pi$ if $\pi \leq \theta_0 \leq 2\pi$, and $\phi = \pi - \phi_0$. Notice that if you are on the equator, then so is your antipodal point.

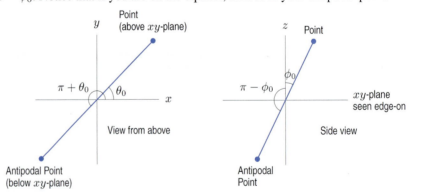

Figure 17.42: Two views of the xyz-coordinate system showing coordinates of antipodal points

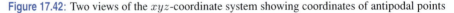

Parameterizing a Sphere Using Spherical Coordinates

The sphere with radius 1 centered at the origin is parameterized by

$$x = \sin \phi \cos \theta, \qquad y = \sin \phi \sin \theta, \qquad z = \cos \phi.$$

Where $0 \leq \theta \leq 2\pi$ and $0 \leq \phi \leq \pi$. (See Figure 17.43.)

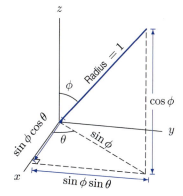

Figure 17.43: The relationship between x, y, z and ϕ, θ on a sphere of radius 1

We can also write these equations in vector form:

$$\vec{r}(\theta, \phi) = \sin\phi\cos\theta\,\vec{i} + \sin\phi\sin\theta\,\vec{j} + \cos\phi\,\vec{k}.$$

Since $x^2 + y^2 + z^2 = \sin^2\phi(\cos^2\theta + \sin^2\theta) + \cos^2\phi = \sin^2\phi + \cos^2\phi = 1$ this verifies that the point with position vector $\vec{r}(\theta, \phi)$ does lie on the sphere of radius 1. Notice that the z-coordinate depends only on the parameter ϕ. Geometrically, this means that all points on the same latitude have the same z-coordinate.

Example 5   Find parametric equations for the following spheres:
(a) Center at the origin and radius 2.
(b) Center at the point $(2, -1, 3)$ and radius 2.

Solution   (a) We must scale the distance from the origin by 2. Thus, we have

$$x = 2\sin\phi\cos\theta, \qquad y = 2\sin\phi\sin\theta, \qquad z = 2\cos\phi,$$

where $0 \le \theta \le 2\pi$ and $0 \le \phi \le \pi$. In vector form, this is written

$$\vec{r}(\theta, \phi) = 2\sin\phi\cos\theta\,\vec{i} + 2\sin\phi\sin\theta\,\vec{j} + 2\cos\phi\,\vec{k}.$$

(b) To shift the center of the sphere from the origin to the point $(2, -1, 3)$, we add the vector parameterization we found in part (a) to the position vector of $(2, -1, 3)$. (See Figure 17.44.) This gives

$$\begin{aligned}
\vec{r}(\theta, \phi) &= 2\vec{i} - \vec{j} + 3\vec{k} + \left(2\sin\phi\cos\theta\,\vec{i} + 2\sin\phi\sin\theta\,\vec{j} + 2\cos\phi\,\vec{k}\right) \\
&= (2 + 2\sin\phi\cos\theta)\vec{i} + (-1 + 2\sin\phi\sin\theta)\vec{j} + (3 + 2\cos\phi)\vec{k},
\end{aligned}$$

where $0 \le \theta \le 2\pi$ and $0 \le \phi \le \pi$. Alternatively

$$x = 2 + 2\sin\phi\cos\theta, \qquad y = -1 + 2\sin\phi\sin\theta, \qquad z = 3 + 2\cos\phi.$$

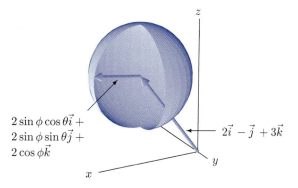

Figure 17.44: Sphere with center at the point $(2, -1, 3)$ and radius 2

Note that the same point can have more than one value for θ or ϕ. For example, points with $\theta = 0$ also have $\theta = 2\pi$, unless we restrict θ to the range $0 \le \theta < 2\pi$. Also, the north pole, at $\phi = 0$, and the south pole, at $\phi = \pi$, can have any value of θ.

Parameterizing Surfaces of Revolution

Many surfaces have an axis of rotational symmetry and circular cross-sections perpendicular to that axis. These surfaces are referred to as *surfaces of revolution*.

Example 6   Find a parameterization of the cone whose base is the circle $x^2 + y^2 = a^2$ in the xy-plane and whose vertex is at a height h above the xy-plane. (See Figure 17.45.)

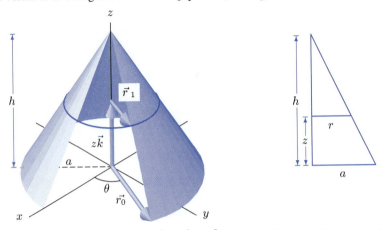

Figure 17.45: The cone whose base is the circle $x^2 + y^2 = a^2$ in the xy-plane and whose vertex is at the point $(0, 0, h)$ and the vertical cross-section through the cone

Solution   We use cylindrical coordinates, r, θ, z. (See Figure 17.45.) In the xy-plane, the radius vector, $\vec{r}_0$, from the z-axis to a point on the cone in the xy-plane is

$$\vec{r}_0 = a \cos \theta \vec{i} + a \sin \theta \vec{j}.$$

Above the xy-plane, the radius of the circular cross section, r, decreases linearly from $r = a$ when $z = 0$ to $r = 0$ when $z = h$. From the similar triangles in Figure 17.45,

$$\frac{a}{h} = \frac{r}{h - z}.$$

Solving for r, we have

$$r = \left(1 - \frac{z}{h}\right) a.$$

The horizontal radius vector, $\vec{r}_1$, at height z has components similar to $\vec{r}_0$, but with a replaced by r:

$$\vec{r}_1 = r \cos \theta \vec{i} + r \sin \theta \vec{j} = \left(1 - \frac{z}{h}\right) a \cos \theta \vec{i} + \left(1 - \frac{z}{h}\right) a \sin \theta \vec{j}.$$

As θ goes from 0 to 2π, the vector $\vec{r}_1$ traces out the horizontal circle in Figure 17.45. We get the position vector, $\vec{r}$, of a point on the cone by adding the vector $z\vec{k}$, so

$$\vec{r} = \vec{r}_1 + z\vec{k} = a\left(1 - \frac{z}{h}\right)\cos\theta\vec{i} + a\left(1 - \frac{z}{h}\right)\sin\theta\vec{j} + z\vec{k}, \quad \text{for } 0 \le z \le h \text{ and } 0 \le \theta \le 2\pi.$$

These equations can be written as

$$x = \left(1 - \frac{z}{h}\right) a \cos\theta, \quad y = \left(1 - \frac{z}{h}\right) a \sin\theta, \quad z = z.$$

The parameters are θ and z.

Example 7   Consider the bell of a trumpet. A model for the radius $z = f(x)$ of the horn (in cm) at a distance x cm from the large open end is given by the function

$$f(x) = \frac{6}{(x+1)^{0.7}}.$$

The bell is obtained by rotating the graph of f about the x-axis. Find a parameterization for the first 24 cm of the bell. (See Figure 17.46.)

Figure 17.46: The bell of a trumpet obtained by rotating the graph of $z = f(x)$ about the x-axis

Solution   At distance x from the large open end of the horn, the cross section parallel to the yz-plane is a circle of radius $f(x)$, with center on the x-axis. Such a circle can be parameterized by $y = f(x)\cos\theta$, $z = f(x)\sin\theta$. Thus we have the parameterization

$$x = x, \quad y = \left(\frac{6}{(x+1)^{0.7}}\right)\cos\theta, \quad z = \left(\frac{6}{(x+1)^{0.7}}\right)\sin\theta, \quad 0 \le x \le 24, \quad 0 \le \theta \le 2\pi.$$

The parameters are x and θ.

Parameter Curves

On a parameterized surface, the curve obtained by setting one of the parameters equal to a constant and letting the other vary is called a *parameter curve*. If the surface is parameterized by

$$\vec{r}(s, t) = f_1(s, t)\vec{i} + f_2(s, t)\vec{j} + f_3(s, t)\vec{k},$$

there are two families of parameter curves on the surface, one family with t constant and the other with s constant.

Example 8   Consider the vertical cylinder

$$x = \cos t, \quad y = \sin t, \quad z = z.$$

(a) Describe the two parameter curves through the point $(0, 1, 1)$.
(b) Describe the family of parameter curves with t constant and the family with z constant.

Solution   (a) Since the point $(0, 1, 1)$ corresponds to the parameter values $t = \pi/2$ and $z = 1$, there are two parameter curves, one with $t = \pi/2$ and the other with $z = 1$. The parameter curve with $t = \pi/2$ has the parametric equations

$$x = \cos\left(\frac{\pi}{2}\right) = 0, \quad y = \sin\left(\frac{\pi}{2}\right) = 1, \quad z = z,$$

with parameter z. This is a line through the point $(0, 1, 1)$ parallel to the z-axis.

The parameter curve with $z = 1$ has the parametric equations

$$x = \cos t, \quad y = \sin t, \quad z = 1,$$

with parameter t. This is a unit circle parallel to and one unit above the xy-plane centered on the z-axis.

(b) First, fix $t = t_0$ for t and let z vary. The curves parameterized by z have equation

$$x = \cos t_0, \quad y = \sin t_0, \quad z = z.$$

These are vertical lines on the cylinder parallel to the z-axis. (See Figure 17.47.)

The other family is obtained by fixing $z = z_0$ and varying t. Curves in this family are parameterized by t and have equation

$$x = \cos t, \quad y = \sin t, \quad z = z_0.$$

They are circles of radius 1 parallel to the xy-plane centered on the z-axis. (See Figure 17.48.)

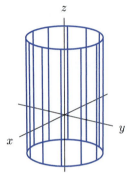

Figure 17.47: The family of parameter curves with $t = t_0$ for the cylinder $x = \cos t, y = \sin t, z = z$

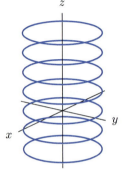

Figure 17.48: The family of parameter curves with $z = z_0$ for the cylinder $x = \cos t, y = \sin t, z = z$

Example 9   Describe the families of parameter curves with $\theta = \theta_0$ and $\phi = \phi_0$ for the sphere

$$x = \sin \phi \cos \theta, \qquad y = \sin \phi \sin \theta, \qquad z = \cos \phi,$$

where $0 \le \theta \le 2\pi, 0 \le \phi \le \pi$.

Solution   Since ϕ measures latitude, the family with ϕ constant consists of the circles of constant latitude. (See Figure 17.49.) Similarly, the family with θ constant consists of the meridians (semicircles) running between the north and south poles. (See Figure 17.50.)

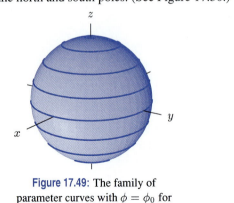

Figure 17.49: The family of parameter curves with $\phi = \phi_0$ for the sphere parameterized by (θ, ϕ)

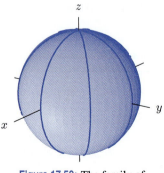

Figure 17.50: The family of parameter curves with $\theta = \theta_0$ for the sphere parameterized by (θ, ϕ)

We have seen parameter curves before in pages 568-570 of Section 12.2: The cross-sections with $x = a$ and $y = b$ on a surface $z = f(x, y)$ are examples of parameter curves. So are the grid lines on a computer sketch of a surface. The small regions shaped like parallelograms surrounded by nearby pairs of parameter curves are called *parameter rectangles*. See Figure 17.51.

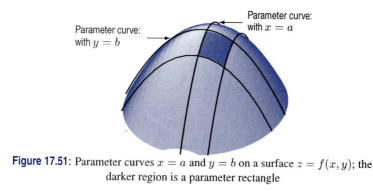

Figure 17.51: Parameter curves $x = a$ and $y = b$ on a surface $z = f(x, y)$; the darker region is a parameter rectangle

Exercises and Problems for Section 17.5

Exercises

Describe in words the objects parameterized by the equations in Exercises 1–8. (Note: r and θ are cylindrical coordinates.)

1. $x = r \cos \theta \qquad y = r \sin \theta \qquad z = 7$
$\quad 0 \le r \le 5 \qquad 0 \le \theta \le 2\pi$

2. $x = 5 \cos \theta \qquad y = 5 \sin \theta \qquad z = 7$
$\quad 0 \le \theta \le 2\pi$

3. $x = 5 \cos \theta \qquad y = 5 \sin \theta \qquad z = z$
$\quad 0 \le \theta \le 2\pi \qquad 0 \le z \le 7$

4. $x = 5 \cos \theta \qquad y = 5 \sin \theta \qquad z = 5\theta$
$\quad 0 \le \theta \le 2\pi$

5. $x = r \cos \theta \qquad y = r \sin \theta \qquad z = r$
$\quad 0 \le r \le 5 \qquad 0 \le \theta \le 2\pi$

6. $x = 2z \cos \theta \qquad y = 2z \sin \theta \qquad z = z$
$\quad 0 \le z \le 7 \qquad 0 \le \theta \le 2\pi$

7. $x = 3 \cos \theta \qquad y = 2 \sin \theta \qquad z = z$
$\quad 0 \le \theta \le 2\pi \qquad 0 \le z \le 7$

8. $x = x \qquad y = x^2 \qquad z = z$
$\quad -5 \le x \le 5 \qquad 0 \le z \le 7$

In Exercises 9–12, for a sphere parameterized using the spherical coordinates θ and ϕ, describe in words the part of the sphere given by the restrictions.

9. $0 \le \theta < 2\pi, \quad 0 \le \phi \le \pi/2$

10. $\pi \le \theta < 2\pi, \quad 0 \le \phi \le \pi$

11. $\pi/4 \le \theta < \pi/3, \quad 0 \le \phi \le \pi$

12. $0 \le \theta \le \pi, \quad \pi/4 \le \phi < \pi/3$

Problems

13. Find a parameterization of a circular cylinder of radius a whose axis is along the z-axis, from $z = 0$ to a height $z = h$. See Figure 17.52.

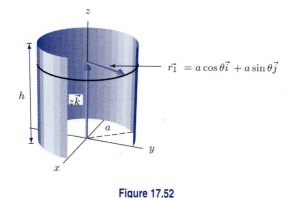

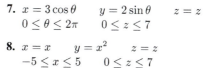

Figure 17.52

14. Find a parameterization of the circle of radius 2 centered at $(0, 1, 0)$ lying in the plane $x + z = 0$.

15. A city is described parametrically by the equation

$$\vec{r} = (x_0 \vec{i} + y_0 \vec{j} + z_0 \vec{k}) + s\vec{v_1} + t\vec{v_2}$$

where $\vec{v}_1 = 2\vec{i} - 3\vec{j} + 2\vec{k}$ and $\vec{v}_2 = \vec{i} + 4\vec{j} + 5\vec{k}$. A city block is a rectangle determined by $\vec{v}_1$ and $\vec{v}_2$. East is in the direction of $\vec{v}_1$ and north is in the direction of $\vec{v}_2$. Starting at the point (x_0, y_0, z_0), you walk 5 blocks east, 4 blocks north, 1 block west and 2 blocks south. What are the parameters of the point where you end up? What are your x, y and z coordinates at that point?

16. Find a parameterization for the plane through $(1, 3, 4)$ and orthogonal to $\vec{n} = 2\vec{i} + \vec{j} - \vec{k}$.

17. Does the plane $\vec{r}\,(s,t) = (2+s)\vec{i} + (3+s+t)\vec{j} + 4t\vec{k}$ contain the following points?
(a) $(4, 8, 12)$ (b) $(1, 2, 3)$

18. Are the following two planes parallel?

$$x = 2 + s + t, \quad y = 4 + s - t, \quad z = 1 + 2s, \quad \text{and}$$

$$x = 2 + s + 2t, \quad y = t, \quad z = s - t.$$

19. You are at a point on the earth with longitude $80°$ West of Greenwich, England, and latitude $40°$ North of the equator.

(a) If your latitude decreases have you moved nearer to or farther from the equator?
(b) If your latitude decreases, have you moved nearer to or farther from the north pole?
(c) If your longitude increases (say, to $90°$ West), have you moved nearer to or farther from Greenwich?

20. Describe in words the curve $\phi = \pi/4$ on the surface of the globe.

21. Describe in words the curve $\theta = \pi/4$ on the surface of the globe.

22. Find parametric equations for the sphere centered at the origin and with radius 5.

23. Find parametric equations for the sphere centered at the point $(2, -1, 3)$ and with radius 5.

24. Find parametric equations for the sphere $(x - a)^2 + (y - b)^2 + (z - c)^2 = d^2$.

25. Adapt the parameterization for the sphere to find a parameterization for the ellipsoid

$$\frac{x^2}{a^2} + \frac{y^2}{b^2} + \frac{z^2}{c^2} = 1.$$

26. Suppose you are standing at a point on the equator of a sphere, parameterized by spherical coordinates θ_0, and ϕ_0. If you go halfway around the equator and halfway up toward the north pole along a longitude, what are your new θ and ϕ coordinates?

27. Find parametric equations for the cone $x^2 + y^2 = z^2$.

28. Parameterize the cone in Example 6 on page 814 in terms of r and θ.

29. Parameterize a cone of height h and maximum radius a with vertex at the origin and opening upward. Do this in two ways, giving the range of values for each parameter in each case: (a) Use r and θ.   (b) Use z and θ.

30. Parameterize the paraboloid $z = x^2 + y^2$ using cylindrical coordinates.

31. Parameterize a vase formed by rotating the curve $z = 10\sqrt{x - 1}$, $1 \le x \le 2$, around the z-axis. Sketch the vase.

For Problems 32–35

(a) Write an equation in x, y, z and identify the parametric surface.
(b) Draw a picture of the surface.

32. $x = 2s \qquad y = s + t \qquad z = 1 + s - t$
$\quad\ 0 \le s \le 1 \qquad 0 \le t \le 1$

33. $x = s + t \qquad y = s - t \qquad z = s^2 + t^2$
$\quad\ 0 \le s \le 1 \qquad 0 \le t \le 1$

34. $x = 3\sin s \qquad y = 3\cos s \qquad z = t + 1$
$\quad\ 0 \le s \le \pi \qquad 0 \le t \le 1$

35. $x = s \qquad y = t \qquad z = \sqrt{1 - s^2 - t^2}$
$\quad\ s^2 + t^2 \le 1 \qquad s, t \ge 0$

36. (a) Describe the surface given parametrically by the equations

$$x = \cos(s - t), \quad y = \sin(s - t), \quad z = s + t.$$

(b) Describe the two families of parameter curves on the surface.

37. Give a parameterization of the circle of radius a centered at the point (x_0, y_0, z_0) and in the plane parallel to two given unit vectors $\vec{u}$ and $\vec{v}$ such that $\vec{u} \cdot \vec{v} = 0$.

38. A torus (doughnut) is constructed by rotating a small circle of radius a in a large circle of radius b about the origin. The small circle is in a (rotating) vertical plane through the origin and the large circle is in the xy-plane. (See Figure 17.53.) Parameterize the torus as follows.

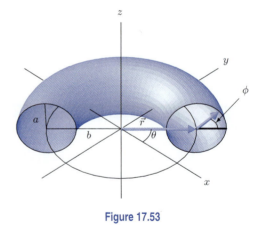

Figure 17.53

(a) Parameterize the large circle.
(b) For a typical point on the large circle, find two unit vectors which are perpendicular to one another and in the plane of the small circle at that point. Use these vectors to parameterize the small circle relative to its center.
(c) Combine your answers to parts (b) and (c) to parameterize the torus.

CHAPTER SUMMARY

- **Parameterized Curves**

    Parameterizations representing motion in 2- and 3-space, change of parameter, vector form of parametric equations, parametric equation of a line.

- **Velocity and Acceleration Vectors**

    Computing velocity and acceleration, uniform circular motion, the length of a parametric curve.

- **Vector Fields**

    Definition of vector field, visualizing fields, gradient fields.

- **Flow Lines**

    Parametric equations of flow lines, approximating flow lines numerically.

- **Parameterized Surfaces** Parametric equations for planes, graphs of functions, spheres, and cylinders. Parameter curves.

REVIEW EXERCISES AND PROBLEMS FOR CHAPTER SEVENTEEN

Exercises

Write a parameterization for the curves in Exercises 1–10.

1. The horizontal line through the point $(0, 5)$.

2. The circle of radius 2 centered at the origin starting at the point $(0, 2)$ when $t = 0$.

3. The circle of radius 4 centered at the point $(4, 4)$ starting on the x-axis when $t = 0$.

4. The circle of radius 1 in the xy-plane centered at the origin, traversed counterclockwise when viewed from above.

5. The line through the points $(2, -1, 4)$ and $(1, 2, 5)$.

6. The line through the point $(1, 3, 2)$ perpendicular to the xz-plane.

7. The line through the point $(1, 1, 1)$ perpendicular to the plane $2x - 3y + 5z = 4$.

8. The circle of radius 2 parallel to the xy-plane, centered at the point $(0, 0, 1)$, and traversed counterclockwise when viewed from below.

9. The circle of radius 3 parallel to the xz-plane, centered at the point $(0, 5, 0)$, and traversed counterclockwise when viewed from $(0, 10, 0)$.

10. The line of intersection of the planes $z = 4 + 2x + 5y$ and $z = 3 + x + 3y$.

11. Are the lines $x = 3 + 2t$, $y = 5 - t$, $z = 7 + 3t$ and $x = 3 + t$, $y = 5 + 2t$, $z = 7 + 2t$ parallel?

12. (a) What is meant by a vector field?
    (b) Suppose $\vec{a} = a_1\vec{i} + a_2\vec{j} + a_3\vec{k}$ is a constant vector. Which of the following are vector fields? Explain.

    (i) $\vec{r} + \vec{a}$ (ii) $\vec{r} \cdot \vec{a}$
    (iii) $x^2\vec{i} + y^2\vec{j} + z^2\vec{k}$ (iv) $x^2 + y^2 + z^2$

Sketch the vector fields in Exercises 13–15.

13. $\vec{F} = \left(\dfrac{y}{\sqrt{x^2 + y^2}}\right)\vec{i} - \left(\dfrac{x}{\sqrt{x^2 + y^2}}\right)\vec{j}$

14. $\vec{F} = \left(\dfrac{y}{x^2 + y^2}\right)\vec{i} - \left(\dfrac{x}{x^2 + y^2}\right)\vec{j}$

15. $\vec{F} = y\vec{i} - x\vec{j}$

Problems

16. A particle travels along a line, with position at time t given by $\vec{r}(t) = (2 + 5t)\vec{i} + (3 + t)\vec{j} + 2t\vec{k}$.

    (a) Where is the particle when $t = 0$?
    (b) At what time does the particle reach the point $(12, 5, 4)$?
    (c) Does the particle ever reach the point $(12, 4, 4)$? Explain.

17. Consider the parametric equations for $0 \le t \le \pi$:

    (I) $\vec{r} = \cos(2t)\vec{i} + \sin(2t)\vec{j}$
    (II) $\vec{r} = 2\cos t\vec{i} + 2\sin t\vec{j}$
    (III) $\vec{r} = \cos(t/2)\vec{i} + \sin(t/2)\vec{j}$
    (IV) $\vec{r} = 2\cos t\vec{i} - 2\sin t\vec{j}$

    (a) Match the equations above with four of the curves C_1, C_2, C_3, C_4, C_5 and C_6 in Figure 17.54. (Each curve is part of a circle.)

    (b) Give parametric equations for the curves which have not been matched, again assuming $0 \le t \le \pi$.

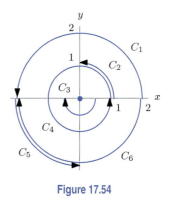

Figure 17.54

18. On a calculator or a computer, plot $x = 2t/(t^2 + 1)$, $y = (t^2 - 1)/(t^2 + 1)$, first for $-50 \le t \le 50$ then for $-5 \le t \le 5$. Explain what you see. Is the curve really a circle?

19. Let $f(x, y) = \dfrac{x^2 - y^2}{x^2 + y^2}$.

    (a) In which direction should you move from the point $(1, 1)$ to obtain the maximum rate of increase of f?
    (b) Find a direction in which the directional derivative at the point $(1, 1)$ is equal to zero.
    (c) Suppose you move along the curve $x = e^{2t}, y = 2t^3 + 6t + 1$. What is df/dt at $t = 0$?

20. Find parametric equations of the line passing through the points $(1, 2, 3)$, $(3, 5, 7)$ and calculate the shortest distance from the line to the origin.

21. An ant, starting at the origin, moves at 2 units/sec along the x-axis to the point $(1, 0)$. The ant then moves counterclockwise along the unit circle to $(0, 1)$ at a speed of $3\pi/2$ units/sec, then straight down to the origin at a speed of 2 units/sec along the y-axis.

    (a) Express the ant's coordinates as a function of time, t, in secs.
    (b) Express the reverse path as a function of time.

22. The motion of a particle is given by $\vec{r}(t) = R\cos(\omega t)\vec{i} + R\sin(\omega t)\vec{j}$, with $R > 0, \omega > 0$.

    (a) Show that the particle moves on a circle and find the radius, direction, and period.
    (b) Determine the velocity vector of the particle and its direction and speed.
    (c) What are the direction and magnitude of the acceleration vector of the particle?

23. A wheel of radius 1 meter rests on the x-axis with its center on the y-axis. There is a spot on the rim at the point $(1, 1)$. See Figure 17.55. At time $t = 0$ the wheel starts rolling on the x-axis in the direction shown at a rate of 1 radian per second.

    (a) Find parametric equations describing the motion of the center of the wheel.
    (b) Find parametric equations describing the motion of the spot on the rim. Plot its path.

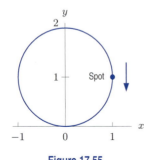

Figure 17.55

24. (a) The ellipse $2x^2 + 3y^2 = 8$ intersects the line of slope t through the point $P = (-2, 0)$ in two points, one of which is P. Compute the coordinate of the other point Q. See Figure 17.56.
    (b) Give a parameterization of the ellipse $2x^2 + 3y^2 = 8$ by rational functions.

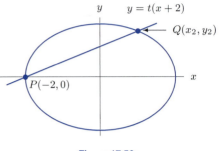

Figure 17.56

25. Let $x(t) = \cos^2 t$, $y(t) = \sin t \cos t$, $z(t) = \sin t$.

    (a) Show that this curve lies on the sphere of radius 1 centered at the origin.
    (b) Find a vector normal to the sphere and the curve at the point where $t = \pi/4$.
    (c) Find a tangent vector to the curve at the point where $t = \pi/4$.

26. A cheerleader has a 0.4 m long baton with a light on one end. She throws the baton in such a way that its center moves along a parabola, and the baton rotates counterclockwise around the center with a constant angular velocity. The baton is initially horizontal and 1.5 m above the ground; its initial velocity is 8 m/sec horizontally and 10 m/sec vertically, and its angular velocity is 2 revolutions per second. Find parametric equations describing the following motions:

    (a) The center of the baton relative to the ground.
    (b) The end of the baton relative to its center.
    (c) The path traced out by the end of the baton relative to the ground.
    (d) Sketch a graph of the motion of the end of the baton.

27. For a and ω positive constants and $t \ge 0$, the position vector of a particle moving in a spiral counterclockwise outward from the origin is given by

$$\vec{r}(t) = at\cos(\omega t)\vec{i} + at\sin(\omega t)\vec{j}.$$

What is the significance of the parameters ω and a?

28. An object is moving on a straight line path. Can you conclude at all times that:

    (a) Its velocity vector is parallel to the line? Justify your answer.
    (b) Its acceleration vector is parallel to the line? Justify your answer.

29. An object is moving in the plane on a circular path. Can you conclude at all times that

   (a) Its velocity vector is tangent to the circle ? Justify your answer.

   (b) Its acceleration vector points toward the center of the circle? Justify your answer.

30. If $\vec{F} = \vec{r}/\|\vec{r}\|^3$, find the following quantities in terms of x, y, z, or t.

   (a) $\|\vec{F}\|$

   (b) $\vec{F} \cdot \vec{r}$

   (c) A unit vector parallel to $\vec{F}$ and pointing in the same direction.

   (d) A unit vector parallel to $\vec{F}$ and pointing in the opposite direction.

   (e) $\vec{F}$ if $\vec{r} = \cos t\vec{i} + \sin t\vec{j} + \vec{k}$

   (f) $\vec{F} \cdot \vec{r}$ if $\vec{r} = \cos t\vec{i} + \sin t\vec{j} + \vec{k}$

For Problems 31–34, find the region of the Gulf stream velocity field in Figure 17.15 on page 799 represented by the given table of velocity vectors (in cm/sec).

31.

$35\vec{i} + 131\vec{j}$	$48\vec{i} + 92\vec{j}$	$47\vec{i} + \vec{j}$
$-32\vec{i} + 132\vec{j}$	$-44\vec{i} + 92\vec{j}$	$-42\vec{i} + \vec{j}$
$-51\vec{i} + 73\vec{j}$	$-119\vec{i} + 84\vec{j}$	$-128\vec{i} + 6\vec{j}$

32.

$10\vec{i} - 3\vec{j}$	$11\vec{i} + 16\vec{j}$	$20\vec{i} + 75\vec{j}$
$53\vec{i} - 7\vec{j}$	$58\vec{i} + 23\vec{j}$	$64\vec{i} + 80\vec{j}$
$119\vec{i} - 8\vec{j}$	$121\vec{i} + 31\vec{j}$	$114\vec{i} + 66\vec{j}$

33.

$97\vec{i} - 41\vec{j}$	$72\vec{i} - 24\vec{j}$	$54\vec{i} - 10\vec{j}$
$134\vec{i} - 49\vec{j}$	$131\vec{i} - 44\vec{j}$	$129\vec{i} - 18\vec{j}$
$103\vec{i} - 36\vec{j}$	$122\vec{i} - 30\vec{j}$	$131\vec{i} - 17\vec{j}$

CAS Challenge Problems

38. Let $\vec{r}_0 = x_0\vec{i} + y_0\vec{j} + z_0\vec{k}$, and let $\vec{e}_1$ and $\vec{e}_2$ be perpendicular unit vectors. A circle of radius R, centered at (x_0, y_0, z_0), and lying the plane parallel to $\vec{e}_1$ and $\vec{e}_2$, is parameterized by $\vec{r}(t) = \vec{r}_0 + R\cos t\vec{e}_1 + R\sin t\vec{e}_2$. We want to parameterize a circle in 3-space with radius 5, centered at $(1, 2, 3)$, and lying in the plane $x+y+z = 6$.

   (a) Let $\vec{e}_1 = a\vec{i} + b\vec{j}$ and $\vec{e}_2 = c\vec{i} + d\vec{j} + e\vec{k}$. Write down conditions on $\vec{e}_1$ and $\vec{e}_2$ that make them unit vectors, perpendicular to each other, and lie in the given plane.

   (b) Solve the equations in part (a) for a, b, c, d, and e and write a parameterization of the circle.

39. Let $\vec{F}(x, y) = -y(1 - y^2)\vec{i} + x(1 - y^2)\vec{j}$.

34.

$-95\vec{i} - 60\vec{j}$	$18\vec{i} - 48\vec{j}$	$82\vec{i} - 22\vec{j}$
$-29\vec{i} + 48\vec{j}$	$76\vec{i} + 63\vec{j}$	$128\vec{i} - 16\vec{j}$
$26\vec{i} + 105\vec{j}$	$49\vec{i} + 119\vec{j}$	$88\vec{i} + 13\vec{j}$

35. Each of the following vector fields represents an ocean current. Sketch the vector field, and sketch the path of an iceberg in this current. Determine the location of an iceberg at time $t = 7$ if it is at the point $(1, 3)$ at time $t = 0$.

   (a) The current everywhere is $\vec{i}$.

   (b) The current at (x, y) is $2x\vec{i} + y\vec{j}$.

   (c) The current at (x, y) is $-y\vec{i} + x\vec{j}$.

36. Wire is stretched taught from the point $P = (7, 12, -10)$ to the point $Q = (-2, -3, 2)$ and from the point $R = (-20, 17, 1)$ to the point $S = (37, 2, 25)$. Spherical balls of diameter 8 cm slide along each cable through holes along an axis through their centers. Can the beads pass each other without touching, regardless of their position?

37. There is a famous way to parameterize a sphere called *stereographic projection*. We work with the sphere $x^2 + y^2 + z^2 = 1$. Draw a line from a point (x, y) in the xy-plane to the north pole $(0, 0, 1)$. This line intersects the sphere in a point (x, y, z). This gives a parameterization of the sphere by points in the plane.

   (a) Which point corresponds to the south pole?

   (b) Which points correspond to the equator?

   (c) Do we get all the points of the sphere by this parameterization?

   (d) Which points correspond to the upper hemisphere?

   (e) Which points correspond to the lower hemisphere?

   (a) Show that $\vec{r} \cdot \vec{F} = 0$. What does this tell you about the shape of the flow lines?

   (b) Show that $\vec{r}(t) = \cos t\vec{i} + \sin t\vec{j}$ has velocity vector parallel to $\vec{F}$ at every point, but is not a flow line.

   (c) Show $\vec{r}(t) = (1/(\sqrt{1 + t^2}))\vec{i} + (t/(\sqrt{1 + t^2}))\vec{j}$ is a flow line for $\vec{F}$. What is the difference between this curve and the one in part (b)?

40. Let $\vec{F}(x, y) = (x + y)\vec{i} + (4x + y)\vec{j}$.

   (a) Show that $\vec{r}(t) = (ae^{3t} + be^{-t})\vec{i} + (2ae^{3t} - 2be^{-t})\vec{j}$, for constant a, b, is a flow line for $\vec{F}$.

   (b) Find the flow line passing through $(1, -2)$ at $t = 0$ and describe its behavior as $t \to \infty$. Do the same for the points $(1, -1.99)$ and $(1, -2.01)$. Compare the behavior of the three flow lines.

41. Two surfaces generally intersect in a curve. For each of the following pair of surfaces $f(x, y, z) = 0$ and $g(x, y, z) = 0$, find a parameterization for the curve of intersection by solving for two of the variables in terms of the third.

(a) $3x - 5y + z = 5, 2x + y + z = 3$,
(b) $3x^2 - 5y + z = 5, 2x + y + z = 3$,
(c) $x^2 + y^2 = 2, 3x - y + z = 5$.

CHECK YOUR UNDERSTANDING

Are the statements in Problems 1–38 true or false? Give reasons for your answer.

1. The parametric curve $x = 3t+2, y = -2t$ for $0 \le t \le 5$ passes through the origin.

2. The parametric curve $x = t^2, y = t^4$ for $0 \le t \le 1$ is a parabola.

3. A parametric curve $x = g(t), y = h(t)$ for $a \le t \le b$ is always the graph of a function $y = f(x)$.

4. The parametric curve $x = (3t + 2)^2, y = (3t + 2)^2 - 1$ for $0 \le t \le 3$ is a line.

5. The parametric curve $x = -\sin t, y = -\cos t$ for $0 \le t \le 2\pi$ traces out a unit circle counterclockwise as t increases.

6. A parameterization of the graph of $y = \ln x$ for $x > 0$ is given by $x = e^t, y = t$ for $-\infty < t < \infty$.

7. Both $x = -t + 1, y = 2t$ and $x = 2s, y = -4s + 2$ describe the same line.

8. The line of intersection of the two planes $z = x + y$ and $z = 1 - x - y$ can be parameterized by $x = t, y = \frac{1}{2} - t, z = \frac{1}{2}$.

9. The two lines given by $x = t, y = 2 + t, z = 3 + t$ and $x = 2s, y = 1 - s, z = s$ do not intersect.

10. The line parameterized by $x = 1, y = 2t, z = 3 + t$ is parallel to the x-axis.

11. The equation $\vec{r}(t) = 3t\vec{i} + (6t + 1)\vec{j}$ parameterizes a line.

12. The lines parameterized by $\vec{r}_1(t) = t\vec{i} + (-2t + 1)\vec{j}$ and $\vec{r}_2(t) = (2t + 5)\vec{i} + (-t)\vec{j}$ are parallel.

13. A particle whose motion in the plane is given by $\vec{r}(t) = t^2\vec{i} + (1-t)\vec{j}$ has the same velocity at $t = 1$ and $t = -1$.

14. A particle whose motion in the plane is given by $\vec{r}(t) = t^2\vec{i} + (1-t)\vec{j}$ has the same speed at $t = 1$ and $t = -1$.

15. If a particle is moving along a parameterized curve $\vec{r}(t)$ then the acceleration vector at any point is always perpendicular to the velocity vector at that point.

16. If a particle is moving along a parameterized curve $\vec{r}(t)$ then the acceleration vector at a point cannot be parallel to the velocity vector at that point.

17. If $\vec{r}(t)$ for $a \le t \le b$ is a parameterized curve, then $\vec{r}(-t)$ for $a \le t \le b$ is the same curve traced backward.

18. If $\vec{r}(t)$ for $a \le t \le b$ is a parameterized curve C and the speed $\|\vec{v}(t)\| = 1$, then the length of C is $b - a$.

19. If a particle moves with motion $\vec{r}(t) = 3t\vec{i} + 2t\vec{j} + t\vec{k}$, then the particle stops at the origin.

20. If a particle moves with constant speed, the path of the particle must be a line.

21. The flow lines for $\vec{F}(x, y) = x\vec{j}$ are parallel to the y-axis.

22. The flow lines of $\vec{F}(x, y) = y\vec{i} - x\vec{j}$ are hyberbolas.

23. The flow lines of $\vec{F}(x, y) = x\vec{i}$ are parabolas.

24. The vector field in Figure 17.57 has a flow line which lies in the first and third quadrants.

Figure 17.57

25. The vector field in Figure 17.57 has a flow line on which both x and y tend to infinity.

26. If $\vec{F}$ is a gradient vector field, $\vec{F}(x, y) = \nabla f(x, y)$, then the flow lines for $\vec{F}$ are the contours for f.

27. If the flow lines for the vector field $\vec{F}(\vec{r})$ are all concentric circles centered at the origin, then $\vec{F}(\vec{r}) \cdot \vec{r} = 0$ for all $\vec{r}$.

28. If the flow lines for the vector field $\vec{F}(x, y)$ are all straight lines parallel to the constant vector $\vec{v} = 3\vec{i} + 5\vec{j}$, then $\vec{F}(x, y) = \vec{v}$.

29. No flow line for the vector field $\vec{F}(x, y) = x\vec{i} + 2\vec{j}$ has a point where the y-coordinate reaches a relative maximum.

30. The vector field $\vec{F}(x, y) = e^x\vec{i} + y\vec{j}$ has a flow line that crosses the x-axis.

31. The vector field $\vec{F}(x, y) = y\vec{i} + \vec{j}$ is a gradient vector field.

32. The equations $x = s + 1, y = t - 2, z = 3$ parameterize a plane.

33. The equations $x = 2s - 1, y = -s + 3, z = 4 + s$ parameterize a plane.

34. If $\vec{r} = \vec{r}(s, t)$ parameterizes the upper hemisphere $x^2 + y^2 + z^2 = 1, z \ge 0$, then $\vec{r} = -\vec{r}(s, t)$ parameterizes the lower hemisphere $x^2 + y^2 + z^2 = 1, z \le 0$.

35. If $\vec{r} = \vec{r}(s, t)$ parameterizes the upper hemisphere $x^2 + y^2 + z^2 = 1, z \ge 0$, then $\vec{r} = \vec{r}(-s, -t)$ parameterizes the lower hemisphere $x^2 + y^2 + z^2 = 1, z \le 0$.

36. If $\vec{r_1}(s,t)$ parameterizes a plane then $\vec{r_2}(s,t) = \vec{r}_1(s,t) + 2\vec{i} - 3\vec{j} + \vec{k}$ parameterizes a parallel plane.

37. Every point on a parameterized surface has a parameter curve passing through it.

38. If $s_0 \neq s_1$, then the parameter curves $\vec{r}(s_0,t)$ and $\vec{r}(s_1,t)$ do not intersect.

For Problems 39–42, decide if the statement is true or false for

all smooth parameterized curves $\vec{r}(t)$ and all values of t for which $\vec{r}\,'(t) \neq \vec{0}$.

39. The vector $\vec{r}\,'(t)$ is tangent to the curve at the point with position vector $\vec{r}(t)$.

40. $\vec{r}\,'(t) \times \vec{r}(t) = \vec{0}$

41. $\vec{r}\,'(t) \cdot \vec{r}(t) = 0$

42. $\vec{r}\,''(t) = -\omega^2 \vec{r}(t)$

PROJECTS FOR CHAPTER SEVENTEEN

1. **Shooting a Basketball**

    A basketball player shoots the ball from 6 feet above the ground towards a basket that is 10 feet above the ground and 15 feet away horizontally.

    (a) Suppose she shoots the ball at an angle of A degrees above the horizontal $(0 < A < \pi/2)$ with an initial speed V. Give the x- and y-coordinates of the position of the basketball at time t. Assume the x-coordinate of the basket is 0 and that the x-coordinate of the shooter is -15. [Hint: There is an acceleration of -32 ft/sec^2 in the y-direction; there is no acceleration in the x-direction. Ignore air resistance.]

    (b) Using the parametric equations you obtained in part (a), experiment with different values for V and A, plotting the path of the ball on a graphing calculator or computer to see how close the ball comes to the basket. (The tick marks on the y-axis can be used to locate the basket.) Find some values of V and A for which the shot goes in.

    (c) Find the angle A that minimizes the velocity needed for the ball to reach the basket. (This is a lengthy computation. First find an equation in V and A that holds if the path of the ball passes through the point 15 feet from the shooter and 10 feet above the ground. Then minimize V.)

2. **Kepler's Second Law**

    The planets do not orbit in circles with the sun at the center, nor does the moon orbit in a circle with the earth at the center. In fact, the moon's distance from the earth varies from 220,000 to 260,000 miles. In the last half of the 16th century the Danish astronomer Tycho Brahe (1546–1601) made measurements of the positions of the planets. Johann Kepler (1571–1630) studied this data and arrived at three laws now known as *Kepler's Laws*:

    I. The orbit of each planet is an ellipse with the sun at one focus. In particular, the orbit lies in a plane containing the sun.

    II. As a planet orbits around the sun, the line segment from the sun to the planet sweeps out equal areas in equal times. See Figure 17.58.

    III. The ratio p^2/d^3 is the same for every planet orbiting around the sun, where p is the period of the orbit (time to complete one revolution) and d is the mean distance of the orbit (average of the shortest and farthest distances from the sun).

    Kepler's Laws, impressive as they are, were purely descriptive; Newton's great achievement was to find an underlying cause for them. In this project, you will derive Kepler's Second Law from Newton's Law of Gravity.

    Consider a coordinate system centered at the sun.[2] Let $\vec{r}$ be the position vector of a planet and let $\vec{v}$ and $\vec{a}$ be the planet's velocity and acceleration, respectively. Define $\vec{L} = \vec{r} \times \vec{v}$. (This is a multiple of the planet's angular momentum.)

    (a) Show that $\dfrac{d\vec{L}}{dt} = \vec{r} \times \vec{a}$.

    (b) Consider the planet moving from $\vec{r}$ to $\vec{r} + \Delta\vec{r}$. Explain why the area ΔA about the origin swept out by the planet is approximately $\frac{1}{2}\|\Delta\vec{r} \times \vec{r}\|$.

    (c) Using part (b), explain why $\dfrac{dA}{dt} = \dfrac{1}{2}\|\vec{L}\|$.

    ---

    [2]We are assuming the center of the sun is the same as the center of mass of the planet/sun system. This is only approximately true.

(d) Newton's Laws imply that the planet's gravitational acceleration, $\vec{a}$, is directed towards the sun. Using this fact and part (a), explain why $\vec{L}$ is constant.

(e) Use parts (c) and (d) to explain Kepler's Second Law.

(f) Using Kepler's Second Law, determine whether a planet is moving most quickly when it is closest to, or farthest from, the sun.

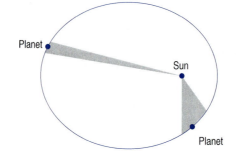

Figure 17.58: The line segment joining a planet to the sun sweeps out equal areas in equal times

3. **Flux Diagrams**

A *flux diagram* uses flow lines to represent a vector field. The arrows drawn on a flow line indicate the direction of the vector field. The flow lines are drawn in such a way that their density is proportional to the magnitude of the vector field at each point. (The density is the number of flow lines per unit length along a curve perpendicular to the vector field.)

Figure 17.59 is a flux diagram for the vector field in 2-space $\vec{F} = \vec{r}/\|\vec{r}\|^2$. Since the field points radially away from the origin, the flow lines are straight lines radiating from the origin. The number of flow lines passing through any circle centered at the origin is a constant k. Therefore, the flow lines passing through a small circle are more densely packed that those passing through a large circle, indicating that the magnitude of the vector field decreases as we move away from the origin. In fact,

$$\text{Density of lines} = \frac{\text{Number of lines passing through circle}}{\text{Circumference of circle}} = \frac{k}{2\pi r} = \frac{k}{2\pi}1/r,$$

so that the density is proportional to $1/r$, the magnitude of the field.

Sometimes we have to start new lines to make the density proportional to the magnitude. For example, the flow lines of $\vec{v} = x\vec{i}$ are horizontal straight lines directed away from the y-axis. However, since the magnitude of $\vec{v}$ increases linearly with x, we have to make the density of lines increase linearly with x. We achieve this by starting new lines at regular intervals (see Figure 17.60).

Draw flux diagrams for the following vector fields:

(a) $\vec{v} = \vec{i}$　　　**(b)** $\vec{v} = -y\vec{i} + x\vec{j}$　**(c)** $\vec{v} = y\vec{i}$　　　**(d)** $\vec{v} = y\vec{j}$

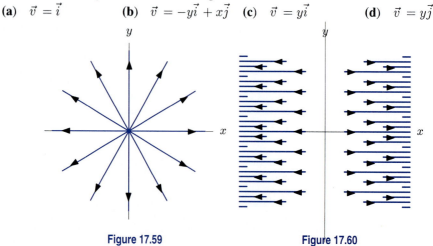

Figure 17.59　　　　　　　　　　**Figure 17.60**

Chapter Eighteen

LINE INTEGRALS

When a constant force, $\vec{F}$, acts on an object as it moves through a displacement, $\vec{d}$, the work done by the force is the dot product, $\vec{F} \cdot \vec{d}$. What if the object moves along a curved path through a variable force field? In this chapter we define a *line integral* that calculates the work in this situation. We also define the *circulation* which is used to measure the strength of eddies in a fluid flow.

Line integrals provide us with an analogue of the Fundamental Theorem of Calculus, which tells us how to recover a function from its derivative. The line integral analogue of the Fundamental Theorem shows how a line integral can be used to recover a multivariable function from its gradient field.

In contrast with the one-variable situation, not all vector fields are gradient fields. The line integral can be used to distinguish those that are, using the concept of a *path-independent* (or *conservative*) vector field. We study path-independent fields, which are central in physics, and end with Green's Theorem.

18.1   THE IDEA OF A LINE INTEGRAL

Imagine that you are rowing in a river with a noticeable current. At times you may be working against the current and at other times you may be moving with it. At the end you have a sense of whether, overall, you were helped or hindered by the current. The line integral, defined in this section, measures the extent to which a curve in a vector field is, overall, going with the vector field or against it.

Orientation of a Curve

A curve can be traced out in two directions, as shown in Figure 18.1. We need to choose one direction before we can define a line integral.

> A curve is said to be **oriented** if we have chosen a direction of travel on it.

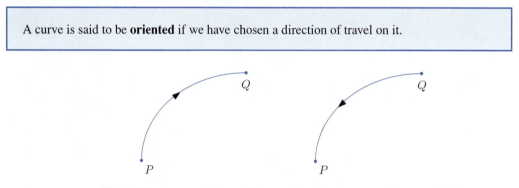

Figure 18.1: A curve with two different orientations represented by arrowheads

Definition of the Line Integral

Consider a vector field $\vec{F}$ and an oriented curve C. We begin by dividing C into n small, almost straight pieces along which $\vec{F}$ is approximately constant. Each piece can be represented by a displacement vector $\Delta\vec{r}_i = \vec{r}_{i+1} - \vec{r}_i$ and the value of $\vec{F}$ at each point of this small piece of C is approximately $\vec{F}(\vec{r}_i)$. See Figures 18.2 and 18.3.

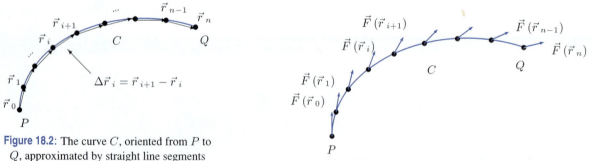

Figure 18.2: The curve C, oriented from P to Q, approximated by straight line segments represented by displacement vectors $\Delta\vec{r}_i = \vec{r}_{i+1} - \vec{r}_i$

Figure 18.3: The vector field $\vec{F}$ evaluated at the points with position vector $\vec{r}_i$ on the curve C oriented from P to Q

For each point with position vector $\vec{r}_i$ on C, we form the dot product $\vec{F}(\vec{r}_i) \cdot \Delta\vec{r}_i$. Summing over all such pieces, we get a Riemann sum:

$$\sum_{i=0}^{n-1} \vec{F}(\vec{r}_i) \cdot \Delta\vec{r}_i.$$

We define the line integral, written $\int_C \vec{F} \cdot d\vec{r}$, by taking the limit as $\|\Delta\vec{r}_i\| \to 0$. Provided the limit exists, we make the following definition:

The **line integral** of a vector field $\vec{F}$ along an oriented curve C is

$$\int_C \vec{F} \cdot d\vec{r} = \lim_{\|\Delta \vec{r}_i\| \to 0} \sum_{i=0}^{n-1} \vec{F}(\vec{r}_i) \cdot \Delta \vec{r}_i.$$

How Does the Limit Defining a Line Integral Work?

The limit in the definition of a line integral exists if $\vec{F}$ is continuous on the curve C and if C is made by joining end to end a finite number of smooth curves. (A vector field $\vec{F} = F_1\vec{i} + F_2\vec{j} + F_3\vec{k}$ is *continuous* if F_1, F_2, and F_3 are continuous, and a *smooth curve* is one that can be parameterized by smooth functions.) We subdivide the curve using a parameterization which goes from one end of the curve to the other, in the forward direction, without retracing any portion of the curve. A subdivision of the parameter interval gives a subdivision of the curve. All the curves we consider in this book are *piecewise smooth* in this sense. Section 18.2 shows how to use a parameterization to compute a line integral.

Example 1   Find the line integral of the constant vector field $\vec{F} = \vec{i} + 2\vec{j}$ along the path from $(1, 1)$ to $(10, 10)$ shown in Figure 18.4.

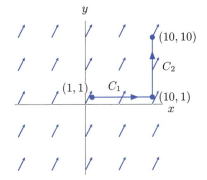

Figure 18.4: The constant vector field $\vec{F} = \vec{i} + 2\vec{j}$ and the path from $(1, 1)$ to $(10, 10)$

Solution   Let C_1 be the horizontal segment of the path going from $(1, 1)$ to $(10, 1)$. When we break this path into pieces, each piece $\Delta \vec{r}$ is horizontal, so $\Delta \vec{r} = \Delta x \vec{i}$ and $\vec{F} \cdot \Delta \vec{r} = (\vec{i} + 2\vec{j}) \cdot \Delta x \vec{i} = \Delta x$. Hence,

$$\int_{C_1} \vec{F} \cdot d\vec{r} = \int_{x=1}^{x=10} dx = 9.$$

Similarly, along the vertical segment C_2, we have $\Delta \vec{r} = \Delta y \vec{j}$ and $\vec{F} \cdot \Delta \vec{r} = (\vec{i} + 2\vec{j}) \cdot \Delta y \vec{j} = 2\Delta y$, so

$$\int_{C_2} \vec{F} \cdot d\vec{r} = \int_{y=1}^{y=10} 2\, dy = 18.$$

Thus,

$$\int_C \vec{F} \cdot d\vec{r} = \int_{C_1} \vec{F} \cdot d\vec{r} + \int_{C_2} \vec{F} \cdot d\vec{r} = 9 + 18 = 27.$$

What Does the Line Integral Tell Us?

Remember that for any two vectors $\vec{u}$ and $\vec{v}$, the dot product $\vec{u} \cdot \vec{v}$ is positive if $\vec{u}$ and $\vec{v}$ point roughly in the same direction (that is, if the angle between them is less than $\pi/2$). The dot product is zero if $\vec{u}$ is perpendicular to $\vec{v}$ and is negative if they point roughly in opposite directions (that is, if the angle between them is greater than $\pi/2$).

The line integral of adds the dot products of $\vec{F}$ and $\Delta \vec{r}$ along the path. If $||\vec{F}||$ is constant, the line integral gives a positive number if $\vec{F}$ is mostly pointing in the same direction as $\Delta \vec{r}$, a negative number if $\vec{F}$ is mostly pointing in the opposite direction. The line integral is zero if $\vec{F}$ is perpendicular to the path at all points or if the positive and negative contributions cancel out. In general, the line integral of a vector field $\vec{F}$ along a curve C measures the extent to which C is going with $\vec{F}$ or against it.

Example 2 The vector field $\vec{F}$ and the oriented curves C_1, C_2, C_3, C_4 are shown in Figure 18.5. The curves C_1 and C_3 are the same length. Which of the line integrals $\int_{C_i} \vec{F} \cdot d\vec{r}$, for $i = 1, 2, 3, 4$, are positive? Which are negative? Arrange these line integrals in ascending order.

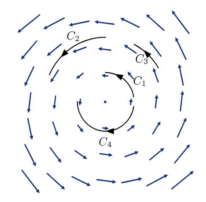

Figure 18.5: Vector field and paths C_1, C_2, C_3, C_4

Solution The vector field $\vec{F}$ and the line segments $\Delta \vec{r}$ are approximately parallel and in the same direction for the curves C_1, C_2, and C_3. So the contributions of each term $\vec{F} \cdot \Delta \vec{r}$ are positive for these curves. Thus, $\int_{C_1} \vec{F} \cdot d\vec{r}$, $\int_{C_2} \vec{F} \cdot d\vec{r}$, and $\int_{C_3} \vec{F} \cdot d\vec{r}$ are each positive. For the curve C_4, the vector field and the line segments are in opposite directions, so each term $\vec{F} \cdot \Delta \vec{r}$ is negative, and therefore the integral $\int_{C_4} \vec{F} \cdot d\vec{r}$ is negative.

Since the magnitude of the vector field is smaller along C_1 than along C_3, and these two curves are the same length, we have

$$\int_{C_1} \vec{F} \cdot d\vec{r} < \int_{C_3} \vec{F} \cdot d\vec{r}.$$

In addition, the magnitude of the vector field is the same along C_2 and C_3, but the curve C_2 is longer than the curve C_3. Thus,

$$\int_{C_3} \vec{F} \cdot d\vec{r} < \int_{C_2} \vec{F} \cdot d\vec{r}.$$

Putting these results together with the fact that $\int_{C_4} \vec{F} \cdot d\vec{r}$ is negative, we have

$$\int_{C_4} \vec{F} \cdot d\vec{r} < \int_{C_1} \vec{F} \cdot d\vec{r} < \int_{C_3} \vec{F} \cdot d\vec{r} < \int_{C_2} \vec{F} \cdot d\vec{r}.$$

Interpretations of the Line Integral

Work

Recall from Section 13.3 that if a constant force $\vec{F}$ acts on an object while it moves along a straight line through a displacement $\vec{d}$, the work done by the force on the object is

$$\text{Work done} = \vec{F} \cdot \vec{d}.$$

Now suppose we want to find the work done by gravity on an object moving far above the surface of the earth. Since the force of gravity varies with distance from the earth and the path may not be straight, we can't use the formula $\vec{F} \cdot \vec{d}$. We approximate the path by line segments which are small enough that the force is approximately constant on each one. Suppose the force at a point with position vector $\vec{r}$ is $\vec{F}(\vec{r})$, as in Figures 18.2 and 18.3. Then

$$\begin{array}{c} \text{Work done by force } \vec{F}(\vec{r}_i) \\ \text{over small displacement } \Delta \vec{r}_i \end{array} \approx \vec{F}(\vec{r}_i) \cdot \Delta \vec{r}_i,$$

and so,

$$\begin{array}{c} \text{Total work done by force} \\ \text{along oriented curve } C \end{array} \approx \sum_i \vec{F}(\vec{r}_i) \cdot \Delta \vec{r}_i.$$

Taking the limit as $\|\Delta \vec{r}_i\| \to 0$, we get

$$\begin{array}{c} \text{Work done by force } \vec{F}(\vec{r}) \\ \text{along curve } C \end{array} = \lim_{\|\Delta \vec{r}_i\| \to 0} \sum_i \vec{F}(\vec{r}_i) \cdot \Delta \vec{r}_i = \int_C \vec{F} \cdot d\vec{r}.$$

Example 3   A mass lying on a flat table is attached to a spring whose other end is fastened to the wall. (See Figure 18.6.) The spring is extended 20 cm beyond its rest position and released. If the axes are as shown in Figure 18.6, when the spring is extended by a distance of x, the force exerted by the spring on the mass is given by

$$\vec{F}(x) = -kx\vec{i},$$

where k is a positive constant that depends on the strength of the spring.

Suppose the mass moves back to the rest position. How much work is done by the force exerted by the spring?

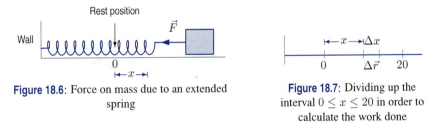

Figure 18.6: Force on mass due to an extended spring

Figure 18.7: Dividing up the interval $0 \le x \le 20$ in order to calculate the work done

Solution   The path from $x = 20$ to $x = 0$ is divided as shown in Figure 18.7, with a typical segment represented by

$$\Delta \vec{r} = \Delta x \vec{i}.$$

Since we are moving from $x = 20$ to $x = 0$, the quantity Δx will be negative. The work done by the force as the mass moves through this segment is approximated by

$$\text{Work done} \approx \vec{F} \cdot \Delta \vec{r} = (-kx\vec{i}) \cdot (\Delta x \vec{i}) = -kx\,\Delta x.$$

Thus, we have

$$\text{Total work done} \approx \sum -kx\,\Delta x.$$

In the limit, as $\|\Delta x\| \to 0$, this sum becomes an ordinary definite integral. Since the path starts at $x = 20$, this is the lower limit of integration; $x = 0$ is the upper limit. Thus, we get

$$\text{Total work done} = \int_{x=20}^{x=0} -kx\,dx = -\frac{kx^2}{2}\bigg|_{20}^{0} = \frac{k(20)^2}{2} = 200k.$$

Note that the work done is positive, since the force acts in the direction of motion.

Example 3 shows how a line integral over a path parallel to the x-axis reduces to a one-variable integral. Section 18.2 shows how to convert *any* line integral into a one-variable integral.

Example 4   A particle with position vector $\vec{r}$ is subject to a force, $\vec{F}$, due to gravity. What is the *sign* of the work done by $\vec{F}$ as the particle moves along the path C_1, a radial line through the center of the earth, starting 8000 km from the center and ending 10,000 km from the center? (See Figure 18.8.)

Solution   We divide the path into small radial segments, $\Delta \vec{r}$, pointing away from the center of the earth and parallel to the gravitational force. The vectors $\vec{F}$ and $\Delta \vec{r}$ point in opposite directions, so each term $\vec{F} \cdot \Delta \vec{r}$ is negative. Adding all these negative quantities and taking the limit results in a negative value for the total work. Thus, the work done by gravity is negative. The negative sign indicates that we would have to do work *against* gravity to move the particle along the path C_1.

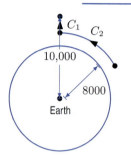

Figure 18.8: The earth

Example 5   Find the sign of the work done by gravity along the curve C_1 in Example 4, but with the opposite orientation.

Solution   Tracing a curve in the opposite direction changes the sign of the line integral because all the segments $\Delta \vec{r}$ change direction, and so every term $\vec{F} \cdot \Delta \vec{r}$ changes sign. Thus, the result will be the negative of the answer found in Example 4. Therefore, the work done by gravity as a particle moves along C_1 toward the center of the earth is positive.

Example 6   Find the work done by gravity as a particle moves along C_2, an arc of a circle 8000 km long at a distance of 8000 km from the center of the earth. (See Figure 18.8.)

Solution   Since C_2 is everywhere perpendicular to the gravitational force, $\vec{F} \cdot \Delta \vec{r} = 0$ for all $\Delta \vec{r}$ along C_2. Thus,

$$\text{Work done} = \int_{C_2} \vec{F} \cdot d\vec{r} = 0,$$

so the work done is zero. This is why satellites can remain in orbit without expending any fuel, once they have attained the correct altitude and velocity.

Circulation

The velocity vector field for the Gulf stream on page 799 shows distinct eddies or regions where the water circulates. We can measure this circulation using a *closed curve*, that is, one that starts and ends at the same point.

> If C is an oriented closed curve, the line integral of a vector field $\vec{F}$ around C is called the **circulation** of $\vec{F}$ around C.

Circulation is a measure of the net tendency of the vector field to point around the curve C. To emphasize that C is closed, the circulation is sometimes denoted $\oint_C \vec{F} \cdot d\vec{r}$, with a small circle on the integral sign.

Example 7   Describe the rotation of the vector fields in Figures 18.9 and 18.10. Find the sign of the circulation of the vector fields around the indicated paths.

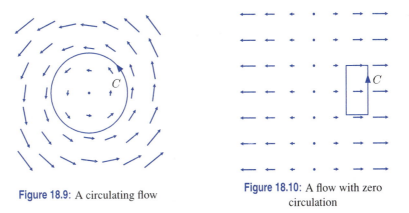

Figure 18.9: A circulating flow

Figure 18.10: A flow with zero circulation

Solution   Consider the vector field in Figure 18.9. If you think of this as representing the velocity of water flowing in a pond, you see that the water is circulating. The line integral around C, measuring the circulation around C, is positive, because the vectors of the field are all pointing in the direction of the path. By way of contrast, look at the vector field in Figure 18.10. Here the line integral around C is zero because the vertical portions of the path are perpendicular to the field and the contributions from the two horizontal portions cancel out. This means that there is no net tendency for the water to circulate around C.

It turns out that the vector field in Figure 18.10 has the property that its circulation around *any* closed path is zero. Water moving according to this vector field has no tendency to circulate around any point and a leaf dropped into the water will not spin. We'll look at such special fields again later when we introduce the notion of the *curl* of a vector field.

Properties of Line Integrals

Line integrals share some basic properties with ordinary one-variable integrals:

> For a scalar constant λ, vector fields $\vec{F}$ and $\vec{G}$, and oriented curves C, C_1, and C_2
>
> 1. $\displaystyle\int_C \lambda\vec{F} \cdot d\vec{r} = \lambda\int_C \vec{F} \cdot d\vec{r}.$ 2. $\displaystyle\int_C (\vec{F} + \vec{G}) \cdot d\vec{r} = \int_C \vec{F} \cdot d\vec{r} + \int_C \vec{G} \cdot d\vec{r}.$
>
> 3. $\displaystyle\int_{-C} \vec{F} \cdot d\vec{r} = -\int_C \vec{F} \cdot d\vec{r}.$ 4. $\displaystyle\int_{C_1+C_2} \vec{F} \cdot d\vec{r} = \int_{C_1} \vec{F} \cdot d\vec{r} + \int_{C_2} \vec{F} \cdot d\vec{r}.$

Properties 3 and 4 are concerned with the curve C over which the line integral is taken. If C is an oriented curve, then $-C$ is the same curve traversed in the opposite direction, that is, with the opposite orientation. (See Figure 18.11 on page 832.) Property 3 holds because if we integrate along $-C$, the vectors $\Delta\vec{r}$ point in the opposite direction and the dot products $\vec{F} \cdot \Delta\vec{r}$ are the negatives of what they were along C.

If C_1 and C_2 are oriented curves with C_1 ending where C_2 begins, we construct a new oriented curve, called $C_1 + C_2$, by joining them together. (See Figure 18.12.) Property 4 is the analogue for line integrals of the property for definite integrals which says that

$$\int_a^b f(x)\,dx = \int_a^c f(x)\,dx + \int_c^b f(x)\,dx.$$

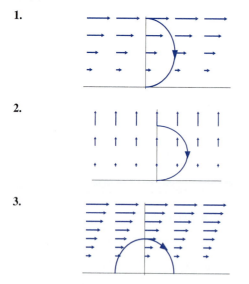

Figure 18.11: A curve, C, and its opposite, $-C$

Figure 18.12: Joining two curves, C_1, and C_2, to make a new one, $C_1 + C_2$

Exercises and Problems for Section 18.1

Exercises

In Exercises 1–4, say whether you expect the line integral of the pictured vector field over the given curve to be positive, negative, or zero.

1.

4.

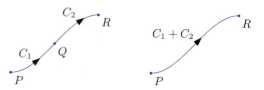

2.

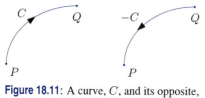

In Exercises 5–10, calculate the line integral of the vector field along the line between the given points.

5. $\vec{F} = x\vec{j}$, from $(1, 0)$ to $(3, 0)$

6. $\vec{F} = x\vec{j}$, from $(2, 0)$ to $(2, 5)$

7. $\vec{F} = x\vec{i}$, from $(2, 0)$ to $(6, 0)$

8. $\vec{F} = x\vec{i} + y\vec{j}$, from $(2, 0)$ to $(6, 0)$

9. $\vec{F} = \vec{r}$, from $(2, 2)$ to $(6, 6)$

10. $\vec{F} = 3\vec{i} + 4\vec{j}$, from $(0, 6)$ to $(0, 13)$

3.

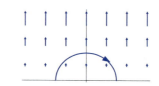

Problems

11. Given the force field $\vec{F}(x, y) = y\vec{i} + x^2\vec{j}$ and the right-angle curve, C, from the points $(0, -1)$ to $(4, -1)$ to $(4, 3)$ shown in Figure 18.13:

(a) Evaluate $\vec{F}$ at the points $(0, -1)$, $(1, -1)$, $(2, -1)$, $(3, -1)$, $(4, -1)$, $(4, 0)$, $(4, 1)$, $(4, 2)$, $(4, 3)$.

(b) Make a sketch showing the force field along C.

(c) Estimate the work done by the indicated force field on an object traversing the curve C.

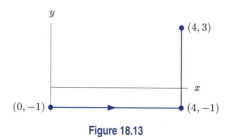

Figure 18.13

12. (a) For each of the vector fields, $\vec{F}$, shown in Figure 18.14, sketch a curve for which the integral $\int_C \vec{F} \cdot d\vec{r}$ is positive.
(b) For which of the vector fields is it possible to make your answer to part (a) a closed curve?

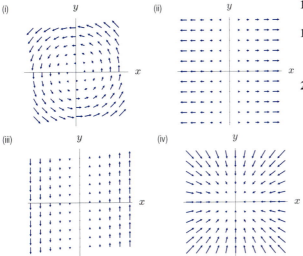

(i) y (ii) y

(iii) y (iv) y

Figure 18.14

13. Consider the vector field $\vec{F}$ shown in Figure 18.15, together with the paths C_1, C_2, and C_3. Arrange the line integrals $\int_{C_1} \vec{F} \cdot d\vec{r}$, $\int_{C_2} \vec{F} \cdot d\vec{r}$ and $\int_{C_3} \vec{F} \cdot d\vec{r}$ in ascending order.

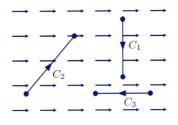

Figure 18.15

For Problems 14–18, say whether you expect the given vector field to have positive, negative, or zero circulation around the closed curve $C = C_1 + C_2 + C_3 + C_4$ in Figure 18.16. The segments C_1 and C_3 are circular arcs centered at the origin; C_2 and C_4 are radial line segments. You may find it helpful to sketch the vector field.

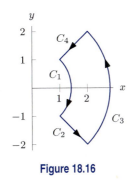

Figure 18.16

14. $\vec{F}(x, y) = x\vec{i} + y\vec{j}$

15. $\vec{F}(x, y) = -y\vec{i} + x\vec{j}$

16. $\vec{F}(x, y) = y\vec{i} - x\vec{j}$

17. $\vec{F}(x, y) = x^2\vec{i}$

18. $\vec{F}(x, y) = -\dfrac{y}{x^2 + y^2}\vec{i} + \dfrac{x}{x^2 + y^2}\vec{j}$

19. Draw an oriented curve C and a vector field $\vec{F}$ along C that is not always perpendicular to C, but for which $\int_C \vec{F} \cdot d\vec{r} = 0$.

20. Let $\vec{F}$ be the constant force field $\vec{j}$ in Figure 18.17. On which of the paths C_1, C_2, C_3 is zero work done by $\vec{F}$? Explain.

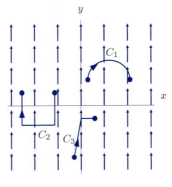

Figure 18.17

21. Explain why the following statement is true: Whenever the line integral of a vector field around every closed curve is zero, the line integral along a curve with fixed endpoints has a constant value independent of the path taken between the endpoints.

22. Explain why the converse to the statement in Problem 21 is also true: Whenever the line integral of a vector field depends only on endpoints and not on paths, the circulation around every closed curve is zero.

23. A square has side 1000 km. A wind blows from the east and decreases in magnitude toward the north at a rate of 6 meter/sec for every 500 km. Compute the circulation of the wind counterclockwise around the square.

In Problems 24–25, use the fact that the force of gravity on a particle of mass m at the point with position vector $\vec{r}$ is

$$\vec{F} = -\frac{GMm\vec{r}}{\|\vec{r}\|^3}$$

where and G is a constant and M is the mass of the earth.

24. Calculate the work done by the force of gravity on a particle of mass m as it moves from 8000 km to 10,000 km from the center of the earth.

25. Calculate the work done by the force of gravity on a particle of mass m as it moves from 8000 km from the center of the earth to infinitely far away.

In Problems 26–29 we explore the notion of the electric potential of an electric field $\vec{E}$. We chose a point P_0 to be the ground, that is, the potential at P_0 is zero. Then the potential $\phi(P)$ at a point P is defined to be the work done moving a particle of charge 1 coulomb from P_0 to P. (The potential does not depend on the path chosen.) Q, E are constants.

26. Write a line integral for $\phi(P)$, using the fact that the electric field acts on a particle with force $\vec{E}$.

27. Let $\vec{E} = \left(\dfrac{Q}{4\pi\epsilon}\right)\dfrac{\vec{r}}{\|\vec{r}\|^3}$, and let P_0 be a point a units from the origin. Describe the set of points with zero potential.

28. An equipotential surface is a surface on which the potential is constant. Describe the equipotential surfaces of $\vec{E} = \dfrac{Q}{4\pi\epsilon}\dfrac{\vec{r}}{\|\vec{r}\|^3}$, where the ground P_0 is chosen to be at distance a from the origin.

29. Let $\vec{E} = \dfrac{Q}{4\pi\epsilon}\dfrac{\vec{r}}{\|\vec{r}\|^3}$ and let the ground P_0 be chosen to be a units from the origin.

(a) Find a formula for ϕ.
(b) Engineers often choose the ground point to be "at infinity." Why?

18.2   COMPUTING LINE INTEGRALS OVER PARAMETERIZED CURVES

The goal of this section is to show how to use a parameterization of a curve to convert a line integral into an ordinary one-variable integral.

Using a Parameterization to Evaluate a Line Integral

Recall the definition of the line integral,

$$\int_C \vec{F} \cdot d\vec{r} = \lim_{\|\Delta\vec{r}_i\|\to 0} \sum \vec{F}(\vec{r}_i) \cdot \Delta\vec{r}_i,$$

where the $\vec{r}_i$ are the position vectors of points subdividing the curve into short pieces. Now suppose we have a smooth parameterization, $\vec{r}(t)$, of C for $a \leq t \leq b$, so that $\vec{r}(a)$ is the position vector of the beginning of the curve and $\vec{r}(b)$ is the position vector of the end. Then we can divide C into n pieces by dividing the interval $a \leq t \leq b$ into n pieces, each of size $\Delta t = (b-a)/n$. See Figures 18.18 and 18.19.

At each point $\vec{r}_i = \vec{r}(t_i)$ we want to compute

$$\vec{F}(\vec{r}_i) \cdot \Delta\vec{r}_i.$$

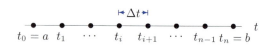

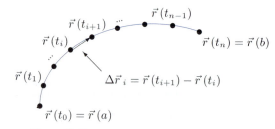

Figure 18.19: Corresponding subdivision of the parameterized path C

Figure 18.18: Subdivision of the interval $a \leq t \leq b$

Since $t_{i+1} = t_i + \Delta t$, the displacement vectors $\Delta\vec{r}_i$ are given by

$$\begin{aligned}
\Delta\vec{r}_i &= \vec{r}(t_{i+1}) - \vec{r}(t_i) \\
&= \vec{r}(t_i + \Delta t) - \vec{r}(t_i) \\
&= \frac{\vec{r}(t_i + \Delta t) - \vec{r}(t_i)}{\Delta t} \cdot \Delta t \\
&\approx \vec{r}'(t_i)\Delta t,
\end{aligned}$$

where we use the facts that Δt is small and that $\vec{r}(t)$ is differentiable to obtain the last approximation.

Therefore,

$$\int_C \vec{F} \cdot d\vec{r} \approx \sum \vec{F}(\vec{r}_i) \cdot \Delta\vec{r}_i \approx \sum \vec{F}(\vec{r}(t_i)) \cdot \vec{r}'(t_i)\,\Delta t.$$

Notice that $\vec{F}(\vec{r}(t_i)) \cdot \vec{r}'(t_i)$ is the value at t_i of a one-variable function of t, so this last sum is really a one-variable Riemann sum. In the limit as $\Delta t \to 0$, we get a definite integral:

$$\lim_{\Delta t \to 0} \sum \vec{F}\left(\vec{r}\left(t_i\right)\right) \cdot \vec{r}\,'(t_i)\,\Delta t = \int_a^b \vec{F}\left(\vec{r}\left(t\right)\right) \cdot \vec{r}\,'(t)\,dt.$$

Thus, we have the following result:

If $\vec{r}\left(t\right)$, for $a \le t \le b$, is a smooth parameterization of an oriented curve C and $\vec{F}$ is a vector field which is continuous on C, then

$$\int_C \vec{F} \cdot d\vec{r} = \int_a^b \vec{F}\left(\vec{r}\left(t\right)\right) \cdot \vec{r}\,'(t)\,dt.$$

In words: To compute the line integral of $\vec{F}$ over C, take the dot product of $\vec{F}$ evaluated on C with the velocity vector, $\vec{r}\,'(t)$, of the parameterization of C, then integrate along the curve.

Even though we assumed that C is smooth, we can use the same formula to compute line integrals over curves which are only *piecewise smooth*, such as the boundary of a rectangle: If C is piecewise smooth, we apply the formula to each one of the smooth pieces and add the results.

Example 1   Compute $\int_C \vec{F} \cdot d\vec{r}$ where $\vec{F} = (x + y)\vec{i} + y\vec{j}$ and C is the quarter unit circle, oriented counterclockwise as shown in Figure 18.20.

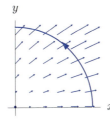

Figure 18.20: The vector field $\vec{F} = (x + y)\vec{i} + y\vec{j}$ and the quarter circle C

Solution   Since all of the vectors in $\vec{F}$ along C point generally in a direction opposite to the orientation of C, we expect our answer to be negative. The first step is to parameterize C by

$$\vec{r}\,(t) = x(t)\vec{i} + y(t)\vec{j} = \cos t\vec{i} + \sin t\vec{j}, \quad 0 \le t \le \frac{\pi}{2}.$$

Substituting the parameterization into $\vec{F}$ we get $\vec{F}\left(x(t), y(t)\right) = (\cos t + \sin t)\vec{i} + \sin t\vec{j}$. The vector $\vec{r}\,'(t) = x'(t)\vec{i} + y'(t)\vec{j} = -\sin t\vec{i} + \cos t\vec{j}$. Then

$$\int_C \vec{F} \cdot d\vec{r} = \int_0^{\pi/2} ((\cos t + \sin t)\vec{i} + \sin t\vec{j}) \cdot (-\sin t\vec{i} + \cos t\vec{j})dt$$

$$= \int_0^{\pi/2} (-\cos t \sin t - \sin^2 t + \sin t \cos t)dt$$

$$= \int_0^{\pi/2} -\sin^2 t\,dt = -\frac{\pi}{4} \approx -0.7854.$$

So the answer is negative, as expected.

Example 2   Consider the vector field $\vec{F} = x\vec{i} + y\vec{j}$.

(a) Suppose C_1 is the line segment joining $(1, 0)$ to $(0, 2)$ and C_2 is a part of a parabola with its vertex at $(0, 2)$, joining the same points in the same order. (See Figure 18.21.) Verify that

$$\int_{C_1} \vec{F} \cdot d\vec{r} = \int_{C_2} \vec{F} \cdot d\vec{r}.$$

(b) If C is the triangle shown in Figure 18.22, show that $\int_C \vec{F} \cdot d\vec{r} = 0$.

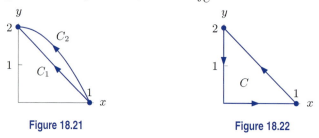

Figure 18.21

Figure 18.22

Solution

(a) We parameterize C_1 by $\vec{r}(t) = (1-t)\vec{i} + 2t\vec{j}$ with $0 \le t \le 1$. Then $\vec{r}'(t) = -\vec{i} + 2\vec{j}$, so

$$\int_{C_1} \vec{F} \cdot d\vec{r} = \int_0^1 \vec{F}(1-t, 2t) \cdot (-\vec{i} + 2\vec{j})\, dt = \int_0^1 ((1-t)\vec{i} + 2t\vec{j}) \cdot (-\vec{i} + 2\vec{j})\, dt$$

$$= \int_0^1 (5t - 1)\, dt = \frac{3}{2}.$$

To parameterize C_2, we use the fact that it is part of a parabola with vertex at $(0, 2)$, so its equation is of the form $y = -kx^2 + 2$ for some k. Since the parabola crosses the x-axis at $(1, 0)$, we find that $k = 2$ and $y = -2x^2 + 2$. Therefore, we use the parameterization $\vec{r}(t) = t\vec{i} + (-2t^2 + 2)\vec{j}$ with $0 \le t \le 1$, which has $\vec{r}' = \vec{i} - 4t\vec{j}$. This traces out C_2 in reverse, since $t = 0$ gives $(0, 2)$, and $t = 1$ gives $(1, 0)$. Thus, we make $t = 0$ the upper limit of integration and $t = 1$ the lower limit:

$$\int_{C_2} \vec{F} \cdot d\vec{r} = \int_1^0 \vec{F}(t, -2t^2 + 2) \cdot (\vec{i} - 4t\vec{j})\, dt = -\int_0^1 (t\vec{i} + (-2t^2 + 2)\vec{j}) \cdot (\vec{i} - 4t\vec{j})\, dt$$

$$= -\int_0^1 (8t^3 - 7t)\, dt = \frac{3}{2}.$$

So the line integrals along C_1 and C_2 have the same value.

(b) We break $\int_C \vec{F} \cdot d\vec{r}$ into three pieces, one of which we have already computed (namely, the piece connecting $(1, 0)$ to $(0, 2)$, where the line integral has value $3/2$). The piece running from $(0, 2)$ to $(0, 0)$ can be parameterized by $\vec{r}(t) = (2 - t)\vec{j}$ with $0 \le t \le 2$. The piece running from $(0, 0)$ to $(1, 0)$ can be parameterized by $\vec{r}(t) = t\vec{i}$ with $0 \le t \le 1$. Then

$$\int_C \vec{F} \cdot d\vec{r} = \frac{3}{2} + \int_0^2 \vec{F}(0, 2 - t) \cdot (-\vec{j})\, dt + \int_0^1 \vec{F}(t, 0) \cdot \vec{i}\, dt$$

$$= \frac{3}{2} + \int_0^2 (2 - t)\vec{j} \cdot (-\vec{j})\, dt + \int_0^1 t\vec{i} \cdot \vec{i}\, dt$$

$$= \frac{3}{2} + \int_0^2 (t - 2)\, dt + \int_0^1 t\, dt = \frac{3}{2} + (-2) + \frac{1}{2} = 0.$$

Example 3   Let C be the closed curve consisting of the upper half-circle of radius 1 and the line forming its diameter along the x-axis, oriented counterclockwise. (See Figure 18.23.) Find $\int_C \vec{F} \cdot d\vec{r}$ where $\vec{F}(x, y) = -y\vec{i} + x\vec{j}$.

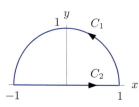

Figure 18.23: The curve $C = C_1 + C_2$ for Example 3

Solution   We write $C = C_1 + C_2$ where C_1 is the half-circle and C_2 is the line, and compute $\int_{C_1} \vec{F} \cdot d\vec{r}$ and $\int_{C_2} \vec{F} \cdot d\vec{r}$ separately. We parameterize C_1 by $\vec{r}(t) = \cos t \vec{i} + \sin t \vec{j}$, with $0 \leq t \leq \pi$. Then

$$\int_{C_1} \vec{F} \cdot d\vec{r} = \int_0^\pi (-\sin t \vec{i} + \cos t \vec{j}) \cdot (-\sin t \vec{i} + \cos t \vec{j}) \, dt$$

$$= \int_0^\pi (\sin^2 t + \cos^2 t) \, dt = \int_0^\pi 1 \, dt = \pi.$$

For C_2, we have $\int_{C_2} \vec{F} \cdot d\vec{r} = 0$, since the vector field $\vec{F}$ has no $\vec{i}$ component along the x-axis (where $y = 0$) and is therefore perpendicular to C_2 at all points.

Finally, we can write

$$\int_C \vec{F} \cdot d\vec{r} = \int_{C_1} \vec{F} \cdot d\vec{r} + \int_{C_2} \vec{F} \cdot d\vec{r} = \pi + 0 = \pi.$$

It is no accident that the result for $\int_{C_1} \vec{F} \cdot d\vec{r}$ is the same as the length of the curve C_1. See Problems 16–17 on page 839.

The next example illustrates the computation of a line integral over a path in 3-space.

Example 4   A particle travels along the helix C given by $\vec{r}(t) = \cos t \vec{i} + \sin t \vec{j} + 2t \vec{k}$ and is subject to a force $\vec{F} = x \vec{i} + z \vec{j} - xy \vec{k}$. Find the total work done on the particle by the force for $0 \leq t \leq 3\pi$.

Solution   The work done is given by a line integral, which we evaluate using the given parameterization:

$$\text{Work done} = \int_C \vec{F} \cdot d\vec{r} = \int_0^{3\pi} \vec{F}(\vec{r}(t)) \cdot \vec{r}'(t) \, dt$$

$$= \int_0^{3\pi} (\cos t \vec{i} + 2t \vec{j} - \cos t \sin t \vec{k}) \cdot (-\sin t \vec{i} + \cos t \vec{j} + 2\vec{k}) \, dt$$

$$= \int_0^{3\pi} (-\cos t \sin t + 2t \cos t - 2 \cos t \sin t) \, dt$$

$$= \int_0^{3\pi} (-3 \cos t \sin t + 2t \cos t) \, dt = -4.$$

The Notation $\int_C P \, dx + Q \, dy + R dz$

There is an alternative notation for line integrals that is quite common. Given functions $P(x, y, z)$ and $Q(x, y, z)$ and $R(x, y, z)$, and an oriented curve C, we consider the vector field $\vec{F} = P \vec{i} + Q \vec{j} + R \vec{k}$. We can write

$$\int_C \vec{F} \cdot d\vec{r} = \int_C P(x, y, z) dx + Q(x, y, z) dy + R(x, y, z) dz,$$

The relation between the two notations can be remembered using $d\vec{r} = dx \vec{i} + dy \vec{j} + dz \vec{k}$.

Example 5   Evaluate $\int_C xy \, dx - y^2 \, dy$ where C is the line segment from $(0, 0)$ to $(2, 6)$.

Solution   We parameterize C by $\vec{r}(t) = x(t)\vec{i} + y(t)\vec{j} = t\vec{i} + 3t\vec{j}$, for $0 \leq t \leq 2$. Thus,

$$\int_C xy \, dx - y^2 \, dy = \int_C (xy\vec{i} - y^2\vec{j}) \cdot d\vec{r} = \int_0^2 (3t^2\vec{i} - 9t^2\vec{j}) \cdot (\vec{i} + 3\vec{j}) \, dt = \int_0^2 (-24t^2) \, dt = -64.$$

Independence of Parameterization

Since there are many different ways of parameterizing a given oriented curve, you may be wondering what happens to the value of a given line integral if you choose another parameterization. The answer is that the choice of parameterization makes no difference. Since we initially defined the line integral without reference to any particular parameterization, this is exactly as we would expect.

Example 6 Consider the oriented path which is a straight line segment L running from $(0,0)$ to $(1,1)$. Calculate the line integral of the vector field $\vec{F} = (3x - y)\vec{i} + x\vec{j}$ along L using each of the parameterizations

(a) $A(t) = (t, t)$, $\quad 0 \le t \le 1$, $\qquad$ (b) $D(t) = (e^t - 1, e^t - 1)$, $\quad 0 \le t \le \ln 2$.

Solution The line L has equation $y = x$. Both $A(t)$ and $D(t)$ give a parameterization of L: each has both coordinates equal and each begins at (0,0) and ends at (1,1). Now let's calculate the line integral of the vector field $\vec{F} = (3x - y)\vec{i} + x\vec{j}$ using each parameterization.

(a) Using $A(t)$, we get

$$\int_L \vec{F} \cdot d\vec{r} = \int_0^1 ((3t - t)\vec{i} + t\vec{j}) \cdot (\vec{i} + \vec{j})\, dt = \int_0^1 3t\, dt = \left.\frac{3t^2}{2}\right|_0^1 = \frac{3}{2}.$$

(b) Using $D(t)$, we get

$$\int_L \vec{F} \cdot d\vec{r} = \int_0^{\ln 2} \left((3(e^t - 1) - (e^t - 1))\, \vec{i} + (e^t - 1)\vec{j}\right) \cdot (e^t\vec{i} + e^t\vec{j})\, dt$$

$$= \int_0^{\ln 2} 3(e^{2t} - e^t)\, dt = 3\left.\left(\frac{e^{2t}}{2} - e^t\right)\right|_0^{\ln 2} = \frac{3}{2}.$$

The fact that both answers are the same illustrates that the value of a line integral is independent of the parameterization of the path. Problems 19–21 at the end of this section give another way of seeing this.

Exercises and Problems for Section 18.2

Exercises

In Exercises 1–12, find $\int_C \vec{F} \cdot d\vec{r}$ for the given $\vec{F}$ and C.

1. $\vec{F} = x\vec{i} + y\vec{j}$ and C is the line from $(0, 0)$ to $(3, 3)$.

2. $\vec{F} = \ln y\vec{i} + \ln x\vec{j}$ and C is the curve given parametrically by $(2t, t^3)$, for $2 \le t \le 4$.

3. $\vec{F} = 2y\vec{i} - (\sin y)\vec{j}$ counterclockwise around the unit circle C starting at the point $(1, 0)$.

4. $\vec{F} = x^2\vec{i} + y^2\vec{j}$ and C is the line from the point $(1, 2)$ to the point $(3, 4)$.

5. $\vec{F} = -y \sin x\vec{i} + \cos x\vec{j}$ and C is the parabola $y = x^2$ between $(0, 0)$ and $(2, 4)$.

6. $\vec{F} = y\vec{i} - x\vec{j}$ and C is the right-hand side of the unit circle, starting at $(0, 1)$.

7. $\vec{F} = e^x\vec{i} + e^y\vec{j}$ and C is the part of the ellipse $x^2 + 4y^2 = 4$ joining the point $(0, 1)$ to the point $(2, 0)$ in the clockwise direction.

8. $\vec{F} = e^y\vec{i} + \ln(x^2 + 1)\vec{j} + \vec{k}$ and C is the circle of radius 2 centered at the origin in the yz-plane in Figure 18.24.

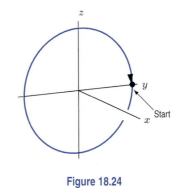

Figure 18.24

9. $\vec{F} = xy\vec{i} + (x-y)\vec{j}$ and C is the triangle joining $(1,0)$, $(0,1)$ and $(-1,0)$ in the clockwise direction.

10. $\vec{F} = x\vec{i} + 2zy\vec{j} + x\vec{k}$ and C is given by $\vec{r} = t\vec{i} + t^2\vec{j} + t^3\vec{k}$ for $1 \le t \le 2$.

11. $\vec{F} = x^3\vec{i} + y^2\vec{j} + z\vec{k}$ and C is the line from the origin to the point $(2,3,4)$.

12. $\vec{F} = -y\vec{i} + x\vec{j} + 5\vec{k}$ and C is the helix $x = \cos t$, $y = \sin t$, $z = t$, for $0 \le t \le 4\pi$.

Problems

13. In Example 6 on page 838 we integrated $\vec{F} = (3x - y)\vec{i} + x\vec{j}$ over two parameterizations of the line from $(0,0)$ to $(1,1)$, getting $3/2$ each time. Now compute the line integral along two different paths with the same endpoints, and show that the answers are different.

  (a) The path (t, t^2), with $0 \le t \le 1$
  (b) The path (t^2, t), with $0 \le t \le 1$

14. A spiral staircase in a building is in the shape of a helix of radius 5 meters. Between two floors of the building, the stairs make one full revolution and climb by 4 meters. A person carries a bag of groceries up two floors. The combined mass of the person and the groceries is 70 kg and the gravitational force is $70g$ downward, where g is the acceleration due to gravity. Calculate the work done by the person against gravity.

15. Suppose C is the line segment from the point $(0,0)$ to the point $(4,12)$ and $\vec{F} = xy\vec{i} + x\vec{j}$.

  (a) Is $\int_C \vec{F} \cdot d\vec{r}$ greater than, less than, or equal to zero? Give a geometric explanation.
  (b) A parameterization of C is $(x(t), y(t)) = (t, 3t)$ for $0 \le t \le 4$. Use this to compute $\int_C \vec{F} \cdot d\vec{r}$.
  (c) Suppose a particle leaves the point $(0,0)$, moves along the line towards the point $(4,12)$, stops before reaching it and backs up, stops again and reverses direction, then completes its journey to the endpoint. All travel takes place along the line segment joining the point $(0,0)$ to the point $(4,12)$. If we call this path C', explain why $\int_{C'} \vec{F} \cdot d\vec{r} = \int_C \vec{F} \cdot d\vec{r}$.
  (d) A parameterization for a path like C' is given, for $0 \le t \le 4$, by
  $$(x(t), y(t)) = \left(\frac{t^3 - 6t^2 + 11t}{3}, (t^3 - 6t^2 + 11t) \right).$$
  Check that this parameterization begins at the point $(0,0)$ and ends at the point $(4,12)$. Check also that all points of C' lie on the line segment connecting the point $(0,0)$ to the point $(4,12)$. What are the values of t at which the particle changes direction?

  (e) Find $\int_{C'} \vec{F} \cdot d\vec{r}$ using the parameterization in part (d). Do you get the same answer as in part (b)?

16. Let $\vec{F} = -y\vec{i} + x\vec{j}$ and let C be the unit circle oriented counterclockwise.

  (a) Show that $\vec{F}$ has a constant magnitude of 1 on C.
  (b) Show that $\vec{F}$ is always tangent to the circle C.
  (c) Show that $\int_C \vec{F} \cdot d\vec{r} = $ Length of C.

17. Along a curve C, a vector field $\vec{F}$ is everywhere tangent to C in the direction of orientation and has constant magnitude $\|\vec{F}\| = m$. Use the definition of the line integral to explain why
$$\int_C \vec{F} \cdot d\vec{r} = m \cdot \text{Length of } C.$$

18. Consider the oriented path which is a straight line segment L running from $(0,0)$ to $(1,1)$. Calculate the line integral of the vector field $\vec{F} = (3x - y)\vec{i} + x\vec{j}$ along L using each of the parameterizations

  (a) $B(t) = (2t, 2t)$, $\quad 0 \le t \le 1/2$,
  (b) $C(t) = \left(\dfrac{t^2 - 1}{3}, \dfrac{t^2 - 1}{3} \right)$, $\quad 1 \le t \le 2$,

In Example 6 on page 838 two parameterizations, $A(t)$, and $D(t)$, are used to convert a line integral into a definite integral. In Problem 18, two other parameterizations, $B(t)$ and $C(t)$, are used on the same line integral. In Problems 19–21 show that two definite integrals corresponding to two of the given parameterizations are equal by finding a substitution which converts one integral to the other. This gives us another way of seeing why changing the parameterization of the curve does not change the value of the line integral.

19. $A(t)$ and $B(t)$

20. $A(t)$ and $C(t)$

21. $A(t)$ and $D(t)$

18.3 GRADIENT FIELDS AND PATH-INDEPENDENT FIELDS

For a function, f, of one variable, the Fundamental Theorem of Calculus tells us that the definite integral of a rate of change, f', gives the total change in f:

$$\int_a^b f'(t)\,dt = f(b) - f(a).$$

What about functions of two or more variables? The quantity that describes the rate of change is the gradient vector field. If we know the gradient of a function f, can we compute the total change in f between two points? The answer is yes, using a line integral.

Finding the Total Change in f from grad f: The Fundamental Theorem

To find the change in f between two points P and Q, we choose a smooth path C from P to Q, then divide the path into many small pieces. See Figure 18.25.

First we estimate the change in f as we move through a displacement $\Delta \vec{r}_i$ from $\vec{r}_i$ to $\vec{r}_{i+1}$. Suppose $\vec{u}$ is a unit vector in the direction of $\Delta \vec{r}_i$. Then the change in f is given by

$$f(\vec{r}_{i+1}) - f(\vec{r}_i) \approx \text{Rate of change of } f \times \text{Distance moved in direction of } \vec{u}$$
$$= f_{\vec{u}}(\vec{r}_i)\|\Delta \vec{r}_i\|$$
$$= \text{grad} f \cdot \vec{u} \,\|\Delta \vec{r}_i\|$$
$$= \text{grad} f \cdot \Delta \vec{r}_i. \qquad \text{since } \Delta \vec{r}_i = \|\Delta \vec{r}_i\|\vec{u}$$

Therefore, summing over all pieces of the path, the total change in f is given by

$$\text{Total change} = f(Q) - f(P) \approx \sum_{i=0}^{n-1} \text{grad} f(\vec{r}_i) \cdot \Delta \vec{r}_i.$$

In the limit as $\|\Delta \vec{r}_i\|$ approaches zero, this suggests the following result:

Theorem 18.1: The Fundamental Theorem of Calculus for Line Integrals

Suppose C is a piecewise smooth oriented path with starting point P and endpoint Q. If f is a function whose gradient is continuous on the path C, then

$$\int_C \text{grad} f \cdot d\vec{r} = f(Q) - f(P).$$

Notice that there are many different paths from P to Q. (See Figure 18.26.) However, the value of the line integral $\int_C \text{grad} f \cdot d\vec{r}$ depends only on the endpoints of C; it does not depend on where C goes in between. Problem 31 on page 847 shows how the Fundamental Theorem for Line Integrals can be derived from the one-variable Fundamental Theorem of Calculus.

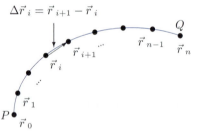

Figure 18.25: Subdivision of the path from P to Q. We estimate the change in f along $\Delta \vec{r}_i$

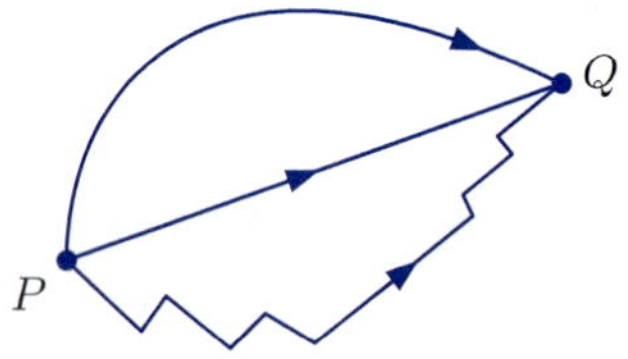

Figure 18.26: There are many different paths from P to Q: all give the same value of $\int_C \text{grad} f \cdot d\vec{r}$

Example 1   Suppose that $\text{grad} f$ is everywhere perpendicular to the curve joining P and Q shown in Figure 18.27.

(a) Explain why you expect the path joining P and Q to be a contour.

(b) Using a line integral, show that $f(P) = f(Q)$.

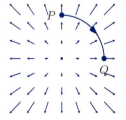

Figure 18.27: The gradient vector field of the function f

Solution

(a) The gradient of f is everywhere perpendicular to the path from P to Q, as you expect along a contour.

(b) Consider the path from P to Q shown in Figure 18.27 and evaluate the line integral

$$\int_C \operatorname{grad} f \cdot d\vec{r} = f(Q) - f(P).$$

Since $\operatorname{grad} f$ is everywhere perpendicular to the path, the line integral is 0. Thus, $f(Q) = f(P)$.

Example 2

Consider the vector field $\vec{F} = x\vec{i} + y\vec{j}$. In Example 2 on page 835 we calculated $\int_{C_1} \vec{F} \cdot d\vec{r}$ and $\int_{C_2} \vec{F} \cdot d\vec{r}$ over the oriented curves shown in Figure 18.28 and found they were the same. Find a scalar function f with $\operatorname{grad} f = \vec{F}$. Hence, find an easy way to calculate the line integrals, and explain how we could have expected them to be the same.

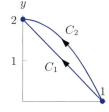

Figure 18.28: Find the line integral of $\vec{F} = x\vec{i} + y\vec{j}$ over the curves C_1 and C_2

Solution

One possibility for f is

$$f(x, y) = \frac{x^2}{2} + \frac{y^2}{2}.$$

You can check that $\operatorname{grad} f = x\vec{i} + y\vec{j}$. Now we can use the Fundamental Theorem to compute the line integral. Since $\vec{F} = \operatorname{grad} f$ we have

$$\int_{C_1} \vec{F} \cdot d\vec{r} = \int_{C_1} \operatorname{grad} f \cdot d\vec{r} = f(0, 2) - f(1, 0) = \frac{3}{2}.$$

Notice that the calculation looks exactly the same for C_2. Since the value of the integral depends only on the value of f at the endpoints, it is the same no matter what path we choose.

Path-Independent, or Conservative, Vector Fields

In the previous example, the line integral was independent of the path taken between the two (fixed) endpoints. We give vector fields whose line integrals have this property a special name.

> A vector field $\vec{F}$ is said to be **path-independent**, or **conservative**, if for any two points P and Q, the line integral $\int_C \vec{F} \cdot d\vec{r}$ has the same value along any piecewise smooth path C from P to Q lying in the domain of $\vec{F}$.

If, on the other hand, the line integral $\int_C \vec{F} \cdot d\vec{r}$ does depend on the path C joining P to Q, then $\vec{F}$ is said to be a *path-dependent* vector field.

Now suppose that $\vec{F}$ is any continuous gradient field, so $\vec{F} = \operatorname{grad} f$. If C is a path from P to Q, the Fundamental Theorem for Line Integrals tells us that

$$\int_C \vec{F} \cdot d\vec{r} = f(Q) - f(P).$$

Since the right-hand side of this equation does not depend on the path, but only on the endpoints of the path, the vector field $\vec{F}$ is path-independent. Thus, we have the following important result:

> If $\vec{F}$ is a continuous gradient vector field, then $\vec{F}$ is path-independent.

Why Do We Care about Path-Independent, or Conservative, Vector Fields?

Many of the fundamental vector fields of nature are path-independent — for example the gravitational field and the electric field of particles at rest. The fact that the gravitational field is path-independent means that the work done by gravity when an object moves depends only on the starting and ending points and not on the path taken. For example the work done by gravity (computed by the line integral) on a bicycle being carried to a sixth floor apartment is the same whether it is carried up the stairs in a zig-zag path or taken straight up in an elevator.

When a vector field is path-independent we can define the *potential energy* of a body. When the body moves to another position, the potential energy changes by an amount equal to the work done by the vector field, which depends only on the starting and ending positions. If the work done had not been path-independent, the potential energy would depend both on the body's current position *and* on how it got there, making it impossible to define a useful potential energy.

Project 1 on page 862 explains why path-independent force vector fields are also called *conservative* vector fields: When a particle moves under the influence of a conservative vector field, the *total energy* of the particle is *conserved*. It turns out that the force field is obtained from the gradient of the potential energy function.

Path-Independent Fields and Gradient Fields

We have seen that every gradient field is path-independent. What about the converse? That is, given a path-independent vector field $\vec{F}$, can we find a function f such that $\vec{F} = \operatorname{grad} f$? The answer is yes, provided that $\vec{F}$ is continuous.

How to Construct f from $\vec{F}$

First, notice that there are many different choices for f, since we can add a constant to f without changing $\operatorname{grad} f$. If we pick a fixed starting point P, then by adding or subtracting a constant to f we can ensure that $f(P) = 0$. For any other point Q, we define $f(Q)$ by the formula

$$f(Q) = \int_C \vec{F} \cdot d\vec{r}, \quad \text{where } C \text{ is any path from } P \text{ to } Q.$$

Since $\vec{F}$ is path-independent, it doesn't matter which path we choose from P to Q. On the other hand, if $\vec{F}$ is not path-independent, then different choices might give different values for $f(Q)$, so f would not be a function (a function has to have a single value at each point).

We still have to show that the gradient of the function f really is $\vec{F}$; we do this on page 844. However, by constructing a function f in this manner, we have the following result:

> **Theorem 18.2: Path-independent Fields are Gradient Fields**
>
> If $\vec{F}$ is a continuous path-independent vector field on an open region R, then $\vec{F} = \operatorname{grad} f$ for some f defined on R.

Combining Theorems 18.1 and 18.2, we have

> A continuous vector field $\vec{F}$ defined on an open region is path-independent if and only if $\vec{F}$ is a gradient vector field.

The function f is sufficiently important that it is given a special name:

> If a vector field $\vec{F}$ is of the form $\vec{F} = \operatorname{grad} f$ for some scalar function f, then f is called a **potential function** for the vector field $\vec{F}$.

Warning

Physicists use the convention that a function ϕ is a potential function for a vector field $\vec{F}$ if $\vec{F} = -\operatorname{grad} \phi$. See Problem 30 on page 847.

Example 3   Show that the vector field $\vec{F}(x, y) = y \cos x \vec{i} + \sin x \vec{j}$ is path-independent.

Solution   If we can find a potential function f, then $\vec{F}$ must be path-independent. We want $\operatorname{grad} f = \vec{F}$. Since

$$\frac{\partial f}{\partial x} = y \cos x,$$

f must be of the form

$$f(x, y) = y \sin x + g(y) \qquad \text{where } g(y) \text{ is a function of } y \text{ only.}$$

In addition, since $\operatorname{grad} f = \vec{F}$, we must have

$$\frac{\partial f}{\partial y} = \sin x,$$

and differentiating $f(x, y) = y \sin x + g(y)$ gives

$$\frac{\partial f}{\partial y} = \sin x + g'(y).$$

Thus, we must have $g'(y) = 0$, so $g(y) = C$ where C is some constant. Thus,

$$f(x, y) = y \sin x + C$$

is a potential function for $\vec{F}$. Therefore $\vec{F}$ is path-independent.

Example 4   The gravitational force field, $\vec{F}$, of an object of mass M, is given by

$$\vec{F} = -\frac{GM}{r^3} \vec{r}.$$

Show that $\vec{F}$ is a gradient field by finding a potential function for $\vec{F}$.

Solution   All the force vectors point in toward the origin. If $\vec{F} = \operatorname{grad} f$, the force vectors must be perpendicular to the level surfaces of f, so the level surfaces of f must be spheres. Also, if $\operatorname{grad} f = \vec{F}$, then $\|\operatorname{grad} f\| = \|\vec{F}\| = GM/r^2$ is the rate of change of f in the direction toward the origin. Now, differentiating with respect to r gives the rate of change in a radially outward direction. Thus, if $w = f(x, y, z)$ we have

$$\frac{dw}{dr} = -\frac{GM}{r^2} = GM\left(-\frac{1}{r^2}\right) = GM \frac{d}{dr}\left(\frac{1}{r}\right).$$

So let's try

$$w = \frac{GM}{r} \qquad \text{or} \quad f(x, y, z) = \frac{GM}{\sqrt{x^2 + y^2 + z^2}}.$$

We calculate

$$f_x = \frac{\partial}{\partial x}\frac{GM}{\sqrt{x^2 + y^2 + z^2}} = \frac{-GMx}{(x^2 + y^2 + z^2)^{3/2}},$$

$$f_y = \frac{\partial}{\partial y}\frac{GM}{\sqrt{x^2 + y^2 + z^2}} = \frac{-GMy}{(x^2 + y^2 + z^2)^{3/2}},$$

$$f_z = \frac{\partial}{\partial z}\frac{GM}{\sqrt{x^2 + y^2 + z^2}} = \frac{-GMz}{(x^2 + y^2 + z^2)^{3/2}}.$$

So

$$\text{grad } f = f_x\vec{i} + f_y\vec{j} + f_z\vec{k} = \frac{-GM}{(x^2 + y^2 + z^2)^{3/2}}(x\vec{i} + y\vec{j} + z\vec{k}) = \frac{-GM}{r^3}\vec{r} = \vec{F}.$$

Our computations show that $\vec{F}$ is a gradient field and that $f = GM/r$ is a potential function for $\vec{F}$.

Why Path-Independent Vector Fields are Gradient Fields: Showing grad $f = \vec{F}$

Suppose $\vec{F}$ is a path-independent vector field. On page 842 we defined the function f, which we hope will satisfy grad $f = \vec{F}$, as follows:

$$f(x_0, y_0) = \int_C \vec{F} \cdot d\vec{r},$$

where C is a path from a fixed starting point P to a point $Q = (x_0, y_0)$. This integral has the same value for any path from P to Q because $\vec{F}$ is path-independent. Now we show why grad $f = \vec{F}$. We consider vector fields in 2-space; the argument in 3-space is essentially the same.

First, we write the line integral in terms of the components $\vec{F}(x, y) = F_1(x, y)\vec{i} + F_2(x, y)\vec{j}$ and the components $d\vec{r} = dx\vec{i} + dy\vec{j}$:

$$f(x_0, y_0) = \int_C F_1(x, y)dx + F_2(x, y)dy.$$

We want to compute the partial derivatives of f, that is, the rate of change of f at (x_0, y_0) parallel to the axes. To do this easily, we choose a path which reaches the point (x_0, y_0) on a horizontal or vertical line segment. Let C' be a path from P which stops short of Q at a fixed point (a, b) and let L_x and L_y be the paths shown in Figure 18.29. Then we can split the line integral into three pieces. Since $d\vec{r} = \vec{j}\,dy$ on L_y and $d\vec{r} = \vec{i}\,dx$ on L_x, we have:

$$f(x_0, y_0) = \int_{C'} \vec{F}\cdot d\vec{r} + \int_{L_y} \vec{F}\cdot d\vec{r} + \int_{L_x} \vec{F}\cdot d\vec{r} = \int_{C'} \vec{F}\cdot d\vec{r} + \int_b^{y_0} F_2(a, y)dy + \int_a^{x_0} F_1(x, y_0)dx.$$

The first two integrals do not involve x_0. Thinking of x_0 as a variable and differentiating with respect to it gives

$$f_{x_0}(x_0, y_0) = \frac{\partial}{\partial x_0}\int_{C'} \vec{F} \cdot d\vec{r} + \frac{\partial}{\partial x_0}\int_b^{y_0} F_2(a, y)dy + \frac{\partial}{\partial x_0}\int_a^{x_0} F_1(x, y_0)dx$$
$$= 0 + 0 + F_1(x_0, y_0) = F_1(x_0, y_0),$$

and thus

$$f_x(x, y) = F_1(x, y).$$

A similar calculation for y using the path from P to Q shown in Figure 18.30 gives

$$f_{y_0}(x_0, y_0) = F_2(x_0, y_0).$$

Therefore, as we claimed

$$\text{grad } f = f_x\vec{i} + f_y\vec{j} = F_1\vec{i} + F_2\vec{j} = \vec{F}.$$

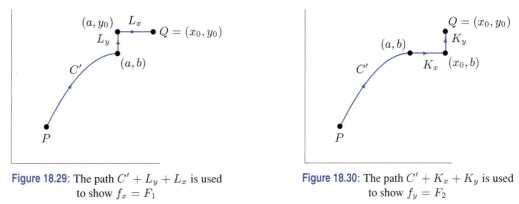

Figure 18.29: The path $C' + L_y + L_x$ is used to show $f_x = F_1$

Figure 18.30: The path $C' + K_x + K_y$ is used to show $f_y = F_2$

Summary

We have studied two apparently different types of vector field: path-independent vector fields and gradient vector fields. It turns out that these are the same. Here is a summary of the definitions and properties of these vector fields:

- **Path-independent vector fields** have the property that for any two points P and Q, the line integral along a path from P to Q is the same no matter what path we choose.

- **Gradient vector fields** are of the form grad f for some scalar function f, called the potential function of the vector field.

- **Gradient vector fields are path-independent** by the Fundamental Theorem of Line Integrals,

$$\int_C \text{grad } f \cdot d\vec{r} = f(Q) - f(P).$$

- **Path-independent vector fields are gradient fields** because we can use a line integral and path independence to construct a potential function.

Exercises and Problems for Section 18.3

Exercises

1. The vector field $\vec{F}(x, y) = x\vec{i} + y\vec{j}$ is path-independent. Compute geometrically the line integrals over the three paths A, B, and C shown in Figure 18.31 from $(1, 0)$ to $(0, 1)$ and check that they are equal. Here A is a portion of a circle, B is a line, and C consists of two line segments meeting at a right angle.

2. The vector field $\vec{F}(x, y) = x\vec{i} + y\vec{j}$ is path-independent. Compute algebraically the line integrals over the three paths A, B, and C shown in Figure 18.32 from $(0, 0)$ to $(1, 1)$ and check that they are equal. Here A is a line segment, B is part of the graph of $f(x) = x^2$, and C consists of two line segments meeting at a right angle.

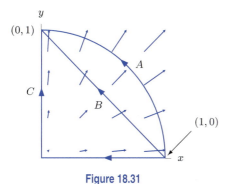

Figure 18.31

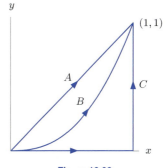

Figure 18.32

In Exercises 3–6, decide whether the vector field could be a gradient vector field. Justify your answer.

3. $\vec{F}(x, y) = x\vec{i}$

4. $\vec{G}(x, y) = (x^2 - y^2)\vec{i} - 2xy\vec{j}$

5. $\vec{F}(x, y, z) = \dfrac{-z}{\sqrt{x^2 + z^2}}\vec{i} + \dfrac{y}{\sqrt{x^2 + z^2}}\vec{j} + \dfrac{x}{\sqrt{x^2 + z^2}}\vec{k}$

6. $\vec{F}(\vec{r}) = \vec{r}/\|\vec{r}\|^3$, where $\vec{r} = x\vec{i} + y\vec{j} + z\vec{k}$

Do the vector fields in Exercises 7–10 appear to be path-independent (conservative)?

7.
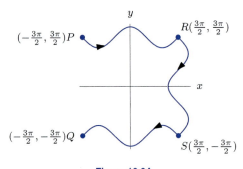

8.

9.
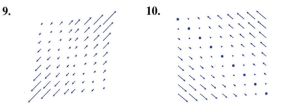

10.

In Exercises 11–14, use the Fundamental Theorem of Line Integrals to calculate $\int_C \vec{F} \cdot d\vec{r}$.

11. $\vec{F} = (x+2)\vec{i} + (2y+3)\vec{j}$ and C is the line from $(1, 0)$ to $(3, 1)$.

12. $\vec{F} = y\sin(xy)\vec{i} + x\sin(xy)\vec{j}$ and C is the parabola $y = 2x^2$ from $(1, 2)$ to $(3, 18)$.

13. $\vec{F} = 2x\vec{i} - 4y\vec{j} + (2z - 3)\vec{k}$ and C is the line from $(1, 1, 1)$ to $(2, 3, -1)$.

14. $\vec{F} = 2xy^2ze^{x^2y^2z}\vec{i} + 2x^2yze^{x^2y^2z}\vec{j} + x^2y^2e^{x^2y^2z}\vec{k}$ and C is the circle of radius 1 in the plane $z = 1$, centered on the z-axis, starting at $(1, 0, 1)$ and oriented counterclockwise viewed from above.

Problems

15. Let $\operatorname{grad} f = 2xe^{x^2}\sin y\vec{i} + e^{x^2}\cos y\vec{j}$. Find the change in f between $(0, 0)$ and $(1, \pi/2)$: **(a)** By computing a line integral   **(b)** By computing f.

16. The line integral of $\vec{F} = (x + y)\vec{i} + x\vec{j}$ along each of the following paths is $3/2$:

  (i) The path (t, t^2), with $0 \le t \le 1$
  (ii) The path (t^2, t), with $0 \le t \le 1$
  (iii) The path (t, t^n), with $n > 0$ and $0 \le t \le 1$

  Show this

  (a) Using the given parameterization to compute the line integral.
  (b) Using the Fundamental Theorem of Calculus for Line Integrals.

For the vector fields in Problems 17–20, find the line integral along the curve C from the origin along the x-axis to the point $(3, 0)$ and then counterclockwise around the circumference of the circle $x^2 + y^2 = 9$ to the point $(3/\sqrt{2}, 3/\sqrt{2})$.

17. $\vec{F} = x\vec{i} + y\vec{j}$

18. $\vec{H} = -y\vec{i} + x\vec{j}$

19. $\vec{F} = y(x + 1)^{-1}\vec{i} + \ln(x + 1)\vec{j}$

20. $\vec{G} = (ye^{xy} + \cos(x + y))\vec{i} + (xe^{xy} + \cos(x + y))\vec{j}$

21. Let $f(x, y, z) = x^2 + 2y^3 + 3z^4$ and $\vec{F} = \operatorname{grad} f$. Find $\int_C \vec{F} \cdot d\vec{r}$ where C consists of four line segments from $(4, 0, 0)$ to $(4, 3, 0)$ to $(0, 3, 0)$ to $(0, 3, 5)$ to $(0, 0, 5)$.

22. Let C be the helix $x = \cos t$, $y = \sin t$, $z = t$ for $0 \le t \le 1.25\pi$. Find $\int_C \vec{F} \cdot d\vec{r}$ exactly for
$$\vec{F} = yz^2e^{xyz^2}\vec{i} + xz^2e^{xyz^2}\vec{j} + 2xyze^{xyz^2}\vec{k}.$$

23. If $\vec{F}(x, y, z) = 2xe^{x^2+yz}\vec{i} + ze^{x^2+yz}\vec{j} + ye^{x^2+yz}\vec{k}$, find exactly the line integral of $\vec{F}$ along the curve consisting of the two half circles in the plane $z = 0$ in Figure 18.33.

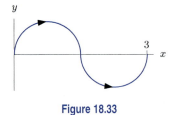

Figure 18.33

24. Calculate the line integral $\int_C \vec{F} \cdot d\vec{r}$ exactly, where C is the curve from P to Q in Figure 18.34 and
$$\vec{F} = \sin\left(\frac{x}{2}\right)\sin\left(\frac{y}{2}\right)\vec{i} - \cos\left(\frac{x}{2}\right)\cos\left(\frac{y}{2}\right)\vec{j}.$$
The curves PR, RS and SQ are trigonometric functions of period 2π and amplitude 1.

Figure 18.34

25. The force exerted by gravity on a refrigerator of mass m is $\vec{F} = -mg\vec{k}$.

   (a) Find the work done against this force in moving from the point $(1, 0, 0)$ to the point $(1, 0, 2\pi)$ along the curve $x = \cos t, y = \sin t, z = t$ by calculating a line integral.

   (b) Is $\vec{F}$ conservative (that is, path independent)? Give a reason for your answer.

26. A particle subject to a force $\vec{F}(x, y) = y\vec{i} - x\vec{j}$ moves clockwise along the arc of the unit circle, centered at the origin, that begins at $(-1, 0)$ and ends at $(0, 1)$.

   (a) Find the work done by $\vec{F}$. Explain the sign of your answer.

   (b) Is $\vec{F}$ path-independent? Explain.

27. Figure 18.35 shows the vector field $\vec{F}(x, y) = x\vec{j}$.

   (a) Find paths C_1, C_2, and C_3 from P to Q such that

$$\int_{C_1} \vec{F} \cdot d\vec{r} = 0, \quad \int_{C_2} \vec{F} \cdot d\vec{r} > 0, \quad \int_{C_3} \vec{F} \cdot d\vec{r} < 0.$$

   (b) Is $\vec{F}$ a gradient field? Explain.

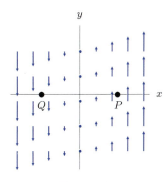

Figure 18.35

28. Consider the vector field $\vec{F}$ shown in Figure 18.36.

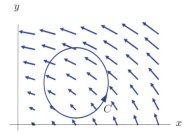

Figure 18.36

   (a) Is $\int_C \vec{F} \cdot d\vec{r}$ positive, negative, or zero?

   (b) From your answer to part (a), can you determine whether or not $\vec{F} = \operatorname{grad} f$ for some function f?

   (c) Which of the following formulas best fits $\vec{F}$?

$$\vec{F}_1 = \frac{x}{x^2 + y^2}\vec{i} + \frac{y}{x^2 + y^2}\vec{j},$$

$$\vec{F}_2 = -y\vec{i} + x\vec{j},$$

$$\vec{F}_3 = \frac{-y}{(x^2 + y^2)^2}\vec{i} + \frac{x}{(x^2 + y^2)^2}\vec{j}.$$

29. Figure 18.37 shows level curves of the function $f(x, y)$.

   (a) Sketch ∇f at P.

   (b) Is the vector ∇f at P longer than, shorter than, or the same length as, ∇f at Q?

   (c) If C is a curve from P to Q, evaluate $\int_C \nabla f \cdot d\vec{r}$.

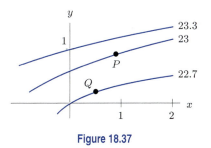

Figure 18.37

30. Let $\vec{F}$ be a path-independent vector field. In physics, the potential function ϕ is usually required to satisfy the equation $\vec{F} = -\nabla\phi$. This problem illustrates the significance of the negative sign.[1]

   (a) Let the xy-plane represent part of the earth's surface with the z-axis pointing upward. (The scale is small enough that a flat plane is a good approximation to the earth's surface.) Let $\vec{r} = x\vec{i} + y\vec{j} + z\vec{k}$, with $z \geq 0$, and x, y, z in meters, be the position vector of a rock of unit mass. The gravitational potential energy function for the rock is $\phi(x, y, z) = gz$, where $g \approx 9.8 \text{ m/sec}^2$. Describe in words the level surfaces of ϕ. Does the potential energy increase or decrease with height above the earth?

   (b) What is the relation between the gravitational vector, $\vec{F}$, and the vector $\nabla\phi$? Explain the significance of the negative sign in the equation $\vec{F} = -\nabla\phi$.

31. In this problem, we see how the Fundamental Theorem for Line Integrals can be derived from the Fundamental Theorem for ordinary definite integrals. Suppose that $(x(t), y(t))$, for $a \leq t \leq b$, is a parameterization of C, with endpoints $P = (x(a), y(a))$ and $Q = (x(b), y(b))$. The values of f along C are given by the single variable function $h(t) = f(x(t), y(t))$.

   (a) Use the chain rule to show that $h'(t) = f_x(x(t), y(t))x'(t) + f_y(x(t), y(t))y'(t)$.

   (b) Use the Fundamental Theorem of Calculus applied to $h(t)$ to show

$$\int_C \operatorname{grad} f \cdot d\vec{r} = f(Q) - f(P).$$

---

[1] Adapted from V.I. Arnold, *Mathematical Methods of Classical Mechanics*, 2nd Edition, Graduate Texts in Mathematics, Springer. Potential energy is also discussed in Project 1 on page 862.

18.4   PATH-DEPENDENT VECTOR FIELDS AND GREEN'S THEOREM

Suppose we are given a vector field but are not told whether it is path-independent. How can we tell if it has a potential function, that is, if it is a gradient field?

How to Tell if a Vector Field is Path-Dependent Using Line Integrals

Example 1   Is the vector field, $\vec{F}$, shown in Figure 18.38 path-independent? At any point $\vec{F}$ has magnitude equal to the distance from the origin and direction perpendicular to the line joining the point to the origin.

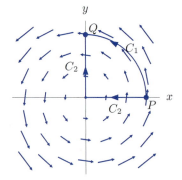

Figure 18.38: Is this vector field
path-independent?

Solution   We choose $P = (1, 0)$ and $Q = (0, 1)$ and two paths between them: C_1, a quarter circle of radius 1, and C_2, formed by parts of the x- and y-axes. (See Figure 18.38.) Along C_1, the line integral $\int_{C_1} \vec{F} \cdot d\vec{r} > 0$, since $\vec{F}$ points in the direction of the curve. Along C_2, however, we have $\int_{C_2} \vec{F} \cdot d\vec{r} = 0$, since $\vec{F}$ is perpendicular to C_2 everywhere. Thus, $\vec{F}$ is not path-independent.

Path-Dependent Fields and Circulation

Notice that the vector field in the previous example has nonzero circulation around the origin. What can we say about the circulation of a general path-independent vector field around a closed curve, C? Suppose C is a *simple* closed curve, that is, a curve which does not cross itself. If P and Q are any two points on the path, then we can think of C (oriented as shown in Figure 18.39) as made up of the path C_1 followed by $-C_2$. Since $\vec{F}$ is path-independent, we know that

$$\int_{C_1} \vec{F} \cdot d\vec{r} = \int_{C_2} \vec{F} \cdot d\vec{r}.$$

Thus, we see that the circulation around C is zero:

$$\int_C \vec{F} \cdot d\vec{r} = \int_{C_1} \vec{F} \cdot d\vec{r} + \int_{-C_2} \vec{F} \cdot d\vec{r} = \int_{C_1} \vec{F} \cdot d\vec{r} - \int_{C_2} \vec{F} \cdot d\vec{r} = 0.$$

If the curve C does cross itself, we break it into simple closed curves as shown in Figure 18.40 and apply the same argument to each one.

Now suppose we know that the line integral around any closed curve is zero. For any two points, P and Q, with two paths, C_1 and C_2, between them, create a closed curve, C, as in Figure 18.39. Since the circulation around this closed curve, C, is zero, the line integrals along the two paths, C_1 and C_2, are equal. Thus, $\vec{F}$ is path-independent.

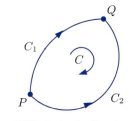

Figure 18.39: A simple closed curve C broken into two pieces, C_1 and C_2

Figure 18.40: A curve C which crosses itself can be broken into simple closed curves

Thus, we have the following result:

> A vector field is path-independent if and only if $\int_C \vec{F} \cdot d\vec{r} = 0$ for every closed curve C.

Hence, to see if a field is *path-dependent*, we look for a closed path with nonzero circulation. For instance, the vector field in Example 1 has nonzero circulation around a circle around the origin, showing it is path-dependent.

How to Tell If a Vector Field is Path-Dependent Algebraically: The Curl

Example 2   Does the vector field $\vec{F} = 2xy\vec{i} + xy\vec{j}$ have a potential function? If so, find it.

Solution   Let's suppose $\vec{F}$ does have a potential function, f, so $\vec{F} = \text{grad } f$. This means that

$$\frac{\partial f}{\partial x} = 2xy \quad \text{and} \quad \frac{\partial f}{\partial y} = xy.$$

Integrating the expression for $\partial f / \partial x$ shows that we must have

$$f(x, y) = x^2 y + C(y) \qquad \text{where } C(y) \text{ is a function of } y.$$

Differentiating this expression for $f(x, y)$ with respect to y and using the fact that $\partial f / \partial y = xy$, we get

$$\frac{\partial f}{\partial y} = x^2 + C'(y) = xy.$$

Thus, we must have

$$C'(y) = xy - x^2.$$

But this expression for $C'(y)$ is impossible because $C'(y)$ is a function of y alone. This argument shows that there is no potential function for the vector field $\vec{F}$.

Is there an easier way to see that a vector field has no potential function, other than by trying to find the potential function and failing? The answer is yes. First we look at a 2-dimensional vector field $\vec{F} = F_1\vec{i} + F_2\vec{j}$. If $\vec{F}$ is a gradient field, then there is a potential function f such that

$$\vec{F} = F_1\vec{i} + F_2\vec{j} = \frac{\partial f}{\partial x}\vec{i} + \frac{\partial f}{\partial y}\vec{j}.$$

Thus,

$$F_1 = \frac{\partial f}{\partial x} \quad \text{and} \quad F_2 = \frac{\partial f}{\partial y}.$$

Let us assume that f has continuous second partial derivatives. Then, by the equality of mixed partial derivatives

$$\frac{\partial F_1}{\partial y} = \frac{\partial^2 f}{\partial y \partial x} = \frac{\partial^2 f}{\partial x \partial y} = \frac{\partial F_2}{\partial x}$$

Thus we have the following result:

If $\vec{F}(x, y) = F_1\vec{i} + F_2\vec{j}$ is a gradient vector field with continuous partial derivatives, then

$$\frac{\partial F_2}{\partial x} - \frac{\partial F_1}{\partial y} = 0.$$

We call $\dfrac{\partial F_2}{\partial x} - \dfrac{\partial F_1}{\partial y}$ the 2-dimensional or scalar **curl** of the vector field $\vec{F}$.

Notice that we now know that if $\vec{F}$ is a gradient field, then its curl is 0. We do not (yet) know whether the converse is true. (That is: If the curl is 0, does $\vec{F}$ have to be a gradient field?) However, the curl already enables us to show that a vector field is *not* a gradient field.

Example 3   Show that $\vec{F} = 2xy\vec{i} + xy\vec{j}$ cannot be a gradient vector field.

Solution   We have $F_1 = 2xy$ and $F_2 = xy$. Since $\partial F_1/\partial y = 2x$ and $\partial F_2/\partial x = y$, in this case

$$\partial F_2/\partial x - \partial F_1/\partial y \neq 0$$

so $\vec{F}$ cannot be a gradient field.

We now have two ways of seeing that a vector field $\vec{F}$ in the plane is path-dependent. We can evaluate $\int_C \vec{F} \cdot d\vec{r}$ for some closed curve and find it is not zero, or we can show that $\partial F_2/\partial x - \partial F_1/\partial y \neq 0$. It's natural to think that

$$\int_C \vec{F} \cdot d\vec{r} \quad \text{and} \quad \frac{\partial F_2}{\partial x} - \frac{\partial F_1}{\partial y}$$

might be related. The relation is called Green's Theorem.

Green's Theorem

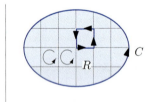

Figure 18.41: Region R bounded by a closed curve C and split into many small regions, ΔR

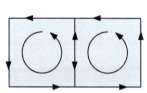

Figure 18.42: Two adjacent small closed curves

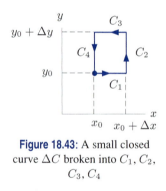

Figure 18.43: A small closed curve ΔC broken into C_1, C_2, C_3, C_4

We obtain the statement of Green's Theorem by splitting a region up into small pieces and looking at the relation between the line integral and the curl on each one.

In 2-space, a simple closed curve C separates the plane into points inside and points outside C. We consider $\int_C \vec{F} \cdot d\vec{r}$, where C is oriented as shown in Figure 18.41. We divide the region R inside C into small pieces, each bounded by a closed curve with the orientation shown. Figure 18.42 shows that if we add the circulation around all of these small closed curves, each common edge of a pair of adjacent curves is counted twice, once in each direction. Hence the integrals along these edges cancel. Thus, the line integrals along all the edges inside the region R cancel, giving

$$\begin{array}{c} \text{Circulation of } \vec{F} \\ \text{around } C \end{array} = \sum_{\Delta C} \begin{array}{c} \text{Circulation of } \vec{F} \\ \text{around small curve, } \Delta C \end{array}$$

Now we estimate the line integral around one of these small closed curves, ΔC. We break ΔC into C_1, C_2, C_3, and C_4, as shown in Figure 18.43. Then we calculate the line integrals along C_1 and C_3, where $\Delta \vec{r}$ is parallel to the x-axis, so

$$\Delta \vec{r} = \Delta x \vec{i}.$$

Thus, on C_1 and C_3,

$$\vec{F} \cdot \Delta \vec{r} = (F_1 \vec{i} + F_2 \vec{j}) \cdot \Delta x \vec{i} = F_1 \, \Delta x.$$

However, the function F_1 is evaluated at (x, y_0) on C_1 and at $(x, y_0 + \Delta y)$ on C_3, so

$$\int_{C_1} \vec{F} \cdot d\vec{r} + \int_{C_3} \vec{F} \cdot d\vec{r} = = \int_{C_1} F_1(x, y_0) \, dx + \int_{C_3} F_1(x, y_0 + \Delta y) \, dx$$

$$= \int_{x=x_0}^{x_0 + \Delta x} F_1(x, y_0) \, dx + \int_{x_0 + \Delta x}^{x=x_0} F_1(x, y_0 + \Delta y) \, dx$$

$$= \int_{x=x_0}^{x_0 + \Delta x} F_1(x, y_0) \, dx - \int_{x=x_0}^{x_0 + \Delta x} F_1(x, y_0 + \Delta y) \, dx$$

$$= \int_{x=x_0}^{x_0 + \Delta x} (F_1(x, y_0) - F_1(x, y_0 + \Delta y)) \, dx.$$

Since F_1 is differentiable and Δy is small,

$$F_1(x, y_0) - F_1(x, y_0 + \Delta y) = -(F_1(x, y_0 + \Delta y) - F_1(x, y_0)) \approx -\frac{\partial F_1}{\partial y}(x, y_0) \, \Delta y,$$

so we have

$$\int_{x=x_0}^{x_0 + \Delta x} (F_1(x, y_0) - F_1(x, y_0 + \Delta y)) \, dx \approx -\left(\int_{x=x_0}^{x_0 + \Delta x} \frac{\partial F_1}{\partial y} \, dx \right) \Delta y.$$

Assuming $\dfrac{\partial F_1}{\partial y}(x, y_0)$ is approximately constant over the interval $x_0 \leq x \leq x_0 + \Delta x$, we get

$$-\left(\int_{x=x_0}^{x_0 + \Delta x} \frac{\partial F_1}{\partial y} \, dx \right) \Delta y \approx -\frac{\partial F_1}{\partial y}(x_0, y_0) \left(\int_{x=x_0}^{x_0 + \Delta x} dx \right) \Delta y = -\frac{\partial F_1}{\partial y}(x_0, y_0) \, \Delta x \, \Delta y.$$

By a similar argument on C_2 and C_4, we have

$$\int_{C_2} \vec{F} \cdot d\vec{r} + \int_{C_4} \vec{F} \cdot d\vec{r} \approx \frac{\partial F_2}{\partial x} \, \Delta x \, \Delta y.$$

Combining these results for C_1, C_2, C_3, and C_4, we get

$$\int_{\Delta C} \vec{F} \cdot d\vec{r} = \int_{C_1} \vec{F} \cdot d\vec{r} + \int_{C_2} \vec{F} \cdot d\vec{r} + \int_{C_3} \vec{F} \cdot d\vec{r} + \int_{C_4} \vec{F} \cdot d\vec{r} \approx \frac{\partial F_2}{\partial x} \, \Delta x \, \Delta y - \frac{\partial F_1}{\partial y} \, \Delta x \, \Delta y.$$

Summing over all small regions ΔC gives

$$\int_C \vec{F} \cdot d\vec{r} \approx \sum_{\Delta C} \int_{\Delta C} \vec{F} \cdot d\vec{r} \approx \sum_{\Delta R} \left(\frac{\partial F_2}{\partial x} - \frac{\partial F_1}{\partial y} \right) \Delta x \, \Delta y.$$

The last sum is a Riemann sum approximating a double integral; taking the limit as Δx, Δy tend to zero, we get

---

Theorem 18.3: Green's Theorem

Suppose C is a piecewise smooth simple closed curve that is the boundary of an open region R in the plane and oriented so that the region is on the left as we move around the curve. Suppose $\vec{F} = F_1\vec{i} + F_2\vec{j}$ is a smooth vector field on an open region containing R and C. Then

$$\int_C \vec{F} \cdot d\vec{r} = \int_R \left(\frac{\partial F_2}{\partial x} - \frac{\partial F_1}{\partial y} \right) dx \, dy.$$

The online supplement contains a proof of Green's Theorem with different, but equivalent, conditions on the region R.

The Curl Test for Vector Fields in the Plane

We already know that if $\vec{F} = F_1\vec{i} + F_2\vec{j}$ is a gradient field with continuous partial derivatives, then

$$\frac{\partial F_2}{\partial x} - \frac{\partial F_1}{\partial y} = 0.$$

Now we show that the converse is true if the domain of $\vec{F}$ has no holes in it. This means that we assume that

$$\frac{\partial F_2}{\partial x} - \frac{\partial F_1}{\partial y} = 0$$

and show that $\vec{F}$ is path-independent. If C is any oriented closed curve in the domain of $\vec{F}$ and R is the region inside C, then

$$\int_R \left(\frac{\partial F_2}{\partial x} - \frac{\partial F_1}{\partial y} \right) dx \, dy = 0$$

since the integrand is identically 0. Therefore, by Green's Theorem

$$\int_C \vec{F} \cdot d\vec{r} = \int_R \left(\frac{\partial F_2}{\partial x} - \frac{\partial F_1}{\partial y} \right) dx dy = 0.$$

Thus, $\vec{F}$ is path-independent and therefore a gradient field. This argument is valid for every closed curve, C, provided the region R is entirely in the domain of $\vec{F}$. Thus we have the following result:

The Curl Test for Vector Fields in 2-Space

Suppose $\vec{F} = F_1\vec{i} + F_2\vec{j}$ is a vector field with continuous partial derivatives, such that
- The domain of $\vec{F}$ has the property that every closed curve in it encircles a region that lies entirely within the domain. In particular, the domain of $\vec{F}$ has no holes.
- $\dfrac{\partial F_2}{\partial x} - \dfrac{\partial F_1}{\partial y} = 0.$

Then $\vec{F}$ is path-independent, so $\vec{F}$ is a gradient field and has a potential function.

Why Are Holes in the Domain of the Vector Field Important?

The reason for assuming that the domain of the vector field $\vec{F}$ has no holes is to ensure that the region R inside C is actually contained in the domain of $\vec{F}$. Otherwise, we cannot apply Green's Theorem. The next two examples show that if $\partial F_2/\partial x - \partial F_1/\partial y = 0$ but the domain of $\vec{F}$ contains a hole, then $\vec{F}$ can either be path-independent or path-dependent.

Example 4   Let $\vec{F}$ be the vector field given by $\vec{F}(x, y) = \dfrac{-y\vec{i} + x\vec{j}}{x^2 + y^2}$.

(a) Calculate $\dfrac{\partial F_2}{\partial x} - \dfrac{\partial F_1}{\partial y}$. Does the curl test imply that $\vec{F}$ is path-independent?

(b) Calculate $\displaystyle\int_C \vec{F} \cdot d\vec{r}$, where C is the unit circle centered at the origin and oriented counterclockwise. Is $\vec{F}$ a path-independent vector field?

(c) Explain why the answers to parts (a) and (b) do not contradict Green's Theorem.

Solution   (a) Taking partial derivatives, we have

$$\frac{\partial F_2}{\partial x} = \frac{\partial}{\partial x}\left(\frac{x}{x^2 + y^2}\right) = \frac{1}{x^2 + y^2} - \frac{x \cdot 2x}{(x^2 + y^2)^2} = \frac{y^2 - x^2}{(x^2 + y^2)^2}.$$

Similarly,

$$\frac{\partial F_1}{\partial y} = \frac{\partial}{\partial y}\left(\frac{-y}{x^2 + y^2}\right) = \frac{-1}{x^2 + y^2} + \frac{y \cdot 2y}{(x^2 + y^2)^2} = \frac{y^2 - x^2}{(x^2 + y^2)^2}.$$

Thus,

$$\frac{\partial F_2}{\partial x} - \frac{\partial F_1}{\partial y} = 0.$$

Since $\vec{F}$ is undefined at the origin, the domain of $\vec{F}$ contains a hole. Therefore, the curl test does not apply.

(b) On the unit circle, $\vec{F}$ is tangent to the circle and $||\vec{F}|| = 1$. Thus,

$$\int_C \vec{F} \cdot d\vec{r} = ||\vec{F}|| \cdot \text{Length of curve} = 1 \cdot 2\pi = 2\pi.$$

Since the line integral around the closed curve C is nonzero, $\vec{F}$ is not path-independent.

(c) The domain of $\vec{F}$ is the "punctured plane," as shown in Figure 18.44. Since $\vec{F}$ is not defined at the origin, which is inside C, Green's Theorem does not apply. In this case

$$2\pi = \int_C \vec{F} \cdot d\vec{r} \neq \int_R \left(\frac{\partial F_2}{\partial x} - \frac{\partial F_1}{\partial y}\right) dx dy = 0.$$

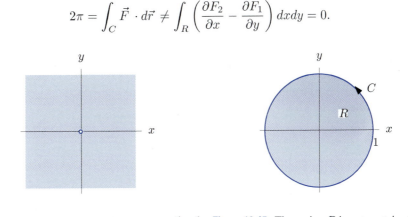

Figure 18.44: The domain of $\vec{F}(x, y) = \frac{-y\vec{i} + x\vec{j}}{x^2 + y^2}$ is the plane minus the origin

Figure 18.45: The region R is *not* contained in the domain of $\vec{F}(x, y) = \frac{-y\vec{i} + x\vec{j}}{x^2 + y^2}$

Although the vector field $\vec{F}$ in the last example was not defined at the origin, this by itself does not prevent the vector field from being path-independent as we see in the following example.

Example 5   Consider the vector field $\vec{F}$ given by $\vec{F}(x, y) = \dfrac{x\vec{i} + y\vec{j}}{x^2 + y^2}$.

(a) Calculate $\dfrac{\partial F_2}{\partial x} - \dfrac{\partial F_1}{\partial y}$. Does the curl test imply that $\vec{F}$ is path-independent?

(b) Explain how we know that $\displaystyle\int_C \vec{F} \cdot d\vec{r} = 0$, where C is the unit circle centered at the origin and oriented counterclockwise. Does this imply that $\vec{F}$ is path-independent?

(c) Check that $f(x, y) = \frac{1}{2} \ln(x^2 + y^2)$ is a potential function for $\vec{F}$. Does this imply that $\vec{F}$ is path-independent?

Solution   (a) Taking partial derivatives, we have

$$\frac{\partial F_2}{\partial x} = \frac{\partial}{\partial x}\left(\frac{y}{x^2 + y^2}\right) = \frac{-2xy}{(x^2 + y^2)^2}, \quad \text{and} \quad \frac{\partial F_1}{\partial y} = \frac{\partial}{\partial y}\left(\frac{x}{x^2 + y^2}\right) = \frac{-2xy}{(x^2 + y^2)^2}.$$

Therefore,

$$\frac{\partial F_2}{\partial x} - \frac{\partial F_1}{\partial y} = 0.$$

This does *not* imply that $\vec{F}$ is path-independent: The domain of $\vec{F}$ contains a hole since $\vec{F}$ is undefined at the origin. Thus, the curl test does not apply.

(b) Since $\vec{F}(x, y) = x\vec{i} + y\vec{j} = \vec{r}$ on the unit circle C, the field $\vec{F}$ is everywhere perpendicular to C. Thus

$$\int_C \vec{F} \cdot d\vec{r} = 0.$$

The fact that $\int_C \vec{F} \cdot d\vec{r} = 0$ when C is the unit circle does *not* imply that $\vec{F}$ is path-independent. To be sure that $\vec{F}$ is path-independent, we would have to show that $\int_C \vec{F} \cdot d\vec{r} = 0$ for *every* closed curve C in the domain of $\vec{F}$, not just the unit circle.

(c) To check that grad $f = \vec{F}$, we differentiate f:

$$f_x = \frac{1}{2}\frac{\partial}{\partial x}\ln(x^2 + y^2) = \frac{1}{2}\frac{2x}{x^2 + y^2} = \frac{x}{x^2 + y^2},$$

and

$$f_y = \frac{1}{2}\frac{\partial}{\partial y}\ln(x^2 + y^2) = \frac{1}{2}\frac{2y}{x^2 + y^2} = \frac{y}{x^2 + y^2},$$

so that

$$\text{grad } f = \frac{x\vec{i} + y\vec{j}}{x^2 + y^2} = \vec{F}.$$

Thus, $\vec{F}$ is a gradient field and therefore is path-independent — even though $\vec{F}$ is undefined at the origin.

The Curl Test for Vector Fields in 3-Space

The curl test is a convenient way of deciding whether a 2-dimensional vector field is path-independent. Fortunately, there is an analogous test for 3-dimensional vector fields, although we cannot justify it until Chapter 20.

If $\vec{F}(x, y, z) = F_1\vec{i} + F_2\vec{i} + F_3\vec{k}$ is a vector field on 3-space we define a new vector field, curl $\vec{F}$, on 3-space by

$$\text{curl } \vec{F} = \left(\frac{\partial F_3}{\partial y} - \frac{\partial F_2}{\partial z}\right)\vec{i} + \left(\frac{\partial F_1}{\partial z} - \frac{\partial F_3}{\partial x}\right)\vec{j} + \left(\frac{\partial F_2}{\partial x} - \frac{\partial F_1}{\partial y}\right)\vec{k}.$$

The vector field curl $\vec{F}$ can be used to determine whether the vector field $\vec{F}$ is path-independent.

The Curl Test for Vector Fields in 3-Space

Suppose $\vec{F}$ is a vector field on 3-space with continuous partial derivatives, such that
- The domain of $\vec{F}$ has the property that every closed curve in it can be contracted to a point in a smooth way, staying at all times within the domain.
- curl $\vec{F} = \vec{0}$.

Then $\vec{F}$ is path-independent, so $\vec{F}$ is a gradient field and has a potential function.

For the 2-dimensional curl test, the domain of $\vec{F}$ must have no holes. This meant that if $\vec{F}$ was defined on a closed curve C, then it was also defined at all points inside C. One way to test for holes is to try to "lasso" them with a closed curve. If every closed curve in the domain can be pulled to a point without hitting a hole, that is, without straying outside the domain, then the domain has no holes. In 3-space, we need the same condition to be satisfied; we must be able to pull every closed curve to a point, like a lasso, without straying outside the domain.

Example 6   Decide if the following vector fields are path-independent and whether or not the curl test applies.

(a) $\vec{F} = \dfrac{x\vec{i} + y\vec{j} + z\vec{k}}{(x^2 + y^2 + z^2)^{3/2}}$

(b) $\vec{G} = \dfrac{-y\vec{i} + x\vec{j}}{x^2 + y^2}$

Solution   (a) Suppose $f = -(x^2+y^2+z^2)^{-1/2}$. Then $f_x = x(x^2+y^2+z^2)^{-3/2}$ and similarly grad $f = \vec{F}$. Thus, $\vec{F}$ is a gradient field and therefore path-independent. Calculations show curl $\vec{F} = \vec{0}$. The domain of $\vec{F}$ is all of 3-space minus the origin, and any closed curve in the domain can be pulled to a point without leaving the domain. Thus, the curl test applies.

(b) Let C be the circle $x^2 + y^2 = 1, z = 0$ traversed counterclockwise when viewed from the positive z-axis. The vector field is everywhere tangent to this curve and of magnitude $1,^2$ so

$$\int_C \vec{G} \cdot d\vec{r} = \|\vec{G}\| \cdot \text{Length of curve} = 1 \cdot 2\pi = 2\pi.$$

Since the line integral around this closed curve is nonzero, $\vec{G}$ is path-dependent. Computations show curl $\vec{G} = \vec{0}$. However, the domain of $\vec{G}$ is all of 3-space minus the z-axis, and it does not satisfy the curl test domain criterion. For example, the circle, C, is lassoed around the z-axis, and cannot be pulled to a point without hitting the z-axis. Thus the curl test does not apply.

Exercises and Problems for Section 18.4

Exercises

1. Find f if grad $f = 2xy\vec{i} + (x^2 + 8y^3)\vec{j}$

2. Find f if grad $f = (yze^{xyz} + z^2\cos(xz^2))\vec{i} + xze^{xyz}\vec{j} + (xye^{xyz} + 2xz\cos(xz^2))\vec{k}$

In Exercises 3–8, decide if the given vector field is the gradient of a function f. If so, find f. If not, explain why not.

3. $y\vec{i} + y\vec{j}$

4. $(x^2 + y^2)\vec{i} + 2xy\vec{j}$

5. $(2xy^3 + y)\vec{i} + (3x^2y^2 + x)\vec{j}$

6. $\dfrac{\vec{i}}{x} + \dfrac{\vec{j}}{y} + \dfrac{\vec{k}}{xy}$

7. $\dfrac{\vec{i}}{x} + \dfrac{\vec{j}}{y} + \dfrac{\vec{k}}{z}$

8. $2x\cos(x^2 + z^2)\vec{i} + \sin(x^2 + z^2)\vec{j} + 2z\cos(x^2 + z^2)\vec{k}$

In Exercises 9–12, use Green's Theorem to calculate the circulation of $\vec{F}$ around the curve, oriented counterclockwise.

9. $\vec{F} = y\vec{i} - x\vec{j}$ around the unit circle.

10. $\vec{F} = xy\vec{j}$ around the square $0 \leq x \leq 1, 0 \leq y \leq 1$.

11. $\vec{F} = (2x^2 + 3y)\vec{i} + (2x + 3y^2)\vec{j}$ around the triangle with vertices $(2, 0), (0, 3), (-2, 0)$.

12. $\vec{F} = 3y\vec{i} + xy\vec{j}$ around the unit circle.

---

^{2}See Problem 17 on page 839.

Problems

13. Example 1 on page 848 showed that the vector field in Figure 18.46 could not be a gradient field by showing that it is not path-independent. Here is another way to see the same thing. Suppose that the vector field were the gradient of a function f. Draw and label a diagram showing what the contours of f would have to look like, and explain why it would not be possible for f to have a single value at any given point.

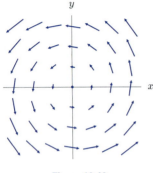

Figure 18.46

14. Repeat Problem 13 for the vector field in Problem 27 on page 847.

15. (a) Sketch $\vec{F} = y\vec{i}$ and hence decide the sign of the circulation of $\vec{F}$ around the unit circle centered at the origin and traversed counterclockwise.
  (b) Use Green's Theorem to compute the circulation in part (a) exactly.

16. Let $\vec{F} = (\sin x)\vec{i} + (x+y)\vec{j}$. Find the line integral of $\vec{F}$ around the perimeter of the rectangle with corners $(3,0)$, $(3,5)$, $(-1,5)$, $(-1,0)$, traversed in that order.

17. (a) Sketch the curves C_1 and C_2:

$$C_1 \text{ is } (x(t), y(t)) = (0, t) \quad \text{for } -1 \le t \le 1$$
$$C_2 \text{ is } (x(t), y(t)) = (\cos t, \sin t) \quad \text{for } \frac{\pi}{2} \le t \le \frac{3\pi}{2}.$$

  (b) Find $\int_C \vec{F} \cdot d\vec{r}$, where $\vec{F} = (x + 3y)\vec{i} + y\vec{j}$ and $C = C_1 + C_2$.

18. Find the line integral of $\vec{F} = (x - y)\vec{i} + x\vec{j}$ around the closed curve in Figure 18.47. (The arc is part of a circle.)

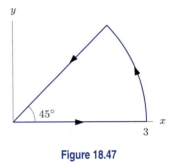

Figure 18.47

19. Show that the line integral of $\vec{F} = x\vec{j}$ around a closed curve in the xy-plane, oriented as in Green's Theorem, measures the area of the region enclosed by the curve.

Use the result of Problem 19 to calculate the area of the region within the parameterized curves in Problems 20–22. In each case, sketch the curve.

20. The ellipse $x^2/a^2 + y^2/b^2 = 1$ parameterized by $x = a\cos t$, $y = b\sin t$, for $0 \le t \le 2\pi$.

21. The hypocycloid $x^{2/3} + y^{2/3} = a^{2/3}$ parameterized by $x = a\cos^3 t$, $y = a\sin^3 t$, $0 \le t \le 2\pi$.

22. The folium of Descartes, $x^3 + y^3 = 3xy$, parameterized by $x = \dfrac{3t}{1+t^3}$, $y = \dfrac{3t^2}{1+t^3}$, for $0 \le t < \infty$.

23. Calculate $\int_C \left((x^2 - y)\vec{i} + (y^2 + x)\vec{j} \right) \cdot d\vec{r}$ if:

  (a) C is the circle $(x - 5)^2 + (y - 4)^2 = 9$ oriented counterclockwise.
  (b) C is the circle $(x - a)^2 + (y - b)^2 = R^2$ in the xy-plane oriented counterclockwise.

24. The following vector field $\vec{F}$ is defined on the disk D of radius 5 centered at the origin in the plane:

$$\vec{F} = (-y^3 + y\sin(xy))\vec{i} + (4x(1-y^2) + x\sin(xy))\vec{j}$$

Consider the line integral $\int_C \vec{F} \cdot d\vec{r}$, where C is some closed curve contained in D. For which C is the value of this integral the largest? [Hint: Assume C is a closed curve, made up of smooth pieces and never crossing itself, and oriented counterclockwise.]

CHAPTER SUMMARY

- **Line Integrals**
  Oriented curves, definition as a limit of a Riemann sum, work interpretation, circulation, algebraic properties, computing line integrals over parameterized curves, independence of parameterization.

- **Gradient Fields**

  Fundamental Theorem for line integrals, path-independent (conservative) fields and their relation to gradient fields, potential functions.

- **Green's Theorem**
  Statement of the theorem, curl test for path-independence.

REVIEW EXERCISES AND PROBLEMS FOR CHAPTER EIGHTEEN

Exercises

1. Figure 18.48 shows a vector field $\vec{F}$ and a curve C. Decide if $\int_C \vec{F} \cdot d\vec{r}$ is positive, zero, or negative.

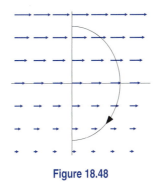

Figure 18.48

For the vector fields in Exercises 2–3, is the line integral positive, negative, or zero along

(a) A?   **(b)** C_1, C_2, C_3, C_4?
(c) C, the closed curve consisting of all the Cs together?

2.

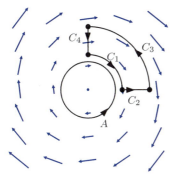

3.

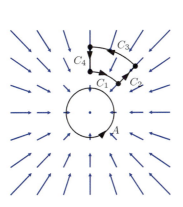

In Exercises 4–9, compute $\int_C \vec{F} \cdot d\vec{r}$ for the given vector fields and paths.

4. $\vec{F} = (x^2 - y)\vec{i} + (y^2 + x)\vec{j}$ and C is the parabola $y = x^2 + 1$ traversed from $(0, 1)$ to $(1, 2)$.

5. $\vec{F} = (3x - 2y)\vec{i} + (y + 2z)\vec{j} - x^2\vec{k}$ and C is the path consisting of the line from $(0, 0, 0)$ to $(1, 1, 1)$.

6. $\vec{F} = (2x - y + 4)\vec{i} + (5y + 3x - 6)\vec{j}$ and C is the triangle with vertices $(0, 0), (3, 0), (3, 2)$ traversed counterclockwise.

7. $\vec{F} = x\vec{i} + y\vec{j}$ and C is the unit circle in xy-plane oriented counterclockwise.

8. $\vec{F} = x\vec{i} + y\vec{j}$ and C is the y-axis from the origin to $(0, 10)$.

9. $\vec{F} = x\vec{i} + y\vec{j} + z\vec{k}$ and C is path consisting of a line from $(2, 3, 0)$ to $(4, 5, 0)$, followed by a line from $(4, 5, 0)$ to $(0, 0, 7)$.

10. Which two of the vector fields (i)-(iv) could represent gradient vector fields on the whole plane? Give reasons for your answer.

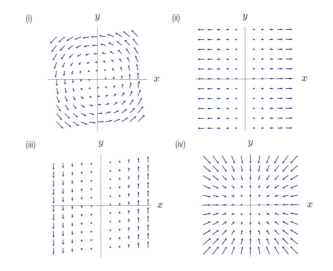

Which of the vector fields in Exercises 11–15 are path independent on all of 3-space?

11. $y\vec{i}$ **12.** $y\vec{j}$

13. $y\vec{i} + x\vec{j}$ **14.** $(x + y)\vec{i}$

15. $yz\vec{i} + zx\vec{j} + xy\vec{k}$

Problems

16. Let $\vec{F}(x, y)$ be the path-independent vector field in Figure 18.49. The vector field $\vec{F}$ associates with each point a unit vector pointing radially outward. The curves $C_1, C_2, \ldots, C_7$ have the directions shown. Consider the line integrals $\int_{C_i} \vec{F} \cdot d\vec{r}$, $i = 1, \ldots, 7$. Without computing any integrals

(a) List all the line integrals which you expect to be zero.

(b) List all the line integrals which you expect to be negative.

(c) Arrange the positive line integrals in ascending order.

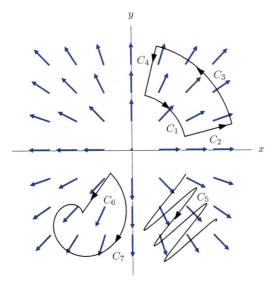

Figure 18.49

17. Let $\vec{F} = x\vec{i} + y\vec{j}$. Find the line integral of $\vec{F}$:

(a) Along the x-axis from the origin to the point $(3, 0)$.

(b) Around the path from A to B to O in Figure 18.50. (The curve is part of a circle centered at the origin.)

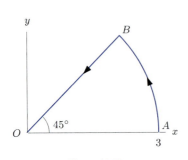

Figure 18.50

18. Calculate the line integral of $\vec{F} = -y\vec{i} + x\vec{j}$ along the following paths in the xy-plane.

(a) Straight line from the origin to the point $(2, 3)$.

(b) Straight line from $(2, 3)$ to $(0, 3)$.

(c) Counterclockwise around a circle of radius 5 centered at the origin, starting from $(5, 0)$ and ending at $(0, -5)$.

(d) Counterclockwise around the perimeter of a triangle of area 7.

19. Let $\vec{F} = (6x + y^2)\vec{i} + 2xy\vec{j}$ and $\vec{G} = (x - y)\vec{i} + (x + y)\vec{j}$. Let C_1 and C_2 be the curves in Figure 18.51. [Note C_1 is made up of line segments and C_2 is part of a circle.] Compute the following line integrals.

(a) $\displaystyle\int_{C_1} \vec{F} \cdot d\vec{r}$ (b) $\displaystyle\int_{C_1} \vec{G} \cdot d\vec{r}$

(c) $\displaystyle\int_{C_2} \vec{F} \cdot d\vec{r}$ (d) $\displaystyle\int_{C_2} \vec{G} \cdot d\vec{r}$

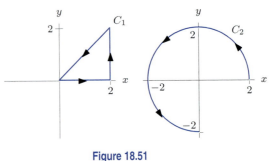

Figure 18.51

20. Find $\int_C \vec{F} \cdot d\vec{r}$ if C is an oriented curve from $(2, -6)$ to $(4, 4)$ and if $\vec{F} = 6\vec{i} - 7\vec{j}$.

21. Let $\vec{F} = x\vec{i} + y\vec{j}$ and let C_1 be the line joining $(1, 0)$ to $(0, 2)$ and let C_2 be the line joining $(0, 2)$ to $(-1, 0)$. Is $\int_{C_1} \vec{F} \cdot d\vec{r} = -\int_{C_2} \vec{F} \cdot d\vec{r}$? Explain.

22. Suppose P and Q both lie on the same contour of f. What can you say about the total change in f from P to Q? Explain your answer in terms of $\int_C \text{grad } f \cdot d\vec{r}$ where C is a part of the contour that goes from P to Q.

23. Consider the line integrals, $\int_{C_i} \vec{F} \cdot d\vec{r}$, for $i = 1, 2, 3, 4$, where C_i is the path from P_i to Q_i shown in Figure 18.52 and $\vec{F} = \text{grad } f$. Level curves of f are also shown in Figure 18.52.

(a) Which of the line integrals is (are) zero?

(b) Arrange the four line integrals in ascending order (from least to greatest).

(c) Two of the nonzero line integrals have equal and opposite values. Which are they? Which is negative and which is positive?

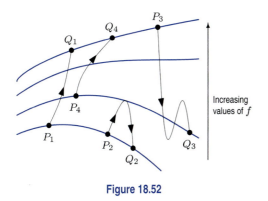

Figure 18.52

24. The vector field $\vec{F}$ has $\|\vec{F}\| \leq 7$ everywhere and C is the circle of radius 1 centered at the origin. What is the largest possible value of $\int_C \vec{F} \cdot d\vec{r}$? The smallest possible value? What conditions lead to these values?

25. The fact that an electric current gives rise to a magnetic field is the basis for some electric motors. Ampère's Law relates the magnetic field $\vec{B}$ to a steady current I. It says

$$\int_C \vec{B} \cdot d\vec{r} = kI$$

where I is the current[3] flowing through a closed curve C and k is a constant. Figure 18.53 shows a rod carrying a current and the magnetic field induced around the rod. If the rod is very long and thin, experiments show that the magnetic field $\vec{B}$ is tangent to every circle that is perpendicular to the rod and has center on the axis of the rod (like C in Figure 18.53). The magnitude of $\vec{B}$ is constant along every such circle. Use Ampère's Law to show that around a circle of radius r, the magnetic field due to a current I has magnitude given by

$$\|\vec{B}\| = \frac{kI}{2\pi r}.$$

(In other words, the strength of the field is inversely proportional to the radial distance from the rod.)

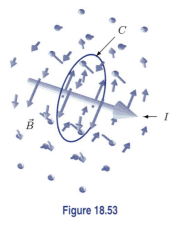

Figure 18.53

26. Figure 18.54 shows the tangential velocity as a function of radius for the tornado that hit Dallas on April 2, 1957. Use it and Problem 28 to estimate K and ω for the Rankine model of this tornado.[4]

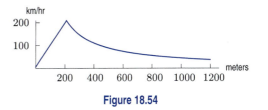

Figure 18.54

27. A *central vector field* is a vector field whose direction is always toward (or away from) a fixed point O (the center) and whose magnitude at a point P is a function only of the distance from P to O. In two dimensions this means that the vector field has constant magnitude on circles centered at O. The gravitational and electrical fields of spherically symmetric sources are both central fields.

(a) Sketch an example of a central vector field.
(b) Suppose that the central field $\vec{F}$ is a gradient field, that is, $\vec{F} = \text{grad } f$. What must be the shape of the contours of f? Sketch some contours for this case.
(c) Is every gradient field a central vector field? Explain.
(d) In Figure 18.55, two paths are shown between the points Q and P. Assuming that the three circles C_1, C_2, and C_3 are centered at O, explain why the work done by a central vector field $\vec{F}$ is the same for either path.
(e) It is in fact true that every central vector field is a gradient field. Use an argument suggested by Figure 18.55 to explain why any central vector field must be path-independent.

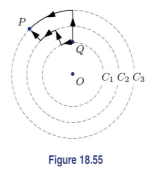

Figure 18.55

---

[3]More precisely, I is the net current through any surface that has C as its boundary.

[4]Adapted from *Encyclopedia Britannica, Macropedia*, Vol. 16, page 477, "Climate and the Weather", Tornados and Waterspouts, 1991.

28. A *free vortex* circulating about the origin in the xy-plane (or about the z-axis in 3-space) has vector field $\vec{v} = K(x^2 + y^2)^{-1}(-y\vec{i} + x\vec{j})$ where K is a constant. The Rankine model of a tornado hypothesizes an inner core that rotates at constant angular velocity, surrounded by a free vortex. Suppose that the inner core has radius 100 meters and that $\|\vec{v}\| = 3 \cdot 10^5$ meters/hr at a distance of 100 meters from the center.

(a) Assuming that the tornado rotates counterclockwise (viewed from above the xy-plane) and that $\vec{v}$ is con-

tinuous, determine ω and K such that

$$\vec{v} = \begin{cases} \omega(-y\vec{i} + x\vec{j}) & \text{if } \sqrt{x^2 + y^2} < 100 \\ K(x^2 + y^2)^{-1}(-y\vec{i} + x\vec{j}) \\ & \text{if } \sqrt{x^2 + y^2} \geq 100. \end{cases}$$

(b) Sketch the vector field $\vec{v}$.
(c) Find the circulation of $\vec{v}$ around the circle of radius r centered at the origin, traversed counterclockwise.

CAS Challenge Problems

29. Let C_a be the circle of radius a, centered at the origin, oriented in the counterclockwise direction, and let

$$\vec{F} = (-y + \frac{2}{3}y^3)\vec{i} + (2x - \frac{x^3}{3} + xy^2)\vec{j}.$$

(a) Evaluate $\int_{C_a} \vec{F} \cdot d\vec{r}$. For what positive value of a does the integral take its maximum value?
(b) Use Green's theorem to convert the integral to a double integral. Without evaluating the double integral, give a geometric explanation of the value of a you found in part (a).

30. If f is a potential function for the two-dimensional vector field $\vec{F}$ then the Fundamental Theorem of Calculus for Line Integrals says that $\int_C \vec{F} \cdot d\vec{r} = f(x, y) - f(0, 0)$

where C is any path from $(0, 0)$ to (x, y). Using this fact and choosing C to be a straight line, find potential functions for the following conservative fields (where a, b, c are constants):

(a) $\vec{F} = ay\vec{i} + ax\vec{j}$
(b) $\vec{F} = abye^{bxy}\vec{i} + (c + abxe^{bxy})\vec{j}$

31. Let $\vec{F} = (ax + by)\vec{i} + (cx + dy)\vec{j}$. Evaluate the line integral of $\vec{F}$ along the paths

$$C_1 : \vec{r}(t) = 2t\vec{i} + t^2\vec{j}, \quad 0 \leq t \leq 3$$
$$C_2 : \vec{r}(t) = 2(3 - t)\vec{i} + (3 - t)^2\vec{j}, \quad 0 \leq t \leq 3$$

Describe and explain the relationship between the two integrals.

CHECK YOUR UNDERSTANDING

In Problems 1–4, each of the statements is *false*. Explain why or give a counterexample.

1. If $\int_C \vec{F} \cdot d\vec{r} = 0$ for one particular closed path C, then $\vec{F}$ is path-independent.

2. $\int_C \vec{F} \cdot d\vec{r}$ is the total change in $\vec{F}$ along C.

3. If the vector fields $\vec{F}$ and $\vec{G}$ have $\int_C \vec{F} \cdot d\vec{r} = \int_C \vec{G} \cdot d\vec{r}$ for a particular path C, then $\vec{F} = \vec{G}$.

4. If the total change of a function f along a curve C is zero, then C must be a contour of f.

Are the statements in Problems 5–8 true or false? Explain why or give a counterexample.

5. $\int_C \vec{F} \cdot d\vec{r}$ is a vector.

6. $\int_C \vec{F} \cdot d\vec{r} = \vec{F}(Q) - \vec{F}(P)$ when P and Q are the endpoints of C.

7. The fact that the line integral of a vector field $\vec{F}$ is zero around the unit circle $x^2 + y^2 = 1$ means that $\vec{F}$ must be a gradient vector field.

8. Suppose C_1 is the unit square joining the points $(0, 0)$, $(1, 0)$, $(1, 1)$, $(0, 1)$ oriented clockwise and C_2 is the same square but traversed twice in the opposite direction. If $\int_{C_1} \vec{F} \cdot d\vec{r} = 3$, then $\int_{C_2} \vec{F} \cdot d\vec{r} = -6$.

Are the statements in Problems 9–41 true or false? Give reasons for your answer.

9. The line integral $\int_C \vec{F} \cdot d\vec{r}$ is a scalar.

10. If C_1 and C_2 are oriented curves, and the length of C_1 is greater than the length of C_2, then $\int_{C_1} \vec{F} \cdot d\vec{r} > \int_{C_2} \vec{F} \cdot d\vec{r}$.

11. If C is an oriented curve and $\int_C \vec{F} \cdot d\vec{r} = 0$, then $\vec{F} = \vec{0}$.

12. If $\vec{F} = \vec{i}$ is a vector field in 2-space, then $\int_C \vec{F} \cdot d\vec{r} > 0$, where C is the oriented line from $(0, 0)$ to $(1, 0)$.

13. If $\vec{F} = \vec{i}$ is a vector field in 2-space, then $\int_C \vec{F} \cdot d\vec{r} > 0$, where C is the oriented line from $(0, 0)$ to $(0, 1)$.

14. If C_1 is the upper semicircle $x^2 + y^2 = 1, y \geq 0$ and C_2 is the lower semicircle $x^2 + y^2 = 1, y \leq 0$, both oriented counterclockwise, then for any vector field $\vec{F}$, we have $\int_{C_1} \vec{F} \cdot d\vec{r} = -\int_{C_2} \vec{F} \cdot d\vec{r}$.

15. If C_1 and C_2 are oriented curves with C_2 beginning where C_1 ends, then $\int_{C_1 + C_2} \vec{F} \cdot d\vec{r} > \int_{C_1} \vec{F} \cdot d\vec{r}$.

16. The work done by the force $\vec{F} = -y\vec{i} + x\vec{j}$ on a particle moving clockwise around the boundary of the square $-1 \leq x \leq 1, -1 \leq y \leq 1$ is positive.

17. The circulation of any vector field $\vec{F}$ around any closed curve C is zero.

18. The line integral $\int_C 4\vec{i} \cdot d\vec{r}$ over the curve C parameterized by $\vec{r}(t) = t\vec{i} + t^2\vec{j}$, for $0 \leq t \leq 2$, is positive.

19. If C_1 is the curve parameterized by $\vec{r}_1(t) = \cos t\vec{i} + \sin t\vec{j}$, with $0 \leq t \leq \pi$, and C_2 is the curve parameterized by $\vec{r}_2(t) = \cos t\vec{i} - \sin t\vec{j}, 0 \leq t \leq \pi$, then for any vector field $\vec{F}$ we have $\int_{C_1} \vec{F} \cdot d\vec{r} = \int_{C_2} \vec{F} \cdot d\vec{r}$.

20. If C_1 is the curve parameterized by $\vec{r}_1(t) = \cos t\vec{i} + \sin t\vec{j}$, with $0 \leq t \leq \pi$, and C_2 is the curve parameterized by $\vec{r}_2(t) = \cos(2t)\vec{i} + \sin(2t)\vec{j}, 0 \leq t \leq \frac{\pi}{2}$, then for any vector field $\vec{F}$ we have $\int_{C_1} \vec{F} \cdot d\vec{r} = \int_{C_2} \vec{F} \cdot d\vec{r}$.

21. If C is the curve parameterized by $\vec{r}(t)$, for $a \leq t \leq b$ with $\vec{r}(a) = \vec{r}(b)$, then $\int_C \vec{F} \cdot d\vec{r} = 0$ for any vector field $\vec{F}$. (Note that C starts and ends at the same place.)

22. If C_1 is the line segment from $(0,0)$ to $(1,0)$ and C_2 is the line segment from $(0,0)$ to $(2,0)$, then for any vector field $\vec{F}$, we have $\int_{C_2} \vec{F} \cdot d\vec{r} = 2\int_{C_1} \vec{F} \cdot d\vec{r}$.

23. If C is a circle of radius a, centered at the origin and oriented counterclockwise, then $\int_C 2x\vec{i} + y\vec{j} \cdot d\vec{r} = 0$.

24. If C is a circle of radius a, centered at the origin and oriented counterclockwise, then $\int_C 2y\vec{i} + x\vec{j} \cdot d\vec{r} = 0$.

25. If C_1 is the curve parameterized by $\vec{r}_1(t) = t\vec{i} + t^2\vec{j}$, with $0 \leq t \leq 2$, and C_2 is the curve parameterized by $\vec{r}_2(t) = (2-t)\vec{i} + (2-t)^2\vec{j}, 0 \leq t \leq 2$, then for any vector field $\vec{F}$ we have $\int_{C_1} \vec{F} \cdot d\vec{r} = -\int_{C_2} \vec{F} \cdot d\vec{r}$.

26. If $\vec{F} = \text{grad } f$, then $\vec{F}$ is path-independent.

27. If $\vec{F}$ is path-independent, then $\int_{C_1} \vec{F} \cdot d\vec{r} = \int_{C_2} \vec{F} \cdot d\vec{r}$, where C_1 and C_2 are any paths.

28. The line integral $\int_C \vec{F} \cdot d\vec{r}$ is the total change of $\vec{F}$ along C.

29. If $\vec{F}$ is path-independent, then there is a potential function for $\vec{F}$.

30. If $f(x,y) = e^{\cos(xy)}$, and C_1 is the upper semicircle $x^2 + y^2 = 1$ from $(-1,0)$ to $(1,0)$, and C_2 is the line from $(-1,0)$ to $(1,0)$, then $\int_{C_1} \text{grad } f \cdot d\vec{r} = \int_{C_2} \text{grad } f \cdot d\vec{r}$.

31. If $\vec{F}$ is path-independent, and C is any closed curve, then $\int_C \vec{F} \cdot d\vec{r} = 0$.

32. The vector field $\vec{F}(x,y) = y^2\vec{i} + k\vec{j}$, where k is constant, is a gradient field.

33. If $\int_C \vec{F} \cdot d\vec{r} = 0$, where C is any circle of the form $x^2 + y^2 = a^2$, then $\vec{F}$ is path-independent.

34. If $f(x)$ and $g(y)$ are continuous one-variable functions, then the vector field $\vec{F} = f(x)\vec{i} + g(y)\vec{j}$ is path-independent.

35. If $\vec{F} = \text{grad } f$, and C is the perimeter of a square of side length a oriented counterclockwise and surrounding the region R, then

$$\int_C \vec{F} \cdot d\vec{r} = \int_R f \, dA.$$

36. If $\vec{F}$ and $\vec{G}$ are both path-independent vector fields, then $\vec{F} + \vec{G}$ is path-independent.

37. If $\vec{F}$ and $\vec{G}$ are both path-dependent vector fields, then $\vec{F} + \vec{G}$ is path-dependent.

38. The vector field $\vec{F}(\vec{r}) = \vec{r}$ in 3-space is path-independent.

39. A constant vector field $\vec{F} = a\vec{i} + b\vec{j}$ is path-independent.

40. If $\vec{F}$ is path-independent and k is a constant, then the vector field $k\vec{F}$ is path-independent.

41. If $\vec{F}$ is path-independent and $h(x,y)$ is a scalar function, then the vector field $h(x,y)\vec{F}$ is path-independent.

PROJECTS FOR CHAPTER EIGHTEEN

1. **Conservation of Energy**

   (a) A particle moves with position vector $\vec{r}(t) = x(t)\vec{i} + y(t)\vec{j} + z(t)\vec{k}$. Let $\vec{v}(t)$ and $\vec{a}(t)$ be its velocity and acceleration vectors. Show that

   $$\frac{1}{2}\frac{d}{dt}\|\vec{v}(t)\|^2 = \vec{a}(t) \cdot \vec{v}(t).$$

   (b) We now derive the principle of Conservation of Energy. The kinetic energy of a particle of mass m moving with speed v is $(1/2)mv^2$. Suppose the particle has potential energy $f(\vec{r})$ at the position $\vec{r}$ due to a force field $\vec{F} = -\nabla f$. If the particle moves with position vector $\vec{r}(t)$ and velocity $\vec{v}(t)$, then the Conservation of Energy principle says that

   $$\text{Total energy} = \text{Kinetic energy} + \text{Potential energy} = \frac{1}{2}m\|\vec{v}(t)\|^2 + f(\vec{r}(t)) = \text{Constant}.$$

Let P and Q be two points in space and let C be a path from P to Q parameterized by $\vec{r}(t)$ for $t_0 \leq t \leq t_1$, where $\vec{r}(t_0) = P$ and $\vec{r}(t_1) = Q$.

(i) Using part (a) and Newton's law $\vec{F} = m\vec{a}$, show

$$\begin{array}{c} \text{Work done by } \vec{F} \\ \text{as particle moves along } C \end{array} = \text{Kinetic energy at } Q - \text{Kinetic energy at } P.$$

(ii) Use the Fundamental Theorem of Calculus for Line Integrals to show that

$$\begin{array}{c} \text{Work done by } \vec{F} \\ \text{as particle moves along } C \end{array} = \text{Potential energy at } P - \text{Potential energy at } Q.$$

(iii) Use parts (a) and (b) to show that the total energy at P is the same as at Q.

This problem explains why force vector fields which are *path-independent* are usually called *conservative* (force) vector fields.

2. **Ampère's Law**

Ampère's Law, introduced in Problem 25 on page 859 , relates the net current, I, flowing through a surface to the magnetic field around the boundary, C, of the surface. The law says that

$$\int_C \vec{B} \cdot d\vec{r} = kI, \quad \text{for some constant } k.$$

The orientation of C is given by the right hand rule; if the fingers of your right hand lie against the surface and curl in the direction of the oriented curve C, then your thumb points in the direction of positive current.

(a) We consider an infinitely long cylindrical wire having a radius r_0, where $r_0 > 0$. Suppose the wire is centered on the z-axis and carries a constant current, I, uniformly distributed across a cross-section of the wire. The magnitude of $\vec{B}$ is constant along any circle which is centered on and perpendicular to the z-axis. The direction of $\vec{B}$ is tangent to such circles. Show that the magnitude of $\vec{B}$, at a distance r from the z-axis, is given by

$$\|\vec{B}\| = \begin{cases} \dfrac{aI}{2\pi r} & \text{for } r \geq r_0 \\ \dfrac{aIr}{2\pi r_0^2} & \text{for } r < r_0. \end{cases}$$

(b) A *torus* is a doughnut-shaped surface obtained by rotating around the z-axis the circle $(x - \beta)^2 + z^2 = \alpha^2$ of radius α and center $(\beta, 0, 0)$, where $0 < \alpha < \beta$. A *toroidal solenoid* is constructed by wrapping a thin wire a large number, say N, times around the torus. Experiments show that if a constant current I flows though the wire, then the magnitude of the magnetic field is constant on circles inside the torus which are centered on and perpendicular to the z-axis. The direction of $\vec{B}$ is tangent to such circles. Explain why, at all points inside the torus,

$$\|\vec{B}\| = \frac{aNI}{2\pi r},$$

and $\|\vec{B}\| = 0$ otherwise. [Hint: Apply Ampère's Law to a suitably chosen surface S with boundary curve C. What is the net current passing through S?]

Chapter Nineteen

FLUX INTEGRALS

In the previous chapter we saw how to integrate vector fields along curves. In this chapter we shall define a new sort of integral, the flux integral, which goes over a surface rather than along a curve. If we view a vector field as representing the velocity of a fluid flow, the flux integral tells us about the rate at which fluid is flowing through the surface. In addition, the flux integral appears in the theory of electricity and magnetism.

19.1 THE IDEA OF A FLUX INTEGRAL

Flow Through a Surface

Imagine water flowing through a fishing net stretched across a stream. Suppose we want to measure the flow rate of water through the net, that is, the volume of fluid that passes through the surface per unit time. (See Figure 19.1.) This flow rate is called the *flux* of the fluid through the surface. We can also compute the flux of vector fields, such as electric and magnetic fields, where no flow is actually taking place.

Figure 19.1: Flux measures rate of flow
through a surface

Orientation of a Surface

Before computing the flux of a vector field through a surface, we need to decide which direction of flow through the surface is the positive direction; this is described as choosing an orientation. [1]

> At each point on a smooth surface there are two unit normals, one in each direction. **Choosing an orientation** means picking one of these normals at every point of the surface in a continuous way. The normal vector in the direction of the orientation is denoted by $\vec{n}$. For a closed surface, we usually choose the outward orientation.

We say the flux through a piece of surface is positive if the flow is in the direction of the orientation and negative if it is in the opposite direction. (See Figure 19.2.)

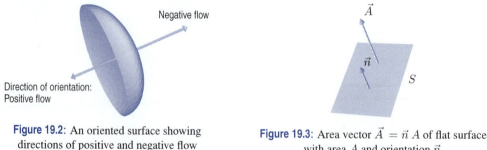

Figure 19.2: An oriented surface showing
directions of positive and negative flow

Figure 19.3: Area vector $\vec{A} = \vec{n}\,A$ of flat surface
with area A and orientation $\vec{n}$

The Area Vector

The flux through a flat surface depends both on the area of the surface and its orientation. Thus, it is useful to represent its area by a vector as shown in Figure 19.3.

---

[1] Although we will not study them, there are a few surfaces for which this cannot be done. See page 869.

> The **area vector** of a flat, oriented surface is a vector $\vec{A}$ such that
> - The magnitude of $\vec{A}$ is the area of the surface.
> - The direction of $\vec{A}$ is the direction of the orientation vector $\vec{n}$.

The Flux of a Constant Vector Field Through a Flat Surface

Suppose the velocity vector field, $\vec{v}$, of a fluid is constant and $\vec{A}$ is the area vector of a flat surface. The flux through this surface is the volume of fluid that flows through in one unit of time. The volume of the skewed box in Figure 19.4 has cross-sectional area $\|\vec{A}\|$ and height $\|\vec{v}\|\cos\theta$, so its volume is $(\|\vec{v}\|\cos\theta)\|\vec{A}\| = \vec{v}\cdot\vec{A}$. Thus we have the following result:

> If $\vec{v}$ is constant and $\vec{A}$ is the area vector of a flat surface, then
>
> $$\text{Flux through surface} = \vec{v}\cdot\vec{A}.$$

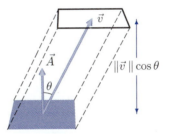

Figure 19.4: Flux of $\vec{v}$ through a surface with area vector $\vec{A}$ is the volume of this skewed box

Example 1   Water is flowing down a cylindrical pipe 2 cm in radius with a velocity of 3 cm/sec. Find the flux of the velocity vector field through the ellipse-shaped region shown in Figure 19.5. The normal to the ellipse makes an angle of θ with the direction of flow and the area of the ellipse is $4\pi/(\cos\theta)$ cm^2.

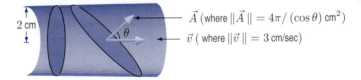

Figure 19.5: Flux through ellipse-shaped region across a cylindrical pipe

Solution   There are two ways to approach this problem. One is to use the formula we just derived which gives
$$\text{Flux through ellipse} = \vec{v}\cdot\vec{A} = \|\vec{v}\|\|\vec{A}\|\cos\theta = 3(\text{Area of ellipse})\cos\theta$$
$$= 3\left(\frac{4\pi}{\cos\theta}\right)\cos\theta = 12\pi \text{ cm}^3/\text{sec}.$$

The second way is to notice that the flux through the ellipse is equal to the flux through the circle perpendicular to the pipe in Figure 19.5. Since the flux is the rate at which water is flowing down the pipe, we have

$$\text{Flux through circle} = \begin{array}{c}\text{Velocity}\\ \text{of water}\end{array} \times \begin{array}{c}\text{Area of}\\ \text{circle}\end{array} = \left(3\,\frac{\text{cm}}{\text{sec}}\right)(\pi 2^2\text{ cm}^2) = 12\pi \text{ cm}^3/\text{sec}.$$

When the vector field is not constant or the surface is not flat, we divide the surface up into small, almost flat pieces such that the vector field is approximately constant on each one, as follows.

The Flux Integral

To calculate the flux of a vector field $\vec{F}$ which is not necessarily constant through a curved, oriented surface S, we divide the surface into a patchwork of small, approximately flat pieces (like a wire-frame representation of the surface) as shown in Figure 19.6. Suppose a particular patch has area ΔA. We pick an orientation vector $\vec{n}$ at a point on the patch and define the area vector of the patch, $\Delta \vec{A}$, as

$$\Delta \vec{A} = \vec{n} \, \Delta A.$$

(See Figure 19.6.) If the patches are small enough, we can assume that $\vec{F}$ is approximately constant on each piece. Then we know that

$$\text{Flux through patch} \approx \vec{F} \cdot \Delta \vec{A}$$

and so

$$\text{Flux through whole surface} \approx \sum \vec{F} \cdot \Delta \vec{A},$$

where the sum adds the fluxes through all the small pieces. As each patch becomes smaller and $\|\Delta \vec{A}\| \to 0$, the approximation gets better and we get

$$\text{Flux through } S = \lim_{\|\Delta \vec{A}\| \to 0} \sum \vec{F} \cdot \Delta \vec{A}.$$

Thus, provided the limit exists, we make the following definition:

> The **flux integral** of the vector field $\vec{F}$ through the oriented surface S is
>
> $$\int_S \vec{F} \cdot d\vec{A} = \lim_{\|\Delta \vec{A}\| \to 0} \sum \vec{F} \cdot \Delta \vec{A}.$$

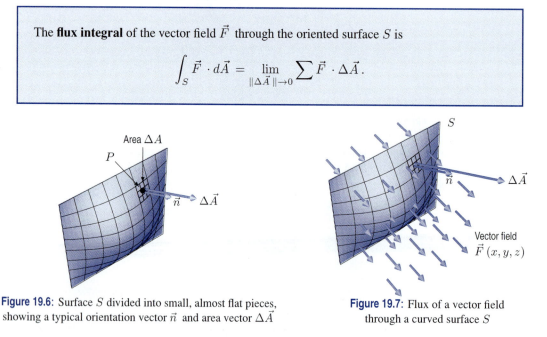

Figure 19.6: Surface S divided into small, almost flat pieces, showing a typical orientation vector $\vec{n}$ and area vector $\Delta \vec{A}$

Figure 19.7: Flux of a vector field through a curved surface S

In computing a flux integral, we have to divide the surface up in a reasonable way, or the limit might not exist. In practice this problem seldom arises; however, one way to avoid it is to define flux integrals by the method used to compute them shown in Section 19.3.

Flux and Fluid Flow

If $\vec{v}$ is the velocity vector field of a fluid, we have

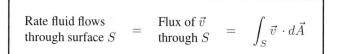

$$\begin{array}{ccc} \text{Rate fluid flows} \\ \text{through surface } S \end{array} = \begin{array}{c} \text{Flux of } \vec{v} \\ \text{through } S \end{array} = \int_S \vec{v} \cdot d\vec{A}$$

The rate of fluid flow is measured in units of volume per unit time.

Example 2   Find the flux of the vector field $\vec{B}(x, y, z)$ shown in Figure 19.8 through the square S of side 2 shown in Figure 19.9, oriented in the $\vec{j}$ direction, where

$$\vec{B}(x, y, z) = \frac{-y\vec{i} + x\vec{j}}{x^2 + y^2}.$$

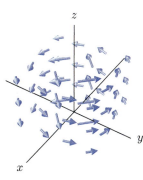

Figure 19.8: The vector field $\vec{B}(x, y, z) = \frac{-y\vec{i} + x\vec{j}}{x^2 + y^2}$

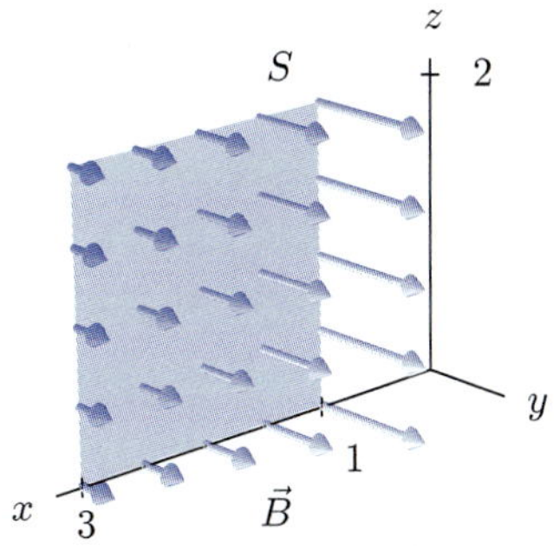

Figure 19.9: Flux of $\vec{B}$ through the square S of side 2 in xy-plane and oriented in $\vec{j}$ direction

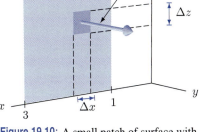

Figure 19.10: A small patch of surface with area $\|\Delta\vec{A}\| = \Delta x \Delta z$

Solution   Consider a small rectangular patch with area vector $\Delta\vec{A}$ in S, with sides Δx and Δz so that $\|\Delta\vec{A}\| = \Delta x \Delta z$. Since $\Delta\vec{A}$ points in the $\vec{j}$ direction we have $\Delta\vec{A} = \vec{j}\,\Delta x \Delta z$. (See Figure 19.10.)

At the point $(x, 0, z)$ in S, substituting $y = 0$ into $\vec{B}$ gives $\vec{B}(x, 0, z) = (1/x)\vec{j}$. Thus, we have

$$\text{Flux through small patch} \approx \vec{B} \cdot \Delta\vec{A} = \left(\frac{1}{x}\vec{j}\right) \cdot (\vec{j}\,\Delta x \Delta z) = \frac{1}{x}\,\Delta x\,\Delta z.$$

Therefore,

$$\text{Flux through surface} = \int_S \vec{B} \cdot d\vec{A} = \lim_{\|\Delta\vec{A}\| \to 0} \sum \vec{B} \cdot \Delta\vec{A} = \lim_{\substack{\Delta x \to 0 \\ \Delta z \to 0}} \sum \frac{1}{x}\,\Delta x\,\Delta z.$$

This last expression is a Riemann sum for the double integral $\int_R \frac{1}{x}\,dA$, where R is the square $1 \le x \le 3$, $0 \le z \le 2$. Thus,

$$\text{Flux through surface} = \int_S \vec{B} \cdot d\vec{A} = \int_R \frac{1}{x}\,dA = \int_0^2 \int_1^3 \frac{1}{x}\,dx\,dz = 2\ln 3.$$

The result is positive since the vector field is passing through the surface in the positive direction.

Example 3   Each of the vector fields in Figure 19.11 consists entirely of vectors parallel to the xy-plane, and is constant in the z direction (that is, the vector field looks the same in any plane parallel to the xy-plane). For each one, say whether you expect the flux through a closed surface surrounding the origin to be positive, negative, or zero. In part (a) the surface is a closed cube with faces parallel to the axes; in parts (b) and (c) the surface is a closed cylinder. In each case we choose the outward orientation. (See Figure 19.12 on page 868.)

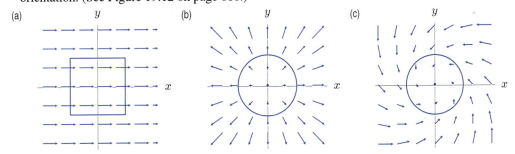

Figure 19.11: Flux of a vector field through the closed surfaces whose cross-sections are shown in the xy-plane

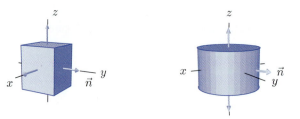

Figure 19.12: The closed cube and closed cylinder, both oriented outward

Solution

(a) Since the vector field appears to be parallel to the faces of the cube which are perpendicular to the y- and z-axes, we expect the flux through these faces to be zero. The fluxes through the two faces perpendicular to the x-axis appear to be equal in magnitude and opposite in sign, so we expect the net flux to be zero.

(b) Since the top and bottom of the cylinder are parallel to the flow, the flux through them is zero. On the round surface of the cylinder, $\vec{v}$ and $\Delta \vec{A}$ appear to be everywhere parallel and in the same direction, so we expect each term $\vec{v} \cdot \Delta \vec{A}$ to be positive, and therefore the flux integral $\int_S \vec{v} \cdot d\vec{A}$ to be positive.

(c) As in part (b), the flux through the top and bottom of the cylinder is zero. In this case $\vec{v}$ and $\Delta \vec{A}$ are not parallel on the round surface of the cylinder, but since the fluid appears to be flowing inwards as well as swirling, we expect each term $\vec{v} \cdot \Delta \vec{A}$ to be negative, and therefore the flux integral to be negative.

Calculating Flux Integrals Using $d\vec{A} = \vec{n}\, dA$

For a small patch of surface ΔS with normal $\vec{n}$ and area ΔA, the area vector is $\Delta \vec{A} = \vec{n} \,\Delta A$. The next example shows how we can use this relationship to compute a flux integral.

Example 4   An electric charge q is placed at the origin in 3-space. The resulting electric field $\vec{E}(\vec{r})$ at the point with position vector $\vec{r}$ is given by

$$\vec{E}(\vec{r}) = q \frac{\vec{r}}{\|\vec{r}\|^3}, \qquad \vec{r} \neq \vec{0}.$$

Find the flux of $\vec{E}$ out of the sphere of radius R centered at the origin. (See Figure 19.13.)

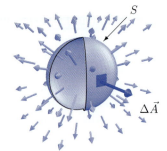

Figure 19.13: Flux of $\vec{E} = q\vec{r}/\|\vec{r}\|^3$ through the surface of a sphere of radius R centered at the origin

Solution

This vector field points radially outward from the origin in the same direction as $\vec{n}$. Thus, since $\vec{n}$ is a unit vector,
$$\vec{E} \cdot \Delta \vec{A} = \vec{E} \cdot \vec{n}\, \Delta A = \|\vec{E}\|\, \Delta A.$$

On the sphere, $\|\vec{E}\| = q/R^2$, so

$$\int_S \vec{E} \cdot d\vec{A} = \lim_{\|\Delta \vec{A}\| \to 0} \sum \vec{E} \cdot \Delta \vec{A} = \lim_{\Delta A \to 0} \sum \frac{q}{R^2}\, \Delta A = \frac{q}{R^2} \lim_{\Delta A \to 0} \sum \Delta A.$$

The last sum approximates the surface area of the sphere. In the limit as the subdivisions get finer we have

$$\lim_{\Delta A \to 0} \sum \Delta A = \text{Surface area of sphere.}$$

Thus, the flux is given by

$$\int_S \vec{E} \cdot d\vec{A} = \frac{q}{R^2} \lim_{\Delta A \to 0} \sum \Delta A = \frac{q}{R^2} \cdot \text{Surface area of sphere} = \frac{q}{R^2}(4\pi R^2) = 4\pi q.$$

This result is known as Gauss's law.

Instead of using Riemann sums, we often write $d\vec{A} = \vec{n}\, dA$, as in the next example.

Example 5   Suppose S is the surface of the cube bounded by the six planes $x = \pm 1$, $y = \pm 1$, and $z = \pm 1$. Compute the flux of the electric field $\vec{E}$ of the previous example outward through S.

Solution   It is enough to compute the flux of $\vec{E}$ through a single face, say the top face S_1 defined by $z = 1$, where $-1 \le x \le 1$ and $-1 \le y \le 1$. By symmetry, the flux of $\vec{E}$ through the other five faces of S must be the same.

On the top face, S_1, we have $d\vec{A} = \vec{n}\, dA = \vec{k}\, dx\, dy$ and

$$\vec{E}(x, y, 1) = q\frac{x\vec{i} + y\vec{j} + \vec{k}}{(x^2 + y^2 + 1)^{3/2}}.$$

The corresponding flux integral is given by

$$\int_{S_1} \vec{E} \cdot d\vec{A} = q\int_{-1}^{1}\int_{-1}^{1} \frac{x\vec{i} + y\vec{j} + \vec{k}}{(x^2 + y^2 + 1)^{3/2}} \cdot \vec{k}\, dx\, dy = q\int_{-1}^{1}\int_{-1}^{1} \frac{1}{(x^2 + y^2 + 1)^{3/2}}\, dx\, dy.$$

Computing this integral numerically shows that

$$\text{Flux through top face} = \int_{S_1} \vec{E} \cdot d\vec{A} \approx 2.0944q.$$

Thus,

$$\text{Total flux of } \vec{E} \text{ out of cube} = \int_S \vec{E} \cdot d\vec{A} \approx 6(2.0944q) = 12.5664q.$$

Example 4 on page 868 showed that the flux of $\vec{E}$ through a sphere of radius R centered at the origin is $4\pi q$. Since $4\pi \approx 12.5664$, Example 5 suggests that

$$\text{Total flux of } \vec{E} \text{ out of cube} = 4\pi q.$$

By computing the flux integral in Example 5 exactly, it is possible to verify that the flux of $\vec{E}$ through the cube and the sphere are exactly equal. When we encounter the Divergence Theorem in Chapter 20 we will see why this is so.

Notes on Orientation

Two difficulties can occur in choosing an orientation. The first is that if the surface is not smooth, it may not have a normal vector at every point. For example, a cube does not have a normal vector along its edges. When we have a surface, such as a cube, which is made of a finite number of smooth pieces, we choose an orientation for each piece separately. The best way to do this is usually clear. For example, on the cube we choose the outward orientation on each face. (See Figure 19.14.)

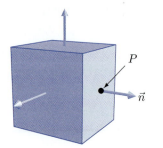

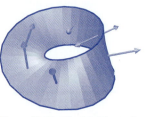

Figure 19.14: The orientation vector field $\vec{n}$ on the cube surface determined by the choice of unit normal vector at the point P

Figure 19.15: The Möbius strip is an example of a non-orientable surface

The second difficulty is that there are some surfaces which cannot be oriented at all, such as the *Möbius strip* in Figure 19.15.

Exercises and Problems for Section 19.1

Exercises

1. Let $\vec{F}(x, y, z) = z\vec{i}$. For each of the surfaces in (a)–(e), say whether the flux of $\vec{F}$ through the surface is positive, negative, or zero. In each case, the orientation of the surface is indicated by the given normal vector.

Compute the flux of the vector field $\vec{v} = 2\vec{i} + 3\vec{j} + 5\vec{k}$ through each of the rectangular regions in Exercises 4–7, assuming each is oriented as shown.

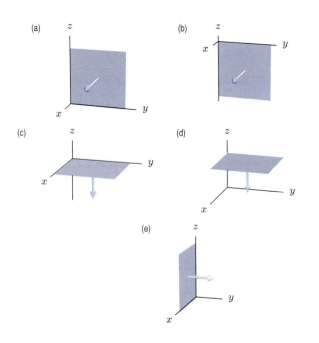

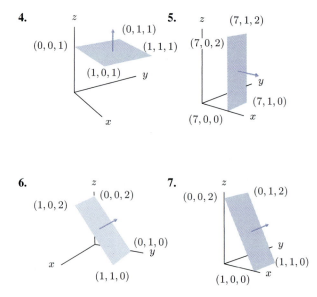

2. Repeat Exercise 1 with $\vec{F}(x, y, z) = -z\vec{i} + x\vec{k}$.

3. Repeat Exercise 1 with the vector field $\vec{F}(\vec{r}) = \vec{r}$.

In Exercises 8–19, calculate the flux of the vector field through the surface.

8. $\vec{F} = x\vec{i}$ through a square of side 3 in the plane $x = -5$. The square and is centered on the x-axis, has sides parallel to the axes, and is oriented in the positive x-direction.

9. $\vec{F} = y\vec{j}$ through the square of side 4 in the plane $y = 5$. The square is centered on the y-axis, has sides parallel to the axes, and is oriented in the positive y-direction.

10. $\vec{F} = (y+3)\vec{j}$ through a square of side 2 in the xz-plane, oriented in the negative y-direction.

11. $\vec{F} = x\vec{k}$ through the square $0 \le x \le 3, 0 \le y \le 3$ in the xy-plane, with sides parallel to the axes, and oriented upward.

12. $\vec{F} = -3\vec{r}$ through the sphere of radius 2 centered at the origin.

13. $\vec{F} = 2\vec{i} + 3\vec{j}$ through a disk of radius 5 in the plane $y = 2$ oriented in the direction of increasing y.

14. $\vec{F} = \vec{i} + 2\vec{j}$ through a sphere of radius 3 centered at the origin.

15. $\vec{F} = \vec{i} - \vec{j} - \vec{k}$ though a cube of side 2 with sides parallel to the axes.

16. $\vec{F} = \vec{i} + 2\vec{j}$ through a square of side 2 lying in the plane $x + y + z = 1$, oriented away from the origin.

17. $\vec{F} = (x^2 + y^2)\vec{k}$ through the disk of radius 3 in the xy-plane and centered at the origin and oriented upward.

18. $\vec{F} = \cos(x^2 + y^2)\vec{k}$ through the disk $x^2 + y^2 \le 9$ oriented upward in the plane $z = 1$.

19. $\vec{F} = e^{y^2 + z^2}\vec{i}$ through the disk of radius 2 in the yz-plane, centered at the origin and oriented in the positive x-direction.

Problems

20. **(a)** What do you think will be the electric flux through the cylindrical surface that is placed as shown in the constant electric field in Figure 19.16? Why?
    (b) What if the cylinder is placed upright, as shown in Figure 19.17? Explain.

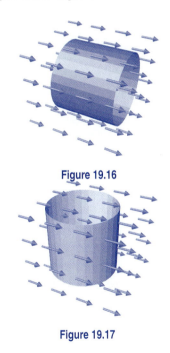

Figure 19.16

Figure 19.17

21. Find the flux of $\vec{F} = 5\vec{i} + 7\vec{i} + z\vec{k}$ out of a closed cylinder of radius 3 centered on the z-axis, with $-2 \le z \le 2$.

22. Find the flux of $\vec{F} = x\vec{i} + y\vec{j} + z\vec{k}$ out of a closed cylinder of radius 2 centered on the y-axis, with $-3 \le y \le 3$.

23. The flux of the constant vector field $a\vec{i} + b\vec{j} + c\vec{k}$ through the square of side 2 in the plane $x = 5$, oriented in the positive x-direction, is 24. Which of the constants a, b, c can be determined from the information given? Give the value(s).

24. Arrange the following flux integrals, $\int_{S_i} \vec{F} \cdot d\vec{A}$, with $i = 1, 2, 3, 4$, in ascending order if $\vec{F} = -\vec{i} - \vec{j} + \vec{k}$ and if the S_i are the following surfaces:

    - S_1 is a horizontal square of side 1 with one corner at $(0, 0, 2)$, above the first quadrant of the xy-plane, oriented upward.
    - S_2 is a horizontal square of side 1 with one corner at $(0, 0, 3)$, above the third quadrant of the xy-plane, oriented upward.
    - S_3 is a square of side $\sqrt{2}$ in the xz-plane with one corner at the origin, one edge along the positive x-axis, one along the negative z-axis, oriented in the negative y-direction.
    - S_4 is a square of side $\sqrt{2}$ with one corner at the origin, one edge along the positive y-axis, one corner at $(1, 0, 1)$, oriented upward.

25. Figure 19.18 shows a cross-section of the earth's magnetic field. Assume that the earth's magnetic and geographic poles coincide. Is the magnetic flux through a horizontal plate, oriented skyward, positive, negative, or zero if the plate is
    (a) At the north pole?   **(b)** At the south pole?
    (c) On the equator?

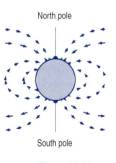

Figure 19.18

26. Let S be the cube with side length 2, faces parallel to the coordinate planes, and centered at the origin.

    (a) Calculate the total flux of the constant vector field $\vec{v} = -\vec{i} + 2\vec{j} + \vec{k}$ out of S by computing the flux through each face separately.
    (b) Calculate the flux out of S for any constant vector field $\vec{v} = a\vec{i} + b\vec{j} + c\vec{k}$.
    (c) Explain why the answers to parts (a) and (b) make sense.

27. Let S be the tetrahedron with vertices at the origin and at $(1, 0, 0)$, $(0, 1, 0)$ and $(0, 0, 1)$.

    (a) Calculate the total flux of the constant vector field $\vec{v} = -\vec{i} + 2\vec{j} + \vec{k}$ out of S by computing the flux through each face separately.
    (b) Calculate the flux out of S in part (a) for any constant vector field $\vec{v}$.
    (c) Explain why the answers to parts (a) and (b) make sense.

28. Explain why if $\vec{F}$ has constant magnitude on S and is everywhere normal to S and in the direction of orientation, then

$$\int_S \vec{F} \cdot d\vec{A} = \|\vec{F}\| \cdot \text{Area of } S.$$

29. The z-axis carries a constant electric charge density of λ units of charge per unit length, with $\lambda > 0$, and that $\vec{E}$ is the resulting electric field.

    (a) Sketch the electric field, $\vec{E}$, in the xy-plane, given

$$\vec{E}(x, y, z) = 2\lambda \frac{x\vec{i} + y\vec{j}}{x^2 + y^2}.$$

    (b) Compute the flux of $\vec{E}$ outward through the cylinder $x^2 + y^2 = R^2$, for $0 \le z \le h$.

30. Let $P(x, y, z)$ be the pressure at the point (x, y, z) in a fluid. Let $\vec{F}(x, y, z) = P(x, y, z)\vec{k}$. Let S be the surface of a body submerged in the fluid. If S is oriented inward, show that $\int_S \vec{F} \cdot d\vec{A}$ is the buoyant force on the body, that is, the force upwards on the body due to the pressure of the fluid surrounding it. [Hint: $\vec{F} \cdot d\vec{A} = P(x, y, z)\vec{k} \cdot d\vec{A} = (P(x, y, z)\, d\vec{A}) \cdot \vec{k}$.]

31. A closed surface S encloses a volume W. The function $\rho(x, y, z)$ gives the electrical charge density at points in space. The vector field $\vec{J}(x, y, z)$ gives the electric current density at any point in space and is defined so that the current through a small area $d\vec{A}$ is given by

$$\text{Current through small area} \approx \vec{J} \cdot d\vec{A}.$$

    (a) What do the following integrals represent, in terms of electricity?

    (i) $\displaystyle\int_W \rho\, dV$ (ii) $\displaystyle\int_S \vec{J} \cdot d\vec{A}$

    (b) Using the fact that an electric current through a surface is the rate at which electric charge passes through the surface per unit time, explain why

$$\int_S \vec{J} \cdot d\vec{A} = -\frac{\partial}{\partial t}\left(\int_W \rho\, dV\right).$$

32. A fluid is flowing along a cylindrical pipe of radius a in the $\vec{i}$ direction. The velocity of the fluid at a radial distance r from the center of the pipe is $\vec{v} = u(1 - r^2/a^2)\vec{i}$.

    (a) What is the significance of the constant u?
    (b) What is the velocity of the fluid at the wall of the pipe?
    (c) Find the flux through a circular cross-section of the pipe.

33. A region of 3-space has a temperature which varies from point to point. Let $T(x, y, z)$ be the temperature at a point (x, y, z). Newton's law of cooling says that $\text{grad}\, T$ is proportional to the heat flow vector field, $\vec{F}$, where $\vec{F}$ points in the direction in which heat is flowing and has magnitude equal to the rate of flow of heat.

    (a) Suppose $\vec{F} = k\, \text{grad}\, T$ for some constant k. What is the sign of k?
    (b) Explain why this form of Newton's law of cooling makes sense.
    (c) Let W be a region of space bounded by the surface S. Explain why

$$\begin{array}{l}\text{Rate of heat} \\ \text{loss from } W\end{array} = k\int_S (\text{grad}\, T) \cdot d\vec{A}.$$

19.2 FLUX INTEGRALS FOR GRAPHS, CYLINDERS, AND SPHERES

In Section 19.1 we computed flux integrals in certain simple cases. In this section we see how to compute flux through surfaces that are graphs of functions, through cylinders, and through spheres.

Flux of a Vector Field through the Graph of $z = f(x, y)$

Suppose S is the graph of the differentiable function $z = f(x, y)$, oriented upward, and that $\vec{F}$ is a smooth vector field. In Section 19.1 we subdivided the surface into small pieces with area vector $\Delta\vec{A}$ and defined the flux of $\vec{F}$ through S as follows:

$$\int_S \vec{F} \cdot d\vec{A} = \lim_{\|\Delta\vec{A}\| \to 0} \sum \vec{F} \cdot \Delta\vec{A}.$$

How do we divide S into small pieces? One way is to use the cross-sections of f with x or y constant and take the patches in a wire frame representation of the surface. So we must calculate the area vector of one of these patches, which is approximately a parallelogram.

The Area Vector of a Parallelogram-Shaped Patch

According to the geometric definition of the cross product on page 630, the vector $\vec{v} \times \vec{w}$ has magnitude equal to the area of the parallelogram formed by $\vec{v}$ and $\vec{w}$ and direction perpendicular to this parallelogram and determined by the right-hand rule. Thus, we have

$$\text{Area vector of parallelogram} = \vec{A} = \vec{v} \times \vec{w}.$$

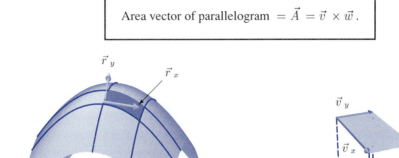

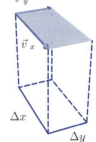

Figure 19.19: Surface showing parameter rectangle and tangent vectors $\vec{r}_x$ and $\vec{r}_y$

Figure 19.20: Parallelogram-shaped patch in the tangent plane to the surface

Consider the patch of surface above the rectangular region with sides Δx and Δy in the xy-plane shown in Figure 19.19. We approximate the area vector, $\Delta \vec{A}$, of this patch by the area vector of the corresponding patch on the tangent plane to the surface. See Figure 19.20 This patch is the parallelogram determined by the vectors $\vec{v}_x$ and $\vec{v}_y$, so its area vector is given by

$$\Delta \vec{A} \approx \vec{v}_x \times \vec{v}_y.$$

To find $\vec{v}_x$ and $\vec{v}_y$, notice that a point on the surface has position vector $\vec{r} = x\vec{i} + y\vec{j} + f(x,y)\vec{k}$. Thus, a cross-section of S with y constant has tangent vector

$$\vec{r}_x = \frac{\partial \vec{r}}{\partial x} = \vec{i} + f_x \vec{k},$$

and a cross-section with x constant has tangent vector

$$\vec{r}_y = \frac{\partial \vec{r}}{\partial y} = \vec{j} + f_y \vec{k}.$$

The vectors $\vec{r}_x$ and $\vec{v}_x$ are parallel because they are both tangent to the surface and in the xz-plane. Since the x-component of $\vec{r}_x$ is $\vec{i}$ and the x-component of $\vec{v}_x$ is $(\Delta x)\vec{i}$, we have $\vec{v}_x = (\Delta x)\vec{r}_x$. Similarly, we have $\vec{v}_y = (\Delta y)\vec{r}_y$. So the upward pointing area vector of the parallelogram is

$$\Delta \vec{A} \approx \vec{v}_x \times \vec{v}_y = (\vec{r}_x \times \vec{r}_y)\, \Delta x\, \Delta y = \left(-f_x \vec{i} - f_y \vec{j} + \vec{k} \right) \Delta x\, \Delta y.$$

This is our approximation for the area vector $\Delta \vec{A}$ on the surface. Replacing $\Delta \vec{A}$, Δx, and Δy by $d\vec{A}$, dx and dy, we write

$$d\vec{A} = \left(-f_x \vec{i} - f_y \vec{j} + \vec{k} \right) dx\, dy.$$

The Flux of $\vec{F}$ through a Surface given by a Graph of $z = f(x, y)$

Suppose the surface S is the part of the graph of $z = f(x, y)$ above a region R in the xy-plane, and suppose S is oriented upward. The flux of $\vec{F}$ through S is

$$\int_S \vec{F} \cdot d\vec{A} = \int_R \vec{F}(x, y, f(x, y)) \cdot \left(-f_x \vec{i} - f_y \vec{j} + \vec{k}\right) dx\, dy.$$

Example 1   Compute $\int_S \vec{F} \cdot d\vec{A}$ where $\vec{F}(x, y, z) = z\vec{k}$ and S is the rectangular plate with corners $(0, 0, 0)$, $(1, 0, 0)$, $(0, 1, 3)$, $(1, 1, 3)$, oriented upwards.

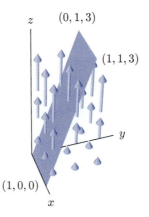

Figure 19.21: The vector field $\vec{F} = z\vec{k}$ on the rectangular surface S

Solution   We find the equation for the plane S in the form $z = f(x, y)$. Since f is linear, with x-slope equal to 0 and y-slope equal to 3, and $f(0, 0) = 0$, we have

$$z = f(x, y) = 0 + 0x + 3y = 3y.$$

Thus, we have

$$d\vec{A} = (-f_x \vec{i} - f_y \vec{j} + \vec{k})\, dx\, dy = (0\vec{i} - 3\vec{j} + \vec{k})\, dx\, dy = (-3\vec{j} + \vec{k})\, dx\, dy.$$

The flux integral is therefore

$$\int_S \vec{F} \cdot d\vec{A} = \int_0^1 \int_0^1 3y\vec{k} \cdot (-3\vec{j} + \vec{k})\, dx\, dy = \int_0^1 \int_0^1 3y\, dx\, dy = 1.5.$$

Flux of a Vector Field through a Cylindrical Surface

Consider the cylinder of radius R centered on the z-axis illustrated in Figure 19.22 and oriented away from the z-axis. The small patch of surface, or parameter rectangle, in Figure 19.23 has surface area given by

$$\Delta A \approx R\, \Delta\theta\, \Delta z.$$

The outward unit normal $\vec{n}$ points in the direction of $x\vec{i} + y\vec{j}$, so

$$\vec{n} = \frac{x\vec{i} + y\vec{j}}{\|x\vec{i} + y\vec{j}\|} = \frac{R\cos\theta\vec{i} + R\sin\theta\vec{j}}{R} = \cos\theta\vec{i} + \sin\theta\vec{j}.$$

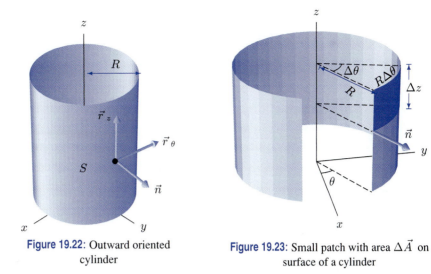

Figure 19.22: Outward oriented
cylinder

Figure 19.23: Small patch with area $\Delta \vec{A}$ on
surface of a cylinder

Therefore, the area vector of the parameter rectangle is approximated by

$$\Delta \vec{A} = \vec{n} \; \Delta A \approx \left(\cos \theta \vec{i} + \sin \theta \vec{j} \right) R \, \Delta z \, \Delta \theta.$$

Replacing $\Delta \vec{A}$, Δz, and $\Delta \theta$ by $d \vec{A}$, dz, and $d\theta$, we write

$$d\vec{A} = \left(\cos \theta \vec{i} + \sin \theta \vec{j} \right) R \, dz \, d\theta.$$

This gives the following result:

The Flux of a Vector Field through a Cylinder

The flux of $\vec{F}$ through the cylindrical surface S, of radius R and oriented away from the z-axis, is given by

$$\int_S \vec{F} \cdot d\vec{A} = \int_T \vec{F}(R, \theta, z) \cdot \left(\cos \theta \vec{i} + \sin \theta \vec{j} \right) R \, dz \, d\theta,$$

where T is the θz-region corresponding to S.

Example 2   Compute $\int_S \vec{F} \cdot d\vec{A}$ where $\vec{F}(x, y, z) = y\vec{j}$ and S is the part of the cylinder of radius 2 centered on the z-axis with $x \geq 0$, $y \geq 0$, and $0 \leq z \leq 3$. The surface is oriented towards the z-axis.

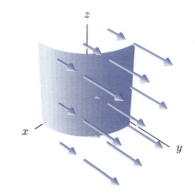

Figure 19.24: The vector field $\vec{F} = y\vec{j}$ on the surface S

Solution
In cylindrical coordinates, we have $R = 2$ and $\vec{F} = y\vec{j} = 2\sin\theta\vec{j}$. Since the orientation of S is toward the z-axis, the flux across S is given by

$$\int_S \vec{F} \cdot d\vec{A} = -\int_T 2\sin\theta\vec{j} \cdot (\cos\theta\vec{i} + \sin\theta\vec{j})2\,dz\,d\theta = -4\int_0^{\pi/2}\int_0^3 \sin^2\theta\,dz\,d\theta = -3\pi.$$

Flux of a Vector Field through a Spherical Surface

Consider the piece of the sphere of radius R centered at the origin, oriented outward illustrated in Figure 19.25. The small parameter rectangle in Figure 19.25 has surface area given by

$$\Delta A \approx R^2 \sin\phi\,\Delta\phi\,\Delta\theta.$$

The outward unit normal $\vec{n}$ points in the direction of $\vec{r} = x\vec{i} + y\vec{j} + z\vec{k}$, so

$$\vec{n} = \frac{\vec{r}}{\|\vec{r}\|} = \sin\phi\cos\theta\vec{i} + \sin\phi\sin\theta\vec{j} + \cos\phi\vec{k}.$$

Therefore, the area vector of the parameter rectangle is approximated by

$$\Delta\vec{A} \approx \vec{n}\,\Delta A = \frac{\vec{r}}{\|\vec{r}\|}\Delta A = \left(\sin\phi\cos\theta\vec{i} + \sin\phi\sin\theta\vec{j} + \cos\phi\vec{k}\right)R^2\sin\phi\,\Delta\phi\,\Delta\theta.$$

Replacing $\Delta\vec{A}$, $\Delta\phi$, and $\Delta\theta$ by $d\vec{A}$, $d\phi$, and $d\theta$, we write

$$d\vec{A} = \frac{\vec{r}}{\|\vec{r}\|}dA = \left(\sin\phi\cos\theta\vec{i} + \sin\phi\sin\theta\vec{j} + \cos\phi\vec{k}\right)R^2\sin\phi\,d\phi\,d\theta.$$

Thus, we obtain the following result:

The Flux of a Vector Field through a Sphere

The flux of $\vec{F}$ through the spherical surface S, with radius R and oriented away from the origin, is given by

$$\int_S \vec{F} \cdot d\vec{A} = \int_S \vec{F} \cdot \frac{\vec{r}}{\|\vec{r}\|}dA$$

$$= \int_T \vec{F}(R, \theta, \phi) \cdot \left(\sin\phi\cos\theta\vec{i} + \sin\phi\sin\theta\vec{j} + \cos\phi\vec{k}\right)R^2\sin\phi\,d\phi\,d\theta,$$

where T is the $\theta\phi$-region corresponding to S.

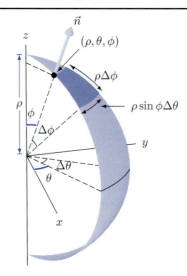

Figure 19.25: Small patch with area $\Delta\vec{A}$ on surface of a sphere

Example 3   Find the flux of $\vec{F} = z\vec{k}$ through S, the upper hemisphere of radius 2 centered at the origin, oriented outward.

Solution   The hemisphere S is parameterized by spherical coordinates θ and ϕ, with $0 \leq \theta \leq 2\pi$ and $0 \leq \phi \leq \pi/2$. Since $R = 2$ and $\vec{F} = z\vec{k} = 2\cos\phi\vec{k}$, the flux is

$$\int_S \vec{F} \cdot d\vec{A} = \int_S 2\cos\phi\vec{k} \cdot (\sin\phi\cos\theta\vec{i} + \sin\phi\sin\theta\vec{j} + \cos\phi\vec{k})4\sin\phi\, d\phi\, d\theta$$

$$= \int_0^{2\pi} \int_0^{\pi/2} 8\sin\phi\cos^2\phi\, d\phi\, d\theta = 2\pi\left(8\left(\frac{-\cos^3\phi}{3}\right)\Big|_{\phi=0}^{\pi/2}\right) = \frac{16\pi}{3}.$$

Example 4   The magnetic field $\vec{B}$ due to an *ideal magnetic dipole*, $\vec{\mu}$, located at the origin is defined to be

$$\vec{B}(\vec{r}) = -\frac{\vec{\mu}}{\|\vec{r}\|^3} + \frac{3(\vec{\mu} \cdot \vec{r})\vec{r}}{\|\vec{r}\|^5}.$$

Figure 19.26 shows a sketch of $\vec{B}$ in the plane $z = 0$ for the dipole $\vec{\mu} = \vec{i}$. Notice that $\vec{B}$ is similar to the magnetic field of a bar magnet with its north pole at the tip of the vector $\vec{i}$ and its south pole at the tail of the vector $\vec{i}$.

Compute the flux of $\vec{B}$ outward through the sphere S with center at the origin and radius R.

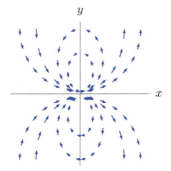

Figure 19.26: The magnetic field of a dipole, $\vec{i}$, at the origin:
$$\vec{B} = \frac{-\vec{i}}{\|\vec{r}\|^3} + \frac{3(\vec{i} \cdot \vec{r})\vec{r}}{\|\vec{r}\|^5}$$

Solution   Since $\vec{i} \cdot \vec{r} = x$ and $\|\vec{r}\| = R$ on the sphere of radius R, we have

$$\int_S \vec{B} \cdot d\vec{A} = \int_S \left(-\frac{\vec{i}}{\|\vec{r}\|^3} + \frac{3(\vec{i} \cdot \vec{r})\vec{r}}{\|\vec{r}\|^5}\right) \cdot \frac{\vec{r}}{\|\vec{r}\|}\, dA = \int_S \left(-\frac{\vec{i} \cdot \vec{r}}{\|\vec{r}\|^4} + \frac{3(\vec{i} \cdot \vec{r})\|\vec{r}\|^2}{\|\vec{r}\|^6}\right)\, dA$$

$$= \int_S \frac{2\vec{i} \cdot \vec{r}}{\|\vec{r}\|^4}\, dA = \int_S \frac{2x}{\|\vec{r}\|^4}\, dA = \frac{2}{R^4}\int_S x\, dA,$$

But the sphere S is centered at the origin. Thus, the contribution to the integral from each positive x- value is canceled by the contribution from the corresponding negative x-value; so $\int_S x\, dA = 0$. Therefore,

$$\int_S \vec{B} \cdot d\vec{A} = \frac{2}{R^4}\int_S x\, dA = 0.$$

Exercises and Problems for Section 19.2

Exercises

In Exercises 1–12 compute the flux of the vector field, $\vec{F}$, through the surface, S.

1. $\vec{F} = (x-y)\vec{i} + z\vec{j} + 3x\vec{k}$ and S is the part of the plane $z = x + y$ above the rectangle $0 \le x \le 2, 0 \le y \le 3$, oriented upward.

2. $\vec{F} = \vec{r}$ and S is the part of the plane $x + y + z = 1$ above the rectangle $0 \le x \le 2, 0 \le y \le 3$, oriented downward.

3. $\vec{F} = \vec{r}$ and S is the part of the surface $z = x^2 + y^2$ above the disk $x^2 + y^2 \le 1$, oriented downward.

4. $\vec{F}(x, y, z) = 2x\vec{j} + y\vec{k}$ and S is the part of the surface $z = -y + 1$ above the square $0 \le x \le 1, 0 \le y \le 1$, oriented upward.

5. $\vec{F} = 3x\vec{i} + y\vec{j} + z\vec{k}$ and S is the part of the surface $z = -2x - 4y + 1$ above the triangle R in the xy-plane with vertices $(0, 0), (0, 2), (1, 0)$, oriented upward.

6. $\vec{F} = x\vec{i} + y\vec{j}$ and S is the part of the surface $z = 25 - (x^2 + y^2)$ above the disk of radius 5 centered at the origin, oriented upward.

7. $\vec{F} = \cos(x^2 + y^2)\vec{k}$ and S is as in Exercise 6.

8. $\vec{F} = -y\vec{j} + z\vec{k}$ and S is the part of the surface $z = y^2 + 5$ over the rectangle $-2 \le x \le 1, 0 \le y \le 1$, oriented upward.

9. $\vec{F}(x, y, z) = -xz\vec{i} - yz\vec{j} + z^2\vec{k}$ and S is the cone $z = \sqrt{x^2 + y^2}$ for $0 \le z \le 6$, oriented upward.

10. $\vec{F} = y\vec{i} + \vec{j} - xz\vec{k}$ and S is the surface $y = x^2 + z^2$, with $x^2 + z^2 \le 1$, oriented in the positive y-direction.

11. $\vec{F} = xz\vec{i} + y\vec{k}$ and S is the hemisphere $x^2 + y^2 + z^2 = 9, z \ge 0$, oriented upward.

12. $\vec{F} = x^2\vec{i} + y^2\vec{j} + z^2\vec{k}$ and S is the oriented triangular surface shown in Figure 19.27.

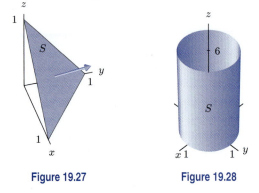

Figure 19.27   Figure 19.28

In Exercises 13–14, compute the flux of the vector field, $\vec{F}$, through the cylindrical surface shown in Figure 19.28, oriented away from the z-axis.

13. $\vec{F} = x\vec{i} + y\vec{j}$

14. $\vec{F} = xz\vec{i} + yz\vec{j} + z^3\vec{k}$

In Exercises 15–16 compute the flux of the vector field, $\vec{F}$, through the given spherical surface, S.

15. $\vec{F} = z^2\vec{k}$ and S is the upper hemisphere of the sphere $x^2 + y^2 + z^2 = 25$, oriented away from the origin.

16. $\vec{F} = x\vec{i} + y\vec{j} + z\vec{k}$ and S is the surface of the sphere $x^2 + y^2 + z^2 = a^2$, oriented outward.

Problems

17. Let $\vec{F} = (xze^{yz})\vec{i} + xz\vec{j} + (5 + x^2 + y^2)\vec{k}$. Calculate the flux of $\vec{F}$ through the disk $x^2 + y^2 \le 1$ in the xy-plane, oriented upward.

18. Let $\vec{H} = (e^{xy} + 3z + 5)\vec{i} + (e^{xy} + 5z + 3)\vec{j} + (3z + e^{xy})\vec{k}$. Calculate the flux of $\vec{H}$ through the square of side 2 with one vertex at the origin, one edge along the positive y-axis, one edge in the xz-plane with $x > 0$, $z > 0$, and the normal $\vec{n} = \vec{i} - \vec{k}$.

19. Electric charge is distributed in space with density (in coulomb/m^3) given in spherical coordinates by

$$\delta(\rho, \phi, \theta) = \begin{cases} \delta_0 \text{ (a constant)} & \rho \le a \\ 0 & \rho > a. \end{cases}$$

(a) Describe the charge distribution in words.
(b) Find the electric field $\vec{E}$ due to δ. Assume that $\vec{E}$ can be written in spherical coordinates as $\vec{E} = E(\rho)\vec{e}_\rho$, where $\vec{e}_\rho$ is the unit outward normal to the

sphere of radius ρ. In addition, $\vec{E}$ satisfies Gauss's Law for any simple closed surface S enclosing a volume W:

$$\int_S \vec{E} \cdot d\vec{A} = k \int_W \delta \, dV, \quad k \text{ a constant.}$$

20. Electric charge is distributed in space with density (in coulomb/m^3) given in cylindrical coordinates by

$$\delta(r, \theta, z) = \begin{cases} \delta_0 \text{ (a constant)} & \text{if } r \le a \\ 0 & \text{if } r > a \end{cases}$$

(a) Describe the charge distribution in words.
(b) Find the electric field $\vec{E}$ due to δ. Assume that $\vec{E}$ can be written in cylindrical coordinates as $\vec{E} = E(r)\vec{e}_r$, where $\vec{e}_r$ is the unit outward vector to the cylinder of radius r, and that $\vec{E}$ satisfies Gauss's Law (see Problem 19).

19.3 FLUX INTEGRALS OVER PARAMETERIZED SURFACES

Most of the flux integrals we are likely to encounter can be computed using the methods of Sections 19.1 and 19.2. In this section we briefly consider the general case: how to compute the flux of a smooth vector field $\vec{F}$ through a smooth oriented surface, S, parameterized by

$$\vec{r} = \vec{r}\,(s,t),$$

for (s,t) in some region R of the parameter space. The method is similar to the one used for graphs in Section 19.2. We consider a parameter rectangle on the surface S corresponding to a rectangular region with sides Δs and Δt in the parameter space. (See Figure 19.29.)

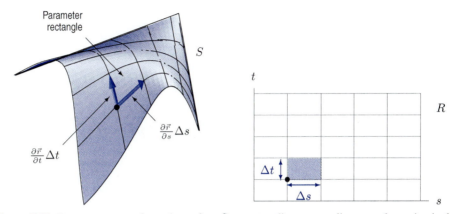

Figure 19.29: Parameter rectangle on the surface S corresponding to a small rectangular region in the parameter space, R

If Δs and Δt are small, the area vector, $\Delta \vec{A}$, of the patch is approximately the area vector of the parallelogram defined by the vectors

$$\vec{r}\,(s+\Delta s, t) - \vec{r}\,(s,t) \approx \frac{\partial \vec{r}}{\partial s} \Delta s, \qquad \text{and} \qquad \vec{r}\,(s, t+\Delta t) - \vec{r}\,(s,t) \approx \frac{\partial \vec{r}}{\partial t} \Delta t.$$

Thus

$$\Delta \vec{A} \approx \frac{\partial \vec{r}}{\partial s} \times \frac{\partial \vec{r}}{\partial t} \Delta s\, \Delta t.$$

We assume that the vector $\partial \vec{r}\,/\partial s \times \partial \vec{r}\,/\partial t$ is never zero and points in the direction of the unit normal orientation vector $\vec{n}$. If the vector $\partial \vec{r}\,/\partial s \times \partial \vec{r}\,/\partial t$ points in the opposite direction to $\vec{n}$, we reverse the order of the cross-product. Replacing $\Delta \vec{A}$, Δs, and Δt by $d\vec{A}$, ds, and dt, we write

$$d\vec{A} = \left(\frac{\partial \vec{r}}{\partial s} \times \frac{\partial \vec{r}}{\partial t} \right) ds\, dt.$$

The Flux of a Vector Field through a Parameterized Surface

The flux of a smooth vector field $\vec{F}$ through a smooth oriented surface S parameterized by $\vec{r} = \vec{r}\,(s,t)$, where (s,t) varies in a parameter region R, is given by

$$\int_S \vec{F} \cdot d\vec{A} = \int_R \vec{F}\,(\vec{r}\,(s,t)) \cdot \left(\frac{\partial \vec{r}}{\partial s} \times \frac{\partial \vec{r}}{\partial t} \right) ds\, dt.$$

We choose the parameterization so that $\partial \vec{r}\,/\partial s \times \partial \vec{r}\,/\partial t$ is never zero and points in the direction of $\vec{n}$ everywhere.

Example 1   Find the flux of the vector field $\vec{F} = x\vec{i} + y\vec{j}$ through the surface S, oriented downward and given by

$$x = 2s, \quad y = s + t, \quad z = 1 + s - t, \qquad \text{where } 0 \le s \le 1, \quad 0 \le t \le 1.$$

Solution   Since S is parameterized by

$$\vec{r}(s, t) = 2s\vec{i} + (s + t)\vec{j} + (1 + s - t)\vec{k},$$

we have

$$\frac{\partial \vec{r}}{\partial s} = 2\vec{i} + \vec{j} + \vec{k} \quad \text{and} \quad \frac{\partial r}{\partial t} = \vec{j} - \vec{k},$$

so

$$\frac{\partial \vec{r}}{\partial s} \times \frac{\partial \vec{r}}{\partial t} = \begin{vmatrix} \vec{i} & \vec{j} & \vec{k} \\ 2 & 1 & 1 \\ 0 & 1 & -1 \end{vmatrix} = -2\vec{i} + 2\vec{j} + 2\vec{k}.$$

Since the vector $-2\vec{i} + 2\vec{j} + 2\vec{k}$ points upward, we use $2\vec{i} - 2\vec{j} - 2\vec{k}$ for downward orientation. Thus, the flux integral is given by

$$\int_S \vec{F} \cdot d\vec{A} = \int_0^1 \int_0^1 (2s\vec{i} + (s + t)\vec{j}) \cdot (2\vec{i} - 2\vec{j} - 2\vec{k}) \, ds \, dt$$

$$= \int_0^1 \int_0^1 (4s - 2s - 2t) \, ds \, dt = \int_0^1 \int_0^1 (2s - 2t) \, ds \, dt$$

$$= \int_0^1 \left(s^2 - 2st \Big|_{s=0}^{s=1} \right) dt = \int_0^1 (1 - 2t) \, dt = t - t^2 \Big|_0^1 = 0.$$

Area of a Parameterized Surface

The area ΔA of a small parameter rectangle is the magnitude of its area vector $\Delta \vec{A}$. Therefore,

$$\text{Area of } S = \sum \Delta A = \sum \|\Delta \vec{A}\| \approx \sum \left\| \frac{\partial \vec{r}}{\partial s} \times \frac{\partial \vec{r}}{\partial t} \right\| \Delta s \, \Delta t.$$

Taking the limit as the area of the parameter rectangles tends to zero, we are led to the following expression for the area of S.

The Area of a Parameterized Surface

The area of a surface S which is parameterized by $\vec{r} = \vec{r}(s, t)$, where (s, t) varies in a parameter region R, is given by

$$\int_S dA = \int_R \left\| \frac{\partial \vec{r}}{\partial s} \times \frac{\partial \vec{r}}{\partial t} \right\| ds \, dt.$$

Example 2   Compute the surface area of a sphere of radius a.

Solution   We take the sphere S of radius a centered at the origin and parameterize it with the spherical coordinates ϕ and θ. The parameterization is

$$x = a \sin \phi \cos \theta, \quad y = a \sin \phi \sin \theta, \quad z = a \cos \phi, \qquad \text{for} \quad 0 \le \theta \le 2\pi, \quad 0 \le \phi \le \pi.$$

We compute

$$\frac{\partial \vec{r}}{\partial \phi} \times \frac{\partial \vec{r}}{\partial \theta} = (a \cos \phi \cos \theta \vec{i} + a \cos \phi \sin \theta \vec{j} - a \sin \phi \vec{k}) \times (-a \sin \phi \sin \theta \vec{i} + a \sin \phi \cos \theta \vec{j})$$

$$= a^2 (\sin^2 \phi \cos \theta \vec{i} + \sin^2 \phi \sin \theta \vec{j} + \sin \phi \cos \phi \vec{k})$$

and so

$$\left\| \frac{\partial \vec{r}}{\partial \phi} \times \frac{\partial \vec{r}}{\partial \theta} \right\| = a^2 \sin \phi.$$

Thus, we see that the surface area of the sphere S is given by

$$\text{Surface area} = \int_S dA = \int_R \left\| \frac{\partial \vec{r}}{\partial \phi} \times \frac{\partial \vec{r}}{\partial \theta} \right\| d\phi d\theta = \int_{\phi=0}^{\pi} \int_{\theta=0}^{2\pi} a^2 \sin \phi \, d\theta \, d\phi = 4\pi a^2.$$

Exercises and Problems for Section 19.3

Exercises

In Exercises 1–5 compute the flux of the vector field, $\vec{F}$, through the parameterized surface, S.

1. $\vec{F} = z\vec{k}$ and S is oriented toward the z-axis and given, for $0 \le s \le 1$, $0 \le t \le 1$, by

$$x = s + t, \quad y = s - t, \quad z = s^2 + t^2.$$

2. $\vec{F} = y\vec{i} + x\vec{j}$ and S is oriented away from the z-axis and given, for $0 \le s \le \pi$, $0 \le t \le 1$, by

$$x = 3 \sin s, \quad y = 3 \cos s, \quad z = t + 1.$$

3. $\vec{F} = z\vec{i} + x\vec{j}$ and S is oriented upward and given, for $0 \le s \le 1$, $1 \le t \le 3$ by

$$x = s^2, \quad y = 2s + t^2, \quad z = 5t.$$

4. $\vec{F} = -\frac{2}{x}\vec{i} + \frac{2}{y}\vec{j}$ and S is oriented upward and parameterized by a and θ, where, for $1 \le a \le 3$, $0 \le \theta \le \pi$,

$$x = a \cos \theta, \quad y = a \sin \theta, \quad z = \sin a^2.$$

5. $\vec{F} = x^2 y^2 z\vec{k}$ and S is the cone $\sqrt{x^2 + y^2} = z$, with $0 \le z \le R$, oriented downward. Parameterize the cone using cylindrical coordinates. (See Figure 19.30.)

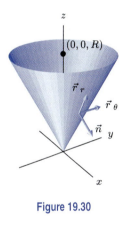

Figure 19.30

Problems

6. Find the area of the ellipse S on the plane $2x + y + z = 2$ cut out by the circular cylinder $x^2 + y^2 = 2x$. (See Figure 19.31.)

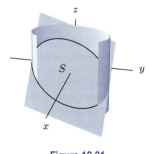

Figure 19.31

7. Evaluate $\int_S \vec{F} \cdot d\vec{A}$, where $\vec{F} = (bx/a)\vec{i} + (ay/b)\vec{j}$ and S is the elliptic cylinder oriented away from the z-axis, and given by $x^2/a^2 + y^2/b^2 = 1$, $|z| \le c$, where a, b, c are positive constants.

8. Consider the surface S formed by rotating the graph of $y = f(x)$ around the x-axis between $x = a$ and $x = b$. Assume that $f(x) \ge 0$ for $a \le x \le b$. Show that the surface area of S is $2\pi \int_a^b f(x)\sqrt{1 + f'(x)^2} \, dx$.

As we remarked in Section 19.1, the limit defining a flux integral might not exist if we subdivide the surface in the wrong way. One way to get around this is to take the formula for a flux integral over a parameterized surface that we have developed in this section and to use it as the *definition* of the flux integral. In Problems 9–12 we explore how this works.

9. Use a parameterization to verify the formula for a flux integral over a surface graph on page 874.

10. Use a parameterization to verify the formula for a flux integral over a cylindrical surface on page 875.

11. Use a parameterization to verify the formula for a flux integral over a spherical surface on page 876.

12. One problem with defining the flux integral using a parameterization is that the integral appears to depend on the choice of parameterization. However, the flux through a surface ought not to depend on how the surface is parameterized. Suppose that the surface S has two parameterizations, $\vec{r} = \vec{r}(s, t)$ for (s, t) in the region R of st-space, and also $\vec{r} = r(u, v)$ for (u, v) in the region T in uv-space, and suppose that the two parameterizations are related by the change of variables

$$u = u(s, t) \quad v = v(s, t).$$

Suppose that the Jacobian determinant $\partial(u, v)/\partial(s, t)$ is positive at every point (s, t) in R. Use the change of variables formula for double integrals on page 775 to show that computing the flux integral using either parameterization gives the same result.

CHAPTER SUMMARY

- **Flux Integrals**
  Oriented surfaces, definition of flux through a surface, definition of flux integral as a limit of a Riemann sum

- **Calculating flux integrals** over surfaces
  Graphs: $d\vec{A} = (-f_x\vec{i} - f_y\vec{j} + \vec{k}) \, dx \, dy$
  Cylinders: $d\vec{A} = (\cos\theta\vec{i} + \sin\theta\vec{j})R \, dz \, d\theta$

  Spheres: $d\vec{A} = (\sin\phi\cos\theta\vec{i} + \sin\phi\sin\theta\vec{j} + \cos\phi\vec{k})R^2 \, d\phi \, d\theta$
  Surfaces parameterized by $\vec{r}(s, t)$: $d\vec{A} = (\partial\vec{r}/\partial s) \times (\partial\vec{r}/\partial t) \, ds \, dt$

- **Area of parameterized surface**
  $dA = \|(\partial\vec{r}/\partial s) \times (\partial\vec{r}/\partial t)\| \, ds \, dt$

REVIEW EXERCISES AND PROBLEMS FOR CHAPTER NINETEEN

Exercises

Find the flux of $\vec{G} = 2\vec{j} + 3\vec{k}$ through the surface in Exercises 1–4.

1. The unit sphere oriented outward.

2. A square of side 7 perpendicular to the y-axis, oriented in the negative y-direction.

3. A circular disk of radius 5 in the plane $z = 2$, centered at $(1, 3, 2)$, and oriented upward.

4. The part of the plane $z = y$ above the square $0 \le x \le 1$, $0 \le y \le 1$, oriented upward.

Find the flux of $\vec{F} = \vec{r} = x\vec{i} + y\vec{j} + z\vec{k}$ through the surfaces in Exercises 5–7.

5. Through a sphere of radius π centered at the origin.

6. Through a square, oriented upward, with corners $(0, 0, 2), (3, 0, 2), (3, 3, 2), (0, 3, 2)$.

7. Through a circular disk of radius 3 in the plane $x = 5$, centered at $(5, 2, 4)$ and oriented in the positive x-direction.

Find the flux of $\vec{F} = 2\vec{i} + 3\vec{j} + z\vec{k}$ through the surfaces in Exercises 8–10.

8. A disk of radius 5 in the plane $x = 4$, oriented away from the origin.

9. A square of side 5 in the plane $z = 4$, oriented upward.

10. A disk of radius 5 in the plane $x + y = 1$, oriented away from the origin.

In Exercises 11–19, calculate the flux of the vector field through the given surface.

11. $\vec{F} = -5\vec{r}$ through the sphere of radius 2 centered at the origin.

12. $\vec{F} = x\vec{i} + y\vec{j} + (z^2 + 3)\vec{k}$ and S is the rectangle $z = 4$, $0 \le x \le 2$, $0 \le y \le 3$, oriented in the positive z-direction.

13. $\vec{F} = z\vec{i} + y\vec{j} + 2x\vec{k}$ and S is the rectangle $z = 4$, $0 \le x \le 2$, $0 \le y \le 3$, oriented in the positive z-direction.

14. $\vec{F} = (x + \cos z)\vec{i} + y\vec{j} + 2x\vec{k}$ and S is the rectangle $x = 2$, $0 \le y \le 3$, $0 \le z \le 4$, oriented in the positive x-direction.

15. $\vec{F} = x^2\vec{i} + (x + e^y)\vec{j} - \vec{k}$, and S is the rectangle $y = -1$, $0 \le x \le 2$, $0 \le z \le 4$, oriented in the negative y-direction.

16. $\vec{F} = (5 + xy)\vec{i} + z\vec{j} + yz\vec{k}$ and S is the 2×2 square plate in the yz-plane centered at the origin, oriented in the positive x-direction.

17. $\vec{F} = x\vec{i} + y\vec{j}$ and S is the surface of a closed cylinder of radius 2 and height 3 centered on the z-axis with its base in the xy-plane.

18. $\vec{F} = -y\vec{i} + x\vec{j} + z\vec{k}$ and S is the surface of a closed cylinder of radius 1 centered on the z-axis with base in the plane $z = -1$ and top in the plane $z = 1$.

19. $\vec{F} = x^2\vec{i} + y^2\vec{j} + z\vec{k}$ and S is the cone $z = \sqrt{x^2 + y^2}$, oriented upward with $x^2 + y^2 \leq 1$, $x \geq 0$, $y \geq 0$.

Problems

20. (a) Describe in words the vector field $\vec{F} = c\vec{r}$, where $\vec{r} = x\vec{i} + y\vec{j} + z\vec{k}$ and $c > 0$ is a constant.
   (b) Explain how you can tell the sign of the flux of $\vec{F}$ out of the sphere of radius 3 centered at the origin without calculating it.
   (c) Calculate the flux described in part (b).

In Problems 21–23, find the flux of the vector field through the surface.

21. $\vec{D}(\vec{r}) = \dfrac{Q}{4\pi\|\vec{r}\|^3}\vec{r}$, where Q is a constant, and S is a sphere of radius R centered at the origin.

22. $\vec{D}(\vec{r}) = \dfrac{\rho_0}{3}\vec{r}$, where ρ_0 is a constant, and S is a sphere of radius R centered at the origin.

23. $\vec{E} = \dfrac{\rho}{2\epsilon_0}(x\vec{i} + y\vec{j})$, where ρ and ϵ_0 are constants, and S is a cylinder (without ends) of height h and radius R centered on the z-axis and oriented outward. Would the answer change if the cylinder included the ends?

24. Water is flowing down a cylindrical pipe of radius 2 cm and that the speed is $(3 - (3/4)r^2)$ cm/sec at a distance r cm from the center of the pipe. Find the flux through the circular cross-section of the pipe, oriented so that the flow is positive.

25. Suppose that $\vec{E}$ is a uniform electric field on 3-space, so $\vec{E}(x, y, z) = a\vec{i} + b\vec{j} + c\vec{k}$, for all points (x, y, z), where a, b, c are constants. Show, with the aid of symmetry, that the flux of $\vec{E}$ through each of the following closed surfaces S is zero:

   (a) S is the cube bounded by the planes $x = \pm 1$, $y = \pm 1$, and $z = \pm 1$
   (b) S is the sphere $x^2 + y^2 + z^2 = 1$
   (c) S is the cylinder bounded by $x^2 + y^2 = 1$, $z = 0$, and $z = 2$

26. According to Coulomb's Law the electrostatic field $\vec{E}$, at the point with position vector $\vec{r}$ in 3-space, due to a charge q at the origin is given by

$$\vec{E}(\vec{r}) = q\frac{\vec{r}}{\|\vec{r}\|^3}.$$

Let S_a be the outward oriented sphere in 3-space of radius $a > 0$ and center at the origin. Show that the flux of the resulting electric field $\vec{E}$ through the surface S_a is equal to $4\pi q$, for any radius a. This is Gauss's Law for a single point charge.

27. An infinitely long straight wire lying along the z-axis carries an electric current I flowing in the $\vec{k}$ direction. Ampère's Law in magnetostatics says that the current gives rise to a magnetic field $\vec{B}$ given by

$$\vec{B}(x, y, z) = \frac{I}{2\pi}\frac{-y\vec{i} + x\vec{j}}{x^2 + y^2}.$$

   (a) Sketch the field $\vec{B}$ in the xy-plane.
   (b) Suppose S_1 is a disk with center at $(0, 0, h)$, radius a, and parallel to the xy-plane, oriented in the $\vec{k}$ direction. What is the flux of $\vec{B}$ through S_1? Is your answer reasonable?
   (c) Suppose S_2 is the rectangle given by $x = 0$, $a \leq y \leq b$, $0 \leq z \leq h$, and oriented in the $-\vec{i}$ direction. What is the flux of $\vec{B}$ through S_2? Does your answer seem reasonable?

28. An ideal electric dipole in electrostatics is characterized by its position in 3-space and its dipole moment vector $\vec{p}$. The electric field $\vec{D}$, at the point with position vector $\vec{r}$, of an ideal electric dipole located at the origin with dipole moment $\vec{p}$ is given by

$$\vec{D}(\vec{r}) = 3\frac{(\vec{r} \cdot \vec{p})\vec{r}}{\|\vec{r}\|^5} - \frac{\vec{p}}{\|\vec{r}\|^3}.$$

Assume $\vec{p} = p\vec{k}$, so the dipole points in the $\vec{k}$ direction and has magnitude p.

   (a) What is the flux of $\vec{D}$ through a sphere S with center at the origin and radius $a > 0$?
   (b) The field $\vec{D}$ is a useful approximation to the electric field $\vec{E}$ produced by two "equal and opposite" charges, q at $\vec{r}_2$ and $-q$ at $\vec{r}_1$, where the distance $\|\vec{r}_2 - \vec{r}_1\|$ is small. The dipole moment of this configuration of charges is defined to be $q(\vec{r}_2 - \vec{r}_1)$. Gauss's Law in electrostatics says that the flux of $\vec{E}$ through S is equal to 4π times the total charge enclosed by S. What is the flux of $\vec{E}$ through S if the charges at $\vec{r}_1$ and $\vec{r}_2$ are enclosed by S? How does this compare with your answer for the flux of $\vec{D}$ through S if $\vec{p} = q(\vec{r}_2 - \vec{r}_1)$?

CAS Challenge Problems

29. Let S be the part of the ellipsoid $x^2 + y^2 + 2z^2 = 1$ lying above the rectangle $-1/2 \leq x \leq 1/2, -1/2 \leq y \leq 1/2$, oriented upward. For each vector field (a)–(c), say whether you expect $\int_S \vec{F} \cdot d\vec{A}$ to be positive, negative, or zero. Then evaluate the integral exactly using a computer algebra system and find numerical approximations for your answers. Describe and explain what you notice.

(a) $\vec{F} = x\vec{i}$
(b) $\vec{F} = (x+1)\vec{i}$
(c) $\vec{F} = y\vec{j}$

30. Let $\vec{F} = (z+4)\vec{k}$, and let S be the surface with normal pointing in the direction of the negative y-axis parameterized, for $0 \leq s \leq 2, 0 \leq t \leq 2\pi$, by

$$\vec{r}(s,t) = s^2 \cos t\vec{i} + s\vec{j} + s^2 \sin t\vec{k}.$$

(a) Sketch S and, without evaluating the integral, say whether $\int_S \vec{F} \cdot d\vec{A}$ is positive, negative, or zero. Give a geometric explanation for your answer.

(b) Evaluate the flux integral. Does it agree with your answer to part (a)?

CHECK YOUR UNDERSTANDING

Are the statements in Problems 1–16 true or false? Give reasons for your answer.

1. The value of a flux integral is a scalar.

2. The area vector $\vec{A}$ of a flat, oriented surface is parallel to the surface.

3. If S is the unit sphere centered at the origin, oriented outward and the flux integral $\int_S \vec{F} \cdot d\vec{A}$ is zero, then $\vec{F} = \vec{0}$.

4. The flux of the vector field $\vec{F} = \vec{i}$ through the plane $x = 0$, with $0 \leq y \leq 1, 0 \leq z \leq 1$, oriented in the $\vec{i}$ direction is positive.

5. If S is the unit sphere centered at the origin, oriented outward and $\vec{F} = x\vec{i} + y\vec{j} + z\vec{k} = \vec{r}$, then the flux integral $\int_S \vec{F} \cdot d\vec{A}$ is positive.

6. If S is the cube bounded by the six planes $x = \pm 1, y = \pm 1, z = \pm 1$, oriented outward, and $\vec{F} = \vec{k}$, then $\int_S \vec{F} \cdot d\vec{A} = 0$.

7. If S is an oriented surface in 3-space, and $-S$ is the same surface, but with the opposite orientation, then $\int_S \vec{F} \cdot d\vec{A} = -\int_{-S} \vec{F} \cdot d\vec{A}$.

8. If S_1 is a rectangle with area 1 and S_2 is a rectangle with area 2, then $2\int_{S_1} \vec{F} \cdot d\vec{A} = \int_{S_2} \vec{F} \cdot d\vec{A}$.

9. If $\vec{F} = 2\vec{G}$, then $\int_S \vec{F} \cdot d\vec{A} = 2\int_S \vec{G} \cdot d\vec{A}$.

10. If $\int_S \vec{F} \cdot d\vec{A} > \int_S \vec{G} \cdot d\vec{A}$ then $||\vec{F}|| > ||\vec{G}||$ at all points on the surface S.

11. If S is parameterized by $\vec{r} = \vec{r}(s,t)$ and $\vec{r} \cdot \vec{F} = 0$ at each point of S, then $\int_S \vec{F} \cdot d\vec{A} = 0$.

12. If S is the part of the graph of f lying above $a \leq x \leq b, c \leq y \leq d$, then S has surface area $\int_a^b \int_c^d \sqrt{f_x^2 + f_y^2 + 1} dx\, dy$.

13. If $\vec{A}(x,y)$ is the area vector for $z = f(x,y)$ oriented upward and $\vec{B}(x,y)$ is the area vector for $z = -f(x,y)$ oriented upward, then $\vec{A}(x,y) = -\vec{B}(x,y)$.

14. If S is the sphere $x^2 + y^2 + z^2 = 1$ oriented outward and $\int_S \vec{F} \cdot d\vec{A} = 0$, then $\vec{F}(x,y,z)$ is perpendicular to $x\vec{i} + y\vec{j} + z\vec{k}$ at every point of S.

15. For any parameterization of the surface $x^2 - y^2 + z^2 = 6$, $d\vec{A}$ at $(1, 2, 3)$ is a multiple of $(2\vec{i} - 4\vec{j} + 6\vec{k})dx\, dy$.

16. If you parameterize the plane $3x + 4y + 5z = 7$, then there is a constant c such that, at any point (x, y, z), $d\vec{A} = c(3\vec{i} + 4\vec{j} + 5\vec{k})dx\, dy$.

PROJECTS FOR CHAPTER NINETEEN

1. Gauss' Law Applied to a Charged Wire and a Charged Sheet

Gauss' Law states that the flux of an electric field through a closed surface, S, is proportional to the quantity of charge, q, enclosed within S. That is,

$$\int_S \vec{E} \cdot d\vec{A} = kq.$$

In this project we use Gauss' Law to calculate the electric field of a uniformly charged wire in part (a) and a flat sheet of charge in part (b).

(a) Consider the electric field due to an infinitely long, straight, uniformly charged wire. (There is no current running through the wire—all charges are fixed.) Assuming that the wire is infinitely long means that we can assume that the electric field is perpendicular to any cylinder that has the wire as an axis and that the magnitude of the field is constant on any such cylinder. Denote by E_r the magnitude of the electric field due to the wire on a cylinder of radius r. (See Figure 19.32.)

Imagine a closed surface S made up of two cylinders, one of radius a and one of larger radius b, both coaxial with the wire, and the two washers that cap the ends. (See Figure 19.33.) The outward orientation of S means that a normal on the outer cylinder points away from the wire and a normal on the inner cylinder points towards the wire.

  (i) Explain why the flux of $\vec{E}$, the electric field, through the washers is 0.

 (ii) Explain why Gauss' Law implies that the flux through the inner cylinder is the same as the flux through the outer cylinder. [Hint: The charge on the wire is not inside the surface S].

(iii) Use part (ii) to show that $E_b/E_a = a/b$.

(iv) Explain why part (iii) shows that the strength of the field due to an infinitely long uniformly charged wire is proportional to $1/r$.

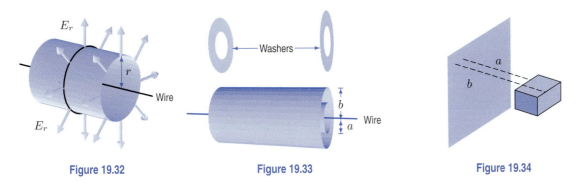

Figure 19.32   Figure 19.33   Figure 19.34

(b) Now consider an infinite flat sheet uniformly covered with charge. As in part (a), symmetry shows that the electric field $\vec{E}$ is perpendicular to the sheet and has the same magnitude at all points that are the same distance from the sheet. Use Gauss' Law to explain why, on any one side of the sheet, the electric field is the same at all points in space off the sheet. [Hint: Consider the flux through the box with sides parallel to the sheet shown in Figure 19.34.]

2. **Flux Across a Cylinder Due to a Point Charge: Obtaining Gauss' Law from Coulomb's Law**

An electric charge q is placed at the origin in 3-space. The induced electric field $\vec{E}(\vec{r})$ at the point with position vector $\vec{r}$ is given by Coulomb's Law, which says

$$\vec{E}(\vec{r}) = q\frac{\vec{r}}{\|\vec{r}\|^3}, \qquad \vec{r} \neq \vec{0}.$$

In this project, Gauss' Law is obtained for a cylinder enclosing a point charge by direct calculation from Coulomb's Law.

(a) Let S be the open cylinder of height $2H$ and radius R given by $x^2 + y^2 = R^2$, $-H \leq z \leq H$, oriented outward.

  (i) Show that the flux of $\vec{E}$, the electric field, through S is given by

$$\int_S \vec{E} \cdot d\vec{A} = 4\pi q \frac{H}{\sqrt{H^2 + R^2}}.$$

 (ii) What are the limits of the flux $\int_S \vec{E} \cdot d\vec{A}$ if
  - $H \to 0$ or $H \to \infty$ when R is fixed?
  - $R \to 0$ or $R \to \infty$ when H is fixed?

(b) Let T be the outward oriented, closed cylinder of height $2H$ and radius R whose curved side is given by $x^2 + y^2 = R^2$, $-H \le z \le H$, whose top is given by $z = H$, $x^2 + y^2 \le R^2$, and bottom by $z = -H$, $x^2 + y^2 \le R^2$. Use part (a) to show that the flux of the electric field, $\vec{E}$, through T is given by

$$\int_T \vec{E} \cdot d\vec{A} = 4\pi q.$$

Notice that this is Gauss' Law. In particular, the flux is independent of both the height, H, and radius, R, of the cylinder.

Chapter Twenty

CALCULUS OF VECTOR FIELDS

We have seen two ways of integrating vector fields in three dimensions: along curves and over surfaces. Now we look at two ways of differentiating them. If we view the vector field as the velocity field of a fluid flow, then one method of differentiation (the divergence) tells us about the net strength of outflow from a point, and the other method (the curl) tells us about the strength of rotation around a point. Each method fits together with one of the ways of integrating to form a vector analogue of the Fundamental Theorem of Calculus: the Divergence Theorem relating the divergence to flux, and Stokes' theorem relating the curl to circulation around a closed path.

20.1   THE DIVERGENCE OF A VECTOR FIELD

Imagine that the vector fields in Figures 20.1 and 20.2 are velocity vector fields describing the flow of a fluid.[1] Figure 20.1 suggests outflow from the origin; for example, it could represent the expanding cloud of matter in the big bang theory of the origin of the universe. We say that the origin is a *source*. Figure 20.2 suggests flow into the origin; in this case we say that the origin is a *sink*.

In this section we will use the flux out of a closed surface surrounding a point to measure the outflow per unit volume there, also called the *divergence*, or *flux density*.

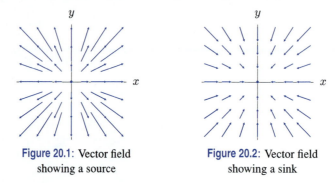

Figure 20.1: Vector field showing a source   Figure 20.2: Vector field showing a sink

Definition of Divergence

To measure the outflow per unit volume of a vector field at a point, we calculate the flux out of a small sphere centered at the point, divide by the volume enclosed by the sphere, then take the limit of this flux to volume ratio as the sphere contracts around the point.

Geometric Definition of Divergence

The **divergence**, or **flux density**, of a smooth vector field $\vec{F}$, written $\textbf{div}\,\vec{F}$, is a scalar-valued function defined by

$$\operatorname{div} \vec{F}(x, y, z) = \lim_{\text{Volume} \to 0} \frac{\int_S \vec{F} \cdot d\vec{A}}{\text{Volume of } S}.$$

Here S is a sphere centered at (x, y, z), oriented outwards, that contracts down to (x, y, z) in the limit.

The limit can be computed using other shapes as well, such as the cubes in Example 2.

Cartesian Coordinate Definition of Divergence

If $\vec{F} = F_1 \vec{i} + F_2 \vec{j} + F_3 \vec{k}$, then

$$\operatorname{div} \vec{F} = \frac{\partial F_1}{\partial x} + \frac{\partial F_2}{\partial y} + \frac{\partial F_3}{\partial z}.$$

Example 1   Calculate the divergence of $\vec{F}(\vec{r}) = \vec{r}$ at the origin

(a) Using the geometric definition.

(b) Using the Cartesian coordinate definition.

---

[1] Although not all vector fields represent physically realistic fluid flows, it is useful to think of them this way.

Solution   (a) Using the method of Example 4 on page 868, you can calculate the flux of $\vec{F}$ out of the sphere of radius a, centered at the origin; it is $4\pi a^3$. So we have

$$\operatorname{div}\vec{F}(0,0,0) = \lim_{a\to 0}\frac{\text{Flux}}{\text{Volume}} = \lim_{a\to 0}\frac{4\pi a^3}{\frac{4}{3}\pi a^3} = \lim_{a\to 0} 3 = 3.$$

(b) In coordinates, $\vec{F}(x,y,z) = x\vec{i} + y\vec{j} + z\vec{k}$, so

$$\operatorname{div}\vec{F} = \frac{\partial}{\partial x}(x) + \frac{\partial}{\partial y}(y) + \frac{\partial}{\partial z}(z) = 1 + 1 + 1 = 3.$$

The next example shows that the divergence can be negative if there is net inflow to a point.

Example 2   (a) Using the geometric definition, find the divergence of $\vec{v} = -x\vec{i}$ at: (i) $(0,0,0)$ (ii) $(2,2,0)$.
(b) Confirm that the coordinate definition gives the same results.

Solution   (a) (i) The vector field $\vec{v} = -x\vec{i}$ is parallel to the x axis, as shown in Figure 20.3. To compute the flux density at $(0,0,0)$, we use a cube S_1, centered at the origin with edges parallel to the axes, of length $2c$. Then the flux through the faces perpendicular to the y- and z-axes is zero (because the vector field is parallel to these faces). On the faces perpendicular to the x-axis, the vector field and the outward normal are parallel but point in opposite directions. On the face at $x = c$, we have

$$\vec{v}\cdot\Delta\vec{A} = -c\,\|\Delta\vec{A}\|.$$

On the face at $x = -c$, the dot product is still negative, and

$$\vec{v}\cdot\Delta\vec{A} = -c\,\|\Delta\vec{A}\|.$$

Therefore, the flux through the cube is given by

$$\int_{S_1}\vec{v}\cdot d\vec{A} = \int_{\text{Face }x=-c}\vec{v}\cdot d\vec{A} + \int_{\text{Face }x=c}\vec{v}\cdot d\vec{A}$$
$$= -c\cdot\text{Area of one face} + (-c)\cdot\text{Area of other face} = -2c(2c)^2 = -8c^3.$$

Thus,

$$\operatorname{div}\vec{v}(0,0,0) = \lim_{\text{Volume}\to 0}\frac{\int_S\vec{v}\cdot d\vec{A}}{\text{Volume of cube}} = \lim_{c\to 0}\left(\frac{-8c^3}{(2c)^3}\right) = -1.$$

Since the vector field points inward towards the yz-plane, it makes sense that the divergence is negative at the origin.

(ii) Take S_2 to be a cube as before, but centered this time at the point $(2,2,0)$. See Figure 20.3. As before, the flux through the faces perpendicular to the y- and z-axes is zero. On the face at $x = 2 + c$,

$$\vec{v}\cdot\Delta\vec{A} = -(2+c)\,\|\Delta\vec{A}\|.$$

On the face at $x = 2 - c$ with outward normal, the dot product is positive, and

$$\vec{v}\cdot\Delta\vec{A} = (2-c)\,\|\Delta\vec{A}\|.$$

Therefore, the flux through the cube is given by

$$\int_{S_2}\vec{v}\cdot d\vec{A} = \int_{\text{Face }x=2-c}\vec{v}\cdot d\vec{A} + \int_{\text{Face }x=2+c}\vec{v}\cdot d\vec{A}$$
$$= (2-c)\cdot\text{Area of one face} - (2+c)\cdot\text{Area of other face} = -2c(2c)^2 = -8c^3.$$

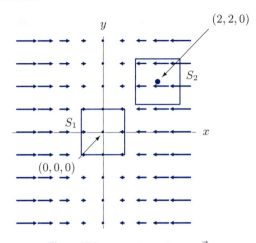

Figure 20.3: Vector field $\vec{v} = -x\vec{i}$

Then, as before,

$$\text{div } \vec{v}\,(2,2,0) = \lim_{\text{Volume}\to 0} \frac{\int_S \vec{v} \cdot d\vec{A}}{\text{Volume of cube}} = \lim_{c\to 0}\left(\frac{-8c^3}{(2c)^3}\right) = -1.$$

Although the vector field is flowing away from the point $(2,2,0)$ on the left, this outflow is smaller in magnitude than the inflow on the right, so the the net outflow is negative.

(b) Since $\vec{v} = -x\vec{i} + 0\vec{j} + 0\vec{k}$, the formula gives

$$\text{div } \vec{v} = \frac{\partial}{\partial x}(-x) + \frac{\partial}{\partial y}(0) + \frac{\partial}{\partial z}(0) = -1 + 0 + 0 = -1.$$

Why Do the Two Definitions of Divergence Give the Same Result?

The geometric definition defines $\text{div }\vec{F}$ as the flux density of $\vec{F}$. To see why the coordinate definition is also the flux density, imagine computing the flux out of a small box-shaped surface S at (x_0, y_0, z_0), with sides of length Δx, Δy, and Δz parallel to the axes. On S_1 (the back face of the box shown in Figure 20.4, where $x = x_0$), the outward normal is in the negative x-direction, so $d\vec{A} = -dy\,dz\,\vec{i}$. Assuming $\vec{F}$ is approximately constant on S_1, we have

$$\int_{S_1} \vec{F} \cdot d\vec{A} = \int_{S_1} \vec{F} \cdot (-\vec{i})\,dy\,dz \approx -F_1(x_0, y_0, z_0) \int_{S_1} dy\,dz$$

$$= -F_1(x_0, y_0, z_0) \cdot \text{Area of } S_1 = -F_1(x_0, y_0, z_0)\,\Delta y\,\Delta z.$$

On S_2, the face where $x = x_0 + \Delta x$, the outward normal points in the positive x-direction, so $d\vec{A} = dy\,dz\,\vec{i}$. Therefore,

$$\int_{S_2} \vec{F} \cdot d\vec{A} = \int_{S_2} \vec{F} \cdot \vec{i}\,dy\,dz \approx F_1(x_0 + \Delta x, y_0, z_0) \int_{S_2} dy\,dz$$

$$= F_1(x_0 + \Delta x, y_0, z_0) \cdot \text{Area of } S_2 = F_1(x_0 + \Delta x, y_0, z_0)\,\Delta y\,\Delta z.$$

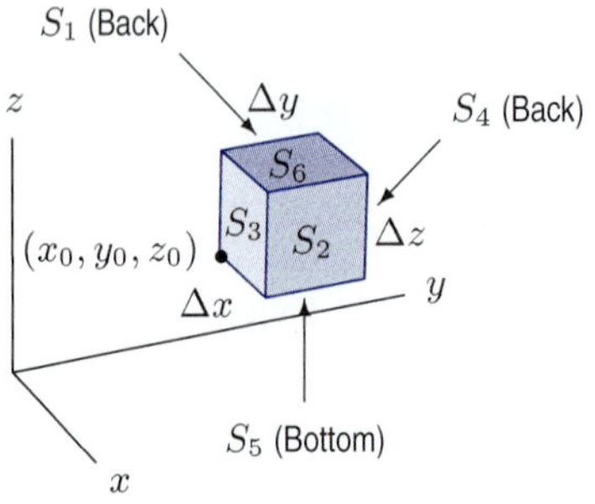

Figure 20.4: Box used to find $\text{div }\vec{F}$ at (x_0, y_0, z_0)

Thus

$$\int_{S_1} \vec{F} \cdot d\vec{A} + \int_{S_2} \vec{F} \cdot d\vec{A} \approx F_1(x_0 + \Delta x, y_0, z_0)\Delta y \Delta z - F_1(x_0, y_0, z_0)\Delta y \Delta z$$

$$= \frac{F_1(x_0 + \Delta x, y_0, z_0) - F_1(x_0, y_0, z_0)}{\Delta x}\Delta x \Delta y \Delta z$$

$$\approx \frac{\partial F_1}{\partial x}\Delta x \Delta y \Delta z.$$

By an analogous argument, the contribution to the flux from S_3 and S_4 (the surfaces perpendicular to the y-axis) is approximately

$$\frac{\partial F_2}{\partial y}\Delta x\,\Delta y\,\Delta z,$$

and the contribution to the flux from S_5 and S_6 is approximately

$$\frac{\partial F_3}{\partial z}\Delta x\,\Delta y\,\Delta z.$$

Thus, adding these contributions, we have

$$\text{Total flux through } S \approx \frac{\partial F_1}{\partial x}\Delta x\,\Delta y\,\Delta z + \frac{\partial F_2}{\partial y}\Delta x\,\Delta y\,\Delta z + \frac{\partial F_3}{\partial z}\Delta x\,\Delta y\,\Delta z.$$

Since the volume of the box is $\Delta x\,\Delta y\,\Delta z$, the flux density is

$$\frac{\text{Total flux through } S}{\text{Volume of box}} \approx \frac{\dfrac{\partial F_1}{\partial x}\Delta x \Delta y \Delta z + \dfrac{\partial F_2}{\partial y}\Delta x \Delta y \Delta z + \dfrac{\partial F_3}{\partial z}\Delta x \Delta y \Delta z}{\Delta x \Delta y \Delta z}$$

$$= \frac{\partial F_1}{\partial x} + \frac{\partial F_2}{\partial y} + \frac{\partial F_3}{\partial z}.$$

Divergence Free Vector Fields

A vector field $\vec{F}$ is said to be *divergence free* or *solenoidal* if $\operatorname{div}\vec{F} = 0$ everywhere that $\vec{F}$ is defined.

Example 3   Figure 20.5 shows, for three values of the constant p, the vector field

$$\vec{E} = \frac{\vec{r}}{\|\vec{r}\|^p} \qquad \vec{r} \neq \vec{0}.$$

(a) Find a formula for $\operatorname{div}\vec{E}$.
(b) Is there a value of p for which $\vec{E}$ is divergence free? If so, find it.

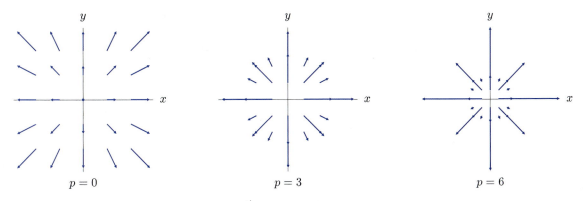

$$p = 0 \qquad\qquad p = 3 \qquad\qquad p = 6$$

Figure 20.5: The vector field $\vec{E}(\vec{r}) = \vec{r}/\|\vec{r}\|^p$ for $p = 0, 3,$ and 6

Solution   (a) The components of $\vec{E}$ are

$$\vec{E} = \frac{x}{(x^2 + y^2 + z^2)^{p/2}}\vec{i} + \frac{y}{(x^2 + y^2 + z^2)^{p/2}}\vec{j} + \frac{z}{(x^2 + y^2 + z^2)^{p/2}}\vec{k}.$$

We compute the partial derivatives

$$\frac{\partial}{\partial x}\left(\frac{x}{(x^2 + y^2 + z^2)^{p/2}}\right) = \frac{1}{(x^2 + y^2 + z^2)^{p/2}} - \frac{px^2}{(x^2 + y^2 + z^2)^{(p/2)+1}}$$

$$\frac{\partial}{\partial y}\left(\frac{y}{(x^2 + y^2 + z^2)^{p/2}}\right) = \frac{1}{(x^2 + y^2 + z^2)^{p/2}} - \frac{py^2}{(x^2 + y^2 + z^2)^{(p/2)+1}}$$

$$\frac{\partial}{\partial z}\left(\frac{z}{(x^2 + y^2 + z^2)^{p/2}}\right) = \frac{1}{(x^2 + y^2 + z^2)^{p/2}} - \frac{pz^2}{(x^2 + y^2 + z^2)^{(p/2)+1}}.$$

So

$$\operatorname{div}\vec{E} = \frac{3}{(x^2 + y^2 + z^2)^{p/2}} - \frac{p(x^2 + y^2 + z^2)}{(x^2 + y^2 + z^2)^{(p/2)+1}}$$

$$= \frac{3 - p}{(x^2 + y^2 + z^2)^{p/2}} = \frac{3 - p}{\|\vec{r}\|^p}.$$

(b) The divergence is zero when $p = 3$, so $\vec{F}(\vec{r}) = \vec{r}/\|\vec{r}\|^3$ is a divergence free vector field.

Magnetic Fields

An important class of divergence free vector fields is the magnetic fields. One of Maxwell's Laws of Electromagnetism is that the magnetic field $\vec{B}$ satisfies

$$\operatorname{div}\vec{B} = 0.$$

Example 4   An infinitesimal current loop, similar to that shown in Figure 20.6, is called a *magnetic dipole*. Its magnitude is described by a constant vector $\vec{\mu}$, called the dipole moment. The magnetic field due to a magnetic dipole with moment $\vec{\mu}$ is

$$\vec{B} = -\frac{\vec{\mu}}{\|\vec{r}\|^3} + \frac{3(\vec{\mu} \cdot \vec{r})\vec{r}}{\|\vec{r}\|^5}, \qquad \vec{r} \neq \vec{0}.$$

Show that $\operatorname{div}\vec{B} = 0$.

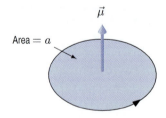

$\vec{\mu}$

Area $= a$

Figure 20.6: A current loop

Solution   To show that $\operatorname{div}\vec{B} = 0$ we can use the following version of the product rule for the divergence: if g is a scalar function and $\vec{F}$ is a vector field, then

$$\operatorname{div}(g\vec{F}) = (\operatorname{grad} g) \cdot \vec{F} + g\operatorname{div}\vec{F}.$$

(See Problem 22 on page 894.) Thus, since $\operatorname{div}\vec{\mu} = 0$, we have

$$\operatorname{div}\left(\frac{\vec{\mu}}{\|\vec{r}\|^3}\right) = \operatorname{div}\left(\frac{1}{\|\vec{r}\|^3}\vec{\mu}\right) = \operatorname{grad}\left(\frac{1}{\|\vec{r}\|^3}\right) \cdot \vec{\mu} + \frac{1}{\|\vec{r}\|^3} \cdot 0$$

and

$$\text{div}\left(\frac{(\vec{\mu}\cdot\vec{r})\vec{r}}{\|\vec{r}\|^5}\right) = \text{grad}(\vec{\mu}\cdot\vec{r})\cdot\frac{\vec{r}}{\|\vec{r}\|^5} + (\vec{\mu}\cdot\vec{r})\,\text{div}\left(\frac{\vec{r}}{\|\vec{r}\|^5}\right).$$

From Problems 34 and 35 on page 672 and Example 3 on page 891 we have

$$\text{grad}\left(\frac{1}{\|\vec{r}\|^3}\right) = \frac{-3\vec{r}}{\|\vec{r}\|^5}, \qquad \text{grad}(\vec{\mu}\cdot\vec{r}) = \vec{\mu}, \qquad \text{div}\left(\frac{\vec{r}}{\|\vec{r}\|^5}\right) = \frac{-2}{\|\vec{r}\|^5}.$$

Putting these results together gives

$$\text{div}\,\vec{B} = -\,\text{grad}\left(\frac{1}{\|\vec{r}\|^3}\right)\cdot\vec{\mu} + 3\,\text{grad}(\vec{\mu}\cdot\vec{r})\cdot\frac{\vec{r}}{\|\vec{r}\|^5} + 3(\vec{\mu}\cdot\vec{r})\,\text{div}\left(\frac{\vec{r}}{\|\vec{r}\|^5}\right)$$

$$= \frac{3\vec{r}\cdot\vec{\mu}}{\|\vec{r}\|^5} + \frac{3\vec{\mu}\cdot\vec{r}}{\|\vec{r}\|^5} - \frac{6\vec{\mu}\cdot\vec{r}}{\|\vec{r}\|^5}$$

$$= 0.$$

Alternative Notation for Divergence

Using $\nabla = \dfrac{\partial}{\partial x}\vec{i} + \dfrac{\partial}{\partial y}\vec{j} + \dfrac{\partial}{\partial z}\vec{k}$, we can write

$$\text{div}\,\vec{F} = \nabla\cdot\vec{F} = \left(\frac{\partial}{\partial x}\vec{i} + \frac{\partial}{\partial y}\vec{j} + \frac{\partial}{\partial z}\vec{k}\right)\cdot(F_1\vec{i} + F_2\vec{j} + F_3\vec{k}) = \frac{\partial F_1}{\partial x} + \frac{\partial F_2}{\partial y} + \frac{\partial F_3}{\partial z}.$$

Exercises and Problems for Section 20.1

Exercises

1. Draw two vector fields that have positive divergence everywhere.

2. Draw two vector fields that have negative divergence everywhere.

3. Draw two vector fields that have zero divergence everywhere.

In Exercises 4–10, find the divergence of the given vector field. (Note: $\vec{r} = x\vec{i} + y\vec{j} + z\vec{k}$.)

4. $\vec{F}(x,y) = -x\vec{i} + y\vec{j}$

5. $\vec{F}(x,y) = -y\vec{i} + x\vec{j}$

6. $\vec{F}(x,y) = (x^2 - y^2)\vec{i} + 2xy\vec{j}$

7. $\vec{F}(\vec{r}) = \vec{a} \times \vec{r}$

8. $\vec{F}(x,y) = \dfrac{-y\vec{i} + x\vec{j}}{x^2 + y^2}$

9. $\vec{F}(\vec{r}) = \dfrac{\vec{r} - \vec{r}_0}{\|\vec{r} - \vec{r}_0\|}, \quad \vec{r} \neq \vec{r}_0$

10. $\vec{F}(x,y,z) = (-x + y)\vec{i} + (y + z)\vec{j} + (-z + x)\vec{k}$

11. Which of the following two vector fields appears to have the greater divergence at the origin? Assume the scales are the same on each.

12. For each of the following vector fields, decide if the divergence is positive, zero, or negative at the indicated point.

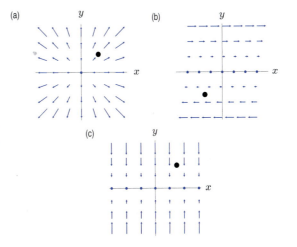

Problems

13. A smooth vector field $\vec{F}$ has div $\vec{F}(1, 2, 3) = 5$. Estimate the flux of $\vec{F}$ out of a small sphere of radius 0.01 centered at the point $(1, 2, 3)$.

14. A vector field $\vec{F}$ has the property that the flux of $\vec{F}$ out of a small cube of side 0.01 centered around the point $(1, 5, 7)$, is 0.0025. Estimate div $\vec{F}$ at the point $(1, 5, 7)$.

15. Let $\vec{F}(x, y, z) = z\vec{k}$.

    (a) Calculate div $\vec{F}$.
    (b) Sketch $\vec{F}$. Does it appear to be diverging? Does this agree with your answer to part (a)?

16. Let $\vec{F}(\vec{r}) = \vec{r}/\|\vec{r}\|^3$ (in 3-space), $\vec{r} \neq \vec{0}$.

    (a) Calculate div $\vec{F}$.
    (b) Sketch $\vec{F}$. Does it appear to be diverging? Does this agree with your answer to part (a)?

For Problems 17–19,

    (a) Find the flux of the given vector field through a cube in the first octant with edge length c, one corner at the origin and edges along the axes.
    (b) Use your answer to part (a) to find div $\vec{F}$ at the origin using the geometric definition.
    (c) Compute div $\vec{F}$ at the origin using partial derivatives.

17. $\vec{F} = x\vec{i}$

18. $\vec{F} = 2\vec{i} + y\vec{j} + 3\vec{k}$

19. $\vec{F} = x\vec{i} + y\vec{j}$

20. The divergence of a magnetic vector field $\vec{B}$ must be zero everywhere. Which of the following vector fields cannot be a magnetic vector field?

    (a) $\vec{B}(x, y, z) = -y\vec{i} + x\vec{j} + (x + y)\vec{k}$
    (b) $\vec{B}(x, y, z) = -z\vec{i} + y\vec{j} + x\vec{k}$
    (c) $\vec{B}(x, y, z) = (x^2 - y^2 - x)\vec{i} + (y - 2xy)\vec{j}$

21. Show that if $\vec{a}$ is a constant vector and $f(x, y, z)$ is a function, then $\text{div}(f\vec{a}) = (\text{grad } f) \cdot \vec{a}$.

22. Show that if $g(x, y, z)$ is a scalar valued function and $\vec{F}(x, y, z)$ is a vector field, then

$$\text{div}(g\vec{F}) = (\text{grad } g) \cdot \vec{F} + g \text{ div } \vec{F}.$$

23. If $f(x, y, z)$ and $g(x, y, z)$ are functions with continuous second partial derivatives, show that

$$\text{div}(\text{grad } f \times \text{grad } g) = 0.$$

24. Let $r = (x^2 + y^2)^{1/2}$. Figure 20.7 shows the vector field $r^A \cdot (x\vec{i} + y\vec{j})$ for $r \neq 0$ and $A = -1, -2, -3$. The vector fields are shown in the xy-plane; they have no z-component and are independent of z.

    (a) Show that $\text{div}(r^A \cdot (x\vec{i} + y\vec{j})) = (2 + A)r^A$ for any constant A.

(b) Using your answer to part (a), find the sign of the divergence of the three vector fields for $A = -1, -2, -3$.

(c) For each value of A, what (if anything) does your answer to part (b) tell you about the sign of the flux of the vector field out of a small sphere centered at $(1, 1, 1)$? At $(0, 0, 0)$?

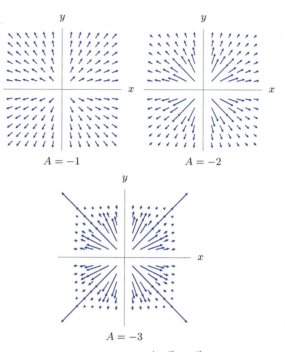

Figure 20.7: The vector field $r^A(x\vec{i} + y\vec{j})$ for the three values of A

25. If $r = (x^2 + y^2)^{1/2}$, show that

$$\text{div}\left(\frac{h(r)}{r^2}(x\vec{i} + y\vec{j})\right) = \frac{h'(r)}{r}.$$

26. Let $\vec{F} = e^{-r}(x\vec{i} + y\vec{j})$, where $r = (x^2 + y^2)^{1/2}$. Use the result of Problem 25 to determine the values of r for which div $\vec{F}$ is positive, negative, or zero.

27. In Problem 33 on page 872 it was shown that the rate of heat loss from a volume V in a region of non-uniform temperature equals $k\int_S(\text{grad } T) \cdot d\vec{A}$, where k is a constant, S is the surface bounding V, and $T(x, y, z)$ is the temperature at the point (x, y, z) in space. By taking the limit as V contracts to a point, show that, at that point,

$$\frac{\partial T}{\partial t} = B \text{ div grad } T$$

where B is a constant with respect to x, y, z, but may depend on time, t.

Problems 28–29 involve electric fields. Electric charge produces a vector field $\vec{E}$, called the electric field, which represents the force on a unit positive charge placed at the point. Two positive or two negative charges repel one another, whereas two charges of opposite sign attract one another. The divergence of $\vec{E}$ is proportional to the density of the electric charge (that is, the charge per unit volume), with a positive constant of proportionality.

28. A certain distribution of electric charge produces the electric field shown in Figure 20.8. Where are the charges that produced this electric field concentrated? Which concentrations are positive and which are negative?

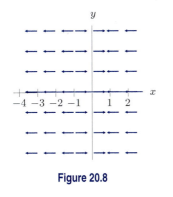

Figure 20.8

29. The electric field at the point $\vec{r}$ as a result of a point charge at the origin is $\vec{E}(\vec{r}) = k\vec{r}/\|\vec{r}\|^3$.

   (a) Calculate div $\vec{E}$ for $\vec{r} \neq \vec{0}$.

   (b) Calculate the limit suggested by the geometric definition of div $\vec{E}$ at the point $(0, 0, 0)$.

   (c) Explain what your answers mean in terms of charge density.

30. A vector field, $\vec{v}$, in the plane is a *point source* at the origin if its direction is away from the origin at every point, its magnitude depends only on the distance from the origin, and its divergence is zero away from the origin.

   (a) Explain why a point source at the origin must be of the form $\vec{v} = \left[f(x^2 + y^2) \right] (x\vec{i} + y\vec{j})$ for some positive function f.

   (b) Show that $\vec{v} = K(x^2 + y^2)^{-1}(x\vec{i} + y\vec{j})$ is a point source at the origin if $K > 0$.

   (c) What is the magnitude $\|\vec{v}\|$ of the source in part (b) as a function of the distance from its center?

   (d) Sketch the vector field $\vec{v} = (x^2 + y^2)^{-1}(x\vec{i} + y\vec{j})$.

   (e) Show that $\phi = \frac{K}{2} \log(x^2 + y^2)$ is a potential function for the source in part (b).

31. A vector field, $\vec{v}$, in the plane is a *point sink* at the origin if its direction is toward the origin at every point, its magnitude depends only on the distance from the origin, and its divergence is zero away from the origin.

   (a) Explain why a point sink at the origin must be of the form $\vec{v} = \left[f(x^2 + y^2) \right] (x\vec{i} + y\vec{j})$ for some negative function f.

   (b) Show that $\vec{v} = K(x^2 + y^2)^{-1}(x\vec{i} + y\vec{j})$ is a point sink at the origin if $K < 0$.

   (c) Determine the magnitude $\|\vec{v}\|$ of the sink in part (b) as a function of the distance from its center.

   (d) Sketch $\vec{v} = -(x^2 + y^2)^{-1}(x\vec{i} + y\vec{j})$.

   (e) Show that $\phi = \frac{K}{2} \log(x^2 + y^2)$ is a potential function for the sink in part (b).

32. Let $\vec{F}(x, y) = u(x, y)\vec{i} + v(x, y)\vec{j}$ be a 2-dimensional vector field. Let $F = \|\vec{F}\|$, and let $\phi(x, y)$ be the angle of $\vec{F}$ with the positive x-axis at the point (x, y), so that $u = F \cos \phi$ and $v = F \sin \phi$. Let $\vec{T}$ be the unit vector in the direction of $\vec{F}$, and let $\vec{N}$ be the unit vector in the direction of $\vec{k} \times \vec{F}$. Show that

$$\operatorname{div} \vec{F} = F\phi_{\vec{N}} + F_{\vec{T}}$$

This problem shows that the divergence of the vector field $\vec{F}$ is the sum of two terms. The first term, $F\phi_{\vec{N}}$, is due to changes in the direction of $\vec{F}$ perpendicular to the flow lines, so it reflects the extent to which flow lines fail to be parallel. The second term, $F_{\vec{T}}$, is due to changes in the magnitude of $\vec{F}$ along the flow lines of $\vec{F}$.

20.2   THE DIVERGENCE THEOREM

The Divergence Theorem is a multivariable analogue of the Fundamental Theorem of Calculus; it says that the integral of the flux density over a solid region equals the flux integral through the boundary of the region.

The Boundary of a Solid Region

A solid region is an open region in 3-space. The boundary of a solid region may be thought of as the skin between the interior of the region and the space around it. For example, the boundary of a solid ball is a spherical surface, the boundary of a solid cube is its six faces, and the boundary of a solid cylinder is a tube sealed at both ends by disks. (See Figure 20.9). A surface which is the boundary of a solid region is called a *closed surface*.

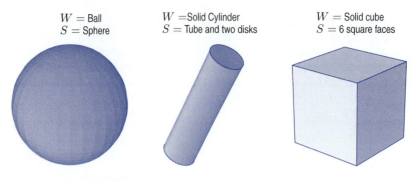

$W = \text{Ball}$
$S = \text{Sphere}$

$W = \text{Solid Cylinder}$
$S = \text{Tube and two disks}$

$W = \text{Solid cube}$
$S = \text{6 square faces}$

Figure 20.9: Several solid regions and their boundaries

Calculating the Flux from the Flux Density

Consider a solid region W in 3-space whose boundary is the closed surface S. There are two ways to find the total flux of a vector field $\vec{F}$ out of W. One is to calculate the flux of $\vec{F}$ through S:

$$\text{Flux out of } W = \int_{S} \vec{F} \cdot d\vec{A}.$$

Another way is to use $\operatorname{div} \vec{F}$, which gives the flux density at any point in W. We subdivide W into small boxes, as shown in Figure 20.10. Then, for a small box of volume ΔV,

$$\text{Flux out of box} \approx \text{Flux density} \cdot \text{Volume} = \operatorname{div} \vec{F} \; \Delta V.$$

What happens when we add the fluxes out of all the boxes? Consider two adjacent boxes, as shown in Figure 20.11. The flux through the shared wall is counted twice, once out of the box on each side. When we add the fluxes, these two contributions cancel, so we get the flux out of the solid region formed by joining the two boxes. Continuing in this way, we find that

$$\text{Flux out of } W = \sum \text{Flux out of small boxes} \approx \sum \operatorname{div} \vec{F} \; \Delta V.$$

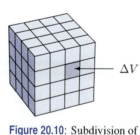

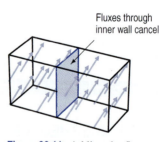

Fluxes through
inner wall cancel

Figure 20.10: Subdivision of
region into small boxes

Figure 20.11: Adding the flux out
of adjacent boxes

We have approximated the flux by a Riemann sum. As the subdivision gets finer, the sum approaches an integral, so

$$\text{Flux out of } W = \int_{W} \operatorname{div} \vec{F} \; dV.$$

We have calculated the flux in two ways, as a flux integral and as a volume integral. Therefore, these two integrals must be equal. This result holds even if W is not a rectangular solid, as shown in Figure 20.10. Thus, we have the following result.[2]

---

[2]A proof of the Divergence Theorem using the coordinate definition of the divergence can be found in the online supplement.

Theorem 20.1: The Divergence Theorem

If W is a solid region whose boundary S is a piecewise smooth surface, and if $\vec{F}$ is a smooth vector field on an open region containing W and S, then

$$\int_S \vec{F} \cdot d\vec{A} = \int_W \operatorname{div} \vec{F} \, dV,$$

where S is given the outward orientation.

Example 1   Use the Divergence Theorem to calculate the flux of the vector field $\vec{F}(\vec{r}) = \vec{r}$ through the sphere of radius a centered at the origin.

Solution   In Example 4 on page 868 we computed the flux integral directly:

$$\int_S \vec{r} \cdot d\vec{A} = 4\pi a^3.$$

Now we use $\operatorname{div} \vec{F} = 3$ and the Divergence Theorem:

$$\int_S \vec{r} \cdot d\vec{A} = \int_W \operatorname{div} \vec{F} \, dV = \int_W 3 \, dV = 3\left(\frac{4}{3}\pi a^3\right) = 4\pi a^3.$$

Example 2   Use the Divergence Theorem to calculate the flux of the vector field

$$\vec{F}(x, y, z) = (x^2 + y^2)\vec{i} + (y^2 + z^2)\vec{j} + (x^2 + z^2)\vec{k}$$

through the cube in Figure 20.12.

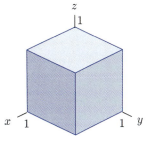

Figure 20.12

Solution   The divergence of $\vec{F}$ is $\operatorname{div} \vec{F} = 2x + 2y + 2z$. Since $\operatorname{div} \vec{F}$ is positive everywhere in the first quadrant, the flux through S is positive. By the Divergence Theorem,

$$\int_S \vec{F} \cdot d\vec{A} = \int_0^1 \int_0^1 \int_0^1 2(x + y + z) \, dx \, dy \, dz = \int_0^1 \int_0^1 x^2 + 2x(y + z)\Big|_0^1 \, dy \, dz$$

$$= \int_0^1 \int_0^1 1 + 2(y + z) \, dy \, dz = \int_0^1 y + y^2 + 2yz\Big|_0^1 \, dz$$

$$= \int_0^1 (2 + 2z) \, dz = 2z + z^2\Big|_0^1 = 3.$$

The Divergence Theorem and Divergence Free Vector Fields

An important application of the Divergence Theorem is the study of divergence free vector fields.

Example 3   In Example 3 on page 891 we saw that the following vector field is divergence free:

$$\vec{F}(\vec{r}) = \frac{\vec{r}}{\|\vec{r}\|^3}, \qquad \vec{r} \neq \vec{0}.$$

Calculate $\int_S \vec{F} \cdot d\vec{A}$, using the Divergence Theorem if possible, for the following surfaces:

(a) S_1 is the sphere of radius a centered at the origin.
(b) S_2 is the sphere of radius a centered at the point $(2a, 0, 0)$.

Solution   (a) We cannot use the Divergence Theorem directly because $\vec{F}$ is not defined everywhere inside the sphere (it is not defined at the origin). Since $\vec{F}$ points outward everywhere on S_1, the flux out of S_1 is positive. On S_1,

$$\vec{F} \cdot d\vec{A} = \|\vec{F}\| dA = \frac{a}{a^3} dA,$$

so

$$\int_{S_1} \vec{F} \cdot d\vec{A} = \frac{1}{a^2} \int_{S_1} dA = \frac{1}{a^2}(\text{Area of } S_1) = \frac{1}{a^2} 4\pi a^2 = 4\pi.$$

Notice that the flux is not zero, although div $\vec{F}$ is zero everywhere it is defined.

(b) Suppose W is the solid region enclosed by S_2. Since div $\vec{F} = 0$ everywhere in W, we can use the Divergence Theorem in this case, giving

$$\int_{S_2} \vec{F} \cdot d\vec{A} = \int_W \text{div} \vec{F} \ dV = \int_W 0 \ dV = 0.$$

The Divergence Theorem applies to any solid region W and its boundary S, even in cases where the boundary consists of two or more surfaces. For example, if W is the solid region between the sphere S_1 of radius 1 and the sphere S_2 of radius 2, both centered at the same point, then the boundary of W consists of both S_1 and S_2. The Divergence Theorem requires the outward orientation, which on S_2 points away from the center and on S_1 points towards the center. (See Figure 20.13.)

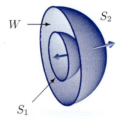

Figure 20.13: Cut-away view of the region W between two spheres, showing orientation vectors

Example 4   Let S_1 be the sphere of radius 1 centered at the origin and let S_2 be the ellipsoid $x^2 + y^2 + 4z^2 = 16$, both oriented outward. For

$$\vec{F}(\vec{r}) = \frac{\vec{r}}{\|\vec{r}\|^3}, \qquad \vec{r} \neq \vec{0},$$

show that

$$\int_{S_1} \vec{F} \cdot d\vec{A} = \int_{S_2} \vec{F} \cdot d\vec{A}.$$

Solution   The ellipsoid contains the sphere; let W be the solid region between them. Since W does not contain the origin, div $\vec{F}$ is defined and equal to zero everywhere in W. Thus, if S is the boundary of W, then

$$\int_S \vec{F} \cdot d\vec{A} = \int_W \text{div} \vec{F} \ dV = 0.$$

But S consists of S_2 oriented outwards and S_1 oriented inwards, so

$$0 = \int_S \vec{F} \cdot d\vec{A} = \int_{S_2} \vec{F} \cdot d\vec{A} - \int_{S_1} \vec{F} \cdot d\vec{A},$$

and thus

$$\int_{S_2} \vec{F} \cdot d\vec{A} = \int_{S_1} \vec{F} \cdot d\vec{A}.$$

In Example 3 we showed that $\int_{S_1} \vec{F} \cdot d\vec{A} = 4\pi$, so $\int_{S_2} \vec{F} \cdot d\vec{A} = 4\pi$ also. Note that it would have been more difficult to compute the integral over the ellipsoid directly.

Electric Fields

The electric field produced by a positive point charge q placed at the origin is

$$\vec{E} = q \frac{\vec{r}}{\|\vec{r}\|^3}.$$

Using Example 3 we see that the flux of the electric field through any sphere centered at the origin is $4\pi q$. In fact, using the idea of Example 4, we can show that the flux of $\vec{E}$ through any simple closed surface containing the origin is $4\pi q$. See Problems 25 and 26 on page 901. This is a special case of Gauss's Law, which states that the flux of an electric field through any closed surface is proportional to the total charge enclosed by the surface. Carl Friedrich Gauss (1777-1855) also discovered the Divergence Theorem, which is sometimes called Gauss's Theorem.

Exercises and Problems for Section 20.2

Exercises

1. Can you evaluate the flux integral in Exercise 16 on page 883 by application of the Divergence Theorem? Why or why not?

For Exercises 2–5, compute the flux integral $\int_S \vec{F} \cdot d\vec{A}$ in two ways, if possible, directly and using the Divergence Theorem. In each case, S is closed and oriented outwards.

2. $\vec{F}(\vec{r}) = \vec{r}$ and S is the cube enclosing the volume $0 \le x \le 2$, $0 \le y \le 2$, and $0 \le z \le 2$.

3. $\vec{F} = x^2\vec{i} + 2y^2\vec{j} + 3z^2\vec{k}$ and S is the surface of the box with faces $x = 1$, $x = 2$, $y = 0$, $y = 1$, $z = 0$, $z = 1$.

4. $\vec{F}(x, y, z) = -z\vec{i} + x\vec{k}$ and S is a square pyramid with height 3 and base on the xy-plane of side length 1.

5. $\vec{F}(x, y, z) = y\vec{j}$ and S is a closed vertical cylinder of height 2, with its base a circle of radius 1 on the xy-plane, centered at the origin.

In Exercises 6–7, use the Divergence Theorem to calculate the flux of the vector field through the surface.

6. $\int_S (x^2\vec{i} + (y-2xy)\vec{j} + 10z\vec{k}) \cdot d\vec{A}$, where S is the sphere of radius 5 centered at the origin, oriented outward.

7. $\vec{F}(x, y, z) = -z\vec{i} + x\vec{k}$ through the sphere of radius a centered at the origin. Give a geometric explanation for your answer.

Problems

8. Suppose $\vec{F}$ is a vector field with div $\vec{F} = 10$. Find the flux of $\vec{F}$ out of a cylinder of height a and radius a, centered on the z-axis and with base in the xy-plane.

9. A cone has its tip at the point $(0, 0, 5)$ and its base the disk D, $x^2 + y^2 \le 1$, in the xy-plane. The surface of the cone is the curved and slanted face, S, oriented upward, and the flat base, D, oriented downward. The flux of the constant vector field $\vec{F} = a\vec{i} + b\vec{j} + c\vec{k}$ through S is given by

$$\int_S \vec{F} \cdot d\vec{A} = 3.22.$$

Is it possible to calculate $\int_D \vec{F} \cdot d\vec{A}$? If so, give the answer. If not, explain what additional information you would need to be able to make this calculation.

10. Find the flux of $\vec{F} = z\vec{i} + y\vec{j} + x\vec{k}$ out of a sphere of radius 3 centered at the origin.

11. Find the flux of $\vec{F} = xy\vec{i} + yz\vec{k} + zx\vec{k}$ out of a sphere of radius 1 centered at origin.

12. A vector field $\vec{F}$ satisfies $\text{div}\,\vec{F} = 0$ everywhere. Show that $\int_S \vec{F} \cdot d\vec{A} = 0$ for every closed surface S.

13. In Figure 20.14, the rectangular solids V_1 and V_2 are in the first quadrant. Both have sides of length 1 parallel to the axes; V_1 has one corner at the origin, while V_2 has the corresponding corner at the point $(1, 0, 0)$. The six-faced surfaces of V_1 and V_2 are S_1 and S_2, respectively. The box-shaped volume, V, consisting of V_1 and V_2 together, has outside surface S. Are the following true or false? Give reasons.

(a) If $\vec{F}$ is a constant vector field, then $\int_S \vec{F} \cdot d\vec{A} = 0$.

(b) If S_1, S_2 and S are all oriented outward and $\vec{F}$ is any vector field, then

$$\int_S \vec{F} \cdot d\vec{A} = \int_{S_1} \vec{F} \cdot d\vec{A} + \int_{S_2} \vec{F} \cdot d\vec{A}.$$

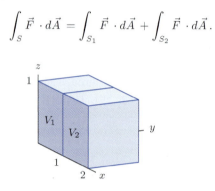

Figure 20.14

14. (a) Let $\vec{F} = \vec{r}/\|\vec{r}\|^3$. Calculate $\text{div}\,\vec{F}$ for $\vec{r} \neq \vec{0}$.
(b) Calculate the flux of $\vec{F}$ out of a box of side a centered at the origin and with edges parallel to the axes.

15. The gravitational field, $\vec{F}$, of a planet of mass m at the origin is given by

$$\vec{F} = -Gm\frac{\vec{r}}{\|\vec{r}\|^3}.$$

Use the Divergence Theorem to show that the flux of the gravitational field through the sphere of radius a is independent of a. [Hint: Consider the region bounded by two concentric spheres.]

16. Suppose $\text{div}\,\vec{F} = xyz^2$.

(a) Find $\text{div}\,\vec{F}$ at the point $(1, 2, 1)$. [Note: You are given $\text{div}\,\vec{F}$, not $\vec{F}$.]
(b) Using your answer to part (a), but no other information about the vector field $\vec{F}$, estimate the flux out of a small box of side 0.2 centered at the point $(1, 2, 1)$ and with edges parallel to the axes.
(c) Without computing the vector field $\vec{F}$, calculate the exact flux out of the box.

17. The divergence of vector field $\vec{F}$ is given by $\text{div}\,\vec{F} = x^2 + y^2 + 3$. Find a surface S such that $\int_S \vec{F} \cdot d\vec{A}$ is negative, or explain why no such surface exists.

18. Suppose $\vec{G}$ is a vector field with the property that $\text{div}\,\vec{G} = 3$ for $2 \leq \|\vec{r}\| \leq 7$ and that the flux of $\vec{G}$ through the sphere of radius 3 centered at the origin is 8π. Find the flux of $\vec{G}$ through the sphere of radius 5 centered at the origin.

19. Let $\vec{a} = a_1\vec{i} + a_2\vec{j} + a_3\vec{k}$ be a constant vector and let $\vec{r} = x\vec{i} + y\vec{j} + z\vec{k}$.

(a) Calculate $\text{div}(\vec{r} \times \vec{a})$.
(b) Calculate the flux of $\vec{r} \times \vec{a}$ out of a cube of side 5 centered at the origin with edges parallel to the axes.

20. (a) Let $\vec{a} = a_1\vec{i} + a_2\vec{j} + a_3\vec{k}$ be a constant vector and let $\vec{r} = x\vec{i} + y\vec{j} + z\vec{k}$. Calculate $\text{div}\,((\vec{a} \cdot \vec{r})\vec{r})$.
(b) Find the flux of the vector field $(\vec{a} \cdot \vec{r})\vec{r}$ out of a sphere of radius 1 centered at the origin.

21. The vector field $\vec{G}$ is given by $\vec{G} = 4r^2\vec{r}$ outside the unit sphere centered at the origin, where $r = \|\vec{r}\|$ and $\vec{r} = x\vec{i} + y\vec{j} + z\vec{k}$. However, $\vec{G}$ is not defined inside the unit sphere.

(a) Calculate the flux of G out of the sphere of radius 2 centered at the origin. Explain your reasoning.
(b) Consider (but do not calculate) the flux of $\vec{G}$ through the box of side 4 centered at the origin and with sides parallel to the axes. Is this flux larger or smaller than the flux you calculated in part (a)? Justify your answer.

22. A basic property of the electric field $\vec{E}$ is that its divergence is zero at points where there is no charge. Suppose that the only charge is along the z-axis, and that the electric field $\vec{E}$ points radially out from the z-axis and its magnitude depends only on the distance r from the z-axis. Use the Divergence Theorem to show that the magnitude of the field is proportional to $1/r$. [Hint: Consider a solid region consisting of a cylinder of finite length whose axis is the z-axis, and with a smaller concentric cylinder removed.]

23. If a surface S is submerged in an incompressible fluid, a force $\vec{F}$ is exerted on one side of the surface by the pressure in the fluid. If the z-axis is vertical, with the positive direction upward and the fluid level at $z = 0$, then the component of force in the direction of a unit vector $\vec{u}$ is given by the following:

$$\vec{F} \cdot \vec{u} = -\int_S z\delta g\vec{u} \cdot d\vec{A},$$

where δ is the density of the fluid (mass/volume), g is the acceleration due to gravity, and the surface is oriented away from the side on which the force is exerted. In this problem we consider a totally submerged closed surface enclosing a volume V. We are interested in the force of the liquid on the external surface, so S is oriented inward. Use the Divergence Theorem to show that:

(a) The force in the $\vec{i}$ and $\vec{j}$ directions is zero.
(b) The force in the $\vec{k}$ direction is δgV, the weight of the volume of fluid with the same volume as V. This is *Archimedes' Principle*.

24. As a result of radioactive decay, heat is generated uniformly throughout the interior of the earth at a rate of 30 watts per cubic kilometer. (A watt is a rate of heat production.) The heat then flows to the earth's surface where it is lost to space. Let $\vec{F}(x, y, z)$ denote the rate of flow of heat measured in watts per square kilometer. By definition, the flux of $\vec{F}$ across a surface is the quantity of heat flowing through the surface per unit of time.

   (a) What is the value of div $\vec{F}$? Include units.
   (b) Assume the heat flows outward symmetrically. Verify that $\vec{F} = \alpha\vec{r}$, where $\vec{r} = x\vec{i} + y\vec{j} + z\vec{k}$ and α is a suitable constant, satisfies the given conditions. Find α.
   (c) Let $T(x, y, z)$ denote the temperature inside the earth. Heat flows according to the equation $\vec{F} = -k\,\text{grad}\,T$, where k is a constant. Explain why this makes sense physically.
   (d) If T is in $°C$, then $k = 30{,}000$ watts/km$°$C. Assuming the earth is a sphere with radius 6400 km and surface temperature $20°C$, what is the temperature at the center?

25. According to Coulomb's Law, the electrostatic field $\vec{E}$ at the point $\vec{r}$ due to a charge q at the origin is given by

$$\vec{E}(\vec{r}) = q\frac{\vec{r}}{\|\vec{r}\|^3}.$$

   (a) Compute div $\vec{E}$.
   (b) Let S_a be the sphere of radius a centered at the origin and oriented outwards. Show that the flux of $\vec{E}$ through S_a is $4\pi q$.
   (c) Could you have used the Divergence Theorem in part (b)? Explain why or why not.
   (d) Let S be an arbitrary, closed, outward-oriented surface surrounding the origin. Show that the flux of $\vec{E}$ through S is again $4\pi q$. [Hint: Apply the Divergence Theorem to the solid region lying between a small sphere S_a and the surface S.]

26. According to Coulomb's Law, the electric field $\vec{E}$ at the point $\vec{r}$ due to a charge q at the point $\vec{r}_0$ is given by

$$\vec{E}(\vec{r}) = q\frac{(\vec{r} - \vec{r}_0)}{\|\vec{r} - \vec{r}_0\|^3}.$$

Suppose S is a closed, outward-oriented surface and that $\vec{r}_0$ does not lie on S. Use Problem 25 to show that

$$\int_S \vec{E} \cdot d\vec{A} = \begin{cases} 4\pi q & \text{if } q \text{ lies inside } S, \\ 0 & \text{if } q \text{ lies outside } S. \end{cases}$$

27. At the point with position vector $\vec{r}$, the electric field $\vec{E}$ of an ideal electric dipole with moment $\vec{p}$ located at the origin is given by

$$\vec{E}(\vec{r}) = 3\frac{(\vec{r} \cdot \vec{p})\vec{r}}{\|\vec{r}\|^5} - \frac{\vec{p}}{\|\vec{r}\|^3}.$$

   (a) What is div $\vec{E}$?
   (b) Suppose S is an outward oriented, closed smooth surface surrounding the origin. Compute the flux of $\vec{E}$ through S. Can you use the Divergence Theorem directly to compute the flux? Explain why or why not. [Hint: First compute the flux of the dipole field $\vec{E}$ through an outward-oriented sphere S_a with center at the origin and radius a. Then apply the Divergence Theorem to the region W lying between S and a small sphere S_a.]

20.3 THE CURL OF A VECTOR FIELD

The divergence of a vector field is a sort of scalar derivative which measures its outflow per unit volume. Now we introduce a vector derivative, the curl, which measures the circulation of a vector field. Imagine holding the paddle-wheel in Figure 20.15 in the flow shown by Figure 20.16. The speed at which the paddle-wheel spins (its angular velocity) measures the strength of circulation. Note that the angular velocity depends on the direction in which the stick is pointing. The paddle-wheel spins one way if the stick is pointing up and the opposite way if it is pointing down. If the stick is pointing horizontally it doesn't spin at all, because the velocity field strikes opposite vanes of the paddle with equal force.

Figure 20.15: A device for measuring circulation

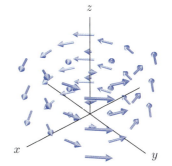

Figure 20.16: A vector field with circulation about the z-axis

Circulation Density

We measure the strength of the circulation using a closed curve. Suppose C is a circle with center (x, y, z) in the plane perpendicular to $\vec{n}$, traversed in the direction determined from $\vec{n}$ by the right-hand rule. (See Figures 20.17 and 20.18.)

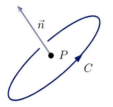

Figure 20.17: Direction of C relates to direction of $\vec{n}$ by the right-hand rule

Figure 20.18: When the thumb points in the direction of $\vec{n}$, the fingers curl in the forward direction around C

We make the following definition:

The **circulation density** of a smooth vector field $\vec{F}$ at (x, y, z) around the direction of the unit vector $\vec{n}$ is defined to be

$$\text{circ}_{\vec{n}}\, \vec{F}\,(x, y, z) = \lim_{\text{Area}\to 0} \frac{\text{Circulation around } C}{\text{Area inside } C} = \lim_{\text{Area}\to 0} \frac{\displaystyle\int_C \vec{F} \cdot d\vec{r}}{\text{Area inside } C},$$

provided the limit exists.

The circulation density determines the angular velocity[3] of the paddle-wheel in Figure 20.15 provided you could make one sufficiently small and light and insert it without disturbing the flow.

Example 1   Consider the vector field $\vec{F}$ in Figure 20.16. Suppose that $\vec{F}$ is parallel to the xy-plane and that at a distance r from the z-axis it has magnitude $2r$. Calculate $\text{circ}_{\vec{n}}\, \vec{F}$ at the origin for
(a) $\vec{n} = \vec{k}$ (b) $\vec{n} = -\vec{k}$ (c) $\vec{n} = \vec{i}$.

Solution   (a) Take a circle C of radius a in the xy-plane, centered at the origin, traversed in a direction determined from $\vec{k}$ by the right hand rule. Then, since $\vec{F}$ is tangent to C everywhere and points in the forward direction around C, we have

$$\text{Circulation around } C = \int_C \vec{F} \cdot d\vec{r} = \|\vec{F}\| \cdot \text{Circumference of } C = 2a(2\pi a) = 4\pi a^2.$$

Thus, the circulation density is

$$\text{circ}_{\vec{k}}\, \vec{F} = \lim_{a\to 0} \frac{\text{Circulation around } C}{\text{Area inside } C} = \lim_{a\to 0} \frac{4\pi a^2}{\pi a^2} = 4.$$

(b) If $\vec{n} = -\vec{k}$ the circle is traversed in the opposite direction, so the line integral changes sign. Thus

$$\text{circ}_{-\vec{k}}\, \vec{F} = -4.$$

---

[3]In fact it is twice the angular velocity. See Example 3 on page 904.

(c) The circulation around $\vec{i}$ is calculated using circles in the yz-plane. Since $\vec{F}$ is everywhere perpendicular to such a circle C,

$$\int_C \vec{F} \cdot d\vec{r} = 0.$$

Thus, we have

$$\text{circ}_{\vec{i}}\,\vec{F} = \lim_{a \to 0} \frac{\int_C \vec{F} \cdot d\vec{r}}{\pi a^2} = \lim_{a \to 0} \frac{0}{\pi a^2} = 0.$$

Definition of the Curl

Example 1 shows that the circulation density of a vector field can be positive, negative, or zero, depending on the direction. We assume that there is one direction in which the circulation density is greatest. Now we define a single vector quantity that incorporates all these different circulation densities.

Geometric Definition of Curl

The curl of a smooth vector field $\vec{F}$, written curl $\vec{F}$, is the vector field with the following properties
- The direction of curl $\vec{F}(x, y, z)$ is the direction $\vec{n}$ for which $\text{circ}_{\vec{n}}(x, y, z)$ is the greatest.
- The magnitude of curl $\vec{F}(x, y, z)$ is the circulation density of $\vec{F}$ around that direction.

If the circulation density is zero around every direction, then we define the curl to be $\vec{0}$.

Cartesian Coordinate Definition of Curl

If $\vec{F} = F_1\vec{i} + F_2\vec{j} + F_3\vec{k}$, then

$$\text{curl}\,\vec{F} = \left(\frac{\partial F_3}{\partial y} - \frac{\partial F_2}{\partial z}\right)\vec{i} + \left(\frac{\partial F_1}{\partial z} - \frac{\partial F_3}{\partial x}\right)\vec{j} + \left(\frac{\partial F_2}{\partial x} - \frac{\partial F_1}{\partial y}\right)\vec{k}.$$

Example 2   For each field in Figure 20.19, use the sketch and the geometric definition to decide whether the curl at the origin points up, down, or is the zero vector. Then check your answer using the coordinate definition of curl. Note that the vector fields have no z-components and are independent of z.

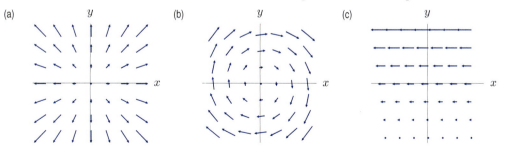

Figure 20.19: Sketches in the xy-plane of (a) $\vec{F} = x\vec{i} + y\vec{j}$ (b) $\vec{F} = y\vec{i} - x\vec{j}$ (c) $\vec{F} = -(y+1)\vec{i}$

Solution   (a) This vector field shows no rotation, and the circulation around any closed curve appears to be zero, so we suspect that the curl is the zero vector. The coordinate definition of curl gives

$$\text{curl}\,\vec{F} = \left(\frac{\partial(0)}{\partial y} - \frac{\partial y}{\partial z}\right)\vec{i} + \left(\frac{\partial x}{\partial z} - \frac{\partial(0)}{\partial x}\right)\vec{j} + \left(\frac{\partial y}{\partial x} - \frac{\partial x}{\partial y}\right)\vec{k} = \vec{0}.$$

(b) This vector field is rotating around the z-axis. By the right-hand rule, the circulation density around $\vec{k}$ is negative, so we expect the z-component of the curl points down. The coordinate definition gives

$$\operatorname{curl} \vec{F} = \left(\frac{\partial(0)}{\partial y} - \frac{\partial(-x)}{\partial z}\right)\vec{i} + \left(\frac{\partial y}{\partial z} - \frac{\partial(0)}{\partial x}\right)\vec{j} + \left(\frac{\partial(-x)}{\partial x} - \frac{\partial y}{\partial y}\right)\vec{k} = -2\vec{k}.$$

(c) At first glance, you might expect this vector field to have zero curl, as all the vectors are parallel to the x-axis. However, if you find the circulation around the curve C in Figure 20.20, the sides contribute nothing (they are perpendicular to the vector field), the bottom contributes a negative quantity (the curve is in the opposite direction to the vector field), and the top contributes a larger positive quantity (the curve is in the same direction as the vector field and the magnitude of the vector field is larger at the top than at the bottom). Thus, the circulation around C is positive and hence we expect the curl to be nonzero and point up. The coordinate definition gives

$$\operatorname{curl} \vec{F} = \left(\frac{\partial(0)}{\partial y} - \frac{\partial(0)}{\partial z}\right)\vec{i} + \left(\frac{\partial(-(y+1))}{\partial z} - \frac{\partial(0)}{\partial x}\right)\vec{j} + \left(\frac{\partial(0)}{\partial x} - \frac{\partial(-(y+1))}{\partial y}\right)\vec{k} = \vec{k}.$$

Another way to see that the curl is nonzero in this case is to imagine the vector field representing the velocity of moving water. A boat sitting in the water tends to rotate, as the water moves faster on one side than the other.

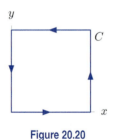

Figure 20.20

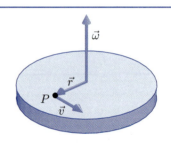

Figure 20.21: Rotating flywheel

Example 3 A flywheel is rotating with angular velocity $\vec{\omega}$ and the velocity of a point P with position vector $\vec{r}$ is given by $\vec{v} = \vec{\omega} \times \vec{r}$. (See Figure 20.21.) Calculate $\operatorname{curl} \vec{v}$.

Solution If $\vec{\omega} = \omega_1\vec{i} + \omega_2\vec{j} + \omega_3\vec{k}$, we have

$$\vec{v} = \vec{\omega} \times \vec{r} = \begin{vmatrix} \vec{i} & \vec{j} & \vec{k} \\ \omega_1 & \omega_2 & \omega_3 \\ x & y & z \end{vmatrix} = (\omega_2 z - \omega_3 y)\vec{i} + (\omega_3 x - \omega_1 z)\vec{j} + (\omega_1 y - \omega_2 x)\vec{k}.$$

Thus,

$$\operatorname{curl} \vec{v} = \begin{vmatrix} \vec{i} & \vec{j} & \vec{k} \\ \frac{\partial}{\partial x} & \frac{\partial}{\partial y} & \frac{\partial}{\partial z} \\ \omega_2 z - \omega_3 y & \omega_3 x - \omega_1 z & \omega_1 y - \omega_2 x \end{vmatrix}$$

$$= \left(\frac{\partial}{\partial y}(\omega_1 y - \omega_2 x) - \frac{\partial}{\partial z}(\omega_3 x - \omega_1 z)\right)\vec{i} + \left(\frac{\partial}{\partial z}(\omega_2 z - \omega_3 y) - \frac{\partial}{\partial x}(\omega_1 y - \omega_2 x)\right)\vec{j}$$

$$+ \left(\frac{\partial}{\partial x}(\omega_3 x - \omega_1 z) - \frac{\partial}{\partial y}(\omega_2 z - \omega_3 y)\right)\vec{k}$$

$$= 2\omega_1\vec{i} + 2\omega_2\vec{j} + 2\omega_3\vec{k} = 2\vec{\omega}.$$

Thus, as we would expect, $\operatorname{curl} \vec{v}$ is parallel to the axis of rotation of the flywheel (namely, the direction of $\vec{\omega}$) and the magnitude of $\operatorname{curl} \vec{v}$ is larger the faster the flywheel is rotating (that is, the larger the magnitude of $\vec{\omega}$).

Why Do the Two Definitions of Curl Give the Same Result?

Using Green's Theorem in Cartesian coordinates, we can show that for curl $\vec{F}$ defined in Cartesian coordinates

$$\text{curl}\,\vec{F} \cdot \vec{n} = \text{circ}_{\vec{n}}\,\vec{F}.$$

This shows that curl $\vec{F}$ defined in Cartesian coordinates satisfies the geometric definition, since the left hand side takes its maximum value when $\vec{n}$ points in the same direction as curl $\vec{F}$, and in that case its value is $\|\,\text{curl}\,\vec{F}\,\|$.

The following example justifies this formula in a specific case. Problem 28 on page 908 shows how to prove curl $\vec{F} \cdot \vec{n} = \text{circ}_{\vec{n}}\,\vec{F}$ in general.

Example 4   Use the definition of curl in Cartesian coordinates and Green's Theorem to show that

$$\left(\text{curl}\,\vec{F}\right) \cdot \vec{k} = \text{circ}_{\vec{k}}\,\vec{F}.$$

Solution   Using the definition of curl in Cartesian coordinates, the left hand side of the formula is

$$\left(\text{curl}\,\vec{F}\right) \cdot \vec{k} = \frac{\partial F_2}{\partial x} - \frac{\partial F_1}{\partial y}.$$

Now let's look at the right hand side. The circulation density around $\vec{k}$ is calculated using circles perpendicular to $\vec{k}$; hence the $\vec{k}$-component of $\vec{F}$ does not contribute to it, that is, the circulation density of $\vec{F}$ around $\vec{k}$ is the same as the circulation density of $F_1\vec{i} + F_2\vec{j}$ around $\vec{k}$. But in any plane perpendicular to $\vec{k}$, z is constant, so in that plane F_1 and F_2 are functions of x and y alone. Thus $F_1\vec{i} + F_2\vec{j}$ can be thought of as a two-dimensional vector field on the horizontal plane through the point (x, y, z) where the circulation density is being calculated. Let C be a circle in this plane, with radius a and centered at (x, y, z), and let R be the region enclosed by C. Green's Theorem says that

$$\int_C (F_1\vec{i} + F_2\vec{j}) \cdot d\vec{r} = \int_R \left(\frac{\partial F_2}{\partial x} - \frac{\partial F_1}{\partial y}\right) dA.$$

When the circle is small, $\partial F_2/\partial x - \partial F_1/\partial y$ is approximately constant on R, so

$$\int_R \left(\frac{\partial F_2}{\partial x} - \frac{\partial F_1}{\partial y}\right) dA \approx \left(\frac{\partial F_2}{\partial x} - \frac{\partial F_1}{\partial y}\right) \cdot \text{Area of } R = \left(\frac{\partial F_2}{\partial x} - \frac{\partial F_1}{\partial y}\right) \pi a^2.$$

Thus, taking a limit as the radius of the circle goes to zero, we have

$$\text{circ}_{\vec{k}}\,\vec{F}\,(x, y, z) = \lim_{a \to 0} \frac{\displaystyle\int_C (F_1\vec{i} + F_2\vec{j}) \cdot d\vec{r}}{\pi a^2} = \lim_{a \to 0} \frac{\displaystyle\int_R \left(\frac{\partial F_2}{\partial x} - \frac{\partial F_1}{\partial y}\right) dA}{\pi a^2} = \frac{\partial F_2}{\partial x} - \frac{\partial F_1}{\partial y}.$$

Curl Free Vector Fields

A vector field is said to be *curl free* or *irrotational* if curl $\vec{F} = \vec{0}$ everywhere that $\vec{F}$ is defined.

Example 5   Figure 20.22 shows the vector field $\vec{B}$ for three values of the constant p, where $\vec{B}$ is defined on 3-space by

$$\vec{B} = \frac{-y\vec{i} + x\vec{j}}{(x^2 + y^2)^{p/2}}.$$

(a) Find a formula for curl $\vec{B}$.

(b) Is there a value of p for which $\vec{B}$ is curl free? If so, find it.

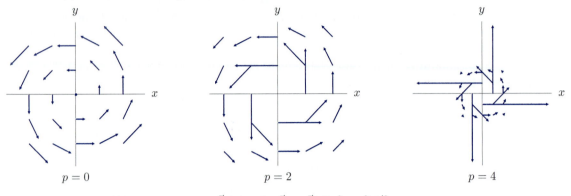

Figure 20.22: The vector field $\vec{B}\,(\vec{r}\,) = (-y\vec{i} + x\vec{j}\,)/(x^2 + y^2)^{p/2}$ for $p = 0, 2,$ and 4

Solution   (a) We can use the following version of the product rule for curl. If ϕ is a scalar function and $\vec{F}$ is a vector field, then

$$\text{curl}(\phi\vec{F}\,) = \phi\,\text{curl}\,\vec{F} + (\text{grad}\,\phi) \times \vec{F}.$$

(See Problem 26 on page 908.) We write $\vec{B} = \phi\vec{F} = \dfrac{1}{(x^2 + y^2)^{p/2}}(-y\vec{i} + x\vec{j}\,)$. Then

$$\text{curl}\,\vec{F} = \text{curl}(-y\vec{i} + x\vec{j}\,) = 2\vec{k}$$

$$\text{grad}\,\phi = \text{grad}\left(\frac{1}{(x^2 + y^2)^{p/2}}\right) = \frac{-p}{(x^2 + y^2)^{(p/2)+1}}(x\vec{i} + y\vec{j}\,).$$

Thus, we have

$$\text{curl}\,\vec{B} = \frac{1}{(x^2 + y^2)^{p/2}}\,\text{curl}(-y\vec{i} + x\vec{j}\,) + \text{grad}\left(\frac{1}{(x^2 + y^2)^{p/2}}\right) \times (-y\vec{i} + x\vec{j}\,)$$

$$= \frac{1}{(x^2 + y^2)^{p/2}}2\vec{k} + \frac{-p}{(x^2 + y^2)^{(p/2)+1}}(x\vec{i} + y\vec{j}\,) \times (-y\vec{i} + x\vec{j}\,)$$

$$= \frac{1}{(x^2 + y^2)^{p/2}}2\vec{k} + \frac{-p}{(x^2 + y^2)^{(p/2)+1}}(x^2 + y^2)\vec{k}$$

$$= \frac{2 - p}{(x^2 + y^2)^{p/2}}\vec{k}.$$

(b) The curl is zero when $p = 2$. Thus, when $p = 2$ the vector field is curl free:

$$\vec{B} = \frac{-y\vec{i} + x\vec{j}}{x^2 + y^2}.$$

Alternative Notation for Curl

Using $\nabla = \dfrac{\partial}{\partial x}\vec{i} + \dfrac{\partial}{\partial y}\vec{j} + \dfrac{\partial}{\partial z}\vec{k}$, we can write

$$\text{curl}\,\vec{F} = \nabla \times \vec{F} = \begin{vmatrix} \vec{i} & \vec{j} & \vec{k} \\ \dfrac{\partial}{\partial x} & \dfrac{\partial}{\partial y} & \dfrac{\partial}{\partial z} \\ F_1 & F_2 & F_3 \end{vmatrix}.$$

Exercises and Problems for Section 20.3

Exercises

Compute the curl of the vector fields in Exercises 1–7.

1. $\vec{F} = (x^2 - y^2)\vec{i} + 2xy\vec{j}$

2. $\vec{F} = x^2\vec{i} + y^3\vec{j} + z^4\vec{k}$

3. $\vec{F} = e^x\vec{i} + \cos y\vec{j} + e^{z^2}\vec{k}$

4. $\vec{F} = 2yz\vec{i} + 3xz\vec{j} + 7xy\vec{k}$

5. $\vec{F} = (-x + y)\vec{i} + (y + z)\vec{j} + (-z + x)\vec{k}$

6. $\vec{F} = (x + yz)\vec{i} + (y^2 + xzy)\vec{j} + (zx^3y^2 + x^7y^6)\vec{k}$

7. $\vec{F}(\vec{r}) = \vec{r}/\|\vec{r}\|$

8. Let $\vec{F} = (xy + z^2)\vec{i} + x^2\vec{j} + (xz - 2)\vec{k}$.

    (a) Find curl $\vec{F}$ at the point $(0, -1, 0)$.
    (b) Is this vector field irrotational?

Decide whether the vector fields in Exercises 9–12 have a nonzero curl at the origin. In each case, the vector field is shown in the xy-plane; assume it has no z-component and is independent of z.

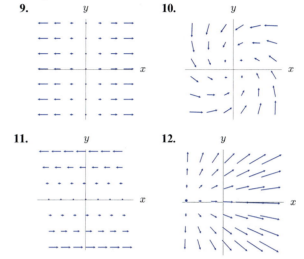

Problems

13. Using your answers to Exercises 2–3, make a conjecture about the value of curl $\vec{F}$ when the vector field $\vec{F}$ has a certain form. (What form?) Show why your conjecture is true.

14. Use the geometric definition to find the curl of the vector field $\vec{F}(\vec{r}) = \vec{r}$. Check your answer using the coordinate definition.

15. The vector field, $\vec{F}(x, y)$ sketched in Figure 20.23 has no z-component and is independent of z

    (a) Do you think that $\vec{F}(x, y)$ is a gradient vector field? Justify your answer using the sketch.
    (b) Do you think curl $\vec{F} \cdot \vec{k}$ is positive, negative, or zero? Explain.
    (c) Give a possible formula for $\vec{F}$.
    (d) Calculate curl $\vec{F}$, using the formula you gave for $\vec{F}$.

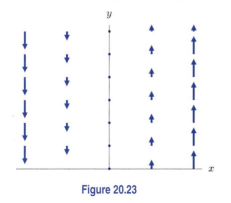

Figure 20.23

16. Let $\vec{F}$ be the vector field in Figure 20.16 on page 901. It is rotating counterclockwise around the z-axis when viewed from above. At a distance r from the z-axis, $\vec{F}$ has magnitude $2r$.

    (a) Find a formula for $\vec{F}$.
    (b) Find curl $\vec{F}$ using the coordinate definition and relate your answer to circulation density.

17. A large fire becomes a fire-storm when the nearby air acquires a circulatory motion. The associated updraft has the effect of bringing more air to the fire, causing it to burn faster. Records show that a fire-storm developed during the Chicago Fire of 1871 and during the Second World War bombing of Hamburg, Germany, but there was no fire-storm during the Great Fire of London in 1666. Explain how a fire-storm could be identified using the curl of a vector field.

18. A tornado is formed when a tube of air circling a horizontal axis is tilted up vertical by the updraft from a thunderstorm. If t is time, this process can be modeled by the wind velocity field

$$\vec{F}(t, x, y, z) = (\cos t\vec{j} + \sin t\vec{k}) \times \vec{r} \quad \text{and} \quad 0 \leq t \leq \frac{\pi}{2}.$$

Determine the direction of curl $\vec{F}$:

    (a) At $t = 0$ (b) At $t = \pi/2$
    (c) For $0 < t < \pi/2$

19. Figure 20.24 gives a sketch of a velocity vector field $\vec{F} = y\vec{i} + x\vec{j}$ in the xy-plane.

(a) What is the direction of rotation of a thin twig placed at the origin along the x-axis?

(b) What is the direction of rotation of a thin twig placed at the origin along the y-axis?

(c) Compute curl $\vec{F}$.

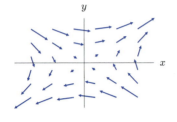

Figure 20.24

20. A smooth vector field $\vec{G}$ has curl $\vec{G}\,(0,0,0) = 2i - 3\vec{j} + 5\vec{k}$. Estimate the circulation around a circle of radius 0.01 in each of the following planes:

(a) xy-plane, oriented counterclockwise when viewed from the positive z-axis.

(b) yz-plane, oriented counterclockwise when viewed from the positive x-axis.

(c) xz-plane, oriented counterclockwise when viewed from the positive y-axis.

21. Three small circles, C_1, C_2, and C_3, each with radius 0.1 and centered at the origin are in the xy-, yz-, and xz-planes, respectively. The circles are oriented counterclockwise when viewed from the positive z-, x-, and y-axes, respectively. A vector field, $\vec{F}$, has circulation around C_1 of 0.02π, around C_2 of 0.5π, and around C_3 of 3π. Estimate curl $\vec{F}$ at the origin.

22. Show that curl $(\vec{F} + \vec{C}) = $ curl $\vec{F}$ for a constant vector field $\vec{C}$.

23. If $\vec{F}$ is any vector field whose components have continuous second partial derivatives, show div curl $\vec{F} = 0$.

24. We have seen that the Fundamental Theorem of Calculus for Line Integrals implies $\int_C \text{grad } f \cdot d\vec{r} = 0$ for any smooth closed path C and any smooth function f.

(a) Use the geometric definition of curl to deduce that curl grad $f = \vec{0}$.

(b) Show that curl grad $f = \vec{0}$ using the coordinate definition.

25. For any constant vector field $\vec{c}$, and any vector field, $\vec{F}$, show that $\text{div}(\vec{F} \times \vec{c}) = \vec{c} \cdot \text{curl } \vec{F}$.

26. Show that curl $(\phi\vec{F}) = \phi\,\text{curl }\vec{F} + (\text{grad }\phi) \times \vec{F}$ for a scalar function ϕ and a vector field $\vec{F}$.

27. Show that if $\vec{F} = f\,\text{grad }g$ for some scalar functions f and g, then curl $\vec{F}$ is everywhere perpendicular to $\vec{F}$.

28. Let $\vec{F}$ be a smooth vector field and let $\vec{u}$ and $\vec{v}$ be constant vectors. Using the definition of curl $\vec{F}$ in Cartesian coordinates, show that

$$\text{grad}(\vec{F} \cdot \vec{v}) \cdot \vec{u} - \text{grad}(\vec{F} \cdot \vec{u}) \cdot \vec{v} = (\text{curl }\vec{F}) \cdot \vec{u} \times \vec{v}.$$

29. A vortex that rotates at constant angular velocity ω about the z-axis has velocity vector field $\vec{v} = \omega(-y\vec{i} + x\vec{j})$.

(a) Sketch the vector field with $\omega = 1$ and the vector field with $\omega = -1$.

(b) Determine the speed $\|\vec{v}\|$ of the vortex as a function of the distance from its center.

(c) Compute div $\vec{v}$ and curl $\vec{v}$.

(d) Compute the circulation of $\vec{v}$ counterclockwise about the circle of radius R in the xy-plane, centered at the origin.

30. Let $r = (x^2 + y^2)^{1/2}$. Figure 20.25 shows the vector field $r^A \cdot (-y\vec{i} + x\vec{j})$ for $r \neq 0$ and $A = -1, -2$, and -3. The vector fields are shown in the xy-plane; they have no z-component and are independent of z.

(a) Show that $\text{curl}(r^A \cdot (-y\vec{i} + x\vec{j})) = (2 + A)r^A\vec{k}$ for any constant A.

(b) Using your answer to part (a), find the direction of the curl of the three vector fields for $A = -1, -2, -3$.

(c) For each value of A, what (if anything) does your answer to part (b) tell you about the sign of the circulation around a small circle oriented counterclockwise when viewed from above, and centered at $(1, 1, 1)$? Centered at $(0, 0, 0)$?

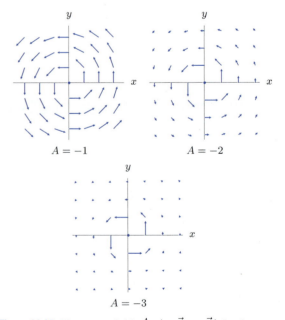

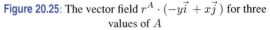

Figure 20.25: The vector field $r^A \cdot (-y\vec{i} + x\vec{j})$ for three values of A

31. Let $\vec{T} = a\vec{i} + b\vec{j}$ be a fixed unit vector, and let $\vec{F} = F(x, y)\vec{T}$ be a vector field everywhere parallel to $\vec{T}$, but of varying magnitude F. Show that curl $\vec{F}$ equals $\vec{k}$ times the directional derivative of F in the direction of $\vec{F} \times \vec{k}$. Do this in two ways:

(a) Graphically, using line integrals
(b) Algebraically

Note that the direction of $\vec{F} \times \vec{k}$ is obtained by rotating $\vec{F}$ through $90°$ clockwise as viewed from above the xy-plane.

32. Express $(3x + 2y)\vec{i} + (4x + 9y)\vec{j}$ as the sum of a curl free vector field and a divergence free vector field.

33. Find a vector field $\vec{F}$ such that curl $\vec{F} = 2\vec{i} - 3\vec{j} + 4\vec{k}$. [Hint: Try $\vec{F} = \vec{v} \times \vec{r}$ for some vector $\vec{v}$.]

20.4 STOKES' THEOREM

The Divergence Theorem says that the integral of the flux density over a solid region is equal to the flux through the surface bounding the region. Similarly, Stokes' Theorem says that the integral of the circulation density over a surface is equal to the circulation around the boundary of the surface.

The Boundary of a Surface

The *boundary* of a surface S is the curve running around the edge of S (like the hem around the edge of a piece of cloth). An orientation of S determines an orientation for its boundary, C, as follows. Pick a positive normal vector $\vec{n}$ on S, near C, and use the right hand rule to determine a direction of travel around $\vec{n}$. This in turn determines a direction of travel around the boundary C. See Figure 20.26. Another way of describing the orientation on C is that someone walking along C in the forward direction, body upright in the direction of the positive normal on S, would have the surface on their left. Notice that the boundary can consist of two or more curves, as the surface on the right in Figure 20.26 shows.

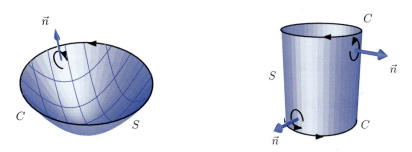

Figure 20.26: Two oriented surfaces and their boundaries

Calculating the Circulation from the Circulation Density

Consider a closed, oriented curve C in 3-space. We can find the circulation of a vector field $\vec{F}$ around C by calculating the line integral:

$$\text{Circulation around } C = \int_C \vec{F} \cdot d\vec{r}.$$

If C is the boundary of an oriented surface S, there is another way to calculate the circulation using curl $\vec{F}$. We subdivide S into pieces as shown on the surface on the left in Figure 20.26. If $\vec{n}$ is a positive unit normal vector to a piece of surface with area ΔA, then $\Delta\vec{A} = \vec{n}\,\Delta A$. In addition, $\text{circ}_{\vec{n}}\,\vec{F}$ is the circulation density of $\vec{F}$ around $\vec{n}$, so

$$\text{Circulation of } \vec{F} \text{ around boundary of the piece} \approx \left(\text{circ}_{\vec{n}}\,\vec{F}\right)\Delta A = ((\text{curl}\,\vec{F}) \cdot \vec{n}\,)\Delta A = (\text{curl}\vec{F}) \cdot \Delta\vec{A}.$$

Next we add up the circulations around all the small pieces. The line integral along the common edge of a pair of adjacent pieces appears with opposite sign in each piece, so it cancels out. (See Figure 20.27.) When we add up all the pieces the internal edges cancel and we are left with the circulation around C, the boundary of the entire surface. Thus,

$$\begin{matrix} \text{Circulation} \\ \text{around } C \end{matrix} = \sum \begin{matrix} \text{Circulation around} \\ \text{boundary of pieces} \end{matrix} \approx \sum \operatorname{curl} \vec{F} \cdot \Delta \vec{A}.$$

Taking the limit as $\Delta A \to 0$, we get

$$\begin{matrix} \text{Circulation} \\ \text{around } C \end{matrix} = \int_{S} \operatorname{curl} \vec{F} \cdot d\vec{A}.$$

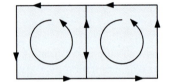

Figure 20.27: Two adjacent pieces of the surface

We have expressed the circulation as a line integral around C and as a flux integral over S; thus, the two integrals must be equal. Hence we have[4]

Theorem 20.2: Stokes' Theorem

If S is a smooth oriented surface with piecewise smooth, oriented boundary C, and if $\vec{F}$ is a smooth vector field on an open region containing S and C, then

$$\int_{C} \vec{F} \cdot d\vec{r} = \int_{S} \operatorname{curl} \vec{F} \cdot d\vec{A}.$$

The orientation of C is determined from the orientation of S according to the right hand rule.

Example 1 Let $\vec{F}(x, y, z) = -2y\vec{i} + 2x\vec{j}$. Use Stokes' Theorem to find $\int_{C} \vec{F} \cdot d\vec{r}$, where C is a circle

(a) Parallel to the yz-plane, of radius a, centered at a point on the x-axis, with either orientation.
(b) Parallel to the xy-plane, of radius a, centered at a point on the z-axis, oriented counterclockwise as viewed from a point on the z-axis above the circle.

Solution We have $\operatorname{curl} \vec{F} = 4\vec{k}$. Figure 20.28 shows sketches of $\vec{F}$ and $\operatorname{curl} \vec{F}$.

(a) Let S be the disk enclosed by C. Since S lies in a vertical plane and $\operatorname{curl} \vec{F}$ points vertically everywhere, the flux of $\operatorname{curl} \vec{F}$ through S is zero. Hence, by Stokes' Theorem,

$$\int_{C} \vec{F} \cdot d\vec{r} = \int_{S} \operatorname{curl} \vec{F} \cdot d\vec{A} = 0.$$

It makes sense that the line integral is zero. If C is parallel to the yz-plane (even if it is not lying in the plane), the symmetry of the vector field means that the line integral of $\vec{F}$ over the top half of the circle cancels the line integral over the bottom half.

---

[4]A proof of Stokes' Theorem using the coordinate definition of curl can be found in the online supplement.

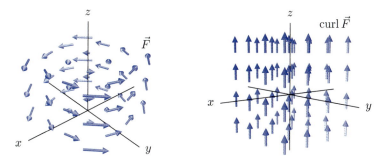

Figure 20.28: The vector fields $\vec{F}$ and curl $\vec{F}$

(b) Let S be the horizontal disk enclosed by C. Since curl $\vec{F}$ is a constant vector field pointing in the direction of $\vec{k}$, we have, by Stokes' Theorem,

$$\int_C \vec{F} \cdot d\vec{r} = \int_S \text{curl}\, \vec{F} \cdot d\vec{A} = \| \text{curl}\, \vec{F} \| \cdot \text{Area of } S = 4\pi a^2.$$

Since $\vec{F}$ is circling around the z-axis in the same direction as C, we expect the line integral to be positive. In fact, in Example 1 on page 902, we computed this line integral directly.

Curl Free Vector Fields

Stokes' Theorem applies to any oriented surface S and its boundary C, even in cases where the boundary consists of two or more curves. This is useful in studying curl free vector fields.

Example 2   A current I flows along the z-axis in the $\vec{k}$ direction. The induced magnetic field $\vec{B}(x, y, z)$ is

$$\vec{B}(x, y, z) = \frac{2I}{c} \left(\frac{-y\vec{i} + x\vec{j}}{x^2 + y^2} \right),$$

where c is the speed of light. In Example 5 on page 906 we showed that curl $\vec{B} = \vec{0}$.

(a) Compute the circulation of $\vec{B}$ around the circle C_1 in the xy-plane of radius a, centered at the origin, and oriented counterclockwise when viewed from above.
(b) Use part (a) and Stokes' Theorem to compute $\int_{C_2} \vec{B} \cdot d\vec{r}$, where C_2 is the ellipse $x^2 + 9y^2 = 9$ in the plane $z = 2$, oriented counterclockwise when viewed from above.

Solution   (a) On the circle C_1, we have $\| \vec{B} \| = 2I/(ca)$. Since $\vec{B}$ is tangent to C_1 everywhere and points in the forward direction around C_1,

$$\int_{C_1} \vec{B} \cdot d\vec{r} = \int_{C_1} \| \vec{B} \| \, dr = \frac{2I}{ca} \cdot \text{Length of } C_1 = \frac{2I}{ca} \cdot 2\pi a = \frac{4\pi I}{c}.$$

(b) Let S be the conical surface extending from C_1 to C_2 in Figure 20.29 on page 912. The boundary of this surface has two pieces, $-C_2$ and C_1. The orientation of C_1 leads to the outward normal on S, which forces us to choose the clockwise orientation on C_2. By Stokes' Theorem,

$$\int_S \text{curl}\, \vec{B} \cdot d\vec{A} = \int_{-C_2} \vec{B} \cdot d\vec{r} + \int_{C_1} \vec{B} \cdot d\vec{r} = -\int_{C_2} \vec{B} \cdot d\vec{r} + \int_{C_1} \vec{B} \cdot d\vec{r}.$$

Since curl $\vec{B} = \vec{0}$, we have $\int_S \text{curl}\, \vec{B} \cdot d\vec{A} = 0$, so the two line integrals must be equal:

$$\int_{C_2} \vec{B} \cdot d\vec{r} = \int_{C_1} \vec{B} \cdot d\vec{r} = \frac{4\pi I}{c}.$$

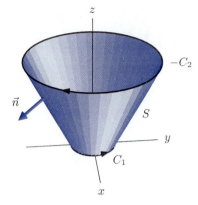

Figure 20.29: Surface joining C_1 to C_2, oriented to satisfy the conditions of Stoke's Theorem

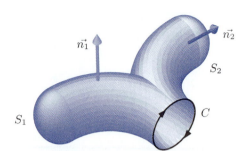

Figure 20.30: The flux of a curl is the same through the two surfaces S_1 and S_2 if they determine the same orientation on the boundary, C

Curl Fields

A vector field $\vec{F}$ is called a *curl field* if $\vec{F} = \text{curl}\,\vec{G}$ for some vector field $\vec{G}$. Recall that if $\vec{F} = \text{grad}\,f$, then f is called a potential function. By analogy, if a vector field $\vec{F} = \text{curl}\,\vec{G}$, then $\vec{G}$ is called a *vector potential* for $\vec{F}$. The following example shows that the flux of a curl field through a surface depends only on the boundary of the surface. This is analogous to the fact that the line integral of a gradient field depends only on the endpoints of the path.

Example 3   Suppose $\vec{F} = \text{curl}\,\vec{G}$. Suppose that S_1 and S_2 are two oriented surfaces with the same boundary C. Show that, if S_1 and S_2 determine the same orientation on C (as in Figure 20.30), then

$$\int_{S_1} \vec{F} \cdot d\vec{A} = \int_{S_2} \vec{F} \cdot d\vec{A}.$$

If S_1 and S_2 determine opposite orientations on C, then

$$\int_{S_1} \vec{F} \cdot d\vec{A} = -\int_{S_2} \vec{F} \cdot d\vec{A}.$$

Solution   Since $\vec{F} = \text{curl}\,\vec{G}$, by Stokes' Theorem we have

$$\int_{S_1} \vec{F} \cdot d\vec{A} = \int_{S_1} \text{curl}\,\vec{G} \cdot d\vec{A} = \int_C \vec{G} \cdot d\vec{r}$$

and

$$\int_{S_2} \vec{F} \cdot d\vec{A} = \int_{S_2} \text{curl}\,\vec{G} \cdot d\vec{A} = \int_C \vec{G} \cdot d\vec{r}.$$

In each case the line integral on the right must be computed using the orientation determined by the surface. Thus, the two flux integrals of $\vec{F}$ are the same if the orientations are the same and they are opposite if the orientations are opposite.

Exercises and Problems for Section 20.4

Exercises

1. Can you use Stokes' Theorem to compute the line integral $\int_C (2x\vec{i} + 2y\vec{j} + 2z\vec{k}) \cdot d\vec{r}$ where C is the straight line from the point $(1, 2, 3)$ to the point $(4, 5, 6)$? Why or why not?

2. At all points in 3-space $\text{curl}\,\vec{F}$ points in the direction of $\vec{i} - \vec{j} - \vec{k}$. Let C be a circle in the yz-plane, oriented clockwise when viewed from the positive x-axis. Is the circulation of $\vec{F}$ around C positive, zero, or negative? Explain.

In Exercises 3–6, use Stokes' Theorem to find the circulation of the vector field around the given paths.

3. $\vec{F} = (z - 2y)\vec{i} + (3x - 4y)\vec{j} + (z + 3y)\vec{k}$ and C is the circle $x^2 + y^2 = 4$, $z = 1$, oriented counterclockwise when viewed from above.

4. $\vec{F} = (2x - y)\vec{i} + (x + 4y)\vec{j}$ and C is a circle of radius 10, centered at the origin.

   (a) In the xy-plane, oriented clockwise as viewed from the positive z-axis.

   (b) In the yz-plane, oriented clockwise as viewed from the positive x-axis.

5. $\vec{F} = \vec{r}/\|\vec{r}\|^3$ and C is the path consisting of line segments from $(1, 0, 1)$ to $(1, 0, 0)$ to $(0, 0, 1)$ to $(1, 0, 1)$.

6. $\vec{F} = xz\vec{i} + (x + yz)\vec{j} + x^2\vec{k}$ and C is the circle $x^2 + y^2 = 1$, $z = 2$, oriented counterclockwise when viewed from above.

In Exercises 7–8, calculate the circulation in two ways, directly and using Stokes' Theorem.

7. Let S be the surface given by $z = 1 - x^2$ for $0 \le x \le 1$ and $-2 \le y \le 2$, oriented upward. Check Stokes' Theorem for $\vec{F} = xy\vec{i} + yz\vec{j} + xz\vec{k}$. Sketch the surface S and the curve C that bounds S.

8. Check Stokes' Theorem for $\vec{F} = y\vec{i} + z\vec{j} + x\vec{k}$ and S, the paraboloid $z = 1 - (x^2 + y^2)$, $z \ge 0$ oriented upward. [Hint: Use polar coordinates.]

Problems

9. The vector fields $\vec{F}$ and $\vec{G}$ are sketched in Figures 20.31 and 20.32. Each vector field has no z-component and is independent of z. All the axes have the same scales.

   (a) What can you say about div $\vec{F}$ and div $\vec{G}$ at the origin?

   (b) What can you say about curl $\vec{F}$ and curl $\vec{G}$ at the origin?

   (c) Is there a closed surface around the origin such that $\vec{F}$ has a nonzero flux through it?

   (d) Repeat part (c) for $\vec{G}$.

   (e) Is there a closed curve around the origin such that $\vec{F}$ has a nonzero circulation around it?

   (f) Repeat part (e) for $\vec{G}$.

10. Suppose that C is a closed curve in the xy-plane, oriented counterclockwise when viewed from above. Show that $\frac{1}{2}\int_C(-y\vec{i} + x\vec{j}) \cdot d\vec{r}$ equals the area of the region R in the xy-plane enclosed by C.

11. Evaluate the circulation of $\vec{G} = xy\vec{i} + z\vec{j} + 3y\vec{k}$ around a square of side 6, centered at the origin, lying in the yz-plane, and oriented counterclockwise viewed from the positive x-axis.

12. Evaluate $\int_C(-z\vec{i} + y\vec{j} + x\vec{k}) \cdot d\vec{r}$, where C is a circle of radius 2, parallel to the xz-plane and around the y-axis with the orientation shown in Figure 20.33.

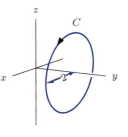

Figure 20.33

13. (a) Let $\vec{F} = (y - z)\vec{i} + (x + z)\vec{j} + xy\vec{k}$, let C be the circle of radius 3 centered at $(2, 1, 0)$ in the xy-plane, oriented counterclockwise when viewed from above. Compute $\int_C \vec{F} \cdot d\vec{r}$.

   (b) Is $\vec{F}$ path-independent (conservative)? Explain.

14. (a) If $\vec{b} = b_1\vec{i} + b_2\vec{j} + b_3\vec{k}$ is a constant vector and $\vec{r} = x\vec{i} + y\vec{j} + z\vec{k}$, show by direct calculation that

$$\text{curl}(\vec{b} \times \vec{r}) = 2\vec{b}.$$

   (b) Let C be the circle of radius 5 in the xy-plane, centered at $(3, 3, 0)$ and oriented counterclockwise when seen from above. Use the result of part (a) to calculate

$$\int_C (\vec{b} \times \vec{r}) \cdot d\vec{r}.$$

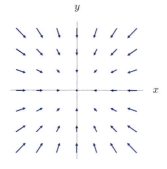

Figure 20.31: Cross-section of $\vec{F}$

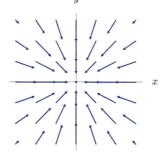

Figure 20.32: Cross-section of $\vec{G}$

15. Let $\vec{F} = -z\vec{j} + y\vec{k}$, let C be the circle of radius a in the yz-plane oriented clockwise as viewed from the positive x-axis, and let S be the disk in the yz-plane enclosed by C, oriented in the positive x-direction. See Figure 20.34.

(a) Evaluate directly $\int_C \vec{F} \cdot d\vec{r}$.

(b) Evaluate directly $\int_S \operatorname{curl} \vec{F} \cdot d\vec{A}$.

(c) The answers in parts (a) and (b) are not equal. Explain why this does not contradict Stokes' Theorem.

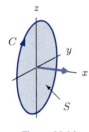

Figure 20.34

16. Let $\vec{F} = -y\vec{i} + x\vec{j} + \cos(xy)z\vec{k}$ and let S be the surface of the lower unit hemisphere $x^2 + y^2 + z^2 = 1, z \leq 0$, oriented with outward pointing normal. Find $\int_S \operatorname{curl} \vec{F} \cdot d\vec{A}$.

17. Let $\vec{F} = (8yz - z)\vec{j} + (3 - 4z^2)\vec{k}$.

(a) Show that $\vec{G} = 4yz^2\vec{i} + 3x\vec{j} + xz\vec{k}$ is a vector potential for $\vec{F}$.

(b) Evaluate $\int_S \vec{F} \cdot d\vec{A}$ where S is the hemisphere of radius 5 shown in Figure 20.35, oriented upwards. [Hint: Use Example 3 on page 912 to simplify the calculation.]

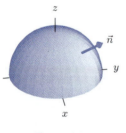

Figure 20.35

18. Use Stokes' Theorem to show that $\int_S \operatorname{curl} \vec{F} \cdot d\vec{A} = 0$ for any surface S which is the boundary surface of a solid region W. Use the geometric definition of the divergence to deduce that $\operatorname{div} \operatorname{curl} \vec{F} = 0$.

19. Show that Green's Theorem is a special case of Stokes' Theorem.

20. A vector field $\vec{F}$ is defined everywhere except on the z-axis and $\operatorname{curl} \vec{F} = \vec{0}$ everywhere where $\vec{F}$ is defined. What can you say about $\int_C \vec{F} \cdot d\vec{r}$ if C is a circle of radius 1 in the xy-plane, and if the center of C is at

(a) The origin?   (b) The point $(2, 0)$?

21. For constants a, b, c, m, consider the vector field

$$\vec{F} = (ax + by + 5z)\vec{i} + (x + cz)\vec{j} + (3y + mx)\vec{k}.$$

(a) Suppose that the flux of $\vec{F}$ through any closed surface is 0. What does this tell you about the values of the constants a, b, c, m?

(b) Suppose instead that the circulation of $\vec{F}$ around any closed curve is 0. What does this tell you about the values of the constants a, b, c, m?

22. Water in a bathtub has velocity vector field near the drain given, for x, y, z in cm, by

$$\vec{F} = -\frac{y + xz}{(z^2 + 1)^2}\vec{i} - \frac{yz - x}{(z^2 + 1)^2}\vec{j} - \frac{1}{z^2 + 1}\vec{k} \text{ cm/sec.}$$

(a) Rewriting $\vec{F}$ as follows, describe in words how the water is moving:

$$\vec{F} = \frac{-y\vec{i} + x\vec{j}}{(z^2 + 1)^2} + \frac{-z(x\vec{i} + y\vec{j})}{(z^2 + 1)^2} - \frac{\vec{k}}{z^2 + 1}.$$

(b) The drain in the bathtub is a disk in the xy-plane with center at the origin and radius 1 cm. Find the rate at which the water is leaving the bathtub. (That is, find the rate at which water is flowing through the disk.) Give units for your answer.

(c) Find the divergence of $\vec{F}$.

(d) Find the flux of the water through the hemisphere of radius 1, centered at the origin, lying below the xy-plane and oriented downward.

(e) Find $\int_C \vec{G} \cdot d\vec{r}$ where C is the edge of the drain, oriented clockwise when viewed from above, and where

$$\vec{G} = \frac{1}{2}\left(\frac{y}{z^2 + 1}\vec{i} - \frac{x}{z^2 + 1}\vec{j} - \frac{x^2 + y^2}{(z^2 + 1)^2}\vec{k}\right).$$

(f) Calculate $\operatorname{curl} \vec{G}$.

(g) Explain why your answers to parts (d) and (e) are equal.

23. Let $\vec{F} = \frac{-y\vec{i} + x\vec{j}}{x^2 + y^2}$.

(a) Calculate $\operatorname{curl} \vec{F}$. What is the domain of $\operatorname{curl} \vec{F}$?

(b) Find the circulation of $\vec{F}$ around the unit circle C_1 in the xy-plane, oriented counterclockwise when viewed from above.

(c) Find the circulation of $\vec{F}$ around the circle C_2 in the plane $z = 4$ and with equation $(x - 3)^2 + y^2 = 1$, oriented counterclockwise when viewed from above.

(d) Find the circulation of $\vec{F}$ around the square S with corners $(2, 2, 0)$, $(-2, 2, 0)$, $(-2, -2, 0)$, $(2, -2, 0)$, oriented counterclockwise when viewed from above.

(e) Using your results to parts (b)–(d) explain how, with no additional calculation, you can find the circulation of this vector field around any simple closed curve in the xy-plane, provided it does not intersect the z-axis. (A simple closed curve does not cross itself.)

20.5 THE THREE FUNDAMENTAL THEOREMS

We have now seen three multivariable versions of the Fundamental Theorem of Calculus. In this section we will examine some consequences of these theorems.

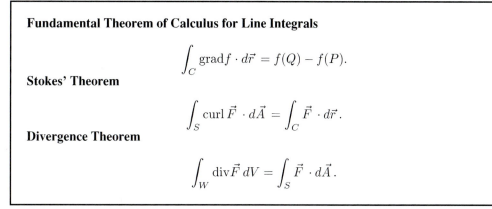

Fundamental Theorem of Calculus for Line Integrals

$$\int_C \operatorname{grad} f \cdot d\vec{r} = f(Q) - f(P).$$

Stokes' Theorem

$$\int_S \operatorname{curl} \vec{F} \cdot d\vec{A} = \int_C \vec{F} \cdot d\vec{r}.$$

Divergence Theorem

$$\int_W \operatorname{div} \vec{F} \, dV = \int_S \vec{F} \cdot d\vec{A}.$$

Notice that, in each case, the region of integration on the right is the boundary of the region on the left (except that for the first theorem we simply evaluate f at the boundary points); the integrand on the left is a sort of derivative of the integrand on the right; see Figure 20.36.

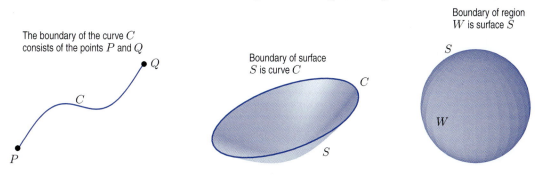

The boundary of the curve C consists of the points P and Q

Boundary of surface S is curve C

Boundary of region W is surface S

Figure 20.36: Regions and their boundaries for the three fundamental theorems

The Gradient and the Curl

Suppose that $\vec{F}$ is a smooth gradient field, so $\vec{F} = \operatorname{grad} f$ for some function f. Using the Fundamental Theorem for Line Integrals, we saw in Chapter 18 that

$$\int_C \vec{F} \cdot d\vec{r} = 0$$

for any closed curve C. Thus, for any unit vector $\vec{n}$

$$\operatorname{circ}_{\vec{n}} \vec{F} = \lim_{\text{Area} \to 0} \frac{\int_C \vec{F} \cdot d\vec{r}}{\text{Area of } C} = \lim_{\text{Area} \to 0} \frac{0}{\text{Area}} = 0,$$

where the limit is taken over circles C in a plane perpendicular to $\vec{n}$, and oriented by the right hand rule. Thus the circulation density of $\vec{F}$ is zero in every direction, so $\operatorname{curl} \vec{F} = \vec{0}$, that is,

$$\boxed{\operatorname{curl} \operatorname{grad} f = \vec{0}.}$$

(This formula can also be verified using the coordinate definition of curl. See Problem 24 on page 908.)

Is the converse true? Is any vector field whose curl is zero a gradient field? Suppose that $\text{curl} \, \vec{F} = \vec{0}$ and let us consider the line integral $\int_C \vec{F} \cdot d\vec{A}$ for a closed curve C contained in the domain of $\vec{F}$. If C is the boundary curve of an oriented surface S that lies wholly in the domain of $\text{curl} \, \vec{F}$, then Stokes' Theorem asserts that

$$\int_C \vec{F} \cdot d\vec{r} = \int_S \text{curl} \, \vec{F} \cdot d\vec{A} = \int_S \vec{0} \cdot d\vec{A} = 0.$$

If we knew that $\int_C \vec{F} \cdot d\vec{r} = 0$ for every closed curve C, then $\vec{F}$ would be path-independent, and hence a gradient field. Thus we need to know whether every closed curve in the domain of $\vec{F}$ is the boundary of an oriented surface contained in the domain. It can be quite difficult to determine if a given curve is the boundary of a surface (suppose, for example, that the curve is knotted in a complicated way). However, if the curve can be contracted smoothly to a point, remaining all the time in the domain of $\vec{F}$, then it is the boundary of a surface, namely, the surface it sweeps through as it contracts. Thus, we have proved the test for a gradient field that we stated in Chapter 18.

The Curl Test for Vector Fields in 3-Space

Suppose $\vec{F}$ is a smooth vector field on 3-space such that
- The domain of $\vec{F}$ has the property that every closed curve in it can be contracted to a point in a smooth way, staying at all times within the domain.
- $\text{curl} \, \vec{F} = \vec{0}$.

Then $\vec{F}$ is path-independent, and thus is a gradient field.

Example 6 on page 855 shows how the curl test is applied.

The Curl and The Divergence

In this section we will use the second two fundamental theorems to get a test for a vector field to be a curl field, that is, a field of the form $\vec{F} = \text{curl} \, \vec{G}$ for some $\vec{G}$.

Example 1   Suppose that $\vec{F}$ is a smooth curl field. Use Stokes' Theorem to show that for any closed surface, S, contained in the domain of $\vec{F}$

$$\int_S \vec{F} \cdot d\vec{A} = 0.$$

Solution   Suppose $\vec{F} = \text{curl} \, \vec{G}$. Draw a closed curve C on the surface S, thus dividing S into two surfaces S_1 and S_2 as shown in Figure 20.37. Pick the orientation for C corresponding to S_1; then the orientation of C corresponding to S_2 is the opposite. Thus, using Stokes' Theorem,

$$\int_{S_1} \vec{F} \cdot d\vec{A} = \int_{S_1} \text{curl} \, \vec{G} \cdot d\vec{A} = \int_C \vec{G} \cdot d\vec{r} = -\int_{S_2} \text{curl} \, \vec{G} \cdot d\vec{A} = -\int_{S_2} \vec{F} \cdot d\vec{A}.$$

Thus, for any closed surface S, we have

$$\int_S \vec{F} \cdot d\vec{A} = \int_{S_1} \vec{F} \cdot d\vec{A} + \int_{S_2} \vec{F} \cdot d\vec{A} = 0.$$

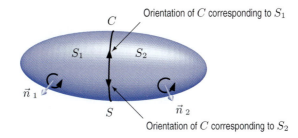

Figure 20.37: The closed surface S divided into two surfaces S_1 and S_2

Thus, if $\vec{F} = \operatorname{curl} \vec{G}$, we use the result of Example 1 to see that

$$\operatorname{div} \vec{F} = \lim_{\text{Volume} \to 0} \frac{\displaystyle\int_S \vec{F} \cdot d\vec{A}}{\text{Volume enclosed by } S} = \lim_{\text{Volume} \to 0} \frac{0}{\text{Volume}} = 0,$$

where the limit is taken over spheres S contracting down to a point. So we conclude that:

$$\boxed{\operatorname{div} \operatorname{curl} \vec{G} = 0.}$$

(This formula can also be verified using coordinates. See Problem 23 on page 908.)

Is every vector field whose divergence is zero a curl field? It turns out that we have the following analogue of the curl test, though we will not prove it.

The Divergence Test for Vector Fields in 3-Space

Suppose $\vec{F}$ is a smooth vector field on 3-space such that
- The domain of $\vec{F}$ has the property that every closed surface in it is the boundary of a solid region completely contained in the domain.
- $\operatorname{div} \vec{F} = 0$.

Then $\vec{F}$ is a curl field.

Example 2   Consider the vector fields $\vec{E} = q\dfrac{\vec{r}}{\|\vec{r}\|^3}$ and $\vec{B} = \dfrac{2I}{c}\left(\dfrac{-y\vec{i} + x\vec{j}}{x^2 + y^2}\right)$.

(a) Calculate $\operatorname{div} \vec{E}$ and $\operatorname{div} \vec{B}$.
(b) Do $\vec{E}$ and $\vec{B}$ satisfy the divergence test?
(c) Is either $\vec{E}$ or $\vec{B}$ a curl field?

Solution   (a) Example 3 on page 891 shows that $\operatorname{div} \vec{E} = 0$. The following calculation shows $\operatorname{div} \vec{B} = 0$ also:

$$\operatorname{div} \vec{B} = \frac{2I}{c}\left(\frac{\partial}{\partial x}\left(\frac{-y}{x^2 + y^2}\right) + \frac{\partial}{\partial y}\left(\frac{x}{x^2 + y^2}\right) + \frac{\partial}{\partial z}(0)\right)$$

$$= \frac{2I}{c}\left(\frac{2xy}{(x^2 + y^2)^2} + \frac{-2yx}{(x^2 + y^2)^2}\right) = 0.$$

(b) The domain of $\vec{E}$ is 3-space minus the origin, so a region is contained in the domain if it misses the origin. Thus the surface of a sphere centered at the origin is contained in the domain of E, but the solid ball inside is not. Hence $\vec{E}$ does not satisfy the divergence test.

The domain of $\vec{B}$ is 3-space minus the z-axis, so a region is contained in the domain if it avoids the z-axis. If S is a surface bounding a solid region W, then the z-axis cannot pierce W without piercing S as well. Hence, if S avoids the z-axis, so does W. Thus $\vec{B}$ satisfies the divergence test.

(c) In Example 3 on page 898 we computed the flux of $\vec{r}/\|\vec{r}\|^3$ through a sphere centered at the origin, and found it was 4π, so the flux of $\vec{E}$ through this sphere is $4\pi q$. Thus, $\vec{E}$ cannot be a curl field, because by Example 1, the flux of a curl field through a closed surface is zero.

On the other hand, $\vec{B}$ satisfies the divergence test, so it must be a curl field. In fact, Problem 5 below shows that

$$\vec{B} = \text{curl}\left(\frac{-I}{c}\ln(x^2 + y^2)\vec{k} \right).$$

Exercises and Problems for Section 20.5

Exercises

Which of the vector fields in Exercises 1–2 is a gradient field?

1. $\vec{F} = yz\vec{i} + (xz + z^2)\vec{j} + (xy + 2yz)\vec{k}$

2. $\vec{G} = -y\vec{i} + x\vec{j}$

3. Let $\vec{B} = b\vec{k}$, for some constant b. Show that the following are all possible vector potentials for $\vec{B}$:

(a) $\vec{A} = -by\vec{i}$ (b) $\vec{A} = bx\vec{j}$

(c) $\vec{A} = \frac{1}{2}\vec{B} \times \vec{r}$.

Problems

4. Find a vector potential for the constant vector field $\vec{B}$ whose value at every point is $\vec{b}$.

5. Show that $\vec{A} = \dfrac{-I}{c}\ln(x^2 + y^2)\vec{k}$ is a vector potential

for $\vec{B} = \dfrac{2I}{c}\left(\dfrac{-y\vec{i} + x\vec{j}}{x^2 + y^2} \right)$.

6. Is there a vector field $\vec{G}$ such that $\text{curl}\,\vec{G} = y\vec{i} + x\vec{j}$? How do you know?

For each vector field in Problems 7–8, determine whether a vector potential exists. If so, find one.

7. $\vec{F} = 2x\vec{i} + (3y - z^2)\vec{j} + (x - 5z)\vec{k}$

8. $\vec{G} = x^2\vec{i} + y^2\vec{j} + z^2\vec{k}$

9. An electric charge q at the origin produces an electric field $\vec{E} = q\vec{r}/\|\vec{r}\|^3$.

(a) Does $\text{curl}\,\vec{E} = \vec{0}$?

(b) Does $\vec{E}$ satisfy the curl test?

(c) Is $\vec{E}$ a gradient field?

10. Suppose c is the speed of light. A thin wire along the z-axis carrying a current I produces a magnetic field

$$\vec{B} = \frac{2I}{c}\left(\frac{-y\vec{i} + x\vec{j}}{x^2 + y^2} \right),$$

(a) Does $\text{curl}\,\vec{B} = \vec{0}$?

(b) Does $\vec{B}$ satisfy the curl test?

(c) Is $\vec{B}$ a gradient field?

11. For constant p, consider the vector field $\vec{E} = \dfrac{\vec{r}}{\|\vec{r}\|^p}$.

(a) Find $\text{curl}\,\vec{E}$.

(b) Find the domain of $\vec{E}$.

(c) For which values of p does $\vec{E}$ satisfy the curl test? For those values of p, find a potential function for $\vec{E}$.

12. The magnetic field, $\vec{B}$, due to a magnetic dipole with moment $\vec{\mu}$ satisfies $\text{div}\,\vec{B} = 0$ and is given by

$$\vec{B} = -\frac{\vec{\mu}}{\|\vec{r}\|^3} + \frac{3(\vec{\mu} \cdot \vec{r})\vec{r}}{\|\vec{r}\|^5}, \qquad \vec{r} \neq \vec{0}.$$

(a) Does $\vec{B}$ satisfy the divergence test?

(b) Show that a vector potential for $\vec{B}$ is given by $\vec{A} = \dfrac{\vec{\mu} \times \vec{r}}{\|\vec{r}\|^3}$.

[Hint: Use Problem 26 on page 908. The identities in Example 3 on page 904, Problem 35 on page 672, and Problem 28 on page 634 might also be useful.]

(c) Does your answer to part (a) contradict your answer to part (b)? Explain.

13. Suppose that $\vec{A}$ is a vector potential for $\vec{B}$.

(a) Show that $\vec{A} + \text{grad}\,\psi$ is also a vector potential for $\vec{B}$, for any function ψ with continuous second-order partial derivatives. (The vector potentials $\vec{A}$ and $\vec{A} + \text{grad}\,\psi$ are called *gauge equivalent* and the transformation, for any ψ, from $\vec{A}$ to $\vec{A} + \text{grad}\,\psi$ is called a *gauge transformation*.)

(b) What is the divergence of $\vec{A} + \text{grad}\,\psi$? How should ψ be chosen such that $\vec{A} + \text{grad}\,\psi$ has zero divergence? (If $\text{div}\,\vec{A} = 0$, the magnetic vector potential $\vec{A}$ is said to be in *Coulomb gauge*.)

CHAPTER SUMMARY

- **Divergence**

  Geometric and coordinate definition of divergence, calculating divergence, interpretation in terms of outflow per unit volume.

- **Curl**

  Geometric and coordinate definition of curl, calculating curl, interpretation in terms of circulation per unit

  area.

- **The Divergence Theorem**

  Statement of the theorem, divergence free fields, harmonic functions.

- **Stokes' Theorem**

  Statement of the theorem, curl free and curl fields.

REVIEW EXERCISES AND PROBLEMS FOR CHAPTER TWENTY

Exercises

1. (a) If $\vec{F} = (\cos x)\vec{i} + e^y\vec{j} + (x - y - z)\vec{k}$, find curl $\vec{F}$.

(b) Find $\int_C \vec{F} \cdot d\vec{r}$ where C is the circle of radius 3 in the plane $x + y + z = 1$, centered at $(1, 0, 0)$ oriented counterclockwise when viewed from above.

2. Use the geometric definition of divergence to find div $\vec{v}$ at the origin, where $\vec{v} = -2\vec{r}$. Check that you get the same result using the definition in Cartesian coordinates.

3. (a) Find the flux of $\vec{F} = 2x\vec{i} - 3y\vec{j} + 5z\vec{k}$ through a box with four of its corners at the points $(a, b, c), (a+w, b, c), (a, b+w, c), (a, b, c+w)$ and edge length w. See Figure 20.38.

(b) Use the geometric definition and part (a) to find div $\vec{F}$ at the point (a, b, c).

(c) Find div $\vec{F}$ using partial derivatives.

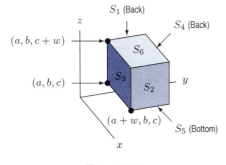

Figure 20.38

4. Suppose $\vec{F} = (3x + 2)\vec{i} + 4x\vec{j} + (5x + 1)\vec{k}$. Use the method of Exercise 3 to find div $\vec{F}$ at the point (a, b, c) by two different methods.

5. Compute the flux integral $\int_S (x^3\vec{i} + 2y\vec{j} + 3\vec{k}) \cdot d\vec{A}$, where S is the $2 \times 2 \times 2$ rectangular surface centered at the origin, oriented outward. Do this in two ways:

(a) Directly

(b) By means of the Divergence Theorem

6. Compute the line integral $\int_C ((yz^2 - y)\vec{i} + (xz^2 + x)\vec{j} + 2xyz\vec{k}) \cdot d\vec{r}$ where C is the circle of radius 3 in the xy-plane, centered at the origin, oriented counterclockwise as viewed from the positive z-axis. Do it two ways:

(a) Directly   **(b)** Using Stokes' Theorem

Exercises 7–9 concern the vector fields in Figure 20.39. In each case, assume that the cross-section is the same in all other planes parallel to the given cross-section.

7. Three of the vector fields have zero curl at each point shown. Which are they? How do you know?

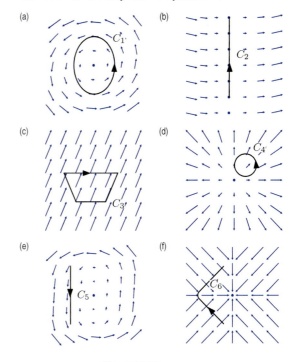

Figure 20.39

8. Three of the vector fields have zero divergence at each point shown. Which are they? How do you know?

9. Four of the line integrals $\int_{C_i} \vec{F} \cdot d\vec{r}$ are zero. Which are they? How do you know?

In Exercises 10–13, calculate div $\vec{F}$ and curl $\vec{F}$. Is $\vec{F}$ solenoidal or irrotational?

10. $\vec{F} = x^2\vec{i} + y^3\vec{j} + z^4\vec{k}$

11. $\vec{F} = xy\vec{i} + yz\vec{j} + zx\vec{k}$

12. $\vec{F} = (\cos x)\vec{i} + e^y\vec{j} + (x + y + z)\vec{k}$

13. $\vec{F} = e^{y+z}\vec{i} + \sin(x + z)\vec{j} + (x^2 + y^2)\vec{k}$

In Exercises 14–17, use the Divergence Theorem to calculate the flux of the vector field out of the given surface.

14. $\vec{F} = (2x+y)\vec{i} + (3y+z)\vec{j} + (4z+x)\vec{k}$ and S is the sphere of radius 5 centered at the origin.

15. $\vec{F} = (e^z + x)\vec{i} + (2y + \sin x)\vec{j} + \left(e^{x^2} - z\right)\vec{k}$ and S is the sphere of radius 1 centered at $(2, 1, 0)$.

16. $\vec{F} = x^2\vec{i} + y^2\vec{j} + e^{xy}\vec{k}$ and S is the cube of side 3 in the first octant with one corner at the origin, and sides parallel to the axes.

17. $\vec{F} = x^3\vec{i} + y^3\vec{j} + \left(x^2 + y^2\right)\vec{k}$ and S is the cylinder $x^2 + y^2 = 4$ with $0 \leq z \leq 5$.

In Exercises 18–21, use Stokes' Theorem to calculate the integral.

18. $\int_C \vec{F} \cdot d\vec{r}$ where $\vec{F} = x^2\vec{i} + y^2\vec{j} + z^2\vec{k}$ and C is the unit circle in the xz-plane, oriented counterclockwise when viewed from the positive y-axis.

19. $\int_C \vec{F} \cdot d\vec{r}$ where $\vec{F} = (y-x)\vec{i} + (z-y)\vec{j} + (x-z)\vec{k}$ and C is the circle $x^2 + y^2 = 5$ in the xy-plane, oriented counterclockwise when viewed from above.

20. $\int_S \text{curl } \vec{F} \cdot d\vec{A}$ where $\vec{F} = -y\vec{i} + x\vec{j} + (xy + \cos z)\vec{k}$ and S is the disk $x^2 + y^2 \leq 9$, oriented upward in the xy-plane.

21. $\int_S \text{curl } \vec{F} \cdot d\vec{A}$ where $\vec{F} = (x+7)\vec{j} + e^{x+y+z}\vec{k}$ and S is the rectangle $0 \leq x \leq 3$, $0 \leq y \leq 2$, $z = 0$, oriented counterclockwise when viewed from above.

Problems

In Problems 22–27, calculate the integral.

22. $\int_S \vec{F} \cdot d\vec{A}$ where $\vec{F} = \left(y^2 + 3x\right)\vec{i} + \left(x^2 - y\right)\vec{j} + 2z\vec{k}$ and S is the unit sphere centered at the origin.

23. $\int_S \text{curl} \vec{F} \cdot d\vec{A}$ where $\vec{F} = (z+y)\vec{i} - (z+x)\vec{j} + (y+x)\vec{k}$ and S is the disk $y^2 + z^2 \leq 3$, $x = 0$, oriented in the positive x-direction.

24. $\int_S \vec{F} \cdot d\vec{A}$ where $\vec{F} = x^3\vec{i} + y^3\vec{j} + z^3\vec{k}$ and S is the sphere of radius 1 centered at the origin

25. $\int_C \vec{F} \cdot d\vec{r}$ where $\vec{F} = (x+y)\vec{i} + (y+2z)\vec{j} + (z+3x)\vec{k}$ and C is a square of side 7 in the xz-plane, oriented counterclockwise when viewed from the positive y-axis.

26. $\int_C \vec{F} \cdot d\vec{r}$ where $\vec{F} = (x - y^3 + z)\vec{i} + (x^3 + y + z)\vec{j} + (x+y+z^3)\vec{k}$ and C is the circle $x^2 + y^2 = 10$, oriented counterclockwise when viewed from above.

27. $\int_S \vec{F} \cdot d\vec{A}$ where $\vec{F} = (y^3 z^3)\vec{i} + y^3\vec{j} + z^3\vec{k}$ and S is the cylinder $y^2 + z^2 = 16$, $-1 \leq x \leq 1$.

28. Are the following vector fields conservative?

(a) $\vec{F}(x, y, z) = y^2 z\vec{i} + 2xyz\vec{j} + xy\vec{k}$
(b) The vector field in Figure 20.40.

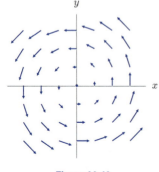

Figure 20.40

29. Figure 20.41 shows the part of a vector field $\vec{E}$ that lies in the xy-plane. Assume that the vector field is independent of z, so that any horizontal cross section looks the same. What can you say about div $\vec{E}$ at the points marked P and Q, assuming it is defined?

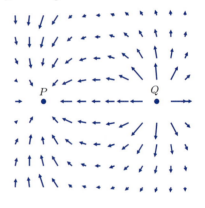

Figure 20.41

30. If V is a volume surrounded by a closed surface S, show that $\frac{1}{3}\int_S \vec{r} \cdot d\vec{A} = V$.

31. Use Problem 30 to compute the volume of a sphere of radius R given that its surface area is $4\pi R^2$.

32. Use Problem 30 to compute the volume of a cone of base radius b and height h. [Hint: Stand the cone with its point downward and its axis along the positive z-axis.]

33. Assume $\vec{r} \neq \vec{0}$. Let $\vec{F} = \vec{r}/\|\vec{r}\|^3$.

(a) Calculate the flux of $\vec{F}$ out of the unit sphere $x^2 + y^2 + z^2 = 1$ oriented outward.
(b) Calculate div $\vec{F}$. Simplify your answer completely.
(c) Use your answer to parts (a) and (b) to calculate the flux out of a box of side 10 centered at the origin and with sides parallel to the coordinate planes. (The box is also oriented outward.)

34. Let $\vec{r} = x\vec{i} + y\vec{j} + z\vec{k}$. Suppose $\vec{F} = \vec{r}/\|\vec{r}\|^3$. Find $\int_S \vec{F} \cdot d\vec{A}$ where S is the ellipsoid $x^2 + 2y^2 + 3z^2 = 6$. Give reasons for your computation.

35. Let $\vec{r} = x\vec{i} + y\vec{j} + z\vec{k}$. Find $\int_S \vec{F} \cdot d\vec{A}$, where $\vec{F} = \vec{r}/\|\vec{r}\|^3$ and S is the cylinder $y^2 + z^2 = 4$, $-2 \le x \le 2$. Give reasons for your computation.

36. Let $\vec{F} = y\vec{i} - x\vec{j}$ and let C be the unit circle in the xy-plane centered at the origin and oriented counterclockwise when viewed from above.

   (a) Calculate $\int_C \vec{F} \cdot d\vec{r}$ by parameterizing the circle.
   (b) Calculate curl $\vec{F}$.
   (c) Calculate $\int_C \vec{F} \cdot d\vec{r}$ using your result from part (b).
   (d) What theorem did you use in part (c)?

In Problems 37–39, find the flux of the vector fields through the upper half of the sphere of radius 5, centered at the origin and oriented upward.

37. $\vec{F} = (x + \cos y)\vec{i} + (y + \sin x)\vec{j} + (z + 3)\vec{k}$

38. $\vec{F} = (\sin z)\vec{i} + e^{xz}\vec{j} + (x^2 + y^2)\vec{k}$

39. $\vec{F} = \text{curl}(-y\vec{i} + x\vec{j})$

40. Suppose $\vec{F}(x, y, z)$ is a vector field with div $\vec{F} = 4$ everywhere. Which of the following quantities can be computed from this information? Give the value of those that can be computed.

   (a) $\int_S \vec{F} \cdot d\vec{A}$, where S is a sphere of radius 2 centered at the origin and oriented outward.
   (b) $\int_C \vec{F} \cdot d\vec{r}$, where C is the unit circle in the xy-plane, oriented counterclockwise viewed from above.
   (c) $\int_S \text{curl } \vec{F} \cdot d\vec{A}$, where S is a sphere of radius 2 centered at the origin and oriented outward.

41. For $\vec{r} = x\vec{i} + y\vec{j} + z\vec{k}$, $\vec{r} \ne \vec{0}$, and constant a, let

$$\vec{F} = \frac{\vec{r}}{\|\vec{r}\|^a}.$$

   (a) Find div $\vec{F}$. Simplify completely. Your answer will contain a.
   (b) Find a value of a such that div $\vec{F} = 0$.
   (c) Using a sketch, explain why you would expect curl $\vec{F} = \vec{0}$ for all a.

42. A central vector field is one of the form $\vec{F} = f(r)\vec{r}$ where f is any function of $r = \|\vec{r}\|$. Show that any central vector field is irrotational.

43. Due to roadwork ahead, the traffic on a highway slows linearly from 55 miles/hour to 15 miles/hour over a 2000 foot stretch of road, then crawls along at 15 miles/hour for 5000 feet, then speeds back up linearly to 55 miles/hour in the next 1000 feet, after which it moves steadily at 55 miles/hour.

   (a) Sketch a velocity vector field for the traffic flow.
   (b) Write a formula for the velocity vector field $\vec{v}$ (miles/hour) as a function of the distance x feet from the initial point of slowdown. (Take the direction of motion to be $\vec{i}$ and consider the various sections of the road separately.)
   (c) Compute div $\vec{v}$ at $x = 1000, 5000, 7500, 10{,}000$. Be sure to include the proper units.

44. The velocity field $\vec{v}$ in Problem 43 does not give a complete description of the traffic flow, for it takes no account of the spacing between vehicles. Let ρ be the density (cars/mile) of highway, where we assume that ρ depends only on x.

   (a) Using your highway experience, arrange in ascending order: $\rho(0), \rho(1000), \rho(5000)$.
   (b) What are the units and interpretation of the vector field $\rho\vec{v}$?
   (c) Would you expect $\rho\vec{v}$ to be constant? Why? What does this mean for div$(\rho\vec{v})$?
   (d) Determine $\rho(x)$ if $\rho(0) = 75$ cars/mile and $\rho\vec{v}$ is constant.
   (e) If the highway has two lanes, find the approximate number of feet between cars at $x = 0, 1000$, and 5000.

45. (a) A river flows across the xy-plane in the positive x-direction and around a circular rock of radius 1 centered at the origin. The velocity of the river can be modeled using the potential function $\phi = x + (x/(x^2 + y^2))$. Compute the velocity vector field, $\vec{v} = \text{grad }\phi$.
   (b) Show that div $\vec{v} = 0$.
   (c) Show that the flow of $\vec{v}$ is tangent to the circle $x^2 + y^2 = 1$. This means that no water crosses the circle. The water on the outside must therefore all flow around the circle.
   (d) Use a computer to sketch the vector field $\vec{v}$ in the region outside the unit circle.

46. A basic property of the magnetic field $\vec{B}$ is that curl $\vec{B} = \vec{0}$ in a region where there is no current. Consider the magnetic field around a long thin wire carrying a constant current. The magnitude of the magnetic field depends only on the distance from the wire and its direction is always tangent to the circle around the wire traversed in a direction related to the direction of the current by the right hand rule. Use Stokes' Theorem to deduce that the magnitude of the magnetic field is proportional to the reciprocal of the distance from the wire. [Hint: Consider an annulus (ring) around the wire. Its boundary has two pieces: an inner and an outer circle.]

47. A vector field is a *point source* at the origin in 3-space if its direction is away from the origin at every point, its magnitude depends only on the distance from the origin, and its divergence is zero except at the origin. (Such a vector field might be used to model the photon flow out of a star or the neutrino flow out of a supernova.)

   (a) Show that $\vec{v} = K(x^2 + y^2 + z^2)^{-3/2}(x\vec{i} + y\vec{j} + z\vec{k})$ is a point source at the origin if $K > 0$.

(b) Determine the magnitude $\|\vec{v}\|$ of the source in part (a) as a function of the distance from its center.

(c) Compute the flux of $\vec{v}$ through a sphere of radius r centered at the origin.

(d) Compute the flux of $\vec{v}$ through a closed surface that does not contain the origin.

Problems 48–49 use the fact that the electric field, $\vec{E}$, is related to the charge density, $\rho(x, y, z)$, in units of charge/volume, by the equation

$$\operatorname{div} \vec{E} = 4\pi\rho.$$

In addition, there is an electric potential, ϕ, whose gradient gives the electric field:

$$\vec{E} = -\operatorname{grad} \phi.$$

48. Calculate and describe in words the electric field and the charge distribution corresponding to the potential function defined as follows:

$$\phi = \begin{cases} x^2 + y^2 + z^2 & \text{for } x^2 + y^2 + z^2 \leq \frac{b^2}{4} \\ \dfrac{b^2}{4} - \dfrac{b^3}{4(x^2 + y^2 + z^2)^{1/2}} & \text{for } \frac{b^2}{4} \leq x^2 + y^2 + z^2 \end{cases}$$

49. A vector field which could possibly represent an electric field is given by

$$\vec{E} = 10xy\vec{i} + (5x^2 - 5y^2)\vec{j}.$$

(a) Calculate the line integral of $\vec{E}$ from the origin to the point (a, b) along the path which runs straight from the origin to the point $(a, 0)$ and then straight from $(a, 0)$ to the point (a, b).

(b) Calculate the line integral of $\vec{E}$ between the same points as in part (a) but via the point $(0, b)$.

(c) Why do your answers to part (a) and (b) suggest that $\vec{E}$ could indeed be an electric field?

(d) Find the electric potential, ϕ, and calculate $\operatorname{grad} \phi$ to confirm that $\vec{E} = -\operatorname{grad} \phi$.

50. The relations between the electric field, $\vec{E}$, the magnetic field, $\vec{B}$, the charge density, ρ, and the current density, $\vec{J}$, at a point in space are described by the equations

$$\operatorname{div} \vec{E} = 4\pi\rho,$$

$$\operatorname{curl} \vec{B} - \frac{1}{c}\frac{\partial \vec{E}}{\partial t} = \frac{4\pi}{c}\vec{J},$$

where c is a constant (the speed of light).

(a) Using the results of Problem 23 on page 908, show that

$$\frac{\partial \rho}{\partial t} + \operatorname{div} \vec{J} = 0.$$

(b) What does the equation in part (a) say about charge and current density? Explain in intuitive terms why this is reasonable.

(c) Why do you think the equation in part (a) is called the charge conservation equation?

51. The speed of a naturally occurring vortex (tornado, waterspout, whirlpool) is a decreasing function of the distance from its center, so the constant angular velocity model of Problem 29 on page 908 is inappropriate. A *free vortex* circulating about the z-axis has vector field $\vec{v} = K(x^2 + y^2)^{-1}(-y\vec{i} + x\vec{j})$ where K is a constant.

(a) Sketch the vector field with $K = 1$ and the vector field with $K = -1$.

(b) Determine the speed $\|\vec{v}\|$ of the vortex as a function of the distance from its center.

(c) Compute $\operatorname{div} \vec{v}$.

(d) Show that $\operatorname{curl} \vec{v} = \vec{0}$.

(e) Compute the circulation of $\vec{v}$ counterclockwise about the circle of radius R at the origin.

(f) The computations in parts (d) and (e) show that $\vec{v}$ has curl $\vec{0}$, but has nonzero circulation around the closed curve in part (e). Explain why this does not contradict Stokes' Theorem.

CAS Challenge Problems

52. (a) Let $\vec{F} = x^3 y\vec{i} + 2xz^3\vec{j} + (z^3 + 4x^2)\vec{k}$. Compute $\operatorname{curl} \vec{F}(1, 2, 1)$.

(b) Consider the family of curves C_a given, for $0 \leq t \leq 2\pi$, by

$$\vec{r}(t) = \vec{i} + (2 + a\cos t)\vec{j} + (1 + a\sin t)\vec{k}.$$

Evaluate the line integral $\int_{C_a} \vec{F} \cdot d\vec{r}$ and compute the limit

$$\lim_{a \to 0} \frac{\int_{C_a} \vec{F} \cdot d\vec{r}}{\pi a^2}.$$

(c) Repeat part (b) for the family D_a given, for $0 \leq t \leq 2\pi$, by

$$\vec{r}(t) = (1 + a\sin t)\vec{i} + 2\vec{j} + (1 + a\cos t)\vec{k}.$$

(d) Repeat part (b) for the family E_a given, for $0 \leq t \leq 2\pi$, by

$$\vec{r}(t) = (1 + a\cos t)\vec{i} + (2 + a\sin t)\vec{j} + \vec{k}.$$

(e) Compare your answers to parts (b)-(d) with part (a) and explain using the geometric definition of curl.

53. Let S be the sphere of radius R centered at the origin with outward orientation and let

$$\vec{F} = (ax^2 + bxz)\vec{i} + (cy^2 + py)\vec{j} + (qz + rx^3)\vec{k}.$$

(a) Use the Divergence Theorem to express the flux integral $\int_S \vec{F} \cdot d\vec{A}$ as a triple integral. Then use symmetry and the volume formula for a sphere to evaluate the triple integral.

(b) Check your answer in part (a) by computing the flux integral directly.

54. Let $\vec{F}(x, y, z) = x^2 y \vec{i} + 2xz\vec{j} + (z^3 + 4x^2)\vec{k}$ and let S_a be the sphere of radius a centered at $(1, 1, 1)$, oriented outwards, parameterized by $\vec{r}(\phi, \theta) = (1 + a\sin\phi\cos\theta)\vec{i} + (1 + a\sin\phi\sin\theta)\vec{j} + (1 + a\cos\phi)\vec{k}$, $0 \le \phi \le \pi, 0 \le \theta \le 2\pi$.

(a) Compute $\operatorname{div}\vec{F}(1, 1, 1)$.

(b) Use the geometric definition of divergence to estimate $\int_{S_a} \vec{F} \cdot d\vec{A}$ for $a = 0.1$.

(c) Evaluate the flux integral $\int_{S_a} \vec{F} \cdot d\vec{A}$. Compare its value for $a = 0.1$ with your answer to part (b). Then compute the limit

$$\lim_{a \to 0} \frac{\int_{S_a} \vec{F} \cdot d\vec{A}}{\text{Volume inside } S_a}$$

and compare your result with part (a). Explain your answer in terms of the geometric definition of the divergence.

CHECK YOUR UNDERSTANDING

1. If $\vec{F}$ is a smooth vector field, defined everywhere in three-space, with $\operatorname{curl}\vec{F}$ everywhere perpendicular to the z-axis, and if C is a circle in the xy-plane, then the circulation of $\vec{F}$ around C is zero. True or false? If true, explain why; if false, give a counterexample.

Are the statements in Problems 2–7 true or false? Assume $\vec{F}$ and $\vec{G}$ are smooth vector fields in 3-space. Explain.

2. $\operatorname{div}(\vec{F} + \vec{G}) = \operatorname{div}\vec{F} + \operatorname{div}\vec{G}$

3. $\operatorname{grad}(\vec{F} \cdot \vec{G}) = \vec{F}(\operatorname{div}\vec{G}) + (\operatorname{div}\vec{F})\vec{G}$

4. $\operatorname{curl}(f\vec{G}) = (\operatorname{grad} f) \times \vec{G} + f(\operatorname{curl}\vec{G})$

5. $\operatorname{div}\vec{F}$ is a scalar whose value can vary from point to point.

6. If $\int_S \vec{F} \cdot d\vec{A} = 12$ and S is a flat disk of area 4π, then $\operatorname{div}\vec{F} = 3/\pi$.

7. If $\vec{F}$ is as shown in Figure 20.42, then $\operatorname{curl}\vec{F} \cdot \vec{j} > 0$.

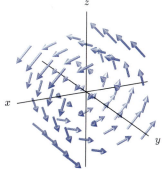

Figure 20.42

Are the statements in Problems 8–41 true or false? Give reasons for your answer.

8. If $\vec{F}$ is a vector field in 3-space, and f is a scalar function, then $\operatorname{div}(f\vec{F}) = f\operatorname{div}\vec{F}$.

9. If $\vec{F}$ is a vector field in 3-space, and $\vec{F} = \operatorname{grad} f$, then $\operatorname{div}\vec{F} = 0$.

10. If $\vec{F}$ is a vector field in 3-space, then $\operatorname{grad}(\operatorname{div}\vec{F}) = \vec{0}$.

11. The field $\vec{F}(\vec{r}) = \vec{r}$ is divergence free.

12. If $f(x, y, z)$ is any given continuous scalar function, then there is at least one vector field $\vec{F}$ such that $\operatorname{div}\vec{F} = f$.

13. If $\vec{F}$ and $\vec{G}$ are vector fields satisfying $\operatorname{div}\vec{F} = \operatorname{div}\vec{G}$ then $\vec{F} = \vec{G}$.

14. If $\vec{F}$ is a vector field in 3-space, and W is a solid region with boundary surface S, then $\int_S \operatorname{div}\vec{F} \cdot d\vec{A} = \int_W \vec{F} \, dV$.

15. If $\vec{F}$ is a divergence free vector field in 3-space, and S is a closed surface oriented inward, then $\int_S \vec{F} \cdot d\vec{A} = 0$.

16. If $\vec{F}$ is a vector field in 3-space satisfying $\operatorname{div}\vec{F} = 1$, and S is a closed surface oriented outward, then $\int_S \vec{F} \cdot d\vec{A}$ is equal to the volume enclosed by S.

17. Let W be the solid region between the sphere S_1 of radius 1 and S_2 of radius 2, both centered at the origin. If $\vec{F}$ is a vector field in 3-space, then $\int_W \operatorname{div}\vec{F} \, dV = \int_{S_2} \vec{F} \cdot d\vec{A} - \int_{S_1} \vec{F} \cdot d\vec{A}$, where both S_1 and S_2 are oriented outward.

18. Let S_1 be the square $0 \le x \le 1, 0 \le y \le 1, z = 0$ oriented downward and let S_2 be the square $0 \le x \le 1, 0 \le y \le 1, z = 1$ oriented upward. If $\vec{F}$ is a vector field, then $\int_W \operatorname{div}\vec{F} \, dV = \int_{S_2} \vec{F} \cdot d\vec{A} + \int_{S_1} \vec{F} \cdot d\vec{A}$, where W is the solid cube $0 \le x \le 1, 0 \le y \le 1, 0 \le z \le 1$.

19. Let S_1 be the square $0 \le x \le 1, 0 \le y \le 1, z = 0$ oriented downward and let S_2 be the square $0 \le x \le 1, 0 \le y \le 1, z = 1$ oriented upward. If $\vec{F} = \cos(xyz)\vec{k}$, then $\int_W \operatorname{div}\vec{F} \, dV = \int_{S_2} \vec{F} \cdot d\vec{A} + \int_{S_1} \vec{F} \cdot d\vec{A}$, where W is the solid cube $0 \le x \le 1, 0 \le y \le 1, 0 \le z \le 1$.

20. If S is a sphere of radius 1, centered at the origin, oriented outward, and $\vec{F}$ is a vector field satisfying $\int_S \vec{F} \cdot d\vec{A} = 0$, then $\operatorname{div}\vec{F} = 0$ at all points inside S.

21. There exist a scalar function f and a vector field $\vec{F}$ satisfying $\operatorname{div}(\operatorname{grad} f) = \operatorname{grad}(\operatorname{div} F)$.

22. Let S_h be the surface consisting of a cylinder of height h, closed at the top. The curved sides are $x^2 + y^2 = 1$, for $0 \leq z \leq h$, and the top $x^2 + y^2 \leq 1$, for $z = h$, oriented outward. If $\vec{F}$ is divergence free, then $\int_{S_h} \vec{F} \cdot d\vec{A}$ is independent of the height h.

23. The circulation density, $\text{circ}_{\vec{n}} \vec{F}(x, y, z)$, is a scalar.

24. If $\vec{F}$ is a vector field with $\text{div} \vec{F} = 0$ and $\text{curl} \vec{F} = \vec{0}$, then $\vec{F} = \vec{0}$.

25. If $\vec{F}$ and $\vec{G}$ are vector fields, then $\text{curl}(\vec{F} + \vec{G}) = \text{curl}\vec{F} + \text{curl}\vec{G}$.

26. If $\vec{F}$ and $\vec{G}$ are vector fields, then $\text{curl}(\vec{F} \cdot \vec{G}) = \text{curl}\vec{F} \cdot \text{curl}\vec{G}$.

27. There exists a vector field $\vec{F}$ with $\text{curl} \vec{F} = \vec{i}$.

28. At all points other than the origin, the vector field $\vec{F} = \frac{-y}{x^2+y^2}\vec{i} + \frac{x}{x^2+y^2}\vec{j}$ has $\text{curl}\vec{F} = \vec{0}$ and $\text{div}\vec{F} = 0$.

29. For any vector field $\vec{F}$, the curl of $\vec{F}$ is perpendicular at every point to $\vec{F}$.

30. If $\vec{F}$ and $\vec{G}$ are vector fields, then $\text{curl}(\vec{F} \times \vec{G}) = (\text{curl}\vec{F}) \times (\text{curl}\vec{G})$.

31. There exists a vector field $\vec{F}$ (whose components have continuous second partial derivatives) satisfying $\text{curl}\vec{F} = x\vec{i}$.

32. If S is the upper unit hemisphere $x^2+y^2+z^2 = 1, z \geq 0$, oriented upward, then the boundary of S used in Stokes' theorem is the circle $x^2+y^2 = 1, z = 0$, with orientation counterclockwise when viewed from the positive z-axis.

33. If C is the boundary of an oriented surface S, oriented by the right hand rule, then $\int_C \text{curl} \vec{F} \cdot d\vec{r} = \int_S \vec{F} \cdot d\vec{A}$.

34. Let S_1 be the disk $x^2 + y^2 \leq 1, z = 0$ and let S_2 be the upper unit hemisphere $x^2 + y^2 + z^2 = 1, z \geq 0$, both oriented upward. If $\vec{F}$ is a vector field then $\int_{S_1} \text{curl}\vec{F} \cdot d\vec{A} = \int_{S_2} \text{curl}\vec{F} \cdot d\vec{A}$.

35. Let S be the closed unit sphere $x^2+y^2+z^2 = 1$, oriented outward. If $\vec{F}$ is a vector field then $\int_S \text{curl}\vec{F} \cdot d\vec{A} = 0$.

36. If $\vec{F}$ and $\vec{G}$ are vector fields satisfying $\text{curl}\vec{F} = \text{curl}\vec{G}$, then $\int_C \vec{F} \cdot d\vec{r} = \int_C \vec{G} \cdot d\vec{r}$, where C is any oriented circle in 3-space.

37. If $\vec{F}$ is a vector field satisfying $\text{curl}\vec{F} = \vec{0}$, then $\int_C \vec{F} \cdot d\vec{r} = 0$, where C is any oriented path around a rectangle in 3-space.

38. If $\vec{F}$ is a gradient field, then $\int_S \text{curl} \vec{F} \cdot d\vec{r} = 0$, for any smooth oriented surface, S, in 3-space.

39. Let S be the cylinder $x^2 + z^2 = 1, 0 \leq y \leq 2$, oriented with inward pointing normal. Then the boundary of S consists of two circles C_1 $(x^2 + z^2 = 1, y = 0)$ and C_2 $(x^2 + z^2 = 1, y = 2)$, both oriented clockwise when viewed from the positive y-axis.

40. Let S be an oriented surface, with oriented boundary C, and suppose that $\vec{F}$ is a vector field such that $\int_C \vec{F} \cdot d\vec{r} = 0$. Then $\text{curl} \vec{F} = \vec{0}$ everywhere on S.

41. Let S be an oriented surface, with oriented boundary C, and suppose that $\vec{F}$ is a vector field such that $\int_S \text{curl}\vec{F} \cdot d\vec{A} = 0$. Then $\vec{F}$ is a gradient field.

PROJECTS FOR CHAPTER TWENTY

1. Divergence of Spherically Symmetric Vector Fields

A vector field is spherically symmetric about the origin if, on every sphere centered at the origin, it has constant magnitude and points either away from or toward the origin. A vector field that is spherically symmetric about the origin can be written in terms of the spherical coordinate $\rho = \|\vec{r}\|$ as $\vec{F} = f(\rho)\vec{e}_\rho$ where f is a function of the distance ρ from the origin, $f(0) = 0$, and $\vec{e}_\rho$ is a unit vector pointing away from the origin.

(a) Show that

$$\text{div} \vec{F} = \frac{1}{\rho^2} \frac{d}{d\rho} \left(\rho^2 f(\rho) \right), \quad \rho \neq 0.$$

(b) Use part (a) to show that if $\vec{F}$ is a spherically symmetric vector field such that $\text{div} \vec{F} = 0$ away from the origin then, for some constant k,

$$\vec{F} = k\frac{1}{\rho^2}\vec{e}_\rho, \quad \rho \neq 0.$$

(c) Use part (a) to confirm the Divergence Theorem for the flux of a spherically symmetric vector field through a sphere centered at the origin.

(d) A form of Gauss's Law for an electric field $\vec{E}$ states that $\text{div} \vec{E}(\vec{r}) = \delta(\vec{r})$, where δ is a scalar-valued function giving the density of electric charge at every point. Use Gauss's Law and part (a) to find the electric field when the charge density is

$$\delta(\vec{r}) = \begin{cases} \delta_0 & \|\vec{r}\| \leq a \\ 0 & \|\vec{r}\| > a, \end{cases}$$

for some nonnegative a. Assume that $\vec{E}$ is spherically symmetric and continuous everywhere.

2. **Divergence of Cylindrically Symmetric Vector Fields**

We say that a vector field is cylindrically symmetric about a line if, on every cylinder centered along that line, it has constant magnitude and points either away from or toward the line. A vector field that is cylindrically symmetric about the z-axis can be written in terms of the cylindrical coordinate r as $\vec{F} = f(r)\vec{e}_r$ where f is a function of the distance r from the z-axis, $f(0) = 0$, and $\vec{e}_r$ is a unit vector pointing away from the z-axis.

(a) Show that

$$\text{div} \, \vec{F} = \frac{1}{r}\frac{d}{dr}(rf(r)), \quad r \neq 0.$$

(b) Use part (a) to show that if $\vec{F}$ is cylindrically symmetric about the z-axis and div $\vec{F} = 0$ away from the z-axis, then, for some constant k,

$$\vec{F} = k\frac{\vec{e}_r}{r}, \quad r \neq 0.$$

(c) Use part (a) to confirm the Divergence Theorem for the flux of a cylindrically symmetric vector field through a cylinder of radius R and height h centered on the z-axis.

(d) A form of Gauss's Law for an electric field $\vec{E}$ states that div $\vec{E}(\vec{r}) = \delta(\vec{r})$, where δ is a scalar-valued function giving the density of electric charge at every point. Use Gauss's Law and part (a) to find the electric field when the charge density is given in cylindrical coordinates by

$$\delta(r, \theta, z) = \begin{cases} \delta_0 & r \leq a \\ 0 & r > a, \end{cases}$$

for some nonnegative constant a. Assume that $\vec{E}$ is cylindrically symmetric and continuous everywhere.

APPENDICES

A   ROOTS, ACCURACY, AND BOUNDS

It is often necessary to find the zeros of a polynomial or the points of intersection of two curves. So far, you have probably used algebraic methods, such as the quadratic formula, to solve such problems. Unfortunately, however, mathematicians' search for similar solutions to more complicated equations, has not been all that successful. The formulas for the solutions to third- and fourth-degree equations are so complicated that you'd never want to use them. Early in the nineteenth century, it was proved that there is no algebraic formula for the solutions to equations of degree 5 and higher. Most nonpolynomial equations cannot be solved using a formula either.

However, we can still find roots of equations, provided we use approximation methods, not formulas. In this section we will discuss three ways to find roots: algebraic, graphical, and numerical. Of these, only the algebraic method gives exact solutions.

First, let's get some terminology straight. Given the equation $x^2 = 4$, we call $x = -2$ and $x = 2$ the *roots*, or *solutions of the equation*. If we are given the function $f(x) = x^2 - 4$, then -2 and 2 are called the *zeros of the function*; that is, the zeros of the function f are the roots of the equation $f(x) = 0$.

The Algebraic Viewpoint: Roots by Factoring

If the product of two numbers is zero, then one or the other or both must be zero, that is, if $AB = 0$, then $A = 0$ or $B = 0$. This observation lies behind finding roots by factoring. You may have spent a lot of time factoring polynomials. Here you will also factor expressions involving trigonometric and exponential functions.

Example 1   Find the roots of $x^2 - 7x = 8$.

Solution   Rewrite the equation as $x^2 - 7x - 8 = 0$. Then factor the left side: $(x + 1)(x - 8) = 0$. By our observation about products, either $x + 1 = 0$ or $x - 8 = 0$, so the roots are $x = -1$ and $x = 8$.

Example 2   Find the roots of $\dfrac{1}{x} - \dfrac{x}{(x + 2)} = 0$.

Solution   Rewrite the left side with a common denominator:

$$\frac{x + 2 - x^2}{x(x + 2)} = 0.$$

Whenever a fraction is zero, the numerator must be zero. Therefore we must have

$$x + 2 - x^2 = (-1)(x^2 - x - 2) = (-1)(x - 2)(x + 1) = 0.$$

We conclude that $x - 2 = 0$ or $x + 1 = 0$, so 2 and -1 are the roots. They can be checked by substitution.

Example 3   Find the roots of $e^{-x} \sin x - e^{-x} \cos x = 0$.

Solution   Factor the left side: $e^{-x}(\sin x - \cos x) = 0$. The factor e^{-x} is never zero; it is impossible to raise e to a power and get zero. Therefore, the only possibility is that $\sin x - \cos x = 0$. This equation is equivalent to $\sin x = \cos x$. If we divide both sides by $\cos x$, we get

$$\frac{\sin x}{\cos x} = \frac{\cos x}{\cos x} \quad \text{so} \quad \tan x = 1.$$

The roots of this equation are

$$\ldots, \frac{-7\pi}{4}, \frac{-3\pi}{4}, \frac{\pi}{4}, \frac{5\pi}{4}, \frac{9\pi}{4}, \frac{13\pi}{4}, \ldots.$$

Warning: Using factoring to solve an equation only works when one side of the equation is 0. It is not true that if, say, $AB = 7$ then $A = 7$ or $B = 7$. For example, you *cannot* solve $x^2 - 4x = 2$ by factoring $x(x - 4) = 2$ and then assuming that either x or $x - 4$ equals 2.

The problem with factoring is that factors are not easy to find. For example, the left side of the quadratic equation $x^2 - 4x - 2 = 0$ does not factor, at least not into "nice" factors with integer coefficients. For the general quadratic equation

$$ax^2 + bx + c = 0,$$

there is the quadratic formula for the roots:

$$x = \frac{-b \pm \sqrt{b^2 - 4ac}}{2a}.$$

Thus the roots of $x^2 - 4x - 2 = 0$ are $(4 \pm \sqrt{24})/2$, or $2 + \sqrt{6}$ and $2 - \sqrt{6}$.

Notice that in each of these examples, we have found the roots exactly.

The Graphical Viewpoint: Roots by Zooming

To find the roots of an equation $f(x) = 0$, it helps to draw the graph of f. The roots of the equation, that is the zeros of f, are *the values of x where the graph of f crosses the x-axis*. Even a very rough sketch of the graph can be useful in determining how many zeros there are and their approximate values. If you have a computer or graphing calculator, then finding solutions by graphing is the easiest method, especially if you use the zoom feature. However, a graph can never tell you the exact value of a root, only an approximate one.

Example 4   Find the roots of $x^3 - 4x - 2 = 0$.

Solution   Attempting to factor the left side with integer coefficients will convince you it cannot be done, so we cannot easily find the roots by algebra. We know the graph of $f(x) = x^3 - 4x - 2$ will have the usual cubic shape; see Figure A.1.

There are clearly three roots: one between $x = -2$ and $x = -1$, another between $x = -1$ and $x = 0$, and a third between $x = 2$ and $x = 3$. Zooming in on the largest root with a graphing calculator or computer shows that it lies in the following interval:

$$2.213 < x < 2.215.$$

Thus, the root is $x = 2.21$, accurate to two decimal places. Zooming in on the other two roots shows them to be $x = -1.68$ and $x = -0.54$, accurate to two decimal places.

Useful trick: Suppose you want to solve the equation $\sin x - \cos x = 0$ graphically. Instead of graphing $f(x) = \sin x - \cos x$ and looking for zeros, you may find it easier to rewrite the equation as $\sin x = \cos x$ and graph $g(x) = \sin x$ and $h(x) = \cos x$. (After all, you already know what these two graphs look like. See Figure A.2.) The roots of the original equation are then precisely the x coordinates of the points of intersection of the graphs of $g(x)$ and $h(x)$.

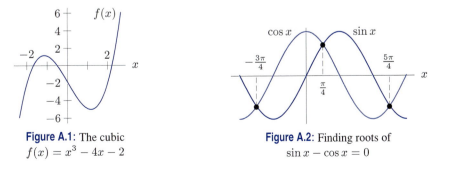

Figure A.1: The cubic
$f(x) = x^3 - 4x - 2$

Figure A.2: Finding roots of
$\sin x - \cos x = 0$

Example 5   Find the roots of $2 \sin x - x = 0$.

Solution   Rewrite the equation as $2 \sin x = x$, and graph both sides. Since $g(x) = 2 \sin x$ is always between -2 and 2, there are no roots of $2 \sin x = x$ for $x > 2$ or for $x < -2$. We need only consider the graphs between -2 and 2 (or between $-\pi$ and π, which makes graphing the sine function easier). Figure A.3 shows the graphs. There are three points of intersection: one appears to be at $x = 0$, one between $x = \pi/2$ and $x = \pi$, and one between $x = -\pi/2$ and $x = -\pi$. You can tell that $x = 0$ is the exact value of one root because it satisfies the original equation exactly. Zooming in shows that there is a second root $x \approx 1.9$, and the third root is $x \approx -1.9$ by symmetry.

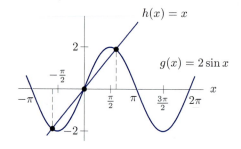

Figure A.3: Finding roots of $2 \sin x - x = 0$

The Numerical Viewpoint: Roots by Bisection

We now look at a numerical method of approximating the solutions to an equation. This method depends on the idea that if the value of a function $f(x)$ changes sign in an interval, and if we believe there is no break in the graph of the function there, then there is a root of the equation $f(x) = 0$ in that interval.

Let's go back to the problem of finding the root of $f(x) = x^3 - 4x - 2 = 0$ between 2 and 3. To locate the root, we close in on it by evaluating the function at the midpoint of the interval, $x = 2.5$. Since $f(2) = -2$, $f(2.5) = 3.625$, and $f(3) = 13$, the function changes sign between $x = 2$ and $x = 2.5$, so the root is between these points. Now we look at $x = 2.25$.

Since $f(2.25) = 0.39$, the function is negative at $x = 2$ and positive at $x = 2.25$, so there is a root between 2 and 2.25. Now we look at 2.125. We find $f(2.125) = -0.90$, so there is a root between 2.125 and 2.25, ... and so on. (You may want to round the decimals as you work.) See Figure A.4. The intervals containing the root are listed in Table A.1 and show that the root is $x = 2.21$ to two decimal places.

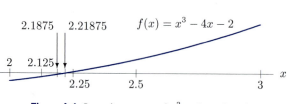

Figure A.4: Locating a root of $x^3 - 4x - 2 = 0$

Table A.1   *Intervals containing root of* $x^3 - 4x - 2 = 0$ *(Note [2, 3] means $2 \leq x \leq 3$)*

$[2, 3]$

$[2, 2.5]$

$[2, 2.25]$

$[2.125, 2.25]$

$[2.1875, 2.25]$ So $x = 2.2$ rounded to one decimal place

$[2.1875, 2.21875]$

$[2.203125, 2.21875]$

$[2.2109375, 2.21875]$

$[2.2109375, 2.2148438]$ So $x = 2.21$ rounded to two decimal places

This method of estimating roots is called the **Bisection Method**:
- To solve an equation $f(x) = 0$ using the bisection method, we need two starting values for x, say, $x = a$ and $x = b$, such that $f(a)$ and $f(b)$ have opposite signs and f is continuous on $[a, b]$.
- Evaluate f at the midpoint of the interval $[a, b]$, and decide in which half-interval the root lies.
- Repeat, using the new half-interval instead of $[a, b]$.

There are some problems with the bisection method:
- The function may not change signs near the root. For example, $f(x) = x^2 - 2x + 1 = 0$ has a root at $x = 1$, but $f(x)$ is never negative because $f(x) = (x - 1)^2$, and a square cannot be negative. (See Figure A.5.)
- The function f must be continuous between the starting values $x = a$ and $x = b$.
- If there is more than one root between the starting values $x = a$ and $x = b$, the method will find only one of the roots. For example, if we had tried to solve $x^3 - 4x - 2 = 0$ starting at $x = -12$ and $x = 10$, the bisection method would zero in on the root between $x = -2$ and $x = -1$, not the root between $x = 2$ and $x = 3$ that we found earlier. (Try it! Then see what happens if you use $x = -10$ instead of $x = -12$.)
- The bisection method is slow and not very efficient. Applying bisection three times in a row only traps the root in an interval $(\frac{1}{2})^3 = \frac{1}{8}$ as large as the starting interval. Thus, if we initially know that a root is between, say, 2 and 3, then we would need to apply the bisection method at least four times to know the first digit after the decimal point.

There are much more powerful methods available for finding roots, such as Newton's method, which are more complicated but which avoid some of these difficulties.

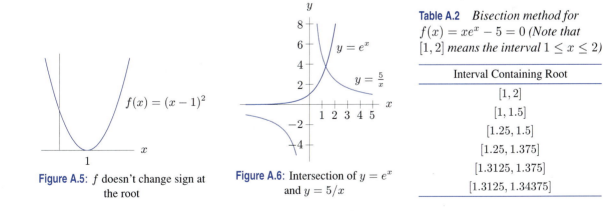

Figure A.5: f doesn't change sign at the root

Figure A.6: Intersection of $y = e^x$ and $y = 5/x$

Table A.2   *Bisection method for $f(x) = xe^x - 5 = 0$ (Note that $[1, 2]$ means the interval $1 \leq x \leq 2$)*

Interval Containing Root
$[1, 2]$
$[1, 1.5]$
$[1.25, 1.5]$
$[1.25, 1.375]$
$[1.3125, 1.375]$
$[1.3125, 1.34375]$

Example 6   Find all the roots of $xe^x = 5$ to at least one decimal place.

Solution   If we rewrite the equation as $e^x = 5/x$ and graph both sides, as in Figure A.6, it is clear that there is exactly one root, and it is somewhere between 1 and 2. Table A.2 shows the intervals obtained by the bisection method. After five iterations, we have the root trapped between 1.3125 and 1.34375, so we can say the root is $x = 1.3$ to one decimal place.

Iteration

Both zooming in and bisection as discussed here are examples of *iterative* methods, in which a sequence of steps is repeated over and over again, using the results of one step as the input for the next. We can use such methods to locate a root to any degree of accuracy. In bisection, each iteration traps the root in an interval that is half the length of the previous one. Each time you zoom in on a calculator, you trap the root in a smaller interval; how much smaller depends on the settings on the calculator.

Accuracy and Error

In the previous discussion, we used the phrase "accurate to 2 decimal places." For an iterative process where we get closer and closer estimates of some quantity, we take a common-sense approach to accuracy: we watch the numbers carefully, and when a digit stays the same for several iterations, we assume it has stabilized and is correct, especially if the digits to the right of that digit also stay the same. For example, suppose 2.21429 and 2.21431 are two successive estimates for a zero of $f(x) = x^3 - 4x - 2$. Since these two estimates agree to the third digit after the decimal point, we probably have at least 3 decimal places correct.

There is a problem with this, however. Suppose we are estimating a root whose true value is 1, and the estimates are converging to the value from below—say, 0.985, 0.991, 0.997 and so on. In this case, not even the first decimal place is "correct," even though the difference between the estimates and the actual answer is very small—much less than 0.1. To avoid this difficulty, we say that an estimate a for some quantity r is *accurate to p decimal places* if the error, which is the absolute value of the difference between a and r, or $|r - a|$, is as follows:

Accuracy to p decimal places	means	Error less than
$p = 1$		0.05
2		0.005
3		0.0005
$\vdots$		$\vdots$
n		$0.\underbrace{000\ldots0}_{n}5$

This is the same as saying that r must lie in an interval of length twice the maximum error, centered on a. For example, if a is accurate to 1 decimal place, r must lie in the following interval:

$$a - 0.05 \qquad a \qquad a + 0.05$$

Since both the graphing calculator and the bisection method give us an interval in which the root is trapped, this definition of decimal accuracy is a natural one for these processes.

Example 7   Suppose the numbers $\sqrt{10}$, $22/7$, and 3.14 are given as approximations to $\pi = 3.1415\ldots$. To how many decimal places is each approximation accurate?

Solution   Using $\sqrt{10} = 3.1622\ldots$,

$$|\sqrt{10} - \pi| = |3.1622\ldots - 3.1415\ldots| = 0.0206\ldots < 0.05,$$

so $\sqrt{10}$ is accurate to one decimal place. Similarly, using $22/7 = 3.1428\ldots$,

$$\left|\frac{22}{7} - \pi\right| = |3.1428\ldots - 3.1415\ldots| = 0.0013\ldots < 0.005,$$

so 22/7 is accurate to two decimal places. Finally,

$$|3.14 - 3.1415\ldots| = 0.0015\ldots < 0.005,$$

so 3.14 is accurate to two decimal places.

Warning:

- Saying that an approximation is accurate to, say, 2 decimal places does *not* guarantee that its first two decimal places are "correct," that is, that the two digits of the approximation are the same as the corresponding two digits in the true value. For example, an approximate value of 5.997 is accurate to 2 decimal places if the true value is 6.001, but neither of the 9s in the approximation agrees with the 0s in the true value (nor does the digit 5 agree with the digit 6).

- When finding a root r of an equation, the number of decimal places of accuracy refers to the number of digits that have stabilized in the root. It does *not* refer to the number of digits of $f(r)$ that are zero. For example, Table A.1 on page 930 shows that $x = 2.2$ is a root of $f(x) = x^3 - 4x - 2 = 0$, accurate to one decimal place. Yet, $f(2.2) = -0.152$, so $f(2.2)$ does not have one zero after the decimal point. Similarly, $x = 2.21$ is the root accurate to two decimal places, but $f(2.21) = -0.046$ does not have two zeros after the decimal point.

Example 8   Is $x = 2.2143$ a zero of $f(x) = x^3 - 4x - 2$ accurate to four decimal places?

Solution   We want to know whether r, the exact value of the zero, lies in the interval

$$2.2143 - 0.00005 < r < 2.2143 + 0.00005$$

which is the same as

$$2.21425 < r < 2.21435.$$

Since $f(2.21425) < 0$ and $f(2.21435) > 0$, the zero does lie in this interval, and so $r = 2.2143$ is accurate to four decimal places.

How to Write a Decimal Answer

The graphing calculator and bisection method naturally give an interval for a root or a zero. However, other numerical techniques do not give a pair of numbers bounding the true value, but rather a single number near the true value. What should you do if you want a single number, rather than an interval, for an answer? In general, averaging the endpoint of the interval is the best solution.

When giving a single number as an answer and interpreting it, be careful about giving rounded answers. For example, suppose you know a root lies in the interval between 0.81 and 0.87. Averaging gives 0.84 as a single number estimating the root. But it would be wrong to round 0.84 to 0.8 and say that the answer is 0.8 accurate to one decimal place; the true value could be 0.86, which is not within 0.05 of 0.8. The right thing to say is that the answer is 0.84 accurate to one decimal place. Similarly, to give an answer accurate to, say, 2 decimal places, you may have to show 3 decimal places in your answer.

Bounds of a Function

Knowing how big or how small a function gets can sometimes be useful, especially when you can't easily find exact values of the function. You can say, for example, that $\sin x$ always stays between -1 and 1 and that $2\sin x + 10$ always stays between 8 and 12. But 2^x is not confined between any two numbers, because 2^x will exceed any number you can name if x is large enough. We say that $\sin x$ and $2\sin x + 10$ are *bounded* functions, and that 2^x is an *unbounded* function.

A function f is **bounded** on an interval if there are numbers L and U such that

$$L \le f(x) \le U$$

for all x in the interval. Otherwise, f is **unbounded** on the interval.

We say that L is a **lower bound** for f on the interval, and that U is an **upper bound** for f on the interval.

Example 9   Use Figures A.7 and A.8 to decide which of the following functions are bounded.

(a) x^3 on $-\infty < x < \infty$; on $0 \le x \le 100$.
(b) $2/x$ on $0 < x < \infty$; on $1 \le x < \infty$.

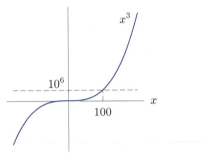

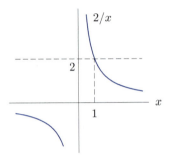

Figure A.7: Is x^3 bounded?

Figure A.8: Is $2/x$ bounded?

Solution

(a) The graph of x^3 in Figure A.7 shows that x^3 will exceed any number, no matter how large, if x is big enough, so x^3 does not have an upper bound on $-\infty < x < \infty$. Therefore, x^3 is unbounded on $-\infty < x < \infty$. But on the interval $0 \le x \le 100$, x^3 stays between 0 (a lower bound) and $100^3 = 1,000,000$ (an upper bound). Therefore, x^3 is bounded on the interval $0 \le x \le 100$. Notice that upper and lower bounds, when they exist, are not unique. For example, -100 is another lower bound and $2,000,000$ another upper bound for x^3 on $0 \le x \le 100$.

(b) $2/x$ is unbounded on $0 < x < \infty$, since it has no upper bound on that interval. But $0 \le 2/x \le 2$ for $1 \le x < \infty$, so $2/x$ is bounded, with lower bound 0 and upper bound 2, on $1 \le x < \infty$. (See Figure A.8.)

Best Possible Bounds

Consider a group of people whose height in feet, h, ranges from 5 feet to 6 feet. Then 5 feet is a lower bound for the people in the group and 6 feet is an upper bound:

$$5 \le h \le 6.$$

But the people in this group are also all between 4 feet and 7 feet, so it is also true that

$$4 \le h \le 7.$$

So, there are many lower bounds and many upper bounds. However, the 5 and the 6 are considered the best bounds because they are the closest together of all the possible pairs of bounds.

The **best possible bounds** for a function, f, over an interval are numbers A and B such that, for all x in the interval,

$$A \le f(x) \le B$$

and where A and B are as close together as possible. A is called the **greatest lower bound** and B is the **least upper bound**.

What Do Bounds Mean Graphically?

Upper and lower bounds can be represented on a graph by horizontal lines. See Figure A.9.

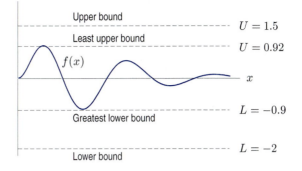

Figure A.9: Upper and lower bounds for the function f

Exercises for Appendix A

1. Use a calculator or computer graph of $f(x) = 13 - 20x - x^2 - 3x^4$ to determine:

   (a) The range of this function;
   (b) The number of zeros of this function.

For Exercises 2–12, determine the roots or points of intersection to an accuracy of one decimal place.

2. (a) The root of $x^3 - 3x + 1 = 0$ between 0 and 1
   (b) The root of $x^3 - 3x + 1 = 0$ between 1 and 2
   (c) The smallest root of $x^3 - 3x + 1 = 0$

3. The root of $x^4 - 5x^3 + 2x - 5 = 0$ between -2 and -1

4. The root of $x^5 + x^2 - 9x - 3 = 0$ between -2 and -1

5. The largest real root of $2x^3 - 4x^2 - 3x + 1 = 0$

6. All real roots of $x^4 - x - 2 = 0$

7. All real roots of $x^5 - 2x^2 + 4 = 0$

8. The smallest positive root of $x \sin x - \cos x = 0$

9. The left-most point of intersection between $y = 2x$ and $y = \cos x$

10. The left-most point of intersection between $y = 1/2^x$ and $y = \sin x$

11. The point of intersection between $y = e^{-x}$ and $y = \ln x$

12. All roots of $\cos t = t^2$

13. Estimate all real zeros of the following polynomials, accurate to 2 decimal places:

    (a) $f(x) = x^3 - 2x^2 - x + 3$
    (b) $f(x) = x^3 - x^2 - 2x + 2$

14. Find the largest zero of

$$f(x) = 10xe^{-x} - 1$$

to two decimal places, using the bisection method. Make sure to demonstrate that your approximation is as good as you claim.

15. (a) Find the smallest positive value of x where the graphs of $f(x) = \sin x$ and $g(x) = 2^{-x}$ intersect.
    (b) Repeat with $f(x) = \sin 2x$ and $g(x) = 2^{-x}$.

16. Use a graphing calculator to sketch $y = 2 \cos x$ and $y = x^3 + x^2 + 1$ on the same set of axes. Find the positive zero of $f(x) = 2 \cos x - x^3 - x^2 - 1$. A friend claims there is one more real zero. Is your friend correct? Explain.

17. Use the table below to investigate the zeros of the function

$$f(\theta) = (\sin 3\theta)(\cos 4\theta) + 0.8$$

in the interval $0 \leq \theta \leq 1.8$.

θ	0	0.2	0.4	0.6	0.8	1.0	1.2	1.4	1.6	1.8
$f(\theta)$	0.80	1.19	0.77	0.08	0.13	0.71	0.76	0.12	-0.19	0.33

   (a) Decide how many zeros the function has in the interval $0 \leq \theta \leq 1.8$.
   (b) Locate each zero, or a small interval containing each zero.
   (c) Are you sure you have found all the zeros in the interval $0 \leq \theta \leq 1.8$? Graph the function on a calculator or computer to decide.

18. (a) Use Table A.3 to locate approximate solution(s) to

$$(\sin 3x)(\cos 4x) = \frac{x^3}{\pi^3}$$

in the interval $1.07 \le x \le 1.15$. Give an interval of length 0.01 in which each solution lies.

Table A.3

x	x^3/π^3	$(\sin 3x)(\cos 4x)$
1.07	0.0395	0.0286
1.08	0.0406	0.0376
1.09	0.0418	0.0442
1.10	0.0429	0.0485
1.11	0.0441	0.0504
1.12	0.0453	0.0499
1.13	0.0465	0.0470
1.14	0.0478	0.0417
1.15	0.0491	0.0340

(b) Make an estimate for each solution accurate to two decimal places.

19. (a) With your calculator in radian mode, take the arctangent of 1 and multiply that number by 4. Now, take the arctangent of the result and multiply it by 4. Continue this process 10 times or so and record each result as in the accompanying table. At each step, you get 4 times the arctangent of the result of the previous step.

$$
\begin{array}{|c|}
\hline
1 \\
3.14159\ldots \\
5.05050\ldots \\
5.50129\ldots \\
\vdots \\
\hline
\end{array}
$$

(b) Your table allows you to find a solution of the equation

$$4 \arctan x = x.$$

Why? What is that solution?

(c) What does your table in part (a) have to do with Figure A.10?
[Hint: The coordinates of P_0 are $(1, 1)$. Find the coordinates of P_1, P_2, P_3,...]

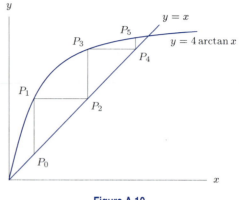

Figure A.10

(d) In part (a), what happens if you start with an initial guess of 10? Of -10? What types of behavior do you observe? (That is, for which initial guesses is the sequence increasing, and for which is it decreasing; does the sequence approach a limit?) Explain your answers graphically, as in part (c).

20. Using radians, apply the iteration method of Problem 19 to the equation

$$\cos x = x.$$

Represent your results graphically, as in Figure A.10.

For Exercises 21–23, draw a graph to decide if the function is bounded on the interval given. Give the best possible upper and lower bounds for any function which is bounded.

21. $f(x) = 4x - x^2$ on $[-1, 4]$

22. $h(\theta) = 5 + 3 \sin \theta$ on $[-2\pi, 2\pi]$

23. $f(t) = \dfrac{\sin t}{t^2}$ on $[-10, 10]$

B POLAR COORDINATES

A point P with Cartesian coordinates (x, y) can be referred to by its *polar coordinates, r* and θ. The number r is the distance between P and the origin, and θ is the angle between the positive x-axis and the line joining P to the origin (with the convention that counterclockwise is positive). Figure B.11 shows the connection between Cartesian and polar coordinates:

Relation between Cartesian and Polar Coordinates

$$x = r \cos \theta \qquad\qquad r = \sqrt{x^2 + y^2}$$

$$y = r \sin \theta \qquad\qquad \tan \theta = \frac{y}{x}$$

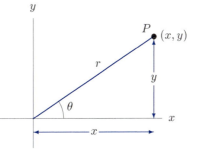

Figure B.11: Cartesian and polar coordinates

Example 1   Convert the point $(x, y) = (-3, 4)$ to polar coordinates.

Solution   The formula for r gives
$$r = \sqrt{x^2 + y^2} = \sqrt{(-3)^2 + 4^2} = 5.$$
The formula involving the angle θ is
$$\tan \theta = \frac{y}{x} = -\frac{4}{3}.$$
This means $\theta = \arctan(-4/3) \approx -0.927$ or $\theta = \arctan(-4/3) + \pi \approx 2.21$. Since the point is in the second quadrant, we choose $\theta \approx 2.21$.

Example 2   Convert the point $(4, \pi/6)$ from polar to rectangular coordinates.

Solution   The formula for x gives
$$x = r \cos \theta = 4 \cos \pi/6 = 2\sqrt{3},$$
and the formula for y gives
$$y = r \sin \theta = 4 \sin \pi/6 = 2.$$

Suppose we have a function $r = f(\theta)$ in polar coordinates. The set of all points (r, θ) satisfying the equation forms the *graph* of the function in polar coordinates. The simplest, and often the most useful, functions are those involving just one of the two variables, as we show in the following example.

Example 3   Sketch the graphs of the polar coordinate curves: (a) $r = 3$ (b) $\theta = \pi/4$.

Solution   (a) Since there is no explicit mention of the variable θ, it can take on any value whatsoever while r must always be equal to 3. The corresponding set of points thus form a circle of radius 3 centered at the origin, as shown in Figure B.12.
(b) Since there is no explicit mention of the variable r, it can take on any value whatsoever, either positive or negative, while θ is always equal to $\pi/4$. The corresponding set of points all lie on a line through the origin inclined at an angle of $\pi/4$, as shown in Figure B.13.

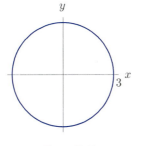

Figure B.12

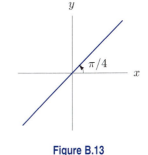

Figure B.13

Exercises for Appendix B

For Exercises 1–8, give polar coordinates for the points with the following Cartesian coordinates. Choose $0 \leq \theta < 2\pi$.

1. $(1, 0)$ **2.** $(0, 2)$ **3.** $(1, 1)$

4. $(-1, 1)$ **5.** $(-3, -3)$ **6.** $(0.2, -0.2)$

7. $(3, 4)$ **8.** $(-3, 1)$

For Exercises 9–15, give Cartesian coordinates for the points with the following polar coordinates (r, θ). The angles are measured in radians.

9. $(1, 0)$ **10.** $(0, 1)$ **11.** $(2, \pi)$

12. $(\sqrt{2}, 5\pi/4)$ **13.** $(5, -\pi/6)$ **14.** $(3, \pi/2)$

15. $(1, 1)$

In Exercises 16–19, sketch the graph of the given function in polar coordinates.

16. $r = 5$ **17.** $r = 2$ **18.** $\theta = \pi/3$ **19.** $\theta = 3\pi/4$

Sketch the regions given in Exercises 20–22.

20. $2 \leq r \leq 3$

21. $r \leq 2, 0 \leq \theta \leq \pi/2$

22. $5 \leq r \leq 10, \pi/3 \leq \theta \leq \pi/2$

In Exercises 23–27, describe the region shown in polar coordinates.

23.   **24.**

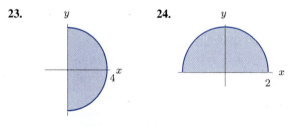

25.   **26.**

27.

In Exercises 28–32, a curve is given in polar coordinates. Transform the equation into xy-coordinates.

28. $r = 1$ **29.** $\theta = \dfrac{\pi}{3}$ **30.** $r = \dfrac{2}{\cos\theta}$

31. $r = 2\cos\theta$ **32.** $r = \dfrac{\sin\theta}{\cos^2\theta}$

33. Graph the function $r = \theta/10$ using a calculator. Explain how the equation relates to the the shape of the graph.

34. Every point in the plane can be represented by some pair of polar coordinates, but are the polar coordinates (r, θ) uniquely determined by the Cartesian coordinates (x, y)? In other words, for each pair of Cartesian coordinates, is there one and only one pair of polar coordinates for that point? Why or why not?

C   COMPLEX NUMBERS

The quadratic equation

$$x^2 - 2x + 2 = 0$$

is not satisfied by any real number x. If you try applying the quadratic formula, you get

$$x = \frac{2 \pm \sqrt{4 - 8}}{2} = 1 \pm \frac{\sqrt{-4}}{2}.$$

Apparently, you need to take a square root of -4. But -4 doesn't have a square root, at least, not one which is a real number. Let's give it a square root.

We define the imaginary number i to be a number such that

$$i^2 = -1.$$

Using this i, we see that $(2i)^2 = -4$, so

$$x = 1 \pm \frac{\sqrt{-4}}{2} = 1 \pm \frac{2i}{2} = 1 \pm i.$$

This solves our quadratic equation. The numbers $1 + i$ and $1 - i$ are examples of complex numbers.

A **complex number** is defined as any number that can be written in the form

$$z = a + bi,$$

where a and b are real numbers and $i^2 = -1$, so we say $i = \sqrt{-1}$.
The *real part* of z is the number a; the *imaginary part* is the number b.

Calling the number i imaginary makes it sound as if i doesn't exist in the same way that real numbers exist. In some cases, it is useful to make such a distinction between real and imaginary numbers. For example, if we measure mass or position, we want our answers to be real numbers. But the imaginary numbers are just as legitimate mathematically as the real numbers are.

As an analogy, consider the distinction between positive and negative numbers. Originally, people thought of numbers only as tools to count with; their concept of "five" or "ten" was not far removed from "five arrows" or "ten stones." They were unaware that negative numbers existed at all. When negative numbers were introduced, they were viewed only as a device for solving equations like $x + 2 = 1$. They were considered "false numbers," or, in Latin, "negative numbers." Thus, even though people started to use negative numbers, they did not view them as existing in the same way that positive numbers did. An early mathematician might have reasoned: "The number 5 exists because I can have 5 dollars in my hand. But how can I have -5 dollars in my hand?" Today we have an answer: "I have -5 dollars" means I owe somebody 5 dollars. We have realized that negative numbers are just as useful as positive ones, and it turns out that complex numbers are useful too. For example, they are used in studying wave motion in electric circuits.

Algebra of Complex Numbers

Numbers such as 0, 1, $\frac{1}{2}$, π, and $\sqrt{2}$ are called *purely real* because they contain no imaginary components. Numbers such as i, $2i$, and $\sqrt{2}i$ are called *purely imaginary* because they contain only the number i multiplied by a nonzero real coefficient.

Two complex numbers are called *conjugates* if their real parts are equal and if their imaginary parts are opposites. The complex conjugate of the complex number $z = a + bi$ is denoted $\overline{z}$, so we have

$$\overline{z} = a - bi.$$

(Note that z is real if and only if $z = \overline{z}$.) Complex conjugates have the following remarkable property: if $f(x)$ is any polynomial with real coefficients ($x^3 + 1$, say) and $f(z) = 0$, then $f(\overline{z}) = 0$. This means that if z is the solution to a polynomial equation with real coefficients, then so is $\overline{z}$.

- Two complex numbers are equal if and only if their real parts are equal and their imaginary parts are equal. Consequently, if $a + bi = c + di$, then $a = c$ and $b = d$.

- Adding two complex numbers is done by adding real and imaginary parts separately:

$$(a + bi) + (c + di) = (a + c) + (b + d)i.$$

- Subtracting is similar:

$$(a + bi) - (c + di) = (a - c) + (b - d)i.$$

- Multiplication works just as for polynomials, using $i^2 = -1$:

$$\begin{aligned}
(a + bi)(c + di) &= a(c + di) + bi(c + di) \\
&= ac + adi + bci + bdi^2 \\
&= ac + adi + bci - bd = (ac - bd) + (ad + bc)i.
\end{aligned}$$

- Powers of i: We know that $i^2 = -1$; then $i^3 = i \cdot i^2 = -i$, and $i^4 = (i^2)^2 = (-1)^2 = 1$. Then $i^5 = i \cdot i^4 = i$, and so on. Thus we have

$$
i^n = \begin{cases}
i & \text{for } n = 1, 5, 9, 13, \ldots \\
-1 & \text{for } n = 2, 6, 10, 14, \ldots \\
-i & \text{for } n = 3, 7, 11, 15, \ldots \\
1 & \text{for } n = 0, 4, 8, 12, 16, \ldots
\end{cases}
$$

- The product of a number and its conjugate is always real and nonnegative:

$$ z \cdot \bar{z} = (a + bi)(a - bi) = a^2 - abi + abi - b^2 i^2 = a^2 + b^2. $$

- Dividing by a nonzero complex number is done by multiplying the denominator by its conjugate, thereby making the denominator real:

$$ \frac{a + bi}{c + di} = \frac{a + bi}{c + di} \cdot \frac{c - di}{c - di} = \frac{ac - adi + bci - bdi^2}{c^2 + d^2} = \frac{ac + bd}{c^2 + d^2} + \frac{bc - ad}{c^2 + d^2}i. $$

Example 1   Compute $(2 + 7i)(4 - 6i) - i$.

Solution $(2 + 7i)(4 - 6i) - i = 8 + 28i - 12i - 42i^2 - i = 8 + 15i + 42 = 50 + 15i$.

Example 2   Compute $\dfrac{2 + 7i}{4 - 6i}$.

Solution

$$ \frac{2 + 7i}{4 - 6i} = \frac{2 + 7i}{4 - 6i} \cdot \frac{4 + 6i}{4 + 6i} = \frac{8 + 12i + 28i + 42i^2}{4^2 + 6^2} = \frac{-34 + 40i}{52} = \frac{-17}{26} + \frac{10}{13}i. $$

You can check by multiplying out that $(-17/26 + 10i/13)(4 - 6i) = 2 + 7i$.

The Complex Plane and Polar Coordinates

It is often useful to picture a complex number $z = x + iy$ in the plane, with x along the horizontal axis and y along the vertical. The xy-plane is then called the *complex plane*. Figure C.14 shows the complex numbers $-2i$, $1 + i$, and $-2 + 3i$.

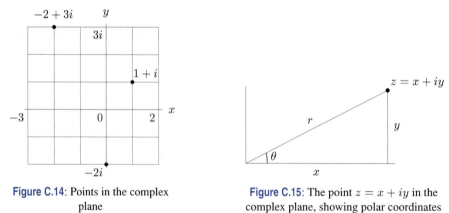

Figure C.14: Points in the complex plane

Figure C.15: The point $z = x + iy$ in the complex plane, showing polar coordinates

The triangle in Figure C.15 shows that a complex number can be written using polar coordinates as follows:

$$ z = x + iy = r\cos\theta + ir\sin\theta. $$

Example 3   Express $z = -2i$ and $z = -2 + 3i$ using polar coordinates. (See Figure C.14.)

Solution   For $z = -2i$, the distance of z from the origin is 2, so $r = 2$. Also, one value for θ is $\theta = 3\pi/2$. Using polar coordinates, $-2i = 2\cos(3\pi/2) + i\,2(\sin 3\pi/2)$.

     For $z = -2 + 3i$, we have $x = -2$, $y = 3$. So $r = \sqrt{(-2)^2 + 3^2} \approx 3.61$, and one solution of $\tan\theta = 3/(-2)$ with θ in quadrant II is $\theta \approx 2.16$. So $-2 + 3i \approx 3.61\cos(2.16) + i\,3.61\sin(2.16)$.

Example 4   Consider the point with polar coordinates $r = 5$ and $\theta = 3\pi/4$. What complex number does this point represent?

Solution   Since $x = r\cos\theta$ and $y = r\sin\theta$ we see that $x = 5\cos 3\pi/4 = -5/\sqrt{2}$, and $y = 5\sin 3\pi/4 = 5/\sqrt{2}$, so $z = -5/\sqrt{2} + i\,5/\sqrt{2}$.

Derivatives and Integrals of Complex-Valued Functions

Suppose $z(t) = x(t) + iy(t)$, where t is real, then we define $z'(t)$ and $\int z(t)\,dt$ by treating i like any other constant:

$$z'(t) = x'(t) + iy'(t)$$

$$\int z(t)\,dt = \int x(t)\,dt + i\int y(t)\,dt.$$

With these definitions, all the usual properties of differentiation and integration hold, such as

$$\int z'(t)\,dt = z(t) + C, \qquad \text{for } C \text{ is a complex constant.}$$

Euler's Formula

Consider the complex number z lying on the unit circle in Figure C.16. Writing z in polar coordinates, and using the fact that $r = 1$, we have

$$z = f(\theta) = \cos\theta + i\sin\theta.$$

It turns out that there is a particularly beautiful and compact way of rewriting $f(\theta)$ using complex exponentials. We take the derivative of f using the fact that $i^2 = -1$:

$$f'(\theta) = -\sin\theta + i\cos\theta = i\cos\theta + i^2\sin\theta.$$

Factoring out an i gives

$$f'(\theta) = i(\cos\theta + i\sin\theta) = i \cdot f(\theta).$$

As you know from Chapter 11, page 497, the only real-valued function whose derivative is proportional to the function itself is the exponential function. In other words, we know that if

$$g'(x) = k \cdot g(x), \quad \text{then} \quad g(x) = Ce^{kx}$$

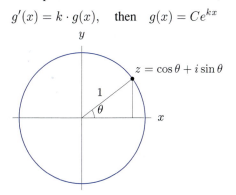

Figure C.16: Complex number represented by a point on the unit circle

for some constant C. If we assume that a similar result holds for complex-valued functions, then we have

$$f'(\theta) = i \cdot f(\theta), \quad \text{so} \quad f(\theta) = Ce^{i\theta}$$

for some constant C. To find C we substitute $\theta = 0$. Now $f(0) = Ce^{i \cdot 0} = C$, and since $f(0) = \cos 0 + i \sin 0 = 1$, we must have $C = 1$. Therefore $f(\theta) = e^{i\theta}$. Thus we have

---

Euler's formula

$$e^{i\theta} = \cos \theta + i \sin \theta$$

This elegant and surprising relationship was discovered by the Swiss mathematician Leonhard Euler in the eighteenth century, and it is particularly useful in solving second-order differential equations. Another way of obtaining Euler's formula (using Taylor series) is given in Problem 41 on page 446. It allows us to write the complex number represented by the point with polar coordinates (r, θ) in the following form:

$$z = r(\cos \theta + i \sin \theta) = re^{i\theta}.$$

Similarly, since $\cos(-\theta) = \cos \theta$ and $\sin(-\theta) = -\sin \theta$, we have

$$re^{-i\theta} = r\left(\cos(-\theta) + i \sin(-\theta)\right) = r(\cos \theta - i \sin \theta).$$

Example 5   Evaluate $e^{i\pi}$.

Solution   Using Euler's formula, $e^{i\pi} = \cos \pi + i \sin \pi = -1$.

Example 6   Express the complex number represented by the point $r = 8$, $\theta = 3\pi/4$ in Cartesian form and polar form, $z = re^{i\theta}$.

Solution   Using Cartesian coordinates, the complex number is

$$z = 8\left(\cos\left(\frac{3\pi}{4}\right) + i \sin\left(\frac{3\pi}{4}\right)\right) = \frac{-8}{\sqrt{2}} + i\frac{8}{\sqrt{2}}.$$

Using polar coordinates, we have

$$z = 8e^{i\,3\pi/4}.$$

The polar form of complex numbers makes finding powers and roots of complex numbers much easier. Writing $z = re^{i\theta}$, we find any power of z as follows:

$$z^p = (re^{i\theta})^p = r^p e^{ip\theta}.$$

To find roots, we let p be a fraction, as in the following example.

Example 7   Find a cube root of the complex number represented by the point with polar coordinates $(8, 3\pi/4)$.

Solution   In Example 6, we saw that this complex number could be written as $z = 8e^{i3\pi/4}$. So,

$$\sqrt[3]{z} = \left(8e^{i\,3\pi/4}\right)^{1/3} = 8^{1/3}e^{i(3\pi/4)\cdot(1/3)} = 2e^{\pi i/4} = 2\left(\cos(\pi/4) + i \sin(\pi/4)\right)$$

$$= 2\left(1/\sqrt{2} + i/\sqrt{2}\right) = \sqrt{2}(1 + i).$$

You can check by multiplying out that $(\sqrt{2}(1 + i))^3 = -(8/\sqrt{2}) + i(8/\sqrt{2}) = z$.

Using Complex Exponentials

Euler's formula, together with the fact that exponential functions are simple to manipulate, allows us to obtain many results about trigonometric functions easily.

The following example uses the fact that for complex z, the function e^z has all the usual algebraic properties of exponents.

Example 8   Use Euler's formula to obtain the double-angle identities

$$\cos 2\theta = \cos^2 \theta - \sin^2 \theta \qquad \text{and} \qquad \sin 2\theta = 2 \cos \theta \sin \theta.$$

Solution   We use the fact that $e^{2i\theta} = e^{i\theta} \cdot e^{i\theta}$. This can be rewritten as

$$\cos 2\theta + i \sin 2\theta = (\cos \theta + i \sin \theta)^2.$$

Multiplying out $(\cos \theta + i \sin \theta)^2$, using the fact that $i^2 = -1$ gives

$$\cos 2\theta + i \sin 2\theta = \cos^2 \theta - \sin^2 \theta + i(2 \cos \theta \sin \theta).$$

Since two complex numbers are equal only if the real and imaginary parts are equal, we must have

$$\cos 2\theta = \cos^2 \theta - \sin^2 \theta \qquad \text{and} \qquad \sin 2\theta = 2 \cos \theta \sin \theta.$$

If we solve $e^{i\theta} = \cos \theta + i \sin \theta$ and $e^{-i\theta} = \cos \theta - i \sin \theta$ for $\sin \theta$ and $\cos \theta$, we obtain

$$\sin \theta = \frac{e^{i\theta} - e^{-i\theta}}{2i} \qquad \text{and} \qquad \cos \theta = \frac{e^{i\theta} + e^{-i\theta}}{2}.$$

By differentiating the formula $e^{ik\theta} = \cos(k\theta) + i \sin(k\theta)$, for θ real and k a real constant, it can be shown that

$$\frac{d}{d\theta}\left(e^{ik\theta}\right) = ike^{i\theta} \qquad \text{and} \qquad \int e^{ik\theta}\, d\theta = \frac{1}{ik}e^{i\theta} + C.$$

Thus complex exponentials are differentiated and integrated just like real exponentials.

Example 9   Use $\cos \theta = \left(e^{i\theta} + e^{-i\theta}\right)/2$ to obtain the derivative formula for $\cos \theta$.

Solution   Differentiating gives

$$\frac{d}{d\theta}(\cos \theta) = \frac{d}{d\theta}\left(\frac{e^{i\theta} + e^{-i\theta}}{2}\right) = \frac{ie^{i\theta} - ie^{-i\theta}}{2} = \frac{i\left(e^{i\theta} - e^{-i\theta}\right)}{2}$$

$$= -\frac{e^{i\theta} - e^{-i\theta}}{2i} = -\sin \theta.$$

The facts that e^z has all the usual properties when z is complex leads to

$$\frac{d}{d\theta}(e^{(a+ib)\theta}) = (a + ib)e^{(a+ib)\theta} \qquad \text{and} \qquad \int e^{(a+ib)\theta}\, d\theta = \frac{1}{a + ib}e^{(a+ib)\theta} + C.$$

Example 10   Use the formula for $\int e^{(a+ib)\theta}\,d\theta$ to obtain formulas for $\int e^{ax}\cos bx\,dx$ and $\int e^{ax}\sin bx\,dx$.

Solution   The formula for $\int e^{(a+ib)\theta}\,d\theta$ allows us to write

$$\int e^{ax}e^{ibx}\,dx = \int e^{(a+ib)x}\,dx = \frac{1}{a+ib}e^{(a+ib)x} + C = \frac{a-ib}{a^2+b^2}e^{ax}e^{ibx} + C.$$

The left-hand side of this equation can be rewritten as

$$\int e^{ax}e^{ibx}\,dx = \int e^{ax}\cos bx\,dx + i\int e^{ax}\sin bx\,dx,$$

The right-hand side can be rewritten as

$$\frac{a-ib}{a^2+b^2}e^{ax}e^{ibx} = \frac{e^{ax}}{a^2+b^2}(a-ib)(\cos bx + i\sin bx),$$

$$= \frac{e^{ax}}{a^2+b^2}\left(a\cos bx + b\sin bx + i\left(a\sin bx - b\cos bx\right)\right).$$

Equating real parts gives

$$\int e^{ax}\cos bx\,dx = \frac{e^{ax}}{a^2+b^2}\left(a\cos bx + b\sin bx\right) + C,$$

and equating imaginary parts gives

$$\int e^{ax}\sin bx\,dx = \frac{e^{ax}}{a^2+b^2}\left(a\sin bx - b\cos bx\right) + C.$$

These two formulas are usually obtained by integrating by parts twice.

Example 11   Using complex exponentials, find a formula for $\int \sin 2x \sin 3x\,dx$.

Solution   Replacing $\sin 2x$ and $\sin 3x$ by their exponential form, we have

$$\int \sin 2x \sin 3x\,dx = \int \frac{\left(e^{2ix} - e^{-2ix}\right)}{2i}\frac{\left(e^{3ix} - e^{-3ix}\right)}{2i}\,dx$$

$$= \frac{1}{(2i)^2}\int \left(e^{5ix} - e^{-ix} - e^{ix} + e^{-5ix}\right)\,dx$$

$$= -\frac{1}{4}\left(\frac{1}{5i}e^{5ix} + \frac{1}{i}e^{-ix} - \frac{1}{i}e^{ix} - \frac{1}{5i}e^{-5ix}\right) + C$$

$$= -\frac{1}{4}\left(\frac{e^{5ix} - e^{-5ix}}{5i} - \frac{e^{ix} - e^{-ix}}{i}\right) + C$$

$$= -\frac{1}{4}\left(\frac{2}{5}\sin 5x - 2\sin x\right) + C$$

$$= -\frac{1}{10}\sin 5x + \frac{1}{2}\sin x + C$$

This result is usually obtained by using a trigonometric identity.

Exercises for Appendix C

For Exercises 1–8, express the given complex number in polar form, $z = re^{i\theta}$.

For Exercises 9–18, perform the indicated calculations. Give your answer in Cartesian form, $z = x + iy$.

1. $2i$

2. -5

3. $1+i$

4. $-3-4i$

9. $(2+3i) + (-5-7i)$

10. $(2+3i)(5+7i)$

5. 0

6. $-i$

7. $-1+3i$

8. $5-12i$

11. $(2+3i)^2$

12. $(1+i)^2 + (1+i)$

13. $(0.5 - i)(1 - i/4)$ **14.** $(2i)^3 - (2i)^2 + 2i - 1$

15. $(e^{i\pi/3})^2$ **16.** $\sqrt{e^{i\pi/3}}$

17. $(5e^{i7\pi/6})^3$ **18.** $\sqrt[4]{10e^{i\pi/2}}$

By writing the complex numbers in polar form, $z = re^{i\theta}$, find a value for the quantities in Exercises 19–28. Give your answer in Cartesian form, $z = x + iy$.

19. $\sqrt{i}$ **20.** $\sqrt{-i}$ **21.** $\sqrt[3]{i}$

22. $\sqrt{7i}$ **23.** $(1 + i)^{100}$ **24.** $(1 + i)^{2/3}$

25. $(-4 + 4i)^{2/3}$ **26.** $(\sqrt{3} + i)^{1/2}$ **27.** $(\sqrt{3}+i)^{-1/2}$

28. $(\sqrt{5} + 2i)^{\sqrt{2}}$

29. Calculate i^n for $n = -1, -2, -3, -4$. What pattern do you observe? What is the value of i^{-36}? Of i^{-41}?

Solve the simultaneous equations in Exercises 30–31 for A_1 and A_2.

30. $A_1 + A_2 = 2$
   $(1 - i)A_1 + (1 + i)A_2 = 3$

31. $A_1 + A_2 = 2$
   $(i - 1)A_1 + (1 + i)A_2 = 0$

32. (a) Calculate a and b if $\dfrac{3 - 4i}{1 + 2i} = a + bi$.
   (b) Check your answer by calculating $(1 + 2i)(a + bi)$.

33. Check that $z = \dfrac{ac + bd}{c^2 + d^2} + \dfrac{bc - ad}{c^2 + d^2}i$ is the quotient $\dfrac{a + bi}{c + di}$ by showing that the product $z \cdot (c + di)$ is $a + bi$.

34. Let $z_1 = -3 - i\sqrt{3}$ and $z_2 = -1 + i\sqrt{3}$.
   (a) Find $z_1 z_2$ and z_1/z_2. Give your answer in Cartesian form, $z = x + iy$.
   (b) Put z_1 and z_2 into polar form, $z = re^{i\theta}$. Find $z_1 z_2$ and z_1/z_2 using the polar form, and verify that you get the same answer as in part (a).

35. Let $z_1 = a_1 + b_1 i$ and $z_2 = a_2 + b_2 i$. Show that $\overline{z_1 z_2} = \bar{z}_1 \bar{z}_2$.

36. If the roots of the equation $x^2 + 2bx + c = 0$ are the complex numbers $p \pm iq$, find expressions for p and q in terms of b and c.

Are the statements in Exercises 37–42 true or false? Explain your answer.

37. Every nonnegative real number has a real square root.

38. For any complex number z, the product $z \cdot \bar{z}$ is a real number.

39. The square of any complex number is a real number.

40. If f is a polynomial, and $f(z) = i$, then $f(\bar{z}) = i$.

41. Every nonzero complex number z can be written in the form $z = e^w$, where w is another complex number.

42. If $z = x+iy$, where x and y are positive, then $z^2 = a+ib$ has a and b positive.

For Exercises 43–47, use Euler's formula to derive the following relationships. (Note that if a, b, c, d are real numbers, $a + bi = c + di$ means that $a = c$ and $b = d$.)

43. $\sin^2 \theta + \cos^2 \theta = 1$ **44.** $\sin 2\theta = 2 \sin \theta \cos \theta$

45. $\cos 2\theta = \cos^2 \theta - \sin^2 \theta$ **46.** $\dfrac{d}{d\theta} \sin \theta = \cos \theta$

47. $\dfrac{d^2}{d\theta^2} \cos \theta = -\cos \theta$

48. Use complex exponentials to show that
$$\sin(-x) = -\sin x.$$

49. Use complex exponentials to show that
$$\sin(x + y) = \sin x \cos y + \cos x \sin y.$$

50. For real t, show that if $z_1(t) = x_1(t) + iy_1(t)$ and $z_2(t) = x_2(t) + iy_2(t)$ then
$$(z_1 + z_2)' = z_1' + z_2' \quad \text{and} \quad (z_1 z_2)' = z_1' z_2 + z_1 z_2'.$$

D NEWTON'S METHOD

Many problems in mathematics involve finding the root of an equation. For example, we might have to locate the zeros of a polynomial, or determine the point of intersection of two curves. Here we will see a numerical method for approximating solutions which cannot be calculated exactly.

One such method, bisection, is described in Appendix A. Although it is very simple, the bisection method has two major drawbacks. First, it cannot locate a root where the curve is tangent to, but does not cross, the x-axis. Second, it is relatively slow in the sense that it requires a considerable number of iterations to achieve a desired level of accuracy. Although speed may not be important in solving a single equation, a practical problem may involve solving thousands of equations as a parameter changes. In such a case, any reduction in the number of steps can be important.

Using Newton's Method

We now consider a powerful root-finding method developed by Newton. Suppose we have a function $y = f(x)$. The equation $f(x) = 0$ has a root at $x = r$, as shown in Figure D.17. We begin with an initial estimate, x_0, for this root. (This can be a guess.) We will now obtain a better estimate x_1. To do this, construct the tangent line to the graph of f at the point $x = x_0$, and extend it until it crosses the x-axis, as shown in Figure D.17. The point where it crosses the axis is usually much closer to r, and we use that point as the next estimate, x_1. Having found x_1, we now repeat the process starting with x_1 instead of x_0. We construct a tangent line to the curve at $x = x_1$, extend it until it crosses the x-axis, use that x-intercept as the next approximation, x_2, and so on. The resulting sequence of x-intercepts usually converges rapidly to the root r.

Let's see how this looks algebraically. We know that the slope of the tangent line at the initial estimate x_0 is $f'(x_0)$, and so the equation of the tangent line is

$$y - f(x_0) = f'(x_0)(x - x_0).$$

At the point where this tangent line crosses the x-axis, we have $y = 0$ and $x = x_1$, so that

$$0 - f(x_0) = f'(x_0)(x_1 - x_0).$$

Solving for x_1, we obtain

$$x_1 = x_0 - \frac{f(x_0)}{f'(x_0)}$$

provided that $f'(x_0)$ is not zero. We now repeat this argument and find that the next approximation is

$$x_2 = x_1 - \frac{f(x_1)}{f'(x_1)}.$$

Summarizing, for any $n = 0, 1, 2, \ldots$, we obtain the following result.

Newton's Method to Solve the Equation $f(x) = 0$

Choose x_0 near a solution and compute the sequence $x_1, x_2, x_3 \ldots$ using the rule

$$x_{n+1} = x_n - \frac{f(x_n)}{f'(x_n)}$$

provided that $f'(x_n)$ is not zero. For large n, the solution is well-approximated by x_n.

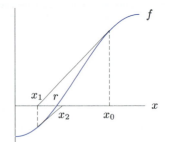

Figure D.17: Newton's method: successive approximations $x_0, x_1, x_2, \ldots$ to the root, r

---

Example 1   Use Newton's method to find the fifth root of 23. (By calculator, this is 1.872171231, correct to nine decimal places.)

Solution   To use Newton's method, we need an equation of the form $f(x) = 0$ having $23^{1/5}$ as a root. Since $23^{1/5}$ is a root of $x^5 = 23$ or $x^5 - 23 = 0$, we take $f(x) = x^5 - 23$. The root of this equation is between 1 and 2 (since $1^5 = 1$ and $2^5 = 32$), so we will choose $x_0 = 2$ as our initial estimate. Now $f'(x) = 5x^4$, so we can set up Newton's method as

$$x_{n+1} = x_n - \frac{x_n^5 - 23}{5x_n^4}.$$

In this case, we can simplify using a common denominator, to obtain

$$x_{n+1} = \frac{4x_n^5 + 23}{5x_n^4}.$$

Therefore, starting with $x_0 = 2$, we find that $x_1 = 1.8875$. This leads to $x_2 = 1.872418193$ and $x_3 = 1.872171296$. These values are in Table D.4. Since $f(1.872171231) > 0$ and $f(1.872171230) < 0$, the root lies between 1.872171230 and 1.872171231. Therefore, in just four iterations of Newton's method, we have achieved eight-decimal accuracy.

Table D.4   Newton's method: $x_0 = 2$

n	x_n	$f(x_n)$
0	2	9
1	1.8875	0.957130661
2	1.872418193	0.015173919
3	1.872171296	0.000004020
4	1.872171231	0.000000027

Table D.5   Newton's method: $x_0 = 10$

n	x_n	n	x_n
0	10	6	2.679422313
1	8.000460000	7	2.232784753
2	6.401419079	8	1.971312452
3	5.123931891	9	1.881654220
4	4.105818871	10	1.872266333
5	3.300841811	11	1.872171240

As a general guideline for Newton's method, once the first correct decimal place is found, each successive iteration approximately doubles the number of correct digits.

What happens if we select a very poor initial estimate? In the preceding example, suppose x_0 were 10 instead of 2. The results are in Table D.5. Notice that even with $x_0 = 10$, the sequence of values moves reasonably quickly toward the solution: we achieve six-decimal place accuracy by the eleventh iteration.

Example 2   Find the first point of intersection of the curves given by $f(x) = \sin x$ and $g(x) = e^{-x}$.

Solution

The graphs in Figure D.18 make it clear that there are an infinite number of points of intersection, all with $x > 0$. In order to find the first one numerically, we consider the function

$$F(x) = f(x) - g(x) = \sin x - e^{-x}$$

whose derivative is $F'(x) = \cos x + e^{-x}$. From the graph, we see that the point we want is fairly close to $x = 0$, so we start with $x_0 = 0$. The values in Table D.6 are approximations to the root. Since $F(0.588532744) > 0$ and $F(0.588532743) < 0$, the root lies between 0.588532743 and 0.588532744. (Remember, your calculator must be set in radians.)

Table D.6   Successive approximations to root of $\sin x = e^{-x}$

n	x_n
0	0
1	0.5
2	0.585643817
3	0.588529413
4	0.588532744
5	0.588532744

Figure D.18: Root of $\sin x = e^{-x}$

When Does Newton's Method Fail?

In most practical situations, Newton's method works well. Occasionally, however, the sequence x_0, x_1, x_2, ... fails to converge or fails to converge to the root you want. Sometimes, for example, the sequence can jump from one root to another. This is particularly likely to happen if the magnitude of the derivative $f'(x_n)$ is small for some x_n. In this case, the tangent line is nearly horizontal and so x_{n+1} will be far from x_n. (See Figure D.19.)

If the equation $f(x) = 0$ has *no* root, then the sequence will not converge. In fact, the sequence obtained by applying Newton's method to $f(x) = 1 + x^2$ is one of the best known examples of *chaotic behavior* and has attracted considerable research interest recently. (See Figure D.20.)

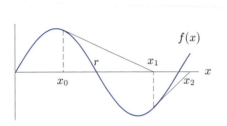

Figure D.19: Problems with Newton's method: Converges to wrong root

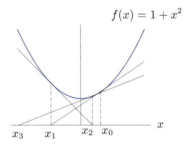

$$f(x) = 1 + x^2$$

Figure D.20: Problems with Newton's method: Chaotic behavior

Exercises for Appendix D

1. Suppose you want to find a solution of the equation

$$x^3 + 3x^2 + 3x - 6 = 0.$$

Consider $f(x) = x^3 + 3x^2 + 3x - 6$.

   (a) Find $f'(x)$, and use it to show that $f(x)$ increases everywhere.
   (b) How many roots does the original equation have?
   (c) For each root, find an interval which contains it.
   (d) Find each root to two decimal places, using Newton's method.

For Exercises 2–4, use Newton's method to find the given quantities to two decimal places:

2. $\sqrt[3]{50}$ 3. $\sqrt[4]{100}$ 4. $10^{-1/3}$

For Exercises 5–8, solve each equation and give each answer to two decimal places:

5. $\sin x = 1 - x$ 6. $\cos x = x$

7. $e^{-x} = \ln x$

8. $e^x \cos x = 1$, for $0 < x < \pi$

9. Find, to two decimal places, all solutions of $\ln x = 1/x$.

10. How many zeros do the following functions have? For each zero, find an upper and a lower bound which differ by no more than 0.1.
   (a) $f(x) = x^3 + x - 1$ (b) $f(x) = \sin x - \frac{2}{3}x$
   (c) $f(x) = 10xe^{-x} - 1$

11. Find the largest zero of

$$f(x) = x^3 + x - 1$$

to six decimal places, using Newton's method. How do you know your approximation is as good as you claim?

12. For any positive number, a, the problem of calculating the square root, $\sqrt{a}$, is often done by applying Newton's method to the function $f(x) = x^2 - a$. Apply the method to obtain an expression for x_{n+1} in terms of x_n. Use this to approximate $\sqrt{a}$ for $a = 2, 10, 1000$, and π, correct to four decimal places, starting at $x_0 = a/2$ in each case.

E   DETERMINANTS

We introduce the determinant of an array of numbers. Each 2 by 2 array of numbers has another number associated with it, called its determinant, which is given by

$$\begin{vmatrix} a_1 & a_2 \\ b_1 & b_2 \end{vmatrix} = a_1 b_2 - a_2 b_1.$$

For example

$$\begin{vmatrix} 2 & 5 \\ -4 & -6 \end{vmatrix} = 2(-6) - 5(-4) = 8.$$

Each 3 by 3 array of numbers also has a number associated with it, also called a determinant, which is defined in terms of 2 by 2 determinants as follows:

$$\begin{vmatrix} a_1 & a_2 & a_3 \\ b_1 & b_2 & b_3 \\ c_1 & c_2 & c_3 \end{vmatrix} = a_1 \begin{vmatrix} b_2 & b_3 \\ c_2 & c_3 \end{vmatrix} - a_2 \begin{vmatrix} b_1 & b_3 \\ c_1 & c_3 \end{vmatrix} + a_3 \begin{vmatrix} b_1 & b_2 \\ c_1 & c_2 \end{vmatrix}.$$

Notice that the determinant of the 2 by 2 array multiplied by a_i is the determinant of the array found by removing the row and column containing a_i. Also, note the minus sign in the second term. An example is given by

$$\begin{vmatrix} 2 & 1 & -3 \\ 0 & 3 & -1 \\ 4 & 0 & 5 \end{vmatrix} = 2 \begin{vmatrix} 3 & -1 \\ 0 & 5 \end{vmatrix} - 1 \begin{vmatrix} 0 & -1 \\ 4 & 5 \end{vmatrix} + (-3) \begin{vmatrix} 0 & 3 \\ 4 & 0 \end{vmatrix} = 2(15 + 0) - 1(0 - (-4)) + (-3)(0 - 12) = 62.$$

Suppose the vectors $\vec{a}$ and $\vec{b}$ have components $\vec{a} = a_1 \vec{i} + a_2 \vec{j} + a_3 \vec{k}$ and $\vec{b} = b_1 \vec{i} + b_2 \vec{j} + b_3 \vec{k}$. Recall that the cross product $\vec{a} \times \vec{b}$ is given by the expression

$$\vec{a} \times \vec{b} = (a_2 b_3 - a_3 b_2)\vec{i} + (a_3 b_1 - a_1 b_3)\vec{j} + (a_1 b_2 - a_2 b_1)\vec{k}.$$

Notice that if we expand the following determinant, we get the cross product:

$$\begin{vmatrix} \vec{i} & \vec{j} & \vec{k} \\ a_1 & a_2 & a_3 \\ b_1 & b_2 & b_3 \end{vmatrix} = \vec{i}(a_2 b_3 - a_3 b_2) - \vec{j}(a_1 b_3 - a_3 b_1) + \vec{k}(a_1 b_2 - a_2 b_1) = \vec{a} \times \vec{b}.$$

Determinants give a useful way of computing cross products.

READY REFERENCE

This section contains a concise summary of the main definitions, theorems, and ideas presented throughout the book.

READY REFERENCE

A Library of Functions

Linear functions (p. 4) have the form $y = f(x) = b + mx$, where m is the **slope**, or rate of change of y with respect to x (p. 5) and b is the **vertical intercept**, or value of y when x is zero (p. 5). The slope is

$$m = \frac{\text{Rise}}{\text{Run}} = \frac{\Delta y}{\Delta x} = \frac{f(x_2) - f(x_1)}{x_2 - x_1} \quad \text{(p. 5).}$$

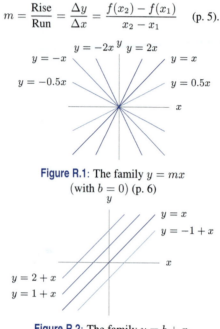

Figure R.1: The family $y = mx$ (with $b = 0$) (p. 6)

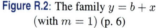

Figure R.2: The family $y = b + x$ (with $m = 1$) (p. 6)

Exponential functions have the form $P = P_0 a^t$ (p. 11) or $P = P_0 e^{kt}$ (p. 13), where P_0 is the initial quantity (p. 11), a is the growth (decay) factor per unit time (p. 11), $|k|$ is the continuous growth (decay) rate (pp. 13, 497), and $r = |a - 1|$ is the growth (decay) rate per unit time (p. 11).

Suppose $P_0 > 0$. If $a > 1$ or $k > 0$, we have **exponential growth**; if $0 < a < 1$ or $k < 0$, we have **exponential decay** (p. 12). The **doubling time** (for growth) is the time required for P to double (p. 12). The **half-life** (for decay) is the time required for P to be reduced by a factor of one half (p. 12). The continuous growth rate $k = \ln(1 + r)$ is slightly less than, but very close to, r, provided r is small (p. 498).

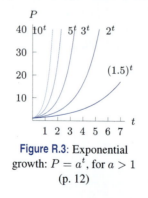

Figure R.3: Exponential growth: $P = a^t$, for $a > 1$ (p. 12)

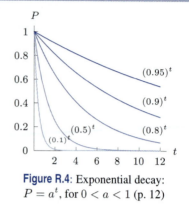

Figure R.4: Exponential decay: $P = a^t$, for $0 < a < 1$ (p. 12)

Common Logarithm and Natural Logarithm

$\log_{10} x = \log x$ = power of 10 that gives x (p. 23)

$\log_{10} x = c$ means $10^c = x$ (p. 23)

$\ln x$ = power of e that gives x (p. 23)

$\ln x = c$ means $e^c = x$ (p. 23)

$\log x$ and $\ln x$ are not defined if x is negative or 0 (p. 24).

Properties of Logarithms (p. 24)

1. $\log(AB) = \log A + \log B$ 4. $\log(10^x) = x$

2. $\log\left(\frac{A}{B}\right) = \log A - \log B$ 5. $10^{\log x} = x$

3. $\log(A^p) = p \log A$ 6. $\log 1 = 0$

The natural logarithm satisfies properties 1, 2, and 3, and $\ln e^x = x$, $e^{\ln x} = x$, $\ln 1 = 0$ (p. 24).

Trigonometric Functions The sine and cosine are defined in Figure R.5 (see also p. 30). The tangent is $\tan t = \frac{\sin t}{\cos t}$ = slope of the line through the origin $(0, 0)$ and P if $\cos t \neq 0$ (p. 33). The period of sin and cos is 2π (p. 30), the period of tan is π (p. 33).

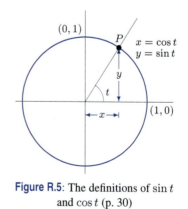

Figure R.5: The definitions of $\sin t$ and $\cos t$ (p. 30)

A **sinusoidal function** has the form $y = C + A\sin(B(t + h))$ or $y = C + A\cos(B(t + h))$ (p. 31). The

amplitude is $|A|$, (p. 31) the **period** is $2\pi/|B|$ (p. 31), and the **phase difference** is $|h|$ (p. 31).

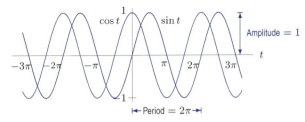

Figure R.6: Graphs of $\cos t$ and $\sin t$ (p. 30)

Trigonometric Identities

$$\sin^2 x + \cos^2 x = 1$$
$$\sin(2x) = 2\sin x \cos x$$
$$\cos(2x) = \cos^2 x - \sin^2 x = 2\cos^2 x - 1 = 1 - 2\sin^2 x$$
$$\cos(a+b) = \cos a \cos b - \sin a \sin b$$
$$\sin(a+b) = \sin a \cos b + \cos a \sin b$$

Inverse trigonometric functions: $\arcsin y = x$ means $\sin x = y$ with $-(\pi/2) \le x \le (\pi/2)$ (p. 34), $\arccos y = x$ means $\cos x = y$ with $0 \le x \le \pi$ (Problem 42, p. 36), $\arctan y = x$ means $\tan x = y$ with $-(\pi/2) < x < (\pi/2)$ (p. 34). The domain of arcsin and arccos is $[-1, 1]$ (p. 34), the domain of arctan is all numbers (p. 34).

Power Functions have the form $f(x) = kx^p$ (p. 37). Graphs for positive powers:

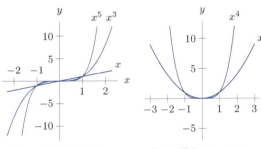

Figure R.7: Odd integer powers of x: "Seat" shaped for $k > 1$ (p. 37)

Figure R.8: Even integer powers of x: $\bigcup$-shaped (p. 37)

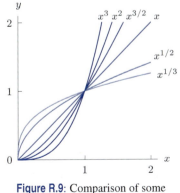

Figure R.9: Comparison of some fractional powers of x

Graphs for zero and negative powers:

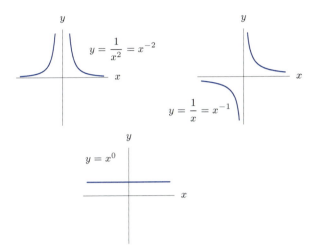

Figure R.10: Comparison of zero and negative powers of x

Polynomials have the form

$$f(x) = a_n x^n + a_{n-1} x^{n-1} + \cdots + a_1 x + a_0, \ a_n \ne 0 \text{ (p. 38)}.$$

The **degree** is n (p. 38) and the **leading coefficient** is a_n (p. 38).

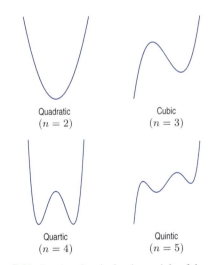

Quadratic $(n = 2)$

Cubic $(n = 3)$

Quartic $(n = 4)$

Quintic $(n = 5)$

Figure R.11: Graphs of typical polynomials of degree n (p. 38)

Rational functions have the form $f(x) = \dfrac{p(x)}{q(x)}$, where p and q are polynomials (p. 41). There is usually a vertical asymptote at $x = a$ if $q(a) = 0$ and a horizontal asymptote at $y = L$ if $\lim_{x \to \infty} f(x) = L$ or $\lim_{x \to -\infty} f(x) = L$ (p. 41).

Hyperbolic functions:

$$\cosh x = \frac{e^x + e^{-x}}{2} \qquad \sinh x = \frac{e^x - e^{-x}}{2} \qquad \text{(p. 203)}.$$

Relative Growth Rates of Functions

Power functions As $x \to \infty$, higher powers of x dominate, as $x \to 0$, smaller powers dominate.

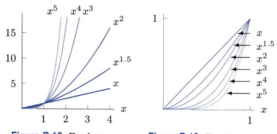

Figure R.12: For large x: Large powers of x dominate

Figure R.13: For $0 \le x \le 1$: Small powers of x dominate

Power functions versus exponential functions Every exponential growth function eventually dominates every power function (p. 37).

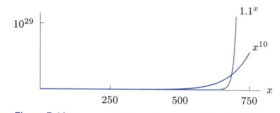

Figure R.14: Exponential function eventually dominates power function

Power functions versus logarithmic functions The power function x^p dominates $A \log x$ for large x for all values of $p > 0$ and $A > 0$.

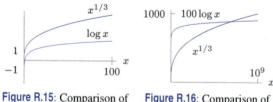

Figure R.15: Comparison of $x^{1/3}$ and $\log x$

Figure R.16: Comparison of $x^{1/3}$ and $100 \log x$

Numerical comparisons of growth rates:

Table R.1 *Comparison of $x^{0.001}$ and $1000 \log x$*

x	$x^{0.001}$	$1000 \log x$
10^{5000}	10^5	$5 \cdot 10^6$
10^{6000}	10^6	$6 \cdot 10^6$
10^{7000}	10^7	$7 \cdot 10^6$

Table R.2 *Comparison of x^{100} and 1.01^x*

x	x^{100}	1.01^x
10^4	10^{400}	$1.6 \cdot 10^{43}$
10^5	10^{500}	$1.4 \cdot 10^{432}$
10^6	10^{600}	$2.4 \cdot 10^{4321}$

Operations on Functions

Shifts, stretches, and composition Multiplying by a constant, c, stretches (if $c > 1$) or shrinks (if $0 < c < 1$) the graph vertically. A negative sign (if $c < 0$) reflects the graph about the x-axis, in addition to shrinking or stretching (p. 17). Replacing y by $(y - k)$ moves a graph up by k (down if k is negative) (p. 17). Replacing x by $(x - h)$ moves a graph to the right by h (to the left if h is negative) (p. 17). The composite of f and g is the function $f(g(x))$; f is the outside function, g the inside function (p. 18).

Symmetry We say f is an **even** function if $f(-x) = f(x)$ (p. 19) and f is an **odd** function if $f(-x) = -f(x)$ (p. 19).

Inverse functions A function f has an inverse if (and only if) its graph intersects any horizontal line at most once (p. 20). If f has an inverse, it is written f^{-1}, and $f^{-1}(x) = y$ means $f(y) = x$ (p. 20). Provided the x and y scales are equal, the graph of f^{-1} is the reflection of the graph of f about the line $y = x$ (p. 21).

Limits and Continuity

Idea of Limit (p. 63) If there is a number L such that $f(x)$ is as close to L as we please whenever x is sufficiently close to c (but $x \ne c$), then $\lim_{x \to c} f(x) = L$.

Definition of Limit (p. 64) If there is a number L such that for any $\epsilon > 0$, there exists a $\delta > 0$ such that if $|x - c| < \delta$ and $x \ne c$, then $|f(x) - L| < \epsilon$, then $\lim_{x \to c} f(x) = L$.

One-sided limits (p. 65) If $f(x)$ approaches L as x approaches c through values greater than c, then $\lim_{x \to c+} f(x) = L$. If $f(x)$ approaches L as x approaches c through values less than c, then $\lim_{x \to c-} f(x) = L$.

Limits at Infinity (p. 67) If $f(x)$ gets as close to L as we please when x gets sufficiently large, then $\lim_{x \to \infty} f(x) = L$. Similarly, if $f(x)$ approaches L as x gets more and more negative, then $\lim_{x \to -\infty} f(x) = L$.

Theorem: Properties of Limits (p. 65) Assuming all the limits on the right hand side exist:

1. If b is a constant, then $\lim_{x \to c} (bf(x)) = b \left(\lim_{x \to c} f(x) \right)$.
2. $\lim_{x \to c} (f(x) + g(x)) = \lim_{x \to c} f(x) + \lim_{x \to c} g(x)$.
3. $\lim_{x \to c} (f(x)g(x)) = \left(\lim_{x \to c} f(x) \right) \left(\lim_{x \to c} g(x) \right)$.
4. $\lim_{x \to c} \dfrac{f(x)}{g(x)} = \dfrac{\lim_{x \to c} f(x)}{\lim_{x \to c} g(x)}$, provided $\lim_{x \to c} g(x) \ne 0$.
5. For any constant k, $\lim_{x \to c} k = k$.
6. $\lim_{x \to c} x = c$.

Idea of Continuity (p. 45) A function is **continuous on an interval** if its graph has no breaks, jumps, or holes in that interval.

Definition of Continuity (p. 95) The function f is **continuous** at $x = c$ if f is defined at $x = c$ and

$$\lim_{x \to c} f(x) = f(c).$$

The function is **continuous on an interval** if it is continuous at every point in the interval.

Theorem: Continuity of Sums, Products, Quotients (p. 95) Suppose that f and g are continuous on an interval and that b is a constant. Then, on that same interval, the following functions are also continuous: $bf(x)$, $f(x)+g(x)$, $f(x)g(x)$. Further, $f(x)/g(x)$ is continuous provided $g(x) \neq 0$ on the interval.

Theorem: Continuity of Composite Functions (p. 95) Suppose f and g are continuous and $f(g(x))$ is defined on an interval. Then on that interval $f(g(x))$ is continuous.

Intermediate Value Theorem (p. 47) Suppose f is continuous on a closed interval $[a, b]$. If k is any number between $f(a)$ and $f(b)$, then there is at least one number c in $[a, b]$ such that $f(c) = k$.

The Extreme Value Theorem (p. 207) If f is continuous on the interval $[a, b]$, then f has a global maximum and a global minimum on that interval.

The Derivative

The slope of the secant line of $f(x)$ over an interval $[a, b]$ gives:

Average rate of change of f over $[a, b]$ $= \dfrac{f(b) - f(a)}{b - a}$ (p. 70).

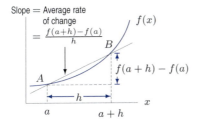

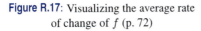

Figure R.17: Visualizing the average rate of change of f (p. 72)

The **derivative** of f at a is the slope of the line tangent to the graph of f at the point $(a, f(a))$:

$$f'(a) = \lim_{h \to 0} \frac{f(a + h) - f(a)}{h},$$

and gives the **instantaneous rate of change** of f at a (p. 71). The **second derivative** of f, denoted f'', is the derivative of f' (p. 89).

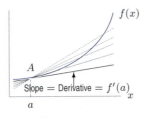

Figure R.18: Visualizing the instantaneous rate of change of f (p. 72)

The **units** of $f'(x)$ are: $\dfrac{\text{Units of } f(x)}{\text{Units of } x}$ (p. 86). If $f' > 0$ on an interval, then f is **increasing** over that interval (p. 80). If $f' < 0$ on an interval, then f is **decreasing** over that interval (p. 80). If $f'' > 0$ on an interval, then f is **concave up** over that interval (p. 89). If $f'' < 0$ on an interval, then f is **concave down** over that interval (p. 89).

The **tangent line** at $(a, f(a))$ is the graph of $y = f(a) + f'(a)(x - a)$ (p. 150). The **tangent line approximation** says that for values of x near a, $f(x) \approx f(a) + f'(a)(x - a)$. The expression $f(a) + f'(a)(x - a)$ is called the **local linearization** of f near $x = a$ (p. 150).

Derivatives of elementary functions

$\dfrac{d}{dx}(x^n) = nx^{n-1}$ (p. 108) $\dfrac{d}{dx}(e^x) = e^x$ (p. 115)

$\dfrac{d}{dx}(a^x) = (\ln a)a^x$ (p. 116) $\dfrac{d}{dx}(\ln x) = \dfrac{1}{x}$ (p. 134)

$\dfrac{d}{dx}(\sin x) = \cos x$ $\dfrac{d}{dx}(\cos x) = -\sin x$ (p. 129)

$\dfrac{d}{dx}(\arctan x) = \dfrac{1}{1 + x^2}$ (p. 135)

$\dfrac{d}{dx}(\arcsin x) = \dfrac{1}{\sqrt{1 - x^2}}$ (p. 135)

Derivatives of sums, differences, and constant multiples

$\dfrac{d}{dx}[f(x) \pm g(x)] = f'(x) \pm g'(x)$ (p. 107)

$\dfrac{d}{dx}[cf(x)] = cf'(x)$ (p. 106)

Product and quotient rules

$(fg)' = f'g + fg'$ (p. 119)

$\left(\dfrac{f}{g}\right)' = \dfrac{f'g - fg'}{g^2}$ (p. 120)

Chain rule

$\dfrac{d}{dx}f(g(x)) = f'(g(x)) \cdot g'(x)$ (p. 123)

Implicit differentiation (p. 138) If y is implicitly defined as a function of x by an equation, then, to find dy/dx, differentiate the equation (remembering to apply the chain rule).

Applications of the Derivative

A function f has a **local maximum** at p if $f(p)$ is greater than or equal to the values of f at points near p, and a **local minimum** at p if $f(p)$ is less than or equal to the values of f at points near p (p. 167). It has a **global maximum** at p if $f(p)$ is greater than or equal to the value of f at any point in the interval, and a **global minimum** at p if $f(p)$ is less than or equal to the value of f at any point in the interval (p. 181).

A **critical point** of a function $f(x)$ is a point p in the domain of f where $f'(p) = 0$ or $f'(p)$ is undefined (p. 167).

Theorem: Local maxima and minima which do not occur at endpoints of the domain occur at critical points (pp. 168, 207).

The First-Derivative Test for Local Maxima and Minima (p. 168):

- If f' changes from negative to positive at p, then f has a local minimum at p.

- If f' changes from positive to negative at p, then f has a local maximum at p.

The Second-Derivative Test for Local Maxima and Minima (p. 170):

- If $f'(p) = 0$ and $f''(p) > 0$ then f has a local minimum at p.

- If $f'(p) = 0$ and $f''(p) < 0$ then f has a local maximum at p.

- If $f'(p) = 0$ and $f''(p) = 0$ then the test tells us nothing.

To find the **global maximum and minimum** of a function on an interval we compare values of f at all critical points in the interval and at the endpoints of the interval (or $\lim_{x \to \pm\infty} f(x)$ if the interval is unbounded) (p. 181).

An **inflection point** of f is a point at which the graph of f changes concavity (p. 170); f'' is zero or undefined at an inflection point (p. 170).

L'Hopital's rule (p. 155) If f and g are continuous, $f(a) = g(a) = 0$, and $g'(a) \neq 0$, then

$$\lim_{x \to a} \frac{f(x)}{g(x)} = \frac{f'(a)}{g'(a)}.$$

Theorems about Derivatives

Theorem: Local Extrema and Critical Points (pp. 168, 207) Suppose f is defined on an interval and has a local maximum or minimum at the point $x = a$, which is not an endpoint of the interval. If f is differentiable at $x = a$, then $f'(a) = 0$.

The Mean Value Theorem (p. 208) If f is continuous on $[a, b]$ and differentiable on (a, b), then there exists a number c, with $a < c < b$, such that

$$f'(c) = \frac{f(b) - f(a)}{b - a}.$$

The Increasing Function Theorem (p. 47) Suppose that f is continuous on $[a, b]$ and differentiable on (a, b).

- If $f'(x) > 0$ on (a, b), then f is increasing on $[a, b]$.

- If $f'(x) \geq 0$ on (a, b), then f is nondecreasing on $[a, b]$.

The Constant Function Theorem (p. 209) Suppose that f is continuous on $[a, b]$ and differentiable on (a, b). If $f'(x) = 0$ on (a, b), then f is constant on $[a, b]$.

The Racetrack Principle (p. 209) Suppose that g and h are continuous on $[a, b]$ and differentiable on (a, b), and that $g'(x) \leq h'(x)$ for $a < x < b$.

- If $g(a) = h(a)$, then $g(x) \leq h(x)$ for $a \leq x \leq b$.

- If $g(b) = h(b)$, then $g(x) \geq h(x)$ for $a \leq x \leq b$.

Theorem: Differentiability and Local Linearity (p. 152) Suppose f is differentiable at $x = a$ and $E(x)$ is the error in the tangent line approximation, that is: $E(x) = f(x) - f(a) - f'(a)(x - a)$. Then $\lim_{x \to a} \dfrac{E(x)}{x - a} = 0$.

Theorem: A Differentiable Function is Continuous (p. 97) If $f(x)$ is differentiable at a point $x = a$, then $f(x)$ is continuous at $x = a$.

The Definite Integral

The **definite integral of f from a to b** (p. 230), denoted $\int_a^b f(x)\, dx$, is the limit of the left and the right sums as the width of the rectangles is shrunk to 0, where

$$\textbf{Left-hand sum} = \sum_{i=0}^{n-1} f(x_i)\Delta x \quad \text{(p. 229)}$$

$$= f(x_0)\Delta x + f(x_1)\Delta x + \cdots + f(x_{n-1})\Delta x$$

$$\textbf{Right-hand sum} = \sum_{i=1}^{n} f(x_i)\Delta x \quad \text{(p. 229)}$$

$$= f(x_1)\Delta x + f(x_2)\Delta x + \cdots + f(x_n)\Delta x$$

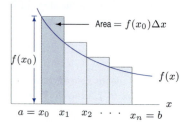

Figure R.19: Left-hand sum (p. 229)

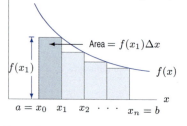

Figure R.20: Right-hand sum (p. 229)

If f is nonnegative, $\int_a^b f(x)\,dx$ represents the area under the curve between $x = a$ and $x = b$ (p. 232). If f has any sign, $\int_a^b f(x)\,dx$ is the sum of the areas above the x-axis, counted positively, and the areas below the x-axis, counted negatively (p. 232). If $F'(t)$ is the rate of change of some quantity $F(t)$, then $\int_a^b F'(t)\,dt$ is the **total change** in $F(t)$ between $t = a$ and $t = b$ (p. 237). The **average value** of f on the interval $[a, b]$ is given by $\dfrac{1}{b-a}\displaystyle\int_a^b f(x)\,dx$ (p. 239).

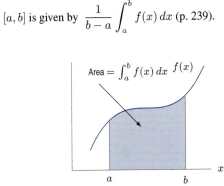

Figure R.21: The definite integral $\int_a^b f(x)\,dx$ (p. 232)

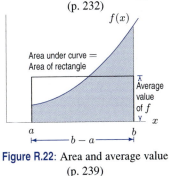

Figure R.22: Area and average value (p. 239)

The **units** of $\int_a^b f(x)\,dx$ are Units of $f(x) \times$ Units of x (p. 238).

The Fundamental Theorem of Calculus If f is continuous on $[a, b]$ and $f(x) = F'(x)$, then

$$\int_a^b f(x)\,dx = F(b) - F(a) \quad \text{(p. 244)}.$$

Properties of Definite Integrals (pp. 246, 248)
If a, b, and c are any numbers and f, g are continuous functions, then

$$\int_b^a f(x)\,dx = -\int_a^b f(x)\,dx$$

$$\int_a^b (f(x) \pm g(x))\,dx = \int_a^b f(x)\,dx \pm \int_a^b g(x)\,dx$$

$$\int_a^c f(x)\,dx + \int_c^b f(x)\,dx = \int_a^b f(x)\,dx$$

$$\int_a^b cf(x)\,dx = c\int_a^b f(x)\,dx$$

Antiderivatives

An **antiderivative** of a function $f(x)$ is a function $F(x)$ such that $F'(x) = f(x)$ (p. 262). There are infinitely many antiderivatives of f since $F(x) + C$ is an antiderivative of f for any constant C, provided $F'(x) = f(x)$ (p. 262). The **indefinite integral** of f is the family of antiderivatives $\int f(x)\,dx = F(x) + C$ (p. 268).

Construction Theorem (Second Fundamental Theorem of Calculus) If f is a continuous function on an interval and a is any number in that interval, then $F(x) = \int_a^x f(t)\,dt$ is an antiderivative of f (p. 279).

Properties of Indefinite Integrals (p. 270)

$$\int (f(x) \pm g(x))\,dx = \int f(x)\,dx \pm \int g(x)\,dx$$

$$\int cf(x)\,dx = c\int f(x)\,dx$$

Some antiderivatives:

$$\int k\,dx = kx + C \quad \text{(p. 268)}$$

$$\int x^n\,dx = \frac{x^{n+1}}{n+1} + C, \quad n \neq -1 \quad \text{(p. 269)}$$

$$\int \frac{1}{x}\,dx = \ln|x| + C \quad \text{(p. 270)}$$

$$\int e^x\,dx = e^x + C \quad \text{(p. 270)}$$

$$\int \cos x\,dx = \sin x + C \quad \text{(p. 270)}$$

$$\int \sin x\,dx = -\cos x + C \quad \text{(p. 270)}$$

$$\int \frac{dx}{1+x^2} = \arctan x + C \quad \text{(p. 314)}$$

$$\int \frac{dx}{\sqrt{1-x^2}} = \arcsin x + C \quad \text{(p. 312)}$$

Substitution (p. 290) For integrals of the form $\int f(g(x))g'(x)\,dx$, let $w = g(x)$. Choose w, find dw/dx and substitute for x and dx. Convert limits of integration for definite integrals.

Integration by Parts (p. 298) Used mainly for products; also for integrating $\ln x$, $\arctan x$, $\arcsin x$.

$$\int uv'\,dx = uv - \int u'v\,dx$$

Partial Fractions (p. 309) To integrate a rational function, $P(x)/Q(x)$, express as a sum of a polynomial and terms of the form $A/(x-c)^n$ and $(Ax+B)/q(x)$, where $q(x)$ is an unfactorable quadratic.

Trigonometric Substitutions (p. 312) To simplify $\sqrt{x^2 - a^2}$, try $x = a\sin\theta$ (p. 313). To simplify $a^2 + x^2$ or $\sqrt{a^2 + x^2}$, try $x = a\tan\theta$ (p. 314).

Numerical approximations for definite integrals (p. 317) Riemann sums (left, right, midpoint) (p. 317), trapezoid rule (p. 318), Simpson's rule (p. 324)

Approximation errors (pp. 319, 320)
f increasing: left underestimates, right overestimates (p. 319)
f decreasing: right underestimates, left overestimates (p. 319)
f concave up: midpoint underestimates, trapezoid overestimates (p. 320)
f concave down: trapezoid underestimates, midpoint overestimates (p. 320)

Evaluating improper integrals (p. 326)

- Infinite limit of integration: $\int_a^\infty f(x)\,dx = \lim_{b\to\infty} \int_a^b f(x)\,dx$ (p. 326)

- For $a < b$, if integrand is unbounded at $x = b$, then: $\int_a^b f(x)\,dx = \lim_{c\to b^-} \int_a^c f(x)\,dx$ (p. 330)

Testing improper integrals for convergence by comparison (p. 335):

- If $0 \le f(x) \le g(x)$ and $\int_a^\infty g(x)\,dx$ converges, then $\int_a^\infty f(x)\,dx$ converges

- If $0 \le g(x) \le f(x)$ and $\int_a^\infty g(x)\,dx$ diverges, then $\int_a^\infty f(x)\,dx$ diverges

Applications of Integration

Total quantities can be approximated by slicing them into small pieces and summing the pieces. The limit of this sum is a definite integral which gives the exact total quantity.

Applications to Geometry
To calculate the volume of a solid, slice the volume into pieces whose volumes you can estimate (pp. 346, 352). Use this method to calculate volumes of revolution (p. 353) and volumes of solids with known cross sectional area (p. 355). Curve $f(x)$ from $x = a$ to $x = b$ has

$$\text{Arc length} = \int_a^b \sqrt{1 + (f'(x))^2}\,dx \quad \text{(p. 356)}.$$

Mass and Center of Mass from Density, δ

$$\text{Total mass} = \int_a^b \delta(x)\,dx \quad \text{(p. 360)}$$

$$\text{Center of mass} = \frac{\int_a^b x\delta(x)\,dx}{\int_a^b \delta(x)\,dx} \quad \text{(p. 364)}$$

To find the center of mass of two- and three-dimensional objects, use the formula separately on each coordinate (p. 365).

Applications to Physics

$$\text{Work done} = \text{Force} \times \text{Distance (p. 368)}$$
$$\text{Pressure} = \text{Density} \times g \times \text{Depth (p. 373)}$$
$$\text{Force} = \text{Pressure} \times \text{Area (p. 373)}$$

Applications to Economics Present and future value of income stream, $P(t)$ (p. 378); consumer and producer surplus (p. 381).

$$\text{Present value} = \int_0^M P(t)e^{-rt}\,dt \quad \text{(p. 378)}$$

$$\text{Future value} = \int_0^M P(t)e^{r(M-t)}\,dt \quad \text{(p. 378)}$$

Series

A **sequence** $S_1, S_2, S_3, \ldots, S_n, \ldots$ has a **limit** L, written $\lim_{n\to\infty} S_n = L$, if we can make S_n as close to L as we please by choosing a sufficiently large n. The sequence **converges** if a limit exists, **diverges** if no limit exists (see p. 412). Limits of sequences satisfy the same properties as limits of functions stated in Theorem 2.1 (p. 65) and

$$\lim_{n\to\infty} x^n = 0 \quad \text{if } |x| < 1 \qquad \lim_{n\to\infty} 1/n = 0 \quad \text{(p. 412)}$$

Theorem: Convergence of an Increasing, Bounded Sequence (p. 412). If a sequence S_n, for $n = 1, 2, 3, \cdots$, is increasing and bounded above, then $\lim_{n\to\infty} S_n$ exists.

A **series** is an infinite sum $\sum a_n = a_1 + a_2 + \cdots$. The nth **partial sum** is $S_n = a_1 + a_2 + \cdots + a_n$ (p. 412). If $S = \lim_{n\to\infty} S_n$ exists, then the series $\sum a_n$ **converges**, and its sum is S (p. 413). If a series does not converge, we say that it **diverges** (p. 413). The sum of a **finite geometric series** is (p. 408):

$$a + ax + ax^2 + \cdots + ax^{n-1} = \frac{a(1-x^n)}{1-x}, \qquad x \ne 1.$$

The sum of an **infinite geometric series** is (p. 408):

$$a + ax + ax^2 + \cdots + ax^n + \cdots = \frac{a}{1-x}, \qquad |x| < 1.$$

The **harmonic series** $\sum 1/n$ diverges (p. 414), the **alternating harmonic series** $\sum (-1)^{n-1}(1/n)$ converges (p. 421).

Convergence Tests

Theorem: Convergence Properties of Series (p. 413)

1. If $\sum a_n$ and $\sum b_n$ converge and if k is a constant, then
   - $\sum (a_n + b_n)$ converges to $\sum a_n + \sum b_n$.
   - $\sum ka_n$ converges to $k \sum a_n$.

2. Changing a finite number of terms in a series does not change whether or not it converges, although it may change its sum if it does converge.

3. If $\sum a_n$ converges, then $\lim_{n\to\infty} a_n = 0$.

Theorem: Integral Test (p. 416) Suppose $c \geq 0$ and $f(x)$ is a decreasing positive function, defined for all $x \geq c$, with $a_n = f(n)$ for all n.

- If $\int_c^\infty f(x)\,dx$ converges, then $\sum a_n$ converges.
- If $\int_c^\infty f(x)\,dx$ diverges, then $\sum a_n$ diverges.

Theorem: Comparison Test (p. 417) Suppose $0 \leq a_n \leq b_n$ for all n.

- If $\sum b_n$ converges, then $\sum a_n$ converges.
- If $\sum a_n$ diverges, then $\sum b_n$ diverges.

Theorem: Convergence of Absolute Values Implies Convergence (p. 418): If $\sum |a_n|$ converges, then so does $\sum a_n$.

Theorem: The Ratio Test (p. 419) For a series $\sum a_n$, suppose the sequence of ratios $|a_{n+1}|/|a_n|$ has a limit:

$$\lim_{n\to\infty} \frac{|a_{n+1}|}{|a_n|} = L.$$

- If $L < 1$, then $\sum a_n$ converges.
- If $L > 1$, or if L is infinite, then $\sum a_n$ diverges.
- If $L = 1$, the test does not tell us anything about the convergence of $\sum a_n$.

Theorem: Alternating Series Test (p. 420) A series of the form

$$\sum_{n=1}^\infty (-1)^{n-1} a_n = a_1 - a_2 + a_3 - a_4 + \cdots + (-1)^{n-1} a_n + \cdots$$

converges if $0 < a_{n+1} < a_n$ for all n and $\lim_{n\to\infty} a_n = 0$.

Theorem: Error Bounds for Alternating Series (p. 421) Let $S_n = \sum_{i=1}^n (-1)^{i-1} a_i$ be the nth partial sum of an alternating series and let $S = \lim_{n\to\infty} S_n$. Suppose that $0 < a_{n+1} < a_n$ for all n and $\lim_{n\to\infty} a_n = 0$. Then $|S - S_n| < a_{n+1}$.

Power Series

$$P(x) = C_0 + C_1(x-a) + C_2(x-a)^2 + \cdots + C_n(x-a)^n + \cdots$$
$$= \sum_{n=0}^\infty C_n(x-a)^n \quad \text{(p. 423)}.$$

The **radius of convergence** is 0 if the series converges only for $x = a$, ∞ if it converges for all x, and the positive number R if it converges for $|x - a| < R$ and diverges for $|x - a| > R$ (p. 424). The **interval of convergence** is the interval between $a - R$ and $a + R$, including any endpoint where the series converges (p. 424).

Theorem: Method for Computing Radius of Convergence (p. 425) To calculate the radius of convergence, R,

for the power series $\sum_{n=0}^\infty C_n(x-a)^n$, use the ratio test with $a_n = C_n(x-a)^n$.

- If $\lim_{n\to\infty} |a_{n+1}|/|a_n|$ is infinite, then $R = 0$.
- If $\lim_{n\to\infty} |a_{n+1}|/|a_n| = 0$, then $R = \infty$.
- If $\lim_{n\to\infty} |a_{n+1}|/|a_n| = K|x - a|$, where K is finite and nonzero, then $R = 1/K$.

Approximations

The **Taylor polynomial of degree** n approximating $f(x)$ for x near a is:

$$P_n(x) = f(a) + f'(a)(x-a) + \frac{f''(a)}{2!}(x-a)^2 + \cdots\cdots$$
$$+ \frac{f^{(n)}(a)}{n!}(x-a)^n \quad \text{(p. 437)}$$

The **Taylor series** approximating $f(x)$ for x near a is:

$$f(x) = f(a) + f'(a)(x-a) + \frac{f''(a)}{2!}(x-a)^2 + \cdots$$
$$+ \frac{f^{(n)}(a)}{n!}(x-a)^n + \cdots \quad \text{(p. 442)}$$

Theorem: Bounding the Error in $P_n(x)$ (p. 454) Suppose f and all its derivatives are continuous. If $P_n(x)$ is the nth Taylor approximation to $f(x)$ about a, then

$$|E_n(x)| = |f(x) - P_n(x)| \leq \frac{M}{(n+1)!}|x-a|^{n+1},$$

where $M = \max f^{(n+1)}$ on the interval between a and x.

Taylor series for $\sin x$, $\cos x$, e^x (p. 442):

$$\sin x = x - \frac{x^3}{3!} + \frac{x^5}{5!} - \frac{x^7}{7!} + \frac{x^9}{9!} - \cdots$$
$$\cos x = 1 - \frac{x^2}{2!} + \frac{x^4}{4!} - \frac{x^6}{6!} + \frac{x^8}{8!} - \cdots$$
$$e^x = 1 + x + \frac{x^2}{2!} + \frac{x^3}{3!} + \frac{x^4}{4!} + \cdots$$

Taylor series for $\ln x$ about $x = 1$ converges for $0 < x \leq 2$ (p. 443):

$$(x-1) - \frac{(x-1)^2}{2} + \frac{(x-1)^3}{3} - \frac{(x-1)^4}{4} + \cdots$$

The Binomial Series for $(1+x)^p$ converges for $-1 < x < 1$ (p. 444):

$$1 + px + \frac{p(p-1)}{2!}x^2 + \frac{p(p-1)(p-2)}{3!}x^3 + \cdots$$

The Fourier series of $f(x)$ is (p. 461)

$$f(x) = a_0 + a_1\cos x + a_2\cos 2x + a_3\cos 3x + \cdots$$
$$+ b_1\sin x + b_2\sin 2x + b_3\sin 3x + \cdots,$$

where

$$a_0 = \frac{1}{2\pi} \int_{-\pi}^{\pi} f(x)\,dx,$$

$$a_k = \frac{1}{\pi} \int_{-\pi}^{\pi} f(x)\cos(kx)\,dx \quad \text{for } k \text{ a positive integer,}$$

$$b_k = \frac{1}{\pi} \int_{-\pi}^{\pi} f(x)\sin(kx)\,dx \quad \text{for } k \text{ a positive integer.}$$

Differential Equations

A **differential equation** for the function $y(x)$ is an equation involving x, y and the derivatives of y (p. 478). The **order** of a differential equation is the order of the highest-order derivative appearing in the equation (p. 480). A **solution** to a differential equation is any function y that satisfies the equation (p. 479). The **general solution** to a differential equation is the **family of functions** that satisfies the equation (p. 479). An **initial value problem** is a differential equation together with an **initial condition**; a solution to an initial value problem is called a **particular solution** (p. 479). An **equilibrium solution** to a differential equation is a particular solution where y is constant and $dy/dx = 0$ (p. 502).

First-order equations: methods of solution. A **slope field** corresponding to a differential equation is a plot in the xy-plane of small line segments with slope given by the differential equation (p. 482). **Euler's method** approximates the solution of an initial value problem with a string of small line segments (p. 488). Differential equations of the form $dy/dx = g(x)f(y)$ can be solved analytically by **separation of variables** (p. 492).

First-order equations: applications. The differential equation for **exponential growth and decay** is of the form

$$\frac{dP}{dt} = kP.$$

The **solution** is of the form $P = P_0 e^{kt}$, where P_0 is the initial value of P, and positive k represents growth while negative k represents decay (p. 497). Applications of growth and decay include **continuously compounded interest (p. 497)**, **population growth (p. 514)**, and **Newton's law of heating and cooling (p. 500)**. The **logistic equation** for population growth is of the form

$$\frac{dP}{dt} = kP\left(1 - \frac{P}{L}\right),$$

where L is the **carrying capacity** of the population (p. 518). The **solution to the logistic equation** is of the form $P = \dfrac{L}{1 + Ae^{-kt}}$, where $A = \dfrac{L - P_0}{P_0}$ (p. 519).

Systems of differential equations. Two interacting populations, w and r, can be modeled by two equations

$$\frac{dw}{dt} = aw - cwr \quad \text{and} \quad \frac{dr}{dt} = -br + kwr \quad \text{(p. 527).}$$

Solutions can be visualized as **trajectories** in the wr-**phase plane** (p. 527). An **equilibrium point** is one at which

$dw/dt = 0$ and $dr/dt = 0$ (p. 528). A **nullcline** is a curve along which $dw/dt = 0$ or $dr/dt = 0$ (p. 533).

Second-order equations. A **linear second-order differential equation with constant coefficients** is of the form

$$\frac{d^2 y}{dt^2} + b\frac{dy}{dt} + cy = 0 \quad \text{(p. 545).}$$

The solution (given on p. 549) depends on the sign of the quantity $b^2 - 4c$ (see pp. 545, 547). Solutions are calculated using the **characteristic equation (p. 545)** $r^2 + br + c = 0$.

- if $b^2 - 4c > 0$, then $y = C_1 e^{r_1 t} + C_2 e^{r_2 t}$, where r_1 and r_2 are the real solutions to the characteristic equation;

- if $b^2 - 4c = 0$, then $y = (C_1 t + C_2)e^{-bt/2}$;

- if $b^2 - 4c < 0$, then $y = C_1 e^{\alpha t}\cos\beta t + C_2 e^{\alpha t}\sin\beta t$, where $\alpha \pm i\beta$ are the complex solutions to the characteristic equation.

The **equation for oscillations of a mass on a spring** is of the form $\dfrac{d^2 s}{dt^2} = -\dfrac{k}{m}s$ (p. 538). The general solution is of the form $s(t) = C_1\cos\omega t + C_2\sin\omega t$, where $\omega = \sqrt{\dfrac{k}{m}}$, and C_1 and C_2 are arbitrary constants (p. 540).

Multivariable Functions

Points in 3-space are represented by a system of **Cartesian coordinates** (p. 562). The **distance** between (x, y, z) and (a, b, c) is $\sqrt{(x - a)^2 + (y - b)^2 + (z - c)^2}$ (p. 565).

Functions of two variables can be represented by **graphs** (p. 568), **contour diagrams** (p. 575), **cross-sections** (p. 355), and **tables** (p. 561).

Functions of three variables can be represented by the family of **level surfaces** $f(x, y, z) = c$ for various values of the constant c (p. 591).

A **linear function** $f(x, y)$ has equation

$$f(x, y) = z_0 + m(x - x_0) + n(y - y_0) \quad \text{(p. 586)}$$
$$= c + mx + ny, \quad \text{where } c = z_0 - mx_0 - ny_0.$$

Its **graph** is a plane with slope m in the x-direction, slope n in the y-direction, through (x_0, y_0, z_0) (p. 586). Its **table of values** has linear rows (of same slope) and linear columns (of same slope) (p. 587). Its **contour diagram** is equally spaced parallel straight lines (p. 587).

The **limit** of f at the point (a, b), written $\lim_{(x,y)\to(a,b)} f(x, y)$, is the number L, if one exists, such that $f(x, y)$ is as close to L as we please whenever the distance from the point (x, y) to the point (a, b) is sufficiently small, but not zero. (p. 597).

A function f is **continuous at the point** (a, b) if $\lim_{(x,y)\to(a,b)} f(x, y) = f(a, b)$. A function is **continuous on a region** R if it is continuous at each point of R (p. 597).

Vectors

A **vector** $\vec{v}$ has **magnitude** (denoted $\|\vec{v}\|$) and **direction**. Examples are **displacement vectors** (p. 606), **velocity and acceleration vectors** (pp. 614, 616). and **force** (p. 373). We can add vectors, and multiply a vector by a scalar (p. 607). Two non-zero vectors, $\vec{v}$ and $\vec{w}$, are **parallel** if one is a scalar multiple of the other (p. 608).

A **unit vector** has magnitude 1. The vectors $\vec{i}$, $\vec{j}$, and $\vec{k}$ are unit vectors in the directions of the coordinate axes. A unit vector in the direction of any nonzero vector $\vec{v}$ is $\vec{u} = \vec{v}/\|\vec{v}\|$ (p. 612). We **resolve** $\vec{v}$ **into components** by writing $\vec{v} = v_1\vec{i} + v_2\vec{j} + v_3\vec{k}$ (p. 609).

If $\vec{v} = v_1\vec{i} + v_2\vec{j} + v_3\vec{k}$ and $\vec{w} = w_1\vec{i} + w_2\vec{j} + w_3\vec{k}$ then

$$\|\vec{v}\| = \sqrt{v_1^2 + v_2^2 + v_3^2} \quad \text{(p. 610)}$$
$$\vec{v} + \vec{w} = (v_1 + w_1)\vec{i} + (v_2 + w_2)\vec{j} + (v_3 + w_3)\vec{k} \quad \text{(p. 611)},$$
$$\lambda\vec{v} = \lambda v_1\vec{i} + \lambda v_2\vec{j} + \lambda v_3\vec{k} \quad \text{(p. 611)}.$$

The **displacement vector** from $P_1 = (x_1, y_1, z_1)$ to $P_2 = (x_2, y_2, z_2)$ is

$$\overrightarrow{P_1 P_2} = (x_2 - x_1)\vec{i} + (y_2 - y_1)\vec{j} + (z_2 - z_1)\vec{k} \quad \text{(p. 610)}.$$

The **position vector** of $P = (x, y, z)$ is $\overrightarrow{OP}$ (p. 610). A **vector in n dimensions** is a string of numbers $\vec{v} = (v_1, v_2, \ldots, v_n)$ (p. 618).

Dot Product (Scalar Product) (p. 620).
Geometric definition: $\vec{v} \cdot \vec{w} = \|\vec{v}\|\|\vec{w}\|\cos\theta$ where θ is the angle between $\vec{v}$ and $\vec{w}$ and $0 \le \theta \le \pi$.
Algebraic definition: $\vec{v} \cdot \vec{w} = v_1 w_1 + v_2 w_2 + v_3 w_3$.

Two nonzero vectors $\vec{v}$ and $\vec{w}$ are **perpendicular** if and only if $\vec{v} \cdot \vec{w} = 0$ (p. 622). Magnitude and dot product are related by $\vec{v} \cdot \vec{v} = \|\vec{v}\|^2$ (p. 622). If $\vec{u} = (u_1, \ldots, u_n)$ and $\vec{v} = (v_1, \ldots, v_n)$ then the dot product of $\vec{u}$ and $\vec{v}$ is $\vec{u} \cdot \vec{v} = u_1 v_1 + \ldots + u_n v_n$ (p. 624).

The **equation of the plane** with **normal vector** $\vec{n} = a\vec{i} + b\vec{j} + c\vec{k}$ and containing the point $P_0 = (x_0, y_0, z_0)$ is $\vec{n} \cdot (\vec{r} - \vec{r}_0) = a(x - x_0) + b(y - y_0) + c(z - z_0) = 0$ or $ax + by + cz = d$, where $d = ax_0 + by_0 + cz_0$ (p. 623).

If $\vec{v}_{\text{parallel}}$ and $\vec{v}_{\text{perp}}$ are components of $\vec{v}$ which are parallel and perpendicular, respectively, to a unit vector $\vec{u}$, then $\vec{v}_{\text{parallel}} = (\vec{v} \cdot \vec{u})\vec{u}$ and $\vec{v}_{\text{perp}} = \vec{v} - \vec{v}_{\text{parallel}}$ (p. 624).

The **work**, W, done by a force $\vec{F}$ acting on an object through a displacement $\vec{d}$ is $W = \vec{F} \cdot \vec{d}$ (p. 368).

Cross Product (Vector Product) (p. 629, 630)
Geometric definition

$$\vec{v} \times \vec{w} = \begin{pmatrix} \text{Area of parallelogram} \\ \text{with edges } \vec{v} \text{ and } \vec{w} \end{pmatrix} \vec{n}$$
$$= (\|\vec{v}\|\|\vec{w}\|\sin\theta)\vec{n},$$

where $0 \le \theta \le \pi$ is the angle between $\vec{v}$ and $\vec{w}$ and $\vec{n}$ is the unit vector perpendicular to $\vec{v}$ and $\vec{w}$ pointing in the direction given by the **right-hand rule**.
Algebraic definition

$$\vec{v} \times \vec{w} = (v_2 w_3 - v_3 w_2)\vec{i} + (v_3 w_1 - v_1 w_3)\vec{j} + (v_1 w_2 - v_2 w_1)\vec{k}$$

$$\vec{v} = v_1\vec{i} + v_2\vec{j} + v_3\vec{k}, \vec{w} = w_1\vec{i} + w_2\vec{j} + w_3\vec{k}.$$

To find the **equation of a plane through three points** that do not lie on a line, determine two vectors in the plane and then find a normal vector using the cross product (p. 632). The **area of a parallelogram** with edges $\vec{v}$ and $\vec{w}$ is $\|\vec{v} \times \vec{w}\|$. The **volume of a parallelepiped** with edges $\vec{a}$, $\vec{b}$, $\vec{c}$ is $\left|(\vec{b} \times \vec{c}) \cdot \vec{a}\right|$ (p. 633).

Differentiation of Multivariable Functions

Partial derivatives of f (p. 641).

$$f_x(a, b) = \begin{array}{c} \text{Rate of change of } f \text{ with respect to } x \\ \text{at the point } (a, b) \end{array}$$
$$= \lim_{h \to 0} \frac{f(a + h, b) - f(a, b)}{h},$$
$$f_y(a, b) = \begin{array}{c} \text{Rate of change of } f \text{ with respect to } y \\ \text{at the point } (a, b) \end{array}$$
$$= \lim_{h \to 0} \frac{f(a, b + h) - f(a, b)}{h}.$$

On the graph of f, the partial derivatives $f_x(a, b)$ and $f_y(a, b)$ give the slope in the x and y directions, respectively (p. 642). The **tangent plane** to $z = f(x, y)$ at (a, b) is

$$z = f(a, b) + f_x(a, b)(x - a) + f_y(a, b)(y - b) \quad \text{(p. 652)}.$$

Partial derivatives can be estimated from a contour diagram or table of values using difference quotients (p. 642), and can be computed algebraically using the same rules of differentiation as for one-variable calculus (p. 647). Partial derivatives for functions of three or more variables are defined and computed in the same way (p 647).

The gradient vector grad f of f is grad $f(a, b) = f_x(a, b)\vec{i} + f_y(a, b)\vec{j}$ (2 variables) (p. 660) or grad $f(a, b, c) = f_x(a, b, c)\vec{i} + f_y(a, b, c)\vec{j} + f_z(a, b, c)\vec{k}$ (3 variables) (p. 667). The gradient vector at P: Points in the direction of increasing f; is perpendicular to the level curve or level surface of f through P; and has magnitude $\|$grad $f\|$ equal to the maximum rate of change of f at P (pp. 662, 667). The magnitude is large when the level curves or surfaces are close together and small when they are far apart.

The **directional derivative** of f at P in the direction of a unit vector $\vec{u}$ is (pp. 659, 661)

$$f_{\vec{u}}(P) = \begin{array}{c} \text{Rate of change} \\ \text{of } f \text{ in direction} \\ \text{of } \vec{u} \text{ at } P \end{array} = \operatorname{grad} f(P) \cdot \vec{u}$$

The **tangent plane approximation** to $f(x, y)$ for (x, y) near the point (a, b) is

$$f(x, y) \approx f(a, b) + f_x(a, b)(x - a) + f_y(a, b)(y - b).$$

The right-hand side is the **local linearization** (p. 653). The **differential of** $z = f(x, y)$ at (a, b) is the linear function of dx and dy

$$df = f_x(a, b)\, dx + f_y(a, b)\, dy \quad \text{(p. 655)}.$$

Local linearity with three or more variables follows the same pattern as for functions of two variables (p. 654).

The **tangent plane to a level surface** of a function of three-variables f at (a, b, c) is (p. 670)

$$f_x(a,b,c)(x-a)+f_y(a,b,c)(y-b)+f_z(a,b,c)(z-c) = 0.$$

The Chain Rule for the partial derivative of one variable with respect to another in a chain of composed functions (p. 675):

- Draw a diagram expressing the relationship between the variables, and label each link in the diagram with the derivative relating the variables at its ends.

- For each path between the two variables, multiply together the derivatives from each step along the path.

- Add the contributions from each path.

If $z = f(x, y)$, and $x = g(t)$, and $y = h(t)$, then

$$\frac{dz}{dt} = \frac{\partial z}{\partial x}\frac{dx}{dt} + \frac{\partial z}{\partial y}\frac{dy}{dt} \quad \text{(p. 673)}.$$

If $z = f(x, y)$, with $x = g(u, v)$ and $y = h(u, v)$, then

$$\frac{\partial z}{\partial u} = \frac{\partial z}{\partial x}\frac{\partial x}{\partial u} + \frac{\partial z}{\partial y}\frac{\partial y}{\partial u},$$

$$\frac{\partial z}{\partial v} = \frac{\partial z}{\partial x}\frac{\partial x}{\partial v} + \frac{\partial z}{\partial y}\frac{\partial y}{\partial v} \quad \text{(p. 675)}.$$

Second-order partial derivatives (p. 681)

$$\frac{\partial^2 z}{\partial x^2} = f_{xx} = (f_x)_x, \qquad \frac{\partial^2 z}{\partial x \partial y} = f_{yx} = (f_y)_x,$$

$$\frac{\partial^2 z}{\partial y \partial x} = f_{xy} = (f_x)_y, \qquad \frac{\partial^2 z}{\partial y^2} = f_{yy} = (f_y)_y.$$

Theorem: Equality of Mixed Partial Derivatives. If f_{xy} and f_{yx} are continuous at (a, b), an interior point of their domain, then $f_{xy}(a, b) = f_{yx}(a, b)$ (p. 682).

Taylor Polynomial of Degree 1 **Approximating** $f(x, y)$ for (x, y) near (a, b) **(p. 685)**

$$f(x, y) \approx L(x, y) = f(a, b) + f_x(a, b)(x-a) + f_y(a, b)(y-b).$$

Taylor Polynomial of Degree 2 **(p. 686)**

$$\begin{aligned}
f(x, y) &\approx Q(x, y) \\
&= f(a, b) + f_x(a, b)(x - a) + f_y(a, b)(y - b) \\
&\quad + \frac{f_{xx}(a, b)}{2}(x - a)^2 + f_{xy}(a, b)(x - a)(y - b) \\
&\quad + \frac{f_{yy}(a, b)}{2}(y - b)^2.
\end{aligned}$$

Definition of Differentiability (p. 689). A function $f(x, y)$ is **differentiable at the point** (a, b) if there is a linear function $L(x, y) = f(a, b) + m(x - a) + n(y - b)$ such that if the **error** $E(x, y)$ is defined by

$$f(x, y) = L(x, y) + E(x, y),$$

and if $h = x - a, k = y - b$, then the **relative error** $E(a + h, b + k)/\sqrt{h^2 + k^2}$ satisfies

$$\lim_{\substack{h \to 0 \\ k \to 0}} \frac{E(a + h, b + k)}{\sqrt{h^2 + k^2}} = 0.$$

Theorem: Continuity of Partial Derivatives Implies Differentiability (p. 692). If the partial derivatives, f_x and f_y, of a function f exist and are continuous on a small disk centered at the point (a, b), then f is differentiable at (a, b).

Optimization

A function f has a **local maximum** at the point P_0 if $f(P_0) \geq f(P)$ for all points P near P_0, and a **local minimum** at the point P_0 if $f(P_0) \leq f(P)$ for all points P near P_0 (p. 702). A **critical point** of a function f is a point where $\operatorname{grad} f$ is either $\vec{0}$ or undefined. If f has a local maximum or minimum at a point P_0, not on the boundary of its domain, then P_0 is a critical point (p. 702). A **quadratic function** $f(x, y) = ax^2 + bxy + cz^2$ generally has one critical point, which can be a local maximum, a local minimum, or a **saddle point** (p. 706).

Second derivative test for functions of two variables (p. 708). Suppose $\operatorname{grad} f(x_0, y_0) = \vec{0}$. Let $D = f_{xx}(x_0, y_0)f_{yy}(x_0, y_0) - (f_{xy}(x_0, y_0))^2$.

- If $D > 0$ and $f_{xx}(x_0, y_0) > 0$, then f has a local minimum at (x_0, y_0).

- If $D > 0$ and $f_{xx}(x_0, y_0) < 0$, then f has a local maximum at (x_0, y_0).

- If $D < 0$, then f has a saddle point at (x_0, y_0).

- If $D = 0$, anything can happen.

Unconstrained optimization

A function f defined on a region R has a **global maximum on** R at the point P_0 if $f(P_0) \geq f(P)$ for all points P in R, and a **global minimum on** R at the point P_0 if $f(P_0) \leq f(P)$ for all points P in R (p. 710). For an **unconstrained optimization problem**, find the critical points and investigate whether the critical points give global maxima or minima (p. 711).

A **closed** region is one which contains its boundary; a **bounded** region is one which does not stretch to infinity in any direction (p. 715).

Extreme Value Theorem for Multivariable Functions. If f is a continuous function on a closed and bounded region R, then f has a global maximum at some point (x_0, y_0) in R and a global minimum at some point (x_1, y_1) in R (p. 716).

Constrained optimization

Suppose P_0 is a point satisfying the constraint $g(x, y) = c$. A function f has a **local maximum** at P_0 **subject to the constraint** if $f(P_0) \geq f(P)$ for all points P near P_0 satisfying the constraint (p. 720). It has a **global maximum** at P_0 **subject to the constraint** if $f(P_0) \geq f(P)$ for all points P satisfying the constraint (p. 720). Local and global minima are defined similarly (p. 720). A local maximum or minimum of $f(x, y)$ subject to a constraint $g(x, y) = c$ occurs at a point where the constraint is tangent to a level curve of f, and thus where grad g is parallel to grad f (p. 720).

To optimize f subject to the constraint $g = c$ (p. 720), find the points satisfying the equations

$$\text{grad } f = \lambda \,\text{grad } g \quad \text{and} \quad g = c.$$

Then compare values of f at these points, at points on the constraint where grad $g = \vec{0}$, and at the endpoints of the constraint. The number λ is called the **Lagrange multiplier**.

To optimize f subject to the constraint $g \leq c$ (p. 722), find all points in the interior $g(x, y) < c$ where grad f is zero or undefined; then use Lagrange multipliers to find the local extrema of f on the boundary $g(x, y) = c$. Evaluate f at the points found and compare the values.

The value of λ is the rate of change of the optimum value of f as c increases (where $g(x, y) = c$) (p. 724). The **Lagrangian function** $\mathcal{L}(x, y, \lambda) = f(x, y) - \lambda(g(x, y) - c)$ can be used to convert a constrained optimization problem for f subject the constraint $g = c$ into an unconstrained problem for $\mathcal{L}$ (p. 724).

Multivariable Integration

The **definite integral** of f, a continuous function of two variables, over R, the rectangle $a \leq x \leq b$, $c \leq y \leq d$, is called a **double integral**, and is a limit of **Riemann sums**

$$\int_R f \, dA = \lim_{\Delta x, \Delta y \to 0} \sum_{i,j} f(u_{ij}, v_{ij}) \Delta x \Delta y \quad \text{(p. 737)}.$$

The Riemann sum is constructed by subdividing R into subrectangles of width Δx and height Δy, and choosing a point (u_{ij}, v_{ij}) in the ij-th rectangle.

A **triple integral** of f, a continuous function of three variables, over W, the box $a \leq x \leq b, c \leq y \leq d, p \leq z \leq q$ in 3-space, is defined in a similar way using three-variable Riemann sums (p. 752).

Interpretations

If $f(x, y)$ is positive, $\int_R f \, dA$ is the **volume** under graph of f above the region R (p. 738). If $f(x, y) = 1$ for all x and y, then the area of R is $\int_R 1 \, dA = \int_R dA$ (p. 739). If $f(x, y)$

is a **density**, then $\int_R f \, dA$ is the **total quantity** in the region R (p. 736). The **average value** of $f(x, y)$ on the region R is $\frac{1}{\text{Area of } R} \int_R f \, dA$ (p. 740). In probability, if $p(x, y)$ is a **joint density function** then $\int_a^b \int_c^d p(x, y) \, dy \, dx$ is the fraction of population with $a \leq x \leq b$ and $c \leq y \leq d$ (p. 770).

Iterated integrals

Double and triple integrals can be written as **iterated integrals**

$$\int_R f \, dA = \int_c^d \int_a^b f(x, y) \, dx dy \quad \text{(p. 744)}$$

$$\int_W f \, dV = \int_p^q \int_c^d \int_a^b f(x, y, z) \, dx \, dy \, dz \quad \text{(p. 752)}$$

Other orders of integration are possible. For iterated integrals over **non-rectangular regions** (p. 746), limits on outer integral are constants and limits on inner integrals involve only the variables in the integrals further out (pp. 747, 754).

Integrals in other coordinate systems

When computing double integrals in polar coordinates, put $dA = r \, dr \, d\theta$ or $dA = r \, d\theta \, dr$ (p. 756). Cylindrical coordinates are given by $x = r \cos\theta, y = r \sin\theta, z = z$, for $0 \leq r < \infty, 0 \leq \theta \leq 2\pi, -\infty < z < \infty$ (p. 760). Spherical coordinates are given by $x = \rho \sin\phi \cos\theta, y = \rho \sin\phi \sin\theta, z = \rho \cos\phi$, for $0 \leq \rho < \infty, 0 \leq \phi \leq \pi, 0 \leq \theta \leq 2\pi$ (p. 763). When computing triple integrals in cylindrical or spherical coordinates, put $dV = r \, dr \, d\theta \, dz$ for cylindrical coordinates (p. 762), $dV = \rho^2 \sin\phi \, d\rho \, d\phi \, d\theta$ for spherical coordinates (p. 764). Other orders of integration are also possible.

For a **change of variables** $x = x(s, t), y = y(s, t)$, the **Jacobian** is

$$\frac{\partial(x, y)}{\partial(s, t)} = \frac{\partial x}{\partial s} \cdot \frac{\partial y}{\partial t} - \frac{\partial x}{\partial t} \cdot \frac{\partial y}{\partial s} = \begin{vmatrix} \frac{\partial x}{\partial s} & \frac{\partial x}{\partial t} \\ \frac{\partial y}{\partial s} & \frac{\partial y}{\partial t} \end{vmatrix} \quad \text{(p. 775)}.$$

To convert an integral from x, y to s, t coordinates (p. 775): Substitute for x and y in terms of s and t, change the xy region R into an st region T, and change the area element by making the substitution $dxdy = \left| \frac{\partial(x, y)}{\partial(s, t)} \right| dsdt$. For triple integrals, there is a similar formula (p. 776).

Parameterizations and Vector Fields

Parameterized curves

The motion of a particle is described by **parametric equations** $x = f(t), y = g(t)$ (2-space) or $x = f(t), y = g(t), z = h(t)$ (3-space). The path of the particle is a **parameterized curve** (p. 784). Parameterizations are also written in **vector form** $\vec{r}(t) = f(t)\vec{i} + g(t)\vec{j} + h(t)\vec{k}$ (p. 785). For a **curve segment** we restrict the parameter to to a closed interval $a \leq t \leq b$ (p. 787). **Parametric equations for the graph** of $y = f(x)$ are $x = t, y = f(t)$.

Parametric equations for a line through (x_0, y_0) in the direction of $\vec{v} = a\vec{i} + b\vec{j}$ are $x = x_0 + at, y = y_0 + bt$.

In 3-space, the line through (x_0, y_0, z_0) in the direction of $\vec{v} = a\vec{i} + b\vec{j} + c\vec{k}$ is $x = x_0 + at$, $y = y_0 + bt$, $z = z_0 + ct$ (p. 785). In vector form, the equation for a line is $\vec{r}(t) = \vec{r}_0 + t\vec{v}$, where $\vec{r}_0 = x_0\vec{i} + y_0\vec{j} + z_0\vec{k}$ (p. 787).

Parametric equations for a circle of radius R in the plane, centered at the origin are $x = R\cos t$, $y = R\sin t$ (counterclockwise), $x = R\cos t$, $y = -R\sin t$ (clockwise).

To find the **intersection points** of a curve $\vec{r}(t) = f(t)\vec{i} + g(t)\vec{j} + h(t)\vec{k}$ with a surface $F(x, y, z) = c$, solve $F(f(t), g(t), h(t)) = c$ for t (p. 787). To find the intersection points of two curves $\vec{r}_1(t)$ and $\vec{r}_2(t)$, solve $\vec{r}_1(t_1) = \vec{r}_2(t_2)$ for t_1 and t_2 (p. 787).

The **length of a curve segment** C given parametrically for $a \leq t \leq b$ with velocity vector $\vec{v}$ is $\int_a^b \|\vec{v}\| dt$ if $\vec{v} \neq \vec{0}$ for $a < t < b$ (p. 796).

The **velocity** and **acceleration** of a moving object with position vector $\vec{r}(t)$ at time t are

$$\vec{v}(t) = \lim_{\Delta t \to 0} \frac{\Delta \vec{r}}{\Delta t} \quad \text{(p. 792)}$$

$$\vec{a}(t) = \lim_{\Delta t \to 0} \frac{\Delta \vec{v}}{\Delta t} \quad \text{(p. 794)}$$

We write $\vec{v} = \dfrac{d\vec{r}}{dt} = \vec{r}'(t)$ and $\vec{a} = \dfrac{d\vec{v}}{dt} = \dfrac{d^2\vec{r}}{dt^2} = \vec{r}''(t)$.

The **components of the velocity and acceleration vectors** are

$$\vec{v}(t) = \frac{dx}{dt}\vec{i} + \frac{dy}{dt}\vec{j} + \frac{dz}{dt}\vec{k} \quad \text{(p. 792)}$$

$$\vec{a}(t) = \frac{d^2x}{dt^2}\vec{i} + \frac{d^2y}{dt^2}\vec{j} + \frac{d^2z}{dt^2}\vec{k} \quad \text{(p. 794)}$$

The **speed** is $\|\vec{v}\| = \sqrt{(dx/dt)^2 + (dy/dt)^2 + (dz/dt)^2}$ (p. 796). Analogous formulas for velocity, speed, and acceleration hold in 2-space.

Uniform Circular Motion (p. 795) For a particle $\vec{r}(t) = R\cos(\omega t)\vec{i} + R\sin(\omega t)\vec{j}$: motion is in a circle of radius R with period $2\pi/\omega$; velocity, $\vec{v}$, is tangent to the circle and speed is constant $\|\vec{v}\| = \omega R$; acceleration, $\vec{a}$, points toward the center of the circle with $\|\vec{a}\| = \|\vec{v}\|^2/R$.

Motion in a Straight Line (p. 796) For a particle $\vec{r}(t) = \vec{r}_0 + f(t)\vec{v}_0$: Motion is along a straight line through the point with position vector $\vec{r}_0$ parallel to $\vec{v}_0$; velocity, $\vec{v}$, and acceleration, $\vec{a}$, are parallel to the line.

Vector fields

A **vector field** in 2-space is a function $\vec{F}(x, y)$ whose value at a point (x, y) is a 2-dimensional vector (p. 799). Similarly, a vector field in 3-space is a function $\vec{F}(x, y, z)$ whose values are 3-dimensional vectors (p. 799). Examples are the **gradient** of a differentiable function f, the **velocity field** of a fluid flow, and **force fields** (p. 799). A **flow line** of a vector field $\vec{v} = \vec{F}(\vec{r})$ is a path $\vec{r}(t)$ whose velocity vector equals $\vec{v}$, thus $\vec{r}'(t) = \vec{v} = \vec{F}(\vec{r}(t))$ (p. 805). The **flow** of a vector field is the family of all of its flow line (p. 805). Flow lines can be approximated numerically using Euler's method (p. 807).

Parameterized surfaces

We **parameterize a surface** with two parameters, $x = f_1(s, t)$, $y = f_2(s, t)$, $z = f_3(s, t)$ (p. 810). We also use

the vector form $\vec{r}(s, t) = f_1(s, t)\vec{i} + f_2(s, t)\vec{j} + f_3(s, t)\vec{k}$ (p. 810). **Parametric equations for the graph of** $z = f(x, y)$ are $x = s$, $y = t$, and $z = f(s, t)$ (p. 811). **Parametric equation for a plane** through the point with position vector $\vec{r}_0$ and containing the two nonparallel vectors $\vec{v}_1$ and $\vec{v}_2$ is $\vec{r}(s, t) = \vec{r}_0 + s\vec{v}_1 + t\vec{v}_2$ (p. 811). **Parametric equation for a sphere** of radius R centered at the origin is $\vec{r}(\theta, \phi) = R\sin\phi\cos\theta\,\vec{i} + R\sin\phi\sin\theta\,\vec{j} + \cos\phi\,\vec{k}$, $0 \leq \theta \leq 2\pi$, $0 \leq \phi \leq \pi$ (p. 812). **Parametric equation for a cylinder** of radius R along the z-axis is $\vec{r}(\theta, z) = R\cos\theta\,\vec{i} + R\sin\theta\,\vec{j} + z\vec{k}$, $0 \leq \theta \leq 2\pi$, $-\infty < z < \infty$ (p. 809). A **parameter curve** is the curve obtained by holding one of the parameters constant and letting the other vary (p. 815).

Line Integrals

The **line integral** of a vector field $\vec{F}$ along an **oriented curve** C (p. 826) is

$$\int_C \vec{F} \cdot d\vec{r} = \lim_{\|\Delta \vec{r}_i\| \to 0} \sum_{i=0}^{n-1} \vec{F}(\vec{r}_i) \cdot \Delta \vec{r}_i,$$

where the direction of $\Delta \vec{r}_i$ is the direction of the orientation (p. 827).

The line integral measures the extent to which C is going with $\vec{F}$ or against it (p. 827). For oriented curves C, C_1, and C_2, $\int_{-C} \vec{F} \cdot d\vec{r} = -\int_C \vec{F} \cdot d\vec{r}$, where $-C$ is the curve C parameterized in the opposite direction, and $\int_{C_1 + C_2} \vec{F} \cdot d\vec{r} = \int_{C_1} \vec{F} \cdot d\vec{r} + \int_{C_2} \vec{F} \cdot d\vec{r}$, where $C_1 + C_2$ is the curve obtained by joining the endpoint of C_1 to the starting point of C_2 (p. 831).

The **work done by a force** $\vec{F}$ along a curve C is $\int_C \vec{F} \cdot d\vec{r}$ (p. 828). The **circulation** of $\vec{F}$ around an oriented closed curve is $\int_C \vec{F} \cdot d\vec{r}$ (p. 830).

Given a parameterization of C, $\vec{r}(t)$, for $a \leq t \leq b$, the line integral can be calculated as

$$\int_C \vec{F} \cdot d\vec{r} = \int_a^b \vec{F}(\vec{r}(t)) \cdot \vec{r}'(t)\, dt \quad \text{(p. 835).}$$

Fundamental Theorem for Line Integrals (p. 840):
Suppose C is a piecewise smooth oriented path with starting point P and endpoint Q. If f is a function whose gradient is continuous on the path C, then

$$\int_C \text{grad}\, f \cdot d\vec{r} = f(Q) - f(P).$$

Path-independent fields and gradient fields
A vector field $\vec{F}$ is said to be **path-independent**, or **conservative**, if for any two points P and Q, the line integral $\int_C \vec{F} \cdot d\vec{r}$ has the same value along any piecewise smooth path C from P to Q lying in the domain of $\vec{F}$ (p. 841). A **gradient field** is a vector field of the form $\vec{F} = \text{grad}\, f$ for some scalar function f, and f is called a **potential function** for the

vector field $\vec{F}$ (p. 843). A vector field $\vec{F}$ is path-independent if and only if $\vec{F}$ is a gradient vector field (p. 842). A vector field $\vec{F}$ is path-independent if and only if $\int_C \vec{F} \cdot d\vec{r} = 0$ for every closed curve C (p. 849). If $\vec{F}$ is a gradient field, then $\frac{\partial F_2}{\partial x} - \frac{\partial F_1}{\partial y} = 0$ (p. 850). The quantity $\frac{\partial F_2}{\partial x} - \frac{\partial F_1}{\partial y}$ is called the 2-dimensional or scalar **curl** of $\vec{F}$.

Green's Theorem (p. 852):
Suppose C is a piecewise smooth simple closed curve that is the boundary of an open region R in the plane and oriented so that the region is on the left as we move around the curve. Suppose $\vec{F} = F_1\vec{i} + F_2\vec{j}$ is a smooth vector field defined at every point of the region R and boundary C. Then

$$\int_C \vec{F} \cdot d\vec{r} = \int_R \left(\frac{\partial F_2}{\partial x} - \frac{\partial F_1}{\partial y} \right) dx\, dy.$$

Curl test for vector fields in 2-space: If $\frac{\partial F_2}{\partial x} - \frac{\partial F_1}{\partial y} = 0$ and the domain of $\vec{F}$ has no holes, then $\vec{F}$ is path-independent, and hence a gradient field (p. 852). The condition that the domain have no holes is important. It is not always true that if the scalar curl of $\vec{F}$ is zero then $\vec{F}$ is a gradient field (p. 852).

Surface Integrals

A surface is **oriented** if a unit normal vector $\vec{n}$ has been chosen at every point on it in a continuous way (p. 864). For a closed surface, we usually choose the outward orientation (p. 864). The **area vector** of a flat, oriented surface is a vector $\vec{A}$ whose magnitude is the area of the surface, and whose direction is the direction of the orientation vector $\vec{n}$ (p. 865). If $\vec{v}$ is the velocity vector of a constant fluid flow and $\vec{A}$ is the area vector of a flat surface, then the total flow through the surface in units of volume per unit time is called the **flux** of $\vec{v}$ through the surface and is given by $\vec{v} \cdot \vec{A}$ (p. 865).

The **surface integral** or **flux integral** of the vector field $\vec{F}$ through the oriented surface S is

$$\int_S \vec{F} \cdot d\vec{A} = \lim_{\|\Delta\vec{A}\| \to 0} \sum \vec{F} \cdot \Delta\vec{A},$$

where the direction of $\Delta\vec{A}$ is the direction of the orientation (p. 866). If $\vec{v}$ is a variable vector field and then $\int_S \vec{v} \cdot d\vec{A}$ is the flux through the surface S (p. 866).

Simple flux integrals can be calculated by putting $d\vec{A} = \vec{n}\, dA$ and using geometry or converting to a double integral (p. 868).

The **flux through a graph** of $z = f(x, y)$ above a region R in the xy-plane, oriented upward, is

$$\int_R \vec{F}(x, y, f(x, y)) \cdot \left(-f_x\vec{i} - f_y\vec{j} + \vec{k} \right) dx\, dy \quad \text{(p. 874).}$$

The **flux through a cylindrical surface** S of radius R and oriented away from the z-axis is

$$\int_T \vec{F}(R, \theta, z) \cdot \left(\cos\theta\vec{i} + \sin\theta\vec{j} \right) R\, dz\, d\theta \quad \text{(p. 875),}$$

where T is the θz-region corresponding to S.

The **flux through a spherical surface** S of radius R and oriented away from the origin is

$$\int_T \vec{F}(R, \theta, \phi) \cdot \left(\sin\phi\cos\theta\vec{i} + \sin\phi\sin\theta\vec{j} + \cos\phi\vec{k} \right)$$
$$R^2 \sin\phi\, d\phi\, d\theta, \quad \text{(p. 876)}$$

where T is the $\theta\phi$-region corresponding to S.

The **flux through a parameterized surface** S, parameterized by $\vec{r} = \vec{r}(s, t)$, where (s, t) varies in a parameter region R, is

$$\int_R \vec{F}(\vec{r}(s, t)) \cdot \left(\frac{\partial\vec{r}}{\partial s} \times \frac{\partial\vec{r}}{\partial t} \right) ds\, dt \quad \text{(p. 879).}$$

We choose the parameterization so that $\partial\vec{r}/\partial s \times \partial\vec{r}/\partial t$ is never zero and points in the direction of $\vec{n}$ everywhere.

The **area of a parameterized surface** S, parameterized by $\vec{r} = \vec{r}(s, t)$, where (s, t) varies in a parameter region R, is

$$\int_S dA = \int_R \left\| \frac{\partial\vec{r}}{\partial s} \times \frac{\partial\vec{r}}{\partial t} \right\| ds\, dt \quad \text{(p. 880).}$$

Divergence and Curl

Divergence

> **Definition of Divergence (p. 888).**
> **Geometric definition**: The **divergence** of $\vec{F}$ is
>
> $$\operatorname{div}\vec{F}(x, y, z) = \lim_{\text{Volume} \to 0} \frac{\int_S \vec{F} \cdot d\vec{A}}{\text{Volume of } S}.$$
>
> Here S is a sphere centered at (x, y, z), oriented outwards, that contracts down to (x, y, z) in the limit.
>
> **Cartesian coordinate definition**: If $\vec{F} = F_1\vec{i} + F_2\vec{j} + F_3\vec{k}$, then
>
> $$\operatorname{div}\vec{F} = \frac{\partial F_1}{\partial x} + \frac{\partial F_2}{\partial y} + \frac{\partial F_3}{\partial z}.$$

The divergence can be thought of as the outflow per unit volume of the vector field. A vector field $\vec{F}$ is said to be **divergence free** or **solenoidal** if $\operatorname{div}\vec{F} = 0$ everywhere that $\vec{F}$ is defined. Magnetic fields are divergence free (p. 891).

The Divergence Theorem (p. 897). If W is a solid region whose boundary S is a piecewise smooth surface, and if $\vec{F}$ is a smooth vector field which is defined everywhere in W and on S, then

$$\int_S \vec{F} \cdot d\vec{A} = \int_W \operatorname{div}\vec{F}\, dV,$$

where S is given the outward orientation. In words, the Divergence Theorem says that the total flux out of a closed surface is the integral of the flux density over the volume it encloses.

Curl

The **circulation density** of a smooth vector field $\vec{F}$ at (x, y, z) around the direction of the unit vector $\vec{n}$ is defined to be

$$\operatorname{circ}_{\vec{n}} \vec{F}(x, y, z) = \lim_{\text{Area}\to 0} \frac{\text{Circulation around } C}{\text{Area inside } C}$$

$$= \lim_{\text{Area}\to 0} \frac{\displaystyle\int_C \vec{F} \cdot d\vec{r}}{\text{Area inside } C} \quad \text{(p. 902).}$$

Circulation density is calculated using the **right-hand rule** (p. 902).

Definition of curl (p. 903).
Geometric definition The curl of $\vec{F}$, written curl $\vec{F}$, is the vector field with the following properties

- The direction of curl $\vec{F}(x, y, z)$ is the direction $\vec{n}$ for which $\operatorname{circ}_{\vec{n}}(x, y, z)$ is greatest.

- The magnitude of curl $\vec{F}(x, y, z)$ is the circulation density of $\vec{F}$ around that direction.

Cartesian coordinate definition If $\vec{F} = F_1\vec{i} + F_2\vec{j} + F_3\vec{k}$, then

$$\operatorname{curl} \vec{F} = \left(\frac{\partial F_3}{\partial y} - \frac{\partial F_2}{\partial z}\right)\vec{i} + \left(\frac{\partial F_1}{\partial z} - \frac{\partial F_3}{\partial x}\right)\vec{j}$$

$$+ \left(\frac{\partial F_2}{\partial x} - \frac{\partial F_1}{\partial y}\right)\vec{k}.$$

Curl and circulation density are related by $\operatorname{circ}_{\vec{n}} \vec{F} = \operatorname{curl} \vec{F} \cdot \vec{n}$ (p. 905). A vector field is said to be **curl free** or **irrotational** if curl $\vec{F} = \vec{0}$ everywhere that $\vec{F}$ is defined (p. 905).

Given an oriented surface S with a boundary curve C we use the right-hand rule to determine the orientation of C (p. 909).

Stokes' Theorem (p. 910). If S is a smooth oriented surface with piecewise smooth, oriented boundary C, and if $\vec{F}$ is a smooth vector field which is defined on S and C, then

$$\int_C \vec{F} \cdot d\vec{r} = \int_S \operatorname{curl} \vec{F} \cdot d\vec{A}.$$

Stokes' Theorem says that the total circulation around C is the integral over S of the circulation density. A **curl field** is a vector field $\vec{F}$ that can be written as $\vec{F} = \operatorname{curl} \vec{G}$ for some vector field $\vec{G}$, called a **vector potential** for $\vec{F}$ (p. 912).

Relation between divergence, gradient, and curl
The curl and gradient are related by $\operatorname{curl} \operatorname{grad} f = 0$ (p. 915). Divergence and curl are related by $\operatorname{div} \operatorname{curl} \vec{F} = 0$ (p. 917).

The curl test for vector fields in 3-space (p. 916) Suppose that curl $\vec{F} = \vec{0}$, and that the domain of $\vec{F}$ has the property that every closed curve in it can be contracted to a point in a smooth way, staying at all times within the domain. Then $\vec{F}$ is path-independent, so $\vec{F}$ is a gradient field and has a potential function.

The divergence test for vector fields in 3-space (p. 917) Suppose that $\operatorname{div} \vec{F} = 0$, and that the domain of $\vec{F}$ has the property that every closed surface in it is the boundary of a solid region completely contained in the domain. Then $\vec{F}$ is a curl field.

ANSWERS TO ODD NUMBERED PROBLEMS

Section 1.1

5 Slope: $-12/7$
  Vertical intercept: $2/7$

7 Slope: 2
  Vertical intercept: $-2/3$

9 $y = (1/2)x + 2$

11 $y = -\frac{1}{5}x + \frac{7}{5}$

13 Parallel: $y = m(x - a) + b$
  Perpendicular:
  $y = (-1/m)(x - a) + b$

15 (a) (V)
  (b) (VI)
  (c) (I)
  (d) (IV)
  (e) (III)
  (f) (II)

17 Domain: $-2 \le x \le 2$
  Range: $-2 \le y \le 2$

19 Domain: all x
  Range: $y \ge 2$

21 $t \ge 4$ or $t \le -4$
  $t = \pm 5$

23 $S = kh^2$

25 $N = k/l^2$

31 (a) \$0.025/cubic foot
  (b) $c = 65 + 0.025w$
    c = cost of water
    w = cubic feet of water
  (c) 2600 cubic feet

33 (a) $\Delta w/\Delta h$ constant
  (b) $w = 5h - 174$; 5 lbs/in
  (c) $h = 0.2w + 34.8$; 0.2 in/lb

35 (a) (i) $q = 320 - (2/5)p$
    (ii) $p = 800 - (5/2)q$

37 (a) $R = k(350 - H)$
    $(k \ge 0)$

Section 1.2

1 Concave up

3 Neither

5 5; 7%

7 3.2; 3% (continuous)

9 (a) $P = 1000 + 50t$
  (b) $P = 1000(1.05)^t$

11 (a) x-interval: D to E, H to I
  (b) x-interval: A to B, E to F
  (c) x-interval: C to D, G to H
  (d) x-interval: B to C, F to G

15 (a) $P = 6(1.013)^t$
  (b) 7.87 billion
  (c) 53.7 years

17 (a) \$3486.78
  (b) Approx 11 years (or 21 years from initial investment)

19 $y = 4(2^{-x})$

21 $y = 4(1 - 2^{-x})$

23 (a) 1.05
  (b) 5%

25 $P = 2(0.61)^t$; decay

27 $P = 7(0.0432)^t$; decay

29 $d = 670(1.096)^{h/1000}$

31 25.5%

33 (a) $k(t)$
  (b) $h(t)$

(c) $g(t)$

35 (a) $Q = Q_0 \left(\frac{1}{2}\right)^{(t/1620)}$
  (b) 80.7%

37 2.3 years

39 (a) $P = 2.5t + 50$
  (b) $P = 50(1.035)^t$
  (d) Exponential

Section 1.3

1 (a) $h^2 + 6h + 11$
  (b) 11
  (c) $h^2 + 6h$

3 (a) $f(n) + g(n) = 3n^2 + n - 1$
  (b) $f(n)g(n) = 3n^3 + 3n^2 - 2n - 2$
  (c) $n \ne -1$
  (d) $f(g(n)) = 3n^2 + 6n + 1$
  (e) $g(f(n)) = 3n^2 - 1$

5 $2z + 1$

7 $2zh - h^2$

11 Length of column of mercury when temperature is $75°$F

13 Not invertible

15 (a) -1

17 $y = (x - 2)^3 - 1$.

19 Neither

21 Not invertible

23 Invertible

25 $f(g(1)) \approx 0.4$

27 $f(f(1)) \approx -0.9$

31 $f(x) = x^3$
  $g(x) = x + 1$

33 $n = 2y^2 - 5y + 3$

35 (a) $k = f(p) = (1/2.2)p$
  (b) $p = 2.2k$; Weight in pounds given mass in kilograms

Section 1.4

1 $A: e^x$
  $B: x^2$
  $C: x^{1/2}$
  $D: \ln x$

3 $(\log 2)/(\log 17) \approx 0.24$

5 $(\log(2/5))/(\log 1.04) \approx -23.4$

7 $(\log(4/7))/(\log(5/3)) \approx -1.1$

9 -3.26

11 6.212

13 0.26

15 1

17 $(\log a)/(\log b)$

19 $(\log Q - \log Q_0)/(n \log a)$

21 $\ln(a/b)$

23 $1/2$

25 $5A^2$

27 $-1 + \ln A + \ln B$

29 $P = 15e^{0.41t}$

31 $P = 174e^{-0.1054t}$

33 $p^{-1}(t) \approx 58.708 \log t$

35 $f^{-1}(t) = e^{t-1}$

37 (a) 10 mg
  (b) 18%
  (c) 3.04 mg
  (d) 11.60 hours

39 About 14.21 years

41 16 kg

43 $B = 78e^{0.555t}$
  55.5%

45 1990

47 (a) 81%
  (b) 32.9 hours

49 96.34 years

Section 1.5

1 Negative
  0
  Undefined

3 Positive
  Positive
  Positive

5 Positive
  Positive
  Positive

7 Positive
  Negative
  Negative

9 Negative
  Positive
  Negative

11 0.588

13 (a) 1
  (b) $2\pi/3$

15 $f(x) = 2 \sin(x/4)$

17 $f(x) = 2 - \sin x$

19 $f(x) = 2 \sin(x/4) + 2$

21 $f(x) = (\sin x) + 2$

23 $f(x) = \sin(2(\pi/5)x)$

25 $f(x) = 2 \cos(5x)$

27 $(\sin^{-1}(2/5))/3 \approx 0.1372$

29 20.94 to 52.36 rad/sec

31 If $f(x) = \sin x$ and
    $g(x) = x^2$ then
    $\sin x^2 = f(g(x))$
    $\sin^2 x = g(f(x))$
    $\sin(\sin x) = f(f(x))$

33 (a) $(2.3, -4.4)$
  (b) Once around circle and back

35 (a) Maximum displacement from equilibrium
  (b) ω

37 US: 156 volts max, 60 cycles/sec
  Eur: 339 volts max, 50 cycles/sec

39 (b) $P = 800 - 100 \cos(\pi t/6)$

41 $hw + h^2/\tan \theta$

Section 1.6

1 $10 \cdot 2^x$

3 As $x \to \infty$, $y \to \infty$.
  As $x \to -\infty$, $y \to -\infty$.

5 (I) Degree ≥ 3, negative
  (II) Degree ≥ 4, positive
  (III) Degree ≥ 4, negative
  (IV) Degree ≥ 5, negative
  (V) Degree ≥ 5, positive

7 $y = \frac{1}{5}(x+2)(x-1)(x-5)$

9 (a) $-\infty, -\infty$
   (b) 3/2, 3/2
   (c) $0, +\infty$

11 $f(x) = kx(x+3)(x-4)$
    $(k < 0)$

13 $f(x) =$
    $k(x+2)(x-2)^2(x-5)$
    $(k < 0)$

15 (a) 1.3 m^2
   (b) 86.8 kg
   (c) $h = 112.6s^{4/3}$

17 (a) (i) $V = 3\pi r^2$
         (ii) $V = \pi r^2 h$

19 $5\pi x^2$

21 44.25 ft and 708 ft

23 (a) 0
   (b) $t = 2v_0/g$
   (c) $t = v_0/g$
   (d) $(v_0)^2/(2g)$

25 (a) (i) $1 = a + b + c$
         (ii) $b = -2a$ and $c = 1 + a$
         (iii) $c = 6$
   (b) $y = 5x^2 - 10x + 6$

27 g is an exponential;
    f is a cubic;
    k is a quadratic

29 1, 2, 3, 4, or 5 roots

31 $g(x) = 2x^2$
    $h(x) = x^2 + k$
    $(k > 0)$

33 $-10^5 \leq x \leq 10^5, \ -10^{15} \leq y \leq 10^{15}$

35 (a) $a(v) = \frac{1}{m}(F_E - kv^2)$
   $(k > 0)$

Section 1.7

1 Yes

3 Yes

5 Yes

7 No

9 No

11 $f(t) = \begin{cases} 6, & 0 < t \leq 7 \\ 12, & 7 < t \end{cases}$
    Not continuous on any interval around 7

13 $k = 6$

15 20

19 $Q = \begin{cases} 1.2t & 0 \leq t \leq 0.5 \\ 0.6e^{0.001}e^{-.002t} & 0.5 < t \end{cases}$

Chapter 1 Review

1 (a) $[0, 7]$
  (b) $[-2, 5]$
  (c) 5
  (d) $(1, 7)$
  (e) Concave up
  (f) 1
  (g) No

5 Amplitude: 2
   Period: $2\pi/5$

7 $y = e^{0.4621x}$

9 $y = -k(x^2 + 5x)$
   $(k > 0)$

11 $z = 1 - \cos\theta$

13 $x = k(y^2 - 4y)$
    $(k > 0)$

15 $y = -(x+5)(x+1)(x-3)^2$

17 Simplest is $y = 1 - e^{-x}$

19 Not continuous

21 Not continuous

23 $f(x) = x^3$
    $g(x) = \ln x$

25 $Q(m) = T + L + Pm$
    $T =$ fuel for take-off
    $L =$ fuel for landing
    $P =$ fuel per mile in the air
    $m =$ length of the trip (miles)

27 13,500 bacteria

29 2010

31 21,153%

33 (a) $S = 2\pi r^2 + 2V/r$
   (b) $S \to \infty$ as $r \to \infty$

35 Depth $= d = 7 + 1.5\sin(\pi t/3)$

37 (a) f is invertible
   (b) $f^{-1}(400) \approx 1979$ is the year in which 400 million motor vehicles were registered in the world.

39 (a) 2π

41 (a) $r(p) = kp(A - p)$
         $(k > 0)$
   (b) $p = A/2$

43 (a) $f(x) = (x - a)(x + a)(x + b)(x - c)$

45 (a) $f(x) \to \infty$ as $x \to \infty$
         $f(x) \to 16$ as $x \to -\infty$
   (b) $(e^x + 1)(e^{2x} - 2)(e^x - 2)(e^{2x} + 2e^x + 4)$
         Two zeros
   (c) $(\ln 2)/2, \ln 2$
         One twice other

47 (a) $p(x) = x^2 + 3x + 9$
         $r(x) = -3, q(x) = x - 3$
   (b) $f(x) \approx -3/(x - 3)$ for x near 3
   (c) $f(x) \approx x^2 + 3x + 9$ as $x \to \pm\infty$

49 (a) $1 - 8\cos^2 x + 8\cos^4 x$
   (b) $1 - 8\sin^2 x + 8\sin^4 x$

Ch. 1 Understanding

1 False

3 True

5 True

7 True

9 False

11 False

13 True

15 False

17 False

19 True

21 True

23 True

25 False

27 Possible answer
    $f(x) = \begin{cases} x & x \leq 3 \\ 2x & x > 3 \end{cases}$

29 Possible answer
    $f(x) = 1/((x-1)(x-2)\cdots(x-17))$

31 $f(x) = 1.5x, g(x) = 1.5x + 3$

33 Impossible

35 Impossible

37 False; $f(x) = \log x$ on $[1, 2]$

39 True; $f(x) = (0.5)^x$

41 False, $f(x) = \begin{cases} 1 & x \leq 3 \\ 2 & x > 3 \end{cases}$

43 (a) Follows
   (b) Does not follow (although true)
   (c) Follows
   (d) Does not follow

Section 2.1

1 265/3 km/hr

5 27

7 1.9...

9 2

11 0.01745...

17 15.47, 57.65, 135.90, 146.35, 158.55 people/min

Section 2.2

1 (a) 3
  (b) 7
  (c) Does not exist
  (d) 8

3 (b) 1
  (d) $-0.0033 < x < 0.0033$, $0.99 < y < 1.01$

5 (b) 0
  (d) $-0.005 < x < 0.005$, $-0.01 < y < 0.01$

7 (b) 2
  (d) $-0.0865 < x < 0.0865$, $1.99 < y < 2.01$

9 (b) 1
  (d) $-0.0198 < x < 0.0198$, $0.99 < y < 1.01$

11 (b) 4
   (d) $1.99 < x < 2.01$, $3.99 < y < 4.01$

13 (b) 0
   (d) $1.55 < x < 1.59$, $-0.01 < y < 0.01$

15 4

17 -2

19 $-1/16$

21 $\lim_{x \to 2+} f(x) = \lim_{x \to 2-} f(x) = \lim_{x \to 2} f(x) = 0$

23 0.05

29 2.71828...

31 1/3

33 2/3

35 3/2

37 4 or -4

39 4

41 Any k

43 0.46, 0.21, 0.09

45 (a) $x = 1/(n\pi)$, $n = 1, 2, 3, \ldots$
   (b) $x = 2/(n\pi)$, $n = 1, 5, 9, \ldots$
   (c) $x = 2/(n\pi)$, $n = 3, 7, 11, \ldots$

Section 2.3

1 12

3 (b) 8.69
  (c) 7.7

5 $f'(1) \approx 0.43427$

7 $f'(d) = 0$, $f'(b) = 0.5$, $f'(c) = 2$, $f'(a) = -0.5$, $f'(e) = -2$

9 (a) $x = 1$ and $x = 3$
   (b) $f(5)$
   (c) $f'(1)$

11 12

13 3

15 $-1/4$

17 $y = 12x + 16$

19 $y = -2x + 3$

21 $(4, 25)$; $(4.2, 25.3)$; $(3.9, 24.85)$

25 From smallest to largest:
    0, $f'(3)$, $f(3) - f(2)$, $f'(2)$

27 (a) $(f(b) - f(a))/(b - a)$
    (b) Slopes same
    (c) Yes

29 (a) $f'(0) \approx 0.01745$

31 $f'(0) \approx -1$
    $f'(1) \approx 3.5$

33 16.0 million people/year
    16.4 million people/year

Section 2.4

11 (a) 3
    (b) Positive: $0 < x < 4$
         Negative: $4 < x < 12$

13 $-1/x^2$

15 $4x$

21 $f'(x)$ positive:
         $4 \le x \le 8$
    $f'(x)$ negative:
         $0 \le x \le 3$
    $f'(x)$ greatest:
         at $x \approx 8$

35 (a) x_3
    (b) x_4
    (c) x_5
    (d) x_3

37 (a) Graph II
    (b) Graph I
    (c) Graph III

39 (a) Periodic: period 1 year
    (b) Max of 4500 on July 1st
         Min of 3500 on Jan 1st
    (c) Growing fastest:
             around April 1st
         Decreasing fastest:
             around Oct 1st
    (d) ≈ 0

Section 2.5

1 (a) Negative
   (b) Degrees/min

3 (a) Costs $350 for 200 gallons
   (b) Costs $1.40 for 201st gallon

5 Dollars/year

7 Dollars/percent; positive

9 Feet/mile; negative

11 (b) Pounds/(Calories/day)

13 1.25 billion people; growing at
    0.0174 billion people per year

17 mpg/mph

19 (a) $f'(a)$ is always positive
    (c) $f'(100) = 2$: more
         $f'(100) = 0.5$: less

21 (a) Liters per centimeter

(b) About 0.042 liters per centimeter
(c) Cannot expand much more

Section 2.6

1 (a) Negative
   (b) Negative
   (c) Positive

3 B

9 $f'(x) = 0$
   $f''(x) = 0$

11 $f'(x) < 0$
    $f''(x) > 0$

13 $f'(x) < 0$
    $f''(x) < 0$

15 $0 \le t \le 1$:
        acceleration = 30 ft/sec^2
    $1 \le t \le 2$:
        acceleration = 22 ft/sec^2

17 (a) $dP/dt > 0$, $d^2P/dt^2 > 0$

    (b) $dP/dt < 0$, $d^2P/dt^2 > 0$
         (but dP/dt is close to zero.)

19 A positive second derivative indicates a successful campaign.
    A negative second derivative indicates an unsuccessful campaign.

23 (a) B and E
    (b) A and D

Section 2.7

1 (a) $x = 1$
   (b) $x = 1, 2, 3$

3 (a) No
   (b) Yes, $x = 0$

5 Yes

7 Yes

9 Yes

11 (a) Yes
    (b) Not at $t = 0$

13 (a) Yes
    (b) No

15 (a) Yes
    (b) No

Chapter 2 Review

7 $10x + 1$

9 (b) 0
   (d) $-0.015 < x < 0.015$,
        $-0.01 < y < 0.01$

11 (b) 0
    (d) $1.570 < x < 1.5715$,
         $-0.01 < y < 0.01$

13 $2a$

15 $-2/a^3$

17 $-1/(2(\sqrt{a})^3)$

19 $\lim_{x \to 0^+} f(x) = \lim_{x \to 0^-} f(x) = \lim_{x \to 0} f(x) = 1$

21 (a) $3, -1$
    (b) $-3, 1$

23 357

25 (a) $f(7) = 3$
    (b) $f'(7) = 4$

31 (a) $f'(0.6) \approx 0.5$
         $f'(0.5) \approx 2$
    (b) $f''(0.6) \approx -15$
    (c) Maximum: near $x = 0.8$
         minimum: near $x = 0.3$

33 0.45, 0.0447, 0.00447

35 (a) At $(0, \sqrt{19})$:
             slope = 0
         At $(\sqrt{19}, 0)$:
             slope is undefined.
    (b) slope $\approx \frac{1}{2}$
    (c) At $(-2, \sqrt{15})$:
             slope $\approx \frac{1}{2}$
         At $(-2, -\sqrt{15})$:
             slope $\approx -\frac{1}{2}$
         At $(2, \sqrt{15})$:
             slope $\approx -\frac{1}{2}$.

37 (a) Period 12 months
    (b) Max of 4500 on June 1st
    (c) Min of 3500 on Feb 1st
    (d) Growing fastest:
             April 1st
         Decreasing fastest:
             July 15 and Dec 15
    (e) About 400 deer/month

39 (a) Concave down
    (b) $120° < T < 140°$
    (c) $135° < T < 140°$
    (d) $45 < t < 50$

41 (a) $f'(0) = 1.00000$
         $f'(0.3) = 1.04534$
         $f'(0.7) = 1.25521$
         $f'(1) = 1.54314$
    (b) They are about the same.

43 (a) $2\cos^2 x - 2\sin^2 x$
         Answers may vary
    (b) $(d/dx)\sin(2x) = 2\cos(2x)$

45 (a) $1/x$
    (b) Graphs same shape, shifted vertically

47 (a) $\cos x$, $-\sin x$, $\cos^2 x - \sin^2 x$
         Form of answers may vary
    (b) No

Ch. 2 Understanding

1 False

3 True

5 True

7 False

9 True

11 False

13 True

15 True

17 True

19 True

21 True; $f(x) = |x - 3|$

23 False; $f(x) = |x|$

25 False

27 True

29 True

31 False

33 True

35 False

37 False

39 (a) Not a counterexample
    (b) Counterexample
    (c) Not a counterexample
    (d) Not a counterexample

Section 3.1

3 $11x^{10}$

5 $11x^{-12}$

7   $-12x^{-13}$

9   $3x^{-1/4}/4$

11   $-4x^{-5}$

13   ex^{e-1}

15   $6t - 4$

17   $2t - k/t^2$

19   $6/w^4 + 3/(2\sqrt{w})$

21   $15t^4 - \frac{5}{2}t^{-1/2} - 7t^{-2}$

23   $2z - \frac{1}{2}z^{-2}$

25   $(z^2 - 1)/3z^2$

27   $1/(2\sqrt{\theta}) + 1/(2\theta^{3/2})$

29   $(8\pi rb)/3$

31   $2ax + b$

33   $-(5x^4 + 2)/2$

35   $5z^4 + 20z^3 - 1$

37   Do not apply

39   $6x$
    (power rule and sum rule)

41   $-2/3z^3$
    (power rule and sum rule)

43   $x \geq 1$ or $x \leq -2$

45   $x > 1$

47   $r = 3\sqrt{2}$

49   $y = 2x - 1$

51   $y = 2x$ and $y = -6x$

53   $n = 4, a = 3/32$

55   Height $= 625$ cm,
    Changing (eroding) at -30 cm/year

57   (a)   15.2 m/sec
    (b)   5.4 m/sec
    (c) -9.8 m/sec^2
    (d)   34.9 m
    (e)   5.2 sec

59   (a) $dT/dl = \pi/\sqrt{gl}$
    (b)   Positive, so the period increases as the length increases

61   $V(r) = 4\pi r^3/3$
    $\frac{dV}{dr} = 4\pi r^2 = $ surface area of a sphere

Section 3.2

1   $2e^x + 2x$

3   $(\ln 5)5^x$

5   $10x + (\ln 2)2^x$

7   $4(\ln 10)10^x - 3x^2$

9   $((\ln 3)3^x)/3 - (33x^{-3/2})/2$

11   e^{1+x}

13   $e^{\theta-1}$

15   $(\ln 4)^2 4^x$

17   $3x^2 + 3^x \ln 3$

19   $\pi^x \ln \pi$

21   $(\ln \pi)\pi^x$

23   $a^x \ln a + ax^{a-1}$

25   $(2\ln 3)z + (\ln 4)e^z$

27   $2x + (\ln 2)2^x$

29   Our rules do not apply here

31   e^{x+5}

33   The methods of the section do not apply here

35   The methods of the section do not apply here

37   ≈ 7.95 ¢/year

39   $\approx 22.5(1.35)^t$

41   (a) $P'(t) = kP(t)$

43   $c = -1/\ln 2$

47   e

Section 3.3

1   $5x^4 + 10x$

3   $e^x(x + 1)$

5   $2^x/(2\sqrt{x}) + \sqrt{x}(\ln 2)2^x$

7   $4s^3 - 1$

9   $(t^3 - 4t^2 - 14t + 1)e^t$

11   $(50x - 25x^2)/(e^x)$

13   $6/(5r + 2)^2$

15   $1/(5t + 2)^2$

17   $(t^2 + 2t + 2)/(t + 1)^2$

19   $2y - 6, \ y \neq 0$

21   $\sqrt{z}(3 - z^{-2})/2$

23   $2r(r + 1)/(2r + 1)^2$

25   $17e^x(1 - \ln 2)/2^x$

27   $\frac{(-4x^2 - 8x - 1)}{(2 + 3x + 4x^2)^2}$

29   $(3t^2 + 5)(t^2 - 7t + 2) + (t^3 + 5t)(2t - 7)$

31   $x < 2$

33   $y = 7x - 5$

35   $f'(x) = 2e^{2x}$

37   $\frac{d}{dx}e^{4x} = 4e^{4x}$

41   (a)   3
    (b)   14
    (c)   13/8

43   $f(x) = x^{10}e^x$

45   $r_2^2/(r_1 + r_2)^2$

47   (a) $g(v) = \frac{1}{f(v)}$
    $g(80) = 20 \frac{\text{km}}{\text{liter}}$
    $g'(80) = -\frac{1}{5}\frac{\text{km}}{\text{liter}}$ for each
    $1 \frac{\text{km}}{\text{hr}}$ increase in speed
    (b) $h(v) = v \cdot f(v)$
    $h(80) = 4 \frac{\text{liters}}{\text{hr}}$
    $h'(80) = 0.09 \frac{\text{liters}}{\text{hr}}$ for
    each $1 \frac{\text{km}}{\text{hr}}$ inc. in speed

49   (a) $f'(x) = $
    $(x - 2) + (x - 1)$
    (b) $f'(x) = $
    $(x - 2)(x - 3) + $
    $(x - 1)(x - 3) + $
    $(x - 1)(x - 2)$
    (c) $f'(x) = $
    $(x - 2)(x - 3)(x - 4) + $
    $(x - 1)(x - 3)(x - 4) + $
    $(x - 1)(x - 2)(x - 4) + $
    $(x - 1)(x - 2)(x - 3)$

51   (a) $(FGH)' = $
    $F'GH + FG'H + FGH'$
    (c) $f_1'f_2f_3 \cdots f_n + $
    $f_1f_2'f_3 \cdots f_n + \cdots + $
    $f_1 \cdots f_{n-1}f_n'$

Section 3.4

1   $99(x + 1)^{98}$

3   $200t(t^2 + 1)^{99}$

5   $50(\sqrt{t} + 1)^{99}/\sqrt{t}$

7   $5(w^4 - 2w)^4(4w^3 - 2)$

9   $\pi e^{\pi x}$

11   $(\ln \pi)\pi^{(x+2)}$

13   $4(x^3 + e^x)^3(3x^2 + e^x)$

15   $(2t - ct^2)e^{-ct}$

17   $6(1 + 3t)e^{(1+3t)^2}$

19   $5 \cdot \ln 2 \cdot 2^{5t-3}$

21   $\frac{3}{2}e^{\frac{3}{2}w}$

23   $3s^2/2\sqrt{s^3 + 1}$

25   $e^{-t^2}(1 - 2t^2)$

27   $e^{-z}/2\sqrt{z} - \sqrt{z}e^{-z}$

29   $e^{5-2t}(1 - 2t)$

31   $\dfrac{\sqrt{x + 3}(x^2 + 6x - 9)}{[2\sqrt{x^2 + 9}(x + 3)^2]}$

33   $-(3e^{3x} + 2x)/(e^{3x} + x^2)^2$

35   $(\ln 2)(3e^{3x})2^{e^{3x}}$

37   $e^{-\theta}/(1 + e^{-\theta})^2$

39   $2we^{w^2}(5w^2 + 8)$

41   $-(\ln 10)(10^{\frac{5}{2} - \frac{y}{2}})/2$

43   $2ye^{[e^{(y^2)} + y^2]}$

45   abe^{bt}

47   $ae^{-bx} - abxe^{-bx}$

49   $y = 3x - 5$

51   $-\frac{1}{\sqrt{2}} < x < \frac{1}{\sqrt{2}}$

53   (a) $H(4) = 1$
    (b) $H'(4) = 30$
    (c) $H(4) = 4$
    (d) $H'(4) = 56$
    (e) $H'(4) = -1$

55   (a) $\pi\sqrt{2}$
    (b) $7e$
    (c) πe

57   Yes

59   $596.73/yr

61   6 billion people,
    0.078 billion people per year,
    6.833 billion people,
    0.089 billion people per year

63   (a) $dH/dt = -60e^{-2t}$
    (b) $dH/dt < 0$
    (c)   At $t = 0$

65   400π cm^2/sec

67   (a)   For $t < 0, I = 0$
    For $t > 0$
    $I = -\frac{Q_0}{RC}e^{-t/RC}$
    (b)   No
    (c)   No

69   56 cm/sec^2

Section 3.5

3   $\cos^2 \theta - \sin^2 \theta = \cos 2\theta$

5   $3\cos(3x)$

7   $3\pi \sin(\pi x)$

9   $2x \cos x - x^2 \sin x$

11   $-\sin x e^{\cos x}$

13   $e^{\cos \theta} - \theta(\sin \theta)e^{\cos \theta}$

15   $\cos(\tan \theta)/\cos^2 \theta$

17   $\sin x/2\sqrt{1 - \cos x}$

19   $\cos x/\cos^2(\sin x)$

21   $2\sin(3x) + 6x \cos(3x)$

23   $e^{-2x}[\cos x - 2\sin x]$

25 $5\sin^4\theta\cos\theta$

27 $-3e^{-3\theta}/\cos^2(e^{-3\theta})$

29 $\cos t - t\sin t + 1/\cos^2 t$

31 $5\sin^4\alpha\cos^4\alpha - 3\sin^6\alpha\cos^2\alpha$

33 $\dfrac{\sqrt{1-\cos x}(1-\cos x - \sin x)}{2\sqrt{1-\sin x}(1-\cos x)^2}$

35 $(\cos y + a + y\sin y)/\left((\cos y + a)^2\right)$

37 $(\ln 2)(2^{2\sin x + e^x})(2\cos x + e^x)$

39 $\theta^2\cos\theta$

41 $-2(\cos w\sin w + w\sin(w^2))$

43 $e^x/\sin x$

45 Decreasing, concave up

47 (a) $dy/dt = -\frac{4.9\pi}{6}\sin\left(\frac{\pi}{6}t\right)$ ft/hr
   (b) Occurs at $t = 6, 12, 18,$ and 24 hrs

49 (a) $t = (\pi/2)(m/k)^{\frac{1}{2}}$;
    $t = 0$;
    $t = (3\pi/2)(m/k)^{\frac{1}{2}}$
   (b) $T = 2\pi(m/k)^{\frac{1}{2}}$
   (c) $dT/dm = \pi/\sqrt{km}$;
      Positive sign means an increase in mass causes the period to increase

51 $k = 7.46, (3\pi/4, 1/\sqrt{2})$

Section 3.6

1 $2t/(t^2 + 1)$

3 2

5 $-1/z(\ln z)^2$

7 $\dfrac{e^{-x}}{1-e^{-x}}$

9 $(e^x)/(e^x + 1)$

11 $ae^{ax}/(e^{ax} + b)$

13 7

15 $-\tan(w - 1)$

17 $2y/\sqrt{1 - y^4}$

19 1

21 $-\sin(\ln t)/t$

23 $\arcsin w + \dfrac{w}{\sqrt{1-w^2}}$

25 $2/\sqrt{1 - 4t^2}$

27 $-3\sin(\arctan 3x)/(1 + 9x^2)$

29 $\ln x/(1 + \ln x)^2$

31 $(\cos x - \sin x)/(\sin x + \cos x)$

33 $1/(1 + 2u + 2u^2)$

35 $-(x + 1)/(\sqrt{1 - (x+1)^2})$

37 $\frac{d}{dx}(\arcsin x) = 1/\sqrt{1 - x^2}$
   $(-1 < x < 1)$

39 (a) $(-0.99, -0.16)$
   (b) $v_x = 0.32$
      $v_y = -1.98$

41 -43.4

43 (a) $f'(x) = 0$
   (b) f is a constant function

45 (a) $y = -x^2/2 + 2x - 3/2$
   (b) From graph, notice that around $x = 1$, the values of $\ln x$ and its approximation are very close
   (c) At $x = 1.1, y \approx 0.095$
      At $x = 2, y = 0.5$

47 (a) $k \approx 0.067$
   (b) $t \approx 10.3$ hours

(c) Formula:
   $T(24) \approx 74.1°$F,
   rule of thumb: $73.6°$F

49 2513.3 cm^3/sec

51 2 cm/sec

53 (a) $z = \sqrt{0.25 + x^2}$
   (b) 0.693 km/min
   (c) 0.4 radians/min

55 (a) $h(t) = 300 - 30t$
      $0 \le t \le 10$
   (b) $\theta = \arctan(\frac{200 - 30t}{150})$
      $d\theta/dt =$
      $-\left(\frac{1}{5}\right)\left(\frac{150^2}{(150^2 + (200 - 30t)^2)}\right)$
   (c) When the elevator is at the level of the observer

Section 3.7

1 $dy/dx = -x/y$

3 $1/25$

5 $(1 - y)/(x - 3)$

7 ax/by

9 $-2xye^{x^2}$

11 $\frac{dy}{dx} = \frac{y(1 - x\ln y)}{x(x + 3y^3)}$

13 $dy/dx = -y^{1/3}/x^{1/3}$

15 $dy/dx = 0$

17 Slope is infinite

19 $-5/4$

21 $y = e^2 x$

23 $y = 0$

25 (a) $dy/dx = -9x/25y$
   (b) The slope is not defined anywhere along the line $y = 0$

27 (a) $\frac{dy}{dx} = \frac{y^2 - 3x^2}{3y^2 - 2xy}$
   (c) $y \approx 1.9945$
   (d) Horizontal:
      $(1.1609, 2.0107)$
      and
      $(-0.8857, 1.5341)$
      Vertical:
      $(1.8039, 1.2026)$
      and
      $(\sqrt[3]{5}, 0)$

29 $(-1/3, 2\sqrt{2}/3)$,
   $(7/3, 4\sqrt{2}/3)$

Section 3.8

1 The particle moves on straight lines from $(0, 1)$ to $(1, 0)$ to $(0, -1)$ to $(-1, 0)$ and back to $(0, 1)$

3 The particle moves on straight lines from $(-1, 1)$ to $(1, 1)$ to $(-1, -1)$ to $(1, -1)$ and back to $(-1, 1)$

5 Clockwise for all t

7 Clockwise: $t < 0$,
   Counter-clockwise: $t > 0$

9 Counterclockwise: $t > 0$

11 $x = 3\cos t, y = -3\sin t, 0 \le t \le 2\pi$

13 $x = 2 + 5\cos t, y = 1 + 5\sin t$,
   $0 \le t \le 2\pi$

15 $x = t, y = -4t + 7$

17 $x = -3\cos t, y = -7\sin t$,
   $0 \le t \le 2\pi$

19 $y = -(4/3)x$

21 Speed $= |t| \cdot \sqrt{(4 + 9t^2)}$,

Particle stops when $t = 0$

23 Speed $= \sqrt{4\sin^2(2t) + \cos^2 t}$,
   Particle stops when
   $t = (2n + 1)\pi/2$, for any integer n

25 $x = 4 + 4(t - 2)$
   $y = 8 + 12(t - 2)$

27 (a) The part of the line with $x < 10$ and $y < 0$
   (b) The line segment between $(10, 0)$ and $(11, 12)$

29 (b) $v \approx 2.2$
   (c) $v = 2.2363$

33 (a) $a = b = 0, k = 5$ or -5
   (b) $a = 0, b = 5, k = 5$ or -5
   (c) $a = 10, b = -10, k = \sqrt{200}$ or $-\sqrt{200}$

35 (b) For example:
   $R = 12, t = \pi$

Section 3.9

1 $\sqrt{1 + x} \approx 1 + x/2$

3 $1/x \approx 2 - x$

7 (b) This estimate is about right
   (c) Below

9 (c) 0

11 (c) -0.135%

13 (a) $\Delta T \approx -T\Delta g/(2g)$
   (b) 0.5% decrease

15 (a) $f'(5)$
   (b) 0
   (c) $f(5) + f'(5)\Delta x$

Section 3.10

1 Negative

3 Negative

5 0

7 0

9 $0.1x^7$

11 $x^{0.2}$

13 0

15 0.909297

17 Does not exist

19 -2

Chapter 3 Review

1 $2e^t + 2te^t + 1/(2t^{3/2})$

3 $3/(y\ln(2y^3))$

5 $kx^{k-1} + k^x\ln k$

7 $-6\cos(3t + 5)\cdot\sin(3t + 5)$

9 $6\cos(3\theta - \pi)\sin(3\theta - \pi)$

11 $-\theta^{-3}(\theta\cos(5 - \theta) + 2\sin(5 - \theta))$

13 $x^{-\frac{1}{2}}/2 - x^{-2} - 3x^{-\frac{5}{2}}/2$

15 $e^{(e^\theta + e^{-\theta})}(e^\theta - e^{-\theta})$

17 $e^{\tan(\sin\alpha)}\cos\alpha/\cos^2(\sin\alpha)$

19 $e(\tan 2 + \tan r)^{e-1}/\cos^2 r$

21 $e^{\tan x} + xe^{\tan x}/\cos^2 x$

23 $6x/\left(9x^4 + 6x^2 + 2\right)$

25 $2^{\sin x}\left((\ln 2)\cos^2 x - \sin x\right)$

27 a

29 $e^{-4kt}(\cos t - 4k \sin t)$

31 $-4a^2 x/(a^2 + x^2)^2$

33 $(-3a^2 s - s^3)/(a^2 + s^2)^{3/2}$

35 $-(\sin\theta\cos\theta)/(\sqrt{a^2 - \sin^2\theta})$

37 $\cos(t/k)/[k\sin(t/k)]$

39 $20w/(a^2 - w^2)^3$

41 $4a/(e^{ax} + e^{-ax})^2$

43 $6(3\theta - \pi)\cos[(3\theta - \pi)^2]$

45 k

47 $(-4 - 6x)(6x^e - 3\pi) +$
    $(2 - 4x - 3x^2)(6ex^{e-1})$

49 0

51 $4x - 2 - 4x^{-2} + 8x^{-3}$

53 2/3

55 $-2xy/(x^2 - 2)$

57 $(y + b\sin(bx))/$
    $(a\cos(ay) - x)$

59 Proportional to r^2

61 (a) $H'(2) = 11$
   (b) $H'(2) = -1/4$
   (c) $H'(2) = r'(1) \cdot 3$
        (we don't know $r'(1)$)
   (d) $H'(2) = -3$

63 (a) $y = 20x - 48$
   (b) $y = 11x/9 - 16/9$
   (c) $y = -4x + 20$
   (d) $y = -24x + 57$
   (e) $y = 8.06x - 15.84$
   (f) $y = -0.94x + 6.27$

65 1.909 radians (109.4°) or 1.231 radians (70.5°)

67 Not perpendicular; $x \approx 1.3$

69 (a) $dg/dr = -2GM/r^3$
   (b) dg/dr is rate of change of acceleration due
        to pull of gravity.
        The further away from the earth's center, the
        weaker gravity's pull
   (c) -3.05×10^{-6}
   (d) It is reasonable because the magnitude of $\frac{dg}{dr}$
        is so small (compared to $g = 9.8$) that for r
        near 6400 km, g is not varying much at all

71 (a) $v(t) = 10e^{t/2}$
   (b) $v(t) = s(t)/2$

73 (a) Falling, 0.38 m/hr
   (b) Rising, 3.76 m/hr
   (c) Rising, 0.75 m/hr
   (d) Falling, 1.12 m/hr

75 (a) $v = -2\pi\omega y_0 \sin(2\pi\omega t)$
        $a = -4\pi^2\omega^2 y_0 \cos(2\pi\omega t)$
   (b) Amplitudes: different
        $(y_0, \; 2\pi\omega y_0, \; 4\pi^2\omega^2 y_0)$
        Periods $= 1/\omega$

77 100π cm^3/sec

79 (a) Angular velocity
   (b) $v = a(d\theta/dt)$

81 $-k/V^2$

83 (a) $y^{(n)} =$
        $(-1)^{n+1}(n-1)!x^{-n}$
   (b) $y^{(n)} = xe^x + ne^x$

85 $(f/g)'/(f/g) = (f'/f) - (g'/g)$

87 (a) 1
   (b) 1
   (c) $\sin(\arcsin x) = x$

89 (a) 0
   (b) 0
   (c) $\ln(1 - 1/t) + \ln(t/(t-1)) = \ln 1$

Ch. 3 Understanding

1 True

3 True

5 True

7 False

9 True

11 False; $f(x) = |x|$

13 False; $\cos t + t^2$

15 False; $f(x) = 6, g(x) = 10$

17 False; $f(x) = 5x + 7, g(x) = x + 2$

19 False; $f(x) = x^2, g(x) = x^2 - 1$

21 False; $f(x) = e^{-x}, g(x) = x^2$

23 (a) Not a counterexample
   (b) Not a counterexample
   (c) Not a counterexample
   (d) Counterexample

Section 4.1

5 (a) Critical point $x \approx 0$;
       Inflection points between -1 and 0
       and between 0 and 1
  (b) Critical point at $x = 0$,
       Inflection points at $x = \pm 1/\sqrt{2}$

7 Local maximum $x = 1$

9 (a) $x \approx 2.5$ (or any $2 < x < 3$)
       $x \approx 6.5$ (or any $6 < x < 7$)
       $x \approx 9.5$ (or any $9 < x < 10$)
  (b) $x \approx 2.5$: local max;
       $x \approx 6.5$: local min;
       $x \approx 9.5$: local max

13 Local max: $(-1.4, 6.7)$
    Local min: $(1.4, -4.7)$

15 Local max: $(-1, 2)$
    Local min: $(1, -2)$
    Crit. pt. (not max/min): $(0,0)$

17 Local min: $(2.3, -13.0)$

19 Local min: $(-0.71, -0.43)$
    Local max: $(0.71, 0.43)$
    Horizontal asymptote: $y = 0$

23 (a) Local maxima at $x \approx 1, x \approx 8$,
        Local minimum at $x \approx 4$
   (b) Local maxima at $x \approx 2.5, x \approx 9.5$
        Local minimum at $x \approx 6.5$,

25 $a = -1/3$

35 (b) At most four zeros.
   (c) Possibly no zeros.
   (d) Two inflection points.
   (e) Degree four
   (f) $\frac{-2}{15}(x+1)(x-1)(x-3)(x-5)$

37 $B = f, A = f', C = f''$

39 III even; I, II odd
    I is f'', II is f, III is f'

Section 4.2

1 $y = -(x-2)^2 + 5$

3 $y = 5(1 - e^{-bx})$

5 $y = -5x/(x-2)$

7 $y = 0.25\sin(\pi x/10) + 1.75$

9 $y = -x^4 + 2x^2 + 3$

11 $y = 3x^{-1/2}\ln x$

13 (a) Local maximum:
        $p(-\sqrt{\frac{a}{3}}) = +\frac{2a\sqrt{a}}{3\sqrt{3}}$
        Local minimum:
        $p(\sqrt{\frac{a}{3}}) = -\frac{2a\sqrt{a}}{3\sqrt{3}}$
   (b) Further from x- and y-axis

15 $a = 200,000$
    $k = -\ln 0.9 \approx 0.105$

19 $(1/b, 1/be)$

21 (a) $f'(x) = abe^{-bx} > 0$
   (b) $f''(x) = -ab^2 e^{-bx} < 0$

23 (a) $x = e^a$
   (c) Max at (e^{a-1}, e^{a-1}) for any a

27 (a) Local max: $x = \frac{1}{b}$
        No local minima
        Inflection point: $x = \frac{2}{b}$
   (b) Varying a stretches or flattens the graph ver-
        tically.
        Incr b shifts critical, inflection points to left;
        lowers max

31 (a) Intercept: $x = a$
        Asymptotes: $x = 0$,
        $U = 0$
   (b) Local min: $(2a, -b/4)$
        Local max: none

33 (a) Zero: $r = B/A$
        Vertical asymptote:
        $r = 0$
        Horizontal asymptote:
        $f(r) = 0$
   (b) Minimum:
        $\left(3B/(2A), -4A^3/(27B^2)\right)$
        Point of inflection:
        $\left(2B/A, -A^3/(8B^2)\right)$

Section 4.3

3 (a) $f(1)$ local minimum;
       $f(0), f(2)$ local maxima
  (b) $f(1)$ global minimum
       $f(2)$ global maximum

5 (a) $f(\frac{2\pi}{3})$ local maximum
       $f(0)$ and $f(\pi)$ local minima
  (b) $f(\frac{2\pi}{3})$ global maximum
       $f(0)$ global minimum

7 44.1 feet

9 $r = \frac{2}{3}R$

11 (a) $0 \le y \le a$
   (b) $y = 0$

13 $r = 3B/(2A)$

15 $x = L/2$

17 Minimum: $0.148mg$ newtons
    Maximum: $1.0mg$ newtons

19 Minimum: -2 amps
    Maximum: 2 amps

21 $0.91 < y \le 1.00$

23 $0 \le y < 1.61$

25 $0 \le y \le 2\pi$

27 (b) 2 hours
   (c) Equal

29 (b) $f(v) = v \cdot a(v)$
   (c) When $a(v) = f'(v)$
   (d) $a(v)$

31 (a) $g(v) = f(v)/v$
   (b) 220 mph
   (c) 300 mph

33 (b) Yes, at $x = 0$
   (c) Max: $x = -2$, Min: $x = 2$
   (d) $5 > g(0) > g(2)$

Section 4.4

1 $5000, $2.40, $4

3 $\pi(q) = 490q - q^2 - 150$
   Max at $q = 245$

5 $1.1 m, 70, \$1.2 m

7 (a) $\pi(q)$ max when
$R(q) > C(q)$ and R and Q are farthest apart
   (b) $C'(q_0) = R'(q_0) = p$

9 (a) No
   (b) Yes

11 (a) 150; \$750
    (b) \$44,500; 130; \$850

13 (b) (i) $N'(x) = 20$
        (ii) $\frac{N(x)}{x} = \frac{100}{x} + 20$

17 (b) $q = [Fa/(K(1-a))]^a$

Section 4.5

1 $2000 - (1200/\sqrt{5})$

3 Min $v = \sqrt{2}k$; no max

5 (a) $V = Ax/4 - x^3/2$
   (c) $(A/6)^{3/2}$

7 $w = 34.64$ cm, $h = 48.99$ cm

9 40 feet by 80 feet

11 $h = \sqrt{50}$ meters

13 $(1, 1)$

15 When the rectangle is a square.

17 15 miles/hour

19 (a) The arithmetic mean unless $a = b$, in which case the two means are equal.
   (b) The arithmetic mean unless $a = b = c$, in which case the two means are equal.

21 0.8 miles from Town 1

23 65.1 meters

25 (a) $T = \sqrt{a^2 + (c-x)^2}/v_1$
         $+\sqrt{b^2 + x^2}/v_2$

Section 4.6

1 $2\sinh(2x)$

3 $\sinh(\sinh t) \cdot \cosh t$

5 $2\cosh t \cdot \sinh t$

7 $3\cosh(3y) \cdot \cosh(\sinh(3y))$

13 $\sinh(2x) = 2\sinh x \cosh x$

15 (a) 0
   (b) Positive for $x > 0$
        Negative for $x < 0$
        Zero for $x = 0$
   (c) Increasing everywhere
   (d) $1, -1$
   (e) Yes; derivative positive everywhere

17 (b) $A = 6.325$
        (stretch factor)
        $c = 0.458$
        (horizontal shift)

19 (b) U-shaped
    (c) Incr $(A > 0)$
         or Decr $(A < 0)$
    (d) Max: $A < 0, B < 0$
         Min: $A > 0, B > 0$

21 $y \approx 715 - 100\cosh(x/100)$

Section 4.7

1 False

3 False

5 False

7 True

9 No; no

11 No; no

19 Racetrack

21 Mean Value

23 $21 \le f(2) \le 25$

Chapter 4 Review

3 (a) Increasing for $x > 0$
        Decreasing for $x < 0$
   (b) Local and global min: $f(0)$

5 (a) Increasing for $0 < x < 4$
        Decreasing for $x < 0$ and $x > 4$
   (b) Local max: $f(4)$
        Local min: $f(0)$

7 (a) $f'(x) = 3x(x-2)$
        $f''(x) = 6(x-1)$
   (b) $x = 0$
        $x = 2$
   (c) Inflection point: $x = 1$
   (d) Endpoints:
        $f(-1) = -4$
        $f(3) = 0$
        Critical Points:
        $f(0) = 0$
        $f(2) = -4$
        Global max:
        $f(0) = 0$ and $f(3) = 0$
        Global min:
        $f(-1) = -4$ and
        $f(2) = -4$
   (e) f increasing:
        for $x < 0$ and $x > 2$
        f decreasing:
        for $0 < x < 2$
        f concave up:
        for $x > 1$
        f concave down:
        for $x < 1$.

9 (a) $f'(x) =$
        $-e^{-x}\sin x + e^{-x}\cos x$
        $f''(x) = -2e^{-x}\cos x$
   (b) Critical points:
        $x = \frac{\pi}{4}$ and $\frac{5\pi}{4}$
   (c) Inflection points:
        $x = \frac{\pi}{2}$ and $\frac{3\pi}{2}$
   (d) Endpoints:
        $f(0) = 0$
        $f(2\pi) = 0$
        Global max:
        $f(\frac{\pi}{4}) = (e^{-\frac{\pi}{4}})(\frac{\sqrt{2}}{2})$
        Global min:
        $f(\frac{5\pi}{4}) = -e^{\frac{-5\pi}{4}}(\frac{\sqrt{2}}{2})$
   (e) f increasing:
        $0 < x < \frac{\pi}{4}$ and
        $\frac{5\pi}{4} < x < 2\pi$
        f decreasing:
        $\frac{\pi}{4} < x < \frac{5\pi}{4}$
        f concave down:
        for $0 \le x < \frac{\pi}{2}$
        and $\frac{3\pi}{2} < x \le 2\pi$
        f concave up:
        for $\frac{\pi}{2} < x < \frac{3\pi}{2}$

11 $\lim\limits_{x \to \infty} f(x) = \infty$
    $\lim\limits_{x \to -\infty} f(x) = -\infty$
   (a) $f'(x) = 6(x-2)(x-1)$
        $f''(x) = 6(2x-3)$
   (b) $x = 1$ and $x = 2$
   (c) $x = \frac{3}{2}$
   (d) Critical points:
        $f(1) = 6, f(2) = 5$
        Local max: $f(1) = 6$
        Local min: $f(2) = 5$
        Global max and min: none
   (e) f increasing: $x < 1$ and $x > 2$
        Decreasing: $1 < x < 2$

f concave up: $x > \frac{3}{2}$
f concave down: $x < \frac{3}{2}$

13 $\lim\limits_{x \to -\infty} f(x) = -\infty$
    $\lim\limits_{x \to \infty} f(x) = 0$
   (a) $f'(x) = (1-x)e^{-x}$
        $f''(x) = (x-2)e^{-x}$
   (b) Only critical point is at $x = 1$.
   (c) Inflection point: $f(2) = \frac{2}{e^2}$
   (d) Global max: $f(1) = \frac{1}{e}$
        Local and global min: none
   (e) f increasing: $x < 1$
        f decreasing: $x > 1$
        f concave up: $x > 2$
        f concave down: $x < 2$

15 Local max: $f(-3)$
    Local min: $f(3)$
    Inflection pts: $x = 0, x = \pm\frac{3}{\sqrt{2}}$
    Global max and min: none

17 Local max: $f(-\frac{2}{5})$
    Global and local min: $f(0)$
    Inflection pts: $x = \frac{-2\pm\sqrt{2}}{5}$
    Global max: none

19 Local maxima: none
    Global and local min:
    $f(0) = 0$
    Inflection pts: $x = \pm\frac{1}{\sqrt{3}}$

21 $a = 1.63, b = -2$

23 (a) x_3
   (b) x_1, x_5
   (c) x_2
   (d) 0

27 $x = y = \sqrt[3]{V}$

29 $r \approx 2.7$ cm

31 (a) $2\sqrt{x^2 + 300^2} + (1000 - x)$
   (b) 173.21 miles south of Boise on the coast

33 $-4.81 \le f(x) \le 1.82$

37 (a) $g(e)$ is a global maximum.
        There is no minimum.
   (b) There are exactly two solutions.
   (c) $x = 5$ and $x \approx 1.75$

39 $a = 363.23$, and $b \approx 4.7665$

43 (a) $1/\sqrt{1 + x^2}$

45 (a) Min at $x = a$
   (c) 4.6477

47 (b) No

Ch. 4 Understanding

1 True

3 False

5 False

7 False

9 True

11 False

13 One possibility:
    $f(x) = ax^2, a \neq 0$

15 Possible answer: $f(x) = |x|$

17 Possible answer
    $$f(x) = \begin{cases} x^2 & \text{if } 0 \le x < 1 \\ 1/2 & \text{if } x = 1 \end{cases}$$

19 (a) Not implied
   (b) Not implied
   (c) Implied

21 Impossible

23 Impossible

25 Impossible

Section 5.1

1 (a) 570 m^3/hr
  (b) Every 2 minutes

3 (a) Lower estimate = 5.25 mi
     Upper estimate = 5.75 mi
  (b) Lower estimate = 11.5 mi
     Upper estimate = 14.5 mi
  (c) Every 30 seconds

5 (a) Car A
  (b) Car A
  (c) Car B

7 (a) 430 ft
  (b) (ii)

9 Upper estimate = 0.75 m;
   Lower estimate = 0.65 m;
   Average = 0.70 m

11 Between 140 and 150 meters

13 (a) Upper estimate = 34.16 m/sec
     Lower estimate = 27.96 m/sec
  (b) 31.06 m/sec;
     It is too high

Section 5.2

1 (a) 224
  (b) 96
  (c) 200
  (d) 136

3 20

5 6375

7 205.5

9 10.0989

11 1.4936

13 limit = 1
   True value is between
   1.00314 and 0.99686.

15 2.545

17 24.7

19 4.39

21 0.0833

23 (a) 78
  (b) 46; underestimate
  (c) 118; overestimate

27 0.80

29 2/125

31 (a) -4
  (b) 0
  (c) 8

33 $a = 2$, $b = 6$, $f(x) = x^2$; other answers
   possible

Section 5.3

1 2

3 15

5 Dollars

7 (a) Car 1: 1031.25 ft
     Car 2: 562.5 ft
  (b) 1.6 minutes

9 (b) Twice
     At each intersection point, the distance be-
     tween the two vehicles is at a local extremum.

11 (a) $\int_0^2 R(t)\, dt$
  (c) lower estimate: 2.81
     upper estimate: 3.38

13 (a) -4

  (b) (ii)

15 $2392.87

17 About $13,800

19 (b) 0.64

21 Units of f

23 $6080

25 $485.80

27 (a) 0.375 thousand/hour
  (b) 1.75 thousands

29 12 newton · meters

31 (a) III
  (b) I
  (c) II and IV

Section 5.4

3 (a) $f(1), f(2)$
  (b) 2, 2.31, 2.80, 2.77

5 $8c$

7 -52

9 8

11 (a) 2
  (b) 12, 2
  (c) 0, 0

15 $f(3) - f(2)$,
   $[f(4) - f(2)]/2$,
   $f(4) - f(3)$.

21 (a) 0.1574
  (b) 0.9759

23 30/7

Chapter 5 Review

1 (a) 260 ft
  (b) Every 0.5 sec

3 $\approx$ 455 feet or 0.086 miles

5 396

7 3.4

9 36.00

11 10.67

13 5.1666

15 486.15 quadrillion BTU
   $\int_0^{30} f(t)dt$

17 65 km from home
   3 hours
   90 km

19 About 13,500 liters

21 (b) Largest to smallest:
     $n = 1$, $n = 3$,
     $n = 4$, and $n = 2$.

23 V < IV < II < III < I
   I, II, III positive
   IV, V negative

25 (a) $\int_0^5 f(x)\, dx -$
     $\frac{1}{2} \int_{-2}^2 f(x)\, dx$
  (b) $\int_{-2}^5 f(x)\, dx -$
     $2 \int_{-2}^0 f(x)\, dx$
  (c) $\frac{1}{2} \left(\int_{-2}^5 f(x)\, dx - \right.$
     $\left. \int_2^5 f(x)\, dx \right)$

27 45.8°C.

29 741.6 liters

31 (a) V
  (b) IV
  (c) III
  (d) II
  (e) III
  (f) I

33 (a) 300 m^3/sec
  (b) 250 m^3/sec
  (c) 1996: 1250 m^3/sec
     1957: 3500 m^3/sec
  (d) 1996: 10 days
     1957: 4 months
  (e) 10^9 meter^3
  (f) $2 \cdot 10^{10}$ meter^3

35 (a) At $t = 17, 23, 27$ seconds
  (b) Right: $t = 10$ seconds
     Left: $t = 40$ seconds
  (c) Right: $t = 17$ seconds
     Left: $t = 40$ seconds
  (d) $t = 10$ to 17 seconds,
     20 to 23 seconds, and
     24 to 27 seconds
  (e) At $t = 0$ and $t = 35$

37 9 years

39 (a) $\sum_{i=1}^n i^5/n^6$
  (b) $(2n^4 + 6n^3 + 5n^2 - 1)/12n^4$
  (c) 1/6

41 (a) $\sum_{i=0}^{n-1} (n+i)^2/n^3$.
  (b) $7/3 + 1/(6n^2) - 3/(2n)$.
  (c) 7/3
  (d) 7/3

43 (a) $\cos(ac)/c - \cos(bc)/c$
  (b) $-\cos(cx)/c$

Ch. 5 Understanding

1 True

3 False

5 False

7 False

9 False

11 True

13 True

15 False

17 True

Section 6.1

5 128, 169, 217

7 (a) -19
  (b) 6

9 (a) $x = 1$, $x = 3$
  (b) Local min at $x = 1$, local max at $x = 3$

11 x_1 local max;
   x_2 inflection pt;
   x_3 local min

13 x_2, x_3 inflection pts.

15 (a) $f(3) = 1$; $f(7) = 0$
  (b) $x = 0, 5.5, 7$

17 Critical points: $(0, 5)$, $(2, 21)$, $(4, 13)$, $(5, 15)$

19 Acceleration is zero at points A and C.

21 (a) (I) volume; (II) flow rate
  (b) (I) is an antiderivative of (II)

23 (a) $f(x)$ is greatest at x_1
  (b) $f(x)$ is least at x_5
  (c) $f'(x)$ is greatest at x_3
  (d) $f'(x)$ is least at x_5
  (e) $f''(x)$ is greatest at x_1
  (f) $f''(x)$ is least at x_5

25 (b) Maximum in July 1993
   Minimum in Jan 1994
   (c) Increasing fastest in May 1993
   Decreasing fastest in Oct 1993

Section 6.2

1 $5x$

3 $x^3/3$

5 $\sin t$

7 $\ln|z|$

9 $-1/2z^2$

11 $-\cos t$

13 $t^4/4 - t^3/6 - t^2/2$

15 $\sin t + \tan t$

17 $-\cos 2\theta$

19 $(t+1)^3/3$

21 $5x^2/2 - 2x^{3/2}/3$

23 $x^4/4 - x^2/2 + C$

25 $t^4/4 + 5t^2/2 - t + C$

27 $2t^{3/2}/3 + C$

29 $x^4 - 7x + C$

31 $2t^{1/2} + C$

33 $-1/x + C$

35 $F(x) = x^2$
   (only possibility)

37 $F(x) = x^2/8$
   (only possibility)

39 $F(x) = \frac{2}{3}x^{3/2}$
   (only possibility)

41 $F(x) = -\cos x + 1$
   (only possibility)

43 $x^4/4 + C$

45 $x^4/4 - 2x + C$

47 $8w^{3/2}/3 + C$

49 $-4/t + C$

51 $\sin\theta + C$

53 $x^2/2 + 2x^{1/2} + C$

55 $\pi x + x^{12}/12 + C$

57 $\sin(x+1) + C$

59 $-e^{-z} + C$

61 36

63 $-(\sqrt{2}/2) + 1 = 0.293$

65 $\frac{609}{4} - 39\pi \approx 29.728$

67 $\ln 2 + \frac{3}{2} \approx 2.193$

69 1

71 $2e - 2 \approx 3.437$

73 $3/(2\ln 2) \approx 2.164$

75 36

77 $e - 1 - \sin 1$

79 $c = 3$

81 $c = 6$

83 $C(x) =$
   $5x^2 + 4000x + 1,000,000$

Section 6.3

1 $x^4/4 + 5x + C$

3 $8t^{3/2}/3 + C$

7 $10e^t + 15$

9 $2z - \cos z + 6$

11 (a) $a(t) = -9.8$ m/sec^2
      $v(t) = -9.8t + 40$ m/sec
      $h(t) = -4.9t^2 + 40t + 25$ m
   (b) 106.6 m; 4.08 sec
   (c) 8.75 sec

13 $y = 2kt^{3/2}/3$

15 19.55 ft/sec^2

17 (c) 200 ft
   (d) 200 ft

19 (a) 6 seconds
   (b) Left sum: 97.5 ft
      overestimate
      Right sum: 82.5 ft
      underestimate
   (c) 90 ft
   (d) $s(t) = 30t - \frac{5}{2}t^2$; $s(6) = 90$ ft
      Distance by antidiff = Average of left and
      right hand sums

21 (a) 4 seconds
   (b) 576 ft
   (c) 10 seconds
   (d) 192 ft/sec downward

23 (a) 80 ft/sec
   (b) 640 ft

25 -33.56 ft/sec^2

Section 6.4

7 $f(x) = 7 + \int_0^x \sin(t^2)\,dt$

9 $f(x) = 2 + \int_0^x \mathrm{Si}(t)\,dt$

11 500

13 -3.905

15 $(1+x)^{200}$

17 $-\cos(t^3)$

19 $(2\sin x^2)/x$

21 (a) 0
   (c) $F(x) \geq 0$ everywhere.
      $F(x) = 0$ only at integer multiples of π.

23 $\mathrm{erf}(x) + \frac{2}{\sqrt{\pi}}xe^{-x^2}$

25 $3x^2 e^{-x^6}$

Section 6.5

1 (a) 1.5 m
   (b) 7 m/sec
   (c) 9.8 m/sec^2 downward

3 $t = 5$; $v = -160$ ft/sec

5 400 feet

9 (a) First second: $-g/2$
      Second: $-3g/2$
      Third: $-5g/2$
      Fourth: $-7g/2$
   (b) Galileo seems to have been correct.

Chapter 6 Review

1 $\frac{5}{2}x^2 + 7x + C$

3 $2t + \sin t + C$

5 $3e^x - 2\cos x + C$

7 $16\sqrt{x} + C$

9 $e^x + 5x + C$

11 $\tan x + C$

13 $(x+1)^3/3 + C$

15 $\frac{1}{10}(x+1)^{10} + C$

17 $\frac{1}{2}x^2 + x + \ln|x| + C$

19 $3\sin x + 7\cos x + C$

21 $2e^x - 8\sin x + C$

23 $\sin x + C$

25 $-\cos x + C$

27 $5\ln|t| + C$

29 $e^x - x + C$

31 $x^4/4 + +2x^3 - 4x + 4.$

33 $e^x + 3$

35 $\sin x + 4$

37 9

39 $\sqrt{3} - \pi/9$

41 (a) 253/12
   (b) $-125/12$

43 $c = 3/4$

45 2

47 (a) $t = 2, t = 5$
   (b) $f(2) \approx 55, f(5) \approx 40$
   (c) -10

51 Global max: at $x = \pi$
   Global min: at $x = \pi/2$

53 (a) $f(x) = 3 + \int_0^x e^{t^2}\,dt$
   (b) $f(x) = 5 + \int_{-1}^x e^{t^2}\,dt$

57 (a) 14,000 revs/min^2
   (b) 180 revolutions

59 (b) Highest pt: $t = 2.5$ sec
      Hits ground: $t = 5$ sec
   (c) Left sum:
      136 ft (an overest.)
      Right sum:
      56 ft (an underest.)
   (d) 100 ft

61 (b) $t = 6$ hours
   (c) $t = 11$ hrs

63 Positive, zero, negative,
   positive, zero

65 (a) $(1/2)e^{2x}, (1/3)e^{3x}, (1/3)e^{3x+5}$
   (b) $(1/a)e^{ax+b}$

67 (a) $x - \ln|x-1|, x - 2\ln|x-1|,$
      $x + \ln|x-2|$
   (b) $x + (b-a)\ln|x-b|$

Ch. 6 Understanding

1 True

3 True

5 False

7 True

9 True

11 True

13 False

15 False

Section 7.1

1 (a) $2x\cos(x^2+1)$;
      $3x^2\cos(x^3+1)$
   (b) (i) $\frac{1}{2}\sin(x^2+1) + C$
      (ii) $\frac{1}{3}\sin(x^3+1) + C$
   (c) (i) $-\frac{1}{2}\cos(x^2+1) + C$
      (ii) $-\frac{1}{3}\cos(x^3+1) + C$

3 $(1/3)e^{3x} + C$

5 $-0.5\cos(2x) + C$

7 $\frac{1}{18}(y^2+5)^9 + C$

9 $\frac{1}{5}y^5 + \frac{1}{2}y^4 + \frac{1}{3}y^3 + C$

11 $\frac{1}{9}(x^2-4)^{9/2} + C$

13. $-2\sqrt{4-x} + C$

15. $\frac{1}{148}(2t-7)^{74} + C$

17. $-\frac{1}{8}(\cos\theta+5)^8 + C$

19. $-\frac{1}{2}e^{-x^2} + C$

21. $\frac{1}{3}e^{x^3+1} + C$

23. $\frac{1}{4}\sin^4\alpha + C$

25. $\frac{1}{3}(\ln z)^3 + C$

27. $\frac{1}{2}\ln(y^2+4) + C$

29. $2e^{\sqrt{y}} + C$

31. $\ln(2+e^x) + C$

33. $\frac{1}{6}\ln(1+3t^2) + C$

35. $t + 2\ln|t| - \frac{1}{t} + C$

37. $\sinh x + C$

39. $e^{\cosh z} + C$

41. $(\pi/4)t^4 + 2t^2 + C$

43. $\sin x^2 + C$

45. $\frac{1}{5}\cos(2-5x) + C$

47. $\frac{1}{2}\ln(x^2+1) + C$

49. 0

51. $1-(1/e)$

53. $3(e^2-e)$

55. $2(\sin 2 - \sin 1)$

57. 40

59. $\ln 3$

61. $14/3$

63. (a) Yes; $-0.5\cos(x^2)+C$
(b) No
(c) No
(d) Yes; $-1/(2(1+x^2))+C$
(e) No
(f) Yes; $-\ln|2+\cos x|+C$

65. $\ln 3$

67. $4\ln 2$

69. $\frac{1}{2}\ln 3$

71. (a) $\frac{\sin^2\theta}{2}+C$
(b) $-\frac{\cos^2\theta}{2}+C$
(c) $-\frac{\cos 2\theta}{4}+C$
(d) Functions differ by a constant

73. (a) 5.3 billion, 6.1 billion
(b) 5.7 billion

75. (a) $E(t)=1.4e^{0.07t}$
(b) $0.2(e^7-1)\approx 219$ million megawatt-hours
(c) 1972
(d) Graph $E(t)$ and estimate t such that $E(t)=219$

77. $-\frac{1}{k}\ln\left(\left(e^{t\sqrt{gk}}+e^{-t\sqrt{gk}}\right)/2\right)+h_0$

Section 7.2

1. $x\cdot\arctan x-\frac{1}{2}\ln(1+x^2)+C$

3. $\frac{1}{5}t^2e^{5t}-\frac{2}{25}te^{5t}+\frac{2}{125}e^{5t}+C$

5. $-t\cos t+\sin t+C$

7. $\frac{x^4}{4}\ln x-\frac{x^4}{16}+C$

9. $-t^2\cos t+2t\sin t+2\cos t+C$

11. $-\frac{1}{2}\sin\theta\cos\theta+\frac{\theta}{2}+C$

13. $\frac{1}{6}q^6\ln 5q-\frac{1}{36}q^6+C$

15. $t(\ln t)^2-2t\ln t+2t+C$

17. $-(\theta+1)\cos(\theta+1)+\sin(\theta+1)+C$

19. $-x^{-1}\ln x-x^{-1}+C$

21. $-2t(5-t)^{1/2}-\frac{4}{3}(5-t)^{3/2}-14(5-t)^{1/2}+C$

23. $w\arcsin w+\sqrt{1-w^2}+C$

25. $\frac{1}{2}x^2\arctan x^2-\frac{1}{4}\ln(1+x^4)+C$

27. $\frac{1}{3}x^3\sin x^3+\frac{1}{3}\cos x^3+C$

29. $\cos 5+5\sin 5-\cos 3-3\sin 3\approx -3.944$

31. $\frac{9}{2}\ln 3-2\approx 2.944$

33. $6\ln 6-5\approx 5.751$

35. $\frac{1}{2}(\frac{\pi}{2}-1)\approx 0.285$

37. π

39. Integration by parts gives: $\frac{1}{2}\sin\theta\cos\theta+\frac{1}{2}\theta+C$ The identity for $\cos^2\theta$ gives: $\frac{1}{2}\theta+\frac{1}{4}\sin 2\theta+C$

41. $\frac{1}{2}e^\theta(\sin\theta+\cos\theta)+C$

43. $\frac{1}{2}\theta e^\theta(\sin\theta+\cos\theta)-\frac{1}{2}e^\theta\sin\theta+C$

45. Integrate by parts choosing $u=x^n, v'=\cos ax$.

47. Integrate by parts choosing $u=\cos^{n-1}x, v'=\cos x$.

49. Approximately 77

51. (a) $-a^2e^{-a}-2ae^{-a}-2e^{-a}+2$
(b) Increasing
(c) Concave up

53. (a) V_0: increases, ω: none, ϕ:none
(b) V_0: increases, ω: increases, ϕ:none
(c) V_0: increases, ω: none, ϕ:none

55. (a) $C_1=\sqrt 2$
(b) $C_n=\sqrt 2$

Section 7.3

1. $\frac{1}{10}e^{(-3\theta)}(\sin\theta-3\cos\theta)+C$

3. $-\frac{1}{5}x^3\cos 5x+\frac{3}{25}x^2\sin 5x+\frac{6}{125}x\cos 5x-\frac{6}{625}\sin 5x+C$

5. $\frac{5}{7}x^7+\frac{5}{2}x^4+25x+C$

7. $-\frac{1}{4}\sin^3 x\cos x-\frac{3}{8}\sin x\cos x+\frac{3}{8}x+C$

9. $\frac{1}{2}\frac{\sin x}{\cos^2 x}+\frac{1}{4}\ln\left|\frac{\sin x+1}{\sin x-1}\right|+C$

11. $\frac{5}{16}\sin 3\theta\sin 5\theta+\frac{3}{16}\cos 3\theta\cos 5\theta+C$

13. $(\frac{1}{3}x^2-\frac{2}{9}x+\frac{2}{27})e^{3x}+C$

15. $(\frac{1}{3}x^4-\frac{4}{9}x^3+\frac{4}{9}x^2-\frac{8}{27}x+\frac{8}{81})e^{3x}+C$

17. $t+\ln|t-1|-\ln|t+1|+C$

19. $\frac{1}{45}(7\cos 2y\sin 7y-2\sin 2y\cos 7y)+C$

21. $\frac{1}{34}e^{5x}(5\sin 3x-3\cos 3x)+C$

23. $-\frac{1}{2\tan 2\theta}+C$

25. $\frac{1}{21}\frac{\tan 7x}{\cos^2 7x}+\frac{2}{21}\tan 7x+C$

27. $\frac{1}{3}\frac{\sin x}{\cos^3 x}-\frac{4}{3}\frac{\sin x}{\cos x}+x+C$

29. $-\frac{1}{4}(\ln|y-2|-\ln|y+2|)+C$

31. $\arctan(y+2)+C$

33. $-\cos x+\frac{1}{3}\cos^3 x+C$

35. $\frac{1}{8}e^{2z^2}(\cos 2z^2+\sin 2z^2)+C$

39. (a) $NA+2\pi B(1-e^{-N})/(1+4\pi^2)$
(b) $A+2\pi B\left(\frac{1-e^{-N}}{N}\right)/(1+4\pi^2)$
(c) A
(d) $R(t)\approx A$ for large t
(e) No, a real oil well will eventually dry out.

Section 7.4

1. $2/(5-x)+2/(5+x)$

3. $-2/y+1/(y-2)+1/(y+2)$

5. $-2\ln|5-x|+2\ln|5+x|+C$

7. $-2\ln|y|+\ln|y-2|+\ln|y+2|+C$

9. (a) Yes; $x=3\sin\theta$
(b) No

11. $-\ln|x|+x^{-1}+\ln|x-1|+K$

13. $x^2/2+\ln|x|+\ln|x+1|+2\ln|x+11|+K$

15. $(1/2)\arcsin(2x/3)+C$

17. $\arcsin(x-2)+C$

19. $x=(4\tan\theta)-3$

21. $x=(\tan\theta)-1$

23. $z=(\sin\theta)+1$

25. $w=(t+2)^2+3$

27. $(\ln|x-5|-\ln|x-3|)/2+C$

29. $\ln|x|-(1/2)\ln\left|x^2+1\right|+\arctan x+K$

31. $-x\sqrt{9-x^2}/2+(9/2)\arcsin(x/3)+C$

33. $-(\sqrt{1+t^2})/t+C$

35. $-\ln|x-1|+2\ln|x-2|+C$

37. $5\ln 2$

39. $\pi/12-\sqrt 3/8$

41. $\ln(1+\sqrt 2)$

43. (b) $-\sqrt{5-y^2}/(5y)+C$

45. (a) $(k/(b-a))\ln|(2b-a)/b|$
(b) $T\to\infty$

Section 7.5

1. (a) Underestimate
(b) Overestimate
(c) Overestimate
(d) Underestimate

3. (a) Underestimate
(b) Overestimate
(c) Underestimate
(d) Overestimate

5. (a) 27
(b) 135
(c) 81
(d) 67.5

7. (a) MID(2)= 24; TRAP(2)= 28
(b) MID(2) underestimate; TRAP(2) overestimate

9. LEFT(6) = 31

RIGHT(6) = 39

TRAP(6) = 35

11 (a) 0.664 = LEFT
0.633 = TRAP
0.632 = MID
0.601 = RIGHT
(b) Between
0.632 and 0.633

13 MID: over; TRAP: under

15 TRAP: over; MID: under

17 (a) TRAP(4); 1027.5
(b) Underestimate

19 (b) LEFT(5) ≈ 1.32350
error ≈ -0.03810
RIGHT(5) ≈ 1.24066
error ≈ 0.04474
TRAP(5) ≈ 1.28208
error ≈ 0.00332
MID(5) ≈ 1.28705
error ≈ -0.001656

25 RIGHT(10) = 5.556
TRAP(10) = 4.356
LEFT(20) = 3.199
RIGHT(20) = 4.399
TRAP(20) = 3.799

Section 7.6

1 72

5 (a) ≈ 53.598
(b) LEFT(2)= 16.778;
error= 36.820
RIGHT(2)= 123.974;
error= -70.376
TRAP(2)= 70.376;
error= -16.778
MID(2)= 45.608;
error= 7.990
SIMP(2)= 53.864;
error= -0.266
(c) LEFT(4)= 31.193;
error=22.405
RIGHT(4)= 84.791;
error=-31.193
TRAP(4)= 57.992;
error=-4.394
MID(4)= 51.428;
error=2.170
SIMP(4)= 53.616;
error=-0.018

7 (a) 4 places: 2 seconds
8 places: ≈ 6 hours
12 places: ≈ 6 years
20 places: ≈ 600 million years
(b) 4 places: 2 seconds
8 places: ≈ 3 minutes
12 places: ≈ 6 hours
20 places: ≈ 6 years

9 0.272

Section 7.7

3 (a) 0.9596, 0.9995, 0.99999996
(b) 1.0

5 Diverges

7 Converges to 1/2

9 1

11 $\ln 2$

13 $\pi/2$

15 Does not converge

17 $\pi/4$

19 Does not converge

21 Does not converge

23 0.01317

25 $2 - 2e^{-\sqrt{\pi}}$

27 $\pi/2$

29 $\frac{1}{2} \ln \frac{5}{3}$

31 Converges; 0

33 1

35 $\sqrt{b\pi}$

37 Converges for $p > -1$
to $pe^{p+1}/(p+1)^2$

39 (b) $t = 2$
(c) 4000 people

Section 7.8

1 Converges; behaves like $1/x^2$

3 Diverges; behaves like $1/x$

5 Diverges; behaves like $1/x$

7 Converges; behaves like $5/x^3$

9 Converges; behaves like $1/x^2$

11 Does not converge

13 Converges

15 Does not converge

17 Converges

19 Converges

21 Converges

23 Converges

25 Does not converge

27 0.606

29 Converges for $p > 1$
Diverges for $p \leq 1$

31 $a = 0.399$

33 (a) $\int_3^\infty e^{-x^2}\,dx \leq \frac{e^{-9}}{3}$
(b) $\int_n^\infty e^{-x^2}\,dx \leq \frac{1}{n}e^{-n^2}$

Chapter 7 Review

1 $-\cos t + C$

3 $(1/5)e^{5z} + C$

5 $(-1/2)\cos 2\theta + C$

7 $2x^{5/2}/5 + 3x^{5/3}/5 + C$

9 $(r+1)^4/4 + C$

11 $x^2/2 + \ln|x| - x^{-1} + C$

13 $\frac{1}{2}e^{t^2} + C$

15 $(\frac{1}{2}x^2 - \frac{1}{2}x + \frac{1}{4})e^{2x} + C$

17 $\frac{1}{2}x^2 \ln x - \frac{1}{4}x^2 + C$

19 $x(\ln x)^2 - 2x \ln x + 2x + C$

21 $-e^{0.5-0.3t}/0.3 + C$

23 $-\frac{1}{3}(4-x^2)^{3/2} + C$

25 $2 \sin \sqrt{y} + C$

27 $\frac{1}{4}\sin 2\theta + \frac{1}{2}\theta + C$

29 $-\frac{1}{2}\ln|\cos(2x-6)| + C$

31 $e - 2 \approx 0.71828$

33 $-11e^{-10} + 1$

35 0

37 $\pi/4$

39 $\ln|t| - 4/t - 2/t^2 + C$

41 $\ln|t| - 1/t + C$

43 $-\ln|\cos\theta| + C$

45 $(1/2)\ln|x^2 + 1| + C$

47 $\frac{1}{2}\arctan 2z + C$

49 $(-1/20)\cos^4 5\theta + C$

51 $(1/12)(t-10)^{12}$
$+ (10/11)(t-10)^{11} + C$

53 $xe^x - e^x + C$

55 $(1/142)(10^{71} - 2^{71})$

57 $(1/3)\arctan(u/3) + C$

59 $-\ln(|\cos(\ln x)|) + C$

61 $(1/2)\arcsin(2x) + C$

63 $y - (1/2)e^{-2y} + C$

65 $\ln|\ln x| + C$

67 $\sin\sqrt{x^2+1} + C$

69 $ue^{ku}/k - e^{ku}/k^2 + C$

71 $(1/\sqrt{2})e^{\sqrt{2}+3} + C$

73 $(1/2)e^{2x} + 2xe^x - 2e^x$
$+ (x^3/3) + C$

75 $\frac{5}{2}\ln|x^2 + 4| + 3\arctan(x/2) + C$

77 $\frac{1}{20}[\ln|r-10| - \ln|r+10|]$
$+ C$

79 $-e^{-ct}(k\cos kt + c\sin kt)/(k^2 + c^2) + C$

81 $x^{\sqrt{k}+1}/(\sqrt{k}+1)$
$+ (\sqrt{k})^x/(\ln\sqrt{k}) + C$

83 $(\ln|x+1| - \ln|x+4|)/3 + C$

85 $x^2/2 - 3x - \ln|x+1|$
$+ 8\ln|x+2| + C$

87 $\frac{1}{b}(\ln|x| - \ln|x+b/a|) + C$

89 $\ln|z| - \ln|z+1| + C$

91 $(1/\ln 2)\ln|2^t + 1| + C$

93 $(1/7)x^7 + 3x^5 + 25x^3 + 125x + C$

95 $\frac{1}{2}x\sqrt{4-x^2} + 2\arcsin(x/2) + C$

97 $-(1/2)\ln|1 + \cos^2 w| + C$

99 $x\tan x + \ln|\cos x| + C$

101 $(2/3)(\sqrt{x+1})^3 - 2\sqrt{x+1} + C$

103 $(1/2)\ln|e^{2y} + 1| + C$

105 $-1/(z-5) - 5/(2(z-5)^2)$
$+ C$

107 $e^{x^2-x} + C$

109 $-\frac{2}{9}(2 + 3\cos x)^{3/2} + C$

111 $\sin^3(2\theta)/6 - \sin^5(2\theta)/10 + C$

113 $(x + \sin x)^4/4 + C$

115 $(\ln|x-2| - \ln|x+2|)/4 + C$

117 $(\ln|x| - \ln|x+5|)/5 + C$

119 $(1/3)\arcsin(3x) + C$

121 $2\ln|x-1| - \ln|x+1| - \ln|x| + C$

123 $\arctan(x+1) + C$

125 $-4\ln|x-1| + 7\ln|x-2| + C$

127 Converges
$\int_4^\infty t^{-3/2}\,dt = 1$

129 Converges
$\int_0^\infty we^{-w}\,dw = 1$

131 Converges
$\int_{-\pi/4}^{\pi/4} \tan\theta\,d\theta = 0$

133 Converges
$\int_{10}^\infty \frac{1}{z^2-4}\,dz = (\ln(3/2))/4$

135 Does not converge

137 Does not converge

139 Does not converge

141 5/6

143 11/3

145 (a)   (i) 0

    (ii) $\frac{2}{\pi}$

    (iii) $\frac{1}{2}$

  (b) Smallest to largest:

    Average value of $f(t)$

    Average value of $k(t)$

    Average value of $g(t)$

147 error for $\mathrm{TRAP}(10)$

    ≈ 0.0078

149 (a) 0.5 ml

  (b) 99.95%

151 (a) $(\ln x)^2/2, (\ln x)^3/3, (\ln x)^4/4$

  (b) $(\ln x)^{n+1}/(n+1)$

153 (a) $(-9\cos x + \cos(3x))/12$

  (b) $(3\sin x - \sin(3x))/4$

155 (a) $x + x/(2(1+x^2)) - (3/2)\arctan x$

  (b) $1 - (x^2/(1+x^2)^2) - 1/(1+x^2)$

Ch. 7 Understanding

1 False

3 False

5 True

7 False

9 False

11 False

13 True

15 False

17 True

19 True

21 False

23 False

25 True

27 False.

Section 8.1

1 15

3 15/2

5 $(5/2)\pi$

7 1/6

9 $\int_0^9 4\pi \, dx = 36\pi$ cm^3

11 $\int_0^5 (4\pi/25)y^2 \, dy = 20\pi/3$ cm^3

13 $\int_0^5 \pi(5^2 - y^2) \, dy = 250\pi/3$ mm^3

15 5 to 1

17 Triangle; $b, h = 1, 3$

19 Quarter circle $r = \sqrt{15}$

21 Hemisphere, $r = 12$

23 Cone, $h = 6, r = 3$

25 $V = (4\pi r^3)/3$

27 (a) $3\Delta x$;

    $\int_0^4 3 \, dx = 12$ cm^3

  (b) $8(1 - h/3)\Delta h$;

    $\int_0^3 8(1 - h/3) \, dh = 12$ cm^3

29 $\int_0^{150} 1400(160 - h) \, dh = 1.785 \cdot 10^7$ m^3

Section 8.2

1 $\pi/5$

3 $\pi(e^2 - e^{-2})/2$

5 $256\pi/15$

7 $\pi^2/4$

9 $\pi((e^6/6) - (e^2/2) + 1/3)$

11 3.526

13 $e - 1$

15 (a) $16\pi/3$

  (b) 1.48

17 $V = (16/7)\pi \approx 7.18$

19 $V = (\pi^2/2) \approx 4.935$

21 $V \approx 42.42$

23 $V = (e^2 - 1/2) \approx 3.195$

25 (a) Volume ≈ 152 in^3

  (b) About 15 apples

27 $40,000LH^{3/2}/(3\sqrt{a})$

29 (a) $dh/dt = -6/\pi$

  (b) $t = \pi/6$

31 (a) $4\int_0^r \sqrt{1 + (-\frac{x}{y})^2} \, dx$

  (b) $2\pi r$

33 $e - e^{-1}$

Section 8.3

1 $1 - e^{-10}$ gm

3 (a) $\sum_{i=1}^{N}(2 + 6x_i)\Delta x$

  (b) 16 grams

5 (b) $\sum_{i=1}^{N}[600 + 300\sin(4\sqrt{x_i + 0.15})]\frac{20}{N}$

  (c) ≈ 11513

7 2 cm to right of origin

9 1 gm

11 (a) $\int_0^5 2\pi r(0.115e^{-2r}) dr$

  (b) 181 cubic meters

13 (a) $\pi r^2 l/2$

  (b) $2klr^3/3$

15 $\int_0^{60} \frac{1}{144}g(t) \, dt$ ft^3

17 $x = 2$

19 $\pi/2$

21 (a) Right

  (b) $2/(1 + 6e - e^2) \approx 0.2$

23 (a) $10/3$ gm

  (b) $\overline{x} = 3/5$ cm; $\overline{y} = 3/8$ cm

25 1.25 cm from center of base

27 (a) $16000\delta/3$ gm

  (b) 2.5 cm above center of base

Section 8.4

1 9/2 joules

3 (a) 1.5 joules, 13.5 joules

  (b) For $x = 4$ to $x = 5$ Force larger

5 $1.489 \cdot 10^{10}$ joules

7 11,000 ft-lbs

9 1,404,000 ft-lb

11 2,822,909.50 ft-lb

13 354,673 ft-lb

15 (a) Force on dam

$\approx \sum_{i=0}^{N-1} 1000(62.4h_i)\Delta h$

  (b) $\int_0^{50} = 1000(62.4h) \, dh = 78{,}000{,}000$ pounds

17 Bottom: 1497.6 lbs

    Front and back: 499.2 lbs

    Both sides: 374.4 lbs

19 (a) 21,840 lb/ft^2; 151.7 lb/in^2

  (b) (i) 546,000 pounds

     (ii) 542,100 pounds

21 $9800 \int_0^{100} h(3600 - 6h) \, dh = 1.6 \cdot 10^{11}$ newtons

23 60 joules

25 $(GmM)/(a(a + l))$

27 $GMmy/(a^2 + y^2)^{3/2}$ toward center

Section 8.5

1 $15,319.30.

3 $8,242, $12,296

7 (a) $5820 per year

  (b) $36,787.94

9 Installments

11 $46,800

13 (a) 10.6 years

  (b) 624.9 million dollars

15 $85,750,000

19 (a) Less

  (b) Can't tell

  (c) Less

Section 8.6

5 pdf; 1/2

7 pdf; 2/3

9 pdf; 2

11 (a) 0.9 m–1.1 m

15 (a) Cumulative distribution increasing

  (b) Vertical 0.2, horizontal 2

17 (a) 22.1%

  (b) 33.0%

  (c) 30.1%

  (d) $C(h) = 1 - e^{-0.4h}$

19 (b) About 3/4

Section 8.7

5 (a) 0.684 : 1

  (b) 1.6 hours

  (c) 1.682 hours

7 (a) $P(t) = $ Fraction of population who survive up to t years after treatment

  (b) $S(t) = e^{-Ct}$

  (c) 0.178

9 (a) $p(x) = \left(e^{-\frac{1}{2}\left(\frac{x-100}{15}\right)^2}\right) / \left(15\sqrt{2\pi}\right)$

  (b) 6.7% of the population

11 (c) μ represents the mean of the distribution, while σ is the standard deviation.

13 (b)

15 (a) $p(r) = 4r^2e^{-2r}$

  (b) Mean: 1.5 Bohr radii

    Median: 1.33 Bohr radii

    Most likely:

     1 Bohr radius

Chapter 8 Review

1 $\int_0^b h\,dx = hb$

3 $\int_0^b \frac{hx}{b}\,dx = \frac{hb}{2}$

5 $\int_0^{12} \pi h\,dh = 72\pi$

7 $\int_0^\pi \sqrt{1+\cos^2 x}\,dx$

9 (b) $\sum_{i=1}^N \pi x_i \Delta x$
   (c) Volume $= \pi/2$

11 $V = \pi$

13 (a) $a = b/l$
   (b) $(1/3)\pi b^2 l$

15 Volume $= 6\pi^2$

17 (a) $2/3$ gm
   (b) Less than $1/2$
   (c) $\overline{x} = \overline{y} = 9/20$ cm

19 1000 ft-lb

21 1,170,000 lbs

23 (a) $\sum_{i=0}^{n-1}(2000-100t_i)$
   $\times e^{-0.1t_i}\Delta t$
   (b) $\int_0^M e^{-0.10t}(2000-100t)\,dt$
   (c) After 20 years
   $\$11,353.35$

25 (a) $\sum_{i=1}^N \pi\left(\frac{3.5\cdot10^5}{\sqrt{h+600}}\right)^2 \Delta h$
   (b) $1.05 \cdot 10^{12}$ cubic feet

27 $\int_0^R 2\pi NrS(r)\,dr$

29 (a) $\pi PR^4/(8\eta l)$

31 (a) $\pi h^2/(2a)$
   (b) $\pi h/a$
   (c) $dh/dt = -k$
   (d) h_0/k

33 The thin spherical shell.

35 (c) $\pi^2 a^3/8$

37 (a) $(1/2)\sqrt{t}\sqrt{1+4t} + (1/4)\mathrm{arcsinh}\,(2\sqrt{t})$
   (b) t

Ch. 8 Understanding

1 True

3 False

5 False

7 False

9 True

11 True

13 False

15 False

17 False

19 True

21 False

23 False

25 False

27 True

29 False

Section 9.1

1 Yes, $a = 1$, ratio $= -1/2$

3 Yes, $a = 5$, ratio $= -2$

5 No

7 Yes, $a = 1$, ratio $= -x$

9 No

11 $y^2/(1-y), |y| < 1$

13 $1/(1+y^2), |y| < 1$

15 $-4/3$

17 $1/54$

19 260.42 mg

21 (a) $P_n =$
   $250(0.04) + 250(0.04)^2$
   $+ 250(0.04)^3 + \cdots$
   $+ 250(0.04)^{n-1}$
   (b) $P_n = 250 \cdot 0.04(1 - (0.04)^{n-1})$
   $/(1 - 0.04)$
   (c) $\lim_{n\to\infty} P_n \approx 10.42$
   We'd expect the difference between them to be 250 mg.

23 (a) $h_n = 10(3/4)^n$
   (b) $D_1 = 10$ feet
   $D_2 = h_0 + 2h_1$
   $= 25$ feet
   $D_3 = h_0 + 2h_1 + 2h_2$
   $= 36.25$ feet
   $D_4 = h_0 + 2h_1 + 2h_2$
   $+ 2h_3 \approx 44.69$ feet
   (c) $D_n =$
   $10 + 60\left(1 - (3/4)^{n-1}\right)$

25 (a) $\$1250$
   (b) 12.50

27 $\$900$ million

Section 9.2

1 0

3 0

5 Does not exist

7 4/7

9 Converges

11 Converges

13 Diverges

15 Diverges

17 Converges

19 Diverges

21 Not convergent

25 For all $\epsilon > 0$, there is an N such that $|S_n - L| < \epsilon$ for all $n \geq N$

31 12

Section 9.3

5 Does not converge

7 Converges

9 Converges

11 Converges

13 Converges

15 Converges

17 Converges

19 Does not converge

21 Converges

23 Diverges

25 Converges

27 Diverges

Section 9.4

1 Yes

3 No

5 $1 \cdot 3 \cdot 5 \cdots (2n-1)x^n/(2^n \cdot n!); n \geq 1$

7 $(-1)^k(x-1)^{2k}/(2k)!; k \geq 0$

9 $(x-a)^n/(2^{n-1} \cdot n!); n \geq 1$

11 1/5

13 2

15 1

17 ∞

19 1/4

21 1

23 Diverges when $x = 1/2$, converges when $x = -1/2$

Chapter 9 Review

3 Converges

5 Converges

7 Diverges

9 Diverges

11 Diverges

13 1/4

15 ∞

17 (a) $\approx 1\%$
   (b) $Q_1 = 250$
   $Q_2 = 250 + 250(0.01)$
   $Q_3 = 250 + 250(0.01)$
   $+ 250(0.01)^2$
   $Q_4 = 250 + 250(0.01)$
   $+ 250(0.01)^2$
   $+ 250(0.01)^3$
   (c) $Q_3 = \frac{250(1-(0.01)^3)}{1-0.01}$
   ≈ 252.5
   $Q_4 = \frac{250(1-(0.01)^4)}{1-0.01}$
   ≈ 252.5
   (d) $Q_n = 250 + 250(0.01)$
   $+ 250(0.01)^2 + \cdots$
   $+ 250(0.01)^{n-1}$
   $= \frac{250(1-(0.01)^n)}{1-0.01}$
   (e) $Q = \lim_{n\to\infty} Q_n = \frac{250}{1-0.01}$
   $= 252.5$

19 $\$25,503$

21 $\$926.40$

23 (a) $\$1000$
   (b) When the interest rate is 5%, the present value equals the principal.
   (c) The value of the bond.
   (d) Because the present value is more than the principal.

27 (b) Coefficient of x stays same coefficient of x^2 stays same after $T_2(S_2(x))$
   (c) Coefficient of x^3 is 31/6
   (d) Coefficient of x^k stays same after $T_k(S_k(x))$

Ch. 9 Understanding

1 True

3 True

5 True

7 True

9 False

11 False

13 True

15 False

17 True

19 True

21 False

23 False

25 False

Section 10.1

1 $P_4(x) = 1 - x + x^2 - x^3$
$\quad + x^4$
$P_6(x) = 1 - x + x^2 - x^3$
$\quad + x^4 - x^5 + x^6$
$P_8(x) = 1 - x + x^2 - x^3$
$\quad + x^4 - x^5 + x^6 - x^7 + x^8$

3 $P_2(x) = 1 + \frac{1}{2}x - \frac{1}{8}x^2$
$P_3(x) = 1 + \frac{1}{2}x - \frac{1}{8}x^2$
$\quad + \frac{1}{16}x^3$
$P_4(x) = 1 + \frac{1}{2}x - \frac{1}{8}x^2$
$\quad + \frac{1}{16}x^3 - \frac{5}{128}x^4$

5 $P_3(x) = P_4(x) = x - \frac{1}{3}x^3$

7 $P_2(x) = 1 - \frac{1}{3}x - \frac{1}{9}x^2$
$P_3(x) = 1 - \frac{1}{3}x - \frac{1}{9}x^2$
$\quad - \frac{5}{81}x^3$
$P_4(x) = 1 - \frac{1}{3}x - \frac{1}{9}x^2$
$\quad - \frac{5}{81}x^3 - \frac{10}{243}x^4$

9 $P_2(x) = 1 - \frac{1}{2}x + \frac{3}{8}x^2$
$P_3(x) = 1 - \frac{1}{2}x + \frac{3}{8}x^2$
$\quad - \frac{5}{16}x^3$
$P_4(x) = 1 - \frac{1}{2}x + \frac{3}{8}x^2$
$\quad - \frac{5}{16}x^3 + \frac{35}{128}x^4$

11 $P_4(x) = 1 - \frac{1}{2!}\left(x - \frac{\pi}{2}\right)^2$
$\quad + \frac{1}{4!}\left(x - \frac{\pi}{2}\right)^4$

13 $P_4(x) = e[1 + (x - 1)$
$\quad + \frac{1}{2}(x - 1)^2 + \frac{1}{6}(x - 1)^3$
$\quad + \frac{1}{24}(x - 1)^4]$

15 $c < 0, b > 0, a > 0$

17 $a < 0, b > 0, c > 0$

19 (a) 0
(b) 3
(c) -24
(d) 0
(e) 3600

21 $P_2(x) = 4x^2 - 7x + 2$
$f(x) = P_2(x)$

23 (a) If $f(x)$ is a polynomial of degree n, then $P_n(x)$, the nth degree Taylor polynomial for $f(x)$ about $x = 0$, is $f(x)$ itself.

27 (a) 3/7
(b) 0

29 (a) $P_4(x) = 1 + x^2 + \frac{1}{2}x^4$
(b) If we substitute x^2 for x in the Taylor polynomial for e^x of degree 2, we will get $P_4(x)$, the Taylor polynomial for e^{x^2} of degree 4.
(c) $P_{20}(x) = 1 + \frac{x^2}{1!} + \frac{x^4}{2!}$
$\quad + \cdots + \frac{x^{20}}{10!}$
(d) $e^{-2x} \approx 1 - 2x + 2x^2$
$\quad - \frac{4}{3}x^3 + \frac{2}{3}x^4 - \frac{4}{15}x^5$

31 (a) $0.94444\cdots$
(b) $0.94611\cdots$

Section 10.2

1 $f(x) = 1 + x + x^2$
$\quad + x^3 + \cdots$

3 $f(x) = 1 + \frac{x}{2} - \frac{x^2}{8}$
$\quad + \frac{x^3}{16} + \cdots$

5 $\sin x = \frac{\sqrt{2}}{2} + \frac{\sqrt{2}}{2}\left(x - \frac{\pi}{4}\right)$
$\quad - \frac{\sqrt{2}}{4}\left(x - \frac{\pi}{4}\right)^2 - \frac{\sqrt{2}}{12}\left(x - \frac{\pi}{4}\right)^3$
$\quad - \cdots$

7 $\sin\theta = -\frac{\sqrt{2}}{2} + \frac{\sqrt{2}}{2}\left(\theta + \frac{\pi}{4}\right)$
$\quad + \frac{\sqrt{2}}{4}\left(\theta + \frac{\pi}{4}\right)^2$
$\quad - \frac{\sqrt{2}}{12}\left(\theta + \frac{\pi}{4}\right)^3 + \cdots$

9 $\frac{1}{x} = 1 - (x - 1) +$
$\quad (x - 1)^2 - (x - 1)^3 + \cdots$

11 $\frac{1}{x} = -1 - (x + 1)$
$\quad - (x + 1)^2 - (x + 1)^3$
$\quad - \cdots$

13 $(-1)^n x^n; n \geq 0$

15 $(-1)^{n-1} x^n / n; n \geq 1$

17 $(-1)^k x^{2k+1}/(2k + 1); k \geq 0$

19 $(-1)^k x^{2k+2}/(2k)!; k \geq 0$

21 (a) $\ln(1 + 2x)$
$\quad = 2x - 2x^2 + \frac{8}{3}x^3 + \cdots$
(b) To get the expression for $\ln(1 + 2x)$ from the series for $\ln(1 + x)$, substitute $2x$ for x in the series.
(c) $-\frac{1}{2} < x < \frac{1}{2}$

23 $-1 < x < 1$

25 1

27 e^2

29 $4/3$

31 $\ln(3/2)$

33 e^3

35 $e^{-0.1}$

37 $e^{0.2} - 1$

39 $\frac{d}{dx}(x^2 e^{x^2})|_{x=0} = 0$
$\frac{d^6}{dx^6}(x^2 e^{x^2})|_{x=0} = \frac{6!}{2} = 360$

Section 10.3

1 $\sqrt{1 - 2x} = 1 - x - \frac{x^2}{2}$
$\quad - \frac{x^3}{2} - \cdots$

3 $e^{-x} = 1 - x + \frac{x^2}{2!}$
$\quad - \frac{x^3}{3!} + \cdots$

5 $\ln(1 - 2y) = -2y - 2y^2$
$\quad - \frac{8}{3}y^3 - 4y^4 - \cdots$

7 $\frac{1}{\sqrt{1-z^2}} = 1 + \frac{1}{2}z^2 + \frac{3}{8}z^4$
$\quad + \frac{5}{16}z^6 + \cdots$

9 $\frac{z}{e z^2} = z - z^3 + \frac{z^5}{2!}$
$\quad - \frac{z^7}{3!} + \cdots$

11 $e^t \cos t = 1 + t - \frac{t^3}{3}$
$\quad - \frac{t^4}{6} + \cdots$

13 $1 + 3x + 3x^2 + x^3$
$0 \cdot x^n$ for $n \geq 4$

15 $1 + \frac{1}{2}y^2 + \frac{3}{8}y^4 + \cdots +$
$\quad \frac{(1/2)(3/2)\cdots(\frac{1}{2}+n-1)y^{2n}}{n!} + \cdots;$
$\quad n \geq 1$

17 $\frac{a}{\sqrt{a^2+x^2}} = 1 - \frac{1}{2}\left(\frac{x}{a}\right)^2 + \frac{3}{8}\left(\frac{x}{a}\right)^4$
$\quad - \frac{5}{16}\left(\frac{x}{a}\right)^6 + \cdots$

19 Smallest: $\cos\theta$
Largest: $1 + \sin\theta$

21 (a) I
(b) IV
(c) III

(d) II

23 (a) $1 + x^2/2! + x^4/4!$
$\quad + x^6/6! + x^8/8!$
(b) 1.54308
(c) $x + x^3/3! + x^5/5!$
$\quad + x^7/7!$

25 $E \approx \frac{kQ}{R^2}\left(\frac{4}{R} + \frac{8}{R^3}\right)$

29 (a) $((l_1 + l_2)/c) \cdot$
$\quad (v^2/c^2 + (5/4)v^4/c^4)$
(b) $v^2, (l_1 + l_2)/c^3$

31 (a) $\frac{GMm}{h^2}\left(1 - \frac{3}{4}\frac{a^2}{h^2}\right)$
(b) $F \approx (GMm)/(h^2)$
(c) 0.03%

33 (a) Left = 0.8076
Right = 0.6812
(b) $1 - x^2 + x^4/2 - x^6/6$
(c) 0.74286

Section 10.4

1 $|E_3| \leq 0.033$

3 $|E_3| \leq 4.375$

5 (a) Underestimate
(b) 1

7 (a) Overestimate:
$\quad 0 < \theta \leq 1$
Underestimate:
$\quad -1 \leq \theta < 0$
(b) $|E_2| \leq 0.17$

9 (a) 4, 0.2
(b) 1
(c) $-4 \leq x \leq 4$

11 Four decimal places: $n = 7$
Six decimal places: $n = 9$

Section 10.5

1 Not a Fourier series

3 Fourier series

5 $F_1(x) = F_2(x) = \frac{4}{\pi}\sin x$
$F_3(x) = \frac{4}{\pi}\sin x + \frac{4}{3\pi}\sin 3x$

7 99.942% of the total energy

9 $H_n(x) =$
$\quad \frac{\pi}{4} + \sum_{i=1}^{n} \frac{(-1)^{i+1}\sin(ix)}{i} +$
$\quad \sum_{i=1}^{[n/2]} \frac{-2}{(2i-1)^2\pi}\cos((2i-1)x),$
where $[n/2]$ denotes the biggest integer smaller than or equal to $n/2$.

11 (a) $F_3(x) =$
$\quad \frac{1}{2} + \frac{2}{\pi}\cos x - \frac{2}{3\pi}\cos 3x.$
(b) There are cosines instead of sines, but the energy spectrum remains the same

13 $F_4(x) = 1 - \frac{4}{\pi}\sin(\pi x)$
$\quad - \frac{2}{\pi}\sin(2\pi x) - \frac{4}{3\pi}\sin(3\pi x)$
$\quad - \frac{1}{\pi}\sin(4\pi x)$

19 (a) 15.9155%, 0.451808%
(b) $(4\sin^2\frac{k}{2})/(k^2\pi^2)$
(c) The constant term and the first five harmonics are needed.
(d) $F_5(x) = \frac{1}{2\pi} + \frac{2\sin(1/2)}{\pi}\cos x$
$\quad + \frac{\sin 1}{\pi}\cos 2x$
$\quad + \frac{2\sin(3/2)}{3\pi}\cos 3x$
$\quad + \frac{\sin 2}{2\pi}\cos 4x$
$\quad + \frac{2\sin(5/2)}{5\pi}\cos 5x$

21 (a) 31.83%, 76.91%
(b) 90.07%
(c) $F_3(x) = \frac{1}{\pi} + \frac{2\sin 1}{\pi}\cos x$
$\quad + \frac{\sin 2}{\pi}\cos 2x + \frac{2\sin 3}{3\pi}\cos 3x$

Chapter 10 Review

1 $e^x \approx 1 + e(x-1) + \frac{e}{2}(x-1)^2$

3 $\sin x \approx -\frac{1}{\sqrt{2}} + \frac{1}{\sqrt{2}}(x + \frac{\pi}{4})$
   $+ \frac{1}{2\sqrt{2}}(x + \frac{\pi}{4})^2$

5 $P_3(x) = 4 + 12(x-1)$
   $+ 10(x-1)^2 + (x-1)^3$

7 $\sin t^2 =$
   $t^2 - \frac{t^6}{3!} + \frac{t^{10}}{5!} - \frac{t^{14}}{7!} + \cdots$

9 $\frac{1}{1-4z^2} =$
   $1 + 4z^2 + 16z^4 + 64z^6 + \cdots$

11 $\sqrt{R-r} =$
   $\sqrt{R}(1 - \frac{1}{2}\frac{r}{R} - \frac{1}{8}\frac{r^2}{R^2} - \frac{1}{16}\frac{r^3}{R^3}$
   $- \cdots)$

13 $3/4$

15 e^{-2}

17 $3e$

19 1.45

21 $-1 < x < 1; R = 1$

23 (a) 2
   (b) Near $\theta = 0$, we make the approximation
        $\frac{\sin 2\theta}{\theta} \approx 2 - \frac{4}{3}\theta^2$.

25 (a) $2 - (x^2/4) - (x^4/64)$
   (b) 1.9135
   (c) $(\sqrt{3}/2) + (\pi/3)$
   (d) Three decimal places

27 (a) $V(x) \approx V(0) + V''(0)x^2/2$
        $V''(0) > 0$
   (b) Force $\approx -V''(0)x$
        $V''(0) > 0$
        Toward origin

29 (a) Set $\frac{dV}{dr} = 0$, solve for r. Check for max or min.
   (b) $V(r) = -V_0$
        $+ 72V_0 r_0^{-2} \cdot (r-r_0)^2 \cdot \frac{1}{2}$
        $+ \cdots$
   (d) $F = 0$ when $r = r_0$

33 (a) $g(x) \approx P_n(x) =$
        $g(0) + \frac{g''(0)}{2!}x^2 + \frac{g'''(0)}{3!}x^3$
        $+ \cdots + \frac{g^{(n)}(0)}{n!}x^n$
   (b) If $g''(0) > 0$:
        0 is a local minimum
        If $g''(0) < 0$:
        0 is a local maximum.

37 (b) If the amplitude of the kth harmonic of f is A_k, then the amplitude of the kth harmonic of f' is kA_k.
   (c) The energy of the kth harmonic of f' is k^2 times the energy of the kth harmonic of f.

41 (a) $P_7(x) = x - \frac{x^3}{6} + \frac{x^5}{120} - \frac{x^7}{5040}$
        $Q_7(x) = x - \frac{2x^3}{3} + \frac{2x^5}{15} - \frac{4x^7}{315}$
   (b) For n odd, ratio of coefficients of x^n is 2^{n-1}

43 (a) $P_{10}(x) = 1 + \frac{x^2}{12} - \frac{x^4}{720} + \frac{x^6}{30240} -$
        $\frac{x^8}{1209600} + \frac{x^{10}}{47900160}$
   (b) All even powers
   (c) f even

Ch. 10 Understanding

1 False

3 False

5 False

7 True

9 False

11 False

13 True

15 False

17 True

19 True

21 False

23 True

Section 11.1

1 (a) (III)
  (b) (V)
  (c) (I)
  (d) (II)
  (e) (IV)

3 Yes.

5 -2

13 (b) 1

15 (a) (IV)
   (b) None
   (c) (V)
   (d) (I), (II)
   (e) (III)

Section 11.2

3 (b) $y = -x - 1$

5 (b) $y = n\pi$

7 (c) Increasing:
       $-1 < y < 2$
       Decreasing:
       $y > 2$ or $y < -1$
       Horizontal:
       $y = 2$ or $y = -1$

9 (a) II
  (b) VI
  (c) IV
  (d) I
  (e) III
  (f) V

Section 11.3

1 (a) $y(0.4) \approx 1.5282$
  (b) $y(0.4) = -1.4$

3 (b) $y(x) = x^4/4$

7 (a) $\Delta x = 0.5, y(1) \approx 1.5$
       $\Delta x = 0.25, y \approx 1.75$
  (b) $y = x^2 + 1$, so $y(1) = 2$
  (c) Yes

9 (a) $B \approx 1050$
  (b) $B \approx 1050.63$
  (c) $B \approx 1050.94$

Section 11.4

1 $P = 20e^{0.02t}$

3 $P(t) = \sqrt{2t+1}$

5 $L(p) = 100e^{p/2}$

7 $m = 5e^{3t-3}$

9 $z = 5e^{5t-5}$

11 $z(y) = e^{y^2/2}$

13 $P = 104e^t - 4$

15 $m = 3000e^{0.1t} - 2000$

17 $y = 200 - 150e^{t/2}$

19 $R(r) = 1 - 0.9e^{1-r}$

21 $z = -\ln(1 - t^2/2)$

23 $y = -2/(t^2 + 2t - 4)$

25 $w = 2/(\cos\theta^2 + 1)$

27 $w = -2/(\sin\psi^2 - 2)$

29 $Q = Ae^{t/k}$

31 $Q = b - Ae^{-t}$

33 $P(t) = (1/a)(Ae^{at} - b)$

35 $y(t) = -1/\left(k(t + t^3/3) + C\right)$

37 $L(x) = b + Ae^{k((1/2)x^2 + ax)}$

39 $x = \arcsin(At^{\ln t + 1})$

41 $y = 2(2^{-e^{-t}})$

43 (c) $y(x) = Ae^{x^2/2}$

45 (c) $y^2 - x^2 = 2C$

Section 11.5

1 (a) (I)
  (b) (IV)
  (c) (III)

3 (a) $y = 2$: stable; $y = -1$: unstable

5 $y = 3$ stable;
   $y = 1$ unstable

7 (a) $dS/dt = kS$
  (b) $S = Ce^{kt}$
  (c) $C = 5$
  (d) $S = 5e^{0.1576t}$

9 (a) 69,300 barrels/year
  (b) 25.9 years

11 (a) $dH/dt = k(68 - H)$
   (b) $H = 68 - Ce^{-kt}$
   (c) $57.8°F$

13 (a) $\frac{dB}{dt} = \frac{r}{100}B$
        Constant $= \frac{r}{100}$
   (b) $B = Ae^{\frac{r}{100}t}$

15 Michigan: 72 years
    Ontario: 18 years

17 (b) $dQ/dt = -0.0187Q$
   (c) 3 days

19 (a) $dT/dt = -k(T - 68)$
   (b) $T = 68 + 22.3e^{-0.06t}$;
        3:45 am.

21 About 2150 B.C.

23 (a) $\frac{dD}{dt} = kD$

Section 11.6

1 $da/dt = -ka/m$
   $a = ge^{-kt/m}$

3 (a) $dB/dt = 0.08B - 2000$
  (b) $B = 25,000 + Ae^{0.08t}$
  (c) (i) $17,540.88
       (ii) $32,549.12

5 $dD/dt = -0.75(D - 4)$
   Equilibrium $= 4$ g/cm^2

7 about 3 days

9 $y = A\sqrt{x}$

11 (a) $dy/dt = ky$
   (b) 0.2486 grams

13 (a) $I = Ae^{-kl}$
   (b) 20 feet: 75%
        25 feet: 82.3%

15 (a) $dx/dt = k(a - x)(b - x)$
   (b) $x = ab(e^{bkt} - e^{akt})/(be^{bkt} - ae^{akt})$

17 (b) $dc/dt = (43.2/35,000) - 0.082c$
   (c) $c = 0.015(1 - e^{-0.082t})$
        As $t \to \infty$,
        $c \to 0.015$ mg/ml

19 (a) $dp/dt = -k(p - p^*)$

(b) $p = p^* + (p_0 - p^*)e^{-kt}$

(d) As $t \to \infty$, $p \to p^*$

21 (a) $\frac{dc}{dt} = \frac{0.0001}{60} - \frac{0.002}{60}c$

(b) $c = 0.05 - 0.05e^{-3 \times 10^{-5}t}$

(c) $c \to 0.05$

23 (a) $\frac{dS}{dt} = 600 - \frac{3S}{100,000}$
(S in grams)

(b) $S(t) =$
$2 \times 10^7 (1 - e^{-\frac{3}{100,000}t})$

(c) As $t \to \infty$
$S(t) \to 2 \times 10^7$

Section 11.7

1 $P = (6.6 \times 10^6)e^{0.002t}$

3 (a) 0.252

(b) All the available land will be used for farming

(c) In 1974

(d) In 1974

5 dP/dt is largest in 1920.
Using this population to estimate the 1990 population, we get 211.4 million.

7 (a) $P = 5000/(1 + 499e^{-1.78t})$

(c) $t \approx 3.5$; $P \approx 2500$

9 (a) $dI/dt = k(M - I)$
$(k > 0)$

(b) $dI/dt = kI(M - I)$
$(k > 0)$

11 (a) $dp/dt = kp(B - p)$
$(k > 0)$

(b) Half of the tin

13 (c) $P = 0$ (stable)
$P = 4$ (unstable)

15 (b) $dP/dt < 0$ for $P < b/a$
$dP/dt > 0$ for $P > b/a$

(c) $P > b/a$: increase
$P < b/a$: extinction

Section 11.8

3 $r = r_0 e^{-t}$, $w = w_0 e^t$

5 Worms decrease, robins increase. Long run: populations oscillate.

9 Robins increase;
Worms constant, then decrease;
Both oscillate in long run

11 x and y increase, about same rate

13 x decreases quickly while y increases more slowly.

15 (a) Symbiosis

(b) Both $\to \infty$ or both $\to 0$

17 (a) Predator-prey

(b) x, y tend to ≈ 1

19 (a) $dy/dx = bx/ay$

21 (b) $dx/dt = -xy$, $dy/dt = -x$

(c) $dy/dx = 1/y$
soln: $y^2/2 = x + C$

(d) If $C > 0$, y wins
If $C < 0$, x wins
If $C = 0$, mutual annihilation

23 (a) $\frac{dy}{dx} = \frac{y(x+3)}{x(y+2)}$
$y^2 e^y = Ax^3 e^x$

(d) $y^2 e^y = x^3 e^{x+4}$

(e) $y = x = e^{-4} \approx 0.0183$

(f) $y \approx e^{-13}$

Section 11.9

1 (a) $\frac{dS}{dt} = 0$ where
$S = 0$ or $I = 0$.

$\frac{dI}{dt} = 0$ where
$I = 0$ or $S = 192$.

(b) Where $S > 192$,
$\frac{dS}{dt} < 0$ and $\frac{dI}{dt} > 0$.
Where $S < 192$,
$\frac{dS}{dt} < 0$ and $\frac{dI}{dt} < 0$.

3 (a) $(0, 0)$; $(5, 2)$

(b) dx/dt and dy/dt not both 0

5 Horizontal nullclines; $x = \frac{1}{2}$, $y = 0$
Vertical nullclines; $y = \frac{1}{3}(2 - x)$, $x = 0$
Equilibrium points; $(0, 0)$, $(\frac{1}{2}, \frac{1}{2})$, $(2, 0)$

7 Horizontal nullclines; $y = 0$, $y = 2(1 - x)$
Vertical nullclines; $x = 0$, $y = 1 - x/3$
Equilibrium points; $(0, 0)$, $(0, 2)$, $(3, 0)$, $(3/5, 4/5)$

9 $\frac{dx}{dt} = 0$ when
$x = 0$ or $y + \frac{x}{2} = 1$.
$\frac{dy}{dt} = 0$ when
$y = 0$ or $x + \frac{y}{3} = 1$.
Equilibrium points:
$(2, 0)$, $(0, 3)$, $(0, 0)$, $(\frac{4}{5}, \frac{3}{5})$

11 (a) $\frac{dP_1}{dt} = 0$ where $P = 0$ or
$P_1 + 3P_2 = 13$.
$\frac{dP_2}{dt} = 0$ where $P = 0$ or
$P_2 + 0.4P_1 = 6$.

13 (a) $dx/dt = 0$ where
$x = 23.3$
$dy/dt = 0$ where
$8.2x - 0.8y - 142 = 0$
Equilibrium point:
$(23.3, 61.7)$

Section 11.10

7 (c) $A = 5$, $\phi = 0.93$

9 $\sqrt{58}$

11 $\omega = 2$, $A = 13$, $\psi = \tan^{-1} 12/5$

13 Highest point, at rest

15 Lowest point, at rest

17 (a) goes with (II)

(b) goes with (I) and (IV)

(c) goes with (III)

(I) $x = 2 \sin 2t$

(II) $x = -\sin t$

(III) $x = \cos 4t$

(IV) $x = -3 \sin 2t$

19 (a) $x = v_0 \sqrt{l/g} \sin \sqrt{g/lt}$

(b) $x = x_0 \cos \sqrt{g/lt}$

21 (a) $x'' + (g/2)x = 0$, $x(0) = 5$, $x'(0) = 0$

(b) $x = 5 \cos((\sqrt{g/2})t)$

23 (a) $A \sin(\omega t + \phi) =$
$(A \sin \phi) \cos \omega t +$
$(A \cos \phi) \sin \omega t$

25 $C = \frac{1}{20}$ farads

Section 11.11

1 $y(t) = C_1 e^{-t} + C_2 e^{-3t}$

3 $y(t) =$
$C_1 e^{-2t} \cos t + C_2 e^{-2t} \sin t$

5 $s(t) =$
$C_1 \cos \sqrt{7}t + C_2 \sin \sqrt{7}t$

7 $z(t) = C_1 e^{-t/2} + C_2 e^{-3t/2}$

9 $p(t) = C_1 e^{-t/2} \cos \frac{\sqrt{3}}{2}t +$
$C_2 e^{-t/2} \sin \frac{\sqrt{3}}{2}t$

11 $y(t) = A + Be^{-2t}$

13 $y(t) = -2e^{-3t} + 3e^{-2t}$

15 $y(t) = \frac{1}{5}e^{4t} + \frac{4}{5}e^{-t}$

17 $y(t) = \frac{5}{4}e^{-t} - \frac{1}{4}e^{-5t}$

19 $y(t) = 2e^{-3t} \sin t$

21 $y(t) = \frac{1}{1-e}e^{-2t} + \frac{-e}{1-e}e^{-3t}$

23 $p(t) = 20e^{(\pi/2)-t} \sin t$

25 (a) (IV)

(b) (II)

(c) (I)

(d) (III)

27 $k = 6$
$y(t) = C_1 e^{2t} + C_2 e^{3t}$

29 (iii)

31 (iv)

33 (ii)

35 Overdamped: $c < 2$
Critically damped: $c = 2$
Underdamped: $c > 2$

37 $z(t) = 3e^{-2t}$

39 (a) $Q(t) = \frac{2}{\sqrt{3}}(e^{(-1+\frac{\sqrt{3}}{2})t}$
$- e^{(-1-\frac{\sqrt{3}}{2})t})$

(b) $Q(t) =$
$\frac{1}{\sqrt{3}}((2 + \sqrt{3})e^{(-1+\frac{\sqrt{3}}{2})t} -$
$(2 - \sqrt{3})e^{(-1-\frac{\sqrt{3}}{2})t})$

41 (a) $Q(t) = 16e^{-\frac{1}{8}t} \sin \frac{t}{8}$

(b) $Q(t) =$
$2e^{-\frac{1}{8}t} \left(\sin \frac{t}{8} + \cos \frac{t}{8} \right)$

(c) Goes from overdamped to underdamped

43 No

Chapter 11 Review

1 (a) Yes (b) No (c) Yes
(d) No (e) Yes (f) Yes
(g) No (h) Yes (i) No
(j) Yes (k) Yes (l) No

3 $y(x) = 40 + Ae^{0.2x}$

5 $H = Ae^{0.5t} - 20$

7 $P = 100Ae^{0.4t}/(1 + Ae^{0.4t})$

9 $P = \frac{40000}{3}(e^{0.03t} - 1)$

11 $y = \sqrt[3]{33 - 6 \cos x}$

13 $\frac{1}{2}y - 4 \ln |y| = 3 \ln |x| - x$
$+ \frac{7}{2} - 4 \ln 5$.

15 $z(t) = 1/(1 - 0.9e^t)$

17 $20 \ln |y| - y =$
$100 \ln |x| - x + 20 \ln 20 - 19$

19 $y = \ln(e^x + e - 1)$

21 $z = \sin(-e^{\cos \theta} + \pi/6 + e)$

23 $y = -\ln(\frac{\ln 2}{3} \sin^2 t \cos t$
$2 \ln 23 \cos t - \frac{2 \ln 2}{3} + 1)/\ln 2$

25 $y(t) = C_1 e^{t/3} + C_2 e^{-t/3}$

27 $y(t) = C_1 e^{-2t} + C_2 e^{-4t}$

29 $y(t) =$
$C_1 e^{-t} \cos 3t + C_2 e^{-t} \sin 3t$

31 (a) $y(1) \approx 3.689$

(b) Overestimate

(c) $y = 5 - 4e^{-x}$
$y(1) = 5 - 4e^{-1} \approx 3.528$

(d) ≈ 3.61

33 Always overdamped

35 2.21 hours

37 5.7 days, 126.32 liters

39 (a) $dB/dt = 0.1B + 1200$
   (b) $B = 12,000(e^{0.1t} - 1)$
   (c) $7784.66

41 $dI/dt = 0.001I(1 - 10(I/M)); M/10$

43 (b) $d^2r/dt^2 = -ABr$
      $r(t) = C_1 \cos \sqrt{AB}t$
         $+ C_2 \sin \sqrt{AB}t$
   (c) $r(t) = \cos \sqrt{AB}t$
      $j(t) = \sqrt{\frac{A}{B}} \sin \sqrt{AB}t$
   (d) One quarter of the time.

45 (a) $P = 0, 1, 2$
   (c) $1/2, 3/2, 0, 2$

Ch. 11 Understanding

1 False

3 False

5 False

7 False

9 True

11 False

13 True

15 True

17 True

19 True

21 False

23 False

25 True

27 False

29 False

31 True

33 True

35 $dy/dx = y - x^2$

37 $dy/dx = x/y$

Section 12.1

7 Increasing function

9 A, B, C

11 $(1, -1, 1)$; Front, left, above.

13 P_2, P_3

15 $(x - 1)^2 + (y - 2)^2$
      $+ (z - 3)^2 = 25$

19 (a) $C = 40d + 0.15m$
   (b) $245

21 $f(x, y) = y^2$

27 Cylinder radius 2 along x-axis

29 $(1.5, 0.5, -0.5)$

31 (a) $z = 7, z = -1$
   (b) $x = 6, x = -2$
   (c) $y = 7, y = -1$

33 h

Section 12.2

1 Horizontal plane 3 units above the xy-plane

3 Sphere, radius 3

5 Plane, x-intercept 6, y-intercept 3, z-intercept 4

7 Upside-down bowl, vertex $(0, 0, 5)$

9 (a) I
   (b) V
   (c) IV
   (d) II
   (e) III

11 (a) IV
   (b) I
   (c) III

13 (a) Bowl
   (b) Neither
   (c) Plate
   (d) Bowl
   (e) Plate

15 IV

21 (a) $y = 1$
   (b) $y = \pi$
   (c) $x = 0$

Section 12.3

11 Table 11 matches (II)
   Table 12.8 matches (III)
   Table 12.9 matches (IV)
   Table 12.10 matches (I)

17 (a) about $137
   (b) about $250

19 (a) A
   (b) B
   (c) A

21 (a) $\pi = 3q_1 + 12q_2 - 4$
      (thousands)

23 (a) (II) (E)
   (b) (I) (D)
   (c) (III) (G)

25 $\alpha + \beta > 1$: increasing
   $\alpha + \beta = 1$: constant
   $\alpha + \beta < 1$: decreasing

27 (a) I
   (b) IV
   (c) II
   (d) III

33 (b) Lines parallel to $y = x$

Section 12.4

1 Linear function

3 Linear function

5 (a) Yes
   (b) Dollars per month
      Dollars per minute
   (c) Charge to connect to service
   (d) $120

7 Not linear

9 Linear

11 (a) $z = -13 + 5x - 3y$

13 $f(x, y) = -95 + 5x - 5y$

15 $f(x, y) = -0.01x + 0.3y + 1$

17 $f(x, y) = 2 - \frac{1}{2} \cdot x - \frac{2}{3} \cdot y$

19 $z = 4 + 3x + y$

21 180 lb person at 8 mph
   120 lb person at 10 mph

23 (a) All $\Delta z = 7$
   (b) All $\Delta z = -5$

29 (a) Linear
   (b) Linear
   (c) Not linear

Section 12.5

1 (a) I
   (b) II

3 $f(x, y) = 2x - (y/2) - 3$
   $g(x, y, z) = 4x - y - 2z = 6$

5 $f(x, y) = -\sqrt{2(1 - x^2 - y^2)}$
   $g(x, y, z) = x^2 + y^2 + z^2/2 = 1$

7 No

9 No

11 $f(x, y, z) = (x - a)^2 + (y - b)^2 + (z - c)^2$

13 $f(x, y) = (1 - x^2 - y)^2$

15 Hyperboloid of two sheets

17 Elliptic paraboloid

19 $f(x, y, z) = x + 2y - z + 1$

21 Elliptical cylinder along y-axis

23 All planes with normal $\vec{i} + \vec{j} + \vec{k}$

25 Parallel planes

27 (a) Graph of f is graph of $x^2 + y^2 + z^2 = 1, z \geq 0$
   (b) $\sqrt{1 - x^2 - y^2} - z = 0$

29 $g(x, y, z) = 1 - \sqrt{x^2 + z^2} - y = 0$

31 Surface of rotation

Section 12.6

1 Not continuous

3 Continuous

5 Not continuous

7 1

9 0

11 1

13

15 $c = 1$

17 (c) No

21 No

Chapter 12 Review

1 Vertical line through $(2, 1, 0)$

5 Lines with slope $3/5$

7 $2x^2 + y^2 = k$

9 $h(x, y) = 4 + 2x - y$

11 $z = f(x, y) = -\sqrt{4 - (x - 3)^2 - y^2}$
   $g(x, y, z) = (x - 3)^2 + y^2 + z^2 = 4$

13 $f(x, y) = x^2 + y^2 + 5$
   $g(x, y, z) = x^2 + y^2 + 5 - z = 0$

15 Could not be true

17 Might be true

19 True

21 $(-2, 3, -6); 7$

25 Infinite families of parallel planes

35 (a) $9 + 3x + 4y, 21 + 7x + 8y$
   (b) $189 + 63x + 64y$

Ch. 12 Understanding

1 True

3 False

5 False

7 True

9 True

11 False

13 False

15 True

17 True

19 False

21 True

23 True

25 True

27 False

29 False

31 True

33 False

35 False

37 False

39 True

41 False

43 True

45 True

47 True

49 False

51 True

53 True

55 True

57 False

59 False

Section 13.1

1 $\sqrt{11}$

3 $-15\vec{i} + 25\vec{j} + 20\vec{k}$

5 $\sqrt{65}$

7 $\vec{i} + 3\vec{j}$

9 $-4.5\vec{i} + 8\vec{j} + 0.5\vec{k}$

11 $\sqrt{11}$

13 5.6

15 $\vec{a} = -2\vec{j}, \vec{b} = 3\vec{i}, \vec{c} = \vec{i} + \vec{j}$,
$\vec{d} = 2\vec{j}, \vec{e} = \vec{i} - 2\vec{j}, \vec{f} = -3\vec{i} - \vec{j}$

17 (b) $\|\vec{v}\| = 8.602$
(c) $54.46°$

19 $\vec{a} = \vec{b} = \vec{c} = 3\vec{k}$
$\vec{d} = 2\vec{i} + 3\vec{k}$
$\vec{e} = \vec{j}$
$\vec{f} = -2\vec{i}$

21 (a) True
(b) False
(c) False
(d) True
(e) True
(f) False

23 $\vec{i} + 4\vec{j}$

25 $-\vec{i}$

27 $(2/\sqrt{6})\vec{i} - (2/\sqrt{6})\vec{j}$
$+ (4/\sqrt{6})\vec{k}$

29 $\vec{v} = 3\vec{i} + 4\vec{j}$ and $\vec{v} = 3\vec{i} - 4\vec{j}$

31 $-40\vec{i} - 30\vec{j}$

Section 13.2

1 Scalar

3 Vector

5 Vector

7 $-37.59\vec{i} - 13.68\vec{j}$

9 $21\vec{j} + 35\vec{k}$

13 (a) $17.93\vec{i} - 7.07\vec{j}$
(b) 19.27 km/hr
(c) 21.52° south of east

15 $-140.8\vec{i} + 140.8\vec{j} + 18\vec{k}$

17 3.4° north of east

19 (79.00,79.33,89.00,68.33,89.33)

21 $-11\vec{i} + 4\vec{j}$

Section 13.3

1 -38

3 14

5 238

7 Many answers possible.
(a) $8\vec{i} + 6\vec{j}$
(b) $-3\vec{i} + 4\vec{j}$

9 $3\vec{i} + 4\vec{j} - \vec{k}$
(multiples of)

11 $\vec{i} + 3\vec{j} + 2\vec{k}$
(multiples of)

13 Parallel:
$3\vec{i} + \sqrt{3}\vec{j}$ and $\sqrt{3}\vec{i} + \vec{j}$
Perpendicular:
$\sqrt{3}\vec{i} + \vec{j}$ and $\vec{i} - \sqrt{3}\vec{j}$
$3\vec{i} + \sqrt{3}\vec{j}$ and $\vec{i} - \sqrt{3}\vec{j}$

15 $\vec{v}_1, \vec{v}_4, \vec{v}_8$ are all parallel
$\vec{v}_3, \vec{v}_5, \vec{v}_7$ are all parallel
$\vec{v}_1, \vec{v}_4, \vec{v}_8$ are perpendicular to $\vec{v}_3, \vec{v}_5, \vec{v}_7$
$\vec{v}_2$ and $\vec{v}_9$ are perpendicular

17 $5x + y - 2z = 3$

19 $2x + 4y - 3z = 5$

21 (a) $\lambda = -2.5$
(b) $a = -6.5$

23 $50.91°, 53.13°$, and $75.96°$

25 $\vec{a} = -\frac{8}{21}\vec{d} + (\frac{79}{21}\vec{i} + \frac{10}{21}\vec{j} - \frac{118}{21}\vec{k})$

27 $38.7°$

29 $710 revenue

33 (a) Same side of plane; opposite side
(b) $ax + by + cz > d, ax + by + cz < d$
where $d = ax_0 + by_0 + cz_0$
(c) Q, R on the same side; P on other side

Section 13.4

1 $-\vec{i}$

3 $\vec{j} - \vec{k}$

5 $\vec{i} + 3\vec{j} + 7\vec{k}$

7 $\vec{a} \times \vec{b} = -2\vec{i} - 7\vec{j} - 13\vec{k}$
$\vec{a} \cdot (\vec{a} \times \vec{b}) = 0$
$\vec{b} \cdot (\vec{a} \times \vec{b}) = 0$

9 $\vec{i} - \vec{j}$

11 $-2\vec{k}$

13 Parallel to the z-axis

15 $x - y - z = -3$

17 (a) 1.5
(b) $y = 1$

19 $4\vec{i} + 26\vec{j} + 14\vec{k}$

21 $4(x - 4) + 26(y - 5) + 14(z - 6) = 0$

Chapter 13 Review

1 $4\vec{i} + \vec{j} + 3\vec{k}$

3 -3

5 $-6\vec{i} - 9\vec{j} + 3\vec{k}$

7 0

9 (a) 4

(b) $-4\vec{i} - 11\vec{j} - 17\vec{k}$
(c) $3.64\vec{i} + 2.43\vec{j} - 2.43\vec{k}$
(d) $79.0°$
(e) 0.784.
(f) $2\vec{i} - 2\vec{j} + \vec{k}$ (many answers possible)
(g) $-4\vec{i} - 11\vec{j} - 17\vec{k}$.

11 $\pm(-\vec{i} + \vec{j} - 2\vec{k})/\sqrt{6}$

13 $-3\vec{i} + 4\vec{j}$

15 $\vec{n} = 4\vec{i} + 6\vec{k}$

17 (a) $(21/5, 0, 0)$
(b) $(0, -21, 0)$ and $(0, 0, 3)$
(for example)
(c) $\vec{n} = 5\vec{i} - \vec{j} + 7\vec{k}$
(for example)
(d) $21\vec{j} + 3\vec{k}$
(for example)

19 (a) $-2\vec{i} + 3\vec{k}$; $-\vec{i} - \vec{j} + \vec{k}$;
Many answers possible
(b) $3\vec{i} - \vec{j} + 2\vec{k}$;
Many answers possible
(c) $3x - y + 2z = 5$

23 (a) $-500\vec{i}$
(b) $(300, 60, 4)$
(c) $-178.7\vec{i} - 89.4\vec{j} - 7.2\vec{k}$

25 228.43 newtons
85.5° south of east

27 (a) $\vec{u}$ and $-\vec{u}$ where
$\vec{u} = \frac{12}{13}\vec{i} - \frac{4}{13}\vec{j}$
$- \frac{3}{13}\vec{k}$
(b) $\theta \approx 49.76°$
(c) 13/2
(d) $13/\sqrt{29}$

29 $9x - 16y + 12z = 5$
0.23

33 $0, \vec{0}$

35 (a) $\vec{c} = -\vec{i} - 4\vec{j} + 3\vec{k}$ or $\vec{c} = \vec{i} + 4\vec{j} - 3\vec{k}$
(b) $\vec{i} + 4\vec{j} - 3\vec{k}$

37 (a) $(tv - sw - ty + wy + sz - vz)\vec{i} +$
$(-tu + rw + tx - wx - rz + uz)\vec{j} +$
$(su - rv - sx + vx + ry - uy)\vec{k}$
(b) $((s(u - x) + vx - uy + r(-v + y))/c)(a\vec{i} + b\vec{j} + c\vec{k})$

Ch. 13 Understanding

1 False

3 True

5 False

7 False

9 False

11 False

13 True

15 True

17 False

19 True

21 True

23 False

25 False

27 True

29 True

Section 14.1

1 $f_x(3, 2) \approx 2, f_y(3, 2) \approx -1$

3 Positive, Negative, 10, 2, -4

5 (a) Negative

7 (a) Concentration/distance
   Rate of change in concentration with distance
   $\partial c/\partial x < 0$
 (b) Concentration/time
   Rate of change in concentration with time
   For small t, $\partial c/\partial t > 0$
   For large t, $\partial c/\partial t < 0$

9 $f_w(10, 25) \approx -1.6$

11 $f_w(5, 20) \approx -2.6$

13 (a) Negative
 (b) Positive

15 (a) Negative
 (b) Negative
 (c) Negative

17 (a) 2.5, 0.02
 (b) 3.33, 0.02
 (c) 3.33, 0.02

19 -1

Section 14.2

1 $f_x = 10xy^3 + 8y^2 - 6x$,
  $f_y = 15x^2y^2 + 16xy$

3 $21x^5y^6 - 96x^4y^2 + 5x$

5 $V_r = \frac{2}{3}\pi rh$

7 $(15x^2y - 3y^2)\cos(5x^3y - 3xy^2)$

9 y

11 g

13 $(1/2)v^2$

15 $2\pi/v$

17 $\partial V/\partial r = \frac{8}{3}\pi rh$,
   $\partial V/\partial h = \frac{4}{3}\pi r^2$

19 Gm_1/r^2

21 $-2\pi r/T^2$

23 $\epsilon_0 E$

25 $\pi/\sqrt{lg}$

27 $c\gamma a_1 b_1 K^{b_1-1}(a_1 K^{b_1} + a_2 L^{b_2})^{\gamma-1}$

29 $-\pi r^{3/2}/(M\sqrt{GM})$

31 $f_x(1, 2) = 15$,
   $f_y(1, 2) = -5$

33 13.6

35 (a) 41.1
 (b) 41.11, yes

37 1.277 m^2, 0.005 m^2/kg, 0.006 m^2/cm

39 $F = 684$ newtons,
   $\partial F/\partial m = 9.77$ newtons/kg,
   $\partial F/\partial r = -0.000214$ newtons/meter

43 $h_x(2, 5) \approx -0.38$ cm/meter
   $h_t(2, 5) \approx 0.76$ cm/second

Section 14.3

1 $z = 6 + 3x + y$

3 $z = -4 + 2x + 4y$

5 $dz = -e^{-x}\cos(y)dx - e^{-x}\sin(y)dy$

7 $dh = e^{-3t}\cos(x + 5t)\,dx + (-3e^{-3t}\sin(x + 5t) + 5e^{-3t}\cos(x + 5t))\,dt$

9 $dg = 4\,dx$

11 $dF = 0.01G\,dm - 0.2G\,dr$

13 (c) $z = 12x - 6y - 7$

15 $df = \frac{1}{3}dx + 2dy$
   $f(1.04, 1.98) \approx 2.973$

17 $P(r, L) \approx 80 + 2.5(r - 8) + 0.02(L - 4000)$, $P(r, L) \approx 120 + 3.33(r - 8) + 0.02(L - 6000)$, $P(r, L) \approx 160 + 3.33(r - 13) + 0.02(L - 7000)$.

21 8246.68

23 (a) $d\rho = -\beta\rho\,dT$
 (b) 0.00015, $\beta \approx 0.0005$

Section 14.4

1 $((15/2)x^4)\vec{i} - ((24/7)y^5)\vec{j}$

3 $\nabla z = e^y\vec{i} + e^y(1 + x + y)\vec{j}$

5 $(x\vec{i} + y\vec{j})/\sqrt{x^2 + y^2}$

7 $\sin\theta\vec{i} + r\cos\theta\vec{j}$

9 $\nabla z = 2x\cos(x^2 + y^2)\vec{i} + 2y\cos(x^2 + y^2)\vec{j}$

11 $\nabla z = \frac{y}{y^2+x^2}\vec{i} - \frac{x}{y^2+x^2}\vec{j}$

13 $\nabla z = \frac{ye^y}{(x+y)^2}\vec{i} + \frac{e^y(x^2+xy-x)}{(x+y)^2}\vec{j}$

15 $50\vec{i} + 96\vec{j}$

17 $10\pi\vec{i} + 4\pi\vec{j}$

19 $(1/2)\vec{i} + (1/2)\vec{j}$

21 Negative

23 Approximately zero

25 Negative

27 $\vec{i}$

29 $-\vec{i}$

31 $\vec{i} + \vec{j}$

33 $-\vec{i} + \vec{j}$

35 5

37 2

39 1

41 2.12

43 0.3

45 0.7

47 0

49 (a) Negative
 (b) Negative

51 (a) $15/\sqrt{10}$
 (b) $9\vec{i} + 12\vec{j}$
 (c) 15

53 (a) $2/\sqrt{13}$
 (b) $1/\sqrt{17}$
 (c) $\vec{i} + \frac{1}{2}\vec{j}$

55 $f_{\vec{u}_3} < f_{\vec{u}_2} < 0 < f_{\vec{u}_4} < f_{\vec{u}_1}$

57 $17.5°$

59 $(3\sqrt{5} - 2\sqrt{2})\vec{i} + (4\sqrt{2} - 3\sqrt{5})\vec{j}$

61 $3\vec{i} + 2\vec{j}$; $3(x - 2) + 2(y - 3) = 0$

63 $-4\vec{i}$; $x = 2$

65 Fourth quadrant

67 (a) Negative
 (b) Positive
 (c) Positive
 (d) Negative

69 (a) P, Q
 (c) $\|\operatorname{grad} f\|$
   $f_{\vec{u}} = \|\operatorname{grad} f\|\cos\theta$

Section 14.5

1 $2x\vec{i} + 3y^2\vec{j} - 4z^3\vec{k}$

3 $-\sin(x + y)\vec{i} + (\cos(y + z) - \sin(x + y))\vec{j} + \cos(y + z)\vec{k}$

5 $-2(x\vec{i} + y\vec{j} + z\vec{k})/(x^2 + y^2 + z^2)^2$

7 $\operatorname{grad} f = (2x_1x_2^3x_3^4)\vec{i} + (3x_1^2x_2^2x_3^4)\vec{j} + (4x_1^2x_2^3x_3^3)\vec{k}$

9 $6\vec{i} + 3\vec{j} + 2\vec{k}$

11 $9/\sqrt{3}$

13 $-1/\sqrt{2}$

15 $-\sqrt{77/2}$

17 $-2\vec{i} - 2\vec{j} + 4\vec{k}$; $-2(x + 1) - 2(y - 1) + 4(z - 2) = 0$

19 $2\vec{i} - 2\vec{j} + \vec{k}$; $2(x+1) - 2(y-1) + (z-2) = 0$

21 $2\vec{j} - 4\vec{k}$; $2(y - 1) - 4(z - 2) = 0$

23 (a) 0
 (b) $24/\sqrt{19}$

25 (a) Spheres centered at origin
 (b) $(-2x\vec{i} - 2y\vec{j} - 2z\vec{k})e^{-(x^2+y^2+z^2)}$
 (c) $-3\sqrt{2}e^{-1}$ degrees/sec

27 (a) $6.33\vec{i} + 0.76\vec{j}$
 (b) -34.69

29 (a) $(x - 2) + 4(y - 3) - 6(z - 1) = 0$
 (b) $z = 1 + (1/6)(x - 2) + (2/3)(y - 3)$

Section 14.6

1 $\frac{dz}{dt} = e^{-t}\sin(t)(2\cos t - \sin t)$

3 $(t^3 - 2)/(t + t^4)$

5 $2e^{1-t^2}(1 - 2t^2)$

7 $\frac{\partial z}{\partial u} = (e^{-v\cos u} - v(\cos u)e^{-u\sin v})\sin v - (-u(\sin v)e^{-v\cos u} + e^{-u\sin v})v\sin u$
   $\frac{\partial z}{\partial v} = (e^{-v\cos u} - v(\cos u)e^{-u\sin v})u\cos v + (-u(\sin v)e^{-v\cos u} + e^{-u\sin v})\cos u$

9 $\frac{\partial z}{\partial u} = e^v/u$
   $\frac{\partial z}{\partial v} = e^v\ln u$

11 $\frac{\partial z}{\partial u} = 2ue^{(u^2-v^2)}(1 + u^2 + v^2)$
   $\frac{\partial z}{\partial v} = 2ve^{(u^2-v^2)}(1 - u^2 - v^2)$

13 $\frac{\partial z}{\partial u} = \frac{1}{vu}\cos\left(\frac{\ln u}{v}\right)$
   $\frac{\partial z}{\partial v} = -\frac{\ln u}{v^2}\cos\left(\frac{\ln u}{v}\right)$

15 $\frac{\partial w}{\partial u} = \frac{\partial w}{\partial x}\frac{\partial x}{\partial u} + \frac{\partial w}{\partial y}\frac{\partial y}{\partial u} + \frac{\partial w}{\partial z}\frac{\partial z}{\partial u}$
   $\frac{\partial w}{\partial v} = \frac{\partial w}{\partial x}\frac{\partial x}{\partial v} + \frac{\partial w}{\partial y}\frac{\partial y}{\partial v} + \frac{\partial w}{\partial z}\frac{\partial z}{\partial v}$

17 -0.6

19 -5 pascal/hour

21 (a) $F_u(x, 3)$
 (b) $F_v(3, x)$
 (c) $F_u(x, x) + F_v(x, x)$
 (d) $F_u(5x, x^2)(5) + F_v(5x, x^2)(2x)$

23 (a) $\frac{\partial z}{\partial r} = \cos\theta\frac{\partial z}{\partial x} + \sin\theta\frac{\partial z}{\partial y}$
   $\frac{\partial z}{\partial\theta} = r(\cos\theta\frac{\partial z}{\partial y} - \sin\theta\frac{\partial z}{\partial x})$
 (b) $\frac{\partial z}{\partial y} = \sin\theta\frac{\partial z}{\partial r} + \frac{\cos\theta}{r}\frac{\partial z}{\partial\theta}$
   $\frac{\partial z}{\partial x} = \cos\theta\frac{\partial z}{\partial r} - \frac{\sin\theta}{r}\frac{\partial z}{\partial\theta}$

25 $(\partial U_3/\partial P)_V$

27 $(\partial U/\partial T)_V = 7/2$
   $(\partial U/\partial V)_T = 11/4$

29 $\left(\frac{\partial U}{\partial P}\right)_T = \left(\frac{\partial U}{\partial V}\right)_T \left(\frac{\partial V}{\partial P}\right)_T$

35 $F(b, x)$

Section 14.7

1 $f_{xx} = 6y$
$f_{xy} = 6x + 15y^2$
$f_{yx} = 6x + 15y^2$
$f_{yy} = 30xy$

3 $f_{xx} = 0$
$f_{xy} = e^y = f_{yx}$
$f_{yy} = xe^y$

5 $f_{xx} = 0$
$f_{yx} = e^y = f_{xy}$
$f_{yy} = e^y(x + 2 + y)$

7 $f_{xx} = \frac{y^2}{(x^2+y^2)^{3/2}}$
$f_{xy} = \frac{-xy}{(x^2+y^2)^{3/2}} = f_{yx}$
$f_{yy} = \frac{x^2}{(x^2+y^2)^{3/2}}$

9 $f_{xx} = -12 \sin 2x \cos 5y$
$f_{xy} = -30 \cos 2x \sin 5y$
$f_{yx} = -30 \cos 2x \sin 5y$
$f_{yy} = -75 \sin 2x \cos 5y$

11 $Q(x, y) = 1 + 2x - 2y + x^2 - 2xy + y^2$

13 $Q(x, y) = 1 - x^2/2 - 3xy - (9/2)y^2$

15 $Q(x, y) = 1 - 2x + y + 4x^2 - 4xy + y^2$

17 $Q(x, y) = 1 + 2x - \frac{1}{2}y^2$

19 (a) Positive
   (b) Zero
   (c) Positive
   (d) Zero
   (e) Zero

21 (a) Negative
   (b) Zero
   (c) Negative
   (d) Zero
   (e) Zero

23 (a) Zero
   (b) Negative
   (c) Zero
   (d) Negative
   (e) Zero

25 (a) Negative
   (b) Negative
   (c) Zero
   (d) Zero
   (e) Zero

27 (a) Negative
   (b) Positive
   (c) Positive
   (d) Positive
   (e) Negative

29 $L(x, y) = 1 + (1/2)(x - 1) + y$
   $Q(x, y) = 1 + (1/2)(x-1) + y - (1/8)(x-1)^2 - (1/2)(x-1)y - (1/2)y^2$
   $L(0.9, 0.2) = 1.15$
   $Q(0.9, 0.2) = 1.13875$
   $f(0.9, 0.2) \approx 1.14018$

31 $L(x, y) = Q(x, y) = x - 1$
   $L(0.9, 0.2) = Q(0.9, 0.2) = -0.1$
   $f(0.9, 0.2) \approx -0.098$

33 (a) $z_{yx} = 4y$
   (b) $z_{xyx} = 0$
   (c) $z_{xyy} = 4$

35 (a) Increasing, decreasing
   (b) $f_x > 0, f_y < 0$
   (c) $f_{xx} > 0, f_{yy} < 0$
   (e) P

37 (a) xy
   $1 - \frac{1}{2}(x - \frac{\pi}{2})^2 - \frac{1}{2}(y - \frac{\pi}{2})^2$

39 (a) $1 + x/2 + y$
   (b) $1 + x/2 + y - x^2/8 - xy/2 - y^2/2$

Section 14.9

1 $(0, 0)$

3 None

5 x-axis and y-axis

7 (b) No
   (c) No
   (d) No
   (e) Exist, not continuous.

9 (b) Yes
   (c) Yes
   (d) No
   (e) Exist, not continuous

11 (b) Yes
   (d) No
   (f) No

13 (c) No
   (e) No

15 (a) No

Chapter 14 Review

1 $\partial z/\partial x = \frac{14x+7}{(x^2+x-y)^{-6}}$
   $\partial z/\partial y = -7(x^2 + x - y)^6$

3 $\partial f/\partial p = (1/q)e^{p/q}$
   $\partial f/\partial q = -(p/q^2)e^{p/q}$

5 $f_{xx} = (2x^2 - y^2)(x^2 + y^2)^{-5/2}$
   $f_{xy} = 3xy(x^2 + y^2)^{-5/2}$

7 $V_{rr} = 2\pi h$, $V_{rh} = 2\pi r$

9 $2(x\vec{i} + y\vec{j})/(x^2 + y^2)$

11 $(3x^2 - yz)\vec{i} - xz\vec{j} + (3z^2 - xy)\vec{k}$

13 $15/\sqrt{2}$

15 -4

17 0

19 $4\vec{i} - 2\vec{j}$

21 $2\vec{i} - 2\vec{j} - 6\vec{k}$

23 $Q(x, y) = 2 + 6x + y + 6x^2 + 3xy$

25 (a) $f_w(2, 2) \approx 2.78$
   $f_z(2, 2) \approx 4.01$
   (b) $f_w(2, 2) \approx 2.773$
   $f_z(2, 2) = 4$

27 (a) $-3\vec{i} + 4\vec{j} - \vec{k}$
   (b) $-3\vec{i} + 4\vec{j}$

31 $15°$C/minute

33 Approx 7.5 at $(1.94, 1.08)$

35 $\partial Q/\partial b < 0$
   $\partial Q/\partial c > 0$

37 $\partial f/\partial P_1 < 0$
   $\partial f/\partial P_2 > 0$

41 $dP \approx 47.6 \, dL + 17.8 \, dK$

45 $4\sqrt{2}$,
   $6\vec{i} + 2\vec{j}$

47 $-2x\vec{i} - 2y\vec{j}$

49 (a) -14, 2.5
   (b) -2.055
   (c) 14.221 in direction
   of $-14\vec{i} + 2.5\vec{j}$.
   (d) $f(x, y) = f(2, 3) = 7.56$.
   (e) For example,
   $\vec{v} = 2.5\vec{i} + 14\vec{j}$,
   (f) -0.32

51 (a) $e^{10} - 2e^{10}x - 6e^{10}y$
   (b) $1 + (x - 1)^2 + (y - 3)^2$

(c) $-2\vec{i} - 6\vec{j}$
(d) $-2e^{10}\vec{i} - 6e^{10}\vec{j} - \vec{k}$

55 (a) $(1/6) + (1/9)x + y - (1/108)x^2 + (2/3)xy + (3/2)y^2$
   (b) $(1/6) + (1/9)x - (1/108)x^2, 1 + 6y + 9y^2, (1/6) + (1/9)x + y - (1/108)x^2 + (2/3)xy + (3/2)y^2$

57 (a) $(A_0 + A_1 + 2A_2 + A_3 + 2A_4 + 4A_5) + (A_1 + 2A_3 + 2A_4)(x - 1) + (A_2 + A_4 + 4A_5)(y - 2) + A_3(x - 1)^2 + A_4(x - 1)(y - 2) + A_5(y - 2)^2$
   (b) Expands to f
   (c) $A_0 - A_3 + A_4 - 4A_5 + (A_1 + 2A_3 + 2A_4)x + (A_2 + A_4 + 4A_5)y$, does not expand to f

Ch. 14 Understanding

1 True
3 True
5 True
7 True
9 True
11 True
13 True
15 False
17 True
19 False
21 False
23 True
25 False
27 False
29 True
31 False
33 True
35 False
37 True
39 True
41 True
43 True
45 False
47 True

Section 15.1

1 A: no
   B: yes, max
   C: yes, saddle

3 Saddle point

5 Saddle pts: $(1, -1), (-1, 1)$
   local max $(-1, -1)$
   local min $(1, 1)$.

7 Local minimum at $(2, 2)$

9 Local maximum at $(-1, 0)$,
   Saddle points at $(1, 0)$ and $(-1, 4)$,
   Local minimum at $(1, 4)$

11 Saddle point: $(0, 0)$.

13 $(0, 0), (1, 1), (-1, -1)$
   saddle point,
   local maximum,
   local maximum

17 Critical points: $(0, 0), (\pm\pi, 0)$,
   $(\pm 2\pi, 0), (\pm 3\pi, 0), \cdots$
   Local minima: $(0, 0)$,
   $(\pm 2\pi, 0), \pm 4\pi, 0), \cdots$
   Saddle points: $(\pm\pi, 0)$,
   $(\pm 3\pi, 0), (\pm 5\pi, 0), \cdots$

19 $y = 0, \pm 2\pi, \pm 4\pi, \ldots$ Local minima

21 (a) $(1, 3)$ is a minimum

23 (a) $(0, 0)$
  (b) $D = -24x^2$
  (c) Saddle point

25 $k = 2$: local minimum;
  $k = 1$: minimum;
  $k = 0$: saddle point;
  $k = -1$: maximum;
  $k = -2$: local maximum

Section 15.2

1 Mississippi:
  $87 - 88$ (max), $83 - 87$ (min)
  Alabama:
  $88 - 89$ (max), $83 - 87$ (min)
  Pennsylvania:
  $89 - 90$ (max), 80 (min)
  New York:
  $81 - 84$ (max), $74 - 76$ (min)
  California:
  $100 - 101$ (max), $65 - 68$ (min)
  Arizona:
  $102 - 107$ (max), $85 - 87$ (min)
  Massachusetts:
  $81 - 84$ (max), 70 (min)

3 Global max $= 0$
  No global min

5 Min $= 0$ at $(0, 0)$
  (not on boundary)
  Max $= 2$ at $(1, 1), (1, -1)$,
  $(-1, -1)$ and $(-1, 1)$
  (on boundary)

7 Max $= 0$ at $(0, 0)$
  (not on boundary)
  Min $= -2$ at $(1, -1), (-1, -1)$,
  $(-1, 1)$ and $(1, 1)$
  (on boundary)

9 Max: 30.5 at $(0, 0)$
  Min: 20.5 at $(2.5, 5)$

11 Max: 11 at $(5.1, 4.9)$
  Min: -1 at $(1, 3.9)$

13 All edges $(32)^{1/3}$ cm.

15 $1/162$

17 $\sqrt{6}$

19 $q_1 = 300, q_2 = 225$.

21 $h = 25\%, t = 25°C$

23 $w = 6.35$ cm
  $l = 6.35$ cm
  $h = 12.7$ cm

25 (a) $L = \left[pA \left(\frac{a}{k} \right)^a \left(\frac{1}{b} \right)^{(a-1)} \right]^{1/(1-a-b)}$
  $K = \frac{la}{kb} L$
  (b) No

29 (b) $f(\sqrt{1/2}, \sqrt{3/5}) = 4\sqrt{2} + 2\sqrt{15} \approx$
  13.403

Section 15.3

1 Min $= -\sqrt{2}$, max $= \sqrt{2}$

3 Min $= -22$, max $= 22$

5 Min $= -2$, max $= 2$

7 no max
  min: 0.866.

9 max: 21, no min

11 Max: $f(0, 2) = f(0, -2) = 8$
  Min: $f(0, 0) = 0$

13 Max: $f(\frac{1}{\sqrt{5}}, \frac{3}{\sqrt{5}}) = 2\sqrt{5}$

Min: $f(-\frac{1}{\sqrt{5}}, -\frac{3}{\sqrt{5}}) = -2\sqrt{5}$

15 Max: 1
  Min: -1.

17 (c) Max of 5 at $(2, 1)$
  Min of -5 at $(-2, -1)$
  (d) Max of 5 at $(2, 1)$
  Min of -5 at $(-2, -1)$

19 (a) $x = 133.33, y = 26.67, z = 25, 219$
  (b) $\lambda = 11.348$

21 (c) $D = 10, N = 20,$
  $V \approx 9,779$
  (d) $\lambda = 14.67$
  (e) \$68, rise

23 $q_1 = 50$ units
  $q_2 = 150$ units

25 $c \left(\frac{aK}{(a+b)P_1} \right)^a \left(\frac{bK}{(a+b)P_2} \right)^b$

27 (b) $s = 1000 - 10l$

29 Along line $x = 2y$

33 (a) Cost of producing quantity u when prices are p, q
  (b) $2\sqrt{pqu}$

Chapter 15 Review

1 $(0, 0), (2, 0)$: Saddle point
  $(1, 1/4)$: Local min

3 $(0, 0)$: Saddle point

5 $(2, 3)$: Local and global max

7 $(\sqrt{2}, -\sqrt{2}/2)$ saddle point

9 Minimum $f(27.907, 23.256) = 1860.484$

11 Maximum $f(10, 12.5) = 250$

13 No global extrema.

15 Maxima: $(-1, 1)$ and $(1, -1)$
  Minimum: $(0, 0)$

19 $K = 20$
  $L = 30$
  $C = \$7,000$

21 $x = 1000, y = 500$

23 (a) (i) Power function
  (ii) Linear function
  (b) $\ln N = 1.20 + 0.32 \ln A$
  Agrees with biological rule

25 (a) Reduce K by $1/2$ unit,
  increase L by 1 unit.

29 (a) $f_1 = \frac{k_1}{k_1 + k_2} mg, f_2 = \frac{k_2}{k_1 + k_2} mg$
  (b) Distance the mass stretches the top spring and compresses the lower spring

31 $-(1/2m^2)(1 + 2(v_1 v_2)^{1/2}/(v_1 + v_2))$
  weeks^2

33 $d \approx 5.37$ m, $w \approx 6.21$ m,
  $\theta = \pi/3$ radians

35 (a) $x = 1/4 - a, y = 1$, saddle point

37 (a) 21.0208
  (b) $g = 20.5, 21.2211; g = 20.2, 21.1008$
  (c) $g = 20.5, 21.2198; g = 20.2, 21.1007$

Ch. 15 Understanding

1 True

3 False

5 True

7 True

9 False

11 False

13 False

15 False

17 False

19 True

21 True

23 True

25 False

27 False

29 True

31 False

33 False

Section 16.1

1 3800, 2400

3 Upper sum $= 46.63$
  Lower sum $= 8$
  Average ≈ 27.3

5 Lower Bound $= 25$
  Upper Bound $= 41$

7 $40/3$

9 $25.2°C$

11 Mosquitos

13 294 mg

15 Positive

17 Zero

19 Zero

21 Zero

23 Zero

25 Positive

27 Positive

29 Zero

31 (b) $(d - c) \int_a^b g(x)\, dx$

Section 16.2

1 26

3 $2/3$

5 $\int_1^4 \int_1^2 f\, dy\, dx$
  or $\int_1^2 \int_1^4 f\, dx\, dy$

7 $\int_0^6 \int_{y/2}^{(15-y)/3} f(x, y)\, dx\, dy$ or
  $\int_0^3 \int_0^{2x} f(x, y)\, dy\, dx + \int_3^5 \int_0^{-3x+15} f(x, y)\, dy\, dx$

9 $\int_1^4 \int_{(x-1)/3}^2 f\, dy\, dx$

11 $(e^4 - 1)(e^2 - 1)e$

13 ≈ -2.68

15 94.5

17 $1/24$

19 2.38176

21 $13/6$

23 15

25 (b) $\int_0^2 \int_0^{-2x+4} g(x, y)\, dy\, dx$

27 $\frac{1}{2}(1 - \cos 1) = 0.23$

29 $\frac{2}{9}(3\sqrt{3} - 2\sqrt{2})$

31 (a) 180
  (b) 144

33 $\int_{-5}^5 \int_{-\sqrt{25-y^2}}^{\sqrt{25-y^2}} (25 - x^2 - y^2)\, dx\, dy$

35 $\int_0^4 \int_{y-4}^{(4-y)/2} (4 - 2x - y)\, dx\, dy$

37 $1/3$

39 $16/3$

41 Volume $= 1/(6abc)$

45 (a) $a + (3/2)b = 20$

47 $k(a^3b + ab^3)/3$

Section 16.3

1 2

3 $a + b + 2c$

15 162

17 $\frac{15}{2}$

19 Positive

21 Zero

23 Positive

25 Zero

27 Positive

29 Positive

31 Zero

33 Zero

35 Negative

37 $\int_{-4}^{4} \int_{-\sqrt{16-x^2}}^{\sqrt{16-x^2}} \int_{3}^{\sqrt{25-x^2-y^2}} dz\,dy\,dx$

39 $m = 2;$
$(\bar{x}, \bar{y}, \bar{z}) = (13/24, 13/24, 25/24)$

41 $5m/3$

Section 16.4

1 $\int_{\pi/4}^{3\pi/4} \int_{0}^{2} f\, r\,dr\,d\theta$

3 $\int_{0}^{2\pi} \int_{0}^{\sqrt{2}} f\, r\,dr\,d\theta$

13 0

15 (b) $\int_{0}^{1} \int_{0}^{3y} f(x, y)\, dx\,dy$

(c) $\int_{\tan^{-1}(1/3)}^{\pi/2} \int_{0}^{1/\sin\theta} f(r\cos\theta, r\sin\theta)r\,dr\,d\theta$

17 6

19 (a) $\int_{\pi/2}^{3\pi/2} \int_{1}^{4} \delta(r, \theta)\, r\,dr\,d\theta$

(b) (i)

(c) About 39,000

21 $32\pi(\sqrt{2} - 1)/3$

23 (a) $\pi(1 - e^{-a^2})$

(b) Volume tends to π

25 (a) $\int_{-\sqrt{3}/2}^{\sqrt{3}/2} \int_{1-\sqrt{1-y^2}}^{\sqrt{1-y^2}} dx\,dy$

(b) $\int_{0}^{1} \int_{-\arccos(r/2)}^{\arccos(r/2)} r\,d\theta\,dr$

Section 16.5

1 $200\pi/3$

3 25π

5 $\int_{0}^{3} \int_{0}^{1} \int_{0}^{5} f\, dz\,dy\,dx$

7 $\int_{0}^{4} \int_{0}^{\pi/2} \int_{0}^{2} f \cdot r\,dr\,d\theta\,dz$

9 $\int_{0}^{2\pi} \int_{0}^{\pi/6} \int_{0}^{3} f \cdot \rho^2 \sin\phi\,d\rho\,d\phi\,d\theta$

11 $\int_{0}^{\pi} \int_{0}^{2} \int_{0}^{r} f\, r\,dz\,dr\,d\theta$

13 (a) $\int_{0}^{2\pi} \int_{\pi/4}^{\pi/2} \int_{0}^{3/\sin\theta} \rho^2 \sin\phi\,d\rho\,d\phi\,d\theta$

(b) 18π

15 π

17 (a) Positive

(b) Zero

19 $\int_{0}^{2\pi} \int_{0}^{5/\sqrt{2}} \int_{r}^{5/\sqrt{2}} r\,dz\,dr\,d\theta$
$= 125\pi/(6\sqrt{2}) = 46.28$ cm^3

21 (a) $\int_{0}^{2\pi} \int_{0}^{\sqrt{2}} \int_{r}^{\sqrt{4-r^2}} r\,dz\,dr\,d\theta$

(b) $\int_{0}^{2\pi} \int_{0}^{\pi/4} \int_{0}^{2} \rho^2 \sin\phi\,d\rho\,d\phi\,d\theta$

23 $\int_{0}^{2\pi} \int_{0}^{4} \int_{3}^{\sqrt{25-r^2}} r\,dz\,dr\,d\theta = 52\pi/3 =$
54.45 cm^3

25 (a) $\int_{0}^{1} \int_{0}^{\sqrt{1-x^2}} \int_{-\sqrt{1-x^2-y^2}}^{0} dz\,dy\,dx$

(b) $\int_{0}^{\pi/2} \int_{0}^{1} \int_{-\sqrt{1-r^2}}^{0} r\,dz\,dr\,d\theta$

(c) $\int_{0}^{\pi/2} \int_{\pi/2}^{\pi} \int_{0}^{1} \rho^2 \sin\phi\,d\rho\,d\phi\,d\theta$

27 (a) $\int_{-1/\sqrt{2}}^{1/\sqrt{2}} \int_{-\sqrt{(1/2)-x^2}}^{\sqrt{(1/2)-x^2}} \int_{\sqrt{x^2+y^2}}^{1} dz\,dy\,dx$

(b) $\int_{0}^{2\pi} \int_{0}^{1/\sqrt{2}} \int_{r}^{1} r\,dz\,dr\,d\theta$

(c) $\int_{0}^{2\pi} \int_{0}^{\pi/4} \int_{0}^{1/(\sqrt{2}\cos\phi)} \rho^2 \sin\phi\,d\rho\,d\phi\,d\theta$

29 27π

31 $3/\sqrt{2}$

33 (a) $\int_{0}^{2\pi} \int_{1}^{5} \int_{-\sqrt{25-r^2}}^{\sqrt{25-r^2}} r\,dz\,dr\,d\theta$

(b) $64\sqrt{6}\pi = 492.5$ mm^3

35 $3/4$

39 $3I = \frac{6}{5}a^2; I = \frac{2}{5}a^2$

41 $2\pi G\delta(H + R - \sqrt{R^2 + H^2})$

43 Total charge $= \frac{k\pi}{2}h^2 R^2$

45 $(q^2/8\pi\epsilon)((1/a) - (1/b))$

Section 16.6

1 Not a joint density function

3 Not a joint density function

5 Is joint density function

7 0

9 1

11 $7/8$

13 $1/16$

15 (a) $20/27$

(b) $199/243$

17 (a) $k = 8$

(b) $1/3$

19 $\int_{65}^{100} \int_{0.8}^{1} f(x, y)\, dx\,dy$

21 (a) $\lambda/(\lambda + \mu)$

Section 16.7

1 1

3 e^{2s}

5 -13

9 $\rho^2 \sin\phi$

11 13.5

13 9

Chapter 16 Review

5 $\int_{0}^{2} \int_{0}^{3} \int_{0}^{5} f\, dx\,dy\,dz$

7 $\int_{0}^{2\pi} \int_{\pi/2}^{\pi} \int_{2}^{5} f\rho^2 \sin\phi\,d\rho\,d\phi\,d\theta$

9 $\int_{0}^{4} \int_{\frac{y}{2}-2}^{-y+4} f(x, y)\, dx\,dy$

or $\int_{-2}^{0} \int_{0}^{2x+4} f(x, y)\, dy\,dx$ $+$
$\int_{0}^{4} \int_{0}^{-x+4} f(x, y)\, dy\,dx.$

11 $\frac{1}{3}(\sin 5 - \sin 2)(\cos 1 - \cos 2)$

13 $(1/20)\sin^5 1$

15 $(\pi/2)(1 - e^{-1})$

19 $162\pi/5$

21 8π

23 Negative

25 Can't tell

27 Zero

29 Zero

31 Zero

33 Negative

35 Positive

37 ≈ 183

39 $pqr/6$

41 $4\pi(a^2 - R^2)^{3/2}/3$

43 $f(x, y) = \frac{30}{\pi}e^{-50(x-5)^2 - 18(y-15)^2}$

45 $8\pi R^5/15$

47 13.95

49 $-1/2 + e/2$

51 $1/2, -1/2$, integrand not continuous

Ch. 16 Understanding

1 False

3 False

5 True

7 True

9 False

11 False

13 True

15 False

17 False

19 True

21 False

23 False

25 True

27 False

29 False

Section 17.1

1 Same direction $-\vec{i} + 4\vec{j} - 2\vec{k}$, point $(3, 3, -1)$ in common

3 $x = 3\cos t, y = 3\sin t, z = 0$

5 $x = 2\cos t, y = 0, z = 2\sin t$

7 $x = 0, y = 3\cos t, z = 2 + 3\sin t$

9 $x = t, y = t^3, z = 0$

11 $x = -3t^2, y = 0, z = t$

13 $x = 3\cos t, y = 2\sin t, z = 0$

15 $x = \frac{3}{2}\cos t, y = 1, z = -2 + \sin t$

17 $x = 3 + t, y = 2t, z = -4 - t.$

19 $x = -3 + 2t, y = 4 + 2t, z = -2 - 3t.$

21 $x = 1, \quad y = 0, \quad z = t.$

23 $x = 1 + 4t, y = 5 - 5t, z = 2 - 3t$

25 $x = 3 - 3t, y = 0, z = -5t.$

27 No

29 $\vec{r}(t) = -\vec{i} - 3\vec{j} + t(6\vec{i} + 5\vec{j}), 0 \le t \le 1$,
$x = -1 + 6t, y = -3 + 5t, 0 \le t \le 1$

31 $x = 0, y = 5\cos t, z = -5\sin t, -\pi/2 \le t \le \pi/2$

33 Yes, no

35 No, no

37 $\vec{r}(t) = (2 + (t/5)10)\vec{i} + (5 + (t/5)4)\vec{j}$

39 $\vec{r}(t) = (2 + ((t - 20)/10)10)\vec{i} + (5 + ((t - 20)/10)4)\vec{j}$

41 $x = t, y = \sqrt{t}, 1 \le t \le 16$.

43 Two arcs:
$\vec{r}(t) = 5\vec{i} + 5(-\cos t\vec{i} + \sin t\vec{j})$,
$0 \le t \le \pi$ or
$\vec{r}(t) = 5\vec{i} + 5(\cos t\vec{i} + \sin t\vec{j})$,
$\pi \le t \le 2\pi$

45 A straight line through the point $(3, 5, 7)$

47 Helix of radius 1 along y-axis

49 $x = \frac{8}{3}, \quad y = t, \quad z = \frac{1}{3} + t$

51 $x = 1 + 2t, \quad y = 2 + 4t, \quad z = 5 - t$

53 Intersection of planes

55 Yes

57 Circle, cosine, sine

59 Many possible answers

  (a) $a = -2, b = 7, c = 4, d = 0$
  (b) $a = -2, b = 7, c = 4, d = 11$
  (c) $a = 7, b = 2, c = 0, d = 41$

61 (a) $x = 1 + 2t, y = 5 + 2t, z = 2 - t$
  (b) $(-8/7, 25/14, 43/14)$

63 (a) Yellow
  (b) Direction $(-0.4 + t)\vec{i} + (-3.8 + t)\vec{j} + (t - 1)\vec{k}$

Section 17.2

1 $\vec{v} = 6t\vec{i} + 3t^2\vec{j}$,
$\|\vec{v}\| = 3|t| \cdot \sqrt{4 + t^2}$,
Stops when $t = 0$

3 $\vec{v} = 6\sin t\cos t\vec{i} - \sin t\vec{j} + 2t\vec{k}$,
$\|\vec{v}\| = \sqrt{36\sin^2 t\cos^2 t + \sin^2 t + 4t^2}$,
Stops when $t = 0$

5 $\vec{v} = -3\sin t\vec{i} + 4\cos t\vec{j}$, $\vec{a} = -3\cos t\vec{i} - 4\sin t\vec{j}$

7 $\vec{v} = 3\vec{i} + \vec{j} - \vec{k}$, $\vec{a} = \vec{0}$

9 $(4, 5), 19\vec{i} + 6\vec{j}, 19.92$

11 $\vec{v} = -6\pi\sin(2\pi t)\vec{i} + 6\pi\cos(2\pi t)\vec{j}$, $\vec{a} = -12\pi^2\cos(2\pi t)\vec{i} - 12\pi^2\sin(2\pi t)\vec{j}$, $\vec{v} \cdot \vec{a} = 0, \|\vec{v}\| = 6\pi, \|\vec{a}\| = 12\pi^2$

13 Line through $(2, 3, 5)$ in direction of $\vec{i} - 2\vec{j} - \vec{k}$,
$\vec{v} = 2t(\vec{i} - 2\vec{j} - \vec{k}), \vec{a} = 2(\vec{i} - 2\vec{j} - \vec{k})$

15 Length $= \sqrt{42}$

17 Length ≈ 24.6

19 (a) $\vec{v}(2) \approx -4\vec{i} + 5\vec{j}$,
    Speed $\approx \sqrt{41}$
  (b) About $t = 1.5$
  (c) About $t = 3$

21 $x = 1 + 2(t - 2), y = 2, z = 4 + 12(t - 2)$

23 Vertical: $t = 3$
  Horizontal: $t = \pm 1$
  As $t \to \infty, x \to \infty, y \to \infty$
  As $t \to -\infty, x \to \infty, y \to -\infty$

25 $x = 5 + 3(t - 7), y = 4 + 1(t - 7)$, $z = 3 + 2(t - 7)$.

27 (a) $x = 2 + 0.6t, y = -1 + 0.8t$,
    $z = 5 - 1.2t$
  (b) $x = 2 + 1.92t, y = -1 + 2.56t$,
    $z = 5 - 3.84t$

29 $x = (3/\sqrt{5})t, y = 5 - (6/\sqrt{5})t$
$x = -(3/\sqrt{5})t, y = 5 + (6/\sqrt{5})t$

31 $\vec{r}(t) = 22.1t\vec{i} + 66.4t\vec{j} + (442.7t - 4.9t^2)\vec{k}$

33 (a) No
  (b) $t = 5$
  (c) $\vec{v}(5) \approx 0.959\vec{i} + 0.284\vec{j} + 2\vec{k}$
  (d) $\vec{r} \approx 0.284\vec{i} - 0.959\vec{j} + 10\vec{k} + (t - 5)(0.959\vec{i} + 0.284\vec{j} + 2\vec{k})$.

35 (a) π m/sec
  (b) 2.45 m
  (c) 3.01 m

37 Same path, B moves twice as fast

41 Counterclockwise

Section 17.3

1 (a) IV
  (b) III
  (c) I
  (d) II

11 $\vec{V} = x\vec{i}$

13 $\vec{V} = x\vec{i} + y\vec{j} = \vec{r}$

15 $\vec{V} = \dfrac{\vec{r}}{\|\vec{r}\|}$

19 $\vec{F}(x, y) = \dfrac{-x\vec{i} - y\vec{j}}{\sqrt{x^2 + y^2}}$
  (for example)

23 (a) II
  (b) III
  (c) IV
  (d) I

25 (a) Circles origin counterclockwise
  (b) Spirals outward counterclockwise around origin
  (c) Spirals inward counterclockwise around origin

Section 17.4

1 $y = $ constant

3 $y = -\frac{2}{3}x + c$

9 (a) III
  (b) I
  (c) II
  (d) V
  (e) VI
  (f) IV

11 (a) Same directions, different magnitudes
  (b) Same curves, different parameterizations

Section 17.5

1 A horizontal disk of radius 5 in the plane $z = 7$.

3 A cylinder of radius 5 around the z-axis, $0 \le z \le 7$.

5 A cone of height and radius 5.

7 Cylinder, elliptical cross-section

9 Top hemisphere

11 Vertical segment

13 $x = a\cos\theta, y = a\sin\theta, z = z$

15 $s = 4, t = 2$
(x, y, z)
$= (x_0 + 10, y_0 - 4, z_0 + 18)$

17 (a) Yes
  (b) No

19 (a) nearer
  (b) farther
  (c) farther

21 Vertical half-circle

23 $x = 2 + 5\sin\phi\cos\theta$
$y = -1 + 5\sin\phi\sin\theta$
$z = 3 + 5\cos\phi$

25 $\begin{cases} x = a\sin\phi\cos\theta & 0 \le \phi \le \pi \\ y = b\sin\phi\sin\theta & 0 \le \theta \le 2\pi \\ z = c\cos\phi \end{cases}$

27 $x = u\cos v, y = u\sin v$, $z = u$ for $0 \le v \le 2\pi$

29 (a) $x = r\cos\theta$,
    $0 \le r \le a$,
    $y = r\sin\theta$,
    $0 \le \theta < 2\pi$,
    $z = \frac{hr}{a}$
  (b) $x = \frac{az}{h}\cos\theta$,
    $0 \le z \le h$,
    $y = \frac{az}{h}\sin\theta$,
    $0 \le \theta < 2\pi$,
    $z = z$

31 $x = ((\frac{z}{10})^2 + 1)\cos\theta$,
$y = ((\frac{z}{10})^2 + 1)\sin\theta$,
$z = z$,
$0 \le \theta \le 2\pi$,
$0 \le z \le 10$.

33 (a) $z = (x^2/2) + (y^2/2)$
    $0 \le x + y \le 2$
    $0 \le x - y \le 2$

35 (a) $x^2 + y^2 + z^2 = 1, x, y, z \ge 0$.

37 $\vec{r}(t) = x_0\vec{i} + y_0\vec{j} + z_0\vec{k} + a\cos t\vec{u} + a\sin t\vec{v}$

Chapter 17 Review

1 $x = t, y = 5$

3 $x = 4 + 4\sin t, y = 4 - 4\cos t$

5 $x = 2 - t, y = -1 + 3t, z = 4 + t$.

7 $x = 1 + 2t, y = 1 - 3t, z = 1 + 5t$.

9 $x = 3\cos t$
$y = 5$
$z = -3\sin t$

11 No

17 (a) $(I) = C_4, (II) = C_1$,
    $(III) = C_2, (IV) = C_6$
  (b) $C_3 : 0.5\cos t\vec{i} - 0.5\sin t\vec{j}$,
    $C_5 : -2\cos(\frac{t}{2})\vec{i} - 2\sin(\frac{t}{2})\vec{j}$

19 (a) In the direction given by the vector: $\vec{i} - \vec{j}$
  (b) Directions given by unit
    vectors:
    $\frac{1}{\sqrt{2}}\vec{i} + \frac{1}{\sqrt{2}}\vec{j}$
    $-\frac{1}{\sqrt{2}}\vec{i} - \frac{1}{\sqrt{2}}\vec{j}$
  (c) -4

21 (a) $(2t, 0)$ when $0 \le t \le \frac{1}{2}$
    $(\cos\frac{3\pi}{2}(t - \frac{1}{2})$,
    $\sin\frac{3\pi}{2}(t - \frac{1}{2}))$
    when $\frac{1}{2} < t \le \frac{5}{6}$
    $(0, -2(t - \frac{4}{3}))$
    when $\frac{5}{6} < t \le \frac{4}{3}$

  (b) $(0, 2t)$ when $0 \le t \le \frac{1}{2}$
    $(\sin\frac{3\pi}{2}(t - \frac{1}{2})$,
    $\cos\frac{3\pi}{2}(t - \frac{1}{2}))$
    when $\frac{1}{2} < t \le \frac{5}{6}$
    $(-2(t - \frac{4}{3}), 0)$
    when $\frac{5}{6} < t \le \frac{4}{3}$

23 (a) $(x, y) = (t, 1)$
   (b) $(x, y) = (t + \cos t, 1 - \sin t)$

25 (b) $\vec{i} + \vec{j} + \sqrt{2}\vec{k}$
   (c) $-\vec{i} + \frac{1}{\sqrt{2}}\vec{k}$

27 ω: Rate of change of polar angle θ of the particle,
   a: Rate of change of particle's distance from the origin

29 (a) Yes
   (b) No

35 Location at $t = 7$ is:
   (a) $x = 8$, $y = 3$
   (b) $x = e^{14}$, $y = 3e^7$
   (c) $x = \cos 7 - 3\sin 7$
      $y = \sin 7 + 3\cos 7$

37 (a) $(0, 0)$
   (b) $x^2 + y^2 = 1$
   (c) No
   (d) $x^2 + y^2 > 1$
   (e) $x^2 + y^2 < 1$

39 (a) Circles centered at origin
   (c) Parameterized at different speeds

41 (a) $x = \frac{20}{13} - \frac{6t}{13}, y = \frac{-1}{13} - \frac{t}{13}, z = t$
   (b) $x = t, y = \frac{1}{6}(-2 - 2t + 3t^2), z = \frac{1}{6}(20 - 10t - 3t^2)$.
   (c) $x = \sqrt{2 - t^2}, y = t, z = 5 + 5t - 3\sqrt{2 - t^2}$
     and $\quad x = -\sqrt{2 - t^2}, y = t, z = 5 + 5t + 3\sqrt{2 - t^2}$

Ch. 17 Understanding

1 False
3 False
5 False
7 True
9 True
11 True
13 False
15 False
17 False
19 False
21 True
23 False
25 True
27 True
29 True
31 False
33 False
35 False
37 True
39 True
41 False

Section 18.1

1 Positive
3 Positive
5 0
7 16
9 32
11 (c) 60
13 $\int_{C_3} \vec{F} \cdot d\vec{r} < \int_{C_1} \vec{F} \cdot d\vec{r} < \int_{C_2} \vec{F} \cdot d\vec{r}$
15 Positive

17 0
23 -1.2×10^7 meter^2/sec
25 $-GMm/8000$
27 Sphere of radius a centered at the origin
29 (a) $\phi(\vec{r}) = -\frac{Q}{4\pi\epsilon}\frac{1}{a} + \frac{Q}{4\pi\epsilon}\frac{1}{||\vec{r}||}$
   (b) Because then $\phi(\vec{r}) = \frac{Q}{4\pi\epsilon}\frac{1}{||\vec{r}||}$

Section 18.2

1 9
3 -2π
5 $4\cos 2$
7 $e^2 - e$
9 -1
11 21
13 (a) 11/6
   (b) 7/6
15 (a) Greater than zero
   (b) 88
   (d) $t = 2 \pm 1/\sqrt{3}$
   (e) 88; yes

Section 18.3

3 Yes
5 No
7 Indep
9 Indep
11 12
13 -7
15 (a) e
   (b) e
17 9/2
19 $\frac{3}{\sqrt{2}}\ln(\frac{3}{\sqrt{2}} + 1)$
21 1859
23 $e^9 - 1$
25 (a) $2\pi mg$
   (b) Yes
27 (b) No
29 (b) Longer
   (c) -0.3

Section 18.4

1 $f(x, y) = x^2y + 2y^4 + K$
   $K = $ constant
3 No
5 Yes, $f = x^2y^3 + xy + C$
7 Yes, $f = \ln A|xyz|$ where A is a positive constant.
9 -2π
11 -6
15 (b) $-\pi$
17 (b) $-3\pi/2$
21 $3\pi a^2/8$
23 (a) 18π
   (b) $2\pi R^2$

Chapter 18 Review

1 Positive
3 (a) Zero
   (b) C_1, C_3: Zero
      C_2: Negative
      C_4: Positive
   (c) Zero

5 5/3
7 0
9 18
11 Not path-independent
13 Path-independent
15 Path-independent
17 (a) 9/2
   (b) $-9/2$
19 (a) 0
   (b) 4
   (c) -12
   (d) 6π
21 Yes
23 (a) $\int_{C_2} \vec{F} \cdot d\vec{r}$
   (b) $\int_{C_3} \vec{F} \cdot d\vec{r} < \int_{C_2} \vec{F} \cdot d\vec{r} < \int_{C_4} \vec{F} \cdot d\vec{r} < \int_{C_1} \vec{F} \cdot d\vec{r}$
   (c) $\int_{C_3} \vec{F} \cdot d\vec{r} = -\int_{C_4} \vec{F} \cdot d\vec{r} < 0$
27 (b) Circles
   (c) No
29 (a) $-(\pi/2)(-6a^2 + a^4), a = \sqrt{3}$
   (b) Integrand $3 - x^2 - y^2$ positive inside disk of radius $\sqrt{3}$, negative outside
31 $18a + 18b + 36c + (81d/2), -18a - 18b - 36c - (81d/2)$, curves go in opposite directions

Ch. 18 Understanding

5 False
7 False
9 True
11 False
13 False
15 False
17 False
19 False
21 False
23 True
25 True
27 False
29 True
31 True
33 False
35 False
37 False
39 True
41 False

Section 19.1

1 (a) Positive
   (b) Negative
   (c) Zero
   (d) Zero
   (e) Zero
3 (a) Zero
   (b) Zero
   (c) Zero
   (d) Negative
   (e) Zero
5 4
7 9
9 80

11 27/2

13 75π

15 0

17 81π/2

19 π(e⁴ − 1) → $\pi(e^4 - 1)$

21 36π

23 $a = 6$
  Cannot say anything about b and c

25 (a) Negative
  (b) Positive
  (c) Zero

27 (a) Zero
  (b) Zero

29 (b) $4\pi\lambda h$

31 (a) (i) Total charge inside W.
  (ii) Total current out of S.

Section 19.2

1 6

3 $\pi/2$

5 7/3

7 $\pi \sin 25$

9 1296π

11 $81\pi/4$

13 12π

15 $625\pi/2$

17 $11\pi/2$

19 (a) Constant inside sphere of radius a

  (b) $\vec{E} = \begin{cases} k\dfrac{\delta_0}{3}\rho\vec{e}_\rho & \rho \le a \\ k\dfrac{\delta_0 a^3}{3r^3}\vec{e}_\rho & \rho > a \end{cases}$

Section 19.3

1 4/3

3 195

5 $-\pi R^7/28$

7 $2\pi c(a^2 + b^2)$

Chapter 19 Review

1 0

3 75π

5 $4\pi^4$

7 45π

9 100

11 -160π

13 12

15 $-8(1 + e^{-1})$

17 24π

19 $(\pi/6) - 1/3$

21 Q

23 $(\rho/\epsilon_0)\pi R^2 h$. No

27 (b) 0
  (c) $Ih\ln|b/a|/2\pi$

29 (a) 0.0958
  (b) 0.0958
  (c) 0.0958

Ch. 19 Understanding

1 True

3 False

5 True

7 True

9 True

11 False

13 False

15 True

Section 20.1

5 0

7 0

9 $2/\|\vec{r} - \vec{r}_0\|$

11 (a)

13 $0.00002\pi/3$

15 (a) 1

17 (a) Flux $= c^3$
  (b) div $\vec{F} = 1$
  (c) div $\vec{F} = 1$

19 (a) $2c^3$
  (b) 2
  (c) 2

29 (a) 0
  (b) Undefined.

Section 20.2

1 No

3 $\int_S \vec{F} \cdot d\vec{A} = 8$

5 2π

9 Yes; -3.22

11 0

17 S closed, oriented inward

19 (a) 0
  (b) 0

21 (a) 512
  (b) Larger

25 (a) 0
  (c) No

27 (a) 0
  (b) 0

Section 20.3

1 $4y\vec{k}$

3 $\vec{0}$

5 $-\vec{i} - \vec{j} - \vec{k}$

7 $\vec{0}$

9 Zero curl

11 Nonzero curl

13 $\text{curl}(F_1(x)\vec{i} + F_2(y)\vec{j} + F_3(z)\vec{k}) = 0$

15 (a) No
  (b) Positive
  (c) $x\vec{j}$
  (d) $\vec{k}$

19 (a) Counterclockwise
  (b) Clockwise
  (c) $\vec{0}$

21 $50\vec{i} + 300\vec{j} + 2\vec{k}$

29 (b) $|\omega| \cdot \sqrt{x^2 + y^2}$
  (c) div $\vec{v} = 0$
   curl $\vec{v} = 2\omega\vec{k}$
  (d) $2\pi\omega R^2$

33 $\vec{F} = (-\frac{3}{2}z - 2y)\vec{i}$
   $+ (2x - z)\vec{j} + (y + \frac{3}{2}x)\vec{k}$

Section 20.4

1 No

3 20π

5 0

9 (a) div $\vec{F} < 0$, div $\vec{G} < 0$
  (b) curl $\vec{F} = 0$, curl $\vec{G} = 0$
  (c) Yes
  (d) Yes
  (e) No
  (f) No

11 72

13 (a) 0
  (b) No

15 (a) $-2\pi a^2$
  (b) $2\pi a^2$
  (c) Orientations not related

17 (b) 75π

21 (a) $a = 0$
  (b) $b = 1, c = 3, m = 5$

23 (a) $\vec{0}$; all except z-axis
  (b) 2π
  (c) 0
  (d) 2π
  (e) $\pm 2\pi$ if encircles z-axis
   0 otherwise

Section 20.5

1 Yes

7 Yes
  $(-xy + 5yz)\vec{i} + (2xy + xz^2)\vec{k}$.

9 (a) Yes
  (b) Yes
  (c) Yes

11 (a) curl $\vec{E} = \vec{0}$
  (b) 3-space minus a point if $p > 0$
   3-space if $p \le 0$.
  (c) Satisfies test for all p.
   $\phi(r) = r^{2-p}$ if $p \ne 2$.
   $\phi(r) = \ln r$ if $p = 2$.

Chapter 20 Review

1 $-6\sqrt{3}\pi$

3 (a) $4w^3$
  (b) 4
  (c) 4

5 (a) 24
  (b) 24

7 (c), (d), (f)

9 C_2, C_3, C_4, C_6,

11 div $\vec{F} = y + z + x$
  curl $\vec{F} = -y\vec{i} - z\vec{j} - x\vec{k}$
  Not solenoidal; not irrotational

13 div $\vec{F} = 0$
  curl $\vec{F} =$
  $(2y - \cos(x + z))\vec{i} - \left(2x - e^{y+z}\right)\vec{j} +$
  $\left(\cos(x + z) - e^{y+z}\right)\vec{k}$
  $\vec{F}$ is solenoidal; not irrotational

15 $8\pi/3$

17 120π

19 -5π

21 6

23 6π

25 -147

27 768π

29 div $\vec{E}(P) \le 0$, div $\vec{E}(Q) \ge 0$.

31 $4\pi R^3/3$

33 (a) 4π
   (b) 0
   (c) 4π

35 4π

37 325π

39 50π

41 (a) $(3-a)/(x^2+y^2+z^2)^{a/2}$
   (b) 3
   (c) No "swirl" or rotation

43 (b) $\vec{v}(x) = (55 - x/50)\vec{i}$ mph
      if $0 \le x < 2000$
      $\vec{v}(x) = 15\vec{i}$ mph
      if $2000 \le x < 7000$
      $\vec{v}(x) = (15 + (x-7000)/25)\vec{i}$ mph
      if $7000 \le x < 8000$
      $\vec{v}(x) = 55\vec{i}$ mph
      if $x \ge 8000$

   (c) div $\vec{v}(1000) = -1/50$
      div $\vec{v}(5000) = 0$
      div $\vec{v}(7500) = 1/25$
      div $\vec{v}(10,000) = 0$
      mph/ft

45 (a) $\vec{v} = (1 + \frac{y^2-x^2}{(x^2+y^2)^2})\vec{i}$
      $+ \frac{-2xy}{(x^2+y^2)^2}\vec{j}$

47 (b) $\|\vec{v}\| = K/r^2$
   (c) Flux $= 4\pi K/3$
   (d) Zero

49 (a) $5a^2b - 5b^3/3$
   (b) $5a^2b - 5b^3/3$
   (c) Line integrals are equal.
   (d) $\frac{5}{3}y^3 - 5x^2y$

51 (b) $|K|/r$
   (c) For $(x,y,z) \ne (0,0,z)$,
      div $\vec{v} = 0$.
   (e) $2\pi K$
   (f) $\vec{v}$ undefined on z-axis

53 (a) $(4(p+q)\pi R^3/3)$
   (b) $(4(p+q)\pi R^3/3)$

Ch. 20 Understanding

1 True

3 False

5 True

7 True

9 False

11 False

---

13 False

15 True

17 True

19 True

21 False

23 True

25 True

27 True

29 False

31 False

33 False

35 True

37 True

39 False

41 False

Appendix A

1 (a) $y \le 30$
   (b) two zeros

3 -1.05

5 2.5

7 $x = 1.05$

9 0.45

11 1.3

13 (a) $x = -1.15$
   (b) $x = 1, x = 1.41$,
      and $x = -1.41$

15 (a) $x \approx 0.7$
   (b) $x \approx 0.4$

17 (a) 4 zeros
   (b) $[0.65, 0.66], [0.72, 0.73]$,
      $[1.43, 1.44], [1.7, 1.71]$.

19 (b) $x \approx 5.573$

21 Bounded $-5 \le f(x) \le 4$

23 Not bounded.

Appendix B

1 $(1, 0)$

3 $(\sqrt{2}, \pi/4)$

5 $(4.2, 5\pi/4)$

7 $(5, 0.92)$

9 $(1, 0)$

11 $(-2, 0)$

13 $(\frac{5\sqrt{3}}{2}, -\frac{5}{2})$

---

15 $(\cos 1, \sin 1)$

17 Circle, radius 2, about origin

19 Line through origin, angle of $3\pi/4$

23 $r \le 4, -\pi/2 \le \theta \le \pi/2$

25 $r \le 1, 0 \le \theta \le \pi/4$

27 $3 \le r \le 7, \pi/3 \le \theta \le 2\pi/3$

29 $y = \sqrt{3}x$

31 $(x-1)^2 + y^2 = 1$

Appendix C

1 $2e^{i\pi/2}$

3 $\sqrt{2}e^{i\pi/4}$

5 $0e^{i\theta}$, for any θ.

7 $\sqrt{10}e^{i\arctan(-3)}$

9 $-3 - 4i$

11 $-5 + 12i$

13 $\frac{1}{4} - \frac{9i}{8}$

15 $-\frac{1}{2} + i\frac{\sqrt{3}}{2}$

17 $-125i$

19 $\frac{\sqrt{2}}{2} + i\frac{\sqrt{2}}{2}$

21 $\frac{\sqrt{3}}{2} + \frac{i}{2}$

23 -2^{50}

25 $2i\sqrt[3]{4}$

27 $\frac{1}{\sqrt{2}}\cos(\frac{-\pi}{12}) + i\frac{1}{\sqrt{2}}\sin(\frac{-\pi}{12})$

29 $-i, -1, i, 1$
   $i^{-36} = 1, i^{-41} = -i$

31 $A_1 = 1 + i$
   $A_2 = 1 - i$

37 True

39 False

41 True

Appendix D

1 (a) $f'(x) = 3x^2 + 6x + 3$
   (b) At most one
   (c) $[0, 1]$
   (d) $x \approx 0.913$

3 $\sqrt[4]{100} \approx 3.162$

5 $x \approx 0.511$

7 $x \approx 1.310$

9 $x \approx 1.763$

11 $x \approx 0.682328$

INDEX